← T-2 VIRUS

BACTERIUM (E. COLI)

EUCARYOTIC CELL

POLLEN GRAIN

LIFE
ON
EARTH

EDWARD O. WILSON   **THOMAS EISNER**   **WINSLOW R. BRIGG**

Harvard University   *Cornell University*   *Carnegie Institution*
of Washington

LIFE ON EARTH

RICHARD E. DICKERSON
California Institute of Technology

ROBERT L. METZENBERG
University of Wisconsin

RICHARD D. O'BRIEN
Cornell University

MILLARD SUSMAN
University of Wisconsin

WILLIAM E. BOGGS

SECOND EDITION

SINAUER ASSOCIATES, INC. • PUBLISHERS
Sunderland, Massachusetts

LIFE ON EARTH
Second Edition

Copyright © 1978 by Sinauer Associates, Inc.
All rights reserved. This book may not be reproduced in
whole or in part, for any purpose whatever, without
permission from the publisher.
For information address Sinauer Associates, Inc.,
Sunderland, Mass. 01375. Printed in U.S.A.

Library of Congress Cataloging in Publication Data

Main entry under title:

Life on earth.

    Bibliography: p.
    Includes index.
    1. Biology.   I. Wilson, Edward Osborne,
1929–
QH308.2.L53   1978   574   78-4656
ISBN 0-87893-936-9

9 8 7 6 5 4 3 2 1

Concise Contents

Full Contents

In the first edition of *Life on Earth* we tried to present biology not just as a set of principles and examples, but as an expanding research effort in which students could easily imagine themselves as participants. In virtually every chapter we pointed out unsolved problems and areas in which present knowledge was incomplete or tentative. In this edition, the growing edge has been added on to the established body of knowledge.

During four years of classroom use we have recorded our own impressions of the strengths and weaknesses of the first edition, and have weighed the criticisms of 18 users who agreed to advise us (*see Acknowledgments*). The second edition is therefore a field-tested version, with some sections modified and new ones added. There are many new drawings and photographs, an enlarged format with bigger type, and an expanded table of contents and index. We have reworked our treatment of the more complex topics to make them more accessible. The principal changes are as follows:

Exobiology. In the introductory chapter we discuss the significance of the data from the latest space probes, especially those to Mars. The negative results from those experiments raise an important question: Is our planet inhabited by the only forms of life that human beings can ever hope to see?

Chemistry. Instructors praised the elegant style of the chapters on the chemistry of life, but some found this material too difficult for their students. We have preserved the style, but have substantially shortened and simplified the contents of these chapters. The rigor of both text and illustrations has been eased to make them easier to grasp by students who lack previous courses in college chemistry.

Photosynthesis. We have expanded and updated the discussion of this very important process.

Genetics. All four genetics chapters have been revised and streamlined. We have added new photo essays on mitosis and meiosis, especially commissioned from Walter Plaut of the University of Wisconsin. Additional text material and boxes cover such pertinent topics as genetic disease, genetic counseling, DNA sequencing and recombinant DNA technology.

Development. We have added new material on this rapidly expanding subject, and have reversed the order of presentation to create a more logical sequence: molecule, cell, organism. We have also expanded the material on animal development.

Physiology. Like the first edition, the second contains parallel discussions of plant and animal physiology, an approach that was favored by most users. Some felt that we overemphasized the diversity of living systems at the expense of the basic mechanisms common to most organisms. In this edition we have tried to redress this imbalance. Several of the physiology chapters have been heavily revised. At the suggestion of many users of the first edition, we have added a separate chapter on hormones.

Immunobiology. We have added a new section on this important topic to Chapter 17 (The Internal Environment).

Behavior. Did we overemphasize animal behavior in the first edition? Some users felt we did, so we have shortened the coverage of the subject, but we have added a separate chapter on the new discipline of sociobiology.

Molecular Evolution. This chapter was one of the most highly praised in the first edition, but some users felt that it was too difficult. We have simplified it to correspond to the revised treatment of biochemistry in the earlier chapters.

The Future of Life. A discussion of the thought-provoking ideas of Garrett Hardin has been added to Chapter 36 (Defeat by Default).

Biology, having replaced physics at the center of the scientific stage, continues to grow at an accelerating pace. We believe that this new edition of *Life on Earth* has kept abreast of this rapid progress, and presents the most important ideas of biology in a way that will stimulate beginning students.

The Authors

ACKNOWLEDGMENTS

We wish to thank Ari Van Tienhoven of Cornell University and Edward S. Golub of Purdue University for their help with the new material on hormones and immunology, respectively. Walter Plaut of the University of Wisconsin deserves special mention for preparing the new micrographs of meiosis and mitosis especially for this edition. Kraig Adler of Cornell University submitted an in-depth review of the entire book which helped us time and again to focus on the needs and interests of the beginning student. The following biologists provided us with detailed suggestions for improvements, based on their use of the First Edition:

William L. Bischoff
University of Toledo

John Tyler Bonner
Princeton University

Richard K. Boohar
University of Nebraska

Clifford Brunk
U.C.L.A.

Mark W. Dubin
University of Colorado

Stephen S. Easter, Jr.
University of Michigan

Stephen D. Hauschka
University of Washington

Brian A. Hazlett
University of Michigan

Lewis J. Kleinsmith
University of Michigan

J. Richard McIntosh
University of Colorado

John M. Palka
University of Washington

Robert W. Poole
Brown University

William K. Purves
Harvey Mudd College

Rudolph A. Raff
Indiana University

David Shappirio
University of Michigan

David S. Woodruff
Purdue University

Charles F. Yokum
University of Michigan

Life on Earth pioneers a new generation of college biology textbooks. It was written according to a plan which we predict will be followed for a long time to come. One man can no longer write a general biology textbook and hope to achieve excellence in more than one or two principal sections of it; the task requires a well balanced and coordinated team. Modern biology is an immensely broad science, spanning an ascending hierarchy of levels from molecules through cells and organisms to populations and societies — from DNA to bands of chimpanzees. It spans several scientific cultures as well. The chemist mapping the structure of an enzyme protein, or the microbial geneticist painstakingly fitting together the molecular steps that lead from the gene to the ribosome, work in a different world from that of the ecologist or the student of animal behavior. Their sense of adventure is equally keen, but the unknown regions they explore are in a sense as different as the surfaces of alien planets. Yet a series of graded steps of knowledge, theory, and technique can be traced as one travels from molecular biology at one end of the science (and this textbook) to sociobiology at the other. Biology is nearing its destiny as a unified science, and it can be presented that way to the beginning student.

The team approach also has its dangers, the chief one being that the resulting book will turn out to be little more than a collection of essays — disorganized, uneven in level, and lacking a unified perspective. To prevent this, three of the authors — Wilson, Eisner and Boggs — also served as editors, coordinating the work of their colleagues and suggesting revisions and shifts of emphasis.

The authors of *Life on Earth* were selected with the expectation that they would enthusiastically celebrate their personal muses — to transmit the folklore and expertise that only specialists can muster. The selection was made with some care

and deliberation, in order that the pieces might come together in a whole more comprehensive and balanced than possible from the pen of a single author. By design, there is no single center of strength, and no one theme was chosen for heavy emphasis. The reader will find that molecular biology, with the introductory chemistry necessary for its comprehension, is about as strong as evolutionary theory, with the population genetics necessary for its comprehension.

Richard E. Dickerson, a protein chemist of wide interests, wrote most of the material on chemistry and biochemistry, including the chapters on the origin of life and molecular evolution. Robert L. Metzenberg and Millard Susman collaborated closely in the sections on cell biology and genetics. Richard D. O'Brien contributed neurobiology and much of the chapter on internal environment; he also collaborated with Thomas Eisner in writing much of the material on animal physiology. Winslow R. Briggs was the team's botanist, drawing on his broad training to contribute a diversity of subjects from photosynthesis and growth to the diversity of microorganisms and higher plants. Thomas Eisner provided numerous photographs, part of the animal physiology, and collaborated with Edward O. Wilson in the chapters on animal behavior and diversity. Wilson also wrote most of the section on evolutionary biology. William E. Boggs, a science writer and editor, helped to plan the book, wrote parts of some chapters and rewrote others, with particular attention to organization, clarity and level.

Life on Earth presents fundamental biology as it actually is. We have avoided distorting the subject to make it seem to be an appendix to the study of man or as a practical art to assist the control of the environment. While working hard to achieve maximum simplicity and clarity, we have also resisted glossing over the harder parts. Biology is indeed the focus of man's hopes, the potential source of knowledge that can solve our most pressing problems. But it can be used only by men and women educated in the fundamentals, prepared to speak with experts about the ideas and facts that really matter and to go on to advanced professional training if their careers require it. The goal of a good liberal arts education is no less demanding. It is to know what science is, to understand how active scientists think and how discoveries are made, and thus to be able to separate the sound and useful from the shoddy and untrue. *Life on Earth* has been written in the confidence that this attitude is shared by the great majority of biology instructors and students.

Edward O. Wilson
Thomas Eisner

ACKNOWLEDGMENTS

The authors wish to thank Abraham S. Flexer of the University of Colorado for his critique of the entire manuscript, and each of the following biologists for his comments on parts of the manuscript:

Charles F. Cleland
Harvard University

Joseph H. Connell
University of California, Santa Barbara

Gary R. Craven
University of Wisconsin

Rowland H. Davis
University of Michigan

John F. Eisenberg
National Zoological Park,
Smithsonian Institute

Paul B. Green
Stanford University

Jack P. Hailman
University of Wisconsin

William J. Hamilton
University of California, Davis

Norman H. Horowitz
California Institute of Technology

Lewis J. Kleinsmith
University of Michigan

James W. Lash
University of Pennsylvania

Herbert W. Levi
Harvard University

J. Richard McIntosh
University of Colorado

William W. Murdoch
University of California, Santa Barbara

Carl S. Pike
Franklin and Marshall College

C. Ladd Prosser
University of Illinois

Carl Sagan
Cornell University

Felix Strumwasser
California Institute of Technology

Ralph W. Wetmore
Harvard University

Fred H. Wilt
University of California, Berkeley

LIFE
ON
EARTH

A year and a half ago when we were coming home on Apollo 8, we looked at the Earth and mentioned that the Earth was really the only place we had to go to. It was the only place that had color. It was the only place that we could see in the universe that had life, that had warmth, that was home to us. —James A. Lovell, Jr.

Very few men have ever left our planet Earth; most of us take the planet that carries us more or less for granted. When the medieval church claimed dominion over *Urbs et Orbis* (City and World) it was claiming just about everything that mattered to mankind. Kepler, Copernicus, and Galileo changed our point of view. By demonstrating that the Earth was not the center of the universe, they implied that other planets might be worlds like ours. Fiction writers from Johannes Kepler to Jules Verne and beyond populated our neighbors around the Sun with life, humanoids, and civilizations. The astronomer Percival Lowell built the Flagstaff Observatory in Arizona in 1894 specifically to study Mars, which he believed to be inhabited by intelligent canal-builders. H. G. Wells skirted the problem of the barren appearance of the lunar surface by postulating a subsurface civilization. Invaders from Mars and Venus became a stock plot line of science fiction.

Today we are more knowledgeable and less optimistic about other life in our solar system. Mercury, the planet nearest the Sun, is a hot cinder turning slowly on its axis. Surface temperatures on the second planet, Venus, are close to that of melting lead. The fifth and sixth planets, Jupiter and Saturn, are surrounded by dense gaseous atmospheres that solar radiation may never penetrate. The three outer planets appear cold and inhospitable. Our own Moon turns out to be an airless, waterless, lifeless gray sphere, the product of four eons of meteoric bombardment and vulcanism (*Figure 1*). Only the fourth planet, Mars, offers slim

EARTHRISE over the moon. Seas of Tranquility and Fertility stretch toward the horizon. Photograph was taken by crew of Apollo 10 as they left lunar orbit, homeward bound.

hope for life of even a primitive kind (a hope that the Viking probes have dimmed but not eliminated). The dawn of space exploration has lent support to the medieval notion that Earth may be unique in the solar system.

A common theme of the Apollo astronauts has been the surprising lack of color anywhere in the visible universe save Earth. James Lovell on Apollo 8 remarked, "The Moon is essentially gray: no color. It looks like plaster of paris or sort of a grayish deep sand. . . . The best way to describe this is really a vastness of black and white — absolutely no color." Charles Conrad, on Apollo 12, was even more blunt, "If I wanted to look at something that looked like the Moon, I'd go out and look at my driveway." The sky above the Moon, Conrad said, appeared ebony black.

The Earth presented a different picture to the first men ever to leave it completely. Lovell described it during the ascent of Apollo 8, "For colors, the waters are all a sort of a royal blue. Clouds, of course, are bright white. . . . The land areas are generally a brownish —sort of dark-brownish to light-brown in texture." Later, he uttered perhaps the most eloquent comment to come out of the space program, "The Earth from here is a grand oasis in the great vastness of space."

What makes Earth such an oasis in an otherwise inhospitable solar system? Of the nine planets, only the Earth has oceans, and they have been the cradle of life. Is life inevitably linked to seas? Or is life a natural stage in the evolution of matter? What features are unique to life wherever it may be found, and what features are accidents of the history of life on this particular planet? We shall try to answer these questions as we study the phenomenon of life. But some questions may not be answered until we find, somewhere, a planet on

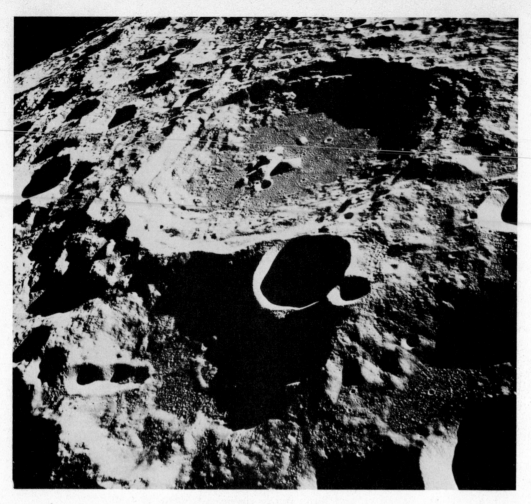

1 BACK SIDE OF MOON was photographed from Apollo II command module in lunar orbit. Large central crater is about 50 miles wide. On moon such craters may persist for billions of years.

which life has appeared independently. Until then, the best we can do is to look carefully at the similarities among the living systems on the planet Earth.

THE EARLY EARTH

Our solar system, including the planet Earth, evolved about 4.7 billion years ago. Early theories about the formation of the Earth spoke of the cooling and condensation of hot gases. Since the early 1950's it has seemed more likely that the planets were built up by the gravitational attraction and aggregation of cold dust and particles into clumps of solid matter. As the Earth grew by this process of cold accretion, the weight of the outer layers compressed the center. This pressure, and energy

from radioactive decay, heated the interior until it finally melted, and the settling of heavier elements led to the fluid iron and nickel core that we have today. Around this core, which has a radius of 2200 miles, lies an 1800 mile thick mantle made of dense silicate minerals. Covering this is a lighter crust, as much as 25 miles thick under the continents but thinning to 3 miles thick under the ocean floor. Only from rare Kimberlite diamond pipes, that erupt to the surface in South Africa and a few other sites, have we ever obtained a specimen of the minerals that comprise the mantle.

The scarcity of helium and other noble gases on Earth compared with their abundance on the Sun suggests that Earth lost its original atmosphere. The Earth was once a sterile rocky ball with neither atmosphere nor oceans. A new atmosphere arose in time from the out-gassing of the mantle and crust. Water vapor from the interior condensed into seas.

The pattern of sea and land in those distant

2

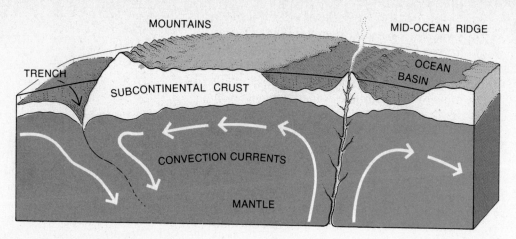

TRENCH

SUBCONTINENTAL CRUST

OCEAN BASIN

CONVECTION CURRENTS

MANTLE

2 CRUST OF EARTH is in constant motion because of planet's liquid core and semifluid mantle. Surface of the planet is shaped by tectonic (crust-deforming) processes. Where currents in the mantle rise beneath an ocean, the ocean floor spreads laterally, separating the bordering continents. Crust may be piled up where it meets sinking currents, producing oceanic trenches, deep earthquake zones and folded mountain ranges like the Andes and Rockies.

times was totally different from now. The crust of the Earth floats on the semifluid mantle, in which convection currents of mantle material flow up from the heated core, expand laterally, and sink again into the interior. One such rising current in the mantle lies beneath the mid-Atlantic ridge, with a flow to the east and west of about a centimeter per year *(Figure 2)*. The Americas are slowly drifting away from Europe and Africa, and one of the side effects of this drift has been the folding and uplift of the Rockies and Andes mountain chains *(Chapter 31)*. Since the beginning of the planet, the Earth's crust has been in constant motion. Sea beds have been uplifted and folded into mountain ranges, and gradually eroded down into plains which have been flooded again. The history of the Earth is one of constant and ceaseless change. Evidence of external accidents, such as impact craters from meteors, have been quickly obscured; today we see only the most recent ones *(Figure 3)*.

By contrast, the Moon and to a lesser degree Mars are essentially static worlds. They have few tectonic processes comparable with the extensive continental drift and crustal folding in Earth, and we think we know why. The Moon, Mars and Earth increase in mass in a regular progression: Mars is 8.7 times the mass of the Moon, and Earth

is 9.4 times the mass of Mars. Both the Moon and Mars were so small that heat produced by collision as they were being formed, or subsequently generated in their interior by gravitational pressure and radioactive decay, could rapidly escape. To use an analogy from nuclear physics, the Moon and Mars were below the critical mass for the temperature buildup required for a permanently molten core. If they melted at all, they probably solidified again as heat was radiated into space from the planetary surface. They became solid balls of rock, and the convection currents that still shape the surface of the Earth either never developed or never lasted. The small size and weak gravitational fields also meant an early loss of what initial atmospheres the Moon and Mars possessed, which meant minimal erosion. One effect of the lack of molten cores in these planets is their absence of magnetic fields.

The oldest rock and soil samples brought back from the Moon by the Apollo 11 and 12 expeditions have been dated by uranium/lead and rubidium/strontium radioisotope methods as 4.5 billion years old. Both expeditions found basaltic lava flows 3.4 to 3.6 billion years old, indicating that there was a widespread outpouring of lava over the mare regions during a relatively narrow time interval. For the first billion years of its history, at least, the Moon was not a static ball of rock. It has been

Of the nine planets, only the Earth has oceans, and they have been the cradle of life.

suggested that Mars and perhaps even the Moon might have been covered early in their histories with shallow seas which were later lost as water molecules escaped the planets' weak gravitational pull and vanished into space. If so, these early moves toward an Earth-like planet were stillborn. The scars of over four billion years of history are visible on the Moon and Mars in the form of heavily cratered surfaces, not erased by subsequent erosion *(Figure 4)*. Earth, by contrast, was large enough to develop differently.

This then, is the stage on which life was to play out its drama —a water-covered planet, with shifting land masses in constant geological turmoil.

3 **BARRINGER METEOR CRATER in Arizona is about three quarters of a mile across and 600 feet deep. It was probably dug by a meteorite about 100 feet in diameter and weighing 100,000 tons, which struck the Earth at a velocity of about 36,000 miles per hour. Geological processes on Earth have erased all but most recent craters. This one is about 25,000 years old.**

Some time around four billion years ago, the first actors appeared on stage. The oldest rocks we can find on Earth, over 3.5 billion years old, already bear traces of living organisms. How and why the first living systems appeared is the subject of Chapter 23. At this point it is enough to say that

The first chemical system that we would describe as "living" appeared about four billion years ago.

they did appear, and spread throughout the seas. The oceans teemed with life, but the exposed land remained nearly as sterile as the Moon. The land was bathed in ultraviolet radiation from the Sun, radiation deadly to living organisms. Life was confined to depths in the seas (below 5–10 meters) that would shield out this radiation. Photosynthesis arose for another reason, yet one effect of photosynthesis was the creation of an ozone layer in the upper atmosphere which blocked the ultraviolet rays. When this happened, the spread of life to shallow coastal waters and eventually to the land itself was only a matter of time.

Photosynthesis evolved in response to a shortage of natural high energy compounds for food. When this new method of tapping solar energy to synthesize sugars developed, life divided into two classes of organisms: those that made their own food (plants) and those that ate the foodmakers (animals). Both plants and animals moved onto the land in their own ways: animals by migration of individuals, and plants by migration of generations, as spores and seeds were carried to new environments.

The invasion of the land cleared the way for the major steps in the history of life that we shall examine in Part III. In the animal kingdom this involved the rise of amphibians, reptiles, mammals, the special type of mammals known as primates, and finally that peculiar primate called Man. With the coming of Man, life crossed another threshold comparable to the synthesis of the first life, the evolution of photosynthesis, and the conquest of land. Man became the first thoroughly cultural animal, by virtue of his ability to communicate by speech and writing. A large part of his heritage became externalized in myths, traditions, custom, and law, rather than being predominately internal and genetic. To use a slightly exaggerated image, Man's genes were supplemented by libraries. There is no clear and obvious break in the thread, however. Stellar evolution is followed by chemical evolution, biological, and even social or cultural evolution. All are stages in the history of life on this planet.

CRATERS ON MARS photographed in July 1976 by the Viking 1 Mars probe. The relatively smooth Plain of Argyre at upper left is an impact basin, surrounded by a ring of many smaller craters. The haze layers, 15 to 25 miles above the Martian horizon at top, are believed to be crystals of carbon dioxide, or dry ice.

WHAT IS LIFE?

The title of this book, *Life on Earth*, can be interpreted two ways. One suggests the eventual appearance of more volumes entitled, *Life on Mars* and *Life on Venus*. The more restrictive interpretation is that life is a phenomenon confined to Earth. Evidence to date, including that from Vikings 1 and 2, supports the second choice, yet few biologists seriously believe that life is peculiar to our own planet. Conditions may not be right for life on the other eight planets of this solar system, but a one-out-of-nine success rate may be pretty good for a phenomenon so tenuous as life. It is a pity that exploration of other solar systems today appears as difficult and futuristic as exploration of other planets in our own solar system seemed to be a century or two ago.

The search for life on other planets has forced us to consider carefully what we are looking for. Martian unicorns went out with Edgar Rice Burroughs and Jules Verne. If life elsewhere evolved independently and differently than here, is it possible that one could look straight at it and not recognize it? What are the generalizations that one can make about all living organisms, no matter where they come from?

Improbable shapes and seemingly voluntary movements of large objects on another planet could be a tip-off that one is looking at life forms. But the two Viking landers revealed no Martian natives staring into the cameras, nor even any changes in the landscape apart from dust storms and seasonal migrations of the ice cap *(Figure 5)*. The challenge is to detect microscopic life, not hulking monsters. What criteria can be used to distinguish between inorganic reactions and true life forms in the Martian soil?

Life must be defined by what it does, not what it is. An alien creature must propagate itself in some way, but the mechanism could be quite strange, and the molecules that store hereditary information might not be DNA. It must extract energy from its environment, but the nutrients need not resemble those of Earth. Some of the life forms on any planet surely would develop means of tapping the energy resources of sunlight, but the photo-receptor molecules could be very different from those in green plants.

All living creatures must METABOLIZE, GROW AND REPRODUCE, PROTECT themselves and their offspring, and EVOLVE in response to long-term changes in their environment. The primary object of one generation is to ensure that another generation follows, by whatever means are necessary. Different life forms, even on Earth, have taken quite different approaches to this problem.

Metabolism is the most obvious hallmark of life. All organisms carry out reactions that release chemical energy. They use some of this energy immediately and store the rest. All life forms break down raw materials ingested from their environment, and use the fragments as building blocks to synthesize their own macromolecules. Extraterrestrial organisms might be imagined using substances other than proteins, carbohydrates, lipids and nucleic acids, but the chemical patterns of breakdown, energy extraction, and synthesis should be recognizable.

Living organisms need enough protection and security to ensure that another generation survives. For some kinds of threat the proper response is the familiar "flight or fight" response.

5 PANORAMA OF MARTIAN LANDSCAPE from the Viking 1 lander, covering an angle of approximately 100°. The large boulder at the left is approximately eight meters from the spacecraft and one by three meters in dimensions. The sand dunes and drifts around boulders indicate that the prevailing wind direction in this region is from upper left to lower right.

Others are protected by armor and shells, and still others by camouflage or obscurity. Some organisms stake their survival on simple fecundity. The salmon, for example, will persist as a species if only a few of its millions of eggs survive. The strategy differs from organism to organism, but the goal is the same: to guarantee the survival of the next generation.

Growth and reproduction are always associated with life. No organism lives forever in a single state. Amoebas and some other simple life forms grow to a certain point and then reproduce asexually by dividing in two. Asexual reproduction can create great numbers of new organisms in a short time, but the offspring are genetically almost identical to their parents. Sexual reproduction, in contrast, mixes the genes of two individuals, creating new — and possibly improved —combinations.

One can find nonliving objects or chemical systems that mimic most of the behavior mentioned so far. Crystals in solution grow and "heal" breaks and imperfections. Automobile engines "metabolize," extracting energy from organic molecules by a series of chemical reactions. Yet life does something that no non-living collections of chemicals do: it evolves.

Reproduction is not simple copying. Occasional mistakes can occur in the copying process, or can result from damage to the genetic material by radiation or chemical agents. These genetic mistakes or mutations are copied in subsequent offspring. In the course of several generations, even the offspring of asexual organisms develop variety as a

result of mutation. In a population of sexual organisms, mating and shuffling of hereditary material means a constant supply of new variations, which makes evolution possible. Individuals identically and perfectly adapted to one set of conditions could be equally (and perhaps lethally) handicapped when these conditions change. But with enough genetic variety, the odds are that some members of the population can survive all but a catastrophic change in the environment. This interaction of variation and selection is the mainspring of evolution.

LIFE ON MARS

How do these four criteria of life—metabolism, growth and reproduction, protection from the environment, and evolution—bear on the recent search for life on Mars? Because evolution takes place very slowly, the designers of the Viking landing vehicles focused on the metabolism, growth, and reproduction of Martian microorganisms. An implicit assumption of their experiments was that Martian life, like that on Earth, would be based on carbon compounds. While exceptions could be imagined, and have served as plots for science fiction stories, the comparative chemistry of carbon and the other elements makes this assumption a good one. The life-detection experiments of the Viking landers asked four questions:

1) If carbon-containing nutrients are added to a Martian soil sample in a sealed chamber containing a sample of Martian atmosphere, do reactions take place that change the gas composition in the chamber? This involved a gas-exchange experiment or GEX.
2) Is there something in the Martian soil that will break down organic nutrient molecules labeled with carbon-14 and release $^{14}CO_2$ as Earth organisms do? This involved a labeled-release experiment or LR.
3) If $^{14}CO_2$, ^{14}CO, and water are supplied to a soil sample along with radiation approximating Martian sunlight, are there reactions that will synthesize organic compounds containing the carbon-14? This involved a pyrolytic-release experiment or PR.
4) Are organic molecules and carbon compounds to be found in the Martian soil itself? This involved a gas chromatograph and mass spectrometer, or GCMS.

The results to date are puzzling. The LR experiment releases carbon dioxide, which can be prevented in a control experiment by sterilizing the soil sample first by heating. The PR experiment shows that CO_2 and CO are incorporated to a small extent into new organic molecules, but the reactions are inhibited, not encouraged, by the presence of moisture. The GEX experiment shows changes in gas composition above the soil that could be attributable to microbial metabolism. But the GCMS runs were completely unable to detect any traces of organic compounds in Martian soil. As one scientist remarked, if there really is life on Mars, then "where are the bodies?". The tentative verdict after six months of study and thought is

that the "positive" results of the first three experiments may be a consequence of the presence, not of living organisms, but of peroxides and other oxidizing inorganic substances in the Martian soil. But the tantalizing question remains: Could we be looking at life, and failing to recognize it because of our Earthbound preconceptions?

LIFE ON EARTH

The search for life elsewhere in the solar system, for the moment, remains inconclusive. Life is a very special phenomenon. We believe that it is a mode of behavior that matter exhibits when it reaches a certain level of organization and complexity, but we cannot yet prove this because we cannot duplicate the phenomenon on Earth, or even examine examples from another planet. We can observe from a distance, and disassemble one part of the whole for study, whether the part is a population, an individual, an organ, or a molecule. We can propose ideas about how living systems work and why they exist as they do, and check the simpler of these ideas experimentally. The problem is complicated because we are part of the system under study —a dilemma that seldom worries a physicist. Still, this makes the study of life an even more absorbing endeavor.

READINGS

F. CRICK, *Of Molecules and Men*, Seattle, University of Washington Press, 1966. A view of life at the molecular level, as seen by the co-discoverer of the structure of DNA.

C. GROBSTEIN, *The Strategy of Life*, 2nd Edition, San Francisco, Freeman, 1974. A little gem of a book. The first four chapters focus on many of the topics discussed here.

F. HOYLE, *Of Men and Galaxies*, Seattle, University of Washington Press, 1964. Three lectures by one of our most noted and articulate astronomers. The last two, "An Astronomer's View of Life", and "Extrapolations into the Future", are especially relevant to this chapter.

C. SAGAN, *The Cosmic Connection*, New York, Dell, 1975. Fact and speculation about the solar system, the galaxy, and the universe, by one of the most enthusiastic pioneers in the search for extraterrestrial life.

E. SCHRÖDINGER, *What is Life?*, New York, Cambridge University Press, 1962 (1st Edition 1944). The inspiration for much of postwar molecular biology. Not as revolutionary as it once was, but still interesting because of the author's eminence as a theoretical physicist.

I. S. SHKLOVSKII AND C. SAGAN, *Intelligent Life in the Universe*, San Francisco, Holden-Day, 1966. A long-distance dialog between a Russian and an American astronomer, both interested in problems of extraterrestrial life.

The results of the Viking 1 and 2 missions to Mars, including the search for life, are reported in detail in the 1 October and 17 December 1976 issues of *Science* (the journal of the American Association for the Advancement of Science), and in the November/December 1976 issue of *American Scientist*.

Part  The cell

One important property of life on Earth is that it is modular. The smallest unit of life is a complex but reasonably standard entity called a cell. A small bit of chemically reactive "soup" is not alive —without the protection of a cell membrane it would rapidly be diluted out of existence. The closed boundaries of a cell permit concentrations and arrangements of chemical components that would be vanishingly improbable in an open solution.

Large organisms do not consist of a single enormous cell but of many small ones. The upper limit on cell size is set by the slow rate at which substances diffuse through liquids. Organisms rely largely on diffusion to move molecules from place to place within the cell. Except for very short distances, the rate of diffusion is simply inadequate to keep pace with biochemical machinery that must carry out thousands of reactions per second. Moreover, a single giant cell would not have enough surface area for adequate exchange of gases, nutrients and waste products. The cytoplasm in the center of the cell would suffocate, starve, or poison itself.

In multicellular organisms a basic cell pattern is laid down in the genetic material, and specialized variations of that pattern emerge as the organism develops from a one-celled fertilized egg to a many-celled adult. One cell of an embryo will eventually give rise to epidermis and another to liver tissue; the exact reason remains a mystery.

A cell is a miniature machine for living —a tiny chemical factory that takes both small and large molecules from the environment and rearranges them into living matter. Part I of this book focuses on the cell and the biochemical processes that occur within it, particularly upon the concept of genetic information, and how this information is encoded within a single molecule of DNA —the master control that regulates the chemistry of life.

Any living cell carries with it the experiences of a billion years of experimentation by its ancestors. You cannot expect to explain so wise an old bird in a few simple words. —Max Delbrück

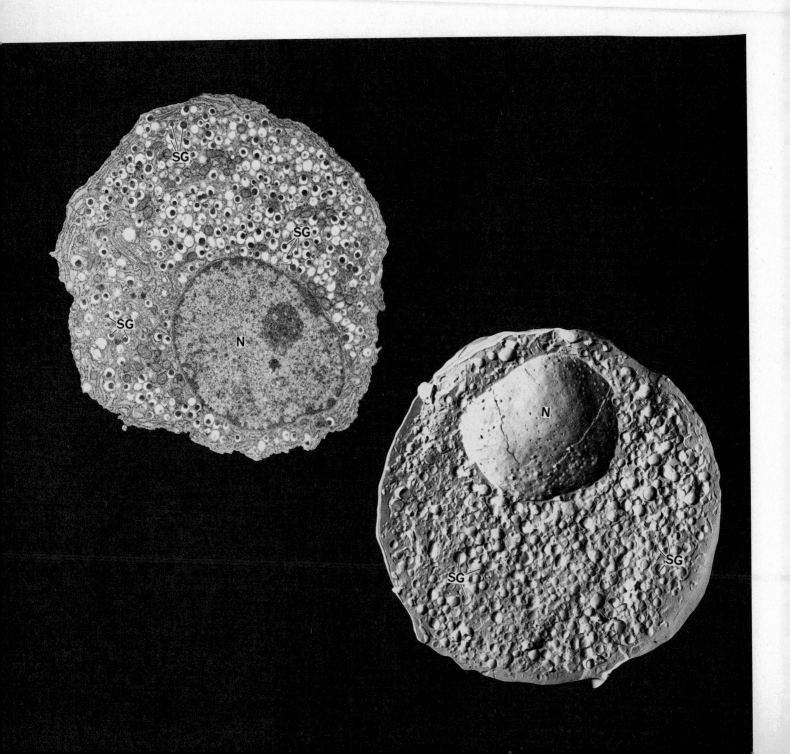

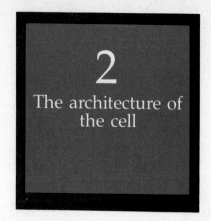

Robert Hooke, a 17th century physicist, natural philosopher, and inventor, is responsible for bringing the word cell into the language of biology. Using the crude microscopes of his day, Hooke looked at a very thin slice of cork and saw a grid which reminded him of a monastery or a prison, with rows of cells to hold the inmates. His comparison of the outer bark of the cork-oak with a network of dungeons was fanciful, but there was insight in his use of the word cells. Much later, in 1839, Matthias Schleiden, a botanist, and Theodore Schwann, a zoologist, promulgated the cell theory—the idea that all living things are composed of cells.

Cells vary greatly in size, from an ostrich egg, with a volume of about 10^{15} cubic micrometers, to parasitic bacteria called Chlamydozoa, with a volume of 10^{-2} cubic micrometers (A micrometer, abbreviated μm, is a unit of length convenient for expressing the sizes of cells. One micrometer equals 10^{-4} cm. The period at the end of this sentence is about 200μm in diameter. *See Box A*). The great majority of cells have a volume of one to 1000 μm³, and there are physical and chemical reasons for this. The volume and mass of a spherical cell increase with the cube of the radius, but the sur-

INSULIN-PRODUCING CELLS from the pancreas of a rat illustrate the power of two different micrographic techniques. Both were made with an electron microscope, but the picture at top is a conventional stained thin section of a cell; the one below was made by the newer freeze-etch technique, which involves splitting a frozen cell and depositing a thin film of platinum and carbon on the exposed faces. Because the cells tend to split along the boundaries of their internal membranes, the freeze-etch technique provides a three-dimensional picture of the surfaces of various membranes. The letter N designates the nucleus, and SG, the secretion granules that contain insulin. Magnification, about 100,000X.

face area available for intake of nutrients, exchange of gases, and excretion of wastes increases only as the square of the radius. By this principle a cow that is twice as long as a calf and about the same shape will yield about four times as much leather and eight times as much beef. Similarly a large cell has a much lower surface-to-volume ratio than a small one, and therefore a smaller absorptive surface per unit of mass. This simple fact sets an upper limit on the size of most cells. The ostrich egg is no exception—most of its bulk consists of stored food; the biologically active part of the egg, which will become the future embryo, is a tiny speck on the surface of the yolk. A single cell the size of a cow would have roughly the same surface area as all the cells of a shrimp. In other words, this imaginary giant cell would have the mass of a cow and the absorptive capacity of a shrimp— biologically an untenable situation, to say the least.

Because all cells face certain common problems, they have architectural features in common. All cells have a surface membrane that regulates molecular traffic into and out of the cell, maintaining an internal environment compatible with the chemistry of life. The control of life processes depends on a complicated set of chemical instructions which is passed from one generation of cells to the next.

All cells store their hereditary information in the form of threadlike molecules of DNA. In some kinds of organisms the hereditary apparatus is segregated from the rest of the cell, enclosed in a membrane-bounded nucleus. In other organisms the DNA is localized within a region called the nucleoid, which is not separated from the rest of the cell by a membrane. This distinction divides all cells into two great classes: eucaryotic cells (from the Greek—true nucleus) and procaryotic cells (primitive nucleus). Bacteria and bluegreen algae are procaryotes. All other organisms are

11

CELL WALL

CELL MEMBRANE

MESOSOME

RIBOSOMES

NUCLEOID

1 PROCARYOTIC CELL lacks a true nucleus and internal organelles. Nucleoid contains the genetic material of the cell. This type of cell is found among bacteria and bluegreen algae.

eucaryotes. Eucaryotic cells are generally much larger and more complicated than procaryotic cells, and they contain several components, collectively called ORGANELLES, that are not found in procaryotes. The evolutionary development of the eucaryotic cell was the breakthrough that made possible the evolution of multicellular organisms, including man. Figures 1 and 2 compare the most important features of a "typical" procaryotic cell

and a "typical" eucaryotic cell. By the end of the book the reader will understand that a typical cell is an over-simplification, rather like the notion of a typical human being, but these illustrations at least indicate what to look for. Notice that both types of cell contain many minute particles called ribosomes. These are the sites of protein synthesis, and all true cells contain them. All procaryote ribosomes are about the same size, and are considerably smaller than eucaryote ribosomes.

Viruses are much simpler than bacteria or bluegreen algae, but they cannot properly be called cells. Viruses do not have ribosomes, and must use the ribosomes of a host cell to synthesize viral proteins. Some biologists classify viruses as procaryotes, but the differences between viruses and bacteria or bluegreen algae are enormous. A virus is simply a packet of hereditary material wrapped in a coat of protein (and in some cases a few additional substances). The coat serves only to protect the hereditary material and to introduce it into a cell of a susceptible host. The coat is not comparable in function with the membrane of a true cell. Biologists have long debated whether viruses are

EUCARYOTIC CELL contains a true nucleus and an assortment of organelles. Function of organelles is discussed in the text. This type of cell is characteristic of all multicellular organisms. **2**

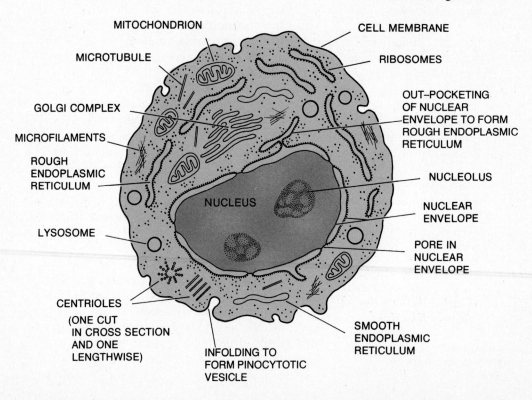

MITOCHONDRION

MICROTUBULE

GOLGI COMPLEX

MICROFILAMENTS

ROUGH ENDOPLASMIC RETICULUM

LYSOSOME

CENTRIOLES (ONE CUT IN CROSS SECTION AND ONE LENGTHWISE)

INFOLDING TO FORM PINOCYTOTIC VESICLE

CELL MEMBRANE

RIBOSOMES

OUT–POCKETING OF NUCLEAR ENVELOPE TO FORM ROUGH ENDOPLASMIC RETICULUM

NUCLEOLUS

NUCLEAR ENVELOPE

PORE IN NUCLEAR ENVELOPE

SMOOTH ENDOPLASMIC RETICULUM

NUCLEUS

A Measuring in the metric system

All the natural sciences, including biology, require frequent measurements using metric units. The reader would be wise to memorize the following divisions and special terms applied to them. Also, we recommend practice in the multiplying and dividing involved in order to get a better feel for the relationships of the units.

Length

1 KILOMETER (abbreviated km) = 1000 meters (also written 10^3 meters) = 0.62 mile

1 METER (abbreviated m) = $\dfrac{1}{1000}$ kilometer (also written 10^{-3} km) = 39.37 inches

1 CENTIMETER (abbreviated cm) = $\dfrac{1}{100}$ meter (also written 10^{-2} m)

1 MILLIMETER (abbreviated mm) = $\dfrac{1}{1000}$ meter (also written 10^{-3} m)

1 MICROMETER (symbolized by μm) = $\dfrac{1}{1000}$ millimeter (also written 10^{-3} mm) = one millionth of a meter (written 10^{-6} m)

1 NANOMETER (abbreviated as nm) = $\dfrac{1}{1000}$ micrometer = one billionth of a meter (written 10^{-9} m)

1 ÅNGSTRÖM (abbreviated Å) = $\dfrac{1}{10}$ nanometer = $\dfrac{1}{10,000}$ micrometer = one ten-billionth of a meter (written 10^{-10} m)

Mass

1 KILOGRAM (abbreviated kg) = 1000 grams (also written 10^3 grams) = 2.20 lbs

1 GRAM (abbreviated g) = $\dfrac{1}{1000}$ kilogram (also written 10^{-3} kilogram)

1 MILLIGRAM (abbreviated mg) = $\dfrac{1}{1000}$ gram (also written 10^{-3} gram)

1 MICROGRAM (abbreviated μg) = $\dfrac{1}{1000}$ milligram (also written 10^{-3} milligram) = one millionth of a gram (written 10^{-6} g)

1 NANOGRAM (abbreviated ng) = $\dfrac{1}{1000}$ microgram (also written 10^{-3} microgram) = one billionth of a gram (written 10^{-9} g)

Volume

1 LITER (abbreviated l) = 1000 milliliters (also written as 10^3 milliliters) = 1.06 liquid quarts

1 MILLILITER (abbreviated ml) = $\dfrac{1}{1000}$ liter (also written 10^{-3} l) = 1 cubic centimeter (abbreviated cc or cm³). One ml (= 1 cc) of water at 4°C weighs 1 gram

1 MICROLITER (abbreviated μl) = $\dfrac{1}{1000}$ milliliter (also written 10^{-3} milliliter) = one millionth of a liter (also written 10^{-6} l). One microliter of water at 4°C weighs 1 milligram.

B Are viruses alive?

Viruses lie in a semantic twilight zone between living and nonliving organisms. They are essentially short segments of genetic material, DNA or RNA, protected by over-coats of protein. They can be crystallized and they will not grow or propagate in isolation, which suggests that they are nonliving *(photo at left)*. They do not carry out any metabolic processes alone. On the other hand viruses are capable of infecting bacteria or higher plants and animals and of seizing control of their metabolic machin-ery. An intestinal bacterium, when infected by a bacteriophage (bacteria eater), shuts down some of its own chemistry and begins copying the phage DNA and making the component parts for new bacteriophage viruses *(photo at right)*. After half an hour or so, the bacterium bursts and several dozen new viruses empty out into the surround-ing medium, ready to infect new bacteria. But from the time such a virus leaves its ''incubator'' until it finds another bacterium to infect, it undergoes no chemical change that we know of. A lone virus can do nothing, but a virus plus a bacterium can do all the things that are required in a generalized living organism. Is the virus alive? Is the virus-plus-bacterium alive? The bacterium would already be considered alive, without the virus. It is probably best to regard the virus as a degenerate parasite which has lost many of the independent functions its ancestors may have had and which it can now make the bacterium do for it. After all, we do the same thing when we eat fruit to obtain Vitamin C, which we can no longer synthesize but which our pre-primate ancestors could. It should be clearly understood, however, that a virus is a degenerate organism which has lost functions. It is not a ''half-way house'' in the evolution of life from nonliving matter.

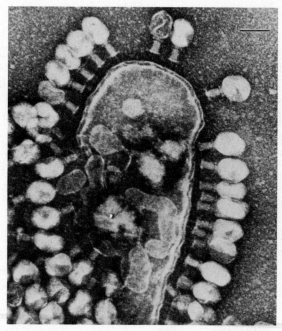

CRYSTALLIZED VIRUS. Tiny rods that constitute this crystal are viruses that cause mosaic disease in tobacco plants.

VIRUSES ATTACKING BACTERIUM. Dozens of bulb-shaped bacteriophage viruses have attached themselves to the cell wall of the sausage-shaped E. coli bacterium.

"alive," but this remains largely a matter of definition (*Box B*).

THE CELL MEMBRANE

This structure, sometimes called the plasma membrane or plasmalemma, regulates the passage of materials into and out of the cell. The cell membrane is not simply a sieve that excludes large molecules and admits small ones. Whether or not a molecule is admitted depends on its size, electric charge, shape, chemical properties, and its relative solubility in water compared with its solubility in fats. Solubility is important because the membrane consists largely of fatlike compounds called phospholipids. Materials soluble in phospholipids are often able to cross the membrane by free diffusion. Water is one of the few exceptions to this rule. Although water is not soluble in fats, it seems to cross the membrane by diffusion, without having to be carried or "pumped." In diffusion the net rate of transport cannot be controlled, because it depends solely on the concentrations of the substance outside and inside the membrane.

To see the cell membrane in any detail, one must examine cells at high magnification and high resolution. The membrane is only 70 to 80 Ångströms thick. The Ångström (abbreviated Å) is a unit generally used to measure the wavelength of light. One Å is 10^{-8} cm. This page is about 70,000 Å thick. The cell membrane is most commonly prepared for microscopy by staining the cell with osmium tetroxide. When thin sections of stained cells are photographed under the electron microscope, the membrane appears as a region of light tone sandwiched between two dark parallel lines (*Figure 3*). The sandwich is called the unit membrane.

Cell membranes are by no means as simple as their appearance under the electron microscope might suggest. Membranes in different cells, and even in different parts of a single cell, are specialized to perform different functions and have demonstrably different properties. These differences between membranes are attributable to the proteins that make up an important part of all biological membranes. A membrane can be visualized as a sandwich composed of two layers of phospholipids. The "wettable" (or hydrophilic) ends of the phospholipid molecules face the aqueous interior and exterior of the cell, and the "greasy" (or hydrophobic) ends lie in the dry interior of the sandwich. The various proteins in the membrane float in the lipid layers and sometimes

The cell membrane is not simply a sieve that excludes large molecules and admits small ones.

may stick out of both surfaces (*see Chapter 4, Figure 4*). Proteins too have hydrophilic and hydrophobic portions, which orient themselves according to the same principles.

TRANSPORT: PASSIVE AND ACTIVE

Compounds that are not fat-soluble must be transported into the cell by specific carrier molecules embedded in the membrane. A carrier can combine with only one compound or at most with a family of closely related compounds. The carrier and compound diffuse across the membrane together; then the compound dissociates from the carrier. This process is called carrier-facilitated diffusion. The carriers are protein molecules, sometimes called permeases because they make the cell permeable to the compound in question. Carrier-facilitated diffusion can work in either direction. If a compound is at higher concentration outside a cell than inside, the net flow will be inward. If the compound is at higher concentration inside, the net flow will be outward. Obviously, carrier-facilitated diffusion cannot be used to create a high concentration of a nutrient inside a cell if the external concentration of that nutrient is low. For this reason, carrier-facilitated diffusion is sometimes called PASSIVE TRANSPORT.

We can distinguish free diffusion from carrier-facilitated diffusion by a careful study of the effect of concentration on the rate of entry of a com-

CELL MEMBRANE photographed by an electron microscope looks like a sandwich viewed edge-on. This illustration depicts the membrane of a red blood cell. The space between the dark lines is about 25 Ångstroms wide.

3

pound into a cell—the number of molecules transported per minute. Imagine that we are starting with cells that contain none of the compound. We may put samples of the cells into solutions containing various concentrations of the compound to be transported. We should not be surprised to find that, in very dilute solutions, the rate of transport will be low, and in more concentrated solutions, the rate of transport will be higher. In the case of free diffusion, the rate of transport of molecules into the cell is directly proportional to the outside concentration, since every molecule striking the outside of the cell membrane is equally likely to get into the cell. In the case of carrier-facilitated diffusion, the molecule to be transported must collide with the carrier. Once contact is made it takes a short but finite time for the carrier to deposit the molecule on the other side of the membrane. During this short time, the carrier cannot react with another molecule. The average time for a carrier to transport a molecule will be the sum of the time it waits for a molecule to collide with the carrier, plus the time of transit across the membrane and back. Therefore, as the external concentration is increased, collisions between molecules and their carriers become more frequent; the rate of transport increases, but only up to a point. When the average time that a carrier is waiting for a passenger becomes small compared to the duration of the trip, increasing the outside concentration further will have almost no effect on the rate of transport. The system will become saturated. The most characteristic difference between these two transport systems is that systems that operate by carrier-facilitated diffusion are saturable, and those that operate by free diffusion are not. Since cells are able to control the number of molecules of each kind of carrier that they build into their membranes, they can control the maximum rate of carrier-facilitated transport of each sort of molecule. However, they cannot control the final internal concentration, because this depends only on the external concentration.

Certain compounds can also be brought into cells by a mechanism known as ACTIVE TRANSPORT, which can move molecules from a region of low concentration to one of high concentration—in the direction opposite from the one dictated by diffusion. This requires the expenditure of energy, which is usually furnished by the hydrolysis of a high-energy compound called ATP *(Chapters 3 and 4)*. Active transport can be thought of as involving a specific carrier molecule coupled to a "pump" driven by the hydrolysis of ATP. Active transport

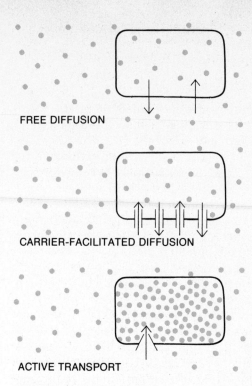

FREE DIFFUSION

CARRIER-FACILITATED DIFFUSION

ACTIVE TRANSPORT

THREE KINDS OF TRANSPORT. In free diffusion of **4** **substances (dots) through the cell membrane, the final concentration of substance inside the cell becomes the same as the concentration outside. The same is true of carrier-facilitated diffusion. In active transport, the final concentration inside the cell will be much higher than concentration outside, because substance is being "pumped" in.**

enables a cell to accumulate certain nutrients in concentrations that are hundreds of times higher than the external concentrations. Conversely, cells can also actively pump out certain substances, such as sodium ions, even though the concentration outside may be higher than that inside. Different cells may transport the same metabolite by different processes. For instance, glucose is actively transported into the body in the intestine, and is actively prevented from being excreted in the kidney, but is transported by carrier-facilitated diffusion in muscle and in liver cells. The mechanisms of the three types of transport are depicted in Figure 4.

The cell membrane creates and maintains a certain constancy of the cell contents in the face of fluctuations in the outside medium. It also creates a degree of biochemical privacy from other cells of the same organism. Frequently in biology we learn a great deal about the normal state from studying aberrations or exceptions. Some organisms, known as coenocytes (pronounced seen-a-sights), consist of a single giant cell that contains many nuclei and other cell components within a single huge cell membrane. The membrane grows without dividing. A fine example of coenocytic organization is presented by the slime mold, *Physarum*. This creature, which resembles Thurber's "blob of glup," oozes about, sometimes covering an area of several square centimeters. Remarkably, the many nuclei in this coenocyte (and other coenocytes) all divide at the same time. The explanation for this is that they are all in the same biochemical environment, all subject to the same signals for nuclear division, and hence are synchronized. When a coenocyte of *Physarum* is shaken vigorously it may break into several pieces. Like soap bubbles, most cell membranes are self-sealing devices that can reform a closed surface after an injury. The several pieces of *Physarum* survive and continue to grow separately. Within each piece, the nuclei divide synchronously, as before. Eventually the different pieces fall out of synchrony with one another. They are not listening to any cosmic metronome, but to an internal one that keeps only fairly good time, and whose biochemical ticking is heard only within the cell membrane.

Organization of the early embryonic stages of

many animals and of the endosperm of plant seeds is coenocytic. For the cells to differentiate (to become different from one another) the coenocyte must first be partitioned into individual cells to provide the biochemical separateness necessary for cells to follow different destinies.

THE CELL WALL

When Hooke examined his slices of cork, he saw that there were no inhabitants in what he called cells. Hooke was actually describing the cell walls laid down by occupants who had long since died. Unlike cell membranes, cell walls are not found in all forms of life. They are characteristic of higher plants, fungi, and bacteria, but are not universally present even in these organisms. Cell walls help to give plants rigidity, a property that is advantageous to a stationary creature, but a handicap to mobile animals. Nevertheless, many bacteria and some cells of green plants manage to propel themselves with hairlike projections that extend through their cell walls.

It is possible to dissolve the cell walls of higher plants, fungi, and bacteria by use of certain enzymes, or to make the cells grow under conditions that inhibit the formation of new cell wall. The resulting cells, which are enclosed only by their cell membranes, are called PROTOPLASTS. Their properties tell us some interesting things about the role of cell walls. First, protoplasts are spherical, though the walled cells from which they originated may be rodlike, cubelike, or more complex in shape. Evidently the cell wall molds the shape of these cells. In the absence of the wall, the protoplast simply assumes the form which has minimum surface area per unit volume (animal cells, which do not normally possess a wall, are not called protoplasts, and their shape is determined by other factors). Second, protoplasts are generally osmotically fragile. This requires a few words of explanation.

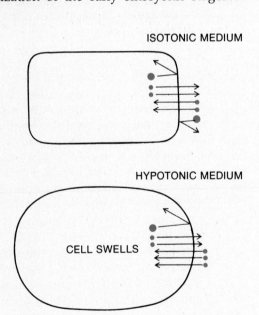

ISOTONIC MEDIUM

HYPOTONIC MEDIUM

CELL SWELLS

HYPERTONIC MEDIUM

CELL SHRINKS

OSMOSIS can cause animal cells to swell or shrink. Cell at top is in osmotic equilibrium with the surrounding medium; concentrations of non-diffusible molecules (large disks) inside and outside the cell are equal. Cell at center swells because it has a higher internal concentration of non-diffusible molecules, causing water molecules (small disks) to diffuse into the cell. Cell at bottom shrinks because higher external concentration of non-diffusible molecules causes water molecules within the cell to diffuse outward.

5

If one fills a membrane, such as a cellophane sausage casing, with the white of a raw egg and ties it closed at the top he will have a crude but useful model for studying osmosis—the movement of water across a cell membrane. Because the egg white contains proteins, salts, and small molecules the concentration of water molecules in egg white is a bit lower than in pure water. If the experimenter now immerses the bag in a tub of pure water, the bag swells, stretches, and may even break. Obviously, extra water has entered the bag. This osmotic flow can be explained on the grounds that each time a water molecule hits the bag it has a certain probability of passing through it. Since the concentration of water inside the bag is less than the concentration outside, more water molecules will pass into the bag than will pass out of it. The ions of salts and small uncharged molecules will also pass through the membrane, but the protein molecules are trapped inside because they are too large to get through the small pores in the cellophane *(Figure 5)*.

Most normal cells of plants, fungi, and bacteria can survive and multiply under hypotonic conditions; that is when the surrounding medium has a lower concentration of dissolved molecules and ions than the inside of the cell. The cells are prevented from bulging and bursting by the strength of the cell wall. The wall limits the tendency of water to flow into the cell by confining the cell to a definite volume. Walled cells can even survive in distilled water. A protoplast, by contrast, swells as the outside medium is made progressively more hypotonic. Ultimately it bursts, killing the cell and releasing its contents into the medium. A protoplast is stable only under isotonic conditions—an osmotic equilibrium in which there is no net flow of water either into, or out of, the cell. This is also true of most animal cells. If any cell is put into a hypertonic medium—one with a higher concentration of solutes (dissolved particles) than the inside of the cell—such as a strong salt or sugar solution, water will diffuse out of the cell. In animal cells, this shrinkage causes a wrinkling of the surface of the cell, called crenation. In plant cells, the mass of protoplasm surrounded by the cell membrane pulls away from the wall, which retains its former dimensions. This process, called plasmolysis, reveals that the cell wall is much more porous than the cell membrane and does not act as a serious barrier to small molecules or ions *(Figure 6)*. In many cases, however, it does impede the diffusion of large molecules, notably proteins.

Walled cells contain enzymes between the wall

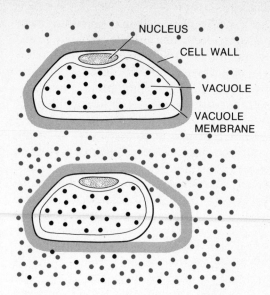

PLANT CELL shows a different response to changes in external osmotic pressure. For simplicity, water molecules are not shown in this drawing. Plant cell consists of a layer of cytoplasm around a central vacuole. In a hypotonic medium (top) cell membrane is pressed firmly against cell wall, which is rigid but permeable. In a hypertonic medium (bottom), cytoplasm shrinks, pulling cell membrane away from the wall. Shrinkage of cell stops when it is in osmotic equilibrium with its environment. 6

and the cell membrane, but not chemically attached to either. Collectively called periplasmic enzymes, they help to transport nutrients into the cell. Some nutrients that cannot cross the highly selective cell membrane may readily diffuse through the cell wall into the periplasmic space. There the enzymes hydrolyze them to smaller, simpler fragments that can be transported across the cell membrane. This process is of crucial importance in the use of sucrose (table sugar) by some yeasts and other fungi. Sucrose readily diffuses through the cell wall but cannot be transported across the cell membrane of yeast. In the periplasmic space it encounters the enzyme invertase which hydrolyzes it to its two constituent sugars, glucose and fructose. Both glucose and fructose are actively transported into the cell. It has been possible to prepare protoplasts of yeast. As might be guessed, these protoplasts make invertase and pour it directly into the growth medium because there is no cell wall to confine it. This points up a third difference between protoplasts and normal cells—protoplasts will be largely or totally lacking in those enzymes that are normally present in the periplasmic space or are part of the cell wall.

Because many kinds of cell never have cell walls, one might suppose that protoplasts maintained in isotonic medium and provided with appropriate nutrients so they do not need periplasmic enzymes could grow and divide indefinitely. Surprisingly, protoplasts of many bacteria can increase in mass under these conditions but cannot divide. Some workers believe that the DNA of the nucleoid is attached at some point to the cell membrane and the cell wall, and that growth of the membrane and wall must separate the genetic machinery into equal portions as a bacterium prepares to divide.

CONTRACTILE VACUOLES

Animal cells lack tough cell walls and cope differently with the problem of osmotic flooding. Many protozoa (one-celled animals) live in fresh water. Water flows in through their cell membranes because the concentration of solutes in the cytoplasm is higher than the concentration of solutes in the water in which they live. They live in a hypotonic environment. This flow would cause a fatal bloat if it were not for the contractile vacuole, an organelle enclosed in a single membrane and serving as the cell's excretory organ (*Figure 7*). The vacuole swells and shrinks in a steady cycle, slowly ballooning as water collects in it, then rapidly contracting as it expels its contents through the cell membrane, then ballooning again. It is not yet known how the

An increase in the number of cells in a plant accounts for only part of the plant's growth; the enlargement of nondividing cells by the intake of water is responsible for much of the increase in size.

cell pumps water into the contractile vacuole, but it is obvious that energy must be expended in a process that moves water from a region where it is less concentrated (inside the cell) into a region where it is more concentrated (outside). It has been suggested that the work done in the cell is not the direct pumping of water across membranes but rather the pumping of solutes to create osmotic gradients between different parts of the cell. This movement of solutes would then lead to the flow of water from one part of the cell to another.

OTHER VACUOLES

In plant cells, as much as 90 per cent of the volume of the cell may be occupied by a large central vacuole (*Figure 13*). Between the vacuole membrane and the plasma membrane lies only a thin layer of cytoplasm pressed against the stiff cell wall. The vacuole seems to serve in part as a cellular sewer into which the cytoplasm flushes certain waste products, such as salts of oxalic acid, that would be toxic if they were kept in the cytoplasmic compartment of the cell. Furthermore, it appears that the vacuole serves as a reservoir for some sugars, amino acids, and even proteins that can be taken up by the cytoplasm as needed. Some of the pigments that give plants their colors are dissolved in the fluid contents of the vacuole. Many of the most interesting plant colors —the reds of Indian war paint, borscht, and chianti, for example —are located in the vacuole.

The high concentration of particles in the central vacuole of plant cells produces a tendency for water to flow into the cell. Since the cell wall does not allow the cell to expand appreciably, a high

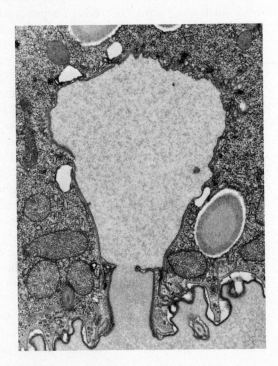

CONTRACTILE VACUOLE serves as excretory organ of one-celled animals, and opens to the outside of the cell (bottom). Another picture of a contractile vacuole appears on the opening page of Chapter 17.

7

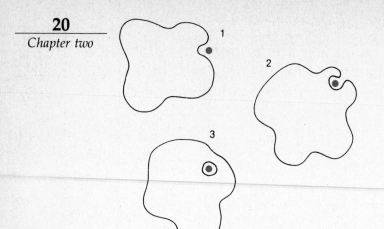

8 ENDOCYTOSIS. This highly schematic diagram depicts a cell engulfing a food particle. Once engulfed, the particle is enclosed within a former segment of the cell membrane.

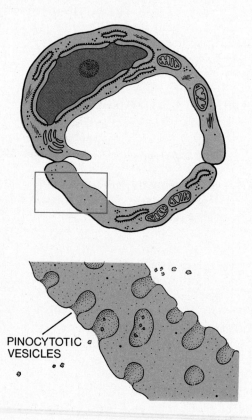

PINOCYTOTIC VESICLES

9 TRANSPORT BY ENDOCYTOSIS. Wall of capillary, shown in cross section in diagram at top, is formed by two cells. Membrane bordering the interior of the capillary takes up fluids by invagination, forming vacuoles. The vacuoles migrate outward and fuse with the outer membrane, releasing their contents to tissues outside the capillary, as shown in the detailed diagram (bottom).

internal pressure, called turgor pressure, results. It is important both in the growth of plants and in maintaining their erect posture. An increase in the number of cells in the plant accounts for only part of the plant's growth; the enlargement of non-dividing cells by the intake of water is responsible for much of the increase in size. Cell division is, for the most part, localized in the tips of stems and roots and in the growing leaves. The dividing cells, containing many small vacuoles, leave behind them as they multiply a zone of nondividing cells in which the vacuoles grow and fuse, producing a tremendous increase in the size of the cells and reducing the cytoplasm to a thin peripheral layer. It is in this zone of elongation that much of the growth of stem and root occurs. Turgor pressure also gives nonwoody plants the stiffness or crispness to hold their stems erect and to spread their leaves out flat under the sun.

Some of the vacuoles in eucaryotic cells originate from a remarkable transport process that moves materials from the outside of the cell to its interior. This process is called PHAGOCYTOSIS (cell eating). Phagocytic cells such as our own white blood cells can ingest solid material into pockets in their cell membranes, then pinch off the pockets, sealing them inside the cell as free-floating vacuoles or vesicles with chunks of solid matter inside. Our white blood cells devour bacteria in this way, thus helping to protect us from bacterial infection (*Figure 8*). Some free-living cells, such as the amoeba, obtain all their food by ingesting bacteria, other one-celled organisms, and bits of organic matter.

Cells also engulf liquid material from their surroundings. This process is called PINOCYTOSIS (cell drinking). It appears, however, that some cells do not simply drink from their fluid surroundings, but rather take in proteins specifically bound to receptor molecules on the cell membrane and, along with these proteins, take a sample of the fluid as well. In higher organisms pinocytosis plays a part in the transport of materials from the circulating blood through the thin cellular walls of capillaries into the surrounding tissues (*Figure 9*). Pinocytosis on the capillary wall facing the interior of the blood vessel leads to the formation of vacuoles of blood plasma which then traverse the cells of the capillary tube and, by reverse pinocytosis, release their contents through the outside wall. This bubble brigade delivers nutrients from the blood to the surrounding tissues and delivers bits of membrane from the inner wall of the capillary cells to the outer wall. It must be supposed, therefore, that the membranes of the surface of the capillary cells are in a constant state of flux to compensate for the one-way transfer of material. Phagocytosis and pinocytosis are such similar processes that they are often lumped together and called ENDOCYTOSIS.

THE ENDOPLASMIC RETICULUM

Eucaryotic cells contain a collection of internal membranes called the endoplasmic reticulum. These membranes are folded in complicated ways to form plump fingerlike projections or flat stacked pockets. The endoplasmic reticulum is not a static structure, but varies from time to time within a cell and shows vastly different appearances from one type of cell to another. The endoplasmic reticulum is capable of continuous reorganization, breaking old connections and forming new ones. The membranes can break up into vesicles and rejoin one another. Under the electron microscope, some of the reticular membranes appear to be heavily peppered with ribosomes *(Figure 10)*. Such membranes are called rough endoplasmic reticulum. Accordingly, membranes that lack ribosomes are called smooth endoplasmic reticulum. It should not be supposed that all ribosomes in eucaryotic cells are bound to the rough endoplasmic reticulum. There are also unattached ribosomes that float free in the cellular fluid. It appears that ribosomes attached to the rough endoplasmic reticulum are involved in the synthesis of proteins for export from the cell. The free ribosomes participate in the synthesis of proteins that will remain inside the cell.

Smooth endoplasmic reticulum rarely takes the form of flattened pockets. It consists mostly of fingerlike or tubular elements. It seems to be the site of synthesis of an important group of compounds called sterols, and of certain other lipids. In liver cells the smooth endoplasmic reticulum is involved in the detoxification of harmful substances. In muscle cells it is given a special name, the sarcoplasmic reticulum, and it helps to regulate muscle contraction. In plant cells the smooth endoplasmic reticulum plays a role in the storage of starch granules, and it has a similar function in animal cells that store glycogen (animal starch).

THE GOLGI COMPLEX

A collection of smooth membranes of characteristic structure near the cell nucleus is called the Golgi complex, after Camillo Golgi, an Italian pathologist who first described the structure in 1898 and who

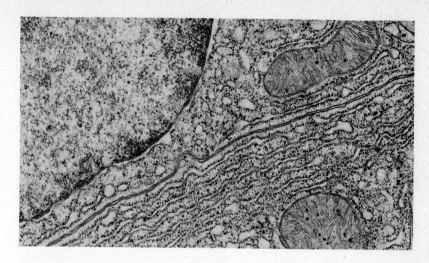

ENDOPLASMIC RETICULUM is clearly visible in this 10 electron micrograph of parts of two cells from the pancreas of a bat. Membrane separating the two cells extends from upper right to bottom left. Parallel and immediately to the right of it lie membranes of the endoplasmic reticulum, richly peppered with ribosomes. Large structure at top left of photo is a cell nucleus. The round structure at the lower right is a mitochondrion.

later won a Nobel prize for his studies on nerve cells. The Golgi complex displays a unique affinity for certain heavy-metal stains and can therefore be seen in stained cells even under the light microscope. For years, however, cytologists debated whether the Golgi complex really existed in living unstained cells or whether it was an artifact, an image produced by some peculiarity of the staining process. A similar question arose when electron microscopists first observed the association between ribosomes and the membranes of the endoplasmic reticulum. Were the ribosomes attached to the membranes in the living cell, or did they become attached during the preparation of the cells for electron microscopy?

Only after the Golgi complex had been seen in living cells with a phase-contrast microscope was its existence accepted. And only after membrane-associated ribosomes had been purified from disrupted cells by physical techniques was the existence of the rough endoplasmic reticulum universally accepted. It is a dilemma of biology that living things must sometimes be treated roughly to study them. An investigator must be skeptical of his findings until he can demonstrate that his results arise from the properties of living things, and not from the nature of the treatment to which he has subjected them *(Box C)*.

C Taking the cell apart

Much can be learned about the function of an organelle by studying the activities of whole cells. But the final proof that an organelle carries out a particular function requires that it be isolated from the cell. Biochemists have devised many ways to break open cells and sort their parts. Needless to say they try to open the cells as gently as possible so that the parts are released undamaged. With animal cells, which have no cell walls, this is relatively easy; with plant cells and bacteria, it is sometimes very difficult.

Sorting the organelles usually involves spinning them in a centrifuge, a device that works more or less like a washing machine spinning in one direction. The centrifuge simply magnifies the effect of gravity. Under enormous centrifugal force, heavy particles settle much more rapidly than lighter ones. If a tube containing a suspension of organelles is centrifuged, the bottom layer of sediment in the tube will contain mostly large particles (such as nuclei) and the top layer will consist almost entirely of small particles (such as ribosomes). Of course the bottom layer will be contaminated with small particles that started at or near the bottom of the tube and had no way of getting to the top. The separation can be improved by resuspending the bottom layer in a fresh solution and repeating the centrifugation.

The diagram shows how the centrifuge can be used to separate particles of two different sizes. By spinning the tube in a centrifuge, an investigator can sediment all of the large particles, leaving most of the small particles in suspension. The suspended particles can then be poured off. By resuspending the mixture and repeating the process, he can obtain a pure sample of the large particles. Using this general approach, and more sophisticated versions of it, it is possible to isolate pure samples of mitochondria, chloroplasts, nuclei, ribosomes, and other organelles.

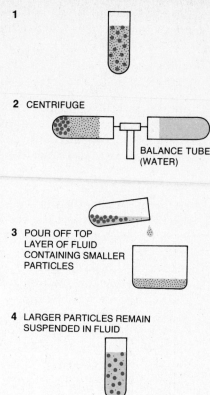

1

2 CENTRIFUGE

BALANCE TUBE (WATER)

3 POUR OFF TOP LAYER OF FLUID CONTAINING SMALLER PARTICLES

4 LARGER PARTICLES REMAIN SUSPENDED IN FLUID

The roles of the endoplasmic reticulum and the Golgi complex in the secretion of cellular products have been thoroughly studied. The rough endoplasmic reticulum is highly developed in secretory cells such as the acinar cells of the pancreas, which produce a variety of digestive enzymes for export through the pancreatic duct to the small intestine. The rough endoplasmic reticulum is also highly developed in the liver cells that produce albumin, the major protein component of blood plasma. After the proteins have been synthesized they are somehow packaged into condensation vacuoles that bud from the rough endoplasmic reticulum. The vacuoles are then transferred to the Golgi complex by a poorly understood process. There the contents of the vacuoles become more and more concentrated until they are ready for export. These membrane-covered packages then migrate to the cell membrane, where they fuse with it and release their contents to the outside by a process that is virtually the reverse of endocytosis. Smooth endoplasmic reticulum is actually continuous with the Golgi complex. Studies with radioactive tracers have indicated that the Golgi complex is also in-

volved in the packaging of products for export from the cell (Box D).

THE NUCLEAR ENVELOPE

The nucleus of the eucaryotic cell is surrounded by an envelope consisting of two unit membranes. The nuclear envelope is perforated by pores. At the walls of these pores, the inner and outer membrane join to form a continuous structure (Figure 11). The pores are not merely openings. In some electron photomicrographs, the pore is covered by a thin plate or by a plug. The way in which nuclei transport ions also shows that the nuclear envelope is not simply a sieve that lets things pass in and out freely. All that is definitely known about the pores is that they exist; their function is obscure. The outer membrane of the nucleus looks very much like the rough endoplasmic reticulum because it contains many ribosomes. Indeed it appears that the rough endoplasmic reticulum is merely an extensive and complicated outpocketing of the outer membrane of the nuclear envelope. The smooth endoplasmic reticulum,

22

however, does not seem to be connected to the nuclear membrane.

MITOCHONDRIA

All eucaryotic cells contain small, double-membrane-bounded organelles called MITOCHONDRIA (singular: mitochondrion). They carry out a variety of functions in different cells, but they always have one function in common. They are the power plants of eucaryotic cells. They perform oxidations that result in the production of ATP (adenosine triphosphate), the common currency of energy conversion in the cell (*Chapter 3*). Mitochondria tend to be most numerous in regions of the cell that consume large amounts of energy. For example, in muscle cells, mitochondria lie close to the contractile elements; in cells that secrete proteins they are closely associated with the rough endoplasmic reticulum; in brown fat cells,

D Radioactive tracers

Chemical pathways in the cell can be traced with the aid of isotopes —atypical atoms that differ from ordinary atoms of the same element only in the weight of their nuclei. Many of the atoms involved in the chemistry of the cell exist in nature in two or more forms, one of which is a radioactive isotope. Hydrogen, for example, exists in three forms. The most abundant isotope of hydrogen has a nucleus consisting of a single proton. About one of every 7000 hydrogen atoms has a nucleus containing both a proton and a neutron. This isotope is deuterium, commonly called heavy hydrogen. One of every 10^{17} hydrogen nuclei contains a proton and two neutrons. This isotope, known as tritium, is radioactive.

The chemical behavior of radioisotopes is almost identical to that of the nonradioactive forms. Cells metabolize radioisotopes identically with the nonradioactive compounds. Amino acids can be labeled by incorporating tritium into them. When cells are provided with labeled amino acids, they will use them to synthesize radioactive proteins. One way to find out where new proteins are being synthesized in the cells is to let the newly-synthesized proteins take their own picture (autoradiography). If a thin slice of tissue containing radioactive proteins is coated with a photographic emulsion, the radiation will expose the emulsion in those areas where the radioisotope is concentrated. The emulsion is then developed just like an ordinary photograph and viewed under a microscope, revealing the location of the radioactive protein within the cell.

One way this technique has been used is to study the sequence of cellular events in the synthesis of proteins for export from the cell. The pancreas is a favorable tissue for such an experiment because it makes certain export proteins, particularly digestive enzymes, much more rapidly than it makes proteins that will remain in the cell. The investigator injects radioactive amino acids into several animals. At sequential times after injection, he kills an animal and removes its pancreas. The pancreas is subjected to a series of treatments that stop protein synthesis and remove amino acids that have not been incorporated into protein. The tissues are stained with appropriate dyes, cut into thin sections, and placed in contact with film.

In studies of this sort, the first part of the cell to become radioactive is the rough endoplasmic reticulum. Here presumably the amino acids are first built into proteins. In animals that are killed a few minutes later, the labeled compounds appear both in the rough endoplasmic reticulum and in the Golgi complex. Later still, the radioisotopes appear in concentrated granules in the Golgi region and in the region of the cell from which the enzymes are released into a duct that empties into the intestine. Apparently the rough endoplasmic reticulum synthesizes digestive enzymes and the Golgi complex packages and concentrates them, then transports them to the cell surface. The actual release of the enzymes is triggered by a hormone, but the nature of this chemical signal is still not completely understood.

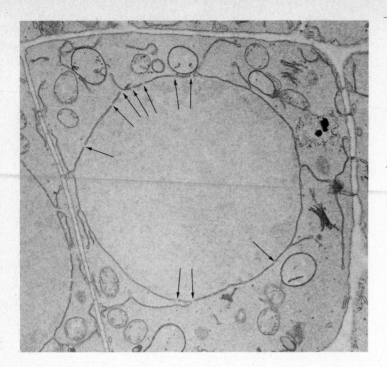

11 NUCLEAR ENVELOPE forms a large circle in the electron micrograph of the cell shown above. Double layers of membrane show up clearly, as do pores in the membrane (arrows). It can also be seen that the outer membrane of the nuclear envelope forms long folded projections into the cytoplasm. These are continuous with the rough endoplasmic reticulum.

12 MITOCHONDRION has an outer and inner membrane. Inner one is folded to form partitions called cristae. Between the cristae lies the matrix, which is in a semifluid state.

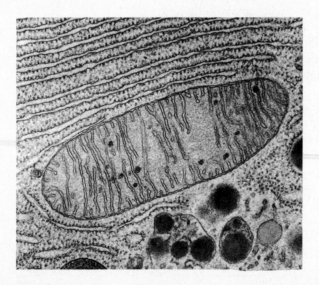

> Some workers believe that mitochondria were once free-living organisms which, at an early stage of evolution, came to live symbiotically within larger cells.

which provide heat for a hibernating animal, they are abundantly distributed throughout the cytoplasm.

The outer of the two membranes of a mitochondrion forms a smooth sack-like covering that does not connect with any of the other membranes of the eucaryotic cell *(Figure 12)*. The inner membrane, quite different in structure and chemical composition, is elaborately infolded to form shelf-like pockets called cristae (singular: crista). Because of the folding, the inner membrane has a much larger area than the outer one. The inner membrane seems to contain the enzymes responsible for many of the most important synthetic activities of the mitochondrion, and it encloses a region called the matrix. The matrix contains a small amount of DNA and the biochemical machinery for duplicating it. It also contains ribosomes very similar to those of procaryotes, which are distinctly smaller than the ribosomes found elsewhere in the eucaryotic cell. Biologists have speculated that mitochondria and chloroplasts were once free-living organisms. In the course of evolution, according to this line of argument, some primitive eucaryotic cells engulfed bacteria and bluegreen algae and eventually formed a symbiotic relationship with them, a relationship that eventually became hereditary. There is a growing body of evidence to support this hypothesis *(Box E)*.

CHLOROPLASTS

The chloroplasts of green plants are the site of photosynthesis —the conversion of carbon dioxide and water to sugar. Chloroplasts also carry out a process called photophosphorylation, which uses the energy of light to synthesize ATP. The mechanism is quite similar to the oxidative phosphorylation that produces ATP in mitochondria. Chloroplasts are larger than mitochondria, but they are basically similar in structure *(Figure 13)*. A chloroplast is bounded by a double membrane, and the smooth outer membrane does not connect

3 CHLOROPLASTS are large dark ovals in this electron micrograph of a plant cell. The light objects within chloroplasts are starch grains. Large vacuole at center is typical of plant cells.

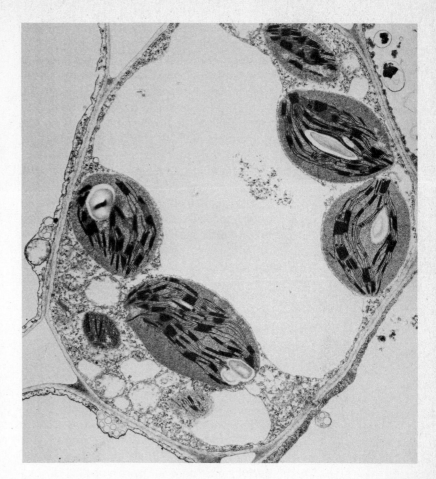

with other membranes of the cell. The inner membrane is even more highly infolded than the inner membrane of the mitochondrion, and the structure of the chloroplast is therefore more complicated. Like mitochondria, chloroplasts contain DNA and the apparatus for replicating it. Chloroplasts also contain ribosomes that are very much like those of mitochondria and of procaryotes.

LYSOSOMES AND MICROBODIES

The cells of plants and animals also contain vesicles called LYSOSOMES that appear to be the functional equivalent of the cyanide capsule that every good spy is supposed to carry. Lysosomes contain enzymes known collectively as acid hydrolases, which under slightly acid conditions can quickly dissolve all of the major molecules that constitute the cell. Like the zymogen granules formed in the pancreas, these enzyme packages are produced in the rough endoplasmic reticulum. Unlike the zymogen granules, lysosomes are kept inside the

E Mitochondria, chloroplasts and evolution

The striking resemblance of mitochondria and chloroplasts to procaryotic cells has encouraged biologists to speculate that these organelles originated by the infection of some primitive eucaryotic cell by procaryotes. In the case of the chloroplast, the inner membrane resembles the cell membrane of a blue-green alga in that it is the site of photosynthesis and photophosphorylation. The inner membrane of the mitochondrion resembles the cell membrane of many bacteria, in that the cell membrane of bacteria is also the site of oxidative phosphorylation. There are many other interesting parallels. For example, many antibiotics inhibit the growth of procaryotes by interfering with their protein synthesis. Almost always, these same antibiotics also interfere with protein synthesis by mitochondria and chloroplasts, but do not inhibit protein synthesis in other parts of the eucaryotic cell, such as in the ribosomes of the rough endoplasmic reticulum.

What about the outer membrane of the mitochondria and chloroplasts? The simplest hypothesis to explain how procaryotes got into a primitive eucaryotic cell in the first place is that they were devoured by the process of endocytosis. Since endocytosis involves surrounding the particle to be ingested with a piece of the cell membrane, it is a reasonable guess that the outer membrane of mitochondria and chloroplasts originated from the cell membrane of the primitive eucaryote. This has given rise to further speculation that the essential innovation that made eucaryote life possible was the invention of endocytosis.

cell. One obvious function of these granules is to accomplish the self-destruction of injured cells or cells that have outlived their usefulness, such as the cells of the tadpole's tail when the tadpole undergoes its metamorphosis into a tailless adult. If the cell is injured or exposed to certain chemicals, including a variety of hormones and dyes, the lysosomal membrane will rupture and empty its lethal contents into the cell.

The lysosome is also involved in the digestion of materials taken into the cell in food vacuoles. When a phagocytic cell engulfs a bacterium, for example, the food vacuole then fuses with a lysosome and the enzymes of the lysosome digest the bacterium into small molecules that can be used as nutrients. Undigested fragments of the bacterium may be kept inside the vacuole for a while or dumped outside the cell.

Lysosomes also appear to play a part in the breakdown of normal cellular products. In humans there are a number of genetic abnormalities called storage diseases that seem to result from an inherited inability to make certain lysosomal enzymes. Deficiencies of these enzymes cause cell products such as mucopolysaccharides or glycogen to accumulate in quantities that become injurious to the cell. It is interesting that enzymes that occur in the lysosomes of eucaryotic cells are very often located in the periplasmic space in procaryotes.

The cytoplasm also contains a number of enzyme-containing vesicles called MICROBODIES. It is not possible to generalize about their functions. Most, but not all, microbodies contain catalase, the enzyme that catalyzes the breakdown of hydrogen peroxide into oxygen and water. The other enzymes packaged in microbodies vary from species to species, although the containers look very much alike. Packaging enzymes with closely-related functions into membrane-bounded compartments is efficient because the product of one enzymatically-catalyzed reaction often serves as the starting material for the next reaction in a chain; a compact bundle of related enzymes can increase the reaction rates because it keeps the local concentrations of reactants high.

For the cell, lysosomes appear to be the functional equivalent of the cyanide capsule that a good spy is supposed to carry.

FIBERS AND TUBULES

Eucaryotes contain a variety of fibrous or cablelike materials that contribute to the shape and rigidity of cells and tissues, or that move parts of the cell in relation to one another. Some of these materials lie outside of cells and others inside. These stringy parts appear, in general, to be assembled from small protein subunits.

Some of the fibrous proteins are simply extracellular structural materials, the rigging that holds the tissue together. Collagen, a tough fibrous substance found between the cells of skin and tendon, is a structural material that strengthens the skin and ties muscle to bone. Elastin is a stretchy fibrous protein that gives bounce to our connective tissues and, in the ligaments, connects the bones to one another without locking them tightly together. Keratin comprises the outermost layer of the skin, and also occurs in hair and horn. Silk is a building material to the caterpillars who use it to spin cocoons and the spiders who weave it into webs.

A number of other fibrous proteins, especially those involved in movement, are found inside cells. The contractile elements of muscle cells are composed of fibrous filaments. In skeletal muscle thick filaments of myosin and thin filaments of actin are arranged in a regular pattern. When the muscle contracts, they slide past one another and overlap (*Chapter 20*). The shortening of muscles is not the result of the shortening of the filaments of actin and myosin, but only the result of the rearrangement of the filaments in space.

Many other kinds of cell contain bundles of networks of threadlike elements called microfilaments, which in some ways resemble actin. Microfilaments facilitate a variety of movements in cells. For example, when a spherical animal cell begins to divide, a furrow forms in the cell membrane and the cell gradually becomes dumbbell shaped; then it separates into two cells. The furrow is formed by the contraction of a ring of microfilaments. The process called cell streaming depends on networks of microfilaments attached to the inside of the cell membrane. By alternately contracting and relaxing, they apparently move the cell membrane back and forth, stirring the contents of the cell. Some cylindrical cells have a ring of microfilaments at one end. When they contract the cell becomes conical. When the "purse-string" microfilaments of the cells in a flat sheet of tissue contract simultaneously, the sheet becomes a hollow bulb. Such cell movements are a common feature of em-

F Cell streaming

Once a molecule has been brought into a cell, it must somehow get from the inner surface of the cell membrane to the site where it will be used. In very small cells such as procaryotes this is no problem, because diffusion can carry a molecule rapidly to any part of the cell. In the larger eucaryotic cell, simple diffusion cannot move nutrients and wastes quickly enough over relatively great distances. It is instructive to compare transport by simple diffusion in a small cell, 1 unit in radius, with that in a large cell, 100 units in radius. The time required for a molecule just inside the cell membrane to diffuse to the center of the large cell will not be 100-fold longer than in the small cell, but 10,000-fold longer. In other words, the time required to achieve a given concentration by diffusion is proportional to the square of the distance that the substance must diffuse. This is not because molecules move more slowly in a big cell than in a little cell, but because they have no sense of direction. Each molecule is like a drunk diffusing away from a lamppost. If he staggers along at a uniform rate of one meter per second, he will be one meter away at the end of one second, but will probably not be 100 meters away at the end of 100 seconds, because he will not stagger in a straight line. If the transport of materials in eucaryotic cells depended completely on diffusion, the cells could not be as large as they are. By keeping the contents of the cell well stirred, protoplasmic streaming solves the problem of intracellular transport.

---

bryological development *(Box F)*.

In addition to microfilaments, eucaryotic cells contain many microtubules. These are not some sort of conduit, as their name implies, because they do not seem to be truly hollow. Microtubules apparently help to maintain cell shape. They are also involved somehow in moving chromosomes during cell division. Like the other fibrous structures in the cell, microtubules are aggregates of small protein subunits.

Cilia and flagella, the hairlike "oars" that propel eucaryotic cells through the water or create currents at the surface of stationary cells, are built around a core of microtubules arranged in a characteristic pattern *(Chapter 20)*. The same pattern is found in centrioles, mysterious organelles that play a key role in cell division.

THE NUCLEUS

Near the center of the cell lies a fibrous tangle of chromosomes, which contain most of the cell's genetic information. In procaryotic cells this region is called the NUCLEOID, and in eucaryotic cells, the NUCLEUS. Strictly speaking, the word chromosome is used mainly to refer to the genetic apparatus of eucaryotes; the corresponding structure in procaryotic cells is usually called a GENOPHORE. The structure of a chromosome is much more complex than the structure of a genophore, but it is not clear that the differences in structure reflect fundamental differences in function and organization.

The genophores of procaryotes are made simply of deoxyribonucleic acid (DNA), a threadlike molecule about 20 Å thick and enormously long. In the bacterium *Escherichia coli*, an inhabitant of the human intestine, the genetic apparatus is a single molecule of DNA about 10 million Å long —the molecule is 500,000 times longer than it is thick. The bacterium itself is about 1 micrometer thick and about 4 micrometers long. Thus, the space into which the long thread of DNA is packed in the bacterial nucleoid is very small relative to the length of the DNA molecule. It is not surprising that the molecule appears in electron photomicrographs as a hopeless tangle of fibers.

When bacterial cells are gently lysed (broken open) to release their contents, the bacterial genophore sometimes comes untangled and spreads out to its full length. Autoradiographs of these molecules have been made by feeding bacteria radioactive thymidine (thymidine is a building block of DNA but of no other cellular molecule), and then collecting radioactively labeled genophores on the surface of a thin artificial membrane. This membrane is then coated with photographic emulsion and stored in the dark while an image is formed as described in Box D.

Many DNA molecules in viruses, procaryotes, mitochondria, and chloroplasts show up as closed rings *(Figure 14)*. It is reasonable to suppose that the ring structure serves some functional purpose, but so far that purpose is a mystery. It might be simply that the ring protects the DNA from certain enzymes that chop up DNA molecules, starting with the free ends. It is also possible that the ring shape is necessary to the duplication of these molecules or to the proper separation of the new DNA molecules once they have been duplicated.

When cells divide, each of the daughter cells needs genetic information to govern its biochemical processes. This requires a mechanism for copying the genetic information and distributing replicas to each daughter cell during cell division. In procaryotes, the separating is relatively simple *(Figure 15)*. The two separating genophores are somehow attached to the new cell wall that begins to form between the dividing halves of the cell. The cell wall grows outward in both directions from the point of attachment, moving the two genophores apart. By the time the cell divides the two genophores have been completely separated.

In eucaryotes cell division is much more complicated. Eucaryotes carry a great deal more genetic information than procaryotes, and they do not carry it in a single piece. The amount of DNA found in the cells of living things is roughly proportional to the complexity of the organisms. The simple virus designated ϕ X 174 contains 2.6×10^{-18} grams of DNA; *Escherichia coli,* a much larger and more complex organism, contains 4.1×10^{-15} grams of DNA; one human cell contains 6.4×10^{-12} grams of DNA. Thus, the ratios of the DNA contents of the three organisms are about 1:1580:2,460,000. In this sense we are over 2 million times more complex than a virus and over a thousand times more complex than a bacterium. The total length of DNA in a cell of *E. coli* is one mm; the total length of DNA in one of our cells is 1.5 meters —about five feet. In order to keep all of that string of information in order within the tiny dimensions of the nucleus, a filing system more

Eucaryotes carry a great deal more genetic information than procaryotes, and they do not carry it in a single piece.

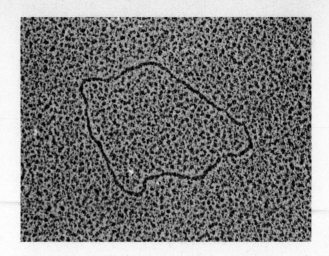

RING-SHAPED DNA appears in many viruses, bluegreen algae and bacteria as well as in mitochondria and chloroplasts. This electron micrograph depicts DNA from polyoma virus.

CELL DIVISION IN PROCARYOTES involves the formation of a new wall between the dividing cells. Genophores are attached to the new wall (2), and as it grows outward in both directions (3), it pulls the points of attachment apart. When cells divide (4), genophores are completely separated.

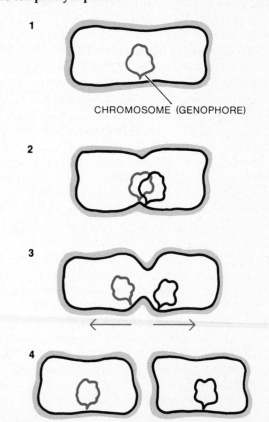

CHROMOSOME (GENOPHORE)

complex than that found in the procaryote had to evolve.

Within the double envelope of the eucaryote nucleus the DNA is organized into several separate, complex structures: the CHROMOSOMES. Chromosome means "colored body," a term that refers to their tendency to absorb certain dyes. In

The total length of DNA in a cell of E. coli is one millimeter; the total length of the DNA in each of our cells is 1.5 meters —about five feet.

The limitations of microscopes

The layman tends to assume that anything in the cell can be seen through the microscope. He might suppose that, if the image in the microscope were not big enough, he could magnify it further by viewing it through a second microscope, and so on. This doesn't work because one needs not only magnification, but also resolution: the power to distinguish detail. Resolving power is limited by the wavelength of the light illuminating the subject. The limit of resolution of a microscope —the smallest detail that it can form into a discrete image —is a dot about half the wavelength of the illuminating light. The shortest wavelength of light that the eye can detect is about 4000 Å and lies toward the violet end of the spectrum. Thus, the smallest object that we can resolve using an ordinary light microscope is about 2000 Å in diameter. This means that plasma membranes (about 75 Å thick), ribosomes (about 200 Å in diameter), microtubles (about 250 Å in diameter), DNA molecules (about 20 Å in diameter), and many of the other cellular structures that we have discussed in this chapter are too small to see or photograph with a light microscope.

The electron microscope illuminates a specimen with a beam of electrons whose wavelength permits a resolution of about 4 to 5 Å. Theoretically the instrument can resolve even a single atom, but the magnetic lenses that focus the electron beam are not perfect, and the "color" of the electron beam (the energy of the electrons in the beam) is not uniform. This means that the finest details visible in the best electron photomicrographs are about 2 to 4 times the size of individual atoms.

A microscope must also distinguish the subject from the background —the image must have contrast. This is a particularly troublesome requirement because most cells and their components are relatively transparent. The observer faces the problem of the cartoon cat trying to catch a mouse who rubs himself with "vanishing cream." The problem for the cat is not resolving the image —the mouse is quite large enough for that. The problem is contrast. The cat throws ink or flour at the mouse in an effort to improve the contrast. For the same reason, the cytologist stains cells with dyes. This almost always kills the cell. The phase contrast microscope uses a clever optical trick to produce contrasty images of unstained living cells, but its resolving power is no greater than that of other light microscopes.

Examination of cells under the electron microscope requires more drastic chemical treatment. First the cells are usually fixed by an embalming process intended to keep them looking in death as they looked in life. Then the cells are stained, usually with a heavy metal, to produce contrast. Finally they must be sliced into thin sections because out-of-focus layers of thick slices complicate the picture. Such harsh techniques for preparing specimens frequently provoke questions about whether a particular structure in a picture is characteristic of a living cell, or whether it is merely an artifact —an accidental byproduct of the preparation process. The problem of artifacts has haunted microscopists since Leeuwenhoek's time, and as instruments become more sophisticated and preparation techniques more rigorous the problem shows no sign of going away.

addition to DNA, chromosomes also contain proteins and some ribonucleic acid (RNA). The proteins of chromosomes are of a characteristic type, called HISTONES, which at cellular pH carry a positive charge. These are attracted to the negatively charged nucleic acids and form a coating of protein over the long strands of DNA. There are also some nonhistone proteins that appear to be involved in turning genes on and off *(Chapter 11)*.

When chromosomes are spread out and examined under the electron microscope, they present a picture that is disappointingly complicated. Depending on how they have been prepared for microscopy, the chromosomes appear as a twisted, knotted webwork of fibers 250 Å thick. The core of each 250 Å fiber is a thread of DNA 20 Å in diameter, covered by a histone coat.

The DNA is apparently coiled like a spring and then further twisted and folded back upon itself to form the 250 Å fiber. This efficient packaging produces a bundle of DNA compact enough to fit into the nucleus, but the complicated folding has confounded all efforts to understand its structure. Does the chromosome contain one long crumpled and folded molecule of DNA? Or does it consist of several short DNA molecules glued end to end or laid side by side and twisted along with their protein covering into a long, multistranded rope? So

far, microscopic studies have not produced an answer, chiefly because of the formidable difficulty of resolving such a tiny structure *(Box G)*.

MITOSIS

Chromosomes undergo a characteristic cycle of condensation and dispersion prior to cell division. In the so-called interphase nucleus (between cell division) the chromosomes are diffuse and fibrous. If cells in this state are treated with stains that specifically color DNA, the interphase nucleus appears in the microscope to contain a collection of exceedingly small grains of stained material. When the chromosomes are in this extended form, access to their genetic content is easier than when they are more compact, and it is in the interphase that the chromosomes are most actively engaged in the business of controlling the biochemistry of the cell *(Chapters 5 and 6)*. During interphase, a cell about to divide makes an additional copy of its DNA and doubles the quantity of everything else. At the end of interphase, it is really two cells inside a single membrane. The process of cell division consists of partitioning the materials into two daughter cells. Organelles such as ribosomes and mitochondria need be distributed only approximately equally between the daughter cells, but the chromosomes must be distributed precisely. The orderly process by which this is accomplished is called MITOSIS.

Mitosis is arbitrarily divided into four stages: PROPHASE, METAPHASE, ANAPHASE, and TELOPHASE *(Figures 16 and 17)*. During prophase the appearance of the chromosomes begins to change. They become visible first as thin fibers. The fibers gradually thicken, and the tangle of chromatin begins to resolve itself into a number of separate, sausage-

16 **MITOSIS in an animal cell, as shown here, is typical of the process in the cells of virtually all eucaryotes. Interphase is the term used to designate growth stages between mitotic divisions. During interphase chromosomes are not readily visible. Mitosis is divided into four stages; prophase, metaphase, anaphase and telophase. Labels on diagram indicate events characteristic of each stage.**

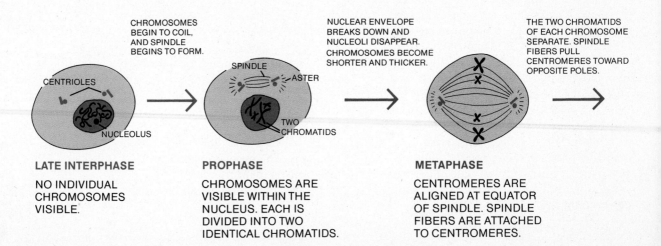

CHROMOSOMES BEGIN TO COIL, AND SPINDLE BEGINS TO FORM.

NUCLEAR ENVELOPE BREAKS DOWN AND NUCLEOLI DISAPPEAR. CHROMOSOMES BECOME SHORTER AND THICKER.

THE TWO CHROMATIDS OF EACH CHROMOSOME SEPARATE. SPINDLE FIBERS PULL CENTROMERES TOWARD OPPOSITE POLES.

CENTRIOLES

SPINDLE · ASTER

NUCLEOLUS

TWO CHROMATIDS

LATE INTERPHASE

NO INDIVIDUAL CHROMOSOMES VISIBLE.

PROPHASE

CHROMOSOMES ARE VISIBLE WITHIN THE NUCLEUS. EACH IS DIVIDED INTO TWO IDENTICAL CHROMATIDS.

METAPHASE

CENTROMERES ARE ALIGNED AT EQUATOR OF SPINDLE. SPINDLE FIBERS ARE ATTACHED TO CENTROMERES.

shaped chromosomes, each composed of two parts called CHROMATIDS. Each is a complete copy of the original chromosome, and the two are linked together at a spot called the CENTROMERE. After jostling about in the nucleus, the chromosomes eventually begin to align themselves so that all the centromeres lie in a single plane at the center of the cell, the METAPHASE PLANE (or metaphase plate). At the end of prophase, the nuclear envelope breaks up and disappears.

The spindle apparatus, which begins to form during prophase, is an array of microtubules that stretches between the poles of the mitotic apparatus. Some tubules run straight across the cell from pole to pole; others connect the poles to the centromeres of the chromosomes. It is thought that the spindle microtubules pull the chromosomes toward the metaphase plane, and later toward the poles. The poles mark the future locations of the nuclei of the two daughter cells.

In animal cells the centriole seems to play some part in the formation of the spindle. For many years, cytologists believed that centrioles were self-replicating structures, but it now appears that new centrioles arise through the assembly of subunit proteins. The new centrioles are usually assembled immediately adjacent to old ones. Centrioles normally occur in pairs. During interphase a pair of centrioles lies near the nuclear envelope. During prophase they double and separate; one pair migrates toward each pole of the cell, with the newly-forming spindle becoming visible behind them. The cells of higher plants, however, do not form centrioles, nor do they have cilia or flagella. The ability to make centrioles, cilia and flagella seems to be correlated.

At metaphase the nuclear envelope has dis-persed, the nucleoli have disappeared, the spindle is complete, and the chromosomes lie in the metaphase plane, their centromeres connected to the poles by spindle fibers. This stage is the easiest time to count the chromosomes and to determine their shapes. Each organism contains a characteristic number of chromosomes, ranging from two to about 600. In the fruit fly, *Drosophila melanogaster*, the number is 8; in humans, it is 46. The number, however, is not a measure of the degree of complexity of an organism. Sugar cane has 80 chromosomes, for example, and the dog has 78. It is frequently possible to identify certain metaphase chromosomes by their sizes and the positions of their centromeres. For example, the 46 human chromosomes can be divided into 7 classes on the basis of their sizes and the location of their centromeres, making it possible sometimes to associate certain genetic anomalies with visible abnormalities in specific chromosomes.

At anaphase, the centromeres seem to split into two, and the chromatids in each pair separate from one another and move quickly apart, one toward each pole, as if drawn by their associated spindle fibers. At telophase, the newly separated chromosomes reach their poles and begin to revert to their previous fibrous appearance. The spindle disappears, except for some microtubules which remain between the separating chromosomes, and the nuclear envelope reforms around the chromosome bundle at each pole. At the center of the cell, along the metaphase plane, the process of cell separation begins. In animal cells, the cell surface begins to constrict, as if a belt were being tightened around it, pinching the old cell into two new ones. Microfilaments play a major role in this process. In plant cells, where a stiff cell wall interferes with this sort

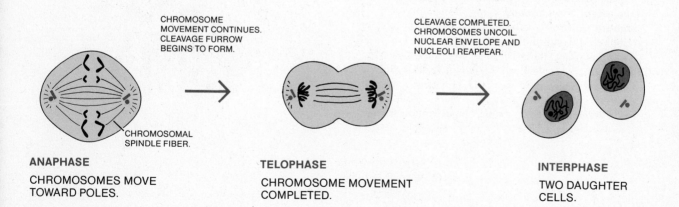

CHROMOSOME MOVEMENT CONTINUES. CLEAVAGE FURROW BEGINS TO FORM.

CLEAVAGE COMPLETED. CHROMOSOMES UNCOIL. NUCLEAR ENVELOPE AND NUCLEOLI REAPPEAR.

CHROMOSOMAL SPINDLE FIBER.

ANAPHASE
CHROMOSOMES MOVE TOWARD POLES.

TELOPHASE
CHROMOSOME MOVEMENT COMPLETED.

INTERPHASE
TWO DAUGHTER CELLS.

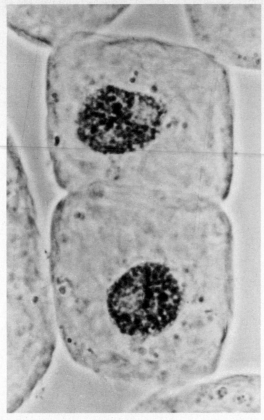

A

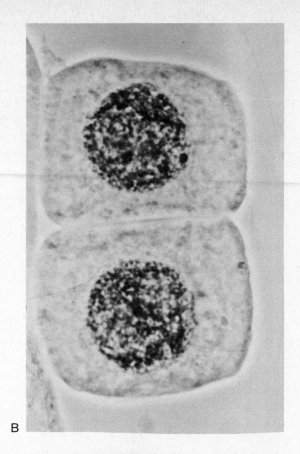

B

17 MITOSIS in the root tip of a bean plant. (A) Two cells early in interphase. (B) Later in interphase, the nuclei are larger; all chromosomes have duplicated in preparation for cell division. (C) Early in prophase, chromosomes begin to condense into distinct fibrous structures. (D) Cell at top is in mid-prophase; cell at bottom, in which condensation has progressed further, is in pro-metaphase. (E) At metaphase centromeres are aligned in the equatorial plane of the cell; at this stage each chromosome consists of two identical sister chromatids. (F) During anaphase the sister chromatids in each metaphase chromosome are pulled toward opposite poles of the mitotic spindle. The two sets of chromosomes being drawn apart are identical; each is a complete diploid set. (G) During telophase the chromosomes reach the division poles. After uncoiling, they will enter the interphase state depicted in (A). New cell walls will form in the equatorial planes of the two cells shown here, separating each into two daughter cells. The daughter cells can then begin the mitotic cycle anew. Magnification varies somewhat from frame to frame, but is about 1500X.

E

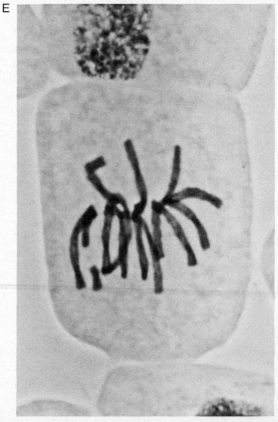

of pinching, new cell membranes form between the two daughter cells along the plane previously defined by the metaphase plate, and a new cell wall is deposited between the two new cell membranes. Biologists think of cell division as two dis-

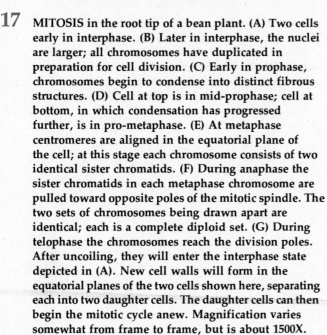

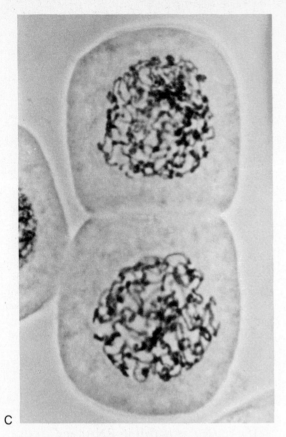

C

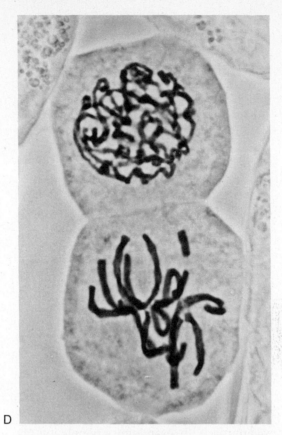

D

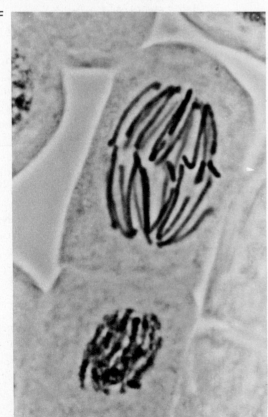

F

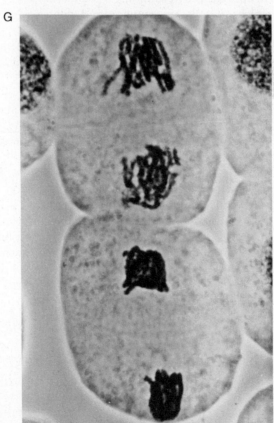

G

tinct processes: the separation of identical copies of the chromosomal content of the parent cell—mitosis or CARYOKINESIS; and the physical separation of the cells —CYTOKINESIS.

The chromosomes duplicate themselves during interphase. The time when the cells are actively engaged in DNA synthesis should correspond to the time when they are making copies of their genetic complement of chromosomes. Using mitosis and DNA synthesis as milestones, one can divide the division cycle of cells into four stages *(Figure 18)*. Mitosis (M) is followed by a gap (G1) during which the new daughter cells make do with the DNA that they inherited from the parent cell. It should not be supposed that the cells are doing nothing during this time —they are simply not making DNA. Next, there is a period of DNA synthesis (S), during which the amount of DNA in the cell must double in preparation for the next cellular division. Between S and mitosis, there is a second gap (G2), during which the cell is preparing itself for mitosis. Measurement of these stages is easy, at least in principle.

One way to measure the amount of time that a cell spends on mitosis is simply to watch a cell with one eye on a clock. Another way is to use a statistical approach. By examining a population of cells in which all cells are growing and dividing and in which division is random in time (so that the cells are not dividing synchronously), one can simply measure the fraction of the cells that are engaged in mitosis.

Experiments have shown that the length of the cell cycle and the relative lengths of the four stages vary from tissue to tissue and from organism to organism. In some tissues —nerve, for example — there is hardly any cell division. In growing tissues, the relative lengths of the various periods and the total length of the division cycle vary, but in a "typical" population of dividing cells, the M phase comprises about 5–10 percent of division cycle; the G1 phase, 30–40 percent; the S phase, 30–50 percent; and the G2 phase, 10–20 percent. Interphase comprises about 90 percent of the division cycle of the cell, and about half of interphase is spent on chromosome duplication.

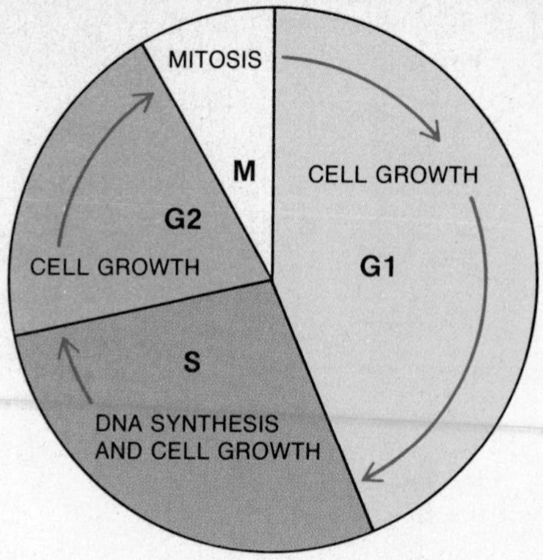

THE NUCLEOLUS

Eucaryotic nuclei contain one or more spherical bodies called NUCLEOLI. They are not surrounded by a membrane. During prophase the nucleolus disappears. When daughter nuclei begin to form at telophase, new nucleoli emerge near specific sites

CELL CYCLE is divided into four stages. The circle graph indicates relative length of each stage. During G1 the cell grows but does not synthesize DNA. During S the amount of DNA in the cell is doubled. G2 is a period of further growth that leads to mitosis, M, and formation of daughter cells.

18

on the unfolding chromosomes. These sites are called nucleolar organizing regions.

The nucleolus is rich in RNA, and electron microscopists have found that its periphery contains granular material similar in size and staining properties to ribosomes. It has been recently demonstrated that the nucleolus contains RNA identical to that extracted from ribosomes. In addition, it contains DNA that seems to represent multiple copies of the chromosomal gene that directs the synthesis of ribosomal RNA. It seems reasonable to suppose that the nucleolus is the site of ribosome synthesis.

There is a limit to what one can learn about cells by looking at them. Many of the most important activities of cells occur at the chemical level —a level that cannot be observed even with the electron microscope. The next four chapters discuss the structure and function of the essential molecules of life.

READINGS

D.W. FAWCETT, *The Cell, Its Organelles and Inclusions: An Atlas of Fine Structure*, Philadelphia, W.B. Saunders Company, 1966. A collection of electron micrographs of many types of cells, with commentary on the activities of the structures shown. A strikingly beautiful book by one of the masters of the art of electron microscopy.

T.L. Lᴇɴᴛᴢ, *Cell Fine Structure: An Atlas of Drawings of Whole-Cell Structure*, Philadelphia, W.B. Saunders Company, 1971. A marvelous collection of ink drawings of cells as they appear in the electron microscope. It is possible in a drawing to do something that is impossible in a photograph; namely, to include in one picture all of the structures found in a given type of cell, even those structures that are hardest to find or that only appear when special fixation or staining techniques are used. One of Dr. Lentz's pictures is the equivalent of a whole chapter of electron micrographs.

A.G. Lᴏᴇᴡʏ ᴀɴᴅ P. Sɪᴇᴋᴇᴠɪᴛᴢ, *Cell Structure and Function*, 2nd Edition, New York, Holt, Rinehart, and Winston, 1969. The emphasis in this fine book is on function. The level is fairly advanced, but the writing is clear and concise.

A.B. Nᴏᴠɪᴋᴏғғ ᴀɴᴅ E. Hᴏʟᴛᴢᴍᴀɴ, *Cells and Organelles*, 2nd Edition, New York, Holt, Rinehart, and Winston, 1976. The emphasis in this book is on the structure of cells, how the structure is studied, and, finally, how the functions of various structures are determined. Much information, economically presented.

L. Oʀᴄɪ ᴀɴᴅ A. Pᴇʀʀᴇʟᴇᴛ, *Freeze-Etch Histology: A Comparison between Thin Sections and Freeze-Etch Replicas*, New York, Springer-Verlag, 1975. A beautiful introduction to a relatively new technique for showing the detailed structure of membranous surfaces. In this book, freeze-etch pictures are paired with "old-fashioned" thin-section electron micrographs, enabling the reader to appreciate and understand these new images of cells.

C.P. Sᴡᴀɴsᴏɴ, *The Cell*, 3rd Edition, Englewood Cliffs, NJ, Prentice-Hall, 1969. A short, well-written introduction to cell structure and function. An excellent starting point for students wishing to embark on the study of cytology.

In my hunt for the secret of life, I started my research in histology. Unsatisfied by the information that cellular morphology could give me about life, I turned to physiology. Finding physiology too complex I took up pharmacology. Still finding the situation too complicated I turned to bacteriology. But bacteria were even too complex, so, I descended to the molecular level, studying chemistry and physical chemistry. After twenty years work, I was led to conclude that to understand life we have to descend to the electronic level, and to the world of wave mechanics. But electrons are just electrons, and have no life at all. Evidently on the way I lost life; it had run out between my fingers. —Albert Szent-Györgyi

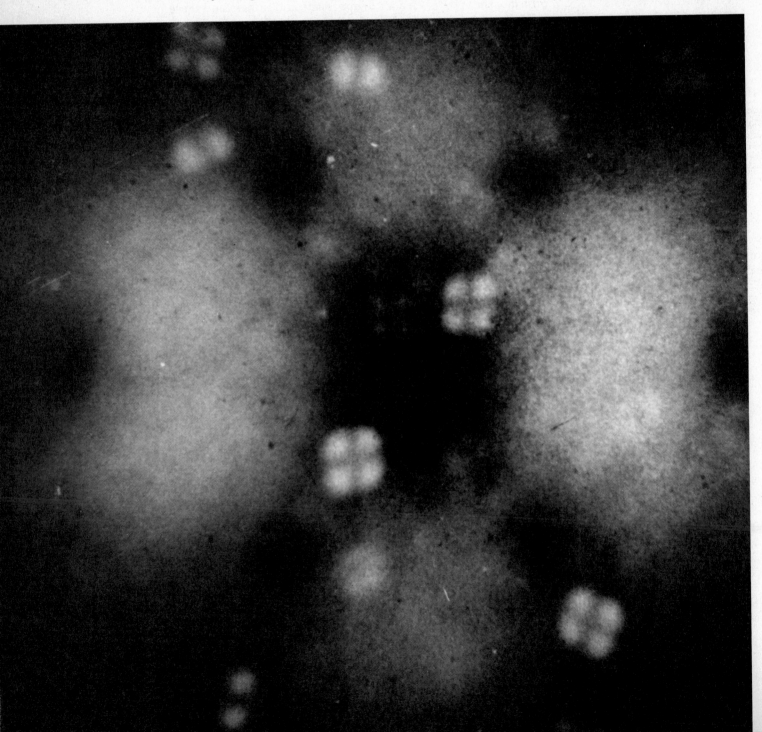

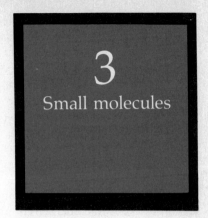

3
Small molecules

Reductionism is an evil word in biology. The fear that life could be reduced to nothing but chemistry and physics troubles many workers who study it. One cannot study life at the level of atoms and molecules, they maintain, because in the act of dissecting out molecules and reactions for examination, one destroys the phenomenon he seeks to study. They and Szent-Györgyi are essentially correct. According to the definition of life in the first chapter, the smallest unit or "quantum" of life is the intact cell. When one dissects the cell and examines its components one is no longer looking at a living system. The organization of cell components is literally as vital as the nature of the components. The secret of life lies not only in the structure of biological molecules, but also in the way they are organized within the cell, an idea expressed with clarity by British astrophysicist Fred Hoyle in his science-fiction novel, *The Black Cloud:* ". . . the distinction between animate and inanimate is more a matter of verbal convenience than anything else. By and large, inanimate matter has a simple structure and comparatively simple properties. Animate or living matter on the other hand has a highly complicated structure and is capable of very involved behavior."

Biochemistry can fairly be defined as the study of the chemical reactions that occur in living organisms. A few overconfident statements by biochemists in the past have led to accusations of reductionism, yet biochemistry has given us an irreplaceable picture of life as a chemical system. Molecular biology, a field that has expanded mightily in the past 20 years, goes one step further.

Molecular biologists do not ignore organization and spatial arrangement, but emphasize them. Their two great triumphs have been the successful explanation at the molecular level of the machinery of heredity and of the action of enzymes. Both of these discoveries have depended on knowing how the chemical groups on a giant molecule are arranged in space. X-ray diffraction studies have revealed the structures of DNA and proteins, and electron microscopy has disclosed the fine structures of many organelles of the cell. There is still a tantalizing no-man's land between the upper resolution limit of x-ray crystallography and the lower limit of electron microscopy that prevents us from seeing the molecular structure of a membrane or the arrangement of enzymes within a mitochondrion. But molecular biology and biochemistry have become thoroughly structural. The phrase, "the molecular architecture of life" is no longer a flight of rhetoric. The modern biologist intends that it be taken literally.

This chapter and the next dissect the atomic and molecular components of the living cell and then reassemble them. Life and life processes will be more understandable if Szent-Györgyi's warning is kept in mind. The first part of this chapter covers the kinds of chemical substances found in living organisms; the second part is a brief review of some fundamental ideas of chemistry that will be needed later. Students who have had a good high school or college course in chemistry might want to study the next section and then merely to skim the rest of the chapter to brush up on what they have forgotten.

SMALL ORGANIC MOLECULES of phthalocyanin appear as four-lobed patterns of light in the photograph on the opposite page, made with a field emission microscope.

THE MOLECULES OF LIVING ORGANISMS

The chemistry of life, at least on this planet, is the chemistry of water and carbon: water, because it is the most abundant molecule in the cell, comprising on the average about 70 percent of its weight *(Figure 1)*; carbon, because about 95 percent of the dry weight of a cell consists of organic (carbon-based) molecules. Life began in the sea, and the properties of water shape the chemistry of all living organisms. Life developed as a liquid-phase phenomenon essentially because reactions in solution are much more rapid than reactions between solids, and because complex and highly-structured molecules can behave in solution in a way that they cannot behave in a gas. Fats, lipids, and the active sites of some enzymes can create nonaqueous microenvironments in which essential reactions proceed, but most reactions of the cell occur in watery surroundings.

If all the water is removed from a cell, the residue contains four main types of large organic molecules, about 100 different small organic molecules, and a few minerals. The large molecules, often called macromolecules, include proteins, carbohydrates, lipids, and nucleic acids. Some of the macromolecules are polymers — compounds formed by the linking together of a number of small subunits. Proteins, for example, are long-chain polymers of amino acids. Proteins serve as structural material (hair, muscle, skin), as carriers of small molecules (such as the protein hemoglobin, which carries oxygen) and, most important, as compounds that speed up chemical reactions in the cell (enzymes). Carbohydrates are simple sugars and their derivatives, including polymers. The simple sugar glucose is a small organic molecule, but its polymer cellulose is a long-chain compound that reinforces the rigid walls of plant cells. Starch, the principal energy storage compound in plants, is a polymer of glucose molecules linked together in a slightly different way. Lipids are actually a mixed bag of compounds. Anything that can be extracted from plant or animal tissues by organic solvents (such as alcohol) is called a lipid. The phospholipids, an important class derived from fats, are widely used along with proteins in building membranes. Other lipids, such as steroids and carotenoids, act as chemical signals that carry information from one

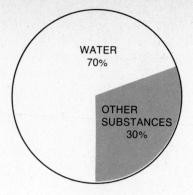

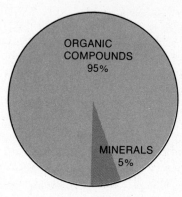

COMPOSITION OF CELLS. Living cells consist mainly **1** **of water (top graph). The dry weight of the cell (bottom) consists mainly of organic compounds.**

part of the organism to another, or as triggering devices activated by outside stimuli such as light. Nucleic acids are enormous molecules that store the hereditary blueprints for the synthesis of proteins.

Many of the small organic molecules in the cell are monomers, the subunits of giant polymers. Some small molecules, such as adenosine triphosphate (ATP), act as storage units for chemical energy. Others carry electrons from one macromolecule to another, or carry small molecules or chemical groups from one place in the cell to another. Still others are bound to the surface of an enzyme and are essential to their action. These are known as cofactors. As with large molecules, some organisms (notably plants and bacteria) can synthesize all of the carriers and cofactors that they need. Most animals cannot, and require an outside source either for the final compound or more commonly for a precursor as raw material for synthesis. Vitamins are small organic molecules

---

The chemistry of life, at least on this planet, is the chemistry of water and carbon.

that must be obtained in the animal diet, and the list of essential molecules varies from one species to another. Ascorbic acid (vitamin C) is required by humans and other primates, but rats and most other mammals can make their own.

At the bottom of the scale of chemical complexity are the ions (charged particles) and minerals that all living organisms need from outside sources. These include sodium, potassium, magnesium, calcium, manganese, iron, copper, zinc, molybdenum, phosphorus, chlorine, and iodine. Some elements, such as calcium, are required in quantity for bones and teeth in vertebrates, but most are required only in minute amounts. Metals, such as iron, copper, and zinc, are essential to the function of enzymes, and are bound to the surface of the enzyme molecule. Other metals, notably sodium and potassium, are important in the process of nerve conduction.

Plants can synthesize all the large molecules they need, given sources of carbon, nitrogen, oxygen, hydrogen, and other elements. An organism that can synthesize its own proteins and carbohydrates is called an AUTOTROPH, from the Greek for self-feeder. Animals have lost much of the capability to manufacture their own food and most obtain the semi-finished raw materials in the food they eat. They are classed as HETEROTROPHS, mixed-feeders. It is more efficient and economical to jettison the genetic and biochemical machinery for making a substance that can easily be obtained from the environment. The fats that a man eats are

---

If all the water is removed from a cell, the residue contains four main types of large organic molecules, about 100 different small organic molecules, and a few minerals.

---

used in part for energy, but also to make lipids. Proteins in the diet are digested to amino acids, and some of these are relinked in a different order to make the proteins of the body. Carbohydrates are degraded to simple sugars like glucose, and stored again in the form of polymers. None of these large molecules is used by animals just as is —all are chopped up into their subunits and the subunits are then reassembled into more complex structures. Most plants and bacteria are auto-

trophs, and animals are familiar examples of heterotrophs.

The remainder of this chapter is a review of the nature of atoms, the periodic table, and simple ideas of chemical bonding. Chapter 4 is a discussion of large molecules and how they operate.

THE BUILDING BLOCKS OF MATTER

A biologist, runs an old definition, is one who thinks that molecules are too small to matter. A physicist is one who thinks that molecules are too large to matter. Anyone who disagrees with both of them is probably a chemist. This definition, fortunately, is at least 15 years out of date. The double-helical structure of DNA has had a profound effect on biology, not only in explaining the genetic mechanism, but in shaping biologists' thinking about problems that cannot yet be explained in molecular terms. Chemists, in turn, have discovered something that biologists knew from the start: that the organization of a system is fully as important as its components in determining its properties. If cells are the brick in the architecture of living organisms, then atoms and molecules are the clay from which the brick is fashioned, and they impose fundamental limits on the mechanisms of life.

Atoms are built from two components: a positively charged nucleus and enough negatively charged electrons to make the entire atom electrically neutral. The nucleus itself is comprised of two kinds of particles: protons and neutrons, which have approximately the same mass, but protons are positively charged while neutrons have no charge. The mass of an individual proton or neutron is quite small: 1.67×10^{-24} grams. To avoid dealing with such small numbers, the mass of a proton or neutron is taken as a standard unit mass in discussing atoms. The atomic mass unit (amu) is 1.66024×10^{-24} grams. The slight discrepancies in mass between one amu and the mass of either a proton or a neutron are important to physicists, but not to biologists; they arise from matters such as the binding energy that holds a nucleus together. The mass of an electron is only 1/1835 that of a proton. In biochemistry electrons too can be ignored when calculating the weight of an atom —only the nucleus matters. The positive charge on a proton is a tiny number, 1.602×10^{-19} coulombs, and this quantity is defined as a *unit* of charge. A proton has a charge of +1 in these units, and an electron a charge of −1. The charge on the neutron, of course, is 0.

The number of protons in the nucleus of an atom is called its ATOMIC NUMBER, and the total number of protons and neutrons is, to a first approximation, its ATOMIC WEIGHT. The atomic number is important in determining the chemical properties of an atom, and all atoms with the same atomic number are classified as the same chemical element. All atoms of gold, for example, have exactly 79 protons in their nuclei. All carbon atoms have six protons, all nitrogen atoms have seven, and all oxygen atoms have eight.

Some atoms of a chemical element are heavier than others because of extra neutrons in the nucleus. These variant atoms are called isotopes; a few are shown in Figure 2. The number of neutrons has little effect on the chemical properties of an element, but the effect on the physical properties is more pronounced. The rate of diffusion of a gas,

for example, depends to some extent on its mass. More important, the presence or absence of a neutron or two makes some nuclei unstable, and they tend to break down by emitting radiation. Biochemists and biologists use radioactive isotopes to label a chemical compound and radiation detectors to trace its pathway through an organism or through a complicated series of reactions.

One of the biochemist's favorite tracers is carbon 14. Of all the atoms of carbon found in nature, 98.9 percent have six protons and six neutrons in their

STRUCTURE OF ATOMS. Electrons are represented as clouds around the central nuclei. The size of the nucleus is exaggerated in this diagram. The electron cloud in a typical atom is 100,000 times the diameter of the nucleus. At top, the three isotopes of hydrogen: hydrogen, deuterium and tritium. Diagrams at bottom depict two isotopes of carbon: carbon 12 and carbon 14. 2

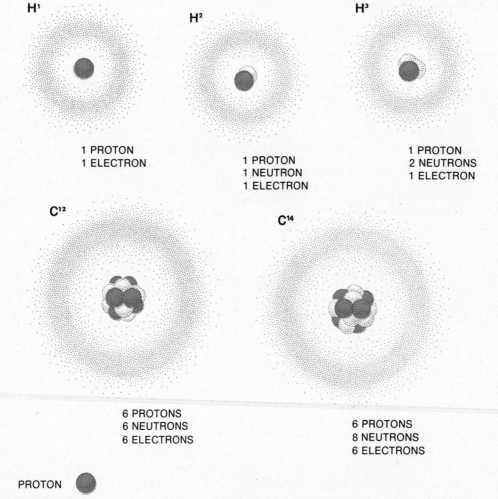

H¹

1 PROTON
1 ELECTRON

H²

1 PROTON
1 NEUTRON
1 ELECTRON

H³

1 PROTON
2 NEUTRONS
1 ELECTRON

C¹²

6 PROTONS
6 NEUTRONS
6 ELECTRONS

C¹⁴

6 PROTONS
8 NEUTRONS
6 ELECTRONS

PROTON

NEUTRON

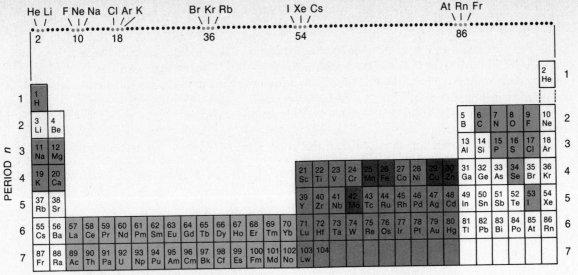

3 **PERIODIC TABLE of the elements. When arranged in order of increasing atomic number, elements of similar chemical properties recur at intervals of 8, 8, 18, 18 and 32, as shown in the dotted row of elements at top. Arranging the elements so that similar ones occur in vertical columns leads to the periodic table below it. Elements can be classified into three categories: representative elements (white blocks), inner transition metals (light gray) and transition metals (darker gray). Elements important in living systems, such as carbon, oxygen, nitrogen and phosphorus, are shown in colored blocks.**

nuclei. This isotope is called carbon 12, and is represented in chemical equations where isotopes matter by a superscript 12 just after (or just before) the symbol for the element: C^{12}. A much smaller number of carbon atoms, 1.1 percent, have seven neutrons instead of six, and a very few have eight neutrons. These isotopes are written C^{13} and C^{14}, respectively. Other isotopes of carbon have been created artificially in nuclear reactors, but do not exist in nature: C^{10}, C^{11}, C^{15}, and C^{16}. All of these artificial isotopes are unstable and break down spontaneously to other elements. Their half lives, the time required for half of any given quantity of the element to disappear, vary from less than a second to more than 20 minutes. Carbon 14 is also unstable and radioactive, but its half life is 5,570 years. Because it is continually being made in the upper atmosphere by the bombardment of nitrogen atoms by cosmic neutrons, a small amount of C^{14} is always present in samples of natural carbon. In effect, the slowly decaying carbon is a built-in

clock that makes it possible to determine the age of ancient organic materials. It has proven an invaluable tool in establishing the time scale of recent evolutionary events *(Box A)*.

THE PERIODIC TABLE

If one arranges the elements in order of increasing atomic number, elements of similar properties recur at intervals of 8, 8, 18, 18, and 32. For example, element 2 is helium (He), an inert gas. Element 10 is neon (Ne), 18 is argon (Ar), 36 is Krypton

---

Carbon, nitrogen, oxygen, hydrogen, phosphorus and sulfur are the basic elements in large biological molecules.

(Kr), and 54 is Xenon (Xe), all the so-called noble gases with little or no tendency to react. Just before each of these noble gases in the atomic number list is a member of the halogen family: fluorine (F), chlorine (Cl), bromine (Br), and iodine (I). Just after them is a light, very reactive silvery metal: lithium (Li), sodium (Na), potassium (K), rubidium (Rb), or cesium (Cs).

If one folds the list so that elements with similar

A Radioisotope dating

Because C^{14} decays with a half life of 5570 years, it can be used to determine the time of death of any organism that has died within the last 10,000 years or so and left carbonaceous remains. The ratio of C^{14} to C^{12} in a living organism is always the same as that in the atmosphere at large, for carbon compounds and CO_2 are constantly being exchanged between the atmosphere and plants and animals. Plants use CO_2 directly in photosynthesis, and animals eat the plants enriched with carbon from the atmosphere. The production of new C^{14} in the upper atmosphere by the reaction of fast neutrons with nitrogen:

$$N^{14} + n^1 \rightarrow C^{14} + H^1$$

just balances the natural radioactive decay of C^{14}, and a steady state exists.

As soon as a tree or any other creature dies, it no longer equilibrates its carbon compounds with the rest of the living world. Its decaying C^{14} is not replenished from outside, and the ratio of C^{14} to C^{12} falls. By measuring how much of the total carbon is C^{14} at a given time, we can calculate how much time has elapsed since death occurred. Wooden pillars, therefore, can date a temple, and roof beams a house. Charcoal fragments can tell us how long ago primitive farmers built a pottery kiln, and human hair from a mummy can verify the date of an Egyptian dynasty. Some radiocarbon dates of archeological objects from various cultures are shown on the graph below.

The radioactive decay law is: $C_t = C_o e^{-bt}$, where C_t is the concentration of C^{14} at a time t, C_o is the initial concentration, and B is a decay constant that can be evaluated from the knowledge that $C_t/C_o = \frac{1}{2}$ at $t = 5570$ years. The decay law can also be written

$$\ln C_t = \ln C_o - Bt.$$

If the logarithm of the C^{14} concentration (expressed in radioactive decay counts per minute per gram of carbon) is plotted against age or time since death, a straight line is obtained.

Other radioactive isotopes with longer half lives are more useful in dating geological and paleontological events in the history of our planet. Potassium-40, K^{40}, decays to argon-40 with a half life of 1.3 billion years. Rubidium-87 decays to strontium-87 with a half life of 47 billion years. Uranium-238 breaks down in a series of steps to lead-206 with a half life of 4.5 billion years, for the slowest, rate-determining step. It gives us confidence that we may know what we are doing, to find that potassium/ argon, rubidium/strontium, and uranium/lead dating methods all give the same answers for ages of terrestrial and lunar rock samples.

---

properties lie in vertical columns, the result is the table in Figure 3. The elements appear in horizontal rows or PERIODS, but these periods have different lengths. The first period has only two elements, hydrogen (H) and helium (He). The second and third periods have 8, the fourth and fifth have 18, and the sixth and probably the seventh have 32. The elements fall naturally into three classes. The representative elements are those one ordinarily thinks of when chemical elements are mentioned. They show a wide variety of properties, from metals like sodium (Na) and aluminum (Al), to non-metallic solids like carbon (C) and sulfur (S), to gases such as oxygen (O) and chlorine (Cl). Metals are at the left of the periodic table, and nonmetals are at the right. Every period after the first has eight representative elements. These are the fundamental building blocks of living organisms. (The names and symbols of the chemical elements are listed in Box B.)

The fourth and later periods each have ten transition metals, which include such metals as iron (Fe), copper (Cu), zinc (Zn), silver (Ag), tungsten (W), and platinum, (Pt). Although they are distinc-

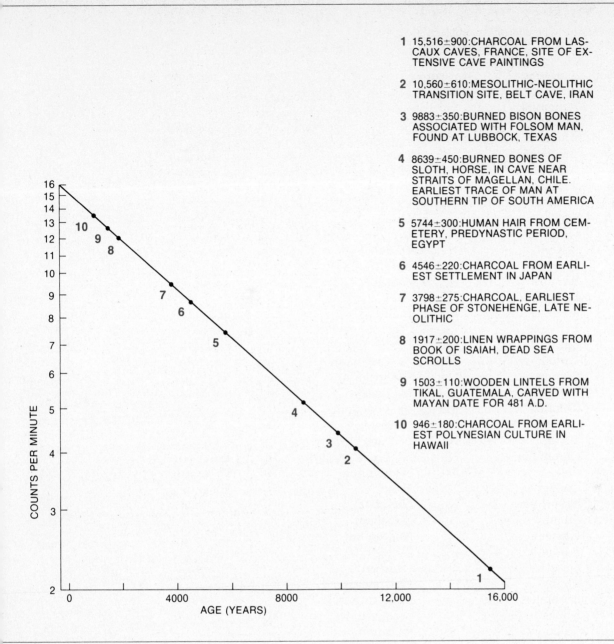

1 15,516±900:CHARCOAL FROM LAS-
   CAUX CAVES, FRANCE, SITE OF EX-
   TENSIVE CAVE PAINTINGS

2 10,560±610:MESOLITHIC-NEOLITHIC
   TRANSITION SITE, BELT CAVE, IRAN

3 9883±350:BURNED BISON BONES
   ASSOCIATED WITH FOLSOM MAN,
   FOUND AT LUBBOCK, TEXAS

4 8639±450:BURNED BONES OF
   SLOTH, HORSE, IN CAVE NEAR
   STRAITS OF MAGELLAN, CHILE.
   EARLIEST TRACE OF MAN AT
   SOUTHERN TIP OF SOUTH AMERICA

5 5744±300:HUMAN HAIR FROM CEM-
   ETERY, PREDYNASTIC PERIOD,
   EGYPT

6 4546±220:CHARCOAL FROM EARLI-
   EST SETTLEMENT IN JAPAN

7 3798±275:CHARCOAL, EARLIEST
   PHASE OF STONEHENGE, LATE NE-
   OLITHIC

8 1917±200:LINEN WRAPPINGS FROM
   BOOK OF ISAIAH, DEAD SEA
   SCROLLS

9 1503±110:WOODEN LINTELS FROM
   TIKAL, GUATEMALA, CARVED WITH
   MAYAN DATE FOR 481 A.D.

10 946±180:CHARCOAL FROM EARLI-
   EST POLYNESIAN CULTURE IN
   HAWAII

tive as metals, they have more similar chemical properties than the representative elements do. Several transition metals are important to living organisms, but usually in small amounts as metal atoms attached to enzymes or other proteins. Iron, for example, is found in the oxygen carrier hemoglobin, the energy-extraction protein cytochrome *c*, and in catalase, peroxidase, and other enzymes. The sixth and seventh periods each have 14 additional elements, the inner transition metals, or rare earths and actinides. None of these has any importance for life (or, at least, did not have until man

began playing with uranium), and they all can be ignored here.

The most important elements in living systems are set off in color in Figure 3 and shown in Table I. Carbon, nitrogen, oxygen, hydrogen, phosphorus, and sulfur are the basic elements in large biological molecules. Sodium and potassium occur mainly as ions in solution, and are important in such mechanisms as the conduction of nerve impulses. Calcium is used for bones and teeth, and the other elements are needed in small amounts in special compounds.

SYMBOL	ELEMENT	SYMBOL	ELEMENT	SYMBOL	ELEMENT
Ac	Actinium	Gd	Gadolinium	Pm	Promethium
Ag	Silver	Ge	Germanium	Po	Polonium
Al	Aluminum	H	Hydrogen	Pr	Praseodymium
Am	Americium	He	Helium	Pt	Platinum
Ar	Argon	Hf	Hafnium	Pu	Plutonium
As	Arsenic	Hg	Mercury	Ra	Radium
At	Astatine	Ho	Holmium	Rb	Rubidium
Au	Gold	I	Iodine	Re	Rhenium
B	Boron	In	Indium	Rh	Rhodium
Ba	Barium	Ir	Iridium	Rn	Radon
Be	Beryllium	K	Potassium	Ru	Ruthenium
Bi	Bismuth	Kr	Krypton	S	Sulfur
Bk	Berkelium	La	Lanthanum	Sb	Antimony
Br	Bromine	Li	Lithium	Sc	Scandium
C	Carbon	Lu	Lutetium	Se	Selenium
Ca	Calcium	Lw	Lawrencium	Si	Silicon
Cd	Cadmium	Md	Mendelevium	Sm	Samarium
Ce	Cerium	Mg	Magnesium	Sn	Tin
Cf	Californium	Mn	Manganese	Sr	Strontium
Cl	Chlorine	Mo	Molybdenum	Ta	Tantalum
Cm	Curium	N	Nitrogen	Tb	Terbium
Co	Cobalt	Na	Sodium	Tc	Technetium
Cr	Chromium	Nb	Niobium	Te	Tellurium
Cs	Cesium	Nd	Neodymium	Th	Thorium
Cu	Copper	Ne	Neon	Ti	Titanium
Dy	Dysprosium	Ni	Nickel	Tl	Thallium
Er	Erbium	No	Nobelium	Tm	Thulium
Es	Einsteinium	Np	Neptunium	U	Uranium
Eu	Europium	O	Oxygen	V	Vanadium
F	Fluorine	Os	Osmium	W	Tungsten
Fe	Iron	P	Phosphorus	Xe	Xenon
Fm	Fermium	Pa	Protactinium	Y	Yttrium
Fr	Francium	Pb	Lead	Yb	Ytterbium
Ga	Gallium	Pd	Palladium	Zn	Zinc
				Zr	Zirconium

ATOMS, ELECTRONS, CHEMICAL PROPERTIES

The periodic table as a summary of chemical behavior has been known since the late nineteenth century, but it was not successfully explained by any theory of atomic structure until quantum mechanics was developed in the 1930's. We shall not be concerned with this theory, but only with some of its results. The most important prediction of quantum mechanics is that an electron associated with an atom cannot have any arbitrary energy. Its energy is "quantized," or restricted to certain discrete values that correspond to a series of SHELLS around the nucleus. The first shell, which lies nearest to the nucleus, corresponds to the lowest energy level. This shell can hold only two electrons. The next two shells hold eight electrons each; the next two, 18 each, and the rest 32. Electrons in an atom must fill these energy levels from the bottom up (at least in the most stable, or ground state, of the atom). To see how electrons are arranged around the nucleus of a given element, we need only imagine that the electrons are fed into the atom one at a time, and see which shells are filled and which are left unfilled by the time we have used up all the available electrons.

A hydrogen atom has one proton in its nucleus and one electron on the outside of the nucleus. The lone electron will occupy the first shell. In helium, the second electron similarly occupies (and fills) the first shell. In lithium (Li), atomic number 3, the first two electrons completely fill the first shell and the third electron has to occupy the lowest remaining unfilled level in the second shell. This shell can hold a total of eight electrons, and the series: beryllium, boron, carbon, nitrogen, oxygen, fluorine, and neon, across the second period of the periodic table in Figure 3, reflects the gradual filling of this shell (*Figure 4*). Atoms whose outer shells contain eight electrons tend to be particularly stable, even when the maximum capacity of the shell is greater. Neon, for example, is chemically inert. Because chemical reactions involve the approaches of atoms to one another and the interactions of their outermost electrons, the configuration of electrons

ELEMENTS FOUND IN LIVING SYSTEMS are listed below, together with their properties. Carbon, nitrogen, oxygen, hydrogen, phosphorus and sulfur are main components of large biological molecules.

ATOMIC NUMBER	SYMBOL	NAME	ATOMIC WEIGHT	PROPERTIES OF PURE ELEMENT
1	H	HYDROGEN	1.01	LIGHT, COLORLESS GAS
6	C	CARBON	12.01	HARD SOLID (DIAMOND, GRAPHITE)
7	N	NITROGEN	14.01	COLORLESS GAS
8	O	OXYGEN	16.00	COLORLESS GAS
9	F	FLUORINE	19.00	PALE GREENISH GAS
11	Na	SODIUM	23.00	REACTIVE SILVER METAL
12	Mg	MAGNESIUM	24.31	LIGHT, SILVERY METAL
15	P	PHOSPHORUS	30.97	WHITE, RED, OR YELLOW NONMETAL
16	S	SULFUR	32.06	YELLOW SOLID
17	CL	CHLORINE	35.45	YELLOW-GREEN GAS
19	K	POTASSIUM	39.10	LIGHT, SILVER-WHITE METAL
20	Ca	CALCIUM	40.08	SOFT, SILVERY METAL
25	Mn	MANGANESE	54.94	HARD, BRITTLE METAL
26	Fe	IRON	55.85	SILVERY GREY METAL
29	Cu	COPPER	63.54	MALLEABLE REDDISH METAL
30	Zn	ZINC	65.37	BLUEISH-WHITE METAL
34	Se	SELENIUM	78.96	RED OR GREY SEMIMETAL
42	Mo	MOLYBDENUM	95.94	TOUGH, SILVERY METAL
53	I	IODINE	126.90	VIOLET CRYSTALLINE NONMETAL

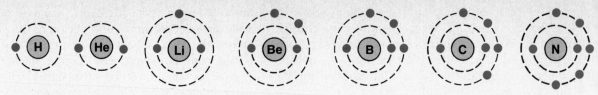

4 ELECTRON SHELLS determine chemical properties of elements. Inner shell (dashed circle) can hold a maximum of two electrons. Next shell holds up to eight. Diagram shows, from left to right, progressive filling of electron shells of elements in first two periods of periodic table: hydrogen and helium (period 1), and lithium, beryllium, boron, carbon, nitrogen, oxygen, fluorine, neon (period 2).

in the outer shell powerfully influences the chemical behavior of the atom.

Because a filled eight-electron outer shell is so stable, an atom having a filled shell plus an extra electron (such as lithium), tends to lose that electron and become a positively charged atom: a POSITIVE ION. Substances that lose electrons easily to form positive ions are called METALS and they appear at the left side of the periodic table. If an atom has one electron less than a closed shell, it has a strong tendency to seize an electron from somewhere else and complete the shell, becoming a NEGATIVE ION in the process. Such compounds are called halogens, and chlorine is a good example. Substances that gain electrons easily to form negative ions are all nonmetals, and appear at the right side of the table.

Metals and nonmetals, lying on opposite sides of the periodic table, can transfer electrons and form IONIC BONDS. Common table salt, sodium chloride, is written NaCl. A more accurate representation, however, is Na^+Cl^-, for each sodium atom has lost an electron to become a positive ion with the closed shell electronic arrangement of the inert gas neon, and each chlorine atom has gained an electron to become a negative ion with the closed shell electronic arrangement of the inert gas argon (*Figure 5*). When sodium chloride dissolves in water

the ions are still present and the salt solution will conduct electricity well because the ions migrate from one electrode to another. If the water evaporates and the salt crystallizes, the Na^+ and Cl^- ions settle down into a regular crystalline lattice in which each ion has ions of opposite charge for nearest neighbors (*Figure 6*). No one sodium ion is associated with a particular chloride ion, but there are equal numbers of both, so the crystal as a whole is electrically neutral. The electrostatic forces that keep the crystal together are nondirectional, and the structure in Figure 6 is dictated by the packing of charged spheres rather than by the arrangements of electrons on the ions.

Metals, at the left of the table, tend to lose enough electrons to attain a closed shell structure, and to become positive ions, or CATIONS. Na^+, K^+, Mg^{2+}, and Ca^{2+} are particularly important cations in living systems. The transition metals do not lose all of their electrons outside of the previous noble gas shell; this would require too much energy. Instead, iron exists commonly in either the ferrous state, Fe^{2+}, or the ferric, Fe^{3+}. Copper can form either cuprous ions, Cu^+, or cupric, Cu^{2+}, and other transition metals have more than one ionic state. The halogens at the far right, and to a lesser extent, O and S, can gain enough electrons to achieve a stable configuration, and form the negative ions or ANIONS: F^-, Cl^-, Br^-, I^-, S^{2-}. What happens with elements nearer the middle of the

IONIC BONDS in sodium chloride. Sodium atom loses **5** **an electron to become a cation (positive ion) and chlorine atom gains an electron to become an anion (negative ion). Chemical shorthand notation at bottom of diagram shows that electrons around chloride ion (right) are grouped into four pairs.**

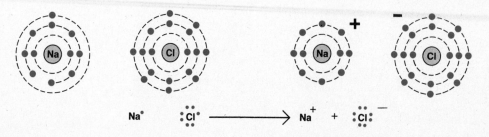

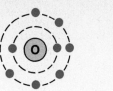

number of atoms or molecules in a mole is AVOGAD-RO'S NUMBER, and this number has been found experimentally to be 6.023×10^{23} molecules/mole. The atomic weight of C^{12} is 12, and the molecular weight of water, H_2O, is $16 + 1 + 1 = 18$, the sum of the weights of its component atoms. One mole of carbon (diamond or graphite) is therefore 12 grams, and to be sure of measuring out as many water molecules as there are carbon atoms in 12 grams of graphite, a chemist must weigh out 18 grams of water.

Atoms which cannot easily either gain or lose enough electrons to produce a stable outer shell can make bonds with other atoms by sharing electrons. A shared pair of electrons is a COVALENT BOND. Carbon can share all four of its outer electrons and can end with four electron pair bonds around it—the desirable eight electrons. This shared-electron bonding is diagrammed for methane in Figure 7.

Methane, CH_4, is a molecular compound in a sense that sodium chloride, NaCl, is not. No one sodium ion is associated with any particular chloride ion, either in solution or in a crystal of rock salt. But the carbon and four hydrogen atoms in methane are tightly bound into a molecule that behaves in most circumstances as a unit. Moreover, the hydrogen atoms are arranged in a specific way around carbon, in four tetrahedral directions *(Figure 8)*. Tetrahedral bonding is the usual geometry of the compounds formed by the most common atoms of living organisms: C, N, and O *(Figure 9)*. Carbon has four outer electrons,

table? Can carbon lose four electrons to produce C^{4+} with the electronic configuration of helium, or gain four electrons to produce C^{4-} with the neon configuration? Do we ever find N^{5-} or N^{3+}?

The answer is that we do not. The price in energy is too high. Elements such as B, C, N, O, Si, and P tend to share electrons rather than donating or accepting them. These shared electrons form what are called COVALENT BONDS, in contrast to the simple ionic forces. But covalent bonds are different in several important ways. They connect specific individual atoms, unlike the forces between ions in a crystal of rock salt or between Na^+ and Cl^- ions in a salt solution. This assemblage of connected atoms is called a molecule. Moreover, the atoms in a molecule are held in specific orientations, which can be predicted from quantum theory. Molecules have shape, and the shape comes from covalent bonds.

IONIC, COVALENT AND HYDROGEN BONDS

In an ionic bond, one atom loses an electron to another, and electrostatic forces between the positive ion (the electron donor) and the negative ion (the acceptor) make the ions attract one another. This attraction is ordinarily in the neighborhood of 90 kilocalories of energy per mole of the two ions.

Two fundamental ideas are introduced in this last sentence. The CALORIE is a measure of heat or energy. One standard calorie is defined as the heat necessary to raise one gram of pure water from 14.5°C to 15.5°C. (This is called the 15° calorie, and the choice of 15° probably represents the ambient temperature in European laboratories in the last century.) The nutritionist's Calorie is 1000 of these small calories, or what we call a KILOCALORIE (kcal).

The concept of the MOLE is one of the most fundamental ideas in practical chemistry. A chemist cannot count or weigh individual atoms, and needs some way of being sure that he can measure out equal numbers of atoms of two different substances. A mole of any substance is an amount whose weight in grams is numerically equal to its atomic or molecular weight, and a mole of any substance has the same number of molecules. The

CRYSTAL STRUCTURE of sodium chloride. Sodium ions (color) and chloride ions (grey) are packed into a cubic lattice in crystals of rock salt. Here atoms have been reduced in size for clarity; actually they touch one another. **6**

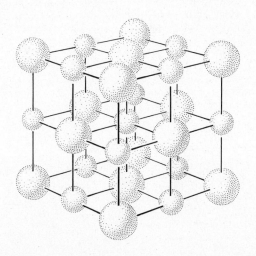

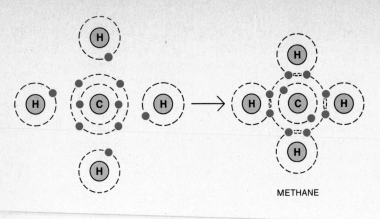

METHANE

H· ·C· ·H ⟶ H :C: H   H—C—H

7 COVALENT BONDS in methane. The carbon atom needs four electrons to complete its outer shell, and the hydrogen atom needs one electron. In forming a molecule of methane the atoms gain the missing electrons by forming four electron pairs. Shorthand notation at bottom shows same reaction. Covalent bond formed by electron pairs can also be represented as straight line between atoms (lower right).

needs four more, and so forms four shared covalent bonds with neighboring atoms. Nitrogen has five electrons, needs three more, and makes three electron pair bonds. These bonds occupy three of the tetrahedral directions, and the unused fourth and fifth electrons of nitrogen pair in the other tetrahedral direction. Oxygen has six outer electrons. Two of these are used to pair with two other electrons from bonded atoms, and the other four occur as two lone pairs. These four electron pairs

around oxygen again point in the four tetrahedral directions. Only when carbon shares two electron pairs with the same neighbor atom in a double bond does the bonding geometry change from tet-

---

Hydrogen bonds are largely responsible for maintaining the elaborately folded structure of the protein molecule.

---

rahedral to trigonal. Because it is difficult to represent tetrahedral geometry on a printed page, schematic drawings such as those on the right side of Figure 9 are often used. They are only shorthand representations, however, and do not give a good idea of the real molecular shapes.

The bond energies of most covalent bonds between carbon, nitrogen, oxygen and hydrogen are of the order of 80 to 100 kcal per mole. Ionic and covalent bonds are therefore roughly of equal strength, although covalent bonds are highly directional whereas ionic bonds are not. A third type of bond, of particular importance in living systems, is the HYDROGEN BOND *(Figure 10)*. Although the energy of such linkages is only around six kcal per mole, hydrogen bonds are important in the cell because they are so common. Hydrogen bonds are largely responsible for keeping protein molecules folded correctly. The coding between base pairs in

BONDS OF CARBON ATOM form a tetrahedron (left). 8 Four shaded lobes pointing outward from each carbon atom indicate probable distribution of electrons forming the bonds, and are called orbitals. Tetrahedral arrangement of carbon bonds can also be indicated by tapered lines (right).

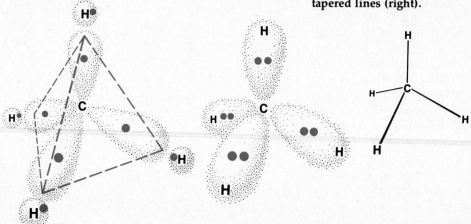

48

DNA is achieved by hydrogen bonds. Even the special property of ice that makes it float in its own liquid, unlike almost every other solid, arises from hydrogen bonding *(Figure 11)*. We shall see later the implications of hydrogen bonding both in proteins and in water.

SIMPLE ORGANIC COMPOUNDS: ACIDS AND BASES

Methane, CH_4, ethane, CH_3—CH_3, propane, CH_3—CH_2—CH_3, and ethylene, CH_2=CH_2, are examples of HYDROCARBONS, compounds of only hydrogen and carbon. Methane, ethane, and propane are called SATURATED hydrocarbons because they contain no carbon-carbon double bonds. Ethylene, in contrast is UNSATURATED, and could add more hydrogen across the double bond:

$$CH_2{=}CH_2 + H_2 \rightarrow CH_3{-}CH_3.$$

Gasolines are hydrocarbons with six to ten carbon atoms, and a typical gasoline fraction, octane, is

9 GEOMETRY OF BONDS around carbon, nitrogen and oxygen atoms is depicted at left. The more conventional diagrams at right do not give a true picture of the geometry of the molecule. Shown here are, from top to bottom, molecules of propane, ethylene, ammonia and water.

10 HYDROGEN BOND is a weak ionic bond. If hydrogen is bonded to another atom, such as oxygen, with a greater affinity for electrons, the electron pair of the bond will shift closer to the oxygen, leaving the hydrogen with a slight positive charge and the oxygen with a negative charge (left). The hydrogen atom will be attracted to another atom having a slight negative charge, such as the oxygen of another water molecule or a carbonyl oxygen in a protein chain (right).

shown in Figure 12a and Figure 12b. Motor oils have 12 to 20 carbon atoms, and longer chain hydrocarbons are waxy semi-solids called paraffin waxes. The ultimate hydrocarbon, with chains thousands of carbon atoms long, is polyethylene plastic.

Hydrocarbons as a family are inflammable, oily, and immiscible with water. Things that dissolve in hydrocarbons ordinarily do not dissolve in water, and vice versa. The heat that is liberated when hydrocarbons are burned with O_2 makes them useful as fuels. It is the hydrocarbon tails of molecules of fats and lipids that make them such an efficient storehouse for energy in the body, and such good foods.

If a hydrogen atom on a hydrocarbon is replaced by a hydroxyl group, —OH, then the compound becomes an ALCOHOL. The common ethyl alcohol or

11 ICE CRYSTALS are held together by hydrogen bonds (dashed lines). When ice melts, this lattice structure collapses, which explains why liquid water at 0°C is denser than solid ice.

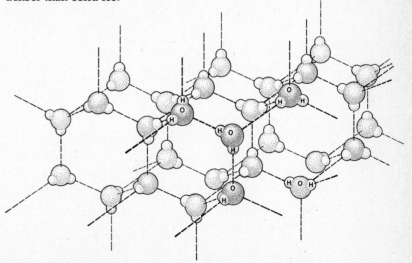

ethanol is shown in Figure 12c and Figure 12d. The smaller alcohols: methanol, CH_3OH, ethanol, C_2H_5OH, propanol, C_3H_7OH, are all soluble in water because of the similarity of the —OH group to water, H—O—H. But in the higher molecular weight alcohols, the long hydrocarbon chain dominates the —OH group, and the alcohols are not soluble in water.

Figure 13 shows a typical carboxylic acid, with the carboxyl group, —COOH. Acetic acid, CH_3COOH, shown here, is the principal constituent of vinegar. Formic acid, HCOOH, is the irritating fluid expelled by many kinds of ants when they defend their nests. The carboxylic acids have the important property of being able to lose a hydrogen ion, or proton, or of IONIZING, as shown in Figure 13. Because the proton has a positive charge, the resulting acetate ion has a negative charge. The double bond that formerly existed between the carbon and one oxygen atom spreads over both oxygen atoms as shown in Figure 13, and the negative charge is also spread over the

12 SHAPE OF SIMPLE ORGANIC COMPOUNDS. In octane, a gasoline (top), bonds about the carbon atom are tetrahedral, as shown by colored outline at right. Tapered lines represent bonds extending out of page toward reader, with the thick end nearest. Dashed lines represent bonds receding into the page. Conventional diagram of octane (center) is easier to write but does not indicate the shape of the molecule. Structure of ethyl alcohol (ethanol) shows tetrahedral bonding of two carbon atoms (lower left). Structure at lower right is the conventional representation of the ethyl alcohol molecule.

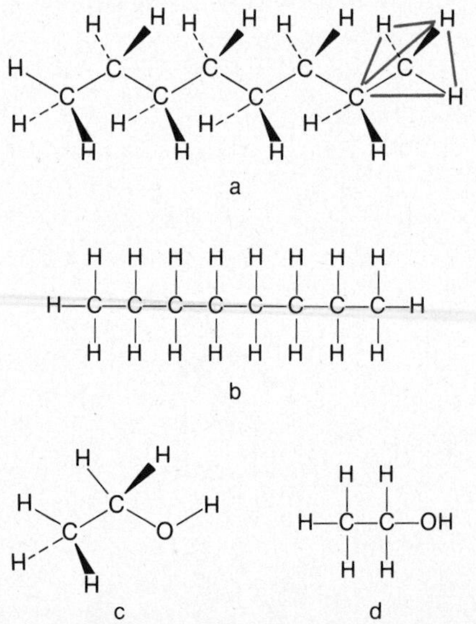

a

b

c

d

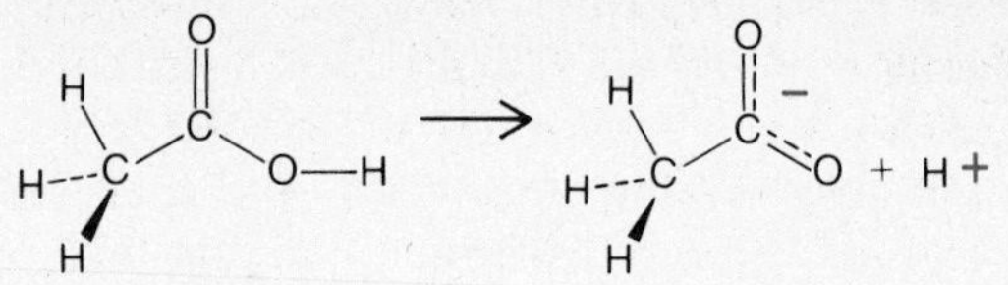

13 IONIZATION OF ACETIC ACID. When a carboxylic acid ionizes and loses a hydrogen ion, the resulting negative charge is spread over both oxygen atoms of the carboxylate ion.

entire carboxyl group, —COO^-. Organic acids, like inorganic acids, have a sharp taste and turn litmus indicator paper red.

Many other substances besides carboxyl groups are capable of ionizing and releasing protons into aqueous solution, and all of these are called ACIDS. Hydrochloric acid is an example:

$$HCl \rightarrow H^+ + Cl^-.$$

When nitric acid, HNO_3, dissociates, the NO_3^- group behaves as if it were a single atom, and does not dissociate further:

$$HNO_3 \rightarrow H^+ + NO_3^-.$$

The three oxygen atoms are bound to the central nitrogen by covalent electron pair bonds of the type we have been examining, and are not easily separated from the nitrogen. The bonding between the hydrogen ion, or proton, and the nitrate group, is electrostatic and more easily disrupted by the solvent. This does not mean that ionic bonds are necessarily weaker than covalent bonds. Both are comparable in absolute strength. But water molecules themselves are polar: The molecule is bent as shown in Figure 14, and the oxygen atom has a slight negative charge while the hydrogen atoms are slightly positive. A water molecule is therefore a miniature dipole. When HCl dissolves, water molecules cluster around the negative chloride ion with their positively charged hydrogen atoms closest to it. This is an electrostatically stable arrangement of negative and positive charge. Each ion therefore regains some of the energy it lost in separating from its partner by making many weaker ionic bonds to water molecules. This is why substances that can break into ions, or that have localized positive and negative charges on their molecular framework, are soluble in water. Hydrocarbons are not soluble in water because they cannot interact this way with the polar water molecules. They gain nothing in energy by dissolving. Substances that have a high affinity for aqueous solution are hydrophilic; those, such as fats, oils, and lipids, which are more stable when separated in their own environment away from water are hydrophobic.

Sulfuric acid, H_2SO_4, is a common acid which can lose two protons in successive steps

$$H_2SO_4 \rightarrow H^+ + HSO_4^-,$$
$$HSO_4^- \rightarrow H^+ + SO_4^{2-}.$$

The four oxygen atoms are bound covalently to the sulfur at the corners of a tetrahedron, like hydrogens about carbon in methane. It would be better to diagram the dissociation process as

$$O=\overset{\displaystyle OH}{\underset{\displaystyle O}{S}}-OH \rightarrow O=\overset{\displaystyle O^-}{\underset{\displaystyle O}{S}}-OH + H^+ \rightarrow O=\overset{\displaystyle O^-}{\underset{\displaystyle O}{S}}-O^- + 2H^+.$$

Sulfur, in the row of the table below oxygen, can use six of its outer electrons to make six covalent bonds with other atoms, and it is said to have a VALENCE of six. Its other common valence state is two, as in hydrogen sulfide, H_2S. Oxygen cannot make six bonds and show a valence of six. This representation of the sulfate ion, SO_4^{2-}, is deceptive in the same way the representation of the carboxyl group is deceptive. All four bonds in the sulfate ion actually have partial double bond character, and the two negative charges are spread uniformly over all the oxygens.

Another inorganic acid of great importance for living systems is phosphoric acid, H_3PO_4. It has three dissociable protons, and its dissociation can be written

$$H_3PO_4 \rightarrow H^+ + H_2PO_4^- \rightarrow 2H^+ + HPO_4^{2-} \rightarrow$$
$$3H^+ + PO_4^{3-}.$$

Again, the oxygen atoms are tetrahedrally arranged around the phosphorus atom, and the dissociations are better represented as

$$O=\overset{\displaystyle OH}{\underset{\displaystyle OH}{P}}-OH \rightarrow O=\overset{\displaystyle OH}{\underset{\displaystyle OH}{P}}-O^- + H^+ \rightarrow$$

$$O=\overset{\displaystyle O^-}{\underset{\displaystyle OH}{P}}-O^- + 2H^+ \rightarrow O=\overset{\displaystyle O^-}{\underset{\displaystyle O^-}{P}}-O^+ + 3H^+.$$

Phosphorus often forms five covalent bonds, but nitrogen cannot.

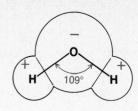

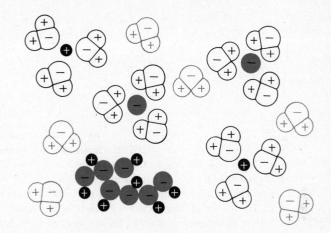

SOLVATION OF IONS by water molecules. Water molecule (top) is a miniature dipole, with the positive pole near the hydrogens and the negative pole near the oxygen. When HCl dissolves, both the positive and the negative ion can interact with water molecules, and this stabilizes the solution. **14**

The amines, with $-NH_2$ groups, are organic BASES. Methylamine, CH_3NH_2, is shown in Figure 15 *(top)*. If acids are substances which increase the hydrogen ion concentration in aqueous solution, bases are substances which decrease the concentration of H^+. Amines do this by combining with hydrogen ions. The positively charged proton is attracted to the lone electron pair on the nitrogen atom, and the two-electron covalent bond formed is in no way different from the two original $N-H$ bonds. The entire $-NH_3^+$ group as a whole possesses the positive charge. Common inorganic bases are sodium hydroxide, NaOH, and potassium hydroxide, KOH. These bases dissociate in solution and the hydroxyl ions produced combine with hydrogen ions already present in the solution to decrease the hydrogen ion concentration or increase the pH *(Box C)*:

$$NaOH \rightarrow Na^+ + OH^-$$
$$OH^- + H^+ \rightarrow H_2O.$$

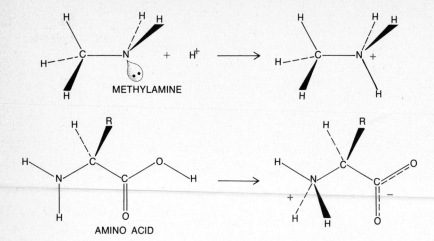

METHYLAMINE

AMINO ACID

15 **AMINES AND AMINO ACIDS.** The nitrogen atom in methylamine has a lone pair of electrons that can attract a proton from solution to form a positive ion. Amino acids have both carboxyl and amine groups attached to same carbon atom. Side groups (R) give each amino acid its individual chemical properties. In aqueous solution, both ends of amino acid are ionized to form a zwitterion.

Solutions of inorganic bases feel slippery to the touch because the hydroxide ion makes soap out of the fats of the skin on your fingertips.

The AMINO ACIDS are an important class of compounds that have both a carboxylic acid group and an amine group attached to the same carbon atom, called the α-carbon. Also attached to the α-carbon are a hydrogen atom and a side chain that gives the amino acid its distinctive chemical properties. Twenty amino acids are the building blocks of the giant protein molecules of living organisms. In aqueous solutions of amino acids, both the amine and carboxyl groups are ionized, as shown in Fig-

ure 15 (bottom). The carboxyl group has lost a proton, and the amine group has gained one. Such a double-ended ion is called a ZWITTERION, meaning a hermaphrodite or hybrid ion.

Whenever a tetrahedral carbon atom has four different atoms or groups attached to it, there will be two different ways of making the attachments, mirror images of one another. Such a carbon atom is called an asymmetric carbon, and the resulting compounds are called optical isomers of one another or enantiomorphs (your right and left hands are enantiomorphs). The α-carbon in an amino acid is an asymmetric carbon, and amino acids can exist in two isomeric forms called D- and L-amino acids (Figure 16). With a few rare exceptions in lower organisms, only L-amino acids are found in living things. There is nothing special about L-amino acids. In test tube reactions not involving other asymmetric compounds, their chemical properties are identical to those of D-amino acids. But as will become clear when we look at the structures of active sites in enzymes, it would be wasteful to set up a protein-synthesizing system that would accommodate both D- and L-amino acids. At an early stage in the development of life, living systems had to commit themselves to one isomer or the other. The choice was presumably random and accidental, but once made, it was preserved throughout the rest of the evolution of life.

The chemical compounds discussed in this chapter are all found in living organisms, but they fall far short of being alive. If we compare a living organism with a digital computer (an unflattering

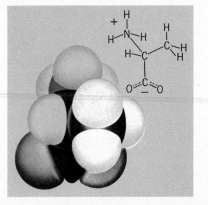

OPTICAL ISOMERS of the amino acid alanine are represented by these two molecular models. Structural formulas are given beside each model. Alanine can exist as one of two mirror images: L-alanine (left) or D-alanine (right). Naturally occurring proteins contain only L-amino acids.

pH and acidity

Substances which increase the concentration of hydrogen ion, H^+, in aqueous solution are acids, and substances which decrease the H^+ concentration are bases. By this definition, water itself is a weak acid, for it partially dissociates

$$H_2O \rightarrow H^+ + OH^-.$$

In pure water, the hydrogen ion concentration is 10^{-7} moles per liter. In 1 molar hydrochloric acid, the hydrogen ion concentration is 1 mole per liter, and in 1 molar sodium hydroxide, it is 10^{-14} moles per liter. When concentrations vary over such a wide range from 1.00 to 10^{-14}, the hydrogen ion concentration itself becomes an inconvenient quantity to handle. It is easier to work with the logarithm of the concentration.

The pH of a solution is defined as the negative logarithm of the hydrogen ion concentration

$$pH = -\log_{10}[H^+].$$

When the concentration is 10^{-7} moles per liter in pure water, the pH is 7. In 1 molar HCl, the pH is the negative logarithm of 1, or 0. In 1 molar NaOH, the pH is the negative logarithm of 10^{-14}, or 14. The proper control of acidity or pH in living organisms is critical. Blood plasma is kept in the pH range 7.3–7.5, or slightly basic. Milk has a pH of about 6.9, black coffee is 5.0, tomato juice is 4.1, orange juice is around 3.0, lemon juice is 2.1, and gastric juice varies between 1.3 and 3.0. On the basic side, a solution of baking soda has a pH of 8.4, and can therefore neutralize a limited amount of acid in the stomach. Household ammonia has a typical pH of 11.9, depending on its strength.

A mixture of a weak acid and its salt, such as acetic acid and sodium acetate, can act as a buffer to keep the pH controlled within limits. If more acid is added to such a buffer mixture, the added hydrogen ions combine with acetate ion from the sodium acetate to produce undissociated acetic acid. If more base is added, removing hydrogen ions, more acetic acid dissociates and restores the hydrogen ion concentration. Carbonic and phosphoric acid buffers are especially important in controlling the pH of fluids in living organisms.

---

simplification for the organism), the compounds we have seen so far are not even analogous to the transistors and capacitors. They are more like the lumps of silicon and germanium which with care and skill can be fashioned into transistors. In the next chapter, we shall climb the next step in the organization of living systems —the arrangement of molecules into macromolecules such as proteins, lipids, carbohydrates, and nucleic acids.

READINGS

T.R. Dickson, *Introduction to Chemistry*, New York, John Wiley & Sons, 1971. A good introductory chemistry textbook at a very simple level.

A.L. Neal, *Chemistry and Biochemistry: A Comprehensive Introduction*, New York, McGraw-Hill, 1971.

J.I. Routh, D.P. Eyman, and D.J. Burton, *Essentials of General, Organic, and Biochemistry*, Philadelphia, W.B. Saunders Company, 1969. Two introductory textbooks of chemistry at a slightly higher level, which blend the biochemical with the purely chemical. Not easy, but perhaps more palatable that way.

R.E. Dickerson and I. Geis, *Chemistry, Matter and the Universe*, Menlo Park, CA, W.A. Benjamin, 1976. A more extended introduction to chemistry. Written with a biological point of view, at about the same level as this chapter.

Real differences between biological and nonbiological molecules first appear at the level just above simple molecules. [The formation of long-chain molecules] occurs in nonbiological systems, but is of spectacular frequency and variety in biological ones. —Clifford Grobstein

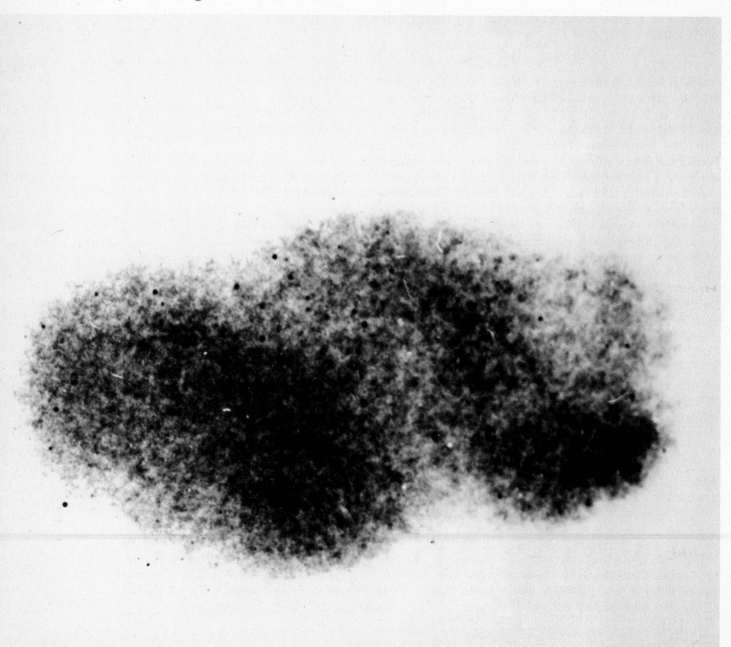

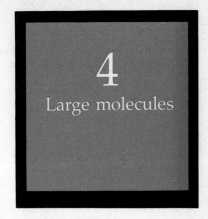

Instructions for making a new organism lie in the DNA (deoxyribonucleic acid) inherited from the parents in the fertilized egg. The chemical machinery of the egg is required to express this information, just as the electronic machinery of a playback deck is required to express music recorded on a tape. Clearly the music is on the tape and not built into the recorder. In exactly the same sense, genetic information lies in the DNA and not in the cell around it. Among the four types of macromolecule in the cell, DNA is the primary information carrier.

The information encoded in DNA governs only the sequence in which amino acids are strung together to make a protein chain. The way in which this chain folds after it has been put together, and the exact positions of important chemical groups at the active site of an enzyme, all are determined by the sequence in which amino acid side groups occur along the chain. There are no three-dimensional templates that guide the folding of a protein molecule or shape the geometry of an enzyme surface. Specifically, the information in the DNA of a cow says only that "The enzyme bovine ribonuclease shall begin with lysine, then continue with glutamic acid, threonine, three alanines, lysine,. . ." and so on. The molecule, once synthesized, assumes its intricate three-dimensional shape automatically.

The DNA molecule is the master control of the cell, but the only processes that it controls directly are its own replication and the synthesis of proteins, including enzymes. Enzymes then direct the

thousands of reactions that occur every second in the cell. Enzymes are biological catalysts: compounds that speed up chemical reactions without being permanently changed in the process. Their power to control cell chemistry is a result of their astonishing efficiency. The proper enzyme can speed up a reaction by as much as 10^8 times —a factor that makes the difference between a reaction time of one second or 3 years! Each enzyme catalyzes only one particular type of reaction, and if the enzyme is absent, the reaction proceeds so slowly that for all practical purposes it does not occur at all.

DNA does not contain direct information for making lipids or carbohydrates. Instead, the enzymes (proteins), that have been made from instructions in DNA, catalyze the particular reactions that put fats and glycerol and phosphate together to make lipids, or snap sugar rings together to make cellulose or starch, both polymers of carbohydrate. Classical geneticists speak of "the gene for blue eyes" in humans or "the gene for short wings" in fruit flies. But no one stretch of DNA is solely responsible for causing eyes to be a particular color. Instead, several sections of DNA make enzymes that control various steps of the chemical reactions by which a pigment is synthesized or not synthesized, or by which growth of wings is affected. The organismic biologist talks about the genes that produce wrinkled seeds in peas. The molecular biologist, if he knew the whole story, might speak instead about the DNA that makes the enzymes that critically affect the seed-wall building process, leading to partial loss of moisture and wrinkling. In the chain of genetic cause and effect, the middle links are always enzymes.

An elaborate apparatus has evolved for translating information from DNA into RNA (ribonucleic acid), and then into protein sequences (*Chapter 7*). If the amino acid sequence is really a translation of

SEGMENT OF DNA on opposite page was photographed with a field ion microscope. Helical structure of molecule is clearly visible. Even though DNA is an enormous molecule, details of its structure lie beyond the resolving power of even the most powerful electron microscopes.

55

Each enzyme catalyzes only one
particular type of reaction, and if the
enzyme is absent, the reaction
proceeds so slowly that for all
practical purposes it does
not occur at all.

what was in the DNA, then both molecules contain information. A biochemist can look at the amino acid sequence of an enzyme and tell that it is chymotrypsin and not trypsin. What is more, he can usually tell the species it came from. The amino acid sequence of a protein is almost as characteristic of a species as fingerprints are of an individual. Because of this special property of DNA and proteins, Emil Zuckerkandl and Linus Pauling have called them SEMANTOPHORES, which simply means information carriers. Lipids and carbohydrates are not semantophores. Some types of lipid are found more often in certain kinds of organisms, but in general the structure of a lipid does not pinpoint what species it came from and exactly what it was doing in that species. Starch and cellulose are about the same no matter what their source. In contrast, nucleic acids and proteins are the really vital components of living organisms, and are the sole true bearers of information. Lipids and carbohydrates, although macromolecules, are passive raw materials to be shaped directly by enzymes, and indirectly by DNA as the synthesizer of enzyme proteins.

FATS, LIPIDS, AND MEMBRANES

Fatty acids are carboxylic acids with long hydrocarbon tails. Palmitic acid, $C_{15}H_{31}COOH$, is common in animal fats and lipids:

$$CH_3-CH_2-CH_2-CH_2-(CH_2)_8-CH_2-CH_2-CH_2-\overset{\overset{\displaystyle O}{\|}}{C}-OH.$$

So is stearic acid, $C_{17}H_{35}COOH$, which has two more carbon atoms:

$$CH_3-CH_2-CH_2-CH_2-(CH_2)_8-CH_2-CH_2-CH_2-CH_2-CH_2-\overset{\overset{\displaystyle O}{\|}}{C}-OH.$$

These are both called SATURATED fatty acids because their hydrocarbon tails contain no double bonds. The common oleic acid is UNSATURATED:

$$CH_3-CH_2-(CH_2)_4-CH_2-CH_2-CH=CH-CH_2-(CH_2)_4-CH_2-CH_2-\overset{\overset{\displaystyle O}{\|}}{C}-OH.$$

It can be hydrogenated under high H_2 pressure with a nickel catalyst to produce the saturated stearic acid. Other fatty acids have several carbon-carbon double bonds and are said to be POLYUNSATURATED.

The fatty acids are schizophrenic molecules. Their carboxyl groups are polar and ionize in water, but their long hydrocarbon tails make them insoluble. The dual personality of the fatty acids gives soaps their cleaning ability. Soaps are the sodium or potassium salts of fatty acids *(Box A)*. One of the most common soaps is sodium stearate, which can be regarded as the salt formed from stearic acid and sodium hydroxide:

$$CH_3-(CH_2)_{16}-COO^-Na^+.$$

Oil and grease, in the presence of water, normally segregate into a separate phase and do not mix with the water. Soaps emulsify greases, as shown in Figure 1, and bring them into solution as small droplets (micelles) that can easily be washed away.

The salts of fatty acids with doubly charged calcium and magnesium ions, Ca^{2+} and Mg^{2+}, are

Fats and fatty acids are the chief
molecules for energy storage in
animals. They have twice the energy
content per gram of carbohydrates or
proteins.

insoluble. Calcium stearate has no detergent action and is not a soap. Even worse, in hard water containing these ions, soap combines with Ca^{2+} and Mg^{2+} to create a film or scum. One can wash in hard water either by adding so much soap that all the calcium and magnesium ions are removed as scum and more soap is left for the grease, or more efficiently by removing the offending ions

The naming of an element

The leachings from wood ash cooking fires, or pot ash, used in primitive soap-making, were called caustic potash because they would burn the skin as they made soap of one's natural oils. The extract was actually a highly impure solution of KOH (potassium hydroxide). The element that was finally separated electrolytically from purified caustic potash by Sir Humphrey Davy in 1807 was therefore named potassium in English and French. The chemical symbol, K, comes from the modern Latin and German name for the element, *kalium*. This ultimately derives from the old Latin *calere*, "to burn." The Arabic alchemists added the definite article, and gave the name, *Al kali*, "the kali," to wood ashes used in soap and glassmaking. From this we have obtained the term alkali for the hydroxides of lithium, sodium, and potassium. From the old original Latin *calere* we have also obtained calorie and calcium, cauldron, and caldera for a volcanic crater.

---

beforehand. Home water softeners are ion exchange resins that bind Ca^{2+} and Mg^{2+} as the hard water is run through, and release harmless Na^+ ions in their place. The resin has to be regenerated periodically by running concentrated NaCl solution through the tank to flush out the Ca^{2+} and Mg^{2+} ions and recharge the resin with Na^+.

The combination of three fatty acid molecules with glycerine is a FAT:

$$CH_3(CH_2)_{16}-\overset{\overset{\textstyle O}{\|}}{C}-OH \quad HO-CH_2$$

$$CH_3(CH_2)_{16}-\overset{\overset{\textstyle O}{\|}}{C}-OH + HO-CH \rightarrow$$

$$CH_3(CH_2)_{16}-\overset{\overset{\textstyle O}{\|}}{C}-OH \quad HO-CH_2$$

STEARIC ACID   GLYCEROL

FATTY ACIDS   GLYCERINE

$$CH_3(CH_2)_{16}-\overset{\overset{\textstyle O}{\|}}{C}-O-CH_2$$

$$CH_3(CH_2)_{16}-\overset{\overset{\textstyle O}{\|}}{C}-O-CH + 3H_2O.$$

$$CH_3(CH_2)_{16}-\overset{\overset{\textstyle O}{\|}}{C}-O-CH_2$$

TRISTEARIN

FAT

The three fatty acids in one molecule do not have to be the same length, nor do they all have to be either saturated or unsaturated. Fats with short chains and unsaturated chains are usually oily liquids. Fats with long and saturated chains are waxy solids. Animal fats such as lard and tallow are usually long-chain virtually saturated solids, with palmitic, stearic, and oleic acids the most common. Hydrocarbon chain lengths range between 10 and 20 carbon atoms. Vegetable fats tend

DETERGENT ACTION of soaps breaks oil or grease into droplets. Soap molecules (small disks) surround the oil droplets (large disks). Hydrocarbon tails of soap molecules are dissolved in the oil while their heads remain in the surrounding water. The grease is dispersed, and the mutual repulsion of negative charges on the soap molecules keeps the droplets apart and suspended in water.   **1**

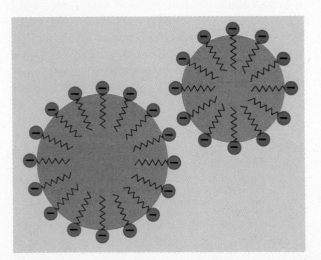

to be more unsaturated, oily liquids. These are hardened in the manufacture of margarine by hydrogenating the double bonds and saturating the chains.

Fats and fatty acids are the chief molecules for energy storage in animals. They contain twice as much energy per gram as carbohydrates or proteins. The hydrocarbon tails of fats are literally the gasolines of the body. One curious feature of fatty acids from animal fats is that they usually have an even number of carbon atoms. For example, there are 16 carbon atoms in palmitic acid, and 18 in stearic and oleic acids. This is really not so mysterious. In the act of storing energy, fatty acids are synthesized in the body by adding acetic acid units to the growing chain, and acetic acid has two carbon atoms. This especially efficient energy storage system is appropriate in mobile creatures, like animals, where weight is at a premium. Stationary plants store energy primarily in carbohydrates: polymers of sugars called starches.

Fats are insoluble in water, for they have lost their polar carboxylic acids. Since most of the chemical reactions of the body take place in aqueous solution, how can fats be brought into solution so their energy can be extracted? This is done in two steps. Bile salts, detergents secreted by the liver into the intestines (*Figure 6e*), emulsify fats in the way that soap lifts grease, producing micelles with large surface areas. Enzymes called LIPASES then attack the fats, breaking them down to fatty acids and glycerol, both of which are much more soluble in water than are fats. The fatty acids are finally degraded systematically to obtain the energy stored in their hydrocarbon chains.

If fats are digested with sodium or potassium hydroxide, the products formed are glycerine and sodium or potassium soaps. The American frontiersman knew that water leachings from wood ashes, when boiled with lard or tallow, produced a very serviceable soap. The beads of glycerine that can be squeezed out of this homemade soap are a valuable byproduct of commercial soap manufacture, valuable enough to defray much of the cost of making the soap.

The reaction by which fats are produced from fatty acids is an example of a general reaction called ESTERIFICATION, in which an acid and an alcohol combine with the elimination of water to form an ester

$$CH_3-\overset{\overset{\displaystyle O}{\|}}{C}-OH + HO-CH_2CH_3 \rightarrow CH_3-\overset{\overset{\displaystyle O}{\|}}{C}-O-CH_2CH_3 + H_2O.$$

ACETIC ACID   ETHANOL   ETHYL ACETATE

Esters can also be formed with inorganic acids

$$HO-\overset{\overset{\displaystyle O}{\|}}{\underset{\underset{\displaystyle OH}{|}}{P}}-OH + HO-C_2H_5 \rightarrow$$

PHOSPHORIC ACID   ETHANOL

$$HO-\overset{\overset{\displaystyle O}{\|}}{\underset{\underset{\displaystyle OH}{|}}{P}}-O-C_2H_5 + H_2O.$$

ETHYL PHOSPHATE

If two alcoholic —OH groups of glycerol are esterified with fatty acids, but the third reacts with phosphoric acid, the product is a PHOSPHOLIPID:

$$CH_3-CH_2-(CH_2)_8-CH_2-\overset{\overset{\displaystyle O}{\|}}{C}-O-CH_2$$

$$CH_3-CH_2-(CH_2)_{10}-CH_2-\overset{\overset{\displaystyle O}{\|}}{C}-O-CH$$

$$H_2C-O-\overset{\overset{\displaystyle O}{\|}}{\underset{\underset{\displaystyle OH}{|}}{P}}-OH.$$

The two protons on the phosphate group can dissociate as in any acid, giving the phospholipid molecule a double negative charge. The molecule now resembles a fatty acid in having a hydrophobic hydrocarbon tail and a charged polar head. The phosphate itself can be esterified again with one of several compounds (usually of nitrogen) to form a PHOSPHATIDE such as lecithin:

$$R'-\overset{\overset{\displaystyle O}{\|}}{C}-O-CH_2$$

$$R''-\overset{\overset{\displaystyle O}{\|}}{C}-O-CH$$

$$H_2C-O-\overset{\overset{\displaystyle O}{\|}}{\underset{\underset{\displaystyle O^-}{|}}{P}}-O-C_2H_4-N^+(CH_3)_3$$

PHOSPHATIDYL CHOLINE (LECITHIN)

or cephalin (*on opposite page*), two compounds found in cell membranes.

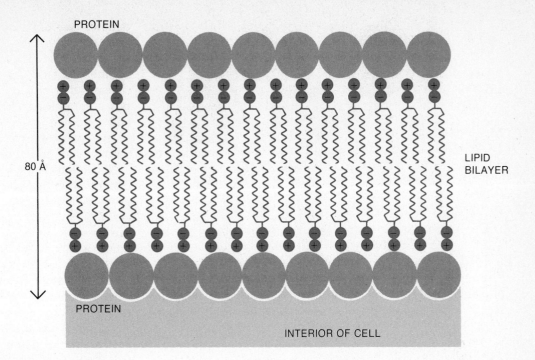

PROTEIN

80 Å

LIPID BILAYER

PROTEIN

INTERIOR OF CELL

2 MODEL OF CELL MEMBRANE shows a bilayer of phospholipid molecules with hydrocarbon tails inside and polar heads outside. Phospholipids are sandwiched between two layers of protein molecules. Thickness of 80 Å agrees with observed thickness of membranes in electron micrographs.

$$R'—\overset{\overset{O}{\|}}{C}—O—CH_2$$
$$R''—\overset{\overset{O}{\|}}{C}—O—CH$$
$$H_2C—O—\overset{\overset{O}{\|}}{\underset{\underset{O^-}{|}}{P}}—O—C_2H_4—NH_3^+.$$

PHOSPHATIDYL ETHANOLAMINE (CEPHALIN)

Here R' and R" stand for "residue," and represent any fatty acid chains. These molecules have two hydrocarbon tails and a doubly charged head that is strongly hydrophilic but electrically neutral, and so does not require any small ions nearby to neutralize the charge.

Lecithin is an important component of cell membranes. Phospholipids added to a beaker of water will form a film on the surface, with their polar heads in the water and hydrophobic tails out of it. If one draws the film down into the water, he can make a lipid bilayer with many of the mechani-

cal and chemical properties of a biological membrane. A simplified but useful model for the structure of cell membranes is shown in Figure 2. It consists of a double layer of lipids such as lecithin, with a covering layer of protein molecules on each side. The dimensions, calculable from the known dimensions of lecithin molecules, are quite close to what we can see as double-layered boundaries in electron micrographs (*Figure 3*). A more realistic model of the structure of a biological membrane, Figure 4, has the protein molecules embedded in the lipid bilayer and capable of limited diffusion within the membrane surface.

The word lipid is really a functional term that

CELL MEMBRANE in this electron micrograph appears **3** as a pale band sandwiched between two dark bands. Membrane shown here is from a red blood cell.

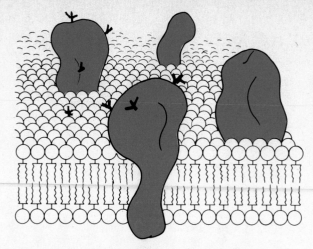

4 MODEL OF CELL MEMBRANE shows large protein molecules (color) embedded in a two-layer mosaic of lipid molecules. Some proteins extend all the way through the lipid layers, and can diffuse, to some extent, from one region of the membrane to another.

5 CAROTENOID PIGMENTS are depicted in a shorthand notation that can be deciphered by comparing the two ends of the molecule of β-carotene, which is symmetric about the double bond printed in color. Crucial double bond in retinene that is triggered by light is detailed at bottom.

β-CAROTENE

VITAMIN A

RETINENE

TRANS

CIS

denotes any substance that is easily extractable from animal or vegetable tissues by organic solvents such as ether, chloroform, or benzene. In addition to fatty acids, fats, and phospholipids, several other organic molecules with special purposes are classed as lipids, including carotenoids and steroids.

The CAROTENOIDS are a family of light-sensitive pigments found in both plants and animals (*Figure 5*). β-carotene is the pigment that activates the process of phototropism —the tendency of plants to grow toward the light. It is the pigment that colors carrots, tomatoes, pumpkins, egg yolk, and butter. We need β-carotene, or its breakdown product vitamin A, to maintain normal body reactions to light.

In the history of evolution, eyes of advanced construction were invented independently by three different types of organism: by mollusks (such as the scallop and the octopus), by insects, and by vertebrates. Each time the visual system evolved, it was based on identical chemical compounds — retinene and special-purpose proteins called opsins, including rhodopsin. When light strikes a retinene molecule it twists one carbon-carbon double bond 180°, from the kinked *cis* configuration on the right in Figure 5, to the extended *trans* geometry on the left. This molecular spasm triggers the production of a nerve impulse that travels along the optic nerve to the brain. Because of the energy barrier that resists twist about a double bond, retinene molecules do not flip over and trigger nerve impulses randomly, but do so only when they are hit by a burst of light energy. They are subsequently restored to the *cis* configuration, and the molecular trap is set for another burst of light. No animal can entirely synthesize retinene or vitamin A; they must be obtained from plant carotenoids. This is one reason why minute amounts of vitamin A are absolutely essential in the diet.

The STEROIDS are a family of organic compounds based on a common multiple-ring system, and are used chiefly as HORMONES, chemical signals that carry messages from one part of the body to another. Testosterone (*Figure 6a*) regulates sexual development in male vertebrates, and the estrogens play a similar role in females. Cortisone (*Figure 6b*) is one of a family of hormones secreted in the adrenal cortex which play a wide variety of regulatory roles. Vitamin D (*Figure 6c*) regulates the absorption of calcium from the intestines, either by altering the permeability of membranes to calcium ions or by active transport. Vitamin D is necessary for proper deposition of calcium in bones, and a

6 IMPORTANT STEROIDS include the hormone cortisone and the male sex hormone testosterone, as well as vitamin D, cholesterol, and the bile salt glycocholic acid.

a TESTOSTERONE

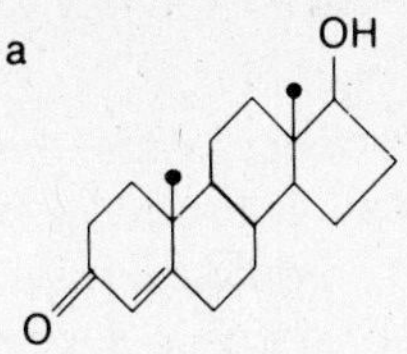

b CORTISONE

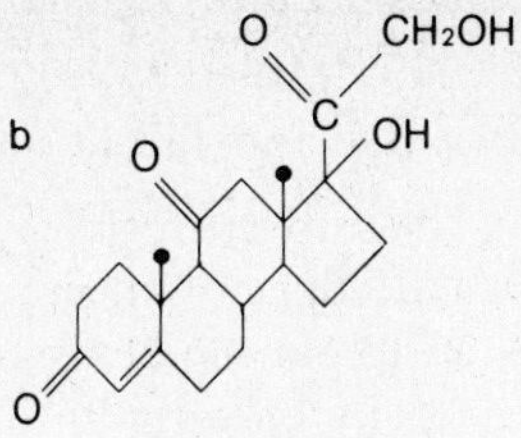

c VITAMIN D

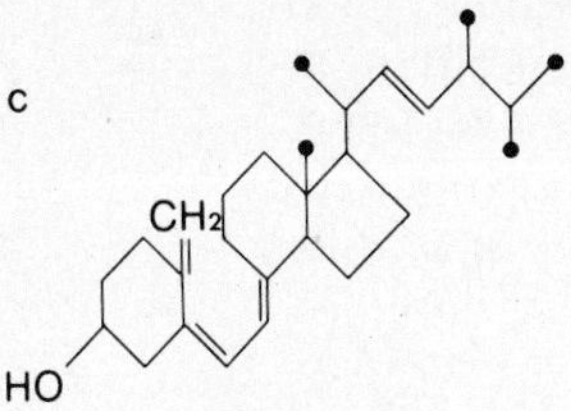

d CHOLESTEROL

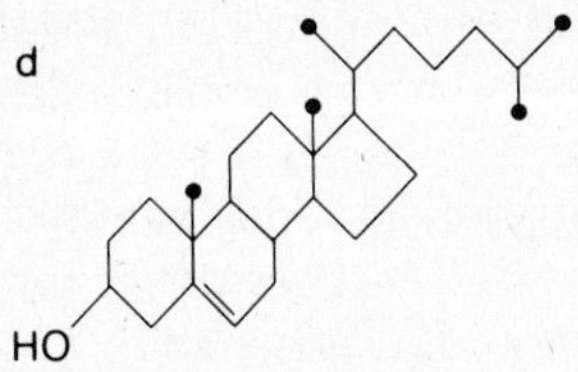

e GLYCOCHOLIC ACID

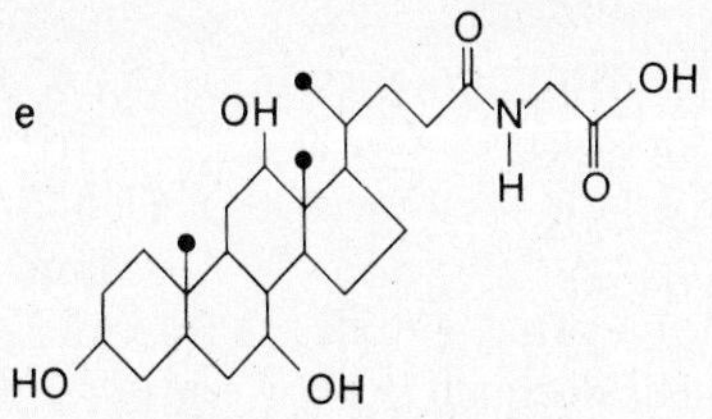

deficiency of vitamin D leads to rickets, a bone-softening disease. Vitamin D is produced when certain other sterols are irradiated with sunlight or ultraviolet light.

Cholesterol *(Figure 6d)* is synthesized in the liver. It is a raw material for testosterone and several other steroid hormones, and for the bile salts that help to get fats into solution for digestion. The salt of glycocholic acid *(Figure 6e)* is one such bile salt, and is a powerful detergent. Too much cholesterol, deposited in the aorta, leads to arteriosclerosis and heart attack in adults. Cholesterol is absorbed from foods such as milk, butter, and fats. Low-cholesterol diets are sometimes recommended to ward off heart trouble. Just enough cholesterol is essential, but too much can be lethal.

In brief, lipids are a mixed bag of compounds with a wide variety of uses: energy storage, digestion, membrane structure, bone formation, photosensitivity for growth and for vision, and chemical signaling. Most lipids can be synthesized in the body, and synthesis and storage of fats is an important means of locking energy away until needed. The few that cannot be synthesized must be obtained in small amounts in the diet. These are the fat-soluble vitamins: A, D, E, and K (essential for blood clotting).

CARBOHYDRATES

Carbohydrates are a class of organic compounds with the general formula: $C_n(H_2O)_m$, where n and m are whole numbers. The name carbohydrate is derived from the old, erroneous idea that they were in some sense compounds of carbon and water. The most important carbohydrates for our purposes are the sugars, starch, glycogen, and cellulose. SUGARS such as glucose are efficient energy yielding molecules, although only half as good as fats on a weight basis. STARCHES and GLYCOGENS are long, branched-chain polymers of glucose, and are the media for carbohydrate energy storage in plants and animals, respectively. CELLULOSE is a slightly different linear polymer of glucose, which performs the same kind of supporting and structural role in plants that the calcium skeleton does in vertebrates. Chitin, a carbohydrate derivative, accomplishes the same purpose in insect exoskeletons, and even the cell walls of bacteria have carbohydrate components. Like the witch in Hansel and Gretel, organisms on this planet seem to have fashioned their bodies from edibles, so that one creature's architecture is another's dinner.

The linkage between sugar units in cellulose is a simple evolutionary trick to defeat the enzymes that digest starch. Very few organisms can digest cellulose. But for cows and termites, whose guts harbor microorganisms that can synthesize the enzyme cellulase, even cellulose becomes a good meal *(Box B)*.

Lipids are a mixed bag of compounds with a wide variety of uses: energy storage, digestion, chemical signaling, membrane structure, bone formation, and photosensitive pigments for growth and vision.

B The rise and fall of nonfood

The only current way in which man can make nutritional use of cellulose, although admittedly not an unpleasant one, is to feed grass to cows or other ruminants, and then to eat the beefsteak. Direct degradation with the enzyme cellulase, which is made by bacteria in the cow's stomach, is successful in the laboratory but is not a commercial proposition. Cows are still cheaper than chemists.

One biochemist, while looking for a better tire cord material at American Viscose, did discover how to make a form of cellulose which was at least edible even if it was nutritionally worthless. O.A. Battista sheared cellulose molecules mechanically in a Waring blender and produced a low-molecular weight cellulose product that could be used in diet foods in place of flour. It was not absorbed in the digestive tract, and had no taste, no calories, no vitamins, no nutritional or food value whatever. Cookies and cakes made with the material would stave off hunger pangs for a few hours, fooling the stomach into equating "full" with "fed." It was hailed by *Life* magazine in 1961 as the answer to the average American's tendency to overeat, but sank into rapid oblivion. At the time there seemed to be a concensus that our food-processing industries produce enough nutritionally worthless foods without bothering with anything that began that way.

More recently, however, "edible" cellulose has found a use in diet foods and as a substitute for milk and ice cream in milk shakes sold in cheap drive-in restaurants. The inventor received a promotion and a pay raise, but as is often the case in industrial laboratories, had signed his patent rights away in advance for the token sum of one dollar, making him ineligible to share in any financial bonanza from the use of his product in diet foods. As *Life* concluded, it was somehow fitting that the inventor of nonfood be paid in nonmoney.

Sugars are classified according to the number of carbon atoms they contain. Three-carbon sugars are called trioses. Sugars with more carbons are tetroses, pentoses, hexoses, and heptoses. To the biologist, perhaps the most important are two pentoses, RIBOSE and DEOXYRIBOSE, and two different hexoses, FRUCTOSE and GLUCOSE. In biological systems these sugars exist as closed rings of atoms. Ribose has the five-membered ring structure shown in Figure 7, which is closed by an oxygen atom. The ring

is almost flat because the interior angle of a regular pentagon is 108°, and the angle between the tetrahedral bonds of carbon is 109.5°. Each carbon in the ring has a hydrogen atom and either a hydroxyl group or —CH₂OH attached to it. Ribose is particularly interesting to biologists because it is incorporated into RIBONUCLEIC ACID or RNA. Deoxyribose (*Figure 7*) lacks the oxygen of the hydroxyl group at C_2. The similar polymer of deoxyribose is DEOXYRIBONUCLEIC ACID, or DNA.

Fructose is a six-carbon hexose, although it has the five-membered ring of the pentoses (*Figure 8*). Glucose is a hexose with the same chemical composition as fructose, $C_6H_{12}O_6$, but with a six-membered ring. The five-membered ring in fructose is flat, but a six-membered ring must be puckered. However, in Figure 8 and subsequent drawings, it will be shown as a flat ring for simplicity. In water solution, the glucose ring opens and closes by breaking and making the bond between the ring oxygen and carbon position 1, so glucose is a mixture of the α and β forms shown in Figure 8, differing in the location of the —OH group on carbon 1. Interchanges of —H and —OH at any other carbon position lead to chemically distinct molecules.

7 TWO EXAMPLES OF PENTOSES, or five-carbon sugars. Ribose is part of the backbone of ribonucleic acid, or RNA. Deoxyribose has one fewer—OH group, and occurs in deoxyribonucleic acid, or DNA.

RIBOSE   DEOXYRIBOSE

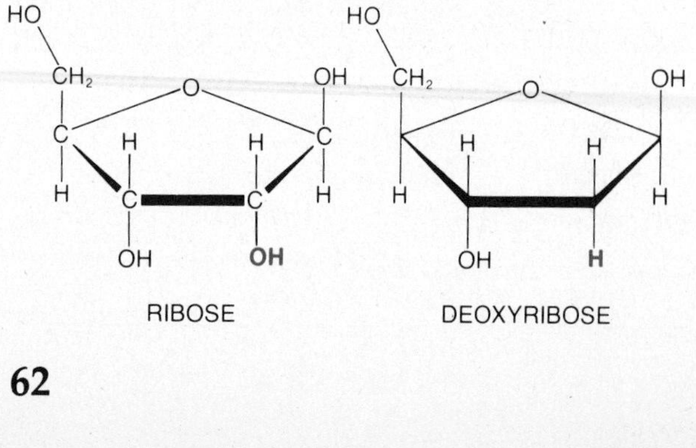

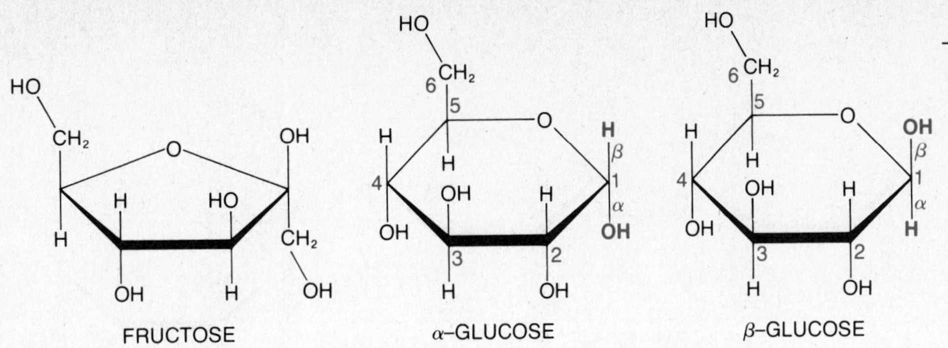

FRUCTOSE α–GLUCOSE β–GLUCOSE

8 **THREE EXAMPLES OF HEXOSES,** or six-carbon sugars, with the overall formula $C_6H_{12}O_6$. Fructose has the five-sided ring structure characteristic of pentoses. Six-sided rings of α- and β-glucose differ only at carbon position 1.

Mannose, with an interchange at carbon position 2, and galactose, with an interchange at position 4, are found in nature (*Figure 9*), but all the other variations that might be imagined are no more than laboratory curiosities.

The sugars depicted so far are monosaccharides: sugars with one ring. Disaccharides (*Figure 10*) are common in nature, especially sucrose, which is found in cane, beets, and a great many other plants. Maltose is the disaccharide breakdown product of starch, and hence is encountered in the brewing process. Lactose is a milk sugar produced only by mammals. Cellobiose is not a natural disaccharide, but is a breakdown product of cellulose, as maltose is for starch. Note that lactose and cellobiose, the exclusive products of a special class of animals and of plants, respectively, differ only by the conformation of −H and −OH around the C_4 carbon atom in one of the two rings. This indicates just how specific are the enzymes that assemble biological molecules.

Maltose and cellobiose are both comprised of

two joined glucose rings, but the maltose units are connected through the α position at carbon 1 whereas cellobiose uses a β connection. If we imagine maltose to be extended indefinitely we obtain STARCH. If we do the same with cellobiose we get CELLULOSE. Such compounds are called polymers —giant molecules assembled from subunits called monomers. Starch and cellulose are both polymers of glucose, but they are very different compounds. Amylases, the enzymes that make and break the α bonds of starch, cannot touch the β bonds of cellulose. The foods and the containers are therefore not confused in plants, although the same basic pieces are used to build both. Cellulose chains are long, straight, and unbranched, with several thousand glucose molecules per chain. The purest source is cotton, which is 90 percent pure cellulose. Cellulose is also the standard building material for woody stalks, fibers, and all types of cell walls in plants. It is by far the most common organic compound on the planet, containing over half the carbon involved in plant life.

Starch chains are branched like a thicket, with tens of thousands of glucose monomer units per branched-chain molecule of starch. A similar branched-chain glucose polymer, GLYCOGEN, is sometimes called animal starch. Glycogen is used for energy storage in liver and muscle. The main energy reserves in animals are stored as fats. In animals a much smaller amount of energy is locked

9 **GLUCOSE, MANNOSE AND GALACTOSE** all have the formula $C_6H_{12}O_6$, and differ only in the arrangement of −H and −OH around carbons 2 and 4.

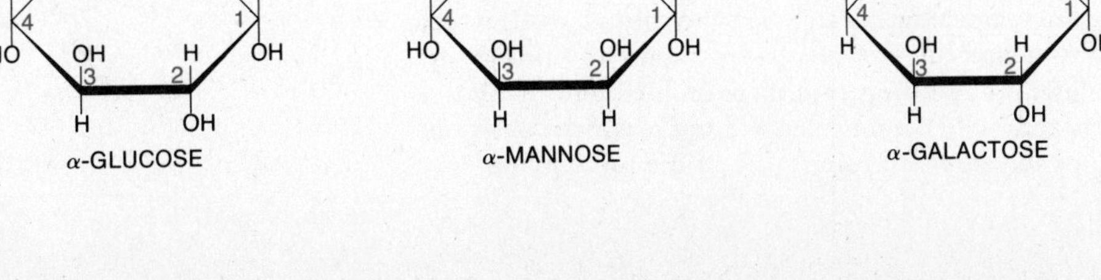

α-GLUCOSE α-MANNOSE α-GALACTOSE

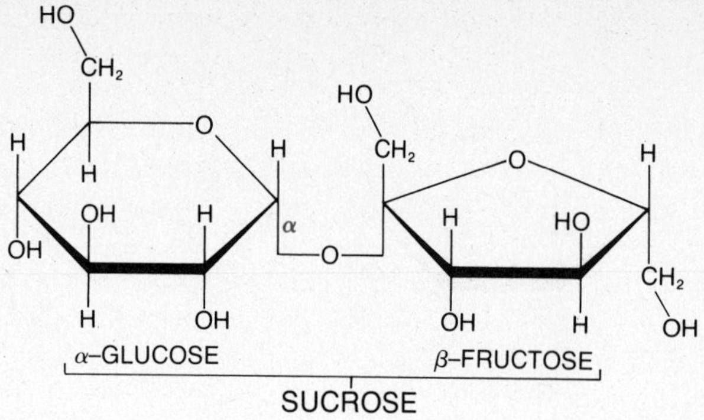

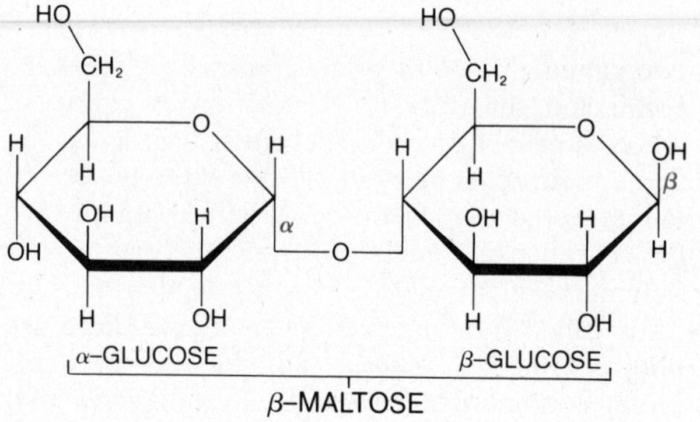

α-GLUCOSE   β-FRUCTOSE

SUCROSE

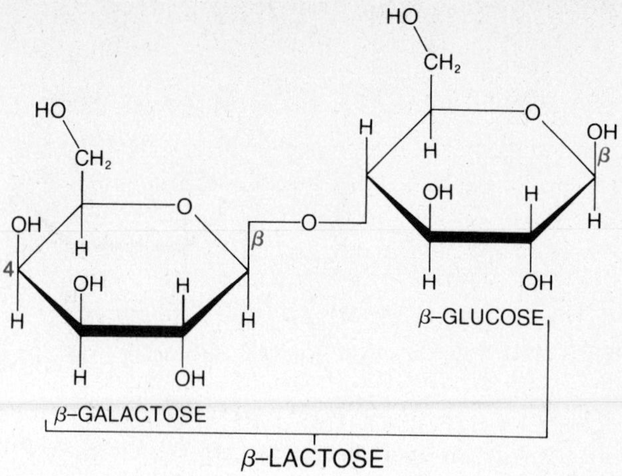

β-GALACTOSE

β-LACTOSE

α-GLUCOSE   β-GLUCOSE

β-MALTOSE

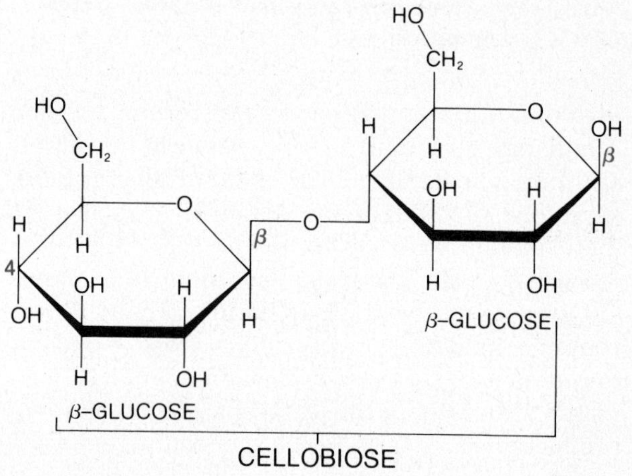

β-GLUCOSE   β-GLUCOSE

CELLOBIOSE

10   **COMMON DISACCHARIDES. Sucrose is an α,1,2 linkage of glucose and fructose. Maltose is an α,1,4 linkage of two glucose molecules. Lactose is a β,1,4 linkage of galactose and glucose. Cellobiose is a similar linkage of two glucose molecules. Maltose linkage also occurs in starch molecules.**

away in glycogen than in fats, but the glycogen reserves are used to stabilize the level of sugar in the blood. It is a rapid access energy store, whereas fats require more intermediate steps for extraction of their energy.

AMINO ACIDS AND PROTEINS

Proteins are long-chain polymers of amino acids. In the polymerization reaction, the carboxyl group from one amino acid is linked to the amine of the next, splitting out water and forming a peptide bond (*Figure 11*). In Figure 12a the peptide bond between C and N is shown as a single bond, and a

lone electron pair is drawn on the N atom. Another electron pair (one electron each from C and O) forms the second bond of the carbon-oxygen double bond. But one could imagine a shift of these two electron pairs so the pair that formerly contributed the second carbon-oxygen bond moved to the oxygen atom and became a lone pair, while the former lone pair on nitrogen moved between C and N to give these two atoms a double bond. This

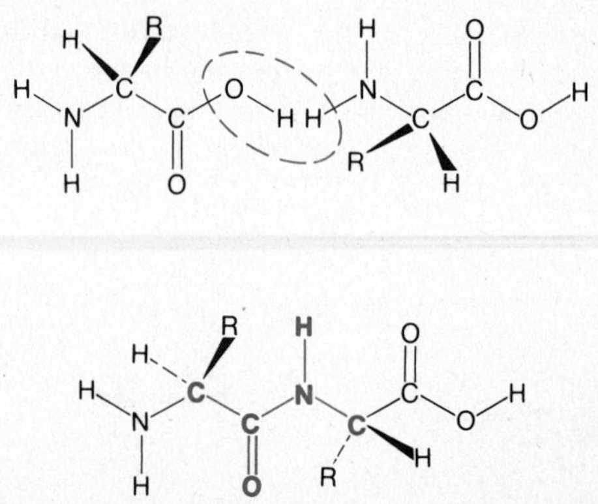

11   **PEPTIDE BOND links amino acids to form a polypeptide chain. At top, two amino acids are linked by the removal of the atoms in the colored oval, which forms a molecule of water. R represents any side group. Below, the resulting peptide or amide bond involves a partial double bond between C and N which constrains all of the atoms in color to lie in the same plane.**

situation, involving only a shift of electrons, is shown in Figure 12b. Such a shift would place a net negative charge on the oxygen and a positive charge on the nitrogen. If the double bond were between carbon and nitrogen, then it would be impossible to twist the chain about this bond without reducing the double bond to a single bond, an operation that would require 64 kilocalories of energy per mole.

The true bond pattern in the peptide group is somewhere between the extremes of Figure 12a and b, and can be represented approximately by Figure 12c. Both the carbon-oxygen and carbon-nitrogen bonds are more than single and less than double. The energy required to twist the chain about the carbon-nitrogen bond is more like 21 kilocalories than 64 kcal. But this is still too high a price to pay. In every protein whose atomic structure is known from x-ray analysis, the peptide groups are always planar. The six-atom cluster outlined by the colored dotted lines in Figure 12c is a rigid, planar unit, and the only degrees of freedom allowed to the protein chain are the swiveling of this unit about the bonds connecting it to the carbons. This is the most important single principle of protein chain folding.

12 PLANARITY OF PEPTIDE BOND is further explained in these diagrams. Partial double bond between C and N arises because N shares its lone pair of electrons with C (center). Repulsion between electron pairs then pushes the electron pair of the C═O double bond nearer to O atom. This weakens the C═O double bond and confers a slight negative charge on the O atom. H atom acquires a slight positive charge. These slight charges are important in hydrogen bonding in proteins. Dotted lines indicate plane of peptide bond (bottom).

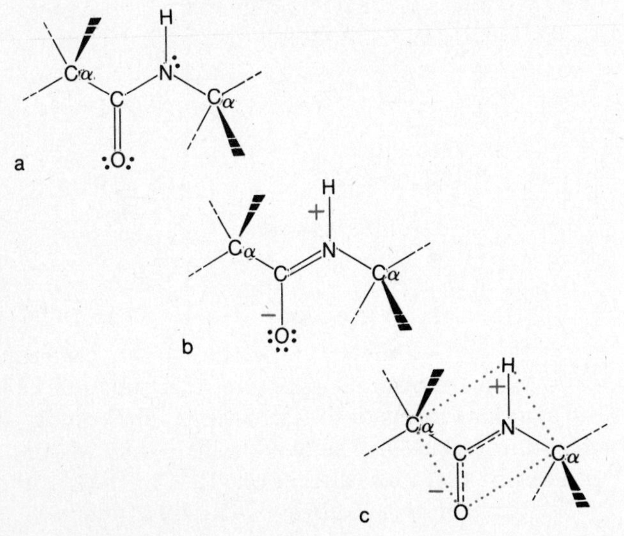

Another effect of this bond structure is that both the carbonyl or C═O oxygen and the amide or N—H hydrogen are particularly well suited for hydrogen bonding—one with a negative charge and the other with a positive. Hydrogen bonds are an important element in holding protein chains together.

As was mentioned earlier, the only information that is coded in the DNA is the order of amino acids in the protein chain. Twenty different amino acids are coded by DNA, and their side chains (R in Figure 11) show a wide variety of chemical properties. The order of amino acids determines how the protein folds. Perhaps the most straightforward classification of amino acids is based on whether they are usually found on the inside of a folded protein molecule, on the surface, or both.

Valine, leucine, isoleucine, phenylalanine, and methionine are all either hydrocarbons or near-hydrocarbons of varying bulk:

Amino acid	Side chain, R, in Figure 11
Valine	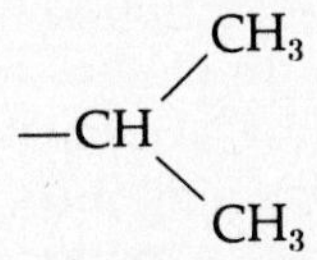 $-CH\begin{smallmatrix}CH_3\\CH_3\end{smallmatrix}$
Leucine	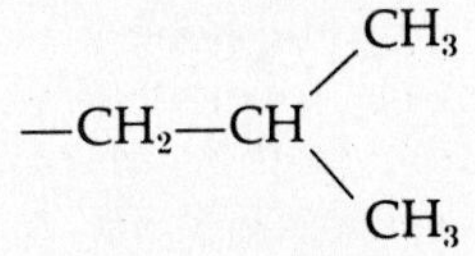 $-CH_2-CH\begin{smallmatrix}CH_3\\CH_3\end{smallmatrix}$
Isoleucine	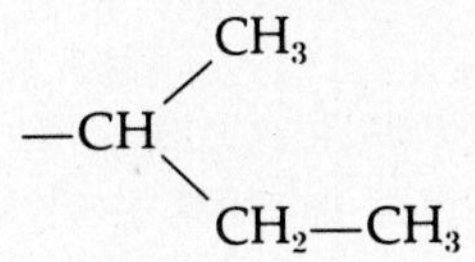 $-CH\begin{smallmatrix}CH_3\\CH_2-CH_3\end{smallmatrix}$
Phenylalanine	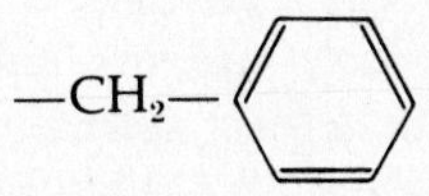 $-CH_2-\bigcirc$
Methionine	$-CH_2-CH_2-S-CH_3$

All of these side chains are hydrophobic—they are more stable when removed from an aqueous environment. Since most enzymes and similar proteins operate in aqueous solution, those parts of the protein chain that bear these hydrophobic amino acids tend to fold into the interior of the molecule. These side chains are the protein analogues of the hydrocarbon tails of phospholipids.

Seven strongly hydrophilic amino acids show behavior just the opposite of the hydrophobic amino acids.

Amino acid	Un-ionized form	Ionized form
Aspartic acid	$-CH_2-COOH \rightarrow$	$-CH_2-COO^- + H^+$
Glutamic acid	$-CH_2-CH_2-COOH \rightarrow$	$-CH_2-CH_2-COO^- + H^+$
Asparagine	$-CH_2-CO-NH_2$	
Glutamine	$-CH_2-CH_2-CO-NH_2$	
Lysine	$-CH_2-CH_2-CH_2-CH_2-NH_2 + H^+ \rightarrow$	$-CH_2-CH_2-CH_2-CH_2-NH_3^+$

Arginine

$$-CH_2-CH_2-CH_2-NH-C\begin{smallmatrix}NH\\ \\NH_2\end{smallmatrix} + H^+ \rightarrow -CH_2-CH_2-CH_2-NH-C\begin{smallmatrix}NH_2^+\\ \\NH_2\end{smallmatrix}$$

Histidine

$$-CH_2-\text{[imidazole ring, N, NH]} + H^+ \rightarrow -CH_2-\text{[imidazole ring, N}^+\text{H, NH]}$$

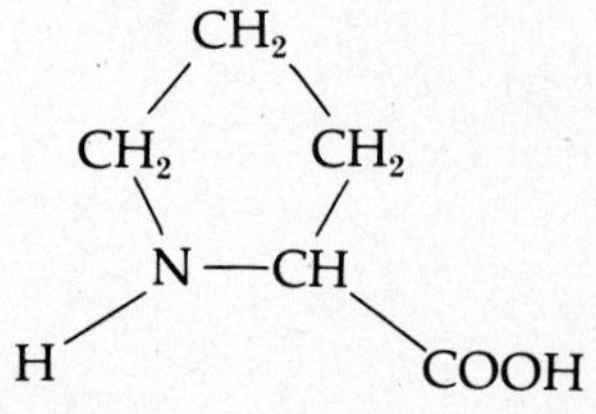

These groups are either charged or so polar that they gain in stability by having water molecules around them. In aqueous solution they tend to orient their neighborhood of the protein chain toward the outside of the molecule. Aspartic and glutamic acid are acidic, with negative charges. Lysine, arginine and histidine are basic, with positive charges. Asparagine and glutamine are neutral but quite polar.

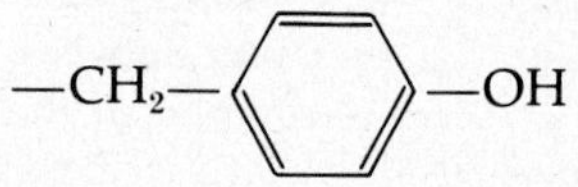

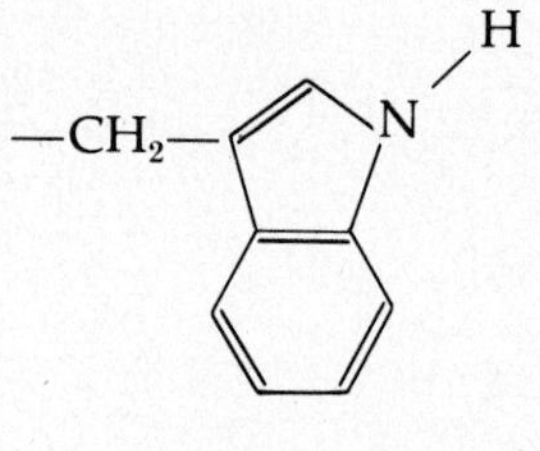

A third class of eight amino acids apparently can occur either on the surface or in the interior with equal ease. These side chains are to the right. The entire amino acid is shown for proline since the side chain curves back to bond to the nitrogen atom. These side chains are moderately polar but are uncharged. Some of them—serine, threonine, tyrosine, and tryptophan—form hydrogen bonds whenever they lie in the interior of a protein molecule. Two cysteine side chains can be oxidized so that their sulfur atoms are joined by a covalent bond in a disulfide bridge *(Figure 13)*. These two types of bond are important in determining how a protein chain folds. The glycine side chain is important because of its absence. Glycines often appear in tight corners in the interior where two chains come together too closely for a side chain to fit. Alanine is technically a hydrocarbon, but its small size tends to minimize its water-repellent properties, and it is frequently found on the surface of protein molecules.

In brief, the raw materials of proteins are a long polypeptide chain with rigid connecting links and a strong tendency to make hydrogen bonds, and an assortment of 20 different side groups with widely different chemical properties. Most of the

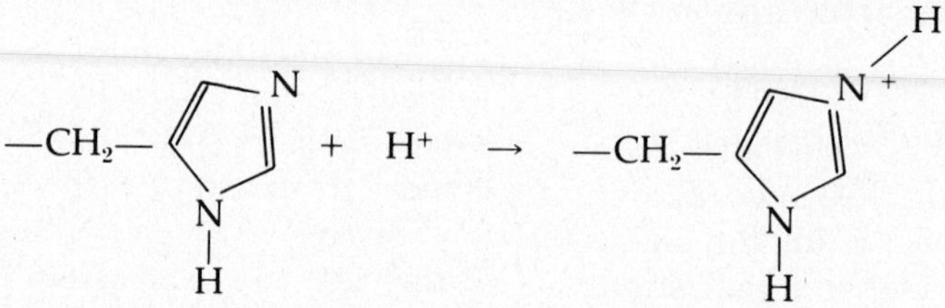

Amino acid	Side chain, R, in Figure 11
Glycine	$-H$
Alanine	$-CH_3$
Cysteine	$-CH_2-SH$
Serine	$-CH_2-OH$
Threonine	$-CH\begin{smallmatrix}OH\\ \\CH_3\end{smallmatrix}$
Tyrosine	$-CH_2-$[benzene ring]$-OH$
Tryptophan	$-CH_2-$[indole ring with N-H]
Proline	[ring: CH_2, CH_2, CH_2, $N-CH$, H, $COOH$]

chemical and catalytic properties of the protein molecule arise from its elaborate structure. One of the standard forms of this structure is the α-helix, shown in Figure 14. The helix is shaped by hydrogen bonds that extend from one turn of the chain to the turns above and below. The bonds serve to

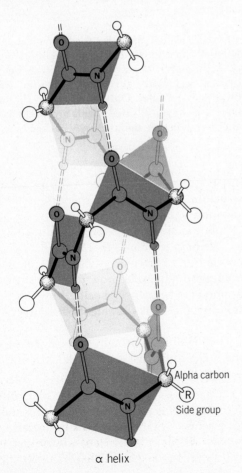

.3 BONDS BETWEEN CHAINS in proteins. Top, a hydrogen bond between a serine group and a main chain carbonyl oxygen. Bottom, a disulfide bridge between two cysteine side groups. Such bridges form after the polypeptide chain is assembled.

.4 ALPHA-HELIX is found in fibrous proteins of hair, wool, skin, fingernails, claws, beaks and muscle. The carbonyl oxygen atom of one amino acid residue is hydrogen bonded to N—H group four residues farther along the chain. Color indicates planes of the polypeptide bonds.

Alpha carbon

Side group

α helix

make a rigid cylinder from a floppy chain. The α-helix occurs in a class of fibrous structural proteins called keratins, which includes most of the protective tissues found in animals, such as nails and claws, skin, hair, and wool. Contractile muscle protein is also made of two protein components, actin and myosin, of which at least the myosin is arranged in an α-helix. Hair can be stretched because this involves only the breaking of hydrogen bonds in an α-helix, and the bonds are remade when the tension is released and the helix reforms.

Silk fibers are based on a second basic structure, the β-pleated sheet. Here the protein chains are almost completely extended and are bound into sheets by hydrogen bonds from one chain to a neighbor. Weaker forces tend to hold the sheets in three-dimensional stacks. The direction of the polypeptide chain (*left to right in Figure 15*) is the direction of the silk thread. Silk is supple for the same reason that a ream of typing paper held at both ends is flexible—the sheets can slip over one another in bending. But silk is not as stretchable as wool. To pull a ream of paper apart in the middle one must rip the sheets instead of flexing them; and to stretch a silk fiber appreciably, the fully extended, covalently bonded protein chains must be snapped. This demands energies in the range of 100 kilocalories per mole of bonds, rather than the one or two kilocalories required to flex the sheets. Silk is therefore flexible, but comparatively resistant to stretching.

The collagen that is found in cartilage, tendons, the underlayers of the skin, and the cornea of the eye, is a protein based on yet another structural pattern—a twisted three-stranded polypeptide chain. The three chains are twisted around one another like the strands of a cable. Hydrogen bonds run from one chain to the other two, producing a structure that is strong, rigid and unstretchable.

Some of the most important proteins are globular. Their polypeptide chains are folded back and forth on themselves to build globular molecules with typical diameters of 25 to 100 Å or more. Most of the chains in globular proteins are 80 to 400 amino acids long, with molecular weights of 9000 to 45,000. Some proteins have molecular weights of several millions, but these molecules are merely aggregates of smaller subunits. Apparently it is inefficient for a biological system to code its DNA for enormous protein chains when aggregates of shorter chains will serve as well.

Myoglobin is a typical example of a globular

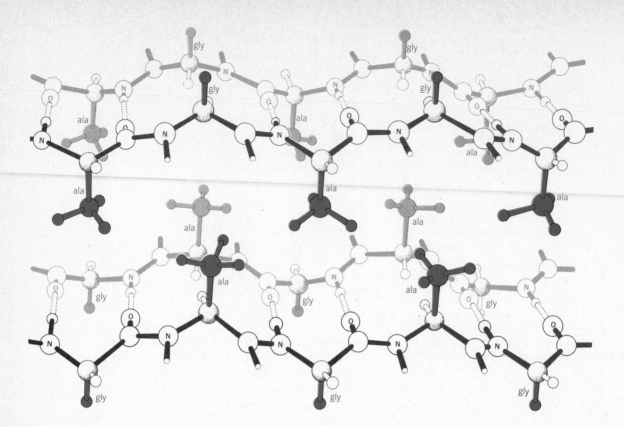

protein *(Figure 16)*. Its function is to store O_2 in the tissue until needed. It has 153 amino acids in one chain, with no disulfide bridges, and is somewhat unusual in consisting almost entirely of α-helices. Its eight helices make a pocket that encloses a

STRUCTURE OF SILK is called a β-pleated sheet. Polypeptide chains extend horizontally in the direction of the silk fiber. Many parallel chains are held together by hydrogen bonds at right angles to the direction of the chain and fiber. Only two parallel chains in each layer are shown. Pleated sheets are stacked atop one another, with side chains closely packed between them. 15

heme group—an iron-containing organic ring structure. In myoglobin the difference between secondary and tertiary structure is clear. Hydrophobic side chains on the inner sides of the helices help to ensure that the helices fold against one another correctly as the molecule is formed. Myoglobin illustrates how primary structure—the amino acid sequence specified by DNA—can determine the three-dimensional folding of a protein.

Hemoglobin, a protein that brings oxygen from the lungs to the tissues and delivers it to myoglobin for storage, illustrates the concept of quater-

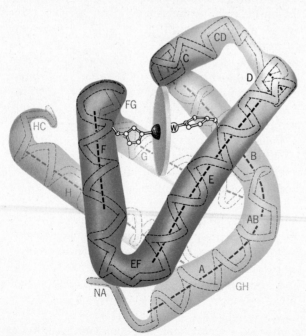

MYOGLOBIN is a globular protein. Tubular elements are α-helices. Disk at center is a heme group seen edge-on. Single letters designate helical regions; double letters, nonhelical regions. 16

17 **HEMOGLOBIN MOLECULE consists of four chains, each similar in length and folded like the myoglobin molecule in the previous figure. A more detailed view of hemoglobin appears in Figure 18 in Chapter 5.**

nary structure *(Figure 17)*. As hemoglobin takes up or releases oxygen its four chains slightly shift their relative positions, changing the quaternary structure. Ionic bonds are broken, exposing buried side chains that enhance the binding of molecular oxygen. Each subunit of hemoglobin is folded exactly like a myoglobin molecule because they are both evolutionary descendants of a common oxygen-binding ancestor protein *(Chapter 29)*. But on the surfaces where subunits come in contact—regions which on the myoglobin molecule are exposed to the aqueous surround-

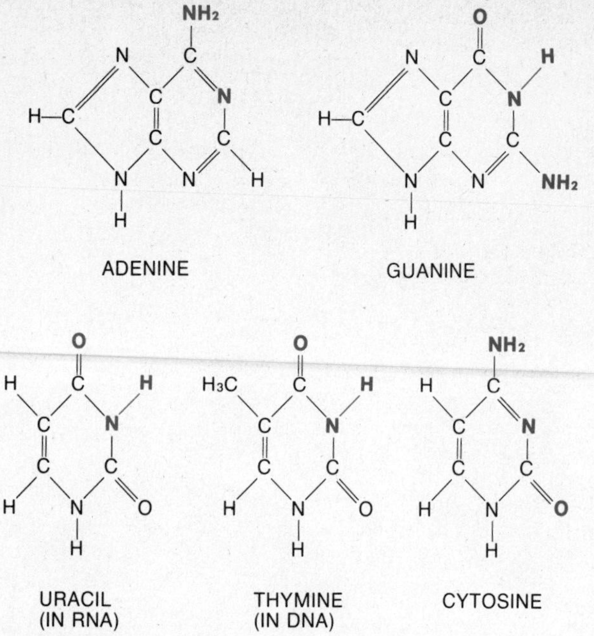

ADENINE

GUANINE

URACIL
(IN RNA)

THYMINE
(IN DNA)

CYTOSINE

18 **ORGANIC BASES of DNA and RNA. Adenine and guanine are purines. Uracil, thymine and cytosine are pyrimidines. In DNA and RNA, guanine on one chain is bonded to cytosine on another chain. Adenine is paired with thymine in DNA and with uracil (minus the CH$_3$ group) in RNA. Atoms involved in hydrogen bonding are printed in color.**

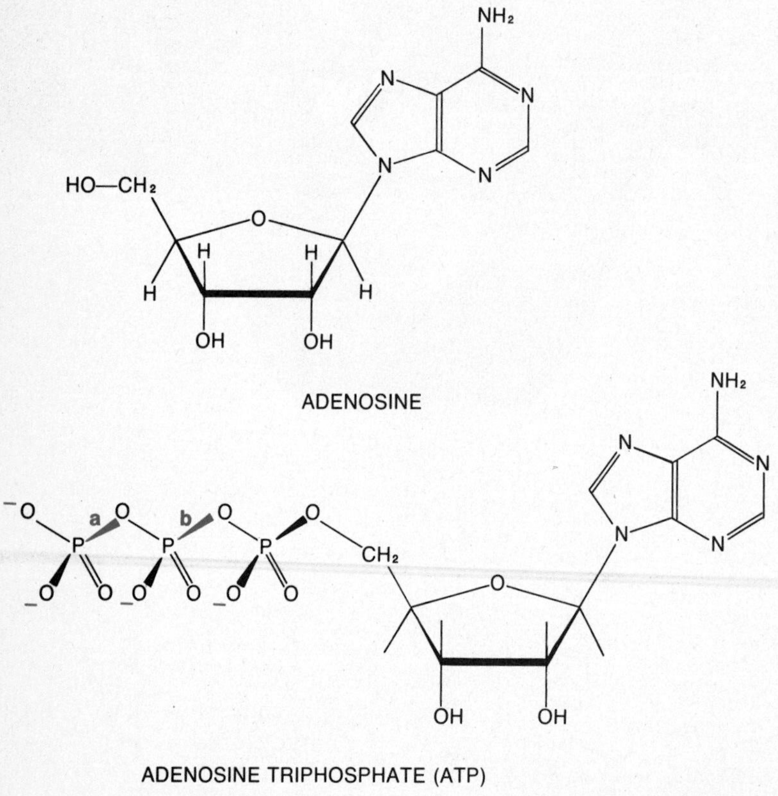

ADENOSINE

ADENOSINE TRIPHOSPHATE (ATP)

ings—hemoglobin has hydrophobic side chains where myoglobin has hydrophilic ones. Again, the chemical nature of the side chains, as coded by DNA, helps in deciding how the molecule will fold and pack in three dimensions.

It is possible, although not particularly flattering to our egos, to describe a human as the best possible local environment for the survival of human DNA.

Globular proteins are generally either carriers or enzymes. Myoglobin and hemoglobin are carriers (or storehouses) of oxygen. The cytochromes carry energy. Other proteins transfer small molecules or chemical groups from one reaction to another or from one part of the cell to another. The gamma globulins are antibodies—two-headed globular proteins whose function is to bind foreign proteins (such as the coat of a virus) into an insoluble clump, out of harm's way. The enzymes control the rates and pathways of all the reactions of an organism (*Chapter 5*).

NUCLEOTIDES AND NUCLEIC ACIDS

A biologist's flip reply to the false paradox; "Which came first, the chicken or the egg?", might be: "A chicken is merely the egg's way of producing another egg." Every living thing organizes matter from its environment into its own molecules. It is possible, although not particularly flattering to our egos, to describe a human as the best possible local environment for the survival of human DNA. This is a gross oversimplification, but contains some element of truth. Not everything in a new organism comes solely from the DNA of its parents, and a DNA molecule without the protein-synthesizing machinery is as useless as a prerecorded tape without a high-fidelity sys-

NUCLEOSIDES AND NUCLEOTIDES. The **19** **combination of a purine or a pyrimidine with ribose is called a nucleoside. Their phosphate esters are nucleotides. Adenosine (top) is a nucleoside, and adenosine triphosphate (bottom) is a nucleotide. Letters a and b designate high-energy bonds in ATP.**

tem. Nevertheless, the information required to construct a new organism comes from the DNA.

The closely related ribonucleic acid, RNA, is a polymer whose links are ribose molecules with basic organic rings attached to carbon atoms. DNA differs from RNA only in having one of its hydroxyl groups replaced by a hydrogen atom, so that the sugar is deoxyribose rather than ribose.

Four different organic bases are found in DNA: the purine rings adenine and guanine, and the pyrimidine rings thymine and cytosine *(Figure 18)*. RNA uses uracil rather than thymine, with one less methyl group. The combination of a purine or pyrimidine with ribose is a NUCLEOSIDE, and if the ring is adenine, for example, the nucleoside is adenosine *(Figure 19)*. DNA and RNA are polymers of nucleosides, in which the ribose (or deoxyribose) monomers are connected to one another by a phosphate group. This sugar-phosphate-sugar-phosphate-sugar-phosphate polymer with purine and pyrimidine side groups on the sugars is diagrammed in Figure 20.

A nucleoside such as adenosine can add a phosphate group to form the molecule known as adenosine monophosphate, AMP. But most importantly for energy storage purposes, the nucleoside can add a second and a third phosphate to build adenosine di- and triphosphate, ADP and ATP. ATP is the main short-term energy storage molecule in cells. It is the source of instant energy for muscle contraction, active transport across membranes, activation of molecules for subsequent reaction, and almost every other uphill (energy-consuming) chemical process in the cell. The synthesis of ATP from ADP and inorganic phosphate is the first step in packaging and storing energy obtained from the breakdown of foods. We shall look more extensively at the role of ATP in Chapter 6.

One of the most essential properties of any set of instructions is that they can be read and copied. A book must be readable by those who intend to use its information. There must also be some way of making additional copies of the book for multiple users. Reading and copying in DNA and RNA is possible because the organic bases "recognize" one another by means of hydrogen bonds between the rings *(Figure 20)*. Adenine can only form a hydrogen bond with thymine (or with uracil in RNA), and guanine can only bond to cytosine. DNA is a double-stranded structure in which every base on one strand is hydrogen bonded and paired with its corresponding partner on the other strand. The secret of information storage is that every three

successive bases along the chain comprise a code word for one amino acid. If there are four different letters in this alphabet, and every word has three letters in it, then obviously there are $4^3 — 64$ different code words for only 20 amino acids. The code is redundant. As we shall see in Chapter 7, some amino acids have several code combinations, and some code words are merely punctuation marks that tell the protein-assembly machinery to stop work. In biological systems redundancy serves the same purpose as it does in the error-checking codes used in computer programming.

The two strands of the DNA ladder run in opposite directions, and the ladder is twisted about its long axis into a helix *(Figure 21)*. The hydrogen-bonded base pairs are stacked parallel to one another like steps in a spiral staircase. The sugar-

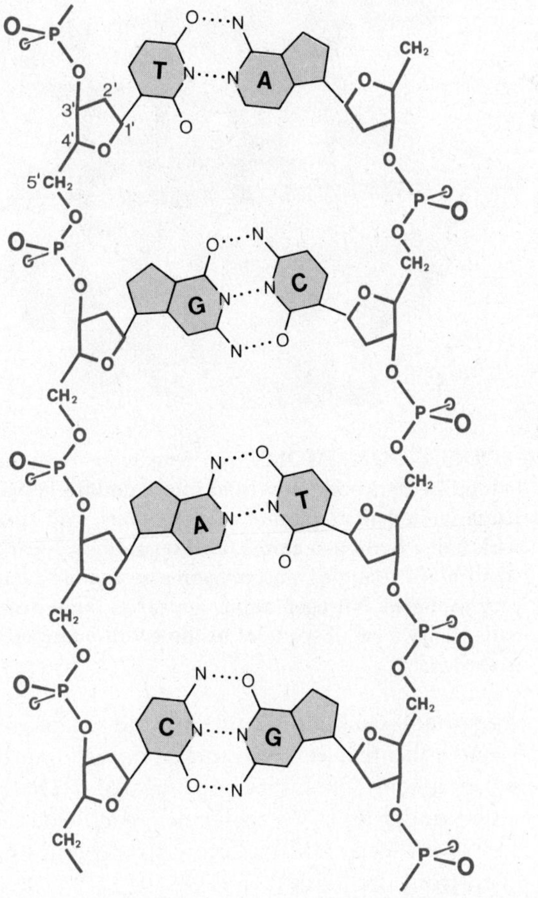

DNA is a two-chain structure. The two backbones of the molecule, built from phosphate groups alternating with deoxyribose molecules, are antiparallel: the chains are oriented in opposite directions. Base pairs connect the backbone chains like rungs of a ladder. Dots represent hydrogen bonds. **20**

21 MODEL OF DNA MOLECULE reveals the way that the ladderlike backbone is twisted into a double helix. Large white balls represent oxygen atoms, and small dark balls, hydrogen atoms. Carbon atoms appear as small black triangles, and phosphorus atoms as larger grey triangles. Nitrogen atoms appear as large dark balls. Only a small segment of the enormous molecule is shown.

phosphate backbone runs around them. When DNA replicates, the two strands unwind and each makes a complementary copy of itself. The result is two daughter DNA molecules, each exactly like the parent, each with one new strand and one derived from the parent.

The structure of DNA was worked out in 1953 by Francis Crick, a British physical chemist, and James Watson, an American biochemistry postdoctoral fellow, as a result of their experiments in the Cavendish Laboratory at Cambridge University. Watson and Crick shared the Nobel Prize in medicine for 1961 with Maurice Wilkins, the London crystallographer whose x-ray photographs provided the evidence for the validity of the double helix. The story of the inspiration and the strife that led to recognition of the DNA structure has been recounted in a remarkable book by James Watson, called, appropriately enough, *The Double Helix*. Because the two strands in DNA are complementary but opposite, molecular geneticists often differentiate between them by calling one "Watson" and the other one "Crick."

We have come a long way in the last two chapters, from protons and electrons to DNA. The progression is symbolic, for over the first one or two billion years of its history, our solar system came along precisely this path. The biologist George Wald has on occasions defined life as a property of molecules in a sufficiently intricate state of organization. We have seen some of these molecules. The rest of this book will be concerned with the consequences of their organization.

READINGS

R.E. DICKERSON AND I. GEIS, *The Structure and Action of Proteins*, Menlo Park, CA, W.A. Benjamin, 1969. A more extended treatment of protein structure, folding and physical and chemical properties, at about the level of this chapter and Chapter 5. Profusely illustrated, including stereo pair drawings of protein molecules.

F.R. JEVONS, *The Biochemical Approach to Life*, London, Allen & Unwin, 1964. An unusually readable introduction to some of the central ideas of biochemistry. Not as structurally or as biologically oriented as *Dickerson & Geis* or *Smith*.

C.U.M. SMITH, *Molecular Biology: A Structural Approach*. Cambridge, MA, M.I.T. Press, 1968. An extremely well-written introduction to the field, with constant emphasis on what is important to the biologist. Highly recommended.

J.D. WATSON, *The Double Helix*, New York, Atheneum, 1968. The view by a young American post-doctoral fellow of the process by which he and his British coworkers solved the structure of DNA and ultimately won the Nobel Prize. Biased and opinionated, but honestly so, and revealing as to the motives that sometimes impel scientists to achieve. Excellent reading.

The following books are more advanced, but easier to follow and understand than many of their competitors:

A.L. LEHNINGER, *Biochemistry*, 2nd Edition, New York, Worth, 1975. Perhaps the best introductory biochemistry text available. Authoritative, yet extremely readable. Strong molecular biological slant.

L. STRYER, *Biochemistry*, San Francisco, W.H. Freeman, 1975. Another clear and readable textbook in elementary biochemistry.

J.D. WATSON, *The Molecular Biology of the Gene*, 3rd Edition, Menlo Park, CA, W.A. Benjamin, 1976. A lucid presentation of molecular genetics by one of the discoverers of the DNA double helix.

A. WHITE, P. HANDLER AND E.L. SMITH, *Principles of Biochemistry*, 5th Edition, New York, McGraw-Hill, 1973. A widely used textbook that takes the trouble to explain things from the beginning. Comprehensive, with special attention to human biochemistry.

The whole system of chemical catalysis and its regulation is so precise that it almost suggests purpose. . . . But, as elsewhere in evolution, the purposefulness is only apparent. —S. E. Luria

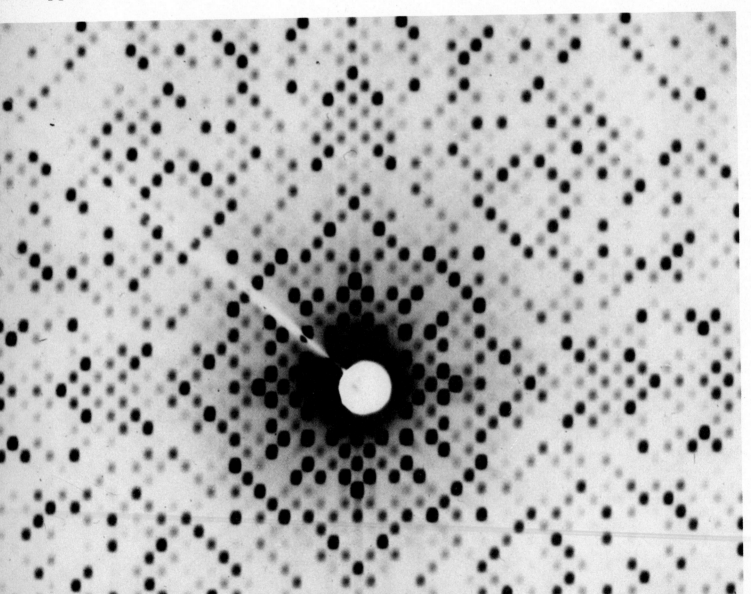

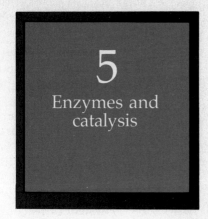

5
Enzymes and catalysis

To sustain the processes of life the cell carries out thousands of chemical reactions per second. In the nonliving world most of these reactions would proceed sluggishly, if at all. Outside a living organism, in fact, many biological reactions run spontaneously in the opposite direction. On a planet with an oxygen atmosphere, large organic molecules are chemically unstable and tend to decompose gradually into carbon dioxide and water.

The chemistry of life is incredibly elaborate in its working details but elegantly simple in principle. Living systems have circumvented the limitations of their environment by devising a fourfold chemical strategy. First, to accelerate the tempo of slow reactions living systems invented enzymes, which are superbly efficient catalysts. Second, to drive reactions uphill—to run them opposite to their spontaneous direction—living systems harness external sources of energy. Plants tap the energy of sunlight, and other organisms rely on the chemical energy of various foods, including plants. Third, living systems have evolved feedback mechanisms to regulate their network of reactions. Here life displays an impressive economy of effort by using enzymes as the instruments of control. Recall that enzymes are highly specific: each catalyzes only one type of reaction. By providing a favorable pathway, enzymes make one reaction go much faster than another. Enzymes "decide" which reactions will be important and which will not. Within the cell enzymes comprise a set of chemical switches that turn reactions on and off, preventing

DIFFRACTION PATTERN on opposite page was produced by x-rays beamed through a crystal of lysozyme, an enzyme that causes the breakdown of the cell walls of bacteria. Computer analysis of diffraction patterns has made it possible to work out the detailed three dimensional structure of complicated enzyme molecules, such as the one depicted in figure 13.

biochemical surpluses and deficits. When the automatic control process blocks the action of an enzyme, the reaction that the enzyme catalyzes is shut down. Fourth, living systems have based much of their chemistry on spontaneous reactions. A chemical system that contains spontaneous processes can do useful work for an organism. If a reaction is not spontaneous, enzymes cannot make it go. But enzymes can couple a highly spontaneous reaction to one that will not go by itself, diverting the energy of the spontaneous reaction to the synthesis of unstable compounds essential to life.

How can the unstable organic molecules of living systems persist for any length of time? The answer lies in reaction rates. Although organic compounds tend to decompose spontaneously, the decomposition reactions are usually very slow, and most require an input of energy to get them started, just as firewood requires a match. An organism's large molecules represent a reservoir of stored chemical energy, and when the occasion arises, enzymes catalyze the breakdown of these molecules under carefully controlled conditions, releasing energy for a variety of biological tasks.

This chapter focuses on two main questions: exactly what is catalysis and how does it speed up a chemical reaction? And how do enzyme molecules function as biological catalysts?

HEATS OF REACTION

Living organisms use food molecules as a source of spare parts for synthesizing materials that they need, and as a source of energy to drive these syntheses and keep themselves going. This energy is carefully saved by using it to make chemical bonds in ATP, fats, starches, and other storage molecules, since any energy that is converted to heat and given off in that form is usually considered as wasted. One can determine how much

75

energy a substance potentially has available during a reaction, however, by running the reaction in isolation (without the energy-storing apparatus) and measuring the heat given off. This heat of a reaction has a special name in thermodynamics, the ENTHALPY, H; but one can think of the symbol, H, as standing simply for "heat". Like other forms of energy, heat or enthalpy is measured in calories, with the calorie being defined as that quantity of heat required to raise the temperature of one gram of water from 14.5°C to 15.5°C. The change in enthalpy during a reaction is represented by: ΔH. If a system of reacting chemicals gives off heat, then the system has lost enthalpy, so ΔH is a negative number. Conversely, if a process (such as the evaporation of water) absorbs heat, then the system gains enthalpy, and ΔH is positive.

As an example, every time one mole of hydrogen gas (two grams) reacts with one half mole of oxygen gas (16 grams) to form a mole of liquid water (18 grams) 68,000 calories or 68 kilocalories (kcal) of heat are given off *(Figure 1)*. Hence ΔH is negative, and one can write the equation:

$$H_2(g) + \tfrac{1}{2} O_2(g) \rightarrow H_2O(l) \qquad \Delta H = -68 \text{ kcal}$$

As Figure 1 indicates, the heat emitted can be thought of as arising because the hydrogen and oxygen atoms fall from a high energy level (hydrogen and oxygen molecules) to a lower energy level (water molecules in a liquid).

To evaporate this liquid water, one must supply heat energy from outside: 10 kcal per mole of water vaporized. This energy is needed to pull the water molecules of the liquid apart from one another and turn them into a gas. Water vapor hence is in a higher energy state than liquid water, by 10 kcal/mole, as shown in the energy level diagram in Figure 1.

Heats of reactions are additive. Part of the heat released when hydrogen and oxygen molecules rearrange themselves into liquid water arises because H_2O molecules are intrinsically more stable, but part comes also from the extra stability and attractions between water molecules in a liquid. If we were to react H_2 and O_2 to form water vapor, the heat of reaction would be less:

$$H_2(g) + \tfrac{1}{2} O_2(g) \rightarrow H_2O(g) \qquad \Delta H = -58 \text{ kcal}$$

This is easy to see from the energy level diagram in Figure 1. If two reactions can be added to give a third, the two heats of reaction can be added to give the heat of the third reaction. This is sometimes called "Hess's Law of Heat Summation", but is really only a form of the First Law of Thermodynamics. For the water reactions that we have just been examining:

$$H_2(g) + \tfrac{1}{2} O_2(g) \rightarrow H_2O(l) \qquad \Delta H = -68 \text{ kcal}$$
$$H_2O(l) \rightarrow H_2O(g) \qquad \Delta H = +10 \text{ kcal}$$
$$\overline{H_2(g) + \tfrac{1}{2} O_2(g) \rightarrow H_2O(g) \qquad \Delta H = -58 \text{ kcal}}$$

Every chemical reaction has its heat or enthalpy change, and this is one measure of the useful energy that can be obtained from the reaction. Some processes are very productive of energy, such as the combustion of gasoline (octane) or of glucose:

$$C_8H_{18}(l) + 12\tfrac{1}{2} O_2(g) \rightarrow 8 CO_2(g) + 9 H_2O(l)$$
octane $\qquad \Delta H = -1303 \text{ kcal}$

$$C_6H_{12}O_6(s) + 6 O_2(g) \rightarrow 6 CO_2(g) + 6 H_2O(l)$$
glucose $\qquad \Delta H = -673 \text{ kcal}$

Other reactions absorb heat, such as the above reactions run in reverse, or the synthesis of ATP (adenosine triphosphate) from ADP (adenosine diphosphate) and inorganic phosphate (P_i):

$$ADP + P_i \rightarrow ATP \qquad \Delta H = +8.2 \text{ kcal}$$

Each of these reactions can be represented on an energy level diagram of the type shown in Figure 2.

Combustion—the oxidation of fuels—always

1 RELATIVE ENERGIES (enthalpies) of a mixture of hydrogen and gases, liquid water, and water vapor, are plotted as energy levels on a vertical scale. Arrows indicate changes from one state to another, and the corresponding enthalpy or heat changes are marked. A negative enthalpy change means that heat is given off or lost by the atoms, and a positive value means that heat is absorbed.

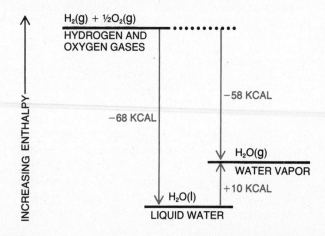

Portable energy sources

Every time a mole of hydrogen gas (2 grams) is burned to produce a mole of liquid water (18 grams), 68.3 kilocalories of heat are given off. This can be written as:

$$H_2 + \tfrac{1}{2} O_2 \rightarrow H_2O \quad \Delta H = -68.32 \text{ kcal/mole } H_2O$$

When glucose or stearic acid is burned, the heat given off is:

$$C_6H_{12}O_6 + 6 O_2 \rightarrow 6 CO_2 + 6 H_2O \qquad \Delta H = -673 \text{ kcal/mole}$$

$$C_{17}H_{35}COOH + 26 O_2 \rightarrow 18 CO_2 + 18 H_2O \quad \Delta H = -2712 \text{ kcal/mole.}$$

By dividing each enthalpy change by the molecular weight of glucose or stearic acid to obtain the heat given off per gram, one can show that fats and fatty acids are twice as efficient at storing energy, on a calories-per-gram basis, than carbohydrates are.

It is interesting to compare these compounds with other potential fuels such as octane (a gasoline) and alanine, a typical amino acid from proteins:

FUEL	CHEMICAL FORMULA	PHYSICAL STATE	MOLECULAR WEIGHT	HEAT OF COMBUSTION, ΔH	
				KCAL/MOLE	KCAL/GRAM
HYDROGEN	H_2	GAS	2.0	−68	−34
N-OCTANE	C_8H_{18}	LIQUID	114.2	−1303	−11.4
STEARIC ACID	$C_{18}H_{36}O_2$	SOLID	284.5	−2712	−9.5
ALANINE	$C_3H_7O_2N$	SOLID	89.1	−388	−4.4
GLUCOSE	$C_6H_{12}O_6$	SOLID	180.2	−673	−3.7

Considering stearic acid, alanine, and glucose as representative of fats, proteins, and carbohydrates, it is easy to see from the column at the right why fats and not carbohydrates are the main energy storage molecules in animals, which must move about and carry their energy with them. Gasoline would be somewhat better yet than fats, but a liquid would be too inconvenient. In terms of energy storage, the animal fat tristearin is essentially a gasoline that has been solidified by making it into a large molecule. Hydrogen gas would be more than three times as efficient an energy carrier on a weight basis, but an animal that stored energy via great sacs of hydrogen gas would be strange indeed. (On second thought, the gas bags might also be useful for locomotion.)

leads to the emission of heat, because the molecules of the products (carbon dioxide gas and liquid water) are more stable than the reactants (fuel and O_2). The most useful fuel to a moving animal is the one that yields the greatest heat per unit of weight. To a stationary plant, however, the weight of the foodstuff is of little importance. The heat-per-gram yield of several biological and non-biological fuels is discussed in Box A.

2 ENERGY-LEVEL DIAGRAMS showing enthalpy or heat changes: (A) the combustion of gasoline, (B) the combustion of glucose, and (C) the storage of energy by synthesizing ATP.

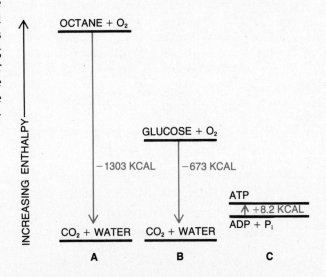

SPONTANEITY AND EQUILIBRIUM

A SPONTANEOUS REACTION is one that will occur by itself, given enough time. This chemical "drive" can be harnessed by a living organism. If a reaction runs spontaneously in one direction, then the reverse reaction will not, and cannot happen without a supply of driving energy from outside. The breakdown of ATP into ADP and phosphate, with the release of 8.2 kcal of energy, is a spontaneous process that can be used to power other reactions; but the synthesis of ATP from ADP and phosphate is a nonspontaneous process that requires energy from the metabolism of foodstuffs. Similarly, the combustion of glucose in oxygen is spontaneous, but the synthesis of glucose from CO_2 and water is a nonspontaneous reaction that must be driven by the energy derived from absorption of light in photosynthesis.

Spontaneity has nothing to do with time. A reaction may be extremely slow, yet spontaneous. The sole criterion is, "Would the reaction take place by itself, given infinite time?" The burning of a newspaper with a match, the slow browning of newsprint in library files, and the gradual erosion of the continent of North America, are all spontaneous processes with different time scales. One can let a mixture of hydrogen and oxygen gases sit for years in a tank without seeing the formation of water, but the explosion that results when a platinum catalyst is dropped into the tank is evidence that the reaction was thermodynamically spontaneous all along.

A spontaneous reaction is one that is moving toward a state of EQUILIBRIUM. All chemical reactions, in principle, can run both forward and backward. One can think of the net observed reaction as being the result of competition between forward and reverse steps. Increasing the amount of reactants speeds up the forward reaction, and increasing the amount of products favors the reverse. Hence a reaction that began by being highly spontaneous will become less so as the reactants are depleted and the products accumulate. Eventually so few reactants and so many products will be present that the forward and reverse steps take place at the same rate. At this point no further change in the system is visible, although individual molecules still are forming and breaking apart. This balance between forward and reverse reactions, with molecular activity but with a stable overall composition, is known as chemical equilibrium. A spontaneous reaction is one that is moving toward equilibrium, using up reactants and making products. Given enough time, every spontaneous process eventually settles down at equilibrium.

How can one decide whether a given reaction is spontaneous, tending to move toward equilibrium on its own, especially if the reaction is also very slow? The criterion for mechanical equilibrium, as with a ball rolling on an uneven tabletop, or a rock tumbling downhill, is the attainment of a state of lowest energy. To find mechanical equilibrium, one should minimize energy. Does the same criterion hold for chemical equilibrium? Should one minimize the energy or enthalpy of a system of reacting molecules? If this were true, then every spontaneous reaction approaching equilibrium would be a heat-emitting process, because the molecules would be moving toward a state of lower enthalpy as in Figure 3.

Is this true? Do all spontaneous reactions give off heat? Most of them do. Obviously explosions and other combustions emit large quantities of heat. Although reactions do not have to be rapid to be spontaneous, if they are rapid, then they are certainly spontaneous.

The trouble with a statement that something "always" happens is that endless verifications cannot prove the law, while one lone exception can disprove it. Examples of spontaneous heat-absorbing reactions are not hard to find if one looks carefully. Ice absorbs heat when it melts, and water when it evaporates, yet melting and evaporation can be spontaneous under the proper conditions. Ammonium chloride crystals, when dissolved in water, absorb so much heat that frost may condense on the outside of the beaker, yet the crystals are not thereby prevented from dissolving.

The missing factor is disorder, or ENTROPY in the language of thermodynamics. Chemical reactions do tend spontaneously toward states of low energy, but they also tend toward maximum disorder. Most energy-releasing processes such as combustions also create more disorder in the

---

All chemical reactions, in principle, can run both forward and backward. Increasing the amount of reactants speeds up the forward reaction, and increasing the amount of products favors the reverse reaction.

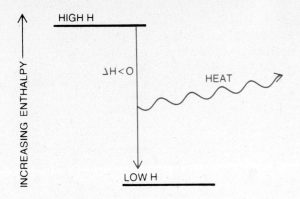

HIGH H

$\Delta H < 0$

HEAT

LOW H

3 **HEAT is given off when a chemical system goes from a high enthalpy state to a lower one.**

for a reaction carried out at a certain temperature, T, the free energy change is:

$$\Delta G = \Delta H - T\Delta S$$

As an example illustrating the relative importance of these factors, in the combustion of one mole of glucose, 673 kcal of heat are given off, and the disorder rises by 43.3 cal/deg. (A cal/deg is sometimes called an entropy unit, e.u.) At 25°C or 298°K, $T\Delta S = 298(43.3)$ cal = 12,903 calories or 12.9 kcal. Both heat and disorder favor the spontaneity of combustion of glucose, and the free energy change is:

$$\Delta G = -673 \text{ kcal} - 13 \text{ kcal} = -686 \text{ kcal}$$

The drive to react is stronger, by 13 kcal, than would have been predicted from heat alone. The entropy or disorder factor is smaller than the heat effect, but is not negligible.

products than had been present in the reactant molecules. In the combustion of glucose:

$$C_6H_{12}O_6(s) + 6 \ O_2(g) \rightarrow 6 \ CO_2(g) + 6 \ H_2O(l)$$

the six moles of CO_2 gas and O_2 gas have about the same degree of disorder or entropy, but six moles of liquid water are more disordered (mixed up) than one mole of solid glucose. Because both heat (enthalpy) and disorder (entropy) favor the burning of glucose, it proceeds with more chemical drive and spontaneity than if heat where the only factor.

The contrary examples, heat-absorbing spontaneous reactions, were carefully chosen so that the creation of disorder and the lowering of energy did not run in parallel. Molecules in water vapor are more disordered than in liquid water, even though energy is required to pull the attracting molecules away from one another during vaporization. Ammonium and chloride ions in solution, NH_4^+ and Cl^-, are more disordered than a perfect crystal of NH_4Cl, even though energy is needed to break up the crystal.

The quantity that should really be minimized to find the conditions of chemical equilibrium, the quantity that is always decreasing during a spontaneous reaction, is the FREE ENERGY, G. It is defined as a combination of heat or enthalpy, H, and disorder or entropy, S:

$$G = H - TS$$

T is the absolute temperature, and its presence in this definition is a recognition that the disorder term is more important at high temperatures than at low. If the changes in free energy, heat, and disorder are represented by ΔG, ΔH, and ΔS, then

REACTION RATES

The simple energy level diagram shown in Figure 4 does not tell the whole story about reaction rates. If one looked closely at the actual mechanism by which oxygen molecules combine with glucose, he would find that intermediate (partially reacted) molecular fragments are present, having higher energies than either the products or the reactants. Molecules must pass through these intermediate or transition states in order to react, as shown in Figure 5. The higher energy of the intermediate states can be thought of as a "barrier" to reaction; the maximum height of the barrier is called the

OXIDATION OF GLUCOSE. When one mole of glucose is burned, 673 kcal of heat are released. Diagram shows only initial and final states, and tells nothing about the pathway of the reaction.

4

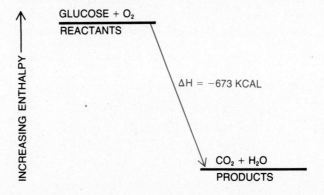

GLUCOSE + O_2
REACTANTS

$\Delta H = -673$ KCAL

$CO_2 + H_2O$
PRODUCTS

INCREASING ENTHALPY

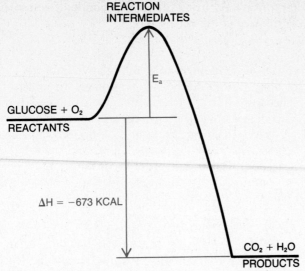

REACTION INTERMEDIATES

E_a

GLUCOSE + O_2
REACTANTS

$\Delta H = -673$ KCAL

$CO_2 + H_2O$
PRODUCTS

5 **INTERMEDIATE COMPOUNDS in the combustion of glucose** have higher energy than either the reactants or the products. The overall heat reaction is still 673 kcal, but the activation energy, E_a, acts as a barrier to the reaction and slows it down.

ACTIVATION ENERGY of the reaction, E_a. All of the activation energy needed to start the reaction is eventually recovered, so the overall drop in enthalpy, ΔH, is unaffected. But the higher the activation barrier to a reaction, the slower the reaction.

The idea of an activation energy barrier is easier to understand with a simpler reaction involving hydrogen and deuterium:

$$H_2 + D_2 \rightarrow 2\ HD.$$

The simplest reaction mechanism would be for a molecule of each gas to collide, the H—H and D—D bonds to stretch and break, and two new H—D bonds to form, as in Figure 6. In the intermediate state, 6c, the H—H and D—D bonds would be longer than normal, and H—D bonds would be in the process of formation. The atomic arrangement in 6c can be thought of as a four-atom transition state or reaction intermediate, at the top

A catalyst does not cause anything to happen that would not occur eventually without it; it merely pushes the reaction toward equilibrium faster.

of the activation energy barrier in a diagram like Figure 5.

How could the reaction be speeded up? Anything that would lower the activation energy barrier, E_a, by providing an easier reaction mechanism, would cause the reaction to go faster. This is the function of a catalyst.

COLLISION MECHANISM for the gas-phase reaction: 6 **$H + D_2 \rightarrow 2\ HD$. The four-atom complex (C) can be termed the transition state for the reaction, and would lie at the top of the E_a energy barrier in Figure 5.**

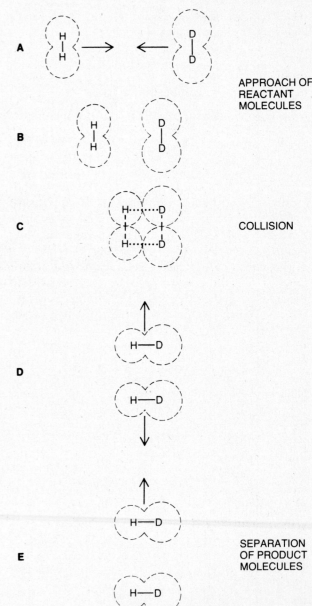

A — APPROACH OF REACTANT MOLECULES

B

C — COLLISION

D

E — SEPARATION OF PRODUCT MOLECULES

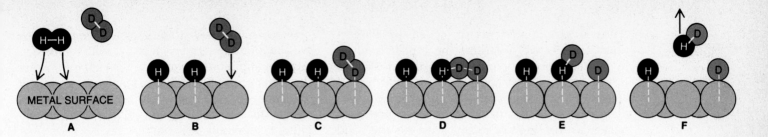

METAL SURFACE

A   B   C   D   E   F

HOW CATALYSTS ACT

If H_2 and D_2 gases are mixed in a tank, it will be a long time before $H—D$ molecules can be detected. But if a small amount of platinum black (clean, finely ground platinum dust) is added to the tank, then HD forms rapidly. Finely divided platinum is a CATALYST for this reaction. The catalyst does not cause anything to happen that would not occur eventually without it; it merely pushes the reaction toward equilibrium faster.

The catalyst works by providing an easier mechanism for reaction, with a lower activation energy, E_a. The actual catalytic mechanism for the $H_2 + D_2$ reaction is outlined in Figure 7. Atoms of metallic platinum are just far enough apart that when an H_2 molecule binds to the surface, the $H—H$ bond is pulled apart, leaving two H atoms bonded to the platinum. (The energy required to break the $H—H$ bond is provided by the making of two temporary $H—Pt$ bonds.) These two H atoms then are more reactive than an H_2 molecule; when a D_2 molecule binds itself to the surface of the platinum as shown in 7c, it readily combines with one of the dissociated H atoms to form a three-atom intermediate complex. The complex can break apart in either of two ways: back to H and D_2 again, or forward to a $H—D$ molecule and a D atom, which then can react with another H_2 molecule. The end result is the same as in the un-catalyzed reaction of gaseous H_2 and D_2, but the mechanism involving the three-atom platinum complex (7d) is simpler than one involving the four-atom complex of gases shown in 6c. The activation energy for the catalyzed process is lower because the necessary energy is supplied in a series of steps, each one requiring less energy. The effect on the activation barrier is diagrammed in Figure 8. With a lower E_a, both the forward and the reverse reactions are speeded up, so the molecules go toward equilibrium faster. (The final equilibrium, it must be emphasized, is the same either with or without a catalyst.)

CATALYTIC MECHANISM for the reaction of H_2 and   **7**
D_2. A) Binding of H_2 molecule to platinum (Pt) surface, B) weakening and breaking of H—H bond by formation of bonds to Pt atoms, C) binding of D_2 molecule, D) formation of H—D—D intermediate complex, E) breakdown of complex to H—D and D atom, and F) H—D molecule diffuses away, leaving H and D atoms on the surface of the catalyst for future reaction.

ENZYMES AND SELECTIVE CATALYSIS

There are several ways to speed up a chemical reaction; for example one can increase the concentration of the reactants or one can raise the temper-

LOWERING OF ACTIVATION ENERGY (E_a) in a   **8**
catalyzed reaction. The initial and final states of uncatalyzed and catalyzed reactions are identical, but intermediate or transition states are different. The square four-atom cluster (the activated complex in the gas-phase reaction) has a higher energy than the strained three-atom complex on the platinum surface in the catalyzed process. Hence the catalyzed reaction goes faster.

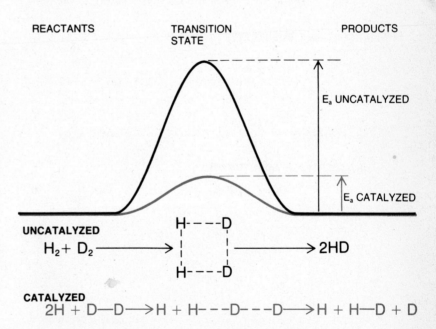

REACTANTS   TRANSITION STATE   PRODUCTS

E_a UNCATALYZED

E_a CATALYZED

UNCATALYZED

$$H_2 + D_2 \longrightarrow \begin{matrix} H\text{----}D \\ | \quad\quad | \\ H\text{----}D \end{matrix} \longrightarrow 2HD$$

CATALYZED

$$2H + D—D \longrightarrow H + H\text{--}D\text{---}D \longrightarrow H + H—D + D$$

81

ature. Raising the temperature speeds up all reactions nonselectively. Adding the proper catalyst, in contrast, speeds up only one particular reaction or class of reactions. Most inorganic catalysts are reasonably nonspecific. Platinum black, for example, will catalyze virtually any reaction involving molecular hydrogen because it weakens the bond between atoms in the H_2 molecule. In contrast, most biological catalysts (enzymes) are highly specific. These molecules that bind themselves to an enzyme and are acted upon catalytically are known as the enzyme's SUBSTRATES; the place where substrate molecules attach themselves to the surface of the enzyme, where catalysis occurs, is called the enzyme's ACTIVE SITE. The enzyme aspartase catalyzes the reversible addition of ammonium ions across the double bond of fumarate (a molecule from the citric acid cycle) to produce aspartic acid (one of the amino acids). This reaction is shown at the top of Figure 9. Fumarate and aspartic acid are the natural substrates of the enzyme aspartase. Relatively minor changes in the fumarate molecule can make it unrecognizable to the enzyme. All of the molecules in the lower half of Figure 9 are unsuitable as substrates, and aspartase will not add ammonium ion to them.

Sometimes a molecule is similar enough to an

enzyme's natural substrate to bind itself to the active site, but different enough that nothing further happens; it inhibits the expected catalytic reaction. These molecular "dogs in the manger" are called REVERSIBLE or COMPETITIVE INHIBITORS. They block the enzyme in a reversible way, that is, by binding without chemically altering the enzyme. If enough

---

Without enzymes that couple spontaneous to nonspontaneous reactions life would be impossible, because the synthesis of organic compounds depends on thermodynamically unfavorable reactions that require an input of energy.

---

of the natural substrate molecules are present, they can compete with the inhibitor and displace it. For example, the enzyme succinate dehydrogenase removes two hydrogen atoms from succinate (another molecule from the citric acid cycle) to produce fumarate, and transfers the hydrogens to a carrier molecule, as shown at the top of Figure 10. The molecules shown at the bottom of Figure 10 are competitive inhibitors of succinate dehydrogenase. They are similar enough to succinate that the enzyme is fooled into binding them, but once this occurs, the enzyme can do nothing more with them because they are the wrong size and shape, or have key chemical groups in the wrong places. The series of molecules of increasing length—oxalate, malonate, succinate, glutarate—reveals that the catalytic action of the enzyme depends on two negatively charged carboxyl groups being a certain specific distance apart; the requirement for simply binding to the active site is less stringent. Even the pyrophosphate ion can bind and block the enzyme.

Some substances can irreversibly modify certain groups at the active site of the enzyme, ruining the enzyme by destroying its capacity to function as a catalyst. The modifying chemical need not resemble the true substrate at all, because it is not necessary to "fool" the enzyme to ruin it. Such an IRREVERSIBLE INHIBITOR for the digestive enzyme trypsin is the organophosphorous compound shown below, Diisopropyl phosphorofluoridate or DFP:

9 ACCEPTABLE SUBSTRATES for the enzyme aspartase (top) are compared with unacceptable ones (below). The normal enzymatic reaction is shown at the top.

ACCEPTABLE SUBSTRATES

FUMARATE   ASPARTIC ACID

UNACCEPTABLE SUBSTRATES

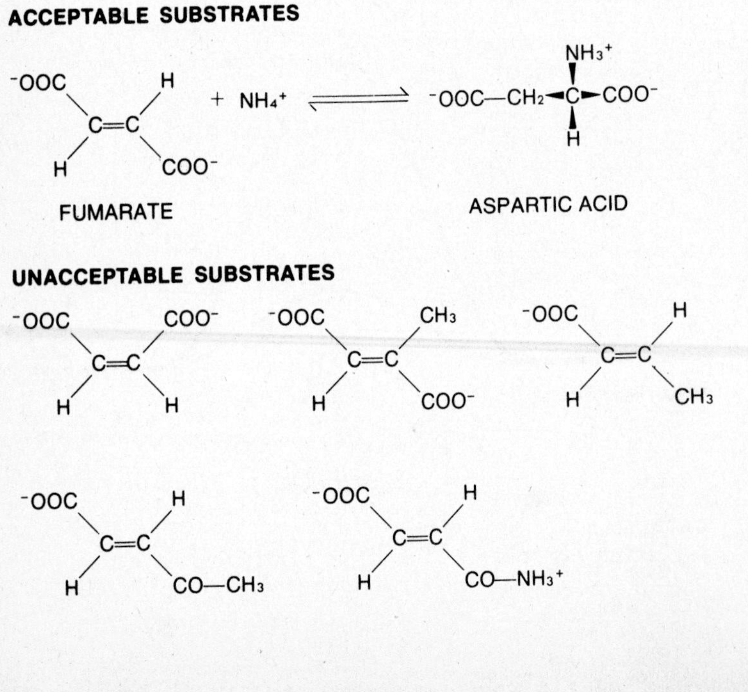

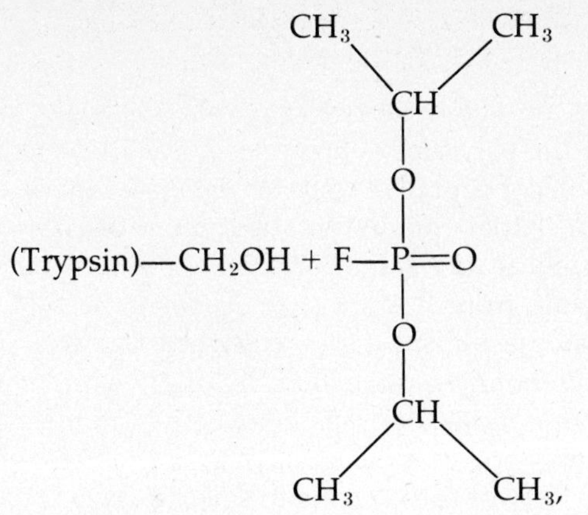

$$CH_3 \quad\quad CH_3$$
$$\backslash\,/$$
$$CH$$
$$|$$
$$O$$
$$|$$
$$(Trypsin)—CH_2OH + F—P{=}O$$
$$|$$
$$O$$
$$|$$
$$CH$$
$$/\,\backslash$$
$$CH_3 \quad\quad CH_3,$$

Diisopropyl phosphorofluoridate (DFP)

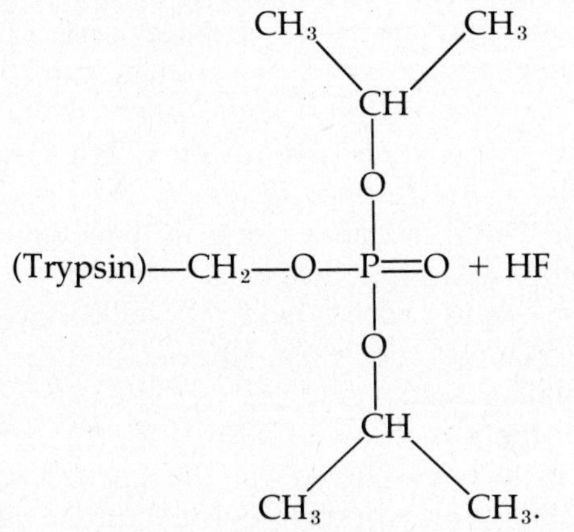

$$CH_3 \quad\quad CH_3$$
$$\backslash\,/$$
$$CH$$
$$|$$
$$O$$
$$|$$
$$(Trypsin)—CH_2—O—P{=}O + HF$$
$$|$$
$$O$$
$$|$$
$$CH$$
$$/\,\backslash$$
$$CH_3 \quad\quad CH_3.$$

Diisopropylphosphoryl enzyme,
(DIP-trypsin)

DFP reacts with a particular serine side chain at the enzyme's active site, preventing the use of this serine in the catalytic mechanism. DFP is also an irreversible inhibitor for many other enzymes whose active sites contain serine. Among these are acetylcholinesterase, which is essential for the propagation of impulses from one nerve cell to another *(Chapter 19).* DFP and other lethal organophosphorous compounds are classified as nerve gases. Many related organophosphorous compounds are lethal to insects but not to vertebrates, and are the major biodegradable substitutes for DDT.

COUPLING OF REACTIONS

Succinate dehydrogenase also illustrates the third aspect of enzymatic activity: the coupling of a spontaneous and a nonspontaneous reaction. The conversion of succinate to fumarate is highly spon-

taneous, with a large drop in free energy. When this reaction takes place in the mitochondria of a cell, the two hydrogen atoms that are removed from succinate are transferred to a molecule of flavin adenine dinucleotide, FAD, shown in Figure 11. The hydrogenation of FAD to $FADH_2$ is highly nonspontaneous, and requires a large input of free energy. $FADH_2$ can release this energy later to another molecule, so it serves as a temporary storage unit for chemical free energy.

Why is it inevitable that a molecule of FAD becomes $FADH_2$ each time that succinate is converted to fumarate? Isn't it more likely that the hydrogen atoms could go somewhere else, and the free energy of the succinate reaction be dissipated as heat? Why is this free energy necessarily stored in the $FADH_2$ molecule? The answer is that the enzyme succinate dehydrogenase links the spontaneous reaction to the nonspontaneous one by ensuring that hydrogen atoms liberated by succinate are used to make $FADH_2$. One site on the enzyme surface binds succinate, and a second site, presumably nearby, binds FAD. It is believed that

SUBSTRATES AND INHIBITORS for the enzyme 10
succinate dehydrogenase. The competitive inhibitors resemble succinate closely enough to fool the enzyme into binding them at the active site, but they cannot undergo the subsequent catalytic steps. Molecule A (top) is FAD, illustrated in Figure 11.

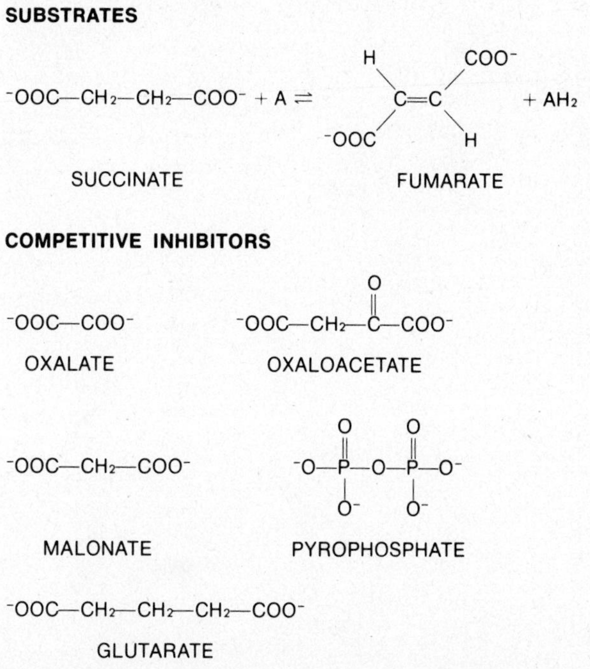

SUBSTRATES

$$^-OOC—CH_2—CH_2—COO^- + A \rightleftharpoons \quad \begin{matrix} H & COO^- \\ \backslash & / \\ C{=}C \\ / & \backslash \\ ^-OOC & H \end{matrix} \quad + AH_2$$

SUCCINATE FUMARATE

COMPETITIVE INHIBITORS

$$^-OOC—COO^-$$
OXALATE

$$^-OOC—CH_2—\overset{\displaystyle O}{\overset{\|}{C}}—COO^-$$
OXALOACETATE

$$^-OOC—CH_2—COO^-$$
MALONATE

$$^-O—\overset{\displaystyle O}{\overset{\|}{P}}—O—\overset{\displaystyle O}{\overset{\|}{P}}—O^-$$
$$\quad\;\;\overset{|}{O^-}\quad\quad\overset{|}{O^-}$$
PYROPHOSPHATE

$$^-OOC—CH_2—CH_2—CH_2—COO^-$$
GLUTARATE

succinate transfers two protons to amino acid side chains on the enzyme, which are then properly oriented to donate protons to the flavin ring of FAD. Thus every time a succinate ion reacts, much of the free energy which would otherwise be released by this highly spontaneous process is trapped and used to synthesize $FADH_2$.

Without enzymatic coupling of spontaneous to nonspontaneous reactions life would be impossible, because the synthesis of organic compounds depends on thermodynamically unfavorable reactions requiring an input of energy. The essence of photosynthesis, for example, is the use of the energy of sunlight to synthesize high-free-energy compounds by reactions that ordinarily run spontaneously in the opposite direction *(Chapter 6)*.

11 FLAVIN ADENINE DINUCLEOTIDE. FAD is a carrier of molecular energy in the form of two hydrogen atoms that reduce FAD to $FADH_2$ (right). The FAD molecule serves as an energy shuttle. Reduced FAD ($FADH_2$) in turn passes its two hydrogen atoms to another molecule, along with the 52 kcal of chemical energy, and is restored to FAD. The vitamin riboflavin furnishes part of the raw material for making the FAD molecule.

MOLECULAR STRUCTURE OF ENZYMES

Until about a decade ago biochemists knew very little about the behavior of enzymes at the molecular level. The concept of surface catalysis was familiar from industrial chemistry: substrate molecules bind themselves to a solid surface before reacting and dissociate from it afterward. It was generally agreed that the substrates of enzymes too were momentarily bound to an active site on the surface of the enzyme molecule, but chemists had no idea of the actual structure of an active site. The remarkable ability of the enzyme to select exactly the right substrate was explained by the assumption that the binding of the substrate to the site depended on a precise interlocking of molecular shapes. In 1894 the great German biochemist Emil Fischer compared the fit between enzyme and substrate to that of a lock and key. Fischer's model persisted for more than half a century with only indirect evidence to support it *(Box B)*.

The first direct evidence came in 1965, when David Phillips and his colleagues at The Royal Institution in London succeeded in crystallizing the enzyme lysozyme and, using the techniques of x-ray crystallography, determining its structure. Since then the structures of several dozen other enzymes have been solved by x-ray diffraction studies *(Box C)*. This work has revealed a great deal about how the enzyme molecule is designed, how it works, and to some extent how it is controlled. Enzymes turn out to be globular proteins, with molecular weights ranging from 10,000 to several million. The smaller enzymes consist of a single folded polypeptide chain; the largest contain several chains, often identical. Many of the active sites contain metal ions that enhance the reaction, particularly by helping to bind the substrate or to withdraw electrons.

Although it is probably an oversimplification to say that if you understand one enzyme molecule you understand them all, the x-ray studies suggest that most enzymes behave in very similar ways. This can be illustrated by comparing the structure and function of four protein-digesting (proteolytic) enzymes secreted into the human intestine by the pancreas: carboxypeptidase, chymotrypsin, trypsin and elastase. All four break the peptide bonds that link amino acids in polypeptide chains, but each attacks a bond having very specific characteristics. Carboxypeptidase snips one amino acid at a time from the carboxyl end of a chain; the other three enzymes cleave a chain in the middle. Tryp-

B **The battle of Sumner's urease**

Long after the discovery of enzymes biochemists refused to accept the idea that such remarkably efficient catalysts could be made of unglamorous materials akin to egg albumen and hair. In 1900 two chemists, Pekelharing and Ringer, crystallized a protein from digestive juices which they claimed was the enzyme pepsin. Seen in retrospect, it probably was. But they and others were unable to repeat the experiment, and their conclusions were rejected. It was then said that enzymes were "neither fats, carbohydrates, nor proteins, but an entirely new and unknown class of compounds." The great biochemist Richard Willstätter of the University of Munich purified many enzymes, including peroxidase, saccharase, lipases and amylase, and shaped the field of protein chemistry in the first three decades of this century. He maintained that enzymes were unknown catalytic substances associated with colloidal protein. (If one substitutes "active sites" for "enzymes," and deletes the word "colloidal," it turns out that he was not too far wrong.) Evidence for this was the observation that catalytic activity persisted even when purification had progressed so far that none of the usual tests for proteins would give a positive reaction. It never occurred to biochemists of the time that these protein enzymes could be so active that the catalytic tests were far more sensitive than the tests for protein.

James B. Sumner, a young chemist at Cornell University, isolated the enzyme urease from jack beans in 1926, and crystallized what he claimed was the pure protein enzyme. The response from Willstätter and his school was immediate and derisory. In a confrontation at a seminar at Cornell, Willstätter sarcastically denied the validity of all of Sumner's work. He maintained that Sumner had only crystallized the "carrier protein," and had let the enzyme slip through his fingers. But Sumner did not wilt under fire. He and others could repeat his work, and over the next few years he demonstrated that the repeatedly recrystallized protein lost none of its enzymatic activity.

When John Northrop, then at the Rockefeller University, crystallized the enzyme pepsin as a protein in 1930, resistance to the new idea began to crumble, although even four years later, diehards were refusing to equate enzyme and catalytic protein. The late 1930's saw the crystallization of many other enzymes—all proteins. This led to the first hesitant steps in what was to be an enormously powerful tool, x-ray analysis of protein crystal structure. A belated recognition of the value of James Sumner's work was the awarding to him and Northrop of the Nobel Prize in Chemistry for 1946. Sumner died in 1955, having had the satisfaction of seeing the heresy for which he was vilified in 1927 become the cornerstone of enzymology.

---

sin cuts a protein chain best at a point next to a basic or positively charged side group; chymotrypsin cuts the chain at a point next to a large hydrophobic residue, and elastase prefers to cut next to a small side chain. (*Box D*). The explanation for this fastidious specificity, which underlies the entire chemical strategy of living organisms, lies in the architecture of the enzyme molecule. As Fischer suggested, the structure of the active site is molded to fit the substrate molecule. The binding of the substrate depends on the same forces that maintain the folded structure of the enzyme protein: hydrogen bonds, the electrostatic attraction and repulsion of charged chemical groups, and the interaction of hydrophobic (water-repelling) groups.

Perhaps the most challenging problem in enzyme chemistry is to devise a molecular explanation for the behavior of regulatory enzymes—the chemical switches that catalyze reactions at the branching points where two or more metabolic pathways diverge. The end product of the pathway, if present in excess, inhibits the activity of the first enzyme in the pathway. Known as end-product inhibition, this feedback mechanism is very common in living cells, and is an example of

C The beginnings of enzyme anatomy

In 1938, Max Perutz, a young Viennese chemist and political refugee at Cambridge University, published x-ray photographs and crystal data for two proteins: hemoglobin and chymotrypsin. This was the beginning of the Cambridge protein structure project, which eventually led to the discovery of the structure of myoglobin in 1959, and of hemoglobin one year later —work that earned the Nobel Prize in 1961 for Perutz and his colleague John Kendrew. Protein crystal structure analyses fortunately no longer take 23 years. More than 75 protein structures are now known, including many enzymes.

For a protein to be studied by this technique it must first be crystallized, a difficult but not impossible task (*see photo below*). A beam of x-rays is passed through the crystal, and the diffraction pattern is collected either photographically or with an automated scintillation counter diffractometer. Heavy atoms such as mercury or platinum are then diffused into the crystals and their effects on the x-ray pattern are measured. Computer analysis of this information leads to a map showing the electron density at each point in the crystal. This map must be interpreted. At high enough resolution, individual atoms would appear as spherical clouds of electron density, but

CRYSTALLIZED PROTEIN. Photograph depicts crystals of ferricytochrome C from the heart muscle of a horse. Crystals are shown here about 100X actual size.

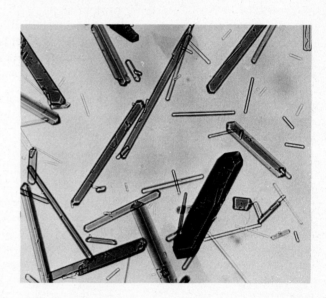

D Enzymology in the service of mankind

Subtilisin is a protein-digesting enzyme from *Bacillus subtilis* which resembles chymotrypsin and trypsin in active site structure, and in catalytic mechanisms. It is remarkably heat-stable, remaining active in hot water when most other enzymes would be destroyed. This is why subtilisin achieved such popularity in recent years as a presoak laundry agent or as a component in washday detergents. Subtilisin will digest protein stains such as chocolate and blood that have traditionally been hard to remove with soap or detergents. It will also digest silk, wool, or human epidermis if given a chance.

This is not the first time that enzymes have been used in a product, as opposed to being used to prepare it, as in brewing or baking. The carbohydrate chemist H. S. Paine achieved immortality of a sort in 1924 for a patent that made him the Father of Liquid Center Chocolates. Have you ever wondered, as you ate a liquid-center cherry

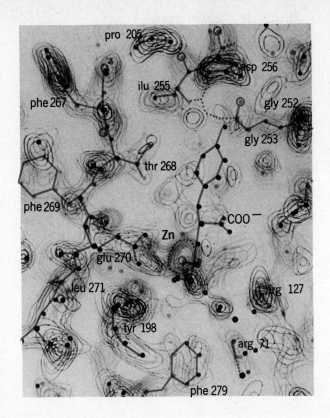

ELECTRON-DENSITY MAP of carboxypeptidase A resembles a contour map of the Earth's surface, except that contours here represent areas of equal electron density. Such maps are made from analysis of x-ray diffraction patterns like the one on the opening page of this chapter. From these maps investigators can deduce the three dimensional structure of the protein. Colored lines superimposed on this map indicate part of the skeleton of the enzyme molecule, and black lines show the position of the substrate. Dots indicate positions of atoms.

it is not customary to take a protein structure analysis that far because of the enormous labor involved. At somewhat lower resolution the main polypeptide chain can be followed and the entire molecular structure is decipherable if the amino acid sequence is known (*see map above*).

The function of computer analysis can be compared to that of the optical system of a microscope. The analogy is mathematically exact, and with a material of a high enough refractive index for x-rays to bend the rays, a true x-ray microscope might be built. X-ray analysis uses electromagnetic wavelengths only 1/5000th as long as light waves, and can "see" atomic details that are far beyond the resolving power of light microscopes.

candy, how the manufacturers managed to wrap the chocolate coating around the center while it is fluid? The answer, thanks to H. S. Paine, is that they don't have to—the center is solid when the candy is made. Paine discovered the trick of mixing a minute quantity of the enzyme invertase with the sugar of the fondant around the cherry. After the candy is chocolate-coated and packaged, the enzyme slowly hydrolyzes the disaccharide sucrose and converts it to the monosaccharides glucose and fructose. The monosaccharides are hygroscopic—they slowly absorb moisture through the chocolate coating until finally they dissolve and the center liquefies. Invertase is so efficient an enzyme that the amount needed to "invert" the sucrose is too small to be tasted or otherwise detected by the customer. This is nearly as clever a trick as finding an enzyme that will grow ships in glass bottles. Paine, like many geniuses, lies unsung and unappreciated, but his work endures.

At last biochemists understand enzymes for what they are: molecular-scale machines.

ACTION OF EXOPEPTIDASES

ACTION OF ENDOPEPTIDASES

superb design logic. By shutting down an entire sequence of reactions at a single stroke, it not only prevents the accumulation of a surplus of end products, but also of all the intermediate compounds.

Regulatory enzymes have two binding sites: an active site and a regulatory site. A small molecule that binds to the regulatory site can switch off the catalytic activity of the active site, or in some regulatory enzymes switch it on. In both cases the transfer of information between the two sites clearly involves some internal rearrangement of the enzyme's polypeptide chains. So far no one has succeeded in mapping the structure of a regulatory enzyme, and the mechanism remains unexplained. Some workers believe that it may resemble the mechanism of certain nonenzymatic proteins such as hemoglobin. In binding oxygen, the four subunits of hemoglobin interact with one another, slightly shifting their relative positions and covering or uncovering various amino acid side chains.

Certainly there is a great deal still to be learned about the enzyme molecule, but there is no longer a reason to suspect that future research will disclose any radical surprises. At last biochemists understand enzymes for what they are: molecular-scale machines.

PORTRAIT OF AN ENZYME

One enzyme whose mechanism has been studied extensively is carboxypeptidase. It belongs to a family of digestive enzymes that attack the polypeptide chains of proteins in food, breaking them into amino acids that can be absorbed through the walls of the intestine. Digestive enzymes HYDROLYZE a peptide bond; that is, they break the chain by adding a water molecule across it, as shown in Figure 12. Carboxypeptidase, as noted earlier, severs one residue at a time from the carboxyl (COOH) end of the chain. Actually there are two carboxypeptidases. Carboxypeptidase A is particularly efficient at removing residues that have aromatic side chains, such as the benzene ring of phenylalanine or tyrosine. Carboxy-

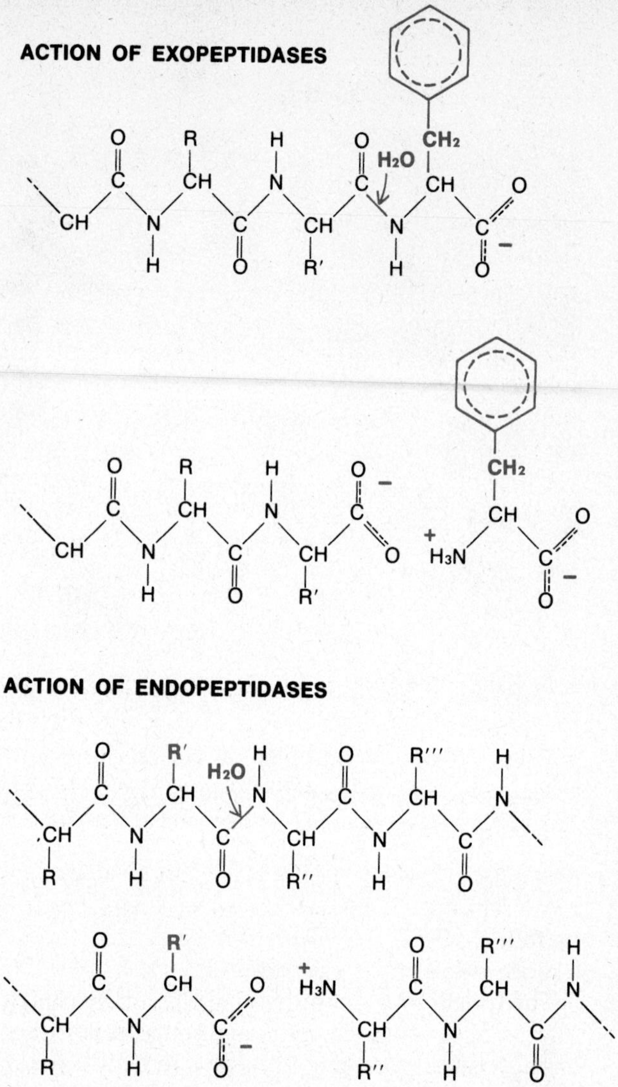

PROTEOLYTIC ENZYMES break a polypeptide chain **1.** and add a water molecule across the severed ends. The process is called hydrolysis, or splitting with water. Exopeptidases cleave off the final residue of a chain (top). Carboxypeptidase, for example, removes the last residue especially well if the residue has a bulky aromatic side group (color). Endopeptidases such as chymotrypsin, trypsin and elastase cleave a polypeptide chain in the middle. The specificity of those three enzymes depends on the nature of the side chain, designated by R' and R".

peptidase B is more efficient at removing residues having basic side chains.

Carboxypeptidase A has a molecular weight of 34,600, a fairly typical size for single-chain enzymes. Its chain is a polymer of 307 amino acids containing about 2500 atoms of C, N, O, and S;

nearly 2000 hydrogen atoms; and one metal atom (zinc). The folding of the chain is depicted schematically in Figure 13, with each of the 307 amino acids represented only by its α-carbon atom, numbered in sequence from the amino end of the chain. The peptide group (—CO—NH—) that connects α-carbons is represented by a straight line bond. Only the most important side chains are drawn. Another drawing of this view of the molecule appears in Figure 14, with only the catalytically important features represented.

The functional heart of the enzyme is its active site, a bowl-shaped depression visible at the upper right of Figure 13, and emphasized in the schematic Figure 14. At the bottom of this depression lies the essential zinc atom. Zinc, like carbon, commonly has four bonds in a tetrahedral orientation. In carboxypeptidase A, three of the four tetrahedral positions around the zinc are occupied by

nitrogens from two histidine side chains and an oxygen from a glutamic acid. The fourth position is vacant and faces the active site. One edge of the active site is folded over to make a pocket (residues 243–251); the interior of the pocket is lined with hydrophobic (water-repelling) side chains. On the opposite side of the active site, the rim is broken and the site blends into a groove running down the side of the molecule (between residues 279–282 and 12–16 on the left, and 125–117 on the right in Figure 13). The polypeptide chain of the substrate

CARBON SKELETON OF CARBOXYPEPTIDASE. **13**
Framework of the molecule is a pleated sheet surrounded by eight α-helices. One disulfide bridge holds more extended regions of chain in place at right. Active site (depicted in detail in the two following illustrations) is a bowl-shaped depression at top. More recent work has indicated that lysine 196 should be histidine 196.

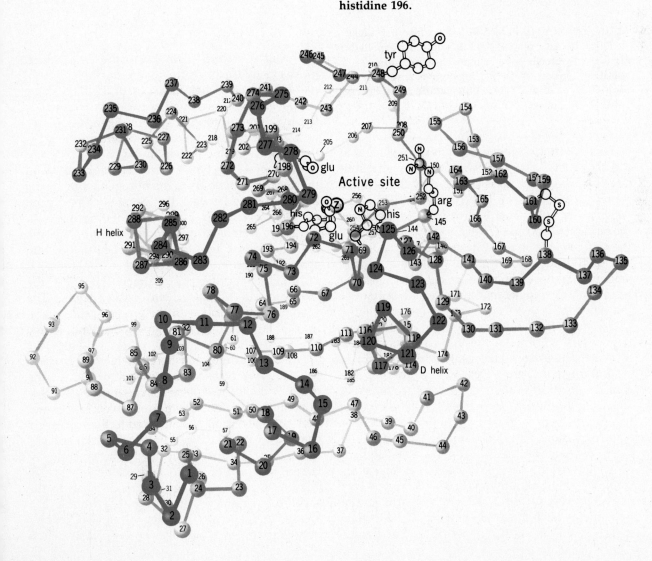

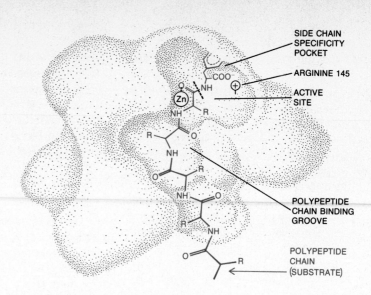

SIDE CHAIN SPECIFICITY POCKET

ARGININE 145

ACTIVE SITE

POLYPEPTIDE CHAIN BINDING GROOVE

POLYPEPTIDE CHAIN (SUBSTRATE)

14   **MAIN CATALYTIC FEATURES of carboxypeptidase A. This schematic rendering of the molecule depicted in Figure 13 emphasizes the active site depression, specificity pocket, chain binding groove, zinc atom and chain-holding arginine 145. Polypeptide chain of substrate is shown in color.**

aligns itself in this groove, with its carboxyl end in the active site. The negative charge on the substrate's carboxyl group is attracted by the positive charge on arginine 145. If the last side chain of the substrate is bulky and hydrophobic, it stabilizes the binding of substrate to enzyme by slipping into the hydrophobic pocket. The carbonyl (C=O) oxygen of the next-to-last amino acid—that is, the amino acid whose bond is to be cut—becomes the fourth coordinating atom for zinc at the bottom of the active site.

The substrate is shown bound to carboxypeptidase in Figure 15; the viewer looks directly into the active site. The catalytic mechanism is diagrammed in Figure 16. As the polypeptide chain binds itself to the enzyme, the arginine-145 side chain of the enzyme moves toward the negative carboxyl end of the polypeptide. This electrostatic attraction helps hold down the end of the chain. Electrons from the carbonyl double bond are drawn toward the zinc atom which is strongly electrophilic. This makes the main-chain carbonyl carbon slightly positive (*Figure 16a*), and attracts the glutamic acid 270 side chain of the enzyme toward it. The peptide bond is now weakened for attack.

The attack comes from tyrosine 248, which moves its hydroxyl group toward the lone electron pair on the main-chain nitrogen of the bond to be cut (*Figure 16a*). As a bond begins to form between the hydroxyl proton and the nitrogen, the bonds between hydrogen and oxygen in tyrosine, and between nitrogen and carbon in the polypeptide, are weakened. As the C—N peptide bond breaks, and the electron pair shifts to the nitrogen atom, a new bond is formed between the carbon and an oxygen of glutamic acid 270, using a pair of electrons from oxygen. At the halfway point (*Figure 16b*) the tyrosine hydroxyl proton has attached to the main chain nitrogen atom and turned it into a —NH$_2$ group; the peptide chain attaches itself to the glutamic-acid-270 side chain of the enzyme. The terminal amino acid with the aromatic side chain, once removed from the polypeptide chain, is free to fall away.

The story cannot end here. So far, the reaction has merely replaced the last amino acid on the chain by a molecule of carboxypeptidase. Restoration of the original conditions involves a water molecule. The proton is attracted to the negatively charged hydroxyl oxygen of tyrosine 248, and the hydroxyl group attacks the newly formed bond between polypeptide chain and glutamic acid 270 (*Figure 16b*). The OH$^-$ group bonds itself to the carbonyl carbon as the bond to glutamic 270 breaks: the glutamic acid retains the electron pair from the broken bond. The polypeptide chain is then free to fall away from the enzyme, glutamic acid 270 once again becomes negatively charged, tyrosine 248 is restored and swings back out of the way (*Figure 15*).

The carboxypeptidase mechanism is typical of many enzymatic reactions. Nothing was done that could not have been accomplished eventually without the enzyme. Hydrolysis of a peptide bond is exothermic and spontaneous, with a free energy drop of about two kilocalories per mole. Fortunately for living organisms, this hydrolysis is also extremely slow because no convenient mechanism exists for a water molecule in neutral solution to attack a peptide bond; otherwise no protein chain

ACTIVE SITE of carboxypeptidase. Shown in dark color and black outline is a substrate-like molecule that binds to the enzyme as the substrate does. Because this molecule is not affected by the enzyme it can be crystallized and studied along with the enzyme. In an actual substrate molecule the hexagonal benzene ring at left would be absent, and the chain would continue to lower left along the groove bounded above by residue 279 and below by residues 124 and 125. For clarity, diagram shows only the α-carbon atoms of the enzyme backbone. The three side chains that move when the substrate binds to the enzyme are shown in their respective positions when the substrate is present (solid outline) and absent (dashed outline). Arrow at center indicates bond to be cleaved in substrate.

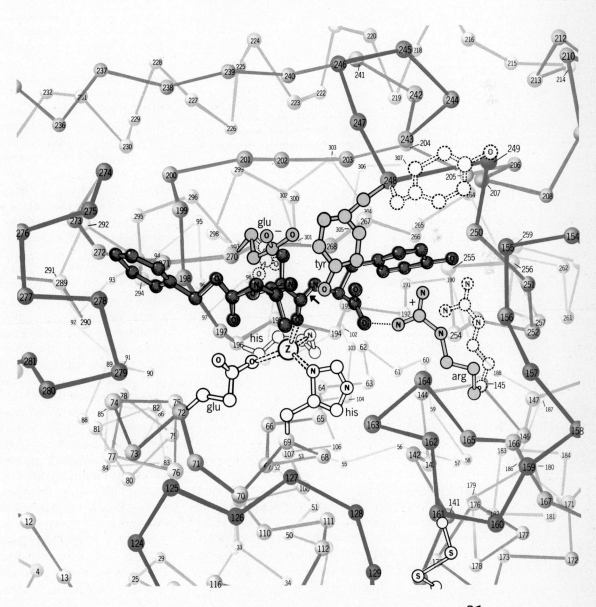

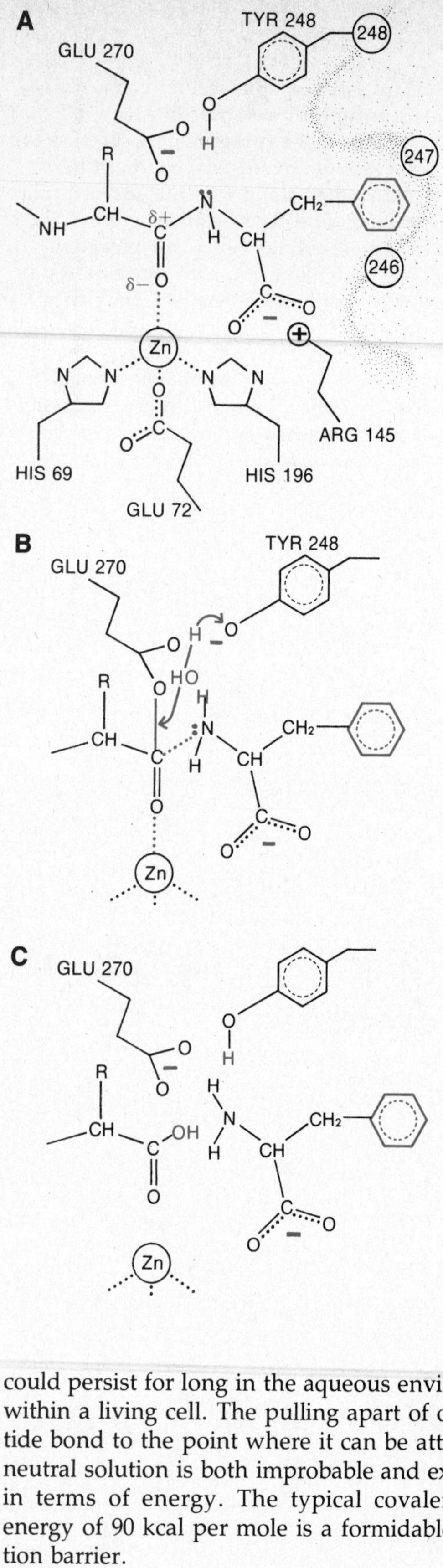

A

GLU 270 TYR 248 (248)

(247)

(246)

Zn

HIS 69 HIS 196

GLU 72

ARG 145

B

GLU 270 TYR 248

Zn

C

GLU 270

Zn

CATALYTIC MECHANISM of carboxypeptidase. (A) The polypeptide chain at center binds to the active site with the aid of (left to right) its carbonyl oxygen, its terminal carboxyl group, and its last aromatic side chain. Tyrosine 248 of the enzyme is poised to donate a proton (colored H) to the polypeptide nitrogen. **(B)** The peptide bond breaks, the terminal amino acid falls free, and the remaining polypeptide chain attaches to glutamic acid residue 270 on the enzyme. A water molecule (color) is split between the enzyme and the substrate. **(C)** Hydroxyl group from water molecule is added to the peptide chain, which then falls away. The proton (colored H) from the water molecule restores tyrosine 248 of the enzyme, leaving the surface of the enzyme in its original state.

Carboxypeptidase assists the early stages of dissociation. The zinc atom weakens the peptide bond by drawing electrons away from the carbon, and the tyrosine hydroxyl weakens it still further by forming a new bond between nitrogen and hydrogen. Glutamic acid 270 is poised to accept the polypeptide carbon when the peptide bond breaks, and other side chains of the enzyme hold the polypeptide in a favorable position for reaction. Carboxypeptidase provides a tunnel through the activation energy barrier, and sharply accelerates the rate of peptide-bond hydrolysis. Because of the specificity of the enzyme, not every peptide bond is hydrolyzed at this rapid rate, but only those carboxyl-terminal bonds next to bulky, hydrophobic side chains.

SUBUNITS, ALLOSTERY AND CONTROL

Most enzymes are much larger than carboxypeptidase, and are built from several subunits. Some of the enzymes of glycolysis and the citric acid cycle discussed in the next chapter, for example, have the subunit structure shown in Table I. Most of these enzymes have two or four subunits of molecular weights 35,000 to 50,000. Two conspicuous exceptions are pyruvate dehydrogenase and α-ketoglutarate dehydrogenase. Both of these are complexes of three different types of enzyme, which are packaged together so that the products of one enzymatic reaction are immediately at hand as the reactants for the next. Considerable economy of effort is inherent in assembling a large enzyme from many identical units, rather than from a single enormous polypeptide chain: less information need be coded in the cell's DNA. The structural principles are the same as those that lead an architect to build a house out of identical and

could persist for long in the aqueous environment within a living cell. The pulling apart of one peptide bond to the point where it can be attacked in neutral solution is both improbable and expensive in terms of energy. The typical covalent bond energy of 90 kcal per mole is a formidable activation barrier.

17 **ALLOSTERIC PROTEIN** in this schematic diagram is comprised of two subunits. When an inhibiting effector molecule (E) binds to a regulatory site on one subunit, the shape or position of that subunit changes in a way that affects the catalytic subunit, preventing its active site from binding a substrate molecule (S). The effector molecule need have no chemical similarity to the enzyme's substrate.

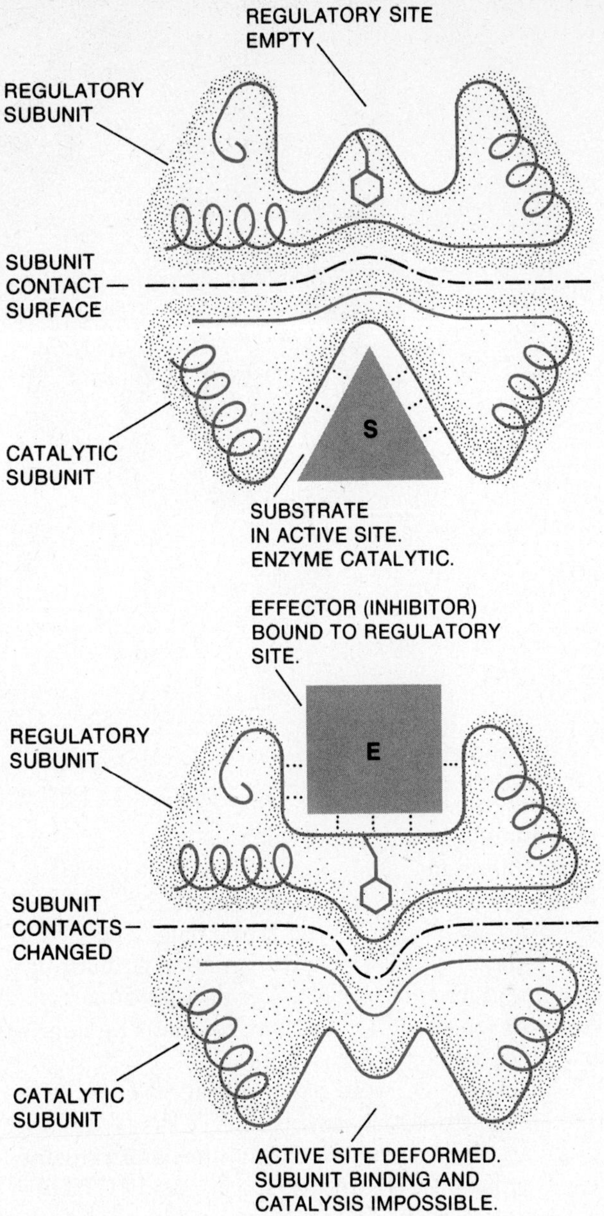

REGULATORY SITE EMPTY

REGULATORY SUBUNIT

SUBUNIT CONTACT SURFACE

CATALYTIC SUBUNIT

S

SUBSTRATE IN ACTIVE SITE. ENZYME CATALYTIC.

EFFECTOR (INHIBITOR) BOUND TO REGULATORY SITE.

REGULATORY SUBUNIT

E

SUBUNIT CONTACTS CHANGED

CATALYTIC SUBUNIT

ACTIVE SITE DEFORMED. SUBUNIT BINDING AND CATALYSIS IMPOSSIBLE.

standardized bricks rather than casting it as a single giant piece of concrete.

The subunits also serve a more important function. They make it practical for the activity of the enzyme to be controlled by outside molecules which have no similarity either to the reactants or the products of the process being catalyzed. These molecules operate by binding themselves to a site other than the catalytic site. Binding at this EFFECTOR SITE (or allosteric site) can enhance or diminish the reactivity at the active site (*Figure 17*). Because of the dissimilarity between the effector and the real enzymatic substrate, this behavior is called ALLOSTERY, meaning "different shape."

Hemoglobin is the best-studied example of an allosteric protein, although strictly speaking it is not an enzyme. It has four chains, each with one iron-containing heme group (*Figure 18*). The binding of oxygen to one heme group assists the binding of O_2 to the other three—a typical allosteric effect. This property enhances the performance of hemoglobin as an oxygen carrier in the blood, particularly in passing oxygen to myoglobin, which stores oxygen within tissues such as muscle. The hemoglobin subunits slightly change their relative positions when oxygen is bound or released—

I **LARGE ENZYMES** often consist of subunits. Table shows the number and molecular weight of subunits that comprise several of the enzymes of glycolysis and citric acid cycle (discussed in Chapter 6).

ENZYME	NUMBER OF SUBUNITS	MOLECULAR WT. OF SUBUNIT	MOLECULAR WT. OF ENZYME
HEXOKINASE	4	24,000	96,000
TRIOSE PHOSPHATE ISOMERASE	2	26,000	52,000
PHOSPHOGLYCEROMUTASE	1	57,000	57,000
LACTATE DEHYDROGENASE	4	35,000	140,000
PYRUVATE DEHYDROGENASE COMPLEX	24 24 24	90,000 36,000 55,000	4,344,000
SUCCINATE DEHYDROGENASE	1	175,000	175,000
FUMARASE	4	50,000	200,000
MALATE DEHYDROGENASE	2	35,000	70,000
α-KETOGLUTARATE DEHYDROGENASE	?	?	2,300,000

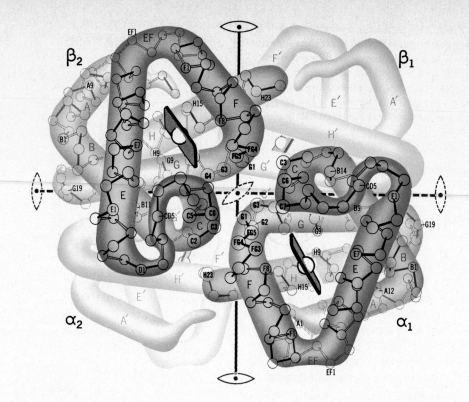

another typical allosteric effect. Because of this movement, the protein releases protons when oxygenated and binds them when deoxygenated. If the tissues lack enough oxygen, lactic acid will build up as an unburned intermediate. This acidic environment will shift the oxygenation equilibrium in hemoglobin in favor of the proton-binding form. In other words, in an acid environment hemoglobin tends to dump its oxygen.

The classic example of an allosteric enzyme is aspartate transcarbamylase (ATC-ase), which catalyzes the first step in the synthesis of cytosine triphosphate (CTP) from aspartic acid. CTP inhibits the activity of the enzyme, although it is eventually produced by a reaction many steps removed from the one catalyzed by ATC-ase and it is totally unlike any of the enzyme's substrates. CTP does not compete with the substrates for the active site; it binds at a completely different site on a different subunit. ATC-ase therefore resembles the two-subunit enzyme shown in Figure 17, but it has 12 subunits: six catalytic subunits with active sites, and six regulatory subunits with allosteric sites. The binding of CTP to the regulator subunits alters the efficiency of the active sites on the catalytic subunits in a way that can be symbolized by Figure 17. One of the unknowns in enzyme chemistry today is how allosteric effects are actually transmitted from one subunit to another.

HEMOGLOBIN is a protein comprised of four subunits, each the size and shape of one myoglobin molecule (Figure 16 in Chapter 4). The binding of oxygen to one of the four heme groups, shown here as rectangles in which iron atoms are embedded, enhances the binding of oxygen to the other three heme groups. Drawing shows fully oxygenated hemoglobin. When oxygen is removed, subunits β_1 and β_2 move apart. Hemoglobin is the best-studied example of allostery in subunit proteins.

FEEDBACK INHIBITION

In terms of control, the subunit structure is less important than the flow of information in a reaction pathway *(Figure 19)*. The end product of CTP synthesis damps the initial steps, keeping the production of CTP within bounds. This is the principle of negative feedback, seen also in flywheel governors on steam engines and thermostats on furnaces. Negative feedback keeps a rising trend within bounds by using the output to inhibit the first stages.

A more involved feedback control network is shown in Figure 20. Aspartic acid is the starting material for the synthesis of several amino acids in the bacterium *E. coli*. The synthetic chain branches into separate pathways for lysine, methionine, and threonine. The methionine-threonine branch occurs at the intermediate compound homoserine. The first step in all of these syntheses is the addition of a phosphate group to aspartic acid by means of the enzyme aspartokinase. Three different allosteric forms of aspartokinase are found in *E. coli,* and are inhibited by, respectively, lysine, homoserine, and threonine. They operate in parallel, so that if a great excess of lysine is present, the production of aspartyl phosphate will be reduced by roughly one-third. This is a relatively sloppy kind of control mechanism, because the excess of lysine ultimately decreases the synthesis of methionine and threonine as well. As an added refinement, however, both lysine and threonine inhibit the first enzymatic reaction past the branch point in their own direction. Up to the appearance of compound A in Figure 20, the synthesis is not committed to a given end product; the end product of the synthesis could ultimately be lysine or any of the products on the right-hand branch of the pathway. But once A is converted to B, a choice

has been made; the end product can only be lysine. The step from A to B is called the committed step. Similarly, the synthesis of homoserine from intermediate A represents a committed step toward one of the other end products, although more branch points will be passed before the commitment to one product becomes final. The committed steps in metabolic pathways are particularly effective points for feedback control. Inhibition of the A to B step in Figure 20 shunts all of the reactants over into the right-hand set of pathways, whereas inhibition of the B to C reaction, one step later,

FEEDBACK INHIBITION of the enzyme aspartate transcarbamylase. Cytosine triphosphate (color) is the end product of a long step synthesis. The first step is catalyzed by the allosteric enzyme aspartate transcarbamylase. Although CTP chemically resembles neither the reactants nor the products of the catalyzed reaction, it can bind to the enzyme and impede its action. An oversupply of CTP therefore shuts down the synthetic machinery and regulates its own synthesis.   19

CYTOSINE TRIPHOSPHATE (CTP)

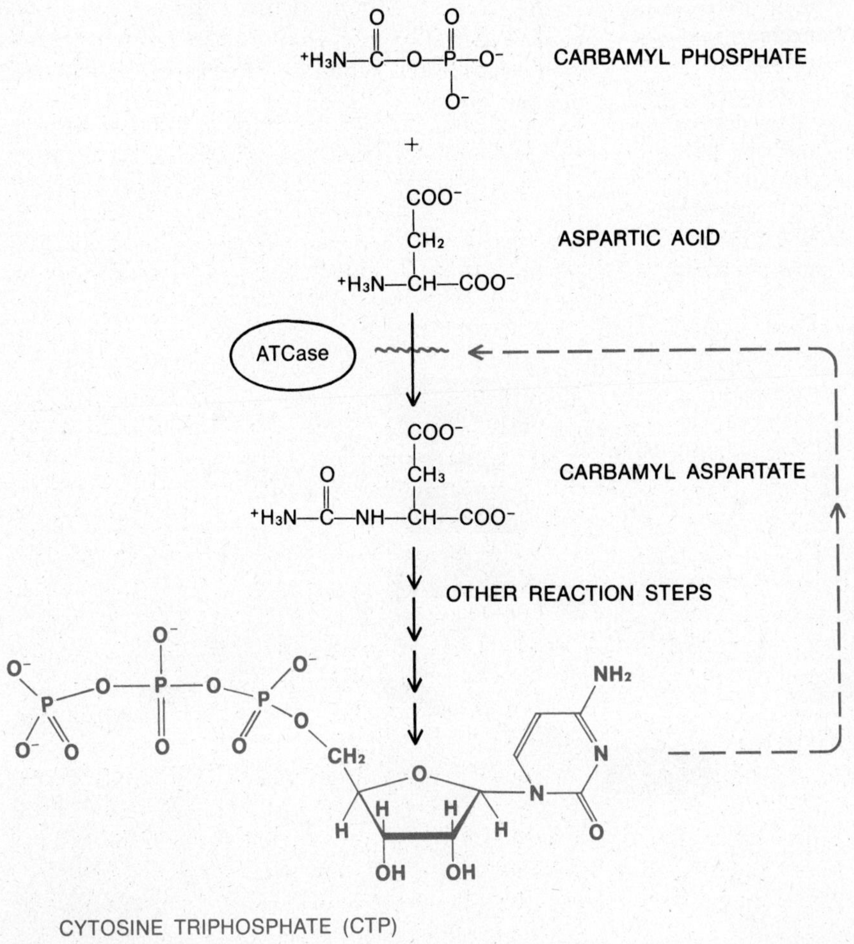

would lead only to a possibly harmful and certainly wasteful buildup of useless substance B. In this network a surplus of lysine decreases the conversion rate of aspartic acid at the beginning, and also shunts whatever aspartic acid is used into the pathway leading to homoserine and then to methionine or threonine. Threonine throws two roadblocks in the way of its own synthesis when it is present in excess, blocking two different committed steps in favor first of lysine and then of methionine.

The operation of this network of pathways is typical of many feedback control systems in metabolism. An excess of one product slows the first step in the synthesis, and also interferes with the first enzyme after a branch point that could lead to different products. All of the inhibitions in Figure 20 appear to be genuine examples of allostery, rather than competitive inhibition at the active site of the enzyme.

20 **FEEDBACK CONTROL governs synthesis of four amino acids (lysine, methionine, threonine and isoleucine) from aspartic acid in the bacterium E. coli. All four amino acids and one intermediate product act as allosteric inhibitors, either at the first step of the synthesis or at critical branch points. Unlabeled ovals and letters A to E represent intermediate compounds. Small colored blocks indicate feedback inhibition of enzymes, and dashed colored lines show source of inhibition.**

Note that there is one design flaw in the feedback control mechanism in Figure 20; an excess of threonine not only shuts down the production of threonine, but also of methionine. Given an oversupply of threonine, a culture of *E. coli* will become deficient in methionine, and will need an outside supply to grow normally. It is not surprising that such flaws occasionally appear in metabolic pathways. Biological control networks are the result of long evolution and natural selection. Since large overdoses of pure threonine are seldom encountered by *E. coli* in human intestines (or in fact, in any place except the laboratory of a zealous biochemist), the flaw in the scheme would pose no disadvantage for the bacterium and hence there would be no selection pressure against the flaw. As the Harvard biologist George Wald has remarked, we are less the product of authorship than of editing. An organism is only as perfect as its environment demands.

The energy-extracting systems of living organisms are under allosteric control that involves both negative and positive feedback *(Chapter 6)*. Some reaction intermediates inhibit earlier stages of the process; others increase the activity of enzymes. An allosteric enzyme can be activated as well as inhibited when an effector binds to a regulatory site.

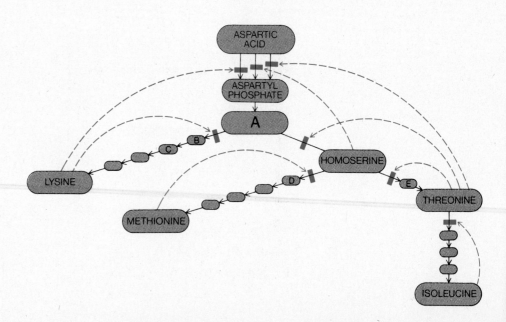

READINGS

R.E. DICKERSON AND I. GEIS, *The Structure and Action of Proteins*, Menlo Park, CA, W. A. Benjamin, 1969. This book continues the treatment of hemoglobin and of enzymes that has been begun in this chapter. It has a fuller treatment of allostery and control, including antibody structure and the serum complement system.

A.L. LEHNINGER, *Biochemistry*, New York, Worth, 2nd Edition, 1975. A remarkably lucid and readable introductory biochemistry textbook. Highly recommended as a reference. Particularly good treatment of glycolysis and respiration, and the enzymes and mitochondrial structure involved.

L. STRYER, *Biochemistry*, San Francisco, W.H. Freeman, 1975. An excellent and very readable textbook with a strong emphasis on molecular structure.

The following articles from *Scientific American*, written for the nonspecialist, are clear and well-illustrated:

J.C. KENDREW, "The Three-Dimensional Structure of a Protein Molecule," December 1961. The structure of myoglobin, the first protein to be solved.

H. NEURATH, "Protein Digesting Enzymes," December 1964. Discussion of chymotrypsin and trypsin based on their amino acid sequences. Predates the x-ray structure analysis.

M.F. PERUTZ, "The Hemoglobin Molecule," November 1964.

D.C. PHILLIPS, "The Three-Dimensional Structure of an Enzyme Molecule," November 1966. The structure of hen egg white lysozyme.

R.M. STROUD, "A Family of Protein-Cutting Proteins," July 1974. The three-dimensional structures of trypsin, chymotrypsin, and elastase.

It may form an interesting intellectual exercise to imagine ways in which life might arise, and having arisen might maintain itself, on a dark planet; but I doubt very much that this has ever happened, or that it can happen.

—George Wald

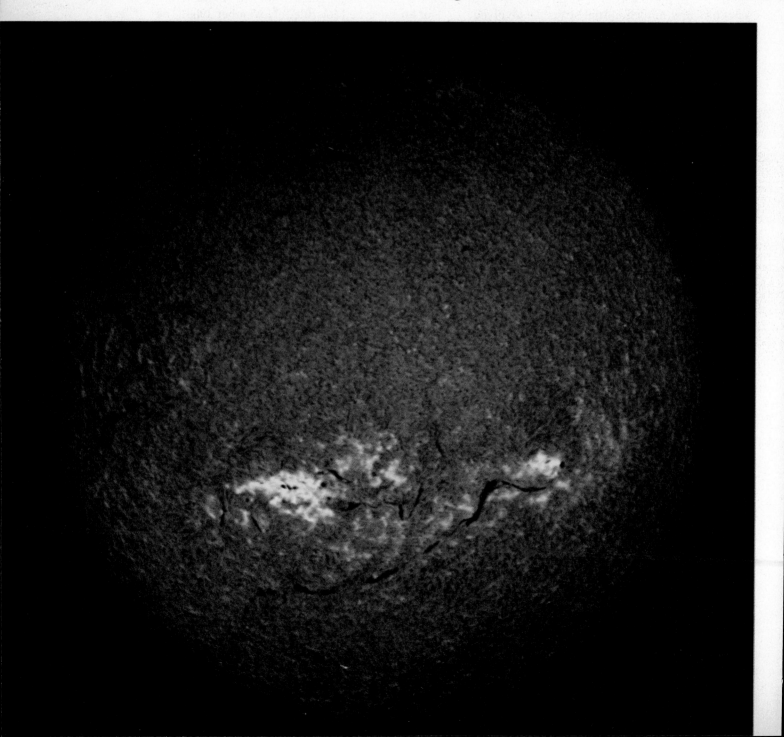

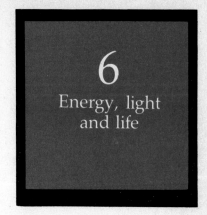
Living systems are basically unstable. They require a continual supply of matter and energy: matter to build new structures and to replace those that have worn out; energy to assemble these structures and to perform biological work. Matter is obtained from the environment, and energy is drawn from three chemical processes; fermentation, respiration, and photosynthesis *(Figure 1)*. These are the biochemical engines that power the machinery of life.

Each process involves a chain of reactions that has developed during countless millenia of trial-and-error evolution. Some links of the energy chain are much older than others. The first organisms on Earth probably were single-celled scavengers that inhabited tidal ponds, ingesting complex energy-rich molecules from the water and digesting them into smaller fragments by fermentation. One of the oldest and commonest fermentation pathways is anaerobic glycolysis—the breakdown of glucose in the absence of oxygen into two- or three-carbon fragments such as ethyl alcohol, acetone, and lactic acid.

The chemical machinery for the fermentation of sugars survives in the cells of all living organisms. It remains the sole source of energy for some microorganisms, such as the soil bacterium *Clostridium botulinum,* which causes the deadliest form of food poisoning *(Figure 2)*. The energy machinery of the Clostridia is a biochemical fossil that has survived for billions of years, dating back to a primitive Earth with an atmosphere devoid of oxygen. To Clostridia oxygen is not only useless but lethal, and they can flourish only in microenvironments devoid of air—buried in soil, in wounds, or in sealed cans of food.

Some microorganisms, such as yeasts and many bacteria, can survive either with or without oxygen. In their cells fermentation comprises the first few links in the energy chain. If oxygen is available, these organisms use the process of respiration to break down further the waste products of fermentation. Respiration not only enables them to burn (oxidize) their biochemical garbage, but to harness the energy of the fire. By oxidizing a sugar molecule all the way to carbon dioxide and water, instead of stopping short at ethyl alcohol, a yeast cell can derive 19 times as much energy per gram of food. If oxygen is not available such cells can live quite well by fermentation alone, although at much lower levels of biochemical efficiency.

Yeasts such as *Saccharomyces* (brewer's yeast) will grow and reproduce rapidly under aerobic conditions *(Figure 3)*. If the oxygen supply is cut off, the yeast simply shuts down its respiratory machinery, slows its rate of reproduction, and maintains itself by fermentation. Wine-growers exploit this metabolic behavior by first aerating crushed grapes to encourage the yeasts to grow, then letting the must (the crush of skins and juice) stand in vats for several days while the yeasts change grape sugar anaerobically into alcohol.

Human cells, and those of all other higher animals and plants, have lost the ability to switch the respiratory machinery on and off. Our cells have no simple way to dump the end product of anaerobic glycolysis, lactic acid. When an athlete exercises vigorously, the quick energy comes from the rapid fermentation of glucose in muscle cells. But as lactic acid accumulates it produces fatigue and eventually muscle cramps. When the muscles have a chance to recover, the slower respiratory process eliminates the lactic acid by oxidizing it to carbon dioxide and water.

THE SUN is ultimate source of all biological energy. In the telescope photograph of the solar disk on opposite page, white areas are solar flares and black strands are filaments of cooler gas.

99

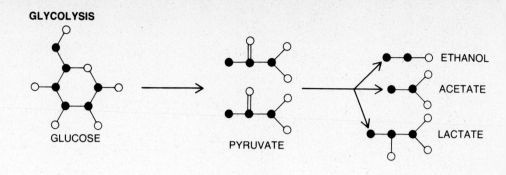

GLYCOLYSIS

GLUCOSE → PYRUVATE → ETHANOL / ACETATE / LACTATE

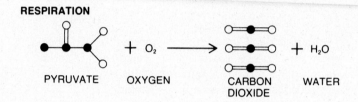

RESPIRATION

PYRUVATE + OXYGEN O_2 → CARBON DIOXIDE + WATER H_2O

PHOTOSYNTHESIS

6 CARBON DIOXIDE MOLECULES + 6 WATER MOLECULES → LIGHT ENERGY → 1 GLUCOSE MOLECULE + 6 OXYGEN MOLECULES

1 **MAIN ENERGY-YIELDING PROCESSES of living organisms.** In glycolysis, glucose molecules are broken down into smaller molecules, yielding relatively little energy. In respiration, pyruvate molecules combine with oxygen to form carbon dioxide and water, with a much higher energy yield. Solar energy is captured by the process of photosynthesis, which makes glucose from carbon dioxide and water. This process releases oxygen into the atmosphere. Black disks are carbon atoms, white ones are oxygen atoms.

More than 20 enzymes control the various reactions in the energy chain. The enzymes for glycolysis are dissolved in the cytoplasm of the cell, and those for respiration are incorporated into the structure of the mitochondrion, the powerhouse of all eucaryotic cells *(Figure 4)*. The reactions take place in a series of steps. At each step there is a small yield of free energy, which the cell uses to synthesize adenosine triphosphate (ATP).

All living cells rely on the ATP molecule for short-term storage of energy. An active cell requires more than two million molecules of ATP per second to drive its biochemical machinery. A fraction of the ATP is diverted to the synthesis of long-term energy storage compounds. Plants synthesize starch, a long-chain polymer of glucose. Animals store energy in glycogen and fats. (Of course, as every nutritionist knows, any large molecule synthesized by the cell—a protein, for example—is a storehouse of energy, but energy storage is not their primary function.) ATP can be considered as the circulating currency of energy exchange in living organisms, starches and fats as a savings account at the energy bank. When ani-

Plants rewind the mainspring of life by creating food, not only for themselves, but also for all other organisms on Earth.

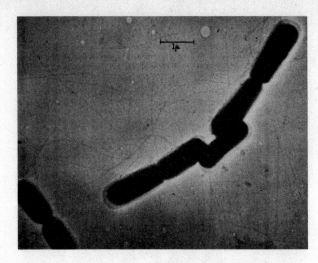

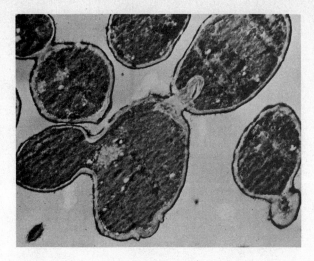

2 CLOSTRIDIUM BOTULINUM is an anaerobic soil bacterium that causes the deadly form of food poisoning known as botulism. This photomicrograph shows seven of these rod-shaped bacteria, each about two micrometers long.

3 BREWER'S YEAST, Saccharomyces cerevisiae, is a fungus that converts sugar into alcohol by the process of fermentation. In this electron micrograph new cells are budding from larger parent cells.

4 MITOCHONDRION contains the respiratory machinery of the cell. These organelles are found in all eucaryotic cells. This electron micrograph depicts mitochondrion from bat pancreas cell.

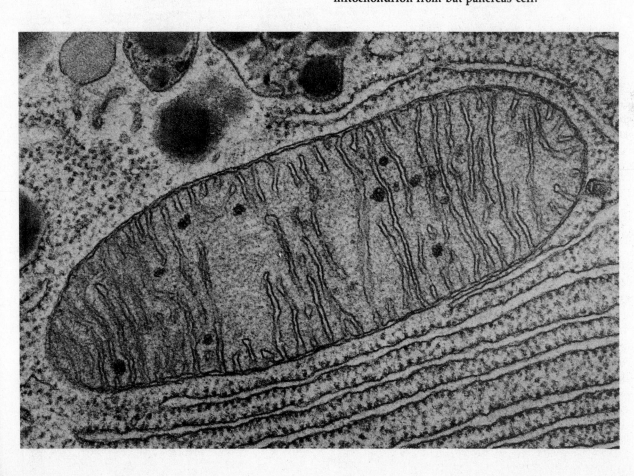

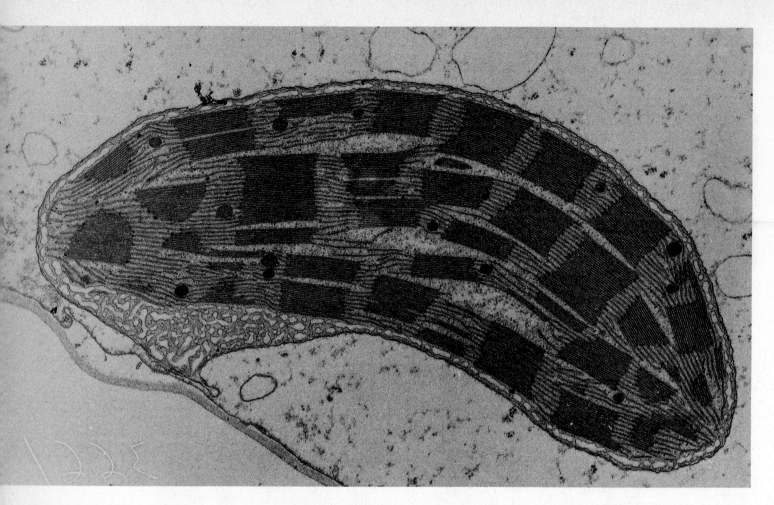

5 **CHLOROPLAST is the site of photosynthesis, the process that converts the energy of sunlight into chemical energy. Parallel bands are membranes that contain photosynthetic pigments.**

mals need energy they draw on their deposits of fat, which they oxidize to carbon dioxide and water. Similarly, plants draw on their deposits of starch, which they convert to glucose and oxidize to carbon dioxide and water.

Unlike animals, green plants can make their own food. The invention of photosynthesis enabled them to tap a virtually unlimited source of free energy—sunlight—to synthesize new molecules of glucose. Within the plant cell photosynthesis occurs in chloroplasts (*Figure 5*), and involves two sets of reactions. The dark reactions, which do not require light, reduce (or "hydrogenate") carbon dioxide to glucose, using water as the ultimate source of hydrogen. The light reactions provide reducing power and a supply of energy, in the form of ATP molecules, to drive the synthesis of glucose.

Sunlight is the ultimate source of nearly all biological energy. Plants rewind the mainspring of life by creating food not only for themselves but for all other organisms on earth. Animals obtain their energy either second-hand by eating plants, or third-hand by eating other animals. Scavengers such as fungi and bacteria complete the cycle by feeding on the energy stored in the large organic molecules of dead plants and animals.

The planet Earth can support life only so long as its supply of sunlight persists. The evolution of life on Earth has led from unorganized chemicals to highly organized creatures. Individually and collectively, living organisms seem to contradict the Second Law of Thermodynamics, which states that spontaneous processes tend toward states of lower energy and greater disorder. How do living systems carry out thousands of spontaneous chemical reactions per second and still maintain their remarkable degree of organization?

Living systems do not violate the Second Law. In the language of the physicist, a living creature is

102

an unstable open thermodynamic system. It can maintain a high level or organization—a state of low entropy—only by extracting free energy from the environment, either in the form of sunlight or of energy-rich molecules, and dumping back into the environment an assortment of low-energy, high-entropy breakdown products such as carbon dioxide, and heat, which is a degraded form of energy.

ANAEROBIC GLYCOLYSIS

Some bacteria ferment organic molecules other than glucose, but for most bacteria and all higher life, the energy chain begins with anaerobic glycolysis. The overall reaction

$$\underset{\text{Glucose}}{C_6H_{12}O_6} \rightarrow 2\ \underset{\text{2 Lactic Acid}}{CH_3-\overset{\displaystyle OH}{\underset{\displaystyle |}{CH}}-COOH,}$$

$$\Delta G = -47.4 \text{ kcal}$$

releases 47.4 kcal of free energy per mole of glucose. This is too much energy to handle at once. If 47.4 kcal were released in one step, most of it would be wasted as heat. An intricate mechanism has evolved to degrade glucose step by step. The cell uses the small packets of free energy released at each step to synthesize ATP.

ATP is a phosphate ester with a free energy of hydrolysis somewhat larger than that for most other esters. If we represent adenine and ribose by A and R, then ATP can be written:

$$A-R-O-\overset{\displaystyle O^-}{\underset{\displaystyle O}{\overset{\displaystyle |}{\underset{\displaystyle \|}{P}}}}-O-\overset{\displaystyle O^-}{\underset{\displaystyle O}{\overset{\displaystyle |}{\underset{\displaystyle \|}{P}}}}-O-\overset{\displaystyle O^-}{\underset{\displaystyle O}{\overset{\displaystyle |}{\underset{\displaystyle \|}{P}}}}-O^-.$$

The hydrolysis of ATP to adenosine diphosphate, ADP, and phosphate is then:

$$A-R-O-\overset{O^-}{\underset{O}{P}}-O-\overset{O^-}{\underset{O}{P}}-O-\overset{O^-}{\underset{O}{P}}-O^- \xrightarrow{H_2O}$$

$$A-R-O-\overset{O^-}{\underset{O}{P}}-O-\overset{O^-}{\underset{O}{P}}-O^- + H-O-\overset{O^-}{\underset{O}{P}}-O^- + H^+,$$

and for this reaction $\Delta G' = -8.1$ kcal mole^{-1}. Hydrolysis of most other phosphate esters produces less energy:

Glucose—6—phosphate + H_2O → glucose + phosphate,

$$\Delta G' = -3.3 \text{ kcal;}$$

Glycerol—1—phosphate + H_2O → glycerol + phosphate,

$$\Delta G' = -2.3 \text{ kcal.}$$

ATP is therefore a vehicle for transferring phosphate groups to other compounds and priming them for later chemical reactions. It is also a useful means of storing energy in packets of eight kilocalories. Part of the unusually large free energy of hydrolysis in ATP comes from the large number of negative charges on its phosphate groups. When ATP is hydrolyzed, the charges are spread over two molecules and the products are more stable. The hydrolysis of ADP to adenosine monophosphate (AMP) and phosphate liberates a similar amount of free energy:

$$A-R-O-\overset{O^-}{\underset{O}{P}}-O-\overset{O^-}{\underset{O}{P}}-O^- + H_2O \rightarrow$$

$$A-R-O-\overset{O^-}{\underset{O}{P}}-O^- + H-O-\overset{O^-}{\underset{O}{P}}-O^- + H^+,$$

$$\Delta G' = -9.5 \text{ kcal.}$$

Hydrolyzing the last phosphate group does not spread out the negative charges any further, so the free energy change is similar to that for any other ester hydrolysis

$$A-R-O-\overset{O^-}{\underset{O}{P}}-O^- + H_2O \rightarrow$$

$$A-R-OH + H-O-\overset{O^-}{\underset{O}{P}}-O^-$$

$$\Delta G' = -2.0 \text{ kcal.}$$

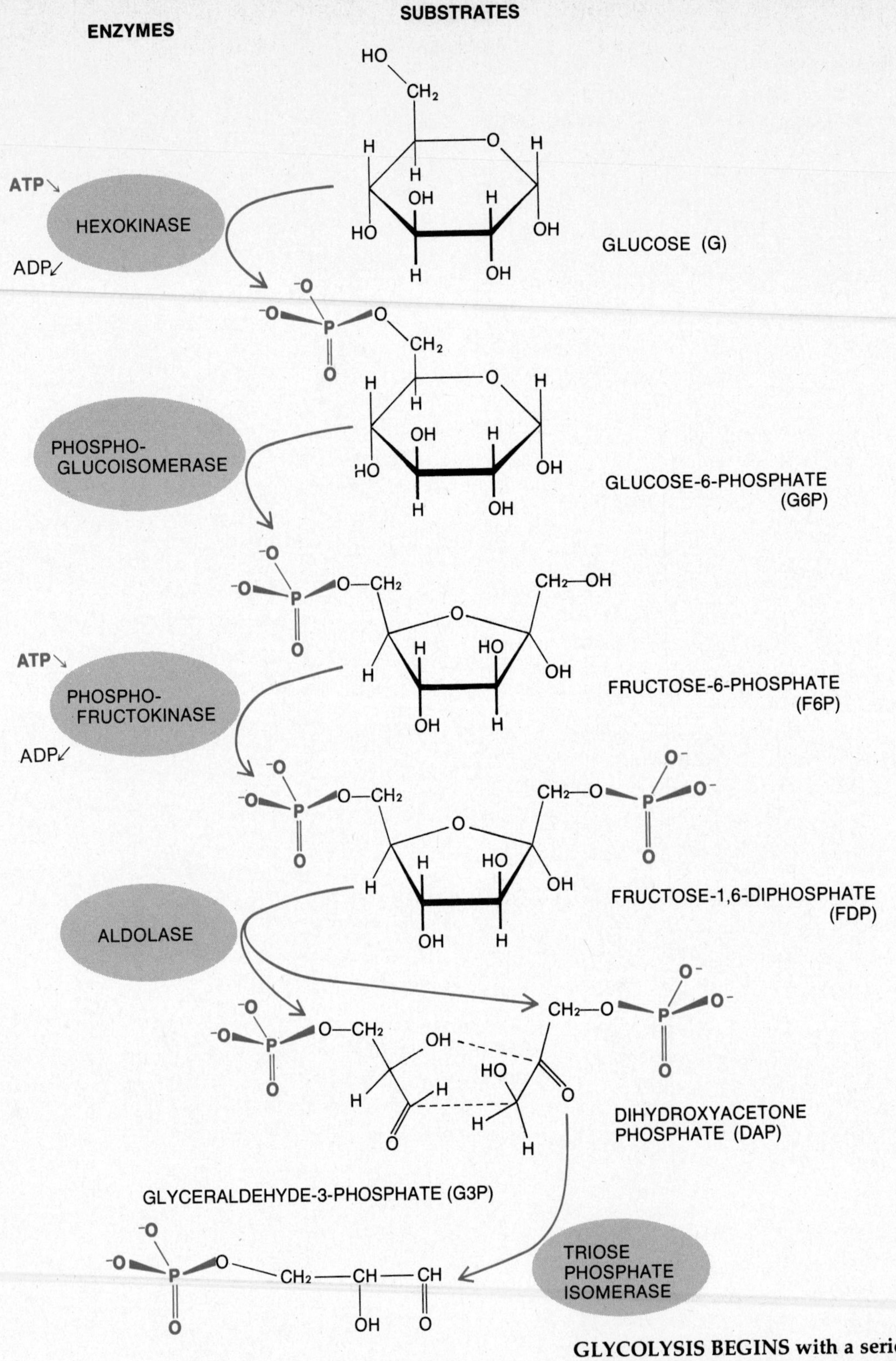

ENZYMES

SUBSTRATES

ATP ↘
ADP ↙

HEXOKINASE

GLUCOSE (G)

PHOSPHO-
GLUCOISOMERASE

GLUCOSE-6-PHOSPHATE
(G6P)

ATP ↘
ADP ↙

PHOSPHO-
FRUCTOKINASE

FRUCTOSE-6-PHOSPHATE
(F6P)

ALDOLASE

FRUCTOSE-1,6-DIPHOSPHATE
(FDP)

DIHYDROXYACETONE
PHOSPHATE (DAP)

GLYCERALDEHYDE-3-PHOSPHATE (G3P)

TRIOSE
PHOSPHATE
ISOMERASE

GLYCOLYSIS BEGINS with a series of pump-priming
reactions (shown above). One molecule of glucose is
split into two molecules of 3-phosphoglyceraldehyde.
Phosphate groups (color) come from two molecules of
ATP. Enzymes that control the process (ovals) are
dissolved in cytoplasm of the cell.

6

ENZYMES   SUBSTRATES

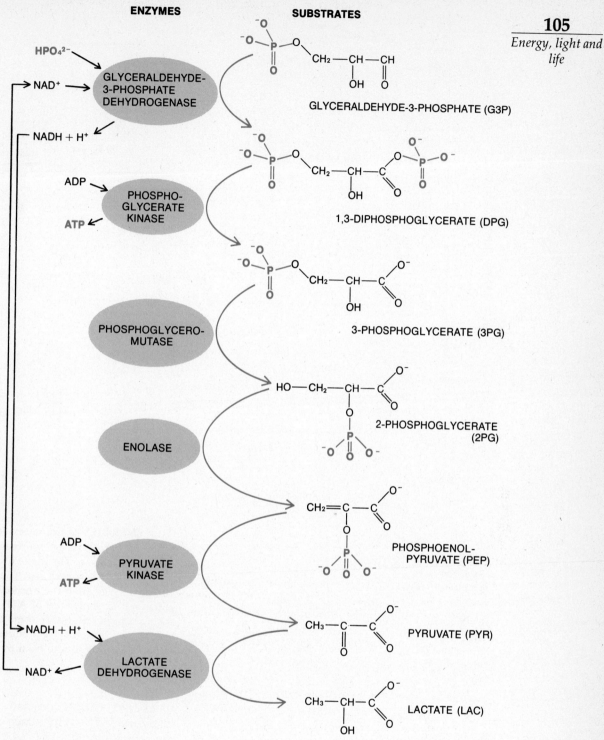

7 GLYCOLYSIS CONTINUES as the two molecules of
3-phosphoglyceraldehyde from Figure 6 are each
primed with a second phosphate group (color). The
molecules eventually transfer both phosphates to ADP
to form two molecules of ATP, the cell's short-term
energy-storage compound. Six different enzymes
(ovals) control the process. Role of NAD is explained in
the text.

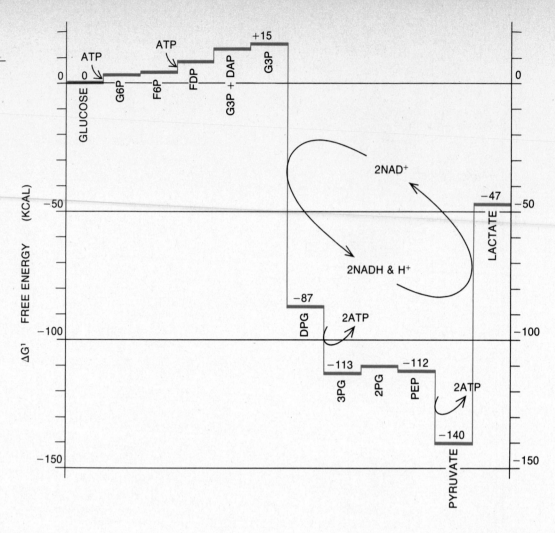

FREE ENERGIES of the intermediate compounds in anaerobic glycolysis are represented as horizontal bars on a vertical scale. The name of the intermediate is abbreviated below each bar. Graph shows total free energy derived from the breakdown of one mole of glucose. Two molecules of ATP are used in the initial steps at upper left, but four molecules are produced at lower right. The NADH produced in one step is used again in making lactate from pyruvate (right).

Because of their larger free energies of hydrolysis, the first and second bonds broken in ATP are sometimes called high-energy bonds, although this refers to the energy of hydrolysis and not to any intrinsic energy of the bond itself. The high-energy bond is sometimes symbolized by: $\sim$, and ATP can be written: $A-R-P\sim P\sim P$, where P now represents an entire phosphate group. The phosphate ion, HPO_4^{2-}, is sometimes abbreviated by P_i, meaning inorganic phosphate.

The eleven steps involved in degrading glucose in small stages to two molecules of lactic acid are outlined in Figures 6 and 7. It is not important to memorize these structures or their reactions, but to grasp the overall strategy of the energy-extraction process. Why so many different steps? At what points and for what kinds of reactions is energy extracted? Why, if the purpose of the machinery is to take energy out, does nature begin by feeding in energy as ATP? Why attach phosphate groups to the sugar?

The first five stages in Figure 6 are essentially pump-priming. They convert one molecule of a six-carbon sugar into two three-carbon molecules, primed with phosphate groups for later reactions. The last six reactions are the real source of energy. There is a net gain of two molecules of ATP for every glucose molecule broken down:

Glucose breakdown:	$C_6H_{12}O_6 \rightarrow 2\ C_3H_6O_3$ (Glucose) (Lactic Acid)	$\Delta G' = -47.4$ kcal
ATP synthesis:	$2ADP + 2P_i \rightarrow 2ATP$	$\Delta G' = +16.2$ kcal
	$C_6H_{12}O_6 + 2ADP + 2P_i \rightarrow 2\ C_3H_6O_3 + 2ATP$	$\Delta G' = -31.2$ kcal

Of the 47 kcal of free energy evolved, 16 kcal are saved and 31 kcal are wasted, for an efficiency of $16.2/47.4 \times 100\% = 34\%$. This is not particularly efficient, and in fact is the sort of mechanism that would have evolved only where organic compounds were abundant.

The free energy changes in these reactions are shown in Figure 8. All of the first five steps require energy. Two molecules of ATP are used to attach two phosphate groups to the sugar and to raise the free energy by 15 kcal per mole. Each of these steps is controlled by its own enzyme. Hexokinase removes a phosphate from ATP and attaches it to glucose. Phosphoglucoisomerase rearranges the six-membered ring to a five-membered fructose ring. Phosphofructokinase attaches a second phosphate group, and aldolase breaks the ring into two different three-carbon molecules. Triose phosphate isomerase rearranges one of them so both molecules are the same, and the halfway point in glycolysis is reached.

The second half involves an important carrier molecule: nicotinamide adenine dinucleotide, NAD, shown in Figure 9. (An older name for NAD is *diphosphopyridine nucleotide*, or DPN.) The sole function of NAD is to carry hydrogen atoms and free energy from compounds being oxidized, and to give up hydrogen atoms and energy to compounds being reduced. The reduction of NAD

$$NAD^+ + 2(H) \rightarrow NADH + H^+$$

can be regarded as occurring with a free energy increase of 52.4 kcal mole^{-1} if O_2 is the final reoxidant:

$$NADH + H^+ + \tfrac{1}{2}O_2 \rightarrow NAD^+ + H_2O$$
$\Delta G' = -52.4$ kcal

Just as ATP can be thought of as a means of packaging free energy in eight-kcal bundles, NAD can

be thought of as a means of packaging 52-kcal bundles. One half of NAD looks very much like a molecule of ATP, and it is easy to imagine that several energy carriers involving adenine, ribose, phosphates, and other groups, evolved in the course of time from a common (and less efficient) precursor. The structure of several other carrier molecules can be described by: adenine — ribose — phosphate — phosphate — Group X. Part of the Group X is often a compound that humans can no longer synthesize for themselves, and so is classed as a vitamin. If Group X is a ribose and a nicotinamide ring, the compound is NAD. Nicotinic acid, or NIACIN, is one of the vitamin B complex. Another member of this same vitamin complex is RIBOFLAVIN, which is involved in carriers

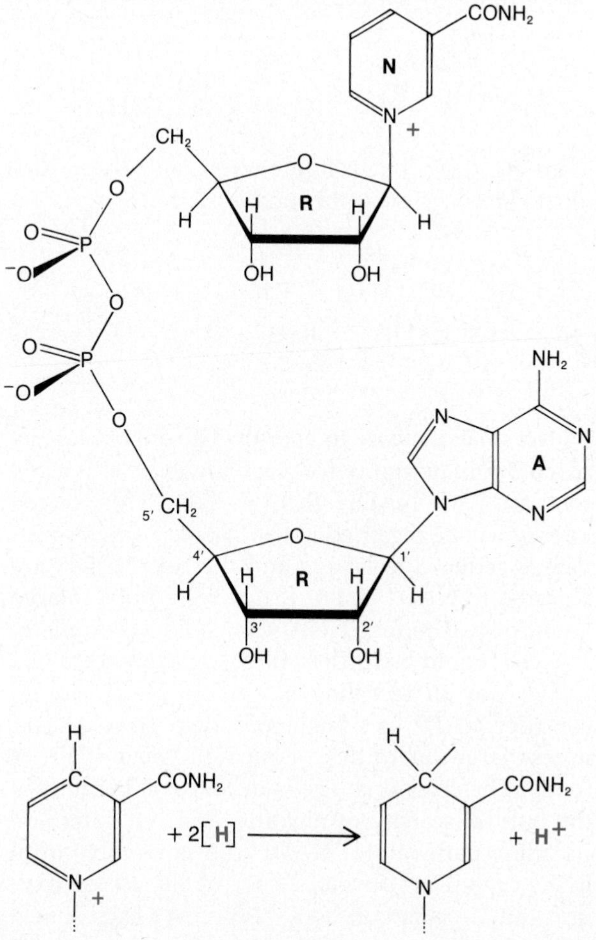

9 NAD$^+$, dicotinamide adenine dinucleotide, is built from adenine (A), ribose (R), two phosphates, and a nicotinamide ring (N). When NAD$^+$ is reduced by two hydrogen atoms from a metabolite, one H atom attaches to a nicotinamide ring carbon, while the other gives its electron to the nitrogen and goes into solution as H$^+$.

that we shall encounter later called FLAVIN MONONUCLEOTIDE, or FMN, and FLAVIN ADENINE DINUCLEOTIDE, or FAD. Pantothenic acid, another member of the vitamin B complex, is used in a carrier of acetate groups called COENZYME A. Other vitamins have similar uses. We need only small amounts, since these carrier molecules are recycled through the metabolic machinery, and need to be replaced only when they are accidentally destroyed.

The first reaction shown in Figure 7 involves the addition of another phosphate group to glyceraldehyde-3-phosphate, yet Figure 8 indicates that the free energy falls by 102 kcal. Why is there a drop in energy on adding a phosphate, when adding phosphate is supposed to be a pump-priming operation that increases the free energy? A close look at the 1,3-diphosphoglycerate molecule in Figure 7, reveals that nature has pulled a fast one: the molecule has also been oxidized. The hydrogen on the aldehyde group at the right of the precursor has been replaced by the oxygen of the phosphate linkage.

One can think of this reaction occurring in two steps: the aldehyde is first oxidized to an acid

$$R-\overset{\overset{\displaystyle O}{\|}}{C}-H + (O) \rightarrow R-\overset{\overset{\displaystyle O}{\|}}{C}-OH$$

with a large drop in free energy, and the acid is then esterified by a phosphate group

$$R-\overset{\overset{\displaystyle O}{\|}}{C}-OH + HPO_4^{2-} \rightarrow R-\overset{\overset{\displaystyle O}{\|}}{C}-O-\overset{\overset{\displaystyle O}{\|}}{\underset{\underset{\displaystyle O^-}{|}}{P}}-O^- + H_2O$$

with a small increase in energy. The energy lost by oxidation is not gone forever, however. It is used to make two NADH from NAD+. This stored energy will be regained in the last step when pyruvate is reduced to lactate and the two NADH are restored to NAD+ again. If this were not so, large quantities of reduced NADH would accumulate. The cell would be forced either to find some alternative way of recycling NADH to NAD+, or to consider NAD+ as a metabolite like glucose, and ingest large quantities of nicotinic acid with its foods. The second choice is impractical because of the natural scarcity of nicotinic acid. An alternate recycling pathway for NADH did not evolve until living organisms developed the ability to use oxygen. But in anaerobic glycolysis, NAD is just a shuttle, picking up energy at one stage of the process and releasing it at another.

The rest of the story is simple. The two phosphate groups of 1,3-diphosphoglycerate are transferred one at a time to ADP, with a rearrangement in between, and 16 kcal of free energy are stored as ATP for every mole of 1,3-diphosphoglycerate broken down. Pyruvate is the end of the chain, except for one final reduction to lactate that recovers the energy held temporarily by NADH and recycles the NAD+. In yeast the process goes one step further. The pyruvate molecule loses a carbon (as CO_2) during the final reduction by NADH, yielding ethyl alcohol as the end product instead of lactate.

This whole process looks complicated at first, but it is hard to see how it could be simplified. Granted that energy is to be stored by transferring a phosphate group to ADP and making ATP out of it, the main problem becomes how to make a compound with a higher phosphate bond energy than ATP, to serve as the phosphate donor. This super-high phosphate bond energy compound is 1,3-diphosphoglycerate. Its phosphate bonds have hydrolysis energies of 11.8 kcal per mole as compared with 8.2 kcal in ATP. The reaction

1,3-diphosphoglycerate + ADP
→ 3-phosphoglycerate + ATP

runs downhill by 3.6 kcal per mole. The enzyme phosphoglycerate kinase acts to ensure that the phosphate group actually is transferred from one molecule to the other. It couples the hydrolysis of one molecule and the esterification of the other. Without the enzyme the reaction might be

1,3-diphosphoglycerate + ADP
→ 3-phosphoglycerate + phosphate + ADP

with all 11.8 kcal of hydrolysis energy lost as heat.

A balance sheet on the entire process would show that two molecules of ATP are used per molecule of glucose, but four are returned —a net

Anaerobic cells use up to 19 times more glucose than do aerobic cells to perform the same work, because they throw out their food when only seven percent of its energy has been extracted.

gain of two ATP molecules. Sixteen kcal of free energy are stored in a process that yields 47 kcal. The other 31 kcal are wasted as heat. This is not very efficient on two counts. Only 34 percent of the energy liberated in the lactate reaction is saved, and even if this percentage were higher, lactate is too rich in energy to throw out as a waste product. Anaerobic cells use up to 19 times more glucose than aerobic cells to perform the same work, because anaerobic cells stop using their food when only 47/686 or seven percent of the possible energy has been extracted. Yeast cells consume glucose and convert it to ethyl alcohol at a prodigious rate, which is fine for the winemaker but a disadvantage to the yeasts. And alcohol, in spite of propaganda about "drinking men's diets," is still a potent source of calories. If ethanol or lactate could be burned to carbon dioxide and water, another 640 kcal would be released per mole of glucose. This is a lot of free energy to throw away. But a more efficient process requires a new chemical substance, oxygen.

RESPIRATION

In anaerobic glycolysis, the largest transfer of energy is from glyceraldehyde-3-phosphate to NAD^+, reducing NAD^+ to NADH and H^+. But this energy is given back again when pyruvate is reduced to lactate, acetate, ethanol, or other molecules that are excreted as waste products. What if this loan of energy did not have to be repaid? What if the 52 kcal of free energy could be used instead to make more ATP? Moreover, what if the cell could keep on oxidizing pyruvate step by step, removing hydrogen atoms to reduce more NAD^+, until finally nothing was left but CO_2? Much more energy then could be obtained from the same amount of glucose.

This metabolic machinery does exist in every living organism that uses oxygen to oxidize its foods. The CITRIC ACID CYCLE (also called the tricarboxylic acid cycle, or the Krebs cycle after its discoverer, Sir Hans Krebs) takes pyruvate from the end of

glycolysis and breaks it down to CO_2, using the hydrogen atoms to reduce NAD^+ and FAD carrier molecules to NADH and $FADH_2$, and passing chemical free energy to these carriers. The RESPIRATORY CHAIN then uses O_2 from the atmosphere to reoxidize and restore the carrier molecules to NAD^+ and FAD so they can be recycled, and saving the chemical free energy by using it to synthesize ATP from ADP and phosphate. An overall flow diagram of glycolysis, the citric acid cycle, and the respiratory chain is shown in Figure 10.

The citric acid cycle itself appears in Figure 11, and the free energy steps are presented in the

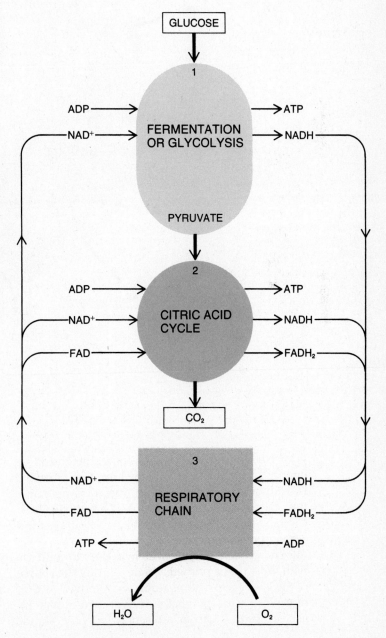

10 GLUCOSE METABOLISM in its most complete form has three parts. Glucose is first fermented or broken down to pyruvate (step 1). Pyruvate is then converted to carbon dioxide in the citric acid cycle (step 2). In both these steps, NAD^+ and FAD carrier molecules are reduced. They are oxidized again in the respiratory chain (step 3), and the energy that was stored in them is used to make ATP.

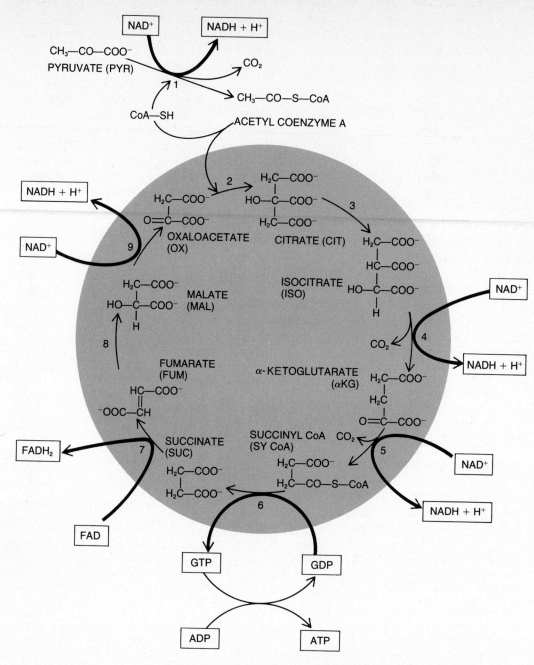

CITRIC ACID CYCLE is the primary energy-extracting process in living organisms. The cycle converts pyruvate and acetate into CO_2 (a waste product) and energy-containing reduced carrier molecules (NADH and $FADH_2$).

graph in Figure 12, a continuation of the glycolysis energy graph of Figure 8. In the first step, pyruvate is oxidized and converted to a primed form of acetate called acetyl coenzyme A. This molecule is combined with oxaloacetate to make citrate, and citrate then is degraded in a series of steps to produce oxaloacetate again, which is ready to combine with more of the primed acetate. During the course of the cycle, two carbon atoms are removed as CO_2 molecules, and four pairs of hydrogen atoms are used to reduce carrier molecules with the simul-

taneous storage of energy. These energy-removing steps, which are the reason for the existence of the cycle, are labeled 4, 5, 7, and 9 in Figure 11.

The compound that enters the cycle, acetyl coenzyme A, is 7.5 kcal higher in energy than simple acetate, and hence is better able to begin the cycle:

$$CH_3COOH + HS\text{-}CoA \rightarrow CH_3\text{-}CO\text{-}S\text{-}CoA + H_2O$$

acetate   coenzyme A   acetyl coenzyme A

$$\Delta G' = +7.5 \text{ kcal/mole}$$

The logic behind this priming step is the same as that for priming glucose to G3P in the early steps of glycolysis. One precycle step is necessary to turn pyruvate into acetyl coenzyme A. This is an oxidation in which three things happen at once: pyruvate is oxidized to acetate with the release of CO_2, part of this energy from oxidation is saved by reducing NAD^+ to NADH, and some of the remaining energy is stored temporarily by adding coenzyme A to the acetate. This same three-for-one reaction occurred in glycolysis when G3P was converted to DPG. In that step, an aldehyde was oxidized to an ester, some of the energy released by oxidation was stored in NADH, and some of the remaining energy was preserved in a second

phosphate bond in the molecule. A good metabolic idea is too valuable not to use more than once, and we shall see it yet a third time in the citric acid cycle.

The energy temporarily stored in acetyl coenzyme A helps to get the citric acid cycle started with a reaction with oxaloacetate to make citrate. When this happens, the coenzyme molecule falls away to be recycled and bound to another acetate. Citrate is rearranged to isocitrate, and then both a CO_2 molecule and two H atoms are removed in converting isocitrate to α-ketoglutarate. As Figure 12 indicates, this step 4 corresponds to a large drop in free energy, and this energy can be recovered later when the NADH carrier molecule is reoxidized with O_2 in the respiratory electron transport chain.

Step 5 is a complex one like the oxidation of pyruvate to acetyl coenzyme A. As in that step, the α-ketoglutarate molecule is oxidized to succinate, CO_2 is given off, some of the oxidation energy is stored in NADH, and some is preserved temporarily by combining succinate with coenzyme A. This latter energy is saved in step 6 by making first guanidine triphosphate (GTP), and then using GTP to make ATP. A smaller amount of free energy is released in step 7 when FAD is reduced,

12   FREE-ENERGY CHANGES in glycolysis plus respiration. The left side of this diagram (light color) is virtually identical with Figure 8 . The endpoint of anaerobic glycolysis is shown by the dashed energy level representing lactate (lac). All of the free-energy drops beyond that point involve the citric acid cycle and the terminal respiratory chain.

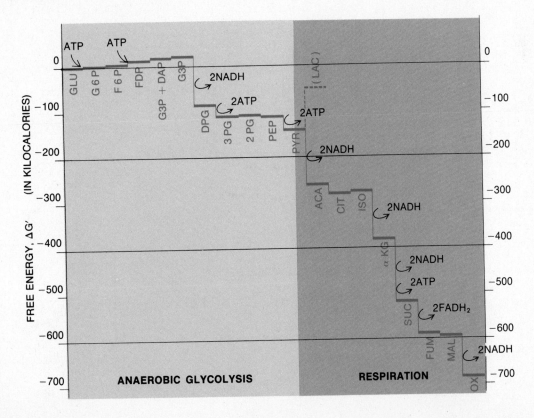

and one more NAD⁺ reduction occurs after a molecular rearrangement. The oxaloacetate that is left over after all these steps is ready to combine with acetyl coenzyme A and go around the cycle once more.

As we shall see in the following section, each NADH ultimately leads to the synthesis of three molecules of ATP, and each FADH₂ yields two ATP. Hence the overall energy yield for glycolysis plus respiration is as shown in the right side of Figure 13: 38 ATP molecules per molecule of glu-

cose starting material. In contrast, the energy yield from glycolysis alone is only 2 molecules of ATP, as the left half of Figure 13 indicates. The advantages of respiration to an organism are obvious.

Several other energy-producing pathways besides glycolysis funnel together and enter the citric acid cycle to produce energy. When fats are used as an energy source, the fatty acids are broken down into two-carbon acetate, and fed into the cycle. During the metabolism of proteins, some amino acids are converted into pyruvate or acetate and thus enter the cycle. Thus the biochemical machinery that probably evolved to make maximum use of the products of glycolysis now is used with many other reactions. Any molecule that can be broken down to pyruvate or acetate can funnel into the citric acid cycle and become a source of energy.

13 **COMPARISON OF ENERGY YIELDS** from anaerobic glycolysis (left) and glycolysis plus respiration (right). Circles represent intermediate compounds, and energy carriers are indicated in color. When the closed NAD loop of anaerobic glycolysis is broken in respiration, the NADH produced after phosphoglyceraldehyde (upper right) becomes available for production of ATP molecules.

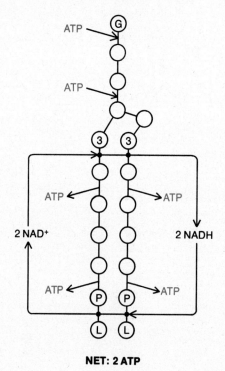

NET: 2 ATP

Ⓖ = GLUCOSE

③ = 3-PHOSPHOGLYCERALDEHYDE

Ⓟ = PYRUVATE

Ⓛ = LACTIC ACID

Ⓐ = ACETYL COENZYME A

Ⓒ = CITRIC ACID

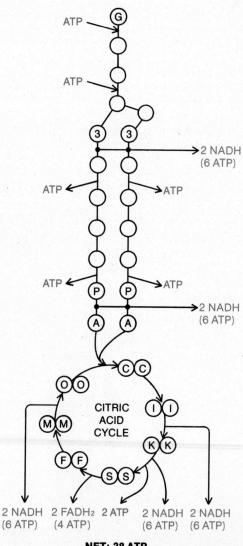

NET: 38 ATP

REOXIDIZING THE CARRIERS

Respiration completes the process begun by glycolysis and the citric acid cycle, by providing a way of reoxidizing the carrier molecules, NADH and $FADH_2$. So far there has been no reason to call these reactions "aerobic", because no oxygen has been involved. The oxidative steps have involved only the transfer of H atoms from the molecules being oxidized to carrier molecules, NAD^+ and FAD. The respiratory chain provides the means of finally linking these reactions to the use of oxygen.

The components of the respiratory electron transport chain are shown schematically in Figure 14. NADH is reoxidized to NAD^+ in the process of reducing the flavin mononucleotide group of an enzyme molecule to $FMNH_2$, and this in turn is reoxidized in the act of reducing a quinone, a small organic molecule indicated by Q in Figure 14. The NAD^+ to FMN step is coupled, by a process that is not fully understood, to a mechanism that uses the free energy of the reaction to synthesize a molecule of ATP. This energy-storing step is known as OXIDATIVE PHOSPHORYLATION —that is, phosphorylation of the ADP molecule in the process of an oxidation-reduction step of the electron transport chain.

Only the electrons from $FMNH_2$ go to the quinone molecules. The protons wander free into solution, and are only brought back into the reaction at the very end of the chain. The remainder of the chain is, indeed, only an electron-transport process. Electrons can also be poured into the "pool" of quinone molecules from another source, the succinate-to-fumarate reaction of the citric acid cycle. The enzyme that catalyzes this reaction has a FAD carrier molecule attached to it, which is reduced to $FADH_2$. As this carrier is reoxidized to FAD, electrons are given to the quinone molecules and protons are released into solution, as shown in Figure 14. But note that no ATP is generated in the succinate-to-FAD branch, and this is why in the previous section the arithmetic of energy storage gave one less ATP-equivalent to $FADH_2$ than to NADH (Figure 13).

Electrons then flow from the quinone, Q, through a series of heme protein molecules known as cytochromes b, c_1, c, a, and a_3. The b-to-c_1 and a-to-a_3 steps have large free energy drops; these, too, are coupled to the oxidative phosphorylation machinery to produce one ATP molecule each. At the end of the electron transport chain, electrons from cytochrome a_3 and protons from solution combine with O_2 molecules to produce water. The process is over. The carrier molecules have been re-oxidized, and the oxidative phosphorylation machinery has synthesized three molecules of ATP (or two, for succinate).

CELLULAR ORGANIZATION OF ENERGY-EXTRACTION MACHINERY

Of the two parts of the energy-extraction machinery, glycolysis and respiration, anaerobic glycolysis is the older and the more primitive. The enzymes for this series of reactions are found free-floating in the cytoplasm of the cell. In contrast, the enzymes of the citric acid cycle and the respiratory electron transport chain are isolated in cellular organelles called mitochondria. A typical mitochondrion from a mammalian cell was shown in Figure 4. It has a relatively smooth outer mem-

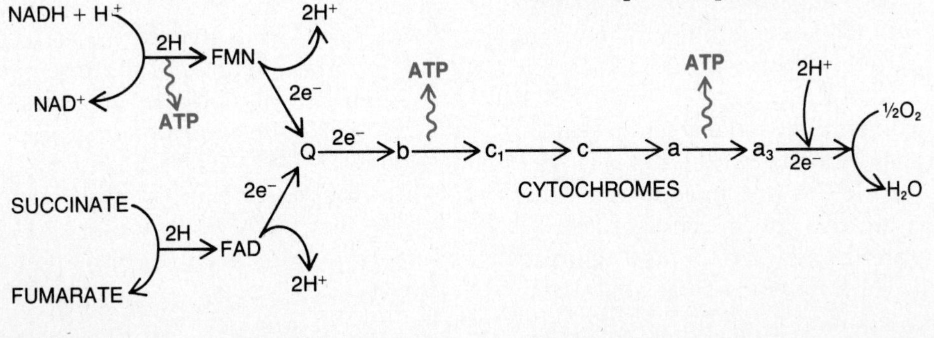

RESPIRATORY ELECTRON-TRANSPORT CHAIN **14** reoxidizes NADH to NAD^+, and uses the energy released to synthesize three moles of ATP per mole of NADH. The ultimate oxidizing power is provided by oxygen from the atmosphere. $FADH_2$ from the succinate-to-fumarate step of the citric acid cycle also can begin the electron transport chain, but with the production of only two moles of ATP. FAD is flavin adenine dinucleotide; FMN is flavin mononucleotide. The different cytochromes differ slightly in the chemical structure of the heme and/or the precise relationship to the protein.

brane, and an inner membrane that is folded back and forth deep into the interior of the mitochondrion, so the inner membrane has enormous surface area compared with the volume that it envelops. Within the inner membrane is a protein-rich fluid called the mitochondrial MATRIX. The cytochromes and enzymes of the electron transport chain are embedded in the inner mitochondrial membrane, and the enzymes of the citric acid cycle are dissolved in the matrix with three exceptions: succinate dehydrogenase, catalyzing step 7 of Figure 11, and the two very large enzyme complexes that catalyze the coenzyme-using steps 1 and 5. These three are buried in the inner membrane along with the electron transport proteins.

Recent research has yielded much new information about the role of electron transport in the synthesis of ATP. Both H^+ ions and electrons apparently move from the inside to the outside of the inner mitochondrial membrane. This migration establishes a substantial difference in pH across the membrane, with the inside being much more alkaline than the outside. The gradient may exceed three pH units, which leads to a strong net influx of H^+ ions back into the inner compartment. The incoming ions flow through special channels lined with ATP-synthesizing enzymes. The idea that this flow of ions drives the synthesis of ATP is called the CHEMIOSMOTIC HYPOTHESIS, and was first proposed in 1961 by Peter Mitchell in England. Perhaps the best supporting evidence comes from experiments in which mitochondria deprived of any carbon source are incubated in a buffer at pH 8, then quickly transferred to one at pH 4, which contains ADP and inorganic phosphate. As hydrogen ions rush into the inner compartment, a burst of ATP synthesis begins.

One correction to the energy arithmetic of the previous sections may have to be made. The inner mitochondrial membrane is permeable to pyruvate, but is a barrier to most other molecules, including NADH. Then how do the NADH molecules from glycolysis get inside the mitochondrion to enter the electron transport chain? The answer appears to be that they reduce a shuttle molecule outside the mitochondrion, and this shuttle goes through the membranes, not the NADH itself. One of the three ATP steps may be lost because of this shuttle, and if recent research is correct, we may have to adjust the net ATP yield per glucose molecule down from 38 to 36.

Mitochondria are the power packs of the cell, taking in pyruvate from glycolysis, and returning ATP to the cell. They are particularly abundant in cells that need large amounts of energy, such as those of the heart and other active muscles. One of the more intiguing hypotheses of biology is the suggestion that respiratory mitochondria and photosynthetic chloroplasts are the descendants of bacteria that once lived in symbiotic partnership with their host cell, and gradually lost their independence.

Mitochondria and chloroplasts are roughly the size of bacteria, and they have similar membrane structure. They each contribute a specific function to the host cell: mitochondria the ability to follow glycolysis by aerobic respiration, and chloroplasts the ability to synthesize glucose with solar energy. It is easy to imagine that some early anaerobic cell solved the problem of gaining more free energy from glucose, not by developing its own respiratory pathways, but by entering into partnership with aerobic bacteria, supplying them with pyruvate and taking some of their ATP for the cell's own needs. Examples of symbiotic cooperation are not rare among organisms alive today. The lichens that grow on rocks and tree trunks are not a single organism but a meshwork of two species: an alga and a fungus. Several kinds of animal, ranging from one-celled paramecia to the reef-building corals of tropical seas, derive at least part of their food from photosynthetic algae that live within their cells.

The bacterial-origin theory was resurrected by the discovery that both mitochondria and chloroplasts contain their own DNA, separate from that in the nucleus of the cell, and that mitochondrial DNA apparently contains the genetic information for some of its inner membrane proteins and some parts of cytochrome *a*. Mitochondria also have their own ribosomes to translate messenger RNA into protein. Even more striking is the observation that the ribosomes of mitochondria, chloroplasts, and bacteria are all of similar size —slightly smaller than the ribosomes in the cytoplasm of the cell. Cytoplasmic ribosomes are rendered inoperative by the compound cycloheximide. Mitochondrial, chloroplast, and bacterial ribosomes are unaffected by cycloheximide but are all poisoned by chloramphenicol. The structures of bacterial and mitochondrial membranes and their permeability to ions and small molecules are quite similar, and much less like the corresponding properties of cell membranes.

Mitochondria are apparently not made anew from nuclear DNA when sperm and egg unite to create a new organism. Instead, they are carried along in the cytoplasm of the egg, and grow and divide autonomously. However, the enzymes of the citric acid cycle and most of the respiratory chain heme proteins are made at cellular ribosomes under the control of the DNA in the cell nucleus, and diffuse into the mitochondria after synthesis. The chloroplast enzyme ribulose-1,5-biphosphate carboxylase, which has two different kinds of subunits, also has a dual origin. The blueprint for one subunit lies in the nuclear DNA, and the blueprint for the other in the DNA of the chloroplast.

If these organelles are truly former bacteria which have lost one biological function after another as they assumed a greater dependence upon their host, then one would expect that the mitochondria of primitive organisms might retain more of their original DNA and the functions it governs. And true enough, the mitochondria of the bread mold *Neurospora crassa* and similar microorganisms contain as much as six to seven times the amount of DNA as do the mitochondria of higher plants and animals.

It now appears as if animals and plants themselves might have arisen from symbiotic relationships between anaerobic, nonphotosynthetic cells, and aerobic or photosynthetic bacteria whose remains can still be seen as mitochondria and chloroplasts. The case is not yet proven, but the evidence is becoming quite persuasive.

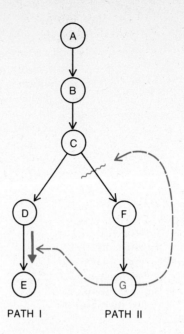

PATH I   PATH II

ALLOSTERIC CONTROL POINTS. Products of reaction path II, if present in excess, can shunt most of the intermediates into path I, either by inhibiting the enzyme that converts compound C into F, or by activating the enzyme that converts D into E. Black arrows represent chemical reactions. Colored arrows and wavy blocking line represent allosteric controls. **15**

FEEDBACK CONTROL

Glycolysis, or anaerobic fermentation of glucose to lactate, produces two molecules of ATP for every glucose molecule used. The addition of the citric acid cycle and respiration increases the number of ATP's to 38 (or perhaps 36). As noted already, a respiring organism obtains 19 times as much energy per mole of glucose. In other words, when yeast switches from respiration to anaerobic fermentation under conditions of low oxygen, it must use glucose 19 times as fast to obtain the same energy. But as soon as respiration begins in yeast, glycolysis suddenly slows down. Only as much glucose is used as is needed for energy production under the conditions being used, anaerobic or aerobic. This is the Pasteur Effect, named for its discoverer. What is the mechanism that turns down glycolysis when the respiratory chain begins to operate?

The mechanism by which glycolysis, the citric acid cycle, and the respiratory chain are regulated is ALLOSTERIC CONTROL of the enzymes involved. Some products of later reactions, if present in oversupply, can suppress the action of enzymes involved with early reactions. Cytosine triphosphate and aspartate transcarbamylase illustrate this kind of control *(Chapter 5)*. Conversely, an excess of the products of one branch of a synthetic chain can speed up reactions in a parallel branch and divert raw materials away from its own synthesis *(Figure 15)*. These positive and negative feedback control mechanisms are used at many points in the energy-extracting processes. They are summarized in Figure 16.

The main control point in glycolysis is the conversion of fructose-6-phosphate (F6P) to fructose-1,6-diphosphate (FDP) by the enzyme phosphofructokinase *(Figure 6)*. This enzyme is allosterically inhibited by ATP, and activated by ADP or AMP *(Figure 16)*. This enzyme is also inhibited by citrate, for reasons that will become clear shortly. As long as anaerobic fermentation proceeds, yielding a relatively small amount of ATP, phosphofructokinase operates at full efficiency. But when respiration begins producing ATP 19 times as rapidly as before, the excess ATP "turns down the burner'' on the conversion of fructose-6-phosphate and the rate of glucose usage drops.

Pyruvate stands at a key position in the network diagrammed in Figure 16. In anaerobic fermentation it is converted to lactate, which can either be returned as pyruvate or be used to resynthesize glucose for storage. Under aerobic conditions pyruvate is changed to acetyl coenzyme A, which enters the citric acid cycle by combining with oxaloacetate. And finally pyruvate can itself be used to produce more oxaloacetate. Which of these pathways it takes depends upon the conditions in the cell.

In the pyruvate-to-lactate conversion, one molecule of NADH must be oxidized to NAD$^+$; the NADH is supplied by the earlier GPDH reaction *(Figure 7)*. But the affinity of the respiratory chain enzymes for NADH is much greater than that of lactate dehydrogenase. If the respiratory chain is operating, it steals the available NADH from lactate dehydrogenase, and the pyruvate-to-lactate conversion does not occur.

A second control point for pyruvate reactions involves acetyl coenzyme A. If enough oxaloacetate is present to keep the citric acid cycle going as fast as acetyl coenzyme A is produced, the acetyl coenzyme A concentration remains low. Pyruvate carboxylase, the enzyme that makes oxaloacetate from pyruvate, cannot work without an appreciable concentration of acetyl coenzyme A present, even though it does not use the coenzyme in the reaction. Acetyl coenzyme A is an allosteric AC-TIVATOR for the enzyme. If for some reason too little oxaloacetate is available, acetyl coenzyme A builds up, pyruvate carboxylase goes to work, and some of the pyruvate is shunted to the left in Figure 16 to make more oxaloacetate. This reaction is also the first step in resynthesizing glucose from pyruvate, another means of storing pyruvate if the cell cannot use it fast enough.

Therefore the concentration of acetyl coenzyme A determines the balance point between two competing reactions: one to use oxaloacetate in turning the citric acid cycle, and the other to make more oxaloacetate if it is in short supply. ADP is an allosteric inhibitor of this same pyruvate carboxylase. If the cycle is inhibited too drastically, the production of ATP falls, ADP accumulates, and the production of oxaloacetate from pyruvate is halted. The cycle is stimulated to get to work again.

The main control point for the citric acid cycle is the conversion of isocitrate to α-ketoglutarate by means of the enzyme isocitrate dehydrogenase. ATP and NADH are feedback inhibitors of this reaction, and ADP and NAD$^+$ are activators. If too much ATP is accumulating, or if NADH is being produced faster than it can be used by the respiratory chain, the isocitrate reaction is blocked and the citric acid cycle is shut down. This would lead to a pileup of large amounts of isocitrate and citrate, except that the conversion of acetyl coenzyme A to citrate is also blocked. The adverse effects of halting the isocitrate reaction are thus

16 **FEEDBACK CONTROL** in glycolysis and in the citric acid cycle. Allosteric effectors are shown in color. Heavy colored arrows represent activation of the enzyme that governs a particular reaction step, and wavy lines represent enzyme inhibition. Colored dashed lines indicate the source of activation or inhibition. The text discusses the effects of these controls.

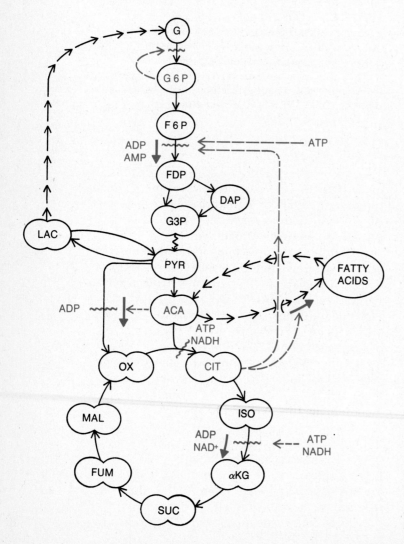

spread backward up the chain of reactions. A certain excess of citrate does accumulate, however, and this excess acts as a negative feedback inhibitor to slow the fructose-6-phosphate reaction early in glycolysis (Figure 16). Thus if the citric acid cycle has been slowed down BECAUSE OF AN EXCESS OF ATP (not because of a lack of oxygen), glycolysis is shut down as well. Both processes resume when the ATP level falls and they are needed.

The last control point in Figure 16 involves a method for storing excess acetyl coenzyme A by using it to synthesize fatty acids. Excess citrate is an allosteric activator for one of the enzymes in the pathway for making fatty acids. If too much ATP is being made, and the citric acid cycle is shut down, the accumulation of citrate switches acetyl coenzyme A to the synthesis of fatty acids for storage.

Allosteric control so elegant as this creates the illusion that it has been designed by a systems analyst. In fact it is one of the most intellectually satisfying examples of the tight logic that can arise by the process of natural selection, when selection pressure favors efficient operation in the fierce competition among organisms for limited resources. Each of the feedback controls regulates the various parts of the glycolysis-oxidation network and keeps them operating in harmony and in balance. It is unnecessary (and therefore inefficient and disadvantageous) to run the glycolytic mechanism too fast if it is supplemented by the more efficient respiratory process.

It is wasteful to produce more acetyl coenzyme A than there is oxaloacetate to handle it in the citric acid cycle. It is also wasteful to shunt too much pyruvate into making oxaloacetate—the gears of the citric acid cycle—and to neglect to produce the fuel, acetyl coenzyme A. Allosteric control maintains the proper balance among the uses of pyruvate by being sensitive to a shortage or an oversupply of acetyl coenzyme A. All of the other allosteric feedback controls have evolved because they make the system more efficient, and hence contribute to the survival of the species that carries them. The control mechanisms do not take into account perils or selection pressures that the organisms have never had to face during their evolution. The breakdown of logic of feedback control in *E. coli* in the presence of massive outside supplies of threonine was mentioned in Chapter 5. Both the threonine and methionine syntheses are turned off and the organism is in trouble unless it has access to external sources of methionine. This design flaw in the control network was never weeded out by natural selection because massive doses of

threonine never occur in nature—only in the laboratory. Allosteric feedback control has proved to be strikingly efficient at keeping metabolic processes in balance in actual living organisms.

WINDING THE MAINSPRING: PHOTOSYNTHESIS

Respiration was probably the third, not the second, great step in the evolution of metabolism. Aerobic respiratory machinery could not have evolved without a supply of oxygen. Mechanisms have been proposed to account for the appearance of O_2 in the atmosphere without the influence of living organisms—for instance, by the dissociation of water vapor in the upper atmosphere—but the primary source of atmospheric oxygen was probably photosynthesis.

Photosynthesis is the process by which plants harness the energy of sunlight to make glucose. The source of carbon atoms for the sugar is carbon dioxide, and the overall reaction is:

$$6CO_2 + 12H_2O \xrightarrow[\text{energy}]{\text{light}} C_6H_{12}O_6 + 6O_2 + 6H_2O.$$

Water appears on both sides of the equation. The 12 water molecules on the left side of the equation are split to yield the six oxygens on the right side. The six water molecules on the right side arise indirectly from some of the hydrogens in the water on the left, plus oxygens from the CO_2.

Some primitive photosynthetic bacteria use hydrogen sources such as H_2S instead of H_2O:

$$6CO_2 + 12H_2S \xrightarrow[\text{energy}]{\text{light}} C_6H_{12}O_6 + 12S + 6H_2O.$$

The strategy of using water as the source of reducing power in photosynthesis evolved quite early, and is universal among all algae and higher green plants. Other mechanisms, such as the use of H_2S, are merely biochemical curiosities.

The liberation of a corrosive and highly reactive gas, molecular oxygen, was probably lethal to many of the anaerobic organisms on the primitive Earth. Obviously any organism that could tolerate

Allosteric control is one of the most intellectually satisfying examples of the tight logic that can arise by the process of natural selection.

oxygen, or even better, use it in some beneficial way, would enjoy a tremendous advantage as oxygen from photosynthesis began to accumulate in the atmosphere. This environment favored the evolution of respiration.

Just as there are two parts of glucose metabolism—glycolysis followed by respiration—so there are two stages of photosynthesis: the dark and light reactions (*Figure 17*). The DARK REACTIONS, so named because they do not require light, synthesize glucose from carbon dioxide. The hydrogen necessary for reduction comes from NADPH (nicotinamide adenine dinucleotide phosphate). This carrier molecule is similar to NADH except for the presence of an extra phosphate group, and provides some of the free energy needed to synthesize glucose; the rest is provided by ATP. Glucose synthesis could proceed without light, so long as a steady supply of NADPH and ATP was available. Plants use the energy of sunlight to manufacture these two molecules, starting with NADP$^+$ and ADP. This is the function of the LIGHT REACTIONS. In the simplest terms, photosynthesis does two things: it synthesizes glucose from carbon dioxide, and it splits water, releasing oxygen gas into the atmosphere. As Figure 17 indicates, CO_2 is captured ("fixed") by the dark reactions. Oxygen is a by-product of the light reactions. The dark and light reactions can be studied separately because, like the citric acid cycle and the respiratory chain, they are connected only by their carrier molecules.

THE DARK REACTIONS

The synthesis of glucose requires energy: 686 kilocalories per mole, to be precise. Hydrogen atoms and some of the energy are provided by NADPH. The only difference between this carrier and NADH is that the $-OH$ on the 2' carbon atom in Figure 9 is replaced by a phosphate. When reduced, NADPH carries virtually the same amount of free energy as NADH: 52 kcal per mole. NAD$^+$ is involved in metabolic breakdown reactions and energy storage, while NADP$^+$ is usually involved in synthetic reactions, which require energy and reducing power. One can think of the extra phosphate in NADP$^+$ as a label that indicates the function of this particular dinucleotide.

The overall equation for the dark reactions can be written:

$$6CO_2 + 12NADPH + 12H^+ + 18ATP \rightarrow$$

$$C_6H_{12}O_6 + 12NADP^+ + 6H_2O + 18ADP + 18P_i$$

with P_i representing inorganic phosphate. As noted above, NADPH provides both reducing power (H atoms) and energy; additional energy is provided by ATP. This simple equation summarizes an elaborate mechanism for converting one-carbon CO_2 molecules to six-carbon glucose molecules. The biochemical tactic of the cell is to combine CO_2 with a five-carbon sugar molecule, split the intermediate product into two three-carbon 3-phosphoglycerate (3PG) molecules, and then to proceed along a synthetic pathway that resembles glycolysis in reverse. Somewhere along the way the five-carbon sugar molecule must be regenerated so that the process can be repeated with further CO_2 molecules.

The machinery for this regeneration lies in the Calvin-Benson cycle (*Figure 18*). In the synthesis of one molecule of glucose, six molecules of ribulose-1,5-biphosphate (RuBP) combine with six molecules of CO_2. The six-carbon products are then split into 12 molecules of 3PG. The enzyme that mediates this reaction is RuBP carboxylase (ribulose-1,5-biphosphate carboxylase). The 12 molecules of 3PG are then pushed up the free-energy ladder with the help of energy from NADPH and ATP. By comparing Figure 18 with Figures 6 and 7, the reader can see that this series of reactions is the reverse of glycolysis. The step at which NADPH is required in the Calvin-Benson

17 TWO HALVES OF PHOTOSYNTHESIS are the light reactions ("photo") and the dark reactions ("synthesis"). They are independent processes, connected only by the NADPH and ATP produced by the light reactions and used by the dark processes. If the dark reactions constitute the glucose factory, the light reactions are the power plant.

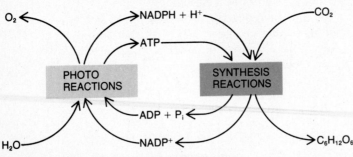

PHOTO REACTIONS (OR LIGHT REACTIONS):
$2H_2O + 2NADP^+ \longrightarrow O_2 + 2NADPH + 2H^+$
$ADP + P_i \longrightarrow ATP$

SYNTHESIS REACTIONS (OR DARK REACTIONS):
$CO_2 + 2NADPH + 2H^+ \longrightarrow \frac{1}{6}C_6H_{12}O_6 + H_2O + 2NADP^+$
$ATP \longrightarrow ADP + P_i$

cycle (from DPG to G3P) is the reverse of the corresponding step in glycolysis.

Only six of the 36 carbon atoms in the 12 3PG molecules end up as glucose. The other 30 are sidetracked to regenerate six RuBP molecules, which are ready to pick up the next six molecules of CO_2. The details of this complicated series of reactions are beyond the scope of this book; suffice to say that the end product is six molecules of ribulose-5-phosphate (Ru5P). The conversion of Ru5P to RuBP requires ATP, and both ATP and NADPH are required to convert 3PG to G3P, as shown in Figure 7. The resynthesis of RuBP closes the loop, turning the reversed-glycolysis pathway into the Calvin-Benson cycle.

Though virtually all carbon incorporated into glucose is processed through the Calvin-Benson cycle, this is not the only possible path for carbon fixation. A number of plants can fix carbon by adding CO_2 to phosphoenolpyruvate (PEP) to form oxaloacetate. The enzyme involved is phosphoenolpyruvate carboxylase (PEP carboxylase). Because this enzyme has a far higher affinity for CO_2 than does RuBP carboxylase, this system is well suited to plants growing in hot dry climates. There the light intensity may be extremely high, but the plant must severely limit diffusion to prevent excessive water loss (Chapter 15). Even when internal CO_2 concentrations drop to a few parts per million, PEP carboxylase, unlike RuBP carboxylase, can still function at almost full capacity. Such plants use the PEP carboxylase system to concentrate CO_2 in certain cells, where it is incorporated into oxaloacetate and then reduced by NADPH to form malic acid. Other cells equipped with the RuBP carboxylase system then break down the malic acid and use the CO_2 to synthesize glucose.

Plants equipped only with the RuBP carboxylase system are usually called C_3 plants, because the first stable product of CO_2 fixation is a three-carbon compound, 3-phosphoglycerate. Those which also have PEP carboxylase are called C_4 plants because they make a four-carbon compound, oxaloacetate, as the first product. Some plants with PEP carboxylase have an even more elegant adaptation to the stressful conditions of desert or semidesert. They accumulate fixed carbon at night, when maximum gas diffusion can proceed with minimum loss of water. During the day they restrict gas exchange to almost zero, but break down the four-carbon oxaloacetates to CO_2, which they incorporate (via the RuBP carboxylase route) into glucose.

RuBP carboxylase has another serious drawback besides its relatively low affinity for CO_2, particularly when rates of photosynthesis are apt to be unusually high. RuBP carboxylase is strongly inhibited by oxygen. Moreover, in the presence of high oxygen and low CO_2 concentrations, it also functions as an oxygenase, binding oxygen to RuBP instead of CO_2. The products are 3PG and glycollic acid, a two-carbon compound that the plant can use only for respiration. There is no net carbon gain—indeed a net loss of fixed carbon—by oxidation of the glycollic acid. This process is called photorespiration, and many biologists interested in increasing agricultural productivity are studying ways to restrict it.

THE LIGHT REACTIONS

Just as cellular respiration takes place within mitochondria, so photosynthesis occurs in chloroplasts (Figure 5). The dark reactions occur in solution within the chloroplasts; the light-trapping reactions occur within stacked platelets called grana. These stacks of grana and their connecting

THE CALVIN-BENSON CYCLE. Schematic diagram **18** illustrates both its cyclic character and the resemblance of the ascending pathway (left) to glycolysis run in reverse. Where glycolysis requires ATP, the Calvin-Benson cycle releases phosphate; and where glycolysis releases ATP or NADH, the Calvin-Benson cycle requires ATP or NADPH from outside sources (meaning the light reactions).

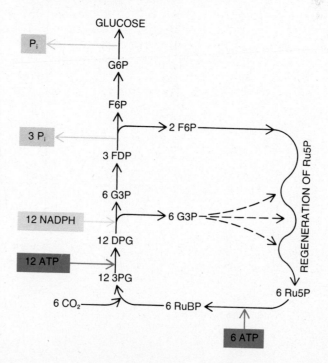

channels are depicted schematically in Figure 19. All of the chlorophylls and other light-absorbing pigments, as well as the enzymes and cytochromes that use this light energy to make ATP and NADPH, are intricately organized in the grana in a way that resembles the probable organization of the terminal respiratory chain in the walls of mitochondria.

Light is trapped in photosynthesis by several pigments: chlorophylls, carotenoids and phycobilins. Table I in Chapter 26 shows the distribution of the different pigments among the various groups of photosynthetic plants Note that all have

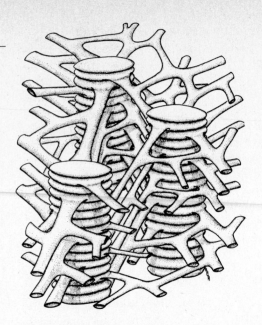

19 GRANA of chloroplasts are the site of the dark reactions of photosynthesis. This drawing indicates the way the grana and their connecting tubules may be stacked within the chloroplasts of green plants.

LIGHT-ABSORBING PIGMENTS trap the energy of sunlight in photosynthesis. Various plant pigments absorb most strongly in different regions of the visible spectrum. β-carotene absorbs only in the blue region of the spectrum, and hence appears yellow (because of the unabsorbed wavelengths). Phycoerythrin in red algae absorbs yellow-green light and hence appears red. Phycocyanin in bluegreen algae absorbs more of the yellow, orange, and red wavelengths and therefore is blue-green. Chlorophylls remove both the blue and red wavelengths, and hence appear green.

20

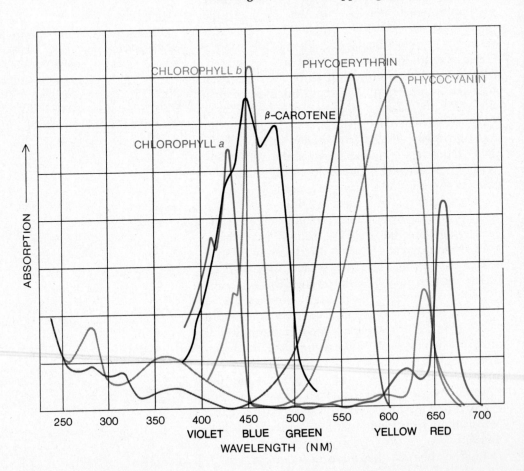

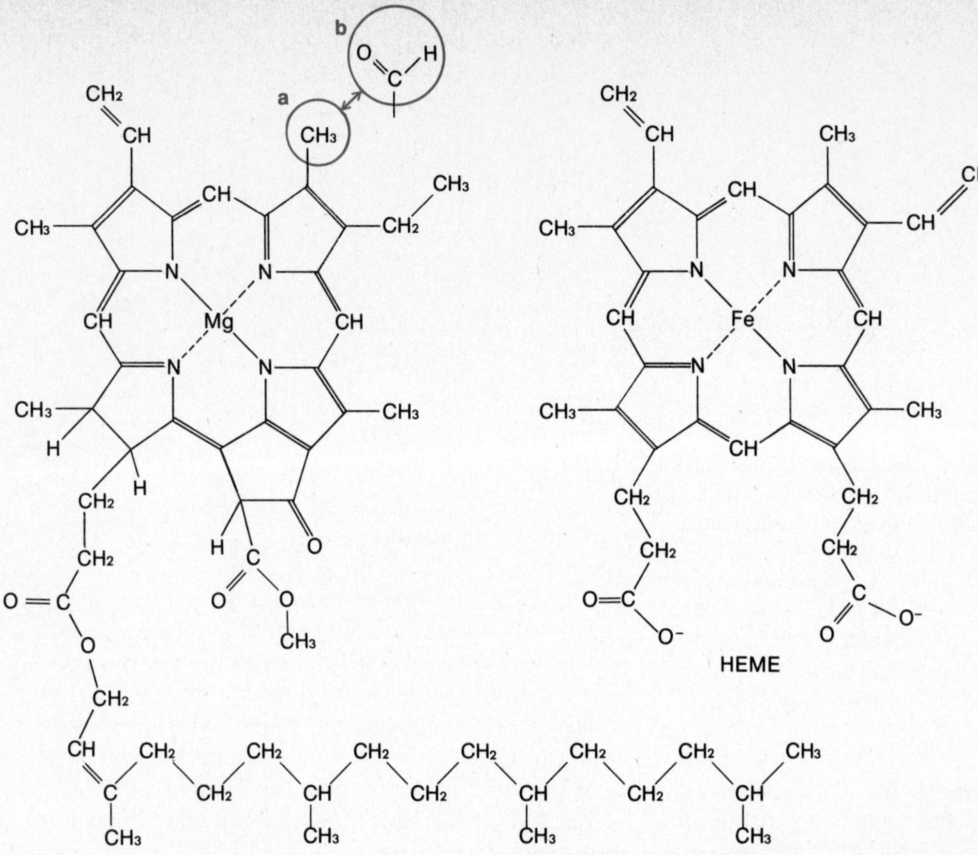

CHLOROPHYLL *a*

CHLOROPHYLL AND HEME have similar structures.
Heme consists of a porphyrin ring centered on an atom
of iron (Fe). Side groups are methyl, vinyl and
proprionic acid. In chlorophyll the porphyrin ring is
centered on a magnesium atom (Mg), and the molecule
has a long tail. Chlorophyll b is identical to chlorophyll
a except for the substitution of —CHO for —CH₃
(colored circles).

21

chlorophyll *a*. Some of the chlorophyll molecules
are located at REACTION CENTERS in the grana, where
the actual separation of negative and positive
charges takes place. Other chlorophyll and various
pigment molecules, spread in an organized pattern
around these reaction centers, act as a light-
gathering antenna system. Together they absorb
light over a range of wavelengths that extend from
the visible to the near-infrared region of the spectrum *(Figure 20)*. The antenna pigments transfer
their excitation energy to the chlorophyll at the
reaction center.

A typical carotenoid, β-carotene, was shown in
Figure 5 of Chapter 4. Carotenoids can also absorb
light because of their extensive double bonding;
their electrons are delocalized to some extent and
the electronic energy levels are close enough together for an electron to be promoted from a lower
to an excited level by visible light. Phycobilins are a
different class of molecule, also with double
bonds. Chlorophyll is a magnesium-porphyrin
compound with a long hydrocarbon tail *(Figure
21)*, related to the heme group found in hemoglobin and cytochromes *(Chapter 4)*. Many of the electrons in the metal atom and in the porphyrin ring

are free to wander over the entire expanse of the
ring. This makes the electronic energy levels particularly close together and makes chlorophylls
particularly good absorbers of visible light. The
electronically excited chlorophyll molecule becomes an excellent reducing agent. It can pass its
excited electron on to a different molecule, thus
reducing it. The second molecule can then reduce
another one, and so on, as the electron cascades
down the energy scale, as in respiration. Only
reaction-center chlorophyll molecules serve directly as reductants. Energy absorbed by other
chlorophylls and by carotenoids or phycobilins
must be transferred to reaction centers to be of use
in electronic charge separation.

Such a process is shown in Figure 22 for CYCLIC
PHOSPHORYLATION. The excited reaction-center
chlorophyll molecule reduces the primary electron

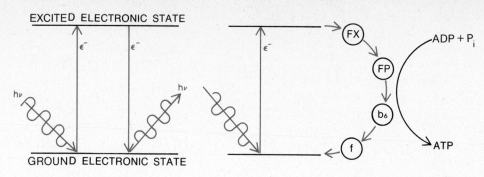

EXCITED ELECTRONIC STATE

GROUND ELECTRONIC STATE

ADP + P$_i$

ATP

22 CYCLIC PHOSPHORYLATION uses the energy of sunlight to synthesize ATP. Chlorophyll absorbs a photon of light (wavy arrow), exciting an electron in the magnesium-porphyrin ring. If the electron falls back to the ground state unimpeded, it loses its energy as fluorescence (left). Plants harness the energy of the excited electron to drive a chain of reactions that leads to the synthesis of ATP (right). FX is ferredoxin, FP flavoprotein, P$_i$ inorganic phosphate, and b$_6$ and f are cytochromes. Energy of photon is hν, where ν is frequency and h is known as Planck's constant.

acceptor, here merely designated X. This molecule reduces ferredoxin, and the ferredoxin in turn reduces a flavoprotein. The flavoprotein is reoxidized as cytochrome b$_6$ is reduced; b$_6$ ultimately reduces cytochrome f, and f passes the electron (via the copper protein plastocyanin) to the reaction center, re-reducing it for another light absorption event. Along the way, the excitation energy that has come from light is used to synthesize ATP. Cyclic phosphorylation resembles the terminal respiratory chain *(Figure 14)*, but with excited reaction-center chlorophyll as the reducing agent instead of NADH, and with the oxidized reaction-center chlorophyll itself the terminal electron acceptor instead of oxygen.

For many photosynthetic bacteria, the ATP produced by cyclic phosphorylation suffices as an energy source. The bacterium *Chromatium* can use hydrogen gas as the source of hydrogen atoms for synthesis. Since NADP$^+$ is reduced by H$_2$ with a standard free energy change of $\Delta G' = -4.1$ kcal, the formation of NADPH is thus spontaneous. This scheme may have been an efficient one when the Earth's atmosphere had large amounts of free H$_2$, but is not a generally usable mechanism now. Organisms need a more dependable supply of hydrogen atoms for producing NADPH.

Purple and green sulfur bacteria live in hot sulfur springs and use H$_2$S as their hydrogen source, either excreting solid sulfur or storing it as a by-product *(Figure 23)*. This system has the disad-

vantage of somewhat restricting the habitable range of the organisms that use it. An obvious and far more widespread source of hydrogen atoms is water, H$_2$O.

The light reactions of photosynthesis in higher plants and algae are diagrammed in Figure 24. Two kinds of reaction centers are involved: pigment systems I and II. Both systems contain chlorophyll *a (Figure 21)*. Other pigments, such as chlorophyll *b* and carotenoids, are usually associated with system II, and may be involved in system I as well.

Excited electrons in system I can tumble back to the ground state through ferredoxin, a flavoprotein, cytochromes b$_6$ and f, and plastocyanin, using the energy of excitation to synthesize molecules of ATP from ADP and inorganic phosphate. This is simply cyclic phosphorylation. But they can do something else — they can reduce NADP$^+$ to

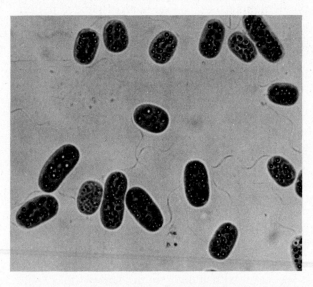

PURPLE SULFUR BACTERIA of the genus Chromatium use hydrogen sulfide as their hydrogen source. Granules of sulfur are visible within their cells.

23

NADPH. This provides a supply of NADPH for the dark photosynthetic reactions, but leaves pigment system I electron-deficient. After one excitation of electrons, the reaction center must either find an alternate supply of electrons or shut down.

The alternate source of electrons is the other kind of reaction center, system II (*Figure 24, left*). These excited electrons are passed to a primary electron acceptor, probably a quinone, but here designated just Q. From Q, the electrons are passed through another cytochrome (b_{559}, because it has strong absorption at 559 nm), another quinone (plastoquinone), cytochrome *f*, the copper protein plastocyanin, and finally to the reaction-center chlorophyll *a*. During this fall from the excited state of system II to the ground (unexcited) state of system I, part of the energy released is again harnessed to synthesize ATP molecules.

But where did the electrons come from that were excited in system II? They came from hydroxyl groups and ultimately from water molecules, by the reaction

$$2 H_2O \rightarrow 4 H^+ + O_2 + 4e^-.$$

The overall result is that water is split into oxygen gas, hydrogen ions, and electrons. The electrons replace those separated from chlorophyll by light in the reaction centers of system II. The oxidized reaction centers of system I are filled by electrons coming through the electron transport chain from system II. Energy released between the two systems is used to synthesize ATP. The protons and electrons from water finally rejoin in the reaction:

$$2 H^+ + 2 e^- + NADP^+ \rightarrow NADPH + H^+.$$

Oxygen gas is released as a waste product just as sulfur is released by green sulfur bacteria.

In brief, the structure of the photosynthetic machinery consists of a set of dark reactions to synthesize glucose from CO_2, and a set of light

reactions that fuel the synthesis by supplying ATP and NADPH. The result is that photosynthetic organisms do not have to depend on other organisms for a source of carbon. All they need is sunlight, carbon dioxide, and a suitable source of hydrogen atoms, such as water.

The chemiosmotic hypothesis for ATP formation applies to chloroplasts just as it did for mitochondria. The flow of electrons from the reducing to the oxidizing end of the photosynthetic electron-transport pathway is accompanied by the movement of hydrogen ions across the thylakoid membrane; this develops a strong pH gradient. The movement of hydrogen ions along this gradient, through channels lined with ATP-synthesizing enzymes, provides the driving force for ATP synthesis. There is an interesting difference, however. In mitochondria, electrons and hydrogen ions move from inside to outside, and the reverse flow drives ATP synthesis; in chloroplasts it is just the opposite: electrons flow from the outside to the inside of the thylakoid, and hydrogen ions accumulate inside. Their return to the outside drives ATP synthesis. Hence one must incubate chloroplasts in acid and transfer them rapidly into base to drive ATP synthesis, rather than the other way around.

4 **LIGHT REACTIONS of photosynthesis. Photons hitting system I (right) produce excited electrons that can generate reducing power by synthesizing NADPH, or can store energy via ATP. If NADPH is generated, then the electron deficiency in system I is made up by electrons from excited system II, which in turn accepts electrons from the decomposition of water. The electron transport chain between systems resembles that of respiration. Q is quinone, PQ is plastoquinone (both small molecules); b_3, b_6 and f are cytochromes (proteins). PC is plastocyanin (a copper protein); FX is ferredoxin and FP is a flavoprotein. X is the presently unidentified primary electron acceptor for system I.**

LIGHT, LIFE AND THERMODYNAMICS

The discussion of the energy extraction and storage machinery of living organisms leads to a topic raised at the beginning of the chapter. Do living things violate the Second Law of Thermodynamics by maintaining a high degree of order? If evolution is described as an historical progression toward states of greater and greater complexity and organization, and if the Second Law says that the entropy or disorder of the universe is steadily increasing, is there not a contradiction or a paradox?

A close look at the Second Law reveals that there is no paradox. The Second Law states that in every spontaneous reaction in a closed system, the entropy of the system rises. No single living organism, however, is a closed thermodynamic system. An animal must have constant access to food, water, and oxygen, and must have some way of eliminating waste products, mainly oxidized compounds of carbon, nitrogen, and hydrogen. These foods cannot be just any compounds of the three elements; they must be compounds with high free energy—in general, highly reduced compounds. A plant can rereduce the carbon and hydrogen by photosynthesis, but cannot obtain nitrogen compounds without the assistance of nitrogen-fixing bacteria or other outside sources. Living organisms convert high-free-energy reduced compounds to low-free-energy oxidized compounds, and use the free energy obtained for their own purposes.

One important purpose for which this energy is used, at least in animals, is mechanical work—locomotion, food gathering, protection, propagation, and all the other actions that ensure that there will be another generation. But a second and very crucial role of free energy is to keep order within the organism. Reactions which lead to a more highly ordered state—a state of lower entropy—are possible in living creatures because the reactions of metabolism lead to an even greater fall in

> Living systems convert high-free-energy reduced compounds to low-free-energy oxidized compounds, and use the extracted free energy for their own purposes.

free energy. Metabolic processes can drive the desirable but thermodynamically unfavorable reactions uphill. If one converts a man into a closed thermodynamic system by placing him in an airtight steel coffin, he will rapidly lose his ability to maintain a state of low entropy. In short he will die.

How much of a man's surroundings must one include to obtain a closed thermodynamic system, to which one can apply the Second Law in the form quoted? A good approximation of a closed thermodynamic system was the Apollo 13 spacecraft. Astronauts Lovell, Swigert, and Haise were sealed into an airtight container with everything needed to keep them alive for a limited period of time. Calculating the conditions for maintenance of life in isolation is no longer the academic exercise it was once. The study of minimum life-support systems has immediate relevance as soon as we try to leave this planet. The Apollo 13 astronauts were provided with food, hydrogen gas in the fuel cells, rocket propellants, and chemical cells capable of delivering electricity. All of these chemical substances are compounds with high free energy, substances which can deliver this energy to suitable receivers on demand. The astronauts were, in effect, given large stores of free energy in various forms.

So long as they remained a closed thermodynamic system, they were doomed to follow the Second Law and run downhill thermodynamically. There was no recycling of wastes on the Apollo 13 mission. Water from the fuel cell reactions was used by the astronauts or vented into space. Urine was stored in bags and vented into space. Solid wastes were stored for later disposal. The rocket exhaust ejected into space. The low free energy compounds in the electrical cells were retained, to be thrown away later. Every time the astronauts drew breath, or used power to radio Houston, or took any action at all, they increased the total entropy of the spacecraft. They steadily approached the time when they would have to convert the craft to an open system or die. And as the entire world listened in suspense, they came very close to doing just that. Had they been unable to return to Earth, and drifted out into their own solar orbit, the Second Law would have caught up with them in a few days.

The simplest way of converting the Apollo 13 craft to an open thermodynamic system would be to return to Earth and open the hatch. The travelers could then throw out all their accumulated low-free-energy waste products, bring in more high-free-energy fuels for man and machinery, and begin over again. Another way, a favorite of science fiction, would be to set up hydroponics tanks on board, feed the chemical wastes to algae, and let them resynthesize carbohydrates and purify the cabin air with oxygen. This would require windows to let the sunlight in so the algae could carry out photosynthesis. The spacecraft would then be an open thermodynamic system in exactly the same way that the planet Earth is an open system.

The sun is the ultimate source of the free energy that enables living organisms to keep their entropy low. Shutting off the sun would have the same effect as enclosing the planet in a steel coffin. All life presently on Earth would rapidly dwindle to extinction. If life has evolved and become more complex, it is only because a virtually infinite source of free energy was at hand. But the Second Law has exacted its price. As the sun consumes its fuel in nuclear reactions, its own entropy is rising. The yearly entropy increase of the sun is greater than the decrease of entropy on Earth. The Earth-sun system obeys the Second Law and is running down. But the energy resources of the sun are so vast that the comparatively small free-energy changes on Earth are negligible perturbations. With a radius of 4000 miles and a distance of 93 million miles from the sun, the Earth intercepts less than one part in 10^{17} of the sun's total radiation. This tiny fraction of the solar output has been enough to transform the planet from a lifeless water-covered sphere to the habitat of millions of species of living creatures.

According to current theories of stellar evolution, the sun should continue to shine at about its present brilliance for tens of millions or hundreds of millions of years. But as it depletes the fuel in its core, its thermonuclear fires will begin to burn its outer shell. The sun will change from a benign golden disk to a swollen red giant, enormously larger and brighter than it is today. Its heat will singe the Earth, boiling away the oceans and destroying all forms of life. Its fuel almost exhausted by this outburst, the dying sun will cool and collapse, first to a white dwarf star, and finally to a dark clinker still circled by the dead planets to which it once gave light and life.

READINGS

S.W. Angrist and L.G. Hepler, *Order and Chaos*, New York, Basic Books, 1967. The best popular introduction to thermodynamics on the market. Nonmathematical, with heavy emphasis on the meaning of thermodynamic ideas and the application of these ideas to everyday life.

R.E. Dickerson, *Molecular Thermodynamics*, Menlo Park, CA, W.A. Benjamin, 1969. Chapter 7, "Thermodynamics and Living Systems," is a restatement of many of the ideas of this chapter, with more emphasis on the thermodynamics and more attention to the citric acid cycle.

I.M. Klotz, *Energy Changes in Biochemical Reactions*, New York, Academic Press, 1967. Rigorous treatment of thermodynamics, less application to metabolic systems than Lehninger. Designed as an elementary introduction to thermodynamics for a biologist with no previous background.

H.A. Krebs and H.L. Kornberg, *Energy Transformations in Living Matter: A Survey*, Berlin, Springer, Verlag, 1957. The source of all of the thermodynamic data in this chapter. Individual free energy values have been questioned since 1957, but this is still the most complete and self-consistent set of data available.

A.L. Lehninger, *Bioenergetics*, 2nd Edition, Menlo Park, CA, W.A. Benjamin, 1971. Thorough, and beautifully clear, but somewhat more detailed than this chapter. Recommended as a follow-up.

L. Margulis, "Symbiosis and Evolution," *Scientific American*, August, 1971.

L. Margulis, *Origin of Eukaryotic Cells*, New Haven, CT, Yale University Press, 1970. The best current statements of the theory that mitrochondria and chloroplasts were once separate but symbiotic organisms.

C.U.M. Smith, *Molecular Biology: A Structural Approach*, Cambridge, MA, M.I.T. Press, 1968. Chapter 10, "Bioenergetics," has a particularly clear elementary introduction to photosynthesis.

The great appeal of the genetic code derives not only from the importance of its role in the cell, but also from man's innate fascination with certain kinds of games and puzzles —chess, logic problems, crossword puzzles, and the like. The matching of the nucleic acid "code words" to the amino acids seemed initially to present this sort of challenge to the scientist. —Carl R. Woese

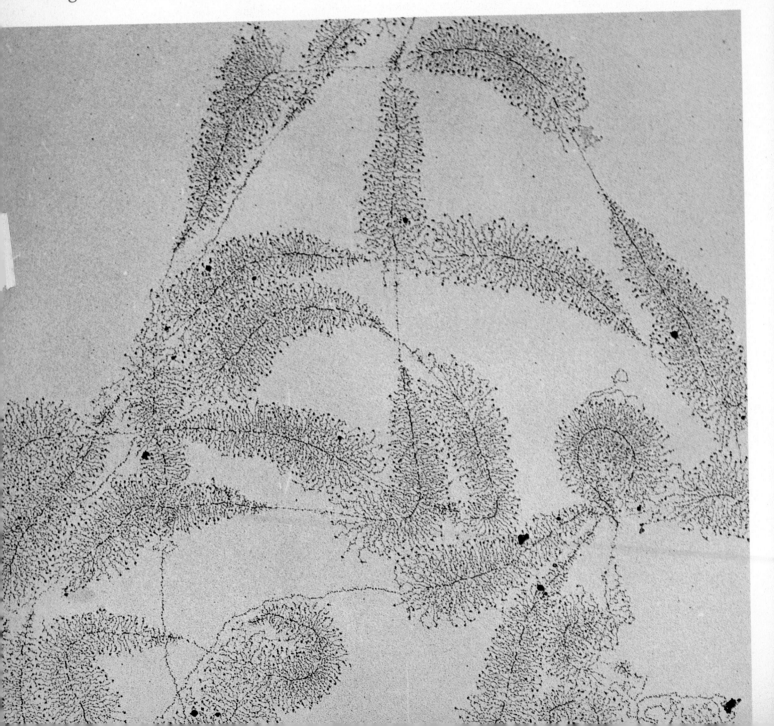

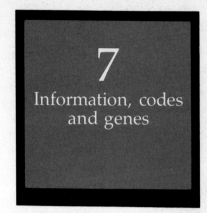

7

Information, codes and genes

The human mind has long been intrigued by the idea of assembling a human being from a collection of parts. A familiar 19th century horror story, made immortal by a series of low-budget movies, describes how an ambitious scientist named Frankenstein created a monster by sewing together parts of corpses, and brought his creation to life with a bolt of lightning. Most biologists believe that if they could assemble an organism, say a human, molecule by molecule, according to nature's blueprint for a human being, this scratch-built working model would AUTOMATICALLY be alive. The bolt of lightning would be completely unnecessary; in fact, it would electrocute him, as it would any other human.

The biological engineer who sets out to create an organism from atomic raw materials would have two problems. He would need a tool that could manipulate individual molecules, and a blueprint that contains a great deal of information. In nature enzymes are the tool for manipulating molecules, and nucleic acids, principally DNA, provide the blueprint.

When this blueprint is passed from one generation to the next it must be reproduced accurately and distributed to the offspring in an orderly way. If it were not, offspring would not resemble their parents. Every living organism is the product of billions of years of natural selection. Any organism

that differed sharply from its parents would almost certainly fail to survive. But if the blueprint were always copied with absolute fidelity, evolutionary progress would be impossible. Errors in copying do occur from time to time, and they are copied with the same fidelity as the original instructions. Known as MUTATIONS, these errors are the only source of new genes. Together with sexual reproduction, which creates new combinations of existing genes, they are the mainspring of evolution *(Chapter 28)*. If the mutation leads to a better-adapted organism, it will be preserved by natural selection and will become part of the new blueprint.

Every cell contains mechanisms for extracting the information in DNA and using it to direct cellular activities. In the main, the instructions that must emerge from the DNA will specify the order of amino acids in each protein chain, how much of each protein is to be made, and when to make it. DNA does not participate directly in the synthesis of protein; it makes disposable copies of appropriate parts of its information in the form of molecules called MESSENGER RNA. The messenger molecules directly control the sequence of amino acids in proteins. In short, there are three kinds of informational macromolecules in cells: DNA, RNA, and proteins. Information flows only from DNA to RNA to protein, never in the reverse direction *(Box A)*.

DNA AND RNA MOLECULES form featherlike structures in this electron micrograph of genetic material from the egg of the spotted newt. The shaft of each "feather" is a long strand of DNA coated with protein. Many RNA molecules extend in clusters from the DNA strand. Transcription of genetic information begins at one end of the gene, with the RNA molecules growing longer as they proceed toward completion. As many as 100 RNA molecules are transcribed simultaneously from each gene.

MOLECULAR INFORMATION

Information is related to probability. If the alphabet contained only one letter —the letter A —and no spaces were allowed (a space would be equivalent to another letter) no information could be conveyed in writing. The probability that the next letter would be A would always be 1.0. The minimum number of symbols that can be used to

A The central dogma

What is semihumorously referred to as The Central Dogma of information transfer is: "DNA makes RNA makes protein." In 1964, Nobel Laureate Howard Temin of the University of Wisconsin studied an RNA virus that causes a cancer of chickens known as the Rous sarcoma. The virus enters the chicken cell and subsequently causes the cell to make a DNA copy of the viral RNA. The afflicted cell does not burst, but changes permanently in shape, metabolism, and growth habits. The new DNA becomes part of the hereditary apparatus of the chicken cell. More recently, Temin and others have shown that the virus carries an enzyme for the manufacture of DNA, using viral RNA as informational template. This discovery does not invalidate the Dogma. In its original and proper form, the Dogma merely stated that information in proteins could not flow back to nucleic acids. The flow of information from RNA to DNA does not contradict this.

convey information is two, like the dot/dash of Morse code, or the on/off of the switching circuits in computers. Of course more than two symbols can be used, such as the ten digits of Arabic numerals, or the 26 letters of the English alphabet. Once we agree upon the coding rules, we can convey a given message using any set of two or more symbols.

DNA contains linear sequences of four different nucleotides, each of which can be considered as a symbol. Any sequence of symbols can carry information. Proteins consist of linear sequences of 20 different amino acids. The four nucleotides in DNA must specify the 20 amino acids in proteins, a problem rather similar to specifying the 26 letters of our alphabet with the dots and dashes of Morse code. The telegrapher's solution is to use various combinations of dots and dashes to construct a code for the letters. In converting information from nucleotide sequences in DNA to amino acid sequences in protein, nature's solution is to use combinations of nucleotides to specify amino acids. If each nucleotide were assigned to code a particular amino acid, only 4 different amino acids could be coded. If PAIRS of nucleotides were used to specify single amino acids, such that AA was one amino acid, AT was another, TA was yet another, etc., 4^2 (or 16) amino acids could be coded. If the DNA code were perfectly analogous to the Morse code, we could specify four of the amino acids by a single nucleotide apiece, and the remaining sixteen by nucleotide doublets (two nucleotides). Nature does not employ a mixed code of this sort, but uses a code in which every amino acid is coded by a triplet of nucleotides. Sequences of three nucleotides can code for 64 amino acids. For a short time after it was first understood that the code

must be based on triplets, biologists thought that the extra 44 triplets might be untranslatable by the cellular decoding machinery. We now know that all 64 possible triplets mean something to the cell. Three of them carry the message: "Don't insert any amino acid; terminate the protein chain." The other 61 all specify one of the 20 amino acids, so that on the average there are about three synonymous triplets to represent each acid. In the language of the branch of mathematics called information theory, this redundancy of the code is called DEGENERACY. The word does not imply that the code had deteriorated. The available degeneracy is not very evenly divided among the 20 amino acids; two of them, methionine and tryptophan, are each represented only by one triplet, while other amino acids are represented by as many as six.

The term degeneracy should not be confused with AMBIGUITY, a word used to describe a single triplet that could specify either of two (or more) different amino acids. Degeneracy in the code for an amino acid (for example, leucine) means that there is more than one unequivocal way to say "put leucine here." Ambiguity would mean that, for a given triplet, there is doubt as to whether to put in leucine or something else. The DNA code is not ambiguous.

A linear sequence of nucleotides in DNA specifies a linear sequence of amino acids in proteins. Because there is no information in proteins that is not also present in DNA, it might seem that DNA, like proteins, could catalyze the chemical reactions of living organisms. It cannot. By its chemical nature, DNA is unable to fold itself into complicated functional shapes that characterize molecules like chymotrypsin and carboxypeptidase. When biologists first began to speculate on the nature of the hereditary material, it was not at all obvious that DNA was a likely candidate, since it seemed to have no biological activity.

THE SEARCH FOR THE HEREDITARY MATERIAL

In 1866, Gregor Mendel, the father of modern genetics, laid down the rules of hereditary transmission of traits (Chapter 9). Meanwhile, chemists were starting to take cells apart and analyze their composition. DNA was discovered in pus cells and in fish sperm by Miescher just three years after Mendel's discovery. Yet it was not until about 1950 that the idea of DNA as hereditary material was generally accepted. Actually, there had been a number of starts in the right direction much earlier, but they had little or no influence on biological thinking of their day. In 1896, for example, E. B. Wilson of Columbia University suggested that the nucleic acid component of the nuclein discovered by Miescher was the chemical form of the genetic material. His conclusion was perhaps more an inspired guess than a closely-reasoned argument, and did not lead directly to any experiments.

Since it was possible to see a relationship between DNA and hereditary material much earlier than 1950, why was such an insight so rare? First, the scientific world remained ignorant of Mendel's monumental discoveries until 1900, partly because he published them in an obscure journal. More important, Mendel's thinking was so far ahead of his time that the few people who did read his work did not understand it. A biologist named Nägeli completely misinterpreted Mendel's findings, and described them incorrectly in a review article. During the last quarter of the 19th century, great advances were made in the field of cytology, and the behavior of the chromosomes during cell division was described. E.B. Wilson and others realized that the chromosomes were behaving as would be expected if they were the bearers of hereditary in-

Because there is no information in proteins that is not present in DNA, it might seem that DNA, like proteins, could catalyze the chemical reactions of living organisms. It cannot.

formation. These changes in the intellectual climate prepared the minds of biologists for the rediscovery of Mendel's work. The decades between 1900 and 1940 saw a great flowering of the field of formal genetics (the rules for the transmission of hereditary information) but the chemical basis of the gene remained a mystery.

Meanwhile, chemists had become interested in enzymes. Throughout most of the 19th century, enzymes had been chemically almost as mysterious as genes. They were will-of-the-wisps that catalyzed certain organic reactions, and were known only by their activity, not by their chemical nature. In 1926 Sumner's assertion that urease was a protein set off a bitter controversy over whether enzymes were or were not proteins. When it finally became accepted that they were, biologists then went overboard and assumed that ALL information-carrying molecules were proteins, including the carriers of genetic information. DNA and RNA, along with protein, were known to be present in chromosomes, but most biologists felt that DNA was merely some sort of skeletal material in the chromosome.

BACTERIAL TRANSFORMATION

It was against this background that Oswald T. Avery and his colleagues at the Rockefeller Institute demonstrated in the 1940's that DNA could carry genetic information. They were investigating a phenomenon discovered by Frederick Griffith in 1928. When injected into mice, one strain of *Pneumococcus* bacterium causes pneumonia and usually death. Another strain of the bacterium is relatively harmless. The infective form always has a capsule (a complicated polysaccharide coating) while the noninfective form does not. Griffith called the encapsulated strain the "S strain" because the colonies looked smooth on a culture plate, and the harmless, unencapsulated one the "R strain" because of its rough colonies (Figure 1). Griffith also found that when large numbers of S bacteria were grown outside of animals, he could obtain a few R bacteria. These proved to be mutants that had lost the ability to make a capsule. Mice injected with R bacteria and simultaneously with a large number of heat-killed S bacteria developed pneumonia and died; live S bacteria were found in their blood. Later work was to show that this process, called TRANSFORMATION, was not a matter of breathing life back into the dead bacteria, but rather that something from the dead bacteria was converting R bacteria into S bacteria.

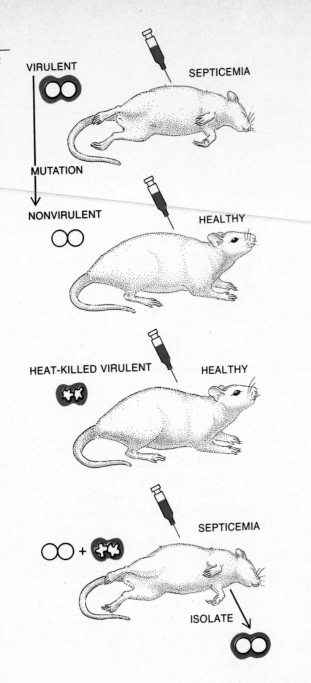

VIRULENT — SEPTICEMIA

MUTATION

NONVIRULENT — HEALTHY

HEAT-KILLED VIRULENT — HEALTHY

⬭⬭ + ✚ — SEPTICEMIA

ISOLATE

GENETIC TRANSFORMATION. Mice injected with live virulent strain of pneumococcus bacteria died of blood poisoning (septicemia). Those injected with nonvirulent or dead bacteria remained healthy. Those injected with a mixture of live nonvirulent and dead virulent bacteria (bottom) died of septicemia. Their blood contained live virulent pneumococci.

did not cause transformation of R bacteria into S bacteria. Finally they found that they had isolated virtually pure DNA. They were able to show something even more impressive. There are different kinds of S *Pneumococci* that form chemically different capsules. If the geneticists extracted DNA from a particular kind of S strain and added it to R bacteria, the type of capsule made by the transformed bacteria was always identical with that of the bacteria that donated the DNA, indicating the hereditary information must be carried by DNA. Geneticists now know that virtually any genetic trait in *Pneumococcus* or in several other genera of bacteria can be passed, via DNA, from one bacterium to another.

The study of transformation showed that some of the traits of a bacterium could be transmitted by DNA, but there was still room for skepticism about whether all of the hereditary information of any organism could be contained in DNA. In 1952, A. D. Hershey and Martha Chase proved this point in their study of the life cycle of the bacteriophage T2. A bacteriophage, or phage, is a virus that grows in a bacterial host cell (*Figure 2*). Bacteriophage T2 is a tadpole-shaped virus that consists of about 50 percent DNA and 50 percent protein. It can infect the bacterium *E. coli* (and a few very close relatives) by attaching by its tail to the bacterial cell wall. About 20 minutes later the bacterium bursts, releasing about 200 new T2 viruses. Hershey and Chase found that the protein coat clings to the outer surface of the bacterium: if the bacteria are put into a blender a few minutes after infection, the virus coats could be sheared off, leaving the DNA core of the viruses inside the bacteria. As before, the bacteria eventually burst, producing a new generation of phage particles. Evidently the protein husk of the phage acts only as a syringe to inject the hereditary material into the bacterium. Hershey and Chase prepared bacteriophages in which the phosphorus of the DNA was labeled with radioactive phosphorus, P^{32}, and others in which the protein portion was labeled with a radioisotope of sulfur, S^{35}. Since DNA contains no sulfur, and proteins generally contain little or no phosphorus,

Griffith, interpreting his findings in terms of the knowledge of his time, felt that protein from the dead bacteria might be the active agent. Others found that the transformation of R bacteria to S bacteria could be accomplished *in vitro* ("in glass," that is, in a test tube) rather than using mice as a culture vessel. The contribution of Avery and his group was to isolate the material that was responsible for the transformation. Instead of adding heat-killed S bacteria, they made an extract of heated S cells and started purifying the material by removing, one after another, those substances that these two isotopes could be used to follow the fate of DNA and protein through the life cycle. The investigators found that if bacteria were infected with these phages, much of the P^{32} turned up in the new phages, but practically none of the S^{35}, demonstrating that the DNA alone was the link between generations.

These experiments imply that DNA is the sole carrier of genetic information. Are there any exceptions? We have already hedged on the DNA story by noting that in some viruses RNA is the carrier of information, and functions very much like the DNA of most other organisms (*Box A*). Perhaps this could be considered as merely a substantial variant on the rule rather than a fundamental exception.

A much more puzzling case is the causative agent of a mysterious disease called scrapie—a neurological disorder that often wipes out whole flocks of sheep. It has proved very difficult to isolate the agent in pure condition and in quantity for chemical studies. The agent appears to be very small in size, and more important, resistant to doses of ultraviolet radiation that would almost surely destroy nucleic acid. The agent is not destroyed by enzymes that degrade DNA and RNA, but is inactivated by enzymes that hydrolyze proteins, and is also destroyed by some reagents that attack carbohydrates. The scrapie "organism" may be the first example of a living thing without any nucleic acid. Moreover, it is probably not unique. There is a similar disease of humans called kuru, which has decimated some populations in the remote New Guinea highlands. Kuru seems to be transmitted by eating the brains of infected victims, and has been on the wane now that laws against cannibalism are being more strictly enforced.

We know nothing about how the scrapie or kuru agents reproduce themselves in their hosts. If it is finally proved that they do not contain any nucleic acid, the discovery will not necessarily contradict the dogma that proteins cannot reproduce themselves. It is possible, for instance, that the scrapie agent is coded by sheep DNA, and that ordinarily the information is never translated into protein. Once the DNA is activated by scrapie agent from another sheep, it starts making more of the agent, causing a fatal infection. The reason that the scrapie agent has attracted so much attention is that it is an exotic exception. Virtually all of the information encoded in both procaryotes and eucaryotes is in the form of DNA. This is not surprising, because DNA is uniquely suited for duplication and storage of information that directs cellular activities.

DNA DUPLICATION

There are two possible ways to duplicate an object, such as a line of type. One is to set an identical line of type, or perhaps carve a copy into a wooden block. The other is to make a mold of the type, and use the mold to cast a second line just like the first. The first process requires a complicated intermediary device, a human typesetter or an engraver. The second process—nature's way of duplicating DNA—can be executed directly by a relatively simple machine. In effect, the cell makes

LIFE CYCLE OF BACTERIOPHAGE T2 and T4 growing on E. coli. Phage attaches itself by its tail fibers to the surface of the bacterial cell (left) and injects virus DNA (color) into the cell. The DNA programs the cell to produce more phage viruses. Eventually the cell bursts, releasing the phages to infect other cells. Empty coat of phage can be shaken loose from cell in a blender. **2**

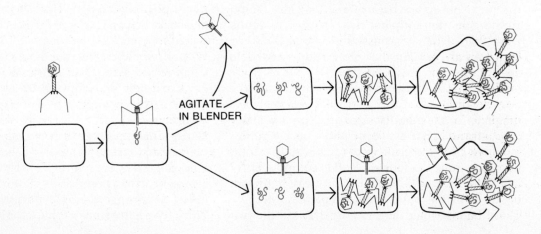

AGITATE
IN BLENDER

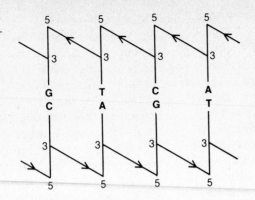

3 ANTIPARALLEL ORIENTATION of the two strands in a double helix of DNA. The direction of a strand depends on the direction of the phosphodiester bonds from the 3′ position of a deoxyribose sugar to the 5′ position of the next one. Phosphodiester bonds are shown as arrows and the sugars are shown as straight vertical lines.

a chemical mold of the sequence of nucleotides in a strand of DNA. The mold itself is made of nucleotides, but differs from the original. When a cast is made from the mold, the resulting nucleotide sequence is identical with the original. In DNA, the original and the mold are wrapped around each other in the form of a double-stranded helix. One of the two complementary strands is whimsically called "Watson," the other "Crick." The nitrogenous bases in the two strands are always paired in such a way that an adenine (A) in one strand always lies opposite a thymine (T) in the other, and a guanine (G) always lies opposite a cytosine (C). Knowing the pairing rules that apply to the bases in the two strands, one can easily write the base sequence of one strand given the base sequence of the other. If the order of bases in Watson is A-A-G-T-C-T-C, for example, the order in the Crick strand must be T-T-C-A-G-A-G, with the sugar-phosphate backbone of the Crick strand running in the antiparallel (opposite) direction from that of the Watson strand (*Figure 3*). Thus each strand contains the information needed to specify the structure of the other.

When the DNA molecule duplicates itself the two strands separate, and each serves as a template for the manufacture of a complementary new strand. If the two strands of a double-stranded DNA molecule were untwisted in a solution containing the four nucleotide components of DNA, the free bases would form weak hydrogen bonds with the bases in the separated strands. These bonds would be so weak that the thermal buffeting of the molecules of the solvent would separate the pairs soon after they formed. But if we could link the free bases together by phosphate ester bonds as they came into position on the template strand, the combined hydrogen bonds of several bases in succession would be sufficient to hold the growing new strand in position next to the template. The nucleotides of a new DNA strand will conform naturally to the shape of the complementary strand.

There are two ways in which a new DNA strand might be assembled. First, all the new nucleotides might line up along the template strand, and when they are all in place, an enzyme might travel down the molecule like the slide of a zipper, forming phosphodiester bonds from one end to the other. Or, the slide of the zipper might need to wait only for the next nucleotide that is to be added to the growing chain. This second model is nature's way. The first would be hopelessly slow, considering the enormous number of nucleotides in a DNA chain.

Though the individual bonds between pairs of complementary nucleotides are quite weak, there are so many of them in a DNA molecule of any substantial length that together they are quite strong. This ensures a stable structure even though nucleotides are added one at a time. The geometry of the nucleotides ensures that the phosphate linkages will be formed only if the base pairs are correct. For example, C can improperly pair with A, but the sugar-phosphate appendage of the unattached nucleotide points in the wrong direction to be connected to the backbone of the growing strand. A-T and G-C pairings produce just the right configurations to lead to the formation of a new bond.

When investigators first started thinking about the duplication of the double helix, it was apparent that there were three general ways in which it might possibly happen. First, a particular Watson and a particular Crick might be destined to stay paired forever. At the time of DNA synthesis, the Watson would synthesize a new Crick, and the Crick would synthesize a new Watson. The new Crick and new Watson would likewise stay paired forever. This possibility was called conservative replication. Second, a particular Watson and a particular Crick might each pair with the new complementary strand whose synthesis it has just guided, so that every double helix of DNA would consist of an old strand and a new strand. This was called semiconservative replication. Third, the Watson and the Crick strands could be somehow

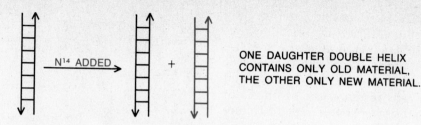

ONE DAUGHTER DOUBLE HELIX
CONTAINS ONLY OLD MATERIAL,
THE OTHER ONLY NEW MATERIAL.

CONSERVATIVE REPLICATION

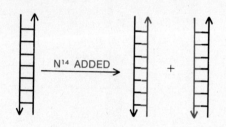

EACH DAUGHTER DOUBLE HELIX
CONTAINS ONE OLD STRAND
AND ONE NEW STRAND.

SEMI-CONSERVATIVE REPLICATION

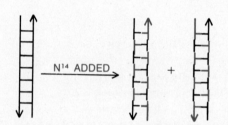

ALL STRANDS CONTAIN
BOTH NEW AND OLD MATERIAL.

DISPERSIVE REPLICATION

4 THREE TYPES OF REPLICATION are possible during the synthesis of new DNA molecules. The original DNA molecule is shown in black and the new material in color. The Meselson-Stahl experiment described in text indicated that the replication of DNA is semi-conservative.

broken into small fragments at the time of DNA synthesis and new materials introduced, so that both strands of every DNA molecule would always consist of a mixture of old and new materials. This was called dispersive replication (*Figure 4*).

THE MESELSON-STAHL EXPERIMENT

DNA is actually duplicated by a semiconservative mechanism. The experiment that demonstrated this fact was done in 1958 by Matthew S. Meselson and Franklin W. Stahl, then at the California Institute of Technology. They used a new technique, called density-gradient centrifugation, which they developed in collaboration with Jerome Vinograd.

The technique enabled them to separate DNA molecules that differed slightly in density (weight per unit volume).

Meselson and Stahl realized that the density of DNA could be modified by incorporating into DNA heavy isotopes of its normal atoms. For example, nitrogen accounts for about 17 percent of the total molecular weight of the DNA in *E. coli*. Replacing all of the N^{14} normally found in *E. coli* DNA with N^{15}, would increase the density of the molecule about 1.2 percent. Meselson and Stahl found that they could grow *E. coli* on a medium in which the only source of nitrogen was ammonium chloride containing N^{15} of high isotopic purity. Cells grown on this medium made "heavy" DNA that could easily be separated from normal DNA by spinning the DNA solution in an ultracentrifuge (*Figure 5*).

After they had grown the bacteria for many generations in the N^{15} medium so that virtually all the

DNA was N^{15} DNA, Meselson and Stahl abruptly transferred the culture of cells to a medium containing N^{14}. All DNA made after the transfer would contain the normal isotope; the average density of the DNA extracted from the bacteria would decrease as the cells resumed the synthesis of normal DNA. Samples were taken from the culture at various times after the change of medium, and

5 ULTRACENTRIFUGE can separate normal DNA from "heavy" DNA containing N^{15}. This schematic diagram depicts the technique known as density-gradient ultracentrifugation, in which the DNA molecules (color) are suspended in a solution of cesium ions (black dots) and spun until they reach an equilibrium position. In effect, this technique provides a very sensitive method of separating DNA molecules of slightly different weights. Results of the technique are depicted in Figure 6.

TUBES OF DNA AND CsCl BEFORE CENTRIFUGATION.

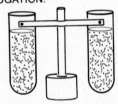

THE SAME TUBES AFTER A BRIEF PERIOD OF CENTRIFUGATION.

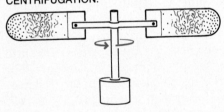

THE SAME TUBES MUCH LATER, AFTER EQUILIBRIUM HAS BEEN REACHED.

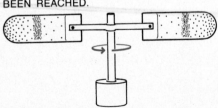

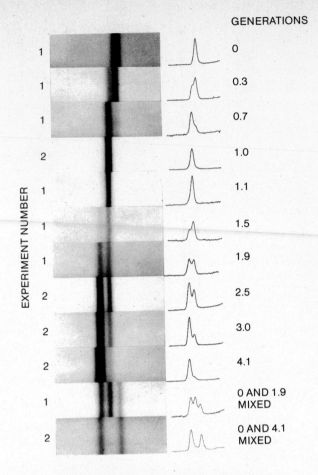

RESULTS OF MESELSON-STAHL EXPERIMENT. **6**
Each photograph shows the equilibrium position of DNA in cesium chloride gradient. The "bottom" of the centrifuge tubes lies at the right. The zero generation is the one in which bacteria were switched from a growth medium containing N^{15} to one containing N^{14}. The photo and the accompanying curve show that at the moment all the DNA is "heavy" N^{15} DNA. Next three frames show progressive formation of hybrid DNA during the first generation after the shift to the "light" N^{14} medium. At the end of the first generation (frame 1.0), all DNA is of hybrid density, consisting of one N^{15} and one N^{14} strand. During second generation (frames 1.1, 1.5, 1.9) pure N^{14} DNA starts to appear, and becomes predominant during third and later generations (frames 2.5, 3.0, 4.1). Curves record amount of DNA in centrifuge tube as measured by a photocell.

DNA extracted from them was spun down in the centrifuge to trace the fate of the heavy N^{15}-labeled DNA. During the first generation of growth following the change of medium, the DNA progressively became lighter until it reached a density midway between that of the heavy DNA and normal DNA *(Figure 6)*. After the next generation, half of the DNA was light and the other half showed an intermediate density corresponding to a hybrid of heavy and normal DNA. In subsequent generations, more and more DNA showed the normal density, but the amount of hybrid DNA remained unchanged. Evidently the heavy parental strands of DNA became associated during synthesis with an exactly equal amount of light DNA.

This conclusion was readily demonstrable by heating DNA to temperatures high enough to "melt" the hydrogen bonds that hold the two strands together. Meselson and Stahl found that when they heated hybrid DNA, they released equal amounts of light and heavy single-stranded DNA. The first duplication of DNA separates the N^{15}-labeled strands so that both enter into DNA of hybrid density. During subsequent cycles of DNA synthesis, the total amount of hybrid DNA re-

mains constant, though it becomes a smaller and smaller fraction of the total DNA. This experiment has been repeated with a variety of organisms, including algae and human cells, and the result has always been the same: the synthesis of DNA is semiconservative.

ENZYMES IN DNA SYNTHESIS

In the mid 1950's, Nobel laureate Arthur Kornberg and his coworkers at Washington University in St. Louis discovered an enzyme that seemed to have the properties required of a biological catalyst for DNA synthesis. Called DNA polymerase, the en-

DEOXYNUCLEOSIDE TRIPHOSPHATES are the building blocks of the DNA molecule. These small molecules are linked together by the enzyme DNA polymerase to form two strands of DNA helix. 7

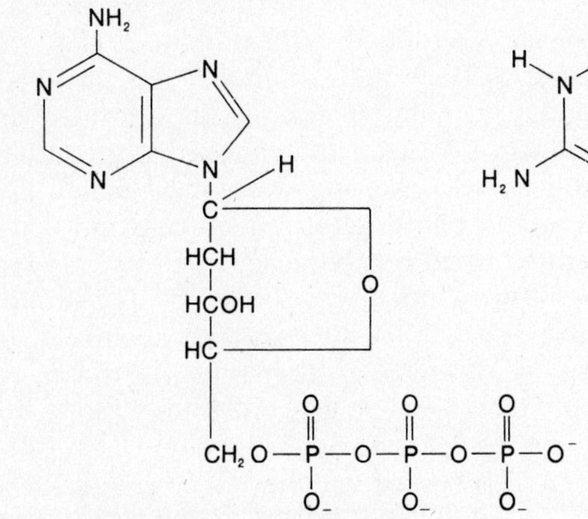

DEOXYADENOSINE TRIPHOSPHATE

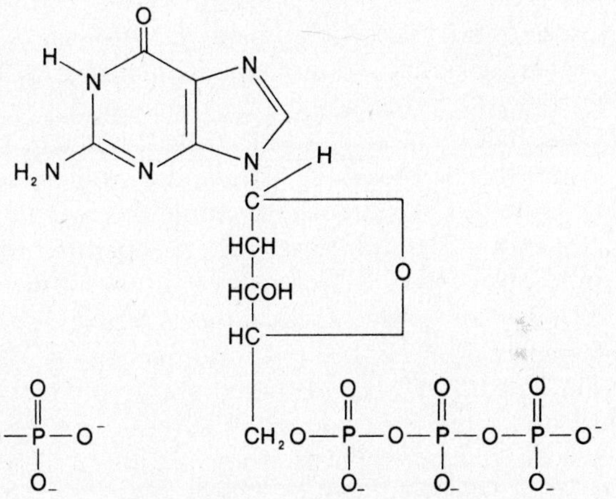

DEOXYGUANOSINE TRIPHOSPHATE

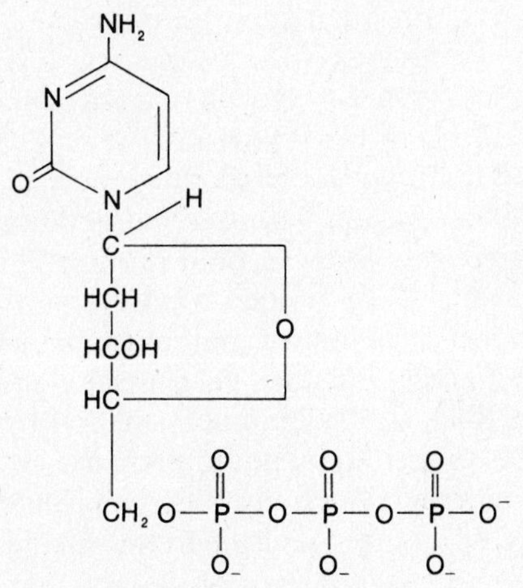

DEOXYCYTIDINE TRIPHOSPHATE

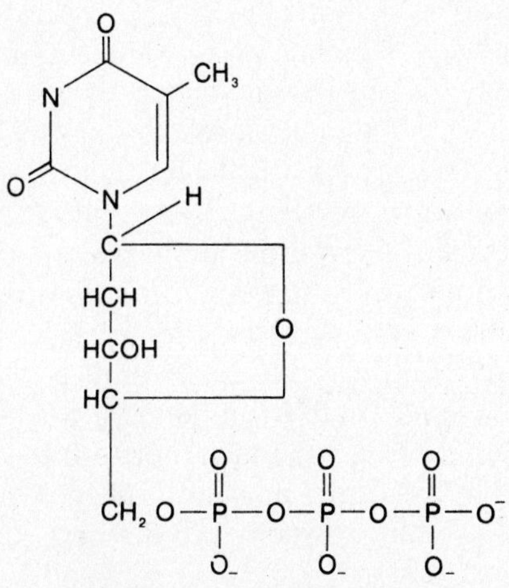

THYMIDINE TRIPHOSPHATE

zyme was capable of building strands of DNA from deoxyribose triphosphates of adenine, guanine, cytosine, and thymine *(Figure 7)*. Like ATP, these triphosphates contain high-energy bonds. Using the energy of these bonds, the enzyme links the hydroxyl group on the C^3 of one nucleotide sugar to the hydroxyl group on the C^5 of the next. The process releases a pyrophosphate (the two terminal phosphates of the triphosphate chain). The addition of nucleotides to the growing chain always occurs at the same end: the end on which the DNA strand has a free hydroxyl group on the C^3 of its

B Nucleic acid in the test tube

In 1967, newspapers announced that three scientists, Mehran Goulian and Arthur Kornberg of Stanford and Robert L. Sinsheimer of Caltech, had "created life" in a test tube. Of course they had not. They had enzymatically copied the DNA molecule of a phage, ϕX 174. This remarkable virus contains SINGLE-STRANDED ring-shaped DNA. The DNA extracted from the virus can infect bacterial spheroplasts (bacteria with part of their cell wall dissolved away) and produce phage. The biological activity of any DNA made in the test tube can be tested by adding it to a culture of spheroplasts. The experiment involved adding DNA polymerase to the ring-shaped Watson strands (they are actually called + strands) that were extracted from the virus. The enzyme proceeded to make Crick (or − strand) complements of the input DNA. This process yields double-stranded rings of DNA with a single gap in the bonding of the new Crick strand. A second enzyme, DNA ligase, was added to the mixture to close that gap and produce a ring-shaped Crick strand. The double-stranded ring was then treated with a nuclease to break one bond, at random, in each double-stranded ring. Thus, half of the rings would consist of a whole Watson and a broken Crick and half would have a broken Watson and a whole Crick. The rings were then heated to separate the two strands, yielding, from each, one single-stranded ring and one single-stranded rod. Rings can be separated from rods because the two kinds of molecules settle at different rates in the ultracentrifuge.

Using the isolated Crick rings as templates, the investigators then added DNA polymerase and DNA ligase and made new Watson rings, then repeated the operation. Next they isolated the wholly synthetic Watson rings, containing none of the DNA isolated from phage in the first place, and showed that the new DNA could infect *E. coli* spheroplasts. They had thus demonstrated that they could enzymatically copy an important biological molecule with astonishing fidelity.

It should be pointed out that Sol Spiegelman and coworkers, at the University of Illinois, had found an enzymatic system that copied infective RNA from the bacteriophage $Q\beta$. In his system, he did not have to go through successive cycles of separation of template and product. They separated themselves, and the result was that there was an enormous net synthesis of copies, such that they far exceeded the number of template molecules ofthat had been added in the first place.

Spiegelman and his collaborators also showed that they could produce molecular evolution in the test tube by stopping the synthetic reaction before it had gone to completion, so that those molecules that could replicate fastest would produce the largest number of progeny in the limited time available. They then used the new molecules that were made as templates for another cycle of limited multiplication, and repeated this competitive growth race through a number of cycles in succession. At the end, the test tubes contained a noninfective RNA molecule, much smaller than the original RNA, and much more efficient than its progenitor at multiplying *in vitro*. The "little monster," as it was called, was no longer a biologically active molecule, but it was eminently suited to the new ecological niche that Spiegelman had created in his test tubes.

Biologists now know that DNA polymerase is not a single enzyme but a small orchestra of different enzymes, each contributing its own special, indispensable functions.

terminal deoxyribose *(Figure 8)*. The synthesis of DNA by DNA polymerase requires the presence of a DNA template. The new strand that is formed by DNA polymerase is the complement of the template, and the two wind around each other to produce a new molecule of double-stranded DNA. The enzyme is remarkably faithful in following the instructions of the DNA template, and it has been shown that the DNA made by the enzyme in the laboratory can be biologically active *(Box B)*.

One of the ground rules of science is that no question is ever considered completely answered. This has certainly been true of the question: what is the exact enzymatic mechanism of DNA synthesis? In 1969, a search by two industrious scientists turned up a mutant of *E. coli* that seemed totally to lack the Kornberg enzyme, yet it grew and manufactured new DNA perfectly well. The obvious conclusion was that the Kornberg enzyme was not needed in the cell for DNA synthesis, and that it might be restricted to some other function — perhaps DNA repair. Still later, however, inves-

tigators found that the mutant did not totally lack the Kornberg enzyme, but showed only one percent of the activity of the parent strain. Subsequently, mutants were found which really did lack the Kornberg enzyme under certain conditions, and under those conditions, they could not make new DNA, and could not reproduce. Hence the argument has come full circle, and the Kornberg enzyme is once again considered to be required for DNA synthesis.

All this would seem to be just an entertaining exercise, except that the existence of mutants showing very low activity of the Kornberg polymerase has allowed a new and important question to be asked: what other enzymes might there be in a cell which are also able to make DNA? To see how such mutants have been useful, imagine an orchestra (the various components of the DNA synthesizing machinery) which is dominated by a large bank of violins (the Kornberg enzyme). The structure of the music being played may absolutely require the woodwinds, brasses, etc., so that the music would collapse without them, yet they

ADDITION OF NUCLEOTIDES to a growing chain of **8**
DNA. Left: upper strand is the template and lower strand is the growing chain. Deoxythymidine triphosphate is about to attack the 3' OH group. Middle: deoxythymidine triphosphate is hydrogen-bonded to template strand. Right: inorganic pyrophosphate has been cleaved off and a phosphodiester bond has formed between deoxythymidine phosphate and growing strand. Reaction is catalyzed by DNA polymerase.

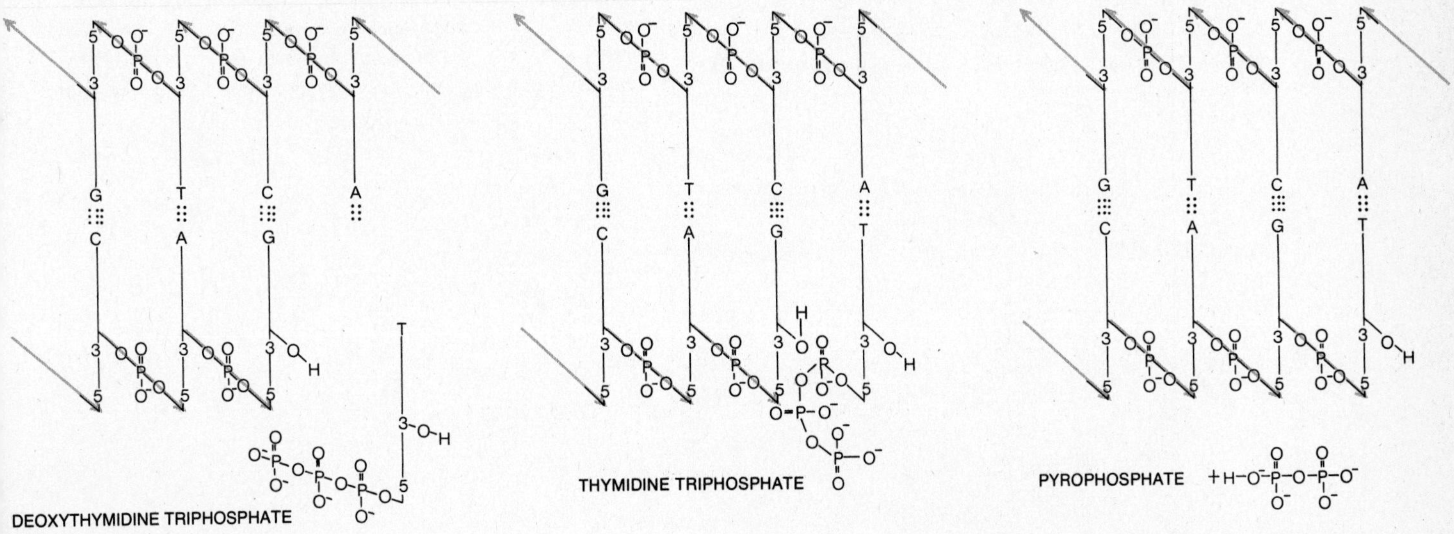

DEOXYTHYMIDINE TRIPHOSPHATE

THYMIDINE TRIPHOSPHATE

PYROPHOSPHATE

are not easily sorted out by the average ear. But if the violins are suddenly muted to one percent of their former loudness, other essential instruments suddenly make themselves known as part of the orchestra. Biologists now know that DNA polymerase is not a single enzyme but a small orchestra of different enzymes, each of which contributes its own special, indispensable functions. Working together, this group of enzymes does the job of DNA synthesis with a speed and accuracy that are almost unimaginable. In *E. coli*, the polymerase complex makes new DNA at a rate in excess of 1000 base pairs per second, and makes mistakes in fewer than one base in a million.

DNA synthesis involves another stubborn paradox. How can enzymatic machinery that can add nucleotides only to ONE END of a growing DNA strand replicate both strands of the parent molecule at the same time? The reader will recall that the two strands are anti-parallel. Both parental strands are replicated at the same time, and in the same direction. It is as if a contractor were required to start in San Francisco, and to convert a two-lane divided highway to New York into two separate highways by adding two new lanes beyond the shoulders of the old ones *(Figure 9)*. But the road-building vehicles are not allowed to violate the one-way signs on the new lanes during construction. The new San Francisco to New York lane could, at least in principle, be paved over in a single uninterrupted effort next to the old New York to San Francisco lane. To keep in step with this effort, and still obey the law, the New York to San Francisco crew would have to leapfrog a few miles east, say to Oakland, California, and pave their way west to San Francisco. Then they would leap frog still further east and pave their way back to Oakland, seaming the two pieces of the road together when they reached their previous con-

struction. Finally, by a series of eastward jumps and westward constructions, they would complete the lane to New York. In just this way, DNA synthesis along one strand (and possibly both strands) is done in a series of short stretches that are promptly seamed together. To push the analogy further, duplication of DNA can be either uni-directional (with construction starting in San Francisco) or bidirectional (with construction starting in Omaha and proceeding simultaneously toward New York and San Francisco). Obviously DNA synthesis is a complicated process, and it is not yet completely understood.

GENOTYPE AND PHENOTYPE

Two organisms are said to have the same GENOTYPE if they have identical hereditary information. They are said to have different genotypes if they have any differences in their hereditary information, even if an observer cannot readily tell them apart. The visible expression of genetic characteristics is the organism's PHENOTYPE. Phenotype is the product of the genotype interacting with the environment. Some expressions of the genotype are very resistant to the effects of the environment, such as blood type, or the color of the eyes. Others, such as stature at maturity, skin color, or athletic ability, are more readily influenced by environmental differences.

When a mutation occurs in the genotype, the

HIGHWAY ANALOGY compares the replication of DNA to the task of converting an old divided highway (black arrows) into two divided highways. Two new lanes (colored arrows) must be constructed along the shoulders of the old highway, without violating two special rules: 1) construction crews must keep the two new lanes finished to about the same point, and 2) they must never break the one-way law, even in the lane that they are currently paving.

9

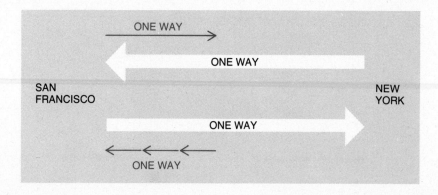

> Two organisms are said to have the same genotype if they have identical hereditary information. The visible expression of genetic characteristics is the organism's phenotype.

change in the phenotype may be complex, such as an alteration in shape or behavior pattern, so that the phenotype gives little indication of the nature of the change in cellular instructions. However, as long as the altered phenotype is heritable and easily characterized, the superficial abnormalities of the affected individual serve as evidence of a change in his genes, even if the molecular anomaly remains unknown.

MUTATIONS

Mutations, heritable changes in the genetic information, are so slight that DNA chemists are usually unable to detect them by chemical analysis. But minute changes in the genetic material often lead to easily observable changes in the outward form and function of the individual. The detection of a mutation depends on the ability to observe its effects. Some mutations are obvious in humans —dwarfism, for example, or the presence of more than five digits on each hand. A mutation may be almost equally obvious in a microorganism —for example, one that results in a change in color, or in nutritional requirements. Other mutations may be virtually unobservable. In humans, for example, there is a mutation that drastically lowers the level of an enzyme called glucose-6-phosphate dehydrogenase, which is present in many tissues, including red blood cells. The red blood cells of a person carrying the mutant gene are abnormally sensitive to an antimalarial drug called primaquine, and when such people are treated with this drug, their red blood cells rupture, causing serious medical complications. Normal people have no such problem with primaquine, and before it came into use, no one was aware that such a mutation existed. Similarly, distinguishing a mutant bacterium from a normal one may be a very subtle matter, and depends on what tools are available.

Mutations are the raw material of evolution, and the working materials of genetics. The only method for "labeling" a gene to discover its func-tion is to find a mutation affecting that gene. In fact, geneticists frequently refer to mutations that alter the function of genes as "markers" because they provide a means for specifically studying the behavior of individual genes.

Some mutations involve extensive chemical changes in the structure of DNA. Others, called POINT MUTATIONS, change only a single nucleotide symbol in the genetic information. Point mutations can generally REVERT —they can mutate back to the original form. Extensive mutations may be REARRANGEMENTS, which change the position or direction of a DNA segment without actually removing any genetic information, or DELETIONS, in which a segment of DNA is irretrievably lost.

TRANSITION & TRANSVERSION MUTATIONS. In **10** a transition mutation a purine is replaced by another purine, or a pyrimidine by another pyrimidine. In a transversion mutation a purine is replaced by a pyrimidine or vice versa. This schematic diagram depicts pyrimidines as disks and purines as squares. Column at right depicts both types of mutation (color). Center column shows how they arise: transition mutations through a mispairing of a purine with the wrong pyrimidine or vice versa, and transversion mutations by a mispairing such as the one shown in brackets. Brackets and question mark indicate that the mechanism is still unknown.

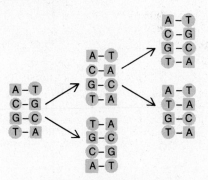

TRANSITION MUTATION

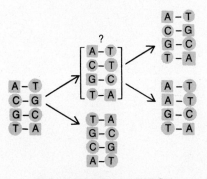

TRANSVERSION MUTATION

All mutations are rare events. The observed frequencies of mutations are different for different organisms and for different genes within a given organism. Usually the frequency of mutation is lower than one mutation per 10^4 genes per DNA duplication, and sometimes the frequency is as low as one mutation per 10^9 genes. Most mutations in nature occur during the copying of the genetic message. The majority are point mutations resulting from the substitution of one nucleotide for another, or from the addition or deletion of a single nucleotide during the synthesis of a new DNA strand.

Nucleotide substitution mutations (base-change mutations) can be either TRANSITION mutations, in which a purine is replaced by another purine (A → G, or G → A) or a pyrimidine is replaced by another pyrimidine (T → C, or C → T), or TRANSVERSION mutations, in which a purine is replaced by a pyrimidine (for instance, A or G → C or T) or vice versa (*Figure 10*). Transition mutations may result from mispairing of bases during semiconservative DNA synthesis. Ordinarily it is not possible for A to pair with C or G to pair with T, but rare alterations can occur in the atomic configurations of the molecules that would make such pairing possible. Two kinds of alterations have been proposed: first, a proton shift within the molecule can move a hydrogen atom from one position to another, producing a tautomeric shift that lines up the hydrogens so that they can participate in an unorthodox pairing: Second, the bases can sometimes ionize, losing a hydrogen atom that had previously stood in the way of the unorthodox pairing (*Figure 11*). Transition mutations may begin with an abnormal pairing during DNA synthesis —for example, a C pairs with an A in the template strand —so that the new DNA molecule contains one noncomplementary C-A base pair. When this molecule replicates again, the C will almost certainly pair with a G from the local supply of free nucleotides, and the A will pair with a T. Of the two daughter molecules produced at the next replication, one will be a mutant with a C-G base pair and the other will have the unchanged configuration with a T-A base pair as in Figure 10.

A number of chemicals can induce mutations. Most of these MUTAGENS induce transition mutations. Among these mutagens are base analogues: purines or pyrimidines not found in natural DNA but enough like the natural bases that the cell can incorporate them into DNA. Base analogues are mutagenic presumably because they are even more likely than the natural DNA bases to undergo the rare molecular alterations that favor mispairing.

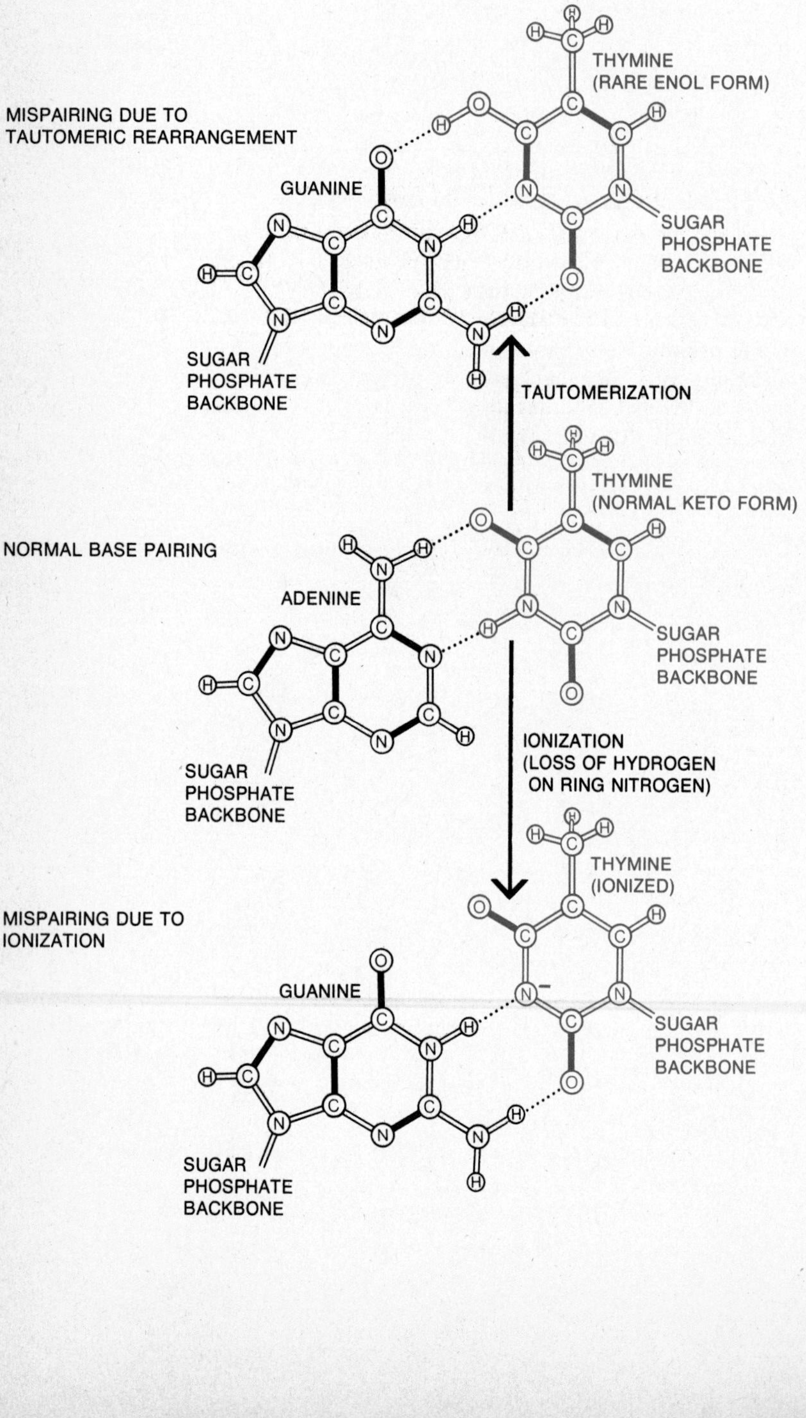

MISPAIRING DUE TO
TAUTOMERIC REARRANGEMENT

THYMINE
(RARE ENOL FORM)

GUANINE

SUGAR
PHOSPHATE
BACKBONE

SUGAR
PHOSPHATE
BACKBONE

TAUTOMERIZATION

NORMAL BASE PAIRING

THYMINE
(NORMAL KETO FORM)

ADENINE

SUGAR
PHOSPHATE
BACKBONE

SUGAR
PHOSPHATE
BACKBONE

IONIZATION
(LOSS OF HYDROGEN
ON RING NITROGEN)

MISPAIRING DUE TO
IONIZATION

THYMINE
(IONIZED)

GUANINE

SUGAR
PHOSPHATE
BACKBONE

SUGAR
PHOSPHATE
BACKBONE

MISPAIRING OF BASES. Normally adenine pairs with thymine (center). Tautomerization produces a form of thymine in which a hydrogen atom shifts from nitrogen atom in ring to a nearby oxygen atom (top). This configuration leads to mispairing with guanine instead of adenine. Ionization (bottom) involves loss of a hydrogen atom from the thymine nitrogen, and also leads to mispairing with guanine. In the ball-and-stick models depicted here, double bonds are printed in color.

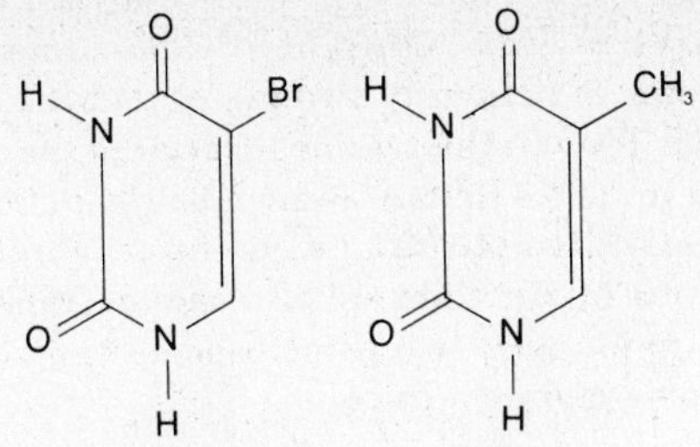

5-BROMOURACIL   THYMINE

12 **MUTAGENS are chemicals that can induce mutations. One example is 5-bromouracil, which the enzymes of DNA synthesis are unable to distinguish from thymine.**

Thus, 5-bromouracil (*Figure 12*) is very similar to thymine, and the cell normally incorporates it into DNA in the place of thymine. But the abnormal base is much more likely than thymine to engage in an abnormal pairing with guanine, and it is therefore a powerful inducer of A-T to G-C and G-C to A-T transitions. Certain other mutagenic agents, such as hydroxylamine or nitrous acid, directly alter the structure of the natural bases in DNA, changing them to other bases that have a tendency to mispair.

The mechanism of transversion mutations is unclear. Apparently they involve the pairing of a purine with a purine (G with A or G with G, for example) or a pyrimidine with a pyrimidine (T with C or T with T). But the purine-purine pair is so long, and the pyrimidine-pyrimidine pair so short, that the sugar-phosphate backbone would be stretched too far out of line for the proper phosphate diester linkages to form. Nevertheless, such mutations do occur, and ultimately investigators will explain their mechanism.

The mechanism by which single base pairs are added to or subtracted from the DNA is another mystery. There is a family of molecules that induces such mutations: the acridine dyes, some of which are used in the treatment of malaria, and are related in structure to primaquine. Mutations resulting from the addition or deletion of single base pairs are known as FRAME-SHIFT mutations, because they interfere with the decoding of the genetic message by throwing the decoding apparatus out of register. Recall that the information in a gene is decoded by a mechanism that starts at one end of the genetic message and reads it as if it were a continuous sequence of three-letter words, each word corresponding to a particular amino acid in

the desired protein product. If a base is added to the message or subtracted from it, the decoding process will work perfectly until it comes to a mutational change. From that point on the three-letter words in the message will be one letter out of register. In other words, such mutations shift the "reading frame" of the genetic message. Frame-shift mutations almost always lead to the production of completely nonfunctional proteins.

A simple analogy can illustrate the concept of frame-shift mutations. If a message contains only three-letter words there is no need to leave spaces between the words, so long as they are read in the right register:

THECOPSAWTHEMANAIMHISGUN.

Now consider the effect of deleting a single letter, analogous to the deletion of a single nucleotide:

THECOPSWTHEMANAIMHISGUN.

It is small wonder that a frame-shift mutation is so disruptive. An organism carrying such a mutant gene can survive only if the gene product affected is not an essential part of the cellular machinery or if the organism also carries another copy of the gene in its normal form. This situation is different from that applying to base-substitution mutations. Often base-substitution mutations change the genetic message so that one amino acid is substituted for another in the protein. Such a substitution may sometimes cause the protein to be completely nonfunctional, but often the effect is only to reduce its functional efficiency. Individuals who carry base-substitution mutations may survive even though the affected protein is absolutely essential to life.

More extensive mutations (non-point mutations) all involve the breakage and rejoining of the genetic strands, with gross disruption of the sequence of genetic information. DELETION MUTATIONS completely remove part of the genetic message. Like frame-shift mutations, they are lethal unless they merely affect unnecessary genes or are masked by the presence, in the same cell, of normal copies of the deleted genes. One mechanism by which such mutations might occur is easy to imagine: two breaks might occur in the DNA molecules, and the end pieces might then rejoin, leaving out the DNA between the breaks.

Another mechanism by which deletion mutations might arise would lead simultaneously to the production of a second kind of mutation: a genetic duplication. This would occur if two identical DNA molecules in the same cell—for example, the

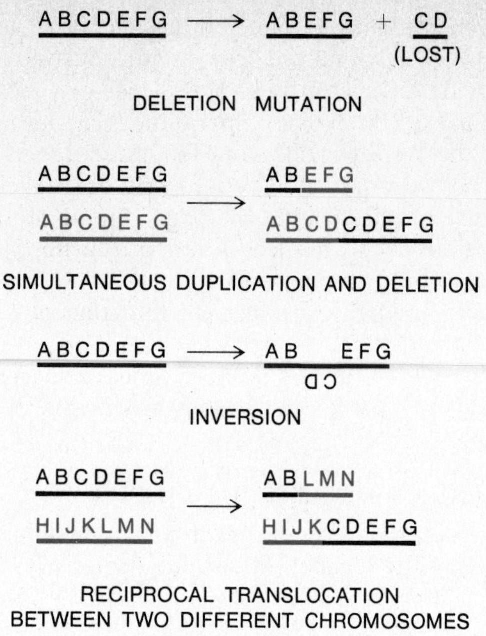

ABCDEFG $\longrightarrow$ ABEFG + CD
(LOST)

DELETION MUTATION

ABCDEFG   ABEFG

$\longrightarrow$

ABCDEFG   ABCDCDEFG

SIMULTANEOUS DUPLICATION AND DELETION

ABCDEFG $\longrightarrow$ AB   EFG
                              CD

INVERSION

ABCDEFG   ABLMN

$\longrightarrow$

HIJKLMN   HIJKCDEFG

RECIPROCAL TRANSLOCATION
BETWEEN TWO DIFFERENT CHROMOSOMES

EXTENSIVE MUTATIONS. In a deletion mutation the chromosome is broken in two places and the segment between is lost. In the rejoined fragment the normal sequence of genes (letters arranged alphabetically) is permanently disrupted in the organism. Three other mechanisms of mutation are also shown. Where mechanism involves two chromosomes, second chromosome is drawn in color.

sister molecules being produced during DNA duplication—were to break at different positions and then reconnect to the wrong partners *(Figure 13)*. One of the two molecules produced by this mechanism would lack a segment of DNA, and the other would have two tandem copies of the information that was deleted from the first *(Box C)*.

Breaking and rejoining of the genetic strands can also lead to rearrangement mutations. The simplest of these, INVERSION, involves the removal of a segment of DNA and its reinsertion in the same location but in the opposite direction. If the breaks leading to an inversion occur within a segment of DNA that codes for a protein, the resulting protein will be drastically altered, and almost certainly nonfunctional. Note that, if a piece of DNA is removed and then put back in the opposite direction, the polarity of the DNA backbone requires that the Watson strand of the inverted segment be joined to the Crick strand of the uninverted part of the molecule.

More complex rearrangements, called TRANSLO-CATIONS, result when a segment of DNA is removed from one position in the genophore or chromosome and is inserted somewhere else. In higher organisms, where there are several chromosomes, this sort of rearrangement might involve the exchange of segments between two chromosomes or the transfer of a bit of one chromosome to another. In procaryotes, which so far as is known always have genetic complements consisting of a single DNA molecule, translocations move segments of the DNA from one part of the molecule to another.

C   The origin of new genes

The great majority of mutations are harmful to the organism that carries them, but once in a while a mutation improves the organism's adaptation to its ecological niche or enables it to invade a new niche. Duplication mutations may be the source of "extra" genes. The more complex creatures living on this planet seem to have more genes than the simplest creatures. Man, for example, has 1,000 times more genetic material than a bacterium. How do new genes arise in evolution? If whole genes are sometimes duplicated by the mechanism described in the text, the bearer of the duplication would have a surplus of genetic information that might be turned to good use. Subsequent mutations in one of the two copies of the gene would have no adverse effect on survival because the other copy of the gene would continue to turn out functional protein. Mutation after mutation could occur in the extra gene without ill effect. If the random accumulation of mutations in the extra gene should produce some useful message, natural selection would take full advantage of it. The result might be a new organism, an organism of greater complexity and versatility than its ancestors.

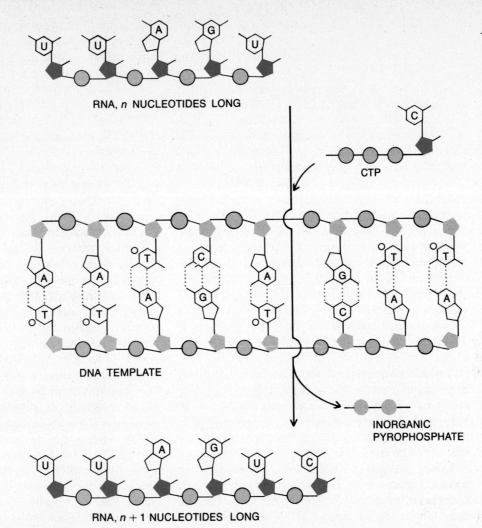

RNA, *n* NUCLEOTIDES LONG

CTP

DNA TEMPLATE

INORGANIC
PYROPHOSPHATE

RNA, *n* + 1 NUCLEOTIDES LONG

14 TRANSCRIPTION BY RNA POLYMERASE.
Polymerase (vertical arrow) moves along DNA
template. Here the upper strand of DNA is being
transcribed. The resulting RNA is an exact copy of
nontranscribed strand of DNA. Structures at top and
bottom show addition of one base (cytosine) to a
growing strand of RNA. Cytosine is originally present
as cytidine triphosphate (CTP).

DECODING THE GENETIC INFORMATION

In double-stranded DNA, all of the genetic information is really present on each strand, but in
complementary form. We may ask whether both
the information from one strand and the complementary information on the other strand are
used for the synthesis of protein. Actually this
would place almost impossible constraints on protein structure. It would be rather like trying to devise a sentence that made good sense in English,
and also made good sense in Russian when read
backward in the mirror. Therefore it is not surprising that only one of the strands provides information for protein synthesis.

Even this one strand is not used directly in the
process of protein synthesis. Its information is
transferred by synthesizing a complementary
molecule of RNA, using the DNA strand as a
template. This RNA is called MESSENGER RNA, abbreviated mRNA. The process of synthesizing it is
called TRANSCRIPTION. This word was first applied to
mRNA synthesis by Sol Spiegelman, who emphasized that the first step in guiding the synthesis
of proteins was merely a change of information
from DNA to RNA. RNA differs from DNA only in
that in contains the sugar, ribose, in place of
deoxyribose and the base, uracil, in place of
thymine. Uracil forms hydrogen bonds with
adenine just as thymine does. Since both are nucleic acids, the change is comparable to a change of
script. Spiegelman contrasted this with the next
step, in which the language itself is changed from

the nucleotides in RNA to the amino acids in proteins, a process he called TRANSLATION. The terms are so vivid and appealing that they have stuck, no doubt permanently.

Superficially the process of transcription is very much like the process of DNA synthesis. An enzyme called RNA polymerase attaches to the DNA molecule it is to copy, somehow choosing the correct strand of the double helix. It then moves along that strand, matching a particular base in the DNA strand (say, G) with the complementary base of a molecule of (ribo)nucleoside triphosphate, namely, cytidine triphosphate (CTP). The two terminal phosphate groups of CTP are split off, while the innermost phosphate becomes joined in an internucleotide (phosphodiester) bond with the 3' hydroxyl of the nucleotide put into the growing chain just ahead of it. This adds a new (ribo)nucleotide to the chain. The enzyme moves one more (deoxyribo)nucleotide down the DNA chain, say to a T. It then pairs the complementary base, A, to it in the form of ATP, and so on (_Figure 14_). There is still much to be learned about transcription: how RNA polymerase picks the right strand of DNA to copy, how it knows where to start and where to stop, and even how the tightly-coiled double helix of DNA unravels enough to permit one of the strands to be copied.

The translation of mRNA into protein is much more complicated. The formation of peptide bonds is thermodynamically an uphill reaction. The cell must invest energy from ATP in the synthesis of proteins. This process, called activation of amino acids, occurs in free solution in the cell. The following step, the decoding of the messenger, occurs on an elaborate workbench —the RIBOSOME. Any ribosome can act as the workbench for the synthesis of any protein, since the instructions are built into the molecular structure of mRNA.

The soluble phase of the cell contains some 20 different activating enzymes, each one specific for one of the 20 kinds of amino acid that must be incorporated into proteins. Each of these enzymes participates in a reaction of the following sort:

$$\text{activating enzyme 1} + \underset{\substack{| \\ NH_2}}{R_1CH} -COOH + \text{A-p-p-p} \rightarrow$$

AMINO ACID   ATP

$$\underset{\substack{| \\ NH_2}}{R_1CH} -CO-p-A \cdots \text{activating enzyme 1} + \text{p-p}$$

AMP-AMINO ACID

A most important feature of this reaction is that the high-energy AMP-amino acid formed does not drift off the activating enzyme, but is firmly associated with it until the next step.

This next step involves a very special sort of RNA, called TRANSFER RNA, abbreviated tRNA. Compared with other kinds of RNA in the cell, each tRNA is quite a small molecule, consisting of a chain of about 75 nucleotides. This chain has the remarkable property of being able to wind up on itself so that large portions of it consist of a double helix of RNA reminiscent of the Watson-Crick structure of DNA. How can a single chain of nucleotides behave in such a manner? The answer is to build the molecule so that one part of it is a mirror image of the other. Figure 15 shows that a simple nucleotide sequence can easily be designed to take on a hairpin shape, and that unpaired regions, free to pair with something else, can be built into the sequence. Actually tRNA molecules have a number of alternating paired and unpaired regions. Drawn in two dimensions, they look more like three-leaf clovers than hairpins, but it is easier to diagram them as hairpins. The structure of tRNA molecules enables them to "recognize" both a specific activating enzyme with its bound AMP-amino acid, and a particular triplet of nucleotides in mRNA. These mRNA triplets are called CODONS. To simplify the picture temporarily, imagine that there are only 20 codons, one for each amino acid. (There are really 64 codons.) There is a separate kind of tRNA for each of these codons, and each of these tRNA's also recognizes one specific activating enzyme, with its bound AMP-amino acid. The second and final reaction for activation of amino acid can be summarized by the equation

$$\underset{\substack{| \\ NH_2}}{R_1} -CH-CO-p-A \cdots \text{activating enzyme} + \text{tRNA}_1 \rightarrow$$

$$\underset{\substack{| \\ NH_2}}{R_1} -CH-CO-\text{tRNA}_1 + \text{activating enzyme} + \text{AMP}$$

"CHARGED" tRNA₁

The amino acid has been attached by an ester bond to the terminal nucleotide of tRNA which bears the free 3'-hydroxyl group. This happens to be an unusually high-energy bond for an ester. It preserves enough energy from the overall conversion of ATP to AMP in the above two reactions to allow the synthesis of a peptide bond.

The mRNA, which attaches itself to the

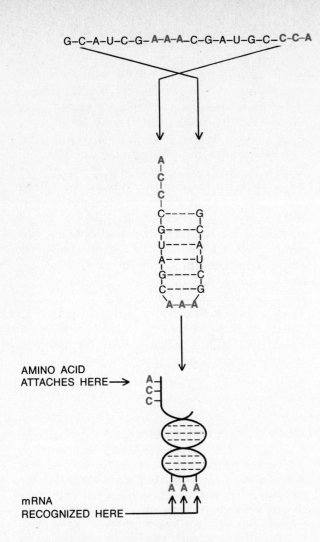

G–C–A–U–C–G–A–A–A–C–G–A–U–G–C–C–C–A

A
C
C
C– – – –G
G– – – –C
U– – – –A
A– – – –U
G– – – –C
C– – – –G
A–A–A

AMINO ACID
ATTACHES HERE→
A
C
C

mRNA
RECOGNIZED HERE
A A A

15 SELF-WINDING HELIX arises as a result of
Watson-Crick pairing. Diagram depicts an imaginary
simplified molecule of tRNA. Color indicates initial
and final position of two triplets.

ribosomal workbench, can be thought of as a series
of codons with no punctuation marks between
them. According to the rules of Watson-Crick pairing, each codon in mRNA will have an attraction
for a certain triplet (called an ANTICODON) at the unpaired hairpin turn of one kind of tRNA. This will
cause that tRNA, charged with its specific amino
acid, to become bound to the messenger. A series
of codons in mRNA can thus arrange amino acids
in a definite order. All that remains is to attach the
amino acids to one another.

The process just described is the key to translation. The crucial molecule is the tRNA. This
molecule must be recognized by the specific ac-

tivating enzyme on one hand, and by the messenger on the other. The situation is analogous to
the way a human and a car ignition recognize each
other, with tRNA being the key. The driver recognizes the key by its octagonal plate at one end, and
the car recognizes the key by the indentations in its
other end. The result is decoding.

How are the amino acids added to a growing
protein chain? Protein synthesis starts at the end
that will ultimately bear the free amino group. The
laying down of the first amino acid is too complicated to discuss here. The mechanism differs from
the one that governs the addition of all the subsequent amino acids. The mechanism also differs
between procaryotes and eucaryotes.

Once the chain has been initiated it will grow,
one amino acid at a time, toward the end that will
ultimately bear the free carboxyl group. At any
moment during growth, the most recently added
amino acid (n) will still be attached to its tRNA.
The tRNA, in turn, will still be paired by its anticodon to the n codon of mRNA. The n + 1 amino
acid, also attached to a tRNA, then approaches,
and the anticodon of this tRNA pairs with the n +
1 codon. Following this, the n amino acid lets go of
its tRNA and attaches to the free amino group of
the n + 1 amino acid, forming a peptide bond. The
tRNA which has lost its amino acid drifts out of the
way. The ribosome then moves along the messenger a distance of three nucleotides, putting the
n + 1 amino acid (still attached to its tRNA) in the
position on the ribosome formerly occupied by the
n amino acid. The machinery is then ready
for the n + 2 amino acid, and the whole cycle repeats *(Figure 16)*. This mechanism has several implications:

1. Each ribosome has binding sites for only two
tRNA molecules at a time. One of the tRNA's bears
the string of amino acids that have been joined in
peptide linkage up to that time, and the other
bears the next amino acid that is to be added.

2. The ribosome must move with respect to the
mRNA (and vice-versa).

3. Since there is no punctuation between codons
in the mRNA, the ribosome must start translating
the message in exactly the right place and move
exactly three nucleotides along the message each
time an amino acid is added to the growing
polypeptide chain. If it occasionally moved only
two nucleotides, or moved four nucleotides, the
result would be a shift in the reading frame. The
protein chain past the point where the mistake was
made would be just as useless as the protein made
from a DNA frame-shift mutant.

The sequence of nucleotides that appears in the anticodons as they line up, one after another, on the ribosome resembles a news release spelled out in moving lights on a marquee. The "news" is the nucleotide sequence in the transcribed strand of DNA.

4. The growing polypeptide or protein chain, with its single tRNA at the growing end, is the aggressor molecule in forming the peptide bond; it contains the next high energy bond that will be used in the synthesis. This is fundamentally different from mRNA synthesis, in which the monomer that is to be added contains the high-energy bond for its condensation with the growing RNA chain.

5. The sequence of nucleotides that appears in the anticodons as they line up, one after another, on the ribosome resembles a news release told in moving lights on a marquee. The news is the nucleotide sequence in the transcribed strand of DNA.

Most messenger molecules are much longer than a ribosome, and therefore a single molecule of mRNA can be read by several ribosomes at once. The first ribosome threaded onto one end of the message will be the first to finish translating the message and to drop off the other end. The assemblage of a thread of mRNA with its beadlike ribosomes and their growing polypeptide chains is called a POLYRIBOSOME, or simply a POLYSOME. Large numbers of polysomes, and relatively fewer free ribosomes or ribosomal subunits, are characteristic of cells that are synthesizing proteins very actively. Figure 17 diagrams a polysome in action. Each ribosome is made up of two subunits of different size, a fact that hints at the mechanism by which mRNA and ribosomes become associated. The ribosomes are not really threaded onto the messenger one at a time. When the ribosome is not engaged in translation, its two subunits are actually separate. To start translation, the smaller subunit first binds to the message at its beginning. Then the first amino acid, attached to its tRNA, is added. Then the larger subunit binds to the smaller subunit-mRNA-amino acid-tRNA assembly, and translation begins.

TRANSLATION OF mRNA. Ribosome slides along a **16** **molecule of mRNA, reading the codons that specify the next amino acid to be added to the end of a growing polypeptide chain. Ribosome (large disk) is shown consisting of two subunits, a small one that binds to the mRNA and a large one on which peptide bonds are formed. Two helices at center are molecules of tRNA, shown with their specific anticodon triplets at one end. The polypeptide chain appears at top, with amino acids represented as circles. N is last amino acid on chain, N + 1 the next to be added, and so on.**

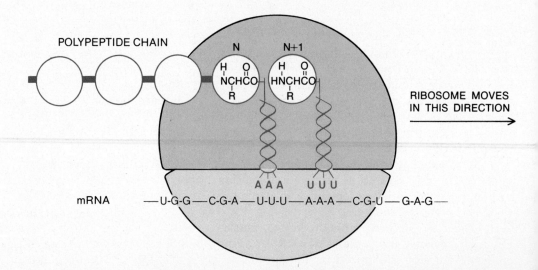

The genetic code involves 64 codons, but there are only 20 amino acids, and apparently 20 activating enzymes. Three codons are signals for terminating the protein chain at its carboxyl end. The other 61 all designate some amino acid. There are more than 20 tRNA types, but some of them can recognize more than one codon. This sounds like a violation of the Watson-Crick pairing rules, and it is. Watson-Crick pairing is not the only way that hydrogen bonds can form between two bases; it is the only way that bonding can occur and still allow the sugar-phosphate portion to be in position to form a new phosphodiester bond. When either DNA duplication or transcription is occurring, the Watson-Crick pairing rules are absolute. However, the binding of anticodon to codon does not involve the formation of new phosphodiester bonds, and the pairing rules are more relaxed. Francis Crick refers to this permissiveness as "wobble." This accounts for the ability of some tRNA types to recognize more than one codon.

PUNCTUATION IN PROTEIN SYNTHESIS

To make a protein of a definite size, the cell not only must put the amino acids in the right order, it must also initiate and terminate the chain in the right places. Since a molecule of mRNA may contain the information to make several proteins, start and stop signals are necessary to avoid synthesizing just a single gigantic protein. The nature of these signals has been described in part. The mRNA region between a start and a stop signal codes for an unbroken polypeptide or protein chain. The DNA corresponding to this stretch of mRNA is called a GENE, or a CISTRON. The word gene is far the older of these two words, and through the years has had a number of less precise meanings. It was for this reason that the word cistron was coined. However, the old word has reasserted itself, and is now used as a synonym for cistron. We shall be using these words a great deal, and it is important that they be understood as the unit which codes for a single polypeptide chain.

UNIVERSALITY OF THE CODE

Figure 18 shows the genetic code of *E. coli* expressed in mRNA language (codons). It was worked out painstakingly by a large number of enzymologists and organic chemists, especially Nirenberg, Khorana, and Ochoa. Work with organisms other than *E. coli* soon showed that the code was not just a local dialect used by that particular organism. The code proved to be identical

17 POLYSOME in this schematic diagram consists of three ribosomes (large colored disks). The track along which they travel is a molecule of mRNA. Twisted-hairpin structures with three prongs at one end are molecules of tRNA with their anticodon triplets. At center a molecule of tRNA carries the next amino acid to be added to the growing polypeptide chain, which looks like a string of beads. Arrows at bottom indicate direction that the ribosomes move. Ribosome at right, which has traveled farthest along the mRNA molecule, has longest polypeptide chain.

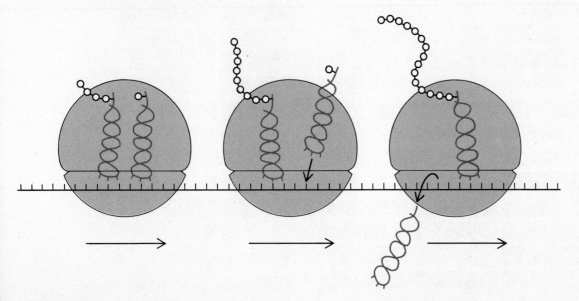

in a eucaryotic bread mold, *Neurospora crassa*, and in yeast. mRNA has been isolated from the immature red blood cells of rabbits; such cells make almost nothing but hemoglobin, and much of their messenger is messenger for hemoglobin. When the rabbit messenger was combined with ribosomes, activating enzymes, and tRNA's from *E. coli*, the *E. coli* machinery proceeded to translate the rabbit mRNA into regular rabbit hemoglobin. This experiment, and analogous experiments in algae and a wide assortment of organisms suggest that all earthly creatures use a similar or probably identical code. Of course, this is not certain; the code has not yet been studied in snapdragons, or armadillos, or in most creatures. But no surprises are expected any more, and the matter is usually considered closed.

How did the genetic code originate? There is nothing chemically obvious about why the codon UUU should mean phenylalanine, for instance, since there are no clear-cut affinities between the codons (or the anticodons) and the amino acids themselves. But such affinities may exist under conditions of pH, temperature, or concentration that have not yet been tried. If so, the code in Figure 18 may not be just one of millions of equally likely codes, and, if life is found in other planetary systems, it may employ a similar or identical code. The fact that, on this planet, the code seems to

GENETIC CODE consists of three-letter words called triplets. Letters represent nucleotide bases uracil, cytosine, adenine and guanine. Each triplet specifies a single amino acid or a single instruction such as "end chain." To decode a triplet, read this grid across (left to right), then down, then across (right to left). Thus the triplet AUG designates the amino acid methionine (dark colored block). Expressed here in the language of mRNA, code is the same in all organisms, from bacteria to man.

FIRST LETTER	SECOND LETTER				THIRD LETTER
	U	C	A	G	
U	PHENYLALANINE	SERINE	TYROSINE	CYSTEINE	U
U	PHENYLALANINE	SERINE	TYROSINE	CYSTEINE	C
U	LEUCINE	SERINE	(END CHAIN)	(END CHAIN)	A
U	LEUCINE	SERINE	(END CHAIN)	TRYPTOPHAN	G
C	LEUCINE	PROLINE	HISTIDINE	ARGININE	U
C	LEUCINE	PROLINE	HISTIDINE	ARGININE	C
C	LEUCINE	PROLINE	GLUTAMINE	ARGININE	A
C	LEUCINE	PROLINE	GLUTAMINE	ARGININE	G
A	ISOLEUCINE	THREONINE	ASPARAGINE	SERINE	U
A	ISOLEUCINE	THREONINE	ASPARAGINE	SERINE	C
A	ISOLEUCINE	THREONINE	LYSINE	ARGININE	A
A	METHIONINE	THREONINE	LYSINE	ARGININE	G
G	VALINE	ALANINE	ASPARTIC ACID	GLYCINE	U
G	VALINE	ALANINE	ASPARTIC ACID	GLYCINE	C
G	VALINE	ALANINE	GLUTAMIC ACID	GLYCINE	A
G	VALINE	ALANINE	GLUTAMIC ACID	GLYCINE	G

have remained the same in various organisms while all other cellular components have diverged through evolution should not, however, be taken as evidence that our code is the only one that can work.

READINGS

B. Lewin, *Gene Expression*, New York, John Wiley and Sons, 1974. A two-volume review of gene expression and gene regulation. The first volume deals with bacteria, the second with eucaryotes. Rich in experimental detail.

G.S. Stent, *Molecular Genetics: An Introductory Narrative*, San Francisco, W.H. Freeman and Company, 1971. By one of the most articulate and witty investigators of molecular biology.

J.D. Watson, *The Molecular Biology of the Gene*, 3rd Edition, Menlo Park, CA, W.A. Benjamin, 1976. A brilliant profile of the gene, written in deceptively simple language by one of the discoverers of the structure of DNA.

C.R. Woese, *The Genetic Code: The Molecular Basis for Genetic Expression*, New York, Harper & Row, 1967. A thorough summary of the subject. Everything from the genesis of the code in the primordial sea to its translation in modern test tubes. Difficult but rewarding reading.

At one time, students of evolution thought that bacteria resembled some primitive forms of life. But a bacterium, with its thousands of genes and enzymes and its remarkably refined adaptations to its environment, is far from primitive. —S. E. Luria

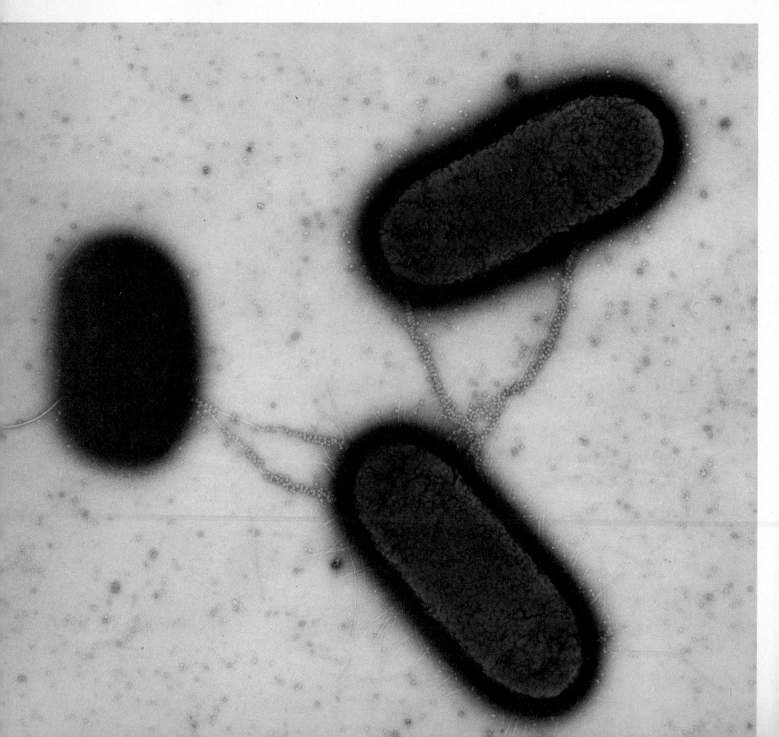

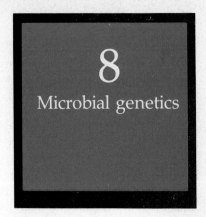

Microorganisms are the darlings of genetic research, and for good reason. First, they are simple. A typical bacterium contains about one thousandth as much DNA as a single human cell, and a typical bacteriophage virus contains perhaps one hundredth as much DNA as a bacterium. Because DNA contains the information for manufacturing all of the molecules in an organism, unraveling the structure of this giant molecule would be a logical first step in working out a complete chemical description of a living system. If the goal of biological research is to understand the processes of life in terms of chemistry and physics, the place to start would not be a human or some other large complex animal or plant, but with a microorganism. Second, microorganisms are small. In many kinds of experiment it is necessary to use extremely large numbers of individual organisms, so that smallness itself becomes a virtue. A single milliliter of medium can contain more than 10^9 E. coli cells, or 10^{11} bacteriophage particles, and costs less than one cent. By contrast, 10^9 mice cost approximately 10^9 dollars, and would require a cage that would cover about three square miles. A third advantage is the short generation time of microorganisms. A culture of E. coli can be grown under conditions that allow it to double about every half-hour. Starting with one bacterium, a researcher can grow in less than a day more bacteria than he can possibly use. Genetic experiments with microorganisms can be done in a day or two, while an experiment

CONJUGATING BACTERIA. Male bacterium at bottom of electron micrograph on opposite page is mating simultaneously with two females. Four long projections called pili connect male to females. Through the pili the male transfers DNA to the females. Tiny beadlike structures along the pili are bacteriophage viruses.

that involves breeding mice requires two months before the progeny can be used for a further experiment. This alone is sufficient to make microbes the organisms of choice for certain types of experiments. Finally, analysis of the inheritance patterns of microorganisms, combined with chemical and physical studies, reveals a sharp picture of the structure of the gene.

Microorganisms lend themselves extremely well to the isolation of mutants that have simple, definable errors in metabolism. Generally speaking, these are mutants that have lost the activity of a single enzyme, and hence fail to make the product of a particular metabolic pathway. If this product is not furnished in the medium, the organisms will not grow. Such nutritional mutants exist in multicellular animals and plants, but they are few and far between. It is difficult to handle large organisms in sufficient numbers to isolate mutants of this sort. Even more important, large organisms almost always have two copies of every gene, so that a mutation in one of the copies is masked by the presence of a normal gene.

True genetic analysis can only be done with organisms that have some form of sexuality. Sexuality, in its broadest sense, means that an individual can have more than one parent. For many years, it was thought that bacteria had no sexual processes, and reproduced only by simple division, so that each bacterium had only one parent. We now know, however, that there are at least three different means by which genetic material from two different bacteria can end up in one individual. CONJUGATION involves actual contact between two bacteria, during which DNA is transferred from one bacterium to the other. TRANSDUCTION involves transfer of DNA from one bacterium to another by a nonlethal bacteriophage. A third process, TRANSFORMATION, was described in Chapter 7. These processes may seem very different from the sexual

151

A Sequencing Macromolecules

To understand exactly how a macromolecule works, one must know the detailed structure of the molecule. What is the sequence of amino acids in a particular protein? Or the sequence of bases in a particular nucleic acid? One way to approach this problem is to lop off one amino acid (or base) at a time from the end of the molecule. The investigator notes which amino acid (or base) comes off first, which comes off second, and so on down the whole length of the molecule. This sequential removal is much easier for small molecules than for large ones. Large molecules are often broken up into fragments which are separated from one another and then sequenced individually. From the sequences of a large number of these small fragments, one can deduce the structure of the original molecules. The process is very much like solving a puzzle, as illustrated by the following anecdote.

A number of guests at a Chinese dinner discovered that they had all received identical messages in their fortune cookies. Feeling cheated and annoyed, they tore up their fortune strips and threw them on the floor. The janitor who swept up the scraps heard what had happened at the dinner, but not what the message said. He examined the scraps of paper. Some had just one word on them; others had two or three. For simplicity, let us consider eleven three-word scraps:

1. cookie says; it
2. not what your
3. your neighbor's cookie
4. the same as
5. Ask not what
6. neighbor's cookie says;
7. says; it says
8. it says the
9. what your neighbor's
10. same as yours.
11. says the same

With a little effort, the janitor was able to reconstruct the original fortune.

The basic requirements for determining the sequence of groups in a giant molecule are comparable to those for solving the fortune-cookie message. First, one must have an unadulterated preparation of molecules. Suppose that 25 percent of the fortunes had read, "Your lap is covered with crumbs; that's the way the fortune cookie crumbles." Obviously the job of reconstructing the most common message —or both messages —would be more difficult because it would not be immediately obvious which fragments came from the same fortune and which came from the two different fortunes. Second, one must be able to separate the fragments from one another. Third, one must be able to read the sequence within each fragment. Fourth, there must be overlapping fragments. If everyone tore his fortune in exactly the same way, there would have been no way to deduce the order of the fragments within the original fortune (except, of course, by looking for arrangements that make sense, a method not available to investigators of amino-acid sequences). Fifth, it is very helpful to be able to identify the first and last fragments. Chemical methods are available for labeling the first and last groups of protein and DNA molecules. In the fortune-cookie message, the capital letter in the first word and the period at the end of the sentence serve as labels.

Recently investigators have devised a new way of determining the sequence of bases in DNA, which works for DNA molecules up to about 100 nucleotides long. One must start with a homogeneous preparation, which may be eigher single- or double-stranded; let us consider only single-stranded DNA. With the help of an enzyme, one end of the molecule is labeled with a radioactive isotope of phosphorus. The labeled molecules are then treated with a reagent that cleaves the chain next to one particular base, such as guanine. The attack by the reagent is deliberately kept very weak; even though a given molecule may contain perhaps 20 guanines, most molecules in the sample will only be broken at one of the 20 possible places, and some will not be broken at all.

A second sample is treated with a reagent that breaks the chain at adenine, a third with one that breaks at cytosine, and a fourth with one that breaks at thymine. The result is four separate and different mixtures of fragments. Each mixture is then subjected to a procedure called gel electrophoresis, which separates the fragments very precisely according to their rate of migration in an electric field. Finally the gel is placed on a sheet of x-ray film; each fragment that contains the original endgroup of the molecule is detected by its radioactivity, which darkens the film underneath it. All other fragments go undetected.

How this works can be seen more easily by sequencing the "cookie crumbles" fortune, as shown in the diagram below. Suppose that the capital Y in "Your . . . ." is the radioactive atom that labels each fragment, and that we treat samples of the sentence with four different reagents. One breaks the sequence at *o*, another at *e*, and so on. (To do a complete job of sequencing a sample of English prose, one would need 26 different reagents.) When the fragments terminating in *o* are examined, we see that they have lengths of 2, 11, 42, 49, and 50 letters. If we number the letters in the fortune sequentially we will see that the letter *o* occupies those positions in the sentence. In a similar way, the letter *e* must occupy positions 13, 15, 34, 40, 47, 53, and 60 in the *e* channel. Notice in the diagram that all four channels show fragments that are 61 letters long. These are sentences that remained uncleaved under the very mild conditions that were used.

This technique has made it possible to determine the sequence of a DNA molecule in a few days. Three or four years ago, most biologists would have guessed that it would take 20 years to determine the sequence of bases in the simplest DNA viruses. That has already been accomplished; right now, no geneticist would bet on how long it will take, for example, to determine the total genome of the fruit fly.

DNA-SEQUENCING TECHNIQUE is illustrated by applying it to the fortune-cookie problem. Here the capital letter Y represents the labeled end group. Dark colored bands at left show starting positions for gel electrophoresis, and light-colored bands show final positions, corresponding to final letters in fragments at top.

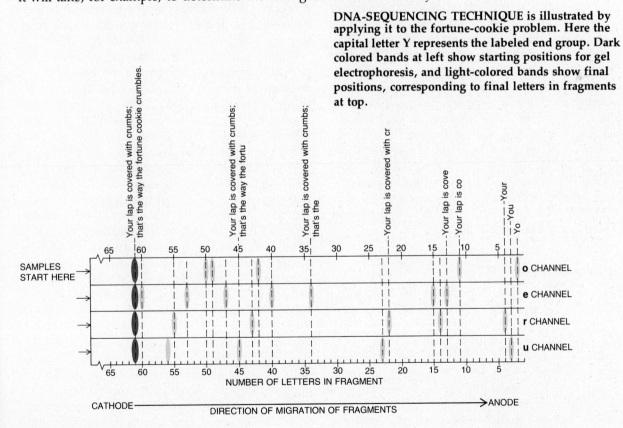

processes of higher organisms, but they accomplish the same purpose, namely, the reshuffling of genetic materials.

CHEMICAL ANALYSIS

In the quest for a complete understanding of the anatomy and physiology of an organism, the first step would be to determine the sequence of nucleotides in its DNA. Protein chemists have developed their art so highly that, starting with a purified protein, they can determine the complete amino acid sequence from one end to the other. Until recently, chemists who tried to determine the nucleotide sequence of DNA molecules were not nearly so successful. It is usually very difficult to isolate even a single kind of DNA molecule, with the exception of viral DNAs. DNA molecules are so large that they usually break when a cell is broken open, or even when DNA solutions are poured from tube to tube. Moreover, DNA contains only four different bases, and they are not as easy to tell apart by specific chemical reagents as are many of the amino acids. The analysis of amino acid sequences in proteins generally involves breaking the molecules at well-defined points, to give medium-sized fragments that can be further analyzed. This fragmentation is accomplished by specific enzymes, for example, trypsin. Conversely, most of the enzymes that snip DNA into fragments can break the molecule at virtually any point, yielding a hopeless mixture of fragments of different size. Recently special enzymes have been found that hydrolyze DNA only at very specific sequences of bases. Called RESTRICTION ENDONUCLEASES, these enzymes have already proved to be very useful for determining nucleotide sequences in viral DNA *(Box A)*. The task of finding the complete blueprint of nucleotide sequences in anything more complex than a virus is still beyond present technical capabilities.

RECOMBINATION

Most of our present knowledge about the organization of DNA in living systems has been obtained not by chemical analysis, but by genetic analysis. The geneticist starts with a "normal" organism, which he calls WILD TYPE. He can then introduce mutations and identify them by looking for individuals with an altered phenotype. These may tell him something about the physiology of the organism, but they do not directly tell him anything about the organization of the DNA. Specifically, he

> The geneticist starts with a normal organism, which he calls wild type. He can introduce mutations and identify them by finding individuals with altered phenotypes.

has no idea where any of the mutations in the DNA molecule have occurred, or how far they are from one another. He would like to be able to MAP these mutations —to find out the order in which they occur on the DNA molecule, and the distances between them.

The way he does this is in principle very simple. He performs a CROSS, or mating, between two individuals that carry two different mutations, and examines the progeny. There are four possible types of offspring. Some will be just like one mutant parent. Some will be just like the other mutant parent. These are called parental-type offspring. Some progeny will carry both mutations, and are called double mutants. Still others will carry neither mutation, and will be identical with the original wild type from which the mutants were derived. The double-mutant and wild-type progeny are called RECOMBINANTS *(Figure 1)*.

The frequency of recombinant progeny compared with that of parental-type progeny is a measure of the distance on the DNA molecule between the two mutations. Why is this so? If two similar DNA molecules are present in the same cell, there is a certain probability that these two molecules will break and exchange parts. The points of exchange will be random, but both DNA molecules are broken at equivalent positions, so that there is no net gain or loss of genetic information in either molecule. If the two DNA molecules bear mutations at different points, there is a certain likelihood that the breakage and exchange will occur between these two points, producing a wild-type recombinant, and a double mutant as offspring. The likelihood that the breakage will occur between the mutations depends on how far apart they are. The process of recombination can be compared to the splicing of tape recordings by a deaf editor. This editor can line up the two tapes side by side so that they correspond to one another, note by note. Here and there, he cuts across both tapes with a scissors and rejoins them with a splice, producing a recombinant tape without ever listening to it. Now suppose that the vio-

3 Recombination

The molecular mechanism of recombination is poorly understood at this time. How cells manage to cut and splice chromosomes so precisely is a mystery. It seems likely that the Watson-Crick pairing rules that govern the precise duplication of genetic information during DNA synthesis and the matching of tRNAs to codons in mRNA are also responsible for the perfect matching of chromosomal fragments when they rejoin during recombination. A highly speculative diagram of this process appears below.

The two sets of Watson-Crick strands shown at top are the parental molecules. The upper molecule contains a mutation (m_1): a G···C has replaced an A···T. The lower molecule contains a different mutation (m_2): a C···G has replaced a T···A. After some unknown intermediate steps, the two molecules are cut, with their Watson-Crick strands broken in different places (middle). The fragments then pair in a new arrangement by forming hydrogen bonds between complementary single strands.

After a few more unknown events, the recombined strands have restored their continuity (bottom). Extra nucleotides have been discarded, missing nucleotides have been inserted, and interruptions in the Watson-Crick strands have been mended. The nucleotides shown in boldface type were not present in either of the parental molecules, but are replacements drawn from the pool of free nucleotides within the cell.

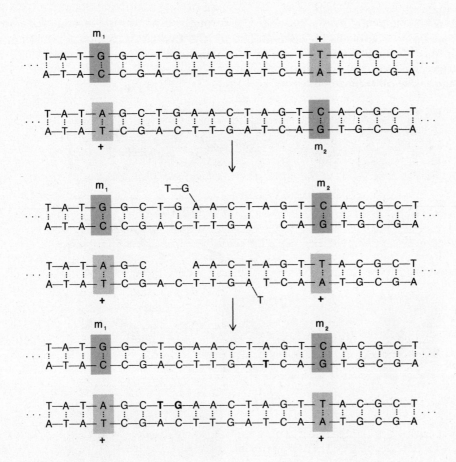

PROGENY

WILD TYPE (NO SPECIAL REQUIREMENTS) HIS+ LEU+ (EVERYTHING ELSE+)	MUTATION 1 → MUTANT PARENT 1 (REQUIRES HISTIDINE) HIS− LEU+ (EVERYTHING ELSE+)
WILD TYPE (NO SPECIAL REQUIREMENTS) HIS+ LEU+ (EVERYTHING ELSE+)	MUTATION 2 → MUTANT PARENT 2 (REQUIRES LEUCINE) HIS+ LEU− (EVERYTHING ELSE+)

CROSS →

PARENTAL TYPE 1
HIS− LEU+ (REQUIRES HISTIDINE)

PARENTAL TYPE 2
HIS+ LEU− (REQUIRES LEUCINE)

WILD TYPE RECOMBINANT
HIS+ LEU+ (NO SPECIAL REQUIREMENTS)

DOUBLE MUTANT TYPE RECOMBINANT
HIS− LEU− (REQUIRES BOTH HISTIDINE AND LEUCINE)

1 TYPICAL CROSS used in genetic analysis involves two mutants, in this case one that requires the amino acid histidine and another that requires the amino acid leucine. The progeny are examined to determine the frequency of the two parental types and of the two recombinant types. The superscript (+) indicates the wild-type allele of a gene, and the superscript (−) indicates the mutant allele.

2 SPLICED TAPE RECORDINGS illustrate the process of genetic recombination. Illustration reproduces the violin solo from Max Bruch's G Minor Violin Concerto. First and second tapes contain errors by the performer (notes in solid color) at different points in the score. By splicing the two tapes together, editor can produce either a perfect tape or one containing both errors.

lin soloist has a borrowed bow on the day that his orchestra records a concerto, and his instrument squawks in the middle of his first solo when the first tape is made. The conductor gamely carries on to the end of the movement, and then the orchestra starts over for a second try. As might be expected, the violinist does it again the next time through. If he muffs the same note on the second try the two tapes cannot possibly be put together to make a perfect performance. But if he muffs

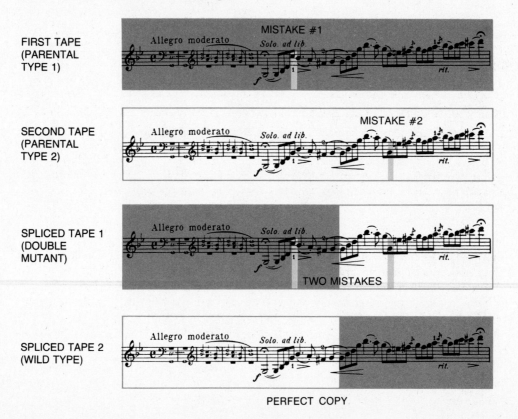

FIRST TAPE (PARENTAL TYPE 1) — Allegro moderato — Solo. ad lib. — MISTAKE #1 — rit.

SECOND TAPE (PARENTAL TYPE 2) — Allegro moderato — Solo. ad lib. — MISTAKE #2 — rit.

SPLICED TAPE 1 (DOUBLE MUTANT) — Allegro moderato — Solo. ad lib. — TWO MISTAKES — rit.

SPLICED TAPE 2 (WILD TYPE) — Allegro moderato — Solo. ad lib. — rit.

PERFECT COPY

different notes on the two tapes, the deaf editor, using his scissors and splicer, will occasionally produce a flawless recording. And, of course, he will at the same time produce the reciprocal product—a recording in which both mistakes are present. How often he will get a flawless recording, and a recording with two mistakes, will depend on how far apart the two mistakes are *(Figure 2)*.

The genophores of viruses and bacteria (and the chromosomes of eucaryotes) seem to recombine in just this way, breaking at perfectly corresponding positions, exchanging pieces of genetic information, and then putting them back together again. We would never know anything had happened if we had not marked the genes with mutations in order to detect the informational exchange *(Box B)*.

PHAGE EXPERIMENTS

The modern concept of the gene and most of the vocabulary in which that concept is expressed grew out of experiments with the bacteriophage T4 *(Figure 3)*. In order to understand the experiments, it is necessary to know how bacteriophages are counted. A high-titer culture of T4 may contain as many as 10^{11} or 10^{12} phages per milliliter. In order to reduce this to a countable number, the culture is carefully diluted, a record being kept of the exact amount of dilution. The diluted phage suspension is then mixed with a very large number of bacteria so that the mixture contains about 100 phage particles and about 10^8 bacteria. The mixture is spread over the surface of sterile nutrient medium gelled with a non-nutrient substance called agar. The bacteria and the phages are now essentially immobilized on the surface of the solid medium so that each bacterium gives rise to a localized colony of bacteria, forming a small bacterial heap on the spot in which the original bacterium was planted. Because there is an immense number of bacteria on the surface of the agar in the dish, this bacterial growth produces a confluent, opaque "lawn" of bacteria over the agar surface. While the bacteria are growing, the phages are growing also. Each of the 100 or so phages in the original mixture infects a bacterium. The infection leads in a short time to the lysis of the bacterium and the release of about 100 progeny phages onto the surface of the agar gel. These progeny phages cannot crawl over the surface—they diffuse slowly until they encounter another bacterium close to the first, then infect it and go through another cycle of infection. The process continues, and about 12 hours later, the

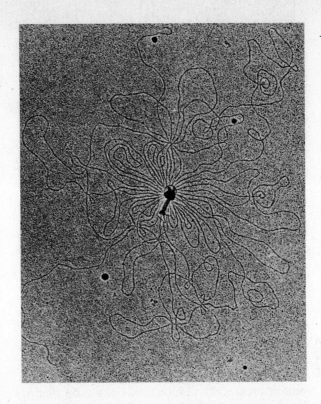

BACTERIOPHAGE VIRUS at the center of this electron 3 micrograph lies tangled in a string of its own DNA. This classic picture gives some idea of the enormous length of a single DNA molecule.

bacteria have formed a lawn that is broken only by small, clear circles called plaques *(Figure 4)*. Each of these plaques is the result of destruction of bacteria by the progeny of a single phage particle. By counting the plaques, the number of phages present in the suspension can be determined. By calculating the dilution factor the number of phages present in the original suspension is discovered.

As soon as the plaque assay technique was invented, it became evident that phage plaques did not all look the same. Compared to the great majority of plaques, some rare plaques were big; some were small; some had smooth edges; some had fuzzy edges; some had haloes; some were turbid; some were clear; and so forth. If phages were isolated from the rare odd-looking plaques and were plated again (that is, spread with bacteria on the agar surface of a petri "plate"), it was found that the abnormalities of the original plaque were usually faithfully reproduced in the plaques formed by the offspring phages. This means that plaque morphology is a heritable characteristic of the phage and that mutations occur that affect plaque morphology in a variety of ways.

One of the kinds of mutation affecting plaque morphology was called the *r* mutation (for rapid lysis). Phages carrying such mutations make large plaques with a characteristic pattern of rings. These *r* mutations were the object of a brilliant study by Seymour Benzer, working at Purdue University in the 1950's. Benzer isolated many independently-produced *r* mutants and found that some of them, the so-called *r*II mutants, had properties that made them particularly favorable for genetic study; namely, they could grow perfectly well on bacteria of one standard strain of *E. coli*, strain B, but could not grow at all on another standard strain of *E. coli*, strain K. The wild type phage grew equally well on both B and K. This provided a selective method for determining whether a phage had the *r*II phenotype or the wild-type phenotype (the notation for wild-type is r^+). The reason that this is called a selective method is that it provides a method for detecting a specific rare type of individual in a mixed population, without even having to look at the common types. One r^+ phage, in a population containing as many as 10^7 or 10^8 *r*II mutant phages, can be detected easily by plating phages on a plate spread with strain K bacteria. The wild-type r^+ phage will form a plaque, and none of the other phages will.

Armed with a large collection of mutations and a selective method for determining the phenotypes of phages, Benzer undertook an intensive analysis of the structural and functional organization of the gene(s) responsible for determination of the *r*II phenotype. The approach was genetical, not chemical. Benzer did not know what the actual molecular function of the gene product(s) might be, and we still do not know. Benzer and his coworkers tried hard to identify the protein product of the gene but never found it. Years later, other workers did solve this puzzle by showing that the product of the *r*II region is a protein that makes up part of the membrane of the bacterium. The fact that Benzer did not need to know this to do his successful analysis shows that one can learn a great deal about the gene without being able to examine anything but a phenotype.

The standard genetic experiment in bac-

teriophages is the phage cross. Bacteria are infected with a mixture of two genetically different phages, using enough phages so that virtually every bacterium is infected with some phages of each genetic type. If there are two *r*II mutant stocks that arose from independent mutations of wild-type phage T4, these can be designated mu-

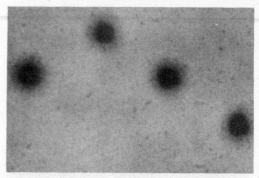

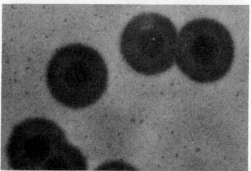

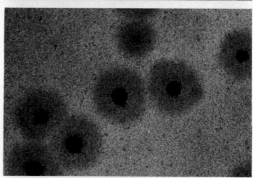

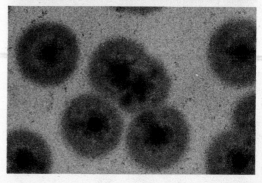

4 PLAQUES on a lawn of bacteria provide researchers with a method for determining both the number and the kind of virus present in a suspension. Shown here are, from top to bottom, plaques of a wild-type T4 bacteriophage virus, an rII mutant, a tu mutant, and a double mutant (rII and tu).

tants r_1 and r_2. E. coli B (the strain in which rII mutants can grow) is then infected with a mixture of the two kinds of mutant phages. The progeny phages that emerge from the mixedly-infected cells after the end of the cycle of phage infection are examined and usually there are some progeny phages that are r^+ in phenotype. That is, they can form plaques on E. coli K bacteria. By special techniques, it can be shown that there are some phages that carry both mutations that were present in the original parents—r_1 and r_2. The number of doubly mutant phages is about the same as the number of wild-type phages. These new types of phages are recombinants. The ability of the two parental phages to give rise to recombinants indicates that the mutations in the two parent phages cannot be located in the same place on the gene. If they were, it would not be possible to break and recombine two mutant genophores in such a way as to produce a nonmutant or doubly-mutant genophore. When a cross of two phage mutants fails to give any wild-type (or doubly-mutant) recombinants, even among a very large number of progeny, we conclude that the two mutants have alterations that affect the same point in their DNA molecules. In general, the test for identical location of two mutations is whether or not wild-type recombinants appear, since there is usually no easy way to select for the double mutants.

Thousands of independently isolated rII mutations were crossed two-by-two, and most pairs did recombine to give wild type. This means that there are hundreds of mutations occupying different positions on the T4 genophore but having identical rII phenotypes. There were, however, pairs of mutations that did not give any wild type when crossed with one another, these presumably being mutations that overlapped.

Among the many mutations examined, some were found that seemed to be extensive. Such mutations failed to recombine with several mutations that were shown by the test to occupy different positions on the genophore. Thus, r_1 and r_2 might recombine with each other to give the wild type, showing that they did not affect the same nucleotide in the DNA molecule. Yet a third mutation, r_3, might fail to give wild-type recombinants with either r_1 or r_2. How could r_3 be identical in position to both of these at once? The answer is that r_3 is a deletion that removes the sites of both r_1 and r_2, so that when it is crossed to either one of them, no perfect copy can be assembled. Extensive deletions virtually never revert to the wild type. Even when very large numbers of phages carrying one of these

suspected deletions were plated on K bacteria they never gave rise to wild-type revertants. Similar examination of the mutations that did not seem to be extensive showed that most of them did give rise to wild-type revertants. By this criterion, Benzer could divide his array of mutants into point mutants and deletion mutants. Some of Benzer's deletion mutations appeared to be very long, failing to recombine with most or all of his point mutations. This suggested that all of the rII mutations occupied a single segment of the DNA molecule that was dispensable, at least for growth on the E. coli B strain.

GENE MAPPING

As has been pointed out, the closer two mutations are to one another on the phage genophore, the less likely they are to produce wild-type recombinant progeny. Suppose two phages of the genotypes $m\ n^+$ and m^+n are crossed. Four kinds of progeny phages are produced in such a cross, the two parental types and the two recombinant types: $m\ n$ and m^+n^+. The RECOMBINATION FREQUENCY in any cross is the number of recombinant offspring (the sum of the two types) divided by the total number of offspring. This frequency differs for different pairs of markers and thus allows summarization of results as a "recombination map." Suppose three mutations, m, n, and p, are crossed in all three pairwise combinations and it is

THREE MARKERS can be mapped by a series of three two-factor crosses. Percentage figures here indicate that the recombination frequency between p and n is slightly less than the sum of the frequencies between p and m and between m and n. 5

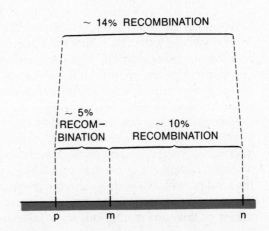

~ 14% RECOMBINATION

~ 5% RECOM-BINATION

~ 10% RECOMBINATION

p   m   n

found that the cross between *m* and *n* gives 10 percent recombinants, that between *m* and *p* gives 5 percent recombinants, and that between *n* and *p* gives 14 percent recombinants. A map of the markers in such a case is shown in Figure 5.

What happens if a series of crosses is done between markers lying farther and farther apart? Will there ever be 100 percent recombinants? The answer is that there will not. It can easily be imagined that the markers might be so far apart that there will virtually always be at least one breakage event between them. When that happens, it is as likely that the rejoining of fragments will put two such markers back together in the recombinant configuration as it is that it will put them back together in the parental configuration. The maximum possible frequency of recombination between two markers would therefore approach 50 percent. Once markers are so far apart that they are almost always separated from one another by at least one break in the genophore, no further increase in the physical distance will measurably increase the frequency of recombination between them. For this reason, this method does not permit mapping of very distant markers. Markers that show frequencies of recombination significantly less than 50 percent are said to be LINKED to each other; markers that show virtually 50 percent recombination are said to be UNLINKED, even if they are really on the same genophore.

Genetic maps are one-dimensional and can be represented as straight lines, very much like a map of the road between two cities.

When Benzer made a map of all his *r*II mutants in this way, he found that all of them were linked, and that all fell within a region that was less than one percent of the recombinational length of the phage genophore. One other very important conclusion has emerged from mapping experiments: that genetic maps are linear, and that distances between mutations are additive. In other words, genetic maps are one-dimensional, and can be represented as straight lines, very much like a map of the road between two cities. This is a research finding, not a mathematical construct. To convince the reader of this, one can try out some fake data that

violate the linearity principle. Imagine three mutations, *m*, *n*, and *p*, with *m* giving ten percent recombination with *p*, *n* giving ten percent recombination with *p*, but *m* giving only five percent recombination with *n*. If one tries to make a map with these data, he ends up, not with a linear map, but with some sort of two-dimensional or branched structure, like the one shown below.

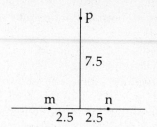

THREE-FACTOR CROSSES

The most reliable way of ordering markers in any organism without having to worry about uncontrolled differences between experiments, is the THREE-FACTOR CROSS. In such a cross, the two parents differ with respect to three mutations. The central marker of the three must be linked to the other two markers, so that breakage separates it from each of them well under 50 percent of the time. Then the probability that the central marker will simultaneously recombine with both flanking markers is roughly the product of the probabilities that it will recombine with each of them singly. Thus, in such a cross, one recombination class is much less frequent than the others, and that is sure to be the double-recombination class. Suppose two phages of genotypes $a\ b\ c$ and $a^+b^+c^+$ are crossed. The progeny from this cross will include eight different genotypic classes. Suppose they occur in the following frequencies:

$a\ b\ c$	28%	
		56%
$a^+b^+c^+$	28%	
$a\ b^+c^+$	12%	
		24%
$a^+b\ c$	12%	
$a\ b\ c^+$	7%	
		14%
a^+b^+c	7%	
$a^+b\ c^+$	3%	
		6%
$a\ b^+c$	3%	

The most common types, totalling 56 percent, are, of course, the progeny that resemble one parent or the other, and require no recombination at all. The two classes that occur with intermediate frequency (24 percent and 14 percent) presumably result from single recombination events. The least frequent class presumably results from double recombination events. These separate the markers at the *b* site from those at the *a* and *c* sites. Therefore it is possible to conclude that *b* was the central marker. The mutations of many species of phages (and of a great many other organisms) have been mapped by this technique.

The *r*II mutations are not the only kinds of mutations in T4 that affect plaque morphology. Other sorts have also been useful, but they have turned out to be widely-separated and to leave large unmarked gaps in the map. Various kinds of mutations, called CONDITIONAL LETHAL MUTATIONS, turned out to be generously distributed about the map and to fill in most of the gaps. One kind of conditional lethal mutation produces a phage whose growth is abnormally temperature-sensitive. Such mutations cause alterations in gene products, so that the affected phages can grow perfectly well on bacteria incubated at 30°C, but cannot grow at all on bacteria incubated at 42°C. This provides a selective method for scoring wild-type phages. A progeny phage population from a cross between two temperature-sensitive mutants is plated, and the plate is placed in an incubator at 42°C. Any plaques that form are wild-type recombinants. By comparing the number of such recombinants to the number of total phages (the plaque count at 30°C incubation), and by assuming that the double-mutant recombinant occurs at the same frequency as the wild type, the frequency of recombinants from the cross can be measured. The *r*II mutations are also conditional lethals, in that they allow the phage to grow in *E. coli* B, but not in *E. coli* K. Mutations such as these, which make possible the selection of very rare wild type recombinants, enable the geneticist to map mutations that are very close together, and to produce a very detailed map. Conditional lethal mutants have also been invaluable in the analysis of gene function.

FUNCTIONAL ANALYSIS

It has been very profitable to look upon a gene as a unit of function. When it is so defined, a gene is more precisely called a CISTRON, a term introduced in Chapter 7 in a different context. If a mutation occurs in a cistron, a single function will be disrupted or destroyed. If two different mutations each disrupt the same function, they lie in the same cistron; if they disrupt two different functions, they lie in different cistrons.

The laboratory procedure that tells us whether two mutations disrupt the same or different functions is called the COMPLEMENTATION (or CIS-TRANS) TEST. It does not depend upon recombinational events, and in fact, can be performed perfectly well in some organisms in which recombination has never been demonstrated. In order to perform the complementation test, one must put two mutant chromosomes into the same cytoplasm under conditions that distinguish wild-type function from mutant function. Once again, Benzer's *r*II system was the source of many ideas. Benzer collected many independent *r*II mutations, each of which was unable to grow in the cytoplasm of *E. coli* K. These did not all occupy identical positions on the genetic map, as demonstrated by recombination analysis. One might now ask whether the *r*II PHENOTYPE always results from disruption of the same function.

First the meaning of wild-type function must be clear. Wild-type function leads to the growth of phage in *E. coli* K, whereas mutant function does not. If wild-type T4 phage are mixed with phage carrying an *r*II mutation and *E. coli* K are infected with the mixture under conditions such that each bacterium is infected with several phages of each type, the result is that the cells do produce phages and that the number of phages of the two types is roughly equal. That is, the presence of wild-type phage in the cell is sufficient to permit the mutant phage to grow as well, and the presence of mutant phage does not interfere with the growth of wild-type phage. This is the genetic definition of DOMINANCE and RECESSIVENESS. When two forms of the gene are present in the same cytoplasm, the course of the infection of the bacterial cell is determined by the dominant phage gene. We say, since the phenotype in this mixed infection is the phenotype of the wild-type phage, that the wild type, *r*⁺, form of the gene is dominant to the *r*II mutant form of the gene and that the *r*II form is recessive. A COMPLEMENTATION TEST WILL GIVE MEANINGFUL RESULTS ONLY WITH TWO RECESSIVE MUTATIONS. This is a very important limitation, and the reader is advised to think about it for a while to be sure to understand why the limitation exists.

The complementation test itself is called the cis-trans test because it consists of two parts. In the "trans" test, two chromosomes (in this discussion "genophore" is included in this term) each carrying one of the two mutations to be tested, are put into the same cytoplasm. One can picture this as follows:

Here, one chromosome carries mutation m_1 and the other carries mutation m_2. They are shown at different positions on the chromosome because we are primarily interested in testing mutations that have been shown to recombine with each other to give wild-type progeny, and such mutations are, by definition, at different places on the chromosome. The + sign on a chromosome indicates that the DNA at the position has the normal, wild-type sequence of nucleotides. The mutations, because they are on opposite chromosomes, are said to be in the trans configuration, a term borrowed from the chemists.

Again, Benzer's experiments with phage T4 will serve well to illustrate these ideas. Benzer tested *r*II mutations that recombined with one another to see whether, in mixed infections of K bacteria (where neither phage could grow by itself), the two phages together were capable of forming

phage progeny. The test here is experimentally simple. The two phages are mixed in sufficiently high concentrations so that when the mixture is spotted on a plate spread with K bacteria, mixed infections of the bacteria are fairly frequent. If the two mutations complement one another—if the two mutant phages can cooperate to produce progeny phages—there will be a clear spot of phage growth on the plate when the plate is examined the next morning. If the two mutations are noncomplementary, there will be no clearing where the phage mixture was spotted. Complementation means that the two mutants, together, possess every function that wild type possesses, and that therefore they must be injured in different functions. To repeat: complementation does not involve recombination to produce a normal chromosome. Both chromosomes remain mutant, but together they give wild-type function. What, then, does it mean if complementation fails to occur? Normally, it means that the two mutants are altered in the same function. This conclusion would be unwarranted if it should turn out that one (or both) of the mutants were actually dominant, so that wild-type function could not possibly be expressed. The control, to make sure that neither of the mutants is dominant, is the second half of the complementation test, or the "cis" test. It is as follows:

6 **COMPLEMENTATION TEST can determine whether two mutations disrupt the same function. The rules of the test, also known as the cis-trans test, are summarized briefly in this diagram.**

If the two chromosomes in the CIS test function together to produce a mutant phenotype, the results of the TRANS test will be meaningless.

If the CIS test gives a wild-type phenotype, continue.

If the two chromosomes in the TRANS test produce a mutant phenotype, the two mutations lie in the same COMPLEMENTATION GROUP or CISTRON. This result means that the two mutations disrupt the same genetic function.

If the two chromosomes in the TRANS TEST function together to produce a WILD-TYPE phenotype, the two mutations lie in different COMPLEMENTATION GROUPS or CISTRONS. This result means that the two mutations affect different genetic functions.

The phenotype of the combined chromosomes in this cis test should be the wild-type phenotype since one of the chromosomes contains no mutations at all. The rules applying to the complementation test are shown in Figure 6.

In the last chapter the cistron was defined as a segment of DNA that gives rise to a single polypeptide. Now it is defined as a unit of function. The unit of function is the unit that produces a single polypeptide. Two defective polypeptides, with defects in different amino acids, usually cannot cooperate to give normal function. Unlike DNA chains, polypeptide chains do not recombine.

A specific example: Benzer found that the mutations in his collection could be assigned to two cistrons. The mutations in cistron A complemented all mutations in cistron B but did not complement any other cistron A mutations. Likewise, the mutations in cistron B complemented all mutations in cistron A, but no other cistron B mutations. When Benzer compared the results of his complementation tests with the results of his recombination tests, he found that all of the cistron A mutations were in one continuous segment of his recombination map and that all of the cistron B mutations were in another. There were no mutations of the B type that lay anywhere in the A segment and no mutations of the A type that lay anywhere in the B segment. There were some deletion mutations that failed to complement either A or B mutations. These deletions were shown to cover portions of both segments.

The complementation experiments were consistent with the following model: There are two phage functions that are required for growth on *E. coli* K and for the formation of plaques with wild-type morphology on lawns of *E. coli* B. Perhaps these functions involve two enzymes that work in a short reaction sequence to produce some small molecule that is essential for growth on K. That is,

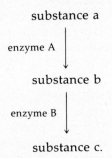

substance a

enzyme A ↓

substance b

enzyme B ↓

substance c.

If substance *c* is required for growth on *E. coli* K, a mutation that destroys the function of either en-

zyme, A or B, will cause a fatal shortage of the required substance. Mutations anywhere in the DNA segment that encodes the A enzyme, for example, may cause the enzyme to be nonfunctional. Thus, two mutations at different positions on the DNA molecule may affect the same enzymatic function. Such mutations will recombine to produce the wild type if crossed in *E. coli* B. However, if the two mutant phages are used to infect *E. coli* K, they will not complement each other because both phages are deficient in the same enzyme. On the other hand, if a K bacterium is infected with a mixture of two phages, one of which is defective in making enzyme A, and the other of which is defective in making enzyme B, the two phages together will grow because each is capable of making the enzyme that the other lacks.

BIOCHEMICAL GENETICS

The principle that one cistron controls the structure of one polypeptide is not a new one. In its older version, this "one cistron, one polypeptide" principle was called the "one gene, one enzyme" hypothesis. The earlier evidence for it did not come from phages, but from mutations in more complex organisms, such as humans.

The first suggestion that enzymes might be subject to genetic alterations was made by A. E. Garrod, an English physician. He suggested in 1909 that alkaptonuria, a human hereditary disease causing darkening of the urine, might be due to a defect in an enzyme that is normally involved in the biological degradation of the amino acid tyrosine. His guess was based on his finding that the urine of affected persons contained a compound that appeared to be derived from tyrosine. This compound could not be found in the urine of normal persons. Garrod postulated that it was actually formed by normal persons, but that they possessed an enzyme that degraded it as fast as it was formed. A few years later, his speculation was confirmed by enzymatic studies on normal and alkaptonuric individuals. One important aspect of Garrod's theory was the implication that by studying mutants one might identify the enzymes engaged in the processes of life. Until the alkaptonuria mutation was subjected to thoughtful analysis, no one had bothered to consider whether there might be an enzyme, or several enzymes, responsible for the biological degradation of tyrosine.

Garrod's speculation bore fruit when George W. Beadle and Edward L. Tatum began in the 1940's to

7   BREAD MOLD, Neurospora crassa, looks like a meshwork of filaments in this scanning electron micrograph. This common mold has become famous as the subject of countless genetic experiments. Magnification, about 750X.

study the salmon-colored bread mold, *Neurospora crassa (Figure 7)*. This is a simple eucaryote that can be grown on a simple, completely defined medium. That is to say, *Neurospora* can grow on a medium in which all of the ingredients are known. All that the mold requires is a few salts, a simple source of nitrogen (such as ammonium chloride), an organic source of energy and carbon (such as glucose), and a single vitamin (biotin). The rest of its cellular components it makes for itself. Beadle and Tatum reasoned that if mutations were capable of altering the functions of enzymes so that the enzymes no longer did their jobs, then mutants of *Neurospora* might be found that were enzymatically unable to make certain cellular compounds. Such mutants would only grow on media to which those compounds were added. Mutants of this type have since been called AUXOTROPHS (increased eaters) as opposed to the wild-type PROTOTROPHS (original eaters) that make up a normal *Neurospora* population. (Auxotrophy is always defined as an extension of the normal nutritional requirements; prototrophy is the wild type, even if the wild-type organism has as complicated a list of nutritional requirements as we do).

Beadle and Tatum isolated from irradiated *Neurospora* a number of nutritional mutants, or auxotrophs. These could grow on a medium enriched with extracts of yeast and malt along with the salts and sugars of their minimal medium, but they would not grow on the defined minimal medium that supported the growth of the original strain. Beadle and Tatum then laboriously tested these mutants to see what were the simplest nutritional supplements that would support their growth. They found among their collection of mutants individual strains that required some specific amino acid or vitamin or purine or pyrimidine. In almost every case, the nutritional requirement turned out to be simple; only a single compound had to be added to the medium to support the growth of any given mutant. This supported the idea that mutations did, in fact, have simple effects, perhaps that each mutation caused a defect in only one enzyme in the series of reactions leading to the synthesis of the required nutrient.

The auxotrophs isolated in this way could be divided into classes on the basis of the nutritional supplements that would support their growth. For example, all mutants that would not grow on minimal medium, but would grow on a minimal medium supplemented with arginine, were classified as *arg* mutants. Other sets of mutants were found that required adenine, or proline, or vitamin B_1, and so forth. The mutants within such sets could be tested against one another in complementation tests to see whether they all affected the same enzymatic step in the relevant biosynthetic reaction sequence. The complementation test with *Neurospora* is in principle identical to that with phage, but it differs in detail. When two mutants of *Neurospora*, each unable to grow on the minimal medium, are plated together on minimal medium, their hyphae (the long strings of cells that form the "body" of the mold) can fuse, and nuclei from the two mutants find themselves in a common cytoplasm. The nuclei themselves do not fuse, but the gene products meet in the cytoplasm so that complementation is possible. The ability of such a heterocaryon (a thing with mixed nuclei) to grow on a medium that will not support the growth of either mutant individually is evidence of complementation. When mutants with the same nutritional phenotype were complementation-tested in this way, it was often found that they fell into different cistrons. For example, it was found that when 15 mutants requiring the amino acid arginine for growth were tested against one another, they fell into seven different complementation groups.

The study of nutritional mutants has taught us even more about biochemistry than about genetics. The study of such mutations during the 1940's and 1950's led to the detailed description of many metabolic pathways and of the enzymes that control the reactions in those pathways. A made-up example will do to illustrate the sort of information that these studies provided. Suppose we have a collection of mutants that require compound d for growth. Suppose, further, that we find that these mutants fall into three cistrons. We hypothesize that the synthesis of compound d must occur by some such mechanism as the following:

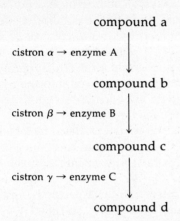

compound a

cistron α → enzyme A

compound b

cistron β → enzyme B

compound c

cistron γ → enzyme C

compound d

The mutants in cistron γ accumulate a compound that, when added to the minimal medium, will support the growth of mutants in cistrons α and β. We conclude, therefore, that cistron γ mutants are trying to make compound d but lack one of the necessary enzymes. Because the enzyme is missing, compound c, the normal substrate for that enzyme, accumulates in the mutant. Since strains mutated in cistrons α or β are able to grow on compound c, we conclude that mutations in cistrons α and β cause defects in enzymes that function earlier in the reaction sequence than the enzyme encoded by cistron γ. By a similar set of experiments, we might also identify compounds a and b and determine which of the cistrons encodes enzyme A and which encodes enzyme B. Sometimes, of course, a mutant will fail to accumulate the compound behind the enzymatic block, either because the compound is unstable or because it is used up by some other enzymatic series of reactions. In such a case, we should have to take an educated guess at the nature of the compound that we wanted to identify. By the time we had completed our analysis of the mutations in our collection, we would have confirmed that biosynthesis of compound d does, in fact, proceed through a sequence of events like that pictured in our hypothetical scheme and we would have identified compounds a, b, and c. Biochemists, using these methods and making good educated guesses when necessary, have unearthed from auxotrophic mu-

tants of *Neurospora,* and other microbes, a rich treasure of information concerning the way in which living things manufacture the molecules of life.

BACTERIAL GENETICS

For many years, bacteriologists thought that there were no sexual processes in bacteria, that all bacterial cells arose from other cells by simple division, and that genetic variation occurred only as a result of mutations. Joshua Lederberg, then a graduate student, and E. L. Tatum, his professor, discovered in 1946 that recombination did actually occur in bacteria. In their original system, however, recombination occurred very rarely. The problem they faced, then, was to show that the rare appearance of new genotypes in mixtures of genetically distinguishable bacteria arose from a mating process rather than from a mutational process. The method they used was to mix cultures of bacteria, each of which carried two mutations for auxotrophy. If a single mutation reverts to wild-type at a frequency of about 10^{-6} mutations per cell per generation, two mutations would simultaneously revert at a frequency of about $10^{-6} \times 10^{-6} = 10^{-12}$ mutations per cell per generation. For example, if a bacterial stock carries two mutations, met^- and bio^-, one causing a requirement for the amino acid methionine, and the other a requirement for the vitamin biotin, it would be expected that a culture of about 10^9 such cells would probably contain no prototrophic revertants at all. Lederberg and Tatum mixed $met^- bio^-$ bacteria with bacteria from another doubly-mutant stock, this one requiring the amino acids threonine and leucine. They found that when they plated these cells on a medium that contained none of the nutritional supplements, some prototrophic cells did appear —about one for every 10^7 parental cells plated. When the two parent bacteria were plated separately on the minimal medium, no prototrophic colonies appeared at all. The conclusion was, therefore, that recombination could occur in bacteria. Further experiments showed that cellular contact was necessary for this process of conjugation to take place.

There are male and female bacteria, and among them maleness is an infective venereal disease.

It was soon discovered by Lederberg and his collaborators, and by William Hayes in England, that bacterial conjugation depended on sexual differentiation of bacteria; there were male and female bacteria! A male is a bacterium that can be killed immediately after mating without in any way affecting the production of recombinants. The female, on the other hand, must survive if recombinants are to be produced. Hayes demonstrated this fact by using the antibiotic, streptomycin, as a specific agent for killing one of the parents. Streptomycin-resistant mutants of *E. coli* can be easily isolated by exposing a culture to the drug and isolating those cells that survive. Hayes isolated resistant mutants of the two parent strains that Lederberg and Tatum had used in their crosses. He found that crosses carried out in the presence of streptomycin produced a normal number of recombinants if one of the two parents, the female parent, was streptomycin-resistant, and no recombinants if the female parent was streptomycin-sensitive. On the other hand, the streptomycin-resistance of the other parent, the male, made almost no difference. This was correctly interpreted by Hayes as meaning that, during conjugation, genetic material is transferred from the male, or donor, bacterium to the female, or recipient, bacterium, and that recombinant prototrophs appeared among the offspring of the female cells, but not among those of the males. Once transfer had occurred, the male was no longer needed for recombinant production. Thus, in bacterial conjugation, genetic transfer occurs only in one direction.

Maleness in bacteria is due to the presence in the male bacterium of a little extra piece of DNA, the F factor, which can replicate itself and persist in the cell population as if it were a genophore independent of the normal bacterial genophore. The F factor may attach to the bacterial genophore. Genes on the F factor direct a number of processes among which is the formation on the surface of the male bacterium of long, thin, hairlike projections called F pili (singular, *pilus* = hair). These seem to be tubes having sticky ends that attach to the surface of female cells. When this attachment of male to female has occurred, DNA may be transferred through the pilus (there is a bit of dispute about this even now) from the male cell into the female cell. In male bacteria in which the F factor is separate from the chromosome, the most likely DNA to be transferred to the female is the F factor itself. When this happens, the female becomes a male. So, in bacteria, maleness is an infective venereal disease (*Figure 8*).

Male bacteria in which the F factor is separate from the bacterial genophore, so-called F⁺ bacteria, transfer F factor to female cells very efficiently, but they rarely, if ever, transfer any of the DNA of the bacterial genophore. It has been possible to isolate from F⁺ bacterial cultures bacteria in which the F factor has become attached to the bacterial genophore. These Hfr (for high frequency of recombination) bacteria do not transfer F factor to the female cells, but they do transfer genophore DNA at high efficiency. It is with these Hfr males that it has been possible to perform the bacterial crosses that yield extensive information about the location of genes on the bacterial genophore. In

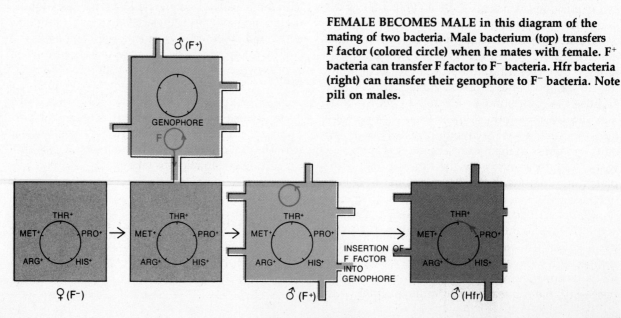

FEMALE BECOMES MALE in this diagram of the mating of two bacteria. Male bacterium (top) transfers F factor (colored circle) when he mates with female. F⁺ bacteria can transfer F factor to F⁻ bacteria. Hfr bacteria (right) can transfer their genophore to F⁻ bacteria. Note pili on males.

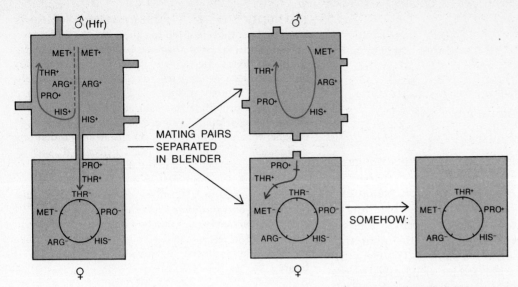

MATING PAIRS
SEPARATED
IN BLENDER

SOMEHOW:

INTERRUPTED MATING is one technique for mapping the bacterial genophore. Conjugating pair is shown at left. The longer they stay together before being separated, the more genes will be transferred to the female. Male synthesizes a new genophore for himself during mating.

any given Hfr strain, genophore transfer begins from a fixed point on the male genophore and proceeds in a fixed direction. This beginning point, the ORIGIN OF TRANSFER, is determined by the spot at which the F factor attaches to the genophore and the direction in which the F factor was pointing when it attached. When two cells form a mating contact, DNA from the donor is transferred to the recipient cell, starting with the DNA at the origin of the Hfr strain being used. After a time, generally before a whole copy of the male genophore has been transferred to the female, the two cells separate. Then, by a process that we still do not completely understand, the transferred DNA and the resident female genophore interact to produce recombinant genophores that contain some information from the donor and some from the recipient. These recombinant genophores appear among the daughter cells of the female bacterium.

A typical bacterial cross involves mixing a culture of Hfr cells and one of female (called F⁻) cells that differ in at least two genetic properties. This mixture is incubated for some time to allow cells to pair and gene transfer to occur. After the mating of the cells, the mixed culture is plated on a medium on which neither parent strain can grow, and the number of recombinant bacteria, having one genetic trait of the Hfr strain and one of the F⁻ strain, is determined.

One way to map mutations in conjugation crosses is the interrupted mating technique *(Figure 9)*. In this sort of experiment, markers are mapped according to how long (in minutes) it takes the donor cell to transfer them to the recipient. The two parent strains of bacteria are mixed, and, at

intervals, samples are removed from the mating mixture and violently agitated in a kitchen blender. This separates the mating cells so that further genetic transfer is impossible. The samples are then plated on selective medium to measure the number of recombinants of a given type. In this sort of experiment, it is found that markers near the origin are transferred soon after mixing the cells. The first recombinants for such markers appear in samples from matings that were interrupted after three to five minutes. However, markers near the end of the chromosome are not transferred until about 90 minutes after mating begins. If the cells are blended at any time before 90 minutes, no recombinants involving these markers are found. For one particular Hfr strain, *thr⁺* is about 8 minutes from the origin, *pro⁺* is about 15 minutes from the origin, and *his⁺* is about 46 minutes from the origin. Thus, the interrupted mating technique permits us to learn the order of the genes on the genophore of *E. coli* and to measure the distances between them. The measure of the distance between two genes is the time elapsing between the insertion of the first gene and the insertion of the second.

Many different Hfr strains have been isolated from F⁺ cells of *E. coli*. The origin varies from one Hfr strain to another, and so does the order of transfer. If one Hfr strain transfers markers in the order,

ABCDEFGHIJKLMNOPQRSTUVWXYZ

other strains have been found that transfer markers in the order

MNOPQRSTUVWXYZABCDEFGHIJKL

and still others in the order

FEDCBAZYXWVUTSRQPONMLKJIHG.

Note that these orders are all circular permutations of the alphabet. This means that we can generate all of these orders by writing the alphabet around the circumference of a circle and then, starting at some randomly chosen point on the circle, reading around the circle in one direction or the other until we come back to where we started. The comparison of the various Hfr strains suggested to the workers who first discovered this phenomenon that the genophore of *E. coli* is in the form of a ring, and that the F factor can attach to the ring at almost any point. When it does so, it creates a transfer origin from which transfer may proceed in one direction or the other, depending on the orientation of the attached F factor. These genetic experiments suggested that the bacterial genophore was ring-shaped years before the prediction was verified by autoradiography of DNA molecules.

Conjugation permits us to map markers that are far apart on the genophore of *E. coli*. Fine-structure mapping of markers that are much closer together —for example, mutations in the same or neighboring cistrons —depends on a different process. In the process of TRANSDUCTION some phages sometimes wrap up a piece of bacterial DNA in their protein coats instead of their own DNA. (For any particular gene, the probability of being wrapped up is about one in a million.) When such phages then infect other bacteria, they inject into them fragments of bacterial DNA from the cells on which they were grown. Cells thus infected often do not die, but continue to grow and divide, sometimes incorporating into their own genophores the fragments of DNA that have been injected by the phage. The amount of DNA that will fit into a phage is rather small compared to the total amount of DNA in a bacterium. Two markers must, therefore, be fairly close to one another on the bacterial genophore if both of them are to fit into the same phage. The farther apart two markers are, the less likely it is that they will be carried together by a single transducing phage. This provides a means of genetic mapping. Suppose, for example, that we have a number of independently isolated mutants that have been shown by complementation tests to

TRANSDUCTION BY VIRUSES enables geneticists to map markers that lie close together. Top half of diagram shows crop of phage viruses grown on a single strain of "donor" bacteria. Most of the viruses contain only virus DNA (white bar), but a few have accidentally enclosed a fragment of bacterial DNA (colored bars) within their protein coats. Bottom half of diagram shows an experiment designed to map the location of two markers: d^r (drug resistance) and nearby mutations designated $e_1{}^+$, $e_2{}^+$ and $e_3{}^+$. A marker close to d^r, such as $e_1{}^+$, will more often accompany d^r in a transducing phage than will a more distant marker such as $e_3{}^+$. Viruses are used to infect three strains of bacteria that differ from one another only in mutations on the e cistron. Geneticists map markers on the basis of recombination frequency. Recipient strain 1 produces greatest number of $d^r e_1{}^+$ transductants, strain 3 the least.

reside in the same cistron, and we want to map these mutations with relation to one another. Suppose, moreover, that the cistron in question is known from conjugation experiments to lie close to another cistron containing markers that can easily be scored selectively. To simplify the example, we shall assume that the nearby marker is a drug-resistance marker and that all of the mutant isolates are sensitive to the drug in question. The mutant bacterial strains, therefore, can be designated $d^s e_1$, $d^s e_2$, $d^s e_3$, and so forth, indicating that they are drug-sensitive and contain some numbered mutation in the cistron concerned (*Figure 10*). The drug-resistant strain is $d^r e^+$, drug-resistant and wild-type for the cistron in question. The transducing phage, P1, is grown on the drug-resistant strain, and the progeny are used to infect mutant strains 1, 2, 3, and so forth. From each of these, we selectively isolate those bacteria that have become drug-resistant because, through the transducing phage, they have received the d^r mutation from the donor strain of bacteria. These selected recombinants are scored to see with what frequency they have inherited the e^+ property along with the drug resistance. Suppose that 90 percent of the d^r cells from strain 1 have become e^+, 70 percent from strain 2, and 50 percent from strain 3. We would conclude that the order of these mutations on the genetic map is $d^r - e_1 - e_2 - e_3$.

This method is not very different from TRANSFORMATION, described earlier, in which DNA fragments are transferred as purified DNA from one bacterium to another. Transformation has, in fact, been used in much this way to order markers on bacterial maps. The P1 phage is really just a

BACTERIUM — $a^+ b^+ c^+ d^r e_1^+ e_2^+ e_3^+ f^+$ — PHAGE

TRANSDUCING PHAGES THAT CARRY DRUG RESISTANCE (d^r):
- $a^+ b^+ c^+ d^r$
- **1** $b^+ c^+ d^r e_1^+$
- **2** $c^+ d^r e_1^+ e_2^+$
- **3** $d^r e_1^+ e_2^+ e_3^+$

"IRRELEVANT" TRANSDUCING PHAGES CARRYING OTHER GENES — $k^+ l^+ m^+ n^+$

ORDINARY NON-TRANSDUCING PHAGES (VAST MAJORITY)

TRANSDUCTION

RECIPIENT STRAIN 1 — $a^+ b^+ c^+ d^s e_1^- e_2^+ e_3^+ f^+$

RECIPIENT STRAIN 2 — $a^+ b^+ c^+ d^s e_1^+ e_2^- e_3^+ f^+$

RECIPIENT STRAIN 3 — $a^+ b^+ c^+ d^s e_1^+ e_2^+ e_3^- f^+$

THESE CAN GIVE d^r, e^+ TRANSDUCTANTS (1, 2, 3)

THESE CAN GIVE d^r, e^+ TRANSDUCTANTS (2, 3)

THIS CAN GIVE d^r, e^+ TRANSDUCTANTS (3)

d^r, e^+ TRANSDUCTANT — $a^+ b^+ c^+ d^r e_1^+ e_2^+ e_3^+ f^+$

d^r, e^+ TRANSDUCTANT — $a^+ b^+ c^+ d^r e_1^+ e_2^+ e_3^+ f^+$

d^r, e^+ TRANSDUCTANT — $a^+ b^+ c^+ d^r e_1^+ e_2^+ e_3^+ f^+$

syringe for carrying the DNA from one bacterium to another. For some reason, *E. coli* resisted for years all efforts to produce transformation. Thus most fine-structure mapping in *E. coli* bacteria has been done by transduction.

GROWING FOREIGN DNA IN BACTERIA

Biologists who are interested in how genes are controlled in higher eucaryotes such as humans have long been stymied by their genetic complexity. Ingenious methods have enabled biochemists and geneticists to get very detailed information on how bacteria and their phages regulate the activity of their genes, and now investigators even know the nucleotide sequence in some regions of DNA, such as promoters and operators. But phages have very few genes —anywhere from three to about 100 — and bacteria have only about 4000, or about four million nucleotide pairs in their total DNA. Sorting out a single sequence of perhaps 100 or 1000 nucleotide pairs from such a genome is no small task, but it can be done. Mammals, as noted in Chapter 2, have about 1000 times more DNA per cell than a typical bacterium, and about 100,000 times more than the best-studied phages. The problem of isolating any one piece of pure DNA from such organisms is truly awesome.

Recently a very successful method has been devised for synthesizing large amounts of various DNAs, not in the laboratory but in *E. coli*. The method is known as RECOMBINANT DNA TECHNOLOGY. It is not terribly complicated, but requires a little explanation. Bacteria often harbor circular pieces of DNA called PLASMIDS, which replicate in the cell independently of the regular bacterial chromosome. Some plasmids carry information for making enzymes that destroy antibiotics such as tetracycline. When a bacterium carrying one of these resistance (R) plasmids is exposed to antibiotics, the number of plasmids may increase from one or two per cell to a thousand. Biochemists have developed methods of purifying the circular plasmids, cutting them open at a specific place with some sort of enzyme, inserting a piece of eucaryo-

tic DNA, and sealing it in place with another enzyme to make a new, enlarged plasmid. With the transformation techniques described in Chapter 7, the investigators can then use this DNA to transform an *E. coli* bacterium that does not carry the plasmid. The piece of eucaryotic DNA is carried into the bacterium along with the rest of the plasmid. If the antibiotic-resistance genes are then multiplied up to 1000 copies per cell, the same will happen to the eucaryotic hitch-hiker.

The cutting enzymes used for recombinant DNA technology are called restriction endonucleases. They are highly specific in their action. A typical example is EcoR1, the restriction enzyme found naturally in many strains of *E. coli*. It cuts only double-stranded DNA, and it cuts both strands. Under the right conditions, it completely ignores a DNA molecule unless the molecule contains the sequence of six bases GAATTC on one strand, and CTTAAG on the other. According to the rules of chance, any particular sequence of six bases would occur about once in every 4^6 nucleotide pairs (about one in 4000). A typical DNA fragment prepared with this enzyme may be 4000 base pairs long —enough DNA to comprise about four genes. When a restriction endonuclease "sees" the right sequence of nucleotide pairs, it cuts each of the strands between G and A. If one uses the letter B to denote all the neighboring bases that are not part of the specific sequence, the structure of the fragments will be

$$----B\ B\ B\ G \qquad A\ A\ T\ T\ C\ B\ B\ B\ B----$$
$$\text{and}$$
$$----B\ B\ B\ C\ T\ T\ A\ A \qquad G\ B\ B\ B\ B----$$

The two single-stranded ends are Watson-Crick complements of each other, and have a weak tendency to link themselves together with hydrogen bonds. These "sticky ends" reassociate only at low temperature, where thermal buffeting by water molecules is small. A pair of such fragments can be joined by keeping them at 0°C for a day or two in the presence of an enzyme called DNA ligase. In short, any two fragments with the same sort of sticky ends can be joined to form a stable double-stranded DNA molecule having continuous chains.

The steps involved in making a plasmid containing a random piece of eucaryotic DNA are illustrated in Figure 11. The first step is to use a restriction enzyme that opens the circular plasmid at a particular place. The same enzyme is then used to cleave a sample of eucaryotic DNA. When the fragments are mixed and DNA ligase is added, a plasmid containing the added fragment is formed.

11 INTRODUCING FOREIGN DNA (color) into a bacterial plasmid. The sequence to be inserted, except for the six nucleotides at each end, is denoted as B'—B'. The rest of the procedure is explained in the text.

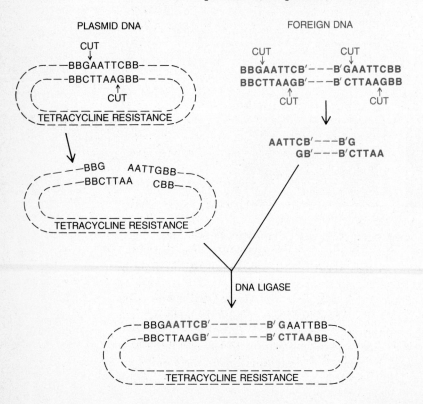

The plasmid is then used to transform an *E. coli* strain which is sensitive to tetracycline, and which cannot destroy the added plasmid because it lacks the gene for the restriction enzyme EcoR1. The bacterium can make very large amounts of the plasmid; a type of DNA which might comprise 0.0001 percent of the genome of a mammal can comprise about 50 percent of the DNA in *E. coli*. Moreover, once this oversized plasmid has been recovered in quantity from the bacteria, the DNA can be cut with the same restriction endonuclease, regenerating the original fragment of eucaryotic DNA. The fragment will be uncontaminated by even a single nucleotide of plasmid DNA.

The desirability of pursuing recombinant DNA research has become a heated issue in recent years, often with political overtones. Proponents cite the value of new knowledge about gene regulation, and the medical advantages of developing ways to produce large quantities of scarce and expensive proteins such as insulin, anti-hemophilic globulin, and pure antibodies against various diseases. Other advocates suggest the possibility of modifying crop plants to fix their own atmospheric nitrogen (like bluegreen algae), thus eliminating the need for nitrogenous fertilizers. Opponents of recombinant DNA research worry about the accidental creation of new pathogens. Others feel that a malevolent government might somehow use recombinant DNA technology to create a race of passive robots or perhaps ultra-militant soldiers programmed for conquest.

Scientists involved in the research generally seem to feel that repressive governments have more than adequate techniques of propaganda and police terror without resorting to exotic biological approaches. In any case, so far there is no obvious way to use recombinant DNA technology for such purposes. Pioneers in the field have generally believed that all bacteria containing artificial recombinant DNA should be grown and handled as if they were potential pathogens, and that they should also be crippled with mutations that would prevent them from surviving outside of the laboratory. Another rule is that certain experiments, such as putting antibiotic-resistant genes into a bacterium in which they do not normally occur, or putting random fragments of human DNA into bacterial hosts, are not to be done under any circumstances. University laboratories in the United States that wish to engage in recombinant DNA work must demonstrate that they have appropriate facilities for physical containment, and that they understand and will adhere to approved procedures. But because everyone does not agree on acceptable criteria of safety, the issue promises to remain a subject of debate for some time to come.

This chapter has introduced the most important operations in the science of genetics —functional analysis and mapping. These techniques permit the geneticist to make a one-dimensional recombination map that is a direct consequence of the one-dimensional organization of information in DNA. Phages and bacteria serve as simple examples, but a genetic map can be made for any sexual organism in which heritable traits can be recognized. Although the rules remain the same in genetic analysis of eucaryotes, their application can be complex. There is more than one chromosome involved, and frequently there is more than one copy of all the genetic information. In the next chapter the basic rules of genetic analysis are applied to "real" organisms —the eucaryotes that the reader is likely to encounter in everyday life.

READINGS

P.E. HARTMAN AND S.R. SUSKIND, *Gene Action*, 2nd Edition, Englewood Cliffs, NJ, Prentice-Hall, 1969. This book concerns itself mainly with regulation mechanisms. At present, these regulation mechanisms are best understood in microorganisms, and the book is largely devoted to those organisms. In passing, the authors do a good job of explaining more general aspects of microbial genetics.

W. HAYES, *The Genetics of Bacteria and Their Viruses*, 2nd Edition, New York, John Wiley & Sons, 1968. A magnificent text. This book covers all of microbial genetics. It is so clearly written that it can be read by a beginning student and so comprehensive that it is an indispensable reference for the expert.

F. JACOB AND E.L. WOLLMAN, *Sexuality and the Genetics of Bacteria*, New York, Academic Press, 1961. A translation of a French classic in microbial genetics. This is a technical book that deals in detail with bacterial sexuality, but the flow of ideas is smooth and orderly so that the book can be read rapidly with pleasure.

F.W. STAHL, *The Mechanics of Inheritance*, 2nd Edition, Englewood Cliffs, NJ, Prentice-Hall, 1969. This is a general genetics text written by a microbial geneticist. The sections on microbial genetics are predictably strong. Challenging problems appear at the ends of the chapters.

It seems to me that among the things we commonly see there are wonders so incomprehensible that they surpass all the perplexity of miracles. What a wonderful thing it is that this drop of seed from which we are produced bears in itself the impressions not only of the bodily form, but also the thoughts and inclinations of our fathers! Where can that drop of fluid contain that infinite variety of forms?

—Michel de Montaigne

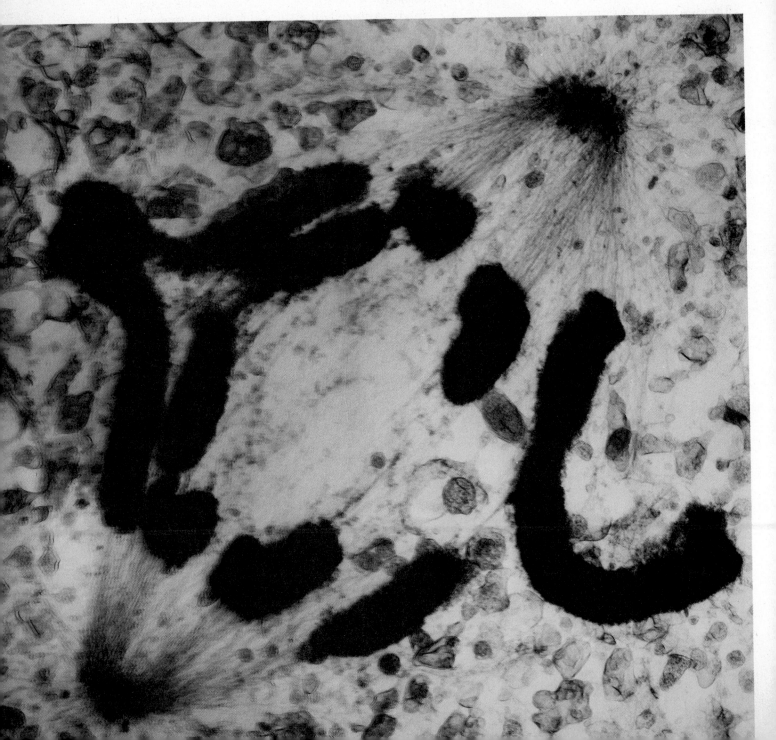

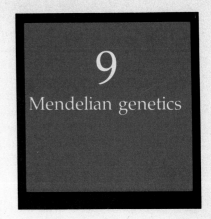
In 1578 Michel de Montaigne was 45 years old and had begun to suffer from a bladder stone. His father too had had such a stone and Montaigne, discomforted and "strangely diswenched" by his ailment, wondered what mechanism could explain the transmission of form and foible from one generation to the next. It would be almost 300 years before Gregor Mendel, a monk at the monastery at Brünn (then Austria, now Czechoslovakia), would discover the laws that govern the transmission of hereditary traits from generation to generation.

What had happened in the intervening centuries that enabled Mendel to discover in the 1860's what men before him could not? The surprising answer is that, in biology, very little happened. Nehemiah Grew, a friend of Robert Hooke and a sharer of his microscope, pointed out in 1676 that the higher plants reproduced sexually and that pollen was the vegetable equivalent of animal semen. But even the ancients knew that sex existed in some plants, particularly the date palm, which (unlike most higher plants) has separate male and female trees. And of course animal sexuality has been recognized from prehistoric times. When Mendel started his work, fertilization in plants was more or less understood and, although it was still debated, there was some evidence that one sperm or one pollen grain would suffice to fertilize one ovum. The study of plant hybrids began early in the 18th century and, by Mendel's time, methods for producing artificial hybrids were well developed. Perhaps most important, Darwin's theory made it clear that an understanding of the laws of heredity was essential not only to explain the recurrence of

CHROMOSOMES of an animal cell during anaphase appear as bold dark strands at the center of the electron micrograph on opposite page. Thin fibers connected to the chromosomes are the microtubules of the mitotic spindle.

bladder stones and baldness within families but also to explain the process of evolution. If Montaigne's discomfort had goaded him to study of the problem of inheritance he could have performed Mendelian experiments with doves, cats, rabbits, or any other domestic animals that were at hand. It was not so much Mendel's knowledge that gave him the scientific edge over Montaigne as it was his outlook. That outlook was the product of the 17th, 18th and 19th centuries.

Until the 17th century science was dominated by the Aristotelian passion for classification. Then came the urge to measure and to experiment. Calculus was invented in the 17th century by Newton and Leibniz, giving birth to modern physics and astronomy. It appeared that the natural world might be understood as well as catalogued. The work of the 18th century elucidated the physical and chemical properties of matter and the nature of electricity. The 19th century, in which most science was called engineering, saw the birth of the steam railways, the friction match, the gas light, the camera, the incandescent bulb, the internal combustion engine, and the phonograph.

Meanwhile biology had discovered experimentation, but the best efforts of biologists were devoted to description and classification. The cell theory and the theory of evolution had been advanced, but biologists, slow to adopt the lessons of their bright colleagues in the other sciences, had scarcely learned to count.

Arithmetic was the basis for Mendel's success in discovering the laws of heredity and for the failure of his colleagues to recognize his discovery. When Mendel presented his results in 1865 to a meeting of the Brünn Natural History Society, his audience perhaps assumed that he was simply trying to overwhelm them with numerical mumbo jumbo. It was not until 1900, when biologists had finally learned the value of numbers, that Mendel's breakthrough was discovered by the biological world and appreciated for what it was.

Perhaps the most significant explanation of Mendel's success is that he was a physicist —or, to be precise, a teacher of physics. His training prepared him to believe that quantitative study of complicated phenomena could yield powerful insights. Genetics was the first of the biological sciences to incorporate the quantitative method. In fact, during the first decade of the 20th century, genetics was the only quantitative biological science. Today they are all quantitative. Much of this trend can be traced to another infusion of physicists into biology in the 1940' and 1950's, led mostly by Max Delbrück, who started his career as a theoretical physicist.

Mendel's success rested upon his invention of a simple model that lent itself to a quantitative test. Historians of science suspect that he anticipated the outcome of the experiments before he started. Mendel's favorite organism was the pea, which is a eucaryote. Eucaryotes in general can be considered to have a mother and a father, and the genetic contributions of the two parents to the makeup of the offspring are essentially equal. This means that an individual must have at least two copies of each gene, one from the mother and one from the father. Mendel based his model on this hypothesis. Because fertilization involves the fusion of two GAMETES (for example, one ovum and one sperm), at some stage in the production of gametes the number of genes must be reduced to half of the number in the rest of the parent's cells. Otherwise the number of genes per cell would double at each generation —a process that would obviously lead to chaos in a few generations. Later in this chapter we shall discuss the specialized process of cell division called MEIOSIS that leads to the production of gametes with half the normal genetic content.

MENDEL'S FIRST LAW

Mendel imagined that during the formation of gametes the parent organism randomly put one copy of a particular gene into each gamete. When egg and sperm united, the new individual would get one copy from each parent. These two gene copies in the new individual might or might not be identical. If they are identical, the individual is said to be HOMOZYGOUS for this gene; if they are different, the individual is said to be HETEROZYGOUS. Different forms of a given gene (or CISTRON) are called ALLELES. For example, all of the different *r*II mutants of the *A* cistron in phage T4 are alleles of one another *(Chapter 8)*. If the new individual is homozygous for the gene, the phenotype will be governed by the allele present in both copies of the gene. If the individual is heterozygous, the phenotype will be determined by the dominant allele. If neither allele is dominant, the phenotype of the heterozygote will show the presence of both alleles. When the heterozygote offspring reproduces it will produce gametes carrying one allele or the other in equal numbers. Mendel called the separation of the two alleles during the formation of gametes SEGREGATION. Mendel's First Law is that alleles segregate from one another during the formation of gametes *(Box A)*.

The consequences of this law can be examined with real organisms, beginning with one that makes an even better textbook case than Mendel's peas: the common snapdragon. Two common kinds of true-breeding snapdragons differ in flower color; one is red, the other is white. True-breeding means that red snapdragons crossed to red give only red offspring generation after generation, and white crossed to white give only white. Red and white are allelic forms of a single gene. When red snapdragons are crossed to white, all of the progeny are pink *(Figure 1)*. The pink progeny are heterozygotes that contain an allele for red and an allele for white. Because the offspring resemble neither parent but have an intermediate phenotype, geneticists say that neither allele is dominant. Alleles can be designated by letters. If *C* stands for color, the red allele can be designated C^r and the white allele C^w. The red homozygous parents have the genotype C^r/C^r. The white homozygous parents have the genotype C^w/C^w.

A   Mendel's laws

1. Alleles segregate from one another during the formation of gametes.
2. Alleles of different genes are assorted independently of one another during the formation of gametes.

---

1 INHERITANCE OF FLOWER COLOR follows the classic Mendelian pattern. Cross between red and white snapdragons produces only pink offspring. This is due to mixing of pigments, not mixing of genes. When pink hybrids are crossed with one another, about half the progeny are pink, one fourth are pure red and one fourth pure white. There are no intermediate gradations of color.

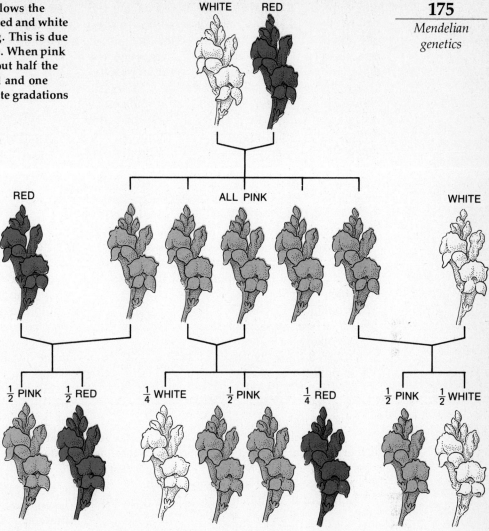

The pink progeny have the genotype C^r/C^w.

Three new sorts of crosses are possible. The test of Mendel's model is whether it correctly predicts the results of these crosses. If white snapdragons are crossed with pink ones, pre-Mendelian biologists would predict that all of the progeny would be a pale pink, about half as dark as the pink parents, since they believed that hereditary traits could be blended. But on the basis of Mendel's First Law, one would predict that the white snapdragons would produce gametes bearing only C^w, and the pink snapdragons would produce equal numbers of two types of gametes: C^w and C^r. These two types would unite with C^w gametes from the white parent to give two kinds of offspring in equal numbers: C^w/C^w, or white, and C^r/C^w, or pink. The pink offspring would be just as pink as the pink parent, and the white offspring would be pure white. The Mendelian prediction is correct.

Another kind of cross, red × pink, works in the same general way. The progeny are not dark pink; rather, half the progeny are regular pink and the other half pure red.

What about a cross of pink × pink? Here each parent produces two kinds of gametes. Therefore, four kinds of fertilization are possible:

1. C^w pollen + C^w ovum → C^w/C^w (white)
2. C^w pollen + C^r ovum → C^w/C^r (pink)
3. C^r pollen + C^w ovum → C^r/C^w (pink)
4. C^r pollen + C^r ovum → C^r/C^r (red).

Since all of these four types of fertilizaton are equally probable, one would expect that in a large number of progeny, the ratio would be about 1 white: 2 pink: 1 red. This is the result depicted in Figure 1.

Peas are not quite so straightforward. They too have red and white flowers, and Mendel used these two allelic forms in some of his experiments. When he crossed red peas to white peas, the progeny were not pink; they were as red as the red parent. In peas the red allele is DOMINANT. When one allele is clearly dominant over the other, geneticists use a capital letter to symbolize the dominant allele and a lower case letter to symbolize the RECESSIVE allele. Mendel's white peas had a genotype designated *r/r*; the homozygous red parents had the genotype *R/R*; and the heterozygous red progeny had the genotype *R/r*. Mendel crossed the heterozygous red progeny to homozygous white plants of the original parental type. Half the progeny were red, and the other half were white. This confirms Mendel's expectation that half the offspring should be *R/r*, and the other half *r/r (Figure 2)*.

When Mendel crossed heterozygous red peas to themselves, he expected that there would be four different types of fertilization, all of them equally probable:

1. *r* pollen + *r* ovum →
   r/r (white homozygote)
2. *r* pollen + *R* ovum →
   r/R (red heterozygote)
3. *R* pollen + *r* ovum →
   R/r (red heterozygote)
4. *R* pollen + *R* ovum →
   R/R (red homozygote).

The ratio then is 3 red: 1 white. Mendel went even further, and let the red offspring from this last cross pollinate themselves. He found that about 2/3 of them gave at least some white offspring, and therefore had to be heterozygous. The rest of the red plants gave only red offspring. Once again, this was in accord with Mendel's hypothesis. He had predicted that 2/3 of the red plants from the *R/r* × *R/r* cross would be heterozygous.

Mendel's most revolutionary discovery was that genes do not blend like cans of red and white paint.

Mendel's most revolutionary finding was that genes do not blend like cans of red and white paint. The results may be red, as in peas, or pink, as in snapdragons, but the genes are easily sorted out in the next generation to give the original colors. This is because there are only a very small number of determinants for each trait —two, to be exact. In all, Mendel did essentially the same experiment with several other traits, tallness of the plant, shape of seed, shape of the ripe pod, color of the unripe pod, and position of the flower. None of these traits showed any blending. It was clear that the genes were discrete entities that could enter into a heterozygote in which their expression

GARDEN PEAS illustrate dominant and recessive inheritance. Cross between a true-breeding white-flowered plant and a true-breeding red-flowered one yields only red-flowered heterozygous offspring (center). Crosses between red homo- and heterozygotes yields all red offspring (lower left). Cross between red heterozygote and white homozygote yields half red, half white offspring (lower right).

2

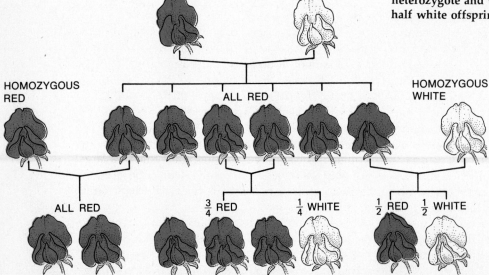

HOMOZYGOUS RED   HOMOZYGOUS WHITE

HOMOZYGOUS RED   ALL RED   HOMOZYGOUS WHITE

ALL RED $\frac{3}{4}$ RED $\frac{1}{4}$ WHITE $\frac{1}{2}$ RED $\frac{1}{2}$ WHITE

might be masked, but they could emerge unchanged when the heterozygote produced gametes. This discovery was of great importance to the understanding of the processes of evolution, because it plugged a gaping hole in the theory of evolution as it had been proposed by Darwin. Darwin realized of course that for natural selection to work at all it had to work on stable heritable traits. Unfortunately the genetic models current at Darwin's time implied that any adaptive heritable change in an individual would be rapidly diluted in the genetic material of later generations. Genetic variability, according to this reasoning, would tend to disappear before natural selection had a chance to alter the characteristics of the population. Mendel's demonstration of the particulate nature of the gene solved this problem and put Darwin's theory on a reasonable basis.

ASSORTMENT OF GENES

The second part of Mendel's investigation dealt with the question of inheritance of multiple genetic differences. What would happen if a cross were made of two parents that differed with respect to two or more genes? When the double heterozygote makes gametes, do the alleles from one of the parents go together to one gamete and those from the other parent go to the other gamete? Or do parental traits form new associations at the time of gamete formation?

Mendel's plants differed with respect to two seed properties: one made spherical yellow peas and the other made dented green ones. The first type can be designated S/S Y/Y, indicating that it is homozygous both for the S allele of the spherical, or "shape" gene and for the Y allele of the yellowness gene. The second doubly-homozygous variety is s/s y/y. The doubly heterozygous offspring from a cross between these two varieties would be expected to be uniformly S/s Y/y.

There are two ways in which these doubly heterozygous plants might produce gametes. If the alleles maintained their original associations during gametogenesis, the double heterozygotes would produce only two kinds of gametes —S Y and s y —and the offspring would be of three types: ¼ S/S Y/Y, ½ S/s Y/y, and ¼ s/s y/y. In other words, the results of crossing the doubly-heterozygous plants to themselves, or to one another, would be exactly like the result of the cross involving flower color. There would be no reason to suppose that seed shape and seed color were really regulated by two different genes, because spherical seeds would always be yellow, and dented seeds green.

Alternatively, the segregation of S from s could be independent of the segregation of Y from y during formation of gametes. This would mean that the doubly-heterozygous plant would produce four kinds of gametes, all in equal numbers: S Y, S y, s Y, and s y. The fusion of two of these gametes at fertilization can give rise to nine different genotypes among the progeny:

1. S/S Y/Y 4. S/S Y/y 7. s/s Y/Y
2. S/s Y/Y 5. S/S y/y 8. s/s Y/y
3. S/s Y/y 6. S/s y/y 9. s/s y/y.

The progeny can thus have any of three possible genotypes for shape (S/S, S/s, or s/s) and any one of three possible genotypes for color (Y/Y, Y/y, or y/y).

The number of different phenotypes cannot be greater than nine (the number of genotypes). It may be even smaller if some of the alleles are fully dominant, because the heterozygotes will be indistinguishable from homozygotes for the dominant gene. This is the case with shape and color of peas. S (spherical) is dominant to s (dented), and Y (yellow) is dominant to y (green). Then genotypes 1, 2, 3, and 4 above will all have the same phenotype: spherical, yellow. Genotypes 5 and 6 will both be spherical, green. Genotypes 7 and 8 will both be dented, yellow. And genotype 9 will be dented, green. Actually there are only four possible phenotypes because one allele for each of the two seed characteristics is dominant. There would be six possible phenotypes if S were dominant but Y were not, so that Y/Y, Y/y, and y/y seeds would form three distinct phenotypic classes.

The frequencies of the nine genotypes listed above are not all the same. Some genotypes can only be produced by one kind of fertilization event, but other genotypes can be produced by several different kinds of fertilization. For example, S/S Y/Y individuals can arise only from fertilization of S Y ova by S Y pollen, whereas S/s Y/y individuals can arise four ways: from fertilization of S Y ova by s y pollen, of s Y ova by S y pollen, of s y ova by S Y pollen, or of S y ova by s Y pollen. A convenient illustration of the possible kinds of fertilization events is given by the so-called Punnett square shown in Figure 3. This type of diagram, fashioned in 1905 by R. C. Punnett, puts Mendel's hypothesis into a form that anyone can understand. Perhaps if Mendel himself had thought of it in 1866, his work would have been readily accepted by his contemporaries.

In the Punnett square the four ovum genotypes are written vertically at the left of the square and

POLLEN GENOTYPES

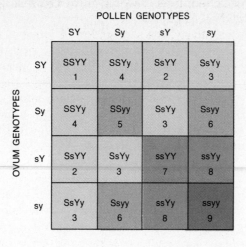

PUNNETT SQUARE shows independent assortment of alleles for seed shape and seed color. Dominant S denotes spherical seed, recessive s a dented seed. Dominant Y stands for yellow seed, recessive y for green seed. Color of squares indicates four phenotypes of offspring.

the four pollen genotypes are written across the top. The offspring resulting from fusion of two gametes have the genotypes shown in the boxes where the appropriate ovum rows and pollen columns intersect. The number shown in each box is the genotypic class to which the offspring belong. Note that Classes 1, 5, 7 and 9, which are homozygous for both genes, are all equally hard to produce. Each of these genotypes appears in only

B Calculating probabilities

The Mendelian formulation allows us to calculate average expectations. Suppose, for example, that we crossed two hybrids for seed color (each was Y/y) and they produced a progeny of 13 seeds. How many seeds would be yellow and how many green? We know from the discussion in the text that on the average ¾ of the seeds should be yellow and ¼ should be green. But that is an AVERAGE that we would expect to measure if we examined a very large number of "families" of 13 sibling seeds. Some sets of 13 would contain ten yellow and three green; some would contain nine yellow and four green; some would be all yellow; and so forth. We can be sure that no set of 13 seeds will contain exactly the expected (average) proportion of yellow and green seeds because 13 is not divisible into integral quarters. It is relatively easy to calculate the probability that a family of 13 seeds will have a given proportion of yellow and green seeds. Such a calculation predicts a FREQUENCY DISTRIBUTION of families of the various possible types. It tells us, for example, that if we examine a large number of families of 13 seeds from hybrid-by-hybrid crosses, the expected frequency of families with nine yellow and four green seeds is 0.210. Let us consider how to make that calculation.

First, we assume that the colors of the various seeds are determined independently of one another. That is, the color of one seed does not influence the probability that some other seed will be a given color. When events are independent of one another, the probability that the events will occur simultaneously is the product of their individual probabilities. In our example, the probability that a given two seeds will both be green is $\frac{1}{4} \times \frac{1}{4} = 1/16$. If we number the 13 seeds in the sample, the probability that seeds 1 through 4 will be green and 5 through 13 will be yellow is $(\frac{1}{4})^4 (\frac{3}{4})^9$. Similarly, the probability that seeds 4, 7, 8, and 12 will be green and the rest will be yellow is $(\frac{1}{4})^4 (\frac{3}{4})^9$. For every possible specific order of four green and nine yellow seeds, the probability is the same. Thus, the probability of four green and nine yellow seeds in a family is $(\frac{1}{4})^4 (\frac{3}{4})^9$ multiplied by the number of possible orders of four green and nine yellow seeds.

Now the problem is to determine the number of possible orders of four green and nine yellow seeds. This is easily done. The 13 numbered seeds can be arranged in many different orders: any one of the 13 can occupy position 1; any of the remaining 12 position 2; any of the remaining 11 position 3; and so on. Mathematically, the number of possible orders of 13 distinctly identifiable seeds is 13!, that is, $13 \times 12 \times 11 \times \ldots \times 2 \times 1 = 13!$. Of these orders, however, a number are indistinguishable from one of the 16 boxes of the Punnett square. Classes 2, 4, 6 and 8 (single heterozygotes) each appear in two boxes. And Class 3 (the double heterozygote) appears in four of the 16 boxes. Thus, the nine genotypic classes appear, respectively, in the relative proportions 1:2:4:2:1:2:1:2:1. Since the capitalized alleles are dominant, the various genotypic classes will give rise to the following phenotypes:

Classes 1, 2, 3 and 4: spherical, yellow
Classes 5 and 6: spherical, green
Classes 7 and 8: dented, yellow
Class 9: dented, green.

The proportion of the four possible phenotypic classes can, therefore, be determined from the Punnett square to be 9:3:3:1. The boxes of the Pun-nett square are tinted to indicate this distribution of phenotypes.

The Punnett square is clear, but it becomes tedious to lay out a square that predicts the outcome of a cross involving alleles of three genes. In such a cross each parent will produce eight different kinds of gametes, and the Punnett square contains 64 boxes. It is easier to think of such a cross as three separate crosses in one, in which the alleles of the three genes segregate independently of one another. Suppose a geneticist crosses two plants of the genotype *A/a B/b C/c*. What proportion of the offspring would be expected to be *A/A B/b C/c*? Of all offspring, $\frac{1}{4}$ are *A/A*, $\frac{1}{2}$ are *B/b*, and $\frac{1}{2}$ are *C/c*. Since the three genes are inherited independently of one another, the frequency of *A/A B/b C/c* indi-

---

one another if we do not number the seeds, but only identify them with respect to color. The four green seeds in our hypothetical sample can be arranged in 4! ways, and the nine yellow seeds can be arranged in 9! ways. Thus, there are 4! × 9! different ways to arrange 13 seeds so that seeds 1 through 4 are green and 5 through 13 are yellow, but all of those orders look identical if we see only the seed colors and not their arbitrarily assigned numbers. Similarly, there are 4! × 9! ways in which we can arrange the 13 seeds so that seeds 4, 7, 8, and 12 will be green and the rest yellow. And all of these arrangements will look alike if we examine only the color of the seeds. This means that 13! overestimates the number of possible color orders of our 13 seeds by a factor of 4! × 9! The number of possible orders of four green and nine yellow seeds is, then

$$\frac{13!}{4!\ 9!}.$$

Putting this into our previous expression, we find that the probability that a family of 13 peas produced by a cross between two seed-color hybrids will contain four green and nine yellow peas is

$$\frac{13!}{4!\ 9!}\left(\frac{1}{4}\right)^4\left(\frac{3}{4}\right)^9 = 0.210.$$

By similar reasoning, we could derive a general expression to calculate the probability that a sample of *n* individuals would contain *s* members with one phenotype and *t* with a second, alternative phenotype. We assume that the phenotypes of the individuals are determined independently and that phenotype 1 occurs with a probability of *p* and phenotype 2 with a probability *q*, where $p + q = 1$, because an individual must have one phenotype or the other, and $s + t = n$. Among all sets of size *n*, we would expect that the frequency with *s* individuals of phenotype 1 and *t* individuals of phenotype 2 would be:

$$\frac{n!}{s!\ t!}(p)^s(q)^t.$$

In applying this formula, it is necessary to accept the definition, $0! = 1$.

This is the *s*th term of the binomial expansion of $(p + q)^n$, and, therefore, we say that we expect the various possible families of size *n* to be distributed according to a binomial distribution, the number of families with *s* offspring of one type and *t* offspring of the other being calculated by the general expression given above.

If there are three possible phenotypes, with probabilities *p*, *q*, and *r*, the above expression becomes:

$$\frac{n!}{s!\ t!\ u!}(p)^s(q)^t(r)^u.$$

viduals would be the product of these three individual frequencies: ¼ × ½ × ½ = 1/16. The frequencies of the other genotypic classes can be calculated in the same way. The method also works for calculating the frequency of phenotypic classes as well as genotypic classes; and it works just as well with crosses involving only two factors as it does with crosses involving three or more. The reader can predict the frequency of the four

phenotypic classes in the cross between two plants of genotype *S/s Y/y*. If he pays attention only to the shape of the seeds, it is clear that ¾ of the offspring will be spherical (the *S/S* and *S/s* classes) and ¼ will be dented (the *s/s* class). Similarly, ¾ will be yellow and ¼ will be green. Therefore, the frequencies of the four possible phenotypes will be:

spherical, yellow: ¾ × ¾ = 9/16

C Problem set 1

Problem 1. Suppose you crossed two pea plants of the following genotypes: *R/r s/s Y/y* and *R/R S/s Y/y*. (Upper case letters indicate dominant alleles.)
  a. How many different kinds of gametes could be produced by the first parent?
  b. How many different kinds of gametes could be produced by the second parent?
  c. How many different genotypes would be found among the progeny of the cross?
  d. How many different phenotypes would be found among the progeny of the cross?

Answers to problem 1.

  a. The *R/r s/s Y/y* parent can produce 4 kinds of gametes: *R s Y, R s y, r s Y,* and *r s y*.
  b. The *R/R S/s Y/y* parent can also produce 4 kinds of gametes: *R S Y, R s Y, R S y,* and *R s y*.
  c. The number of genotypes found among the progeny will be 12. Consider the possibilities for one gene at a time.
    1. The offspring can be *R/R* or *R/r*, and the frequencies of these two types would be expected to be equal.
    2. The offspring may be either *S/s* or *s/s*. Again, the two types would be expected to be equally frequent.
    3. The offspring may be *Y/Y, Y/y,* or *y/y*. The frequencies of the three types would be expected to be 1:2:1.
  The twelve possible genotypes, and their expected frequencies among the progeny, are as follows:

R/R S/s Y/Y	½·½·¼ =	¹⁄₁₆
R/r S/s Y/Y	½·½·¼ =	¹⁄₁₆
R/R s/s Y/Y	½·½·¼ =	¹⁄₁₆
R/r s/s Y/Y	½·½·¼ =	¹⁄₁₆
R/R S/s Y/y	½·½·½ = ⅛	= ²⁄₁₆
R/r S/s Y/y	½·½·½ = ⅛	= ²⁄₁₆
R/R s/s Y/y	½·½·½ = ⅛	= ²⁄₁₆
R/r s/s Y/y	½·½·½ = ⅛	= ²⁄₁₆
R/R S/s y/y	½·½·¼ =	¹⁄₁₆
R/r S/s y/y	½·½·¼ =	¹⁄₁₆
R/R s/s y/y	½·½·¼ =	¹⁄₁₆
R/r s/s y/y	½·½·¼ =	¹⁄₁₆

Total = ¹⁶⁄₁₆ = 1

  d. There are only 4 possible phenotypes to be found among the progeny. Again, consider one gene at a time.

spherical, green: $\frac{3}{4} \times \frac{1}{4} = 3/16$
dented, yellow: $\frac{1}{4} \times \frac{3}{4} = 3/16$
dented, green: $\frac{1}{4} \times \frac{1}{4} = 1/16$.

This is the 9:3:3:1 ratio established earlier with a Punnett square *(Box B)*.

Mendel's actual results were in excellent agreement with the predicted ratio. They caused him to formulate what is now known as Mendel's Second

Law: that alleles of different genes are ASSORTED independently of one another during gamete formation. This second law is not as universal as the first. It applies to genes that lie on separate chromosomes, but not to those that lie on the same chromosome. It becomes a true law if one says that CHROMOSOMES assort independently during formation of gametes *(Box C)*.

---

1. *R* is dominant to *r*. Therefore, since all of the offspring carry at least one copy of the *R* allele, all will have red flowers.
2. The *S/s* offspring, constituting ½ of all the progeny, will have spherical seeds. The *s/s* offspring, constituting the other ½, will have dented seeds.
3. The *Y/Y* and *Y/y* offspring together make up ¾ of all offspring and have the same phenotype; namely, yellow seeds. The *y/y* offspring, the other ¼, will have green seeds.

The possible phenotypes, with their expected frequencies, are as follows:

red flower, spherical yellow seed	$1 \cdot \frac{1}{2} \cdot \frac{3}{4} = \frac{3}{8}$
red flower, spherical green seed	$1 \cdot \frac{1}{2} \cdot \frac{1}{4} = \frac{1}{8}$
red flower, dented yellow seed	$1 \cdot \frac{1}{2} \cdot \frac{3}{4} = \frac{3}{8}$
red flower, dented green seed	$1 \cdot \frac{1}{2} \cdot \frac{1}{4} = \frac{1}{8}$
	Total = 1

Problem 2. Dominance relations are not always as simple as those that we have described for snapdragons and for Mendel's peas. Consider the following examples:

a. Early in this century, Bateson and Punnett performed a cross between two true-breeding white varieties of sweet pea. All of the offspring of this cross had colored flowers! When the colored flowers were crossed to one another, they produced colored and white offspring in a 9:7 ratio. Can you find a rule relating genotype to phenotype in such a way that Mendel's 9:3:3:1 ratio of phenotypes is converted into a simple 9:7 because the last three classes are not distinguishable?

b. Mendel himself did a preliminary experiment in which true-breeding white-flowered bean plants were crossed to true-breeding colored-flowered plants. All of the progeny had colored flowers. When the colored offspring were crossed to one another, it appeared that only 1/16 of their offspring had white flowers and the other 15/16 had colored flowers. Can you account for that?

Answers to problem 2.

a. The cross involves two genes, *A* and *B*, each playing an essential role in the determination of plant color. The flower is white if it is homozygous for the recessive mutation *a* or for the recessive mutation *b*. Thus, *A/A b/b* individuals are white, and *a/a B/B* individuals are white. If these two true-breeding white varieties are crossed, they yield colored offspring of genotype *A/a B/b*.

When *A/a B/b* individuals are crossed with others of the same genotype, 9/16 of the offspring have at least one dominant allele of each gene, and will be colored. The other 7/16 will be homozygous for *a* or *b* or both *a* and *b*, and these will be white.

b. Mendel presumably crossed an *A/A B/B* colored plant to an *a/a b/b* white plant

---

(Continued on next page)

MEIOSIS

Mendel never saw chromosomes, and probably never imagined that cytologists would soon witness under the microscope the actual process that leads to the segregation of alleles. The process of segregation and assortment in eucaryotes is called MEIOSIS (from the Greek for "diminution"). In meiosis, two cell divisions occur in rapid succession without intervening duplication of chromosomes. Four daughter cells result, each with only one copy of each of its chromosomes, instead of two copies, as in the parent cell.

The sequence of events in meiosis is depicted in Figures 4, 5 and 6. The process begins like mitosis. Chromosomes that have already duplicated in the

---

C *(Continued)*

that was homozygous for two flower-color mutations. This produced $A/a\ B/b$ colored offspring. When he crossed these $A/a\ B/b$ plants to one another, he got only 1/16 white offspring. This suggests that either one A allele or one B allele is sufficient to produce color and that plants will be white only if they are of genotype $a/a\ b/b$.

Problem 3. There is a dominant mutation, Cy, in the fruit fly, *Drosophila melanogaster*, that causes the wings to curl. When a heterozygous curly fly is crossed to the wild type, ½ of the offspring are curly and ½ are of the wild type. When two heterozygous curly flies are crossed to one another, ⅔ of the offspring are curly, and ⅓ are of the wild type. Can you explain this deviation from the expected ratio of 3:1?

Answer to problem 3. The real ratio of offspring is 1 cy/cy wild type: 2 Cy/cy curly heterozygotes: 1 Cy/Cy curly homozygote. This last type, however, dies in an embryonic stage and goes unnoticed. Of the survivors, ⅓ are cy/cy and ⅔ are Cy/cy.

Problem 4. A mutation in humans gives rise to a disease in which one of the chains of hemoglobin (the β chain) is made in reduced amounts. People heterozygous for this mutation are said to have *thalassemia minor*, and have a mild but detectable anemia which does not seriously affect their health. Homozygotes are said to have *thalessemia major*, are severely anemic, and frequently die in childhood or adolescence. If a couple that knows they are both heterozygous for the thalassemia mutation nevertheless decides to have a family of three children, what are the chances that:

  a. They will have exactly one child with thalassemia major.
  b. They will have *at least* one child with thalassemia major.
  c. If they have already had one child with thalassemia major and then have three more children, what is the chance that they will have exactly three total children with thalessemia major?

Answers to problem 4.

  a. The probability is $(3!/2!1!)\ (¼)^1(¾)^2 = 27/64$.
  b. The "brute force" way to calculate this is to sum up the probability of having exactly one, plus that of having exactly two, plus that of having exactly three. This is $(27/64) + (9/64) + (1/64) = 37/64$.

  An easier way is to recognize that the probability of their having at least one is 1.00 minus the probability of their not having any, or $1.00 - (3!/3!0!)\ (¾)^3 = 37/64$.
  c. The probability that they will have three total is the same as the probability that they will have two more, or 9/64.

interphase stage begin to shorten in the MEIOTIC PROPHASE to form condensed, easily visible chromosomes, each consisting of two identical chromatids joined together at the centromere. It is during this condensation of the chromosomes that the first unique event of meiosis occurs: the SYNAPSIS of homologous chromosomes. Recall that the cell contains two copies—homologues—of each chromosome, one from each parent. At the first meiotic division the two chromosomes in each homologous pair, each made up of two chromatids, line up next to each other and migrate together to the metaphase plate so that the centromere of one homologue lies just above the plate and the centromere of the other lies just below it. The paired homologous chromosomes go by two names, one referring to the chromosome content of the pair and the other to the chromatid content. Thus, the paired homologues are a chromosome BIVALENT or a chromatid TETRAD. As the chromatids become more and more condensed, it becomes possible to see that there appear to be points at which the homologous chromatids connect to one another, forming cross-shaped patterns called CHIASMATA (singular, chiasma: a "crosspiece"). These appear to be the visible manifestations of recombination between the homologous chromatids. It appears that at any one position on the chromosome, only one pair of homologous chromatids is involved in the formation of the chiasma.

At the meiotic METAPHASE the chromosomes have reached their maximum state of condensation, and the nuclear envelope and nucleoli have disappeared. This is followed by the FIRST MEIOTIC ANAPHASE, at which the chromosomes separate from their homologues without separation of the sister chromatids, which remain attached at the centromere. As the two chromosomes, each retaining its double structure, move apart, connections at the

chiasmata move toward the end of the chromosomes until they run off the end, like a twist in two pieces of string as the strings are pulled apart. The chromosomes move to the two poles of the meiotic spindle. At TELOPHASE, when they have reached the two poles, they begin to grow less condensed and to enter into a brief MEIOTIC INTERPHASE, in which the nuclear envelope forms again. The chromosomes grow diffuse but do not replicate. The existence of this interphase is not a universal feature of meiosis. In some organisms, the chromosomes do not uncoil and the nuclear envelope does not reform at this point. In such organisms, the second meiotic metaphase and anaphase take place without further ado.

A second prophase follows rapidly. The chromosomes again condense. In this division there is no synapsis, because the homologous chromosomes were separated at the first meiotic anaphase, one going to each pole. Here the chromosomes are called UNIVALENTS and the chromatids DYADS. The chromosomes migrate to the metaphase plate once again and the nuclear envelope disappears. The chromatids then separate at the centromere so that, at the SECOND MEIOTIC ANAPHASE, the sister chromatids detach from each

MEIOSIS, depicted schematically below, begins like mitosis. Very early in prophase chromosomes begin to condense into easily visible form. Chromatids shown in color are maternal; black ones, paternal. By metaphase the nuclear envelope has disappeared. Homologous chromosomes have lined up next to each other and joined to form X-shaped chiasmata at point of contact. After anaphase I the cells divide, halving the number of chromosomes. After anaphase II the two haploid cells also divide. The second meiotic division resembles mitosis. The overall result of the process is the production of four daughter cells (gametes) from one parent cell. Each gamete has a haploid number of chromosomes. This diagram is an abbreviated version of meiosis. For a more complete depiction see Figures 5 and 6. **4**

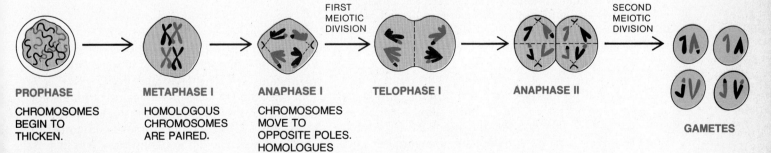

PROPHASE

CHROMOSOMES
BEGIN TO
THICKEN.

METAPHASE I

HOMOLOGOUS
CHROMOSOMES
ARE PAIRED.

ANAPHASE I

CHROMOSOMES
MOVE TO
OPPOSITE POLES.
HOMOLOGUES
SEPARATE.

FIRST
MEIOTIC
DIVISION

TELOPHASE I

ANAPHASE II

SECOND
MEIOTIC
DIVISION

GAMETES

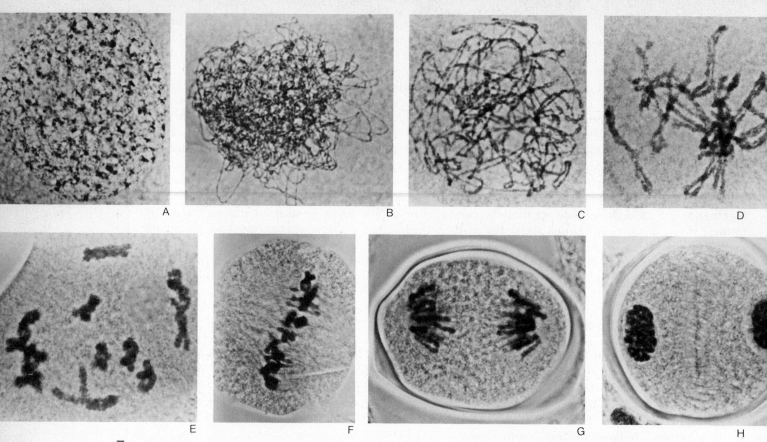

A   B   C   D

E   F   G   H

5 FIRST DIVISION OF MEIOSIS begins with prophase (A). The sequence of micrographs in this figure and the next depicts the stages of meiosis in the anther of a lily. (B) Later in prophase the chromosomes have condensed into threads. (C) Still later in prophase the chromosomes have paired, forming four-stranded structures consisting of two homologous chromosomes, each consisting of two sister chromatids. The paired chromosomes thicken (D) and become compact (E). Cross-shaped chiasmata are visible at this stage. During the first metaphase (F) the condensed chromosomes are lined up, with the centromeres aligned on the metaphase plate. During first meiotic anaphase (G) chromosomes are pulled apart, with homologous centromeres moving toward opposite poles of the cell. (H) At telophase the chromosomes have reached the poles and new cell wall begins to form at middle of cell. (I) The next stage is a brief interphase, during which chromosomes decondense but do not replicate; new cell wall is almost complete. Sequence is continued in the next illustration.

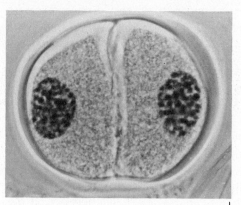

I

other, one going to each pole. At this point, when they are no longer connected at the centromere, they are once again called chromosomes. After this second meiotic division the number of chromosomes at each pole is half the original number, and reduction is complete.

The process of meiosis started with a single cell that had already doubled its chromosomes, so that there were four copies of each chromosome. The two meiotic cell divisions then produce four cells from one. But because there was no chromosome doubling between the two divisions, there was necessarily a reduction in chromosome number by a factor of two. The different appearance of meiotic and mitotic cells is illustrated in Figure 7.

184

ALTERNATION OF GENERATIONS

Before meiotic cell division the cell has two copies of each chromosome; it is said to be DIPLOID (from the Greek for "double") and to have $2n$ chromosomes. After the meiotic divisions the cell is HAPLOID (from the Greek for "single") and has n chromosomes. Fertilization reverses the effects of meiosis, fusing two haploid cells to produce a diploid cell called a ZYGOTE. There is considerable variation in living things in the number of cell divisions separating meiosis and fertilization. In any sexual organism there are two genetic forms, the haploid and the diploid, which alternate from generation to generation. The diploid generation gives rise to the haploid generation by meiosis, and the haploid generation gives rise to the diploid generation by fertilization, the fusion of two cells. This

> In any sexual organism there are two genetic forms, the haploid and the diploid, which alternate from generation to generation.

point is best illustrated by comparing a few representative life cycles: *Ulva* (sea lettuce), corn, animals, and bread mold.

ULVA

Ulva stenophylla is a green edible broadleafed marine alga that exhibits ALTERNATION OF GENERATIONS in its purest form (*Figure 8*). The organism exists in two forms, one diploid, the other haploid. The "leaves" of these two forms are indistinguishable to the unaided eye, though one has twice as many chromosomes per cell as the other. The haploid form is called a GAMETOPHYTE because it produces gametes. These are produced by a mitotic process that does not change the number of chromosomes per nucleus. Gametophytes can be of two different kinds, called MATING TYPES. The two types look exactly alike, and their gametes also look exactly alike, but are in fact different. Two gametes will fuse only if they belong to different mating types. (Box D explains the biological distinction between mating type and sex.) When two gametes of different mating types do meet and fuse, they form a diploid zygote which carries one gene for each mating type. This settles to the bot-

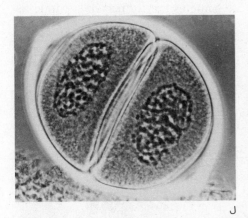

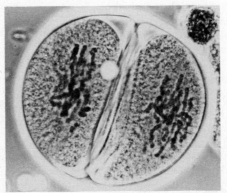

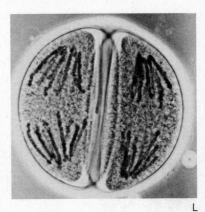

J　　　　　K　　　　　L

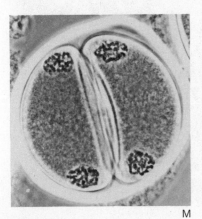

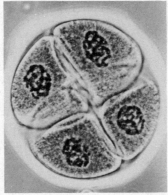

M　　　　　N

DIVISION TWO OF MEIOSIS begins with a second **6** **meiotic prophase (J), during which the chromosomes recondense. (K) During second meiotic metaphase, chromosomes once again line up with their centromeres on the metaphase plate. (L) At second meiotic anaphase, sister centromeres are pulled apart. (M) During second meiotic telophase, newly separated chromosomes gather at the poles of the cells. (N) Cell division is now complete, producing four haploid cells. Each of these will mature into a pollen grain.**

tom and grows into a diploid "leaf" or SPOROPHYTE, whose cells can undergo meiosis to give asexual haploid spores. These spores swim about for a time and then settle to the bottom. Without fertilization they grow into haploid gametophytes of one mating type or the other.

In *Ulva* the two generations are equally prominent in the life cycle of the organism. This is by no means common; most organisms have greatly emphasized either the diploid or the haploid generation. In animals and in higher plants the haploid generation is short and inconspicuous. Conversely, in mosses and fungi the diploid generation is greatly reduced.

7 DIFFERENCE BETWEEN MEIOTIC AND MITOTIC CELLS is evident in these three photomicrographs of lily anther cells. Picture at left depicts one cell during meiotic prophase (top) and another during mitotic prophase (bottom). Picture at center depicts mitotic metaphase. Although this cell contains same number of chromosomes as cell undergoing first meiotic division (Figure 5F), it appears to have twice as many because in a mitotic cell, homologous chromosomes do not pair; each chromosome is seen individually. Picture at right depicts mitotic anaphase. Number of visible chromosomes is again greater than at first meiotic anaphase (Figure 5G). Chromosomes seen here are single structures, while those of meiotic anaphase are double, consisting of two sister chromatids.

CORN

The maize plant commonly seen growing in the corn field belongs to the sporophyte generation. It is diploid and bisexual. By the process of meiosis, the male tassels of the plant produce pollen cells and the female tissues produce ova. In pollen formation, a diploid cell undergoes meiosis to produce four haploid pollen cells which are really small, asexual spores *(Figure 9)*. When these land on the female flower, they grow into male gametophytes. The haploid nucleus of each spore undergoes mitosis to produce two identical haploid nuclei, and one of these divides again by mitosis to give a total of three nuclei. One of these three is the pollen tube nucleus, an integral part of the long, one-celled POLLEN TUBE that grows through the corn-silk from the tip, where the pollen grain lands, to the ovum at the base of the silk. The other two nuclei of the male gametophyte enter the ovum and act as gametes. One fertilizes the female gamete and ultimately gives rise to an EMBRYO. The other nucleus fuses with female nuclei to give rise to a nutrient storage tissue called ENDOSPERM. The whole male haploid generation consists only of small gametophytes containing three nuclei.

The female gametes are formed in separate tissues of the diploid sporophyte. A diploid cell undergoes meiosis to produce four haploid nuclei in-

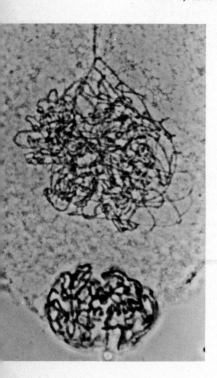

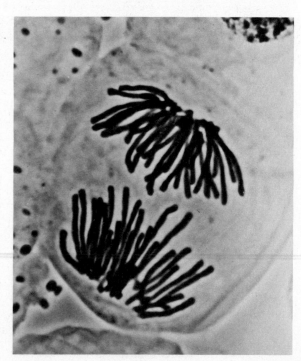

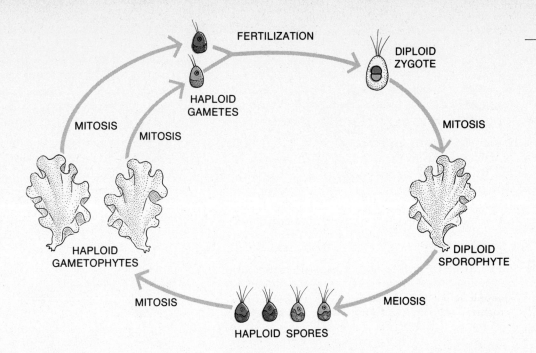

8

side a single cell membrane. Three of these nuclei degenerate and the fourth is an asexual spore that divides three times mitotically to produce eight identical haploid nuclei, still enclosed in the membrane. This is the female gametophyte. One of the eight nuclei is the egg nucleus which unites with the male gamete nucleus to form the diploid embryo of the new sporophyte. Two of the other female gametophyte nuclei are called polar nuclei. These fuse with the second male gamete nucleus to

LIFE CYCLE OF SEA LETTUCE, Ulva stenophylla, shows alternation of generations. Diploid and haploid forms are almost indistinguishable. Color indicates differences in mating types.

form a triploid nucleus—one containing three chromosomes of each type. This triploid nucleus divides repeatedly by mitosis to form the endosperm of the seed. In maize, then, both the male and the female gametophyte generations are re-

D Mating type versus sex

Biologists define the female gamete as the one that furnishes the zygote with most of its cytoplasm. The male gamete furnishes a smaller amount or none at all. In an organism such as the bread mold *Neurospora,* a single individual can act as either sex. There is no absolute requirement that the sex of an organism must be determined by its genes.

Mating type, on the other hand, is determined by genes. Two genetically different cells can fuse and form a zygote; two genetically identical cells cannot. In other words, the two gametes need only be different genetically, not complementary in any specifically sexual way. In some organisms there are hundreds of mating types, and each type can fertilize any of the others.

Sex and mating type are not mutually exclusive. Some organisms, such as *Neurospora,* exhibit both. *Ulva stenophylla (Figure 8)* has two mating types, but because their gametes are the same size, and thus furnish the zygote with equal amounts of cytoplasm, biologists do not consider the mating types as belonging to different sexes. Many higher plants, such as corn and peas, have sex but not mating type, because the pollen of a given individual can fertilize a genetically identical ovum.

duced to minute haploid organisms. They undergo only a few cell divisions within the shelter of the much larger diploid sporophyte plant.

ANIMALS

In animals, including man, the haploid stage is generally represented only by the gametes. Gametogenesis in the two sexes is different in detail. In males, a diploid spermatocyte undergoes meiosis to produce four haploid spermatids, which then mature into SPERMATOZOA. These spermatozoa are all alike in appearance. The female produces a large egg cell containing a sufficient store of nutrients to support the early stages in the development of the embryo. The process begins with a diploid cell, the oocyte, which undergoes two meiotic divisions in which the cytoplasm is divided asymmetrically between daughter cells *(Figure 10)*. Attached to the surface of the haploid egg are three tiny haploid POLAR BODIES, which are not functional gametes. They contain very little of the precious nutrient cytoplasm of the egg cell and are simply waste products of meiosis.

9 **LIFE CYCLE OF CORN** also involves alternation of generations. Most prominent phase is the diploid sporophyte (center). Haploid gametophytes are tiny structures nourished and protected by sporophyte.

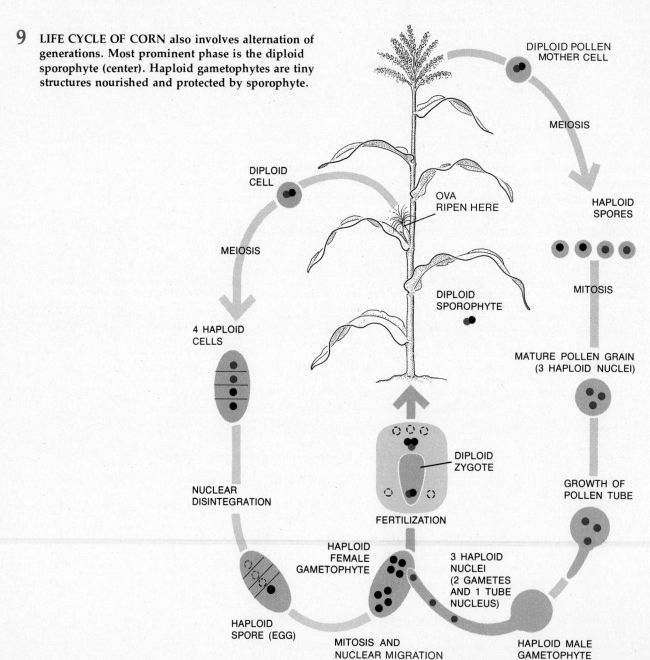

DIPLOID POLLEN
MOTHER CELL

MEIOSIS

DIPLOID
CELL

OVA
RIPEN HERE

HAPLOID
SPORES

MEIOSIS

DIPLOID
SPOROPHYTE

MITOSIS

4 HAPLOID
CELLS

MATURE POLLEN GRAIN
(3 HAPLOID NUCLEI)

NUCLEAR
DISINTEGRATION

DIPLOID
ZYGOTE

GROWTH OF
POLLEN TUBE

FERTILIZATION

HAPLOID
FEMALE
GAMETOPHYTE

3 HAPLOID
NUCLEI
(2 GAMETES
AND 1 TUBE
NUCLEUS)

HAPLOID
SPORE (EGG)

MITOSIS AND
NUCLEAR MIGRATION

HAPLOID MALE
GAMETOPHYTE

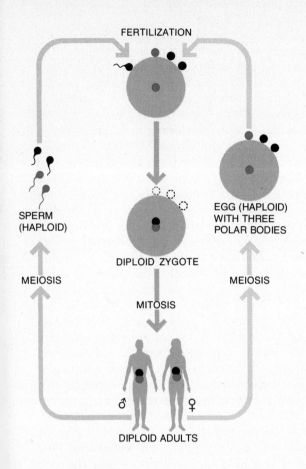

FERTILIZATION

SPERM
(HAPLOID)

MEIOSIS

EGG (HAPLOID)
WITH THREE
POLAR BODIES

DIPLOID ZYGOTE

MEIOSIS

MITOSIS

♂   ♀

DIPLOID ADULTS

10

LIFE CYCLE OF MAN is typical of most mammals. As in corn, diploid phase is most prominent. Haploid phase is reduced to a single cell (egg or sperm). Unlike corn, each mammal belongs to one sex or the other. Light- and dark-colored gametes show segregation for a specific trait.

NEUROSPORA

Conversely, in the commonplace salmon-colored bread mold *Neurospora* the diploid generation is reduced. In fact, meiosis immediately follows fertilization; the zygote itself comprises the entire diploid generation *(Figure 11)*. *Neurospora* grows from haploid spores that divide mitotically to produce a feltlike mat of long strands, or HYPHAE. Hyphae are not made up of cells in the usual sense, but consist of a mass of cytoplasm containing many nuclei. The hyphae readily give rise to haploid spores, called conidia, by a simple asexual process that does not involve meiosis. If a conidium lands on a suitable medium, it can grow into a new fungal

LIFE CYCLE OF NEUROSPORA differs from cycles of
corn and man in that haploid phase is more prominent. Diploid phase is reduced to a single cell. Spores produced by meiosis are enclosed in a sack called an ascus. Many of these asci are visible in the photomicrograph. White spores are due to the presence of a genetic mutation. Wild type spores are black, except that very immature ones are white.

HAPLOID CELL
IN A PROTOPERITHECIUM

FERTILIZATION

DIPLOID
ZYGOTE

HAPLOID NUCLEUS
FROM A CONIDIUM

PROTOPERITHECIUM

MEIOSIS

MITOSIS

MITOSIS

MITOSIS

MITOSIS

MITOSIS   MITOSIS

HAPLOID
MYCELIUM

HAPLOID
MYCELIUM

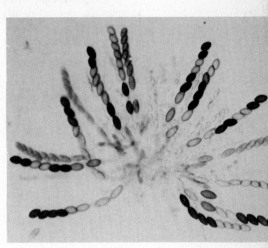

mat. Under different conditions a conidium may play a role in the sexual cycle of this fungus.

Neurospora, like *Ulva*, occurs in two mating types, called *A* and *a*, which are indistinguishable to the eye. But *Neurospora* forms different male and female structures. Each mating type can function either as a male or as a female. If one of the mating types, say *A*, is grown on a suitable medium, it will, after about four days, form differentiated female structures called protoperithecia. If the protoperithecia are dusted with conidia of the opposite mating type, *a*, a single nucleus from a conidium will be conducted into each protoperithecium. There it will divide a number of times by simple mitosis to give more haploid nuclei of mating type *a*. In other words, at this stage the cytoplasm contains two different types of nucleus. Nuclei of mating type *a* then fuse with those of *A* to form a number of diploid zygote nuclei. As soon as a diploid nucleus is formed, it enters into meiosis. Sex is determined by an accident of priority. The first mating type to colonize an area becomes the female. The conidia that arrive later function as the male.

Meiosis in *Neurospora* is a particularly tidy process that packages all of the nuclei produced by the divisions of a single zygote in a long, thin sac called an ASCUS. The two meiotic divisions occur as usual, producing four haploid nuclei which divide once again by mitosis. The eight nuclei produced by this sequence of events are incorporated into eight spores, all neatly lined up within the narrow ascus. Because the ascus is so narrow, the nuclei cannot pass one another as the divisions proceed, so the pairs of spores can easily be identified with the meiotic division that produced them. This makes *Neurospora* an especially useful organism in which to study segregation, assortment, and recombination of genetic markers.

SEX DETERMINATION

In maize, every diploid sporophyte gives rise to both male and female structures. These two types of tissue are genetically identical, just as roots and leaves are genetically identical. Plants such as maize, and animals such as earthworms, which produce both male and female gametes in the same organism, are said to be MONOECIOUS (Greek: "single house"). Some higher plants, notably the date palm, and most animals are DIOECIOUS —male and female gametes are produced in separate organisms. In most dioecious creatures, the sex is determined by differences in the chromosomes.

But even this mechanism operates in a bewildering variety of ways *(Figure 12)*.

The sex of a honeybee depends on whether it developed from a fertilized or an unfertilized egg. A fertilized egg is diploid, and it gives rise to a female bee —either a worker or a queen, depending on its diet during larval life. An unfertilized egg is haploid, and gives rise to a male drone. In many other animals, including man, sex is determined by a single chromosome or pair of chromosomes. The classical example is found in an insect called *Pyrrocoris*. Two copies of a sex chromosome are present in the female, only one copy in the male. The rest of the chromosomes are all present in two copies in both male and female. The chromosomes that are equally present in both sexes are called AUTOSOMES. The one that is present in different amounts is called the X chromosome. Females form eggs that contain one copy of each autosome, and one X chromosome. Males, on the other hand, form two types of sperm. Half contain an X chromosome and one copy of each autosome, and the other half contain only the autosomes. Females are referred to as XX (ignoring the autosomes) and males as XO (pronounced X-oh). When an X-bearing sperm fertilizes an egg the zygote will be XX and will develop into a female. When a sperm without an X fertilizes an egg the zygote will develop into an XO male. This chromosomal mechanism insures that the two sexes are produced in equal numbers. No such mechanism exists in the diploid-haploid system of bees, nor in the "squatters' rights" system of *Neurospora*.

In the fruit fly *Drosophila*, females have two X chromosomes and males have one. But males in turn have a kind of chromosome that is not found in females: the Y chromosome. The females can be represented as XX and the males as XY. The males form two kinds of gametes that differ with respect to their sex chromosomes. Half carry an X chromosome and the other half carry a Y. When an X sperm fertilizes an egg, the zygote will develop into a female; when a Y sperm fertilizes an egg, the zygote will become a male.

Humans and other mammals follow the same pattern as *Drosophila:* females are XX and males are XY. But there are some subtle but very important differences which show up clearly in individuals with abnormal chromosome constitutions. In both *Drosophila* and in humans, XO individuals sometimes appear. In *Drosophila* these XO individuals are males, almost indistinguishable from normal XY males except that they are sterile. In humans

XO individuals are moderately abnormal females who are almost always sterile. In mice, XO individuals are fertile females that are virtually normal. XXY *Drosophila* occasionally occur. They are normal, fertile females. XXY humans are decidedly abnormal and always sterile. In general they look somewhat more like males than like females. In brief, sex in *Drosophila* is determined mostly by the

SEX IS DETERMINED by different chromosomal mechanisms in the four types of organism depicted here. In humans two X chromosomes are characteristic of normal females, X and Y of normal males. Bees and other social insects have no sex chromosomes. Generally females, which develop from fertilized eggs, are diploid while males, which develop from unfertilized eggs, are haploid. In birds, two X chromosomes produce a male, XY a female. Pattern in Drosophila resembles that of man except for abnormal genotypes. Human with one X is abnormal female, but in Drosophila is a sterile but normal male. XXY human is an abnormal male, but in Drosophila is a fertile, virtually normal female.

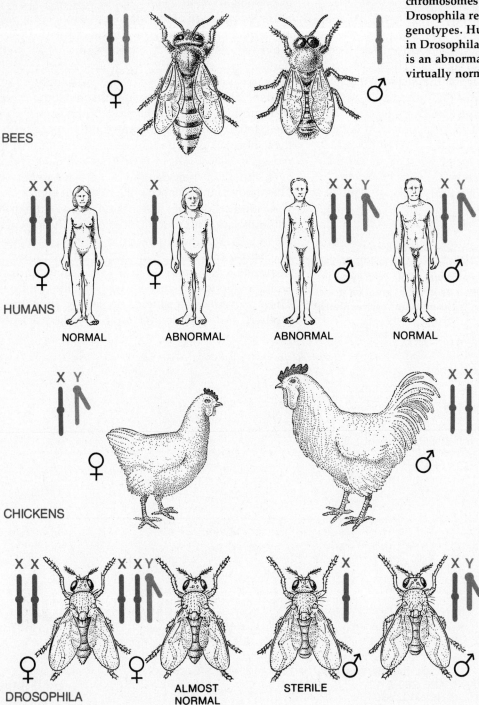

BEES

HUMANS

NORMAL   ABNORMAL   ABNORMAL   NORMAL

CHICKENS

DROSOPHILA   ALMOST NORMAL   STERILE

number of X chromosomes, with the Y chromosome needed only for male fertility. In humans the absence of Y leads to femaleness, and the presence of Y has a definite masculinizing effect. In birds, moths and butterflies, males are XX and females are XY. In these organisms it is the female that produces two types of gametes, and the sex of the offspring is hence determined by the egg, not by the sperm as in humans.

SEX LINKAGE

In *Drosophila* and in man the Y chromosome is almost devoid of genetic content. This leads to an important deviation from the usual Mendelian laws when one examines the inheritance of genes located on the X chromosome. Any gene on the X chromosome is carried in two copies by females, but in only one copy by males. Therefore, females may be heterozygous for genes that are on the X chromosome, but males will always be HEMIZYGOUS for these genes. The first and still one of the best

examples of sex-linked inheritance is the eye color in *Drosophila*. Their normal eye color is red, and the gene is carried on the X chromosome. In 1910, Thomas Hunt Morgan discovered a mutation that causes white eyes. When red-eyed, homozygous females were crossed to white-eyed males, all of the sons and daughters had red eyes because red is dominant over white, and all of the progeny inherited a normal X chromosome from their mother. But when a white-eyed female was mated to a red-eyed male, all of the sons were white-eyed and all of the daughters red-eyed, because the sons inherited their only X chromosome from their mother, and the Y inherited from the father does not carry any gene for eye color. The daughters, of course, get a white-bearing chromosome from their mother, and a red-bearing chromosome from their father, and are therefore heterozygotes. If these heterozygous daughters are mated in turn to red-eyed males, half of their sons will have white eyes, but all of their daughters will have red eyes. This mode of inheritance is illustrated in Figure 13.

A number of genes are located on the human X chromosome, and mutations affecting these genes are inherited in exactly the same way as white eyes in *Drosophila*. Hemophilia is a good example. A hemophilic man married to a homozygous normal woman will not produce any hemophilic children. The sons inherit a single, normal X from their mother, and will neither have the disease, nor

13 SEX LINKAGE in Drosophila is depicted in the diagram below. The gene that governs eye color lies on the X chromosome. The wild type allele (red eyes) is dominant. Only females can be heterozygous for genes on X chromosome because Y chromosome carries almost no known genes.

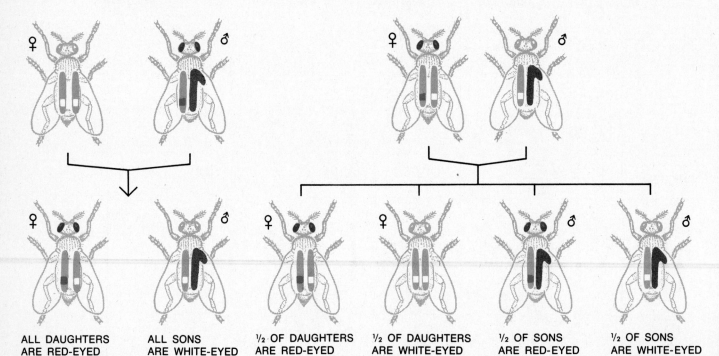

| ALL DAUGHTERS ARE RED-EYED HETEROZYGOTES | ALL SONS ARE WHITE-EYED HEMIZYGOTES | ½ OF DAUGHTERS ARE RED-EYED HETEROZYGOTES | ½ OF DAUGHTERS ARE WHITE-EYED HOMOZYGOTES | ½ OF SONS ARE RED-EYED HEMIZYGOTES | ½ OF SONS ARE WHITE-EYED HEMIZYGOTES |

transmit it to their children. The daughters will get a normal X chromosome from their mother and hemophilia-bearing chromosome from their father. Since hemophilia is recessive, the daughters will not be hemophilic. They will, however, be heterozygous carriers, and will transmit the disease to half their sons, and the carrier role to half their daughters. What would be needed to produce a female hemophiliac analogous to the white-eyed female fly? Her father would have to be a hemophiliac and her mother a carrier. Since hemophilia is quite rare, such a couple would be unlikely to meet. Moreover, hemophilic males rarely survive long enough to reproduce. One might expect hemophilic females to be extremely rare, and in fact very few have ever been found.

The gene for glucose-6-phosphate dehydrogenase deficiency also lies on the X chromosome in humans, and, like the gene for hemophilia, is recessive to the normal allele. As noted in Chapter 7, a deficiency of this enzyme causes an abnormal sensitivity to primaquine, an antimalarial drug. This condition is much more common than hemophilia, and the disease is not commonly fatal. As one might expect, women who are homozygous for primaquine sensitivity are rather rare, but they have been found. All of their sons are primaquine-sensitive; all of their daughters will be heterozygous, provided the father is normal.

RECOMBINATION IN EUCARYOTES

It is obvious that the number of genes in a cell far exceeds the number of chromosomes, for each chromosome contains many genes. If chromosomes were not capable of pairing and recombining, the markers on a given chromosome would all segregate together, and the genes in an organism would fall into several segregation groups, one for each chromosome. The different segregation groups would segregate independently of one another, but within each group all markers would behave as a unit. A geneticist would have no way of telling that they were actually different genes. Mendel's Second Law (independent assortment of alleles of different genes) would apply only to genes that were on different chromosomes.

The actual situation is more complex and therefore more interesting. Mutations located at different places on the same chromosome do separate from one another as the result of recombination, and the frequency with which they separate from one another depends on the distance between them on the chromosome1 With eucaryotes, as with the procaryotes, geneticists can use recombination frequencies to generate genetic maps that indicate the actual arrangement of genes along the chromosome *(Box E)*. Because meiosis in eucaryotes is a more orderly process than the pairing and recombination of the phages and bacteria, geneticists know much more about the formal details of recombination in eucaryotes than in procaryotes. But because the chromosomes of eucaryotes are much more complex than the simple DNA molecules of phages and bacteria, less is known about the molecular details of the recombination process in higher organisms than in viruses and procaryotes.

E Gene Mapping in Eucaryotes

Neurospora is an excellent organism for illustrating the principles of mapping. Most genetic analysis in *Neurospora* does not require that the products of single meiotic events be held together in an ascus. It is usually more convenient to examine a mixed population of spores or "random spores" that have been ejected from the fruiting body. Suppose two strains of *Neurospora* are crossed, carrying the markers *A B* and *a b*. Examination of the random spores that emerge from the cross indicates the following frequencies: *A B* 40 per cent; *a b* 40 per cent; *A b* 10 per cent; *a B* 10 per cent.

Of all the spores, 40 + 40 = 80 per cent are of parental genotype, and 10 + 10 = 20 per cent are recombinant. The frequency of recombination between *A* and *B* is 20 per cent.

As with phages, the ordering of markers is best accomplished by doing a three-factor cross. If strains that are *A B C* and *a b c* are crossed, the following might result:

1. *A B C* 38.0 per cent
2. *a b c* 40.2 per cent
3. *A b c* 7.2 per cent
4. *a B C* 6.6 per cent
5. *A B c* 3.1 per cent
6. *a b C* 3.7 per cent
7. *A b C* 0.5 per cent
8. *a B c* 0.7 per cent.

(Continued on next page)

Classes 1 and 2 are parental types. Classes 3 and 4 are single recombinants between *A* and *B*. Classes 5 and 6 are single recombinants between *B* and *C*. Note that all four of the classes 3, 4, 5, and 6, are recombinants between *A* and *C*. Classes 7 and 8 are the least frequent classes and therefore represent the double crossover types. They are recombinant between *A* and *B* and between *B* and *C*, but not between *A* and *C*. The two crossovers cancel each other with respect to the *A-C* recombination and leave the two markers in the parental arrangement relative to one another. If marker *B* were absent from the cross, we would score these doubles as parental types with respect to *A* and *C*.

To compute recombination distances, we add up the frequencies of all of the classes that are recombinant between a given pair of markers: The "distance" between *A* and *B* is the sum of classes $3 + 4 + 7 + 8 = 7.2 + 6.6 + 0.5 + 0.7 = 15.0$ per cent recombination. The "distance" between *B* and *C* is the sum of classes $5 + 6 + 7 + 8 = 3.1 + 3.7 + 0.5 + 0.7 = 8.0$ per cent recombination. The "distance" between *A* and *C* is the sum of classes $3 + 4 + 5 + 6 = 7.2 + 6.6 + 3.1 + 3.7 = 20.6$ per cent recombination. We can, therefore, draw a map of the three markers, showing their recombination "distances," as follows:

Why do the recombination frequencies between *A* and *B* and between *B* and *C* not add up to the recombination frequency between *A* and *C*? It is because the double crossovers between *A* and *C* contribute to the recombination frequency between *A* and *B* and between *B* and *C*, but not to that between *A* and *C*. The deviation that this produces from perfect additivity of the distances can easily be calculated. It is expected that the frequency of double crossover events will be the product of the frequencies of the single crossover events, provided that the occurrence of a crossover between *A* and *B* is independent of the occurrence of a crossover between *B* and *C*. It can be predicted, therefore, that the frequency of double crossover events will be 15.0 per cent $\times$ 8.0 per cent, or 1.2 per cent. In this example, the frequency of doubles is, in fact, $0.5 + 0.7 = 1.2$ per cent of the total progeny. Even if only the frequency of recombination between *A* and *B* (R_{AB}) and the frequency of recombination between *B* and *C* (R_{BC}) were known, the expected recombination frequency of *A* and *C* (R_{AC}) could be calculated:

$$R_{AC} = R_{AB} + R_{BC} - 2R_{AB}R_{BC}.$$

In the present example,

$$R_{AC} = 0.15 + 0.08 - 2(0.15)(0.08) = 0.206, \text{ or } 20.6 \text{ per cent.}$$

The reason the frequency of double crossovers must be multiplied by two before being subtracted from the sum of the single crossovers is that the double crossovers are counted in both classes of single crossovers.

It is obvious that the observed frequency of recombination between two markers is a complicated function of the probability that one, two, three, or more crossovers will occur in the chromosomal segment between the two markers. As two markers get farther and farther apart, so that recombination arranges them at random with respect to one another, the frequency of recombination approaches 50 per cent. A mathematical expression that predicts the observed frequency of recombinants between two markers as a function of a map distance between the two markers is called a MAPPING FUNCTION. A derivation of such a function is beyond the scope of this book, but the illustration in this box shows how one mapping function generates recombination frequencies as a function of marker distance. It is clear from the figure that recombination frequency shows good proportionality to map distance for markers that are close together and becomes less and less sensitive to increases in map distance for markers that are farther apart.

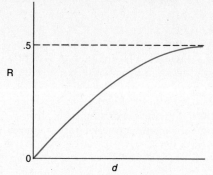

MAPPING FUNCTION plots recombination frequencies (R) versus the distance between markers (d). For short map distances, recombination frequency closely approximates "true" map distance. As distance between two markers increases, frequency of recombination approaches a limiting value of 0.5.

If there were no crossing over, all of the markers on a particular chromosome would behave as a simple segregation group and would always stay together during meiosis. Now that the principles of mapping are understood, a more useful concept can be defined. The LINKAGE GROUP is the genetic abstraction that corresponds to the physical structure called a chromosome. Exhaustive mapping of a variety of different organisms shows that the number of linkage groups is always equal to the number of chromosomes in the gamete, or haploid form of the organism. Of course, many markers on each chromosome must be observed before it is certain how many linkage groups there are. Markers at opposite ends of a given chromosome will often give very nearly 50 per cent recombination —so nearly that the number cannot be distinguished from 50 per cent with any degree of confidence. Such markers can be shown to be in the same linkage group, however, if each of them gives distinctly less than 50 per cent recombination when crossed to a marker that lies between them. If there are enough well-spaced markers on a chromosome —even a chromosome that is hundreds of map units long —all the markers on that chromosome can be shown to belong to the same linkage group.

In the mapping of diploid organisms the methods used are virtually identical to those described for analysis of haploid spores in *Neurospora*, except the products of meiosis cannot be directly examined. Crosses are done, therefore, in such a way that the genotype of the gametes to be tested can be directly determined by examining the phenotypes of the offspring. For example, suppose that the distance between two markers on the same chromosome in a diploid organism is to be measured. Call the markers *A* and *B*, and indicate the chromosome as a line with markers over it, for example, $\underline{A\ B}$ or $\underline{a\ b}$. First cross a diploid that has the genotype $\underline{A\ B}/\underline{A\ B}$ to a diploid that has the genotype $\underline{a\ b}/\underline{a\ b}$. Such a cross produces 100 per cent heterozygotes of the genotype $\underline{A\ B}/\underline{a\ b}$. To detect crossing over, it would be desirable to look directly at the gametes formed by these heterozygotes to see how many of them are $\underline{A\ b}$ or $\underline{a\ B}$. Unfortunately, it is hardly ever possible to tell anything about the genes a gamete is carrying by examining it directly. What does a blue-eyed sperm look like, or a hemophilic egg? The way to find out what genes a gamete is carrying is to allow it to unite with another gamete carrying the recessive alleles of all markers used in the cross, and examine the resulting individual. In practice this is done by mating the heterozygote to a recessive homozygote. In the present case, $\underline{A\ B}/\underline{a\ b}$ would be mated to $\underline{a\ b}/\underline{a\ b}$. In this cross the genetic constitution of the gametes produced by the heterozygous parent will be directly revealed in the phenotypes of the progeny *(see problems in Box F)*.

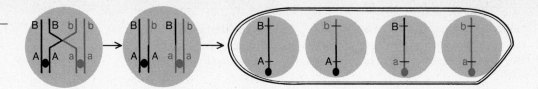

14 CROSSOVER occurs during the four-strand stage of meiosis. Diagram of meiosis in Neurospora proves this assertion. Crossover between two strands can produce only two types of spore in any one ascus. Crossover in four-strand stage can produce four different kinds of spore, as often found in nature. For simplicity, this diagram shows only one member of each pair of spores in the ascus.

15 DOUBLE CROSSOVER. Crossovers in the top diagram involve only two strands, and produce no recombinants between A and B. The middle diagram depicts two crossovers between three strands. Half of the resulting spores are recombinants and half are the parental type. The bottom diagram shows a crossover that involves all four strands of the tetrad. All of the resulting spores are recombinants.

Recombination between genetic markers on the same chromosome occurs by a process called CROSSING OVER. This presumably results from the physical exchange of corresponding genetic segments between two homologous chromosomes. Crossing over occurs during meiosis between the time when the chromosomes double before the first meiotic division and the time when the chromosomes separate from one another at the first meiotic anaphase. That is, recombination occurs when there are four homologous chromatids representing each chromosome pair. In the language of the geneticist, crossing over occurs at the four-strand stage. The exchange event at any point on the length of the chromosome involves only two of the four chromatids present in the cell. The

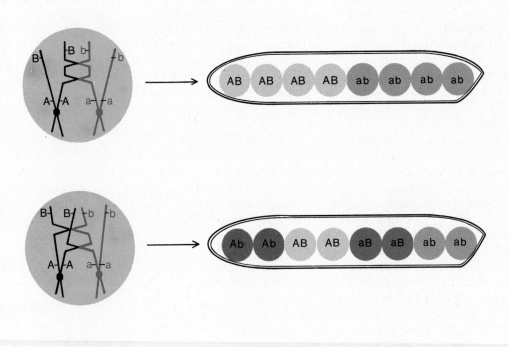

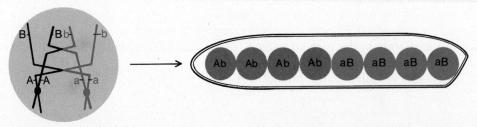

material is exchanged reciprocally, so that both chromatids involved in the crossing over process are recombinant, and there are no markers created or destroyed. The points at which the chromatids break in the exchange seem to correspond perfectly, so that no inequality exists in the amounts of material exchanged (*Figure 14*).

Crossing over may involve two, three, or all four chromatids in a tetrad. At any point along the chromosome only two chromatids participate in crossover, although other crossovers may occur at other points. These crossovers may involve the same pair of chromatids or any other possible pair of homologous chromatids. The probability that more than one crossover will occur in the genetic segment between two mutational markers will depend on the distance between the markers (*Figure 15*).

CYTOGENETICS

By making experimental test crosses, and calculating the recombination frequencies, geneticists can map the arrangement of genes along a chromosome (*Box F*). Such a map is a mathematical abstraction. To what extent does it correspond with the physical structure of the chromosome as seen under the microscope? To establish a relationship between a genetic linkage group and a chromosome, the geneticist tries to find a mutant whose normal linkage relationships are changed by some extensive mutation. He then examines its cells under the microscope to find a corresponding visible change in one or more chromosomes.

In the tissues of most organisms the chromosomes are too small for an observer to see any except the grossest sort of change. One exception,

---

F Problem set 2

Problem 1. Two true-breeding varieties of unicorn differ with respect to three traits. The first has a straight horn, a yellow coat, and a straight mane. The other variety has a twisted horn, a blue coat, and a curly mane. When a male of either variety is crossed with a female of the other, all of the progeny have straight horns, green coats, and a straight mane. These hybrid offspring are then crossed to unicorns of the second parental variety (twisted, blue, curly). This cross produces offspring in the following numbers:

straight horn, green coat, straight mane	855
twisted horn, blue coat, curly mane	855
twisted horn, green coat, straight mane	95
straight horn, blue coat, curly mane	95
straight horn, green coat, curly mane	5
twisted horn, blue coat, straight mane	5
straight horn, blue coat, straight mane	45
twisted horn, green coat, curly mane	45
	2000

Construct a linkage map for the three genes involved in this cross.

Answer to Problem 1.

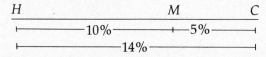

H is the gene controlling shape of horn, *M* controls mane straightness, and *C* controls coat color. Straight horn is dominant to twisted, and straight mane is dominant to curly. The homozygotes for the coat-color alleles are yellow or blue; the heterozygote is green.

(*Continued on next page*)

however, is the giant chromosomes in the salivary glands of the larvae of *Drosophila*. Called POLYTENE (multistranded) CHROMOSOMES, they are much longer than the chromosomes of other cells in the larva and have replicated so many times that the many copies lying side by side form thick, snakelike structures that can be seen clearly even with a low-magnification lens (*Figure 16*). The condensed thickenings along the interphase chromosomes (chromomeres) are all paired with sister chromomeres on the parallel strands of the polytene chromosome so that the chromosomes appear to have a pattern of transverse bands, each chromosome having a characteristic band pattern. The two homologous polytene chromosomes are closely synapsed. The patterns of bands —their thickness, spacing, sharpness or diffuseness and so on—are so characteristic for each chromosome that an experienced cytogeneticist can tell at a glance if the order or position of a group of bands has been changed.

A fairly common type of extensive mutation is

F (Continued)

Problem 2. Two true-breeding stocks of *Drosophila* are available. One of them is the wild type, with red eyes and normal wings. The other has vermilion eyes (*v*) and abnormal wings called crossveinless (*cv*). When wild-type females are crossed to vermilion, crossveinless males, all of the progeny are of the wild type in appearance. When the cross is done the other way around —a vermilion, crossveinless female is crossed to a wild-type male, all of the sons are vermilion, crossveinless, and all of the daughters are of the wild-type. If these hybrid daughters are crossed to their vermilion, crossveinless brothers, offspring of four different phenotypes are produced:

vermilion, crossveinless	35 per cent
normal eyes, normal wings	35 per cent
normal eyes, crossveinless	15 per cent
vermilion, normal wings	15 per cent.

All of these classes are composed of 50 per cent males and 50 per cent females. On what chromosome are the genes that control the vermilion eye-color trait and the crossveinless wing condition? Which alleles are dominant, the mutant or the wild type? How far apart are the two genes on the chromosome?

Answer to Problem 2. These genes must be on the X chromosome. This is shown by the results of the first two crosses. In the first cross, the daughters must have received an X chromosome from their fathers, yet they are not VERMILION, or CROSSVEINLESS. Therefore the wild-type alleles that came from the mother must be dominant. The total percentage of recombinants emerging from the third and last cross is 30 per cent. The recombination frequency can be related to a map distance by a mapping function that corrects for multiple crossovers, as shown by the illustration in Box E.

Problem 3. In mice, the sex-linked mutation *spf* (sparse fur) causes delayed, sparse growth of the first coat. Suppose that a sparse-furred male is crossed to an XO female with normal fur. What kinds of offspring would they be expected to have and in what proportions?

Answer to Problem 3. The female will produce two kinds of eggs, half with a normal (*spf*⁺) X chromosome and half with no X chromosome at all. The male will produce two kinds of spermatozoa, half with a mutant (*spf*) X chromosome and half with a Y. Four kinds of zygotes will be produced: X*spf*⁺/X*spf*, X*spf*⁺/Y, X*spf*/O, and Y/O. The first class will be normal-furred females; the second will be normal-furred males; the third will be sparse-furred females; and the fourth will die since it has no X chromosome at all. Thus, the living offspring of the cross are expected to be phenotypically normal females, normal males, and sparse-furred females in a 1:1:1 ratio.

bands in one of the chromosomes appears reversed. A linkage group with an inversion in it can be correlated with a chromosome whose bands are inverted, and the geneticist can assume that certain genes are located in certain regions of the visible structure of the chromosome.

Other extensive mutations, such as translocations and deletions, can be useful in correlating the map with the physical chromosome. Where there is genetic evidence of a translocation, examination of the cell shows that a group of bands has been moved to a completely different chromosome. Where there is genetic evidence of a deletion, a small group of bands, or even a single band, often appears to be missing from one of the chromosomes. The study of a great many extensive mutations in *Drosophila* has yielded the picture shown in Figure 17. This figure shows genetic and cytological maps of the X chromosomes of *Drosophila* drawn side by side. Notice that the order of genes deduced by the two methods is in agreement, but the distances are not. It is not known which map reflects more accurately the physical spacing of genetic information. Because of the exceedingly complex folding of DNA in a eucaryote chromosome one can never be sure that the cytological length of a segment of chromosome is a good reflection of the length of DNA in that segment. Nor is the recombinational length a reliable reflection of the length of DNA. Any region in which there is an increased frequency of crossing over per unit length of DNA will have a longer genetic length than a similar stretch of DNA with a lower frequency of crossovers. Despite these limitations, the combination of crossover analysis and cytogenetics has been very successful in probing the chromosomes of eucaryotes.

GENETIC DISEASE

One of the long-term priorities of genetic research is to minimize the effects of genetic diseases. There is no obvious borderline between genetic and nongenetic afflictions. Is allergy to ragweed (a variety of hay fever) a nongenetic disease? It is caused by an environmental agent, but not everyone exposed to a particular amount of ragweed pollen suffers equally. The variation in susceptibility to this allergen is due largely to genetic differences.

Of course one can argue that any disease is the result of an interaction between genotype and environment. It might be more useful to say that a disease is clearly genetic if people exposed to a

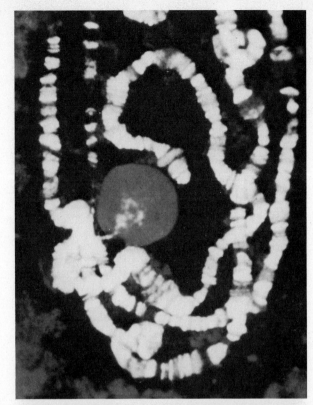

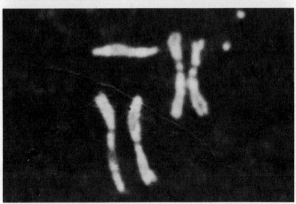

6 **POLYTENE CHROMOSOMES from the salivary gland of Drosophila, stained with fluorescent dye, appear as thick bright strands in photomicrograph at top. Sphere at center of top photo is nucleolus. Bottom cell shows normal Drosophila chromosomes from another type of cell.**

an inversion. Such mutations can be detected by genetic mapping as a change in the linkage relations of markers on a chromosome; if the normal order of markers on a chromosome is ABCDEFGHI, an inversion may have the order ABCDGFEHI. When the polytene chromosomes of such mutants are examined, the order of a group of

normal environment become ill because of their genotypes. Conversely nongenetic diseases affect different people similarly, regardless of their genotypes. One hesitates to cite specific examples of nongenetic diseases, however, because the failure to appreciate the importance of genotype in determining the severity of a particular disease might simply reflect medical ignorance.

Even if one uses the most conservative definition of genetic disease, the amount of human suffering attributable to it is enormous. Approximately ⅓ of the children in hospitals at any given time are there because of genetic diseases. Between the cradle and the grave, at least one person in 20 requires treatment for a serious genetic disease. Because genetic diseases are often disabling and chronic, the cost for treatment —whether measured in time, money or pain —is especially high.

There is no satisfactory treatment for many genetic diseases; the best that can be done is to maintain life and minimize discomfort. Specific treatments have been found for some, such as phenylketonuria (PKU), diabetes and hemophilia. Phenylketonuria, which causes mental retardation and other disabilities, can be treated by providing affected infants with a diet low in the amino acid phenylalanine. Diabetes, a disease in which genes clearly play a significant part, is treated by control of diet and often by insulin injections. Hemophilia is treated by introducing clotting factors into the bloodstream of the patient. In some cases the treatment of a genetic disease requires the identification of a substance that is missing in the affected

individual, such as a hormone or an enzyme, followed by replacement of the missing molecule. In other cases, treatment requires the identification of some environmental factor —a component of food, for example, or an allergen —that provokes a pathological response in persons having a particular genotype. Treatment simply involves the elimination of this agent from the sufferer's environment. A major goal of research in medical genetics is the discovery of the chemical causes of genetic diseases, and the development of effective treatments.

> Even if one uses the most conservative definition of genetic disease, the amount of human suffering attributable to it is enormous. Approximately ⅓ of the children in hospitals at any given time are there because of genetic diseases.

A clinical approach to genetic disease is aimed at prevention through counseling. To do this, it is necessary to identify prospective parents who have a high probability of giving birth to affected children and to provide them with information on how they might avert such misfortune. High-risk parents can be identified in several ways. Individuals who are affected by a genetic disease, or whose relatives are affected, are obvious candidates for genetic counseling. Those who already have had one affected child will also want advice on the likelihood of recurrence. Members of certain distinguishable segments of the population have a relatively high risk of carrying certain deleterious

17 GENETIC AND CYTOLOGICAL MAPS of the X chromosome of Drosophila are compared in this diagram. Position of centromere is indicated by a star. Numbers below chromosome refer to the banding pattern observed under the microscope. Lines and numbers above chromosome indicate the genetic map worked out from recombination frequencies. Letters denote marker mutations. Photomicrograph at upper left depicts the tip of the X chromosome.

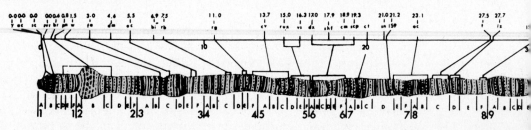

genes. For example, the recessive gene that causes sickle cell anemia is common among West African and American blacks; the recessive gene that causes Tay-Sachs disease (which is fatal in early childhood) is rather high among Jews. Heterozygous carriers of these recessive genes can be identified by analysis of blood samples. In many states voluntary screening programs have been made available to members of these populations who want to determine whether they are in danger of having afflicted offspring. Genetics laboratories are busy trying to develop reliable and inexpensive tests to detect carriers of other serious genetic diseases such as cystic fibrosis and muscular dystrophy. Women over 40 face a rather high risk of bearing a child with Down's syndrome, a congenital disease caused by an incorrect number of chromosomes *(Chapter 10)*. Down's syndrome, commonly called Mongolism, occurs in fewer than one in 600 infants, but the incidence rises to one in 50 born to mothers over 45 years old.

What can high-risk parents do? Several alternatives are available. For example, a couple can decide to adopt children rather than having any of their own. When both partners are heterozygous carriers of a recessive gene (which leads to the probability that ¼ of their children will be afflicted), the couple may choose to have the woman artificially inseminated with sperm from an anonymous non-carrier male. A technique known as amniocentesis enables prospective parents to learn fairly early in pregnancy whether or not a fetus is affected by certain genetic diseases. A sample of fluid (and free-floating fetal cells) is withdrawn from the embryo sac by a needle inserted through the abdomen of the mother. The technique does not endanger the fetus, and it provides a diagnosis in time for the couple to decide to abort an abnormal fetus and try again. As a final alternative, high-risk parents can simply choose to ignore the odds and take their chances, realizing that they might produce a diseased child who will

be a burden to them, to himself, and to society, which pays a large share of the extraordinary costs of chronic and incurable disease.

Medical genetics has provoked considerable debate about ethics. The knowledge gained by the medical geneticist forces prospective parents to make difficult choices: to conceive or not, to abort or not. Some moralists argue that a person should not be told so much about himself that he is forced to make such difficult decisions; they believe that genetic screening creates more problems than it cures. Others argue that the more one knows about himself, the more intelligently and humanely he can conduct his life. There is no doubt that as the technology of medical genetics becomes more powerful, the debate will become wider and louder.

NON-MENDELIAN INHERITANCE

The essence of Mendelian inheritance is that a very few copies of chromosomal information are partitioned with great precision during meiosis. But there are self-reproducing entities within eucaryote cells other than the chromosomes of the nucleus. Mitochondria, chloroplasts, and certain other cytoplasmic organelles also appear to carry some genetic information. The DNA of these organelles is subject to mutation just as is chromosomal DNA, and therefore these organelles can carry "markers." But these markers are not inherited in the same way as nuclear chromosomal markers because the amounts of cytoplasm contributed by the maternal and paternal gametes are grossly unequal. Generally speaking, all of the mitochondria (and, in eucaryote plants, chloroplasts) in a zygote come from its mother, even though half of its chromosomes come from its father. Hence any particle that is inherited through the cytoplasm is sometimes said to be maternally inherited.

A good example is furnished by a mutation in

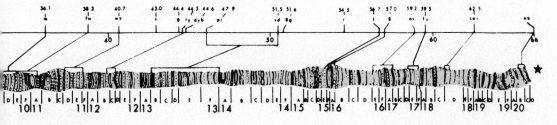

SCALE ⟵ 10μ ⟶

Neurospora that results in abnormal mitochondria. The mutant is called *poky* because it grows slowly. If protoperithecia of *poky* are fertilized with normal, wild-type conidia all the progeny are *poky*. If wild-type protoperithecia are fertilized by *poky* conidia, all the progeny are wild type. Thus, all the progeny resemble their mother. This might resemble sex linkage, but it is not. There is no sex chromosome in *Neurospora,* and all of the progeny are haploid.

Maternal inheritance can easily be confused with infections that are passed from mother to young. The transmission of mouse cancer was originally thought to occur by cytoplasmic inheritance. One strain of mice has a very high incidence of mammary cancer. The females of this strain transmit the trait to all their offspring, but the male does not. At first it seemed that this might be due to a "gene" that was passed only through the cytoplasm of the ovum. But further investigation showed that if baby mice were taken from the high-cancer mothers at birth and suckled by a substitute mother from a low-cancer strain, they did not develop cancer. The high-cancer trait is passed from mother to progeny by way of viruslike particles in the milk. Many other traits that masquerade as maternal inheritance can be understood as an infection by a virus or an intracellular bacterium that is transmitted through intimate contact between mother and offspring.

Certain patterns of cellular architecture are inherited in a way that may not involve DNA at all. This sort of "pattern inheritance" seems to be especially important in protozoa such as *Paramecium;* but perhaps we merely notice it more in protozoa because they are large, complex cells with distinctive external anatomy. One of the most bizarre examples of pattern inheritance is found in the protozoan, *Difflugia corona,* which was studied by H. S. Jennings in 1937. Jennings' work was eloquently summarized by David Nanney, writing in *Science:* "The organism constructs a shell by cementing sand grains together with a cellular secretion. The ventral surface of the shell possesses an opening, the 'mouth,' though which the cell communicates with the outside world. The edges of the openings are surrounded by a symmetrical array of 'teeth.' Jennings observed that the numbers of teeth varied among individuals, and he explored the question of their heredity by the only means available; he isolated individuals, allowed clones to develop, and inquired into clonal uniformity. Tooth number remained constant within a clone; differences in tooth number were hereditary. He was not able to conduct a breeding analysis, but he noted that, when the cell body divided, one of the daughter cells was extruded naked through the mouth and, while still in contact with its sister, began to construct its own sand castle, beginning in the region of contact *(Figure 18).* The new mouth structures were therefore constructed in direct contact with structures of the old mouth. This observation suggested to him that the old mouth might serve as a template to guide the organization of the new one —that new teeth were initiated in the interstices between the old teeth. This curious speculation might have remained just that had Jennings not tried his hand at oral surgery. He broke out denticles with a glass needle and examined the consequences of mutilating the parental template. Modified parents produced modified progeny, and new lineages were established with new tooth numbers. A few generations were required for symmetry to be achieved, but

18 **PATTERN INHERITANCE governs the formation of "teeth" in the protozoan Difflugia corona. The teeth are made of sand. Cell with 13 teeth produces offspring with same number of teeth, because new teeth are formed while offspring is still in contact with the mouth of the parent cell (left center). If four teeth are removed by microsurgery, parent cell produces offspring with nine teeth (far right).**

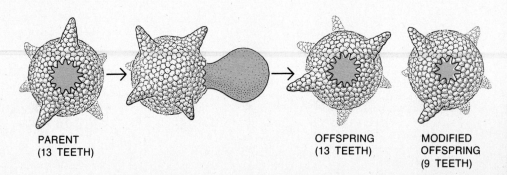

PARENT
(13 TEETH)

OFFSPRING
(13 TEETH)

MODIFIED
OFFSPRING
(9 TEETH)

once that had been accomplished the tooth number stabilized and a new hereditary state was achieved. After considering these studies, one of my colleagues concluded ruefully that genetic specificity may be based on two different structural foundations —nucleic acid and sand. In view of the notorious instability of structures built upon sand, this conclusion is peculiarly disturbing."

READINGS

A.M. SRB, R.D. OWEN AND R.S. EDGAR, *General Genetics*, 2nd Edition, San Francisco, W.H. Freeman and Company, 1965. One of the favorite texts used in general genetics courses, this book is written in lucid and interesting language, and every chapter contains an instructive problem set.

C. STERN AND E.R. SHERWOOD, *eds, The Origin of Genetics: A Mendel Source Book*, San Francisco, W.H. Freeman and Company, 1966. This is a collection of original writings of the founders of genetics. It includes readable translations of Mendel's original paper, of his letters to Nägeli, and of papers by de Vries and Correns, who in 1900 independently rediscovered Mendel's work. In the last two articles in the book, two of the masters of mathematical genetics discuss the likelihood that Mendel fudged his data.

M.W. STRICKBERGER, *Genetics*, 2nd Edition, New York, The Macmillan Company, 1976. A fine, up-to-date text that deals in considerable depth with a broad range of genetic topics. Also contains useful problems.

A.H. STURTEVANT, *A History of Genetics*, New York, Harper & Row, 1965. An eye-witness account by one of the pioneers of genetics. The book begins with Mendel and ends at the dawning of molecular genetics. A fascinating and charming book.

A.H. STURTEVANT AND G.W. BEADLE, *An Introduction to Genetics*, New York, Dover Publications, 1962 (reprint of a book first published by W.B. Saunders Company in 1939). The text is old, but the years have treated it well. This is still one of the best introductions to "classical" genetics.

The ultimate aim of biochemistry is to gain complete insight into the unending series of changes which attend plant and animal metabolism. —Emil Fischer

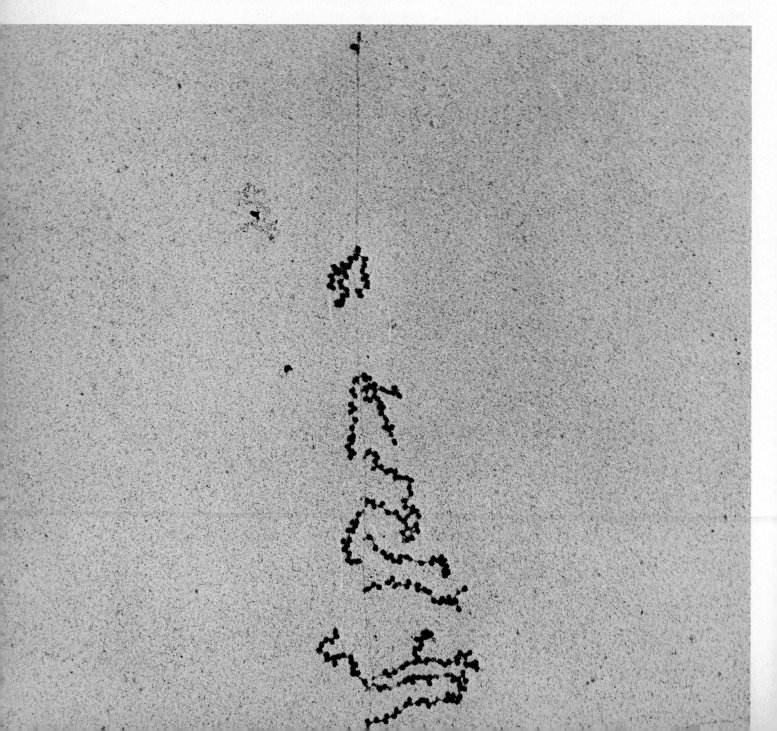

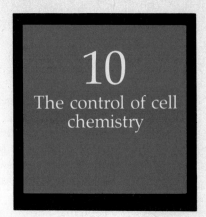

If the cell is thought of as a miniature chemical plant designed to carry out all the processes of life, then the nucleus can be compared to a central computer that controls a network of sophisticated and highly automated biochemical machinery. The genes comprise a library of programs stored in the computer's memory banks, programs that specify the precise nature of each protein synthesized by the cell. The computer monitors changing conditions both within the cell and in the external environment, and responds to this input of information by selecting and activating the appropriate genetic programs.

Control of enzyme synthesis is the key to controlling the chemical pathways of the cell. At any given moment most of the genes in a cell are inactive, or, in the language of the geneticist, they are not being EXPRESSED. By switching particular genes on and off, the cell controls not only the kinds of enzymes that it produces, but also the amounts. Both qualitative and quantitative control of enzyme synthesis are crucial to the proper functioning of the cell, for three main reasons.

First, some enzymes are needed in much larger amounts than others. For example, most cells burn large quantities of glucose, and it seems logical that a relatively large number of enzyme molecules are required to break down this sugar. Conversely, other reactions may require only a few molecules of certain enzyme cofactors, and it seems logical that the enzymes for cofactor synthesis would be made in correspondingly tiny amounts. Sharp differences in abundance between major and minor

POLYSOMES appear as strings of beads in the electron micrograph on opposite page. A strand of DNA from E. coli stretches like a thin wire from top to bottom of picture. Dark spot at extreme top appears to be the promoter site with a molecule of RNA polymerase attached to it.

enzymes persist even in cells maintained in a constant environment.

Second, all cells must be able to adapt at least to some degree to short-term changes in their environment. They do this by switching various genes on or off. The concentration of some enzymes within a cell can be changed more than a thousandfold in response to environmental changes.

Third, the cells of multicellular organisms are highly specialized in their functions, and they may differ tremendously in their enzymatic equipment. This is not because they have different genes from the other cells in the organism, but because they are expressing a different set of genes. The whole mystery of the development of a fertilized egg into an adult, or of a normal cell into a cancer cell, can be thought of in terms of differences in gene expression.

The mechanisms that control gene expression are not nearly as well understood as those that underlie the coding, replication and transmission of genetic information, but at least the overall strategies of the control mechanisms have become clear. Faced with the problem of decreasing the flow of a metabolite through a biochemical pathway, a cell could simply decrease the activity of the existing enzyme molecules that catalyze that pathway. This control strategy, called FEEDBACK INHIBITION, is common in both bacteria and higher organisms (Chapter 5). Another way to reduce the flow is to decrease the synthesis of the unneeded enzymes by switching off the corresponding genes. This strategy is more efficient than making the enzymes and then deactivating them, and the mechanism for it is built into the design of all living cells.

Conversely, to increase the output of a metabolic pathway within a cell, nature has a choice of at least five alternative mechanisms: 1, to increase the number of structural genes for the enzymes in that

205

pathway without changing any of the other genes; 2, to leave the number of genes constant, but speed up the rate of transcription to increase the number of messenger RNA molecules involved in the synthesis of the necessary enzymes; 3, to leave the rate of transcription constant, but decrease sharply the rate of breakdown on the mRNA, thus increasing the available pool of these mRNA molecules; 4, to leave both the rate of synthesis and the breakdown of mRNA constant, but speed up the rate of translation of the message, enabling the ribosomes to synthesize more of the enzyme proteins; and 5, to decrease the rate at which the enzyme molecules themselves are broken down within the cell, thus increasing the pool of enzyme molecules. Nature seems to use each of these mechanisms in one situation or another. Some of them are better understood than others. One of the most important and best understood is the control of transcription.

CONTROL OF TRANSCRIPTION

As a normal inhabitant of the mammalian gut, the bacterium *E. coli* has to adjust to sudden changes in environment. Its host may present it with one foodstuff one hour, another the next. For example, the bacterium may suddenly be deluged with milk, the main carbohydrate of which is lactose. Before this sugar can be of any use to the bacteria it must first be taken into their cells by an enzyme called β-galactoside permease. Then it must be hydrolyzed to glucose and galactose by another enzyme, β-galactosidase. A third enzyme, called β-galactoside transacetylase, whose function is

> The whole mystery of the development of a fertilized egg into an adult, or of a normal cell into a cancer cell, can be thought of in terms of differences in gene expression.

unclear, is also involved in lactose metabolism. When *E. coli* is grown in a medium that does not contain lactose or β-galactosides, the level of all three of these enzymes within the bacterial cell is very low. If lactose is added, the synthesis of these three enzymes begins promptly, and their levels may rise more than a thousandfold. Compounds that evoke the synthesis of an enzyme are called INDUCERS, and the system that is evoked is called an INDUCIBLE system. (Enzymes that are made all the time at a constant rate are called CONSTITUTIVE ENZYMES.) If the lactose is removed from the medium, the synthesis of the three enzymes stops almost immediately. The enzyme molecules that have already been formed do not disappear; they are merely diluted during subsequent growth and reproduction until their concentration falls to the original low level per bacterium.

The blueprints for the synthesis of these three enzymes are called STRUCTURAL GENES, indicating that they specify the primary amino acid sequence of protein molecules. When mutations in these structural genes were mapped crudely by conjugation experiments, it was found that all three genes

A Nonsense mutations

The evidence for the nature of chain-terminating codons did not come from a study of the natural terminators at the ends of cistrons, but from examining mutants in which termination occurred at abnormal positions. Figure 18 in Chapter 7 illustrates that the codons for a good many amino acids can be changed by a single mutation in DNA to one of these chain terminators. For example, the codon UGG (tryptophan) can mutate to UAG (chain terminator). If tryptophan normally occurs at a certain place in a protein, but the codon is changed by mutation as above, the translation does not proceed normally; instead the protein is terminated to give a carboxyl group where tryptophan would normally have been inserted, giving a shorter-than-normal protein. Such proteins are rarely functional, and comprise a large share of the clearcut auxotrophic mutants that can be isolated in microorganisms. Such misplaced punctuation can occur in any gene, and in many different places in each gene. For historical reasons they have been called NONSENSE MUTATIONS, as well as CHAIN TERMINATOR MUTATIONS, and they have proved enormously useful in biochemical genetics.

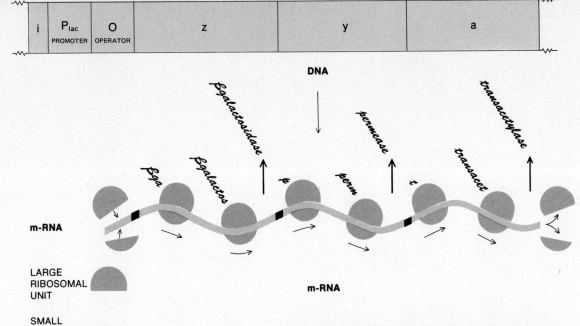

| i | P$_{lac}$ PROMOTER | O OPERATOR | z | y | a |

DNA

m-RNA

LARGE RIBOSOMAL UNIT

SMALL RIBOSOMAL UNIT

m-RNA

TRANSLATION of a polycistronic message from the lactose operon. The lac region of bacterial chromosome is shown at top. The z, y and a genes are the structural genes for β-galactosidase, permease and transacetylase. They are transcribed to give a single mRNA molecule. Ribosomes attach to the messenger at left and move toward right, translating as they go. Finally they drop off the messenger.

1

lie quite close together in a region that covers only about one percent of the bacterial genophore.

It is no coincidence that these three genes lie next to one another. The information from them is transcribed into a single continuous molecule of mRNA. Such a molecule is called a POLYCISTRONIC messenger, and it governs the synthesis of all three enzymes. Provided that the structural genes for all of the enzymes are intact, either all of the enzymes will be made, or none of them.

How can a single messenger RNA molecule make three different polypeptides? The answer is that the mRNA contains punctuation marks that specify the ending of one polypeptide chain and the beginning of the next. The initiation of translation is beyond the scope of this book, but once begun, the termination of translation is much simpler.

How can a single mRNA molecule make three different polypeptides? The answer is that the mRNA contains punctuation marks that specify the end of one polypeptide chain and the start of the next.

A molecule of tRNA is always attached to a growing polypeptide chain, but a finished molecule of protein does not contain any tRNA. This indicates that the last step in protein synthesis must involve not only termination of the polypeptide chain, but also the removal of the terminal tRNA. The termination signal is encoded in the messenger RNA. Three codons of the genetic code (UAA, UAG, and UGA) mean "terminate translation," and one of them must be present at the end of each cistron (Box A). It is easy to see how a polycistronic messenger gives rise to one polypeptide for each cistron. A ribosome begins at one end of the message, translates until it comes to the end of the first cistron, and then releases the first polypeptide. The ribosome then starts translating the second cistron and, when it finishes, releases the second polypeptide, and so on. When the ribosome has translated all the cistrons of the messenger, it drops off the far end of the polynucleotide chain (Figure 1).

Some genes are transcribed more often than others. In Chapter 7 it was stated that RNA polymerase attaches to DNA and starts transcrib-

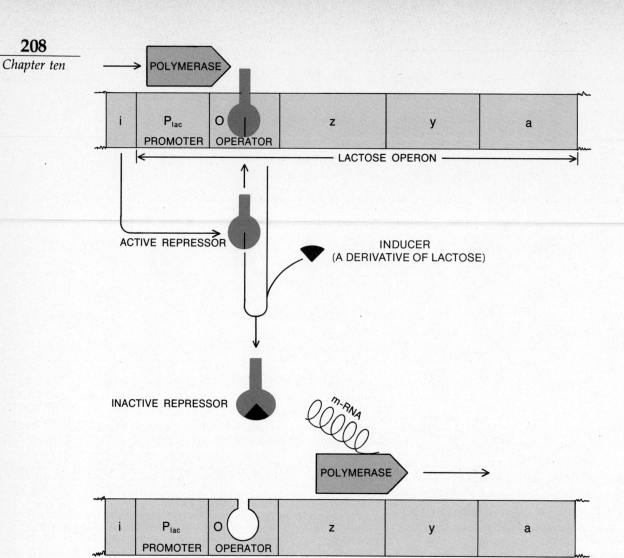

MODEL OF LACTOSE OPERON depicts the induction of enzyme synthesis. Repressor molecule, shown here as a flask-shaped object, binds to the operator (top) and prevents transcription by RNA polymerase. Small inducer molecule can change the shape of the repressor so that it can no longer bind to the operator. Removal of repressor allows transcription to proceed (bottom).

ing, but nothing was mentioned about where it attaches. Clearly the polymerase does not attach itself randomly; special regions for attachment are built into the DNA molecule. These regions are called PROMOTERS. There is one promoter for each cistron or set of cistrons to be encoded into messenger RNA. Promoters serve as punctuation, telling the RNA polymerase where to start and which strand of DNA to read. A promoter and one or more structural cistrons are enough to specify the synthesis of an mRNA molecule.

Not all promoters are identical. One may bind RNA polymerase very effectively, and will therefore trigger frequent transcription of its family of cistrons; in other words, it competes effectively for the available RNA polymerase. Another promoter may bind RNA polymerase poorly, and its family of cistrons will rarely be transcribed. The nature of the promoter sets a limit on how often each family of cistrons can be transcribed. An enzyme that will always be needed in very large amounts is coded by a cistron whose promoter is efficient, and one that will always be needed in tiny amounts is coded by a cistron whose promoter is inefficient.

THE OPERON THEORY

What about enzymes (such as those of lactose metabolism) which are needed in large amounts at

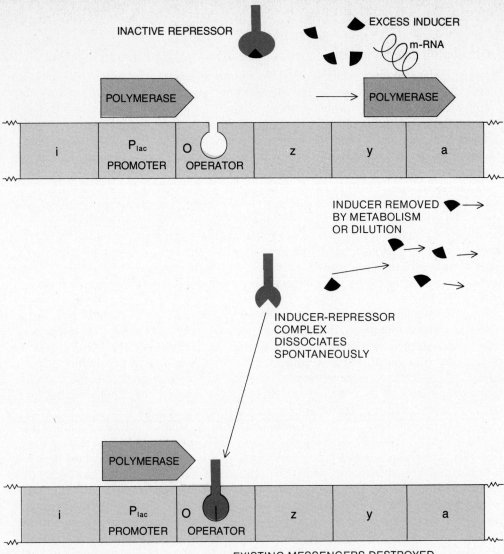

3 **TRANSCRIPTION STOPS when inducer is removed. Symbols are the same as those in Figure 2. Inducer-repressor complex has a tendency to dissociate. If inducer is used up or removed from cell, repressor molecules then bind to the operator, preventing transcription and stopping the synthesis of enzymes coded by z, y and a.**

some times, and not at all at others? Obviously the corresponding genes must contain a very efficient promoter, so that the maximum rate of messenger synthesis can be high. Call this promoter p_{lac}. There must also be a way to shut down mRNA synthesis when the enzymes are no longer needed. Nature does this by placing an obstacle between the promoter and its structural cistrons. The obstacle is a special sort of protein molecule called a REPRESSOR, which can attach to the DNA molecule at a site called the OPERATOR. When the repressor is bound to the operator, it blocks the transcription of the mRNA for the (lactose) enzymes. When the repressor is not attached to the operator, messenger synthesis proceeds rapidly. A controllable unit of transcription of this sort is called an OPERON.

An operon always consists of a binding site for RNA polymerase (a promoter), a binding site for a specific repressor (an operator), and one or more structural cistrons. A representation of the lactose operon is shown in Figure 2. At first glance, it might seem that a reasonable way to switch on an operon would be for the cell to stop making the

B Deducing the operon model

If the model is correct, we might expect some mutations in the i gene to destroy the ability of the repressor to recognize the operator. Such strains (i^- mutants) would synthesize the enzymes for lactose metabolism even in the absence of any inducer, because the repressor would never be bound to the operator. These mutants, called i^- constitutives, are easily isolated. Geneticists have found that it is easy to prepare bacterial strains that contain one copy of the lactose region on the bacterial genophore and another copy on an F' factor *(Chapter 8)*. The i gene is so close to the lactose operon that a single F' factor can carry both of them. Strains have been prepared that carry both an i^- gene and an i^+ gene in the same bacterium. Would we expect these strains to synthesize the enzymes of the lactose operon all the time, that is, to be constitutive? Or would we expect them to make the enzymes only when an inducer is present, that is, to be inducible? The answer is that they should be inducible. The i^+ gene should be dominant over i^-, because the i^+ makes repressor normally, and the repressor can diffuse from place to place within the cell and turn off any normal lactose operator.

It is also possible to imagine an i gene that has mutated so that it still recognizes the operator, but no longer can bind inducer. Such mutants will have their operators irreversibly blocked by repressor and will not produce any of the three enzymes of the operon, even in the presence of inducer. Such mutants are called i^s, where the superscript stands for "superrepressor." If such an i^s gene is present in a bacterium that also contains an i^+ or an i^- gene, the i^s phenotype will be dominant, because superrepressor will permanently block all normal lactose operators.

Mutants of the i^- sort are not the only constitutive mutants one can isolate. Mutations in the operator gene might also be found. Because the only function of the operator is to recognize and bind the repressor, it is not surprising to find that it is possible to isolate mutants in which the operator is altered so that it has a greatly reduced affinity for the repressor. These mutants, called OPERATOR CONSTITUTIVES (o^c) cannot be repressed, even by an i^s gene product. As might also be expected from the model of the lactose operon, the condition of the o gene will affect only those cistrons that are on the same piece of DNA. Therefore, if one puts an o^c and a normal operator (o^+) into the same bacterium, the cistrons attached to the o^c will be transcribed constitutively, while those attached to the o^+ will be transcribed only in the presence of inducer, provided that there is normal repressor in the cell. The difference between the two types of constitutive mutations, i^- and o^c, is revealed only when they are present as diploids with their wildtype alleles, i^+ and o^+. As noted above, the i^-/i^+ diploid is not constitutive, since the repressor made by the i^+ gene is sufficient to bind to both operators. The o^c/o^+ diploid is constitutive, since one of the operators does not bind repressor, and the cistrons on that piece of DNA will be transcribed. Adding inducer to such a diploid will release the repressor even from the o^+ operator, and will further increase the production of the lactose enzymes. The test that is usually done is not quantitative enough to distinguish a substantial amount of enzyme from twice that amount, so that the diploid is called constitutive, even though inducer does somewhat increase the production of the enzymes. Promoter mutations have also been obtained. As you might expect, these affect only the maximum rate of transcription, and affect only the cistrons that are on the same piece of DNA.

repressor. One could, of course, postulate a repressor of the repressor, and a repressor of the repressor of the repressor, but the buck cannot be passed forever. The model postulates a different mechanism. It has become clear that the repressor is able to bind not only to its specific operator, but also to inducers—in this case, lactose and compounds related to lactose. The binding of the inducer changes the shape of the repressor (by allosteric modification) virtually destroying the repressor's affinity for the operator. When an inducer is added to a culture of *E. coli*, the inducer enters the cell and promptly combines with the repressor, changing its shape and detaching it from the operator. RNA polymerase bound to the promoter p_{lac} starts transcribing the structural cistrons of the operon. Messenger RNA transcribed from these cistrons is translated by ribosomes, which synthesize the three enzymes of lactose metabolism. If the inducer is used up or removed from the culture, the inducer molecules bound to the repressor dissociate, the repressor quickly becomes bound to the operator, and transcription of the operon stops. The messenger that is already present is degraded over a period of a few minutes, and translation rapidly comes to an end (*Figure 3*).

Repressor proteins are coded by REGULATOR GENES. The one that codes for the repressor of the lactose operon is called the *i* gene (for "inducibility"). The *i* gene happens to lie close to the operon that it controls, but many other regulatory genes are distant from their operons. Like all genes, the *i* gene has a promoter, which can be designated p_i. It is a very inefficient promoter, and makes just enough messenger to synthesize about ten molecules of repressor per cell generation. There is no operator between p_i and the *i* cistron. Therefore the repressor is constitutive, that is, it is made at a constant rate not subject to environmental control.

The model implies that the natural condition of the lactose operon is being turned on, and that control is exerted by a regulatory protein, the repressor, which turns the operon off. Thus, the control mechanism is negative. A second important implication is that some genes, such as *i*, produce proteins whose sole function is to regulate the expression of other genes, and that certain other genes, namely operators and promoters, very probably do not code for any proteins at all, and perhaps are not even transcribed. Most of the properties of promoters, operators, and repressors were deduced from the behavior of organisms in which these genes were altered by mutation. Once you understand the model that resulted from

studying these mutants, you should be able to follow the logic by which the model was deduced (*Box B*).

REPRESSIBLE SYSTEMS

Inducible systems like the one for lactose metabolism are of adaptive value to an organism that must switch on enzyme synthesis in response to the presence of an outside agent such as lactose. It is equally valuable to an organism to be able to switch off the synthesis of certain enzymes. For example, if tryptophan is present in the medium it is advantageous to be able to stop making all of the enzymes that are involved in tryptophan biosynthesis. When the formation of an enzyme is turned off in response to a biochemical cue, the enzyme is said to be REPRESSIBLE. Repressible systems, such as the one for tryptophan synthesis, work by mechanisms quite similar to those of inducible systems. The regulatory macromolecule (the repressor) cannot shut off its operon unless it unites with a COREPRESSOR, which may be either the nutrient itself (in this case, tryptophan) or a compound derived from it. If the nutrient is absent the operon is transcribed at a maximum rate. If the nutrient is present, the operon is turned off (*Figure 4*). The repressible system is another example of negative control of transcription. In both the lactose and tryptophan systems the function of the normal regulatory macromolecule is to prevent transcription. As in the case of the lactose operon, the tryptophan system has a regulator gene and an operator. Mutations in the regulator gene can change the repressor so that it never binds to the operator, even when corepressor is present. In such mutants the enzymes are made constitutively—their synthesis cannot be repressed.

The first few control systems studied all involved negative control. This initially convinced many biologists that all control of enzyme synthesis is negative. More recently it has become clear that there are also many systems that involve positive control. In such systems, failure to synthesize a regulatory protein in turn blocks the synthesis of a whole family of enzymes. The most thoroughly studied case of positive control is that of the *E. coli* arabinose operon. When arabinose is added to a culture of *E. coli*, the bacteria promptly start making enzymes for arabinose metabolism. The regulatory macromolecule is coded by a gene outside of the operon. If this gene is damaged by a nonsense mutation, synthesis of the arabinose enzymes does not become constitutive, as would be

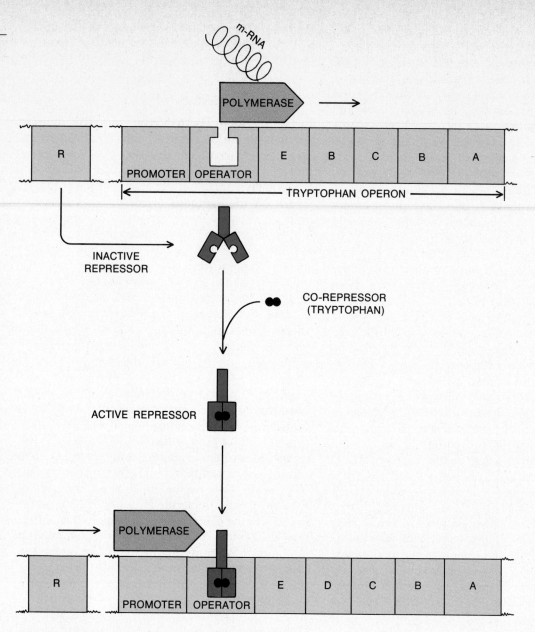

4 **TRYPTOPHAN OPERON is a repressible system.**
Letters on operon denote structural genes controlled by
promoter and operator. Repressor is activated by
binding to a small co-repressor molecule, either
tryptophan or a derivative of it. Active repressor binds
to operator and stops transcription.

the case with the lactose operon. Instead the arabinose enzymes cannot be made at all, even in the presence of arabinose. The regulatory protein plus arabinose is required for transcription of the genes coding for these enzymes.

CATABOLITE REPRESSION

A great many enzymes function directly or indirectly to furnish the cell with glucose. Some of them produce glucose from various carbohydrates; others, from certain amino acids. When glucose or some other excellent carbon source is abundant, the cell curtails or abolishes the synthesis of these enzymes. This type of transcriptional control is often called CATABOLITE REPRESSION, but its mechanism is very different from the kind of

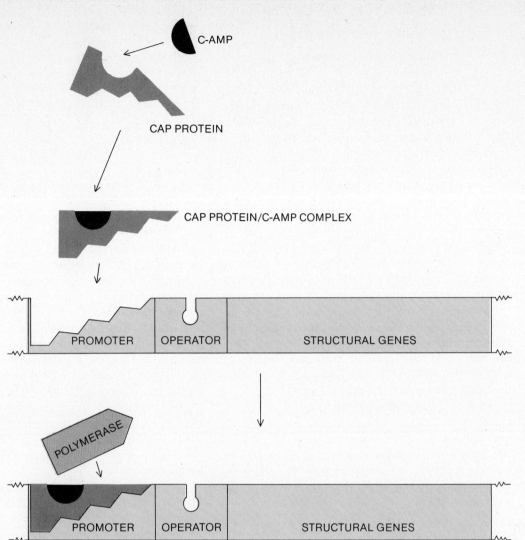

C-AMP

CAP PROTEIN

CAP PROTEIN/C-AMP COMPLEX

PROMOTER OPERATOR STRUCTURAL GENES

POLYMERASE

PROMOTER OPERATOR STRUCTURAL GENES

5 **CATABOLITE REPRESSION is a positive control system that regulates many operons. Promoter does not bind RNA polymerase unless it is covered by the protein CAP. In turn, the ability of CAP to attach to promoter depends on presence of cyclic AMP (c-AMP) bound to CAP. When glucose level in cell is high, c-AMP disappears and CAP cannot bind to promoter, thus preventing transcription of structural genes. Here positive control refers to regulatory macromolecule required for enzyme synthesis.**

operator-repressor mechanism that controls the lactose operon. Catabolite repression is a positive control process, and it involves promoters, not operators. The promoters bind RNA polymerase in a series of steps *(Figure 5):*

1. A special protein, abbreviated CAP, binds a low-molecular weight compound called 3′,5′-cyclic AMP (abbreviated cAMP).

2. The CAP-cAMP complex binds to the corresponding promoters.

3. The promoter-CAP-cAMP complex binds RNA polymerase.

Only then does transcription of the operon become possible. Glucose causes catabolite repression by lowering the level of cAMP, although the mechanism of this process remains unknown. When glucose levels fall, the level of cAMP rises again, and the promoters of many genes involved in sugar metabolism are activated. cAMP is also present in eucaryotes, where it is involved in the action of many hormones *(Chapter 18).* In eucaryotes it seems to work by a mechanism very different from that in procaryotes.

EUCARYOTES AND POSITIVE CONTROL

The operon theory has elucidated some important differences between the control mechanisms of procaryotes and eucaryotes. At least 70 percent of the genes in *E. coli* are thought to occur as clusters of two or more functionally related cistrons, but in eucaryotes clusters seem to be relatively uncommon. Moreover, there is not a single indisputable example of an operator-constitutive mutation in any eucaryote. This may be because positive control is much more common in eucaryotes than in procaryotes; destruction or alteration of a target region next to a structural gene will stop the expression of that gene, rather than making it constitutive, since the positive regulatory element would no longer be able to turn on the transcription of the gene.

Biochemists have a family of candidates for the positive regulatory elements that appear to activate transcription of specific genes in eucaryotes. They are the so-called "acidic chromosomal proteins," or "non-histone chromosomal proteins." Unlike the basic histones, which also comprise part of the structure of all eucaryotic chromosomes, the acidic chromosomal proteins are very diverse in a given organism, and within an organism are very different from one tissue to another. This is what one would expect of regulatory proteins. The evidence also suggests that these proteins do indeed stimulate transcription.

CONTROL BY DEGRADATION

The amount of an enzyme within the cell depends not only on its rate of synthesis, but also on its rate of degradation *(Box C)*. The cell can synthesize two enzymes at the same rate, but breaks them down at different rates. The one broken down more slowly will be present at a higher steady-state level. Different proteins in the same organ (say, the liver) turn over at different rates. On the average, half of the liver protein of a rat turns over every six days, but some liver enzymes have a half-life as short as 90 minutes. This means that half of the molecules synthesized at a given moment will have been destroyed 90 minutes later. In some cases an enzyme is stabilized by one of its substrates. A high substrate concentration tends to decrease the degradation rate without affecting the rate of synthesis, so that the steady-state level of the enzyme rises. The elevated level of enzyme, in turn, causes the substrate to be used up more rapidly, so that its concentration drops back toward normal levels. This self-adjusting system tends to smooth out fluctuations in the concentration of a substrate. Moreover, both the rate of synthesis and the rate of degradation of each enzyme seem to be under genetic control, enabling higher organisms to control their enzyme levels in many different ways.

CONTROL OF TRANSLATION

Many cells, especially those of higher organisms, control enzyme synthesis by shifting, not the rate at which genetic information is transcribed into mRNA, but the rate at which stored mRNA is translated by the ribosomes. Eggs of the sea urchin furnish a good example. In the unfertilized egg very little protein synthesis occurs once the egg is "ripe," even though it contains ribosomes, messenger RNA, and compounds necessary for activation of amino acids. After fertilization, polysomes are quickly formed and protein synthesis increases many-fold without any appreciable synthesis of RNA. This happens even in the presence of an antibiotic (actinomycin D) which blocks all RNA synthesis. Apparently mRNA is stored in the unfertilized egg, but in a masked form that the ribosomes cannot translate. Fertilization unmasks it, allowing the ribosomes to begin translation. Similar results have been obtained with many other creatures.

One of the most impressive demonstrations of the importance of stored mRNA comes from a combination of cell surgery and biochemistry when applied to an enormous one-celled alga, *Acetabularia mediterranea*. This complicated cell begins life as an irregular blob of protoplasm with a single nucleus, and evolves into a rootlike RHIZOID, which retains the nucleus, and a stalk that may grow to several centimeters in length. Ultimately the stalk develops an umbrellalike cap, and the nucleus starts dividing. The daughter nuclei migrate into the cap, filling it with nucleated cysts that later become engorged with gametes. If the stalk is amputated with a razor blade, the rhizoid will regenerate a new stalk, and finally a cap full of cysts. If the amputated stalk is reasonably mature, it will continue to photosynthesize and metabolize, even though it has no nucleus and cannot make mRNA. Eventually it will even regenerate a perfect cap, but the cap will be devoid of nucleated cysts. The German biologist Joachim Hämmerling, who devoted many years to the study of *Acetabularia*, theorized that the maturing stalk must receive from the nucleus a "morphogenetic substance" that controls the formation of the cap. It now

Turnover of proteins

Biologists of the late 19th century were influenced as much as were their lay contemporaries by the spirit of the Industrial Revolution, and tended to view the living animal as a sort of steam engine. Early in the 20th century, it became widely accepted that proteins were the permanent machinery of the human engine, and that carbohydrates and fats were the fuel shoveled into the firebox. In this view, an adult, nongrowing animal needed a little dietary protein to repair wear and tear on the machine; if more protein was ingested, the excess amino acids were thrown into the firebox along with the carbohydrates and fats.

Evidence that flesh-and-blood creatures differed from industrial machinery mounted rapidly in the late 1930's when isotopically labeled compounds became available. Rudolf Schoenheimer and David Rittenberg, working at Columbia University, found that if adult rats were given a dose of N^{15}-labeled amino acid, about $\frac{1}{2}$ of the isotopic label showed up within a few days in the body proteins, about $\frac{1}{4}$ remained in the body in compounds other than proteins, and about $\frac{1}{4}$ was promptly excreted. The nitrogen atoms that were entering the body, by and large, were not the same ones that were leaving. It was soon found that proteins in virtually all organs of the body were in a constant state of flux, and that under the placid surface, the proteins of mammals, at least, are in a dynamic steady state. During the 1950's, this view was in turn challenged by several groups, principally that of Jacques Monod in Paris. Working with β-galactosidase in *E. coli*, Monod's group found that this enzyme does not turn over. If β-galactosidase is induced and then inducer is removed but growth is allowed to continue, the enzyme that was formed during the induction period is not scrapped, but is simply passed on to the daughter cells at each division. The total amount of enzyme remains the same, but there is progressively less and less per bacterium as growth proceeds. Who was right? Or were Schoenheimer and Rittenberg right about animals and Monod right about *E. coli*? It now seems likely that both groups were partly right. In 1958, Joel Mandelstam reported that although proteins in rapidly-growing *E. coli* are metabolically stable, nondividing cells which are being starved of a required amino acid will break down (and resynthesize) as much as 5 percent of their protein per hour. With the benefit of hindsight, we can understand why it is advantageous for starving bacteria, but not for growing bacteria, to turn over their existing proteins. Breaking down proteins only to then resynthesize the same proteins is metabolically wasteful, and it is not likely to be necessary for growing bacteria, which are successfully using externally available nutrients to synthesize all their enzymes. On the other hand, starving cells are, by definition, already in trouble, and can only survive if they liquidate some of their capital by converting existing enzymes into free amino acids. The amino acids can then be used to make kinds of enzymes that may previously have been repressed. These new enzymes may be thought of as a last-ditch attempt to escape from the metabolic impasse that caused the cessation of growth.

Most cells in an adult animal are, in some respects, quite analogous to a starving bacterial cell. They are not dividing or increasing in mass. Therefore, they have no way to adjust their content of various enzymes except by breaking down existing proteins. This is true even in a well-fed animal, provided it is not growing. The situation is even more dramatic when an organism is in a state of starvation. For example, during the cocoon-bound metamorphosis of a caterpillar into a moth, the insect cannot eat . Similarly a tadpole fasts as it turns into a frog. Certain bacteria, when they are starved, develop spores that can survive for years in the dormant state. In each of these cases, there is no way to change over to a new set of proteins except to break down the existing ones.

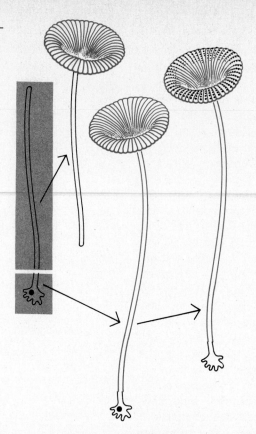

6 ACETABULARIA EXPERIMENT demonstrated that mRNA from nucleus controls formation of the umbrella-like cap. If rootlike rhizoid containing nucleus is cut from a maturing stalk, the stalk will form a cap whose gills lack gamete-forming cysts (top left). Amputated rhizoid (bottom left) regenerates a normal fertile organism (right).

seems certain that the substance is mRNA stored in the stalk, and translated into cap protein according to a prescribed program *(Figure 6)*.

In bacteria, it can be very difficult to distinguish transcriptional control from translational control because the two processes are not completely separated in time and space. Translation of the messenger by ribosomes begins at one end of the molecule while the other end of the molecule is still being transcribed. Translation is necessary to pull the mRNA away from the DNA, and most inhibitors of translation also bring transcription to a halt. Since transcription and translation occur together in bacteria, one might hope to be able to see polysomes attached to the genophore. The remarkable electron micrograph that opens this chapter shows clearly a region of transcription (perhaps an operon) of the *E. coli* genophore, with polysomes protruding from a strand of DNA.

MULTIPLE COPIES OF GENES

One obvious way for a cell to make more of one enzyme than of another would be to have more genes of one kind than of another, and to transcribe them all. This would require having more DNA per cell than would be necessary if there were only one gene of each kind. In Chapter 2 it was pointed out that humans have about 1000 times as much DNA per cell as a typical bacterium. It pleases us to feel that we contain so much genetic information because we represent the zenith of evolutionary complexity. But the cells of a number of amphibians and fishes contain from 10 to 100 times more DNA than do ours. The largest known DNA content per cell is found in the Congo eel *Amphiuma*, a snake-like salamander. Although there is a rough correlation between the amount of DNA per cell and the complexity of an organism, the idea cannot be pushed very far. For example, some amphibians that are obviously closely related and about equally complex, differ many-fold in the amount of DNA per cell. It seems unlikely that one of them makes vastly more different kinds of enzymes than another. This raises the question of whether more DNA really means more genetic information in the form of unique sequences of nucleotides, or whether it merely means that there are many copies of some or most of the genes. The answer is unknown, but there is some evidence of repetitive sequences of DNA in the chromosomes of eucaryotes.

THE NUCLEOLAR ORGANIZER

In at least one case there is more than one copy of a gene per haploid set of chromosomes. Every eucaryotic cell contains in the nucleolar organizer many copies of the DNA that codes for ribosomal RNA *(Chapter 2)*. Moreover, the number of copies of this gene can be adjusted from cell to cell accord-

---

It pleases us to feel that we contain so much genetic information because we represent the zenith of evolutionary complexity. But the cells of a number of amphibians and fishes contain from ten to 100 times more DNA than ours.

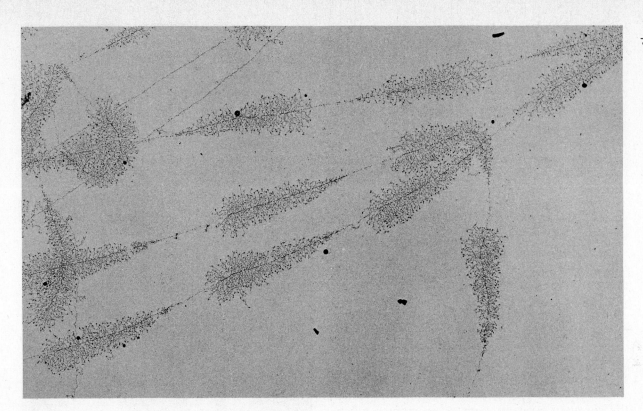

STRANDS OF DNA look like long thin wires in this electron micrograph of the nucleolus of the oocyte of the spotted newt. Fuzzy triangles consist of molecules of rRNA. A similar electron micrograph, reproduced to a much larger scale, appears on the opening page of Chapter 7.

ing to need. The multiple copies are always arranged in a series, one after another, which biologists call a tandemly repetitive region.

In most cells this region appears to contain enough gene copies to produce rRNA as fast as it is needed. But in some cells even these multiple genes are apparently not enough. In the oocytes of the African clawed toad *Xenopus laevis*, the whole nucleolar organizer region of the chromosome multiplies until there are several hundred to a thousand nucleoli floating free in the nuclear sap. Each of these nucleoli consists of DNA with repeating segments corresponding to rRNA, and each is transcribed to furnish the tremendous amount of rRNA that is needed in a mature oocyte. The process of actually creating more genes to enhance transcription is called GENE AMPLIFICATION. Evidently no genes except those responsible for rRNA synthesis become amplified. It is thought that amplification occurs in the oocytes of many animals and, under certain conditions, in higher plants.

Figure 7 shows part of the DNA-containing core of a single, nonchromosomal nucleolus of another amphibian, the spotted newt *Notophthalmus viridescens*. The axial strand is the DNA of the nucleolar organizer. The fuzzy-looking triangles attached to the DNA are composed of many strands of rRNA in the making. Many molecules of RNA polymerase are transcribing the DNA at the same time. The apex of each triangle is the point at which RNA synthesis starts, so the RNA strands protruding from this region are very short. As transcription proceeds along the DNA, the RNA strands become longer and longer, forming the "base" of the fuzzy triangles. The repeated, or "tandem," occurrence of these fuzzy triangles along the DNA of the nucleolar organizer indicates that the DNA itself is repetitive. Notice that there is quite a bit of DNA between the fuzzy triangles that does not seem to be transcribed. Its function is unknown.

CHROMOSOME INACTIVATION

For centuries physicians were puzzled by the human disease called mongolism. It is inborn, but not inherited, at least in the usual sense. Its symptoms are a drastically impaired intelligence and widespread changes in body chemistry and in gross architecture: lowered and somewhat rotated ears, abnormal position of creases in the palms of the hands, changed appearance of the eyes, and other characteristic deformities *(Figure 8)*. The structure of the eyes suggested the name mongolism to Caucasians, but the disease has nothing to do with ethnic origins, and it is better called by its proper name —DOWN'S SYNDROME —to avoid a misleading racial slur. In 1959 Jerome Lejeune and his coworkers in France made the astonishing discovery that sufferers from Down's Syndrome have 47 chromosomes instead of the normal human number of 46. The extra is a third copy of chromosome 21, one of the smallest in the genome, which presumably carries a correspondingly small proportion of the genetic information. The many abnormalities in Down's Syndrome must be attributed to a 50 percent increase in the hereditary dose of this information.

People have long been fascinated or dismayed by the difference between men and women, or perhaps even denied that differences exist, but whatever one's stance, Down's Syndrome raises the question: Why are men and women so much alike? The normal female has two X chromosomes, and the normal male has one X and one Y. The Y chromosome in a male has few if any identifiable genes that are also present on the X chromosome, and appears to be genetically almost inactive. Hence there is a 100 percent difference between women and men in the dosage of X-chromosome genes, and the X chromosome, unlike chromosome 21, is one of the largest in the human genome. Why then is not one sex or the other grossly deformed or completely inviable?

The answer was found in 1961 by two scientists working separately, Mary Lyon and Liane Russell. Lyon suggested that in any cell of a normal female one of the X chromosomes is inactivated early in embryonic life and remains inactive ever after. The choice as to which X in an XX female will remain active is random, but since the female embryo already consists of tens or hundreds of cells by the time the choice is made, virtually all women —in fact, most female mammals—contain patches of tissue in which one or the other X is active. The evidence for this comes from the study of females that are heterozygous for certain genes on the X chromosome. A female who is heterozygous for X chromosome genes that govern a visible trait, such as color of fur in the mouse or cat, will have a mottled or tortoiseshell coat. The coat of the hemizygous male will be a solid color. Biochemical

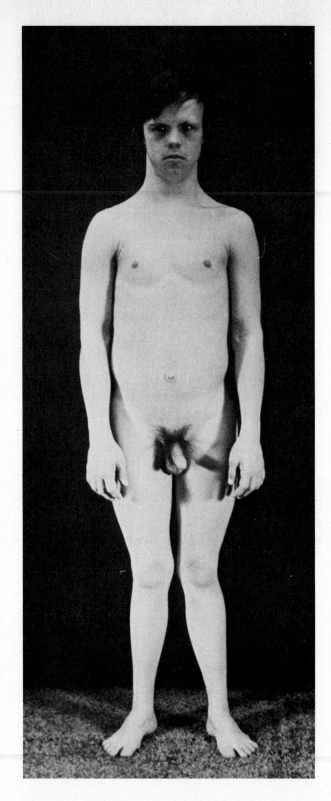

DOWN'S SYNDROME, or mongolism, is the result of having three copies of chromosome 21. Affected individuals such as the one depicted here have very low intelligence, ''Oriental'' eye folds, and short bodies with stubby hands and feet.

studies have confirmed this interpretation. Many diseases are known in humans which are caused by defects in an X chromosome gene, and some of these are associated with absence or abnormality of an enzyme which is easily assayed. Cells can be isolated from a woman who is known to be a heterozygous carrier of such a disease, and a separate culture can be grown from each cell. This produces two kinds of cultures: those with normal enzyme, and those with abnormal or inactive enzyme. No culture contains both normal and abnormal enzyme. The reason women who are heterozygous for such a disease —hemophilia, for example —are not usually clinically ill is that there is enough tissue in which the normal X is turned on that the normal protein (in this case, antihemophilic globulin) can be made in sufficient amounts to supply the entire organism.

The results of cytological studies provide additional confirmation of these genetic and biochemical findings, and allow geneticists to refine the theory of active and inactive X chromosomes. It has long been clear that interphase cells of normal females have a single stainable nuclear body (called a BARR BODY, after its discoverer) which is not present in males, and which represents condensed chromatin, called heterochromatin. More recently it has been shown that heterochromatin does not produce mRNA, and is metabolically inactive. Most interestingly, certain moderately abnormal women were found who had only one X chromosome and no Y chromosome. The cells of these women contained no Barr bodies at all. Other women were found who had a chromosome constitution of XXX, and their cells contained two Barr bodies. Some very abnormal men have been found with the bizarre genetic constitution, XXXXY. As might be guessed, these men had three Barr bodies in each interphase cell. The Lyon-Russell findings are now stated in the Single Active X rule: all but one of the X chromosomes in a cell are turned off. This explains why humans can survive with one X chromosome, with two, and with even more than two. The rule even explains, perhaps to the dismay of everyone, why men and women are so similar.

READINGS

J.D. WATSON, *The Molecular Biology of the Gene*, 3rd Edition, Menlo Park, CA, W.A. Benjamin, 1976. This is an unusually clear presentation of "The Central Dogma," by one of molecular biology's best-known theologians.

1961 was a vintage year in our understanding of the code, of transcription, and of regulation, and some articles written in that year are still the best. We particularly recommend three articles from the *Cold Spring Harbor Symposium on Quantitative Biology*, Vol.26, *Cellular Regulatory Mechanisms*, Baltimore, Waverly Press, Inc., 1961. Especially recommended are the articles by B.D. Davis (pp. 1–10), F. Jacob and J. Monod (pp. 193–212) and by J. Monod and F. Jacob (pp. 389–401). These are not very recent, but they contain the main ideas, and are extraordinarily well-written. The following articles are also especially interesting:

A. GIBOR, "Acetabularia: A Useful Giant Cell," *Scientific American*, November, 1966.

M. PTASHNE AND W. GILBERT, "Genetic Repressors," *Scientific American*, June, 1970.

Part **2**   Multicellular life

Single-celled organisms are one of the great success stories
of evolution. They probably comprise more than half the
total biomass on Earth, and have successfully colonized
even its harshest environments. Bacteria flourish in scald-
ing springs and in the frozen soil of Antarctica; they survive the aridity of deserts and the crushing pressures of the
ocean floor; they float freely in the atmosphere. Biochemi-
cally many microorganisms are far more versatile than
man, being able to synthesize virtually everything they
need from a few simple nutrients. And unlike man, many
of them are potentially immortal, or at least ageless: when a
bacterium reproduces by dividing in two, its daughter cells
are equally young. A whole new life stretches before them.
Although a bacterium can of course die of starvation or of
exposure to toxic chemicals, or be eaten by another organism, it will never die of old age. It would seem that in
the bacteria and their unicellular cousins nature has come
close to creating the ideal organism —an observation that
poses one of the profound enigmas of biology: Why, given
the spectacular success of unicellular organisms, is the
main thrust of evolution directed toward ever higher levels
of organization?

  This thrust has been marked by three great breakthroughs. The first was the invention of the eucaryotic cell,
which according to one theory arose as a commune of previously free-living microorganisms. Eucaryotic cells are not
only tens to hundreds of times larger than procaryotes, but
they have a significant new property: the ability to aggregate into multicellular communes. The emergence of multicellular organisms was the second breakthrough, because
unlike a colony of bacteria, they were more than a mere
heap of cells. Their cells became specialized for a variety of
functions, and interacted in a way that made the organism

221

more than the sum of its parts. Multicells embodied a new level of organization, a whole new order of complexity. The first two breakthroughs occurred billions of years ago, but the third was relatively recent: the emergence of conscious intelligence. Although man is assembled from the same type of cells and organs as his biological relatives, his intelligence makes him a creature of a higher order.

The consequences of the third breakthrough are explored in Part IV; Part II focuses on the first and second. An immense gap separates the great majority of unicellular from multicellular organisms, a gap so large as to seem almost unbridgeable. Yet it was bridged long ago, and each multicellular organism, in its development from fertilized egg to mature organism, bridges the gap in a single lifetime. The basic question remains: What biological advantages favored this leap of evolutionary invention? Even with the almost explosive growth of biological knowledge in the past few decades, a really satisfying answer remains elusive. The classic argument is that multicellular organisms are somehow better adapted to flourish in certain environments, or can fill ecological niches that are totally inaccessible to unicellular forms.

One such environment is land. Life began in water, and for perhaps two billion years was confined to water, especially to the margins of the primitive seas *(Chapter 23)*. Plants, the descendants of ancient marine algae, were probably the first organisms to colonize the land. The unicellular algae that flourish in oceans and lakes today are not very different from their ancient ancestors. Water bathes them with a solution of mineral nutrients and buoys them near the surface, where sunlight for photosynthesis is plentiful. Nevertheless, even greater rewards are available for plants on land: higher light intensity and richer concentrations of minerals.

The first plants to invade the land enjoyed the advantage of diminished competition with other organisms. But terrestrial life imposed a whole new set of demands. Fossil remains of the first land plants have not yet been discovered, but it is reasonable to suppose that they were not unicells, or at least not procaryotic unicells like bluegreen algae. On land, water and nutrients are mostly trapped in the darkness of the soil, requiring plants to inhabit two environments at the same time: one end in the ground, the other in sunlight. No procaryotic cell is long enough to fulfill this requirement. It might seem that a long thin eucaryotic cell would serve the purpose, but as Chapter 2 pointed out, cells depend largely on diffusion for the internal transport of nutrients and other substances, and this imposes strict limits on their size. The eucaryotic marine alga *Acetabularia,* two or three centimeters high, is a giant among unicellular plants and probably lies near the upper limit of cell size. Although it is larger than many multicellular land plants it lacks the specialized equipment necessary for survival on land.

The invention of the eucaryotic cell was life's first successful attempt to exploit the advantages of greater size and complexity. Size in itself was an important new property. Not only did it help plants to colonize the land, but it enabled animals to feed on smaller organisms by overpowering and engulfing them. Even more important, the eucaryotic cell opened the way for the next breakthrough: the emergence of multicellular organisms. For reasons that are not at all clear, all multicellular organisms are comprised of eucaryotic cells. The main advantage of multicellularity is cell specialization. A multicellular organism can delegate highly specialized functions to particular groups of cells. For example, it can encase itself in a waterproof skin fashioned from one population of its cells, creating a wet internal environment favorable to the growth of other types of specialized cells. Clearly an organism capable of creating and maintaining its own internal environment is suited to pioneering new environments in a way that bacteria and other unicellular forms are not.

True, microorganisms are now found virtually everywhere on Earth, but they are there only because other organisms were there first. Parasitic bacteria, for example, rely on their hosts to carry them into new regions; many other microorganisms are saprophytes that feed on decaying material from higher organisms. Lastly, most of them depend on a supply of atmospheric oxygen created by plants. If microorganisms were forced to earn a living solely from one another they would comprise a much smaller percentage of the biomass. In the language of the ecologist, the presence of multicellular organisms increases the overall productivity of an ecosystem —that is, they not only carve out a niche for themselves, but they create new opportunities for other forms of life.

Multicellularity also carries the potential for diversity: millions of different shapes, specialized organ systems, and patterns of behavior. This potential greatly increased the kinds of environment that organisms could exploit, and thus constituted a powerful driving force toward the evolution of new and diverse adaptations. Once begun, the trend toward diversity reinforced itself. The existence of land plants created opportunities for land animals. Herbivores that fed on plants and carnivores that fed on herbivores are the most obvious examples, and their presence in turn created opportunities for a vast array of microorganisms. The wondrous assortment of plants and animals that inhabit the Earth today is the product of billions of years of interaction between life and Earth. The planet has provided the environment and the raw materials, and life has responded with dazzling inventiveness, fashioning organisms that can exploit every conceivable ecological niche. After glancing at the photographs of organisms in Chapters 25, 26 and 27 one is tempted to ask, is there any limit to the potential diversity of living things?

The question applies both to whole organisms and to the systems that comprise them: digestive, sensory, reproductive and so on. The answer depends on what one means by diversity. The possible combinations of known types of cell are virtually unlimited. These combinations can be "packaged" in a vast but nonetheless limited number of sizes and shapes. The basic restrictions on the design of living creatures are imposed by the conditions on the planet, and by the laws that govern the behavior of matter. These restrictions pose a set of problems, and each species of organism represents one successful solution. For example, there are both lower and upper limits on the size of an organism.

The extremes of size range from viruses and bacteria at one end of the scale to the elephant, the sequoia and the blue whale at the other. The lower limit is imposed by chemistry and the upper by physics. A single-celled organism must be large enough to contain the genetic and metabolic machinery required for an independent existence —usually a few cubic micrometers. The elephant and the giant sequoia tree approach the theoretical upper limit to which land organisms can grow. As Galileo pointed out in the 17th century, the limitations are largely due to mechanical factors, particularly the bending moment: the gravitational effect that causes a beam supported at both ends to sag in the middle. A five-centimeter matchstick maintains its rigidity if it is held horizontally at one end, but a matchstick one meter long would droop. To maintain rigidity, the cross-sectional area of a beam must be increased in proportion to its length. This rule accounts for the difference in diameter among the leg bones of a mouse, a deer and an elephant; the slimness of young trees, the fatness of tall ones. In the sea the buoyancy of water eases the engineering problems somewhat. The largest marine creature is the blue whale, which can attain a length of 30 meters and weigh 150 tons. Now hunted to the verge of extinction, it may be the largest animal that has ever lived on Earth.

In the arthropods (primarily insects and crustaceans) a rigid exoskeleton serves in place of the endoskeleton of our own bodies. Its size is limited by the mechanics of hollow objects. A small spherical shell can be remarkably strong, but as the diameter increases the shell becomes more fragile unless it is proportionately thickened. This is one reason for the absence of truly gigantic insects or crabs. Apparent giants (such as the Alaskan king crab, almost two meters across) achieve size not by an enormous body shell but by having long thin legs whose strength lies in their tubular structure. Few insects are longer than a few centimeters, and few are shorter than 0.2 millimeters; at the lower size limit the need for a rigid exoskeleton disappears.

Another factor that limits size is the disproportion between surface area and volume, which was mentioned in Chapter 1. The larger the diameter of an organism, the smaller the surface area per unit volume. The consequences are particularly clear in warm-blooded organisms. Heat loss depends on surface area. Very small animals have such a large surface area per gram of tissue that they suffer very high heat losses to the environment, and need to consume disproportionate quantities of food to maintain their body temperature. A mouse may eat its own weight of food in a day, while we eat about two percent of our weight per day. Large animals suffer the opposite problem. Their need to dispose of metabolic waste heat requires a special heat disposal mechanism, usually an elaborately controlled circulatory system that transports overheated blood to the body surfaces. The insect lacks such a controlled circulation, another reason that giant insects are an impossibility. Other solutions to the heat disposal problem among large animals include sluggish behavior, which generates less heat (elephant), or a partially aquatic lifestyle (hippopotamus).

In the course of billions of years of evolutionary experiments, life has found successful solutions to each of the problems posed by existence on Earth. It is logical to ask why nature has not combined these solutions into an ageless superorganism. One answer is that nature is less concerned with the survival of the individual than with the survival of DNA. It is ironic that the death of the individual is often the best strategy to ensure the survival of the species. Among multicellular organisms, the production of offspring is a better strategy for perpetuating DNA than the mere tenacious survival of individual organisms.

Agelessness has three main drawbacks as a survival strategy. First, on a changing planet the problems and opportunities for living organisms vary over the years in unpredictable ways. Climate shifts, dry land becomes flooded, new predators evolve. Such unforeseen threats can be answered only by providing opportunities for trial changes in the organism. These changes arise from sexual reproduction, which confers on the next generation a new assortment of genetic combinations and mutations. Among the offspring the failures die out, but those that succeed in the new environment survive to reproduce their better adapted genes. Second, ageless parents would compete for food and mates with their offspring, which would jeopardize the survival of future generations. A third drawback of agelessness is that living matter deteriorates. In particular, the genes are under the constant bombardment of background radiation from space, and the resulting damage accumulates over a period of years. The probability of birth defects in humans increases with the age of the mother. Exposure to 20 extra years of radiation distinguishes the 40-year-old mother from the 20-year-old, and often results in damaged genes and inferior offspring. A population of immortal mothers would give birth to an appalling number of deformed children and genetic misfits. There are still other factors that favor the evolution of senescence and death, which we shall discuss in the more detailed treatment of evolutionary theory in Part IV.

The next 12 chapters explain the workings of multicellular organisms, beginning with development and growth: the journey from a one-celled fertilized egg to a mature organism consisting of thousands or even trillions of cells. The emphasis is on the diversity of systems that life has devised to propagate itself, to acquire raw materials and energy, to transport substances from one part of the organism to another, and to integrate and control its activities, including behavior. The strict design limitations imposed by conditions on Earth will become clear, and the reader will note how similar are the problems that each organism faces, and how different the solutions.

The organism is a self-constructing machine. Its macro-scopic structure is not imposed on it by outside forces. It shapes itself autonomously by dint of constructive internal interactions. —Jacques Monod

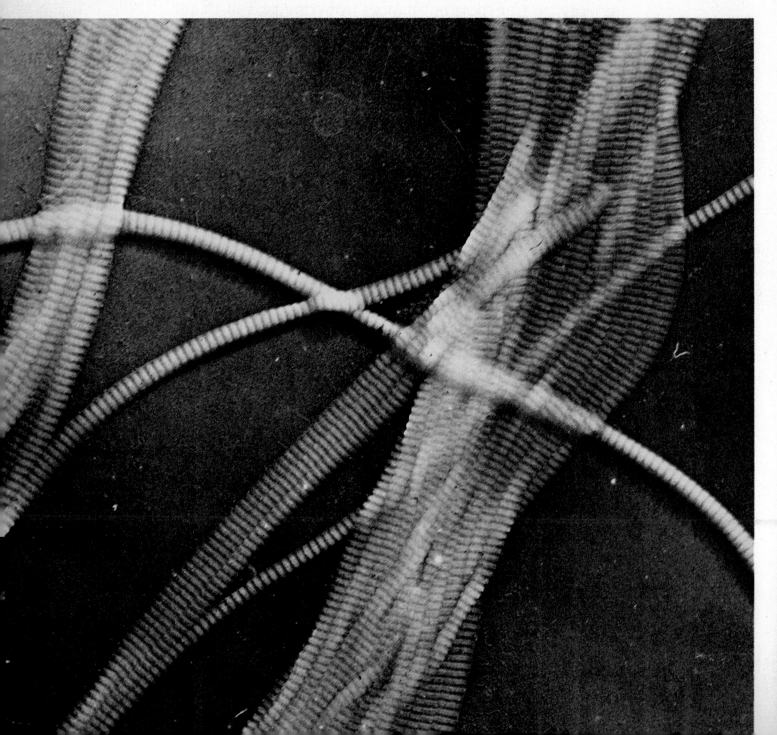

The entire life history of an organism, from birth through maturity to death, can be summarized as the emergence of order, the maintenance of order, and finally the disappearance of order. Development corresponds to the progressive appearance of order. At the cellular level and above there are sharp differences between the development of plants and animals, but at the molecular level the mechanisms of development are remarkably similar. Both plant cells and animal cells —all cells in fact —are, to some extent, self-assembling systems. Polypeptide chains arrange themselves into protein molecules; giant molecules aggregate to form structures like membranes and chloroplasts; and these components assemble themselves into cells. Some steps in the assembly process are spontaneous, relying upon the formation of hydrogen bonds between opposed surfaces of macromolecules with complementary shape or perhaps relying in addition upon the formation of disulfide bridges. Such associations are affected by any environmental change that alters the shape of the molecule, the distribution of charges on it, or the redox potential. Other steps are controlled directly by DNA or RNA within the developing cell, or indirectly by hormones sent from other parts of the organism. Like all biological systems, the cell assembly line has feedback loops and other control mechanisms that make the assembly process largely self-regulating.

Although occasionally this chapter and the next will refer to development in unicellular organisms (and even viruses), the emphasis will be primarily on multicellular organisms, particularly the higher plants and animals. The present chapter will look closely at molecular aspects of development, from self-assembly processes to the role of genes and the regulation of gene products. Chapter 12 will trace development of the fertilized egg (ZYGOTE) from its formation to the appearance of a recognizable multicellular organism.

SELF-ASSEMBLING MOLECULES

Many enzymes and other proteins are comprised of several polypeptide chains. The chains become folded into three-dimensional structures whose precise shape is determined by their amino acid sequence. In some proteins these chains or subunits are identical, while others contain two or more chains of different length and amino acid composition. The enzyme catalase, which facilitates the breakdown of hydrogen peroxide (H_2O_2) into oxygen and water, consists of four identical subunits. The blood pigment hemoglobin consists of two each of two different chains (called alpha and beta). Immunoglobulin, an antibody protein, consists of two long and two short chains. The subunits themselves can rarely perform the chemical functions of the assembled molecule. But once the chains are synthesized, they aggregate spontaneously into functional proteins. The influence of the amino acid composition and sequence on the properties of the assembled protein are dramatically illustrated by the case of sickle-cell anemia. Substitution of a single amino acid (one glutamic acid is replaced by valine) in the beta chain yields a defective hemoglobin which not only has an altered affinity for oxygen, but also, under certain conditions, may even affect the shape of the red blood cells containing it.

The subunits of a functional protein are held together in a very precise manner, sometimes by hydrogen bonds, and sometimes by covalent

COLLAGEN FIBRILS in the electron micrograph on the opposite page assemble themselves outside the cell. Bundles of such fibrils form structures of impressive strength, such as tendons.

229

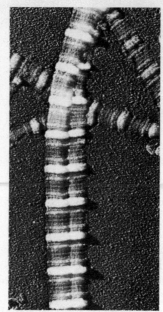

1 SELF ASSEMBLY OF TROPOCOLLAGEN into collagen rods or fibrils depends on the nature of the chemical environment. Collagen was dissolved in acetic acid, then reprecipitated in three different solutions. Electron micrograph at left shows fibrils that formed in salt solution. Banded fibrils (center) formed in acid glycoprotein solution. Short rods (right) formed in solution containing ATP.

sulfur-sulfur or disulfide bonds. In the aqueous interior of a cell the concentration of monovalent or divalent cations, the pH of the medium, the oxidation-reduction potential of the subunits, and the concentration of the subunits can regulate the assembly of a protein. Many enzymes can be completely disaggregated by changing one of these parameters, but when the original conditions are restored, the subunits become reassembled into functional molecules. Under the proper conditions, the whole molecule is thermodynamically more stable than a collection of its separate components.

A good example of self-assembly of a functional protein is COLLAGEN, the fibrous protein that strengthens most animal connective tissue, including cartilage and bone. The fundamental protein of collagen, TROPOCOLLAGEN, consists of three polypeptide subunits coiled helically around each other to form short rods. Synthesized within the cell, the rods are transported outside the cell membrane where they assemble into collagen fibrils (*chapter-opening photograph and Figure 1*). The fibrils then orient themselves quite precisely, forming insolu-

ble structures of enormous strength, such as tendons.

Tropocollagen molecules will aggregate into fibrillar structures in the test tube, and the pattern in which they aggregate can be changed by altering some other component of the solution. The presence of proteins from blood serum results in one type of aggregation, ATP a second, and different salt concentrations still another. In 1956 Jerome Gross suggested that one end of each rod differed from the other. He showed that different patterns could arise depending upon whether the long chains of rods were aligned parallel, antiparallel, or parallel, but with the ends of adjacent rods offset by one-fourth of a rod length. Though to date little is known about regulation of tropocollagen aggregation within the living animal, it is clear that small changes in the animal's internal environment can shape the resulting structure.

Another excellent example of a simple self assembly system is the protein of bacterial flagella, FLAGELLIN. To form a flagellum, a single kind of subunit becomes aggregated into a structure consisting of eight vertical rows of subunits arranged in a cylinder. The rows are arranged so the subunits are slightly offset, giving a helical appearance to flagella in electron micrographs. As with collagen, self assembly will occur in the test tube, and flagella so formed are indistinguishable from those isolated directly from bacteria. Subunits are added to one end only, indicating the presence of polarity; even minor alterations of amino acid sequences can yield flagella of dramatically different shape.

The bacterial virus bacteriophage T4 presents a more complex example of self-assembly. The T4 bacteriophage is one of the most elaborate of all viruses, consisting of a head, neck and collar, a sheath and core, and an endplate with tail fibers (*Figure 2*). This virus is constructed from only two types of macromolecule, proteins and (within the

---

Biologists have little understanding of the program that governs development.

---

head) some tightly wound DNA that contains the information for eventual synthesis of the various proteins. When T4 infects a bacterium only the DNA enters the host cell, replicating itself and directing the protein-synthesizing machinery of the bacterium to construct new viral protein. William Wood and Robert Edgar have isolated numerous T4 mutants, including one that produced no head protein, and one that produced no tail fibers. Spontaneous self-assembly of virus particles was clearly demonstrated when, by mixing headless and tail-less viruses in a test tube, Wood and Edgar obtained completely normal viruses.

Subsequent work by Wood, Edgar and many others has revealed further details of the interesting process of viral self-assembly as exemplified by T4. The assembly of the endplate is particularly fascinating. Seventeen different proteins are required, and all are synthesized in soluble and nonaggregating form. The endplate is formed by the step-by-step addition of each protein in a specific sequence. Evidently the binding of one protein to the developing endplate alters its structure, probably in allosteric fashion, enabling it to add the next protein in the sequence. Not all of the required proteins are structural components. In the formation of tail fibers, four proteins serve simply to bring structural subunits together and to assist in the formation of chemical bonds between them.

The self-assembly of viral proteins yields an extraordinary degree of order and structure at the molecular level. And, like the proteins just mentioned, under appropriate conditions the whole virus is more stable than the sum of its parts. Self-

assembly processes like those of T4 may well play a general role in the assembly of complex cellular structures such as the chloroplast, although so far there is no specific evidence for this.

ASSEMBLY OF AN ORGANELLE

A mature chloroplast is far more complex than a virus. Enclosed by a double membrane, it contains elaborate internal stacks of membranes, precisely arranged. It is made up of proteins, a special type of DNA, and lipids. Although the self-assembly of molecules and membranes may be important in chloroplast development, they are not the whole story. The synthesis of these structures requires an input of energy and information.

Chloroplasts can arise by division of previously existing chloroplasts in already green tissue, but in development, we are interested only in the generation of chloroplasts in nonphotosynthetic tissue, which happens when a seed germinates. Chloroplasts first arise as small inpocketings of the cyto-

MODEL OF T4 BACTERIOPHAGE VIRUS reveals the **2** **elaborate structure of this self-assembling molecular system. Head, neck, collar, sheath and tail fibers are constructed from numerous different proteins. Head contains core of tightly wound DNA.**

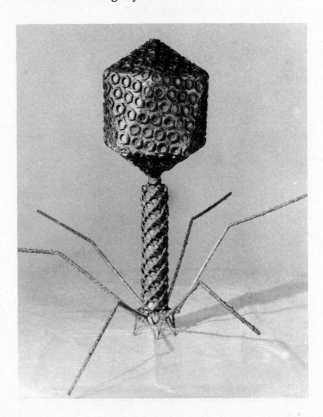

plasmic membrane. When exposed to light, these inpocketings increase enormously in size to form the mature organelles *(Figure 3)*. In the dark, however, their development is arrested. One finds only very small PROPLASTIDS, vesicles bounded by a double membrane and containing in their centers one or more small crystalline structures called PROLAMELLAR GRANULES *(Figure 4)*. The granules contain a few pigments, including carotenoids and minute amounts of protochlorophyll *a*. In some unknown way protochlorophyll inhibits its own synthesis. When the plant embryo is exposed to light, the protochlorophyll is converted to chlorophyll *a* simply by the reduction of one double bond, a process requiring the absorption of light by protochlorophyll itself. (The structure of chlorophyll is illustrated in Figure 21 of Chapter 6.) This step removes the feedback inhibition of protochlorophyll on its own biosynthesis, and so long as the light shines protochlorophyll is synthesized and converted to chlorophyll. Many structural changes accompany these reactions. The prolamellar granules loosen up and disaggregate into small

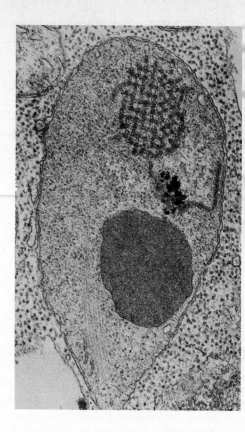

3 MATURE CHLOROPLASTS appear as swollen baglike structures in this photograph of a leaf cell from the pitcher plant Sarracenia flava. Picture was made with the scanning electron microscope.

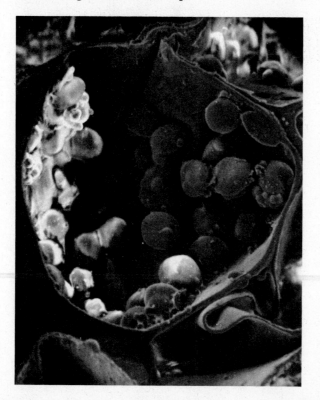

vesicles which migrate toward the periphery of the chloroplast and finally become flattened into stacks of discs. Protein and membrane synthesis begins on a massive scale. The organelle increases manyfold in size, and ultimately becomes an elaborate complex of soluble enzymes, membranes, membrane-bound proteins, and pigments.

The complete development and assembly of a chloroplast cannot occur in a test tube because a proplastid does not contain the information needed to synthesize all the molecular components of a mature chloroplast. The chloroplast DNA in the proplastid contains the genetic code for some of the chloroplast proteins, and they can be synthesized on the chloroplast's own ribosomes. But the codes for many other chloroplast proteins reside in the DNA of the cell nucleus, and protein must be synthesized on cellular, not chloroplast ribosomes. Only then are they imported into the chloroplast. A particularly interesting example, already mentioned in Chapter 6, is the enzyme ribulose-1,5-bisphosphate carboxylase, which is responsible for the fixation of CO_2 in photosynthesis. The functional enzyme consists of two different kinds of subunits, one encoded in nuclear DNA and the other encoded in chloroplast DNA.

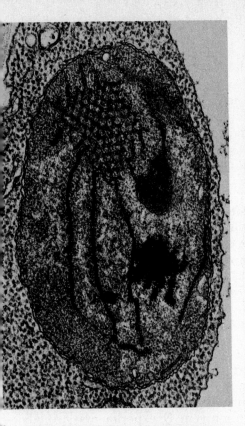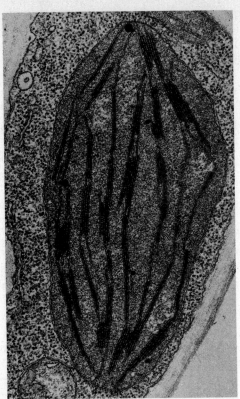

DEVELOPMENT OF CHLOROPLAST is depicted in four electron micrographs above. Proplastid at far left contains crystalline prolamellar granule (top). In the presence of light the granule begins to break up into elongated vesicles (left center). Vesicles flatten into stacks of disks, seen edge-on at right center. Mature chloroplast is shown at far right.

Interaction among organelles is required for the development of many components of the cell. As with the self-assembly of molecules, the assembly of organelles depends on the right environment. In the development of chloroplasts in higher plants, the requirement for light is absolute. Otherwise feedback inhibition limits chlorophyll synthesis.

CELL DEVELOPMENT

Although the developmental biologist can now speak with some confidence about the assembly of molecules and organelles, the molecular processes that underlie the development of the complete cell, and particularly the differentiation of eucaryotic cells, remain largely a mystery. We have little understanding about the nature of the program that governs development. It is not enough to say that the program must somehow be encoded in the genes, because that does not explain why two ad-

jacent embryonic cells with identical genes can follow two completely different pathways of development. It is clear today that cellular differentiation is based on sequential changes in the amount and kind of proteins synthesized by the cell during its development. As the synthesis of one group of proteins slows down or stops, the synthesis of others begins. In the process the cell becomes progressively more specialized. Both the temporal sequence and the final products vary from one type of cell to another.

The replication of DNA is an important step in the development of both eucaryotic and procaryotic cells. As pointed out in Chapter 2, eucaryotic cells go through four distinct phases from one mitosis to the next. The growth period, G_1, is marked by extensive RNA and protein synthesis but not by the replication of DNA, which occurs during the S period. Finally the cell enters mitosis, the M period. The length of the cycle varies from minutes to days. In any given organism, however, the length of the cycle is usually determined by the length of the G_1 period; the total time required for the S + G_2 + M periods remains relatively constant.

The continuation of DNA synthesis and mitosis

233

are not, however, prerequisites for continued differentiation. Substantial development and differentiation can occur with little or no DNA synthesis and long after mitosis has ceased. A striking example is the strange green alga *Acetabularia*, which was discussed in Chapter 10. During most of its development this giant unicell has only a single nucleus. Nevertheless, different parts of the cell grow and differentiate into a basal rhizoid (containing the nucleus), a stalk, and finally an umbrella-shaped cap. Neither nuclear DNA synthesis nor mitosis occurs until the cap begins to produce reproductive cysts, which are released to start new plants.

An even more striking example comes from some experiments performed by Ethel B. Harvey in 1940 at Woods Hole, Massachusetts. She developed an ingenious technique for separating unfertilized sea urchin eggs into two halves, one containing the nucleus, and the other only egg cytoplasm. One can trick sea urchin eggs into starting to develop (as though they had been fertilized) by soaking them for a few minutes in a hypertonic salt solution. This early development, discussed more fully in Chapter 12, consists of cell division with little growth to produce a hollow ball of cells surrounding a central cavity. The embryo at this stage is called a BLASTULA, and the cavity a BLASTOCOEL. Harvey simply treated egg fragments lacking nuclei with hypertonic salt solution and watched them. Astonishingly they began to cleave into separate cells within three hours, with some of them forming perfectly respectable blastulae in just over one day! Here is development occurring in the absence not only of mitosis and DNA synthesis, but of any RNA synthesis as well. Harvey's experiments lead us directly to a consideration of the role of RNA in development.

THE ROLE OF RNA

The central role of messenger, transfer, and ribosomal RNA in protein synthesis *(Chapter 7)* suggests that the regulation of RNA synthesis may be an important link in the control of differentiation. This hypothesis is supported by studies of several systems in which differentiation is expressed by the massive production of one, or at most a few, types of protein. In the human eye, for example, the cells of the developing lens produce massive quantities of the protein crystallin and almost nothing else. Red blood cells produce almost nothing except hemoglobin. The appearance of these proteins depends on the synthesis of the messenger RNA coded for them, as was demonstrated by the experimental inhibition of RNA synthesis by the antibiotic actinomycin D. In all of these systems actinomycin D shows the same effects. If the inhibitor is given to the cells a few hours prior to the scheduled appearance of the characteristic protein, the protein is not synthesized. But if the antibiotic treatment is delayed until the first detectable appearance of the protein, it has almost no effect on subsequent synthesis of the protein.

Because the messenger RNA must be synthesized before it can be translated into protein at the ribosomes, the sequence of events seems obvious. The appearance of a new protein is preceded by: 1, the activation of a previously blocked gene; and 2, the synthesis of a new kind of messenger RNA. Once the messenger has been synthesized, actinomycin D cannot block the subsequent synthesis of the protein. This interpretation is probably correct, but a note of caution is in order. Although actinomycin D does indeed inhibit RNA synthesis in plants and animals, that may not be the only thing it does. Moreover, at least in bacteria, one of the characteristics of messenger RNA is its instability. The actinomycin D experiments strongly suggest that messenger RNA has an enormously longer lifetime in eucaryotic cells than in bacteria. Ethel Harvey's experiments, though from a premolecular biology era, are difficult to interpret on any other basis.

Experiments with a wide variety of developing organisms generally support the existence of long-lived mRNA in developing systems. One example is the red blood cells of mammals, which, unlike those of most other animals, lack a nucleus when mature. Some time after hemoglobin synthesis begins the nucleus disappears, but hemoglobin synthesis may continue for hours, suggesting that the mRNA for hemoglobin must still be functional. It is even possible to isolate ribosomes with mRNA from the mature nonnucleated cells, provide them with amino acids and the appropriate enzymes, and obtain hemoglobin synthesis in the test tube. The mRNA for hemoglobin is clearly quite stable.

The sea urchin egg, already mentioned, provides particularly elegant evidence for stable messenger RNA. Paul R. Gross and his colleagues at MIT provided fertilized sea urchin eggs with a radioactive amino acid to monitor protein synthesis and radioactive uridine to monitor RNA synthesis. They then added enough actinomycin D to inhibit messenger RNA synthesis —that is, to block completely the incorporation of uridine. Normal

In the early stages of development, protein synthesis is controlled not by switching on new sets of genes in the DNA, but by making the existing RNA available for translation.

---

protein synthesis nevertheless continued throughout blastula development. After that stage the zygotes failed to develop further, suggesting that messenger RNA for proteins needed beyond the blastula must be synthesized during blastula development, and did not pre-exist in the fertilized egg. Other workers ultimately were able to isolate stable messenger RNA from unfertilized eggs and to use it to synthesize characteristic sea urchin proteins in the laboratory.

All of these experiments, plus those of Harvey, lead to the inescapable conclusion that the sea urchin egg contains long-lived messenger. Moreover, they suggest that before fertilization (or activation) this mRNA is somehow prevented from participating in protein synthesis; in other words, it is masked. Fertilization leads to unmasking, permitting the start of the protein synthesis required for development to the blastula stage. The molecular mechanism of the masking and unmasking processes remains obscure.

CONTROL OF RNA SYNTHESIS AND TRANSLATION

Blocks to the translation of preformed mRNA are now well documented in both plants and animals. As noted above, fertilization somehow unblocks translation. There are many other ways to regulate protein synthesis during development, at points ranging from transcription to the final stages of synthesis. Even a finished product can inhibit its own synthesis, as in the case of protochlorophyll. The processes of feedback inhibition and the repression of enzyme synthesis were discussed in detail in Chapter 5.

The synthesis of RNA itself is regulated during development by a number of different mechanisms discussed in Chapter 10, particularly gene amplification and gene activation. The South African clawed toad *Xenopus laevis* was cited as an organism whose cells rely on the strategy of gene amplification. The process involves greater replication of certain genes. An ordinary diploid cell of

Xenopus contains two nucleoli, both of which are known to contain 450 copies of the ribosomal genes. The developing oocyte, on the other hand, may contain as many as 1,000 nucleoli—apparently as a result of the disproportionate replication of the DNA that codes for ribosomal RNA. Thus instead of 450 copies, the oocyte contains almost 450,000. This tremendous amplification of ribosomal genes permits the developing oocyte to produce billions of ribosomal RNA transcripts in a very short time. With its enormous store of ribosomes and its long-lived mRNA, the egg can undergo substantial development even if RNA synthesis is blocked by an inhibitor or, as in Harvey's experiments, by the absence of a nucleus. At the moment, little is known about the causes or regulation of differential gene replication. To date, it has only been demonstrated during oogenesis.

At different times during development, different sets of genes are active in RNA synthesis. Furthermore, in a multicellular eucaryote with a multitude of different tissue types the same gene may become active in different tissues at different times in development. The fruit fly *Drosophila melanogaster* is an excellent example. One of the genes controlling pigment formation in the eye, testes, and excretory tissue was studied in detail by Thomas Grigliatti and David Suzuki at the University of British Columbia. In the future excretory tissue the gene was active in eggs and very young larva, but the same gene did not become active in future eye and testis tissue until the larva was well on its way to becoming an adult fly *(see discussion of metamorphosis in Chapter 18).* Development evidently involves mechanisms for turning genes on or off—in other words, for controlling transcription. This mechanism, called gene activation, is the second major means of controlling the amount and kind of proteins that appear or disappear during development.

There is a great temptation to invoke the operon hypothesis of Jacob and Monod, described in detail in Chapter 10. In their model, a specific product of a regulator gene acts as a repressor, which can combine with an operator gene whose role is to permit or induce transcription of one or more structural genes. The structural genes specify the mRNA for the polypeptide chains of proteins. The presence of repressor on an operator gene simply prevents transcription of the structural genes. To date, two repressors have been isolated and characterized, one from *E. coli,* and the other from bacteriophage λ. Both are small proteins, so the regulator gene must contain the code for their

> It is an axiom of genetics that virtually every cell in a multicellular organism contains a complete set of genes —all the information necessary to create a complete organism.

messenger. The Jacob-Monod model further postulates that certain small molecules called activators in the cytoplasm can form specific complexes with different repressors. When such complexing occurs, the repressor can no longer bind to the operator gene; the structural gene is derepressed and transcription proceeds.

The operon hypothesis has been cautiously invoked time and again by developmental biologists probing cell differentiation in higher plants and animals. The caution arises from the fact that the operon model is based entirely on work with bacteria, and the chromosomes of eucaryotic organisms are far more complicated —a complex of DNA, RNA, and protein collectively called CHROMATIN. Characteristically chromatin can occur either in an extended or in a condensed state. The metaphase chromosome seen in mitosis is maximally condensed. The puffs seen on the giant salivary gland chromosomes of certain insects represent maximally extended chromatin *(Figure 5)*. Only chromatin in its extended state is transcribed, and different regions of the chromosome are puffed at different times in development. Among the components of chromatin are highly basic proteins called histones. Because they contain large amounts of the basic amino acids lysine and argenine, the proteins interact ionically with the acidic phosphate groups of DNA. Some workers believe that histones are involved in regulation of gene transcription in eucaryotes, but their mode of action is at present unresolved.

It would be a mistake to conclude that the sole basis of development is the regulated and programmed appearance of different gene products (whatever the regulatory mechanism). This was demonstrated by experiments on a water mold known as *Blastocladiella*, which occasionally reproduces itself by producing small motile cells with a single flagellum. These swimming cells (called ZOOSPORES) eventually land on a favorable substrate and germinate. The process of germination involves remarkable intracellular reorganization: a vesicle filled with ribosomes opens up and re-

leases them, the single mitochondrion elongates and multiplies, the flagellar apparatus disappears, a cell wall forms, and finally a small filament emerges from the cell. In experimental studies, only the disappearance of the flagellar apparatus and the outgrowth of the filament were blocked by treatment with compounds that completely stopped protein synthesis. Clearly mechanisms altering preformed structures must supplement the programming of the transcription-translation machinery.

IS DIFFERENTIATION REVERSIBLE?

It is an axiom of genetics that virtually every cell in a multicellular organism contains a complete set of genes —all the genetic information necessary to create a complete organism. A corollary of this is that any specialized cell in a multicellular organism

GIANT CHROMOSOME from salivary gland of the larva of the midge Chironomus tentans. Three chromosome puffs are visible as swellings at top, center and bottom. Puff at bottom was induced by treating the larva with ecdysone.

5

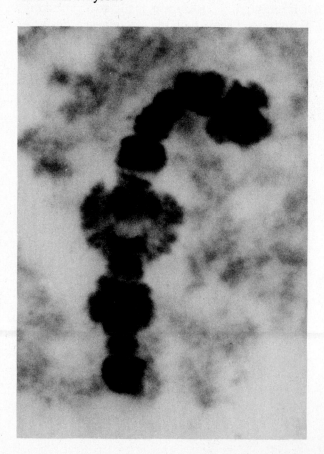

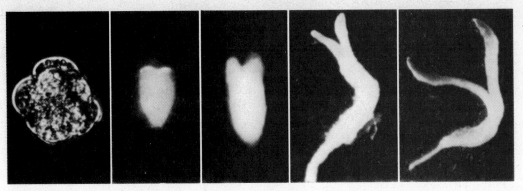

6 EMBRYOIDS obtained from carrot cells grown in tissue culture pass through all of the stages typical of the development of a normal embryo. Globular embryoid (far left), a fraction of a millimeter long, develops first into heart-shaped cell, elongates into torpedo shape (center), then sprouts as a young carrot plant several millimeters long (right).

retains the genetic equipment necessary to carry out all the functions of any other cell in the organism. A nerve cell, for example, has the same set of genes as a liver cell; they carry out different functions only because they are expressing different sets of genes. In theory it should be possible to transform a nerve cell into a liver cell, or to regenerate a missing finger from a muscle cell, or to grow an oak tree from a leaf cell. But in nature things don't work out that way, which in the past has led some biologists to speculate that differentiation is irreversible. The tentative explanation was that once development had run its course, most of the genes of the cell were permanently switched off.

In certain types of cell, differentiation is clearly irreversible. The mammalian red blood cell, which loses its nucleus during development, is one example. Another is the xylem tracheid, a water-conducting cell in higher plants. The development of the tracheid culminates in the death of the cell, leaving only the thickened and pitted cell walls that were fashioned while the cell was alive (*Chapter 16*). In these two extreme cases the irreversibility of differentiation can be explained by the absence of a nucleus. But it is much harder to generalize about mature cells that retain normal nuclei. Most botanists tend to think of differentiation as reversible, while zoologists tend to think of it as permanent, but this is not a hard and fast rule. A lobster can regenerate a missing claw, but a cat cannot. Why is differentiation reversible in some cells but not in others? At some stage of develop-

ment do changes within the nucleus permanently commit a cell to specialization?

At present there are no answers to these questions, but promising avenues for further research have been opened by the pioneering experiments of F. C. Steward at Cornell, Robert Briggs and Thomas King of the Institute for Cancer Research, Philadelphia, and J. B. Gurdon and his associates at Oxford. Steward and his colleagues cultured suspensions of single cells and small clusters of cells from the roots of carrots. Originally a small piece of carrot was grown on a nutrient medium solidified with agar. Under the right conditions, such pieces of tissue began relatively random cell division to form an undifferentiated mass of tissue called CALLUS. The callus was transferred to special

---

Most botanists tend to think of differentiation as reversible, while zoologists tend to think of it as permanent; but this is not a hard and fast rule. A lobster can regenerate a missing claw, but a cat cannot.

---

rotating flasks containing liquid nutrient medium. Although the cells continued to divide, the constant tumbling and agitation broke up the callus tissue and prevented the formation of masses of more than a few cells. When a suspension of such cells was plated on nutrient agar, the cells developed first into globular EMBRYOIDS (*Figure 6*) and eventually into mature carrot plants. But the experimental technique made it impossible to follow the fate of any one individual cell in the culture,

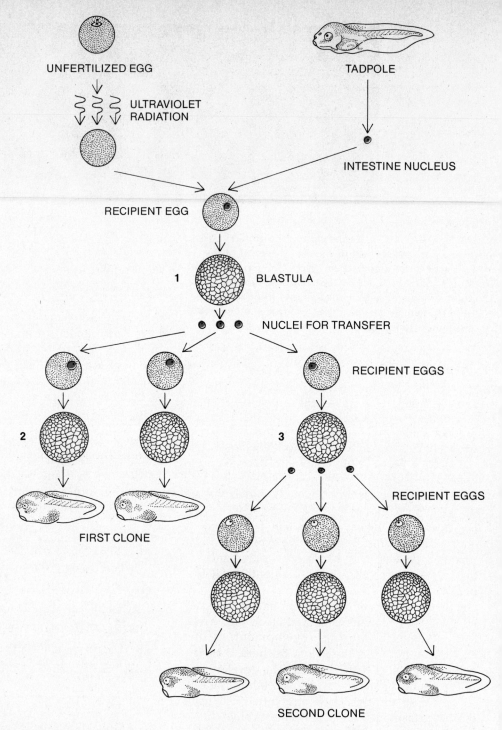

UNFERTILIZED EGG

TADPOLE

ULTRAVIOLET RADIATION

INTESTINE NUCLEUS

RECIPIENT EGG

1 BLASTULA

NUCLEI FOR TRANSFER

RECIPIENT EGGS

2 3

FIRST CLONE

RECIPIENT EGGS

SECOND CLONE

and hence to say for certain that the entire plant had indeed arisen from a single cell. In a later experiment Vasil and Hildebrandt managed to accomplish this task. They isolated single callus cells from tobacco plants and suspended them in a nutrient medium. After several days the single cell had divided repeatedly, forming a tiny mass of callus. When the minute callus was transferred to solid nutrient medium containing appropriate con- centrations of the plant growth hormones auxin and cytokinin, shoots and roots appeared *(Chapter 18; also see Chapter 12 for a discussion of the interaction of these substances in organ formation)*. Eventually the small plantlets were transplanted to soil, where they grew to maturity and flowered. Clearly the nucleus of the callus cell had not differentiated ir- reversibly, and could still express the full tobacco genome.

7

GURDON EXPERIMENT demonstrated that mature cells from a toad tadpole contain all the genes necessary to guide the development of an egg into a mature adult. Nucleus from intestinal cell of Xenopus tadpole was surgically transplanted into unfertilized egg. When egg developed to blastula stage (1), nuclei from blastula were in turn transplanted into eggs whose nuclei had been removed. These eggs developed into normal blastulas (2), then into tadpoles and eventually into normal toads. Nuclei from one blastula were transplanted into new eggs, giving rise to another generation of toads. Known as serial transplanting, technique can be repeated indefinitely to generate thousands of genetically identical toads.

The ready formation of callus by many mature plant tissues, and the capacity of these calluses to differentiate into active roots and shoots, certainly supports the argument that as long as a nucleus retains its normal complement of chromosomes, it still possesses all of the information needed to produce an entire organism. Animal cells have proved more refractory. Isolated animal cells grown in tissue culture never develop into complete organisms or even into relatively simple organs. As will be shown in the following chapter, organ formation in animal tissue culture normally requires interaction between two or more cell types. Perhaps differentiation in animals does involve permanent changes within the nucleus.

The first indication that this statement should be qualified came from the elegant experiments of Briggs and King. They activated a frog egg to begin development (in this case by puncturing it with a glass needle), then carefully removed the egg nucleus. Into the enucleated egg they transplanted a nucleus from one of the cells of an older embryo. If the donor cell came from a blastula, each recipient egg developed into a normal embryo. If the donor cells came from a somewhat later stage of development, say a late gastrula, the resulting embryos were usually defective, or failed to develop at all. The experiment demonstrated that at least through blastula formation, the frog embryo nucleus contains (and can express) all of the information needed to make a whole frog.

The work of Briggs and King emphasized the progressive restriction imposed on nuclei during development. They concluded that differentiation irreversibly restricts the capacity of nuclei to express all of their genes. Gurdon and his colleagues took quite the opposite view. They were actually able to go considerably further with the toad *Xenopus* than were Briggs and King with the frog. In 1962 they succeeded in transplanting nuclei from the intestines of tadpoles into enucleated eggs, and obtaining tadpoles and a small number of mature toads *(Figure 7)*. In later experiments they successfully transplanted nuclei from kidney, heart and skin cells; in these cases, however, the resulting tadpoles failed to develop into toads. More recently L. D. Smith compared the capacity of nuclei from closely adjacent cells —primordial germ cells, destined to become eggs or sperm, and nearby cells with the same origin as the germ cells but already recognizably distinct from them —to produce normal embryos. With germ cell nuclei he had 40 percent success, but with nuclei from adjacent tissue, the success rate was less than 20 percent. These experiments collectively suggest that the capacity of a nucleus to express its entire genome in the orderly way required to progress from egg to adult probably becomes restricted in some nuclei but not all. The nature of the restriction, however, remains unknown.

This chapter has described molecular events which play important and varied roles in the flow of development. The next chapter will consider the ways in which cells become organized into tissues, tissues into organs, and organs into organisms.

READINGS

J.D. EBERT AND I.M. SUSSEX, *Interacting Systems in Development*, 2nd Edition, New York, Holt, Rinehart and Winston, 1970. A good treatment both of descriptive and experimental aspects of development, with a good introduction to molecular aspects. Emphasis is on animals, with only a few plant chapters.

C. FULTON AND A.O. KLEIN, *Explorations in Developmental Biology*, Cambridge, MA, Harvard University Press, 1976. An exceptional collection of reprints of important papers with excellent text between; the emphasis is on molecular aspects of development and animal development. Contains original papers on many of the experiments described in this chapter.

F.B. SALISBURY AND C. ROSS, *Plant Physiology*, 2nd Edition, Belmont, CA, Wadsworth, 1978. A comprehensive textbook of plant physiology with a good treatment of cellular and molecular aspects of plant development.

And what a glorious society we would have if men could regulate their affairs as do the millions of cells in the developing embryo. —R. M. Eakin

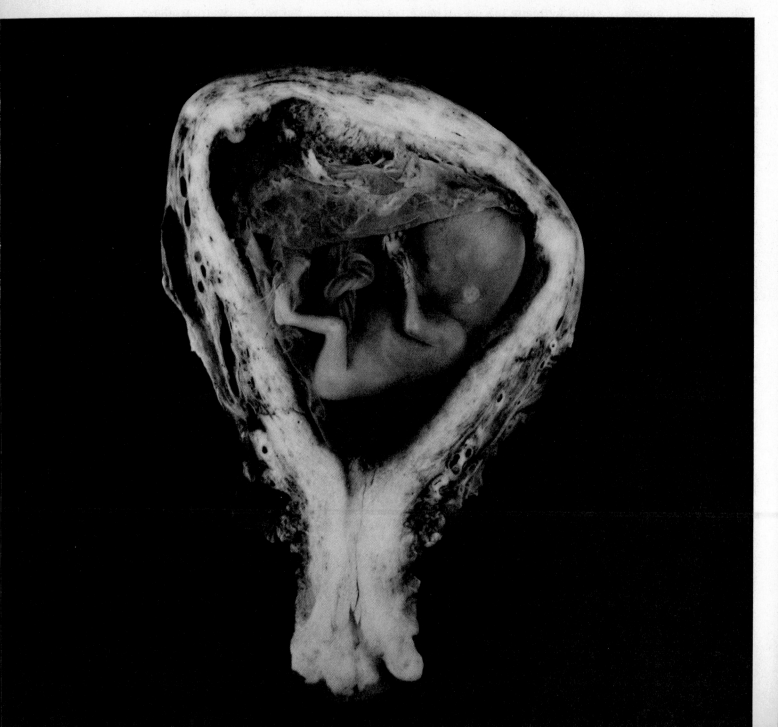

The preceding chapters clearly demonstrate that cells are not static structures; they change with time. The life cycle of a multicellular organism is the story of changes in its cells. Taken together, these changes are called development. The basic problem facing the developmental biologist is to determine precisely how these changes are brought about—how a fertilized one-celled egg becomes an adult organism constructed of many different types of cells.

The story begins with the fusion of egg and sperm to form a ZYGOTE. Cell division of the zygote and its mitotic products leads ultimately to the aggregation of cells into tissues, tissues into organs, and organs into organisms. In one sense, development is best understood in terms of geometry, particularly topology, or "rubber-sheet geometry." Animal zygotes quickly develop into a hollow ball of cells. As the cells multiply and migrate, the embryo attains its final shape as a result of a complicated series of bulgings, foldings, and in- and out-pocketings of its layers of tissue. In plant embryos the rigid cell walls prevent cell migration, so the embryo attains its shape by differential cell division and cell enlargement: some parts of the embryo grow much more rapidly than others, forming new leaves and roots. The study of development is also an exercise in the mastery of an unfamiliar vocabulary. This chapter contains many new terms, and it is appropriate that we begin with precise definitions of four of the most important: growth, determination, differentiation, and morphogenesis (Box A).

TEN-WEEK HUMAN FETUS within the uterus. Part of uterus and fetal membranes have been dissected away to reveal fetus. Picture shows embryo somewhat larger than actual size at this age.

FERTILIZATION IN ANIMALS

Before looking closely at the events which follow the fusion of egg and sperm, the origins of these two remarkable cells in animals should be briefly summarized. In both males and females certain PRIMORDIAL GERM CELLS, which will eventually give rise to the functional gametes, are recognizable in extremely young embryos. One usually thinks of the gametes as being products of the GONADS: the male testis for sperm and the female ovary for eggs. But the primordial germ cells are evident long before the gonads themselves have appeared. With the eventual development of the gonads, germ cells migrate into the testis or ovary and continue their course of development.

The primordial egg cells go through a period of cell division followed by extensive growth and then enter meiosis. Of the four nuclear products of meiosis, three are simply extruded as POLAR BODIES. The fourth will serve as the egg nucleus. Depending on species, a small or large amount of yolk is deposited within the egg. Yolk consist of droplets or platelets of protein, phospholipid, and fat. In eggs with a large amount of yolk, the yolk will nourish the developing embryo for up to a few weeks (Figure 1). Much, if not all, of the yolk material is synthesized elsewhere in the organism, transported to the developing egg, and taken up by invagination (in-pocketing) of the egg membrane. In some insects, even ribosomes may be donated to the egg by other cells in this manner.

The primordial sperm cells find their way into the testis, proliferate, and eventually go through meiosis, with each meiotic cell developing into a characteristic sperm (Figure 2). There is a head region, consisting most anteriorly of a head cap and an underlying ACROSOME, frequently a vesicle or granule, and then the haploid nucleus. A middle

241

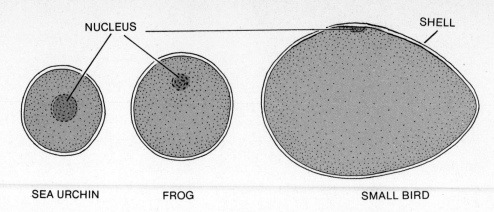

NUCLEUS

SHELL

SEA URCHIN   FROG   SMALL BIRD

1 **EGGS OF THREE SPECIES contain different proportions of yolk (gray). Yolk of sea urchin and frog egg can subdivide into daughter cells by cleavage; yolk of bird egg cannot. Amount of yolk determines how long the egg can nourish developing embryo. Eggs shown here are not drawn to the same scale. The three nuclei are actually about the same size.**

section contains mitochondria, presumably to supply the power for swimming, plus a usual array of microtubules within. In some lower plant species with swimming sperm, the middle section may also contain a chloroplast. The acrosome, as we shall see in a moment, plays an important role in fertilization.

There is a dramatic difference between the contribution of cytoplasm from the egg to the zygote and that from the sperm. Chapter 11 described experiments that revealed the remarkable developmental capacity of the zygote. Recall that the zygote could develop as far as the blastula stage without any RNA synthesis. Recall also that the unfertilized egg contained messenger RNA in a masked form. And finally recall Ethel Harvey's description of development in egg fragments lacking any nuclei. Virtually all of the zygote cytoplasm comes from the egg, not only informational RNA and protein-synthesizing machinery, but also mitochondria and all manner of other organelles. The sperm provides genetic information in the form of DNA, and virtually nothing else. It is of little biochemical assistance in the early stages of development.

The various mechanisms by which egg and sperm are brought together in different groups of

A Four important words

GROWTH. The definition of growth is deceptively simple —it is simply irreversible increase in mass. But does growth involve cell division? Does it require protein synthesis? Though the answer to these questions is normally yes, the literature abounds with examples of growth in the absence of one or both of these processes.

DETERMINATION in a narrow sense is the process whereby a group of cells, a single cell, or even a part of a cell, becomes committed to some predictable pathway of development. In certain fertilized eggs, an observer can even pinpoint the part of the outer cytoplasm that will become the epidermis, the lining of the gut, and so on.

DIFFERENTIATION is the actual expression of determination. A cell differentiates both from its neighbors and from its past. A common definition of differentiation is: change leading to modification of structure and/or function. The fundamental processes responsible for determination and differentiation are among the most important unsolved puzzles in modern biology.

MORPHOGENESIS is the appearance of shape or form. It is the result of coordinated growth and differentiation —the sum of the developmental changes that lead ultimately to the formation of a leaf, a hand, or a heart. Usually several different differentiating systems, acting with remarkable coordination, mutual interaction and growth, lead to morphogenesis.

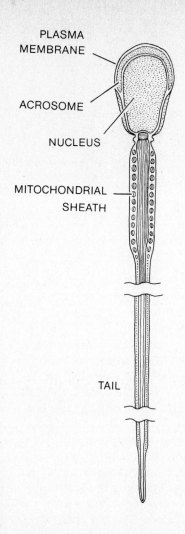

PLASMA
MEMBRANE

ACROSOME

NUCLEUS

MITOCHONDRIAL
SHEATH

TAIL

2 **SPERM CELL consists of head, mitochondrial sheath and tail. Genetic material is packaged in the nucleus. Broken areas in tail indicate longer length. Diagram depicts sperm of the rabbit.**

from some intracellular depots. Small projections arise from the egg plasma membrane, curl over the attached sperm, and draw it into the egg. Meanwhile, a series of small membrane-bound granules just under the plasma membrane move out, fuse with the plasma membrane, and discharge their protein contents outside. All of the events so far described occur within the first 20 seconds. One protein from the granules destroys other sperm receptor sites, another in some manner raises the jelly-like outer layer off the plasma membrane, and still other proteins transform the detached layer into a protective envelope called the fertilization membrane. By this time only 60 seconds have passed. Within five minutes or so one begins to see an increase in protein synthesis. The egg and sperm nuclei do not become fused completely until 20 minutes after binding; then DNA synthesis begins. The first cell division occurs about an hour and a half after fertilization. The raising of the fertilization membrane and the destruction of sperm receptor sites assures that the egg is penetrated by only a single sperm.

During most of the blastula stage, the embryo develops without any real increase in size, the original cytoplasm merely being partitioned into successively smaller cells.

THE ANIMAL ZYGOTE

The course of events following fertilization is determined to a substantial extent by the amount of yolk present in the egg. Three cases show the range of variation: the sea urchin, with little yolk evenly distributed; amphibia, with substantial yolk toward one end of the egg; and birds, with enormous amounts of yolk to support the extensive development that occurs within the egg before hatching.

By far the simplest pattern is that illustrated by the sea urchin. The zygote divides, cleaving the cytoplasm into two equal portions, each with a nucleus. Cleavage is repeated until the embryo becomes a small hollow ball of cells called a BLASTULA, surrounding a central cavity or BLASTOCOEL *(Figure 3)*. During most of the blastula stage, the embryo develops without any real increase in the size, the

animals, and the evolution of reproductive systems in the animal kingdom, are discussed in Chapter 13. As the small sperm approaches the much larger egg the acrosome makes the first physical contact. Most eggs secrete a jellylike protein-polysaccharide mixture that covers their plasma membrane, and the acrosome contains enzymes that digest this material. The most detailed knowledge of subsequent events comes from extensive studies of fertilization of the sea urchin egg. The tip of the acrosome makes contact with a sperm receptor site on the surface of the egg's plasma membrane, initiating a series of extremely rapid events. Two seconds after contact and binding, there is a small influx of sodium followed (at eight seconds) by a massive release of calcium ions

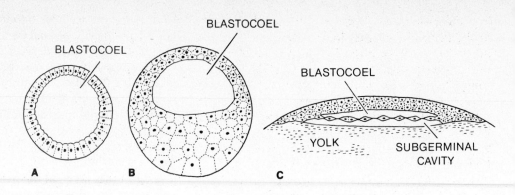

BLASTOCOEL

BLASTOCOEL

BLASTOCOEL

YOLK

SUBGERMINAL
CAVITY

A   B   C

3 **BLASTULAS of a sea urchin (A), a frog (B) and a bird (C). In sea urchin and frog, blastula is a hollow ball of cells. In birds it is a disk of cells that rests on the surface of the yolk.**

original cytoplasm merely being partitioned into successively smaller units. Development occurs in the absence of growth. And, as noted in the last chapter, neither DNA synthesis nor RNA synthesis are required for the cell cleavages which lead to blastula formation. In some animals, like the frog, the cleavage pattern is somewhat different. The cells formed at the end nearest the nucleus are small and rich in cytoplasm, while those at the other end are large and yolk-filled. The smaller cells will produce most of the embryo; this end of

the egg is frequently called the ANIMAL POLE. The end containing most of the yolk is called the VEGETAL POLE. Figure 4 is a striking illustration of an early frog blastula at the sixteen cell stage; the animal pole points over the reader's right shoulder.

In birds, the entire egg does not cleave; instead, mitosis and cytokinesis of the zygote produces a flat plate of cells, the BLASTODISC, perched on the massive yolk. A thin layer of cells next to the yolk separates from the cells above to form the blastocoel.

The next stages of development involve the formation of a hollow three-ply embryo, most easily visualized as three concentric tubes surrounding a central cavity. With the beginning of these stages, a major difference between plant and animal development becomes obvious. Animal cells, unrestricted by a tough cellulose wall, migrate extensively during development. The cells may move by cytoplasmic flowing, as amoebas, or by extending long filamentous processes that attach to other cells and then contract, pulling the cells together. Within the cytoplasm minute microfilaments in some way mediate the changes in cell shape that cause these movements.

The first of these MORPHOGENETIC MOVEMENTS appear in the process of GASTRULATION. Again the sea urchin is the simplest case. Though the urchin egg may look symmetrical it has an inherent polarity, like the frog egg, with distinct animal and vegetal halves. The animal half will become the anterior region of the embryo, and the vegetal half the posterior. In the blastula the cells at the vegetal end form a small pocket as they begin migration

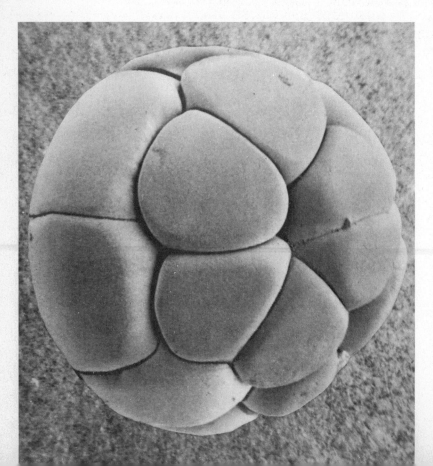

BLASTULA OF FROG after four rounds of cell division is a 16-celled sphere. Picture was made with a scanning electron microscope, and is reproduced at a magnification of about 60X. **4**

into the blastocoel *(Figure 5)*. The whole embryo finally begins to increase in size. The new cavity is the ARCHENTERON or primitive gut, which opens to the outside through the BLASTOPORE, destined to become the anus. By now the zygote has become a two-ply embryo. The outermost layer of cells is the ECTODERM, and the innermost layer the ENDODERM. What is the origin of the third layer? Close to the blastopore, near the junction of ectoderm and endoderm, certain cells begin to migrate away from the two primary cell layers and move into the blastocoel. These cells are the primary MESENCHYME, and will form the middle layer or MESODERM of the three-ply embryo. Other mesenchyme cells become detached from the advancing tip of the primitive gut, and will also contribute to the developing mesoderm. At this stage the blastula has become a GASTRULA. At the end of gastrulation, the inpocketing cells merge with those on the far side of the embryo, and the archenteron opens at its other end to form a tube. The new opening will become the mouth.

In eggs with more yolk, such as those of amphibians, gastrulation proceeds somewhat differently. The large yolk-filled vegetal cells simply don't migrate. Instead, cells from the animal half migrate down, turn in, and then move up into the blastocoel, as shown in Figure 6. Most of the activity occurs at the upper, or DORSAL, lip of the blastopore. Nevertheless, an archenteron is formed. The cells migrating inward segregate in a coordinated way to form both endoderm and a mesenchyme layer destined to become mesoderm. The first mesenchyme cells appear between ectoderm and endoderm just at the dorsal lip. By a combination of proliferation and cell migration the mesenchyme moves forward (remember that the blastopore will become the anus) to form a rod of cells whose center will form the NOTOCHORD: a cartilaginous structure found under the spine of all animals with backbones. As with the sea urchin, the theme is coordinated cell movement that leads to the construction of a three-ply embryo.

In eggs with massive yolks the three-ply embryo must be constructed from the flat blastodisc; gastrulation as described above is impossible. Instead a thin layer of cells against the yolk becomes separated from the rest, forming the blastocoel, and extensive cell movement begins. The blastodisc has a clear head-to-tail polarity; cells on both sides of it begin to migrate toward the center line and then forward, forming a structure called the PRIMITIVE STREAK *(Figure 7)*. The cells toward the forward end begin to move inward and enter the blas-

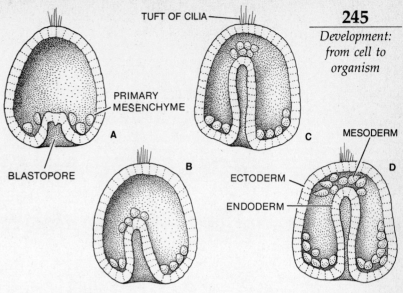

GASTRULATION IN SEA URCHIN. Letters indicate progressive stages of development from blastula (A) to gastrula (D). Invagination produces the primitive gut. Blastopore will become the anus. 5

tocoel. All of these cells are destined to become mesoderm. Although the spatial relationships are quite different from those in the sea urchin, the end result is the same: a three-layered embryo.

Although an introductory text is not the place to catalogue all of the complex events that follow gastrulation and lead to formation of the characteristic tissues and organs which comprise the adult animal, we shall nevertheless follow early development through one more process called NEURULATION *(Figure 8)*. The frog embryo is a good example. At the end of gastrulation, a distinct elongate plate of cells appears in the ectoderm overlying the notochord mesoderm. On either side of this plate —called the NEURAL PLATE —there arise ridges which arch up, fold inward, and fuse to form the NEURAL TUBE, completely covered by a thin layer of ectoderm. The tube then pinches away from the overlying ectoderm to form what will become the spinal cord of the nervous system. Meanwhile, the mesodermal cells on either side of the notochord become organized into flat plates of tissue extending laterally from the notochord. Ultimately the mesodermal tissues become segmented fore and aft into SOMITES. The somites establish the basic segmentation pattern of the mature animal, best exemplified by the sequential arrangement of the muscles and vertebrae of the spinal column. The lateral margins of the three tissue layers even-

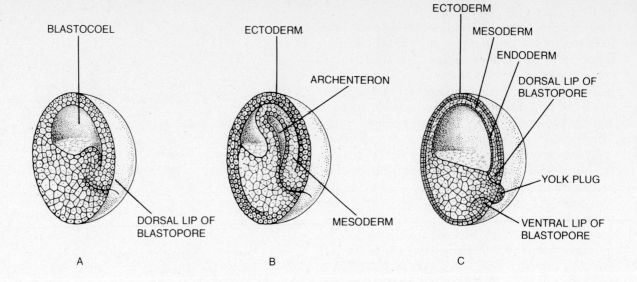

BLASTOCOEL

DORSAL LIP OF
BLASTOPORE

A

ECTODERM

ARCHENTERON

MESODERM

B

ECTODERM

MESODERM

ENDODERM

DORSAL LIP OF
BLASTOPORE

YOLK PLUG

VENTRAL LIP OF
BLASTOPORE

C

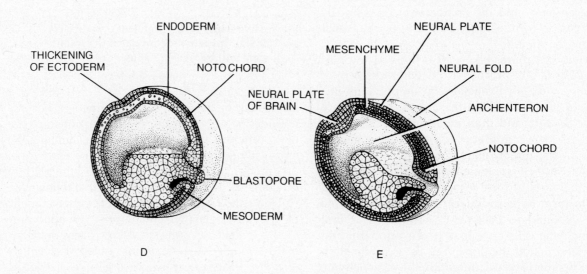

THICKENING
OF ECTODERM

ENDODERM

NOTOCHORD

BLASTOPORE

MESODERM

D

MESENCHYME

NEURAL PLATE
OF BRAIN

NEURAL PLATE

NEURAL FOLD

ARCHENTERON

NOTOCHORD

E

GASTRULATION IN FROG. In gastrula stages (A through E) cells migrate from animal pole, moving downward and inward to form the three primary germ layers and the primitive gut (archenteron). Blastocoel diminishes and a yolk-plug forms from the vegetal cells.

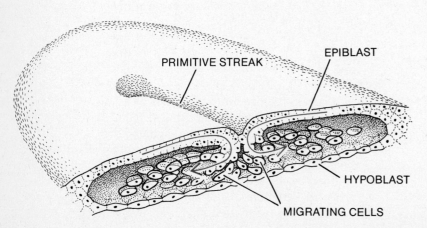

PRIMITIVE STREAK

EPIBLAST

HYPOBLAST

MIGRATING CELLS

GASTRULATION OF CHICK BLASTODISC. Arrows indicate direction of cell migration into and through the primitive streak to form mesoderm. Result is formation of thin three-layered embryo.

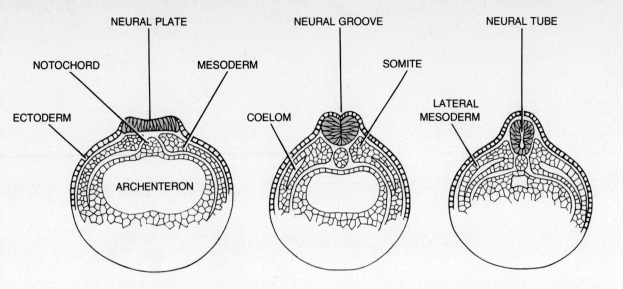

Each of the three cell layers of the gastrula develops into a specific group of tissues in the adult. Ectoderm gives rise to the nervous system and skin. Endoderm forms the entire digestive tract and related organs, including lungs, liver, pancreas and bladder. Mesoderm becomes bone, muscle, blood, heart and the rest of the circulatory system.

FORMATION OF NEURAL TUBE in the frog embryo. **8**
After gastrulation, ectodermal ridges (color) on each side of primitive streak begin to form tube. Note appearance of coelom in lateral mesoderm.

DEVELOPMENT OF MAMMALIAN EMBRYO from **9**
fertilization (upper left) until implantation in the uterine wall (upper right) is depicted two-dimensionally in this schematic diagram.

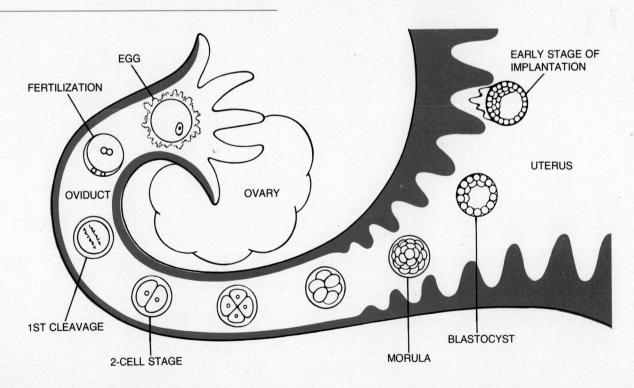

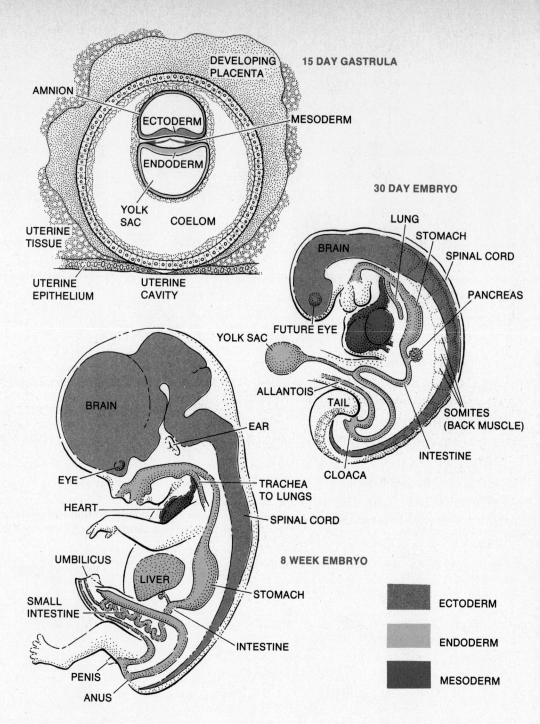

15 DAY GASTRULA

DEVELOPING PLACENTA

AMNION

ECTODERM

MESODERM

ENDODERM

YOLK SAC   COELOM

UTERINE TISSUE

UTERINE EPITHELIUM

UTERINE CAVITY

30 DAY EMBRYO

LUNG

STOMACH

SPINAL CORD

BRAIN

PANCREAS

FUTURE EYE

YOLK SAC

ALLANTOIS   TAIL

SOMITES (BACK MUSCLE)

INTESTINE

CLOACA

BRAIN

EAR

EYE

TRACHEA TO LUNGS

HEART

SPINAL CORD

UMBILICUS

8 WEEK EMBRYO

LIVER

SMALL INTESTINE

STOMACH

INTESTINE

PENIS

ANUS

ECTODERM

ENDODERM

MESODERM

10 **HUMAN DEVELOPMENT** follows the pattern typical of most mammals. At gastrula stage (top) embryo is a hollow ball of cells, folded in at the center. Colors indicate the three germ layers: ectoderm, endoderm and mesoderm. Mesoderm has just begun to migrate from its point of origin. Diagram at right is a longitudinal cross-section through a 30-day embryo. By this time the three germ layers have begun to differentiate into primitive tissues and organs. The major structures derived from each germ layer are indicated by the same colors in all three diagrams. By the 8th week the embryo has developed recognizably human form (left).

tually move inward and fuse, forming the three concentric tubes of cells surrounding an archenteron. A separation of mesodermal tissue opens a cavity called the COELOM which will eventually enlarge to accommodate the internal organs of the animal.

Each of the three layers of the completed gastrula develops into a specific group of tissues in the adult. Ectoderm gives rise to the neural tube, which will produce the nervous system; and to the outermost layer of skin, the EPIDERMIS, and its derivatives, such as nails, hair, feathers, the lens of the eye, and the linings of the mouth and anus. The tip of the archenteron eventually fuses with the opposite layer of ectoderm, and cell migration then opens the end of the tube to the outside, forming the mouth. As a consequence teeth have both ectodermal and endodermal components. The endoderm forms the entire digestive tract and all related structures, such as the lungs and respiratory system, liver, pancreas, thyroid gland, and bladder. These structures all originate as outpocketings of the archenteron. Starting with the formation of a segmented spinal column around the notochord, the mesoderm gives rise to a vast amount of tissue: bone, muscle, blood, and the entire circulatory system including the heart.

VERTEBRATE DEVELOPMENT

The early stages of vertebrate development, with the embryo going through extensive development within the mother and relying exclusively on her for nourishment, are slightly different from those described so far. The egg, released by the ovary, finds its way to the inner end of a tube called the OVIDUCT, which connects directly to the UTERUS. When a sperm enters from the vagina, it must traverse the uterus and make its way up the oviduct to effect fertilization. The events following fertilization are depicted in Figure 9. The zygote cleaves as it is moved down the oviduct; by the time it reaches the uterus it is a small solid ball of cells called a MORULA. Eventually the morula opens to form a fluid-filled vesicle called the BLASTOCYST, which implants itself in the wall of the uterus. After implantation, the outer layer of the blastocyst becomes almost entirely dedicated to forming an elaborate envelope of extra-embryonic membranes, including the placental tissue which will absorb nutrients from the mother and transfer them to the embryo through the umbilical cord. The embryo itself arises from a tiny inner group of cells. Subsequent development involves the formation of a blastula, gastrulation, neurulation, and the formation of somites, as in frogs and birds. The human embryo at various stages of development is shown in two photographs *(one on the opening page of this chapter, one in Figure 11)* and in one diagram *(Figure 10)*.

One group of cells in the young embryo deserves special mention, those of the NEURAL CREST. These cells appear during the ectodermal folding that produces the neural tube. The process is dia-

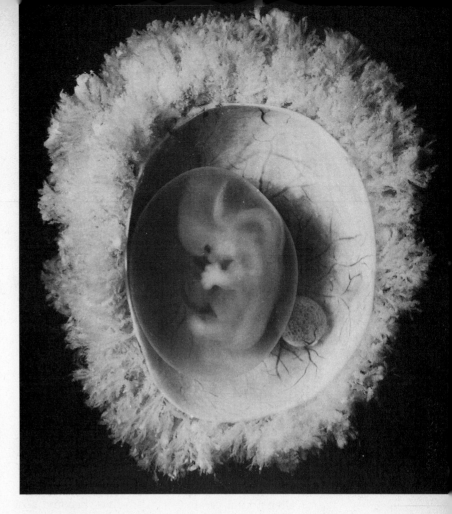

HUMAN FETUS depicted in photograph above is **11** six weeks old and about half an inch long. Fetus floats within the amniotic sac. Yolk sac lies at lower right center. Enclosing them is the placenta, half of it dissected away to reveal fetus.

grammed in Figure 12. The cells of the neural crest become detached from the neural tube and migrate enormous distances, becoming differentiated only when they reach their final destination. Some form sheaths around the motor nerve cells which are

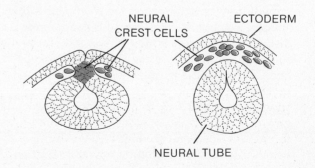

NEURAL CREST cells arise from ectoderm during **12** formation of neural tube.

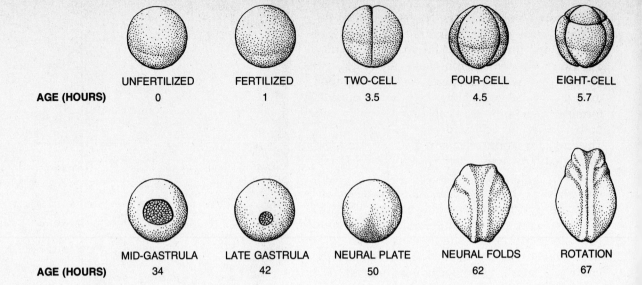

	UNFERTILIZED	FERTILIZED	TWO-CELL	FOUR-CELL	EIGHT-CELL
AGE (HOURS)	0	1	3.5	4.5	5.7

	MID-GASTRULA	LATE GASTRULA	NEURAL PLATE	NEURAL FOLDS	ROTATION
AGE (HOURS)	34	42	50	62	67

13 EARLY DEVELOPMENT of a frog embryo at 18°C from fertilization (upper left) through neural tube formation to the appearance of the tail bud (lower right). All drawings are made to the same scale. Unfertilized egg is 1.6 mm in diameter.

growing into the rapidly developing musculature. Other neural crest cells themselves become sensory nerve cells. Still others may become cartilage, or even part of the adrenal gland. The neural crest cells surpass all other embryonic cells in the extent of their wandering.

The complete development of the frog embryo from fertilization to a swimming tadpole is illustrated in Figures 13 and 14. The series takes the animal from a single cell to a stage containing roughly one million cells—through a remarkable series of coordinated events. Several points are illustrated here. First, the unfertilized egg is roughly 1.6 mm in diameter. An enormous amount of development takes place without any significant growth; not until the neural ridges appear is there an increase in size. Following neurulation there is a distinct rotation of tissues forward to form the rudimentary head, and the simultaneous appearance of a tail bud. Figure 14 shows the appearance of various features in succession—muscle response, heartbeat and so on—during these early growth stages. Only at an age of 162 hours does the mouth open and feeding begin. All prior development was supported by nutrients contained in the yolk.

Animal development is characterized by cell migration. Later in this chapter, we shall consider

how this movement is regulated and coordinated. But first we shall consider the case in which cell migration is impossible because the cells are encased in a straightjacket of cellulose: the development of higher plants.

MORPHOGENESIS IN PLANTS

Almost all plant cells are encased in a cell wall. The principal structural element of the wall is cellulose, a long chain of glucose molecules joined by beta linkages. Packaged into discrete micelles held together by hydrogen bonding, the cellulose molecules are organized into microfibrils of enormous strength *(Figures 15 and 16)*, clearly visible in the electron microscope. The rigidity of the wall depends on an extensive network of such microfibrils embedded in a gelatinous matrix of pectins and hemicelluloses, both polymers of various sugars and sugar acids. The wall restricts the size of the plant cell and prevents both amoeboid movement and the extension of filamentous processes.

Cell division is different in plants and animals. In animal cells, after mitosis, the membrane between the two daughter nuclei becomes constricted and the cytoplasmic connection between the two daughter cells is finally pinched shut. The cytoplasmic filaments mentioned in connection with morphogenetic movements play an important role in the process of constriction. Once cell division is complete, the two cells are free to wander their respective ways, and frequently do. In the plant cell following mitosis small vesicles filled

SIXTEEN-CELL
6.5

THIRTY-TWO CELL
7.5

MID-CLEAVAGE
16

LATE CLEAVAGE
21

DORSAL LIP
26

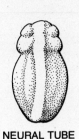

NEURAL TUBE
72

TAIL BUD
84

with matrix material appear in a plate that forms across the center of the mitotic axis. The vesicles coalesce, new ones appear at the plate margins, and eventually the plate divides the entire cytoplasm in two *(Figure 17)*. Meanwhile new cellulose microfibrils begin appearing on both surfaces of the plate intermixed with matrix materials, and two new opposing walls are formed, cemented together by the original plate material, the MIDDLE LAMELLA. Opposite each other in the new walls there are almost always thin areas through which slender cytoplasmic connections extend between the two cells. These areas, called PRIMARY PIT FIELDS, are important regions for intercellular communication. The membranes of animal cells can come into intimate contact. Plants must rely on these cytoplasmic connections or PLASMODESMATA. The wall we have just described is the PRIMARY CELL WALL.

Following cell division, the daughter cells increase in size. In animal cells extensive protein synthesis combined with water uptake and the manufacture of more membrane material is sufficient. Small molecules readily enter the cell, and larger molecules such as proteins, and even solid particulate matter, can be taken up by invagination of the cell membrane. The small vesicle thus formed becomes pinched off inside the cell, and dissolution of its membrane releases its contents to the cytoplasm.

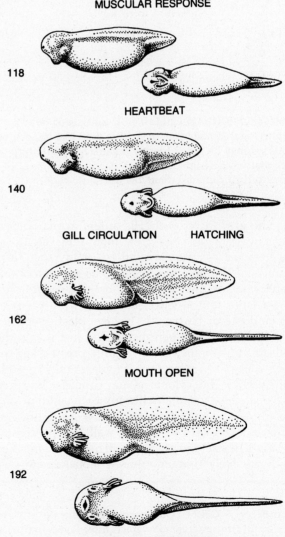

AGE (HOURS)   LATERAL VIEW   VENTRAL VIEW

96

MUSCULAR RESPONSE

118

HEARTBEAT

140

GILL CIRCULATION   **HATCHING**

162

MOUTH OPEN

192

TAIL FIN CIRCULATION

.4 LATER DEVELOPMENT of a frog embryo at 18°C, from late tail-bud stage (top) to swimming and feeding tadpole (bottom).

HIGHLY CRYSTALLINE
ARRANGEMENT
OF CELLULOSE
MOLECULES
INTO A MICELLE

INDIVIDUAL
CELLULOSE CHAINS

POORLY CRYSTALLINE
ARRAY OF CELLULOSE
MOLECULES

15 **MICROFIBRIL OF CELLULOSE consists of both individual strands and crystalline arrays of molecules. Arrays may be either loosely organized or highly crystalline packages (micelles). A cross-section of a microfibril contains about 250 micelles.**

The growth of plant cells differs in two important ways. First, they grow mainly by taking up water. Far less protein synthesis is involved, and much of the entering water, plus inorganic ions and small organic molecules, become sequestered in a large central vacuole, limited by a vacuolar membrane or TONOPLAST. The cytoplasm becomes restricted to a thin peripheral layer just within the plasma membrane, usually tightly pressed against the wall. A few strands of cytoplasm may traverse the vacuole, but cytoplasm may comprise less than 10 percent of the total cell volume. Second, during plant cell growth, there must be a substantial increase in the area of the cell wall. As the cell increases in size, the existing wall becomes stretched progressively thinner. Simultaneously new wall

16 **FIBRILS OF CELLULOSE strengthen the walls of plant cells. Electron micrograph depicts three layers of fibrils in wall of an algal cell, reproduced at a magnification of 24,000 X.**

material is deposited —hemicelluloses and pectic compounds throughout the wall, and new cellulose microfibrils against the inner surface.

The absence of coordinated cell migration imposes a strict requirement for coordinated cell enlargement. Adjacent cells are cemented together rigidly, and interconnected by plasmodesmata. Thus two adjacent walls must increase in area in precisely the same manner. If the cells slipped past each other, the plasmodesmata would be sheared, destroying the continuity of the cytoplasm.

With cell enlargement and cell movement thus restricted in plants, what determines cell shape, and ultimately organ shape? Plants don't increase evenly in size in all directions. The answer is to be found in the arrangement of the cellulose microfibrils. They are not oriented at random. In a rapidly elongating stem, for instance, the orientation of the microfibrils is predominantly horizontal (across the long axis of the stem). As entering water forces the cell contents against the wall, the wall can bulge very little at the sides (along the axis of the microfibrils), and instead becomes stretched predominantly in a longitudinal direction, the microfibrils becoming separated from one another (*Figure 18*). The horizontal orientation is not perfect; some of the microfibrils do become tilted, but the new ones being added are still deposited in the roughly horizontal orientation. Eventually microfibrils brought into a vertical orientation by this cell elongation may terminate growth. Thus wall thickness, determined by the number of fibrils and the orientation of the fibrils, regulates the rate and direction of plant cell growth, and extensive depo-

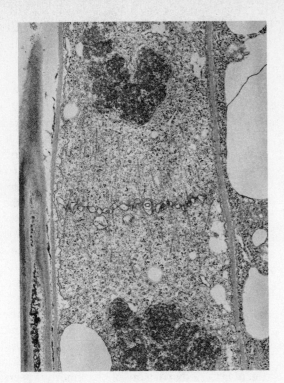

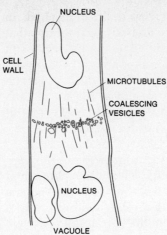

FORMATION OF CELL PLATE marks the last stage of **17**
cell division in plants. Electron micrograph of two cells
of a seedling of soft maple (Acer sacharinum) reveals a
horizontal row of vesicles against which new cellulose
will be deposited to form plate that divides the two
cells.

sition of fibrils eventually terminates the growth.
The factors determining orientation of the microfibrils are not known, although various workers
have found cytoplasmic microtubules similarly
oriented just under the microfibrils. But this discovery merely means that biologists now must determine what orients the microtubules, and how
their orientation orients the microfibrils.

Absence of morphogenetic movement in plant
cells should not be construed as absence of the fine
microfilaments that direct such movements. (Microfilaments should not be confused with microfibrils or microtubules; the filaments consist of an
actin-like protein, similar to one of the contractile
proteins in muscle.) In plant cells a microfilament
system causes extensive streaming of the cytoplasm. Streaming is essential for thorough mixing
of materials in large cells, but the wall prevents the
flowing cytoplasm from causing the membrane deformation essential for cell movement. And for
reasons lost in evolutionary history, plant cells do
not divide by constriction of the cell membrane,
but rather by cell plate formation.

Plants also differ from animals in their solution
to the topological problem of forming organs of
specific shape and size from undifferentiated
groups of cells. In animals cell migration and extension of filamentous cell processes (as in nerve
cells) accomplish this end. Cell proliferation is

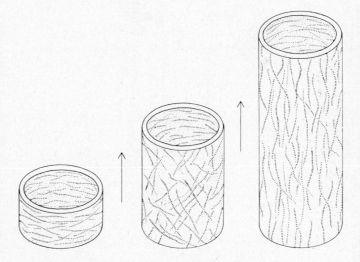

REORIENTATION OF CELLULOSE FIBRILS occurs **18**
during elongation of plant cell. At the start of
elongation (left) microfibrils of cellulose in the cell wall
show roughly horizontal orientation. As the cell grows,
the original fibers are displaced vertically (center),
while new horizontal fibers are added to inner wall. At
a late stage of growth (right) the original fibrils are
almost vertical, but recently added innermost fibrils are
still horizontal. Cross-ply orientation strengthens cell
wall.

necessary simply to provide the raw material. In plants the absence of cell movement requires that cell proliferation be confined to specific regions of the tissue. These local centers of mitotic activity are called MERISTEMS. At the tip of each shoot or branch is a SHOOT APICAL MERISTEM, and at the tip of each root, an analogous ROOT APICAL MERISTEM *(Figure 19)*. Leaves arise from a LEAF PRIMORDIUM formed by particularly active cell division in localized areas on the flanks of the shoot apical meristem. They appear first as small mounds; cell division along their sides (which forms a MARGINAL MERISTEM) yields a characteristic flattened leaf. Deep within the root tissues and far back from the root apical meristem

arise small meristematic areas which produce the ROOT PRIMORDIA, which will become the lateral roots. Root and shoot apical meristems are discussed in more detail later in this chapter.

A number of plants develop what we commonly refer to as wood and bark. These tissues are derived from another meristem, a highly specialized one called a VASCULAR CAMBIUM. It forms a cylinder surrounding the wood, and is the place of least resistance when one peels the bark off of many species. Vertically elongated cells divide to produce derivatives both to the inside and to the outside (occasionally laterally, since obviously the cylinder the cambium surrounds is increasing in size). These cells are called FUSIFORM INITIALS. The outer derivatives of the fusiform initials become the conducting cells of the PHLOEM tissue, through which the products of photosynthesis are transported from leaves to other parts of the plant. The inner derivatives become the conducting and support cells of the XYLEM, the conducting system for water and minerals from the roots to aerial portions of the plant. A discussion of the highly specialized conducting cells found in the xylem and the phloem is deferred to Chapter 16, as is a consideration of transport mechanisms.

Shorter cells, called ray initials, occur in groups that produce storage cells. The storage cells comprise the RAYS of xylem and phloem in wood and bark respectively. Perhaps the best known example of storage by xylem rays is found in the sugar maple. As the weather gets colder, large amounts of sugar are transported through the phloem, across the cambium, and into the xylem rays. In the early spring, hydrolysis of this sugar and its transport into the xylem makes a highly sweetened sap which is tapped and concentrated to yield maple syrup and maple sugar. The vascular cambium of the apple *(Malus sylvestris)* is shown in Figure 20.

Where cambial activity is extensive, additional meristems appear in the phloem tissue. Such a meristem, which may function only for a single season or even a part of it, is called a CORK CAMBIUM. It (and quite obviously everything else functional outside of the vascular cambium) must continuously be replaced because the activity of the vascular cambium is increasing the size of the wood cylinder and sloughing off the outside. A cork cambium produces new cells mainly in an outward direction. Their walls become impregnated with a waxy substance called suberin, which provides water-proofing for the stem and protects the cambium (both vascular and cork) from drying out.

19 MERISTEMS are centers of mitotic activity and growth in plants. Diagram shows their locations in shoot, stem and roots. Girth of plant is controlled by cylindrical meristems: the vascular and cork cambium (shown in cutaway view at center).

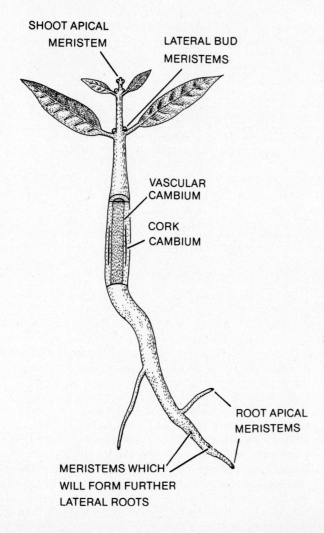

SHOOT APICAL MERISTEM

LATERAL BUD MERISTEMS

VASCULAR CAMBIUM

CORK CAMBIUM

ROOT APICAL MERISTEMS

MERISTEMS WHICH WILL FORM FURTHER LATERAL ROOTS

> Lacking yolk, the plant egg is normally surrounded by cells that actively synthesize nutrients for the developing embryo. The embryo does not become independent until the dormant seed germinates, which may be years after fertilization.

In brief, plant development and morphogenesis depend directly upon the activity of a number of different meristems. The shape of these meristems and the distribution of cell divisions within them determine the final shape of plant tissues and organs.

PLANT EMBRYOS

As will be discussed in Chapter 13, higher plants have entirely dispensed with the swimming male gamete so characteristic of animal reproduction (and common in lower plants); instead they rely on the mechanism of POLLINATION, which was discussed briefly in Chapter 9. The egg is deeply buried within the parent tissues. A pollen grain adheres to the sticky surface of a female portion of a flower and germinates, forming a POLLEN TUBE which may grow several centimeters, digesting its way through female tissue to reach the egg. The tip of the pollen tube then ruptures, releasing a sperm nucleus that fuses with the egg nucleus to form the zygote. The zygote resides in a specialized structure called the OVULE, which will eventually develop into a dormant seed. The seed serves two functions: it harbors the young plant embryo and permits its dispersal. The life cycle of the flowering plant, as well as the structure and function of the various flower parts, will be treated in Chapter 26.

Lacking yolk, the plant egg is normally surrounded by cells which actively synthesize nutrients for the developing embryo. The embryo does not become independent of the surrounding tissue until the dormant seed germinates (sprouts), an event that may come months or years after fertilization.

Figure 21 shows the sequence of events following fertilization in the flowering plant *Capsella*, known as Shepherd's Purse. The zygote begins development by forming a short filament of cells. One end of the filament then begins cell division in

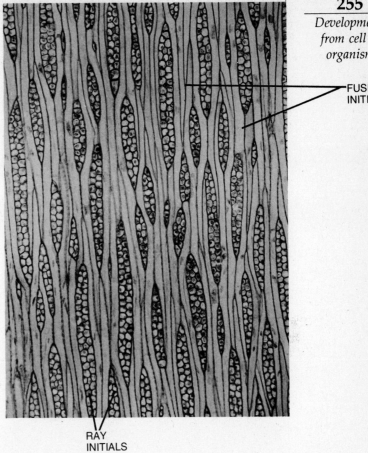

FUSIFORM
INITIALS

RAY
INITIALS

VASCULAR CAMBIUM from an apple tree (Malus sylvestris). Illustration is a tangential view, seen as if the observer had carefully removed the bark and examined the exposed surface under low magnification. Area depicted here is about 0.5 mm wide.

20

three dimensions, forming a small globular mass of cells away from the point of entry of the pollen tube. This ball of cells only superficially resembles a blastula, since further development of organs must depend upon localization of cell division to specific regions, plus regulation of the manner in which the cells then enlarge.

The slender remaining filament, called the SUSPENSOR, then elongates, pushing the globular embryo into the surrounding nutrient tissues. Eventually, localization of cell division produces two small lobes, making the embryo appear heart-shaped. These lobes are the so-called seed leaves or COTYLEDONS. (All flowering plants belong to one of two different groups: those like the *Capsella* illustrated here, which have two cotyledons, are called DICOTS; those with only a single cotyledon

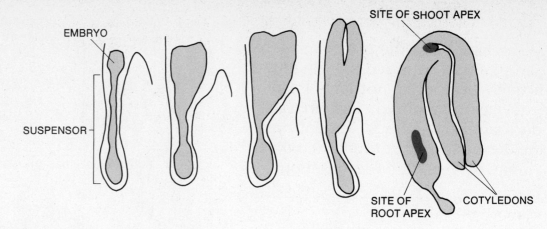

EMBRYO

SUSPENSOR

SITE OF SHOOT APEX

SITE OF
ROOT APEX

COTYLEDONS

21 EMBRYONIC DEVELOPMENT OF PLANTS proceeds during formation of a seed. Sequence of drawings illustrates events that follow fertilization in the flowering plant Capsella, a familiar weed known as Shepherd's Purse. Suspensor elongates (far left), pushing embryo into surrounding nutrient tissues. Embryo eventually develops two small lobes (cotyledons) and below them a stemlike hypocotyl (right center). With continued differentiation embryo bends back upon itself and meristems appear (far right). Meristems are rendered in color in drawing at far right.

are called MONOCOTS. Other differences between these two groups appear later in this chapter and in Chapter 26.)

As development proceeds, continued cell division and enlargement elongate the basal region of the embryo into the stemlike HYPOCOTYL (Greek for below the cotyledons) and elongate the cotyledons themselves, producing a torpedo-shaped embryo. Continued development then causes the whole embryo to be bent back on itself, as shown in Figure 21. Meanwhile the surrounding tissues also undergo substantial development, finally producing a mature seed, consisting of a seed coat, stored food material (either in the cotyledons or in a specialized tissue called ENDOSPERM), and the embryo itself. At the same time, cellular differentiation produces strands of elongated cells, within both the hypocotyl and cotyledons, which will develop into the first functional xylem and phloem. These cells are called the PROCAMBIUM. The root apical meristem arises fairly deep within the lower end of the hypocotyl, while the shoot apical meristem appears as an insignificant mound of tissue between the cotyledons.

When the seed germinates, both root and shoot apical meristems begin the rapid and organized cell division which produces the young seedling. Here another major difference between plant and animal development emerges. Following gastrula-

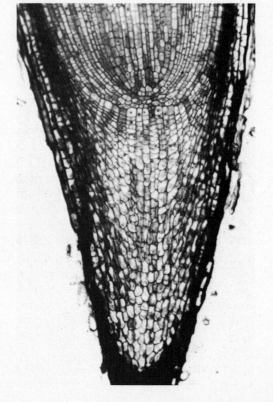

APICAL INITIALS
OF ROOT PROPER

QUIESCENT
CENTER

ROOT CAP
INITIALS

APEX OF PEA ROOT, shown in longitudinal section in photomicrograph (left), contains meristem that gives rise to all the tissues of the root. Accompanying diagram shows undifferentiated meristem cells (called initial cells) that will develop into the root cap and into the structures of the root proper: epidermis, endodermis, cortex, pericycle, xylem and phloem.

tion and neurulation, the animal embryo goes on to elaborate all of the various tissues and organs of the mature individual, at least in embryonic form (the only exceptions are animals that undergo METAMORPHOSIS, which will be considered in Chapter 18). Following this embryonic period the embryonic structures grow extensively and increase in complexity as the animal matures. By contrast the plant embryo, even at the mature seed stage, is singularly undifferentiated. None of the many leaves, branches, or roots of the mature plant is represented by primordial cells or organs. Only the rather unimpressive primary root and shoot apical meristems suggest the events to come. These two meristems (or meristems derived from them) remain active throughout the life of the plant, maintaining the presence of embryonic and developing regions.

ROOT AND SHOOT MERISTEMS

At the lower end of the hypocotyl, but still beneath the embryo surface, a center of active cell division appears: the primary root apical meristem (*Figure 22*). The products of this meristem radiate out in all directions. Those below form the root cap, a structure which among other things protects the delicate root meristem as it advances through the soil. Those behind produce the root EPIDERMIS, from which absorptive root hairs will soon emerge; a CORTEX of thin walled cells; an ENDODERMIS which eventually will serve to regulate entry and exit of materials from the root vascular system; a region of very small cells called the PERICYCLE; and finally the PROCAMBIUM which will form the vascular xylem and phloem. Both the cortex and the pericycle are relatively undifferentiated at maturity, consisting of vacuolated and thin-walled cells called PARENCHYMA. Endodermal cells develop a highly specialized and thickened wall on their radial surfaces, designed to seal the central vascular cylinder from the outside and to restrict movement of materials across the endodermis. In cross section, the xylem of a dicot appears as a solid star-shaped mass with phloem found between the points of the star. The xylem in monocots appears as a ring of isolated strands on alternate radii with isolated strands of phloem, all surrounding a central tissue of living cells called PITH. Figure 5 in Chapter 16 illustrates these two types of roots, and Figures 6 and 7 in that chapter illustrate the cellular details of a dicot root. Both xylem and phloem cells are much elongated and specialized for transport. If a cambium arises, it appears between the xylem and

phloem, spreading from the bays laterally to surround the points of the star, forming a complete cylinder.

Following the seedling stage of development, during which the central region of the root meristem is highly active, cell division becomes more and more restricted to the outer surfaces of the meristem, and a central QUIESCENT ZONE is formed. Mitosis is infrequent, DNA synthesis extremely slow, and RNA and protein syntheses are also severely limited. But quiescence does not necessarily mean absence of physiological activity, as will become clear in the discussion of the shoot apex.

Figure 23 is a photomicrograph of the shoot apex of a flowering plant, the herb *Coleus*. The superficial layer of cells, the TUNICA, gives rise only to epidermis, and the underlying CORPUS produces the rest of the stem tissues. No aerial structure in the plant is comparable to the root cap. Close to its tip lie small mounds of tissue, the LEAF PRIMORDIA, which develop into the mature leaves. Their

SHOOT APEX of the herb Coleus. Photomicrograph depicts developmental structures characteristic of flowering plants: leaf primordia, lateral bud primordia, and shoot apical meristem. 23

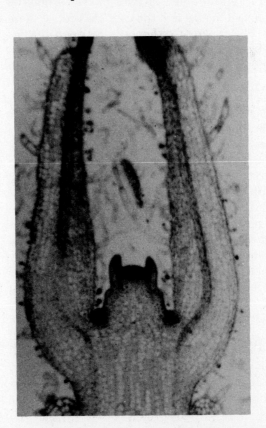

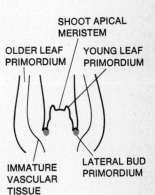

SHOOT APICAL MERISTEM

OLDER LEAF PRIMORDIUM

YOUNG LEAF PRIMORDIUM

IMMATURE VASCULAR TISSUE

LATERAL BUD PRIMORDIUM

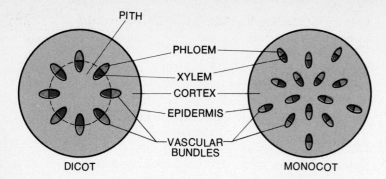

PITH
PHLOEM
XYLEM
CORTEX
EPIDERMIS
VASCULAR
BUNDLES
DICOT
MONOCOT

24 **CROSS-SECTIONS OF STEMS reveal differences between dicot and monocot. Dicot has distinct ring of vascular bundles and well-defined pith. Monocot has scattered bundles. In both types of plant xylem and phloem are found together, with the phloem toward the outside. Dotted line indicates position of vascular cambium in some dicots.**

epidermis is derived from the tunica, with the remainder of their tissue derived from the underlying corpus. Beneath the corpus, longitudinal files of cells first form cortex on the outside, then elongate strands of procambial cells which will form the stem xylem and phloem, and finally, in the center, very large parenchyma cells which will form the central tissue or pith. Cross sections of stems reveal that as was the case with roots, there are readily apparent differences between the organization of the vascular system in dicots and that in monocots. These differences are diagrammed in Figure 24. A dicot stem has a ring of vascular bundles surrounding a clearly defined pith, with each bundle including phloem outside and xylem inside. In monocots, the vascular bundles are scattered at various distances in from the epidermis, and the distinction between pith and cortex is difficult to make, though normally only the outer cortex is photosynthetic.

The shoot apical meristem, like the root meristem, possesses a central quiescent region comprising both tunica and corpus. Unlike the root quiescent center, that of the shoot apex is readily accessible for experimental manipulation. If it is destroyed, shoot growth at the apex ceases almost immediately. Growth may be resumed eventually if marginal regions of the meristem begin mitosis, eventually forming their own quiescent centers. If a vascular cambium arises (this occurs only in dicots), it originates with divisions of cells between the xylem and phloem, and then spreads out laterally along a line separating cortex from pith (*Figure 24*).

In addition to the various features mentioned above, there is an additional major difference between root and shoot meristems. The shoot apex produces superficial lateral structures in a defined sequence —structures which will become leaves or flower parts. In the angle between the leaf primordium and the apex (the LEAF AXIL) a cluster of small cells forms a BUD PRIMORDIUM. Under suitable conditions, bud primordia will resume meristematic activity, become organized as shoot apical meristems, and produce lateral branches. The root apex, by contrast, produces no lateral structures except tiny root hairs. Lateral roots arise from more mature regions of the primary root. Opposite the points of the xylem star, one finds nests of pericycle cells which begin mitosis, eventually becoming organized into a root apex. This root apex must digest its way through the endodermis and cortex to reach the substrate outside. Thus not only internal organization but also the origin of lateral structures are quite different in the two meristems. Yet both can remain meristematic for extremely long periods of time, in marked contrast to the sharply limited embryonic period of animal development.

PHYTOCHROME

Many of the developmental processes in plants described above are regulated by developmental hormones (discussed in some detail in Chapter 18). In addition, some of them are regulated in a remarkable way by red light. For example, if certain varieties of dry and dormant lettuce seeds are soaked in water in complete darkness and then set out on moist filter paper, very few will germinate. But if after soaking they are given just a few minutes' exposure to red light and then returned to darkness, virtually all will germinate. Red light breaks their dormancy. The most effective wavelengths are near 660 nm. If the red treatment is followed by brief exposure to light of about 730 nm (far-red light, just beyond the limit of vision), the seeds behave as though they have never been exposed to light at all. The far-red light cancels the effect of red. The seeds clearly contain some sort of plant photoswitch turned on by red light and **turned off by far-red.** One can in fact alternate red and far-red treatments many times, and whether or not the seeds germinate depends entirely on the final exposure. If it is red the switch is on and the seeds germinate. If it is far-red the switch if off and the seeds stay dormant.

H. A. Borthwick and S. B. Hendricks, of the U.S. Department of Agriculture, pioneered this field and guided it for almost 35 years. They and their colleagues soon found that the red/far-red effect was universal in the development of higher plants, and probably in most lower plants as well, even including certain algae. (It is apparently lacking in the fungi.) Once the seeds germinate, the elongation of the stem, the expansion of leaves, the induction of flower formation, the development of fruits, and even fruit pigmentation, can all be modified by a brief exposure to red light during a normal dark period. In every case, the modulation can be cancelled by exposure to far-red light soon after the red. Even our incomplete list of effects reveals a range of responses as varied and fundamental to integrated plant development as, say, responses to growth hormones (*Chapter 18*). The effects can be positive (promotion of leaf expansion) or negative (inhibition of stem elongation), depending on the target cells or tissues.

The action spectra for these phenomena are all similar, which led to an intensive biochemical search for the photoreceptor pigment. It turned out to be PHYTOCHROME, a protein with a molecular weight near 240,000, conjugated with one or more molecules called linear tetrapyrroles, which absorb visible light. The whole molecule can exist in either of two relatively stable forms, with quite different absorption spectra. Exposure to red light causes a molecular rearrangement reminiscent of the behavior of the visual pigment, rhodopsin, following its excitation by light. Far-red light reconverts the pigment to the red-absorbing form.

The most provocative studies on the mechanism of action of phytochrome come, not from a study of developmental responses, but rather from the regulation by phytochrome of plant "sleep movements." In some plants such as the shamrock *Oxalis*, or the tropical tree *Albizzia*, the leaves are normally wide open in the light, but fold and droop when darkness falls (*Chapter 17*). A special group of cells at the base of each leaf or leaflet is responsible for the movement. Collectively called the PULVINUS, these cells are normally swollen with water in the light. The onset of darkness brings

The fate of an embryonic cell is determined not merely by its own genes, but also by interactions with neighboring cells.

about rapid water loss, the cells collapse, and the leaf folds inward. The exact course of folding is determined by the location of the pulvinar cells, usually on the upper side of the base of the leaf. If just before transfer to darkness the leaves are briefly exposed to far-red light (daylight contains more red than far-red) closure is at most incomplete. Red light will then cause complete closure within as little as a half hour. Repeated photoreversibility can easily be shown, and the system obeys all of the rules for phytochrome-mediated processes. Though many phytochrome-mediated responses undoubtedly involve eventual changes in protein and RNA synthesis, it is clear that one of the earliest responses to red light must be some change in membrane properties—a consequence of a swift interaction between the 730-nm-absorbing form of the pigment and some as yet unidentified cell membrane. In the light, pulvinar cells contain an abnormally high concentration of potassium. Osmotic principles (*Chapter 15*) dictate that the cells should be fully turgid, since their total solute concentration is far higher than that of surrounding cells. Closure is accompanied by a massive efflux of potassium from the cells, accompanied by rapid water loss. Reopening in continuous light involves gradual reaccumulation of potassium, followed by water uptake.

INDUCTION AND TISSUE INTERACTION

The fate of an embryonic cell is determined not merely by its own genes, but also by interactions with neighboring cells. In 1924 Hans Spemann and Hilde Mangold published the first results of a brilliant series of experiments on amphibian embryos which eventually earned Spemann a Nobel prize. They excised the dorsal lip of the blastopore of a gastrulating embryo and grafted it within the blastocoel of a similar embryo. The results of this exacting microsurgery were startling. The dorsal lip developed into a normal neural tube, and the underlying host tissue formed a spinal cord and other associated structures. In some experiments, twin embryos developed attached belly to belly (*Figure 25*). The relationship between the grafted and host cells was clearly a partnership. Development of host tissue was induced by the grafted piece, and development of the grafted piece was induced by the host.

Another elegant example of tissue interaction is the development of the vertebrate eye. An outgrowth at the anterior end of the neural tube forms the optic cup, which comes into intimate contact

B Animal tissue culture

The desire to grow living cells in the laboratory arose during the pioneering work in experimental embryology toward the end of the 19th century. Experimenters wanted to isolate parts of living embryos by microdissection and follow the interaction of living cells away from the influence of the rest of the organism. Early in the 20th century Ross G. Harrison of Yale succeeded in isolating nerve cells from frog embryos and watching their axons grow. He grew them alive in a drop of lymph fluid sealed between a cover slip and a glass slide. His techniques were quickly applied by other workers who kept cells alive in a few drops of saline solution enriched with glucose, phosphates and other mineral salts. Later workers gradually developed more complex media capable of sustaining a wide variety of cells. Some of the media in use today contain not only glucose and saline solution, but also amino acids, enzymes and vitamins. Culture vessels too have become more elaborate, and now there are dozens of slides, dishes and flasks especially designed for tissue culture.

The problems encountered in transferring animal tissues and later whole organs from the intact organism (in vivo) to an artificial culture environment (in vitro) are enormous. Claude Bernard, the famous French physiologist, stressed the importance of the *milieu interieur:* the body fluids surrounding and bathing all the tissues of the body. The basic aim of *in vitro* studies is to mimic as closely as possible the internal environment from which the tissues or cells are isolated. The chemical composition and physicochemical environment (pH, ionic concentrations, nutrients, energy sources and so on) were found to vary not only from organism to organism but also from tissue to tissue within the same embryo. The internal environment even varies at different stages of development in the same tissue.

Despite these difficulties the science of cell and tissue culture has made great progress and is yielding a wealth of new information about the development of plant and animal tissues, particularly in the fields of virology, cancer research, immunology and molecular biology. Although some of the early research in tissue culture must be re-evaluated as the culture media are improved, the methods introduced at the beginning of the century are among the most powerful tools of contemporary biology.

25 TRANSPLANTATION EXPERIMENT demonstrated the role of tissue interactions in the development of frog embryos. Part of the lip of the blastopore of one gastrula was grafted into the blastocoel of another gastrula (left). Grafted gastrulas developed two neural tubes instead of one (center), and a few of the gastrulas survived to develop into twin tadpoles joined at the belly (right).

with the overlying ectoderm. The ectoderm reorients itself, with one group of cells synthesizing massive amounts of CRYSTALLINS, the proteins of the transparent lens. A piece of cellophane placed experimentally between the optic cup and the ectoderm prevents any manifestation of lens formation. If the optic cup is surgically transplanted beneath some other region of ectoderm, a lens will form at the new site. Conversely, the ectoderm over the optic cup can be replaced with ectoderm from the belly region, and one still obtains a lens from the alien ectoderm. The relationship, as in the case of the Spemann-Mangold experiments, is reciprocal. The developing lens also clearly influences optic cup development.

The two cases just described and many others are elegant examples of tissue interaction. Is such interaction essential? The answer is usually yes, but not always. Small pieces of stem tissue excised from a number of plants can be placed in a nutrient

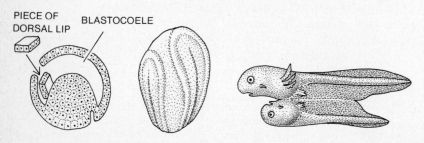

PIECE OF
DORSAL LIP   BLASTOCOELE

agar containing only mineral salts and a carbon source such as sucrose. (Applications of this research technique, known as tissue culture, are described in Box B.) As mentioned in Chapter 11, such stem (or other plant) tissue will initiate cell division to form an undifferentiated mass of cells called a callus. R. H. Wetmore and his students found that an excised shoot apex could be grafted into a cut on top of callus tissue. The shoot apex first established a graft bridge where plasmodesmata from the excised apex fused with those in callus cells, and then began normal development, producing new leaves and buds. Simultaneously a series of isolated xylem (and in some cases phloem) strands appeared in the callus tissue (Figure 26). These results raised two questions: first, does the callus produce something special required for the development of the grafted shoot apex? Second, what is produced by the shoot apex which induces xylem differentiation?

It turned out that development of the shoot apex did not require the presence of callus at all. The apex itself grew normally if placed directly on the nutrient medium. Moreover, a small amount of agar containing the plant growth hormone indoleacetic acid (Chapter 18) could readily substitute for the shoot apex in induction of xylem differentiation. Finally, the position of the new xylem strands depended on the hormone concentration. The higher the concentration, the further away from the hormone source were the xylem strands. Not only was the hormone responsible for the cellular differentiation, but its concentration was critical in determining the position of the differentiated cells.

In animals there is so far no such clear evidence either on the nature of the inducing substance or on the stringency of the requirement for interaction for normal development.

Following the exciting early work on the induction of differentiation in animal systems came a whole series of apparently exciting advances. In 1933 the experiments of C. H. Waddington with Joseph and Dorothy Needham demonstrated that cell differentiation could be induced by a cell-free extract from the dorsal lip of a gastrulating amphibian embryo. The next step was induction in pieces of gastrula ectoderm, maintained in tissue culture, by pieces of other tissue or by cell extracts. Unfortunately the induction of differentiation, at least in these amphibian systems, proved to be a relatively unspecific effect. Many things, including the distinctly unbiological dye, methylene blue, can induce differentiation. Clearly much remains to be learned about this fascinating but still enigmatic process.

In the study of tissue interactions, the mobility of animal cells presents a special problem. What factors direct the migration of specific cell types in such a precise manner? Why, for example, are only liver cells found in liver, and in addition, why are they never found elsewhere? Shortly after the turn of the century, H. V. Wilson discovered that by pushing a sponge through a fine mesh, he could separate it into individual cells and clusters of cells. If the disaggregated cells were then allowed to stand in a bath of seawater, they reaggregated themselves into a perfectly respectable sponge! Wilson and later workers then showed that if one mixed such suspensions from two different species of sponge, the cells reaggregated strictly according to species —not into hybrid sponges. The experiments suggested that a species-specific aggregation factor must be involved.

After the problem had remained dormant for almost five decades, Philip L. Townes and Johannes Holtfreter attacked it anew. They used high pH to disaggregate both neural plate tissue and lateral ectoderm from amphibian embryos. Reaggregation was studied at a more physiological pH. After some random migration, the neural plate cells moved centrally, and were surrounded

XYLEM DIFFERENTIATION can be induced in lilac callus either by a grafted shoot apex or simply by a piece of agar containing an auxin (indoleacetic acid). Grafted callus develops strands of xylem (lower right). Colored line in that diagram shows plane of cross-section depicted above.

26

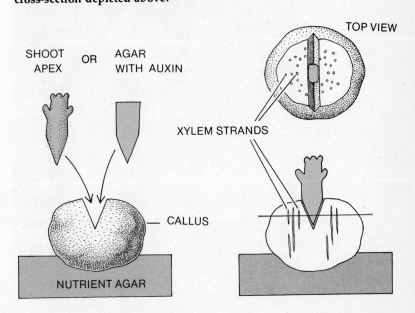

by lateral ectodermal cells. Unlike the sponges, the various vertebrate embryos studied did not show species-specific reaggregation. A mixture of embryonic cells from a chicken and a mouse will produce differentiated structures containing cells from both animals. If embryonic cartilage and liver cells —both kinds from each animal—are mixed, the cartilage cells move into the center, to be surrounded by liver cells. In this case, the reaggregation is tissue- and not species-specific. Conversely a mixture of mouse and chicken heart cells will become completely separated in culture into pure mouse and pure chicken tissues.

Like the mechanism of tissue interactions, the mechanism of cell sorting is still unknown. Cell recognition sites clearly lie on the cell surfaces, and most probably involve protein-polysaccharide complexes.

In one well-known aggregating system, aggregation has indeed been related to a single chemical substance. The cellular slime mold, *Dictyostelium*

27 FRUITING BODIES of the cellular slime mold Dictyostelium discoideum are sacs of spores supported by stalks. About half an inch high, the entire structure is an aggregation of thousands of single-celled social amoebas.

discoideum, a remarkable fungus, begins its life cycle as a group of spores. The spores germinate under favorable conditions, each producing an amoeboid cell or myxamoeba which ranges about freely, feeding on bacteria and dividing in two. When the food supply dwindles, however, thousands of myxamoebae begin to aggregate to form a sluglike "organism." After some aimless wandering, during which each cell retains its separate identity, the slug develops into a stalk capped by a globular fruiting body *(Figure 27)*. Each cell in the globe becomes a spore, the whole structure dries, and the spores are shed to be wafted to some new favorable location for germination.

The *Dictyostelium* system has recently provided elegant material for studies of cell aggregation. First, Maurice Sussman and his colleagues have shown that as the slug forms and develops into a fruiting structure, several enzymes are synthesized in a highly programmed sequence. Disaggregation of the slug abruptly turned the program off. Then, on reaggregation, it started all over again. The cells could be disaggregated and allowed to aggregate a third time with similar results. On the second and third tries the program was speeded up, but otherwise unchanged; all three times gene transcription was required. Clearly cell aggregation in some way influences developmental programming at the gene level.

It is clear that the accumulation of some substance in the medium during food depletion is responsible for aggregation of the myxamoebae. It has now been clearly identified as cyclic AMP, which was mentioned in Chapter 10.

The regulation of cell migration is a fascinating area for further study. There is evidence (for example from *Dictyostelium*) for processes leading to mutual cell attraction. There is evidence (from studies of salamander pigment-cell migration) of processes leading to mutual cell repulsion. When cells come together to form tissues, a process called contact inhibition stops their movement and they form a stable tissue. Moreover, as cells become more densely packed, their density progressively inhibits growth. Cells unaffected by density-dependent growth inhibition can form tumors, and if they are also unaffected by contact inhibition, the tumorous cells can readily invade healthy tissues.

Despite spectacular advances in the past 20 years, developmental biology is in many ways still in its infancy. Understanding of normal (and abnormal) development presents some of the greatest challenges in modern experimental biology.

READINGS

B.I. BALINSKY, *An Introduction to Embryology*, 4th Edition, Philadelphia, W. B. Saunders, Co., 1975. An excellent and complete animal embryology text, particularly strong in its treatment of cellular aspects of development and morphogenesis.

J.D. EBERT AND I.M. SUSSEX, *Interacting Systems in Development*, 2nd Edition, New York, Holt, Rinehart and Winston, 1970. A good modern treatment both of descriptive and experimental aspects of development. Emphasis is on animals, with only a few plant chapters.

D. EPEL, "The Program of Fertilization," *Scientific American*, November, 1977. An excellent account of the recent advances in the study of fertilization, written by the man who has done a great deal of the research.

C. FULTON AND A.O. KLEIN, *Explorations in Developmental Biology*, Cambridge, MA, Harvard University Press, 1976. An exceptional collection of reprints of important papers with excellent text between; particular emphasis, molecular aspects of development and animal development. Contains original papers on many of the experiments described in this chapter.

J. LASH AND J.R. WHITTAKER (Editors), *Concepts of Development*, Sunderland, MA, Sinauer Associates, 1974. A detailed, up-to-date collection of essays by various specialists on a wide range of topics in animal development.

F.B. SALISBURY AND C. ROSS, *Plant Physiology*, 2nd Edition, Belmont, CA, Wadsworth, 1978. A comprehensive textbook of plant physiology with a good treatment of cellular and molecular aspects of plant development.

T.A. STEEVES AND I.M. SUSSEX, *Patterns in Plant Development*, Englewood Cliffs, NJ, Prentice-Hall, 1972. A thorough and excellent treatment of cellular aspects of plant development.

V.C. TWITTY, *Of Scientists and Salamanders*, San Francisco, Freeman, 1966. A delightful account of a highly productive career in experimental embryology. Includes definitive studies on pigment-cell migration and tissue interaction.

The Earth, after having brought forth the first plants and animals at the beginning . . . has never since produced any kinds of plants or animals, either perfect or imperfect; and everything which we know in present times she has produced, came solely from the true seeds of the plants and animals themselves. . . . —Francesco Redi (1688)

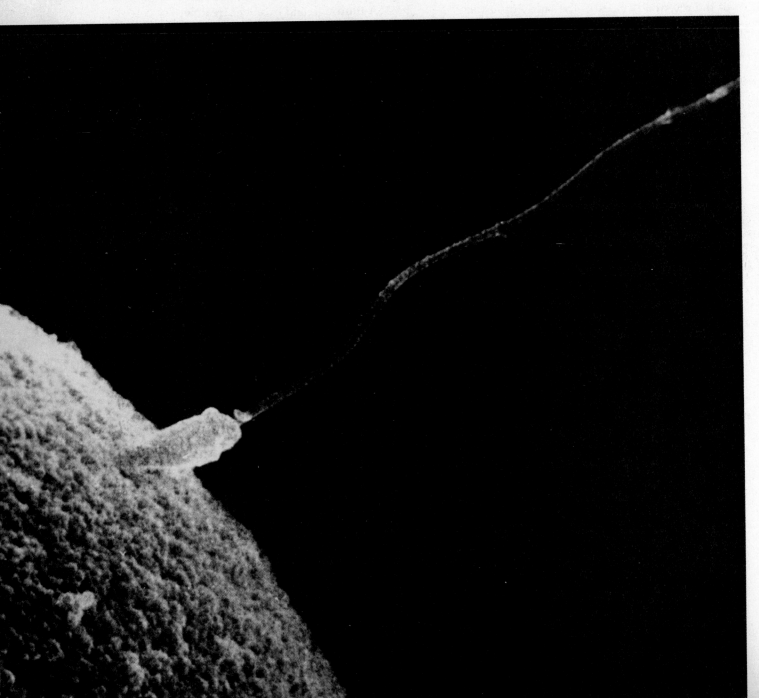

If every organism on Earth were immortal there would be no reproduction. The amount of life on Earth remains approximately constant. If no organism ever died, there would be no room for new ones to come into existence. Reproduction is the means by which the mortal organism increases the probability that its genes will continue to exist after it has perished. The more likely a given kind of organism is to die —in other words the shorter its average life span —the higher is its rate of reproduction. Insects have short lives and high reproduction rates; elephants have very long lives and low reproduction rates.

Another way to look at the relationship between death and reproduction is to regard the time and energy of an organism as resources available for investment. If the organism invests a large amount of its biological resources into reproduction, it will create offspring rapidly but reduce its own chances for survival. Or it can conserve its resources, prolonging its own survival but lowering the reproduction rate. How much time and energy should a species commit to reproduction? The answer is different for each kind of organism. The average bacterium is killed quickly, so the most successful are those that reproduce as fast as their biochemical machinery allows. The average elephant has a long life guaranteed. The female need go through the arduous and even dangerous procedure of reproduction only a few times to reach her optimum life-time production.

SPERM PENETRATES EGG of sea urchin in the photograph on opposite page. Picture was made with a scanning electron microscope.

ASEXUAL VS. SEXUAL REPRODUCTION

The fastest way to reproduce is asexually. Single-celled organisms such as *Amoeba* simply DIVIDE to produce two genetically identical daughter organisms. A variety of both unicellular and multicellular organisms, from parasitic protozoans to fungi, convert one or more of their cells into numerous SPORES. These tiny units, each containing at least one complete set of DNA, are especially adapted for wide dispersal. Fungal spores, for example, are carried like dust particles for long distances —in extreme cases all the way around the Earth in the winds of the upper atmosphere. A third principal method of asexual reproduction in multicellular organisms is BUDDING (*Figure 1*). The parent organism simply spouts an offspring like an appendage from part of its body. Familiar forms of VEGETATIVE REPRODUCTION are employed by many plants. The parent plant may simply become fragmented, with each fragment developing into a new plant independently. Special organs may serve the same purpose —for example the tubers of potatoes or the runners of strawberries or certain grasses. A fifth mode of asexual reproduction is PARTHENOGENESIS, the growth of an organism from an unfertilized egg.

If asexual reproduction is so simple and fast, why has it not become universal? What is the function of sex? Sexual reproduction —the fusing of genetic material from two or more organisms — creates variety among the offspring. This becomes clear when one compares the offspring of two organisms having the genotype AaBb. In an organism that reproduces asexually, all of the offspring will be AaBb unless by rare chance one or more of them mutates to a new kind of allele. But if two sexually reproducing organisms with AaBb

265

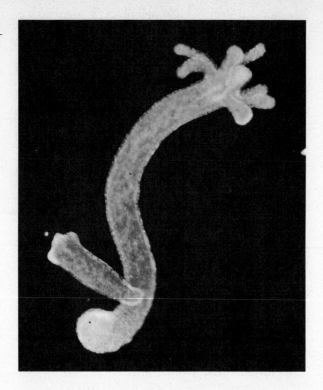

1 BUDDING is one means of asexual reproduction. Here a young hydra develops near the base of the parent animal's stalk (lower left).

genotypes mate with each other, the offspring could be AABB, AaBB, aaBB, AABb, AAbb, AaBb, aaBb, Aabb or aabb. In short, while the sexual organism spreads its investments, the asexual organism puts all its eggs in one hereditary basket.

The environment of most organisms changes constantly, from hour to hour and year to year. Circumstances favor the species that faces the environment with variable offspring. This explains why sexual reproduction occurs almost universally, in groups from viruses and bacteria to man and elephants. The exceptions are instructive. Some organisms, like fungi, persist by scattering huge numbers of offspring (spores) into the environment. The vast majority of these offspring land in places where they cannot survive. But some succeed in reaching the right habitats, however scarce such places may be. Asexual reproduction can also be found in species which at least for a time have a favorable environment and need to breed rapidly to exploit it to the fullest. When a female aphid lands on a suitable plant in the early spring she has a very favorable home and a source of food that she must exploit rapidly before predators eat her or other aphids arrive to crowd her

out; she reproduces swiftly by means of parthenogenesis. Many kinds of plants and animals, including fungi and aphids, alternate sexual and asexual phases in their life cycle and thus enjoy the benefits of both kinds of reproduction. Sexual reproduction has many profound implications in evolution *(Chapters 28 to 34)*.

FERTILIZATION AND SEXUALITY

In the most primitive single-celled organisms sexual union takes place in the water. Two cells conjugate, pressing closely together along their cell borders and exchanging genetic material directly through the cytoplasm *(see Figure 25 in Chapter 25)*. In the bacterium *Escherichia coli* this process appears in perhaps its most elementary form *(see photograph on opening page of Chapter 8)*. By observing genetic traits such as the ability or inability to synthesize particular sugars and amino acids, the French geneticists F. Jacob and E. Wollman showed that genes are exchanged between conjugating bacteria in a definite order. After 8 minutes, for example, gene A has been exchanged but not B or C; after 11 minutes B has also crossed over but not C; after 15 minutes C has finally been added, and so on. In more evolved organisms the genes are organized into chromosomes and can be exchanged in groups. In some single-celled organisms, such as the green alga *Chlamydomonas* and the protozoan *Paramecium*, there are often many sexes: strains of cells that can conjugate with cells belonging to strains other than their own.

Multicellular organisms usually have only two sexes. The sex cells are specialized as GAMETES whose only function is to achieve fertilization and commence the development of the new organism. In some algae and fungi, male and female gametes are of identical appearance, a situation known as ISOGAMY. In other algae and fungi, and in both

If asexual reproduction is so simple and fast, why has it not become universal? Because sexual reproduction —the fusing of genetic material from two or more organisms —creates diversity among the offspring.

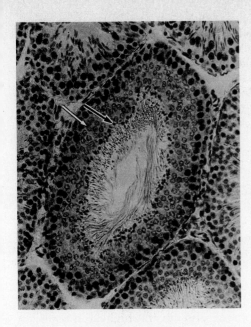

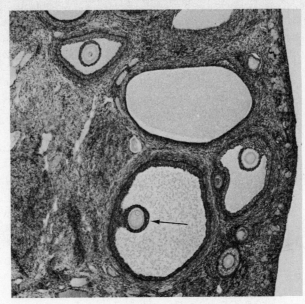

higher plants and animals, they differ greatly in size (ANISOGAMY). The female gamete, the EGG, is ordinarily the larger of the two. It is typically laden with nutrients that later supply the developing embryo and is normally incapable of movement. The male gamete, the SPERM (in plants, motile male gametes are also called antherizoids) is usually tiny, stripped down to the bare essentials: a nucleus to carry the genetic information, a flagellum to propel it to the egg, and a mitochondrion to provide the ATP to drive the flagellum (*Chapter 12*).

The sex cells are produced in animals by a series of cell divisions called GAMETOGENESIS: oogenesis in the female, spermatogenesis in the male (*Figure 2*). The crucial step is meiosis, during which the chromosomes may exchange fragments by crossing over and the two sets of chromosomes in the cell are reduced to one (*Chapter 9*). The diploid cells that produce the haploid gametes are called the GAMETOCYTES —notably the oocyte in the female and the spermatocyte in the male. Each of these two kinds of cell goes through meiosis to create four derived cells. In the male, all four become sperm. But in the female, only one becomes an egg; the other three, called POLAR BODIES, shrink and eventually die. The egg grows at the expense of the polar bodies, acquiring the extra nutrient supplies needed for future embryonic development. The amount of growth can be impressive. The oocyte of the frog (*Rana pipiens*), for example, starts with a diameter of only 50 micrometers, but the final egg shed into the water for fertilization is 1,500 micrometers in diameter—a 27,000-fold increase in

FORMATION OF SPERM AND EGGS. Photomicrograph at left is a cross-section of tubule from testis of a rat. Spermatogenesis proceeds from edge toward center of the tubule. In earliest stage primary spermatocytes (white arrow) arise by mitosis from prespermatocytes that lie near outside edge of tubule. Black arrow indicates spermatids: young tail-less sperm that arise by meiosis from primary spermatocytes. Center of tubule is lined with mature sperm ready to depart. Tails of sperm extend toward center of tubule. Photomicrograph at right depicts three eggs (arrow points to one) developing in follicles within the ovary of a cat. Photomicrographs are not reproduced to same scale.

2

volume. In plants, gametes are produced from a variety of structures but they almost always originate by mitotic division of an already haploid cell. Only very rarely are gametes the direct products of meiosis.

In the higher animals, which have extremely differentiated organ systems, the gametes are produced in special organs, the female OVARIES (singular, ovary) and the male TESTES (singular, testis). These organs also include special tubes and other devices for conveying gametes to the outside. In some cases, including man and other vertebrates, these organs also produce hormones that stimulate the beginning of sexual maturity.

Why are there only two sexes in higher organisms? The answer seems to be simply that two are enough to accomplish the purpose of genetic recombination. Three or more would create complications in the genetic basis of sex determination, in development of the complex sex organs, and in the often complicated behavior patterns that lead to fertilization. The two-sex rule, however, does not preclude the existence of HERMAPHRODITISM, the coexistence of both sex organs in the same organism. The majority of flowering plants are hermaphroditic, or PERFECT, which in a strict botanical sense means that male and female organs occur in at least some of the same flowers. Other kinds of plants are MONOECIOUS, a term that is translated literally from the Greek as "one house," or one organism housing both sexes. This means that male and female organs occur in different flowers but still on the same plant. Some plants are DIOECIOUS; as the term ("two houses") implies, the male and

---

The advantage of having distinct sexes in animals is that it permits a division of labor. The female specializes in making eggs, finding the right places to deposit them, and in the highest animals, nursing and protecting the young. The male specializes in finding and fertilizing females.

---

female organs occur not only in separate flowers but also on separate plants. Most hermaphroditic (including monoecious) plants have elaborate devices to insure that CROSS-POLLINATION occurs, in other words that the sperm-bearing pollen grains fertilize the ovaries of another plant and not their own. One of the commonest devices to insure cross-pollination is the maturing and release of the pollen by the male parts of the flower before the female part of the flower fully develops.

Hermaphroditism is rarer in animals. It occurs chiefly in sessile forms, notably sponges and some mollusks, and in parasites such as tapeworms and flukes, which are unable to seek out one another. Under these circumstances it is advantageous for each individual to breed to its maximum extent. Tapeworms can fertilize themselves, but in most other cases hermaphroditism is accompanied by invariable cross-fertilization between anatomically identical animals, and therefore is closely parallel to the condition of the monoecious plants.

Why aren't all sexually reproducing organisms

SEXUAL AND ASEXUAL GENERATIONS of the fluke Schistosoma mansoni, a parasite that infests the human liver. Eggs hatch into larvae called miracidia, which reproduce asexually in snails. Resulting cercaria larvae then enter humans, usually through skin of foot, and work their way to liver.

hermaphroditic? Why are there separate sexes? Hermaphroditism accompanied by cross-fertilization, as in the earthworms and most flowering plants, facilitates very fast reproduction. Each individual can produce fertilized eggs, and the resulting offspring show the same genetic diversity as those created by separate sexes. The advantage to having distinct sexes in animals is that it permits a division of labor. In general, the female specializes in making eggs, finding the right places to deposit them, and, in the highest animals, in nursing and protecting the young. The male simply specializes in finding and fertilizing females. In many species he also locates and protects the territory in which the young will be born, and sometimes assists the female in rearing the young. Sexual division of labor is impressively diverse among animals. Extreme examples are found among insects, particularly mayflies, mosquitoes and moths. In these species the physiology and behavior of the male equips him to do little more than find and mate with females.

STRATEGIES OF REPRODUCTION

The familiar life cycle of most vertebrates is a very simplified one, committed to a sexual strategy of reproduction. A vertebrate manufactures sex cells by gametogenesis and combines them with those of another individual to create a new organism, completing the cycle. *Amoeba*, in contrast, is committed to an asexual strategy: an individual creates two individuals by dividing, which also completes the cycle. The plants and animals that have committed themselves to a mixed strategy (both sexual and asexual reproduction) usually accomplish this by ALTERNATION OF GENERATIONS: one life form reproduces asexually to create a second life form, which in turn reproduces sexually to create the first *(Figure 3)*. The period of asexual reproduction is usu-

LIFE CYCLE OF ULVA, a green alga commonly called sea lettuce, involves diploid and haploid forms that are almost indistinguishable. Color indicates differences in mating type. 4

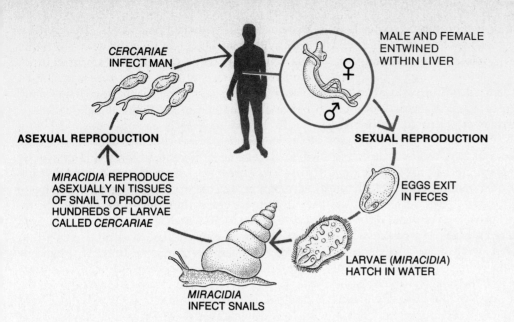

CERCARIAE INFECT MAN

ASEXUAL REPRODUCTION

MIRACIDIA REPRODUCE ASEXUALLY IN TISSUES OF SNAIL TO PRODUCE HUNDREDS OF LARVAE CALLED *CERCARIAE*

MIRACIDIA INFECT SNAILS

MALE AND FEMALE ENTWINED WITHIN LIVER

SEXUAL REPRODUCTION

EGGS EXIT IN FECES

LARVAE (*MIRACIDIA*) HATCH IN WATER

ally devoted to multiplying rapidly and creating offspring that can be widely disseminated; the period of sexual reproduction typically produces more resistant, genetically variable offspring capable of coping with changes in the environment.

REPRODUCTION IN PLANTS

As mentioned above, a major difference between plants and higher animals lies in the origin of the

gametes. In animals, the gametes are produced directly by meiosis of diploid gametocytes. In plants, gametes only rarely arise directly from meiosis. Instead, each meiosis produces four spores that eventually germinate and divide mitotically to produce a multicellular haploid plant, the GAMETOPHYTE. The gametophyte then produces gametes by mitosis, not meiosis. These gametes subsequently fuse to produce a zygote that divides mitotically to produce a diploid plant, the

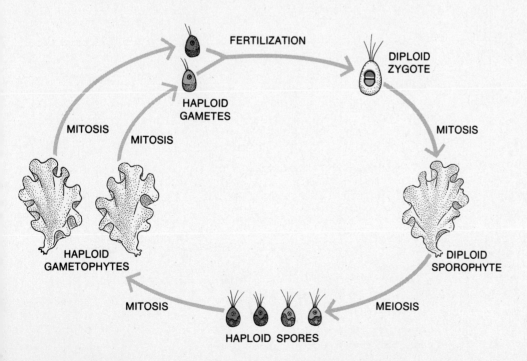

FERTILIZATION

HAPLOID GAMETES

MITOSIS

MITOSIS

HAPLOID GAMETOPHYTES

DIPLOID ZYGOTE

MITOSIS

DIPLOID SPOROPHYTE

MEIOSIS

MITOSIS

HAPLOID SPORES

SPOROPHYTE. At some later time specialized cells of the sporophyte, called SPOROCYTES, undergo meiosis and produce haploid spores, starting the cycle anew.

The alternation of diploid-asexual with haploid-sexual generations is characteristic of multicellular plants. However, particularly in the lower plants —the algae and fungi —there is a wide variation in this basic format. The life cycle of the sea lettuce *Ulva* is one example *(Figure 4)*. The diploid sporophyte of this common alga is a "leaf" two cells thick and a few centimeters across. Specialized cells, the sporocytes, become differentiated and then undergo meiosis to produce motile haploid spores. These swim away, propelled by four flagella, and some eventually find a favorable substrate. They then lose their flagella and begin to divide mitotically, producing first a short filament of cells and ultimately a broad thin sheet. These haploid gametophytes are indistinguishable from sporophytes in gross morphology. The motile spores carry genetic information for just one sex. Any particular gametophyte can produce only

male or female gametes, never both. The gametes arise within single cells called GAMETANGIA, rather than within any specialized multicellular gametophytic structure, as is the case with higher plants. Both types of gametes bear two flagella and hence are motile. However, the female gamete is clearly larger, so *Ulva* is anisogamous. Male and female gametes come together and fuse, losing their flagella as the zygote is formed. The zygote, after a brief resting period, begins mitotic division to form a new sporophyte. Any gametes that fail to find partners can settle down on a favorable substrate, lose their flagella, undergo mitosis, and produce new gametophytes directly; in other words, the gametes can also function as motile spores or ZOOSPORES. Zoospores provide a common method for vegetative reproduction both at the haploid and diploid levels in many algae and fungi. Motile gametes which can also function as zoospores, however, are much less common.

The basic pattern exemplified by *Ulva* is ISOMORPHIC —sporophyte and gametophyte generations are morphologically identical. In the fungi

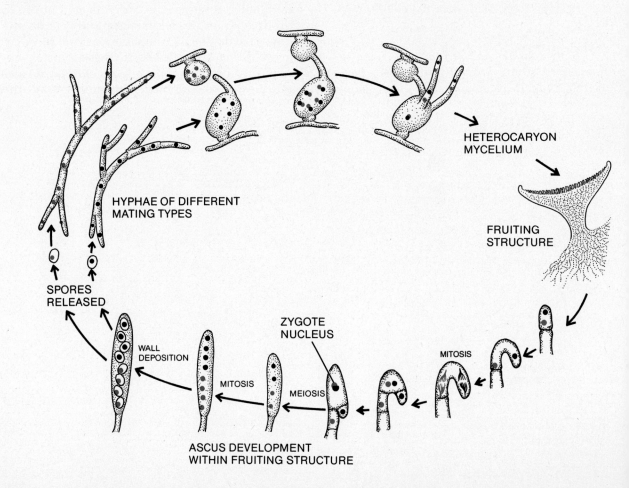

HETEROCARYON MYCELIUM

FRUITING STRUCTURE

HYPHAE OF DIFFERENT MATING TYPES

SPORES RELEASED

WALL DEPOSITION

MITOSIS

ZYGOTE NUCLEUS

MEIOSIS

MITOSIS

ASCUS DEVELOPMENT WITHIN FRUITING STRUCTURE

and algae, one finds variation in two directions producing HETEROMORPHIC generations. First, certain gametophytes produce gametes that fuse to form a zygote. The zygote, with or without a resting period, then functions directly as a sporocyte, undergoing meiosis to produce spores, which in turn produce new gametophytes. In the entire life cycle only one cell, the zygote, is diploid. Other algae go through a life cycle typical of higher animals. Meiosis of sporocytes produces gametes directly; they fuse to form a zygote, and the zygote divides mitotically to form a new sporophyte. In these organisms every cell except the gametes is diploid. Between these two extremes one finds algae whose gametophyte and sporophyte generations are both multicellular, but one phase (usually the sporophyte) is much larger and more prominent than the other.

Among the higher fungi a new wrinkle has evolved. The cytoplasm of opposite sexes fuses long before the nuclei. This is well displayed in a group of higher fungi called ASCOMYCETES (*Chapter 26*). The bread mold *Neurospora* provides an excellent example, already briefly mentioned in Chapter 8. Gametophytes can normally be classified into various mating types. A given gametophyte, consisting of filaments (HYPHAE) woven into a mat (MYCELIUM), will cross only with a gametophyte of a different compatible mating type. Hyphae of different mating types grow toward one another. Where two hyphae touch, the cell walls are broken down by enzymes (*Figure 5*). Cytoplasm and nuclei from one mating type invade the filament of the other, and the invading nuclei begin to divide, forming a special type of filament called a HETEROCARYON ("different nuclei"). A heterocaryon mycelium develops and eventually forms some sort of organized fruiting structure. Within this structure, many pod-shaped asci are produced, and pairs of dissimilar nuclei enter each ascus. Eventually the paired nuclei fuse to form a diploid zygote nucleus. Zygote nuclei then divide meiotically to produce the usual four cells, each of which undergoes one more mitosis to form a total of

5 LIFE CYCLE OF HIGHER FUNGI is exemplified by Neurospora. Hyphae of different mating types grow toward one another and fuse (top). Cytoplasm and nuclei from one type invade the other. Invading nuclei divide to form a heterocaryon, which in turn forms a fruiting structure (right). Within asci of fruiting structure paired nuclei form zygotes that develop into spores (bottom and left).

eight spores. The spores are eventually shed and germinate to form new gametophyte mycelia, each of a particular mating type.

The reproduction of these fungi displays several unusual features. First, there are no gamete cells as such, only gamete nuclei. Second, there is never any true diploid tissue, although for a long period of development the genes of both parents are present in the heterocaryon and can be expressed. Finally, although these plants grow in moist places the gamete nuclei are not motile, and hence water in liquid form is not required for fertilization.

The fungi, like the algae, display a tremendous variation in life cycles. One famous fungus is *Pucinia graminis*, a member of the major group called basidiomycetes. This organism causes the devastating disease known as wheat rust. Different stages in its life cycle occur on two different host plants, wheat (*Triticum*) and barberry (*Berberis*). At one time or another, five different types of spores are formed.

It should be clear by this time that there is really no such thing as a typical algal or fungal life cycle. The array of variations on the basic alternation of generations is bewildering. But above these rather primitive plants, the plant kingdom abruptly begins to exhibit consistent life cycles and some obvious evolutionary trends. The trends include progressively more protection for the developing gametes, a gradual escape from water-based fertilization, a dramatic shift from emphasis on the gametophyte generation to emphasis on the sporophyte, and increased protection for the embryonic sporophyte. All of the changes represent adaptations to life on dry land and a growing independence from a watery environment. This is the single most important theme in the evolution of the plants as a whole.

In mosses, members of the division Bryophyta, the conspicuous plant is the small leafy gametophyte (*Figure 6*). The tips of its prominent leafy branches bear specialized sex organs: the ANTHERIDIA, which produce the motile male gametes, and the ARCHEGONIA, which produce a single immobile egg cell. For the first time in the evolutionary progression, developing gametes of both sexes are protected by an outer layer of sterile tissue. The sperm are eventually released to swim to the archegonia. They pass down the neck of this organ, reach the egg, and cytoplasm and nuclei fuse to form the zygote. The zygote eventually begins division by mitosis, forming a sporophyte embryo. The sporophyte remains attached to the gametophyte throughout its life, forming 1) an ab-

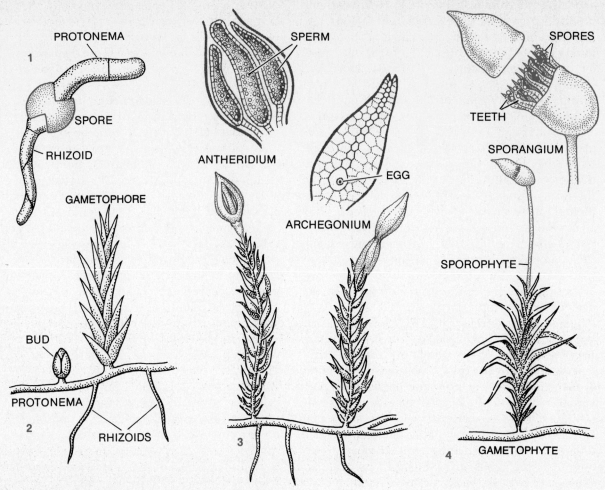

6 **LIFE CYCLE OF MOSSES** begins with the sprouting of a spore (1). Shoots of leafy gametophyte plant develop sex organs at their tips, either male antheridia or female archegonia (3). In some species both sex organs occur on the same plant; in other species, on separate plants. After fertilization the egg develops into a sporophyte plant, which is always attached to the tip of the gametophyte. The sporophyte forms a capsule-like sporangium (4), which eventually bursts. The cap disintegrates, and curved "teeth" around the rim fling the spores outwards.

sorptive foot embedded in gametophyte tissue, 2) a stalk, and 3) a capsule in which sporocytes eventually undergo meiosis to form haploid spores. The spores are shed as in the fungi, and they germinate to form first a filamentous and then an independent new leafy gametophyte. This life cycle is monotonously consistent throughout the bryophytes *(Chapter 26)*.

The most primitive plants that possess true vascular tissue for long distance transport of materials are the ferns and related groups *(see the section on xylem and phloem in Chapter 16)*. In these plants the sporophyte generation is dominant and is frequently enormous in size, as exemplified by certain tree ferns *(Figure 7)*. The undersides of the fern leaves carry specialized SPORANGIA *(see Figure 31 in Chapter 26)* in which sporocytes undergo meiosis to form haploid spores *(Figure 8)*. Once shed, the spores often travel great distances and eventually germinate to form very small and inconspicuous independent gametophytes *(Figure 9)*. The gametophytes produce both antheridia and archegonia, although not necessarily at the same time. Fertilization is accomplished by swimming sperm, as in the bryophytes, and the zygote develops into the new embryo sporophyte. The young sporophyte eventually sprouts a root that gives it the capacity to grow independently. The gametophytes are small, delicate and short lived, but the sporophytes can be very large and can

sometimes survive for hundreds of years. This life cycle is discussed in more detail in Chapter 26 (*Figures 28, 29, 30*).

Above the ferns on the evolutionary scale the gametophyte generation is even further reduced. The most advanced plants are the seed plants, a group comprised of GYMNOSPERMS (pines and their relatives) and ANGIOSPERMS (flowering plants, including most trees). Among seed plants the life cycle is just the reverse of that found in bryophytes: the gametophyte develops partly or entirely while attached to and nutritionally dependent on the sporophyte. And except in the most primitive gymnosperms, these plants have no swimming sperm—an evolutionary advance that enables them to colonize portions of the land habitat that remain closed to most lower plants. In short, no only have these higher plants escaped the aquatic environment, they have also escaped the swamp.

The escape of the gymnosperms and angiosperms was facilitated by a new reproductive strategy. On modified leaves (in cones or flowers) they develop separate male and female sporangia. Within the sporangia, male and female sporocytes undergo meiosis and become spores, but the spores are not shed. Instead the gametophytes begin development within the sporangia and are dependent on them for food. In the case of the female sporangium, usually only one meiotic cell from a given sporocyte survives. Its nucleus divides and the products divide again to produce a multicellular female gametophyte—in the angiosperms normally not more than eight nuclei in all. Meanwhile, within the male sporangium, meiotic division of male sporocytes produces male spores which undergo one or a few divisions to form the

---

Most advanced plants have no swimming sperm—an evolutionary advance that enables them to colonize portions of the land habitat that remain closed to most lower plants.

FERN SPORE is a haploid structure. Scanning electron **8** micrograph depicts spore of the Massachusetts fern (Thelypteris simulata) magnified 1200X.

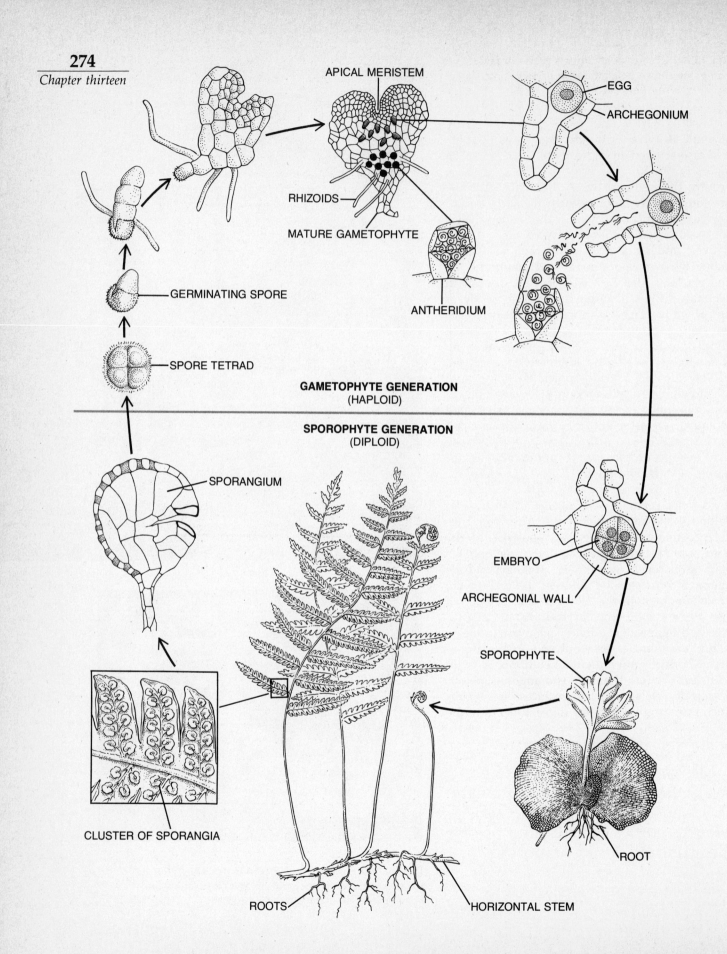

APICAL MERISTEM

EGG

ARCHEGONIUM

RHIZOIDS

MATURE GAMETOPHYTE

ANTHERIDIUM

GERMINATING SPORE

SPORE TETRAD

GAMETOPHYTE GENERATION
(HAPLOID)

SPOROPHYTE GENERATION
(DIPLOID)

SPORANGIUM

EMBRYO

ARCHEGONIAL WALL

SPOROPHYTE

CLUSTER OF SPORANGIA

ROOT

ROOTS

HORIZONTAL STEM

The evolutionary story of animal reproduction, like that of plants, involves the progressive emancipation of gametes from the water, and the development of mechanisms to introduce sperm directly into the body of the female.

---

male gametophytes: the familiar pollen grains. Distributed by the wind, an insect, a hummingbird, or by a plant breeder's brush, a pollen grain which reaches the appropriate surface of a sporophyte, near the female gametophyte, begins further development. The pollen grain sends out a slender tube which digests its way through the sporophyte tissue toward the female gametophyte. The process is accompanied by one or a few mitotic divisions. When the tip of the pollen tube reaches the female gametophyte, it releases a sperm nucleus which fuses with the female (egg) nucleus to form the zygote. (Instead of a sperm nucleus, primitive gymnosperms release a motile sperm that swims up an archegonial neck to reach the egg and fertilize it.) The young sporophyte develops to some embryonic stage, and then the entire system goes dormant; the end product is a seed *(Figure 10)*. A seed may contain tissues from three generations: The seed coat is the original sporangial wall of the parent sporophyte; within the seed coat is a layer of female gametophyte tissue (which may be fairly extensive in gymnosperms, but absent in angiosperms); and in the center of the package lies the embryo of the new sporophyte.

In lower plants spores are usually one-celled structures. These may be rather delicate; despite an extensive cell wall, well impregnated with some waterproofing material, spores often dry out in a short time. In contrast, the multicellular seed is well protected. Many layers of cells enclose the

dormant embryo, and the seed may remain viable for years. The seed habit and the escape from aquatic fertilization are two major reasons for the enormous evolutionary success of seed plants, the dominant elements of the modern flora.

This brief survey of the reproductive cycles in plants has deliberately omitted many of the adaptations and specialized structures (such as flowers) associated with plant reproduction. The effect of day length on flowering is discussed in Chapter 17; the nature of the specialized reproductive structures of both lower and higher plants is discussed in Chapter 26.

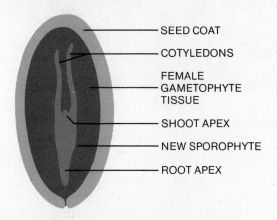

SEED COAT
COTYLEDONS
FEMALE GAMETOPHYTE TISSUE
SHOOT APEX
NEW SPOROPHYTE
ROOT APEX

MATURE PINE SEED contains tissues of three **10** **generations.** Embryo (light color), the new sporophyte generation, is surrounded by nutrient tissue from female gametophyte. Seed coat (gray) is from parent sporophyte. Although only two cotyledons are shown, pine embryo usually has many more.

REPRODUCTION IN ANIMALS

The evolutionary story of animal reproduction, like that of plants, involves the progressive emancipation of gametes from the water, and the development of mechanisms to introduce the sperm directly into the body of the female. But throughout this lengthy sequence, which stretches from the lowest protozoans to the highest vertebrates, animal gametes remain dependent on an aqueous medium; water has simply been replaced with specially secreted body fluids.

The most elementary form of mating in multicellular animals is the release of the sex cells into the water. This method, called SPAWNING *(Figure 11)*, requires an exact synchronization of the behavior of both sexes. If the timing is off, the short-lived

9 **LIFE CYCLE OF FERN involves a large sporophyte plant and an inconspicuous gametophyte.** Swimming sperm shed by antheridia of one gametophyte fertilize eggs within archegonia of another. Zygote develops into embryonic sporophyte within the archegonium and eventually sprouts leaves and roots. Structure of antheridia and archegonia is illustrated in more detail in Chapter 26.

11 SPAWNING OYSTER, seen here edge-on, releases a milky cloud of gametes into the water.

eggs will not be fertilized. One way to achieve such synchronization is by direct communication between individuals. The common quahog (*Venus mercenaria*) of the East Coast, for example, lives tightly packed in "clam beds." When the animals are sexually mature they hold their gametes within their bodies until they can "smell" the seminal fluid of another quahog. Then they discharge a white cloud of their own fluid and gametes. A chain reaction is set in motion among all the sexually mature quahogs in the vicinity, and the water

sometimes becomes opaque with the dense concentrations of their fusing gametes. External cues from the environment can be used to achieve the same effect. The palolo worm (*Leodice*) lives submerged among coral reefs in the South Pacific. As the October-November moon rises for the first time in its last quarter, the worms pinch off the posterior parts of their bodies containing the gametes. These parts rise to the surface of the water in huge numbers, where spawning and fertilization immediately ensue. Meanwhile the anterior part of the worm remains alive down in the reef. During the following year it regenerates a new posterior fragment and awaits the coming of the next spawning time.

A more advanced stage in evolution is spawning combined with pairing. In many kinds of fishes, one or both sexes prepare a nest into which the gametes are simultaneously discharged. Frogs and toads engage in AMPLEXUS: the male seizes the female from above and both discharge their gametes simultaneously into the water (*Figure 12*). In the fishes, frogs, and toads, the pairs come together only after very particular courtship signals have been exchanged. The result is an increase in the degree of certainty that the eggs will be fertilized by males of the same species.

All of the animals mentioned to this point expose their gametes to the perils of an unsheltered existence. The next logical step in evolution is INTERNAL FERTILIZATION, the direct transfer of the sperm into the reproductive tract of the female. This advance is an absolute prerequisite for living entirely on the land. Frogs and toads must return to the water to breed, but reptiles, mammals, and insects, having perfected internal fertilization, can spend their entire lives on land. The simplest method of internal fertilization is the transfer of a SPERMATOPHORE, a package of sperm which the female can pick up and insert in her own reproductive tract. Several animals have evolved variations on this technique, some of them quite bizarre. The male spider spins a small web, deposits a drop of semen (liquid containing sperm) on the web, picks up the drop with one of its palps (a specialized mouth part), and finally places it into the reproductive organ of the female (*Figure 13*). Leeches engage in what has been called "hypodermic" im-

AMPLEXUS in frogs. Male (on top) holds female and both discharge their gametes simultaneously into water. Mass of eggs is visible at left. **12**

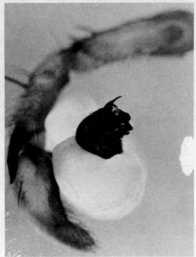

MALE SPIDER deposits semen on web, scoops it into a special receptacle on one of its palps and places it within reproductive organ of the female. At left, courtship of black widow spider, a ritual in which the smaller male is both the entertainment and the dessert. Picture at right depicts the sperm receptacle of the male Micrommata, a giant crab-spider from Europe. 13

pregnation. Although hermaphroditic, only one individual plays the male role in a given mating. "He" deposits spermatophores along the back of the partner. In addition to sperm, each spermatophore contains an enzyme that corrodes a hole through the body wall of the partner. When the breach is made, the sperm penetrate the opening and swim through the body fluids to the ovaries. Such methods represent a kind of experimentation in evolution that has been attempted by only a very few groups. By far the most widespread form of internal fertilization involves INTROMISSION. The male penis is inserted into the female vagina (or an equivalent organ of reception) and the semen is discharged. The sperm are then able to make their way up the fluid-filled canals of the female system and fertilize the eggs internally.

THE SELF-SUFFICIENT EGG

The reader will recall that even in primitive animals the egg is typically larger than the sperm, most of the bulk being due to the yolk. This trend toward providing for offspring in advance is carried to extreme lengths by several groups of higher animals. An insect encloses its egg in a tough, waterproof covering that permits it to remain in the dry open air for weeks or months as the embryo develops inside. Insect exterminators are aware that the egg is the hardest life stage to kill. The most elaborately constructed egg is the AMNIOTIC EGG of birds and reptiles *(Figure 14).* The chicken's egg is a typical example. The shell of the egg is brittle but porous, permitting gases to be exchanged between the embryo and the outside air.

But the shell is also waterproof, holding its liquid contents like a sealed container. Inside, the embryo is attached to the yolk sac, the contents of which it gradually absorbs during its development. Its body floats in the sheltering amniotic cavity, which in turn is enclosed by the amnion, a special membrane formed as an early outgrowth of embryonic tissue. Two other membranes, the allantois and chorion, aid in respiration, the absorption of nutrients within the egg fluids, and the storage of waste materials where they can do no harm. The chorion also supplements the shell in waterproofing the contents of the egg.

AMNIOTIC EGG characteristic of birds and reptiles is the most elaborate in the animal kingdom. 14

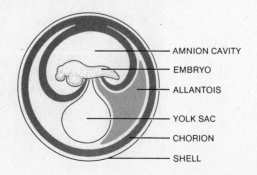

AMNION CAVITY
EMBRYO
ALLANTOIS
YOLK SAC
CHORION
SHELL

NURTURE OF THE YOUNG

In the higher mammals, certain insects and other invertebrate animals, the embryo is retained within the body of the mother during part of its development. In the course of evolution, mammals have enlarged and thickened parts of the tubes that lead from the ovaries to the outside. This modified structure, the UTERUS, is especially adapted for holding the developing embryo (in man and certain other higher mammals the uterus is also called the womb). Marsupial mammals such as kangaroos and opossums have a uterus that merely protects the embryo, but does not nourish it. The young marsupial is born at a very immature stage. It crawls into the marsupium, a pouch on the mother's belly, and attaches itself firmly to a nipple while it completes its development *(Figure 15)*. In placental mammals, including man and most other mammals, the membranes of the embryo come into intimate contact with the walls of the uterus.

The embryonic and maternal tissues intermingle in a complex growth, the PLACENTA, through which the mother provides oxygen and dissolved foodstuffs and receives back carbon dioxide and other waste materials. This direct exchange of liquids and gases is made possible by the fact that the blood capillaries of the mother and the young are intertwined.

Some insects have come close to duplicating the mammalian feat. The females of certain species of cockroach retain the eggs in a modified part of the reproductive tract (also called the uterus) until they hatch.

Many animals give varying degrees of care to their offspring after they are born. Some prepare nests for the eggs or carry them on their bodies after they are laid. Others carry the young, active animals after they hatch *(Figure 16)*. Feeding and grooming, which require intricate forms of communication between parents and young, have been evolved in both vertebrates, particularly birds and mammals, and social insects *(Chapter 22)*.

15 **NEWBORN KANGAROO is depicted beside the urogenital opening of its mother (left). At right, seven-week-old kangaroo sucks on nipple within pouch. Baby kangaroo, born 33 days after fertilization, is about two centimeters long. It must climb six inches to pouch, grasping mother's fur with its claws.**

DISPERSAL OF THE YOUNG

An important function of reproduction is to disperse the offspring. By investing in the colonization of new sites, the organism increases the likeli-

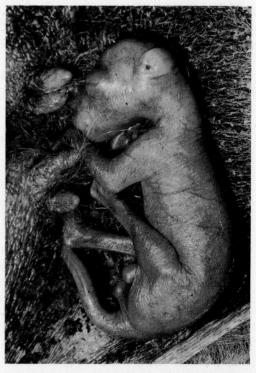

6 FEMALE SCORPION carries her young from place to place on her back. Scorpion depicted here belongs to the species Vejovis spincicrus. Some spiders also carry their young in this fashion.

7 A SPIDER BALLOONING. Spider ejects a long silken thread that is caught by air currents and blown aloft like the string of a kite. The spider then releases its grip on leaf and is carried along by the wind. The thread serves the same function as a balloon carrying a human passenger. Widespread among spiders, ballooning is their single most important means of long-distance dispersal.

hood that its genes will be perpetuated in the next generation. Often the mother accomplishes the dissemination herself. The females of butterflies and moths, for example, fly from plant to plant to lay their eggs. As a result the young caterpillars that hatch later are usually not overcrowded and find ample leaves to eat. When animals disperse on their own, it is ordinarily at an early age, either soon after hatching from the egg, as in most fishes and marine invertebrates, or during early adulthood, as in many mammals, birds, and insects. The means of dispersal are extraordinarily diverse and often elaborate. Two extreme examples—a ballooning spider and an exploding fungus—are illustrated in Figures 17 and 18.

HUMAN REPRODUCTION

With the exception of certain differences in the female cycle, the reproductive biology of human beings is typical of the mammals generally. The two testes of the male are lodged outside the main body cavity in a pouch of skin, the scrotum (*Figure 19*). They contain large numbers of seminiferous tubules, the sites of spermatogenesis. The microscopic sperm are produced by the hundreds of millions and stored in the epididymis, a mass of coiled tubes at the back of the testis. Prior to ejaculation they ascend the sperm duct where they are mixed with the secretions of the seminal vesicle and prostate glands to form the SEMEN. During sexual excitement the penis becomes stiff and erect due to the engorgement of blood in the spaces of the "cavernous bodies" that make up most of its bulk. Upon insertion into the vagina, the penis is stimulated by friction until it ejaculates by a reflex action, discharging the semen into the vagina. The sperm are then able to swim under their own power upward through the uterus to the Fallopian tubes. If they encounter an egg that descended from one of the ovaries, fertilization is likely to occur.

PILOBOLUS is a fungus that grows on dung. Sac of **18** spores (sporangium) caps the stalked sporangiophore. Fluid pressure builds up within sporangiophore and eventually bursts it, blasting the sporangium into a three- to eight-foot trajectory.

Within a woman's body, the vagina receives the semen, the ovaries produce the eggs, and the uterus (womb) carries the developing embryo. All of the cells capable of becoming eggs are already present in the ovaries of an infant girl. There are approximately 500,000 of these cells, called FOLLICLES. During the period of a woman's fertility, which ranges from puberty (12 to 15 years) to menopause (40 to 50 years), no more than 400 of the follicles are fully developed and released to attempt their rendezvous with sperm in the uterus. This process of ovulation does not occur haphazardly. It is the culminating event in an ESTRUS CYCLE that is under the precise control of hormones and lasts approximately 28 days. During the first part of the cycle the wall of the uterus is greatly enlarged. It becomes engorged with blood vessels and is thus prepared to receive and nurture the embryo in the event one is created. The egg is released on about the fourteenth day of the cycle. On all but a very few occasions it is destined to make its journey without being fertilized. The "disappointed womb" then undergoes a breakdown on about the 28th day, accompanied by bleeding, which terminates the cycle and sets the stage for the beginning of the next one. If an egg is fertilized and becomes implanted in the uterine wall, the uterus maintains its functional condition steadily until the birth of the child nine months later. For a discussion of the hormonal control of the menstrual cycle and pregnancy, see Chapter 18.

Because the human estrus cycle is ended so conspicuously by the menses (literally, "month"), the period of uterine breakdown and bleeding, it is often referred to as the MENSTRUAL CYCLE. In other mammals the cycle usually has a different duration, and it is seldom marked by noticeable bleeding. The most striking change comes during the time of ovulation at or near the middle of their cycle. At this time the female becomes sexually receptive and even aggressive. She is said to be IN HEAT; in fact, the word estrus refers to heat. During the remainder of the cycle she is unreceptive, and seldom if ever attempts to mate. In many species ovulation occurs only during a certain breeding season. The behavior of the human female is thus atypical. With the possible exception of her menstrual period, she is sexually receptive throughout the cycle and at all seasons of the year. In other words, depending on how one looks at it, the human female is either always or never "in heat." The evolutionary significance of this trait is discussed in Chapter 34.

HUMAN SEXUALITY

The sex drive in man, as in other animals, functions as a powerful lure that entices individuals to reproduce. Improved contraceptive techniques have enabled man to uncouple sex from its reproductive function, and there has been an increasing tendency to view sex as an activity that is pleasurable in its own right. This view has been strongly reinforced by psychologists, who see sexual activity as an important prerequisite of mental health, and by demographers, who view the population explosion as a phenomenon that is fast approaching the proportions of a natural disaster.

Every human society has felt the need to regulate the sexual activity of its members, probably because sex has long been recognized as a force that can disrupt the fabric of family and society. Probably no aspect of human behavior has been so circumscribed by hypocrisy, myth and taboo.

In the U.S. the sexual revolution that began about two decades ago has led to a less hypocritical approach to the sexual nature of man, to a debunking of long-standing myths, and to a tendency to relax harsh moral judgments about what is "normal" in sexual behavior. But it is not easy to rewrite centuries of cultural tradition in a single generation. At present our laws and beliefs about sex are a patchwork of contradiction and inconsistency. Ideas about virtually every aspect of sexual physiology and behavior are now being reexamined, particularly the sexual response patterns of men and women, homosexuality, the role of marriage, and birth control.

FEMALE SEXUAL RESPONSE

The rise of the women's liberation movement and the physiological studies of Masters and Johnson have led to increased awareness of the sexual responses of the human female. Some writers in the movement have drawn a distinction between clitoral and vaginal orgasms; that is, between orgasms caused by stimulation of the clitoris and those caused by friction of the penis against the walls of the vagina during coitus. Classical Freudian theory argued that normal female sexual development was marked by a gradual change from clitoral to vaginal response. The Masters and Johnson data do not support these views. Subjectively the sensations of orgasm attained through masturbation may differ from those attained through coitus, but the differences are psychological. Physiologically there is only one kind of female

MALE

URINARY BLADDER

URETER

RECTUM

SEMINAL VESICLE

EJACULATORY DUCT

PROSTATE

PUBIS

CAVERNOUS BODIES

URETHRA

GLANS PENIS

COWPER'S GLAND

EPIDIDYMUS

SCROTUM

TESTIS

VAS DEFERENS

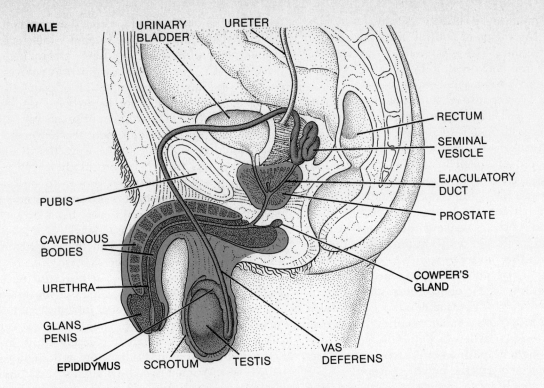

FEMALE

OVARY

URETER

FALLOPIAN TUBE

ABDOMINAL CAVITY

RECTUM

UTERUS

URINARY BLADDER

PUBIS

CERVIX

URETHRA

CLITORIS

LABIUM MAJUS

LABIUM MINUS

VAGINA

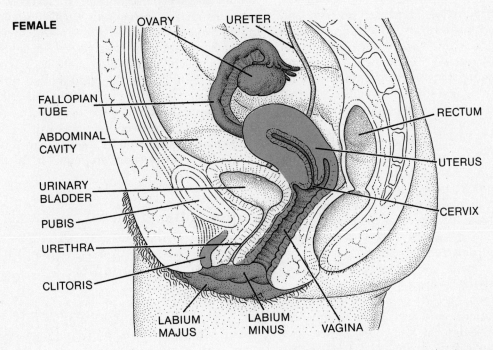

HUMAN REPRODUCTIVE SYSTEM. In the male, sperm produced in testis and stored in the epididymis ascend the vas deferens and mix with secretions from seminal vesicle, prostate gland and Cowper's gland to form semen. Penis becomes erect during sexual excitement as cavernous bodies fill with blood. Ejaculation discharges sperm into female vagina. Sperm swim into Fallopian tubes, where they can fertilize egg descending from ovary.

19

orgasm, and it can be elicited by stimulation of the clitoris or the vagina or both. It might be noted in passing that the clitoris is a remarkable and unique structure —the only organ found in nature whose sole function is to provide sexual pleasure.

The sexual response cycle of both women and men can be divided into four phases: excitement, plateau, orgasm and resolution *(Figure 20)*. In a woman, as sexual excitement begins her heart rate and blood pressure begin to rise, her muscular tension increases, her breasts swell and her nipples become erect. Her clitoris and the inner lips (labia minora) of her genitals swell as they become filled with blood, and the walls of the vagina become moist with lubricating fluid. In times past the appearance of this fluid was regarded as the female equivalent of ejaculation or as evidence of orgasm, but neither is true. Victorian erotica such as Frank Harris's *My Life and Loves* abounds with accounts of women "spending" after a minute or two of genital caresses. Harris was flattering himself that he had brought his partners to orgasm when in fact he had merely begun the preliminaries.

As a woman's sexual excitement increases, she enters the plateau phase. Her blood pressure and heart rate rise further, her breathing becomes rapid, and the head and shaft of the clitoris begin to retract. The greater her excitement, the greater the retraction. The sensitivity that once was restricted to the clitoris spreads over her external genitals, and the clitoris itself becomes so sensitive that touching it can be painful.

Orgasm begins with a long (two- to four-second) contraction of the outer third of the vagina, followed by shorter contractions about a second apart. Orgasm may last as long as a few minutes. Both the intensity and frequency of orgasm vary. Unlike men, some women can experience several orgasms in rapid succession. Five to ten minutes after orgasm a woman's physiology has largely returned to the normal precoital level. This descent from orgasm is the resolution phase. If she does not attain orgasm, return to normal conditions may take 30 minutes or longer. Among women as among men, the descent from orgasm can be a swoop or a gentle slide. Sometimes it leaves her relaxed and ready for sleep; other times it leaves her ready for more.

The expectation that a woman should always attain orgasm during lovemaking is a relatively recent idea that has proved unsettling to both women and men. To women, because they feel that there is something wrong with them if they don't have an orgasm, and to men because they feel inadequate if they fail to bring their partner to a climax. Two thoughts should temper the current overemphasis on sheer sexual "performance." First, that it is possible for a woman to enjoy sex without reaching orgasm, although of course it is vastly more pleasurable for her when she does. Second, that the full attainment of sexual pleasure depends on communication between partners. A man should be sensitive to the ways that female sexuality differs from his, particularly to the fact that a woman is generally slower to become aroused and her excitement is slower to subside. A woman in turn should not expect her partner to guess how she would like him to make love to her. In short, if he's not doing something right, ask him. The bedroom is no place to be shy.

Cultural conditioning plays a powerful role in sexual response. Our sexual responses are generally what we expect them to be. In some cultures, such as that of Victorian England, females were taught that coitus was an unpleasant duty, and that well-bred women should show no sign of sexual passion. Women raised with that sort of conditioning rarely achieve orgasm. In sexually liberal soicieties such as those of the Pacific islands, both males and females are encouraged to regard lovemaking as one of life's greatest delights, and women in such cultures almost always achieve orgasm. In the first case the mind inhibits a normal physiological response and in the second enhances it. The role of the mind in sex should not be underestimated. Most cases of frigidity in women (inability to attain orgasm) and impotence in men (inability to achieve an erection) are rooted in the mind.

Of the entire range of human sexual behavior, only two aspects appear to have no counterpart in the behavior of other animals: the human female has no period of "heat," and female orgasm —apparently a late evolutionary invention —has been conclusively demonstrated in no other species.

THE MALE RESPONSE

The similarities between the male and female cycles of sexual response are greater than the differ-

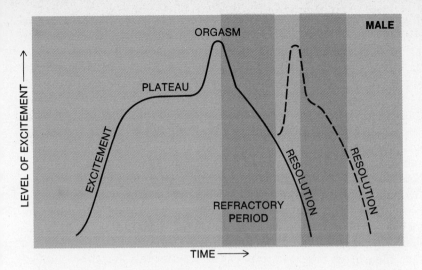

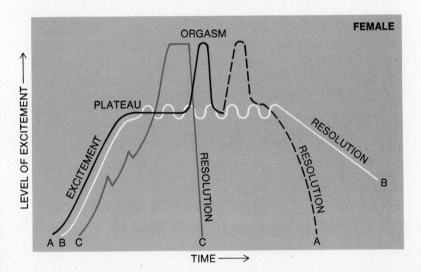

ences. The excitement phase of the male cycle is marked by an increase in blood pressure, heart rate and muscular tension. The penis, engorged with blood, becomes rigid and swells to as much as double its normal length. In the plateau phase breathing becomes rapid, the diameter of the glans (the head of the penis) increases, and a clear lubricating fluid secreted by Cowper's gland oozes from the urethra. The testes swell to half again their normal size, and the scrotum tightens. Pressure and friction against the nerve endings in the glans and in the skin along the shaft of the penis eventually triggers orgasm. Massive spasms of the muscles in the genital area, as well as contractions of the accessory reproductive organs—the prostate, seminal vesicles and ejaculatory duct—result in ejaculation: the spurting of semen from the urethra. Within a few minutes after ejaculation the

CYCLES OF SEXUAL RESPONSE are basically similar in males and females. Differences include the presence of a refractory period in the male cycle and a greater variety of patterns in the female cycle. Female may show any of three patterns on different occasions. Commonest female pattern (curve A) resembles that of male. Alternatively, females may experience sustained multiple orgasms (curve B), or skip the plateau phase in a surge toward very intense orgasm (curve C). Dashed black lines indicate possibility of second orgasm in both sexes. Graphs are based on data of Masters and Johnson. **20**

penis shrinks to its normal size and body physiology drops to normal precoital levels. One of the differences between the male and female cycle is presence of a refractory period immediately after orgasm, as shown in Figure 20. During this period, which may last 20 minutes or longer, a man cannot achieve a full erection or another orgasm, regardless of the intensity of sexual stimulation.

Erotic literature and folklore have propagated a number of myths about male sexuality. Foremost among them is the notion that a giant penis is much more stimulating to a woman than one of average or small size. Pornographic books always endow their heroes with penises of truly awesome dimensions, and purveyors of sexual paraphernalia exploit male anxieties about penis size by selling weighted devices to stretch the penis to greater length. The average penis is three to four inches long when limp and about six inches long when erect, with a diameter of about 1¼ inches. Penises can be considerably larger or smaller. Although medical literature records some that measured over 13 inches when erect, most of this length is wasted because the average woman's vagina is only about three inches deep. Contrary to popular belief, there is no relation between the size of a man's penis and his body build, race, or his ability to give and receive sexual pleasure.

SEXUAL BEHAVIOR

Standards of sexual behavior remain controversial, posing problems of what is normal, what is healthy, what is moral, what is legal. Frequently the answers to these questions are contradictory. Our society has traditionally condoned only sexual intercourse between married adults, and even in marriage certain kinds of sexual activity have long been regarded as taboo. Oral-genital contacts, for example, have been described as "unnatural acts," and they are felonies in many states. The same holds true for homosexual behavior. It is enlightening to examine the presumably more "natural" behavior of man's primate relatives. Nuzzling and licking of the female genitals has been observed among wild male chimpanzees and gorillas, and the females respond with gratitude. Homosexual behavior—males mounting males and females mounting females—is common among primates and other mammals. Of the entire range of human sexual behavior, only two aspects appear to have no counterpart in the behavior of other animals: the human female has no period of "heat" or heightened sexual receptivity, and female orgasm—apparently a late evolutionary invention—has been conclusively demonstrated in no other species.

In the late 1940's the publication of the Kinsey Reports confronted Americans with the fact that the professed standards and laws of the society did not correspond with what people were doing in their bedrooms. Kinsey's sampling techniques had some limitations and the data are by now over two decades old, but they are still cited because they are the best available. When the reports were published many people with no contradictory evidence nevertheless denied their validity. Many of Kinsey's opponents no doubt realized that his statistics would sharply revise public attitudes about what constituted "normal" behavior, and thus would encourage more permissive attitudes toward sex. Indeed the definition of normal behavior is basically a statistical one: it is what the majority does, not what it should or ought to do. The impact of Kinsey's pioneering work no doubt helped to pave the way for the sexual revolution.

Specifically, Kinsey found that masturbation was practiced by about 90 percent of males and almost 60 percent of females; that virtually all males and half of females experienced coitus before marriage; that almost half of college-educated married couples engaged in oral-genital sex; that about half the married men and a fourth of married women admitted that by age 40 they had been involved in at least one extramarital affair; and that by age 40 more than one man in three and one

---

George Bernard Shaw once described marriage as a social contrivance designed to produce the greatest number of offspring and the closest possible care of them.

---

woman in eight had participated in at least one homosexual encounter leading to orgasm. Concerning homosexuality, Kinsey wrote: "Males do not represent two discrete populations, heterosexual and homosexual. . . . It is a fundamental of taxonomy that nature rarely deals with discrete categories. Only the human mind invents categories and tries to force facts into separated pigeon-holes. The living world is a continuum in each and every one of its aspects. The sooner we learn this concerning human sexual behavior the sooner we shall reach a sound understanding of the realities of sex." Recent estimates classify only about two percent of American men and one percent of women as exclusively homosexual, but this corresponds to several million individuals. Among psychiatrists there is considerable debate about whether homosexuality is an illness or a preference, with the trend being to consider it as a preference.

There is great diversity in the human appetite for sex. Beyond a certain point sexual appetite tends to diminish with age. Kinsey estimated the total frequency of sexual outlets both in and out of marriage. He defined outlets as any sexual activity leading to orgasm. The data are presented in Figure 21, but the differences between male and female activity should be viewed with some care, mainly because frequency of orgasm is a better measure of male than of female sexuality. Frequency of coitus is a better index for women. Kinsey found that the average (median) frequency of coitus for young married couples was three times a week. This figure dropped to twice a week by age 30, once every four days by age 40, and once every 12 days by age 60. Because these figures are averages they tend to obscure the wide range of variation from couple to couple. Some couples make love several times a day, some every night, and some hardly at all.

IS MARRIAGE OBSOLETE?

George Bernard Shaw once described marriage as a social contrivance designed to produce the greatest number of children and the closest possible care of them. At a time when a growing population is no longer seen as an asset but as a liability to a society, and ever larger numbers of women are seeking to fulfill themselves in careers rather than in childrearing, the traditional role of marriage is being severely questioned. This reappraisal is hardly surprising. The current failure rate of marriage is appalling. In the U.S. many marriages last about as long as the family car. The median (average) length of marriages ending in divorce is less than seven years.

If one accepts the idea, widely held by social scientists, that trends in California today will become the national norm tomorrow, the long-term outlook is truly bleak. In the state as a whole more than half the marriages end in divorce, and in urban areas where people are better educated and more attuned to the tempo of the times, the failure rate is three marriages out of four and apparently rising. A comparable failure rate in business — three out of four going bankrupt within seven years — would rightly be regarded as a national catastrophe.

Many psychologists attribute the high failure rate of marriage to errors in selecting mates and to marrying for the wrong reasons. William J. Lederer of the Mental Research Institute in Palo Alto argues that "people marry because they are in excessive sexual heat; or they are lonely; or they need an excuse for getting out of their original family home; or they want to be supported; or they have a neurosis which needs to be fed; or society pressures them into marriage. Love is rarely a factor."

AVERAGE FREQUENCY OF ORGASM in both sexes **21** shows a gradual decline after age 30–35, although patterns before age 30 differ. Curves are based on Kinsey data, and do not reflect recent rise in sexual activity of teenage girls. Some individuals are considerably more active than average.

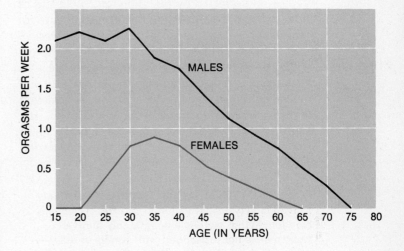

A number of authors have urged that marital counseling should precede marriage rather than being sought only when the marriage is faltering. Anthropologist Margaret Mead has proposed a form of childless marriage based on a renewable marriage contract. After a period of say three years, the marriage would automatically be dissolved unless renewed by mutual consent. At renewal time the couple would have the option of extending their marriage contract for the same period or committing themselves to remain together long enough to raise a family. Other alternatives to conventional marriage are discussed in Carl Rogers' book, *Becoming Partners,* cited in the readings at the end of this chapter. The consensus among students of human behavior seems to be that marriage as an institution will survive for quite a while, but not in its present form.

BIRTH CONTROL

Birth control, also known as contraception, is practiced in one form or another by members of virtually every culture on Earth, and is sanctioned or at least tolerated by most of the major religions. Historically the most common way to regulate family size, even in culturally advanced societies, has been infanticide. In classical Greece, for example, unwanted infants were abandoned on hillsides to die of exposure or to be eaten by animals. In China female infants, traditionally regarded as less desirable than males, were often drowned or abandoned, a practice that persisted well into the 20th century. Modern sensibility recoils at infanticide, but ethical issues raised by contraception and particularly by abortion are still the subject of controversy. On a planet already overburdened with people and afflicted by pollution and shortages of nonrenewable resources, the ethical issues of population control transcend private morality.

One of the oldest and undoubtedly simplest methods of contraception mentioned in the Old Testament is *coitus interruptus*, withdrawal of the penis before ejaculation. If ejaculation is achieved in the moments just following withdrawal, the "seed is spilled upon the ground" —a grave sin in the religion of the Israelites, who were concerned not with birth control but with a rapid increase in their own population to stave off their numerous enemies. (Throughout history many governments have discouraged birth control because a small population makes it hard to raise a large army.) A second ancient and still less pleasurable method of birth control is abstention from sexual relations.

In modern times a variety of sophisticated techniques of contraception have been invented that interfere slightly or not at all with the full act of sexual intercourse. These include condoms, diaphragms, spermicides and "the pill." The condom is a very thin sheath, usually made of rubber, which is pulled tightly over the penis and retains all of the semen after ejaculation. The condom seldom fails in its primary function, and in addition it greatly reduces the likelihood of transmitting venereal diseases, particularly gonorrhea and syphilis. Its chief disadvantage is the loss of some sensitivity in the penis and the necessity of interrupting foreplay in order to fit it on the penis.

The female equivalent of the condom is the diaphragm, a flexible rubber cup which before intercourse is coated on the edge and undersurface with a spermicidal (sperm-killing) jelly or cream and fitted over the cervix. It must be left in position for several hours after intercourse. The diaphragm is an especially effective device *(Table I)*, and it does not interfere with either partner's enjoyment of sexual intercourse. But because it must be inserted into the vagina and fitted exactly each time, the woman needs a prescription and special instruction from a physician. A similar and equally

I FAILURE RATES of various contraceptive methods depend largely on motivation. High effectiveness given here for rhythm method among highly motivated women does not correspond with results of other studies. Table is based on data gathered by Bernard Berelson and his colleagues.

METHOD	NUMBER OF PREGNANCIES RESULTING WHEN 100 WOMEN USE THE METHOD FOR ONE YEAR EACH	
	LOW MOTIVATION FOR BIRTH CONTROL	HIGH MOTIVATION FOR BIRTH CONTROL
NO CONTRACEPTION	80	80
AEROSOL FOAM	—	29
FOAM TABLETS	43	12
SUPPOSITORIES	42	4
JELLY OR CREAM	38	4
DOUCHE	41	21
DIAPHRAGM WITH JELLY	35	4
CONDOM	28	7
COITUS INTERRUPTUS	38	10
RHYTHM	38	0
THE "PILL" (STEROID CONTRACEPTION)	3	0
IUD'S ("LIPPES LOOPS," LARGE SIZE; MOTIVATION NOT DISTINGUISHED)		
0–12 MONTHS	2	
12–24 MONTHS	1	

effective device is the cervical cap, a plastic or metal covering which is applied tightly over the cervix and must be removed only for menstruation.

A woman also has at her disposal a variety of spermicidal jellies, creams, foam tablets, aerosols, and suppositories that can be spread through the upper vagina with special applicators. Or she may use an intrauterine device, or IUD as it is usually abbreviated. The IUD is a small plastic or metallic object in the form of a spiral, an S-shaped loop, a ring, or any one of a number of other designs; it is inserted inside the uterus and left in place. It operates on the general principle that any foreign body in the uterus acts to prevent pregnancy. The greater the surface that comes into contact with the uterine lining, the more effective the contraception—hence the elaborate geometric shapes often given the IUD's. These devices have the great advantage of being very cheap—they cost only a few cents each—and very long-lasting. In most cases they persist unfelt by either partner through an indefinite number of menstrual periods. On the other hand, about ten percent of women expel them spontaneously, often without realizing that this has happened. Also, a small minority of women who retain the IUD's suffer minor bleeding and even pelvic inflammation as side effects.

The "pill" is a radically different birth control device available to women, and one of the most effective and widely used in technologically advanced societies. It consists of a small, orally administered dose of the female hormone estrogen mixed with progestin, a synthetic chemical similar to the natural hormone progesterone, which is secreted by the ovaries. These two substances apparently act together to suppress ovulation, thus controlling the reproductive process at its very source. The pill must be taken daily for 20 or 21 days of the 28-day menstrual cycle, starting with the fifth day after the beginning of menstruation. When taken each day without fail, it is virtually 100 percent effective. The chief disadvantage of the pill, other than the self-discipline required to use it correctly, is its undesirable side-effects. Up to 25 percent of women develop some of the outward symptoms of pregnancy: swelling and tenderness of the breasts, nausea, headaches, changes in complexion, irritability, and so on. A very tiny percentage of pill users also develop increased susceptibility to thrombophlebitis (inflammation of the veins combined with blood clotting) and pulmonary embolism (development of blood clots in the lungs).

Even in these extreme cases, however, the risk of death is considerably less than that caused by complications during pregnancy and childbirth.

The rhythm method, or periodic abstention from sex, is the only technique of birth control now approved by the hierarchy of the Roman Catholic Church. (Surveys of U.S. Catholics reveal that laymen are considerably more liberal than their bishops: there is very little difference between the contraceptive practices of Catholic and non-Catholic couples.) The idea of the rhythm method is simple: a woman must abstain from intercourse during the several days of each menstrual cycle in which she is capable of conceiving. More exactly, she must not be inseminated at any time from two days before ovulation to a half-day following ovulation. The rhythm method is theoretically sound, but it has some serious practical difficulties. First, the time of ovulation can only be determined by keeping careful records of the times of menstruation and changes in body temperature over a period of at least several months. On the day of ovulation the woman's temperature rises half a degree Fahrenheit, and it remains elevated until the onset of menstruation. Once this date in the monthly cycle has been calculated, the couple must add several additional days in advance of and following the estimated ovulation time to give the method an acceptable safety margin. The week or more of sexual abstention can disrupt the marital relationship. Even more serious is the fact that one woman in six has a cycle too irregular for the rhythm method to work.

Sterilization is an increasingly popular step for men and women who have completed their families and wish to enjoy a sexual life wholly free from worry. For the male, sterilization means a vasectomy, the cutting and tying off of the vas deferens. This simple operation, which is relatively painless and requires only about an hour in a doctor's office, ensures that sperm will no longer be included in the ejaculate. It has no effect on the sexual performance of the man or the pleasure experienced by either partner during intercourse. In fact, the only outward sign that a vasectomy has been performed is the absence of sperm in the ejaculate—which requires examination under a microscrope to establish. The sterilization operation for the woman, called a salpingectomy or tubal ligation, is a more complicated form of surgery requiring anesthesia. The abdomen is opened and the fallopian tubes cut and tied, thus preventing the passage of the ovum from the ovaries to the uterus.

ABORTION

Abortion is the most effective of all methods of population control. Almost everyone will agree that it is also the least desirable, since it requires the destruction of a fetus that is already on the developmental path toward the formation of a baby. At what point a fetus becomes a human child is a controversial biological and ethical question. The moral dilemma is further complicated by the knowledge that in many cases a particular fetus allowed to go to term will be seriously defective, or unwanted by its parents. Born with such a handicap, a child is likely to lead a troubled life and to add a heavy burden on an already overpopulated society. Thus the choice must be made between the fetus —which may or may not be regarded as a formed human being—and the adults who must commit two decades of their lives to caring for it if it is allowed to go to term. At one extreme are those who label abortion as murder, even if the embryo consists of no more than a few microscopic cells. Opposing them are those who believe that a woman should have the right to decide for herself whether or not she will carry the fetus to term. Most biologists, including the authors of this book, concur with the 1973 decision of the U.S. Supreme Court confirming this right. We view the early fetus as a developing POTENTIAL human being, still far removed from a newborn infant. The Supreme Court decision does not say that every woman who becomes pregnant must have an abortion; it simply states that she is free to have one if she wants one. We feel it is hard to argue against that logic.

Abortion, when performed before the twelfth week of pregnancy by a physician under proper clinical conditions, is a relatively simple and safe operation. Statistically, in fact, it is much safer than childbirth. Early abortion is usually performed by SUCTION CURETTAGE. The cervix is dilated (spread open) and the contents of the uterus removed by a special tube attached to a vacuum pump. This is generally followed by scraping the uterus with a curved scalpel called a curette to make sure that no fragments of the placenta remain. The suction method simplifies and shortens the operation, and minimizes blood loss and other risks. The operation can be performed under either local or general anesthesia, according to the preference of the patient.

Pregnancies that have advanced to 16 weeks or more are usually terminated by a procedure called SALTING OUT. With a hypodermic needle some of the fluid surrounding the fetus is removed from the uterus and replaced with a salt solution that induces a miscarriage, usually within 24 to 48 hours. This type of abortion is usually performed in a hospital.

Laws against abortions did not stop them, but merely made them harder to get, riskier, and more expensive. Illegal abortionists still flourish in countries where abortion is prohibited or discouraged by law. Abortion has always been one of the most popular methods of birth control throughout the world. In Italy a few years ago, when all forms of contraception except the rhythm method were illegal, the abortion rate was nonetheless estimated to be almost equal to the birth rate. Such illegal abortions, usually performed by sympathetic friends or medical quacks, are of course quite dangerous. The death rate from them is higher than the combined death rate from all legal abortions and all complications arising in pregnancy and childbirth.

READINGS

C.H. BEST AND N.B. TAYLOR, *The Physiological Basis of Medical Practice*, 8th Edition, Baltimore, Williams and Wilkins, 1966. A more advanced treatment of various aspects of human reproductive physiology.

H.A. KATCHADOURIAN AND D.T. LUNDE, *Fundamentals of Human Sexuality*, New York, Holt, Rinehart and Winston, 1972. A comprehensive and clearly written book developed for an undergraduate course in Human Sexuality at Stanford University.

A.C. KINSEY, W.B. POMEROY, AND C.E. MARTIN, *Sexual Behavior in the Human Male*, Philadelphia, W. B. Saunders Company, 1948. Together with the companion volume by the same authors on *Sexual Behavior in the Human Female*, published in 1953, this book is the most complete report on the sexual practices of Americans. Heavy reading, but despite the fact that the data are more than two decades old, the studies remain a landmark.

W.H. MASTERS AND V.E. JOHNSON, *Human Sexual Response*, Boston, Little, Brown, 1966. The definitive work on the subject, based on years of careful laboratory research.

J. PEEL AND M. POTTS, *Textbook of Contraceptive Practice*, New York, Cambridge University Press, 1969. This generally excellent book includes detailed accounts of the reproductive physiology of humans and the ways it can be influenced to achieve birth control.

C.L. Prosser , *Comparative Animal Physiology*, 3rd Edition, Philadelphia, W. B. Saunders Company, 1973. Reproductive biology presented in the setting of a full-dress review of physiological evolution.

P.H. Raven, R.F. Evert and H. Curtis, *Biology of Plants*, 2nd Edition, New York, Worth, 1976. A beautifully illustrated modern botany text with an excellent treatment of plant reproduction.

C.R. Rogers, *Becoming Partners: Marriage and its Alternatives*, New York, Delacorte Press, 1972. An inquiry into the future of marriage. Rogers is a distinguished psychologist, and his book tells the story of several experiments in new forms of marriage, mostly in the words of the couples themselves.

F.B. Salisbury and C. Ross, *Plant Physiology*, 2nd Edition, Belmont, CA, Wadsworth, 1978. A detailed plant physiology text with strong chapters on the reproductive physiology of higher plants.

C.D. Turner and J.T. Bagnara, *General Endocrinology*, 6th Edition, Philadelphia, W. B. Saunders Company, 1976. A comprehensive treatment of all aspects of the relationship of hormones to reproductive physiology in both vertebrates and invertebrates.

In the imperfect development of cohabitation on a crowded planet, the habit of eating one another —dead and alive — has become a general custom. —Hans Zinsser

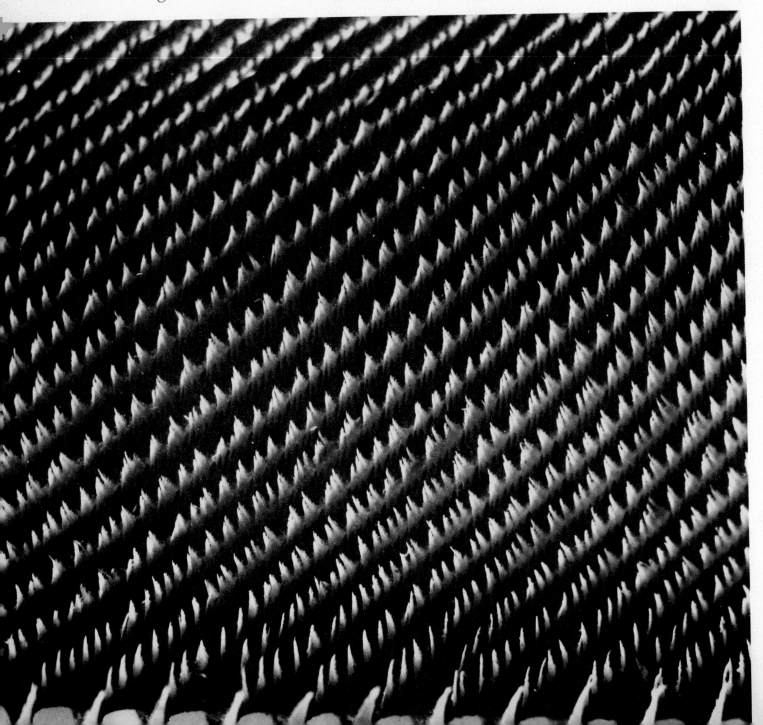

14
Carbon and nitrogen sources

Throughout their lives all organisms require a source of raw materials for synthesis and a source of energy to drive their biochemical machinery. As humans we tend to take it for granted that both come from food, but not all organisms require food in the human sense of the word. Plants and many microorganisms obtain raw materials from the air and the soil, and energy from sunlight. It is more accurate to say that organisms obtain their raw materials from NUTRIENTS, a term that includes any substance used in metabolism.

Among the nutrients required by all organisms are the elements carbon, nitrogen, oxygen and hydrogen. A number of other elements such as phosphorus, sulfur, potassium, calcium, magnesium and iron are required by most organisms, as well as trace quantities of molybdenum, boron, manganese, zinc and copper. Each of these elements frequently fulfills several different roles. During the long process of chemical evolution many organisms have developed their own special requirements. Certain algae require vanadium or cobalt; higher animals require substantial amounts of sodium and chloride, and mammals require iodine. Conversely sodium is not essential to plants, nor are large amounts of chloride. The inorganic elements, sometimes loosely called minerals, may play both structural and catalytic roles in the cell, and even the simplest organisms cannot survive without some of them.

Virtually all living organisms on Earth ultimately obtain two of the four crucial elements, oxygen and hydrogen, from the atmosphere or from wa-

FILTER OF A FLAMINGO (Phoeniconaias minor) is shown in the photograph on opposite page. Fringed platelets in bird's upper jaw strain food, primarily bluegreen algae and diatoms, from water scooped up by the bill.

ter. But the sources of carbon and nitrogen vary enormously from one type of organism to another. The diverse strategies for obtaining these two elements have shaped the anatomy, the lifestyle and the biological destiny of every type of organism on the planet.

All nutrients pass through the cell membrane by diffusion or active transport, although some cells can also ingest particulate matter by pinocytosis. But the ways that nutrients reach the cells of multicellular organisms are remarkably dissimilar. Inorganic elements including nitrogen enter higher plants with water through root systems; carbon in the form of CO_2 enters through the aerial parts of the plant, particularly the leaves. Oxygen may enter through both roots and leaves. All of the essential elements except oxygen enter a mammal via the digestive tract; oxygen enters through the lungs. Amphibians such as frogs can take up inorganic ions and oxygen through their skin, and the gill surfaces of fish serve for exchange of dissolved gases, ion uptake and excretion of soluble waste.

The basic biochemical machinery for building, operating and replicating cells is remarkably similar from the simplest to the most complex living creatures, but lifestyles and basic chemical needs vary enormously. Autotrophs (Chapter 3) manufacture their own organic food from simple inorganic nutrients: carbon dioxide, water, nitrate or ammonium ions, and a few soluble minerals. Most autotrophs are photosynthetic, but a few are chemosynthetic, meaning that they derive energy not from light but from simple oxidizable substances available in solution around them. All other organisms are heterotrophs whose nutritional requirements include one or more organic compounds synthesized by some other organism. Versatile heterotrophs such as bacteria may require merely an oxidizable carbon source like glucose (synthesized by a plant); an exacting heterotroph

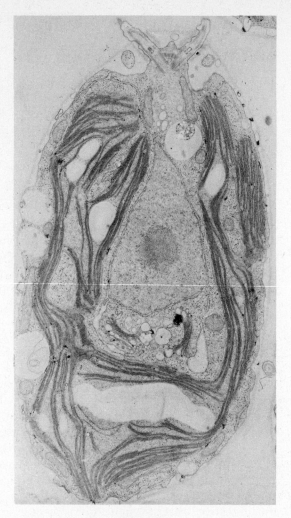

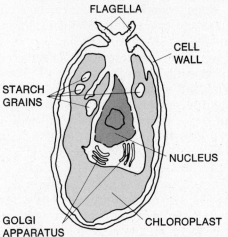

FLAGELLA

CELL
WALL

STARCH
GRAINS

NUCLEUS

GOLGI
APPARATUS

CHLOROPLAST

1 CHLAMYDOMONAS is a unicellular green flagellate that makes its own food by photosynthesis. In the electron micrograph the organism seems to consist chiefly of a chloroplast and a nucleus. Only the stumps of its two flagella are visible. The accompanying drawing indicates the location of the principal structures.

like man requires an external supply of not only oxidizable carbon compounds but also eight of the 20 amino acids, certain fatty acids and an assortment of vitamins.

PHOTOSYNTHESIS

Of the four principal elements of biochemistry, only oxygen occurs abundantly in nature in a readily reactive state. Large amounts of carbon, bound in insoluble carbonates in the earth's crust, are unavailable to organisms. The carbon which is available to organisms exists as CO_2, with low free energy. Nitrogen, about four-fifths of the atmosphere, exists as a relatively inert gas, N_2, which requires an inordinate amount of energy to break the triple bond linking the two atoms and obtain a reasonably reactive form for synthesis of amino acids and other nitrogen-containing organic molecules. Hydrogen is present in vast quantity as water but obtaining it or its electrons for use as reductants also requires a large amount of energy.

The photosynthetic autotrophs have evolved elegantly organized pigment systems *(Figure 1)* to capture the energy of sunlight *(Chapter 6)*. They use the energy to raise an electron to an excited state and then trap it in such a way that a portion of its energy is extracted in usable form as the electron returns to its ground state. What is essentially a light-induced separation of electric charges is sufficiently energetic to split water into reducing hydrogen and oxygen gas. It is also sufficient to generate ATP, which is required (with the hydrogen) to reduce or "fix" carbon dioxide into biologically useful compounds with higher free energy *(Chapter 6)*. Photosynthesis serves two vital functions: it is the ultimate source of carbon compounds for all organisms, and it maintains a continual supply of oxygen gas in the atmosphere. Without this gas, the oxidative processes on which all aerobic organisms depend for energy would cease.

Although photosynthesis is a process unique to green plants and algae, a few unicellular flagellates and two groups of bacteria, the sheer magnitude of the process is extremely impressive. Though estimates vary, there are at least 1.6×10^{10} tons of carbon fixed per year on land, and between 2.0 and 13.5×10^{10} tons per year fixed in the oceans. An amount of oxygen equivalent to that present in the atmosphere is produced by photosynthesis every two years. The debt of aerobic heterotrophs to photosynthesis is enormous.

A unicellular photosynthetic autotroph in water has ready access to carbon as the bicarbonate

An amount of oxygen equivalent to that in the atmosphere is produced by photosynthesis every two years.

formed when CO_2 dissolves in water. Movement of water and diffusion are normally ample to satisfy the carbon requirement. Each cell of single-celled organisms such as photosynthetic bacteria or many algae is in immediate contact with the primary source of carbon. Even enormous algae such as *Macrocystis*, a kelp which on the California coast may reach lengths of 60 meters, or the smaller kelp, *Laminaria,* have ready access to bicarbonate since photosynthesis occurs in limited layers of cells close to the plant surface, and therefore close to the surrounding medium *(Figure 2)*.

Land plants, growing in an environment far more rigorous than the ocean, have no such simple arrangement for obtaining carbon. CO_2 must be in solution to reach the chloroplasts within the cell, requiring that moist surfaces be exposed to the gaseous environment. Unfortunately a moist surface appropriate for dissolving CO_2 also enhances the evaporation of water. In all but the most humid environments plants without some kind of specialized water-conserving adaptation could hardly survive.

An herbaceous plant, one that forms little or no woody tissue in the course of its development, has three major organ systems—leaves, stems, and roots. The parts of the plant directly exposed to a potentially dessicating atmosphere are normally covered with a layer of CUTIN, a waxlike hydrocarbon-containing mixture of substances that provides excellent waterproofing. But cutin is just as impervious to CO_2 as it is to water, and the solution to one problem simply raises another.

The compromise solution is found in the architecture of the leaf. A cross section of a typical leaf from a temperate zone plant is shown in Figure 3. The upper surface is coated with the layer of cutin secreted by the outermost layer of cells, the UPPER EPIDERMIS. The few nonphotosynthetic epidermal cells produce the cutin. The greater portion of the photosynthesis of the leaf occurs in the PALISADE PARENCHYMA, the layer of elongated cells beneath the epidermis. There are small and infrequent intercellular spaces through which gases such as CO_2 might diffuse. Beneath this layer is a layer of irregularly shaped cells with frequent and relatively enormous intercellular spaces. This tissue, the SPONGY PARENCHYMA, is also photosynthetic. Gases can diffuse readily throughout the spongy

BROWN KELP (Laminaria stenophylla) is a large alga that grows along New England coastline. Size of this "seaweed" is indicated by spectacles at left center. Kelp's holdfast lies at far left. 2

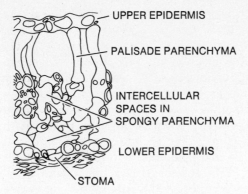

UPPER EPIDERMIS

PALISADE PARENCHYMA

INTERCELLULAR
SPACES IN
SPONGY PARENCHYMA

LOWER EPIDERMIS

STOMA

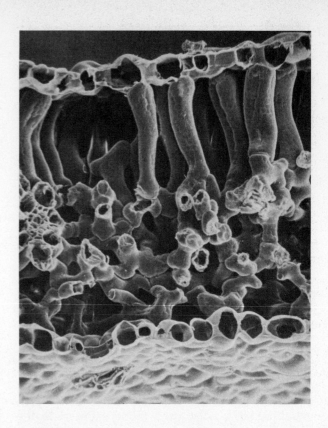

3 CROSS-SECTION OF BEAN LEAF reveals structural compromises that adapt plants to retain water vapor while remaining permeable to the inward diffusion of carbon dioxide. A single stoma is visible at lower left. Gas diffuses readily through empty intercellular spaces. Chloroplasts lie in the parenchymal cells but not in epidermis. Drawing identifies structures in photograph.

parenchyma and the lower surfaces of the palisade parenchyma cells. Since cytoplasmic streaming is occurring in all of these cells, CO_2 that enters a cell through any surface is rapidly mixed throughout the cytoplasm, and in fact can readily be passed to another cell which is not next to any intercellular space. Beneath the spongy parenchyma is the LOWER EPIDERMIS which, like the upper, is non-photosynthetic and secretes a layer of cutin.

The entrance or exit of gases in the seemingly closed system takes place between specialized pairs of epidermal cells called GUARD CELLS, each pair of which is called a STOMA (plural: stomata) (See Figures 5 and 6 in Chapter 15). The guard cells, unlike the other epidermal cells, are photosynthetic. By mechanisms considered in Chapter 16, the facing walls of the guard cells can be drawn apart to form a pore linking the intercellular spaces of the spongy parenchyma with the outside atmosphere. Stomata are normally more frequent in the lower epidermis. The stomata are usually open during the daytime and closed at night, and while open, form a relatively efficient system for gas exchange.

Thus water loss can be regulated under conditions in which adequate CO_2 can still reach the photosynthesizing cells.

To be a functional and effective organ for the whole plant the leaf must have tissue through which it can obtain water and essential ions, and it must also have tissue for export of fixed carbon to parts of the plant in which photosynthesis is either marginal or completely absent (as in roots). Thus a finely branched system of vascular tissue extends throughout the leaf. The veins, the main branches of the system, are obvious, but the system is sufficiently elaborate that no photosynthesizing cell is more than a few cells removed from at least one of the smaller branches (Figure 4). The mechanisms

4 PATTERN OF VEINS in two types of leaf. Linden (left) has netlike pattern. Solomon's seal (Smilacaena) at right shows parallel veins. Veins are comprised of strands of xylem and phloem.

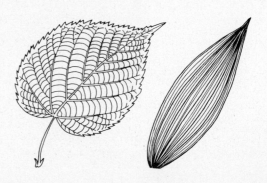

by which water and ions reach the leaf through the xylem, and by which photosynthetic products are exported, sometimes over enormous distances through the phloem, are considered in Chapter 16.

Needless to say, not all plants grow in a reasonably humid and temperate environment, and a wide variety of structural adaptations can be found if one simply explores more extreme habitats. Plants growing in desert regions have evolved a variety of ways to conserve water while still allowing a reasonable level of photosynthesis to take place. They may reduce the number of stomata per unit surface area and greatly increase the thickness of the layer of cutin. They may produce from the epidermis dense layers of hairs which in effect create a layer of still air from which water vapor is not so rapidly swept away by air currents (*Figure 5*). There is evidence that the dense mat of hairs on some desert plants actually reflects back over half of the incoming solar radiation, preventing overheating to lethal temperatures. Plants may assume

shapes with minimal surface-volume ratios, another method of conserving water. The cacti are the best examples of this sort of adaptation (*Figure 6*). Leaves are reduced to nonfunctional thorns and photosynthesis is carried out by subsurface layers of cells in very large and fleshy stems. Obviously any adaptation to reduce water loss also reduces the possible entry of CO_2. Thus it is not surprising that such desert plants commonly show a very small increase in biomass per year in contrast to their counterparts in moister regions.

As mentioned in Chapter 6, there are also important biochemical adaptations against hot arid conditions. C_4 photosynthesis, using the enzyme phosphoenolpyruvate carboxylase, can fix CO_2 into four-carbon acids at vanishingly small CO_2 concentrations —for instance at midday when partially closed stomata and an extremely high rate of photosynthesis have reduced the CO_2 in the intercellular spaces to a tenth of its atmospheric concentration. Also found in some plants is dark fixa-

LEAF HAIRS create a layer of still air over the epidermis. In this scanning electron micrograph, the epidermis is visible through the hairs. The hairs may, however, completely cover the surface.

5

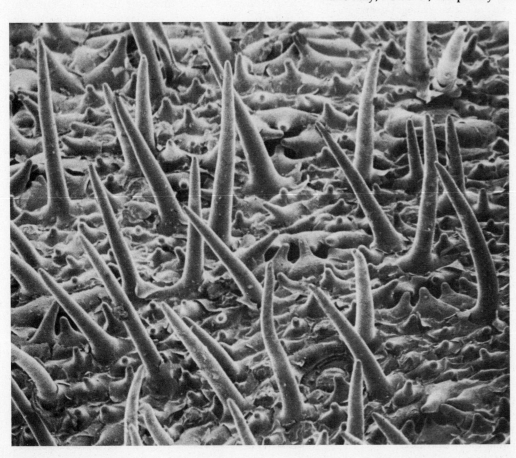

6 BARREL CACTUS (Echinocactus visnaga) grows in Mexico and shows extreme adaptations to prevent water loss. Leaves are reduced to thorns and photosynthesis takes place within the fleshy stem.

tion of CO_2 at night into four-carbon acids while the stomata are open, and then its release and re-fixation through the Calvin cycle during the day-time while the stomata are closed.

At the other extreme, in the humid tropics one finds quite different adaptations. Since potential water loss is not a severe problem, anything providing an increase in surface-volume relationships is usually advantageous. Thus one may find enormous leaves —those of *Raphia*, the Amazon bamboo palm may be 20 meters long —and frequently leaves that are very finely dissected *(Figure 7)*. A very thin layer of cutin and a very large number of stomata per unit of leaf surface area allow extensive gas exchange. Thus limitation in water supply or morphological adaptations maximizing its loss do not apply significant selection pressure in such environments. The advan-

7 LEAF OF TROPICAL TREE FERN is thin and elaborately branched, adaptations that greatly increase its surface-to-volume ratio and enhance gas exchange. Such leaves are common in moist tropics.

tages of rapid growth in a dense tropical rain forest are obvious, and would favor adaptations that would endanger a plant growing in a dry climate.

A number of terrestrial plants have evolved so that they can live partially or even completely submerged in fresh water, and the leaves of these plants are appropriately specialized. For a sub-merged leaf, water, ions, and CO_2 are as accessible as for an alga. There is no cutin, the cells are bathed in the medium containing their elemental needs, and specialized tissue for water transport is superfluous and normally reduced. One frequently finds submerged leaves exquisitely branched and dissected, as surface-volume ratios are maximized. Many plants have both submerged and aerial leaves *(Figure 8)* that are hardly recognizable as belonging to the same plant family or genus, let alone the same individual. The aquatic leaves are elegant examples of adaptation to the fresh water aquatic environment, while the aerial leaves are very simple in shape, cutinized, possess stomata, and resemble the "typical" leaves described for temperate regions. Some plants, for example *Nymphaea*, a fragrant water lily, split the difference with floating leaves. The upper surfaces of these leaves are cutinized and possess stomata, while the lower surfaces have neither cutin nor

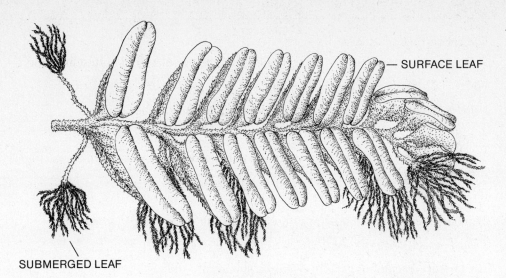

— SURFACE LEAF

SUBMERGED LEAF

8

TWO TYPES OF LEAF appear on the aquatic fern *Salvinia*. **Submerged leaves are finely branched filaments. Flat surface leaves are covered with cutin and have stomata on their upper surfaces.**

stomata. There are large intercellular spaces in the spongy parenchyma and some of the oxygen produced by photosynthesis is driven out of solution, forming pockets of gas which provide flotation.

The acquisition of carbon in plants is closely linked to serious problems of water conservation. Structural adaptations compromise between two conflicting requirements: plants receive carbon as the gas, CO_2, but they must regulate water loss as water vapor. The physiological machinery by which fine tuning is imposed on this gas exchange system during short term changes in environmental conditions will be considered in Chapter 15.

CHEMOSYNTHESIS

Of all of the various forms of free energy available to the biosphere, living organisms can convert only light or chemical energy into a biologically useful form. The photosynthetic autotrophs dominate the biosphere both in terms of biomass and in terms of the tremendous energy conversion. There are, however, a large number of bacteria, relatively small in biomass, which can extract usable energy from simple inorganic substances *(Box A)*. These are the chemosynthetic autotrophs.

The list of reactions in Box A is by no means exhaustive. The purpose of noting all of these wildly different equations for energy production is to emphasize one of the most unifying facts about the energy of life. In every case the free energy of an electron moving from an electronegative to an electropositive environment is trapped and used to provide the chemical bond energy of ATP, and to

provide reducing power. Thus the final outcome of the reactions is the same as photosynthesis, the only difference being the origin of the electron.

Although these chemosynthetic bacteria are autotrophs, they are really dependent on other organisms. For example, the two so-called nitrifying bacteria, *Nitrosomonas* and *Nitrobacter*, depend upon a supply of reduced nitrogen for their chemosynthetic activities. These bacteria are absolutely dependent first upon the nitrogen fixers considered below, second upon other autotrophs and heterotrophs which incorporate the fixed nitrogen, and third upon the many organisms which break down the organic matter of other dead or dying organisms into relatively simple organic compounds, and finally upon certain ammonifying bacteria and fungi which deaminate amino acids to liberate ammonia (as ammonium ion) into the environment. Thus the nitrifying bacteria are a critical link in the whole nitrogen cycle, which is discussed later in the chapter.

A similar point can be made with the bacterium *Hydrogenomonas*. Molecular hydrogen is a pretty rare commodity in nature and one is justified in asking where this chemosynthetic autotroph obtains it. As it turns out, other microorganisms are once again the source. Some of the photosynthetic bacteria and unicellular flagellates have the genetic information to construct an appropriate enzyme, a hydrogenase. If sufficiently electronegative electrons are available, as they are from the photo-

A Chemosynthetic bacteria

The bacterium *Nitrosomonas* obtains energy by the oxidation of ammonium ion:

$NH_4^+ + {}^3/_2 O_2 \rightarrow NO_2^- + H_2O$ + energy.

Nitrobacter can then oxidize the nitrite ion to nitrate to obtain energy:

$NO_2^- + \frac{1}{2} O_2 \rightarrow NO_3^-$ + energy.

The bacterium *Hydrogenomonas* can use energy from the oxidation of molecular hydrogen:

$H_2 + \frac{1}{2} O_2 \rightarrow H_2O$ + energy.

The bacterium *Ferrobacillus* traps energy from the oxidation of ferrous ions:

$2 Fe^{++} + 2H^+ + \frac{1}{2} O_2 \rightarrow 2 Fe^{+++} + H_2O$ + energy.

An obscure bacterium called *Leptothrix* apparently utilizes the oxidation of manganous oxide:

$2 MnO + \frac{1}{2} O_2 \rightarrow Mn_2O_3$ + energy.

In an oxidizing environment, the bacterium *Beggiatoa* can oxidize sulfide ions through elemental sulfur to give sulfate ions:

$S^{--} + 2 O_2 \rightarrow SO_4^{--} + 4 H_2O$ + energy.

If an oxidized product such as sulfate finds its way into a reducing environment that is rich in molecular hydrogen or organic compounds, free energy may be obtained by reducing the sulfate. Such a reaction is the staff of life for the bacterium *Desulfovibrio*.

$SO_4^{--} + 4 H_2 \rightarrow S^{--} + 4 H_2O$ + energy.

---

synthetic pigment systems, the hydrogenase enzyme can use the electron to reduce hydrogen ion:

$$2 H^+ + 2 \text{ electrons} \rightarrow H_2 .$$

Since the process is throwing away reducing potential, it is difficult to conceive how it evolved. One suggestion is that in the nitrogen fixing organisms (*see below*) the process may operate in the reverse direction to scavenge hydrogen produced as a byproduct of the fixation reaction. The evidence, however, is scant.

There are very few nitrogen fixers but their impact on other organisms should not be underestimated; without them no other known organism could survive. . . . About 90 million tons of atmospheric nitrogen per year are fixed by microorganisms.

In considering autotrophy one might make the same distinction between versatile and exacting autotrophs. The exacting autotrophs depend ultimately upon the activities of other organisms to provide their source of oxidizable substrate for energy extraction. The versatile autotrophs have no such limitation: they may require only CO_2, H_2O, essential inorganic elements in soluble form, and nitrogen gas. Though some are aerobic and require that oxygen be replenished by photosynthesizing organisms, some are in fact anaerobic and do splendidly without oxygen. Such organisms are in the majority among the nitrogen fixers.

NITROGEN FIXATION

As was mentioned earlier, nitrogen gas, $N \equiv N$, is extremely unreactive. The covalent triple bond linking the two atoms must be broken to make nitrogen of any use whatsoever to organisms, and most organisms cannot perform the reaction. Fortunately there are a select few which can: the nitrogen fixers. All are procaryotic, although some of

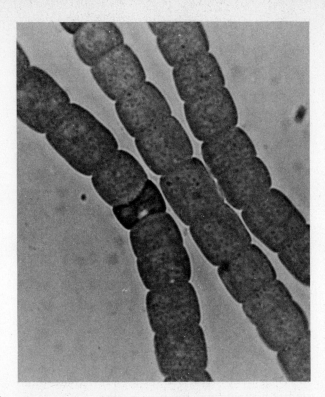

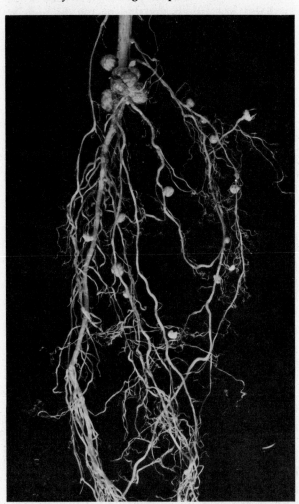

The early Romans, Greeks, and Chinese, and probably many other early civilizations, were aware that higher plants such as clover, alfalfa, and peas were soil improvers. But a pair of German chemists, Hellriegel and Wilfarth, first showed in 1888 that the root nodules which one found on these plants were caused by bacteria, and were in fact sites of nitrogen fixation. These bacteria all belong to the genus *Rhizobium*, and show a fairly high specificity for the particular legume which they nodulate *(Figure 10)*. At almost the same time, the free-living, anaerobic, nitrogen-fixing bacterium *Clostridium* was isolated. Workers later discovered the nitrogen-fixing properties of bluegreen algae and the photosynthetic bacteria in turn, and in the late 1950s and early 1960s another group of higher plants with nitrogen-fixing root

ROOT NODULES are abundant on this 28-day-old soybean plant. Plant was grown on nitrogen-free nutrient solution and had to rely on fixation by nodules for virtually all its nitrogen requirements. 10

9 BLUEGREEN ALGA Fremyella diplosiphon consists of microscopic filaments composed of many cells. Dark cell at left is dying. Filament will break at that point and new cells will proliferate from it.

them must live in intimate association with a particular eucaryotic plant before they develop functional nitrogen-fixing machinery. There are very few nitrogen fixers and what few there are have a small relative biomass. Their impact on other organisms, however, should not be underestimated: without the nitrogen fixers, no other known organism would survive. This small group of procaryotes is in its own way just as essential in the biosphere as the photosynthetic autotrophs.

The list of nitrogen fixers in nature is a relatively simple one. It includes the photosynthetic bacteria; members of at least nine genera of bluegreen algae *(Figure 9)*, the best known of which are *Nostoc* and *Anabaena;* aerobic heterotrophic bacteria such as *Azotobacter;* anaerobic heterotrophic bacteria such as *Clostridium;* and a group of microorganisms which fix nitrogen only in close association with the roots of higher plants, the best known of which belong to the bacterial genus *Rhizobium.* This latter group is found free-living in the soil (where it does not fix nitrogen) or in nodules of the roots of plants from the family Leguminosae, including peas, beans, alfalfa, and a large number of shrubs and trees *(Figure 10)*.

nodules was found. These plants, all unrelated to the legumes, and for that matter mostly unrelated to each other, are all shrubs or trees. At this writing, the nodule-inducing microorganism has just been identified as a member of a group of filamentous bacteria called ACTINOMYCETES *(Chapter 25).*

There is now reasonable agreement that the biological fixation of nitrogen is a progressive reduction of $N \equiv N$ (N_2) through $NH = NH$, $NH_2 - NH_2$, to $2 NH_3$. All nitrogen-fixing systems studied to date require the same components: a strong reductant, for instance ferredoxin; nitrogenase, a complex enzyme using both iron and molybdenum; and ATP. The primary source of electrons for the reductant and of ATP may be photosynthetic or respiratory. The only unique ingredient is the nitrogenase. Fortunately, the nitrogenase is not as rare in nature as the hydrogenase mentioned earlier in this chapter. Evidently all of the reductive steps occur while nitrogen is firmly bound to the enzyme, since first, the partially reduced intermediates are either unstable or highly toxic or both, and second, a heavy isotope of nitrogen, ^{15}N, only appears in the ammonia if it is fed as N_2, never if it is fed as any of the proposed intermediates. The nitrogenase will only function under anaerobic conditions, and is rapidly inactivated by atmospheric oxygen.

It is estimated that approximately 90,000,000 tons of atmospheric nitrogen per year are fixed by biological means. A very small amount of nitrogen may be fixed in the atmosphere by a variety of cataclysms such as lightning, volcanic eruption, and forest fire; the ammonia formed is brought down by rain-water. But the total is trivial compared with what fixers produce. In the oceans, nitrogen is fixed by the photosynthetic bacteria and bluegreen algae, with the latter playing the same role in fresh water. On land, the free-living soil bacteria make some contribution, but it is the root nodules of higher plants that produce most of the fixed nitrogen. Unlike the various free-living procaryotic nitrogen fixers which fix what they need for their own uses and release the fixed nitrogen only upon their death, the root nodules may actually excrete some amino acids into the soil, making the nitrogen immediately available to other organisms.

The pioneering role of some of those nonleguminous plants with nodules is an extremely important one. Shrubs such as alder *(Alnus)* thrive in mountainous areas with their roots grasping chunks of rock in the talus slopes below cliffs. The western mountain lilac *Ceanothus* grows well in extremely gravelly soils where little or no organic matter is present (and hence no fixed nitrogen). The eastern sweet gale *(Myrica)* flourishes on almost pure sand. These plants (with their bacterial partners) are excellent nitrogen fixers, and clearly play a vital role in making otherwise barren habitats available to other plants and the animals that depend on them.

The nodule of the legume represents an excellent example of SYMBIOSIS, a situation in which two different organisms live in close association for their own mutual benefit and in fact do things that neither organism does separately. Neither free-living *Rhizobium* nor an uninfected legume can fix nitrogen. Only when the two are in the close association found in the root nodule does the reaction take place. (The bacteria will fix nitrogen alone, but under very specialized culture conditions.)

The bacterium normally starts the infection process through a root hair, an epidermal outgrowth of very young root tissue *(Figure 11)*. An invagination is formed, and the invagination proceeds inward passing through several cells, while the bacterium divides rather slowly. The bacteria at this stage are still topologically outside the plant which has sheathed the in-growing thread with cellulose and other typical cell wall materials. Surrounding cells begin fairly rapid proliferation. Finally a vesicle forms on the thread, bursts, and releases the bacteria into the cytoplasm of host cells. Then the bacteria undergo a remarkable transformation. They increase about tenfold in size, develop an outside membranous envelope, and form an elaborately folded internal membrane. At this stage the bacterial structures are called bacteroids. Most remarkable is the final event before fixation of nitrogen: the bacteroid becomes surrounded with hemoglobin, an oxygen-carrying pigment that one hardly associates with plants. Some nodules normally contain enough hemoglobin to be bright pink in cross section. The role of hemoglobin is obscure, but it may simply act as an oxygen trap to keep the bacteroids anaerobic. Isolated bacteroids *(Figure 12)* will fix nitrogen only when they are kept anaerobic, since their nitrogenase shows the same oxygen sensitivity as that from other nitrogen-fixing systems.

Any discussion of nitrogen fixation would be misleading and incomplete without mention of the opposite process —denitrification. There are a number of normally aerobic bacteria, mostly species of *Bacillus* and *Pseudomonas*, which can use

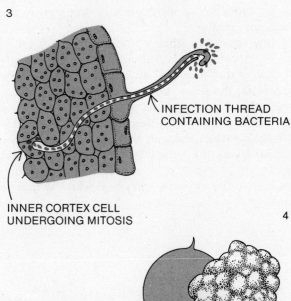

1

← SOIL PARTICLES

ROOT HAIR

← RHIZOBIA
BACTERIA

CORTEX CELL

2

BACTERIA

3

INFECTION THREAD
CONTAINING BACTERIA

INNER CORTEX CELL
UNDERGOING MITOSIS

4

FORMATION OF A NODULE begins when root hairs release a substance that stimulates Rhizobium bacteria to divide (1). Bacteria multiply (2), enter the root hairs and invade the cortex (3), stimulating the cortical cells to divide rapidly. Schematic cross-section of root (4) shows mature nodule formed by proliferation of cortex cells.

HETEROTROPHS

All animals and parasites, as well as carnivorous and saprophytic plants, are heterotrophs — organisms that depend on organic materials built by others. The use of prefabricated goods is obviously advantageous because it frees an organism from the need to synthesize everything itself from simple precursors; but this strategy also has some inevitable consequences. Complex materials do not usually diffuse spontaneously to an organism, nor do they present themselves on demand. They must be hunted, harvested, or otherwise sought, and this requires adaptive specialization, as well as

BACTEROIDS appear as dark structures embedded in **12** the cytoplasm of this cell from a soybean nodule. Nucleus fills lower third of picture. Light blobs within bacteroids are stored food.

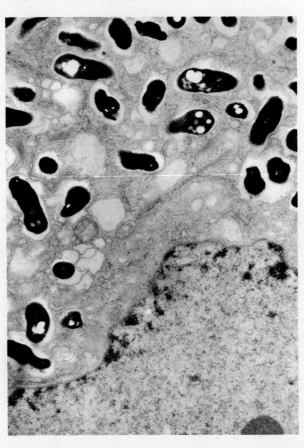

nitrate as a terminal electron acceptor if they are under anaerobic conditions:

$$2\ NO_3^- + 10\ e^- + 12\ H^+ \rightarrow N_2 + 6\ H_2O.$$

These bacteria are extremely common and return nitrogen to the atmosphere as the inert gas that it originally was. Without these organisms, atmospheric nitrogen would steadily decrease as fixation, both biological and industrial, continues on a global scale.

expenditure of time, energy, or both. Moreover, complex materials are usually not utilizable as such, since they are constituted largely of macromolecules that need to be reduced to smaller size before they can be absorbed. Heterotrophs depend on digestion (which is always effected by enzymes) to reduce macromolecules in size.

FEEDING AND DIGESTION IN ANIMALS

Among animals, heterotrophic nutrition is illustrated in its simplest form by those parasites that live literally suspended in nutritious broth. The food of tapeworms *(Figure 13)* for example, which live in the intestine of their host, is predigested, soluble, and available, and can be absorbed without being enzymatically pretreated by the parasite itself. These animals need (and have) no mouth, gut or other feeding devices. Their only problem is to get to the host in the first place, a matter that we shall not consider here.

Free-living animals must not only locate their food and obtain enough of it, but must also provide for its subsequent digestion and absorption. How they do this differs enormously in the various groups, although the principles involved are very much the same in all cases.

13 TAPEWORM is one of the simplest heterotrophs. Head (left) anchors the worm to the intestinal wall of its host by means of hooks, suckers, or both. Photograph at right depicts a few of the many body segments. Tapeworm has no trace of a digestive system. Each segment of body is little more than an envelope of reproductive organs. Worm can grow to several feet long.

Most food is ingested as macromolecules, which poses two problems: the giant molecules are too large to be absorbed by the animal, and their composition is very seldom (except in cannibalism) precisely right for the feeder. Both problems are solved by DIGESTING the macromolecules, that is, by reducing them to small diffusible molecules such as amino acids, fatty acids, monosaccharides and nucleotides; these can later be assembled into macromolecules of the appropriate composition.

Digestion is accomplished by hydrolytic enzymes, which split macromolecules by the addition of water across a chemical bond *(Chapter 5)*. The reader will recall that enzymes are classified according to the substances that they hydrolyze: carbohydrases hydrolyze carbohydrates, proteases hydrolyze proteins, lipases hydrolyze fats, and so on. The prefixes exo- and endo- indicate the site of cleavage. Thus an endoprotease (more commonly called an endopeptidase) hydrolyzes a protein at an internal site along the polypeptide chain, while an exoprotease (or exopeptidase) snips away the terminal residues.

How can an organism produce enzymes to digest such compounds without digesting itself? The answer is that digestion always occurs OUTSIDE the organism; the term "outside" is used in the same sense that a pebble clenched by a hand is outside the hand, rather than being a part of it. The gut is simply a tunnel through the animal; food in the central space, the lumen, is outside the body, and hence can be exposed to extreme conditions (such as high acidity or potent enzymes) which would be intolerable within a cell.

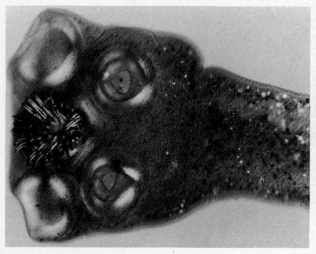
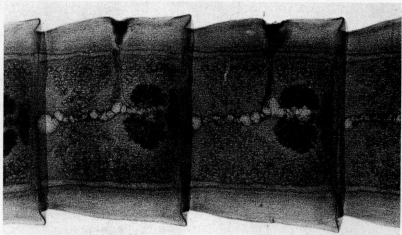

Ⓥ Ⓐ Ⓐ Ⓐ Ⓐ Ⓛ Ⓘ Ⓥ Ⓖ

TRYPSINOGEN
(INACTIVE)

ENTEROKINASE

ACTIVE
CENTER — Ⓘ Ⓥ Ⓖ / Ⓢ

TRYPSIN
(ACTIVE)

14 **CONVERSION OF A ZYMOGEN to an active enzyme.
The hexapeptide (black circles at top) masks the
essential IVG sequence from S, preventing the
formation of the active site. Hexapeptide is linked by
hydrogen or electrostatic bonds to a distant region of
the peptide chain. In this example enterokinase
removes the mask, converting inactive trypsinogen to
active trypsin.**

Many digestive enzymes become active only
when they reach the outside. Within the cell they
are produced in an inactive form, known as a
ZYMOGEN. When secreted into the lumen of the gut,
the zymogens are activated, sometimes by exposure to a different pH, but more often by the action
of another enzyme. Trypsin, for example, is produced by the vertebrate pancreas as inactive trypsinogen. Once it passes into the small intestine it
encounters another enzyme called enterokinase,
which is secreted by the wall of the intestine. Enterokinase converts trypsinogen to trypsin, the active protease, by removing its protective "mask" of
six amino acids *(Figure 14)*.

Special cells in the digestive tracts of many animals secrete mucus, a slimy material that lubricates
the intestinal surface and protects it from abrasion.
Mucus also shields the gut lining from being digested by the activated enzymes. Insects rely on a
different trick: within the gut they secrete a thin
tube of chitin that encloses the food and enzymes
and protects the lining against abrasion and self-
digestion.

There are many other features common to
digestive-absorptive systems. Some chopping and
churning of the food may be needed to mix it with
digestive enzymes; some device will be needed to

conserve the enzymes (by putting them with the
food in a vacuole, or a blind sac, or a tubular gut);
and suitable conditions of acidity and ionic composition are needed for optimal digestion.

Nonparasitic protozoa display a somewhat deceptively simple digestive-absorptive system. For
example, an amoeba engulfs pieces of food by embracing it with protoplasmic extensions.
Paramecium washes bacteria and other particulate
matter down its oral groove (a mouth-like region)
by creating a current with the cilia that line it.
Under pressure from the current, food and fluid
are pushed into the cytoplasm at the base of the
groove, and a food vacuole is formed. The vacuole
breaks away and circulates around the cell,
dwindling in size until it is finally ejected at a spot
called the anal pore *(Figure 15)*. As the vacuole circulates, enzymes are secreted into it and digested
products are absorbed from it. Since the vacuole is
essentially an intracellular device, such digestion is
said to be intracellular.

FOOD VACUOLE OF PARAMECIUM forms at the **15**
**base of the oral groove. Vacuole circulates through the
organism, growing gradually smaller as bacteria within
it are digested and absorbed. Bacteria are intact at first
(A), but have virtually disappeared (B) by the time that
vacuole is ready for expulsion through the anal pore.**

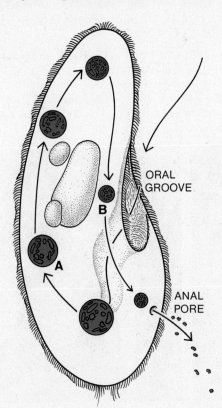

ORAL
GROOVE

B

A

ANAL
PORE

The coelenterates, of which *Hydra* is an example, also have a simple system. The gastrovascular cavity, the single cavity in the coelenterate's body, functions as the digestive and circulatory systems. *Hydra* are carnivores and sting their prey with poisoned daggers called nematocysts, many of which are scattered along the tentacles. The paralyzed prey is then thrust into the gastrovascular cavity where it is subjected first to extracellular digestion by enzymes secreted by gland cells in the cavity's lining. After partial digestion, phagocytic cells ingest the fragments and complete the digestive process by forming a digestive vacuole, in the same way as protozoa *(Figure 16)*.

In higher animals we find what can truly be called a gut, a tubular structure with a mouth at one end for ingestion and an anus at the other for ejection. This use of a tube instead of a vacuole or sac offers sequential treatment of food as one of many advantages. Food can be crushed in the first part, treated with an enzyme at alkaline pH, then passed to an acid area where different enzymes digest out proteins, for instance, and thus help to reduce the particles to diffusible forms that the body can absorb. Before expelling the undigestible material as feces, the body can recover salts and water which had to be added during processing, but which are too important to waste.

There are other features common to all guts. Because they all must absorb digested matter, and

Five kinds of feeder have been recognized among animals, each with its own way of acquiring and handling food. Omnivores and herbivores together comprise the largest group; the others are carnivores, deposit feeders, filter feeders and fluid feeders.

because absorption occurs at the surface of the gut's lumen, many animals increase the surface area of the lumen. The earthworm, for instance, developed a dorsal infolding or TYPHLOSOLE and the shark developed a spiral valve. It is also common for the wall of the gut to be folded many times, with the individual folds bearing many tiny finger-like projections called villi. The villi may in turn be subfolded on their own surface into microvilli *(Figure 17)*.

Guts frequently contain microorganisms that perform functions that the host may be incapable of performing itself. The leech *Hirudo* produces no proteases to digest the proteins in the blood that it sucks from vertebrates. The enzymes are produced in its gut by a bacterium which is found nowhere else. Many animals obtain vitamins from bacteria in their intestines, and herbivores often depend on microorganisms in various parts of their guts for digestion of cellulose.

The gut usually consists of a series of compartments, with a mouth and buccal cavity (a sort of hallway) at one end, and an anus at the other. The mouth may bear mechanical devices, such as the teeth of vertebrates and the mandibles of insects, for grasping or shredding the food; or there may be a special organ farther back, such as the gizzards of birds and some invertebrates, where grinding occurs. The grinding is enhanced by small stones that become lodged in the gizzard. Food and stones are churned by massive gizzard muscles. Other animals simply ingest food that needs no fragmentation.

The importance, relative size, and distinctness of the gut compartments varies considerably *(Figure 18)*. The stomach or crop functions principally as a storage chamber which enables the animal to ingest food quickly and to digest it at leisure. Intermittent rather than continuous feeding frees the animal to devote more time to other activities such

16   **HYDRA CAPTURES DAPHNIA and paralyzes it with stinging tentacles (left). Crustacean is then stuffed into the hydra's gastrovascular cavity (right). Cells lining cavity secrete digestive enzymes and absorb digested food. In addition, the cells engulf food vacuoles for intracellular digestion.**

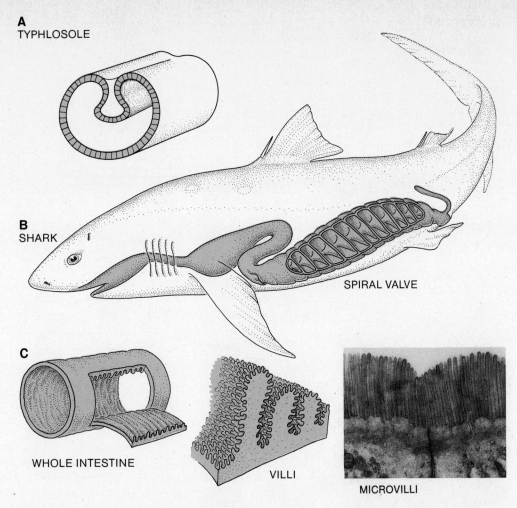

A
TYPHLOSOLE

B
SHARK

SPIRAL VALVE

C
WHOLE INTESTINE

VILLI

MICROVILLI

SURFACE AREA OF GUT is increased by a variety of strategies. Gut of earthworm (A) is infolded along its dorsal length, forming structure called the typhlosole. Intestine of shark contains spiral fold that forces food to travel a longer path, exposing it to an increased absorptive surface. In most vertebrates, including man, intestines are internally folded (C) and the folds are covered with tiny fingerlike projections called villi. Surfaces of villi in turn are folded into microvilli. 17

as courtship and the construction of shelter. Digestion and absorption may or may not occur in the storage chamber, depending on the species. If this compartment is acidic, as it frequently is, it is supplemented by a midgut or intestine. Food delivered into the intestine is well minced and mixed. This zone is usually of neutral pH, and most digestion and absorption occurs here. Enzymes derived from specialized glands (such as the pancreas of vertebrates) are commonly poured into the intestine, and the gut wall secretes other digestive enzymes. The hindgut, which often includes a muscular rectum, recovers water and salts. These materials are especially important to terrestrial animals. The "farms" of microorganisms described above are located in the hindgut of man and many other animals. Sometimes there is a special intestinal diverticulum, or caecum, that houses the mi croorganisms, as for example in rodents.

Although all guts have common features, ther are distinct differences from one type of animal t

another, and the differences depend less on the taxonomic category of the organism than on the kind of food that its gut is called upon to process. Five kinds of feeder have been recognized, each with its own way of acquiring and handling food. OMNIVORES (eaters of everything) and HERBIVORES (plant-eaters) together comprise the largest group feeders. Deer and grasshoppers are both herbi res; men, pigs and cockroaches are all omni es. CARNIVORES (flesh-eaters) comprise another e group, which includes species as diverse as cheetah and the hydra. DEPOSIT FEEDERS take in t quantities of mud or similar deposits and

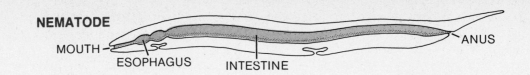

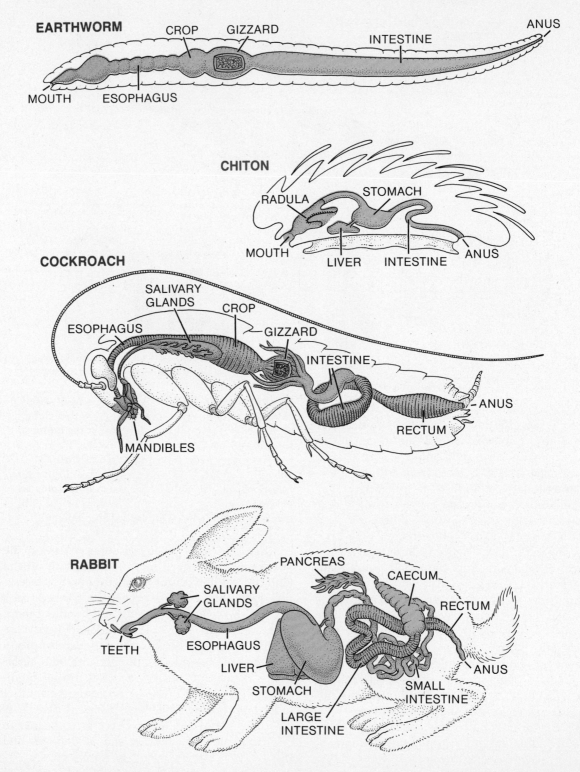

DIGESTIVE TRACTS OF ANIMALS. Gut of nematode worm, a fluid feeder, is little more than an unspecialized tube. Gut of earthworm contains a crop for storage and a pebble-filled gizzard for grinding soil, decayed plant matter and bodies of small animals. Gut of chiton includes a rasping tongue (radula) and a large stomach. Chiton feeds on algae. Cockroach chews food with mandibles, mixes it with saliva, and passes it first into crop, then through grinding gizzard into intestine for digestion and absorption. As in all terrestrial animals, roach's rectum is specialized for reabsorption of water. Gut of rabbit is typical of vertebrates, well equipped with digestive glands, and has a large caecum for farming symbiotic microorganisms.

utilize only the small fraction of organic matter present in it, which may be only four percent. The most familiar examples are the earthworm on land and lugworms and sea-cucumbers in the sea. FILTER FEEDERS, which strain out particles or small organisms suspended in water, include most of the stationary aquatic animals such as sponges, corals, barnacles and bivalve mollusks. FLUID FEEDERS suck up a variety of liquids. Aphids suck the sap of plants; mosquitos, leeches and vampire bats suck blood; and spiders suck the digested contents of insect bodies.

OMNIVORES

The human digestive tract (*Figure 19*) is an example of the gut of an omnivore. Food taken into the mouth is chewed into smaller, more manageable pieces. Human teeth are specialized for various functions: incisors for biting, canines for tearing, and molars and premolars for crushing and grinding. As the chewing proceeds, the tongue manipulates the food into a mass called a BOLUS, mixing it with the secretion that flows into the mouth from three pairs of salivary glands. Saliva contains only one type of digestive enzyme, a carbohydrase called AMYLASE, which initiates the hydrolysis of starch. Saliva also contains a certain amount of mucus, which helps to compact the food into the bolus that is eventually swallowed.

Swallowing is a partly involuntary act. It is initiated voluntarily with the tongue, which is pressed upward against the palate, forcing the food backward through the pharynx into the esophagus. Involuntary contractions of muscles along the middle and lower reaches of the esophagus then take over; they proceed as propagated (peristaltic) waves along the length of the tube, pushing the food downward into the stomach. Swallowing is accompanied by an au-

tomatic closure of the GLOTTIS, the opening through which inhaled air (which must also make its way through the pharynx) ordinarily enters the trachea. As anyone knows who has choked while eating, food does occasionally slip into the trachea when this closure mechanism fails.

The stomach is an altogether remarkable organ. With a capacity of two to four liters, it can accommodate and temporarily store all of a heavy, quickly devoured meal. The major enzyme produced by the stomach is an endopeptidase called PEPSIN. This is secreted as a zymogen (PEPSINOGEN) by special gland cells in the stomach lining. Other gland cells in the lining produce hydrochloric acid, and still others secrete mucus. The hydrochloric acid, which at its normal concentration of 0.5 percent maintains the stomach fluid (the gastic juice) at pH

HUMAN DIGESTIVE TRACT is suited for an omnivorous diet. Chewing reduces food to small bits and increases its surface area for attack by enzymes. Enzymatic digestion begins in the mouth and continues in the stomach and intestines. Absorption begins in the jejunum (small intestine) and continues in the large intestine and rectum.

19

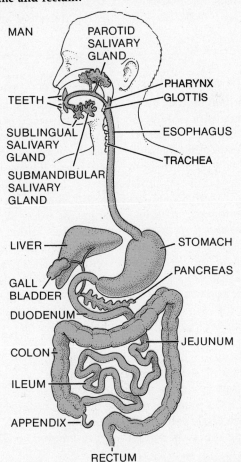

MAN
PAROTID SALIVARY GLAND
PHARYNX
TEETH
GLOTTIS
SUBLINGUAL SALIVARY GLAND
ESOPHAGUS
TRACHEA
SUBMANDIBULAR SALIVARY GLAND
LIVER
STOMACH
PANCREAS
GALL BLADDER
DUODENUM
COLON
JEJUNUM
ILEUM
APPENDIX
RECTUM

1–3, is essential for two reasons. First, it triggers the activation of pepsin —that is, the conversion of pepsinogen to pepsin. Actually, hydrochloric acid merely initiates this activation. Once there is some active pepsin present in the gastric juice, the pepsin itself promotes the conversion of the remaining pepsinogen to pepsin. Second, it provides optimal conditions for pepsin, an enzyme that effects its proteolysis best in an acid medium. The mucus serves the usual lubricating role, and protects the stomach lining from self-digestion by pepsin and from the eroding effect of the hydrochloric acid. The bulk of the stomach wall is composed of powerful compressor muscles. After a meal they become increasingly active, and their periodic contractions churn the food and keep it properly mixed with the gastric juice. Very little of the meal is absorbed in the stomach. Even the few substances that are absorbed, such as alcohol and aspirin, are absorbed in relatively small quantities.

The major events of digestion and absorption take place after the food passes from the stomach into the small intestine. The stomach empties itself gradually, over a period of about four hours. This enables the small intestine to work on a little material at a time, and prolongs the digestive and absorptive processes throughout most of the period between meals. The small intestine, so named because of its diameter, is actually an impressively large organ. More than six meters long, its coils fill much of the lower abdominal cavity. As a consequence of its length, and because of the folds, villi and microvilli of its lining, its inner surface is enormous: an estimated 550 square meters. This area attests to the very impressive chemical traffic, both secretory and absorptive, that must pass across its surface.

The small intestine consists of three sections. The short initial section, the DUODENUM, is where most digestion occurs. Two additional sections, the JEJUNUM and the ILEUM, carry out 90 percent of the absorption of nutrients. Virtually every conceivable digestive process occurs in the small intestine. It contains proteases, lipases, carbohydrases and nucleases. Not all of the enzymes and other glandular products present in the small intestine are secreted by the intestinal wall. The pancreas, a large gland beneath the stomach, empties its products into the duodenum via the pancreatic duct. Bile, a complex fluid produced by the liver and stored in the gall bladder, also empties into the duodenum through a special duct.

B Early experiments on digestion

Excellent studies on the digestive process have been made for hundreds of years. Spallanzani in 1780 obtained gastric juice from live owls by putting sponges down their throats and squeezing out the juice after the sponges had been thrown up. He observed the digestion of calves' intestines by the juices. His quantitative approach and careful controls are noteworthy: "Forty-six grains were immersed in some recent gastric fluid; and at the same time an equal quantity of the same intestine was put in a phial exactly like the former, and an equal quantity of water was poured upon it. . . ." He showed in this way that gastric juices digested proteins but not carbohydrates.

He also puts tubes containing meat and cereals and vegetables into the stomach of owls, hawks, and eagles and showed that only the meat was digested. He also showed that no digestion occurred in the crop or craw as it was called then. For these last experiments, the eagle posed some special problems. "Had this bird been of the same gentle and peaceful disposition as gallinaceous fowls . . . I should have had only to press the portion of the flesh that lay highest in the craw upwards with thumb and forefinger, and . . . should have brought it out at the mouth . . . but the strength and ferociousness of the eagle altered the case totally. After much reflection, I thought of an artifice. . . . I gave my eagle only three or four pieces of flesh, of which the last was tied in the shape of a cross with a fine pack-thread three or four feet long. The eagle, pressed by hunger, devoured the flesh greedily without regarding the string, of which the greater part hung out of its mouth. . . . When I thought it time to examine the piece of flesh I pulled the string forcibly, and the eagle, without growing enraged, opened its beak and allowed me more room."

Protein digestion in the small intestine occurs through the action of several proteases; the endopeptidases TRYPSIN and CHYMOTRYPSIN are of major importance. Why should there be two endopeptidases in the intestine when there is already one (pepsin) in the stomach? These enzymes are not strictly equal in their action. Each is specialized to split only certain types of linkages in a polypeptide chain; there is really little or no duplication in their activities. Pepsin, for example, splits peptide bonds adjacent to certain aromatic amino acids such as phenylalanine and tyrosine, something that trypsin cannot do. The small intestine also contains exopeptidases which are specific in their action. Carboxypeptidase splits the terminal amino acid from the end of a peptide chain bearing the free carboxy group, while aminopeptidase does the same at the opposite end of a chain where there is the free amino group. Dipeptidases split dipeptides.

While aminopeptidase and dipeptidase are produced by glands in the intestinal wall itself, trypsin, chymotrypsin and carboxypeptidase are supplied by the pancreas. All are initially produced as zymogens and activated only after secretion. The activation of trypsinogen and chymotrypsinogen occurs in an interesting way. Trypsinogen, as already noted, is transformed into active trypsin by enterokinase, an enzyme produced by the duodenum. Chymotrypsinogen, in turn, is converted into chymotrypsin by the newly activated trypsin.

The small intestine completes the digestion of carbohydrates, which was initiated in the mouth by salivary amylase. Additional amylase from the pancreas splits starch into maltose units. Maltose is a disaccharide which can be hydrolyzed into glu-

cose, as noted in Chapter 4. The small intestine produces MALTASE specifically for that purpose. It also produces SUCRASE and LACTASE for the hydrolysis of the common dietary disaccharides sucrose and lactose. Oddly enough, lactase is absent in most non-Caucasians.

Fats are digested by pancreatic LIPASE, which hydrolyzes them into glycerol and fatty acids. Because fats are insoluble in water, their digestion tends to proceed slowly. The process is greatly aided by bile, which contains salts that act as detergents, breaking the fat into tiny droplets. The vastly increased total surface area of the droplets enhances the action of lipase. Bile also serves as the vehicle for excretion of cholesterol and bile pigments. The pigments are derived from the breakdown of red blood cells in the liver, and are responsible for the characteristic brown color of feces. The source and function of the major digestive enzymes of humans are listed in Table I.

The digestive enzymes in the small intestine function most efficiently at an alkaline pH. To achieve such alkalinity, and to overcome the acidity of the stomach contents as they are passed into the duodenum, the wall of the small intestine secretes a considerable amount of bicarbonate. Additional bicarbonate is provided by the bile and pancreatic juice.

The products of digestion, such as amino acids, glucose and fatty acids, pass across the cellular lining of the small intestine and into the blood capillaries and lymphatic vessels that lie just inside the lining. Although such small molecules can cross

MAJOR DIGESTIVE ENZYMES of humans are listed **I** in column at left. Column at right center indicates the substrate and the end product of the enzymatic reaction.

ENZYME	SOURCE	ACTION	SITE OF ACTION
SALIVARY AMYLASE	SALIVARY GLANDS	STARCH → MALTOSE	MOUTH
PEPSIN	STOMACH	PROTEINS → PEPTIDES	STOMACH
PANCREATIC AMYLASE	PANCREAS	STARCH → MALTOSE	SMALL INTESTINE
LIPASE	PANCREAS	FATS → FATTY ACIDS AND GLYCEROL	SMALL INTESTINE
NUCLEASE	PANCREAS	NUCLEIC ACIDS → MONONUCLEOTIDES	SMALL INTESTINE
TRYPSIN	PANCREAS	PROTEINS → PEPTIDES	SMALL INTESTINE
CHYMOTRYPSIN	PANCREAS	PROTEINS → PEPTIDES	SMALL INTESTINE
CARBOXYPEPTIDASE	PANCREAS	PEPTIDES → PEPTIDES AND AMINO ACIDS	SMALL INTESTINE
AMINOPEPTIDASE	SMALL INTESTINE	PEPTIDES → PEPTIDES AND AMINO ACIDS	SMALL INTESTINE
DIPEPTIDASE	SMALL INTESTINE	DIPEPTIDES → AMINO ACIDS	SMALL INTESTINE
ENTEROKINASE	SMALL INTESTINE	TRYPSINOGEN → TRYPSIN	SMALL INTESTINE
NUCLEASE	SMALL INTESTINE	NUCLEIC ACIDS → MONONUCLEOTIDES	SMALL INTESTINE
MALTASE	SMALL INTESTINE	MALTOSE → GLUCOSE	SMALL INTESTINE
LACTASE	SMALL INTESTINE	LACTOSE → GALACTOSE AND GLUCOSE	SMALL INTESTINE
SUCRASE	SMALL INTESTINE	SUCROSE → FRUCTOSE AND GLUCOSE	SMALL INTESTINE

cell barriers passively by diffusion, they are usually absorbed against concentration gradients by the process of active transport *(Chapter 2)*. Active transport, the reader will recall, uses ATP to push materials "uphill" from an area of relatively low concentration (such as the gut lumen) to one of relatively high concentration (such as an epithelial cell in contact with the lumen). Such transport systems may be highly specific or relatively nonspecific. The one that moves glucose, for example, may exclude fructose, while the one that moves amino acids may move several indiscriminately. Once absorbed by the epithelial cells of the gut, monosaccharides and amino acids are passed directly into the bloodstream. Fatty acids and glycerol, however, are resynthesized into fats after absorption, and then enter the lymphatic vessels of the intestinal wall as tiny droplets. The lymph eventually enters the blood, together with the droplets. After a meal heavy in fats, the droplets are so numerous that they give the blood a milky appearance.

Like all parts of the digestive tract, the small intestine is surrounded by muscles. Peristaltic waves of contraction push the food toward the large intestine. Its principal function is to process the material that remains after digestion and absorption have run their course in the small intestine. It creates feces by recovering much of the water that had been secreted into the small intestine, leaving behind a more solid residue. Minerals such as bicarbonate and bile salts are also salvaged by reabsorption here. An immense population of bacteria lives within the large intestine; one of the dominant forms is *E. coli*, which has become prominent in research on molecular biology and genetics *(Chapter 8)*. The bacteria subsist on nutrients indigestible by humans. In a sense they are symbiotic, since they also synthesize several vitamins (for example vitamin K, biotin) which are essential to the human host and are absorbed in the large intestine. Prolonged oral intake of antibiotics may affect these bacteria as well as the pathogens that are the targets of the therapy, and can therefore lead to vitamin deficiencies. The bacteria also produce gases, such as methane and hydrogen sulfide, which are unpleasant byproducts of the final stages of digestion. The large intestine has a small diverticular pouch, the appendix, best known for the trouble it can bring when it becomes infected. It appears to play no essential role, and the surgical removal that is performed following infection is without aftereffects.

Digestion in human beings is a complex

assembly-line operation. Like industrial assembly lines, it is coordinated by a variety of control mechanisms. Such controls may involve nerves, hormones, or an interplay of both. Three control mechanisms that involve hormones are of particular interest. When the stomach receives food and becomes active it soon becomes "aware" of its impending digestive "obligations" and it passes this information to other portions of the digestive system. Stimulated by the acidity of food coming from the stomach, the duodenal lining releases SECRETIN, a hormone that travels through the blood, inducing the pancreas to secrete bicarbonate and the gall bladder to release bile. A second hormone, called CHOLECYSTOKININ, also released by the duodenum, concurrently induces the pancreas to secrete enzymes and the gall bladder to empty itself into the intestine. A third hormone, ENTEROGASTRONE, is produced by the duodenum. Enterogastrone secretion is induced by the presence of fatty acids (the products of fat digestion), and it causes the stomach to "slow down" its rate of acid secretion and its rate of churning. Since fat digestion takes longer than the digestion of other nutrients, the duodenum seems to employ this hormone to tell the stomach to "ease off and let me receive food at a more leisurely pace." Stomach-ulcer patients, who must reduce their output of gastric acid, are often urged to adopt a diet rich in milk. Perhaps because the digestion of milk fats stimulates the production of enterogastrone, a milk-rich diet inhibits the secretion of stomach acid.

HERBIVORES

The principal organic compound ingested by a herbivore is cellulose, the carbohydrate of the plant cell wall. Ironically, and for reasons not properly understood, most herbivores are unable to produce the CELLULASES needed to digest this substance. Exceptions include the little insects called silverfish and the wood-boring mollusk *Teredo* (the notorious ship-worm). Other species rely on microorganisms to digest cellulose for them. These microorganisms inhabit various portions of the digestive tract, where they may be present by the billions. Most commonly they are bacteria, but they may also be fungi (yeasts) or protozoans.

The role of microorganisms in digestion has been well studied in ruminants such as cattle, goats, and sheep. Ruminants have a large four-chambered organ in lieu of the conventional stomach *(Figure 20)*. Actually the fourth of these chambers, the ABOMASUM, is a true vertebrate stomach; the other three chambers have evolved by elaboration of the lower portion of the esophagus. The first and largest chamber is the RUMEN. It is packed with anaerobic microorganisms (bacteria and ciliate protozoans) and serves as a cellulose fermentation vat. The microorganisms break down cellulose and other nutrients to simple fatty acids. To facilitate the digestive process, ruminants periodically regurgitate and re-chew the food in the rumen, re-swallowing it after the vegetable fibers have been more thoroughly ground up; this is the familiar "chewing of the cud." As fermentation proceeds, the fatty acids build up in the rumen and are absorbed from there by the host. To buffer the acids in the "fermentation vat," ruminants swallow large quantities of alkaline saliva, which contains bicarbonate. Byproducts such as carbon dioxide, as well as methane and other gases that are also formed in the rumen, are musically belched forth by the animal. Although the fatty acids are the major nutrients derived by the host from this microbial action, the microorganisms themselves are an important source of protein. The vegetable matter ingested by a ruminant is a relatively poor source of protein, but it contains inorganic nitrogen, which the microorganisms use to synthesize their cellular amino acids and proteins. As the food with its complement of microorganisms leaves the rumen, it first passes through the OMASUM where it is concentrated by water reabsorption, and then into the ABOMASUM, which contains hydrochloric acid and proteolytic enzymes. The microorganisms are killed by the acid and digested by the enzymes; the products of digestion are absorbed when they pass into the intestine. It has been estimated that a cow can derive upward of 100 g of protein per day from microorganisms. The propagation rate of the microorganisms in the rumen is high enough to survive the loss. Their symbiotic relationship with the host is evidently well-balanced.

Many mammalian herbivores other than ruminants also have microbial "farms" in their digestive tracts, but they usually maintain them in the CAECUM, a diverticulum of the large intestine. Such is the case in rodents, as well as rabbits (see Figure 18). The digestive processes mediated by microorganisms in the caecum are essentially the same as they are in the rumen. One difference is that there is little nutrient absorption in the caecum. This poses a problem for the animal since there is no small intestine after the caecum to which the role of absorption can be relegated.

Some rodents solve this problem by COPROPHAGY, the ingestion of their own feces. Interestingly, such animals may produce two types of feces: the ordinary type, which they discard (the familiar brown pellets dropped by rabbits), and a softer type, consisting of virtually pure caecal contents, which they ingest directly from the anus. By being taken through the stomach and small intestine, the contents of the caecum can be subjected to additional digestion and to the necessary absorption.

Some invertebrates also rely on microorganisms for cellulose digestion. Interesting studies have been made on termites, most of which subsist exclusively on wood, and which have a dense population of flagellate protozoans in the hindgut (Figure 21). Because the flagellates are anaerobes, they are sensitive to oxygen, and can be killed simply by subjecting the termites to high oxygen concentrations. Termites thus treated continue eating wood, but the cellulose passes unchanged through the gut, and they eventually starve. If artificially reinfested with the protozoans, the termite can recover. These classical experiments, done over 50 years ago by L. R. Cleveland at Harvard, provided the evidence that the protozoans are directly responsible for the digestion of cellulose.

STOMACH OF A RUMINANT contains four chambers: **20** **the rumen (where much of the fermentation takes place), the reticulum, the omasum and the abomasum. The abomasum corresponds to the true vertebrate stomach.**

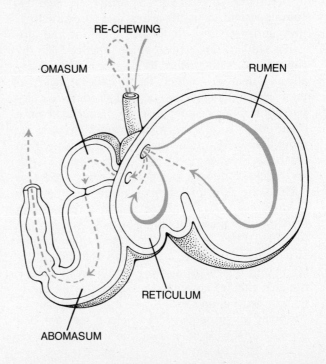

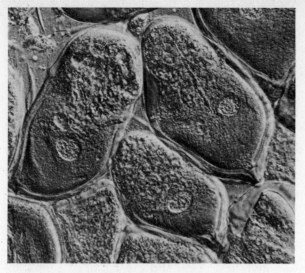

21 CELLULOSE-DIGESTING PROTOZOA live symbiotically in gut of wood-eating termites, and produce enzymes that digest cellulose for the insect. If protozoa are killed, host continues eating but eventually starves for lack of absorbable food.

CARNIVORES

The food of a carnivore has a higher nutritive content than that of a herbivore; it is also more readily digestible, because animal cells are not packaged in impervious cell walls. Carnivores may therefore have shorter intestines than herbivores, and have no special chambers for the accommodation of cellulose-digesting symbionts. Except for these differences, their digestive tracts are similar to those of herbivores or omnivores. Carnivores have become specialized in the ways they capture their prey and prepare it for intake. Catching may involve stalking and chasing, as practiced by the cheetah, creeping toward its intended victim (usually an antelope) and then dashing after it, overtaking it at speeds that have been clocked at 70 miles per hour. Stalking may also be practiced from a position of concealment, and need involve no more than grasping the prey when it strays within reach (as mantids do) or retrieving it with a long and sticky tongue (chameleon). Other predators such as the alligator snapping turtle lure the prey to their jaws *(Figure 22)* or trap it (orb-weaving spiders). The means of killing and swallowing the prey also differ. Snakes can take prey of enormous size, which they often immobilize beforehand by biting and injecting poison. Feeding on oversized morsels enables an animal to space out its meals when need be, and snakes are indeed capable of surviving on infrequent meals (a three-foot boa constrictor can be maintained on one or two rats

ALLIGATOR SNAPPING TURTLE is a carnivore whose wiggly tongue lures prey into its yawning jaws. In the sequence of photographs shown here an inquisitive small fish becomes turtlefood. 22

per week). Some deep sea fish also have enormous jaws; life is scarce where they reside, and they must be prepared to cope with whatever comes, of whatever size. Tentacles like those of *Hydra* have also evolved elsewhere. Cephalopod mollusks, including squid, octopus, and cuttlefish, have suckered tentacles and in addition use salivary toxins to quiet their quarry and parrot-like beaks to tear it up before ingestion. It is also not unusual for some digestion to be accomplished before the prey is actually swallowed. Starfish feed on static bivalve mollusks such as clams. They use their multiple arms, with their hundreds of suckered tube feet, to wrench the shell open; then they evert their stomach and disgorge its contents into the shell. After partially digesting the clam, they withdraw their stomach again, now filled with clam and ready for further digestion. Some carnivores do most of their digestion externally: a blowfly larva excretes proteolytic enzymes to digest the meat in which it lies, then swallows the resulting "soup."

DEPOSIT FEEDERS

The gut of deposit feeders is reduced to the crudest possible system. The low nutritive content dictates the need for little storage and fast throughput. Usually food passes through them in less than an hour. No attempt is made to sort out the material passed through the gut. In an earthworm *(Figure 18)* there is a small anterior crop and gizzard but the remainder of the gut is a long unmodified tube, without clear distinction between the portion where digestion is primary and where water uptake is predominant.

FILTER FEEDERS

Filter feeders include not only the stationary marine animals but also mobile animals as diverse as whales and birds, many tadpoles, and mosquito larvae. Common features include devices for passing great volumes of water through some part of the front end of the gut *(Figure 23)*. Humpback whales swim actively and filter off tiny shrimp-like crustacea called krill from copious water which they strain through the baleen, a set of several hundred closely spaced plates in their mouth. Mosquito larvae have comparable strainers in the pharynx, with which they filter out bacteria from water swept into the mouth by a pair of revolving brushes. Flamingoes sift through water and mud, straining out insect larvae, worms, seeds and other matter, using filtration devices in the bill. One

species having a particularly fine filter feeds predominantly on microscopic algae.

Stationary filter feeders often employ mucus to extract particles from water. Oysters, clams, and other bivalves which draw water through their shells and retain particulate matter on the mucus covering their gills belong to this group. An oyster can sample 30 times its own volume of water per hour. The gills are beset with cilia, which convey the mucus and its trapped particles toward the mouth, where the labial palps reject coarse matter and allow the rest to enter. Mucus is drawn as a continuous cord down the short esophagus and into the stomach by the winding action of a rod called the STYLE, which rubs the cord, as it is drawn in, against the grooved and ciliated stomach lining. More sorting occurs here, with larger pieces being wafted on, by more cilia, into the intestine, and smaller pieces being digested in the stomach, in part by amylase from the style. This surprising structure is continuously generated, and continuously dissolved, with the consequent release of the enzyme amylase.

FLUID FEEDERS

Fluid feeders share with carnivores the problem that although their food is relatively easy to digest, the major problem is locating and contacting it. Some essential nutrients may be very dilute, however, and there is then the problem of passing through great quantities of fluid and processing it economically. A few fluid feeders are illustrated in Figure 24.

The medicinal leech *Hirudo* shows the extreme simplification possible with an ideal diet. Its mouth is armed by sharp teeth and its saliva contains anticoagulant, needed by all bloodsuckers to prevent clotting. When it is fortunate enough to encounter a host, it takes in up to ten times its own volume of blood in a single meal. This giant meal is retained in the great stomach, a much-branched sac, for periods of many weeks. Both digestion and absorption occur in the stomach; the intestine is reduced to an almost trivial appendage.

Animals that suck the sap of plants must cope with a diet that is largely aqueous, rich in carbohydrates, and poor in proteins and vitamins. Aphids, for example, feed almost constantly in order to achieve a throughput containing enough proteins and vitamins. Excess water and carbohydrate is discharged from the anus as a sugary liquid called honeydew. This fluid does not go to waste. It is often drunk by ants which mill about aphid "herds." In exchange for this food the ants provide the aphids with protection against predators and other enemies.

DEPENDENCE

One of the themes of this chapter is the interdependence of organisms on one another for nutrients. The closest thing to an exception is a photosynthetic anaerobic bacterium, which does not need oxygen, can fix both CO_2 and N_2, and needs only a little light and a few ions to survive. At the other extreme lie the heterotrophs, whose

23   **FILTER FEEDERS strain their food from water. Filtering devices include baleen of humpback whale, combs of mosquito larva (color), and platelets in bill of flamingo (see opening page of chapter).**

nutritional requirements often involve compounds synthesized by several other organisms.

A second theme concerns interconversion of nitrogen compounds, starting with N_2. The nitrogen fixers produce ammonia which can either be taken up directly by plants as ammonium ions, or oxidized first by the nitrifying bacteria to NO_3^- (the more common occurrence), also readily taken up by plants. With the expenditure of a bit of ATP and reducing power, and using the enzyme nitrate reductase, plants simply make ammonium again from nitrate, for use in amination reactions. The resulting proteins, nucleic acids, and so forth are already familiar to you. Herbivorous animals can then obtain from the plants the amino acids and whatever else they cannot themselves synthesize. However, when an organism dies many different

HUMPBACK WHALE

MOSQUITO LARVA

FLAMINGO

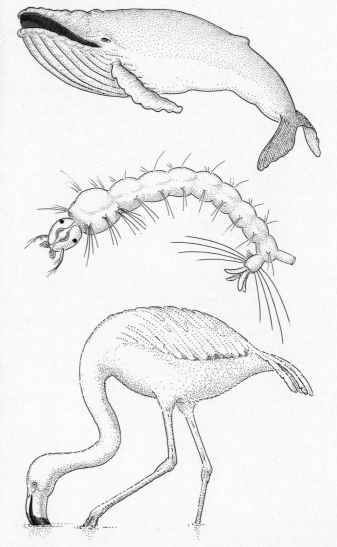

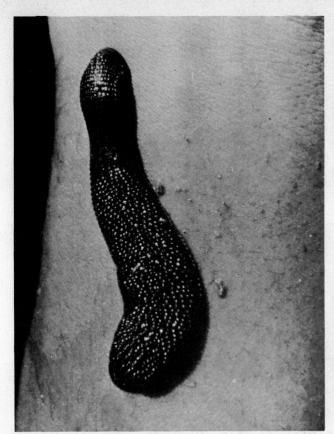

FLUID FEEDERS include leech, vampire bat and skin-piercing moth (*Calpe eustrigata*) from Nepal. All subsist on diet of blood. Other fluid feeders live as parasites in animals or suck sap of plants.

24

kinds of heterotrophs complete the nitrogen cycle, simultaneously obtaining their own carbon and nitrogen.

Heterotrophs have various sizes and shapes, and varying appetites. Among the higher plants, bacteria, and fungi, one finds saprophytes, which live harmlessly on decaying plant and animal detritus; parasites, which infect living plants and animals; there are actually some plants which are carnivores, which trap and digest small insects. It is the saprophytes, by and large, which complete the nitrogen cycle.

Saprophytic fungi and bacteria prevent the world from becoming one vast mass of dead organisms. These saprophytes are remarkably diverse in their appetites. They can use, for example, plastics, paint, vinegar, wax, fingerprints, paper, cloth, dung, leather, paper, kerosene, corpses, or any of a vast number of other carbon sources. They can break these down systematically, using the carbon both for synthesis and for respiration, ultimately returning it to the air as CO_2. Many saprophytes release as ammonium ion any organic nitrogen that they do not use to meet their own nitrogen requirement —a process called ammonification.

The saprophytes are essential organisms in that they close not only the nitrogen cycle but the carbon cycle as well, returning large amounts of CO_2 to the atmosphere for reuse. Both cycles are treated again in Chapter 33. A thin line, separates the saprophytes from the parasites: the organic matter that provides carbon and nitrogen for the parasites is not yet dead. Thus parasitic bacteria and fungi are responsible for a large number of diseases, both minor and devastating, of both plants and animals.

The few carnivorous plants have remarkable adaptations for allowing them to capture small insects or nematodes which they then use as a source of available nitrogen. The pitcher plant *Sarracenia* produces pitcher-shaped leaves which collect small amounts of rainwater. Insects are attracted into the pitchers either by the bright colors or by nectar, and find that they are trapped by stiff downward-pointing hairs. They ultimately die, and a combination of digestive enzymes and saprophytic bacteria in the water do the rest of the job.

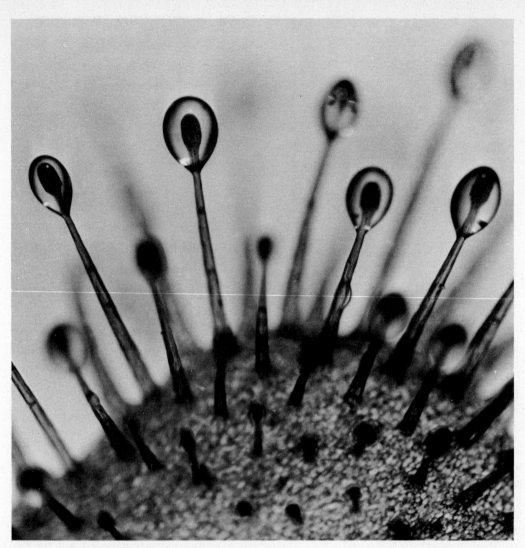

25 **SUNDEW PLANT (Drosera) is a small carnivore. Hairs on specialized leaves are tipped with droplets of a sticky liquid that traps insects. When insect becomes stuck on droplet other tentacles curve over and hold him fast. Plant digests insect by secreting enzymes and absorbs nutrients through leaf.**

The sundew _Drosera_ has leaves covered with hairs which secrete a clear sticky liquid, fairly high in sugar _(Figure 25)_. An insect which touches the leaf becomes stuck, and in fact other of these hairs or "tentacles" will curve over and stick to the insect also. Digestion is taken care of by the secretion of appropriate enzymes, and carbon and nitrogen are taken up. The venus fly trap _Dionaea_ springs an ingenious mechanical trap, triggered by three hairs in the center of a partially closed leaf lobe _(Chapter 21)_. The margins of the leaves have spiny outgrowths, and when the trap is triggered, the two halves of the leaves close with these spines inter-locking to make a prison. These plants are normally found in boggy regions where the pH is well on the acid side. Since most saprophytic organisms require a more neutral pH, decay is slow, and therefore available nitrogen is in short supply. These plants have simply developed adaptations allowing them to pick up a little extra nitrogen on the side. Finally, there are several fungi which have developed ingenious lethal devices for trapping nematodes, very small worms common to the soil and water. One literally forms a noose which contracts when the inside surface is touched by the animal, and another puts forth what W. H. Weston calls lethal lollipops—small projections with adhesive knobs on the end. In both cases, fungal filaments (hyphae) invade the cells of the prey, secrete appropriate enzymes, and carbon and nitrogen become available.

READINGS

J. MORTON, *Guts*, New York, St. Martin's Press, 1967. A paperback whose brevity is anticipated in its title. In 58 pages it offers a readable and well-illustrated survey of what the author calls the first and most obtrusive of the great systems the student finds in dissecting.

V.B. MOUNTCASTLE, *Medical Physiology*, St. Louis, MO, Mosby, 1974. A scholarly treatise on human physiology, written by various specialists. Part IX deals with the physiology of the digestive system.

C.L. PROSSER, *Comparative Animal Physiology*, 3rd Edition, Philadelphia, W.B. Saunders Company, 1973. A basic reference work on the physiology of animals. Nutrition and digestion are treated in Chapters 4 and 5.

F.B. SALISBURY AND C. ROSS, *Plant Physiology*, 2nd Edition, Belmont, CA, Wadsworth, 1978. A detailed plant physiology text with excellent treatment of both carbon and nitrogen relationships in plants.

In respiration is there a sort of digestion of the air? That is to say, a kind of phenomenon that fixes oxygen by means of a structure or tissue avid for oxygen? —Claude Bernard

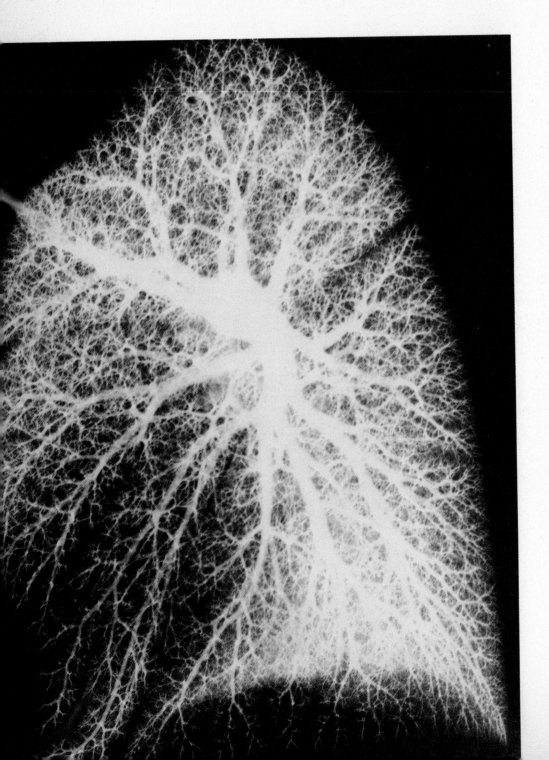

The answer to the outstanding 19th-century physiologist's question is yes. One of the major uses of the food synthesized by plants or ingested by animals is to provide the organism with energy. The commonest strategy for obtaining this energy is to oxidize the food to CO_2 and water, in the process of respiration. Most organisms, including plants, animals and aerobic microorganisms, have developed mechanisms for the continuous acquisition of oxygen and the discharge of CO_2. Photosynthetic organisms also carry out the reverse process, acquiring CO_2 and discharging oxygen. It is important to note that plants carry out both processes, not just photosynthesis alone. Indeed there would be no profit in the photosynthesis of sugars and starches if the plant could not unlock the energy of those compounds by oxidizing them later.

Almost all organisms exchange gases with their environment. Even a virus may have to rely on the gas-exchange properties of the host cell that it infects, and a saprophytic bacterium may have to rid itself of unwanted CO_2, hydrogen sulfide, or some other gaseous waste product.

Because cell chemistry is solution chemistry, a gas must be in aqueous solution to be used by an organism. Likewise if an organism is liberating a gas—be it CO_2 from respiration or O_2 from photosynthesis—it is first produced in solution. Somewhere in the organism there must be a liquid-air interface where gases can efficiently enter or leave solution.

Unicellular aquatic organisms can cope with gas exchange very easily. For them the gas-liquid in-terface is the surface of the water in which they grow. So long as the water is in motion gaseous products are readily removed to be released at this interface, and their gaseous requirements are readily met by the reverse process. Even in water that is relatively stagnant, diffusion of dissolved gases may be rapid enough to meet the needs of at least some organisms. Diffusion is also sufficient to keep the appropriate gas available throughout the organism.

Larger aquatic organisms, however, may face two new problems. First, they may have some sort of "skin" that severely restricts the exchange of dissolved materials, thereby necessitating a special adaptation to provide a liquid-liquid interface between the organism and the solution in which it is living. Second, increase in size means decrease in surface-volume ratio. Above a certain size, diffusion of solutes is simply inadequate to provide required gases in adequate amount or to prevent accumulation of toxic quantities of waste gases.

Terrestrial organisms face yet another problem. Since gases must be in solution to enter or leave the cell's biochemistry, there must be a liquid-to-air interface on the organism. So long as rainfall is adequate, gradual evaporation of water from a lake surface is of little consequence to the aquatic organism beneath the surface. But for a terrestrial organism evaporation of water from its primary liquid-gas interface can easily be fatal. Clearly special adaptations for gas exchange are essential. The nature of these adaptations will be determined by two factors. The first is the rate of gas exchange required. A running jack rabbit obviously requires far more oxygen and must get rid of far more CO_2 than even the most rapidly growing mushroom. The second factor is the use to which gas exchange is put. While excessive loss of water as gaseous vapor would be harmful to a tree, the tree would be equally in trouble if no such loss occurred, since

BLOOD VESSELS OF HUMAN LUNG appear in white in the x-ray photograph on opposite page. Large vessel at center is pulmonary artery, which arises directly from the heart.

319

evaporation of water at the plant's interface with the air provides the driving force for raising water and essential ions from the roots to the aerial parts *(Chapter 16)*. Plants and animals have evolved quite different solutions to the problem of gas exchange.

GAS EXCHANGE IN PLANTS

Aquatic plants need no special adaptations for water conservation so long as they remain submerged (the special osmotic problems of organisms in salt water are considered in Chapter 17). Even when they are quite large, they utilize extensive branching or make very thin organs to maximize their surface-volume ratios and keep actively metabolizing or photosynthesizing cells in close contact with the environment. Gases need never diffuse through more than a few cells. The flatworms are an interesting parallel in the animal kingdom: mold a flatworm into a sphere and its innermost cells would asphyxiate.

The lower land plants such as mosses, liverworts, and hornworts *(Chapter 26)* follow essentially the same principles *(Figure 1)*. Plant parts are thin; flattened, photosynthetic organs are frequently small and numerous; and requirements for long distance transport of any kind are minimized. But such plants are restricted severely in their distribution on land; they must grow where water is plentiful. They are usually not heavily cutinized, and they rapidly dry out in the absence of continual moisture.

> Because aquatic plants have thin or finely branched leaves, gases need never diffuse through more than a few cells. The same principle applies to certain animals; mold a flatworm into a sphere and its innermost cells would suffocate.

Maximizing surface area for gas exchange without elaborate modification to prevent water loss does not necessarily mean that these plants are always restricted to permanent swamps. On a hot summer day in the dry season in the San Joaquin valley of California, one can find rocks too hot to touch. On these rocks are dried brown mosses which crumble readily in your hand. If they are crumbled onto a moist surface, however, intact cells or tissue fragments turn green and are actively photosynthesizing within a few minutes. These plants become dormant rapidly upon desiccation, and begin growth rapidly upon rehydration. They can go through their entire life cycle in a few weeks. Their solution to the gas-exchange-desiccation dilemma is temporal, not structural.

Certain lower plants do have adaptations that permit restriction of gas exchange during relatively dry periods *(Figure 2)*. For example the liverwort *Marchantia*, a plant which grows flat against its substrate, has an internal structure analogous to that of a leaf. A thin layer of nonphotosynthetic cells covers the upper surface, beneath which is an elaborate chamber system. Plates and filaments of photosynthetic cells project up into these chambers, and gain access to the outside environment through pores in the surface layer, the epidermis. The pores open and close in response to humidty changes and a degree of regulation of gas exchange is obtained. Several other lower plants produce

1 THIN LEAVES of many terrestrial plants enhance both gas exchange and water loss. Photograph depicts the leafy moss Dendroalsia covering the trunk of a tanbark oak in coastal forest of California.

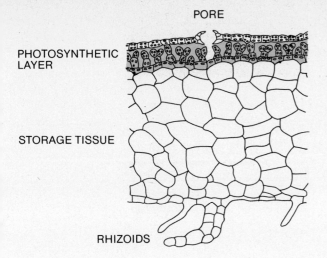

PORE

PHOTOSYNTHETIC LAYER

STORAGE TISSUE

RHIZOIDS

CROSS-SECTION OF LIVERWORT shows adaptations that restrict gas exchange and water loss during dry periods. Gametophyte shown here has pores that admit gases to photosynthetic layer (color) that is only a few cells thick. Inactive storage tissue lies below it. Rhizoids attach plant to its substrate.

sporophytes which project many centimeters away from the substrate. They have a cutinized epidermis with the typical stomata of higher plants, as noted in the next section of this chapter.

As mentioned in Chapter 13, in some higher plants such as ferns both the sporophyte and gametophyte portions of the life cycle are independent and free-living. In these cases the sporophytes, the large and obvious plants of the life cycle, have the typical cutinized epidermis with stomata for gas exchange. The gametophytes, however, often rely on a more primitive solution: a small flat plate of cells appressed against a moist substrate. The gametophyte phase is rapid and can be completed during brief periods of plentiful moisture in an otherwise arid climate, and gas exchange is not regulated.

For practical purposes, stomata are simply pores through a thin sheet of water- or gas-proof material, leading from the outside atmosphere to the

---

Diffusion through pores, in terms of unit area, is almost 50 times more efficient than diffusion from an uncovered surface.

---

intercellular spaces around the moist surfaces of the spongy and palisade parenchyma cells (*see Figure 3 in Chapter 14*). A pair of specialized GUARD CELLS surrounds each pore. One is justified in asking how the gas exchange needs of a rapidly growing plant can possibly be met by what superficially appears to be a rather inefficient system. The an-

swer lies in the nature of diffusion. A diffusible substance in a fluid medium, whether liquid or gas, will diffuse from a region of high concentration to one of low. Ultimately, if the system is closed and left undisturbed, the molecules of the substance will become randomly distributed throughout the fluid. The rate of diffusion depends on the concentration gradient; the steeper the gradient, the more net movement of gas from the region of high concentration to that of low.

LEAVES AND STOMATA

Some 70 years ago Brown and Escombe showed very clearly how these principles apply to diffusion of gas through small pores such as stomata. A pan filled with water is covered with foil perforated with holes of various sizes and patterns of distribution. If the holes are very large the concentration gradient of water vapor is essentially at right angles to the surface and is relatively shallow (*Figure 3*). Likewise if the holes are very small but numerous and close together the diffusion gradient is again almost perpendicular to the surface (except at the edge of the region) and the concentration gradient is again shallow. But if the pores are small and widely spaced, diffusion occurs three-dimensionally into a hemisphere, starting from a point source, and the concentration gradient can be lateral as well as perpendicular. For a given amount of water vapor right at the pore, the concentration gradient is very steep, and the rate of vapor loss therefore great. Brown and Escombe showed that small and well-spaced holes in the foil, exposing only about 0.3 percent of the total available surface area for gas diffusion, allowed in fact almost 15 percent of the rate of gas diffusion from the uncovered pan. Diffusion through pores, expressed on a unit area basis, was almost 50 times as efficient as from the uncovered surface! Ting and Loomis, of the University of California at Riverside, have recently shown empirically that a leaf epidermis with its neatly spaced stomata, between five and ten stoma diameters apart from one another, is indeed an efficient system for gas exchange (*Figure 4*).

321

3 DIFFUSION OF GASES from open tray (top) was compared experimentally with diffusion from tray covered with foil containing large holes (middle) and with foil containing small pores (bottom). Colored contour lines indicate concentration of water vapor above trays, and arrows indicate paths of evaporating water molecules. Small widely spaced pores produce a steep hemispherical concentration gradient which leads to fast efficient diffusion.

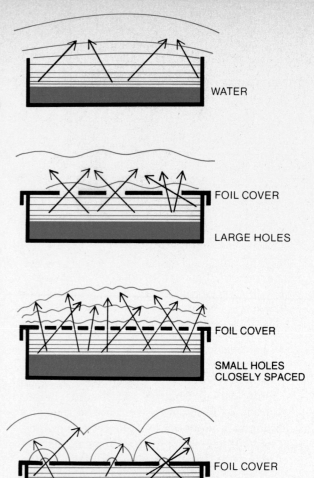

WATER

FOIL COVER

LARGE HOLES

FOIL COVER

SMALL HOLES CLOSELY SPACED

FOIL COVER

SMALL HOLES FAR APART

Such a system, however, is of little value to an organism which must exist in a continually changing environment, unless it is able to adjust itself. Not surprisingly, therefore, stomata open and close in response to a variety of environmental stimuli. The mechanism looks deceptively simple, and operates according to the osmotic principles discussed in Chapter 2. The stomatal guard cells are surrounded by a relatively rigid cell wall against which the cytoplasm presses. If water enters a cell, the cytoplasm increases in volume, stretching the wall and exerting TURGOR PRESSURE against the inside of the wall. Resistance to this pressure is called WALL PRESSURE. If there is a further intake of water, turgor pressure increases, stretching the wall until the resisting wall pressure equalizes the new turgor pressure, and the system is again in equilibrium.

How does such a system operate in opening or closing stomata? Perhaps the best analogy is an almost empty inner tube lying at rest with its opposite inner surfaces touching. If one pumps air

4 TWO STOMATA are visible in this scanning electron micrograph of the underside of a leaf from the spiderwort Tradescantia. The stoma at left is partly open; the one at right is closed.

into it, the inner surfaces are forced apart as the tube assumes its normal expanded shape (*Figure 5*). Let the air out again and it collapses. If it had been designed so that the inner surfaces were always pressed together when it was empty, it would be a perfect mechanical model of a guard cell pair. Each guard cell is one half of the inner tube. As it happens, the direction in which a guard cell pair expands is frequently determined not just by intrinsic shape —a doughnut at high turgor and a pair of limp string beans at low —but also by specialized wall thickenings and cutinization of certain wall regions. The underlying principle, however, remains the same.

Stomata are open when water is plentiful and closed when it is scarce. But many plants photosynthesize and exchange gases effectively under rather dry conditions. It is apparent then that light and CO_2 concentration will have an effect on stomatal aperature size. Stomata are normally closed in the darkness (see the discussion of photosynthesis in Chapter 6 for an interesting ex-

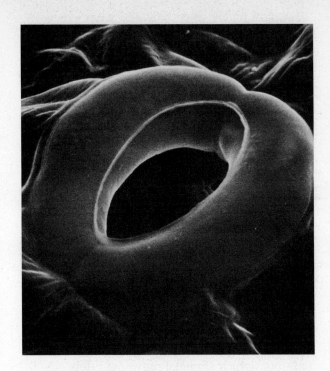

CLOSE-UP OF OPEN STOMA shows the two guard cells surrounding the pore in a cucumber leaf. This picture, made with the scanning electron microscope, is reproduced at a magnification of about 3100X.

higher the concentration of these solutes, the higher the osmotic potential of the guard cell cytoplasm. Water therefore enters the cell following its own diffusion gradient, causing the stomata to open. At night, exactly the reverse process takes place. In the absence of photosynthesis, respiratory CO_2 accumulates as carbonic acid, the pH drops, and conditions for conversion of soluble, and therefore osmotically active sugar to insoluble, and osmotically inactive starch are now optimal. The osmotic potential of the cytoplasm, which is the physical basis for turgor pressure, drops; wall pressure is greater than turgor pressure, water is forced out, and the stomata close.

This elegant photosynthetic theory may account at least partially for stomatal movement, particularly since amylases and phosphorylases with the appropriate pH optima are known. But it cannot be the whole story for several reasons. For example, opening and closing of stomata can be observed in the absence of any starch at all, either in light or in darkness. Also, inhibitors of aerobic respiration such as cyanide or azide, prevent opening in the light. There is no place in Sayre's theory for these observations.

In 1963, Walker and Zelitch developed a clever technique for continuous monitoring of stomatal behavior. These two workers simply smeared a little silicone rubber on the leaf surface, allowed it to set, and then peeled it off again. They then coated the rubber surface with fingernail polish, allowed it to dry, peeled it off, and mounted it on a microscope slide. The fingernail polish was a reasonable replicate of the leaf surface, and one could measure stomatal apertures at leisure. The technique did not seriously disrupt the guard cells, and it was possible to take more than one impression from the same place.

Walker and Zelitch first reinvestigated the influence of various metabolic inhibitors, using leaf discs floating on water. They quickly confirmed that cyanide and azide inhibited light-induced opening, and could also induce stomata already open in the light to close. They next found that light-induced opening could occur only in the presence of oxygen, but that closing could occur anaerobically (Figure 6). Finally, they showed that if one inhibited both aerobic and anaerobic respiration, both opening and closing could be prevented.

ception). If they are exposed to light, they will open within an hour or less; the reverse is true for leaves transferred from light to darkness. However, alteration in CO_2 concentration affects the stomata more than light. High concentrations of CO_2 cause stomata to close whether they are in the light or not, and low CO_2 concentrations allow them to stay at least partially open even in the dark.

There are further important observations necessary to understand the control mechanisms. First, the guard cells are normally the only epidermal cells to have chloroplasts. Furthermore, the light-induced opening suggests a photosynthetic role at least in the opening process. Finally, starch, the principal storage product from photosynthesis, accumulates in granules in the guard cells at night, but not in the daytime — the reverse of the usual photosynthetic cell.

In 1923, Sayre put most of this information together into a theory for stomatal movement. During the daytime, photosynthesis proceeds rapidly, manufacturing carbohydrate. The pH of the guard cell, however, is relatively high because of photosynthetic depletion of CO_2. Under these conditions, the phosphorylase enzymes responsible for starch synthesis are not at their optimal pH, but the amylases, which hydrolyse the insoluble starch back to soluble sugar, are. Therefore, during photosynthetic production of glucose, soluble sugars accumulate in the cell. Obviously, the

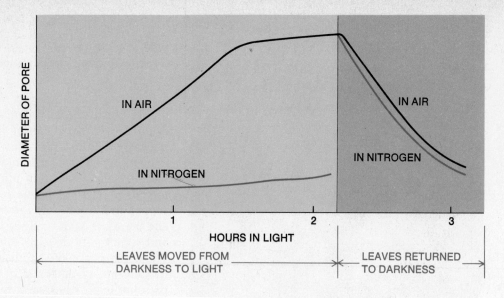

DIAMETER OF PORE

IN AIR

IN NITROGEN

IN AIR

IN NITROGEN

1   2   3

HOURS IN LIGHT

LEAVES MOVED FROM
DARKNESS TO LIGHT

LEAVES RETURNED
TO DARKNESS

6 EFFECT OF LIGHT AND OXYGEN on the action of
stomata. In the presence of oxygen, stomata open
gradually when plants are moved from darkness into
light, and close rapidly when plants are returned to
darkness. Nitrogen atmosphere inhibits opening of
stomata but does not inhibit their closing.

Thus both opening and closing had metabolic
components. Closing could occur with the relatively small amount of energy available from
glycolysis, but opening required the higher efficiency of aerobic respiration *(Chapter 6)*.

In the absence of rapid changes in sugar concentration as a result of photosynthetic or respira-

7 UPTAKE OF POTASSIUM ION accompanies the
opening of a stoma. In these two radiographs potassium
ions appear as bright dots. Picture at left shows
distribution of potassium ions near a closed stoma.
Picture at right shows same location, with potassium
concentrated in guard cells of open stoma.

tory activity, there is one other obvious way in
which to bring about the changes of osmotic potential upon which stomatal movement depends: the
swift uptake of osmotically active solute for opening. Potassium ion is directly involved in at least
one other rapid plant movement, NYCTINASTY or the
night closure of certain leaves *(Chapter 17)*. This
so-called sleep movement also depends upon
rapid turgor changes in certain specialized cells.
Closing involves loss of turgor in the motor cells,
and the loss of potassium has also been observed.
Likewise, opening involves potassium uptake by
the motor cells. The same ion plays a leading role
in stomatal movements. Humble and Raschke at
Michigan State University have followed potassium movement during stomatal opening and closing, using an electron microprobe technique that
causes the potassium ion to photograph itself by
emitting an x-ray *(Figure 7)*. The commonest ac-

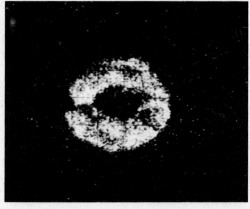

companying anion is chloride, although the organic anions malate and aspartate, among others, may play a role in some plants.

Raschke developed a model of the stomatal system as a series of feedback loops. If, for example, one alters the CO_2 concentration within the leaf, the system readjusts itself to restore the original CO_2 concentration. Since feedback systems may show both an overshoot and then damped oscillations as they shift to a new equilibrium situation, it is not surprising that Raschke can detect both in stomata. One might expect the same kind of response from a rapid change in humidity, and one will probably have to turn to models such as Raschke's for an understanding of the entire system. Probably both the photosynthetic components proposed by Sayre and an active secretion-uptake system are involved.

It is fair to wonder why more progress has not been made in understanding this important regulatory system. After all, we know all about photosynthesis (or do we?) and neurophysiologists have for years monitored much more rapid ionic fluxes across membranes than those proposed for stomatal movement. If there is some guard cell phosphorylase or amylase with an unusual pH optimum, why hasn't anybody found it? Probably because guard cells represent a tiny fraction of the total cells of a leaf and they are well dispersed. The study of guard cell biochemistry has so far been almost impossible, but recently Zeiger and Hepler at Stanford University have made some exciting progress. Using enzymes to digest cell wall material, they isolated guard cell protoplasts from the epidermis of onion cotyledons. These protoplasts increased in volume by from 35 to 60 percent when illuminated. The swelling occurred only when potassium was present in the bathing medium — elegant confirmation of Humble and Raschke's work. Only blue light, not red, was effective, indicating that this particular light response was not photosynthetic. These techniques, combined with those of modern biochemistry, should rapidly increase our knowledge of the mechanism that regulates the size of the stomatal aperture.

ROOTS

Roots too play a role in gas exchange. Roots have a large surface-volume ratio because they branch extensively, and because the surface of recently matured regions is covered with root hairs. Roots can thus obtain gases directly from soil water, or from air spaces in the soil if it is fairly dry. For them, the limiting factors may depend very much on the physical properties of the soil, and on what else is growing there. Gases, aided by wind and air currents, move far faster through open air than through soil. Furthermore, if a soil is saturated with water, the solutions rather than gases dictate the diffusion rates. If a soil is filled with aerobic microorganisms, its oxygen tension may be low and its CO_2 concentration extremely high. Thus roots live a somewhat more precarious existence than leaves, and the physical and biological composition of soil is frequently the factor determining whether a plant will or will not grow on a particular site.

Certain large woody plants live in swampy environments in which water movement is extremely sluggish, and oxygen tension extremely low because of the activities of large numbers of aerobic saprophytes. The bald cypress is a familiar example, as are many of the mangrove species. The roots of these plants frequently develop aerial projections, or pneumatophores, which emerge above

PNEUMATOPHORES of the bald cypress (Taxodium distichum) project above swampy ground. Sometimes called "knees," pneumatophores rise from roots and provide path for diffusion of oxygen. 8

the water. These may be "knees" as in the cypress *(Figure 8)* or they may be the ends of specialized roots. In either case, their function seems to be to provide a path for diffusion of oxygen to the otherwise anaerobic roots.

The stem of an herbaceous plant has an epidermis with stomata just as a leaf has. However, a woody stem is another matter *(Chapter 16)*. Within the cortex there develops a cork cambium *(Chapter 12)*, and the cork cells produced to the outside are far more effective waterproofing—and hence gasproofing—than the simple layer of cutin on the leaf. Yet within the layer of cork one finds numerous living cells, the most important of which are those of the vascular cambium, laying down new xylem and phloem, and ray cells both in the phloem and in the xylem. Thus there is at least a modest requirement for gas exchange. As with the epidermis, there are specialized regions for gas exchange, the LENTICELS *(Figure 9)*. In the cortex be-

9 **LENTICEL on the stem of an elderberry. Gases pass through break in cork and diffuse freely through intercellular spaces of underlying spongy bark. Diagram is key to photomicrograph. Magnification, about 55X.**

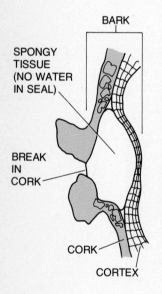

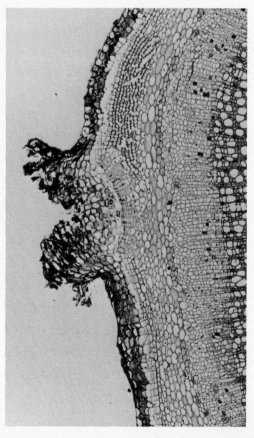

BARK

SPONGY TISSUE (NO WATER IN SEAL)

BREAK IN CORK

CORK

CORTEX

neath a stoma a group of rather spongy cells is produced. The cork cambium beneath these produces a spongy but nonphotosynthetic parenchyma instead of cork, and as the whole system expands with the stem, this production of spongy tissue continues. Gas can readily diffuse through the extensive intercellular spaces of this tissue. Thus even in fairly thick bark, there are small "windows" through which oxygen can enter and CO_2 can exit.

Plants, by and large, are rather sluggish individuals. Their demands for gas exchange are modest for their size and are met locally. There is no need for a special system to move gases rapidly from one part of an organism to another, and diffusion is sufficient. As will shortly become clear, however, active and fast-living animals have gas exchange requirements hundreds or thousands of times higher than plants.

GAS EXCHANGE IN ANIMALS

Animals share with plants the need to provide an extensive wet surface for diffusional gas exchange, and those that live on land must make provision, like plants, to protect the exchange surface from the desiccating effect of air. Animals differ from plants, however, in that their gas exchange is strictly respiratory. They take in O_2 and give off CO_2, but do not also practice the reverse, as do photosynthesizing plants. Animals differ a good deal in the ways and means by which they came to meet their respiratory demands. Three decisive factors influenced the evolution of respiratory mechanisms: the extent of the metabolic need for O_2, the environmental availability of O_2, and the size of the animal.

OXYGEN NEED

Oxygen need is in most cases a precise measure of energy needs. We will express needs (as judged by consumption) as microliters of oxygen per gram of body weight per hour. (A microliter, abbreviated μl, is a millionth of a liter, about the volume of a pinhead. A gram is about half the weight of a dime.) As will become apparent, the values cut completely across phylogenetic lines, and relate exclusively to the activity and the size of the organism.

Growth and movement make large energy demands. Consequently, slowly moving and nongrowing organisms or tissues use relatively little oxygen. A python *(Python molurus)* uses 6.2 μl, a carrot root cell 25 μl, a mussel *(Mytilus)* 22 μl, a dry barley grain 0.06 μl. Organisms with a more active lifestyle require, even at rest, substantially more: man needs 200 μl, the butterfly *Vanessa* 600 μl, a chicken 497 μl. These figures increase greatly when such organisms become highly active: man's need climbs 20-fold to 4000 μl, *Vanessa's* need increases an amazing 170-fold to 100,000 μl. Active forms of plants also have large needs: the antherozoids of the alga *Fucus vesiculosus* need 2550 μl. A germinating barley seed uses 108 μl, which is 1800 times more than the resting seed needs. Obviously growth, like movement, imposes large energy demands.

OXYGEN AVAILABILITY

The quantity of oxygen available in water is far less than in air. Moreover, the amount of oxygen in both water and air can vary considerably. A liter of air contains 210 ml of oxygen, a liter of fully aerated water at 15°C contains only 7 ml (fresh water) or 5 ml (sea water). The oxygen in water falls as the temperature rises (which is why fish leave lake shores as the days warm up and seek cooler, deeper places); at 35°C, fresh water has only 5 ml per liter. These are the levels of oxygen in fully aerated water, and water is often not fully aerated. For instance, unless water movement is excellent (and it rarely is) there will be oxygen-poor zones where organic matter is being oxidized by bacteria. This is a major reason why pouring sewage into rivers can kill massive numbers of fish.

The low oxygen content of water sets severe limits on the possible activity of aquatic animals. One consequence is that all warm-blooded animals, even those such as whales whose life is entirely aquatic, rely upon air for an oxygen source. Such air-breathing aquatic animals commonly have adaptations that enable them to dive for long periods of time *(Box A)*. Additional problems arise whenever water is warm and stagnant. Such conditions frequently prevail in fresh waters in the tropics, where the decay rate of organic matter is high, and where temperatures are relatively constant throughout day and night so that there is little thermal convection to bring oxygen-rich surface waters to the deeper waters. The South

---

A Underwater survival

If a swimmer remains submerged long enough he will die. But what is long enough? Four to six minutes used to be thought the maximum for humans, and this still appears to hold, providing the water is relatively warm. Mere survival, however, is not the whole story, because the brain is often irreparably damaged by lack of oxygen. Humans can survive much longer if the water is chilly (colder than 21°C). In March 1975 a college student from Michigan plunged with his car through the ice of a pond. When pulled out 38 minutes later his body was blue, his breathing had stopped, he had no pulse, and his eyes were fixed in a stare. By conventional criteria he was dead; his rescuers probably would have made no effort to revive him, except that he suddenly belched. Rushed to a hospital and subjected to cardiovascular resuscitation he eventually came to, and recovered completely. (According to one press report he went on to finish the semester with a 3.2 average.)

Although the exceedingly long submersion makes this case exceptional, many individuals have survived submersion of over four minutes in cool water without incurring brain damage. What is the explanation? One theory is that when humans are forced to hold their breath under cold water, a physiological response is triggered in their bodies comparable to the so-called "diving reflex" of mammals such as seals and whales. In essence, the reflex involves a slowing of the heart beat, a reduction of blood flow to tissues such as muscles and skin (which are relatively insensitive to oxygen deprivation), and an increase in flow to the more sensitive brain and heart. Why should a diving reflex occur in a non-diving mammal? A speculative answer is that the reflex persists in all mammals because it plays a role at birth: it may help the infant survive when its oxygen supply is temporarily cut off as it emerges from the birth canal.

American electric eel, *Electrophorus*, has overcome the problem by becoming an air breather; it rises to the surface regularly to gulp air, and absorbs the oxygen through a network of blood vessels in its oral cavity. Many other fish surface for air and some, such as lung-fish, can store air in a special sac-like diverticulum of the gut. In the South American lung-fish, *Lepidosiren*, upward of 90 percent of the total oxygen uptake may occur in this lung. The gill of this fish plays only a minor role in oxygen uptake.

Oxygen insufficiency occurs less in terrestrial animals because of the relatively high oxygen content and the rapid mixing of air. But oxygen content drops proportionately to air pressure, so that at high altitudes it is much reduced. At 17,500 feet air contains about half as much oxygen as at sea level. Animals exposed to such conditions may undergo "acclimation" changes. In humans, lung capacity, number of red blood cells and cardiovascular output all increase notably at high altitudes.

10 **GAS EXCHANGE IN ANIMALS can occur either through the integument (skin), gills, lungs or tracheae. Gills are outfolded structures that increase the exchange surface in aquatic animals. Lungs are infoldings that serve the same purpose in terrestrial animals; the exchange surface is internal and thus protected from drying out. Tracheae are branched internal respiratory tubes; exchange occurs in tiny end tubes (tracheoles). Tracheal systems are characteristic of insects and some other arthropods.**

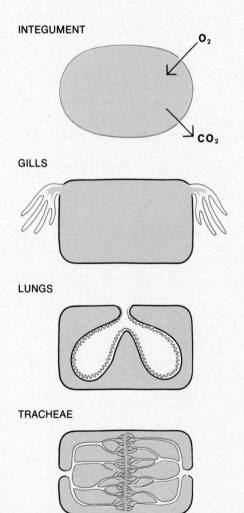

INTEGUMENT

O_2

CO_2

GILLS

LUNGS

TRACHEAE

Plants are rather sluggish individuals. Their demands for gas exchange are modest for their size. Fast-living animals have requirements hundreds or thousands of times higher. A resting elephant needs 148 microliters of oxygen per gram of body weight per hour, a resting man 200, and a resting mouse 2500.

SIZE

Size influences the oxygen needs of warm-blooded animals because of the increase of relative surface area (and therefore of relative heat loss) with decreasing size. The resting elephant needs only 148 μl of oxygen per gram of body weight per hour, the resting man 200 μl, and the resting mouse 2500 μl. Simple diffusion of oxygen through the body wall is adequate only for very small animals such as Protozoa, or those with very modest oxygen requirements. Moreover the distance from the center of the organism to the body surface increases with increasing size, so that even if the required oxygen could be absorbed at the surface, a large organism could suffocate at its center. It was pointed out earlier in the chapter that one way to circumvent this problem is to maintain a flat body shape, so that every part of the body is near a surface, as in the flatworms.

Larger organisms have had to develop special devices to cope with the inadequacy of simple surface diffusion. The physiological devices as-

sociated with various solutions to the respiration problems have some features in common *(Figure 10)*. They all provide large interfaces between the animal and its oxygen source. Some portion of the body usually becomes specialized to provide a vast surface area to speed up gas exchange, as in gills and lungs. The interface must always be wet because the permeability of dry tissues is inadequate. In terrestrial animals special precautions are needed to keep a wet surface, yet not lose excessive amounts of water from it. And except when needs are small, the oxygenated medium in contact with the exchange surface, whether it is air or water, must be continuously renewed to avoid depletion of its oxygen or unwarranted accumulation of carbon dioxide.

SPECIAL RESPIRATORY MECHANISMS

Life began in water, and it is there one should look for the simplest modifications of systems that de-

AXOLOTL is an amphibian with gills, which can be seen protruding from its neck. Such gills are essentially outpocketings of the body wall; fringe-like branches increase their surface area. **11**

pend on free diffusion over the whole body surface, which are adequate for very small organisms. The fundamental mechanism of diffusion across a wet membrane is found at every level, but in animals of increasing complexity various specializations of the wet membranes have evolved. In the simplest type, a portion of the body wall protrudes and increases its surface area by branching to form a GILL. Axolotls are among the many amphibians

GILLS OF CRAYFISH are covered by shell. In drawing at right, shell has been cut away and gills are rendered in color. Each gill is set with hundreds of respiratory filaments which are hollow and filled with blood. Blood cells appear as dark spots in photomicrograph of filaments (left). **12**

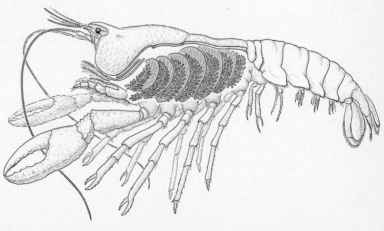

using this device *(Figure 11)*. Many crustaceans also use gills of this type *(Figure 12)*, but because they are covered (and so protected) by the shell, water has to be circulated over them by miniature paddles situated in front of the gills. In a similar fashion, clams circulate water over their enclosed gills by the beating of millions of cilia, and the

octopus actively pumps water in and out of its gill cavity.

Fishes have developed more elaborate gills, forming richly branched fringes around holes (gill slits) in the alimentary canal. Water is brought continuously into the mouth and passes through these slits. Many fish actively force water through the gill slits by a muscular pumping action of the buccal cavity and the operculum, a flap-like shield on each side of the head that covers the gills. Other fast-swimming fish such as the mackerel rely on their movement through the water to keep up

13   **ACTION OF A GILL. Cutaway drawing of mackerel (1) shows position of gills. Broad arrows in (2) indicate flow of water past gill filaments. Diagram (3) shows circulation within filaments. Small arrows indicate that blood flows through transverse capillaries in opposite direction from incoming water. This countercurrent flow maximizes gas exchange.**

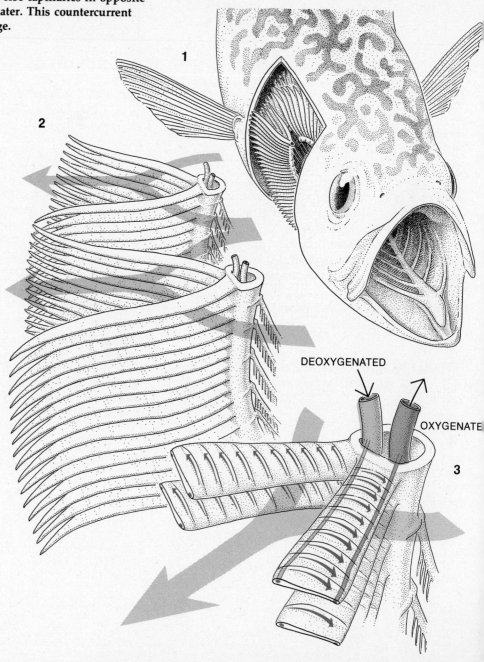

DEOXYGENATED

OXYGENATE

water flow; they cannot stay still or they will die. The efficiency of the fish gill is further improved by a counter-current device *(Figure 13)*. As the blood becomes progressively oxygenated in its passage through gills, it becomes progressively harder to get more oxygen into it. By having the most oxygenated blood meet with the least deoxygenated water, gas exchange is maximized. Comparable counter-current devices are found in both biological and man-made devices designed for ultra-high efficiency extractions. The kidney uses the principle for extracting valuable water from urine before voiding the urine *(Chapter 17)*, and engineers use the principle in designing heat exchangers.

LUNGS

When animals evolved to life on land, one advantage was greatly increased oxygen supply. Gill-like devices in principle would be fully adequate to exchange the oxygen and CO_2. But in practice, gills suffer from a mechanical drawback: being loose and much branched organs, they collapse when not supported by water, and their exchange surface is then too small. Moreover the exposure of a large and wet surface to the outside would lead to excessive water loss, one of the problems of life on

The total gas-exchange area in man may be 100 square meters, which is the surface area of a sphere about six meters in diameter.

land. One solution was to put the large wet surface inside as a LUNG, draw in pulses of air, and reduce one's water-loss to the amount that would saturate this relatively small pulse. The lungs of such different animals as reptiles, amphibians, mammals and birds all consist of elastic bags with linings only about 1 μ thick, across which exchange occurs between blood (in a net of capillaries that is densely appressed to the lining) and the air within the chamber.

In amphibians the skin has an importance for gas exchange comparable to the lung; in fact most terrestrial salamanders are lungless and use only the skin. Frogs respire through their skin and their buccal cavity as well as through their lungs. They

HUMAN RESPIRATORY SYSTEM. Trachea divides into two bronchi, one for each lung. Bronchi in turn branch into bronchioles that terminate in clusters of tiny alveoli, shown in detail at right. **14**

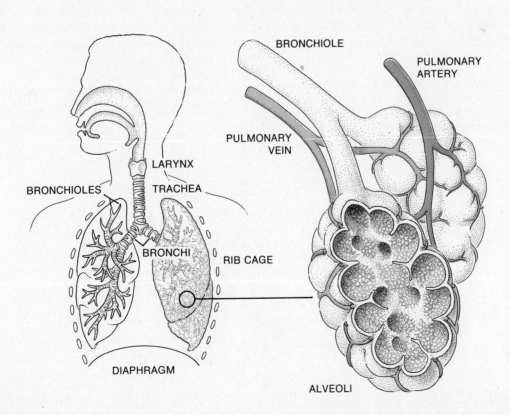

BRONCHIOLE

PULMONARY ARTERY

PULMONARY VEIN

LARYNX

BRONCHIOLES   TRACHEA

BRONCHI   RIB CAGE

DIAPHRAGM

ALVEOLI

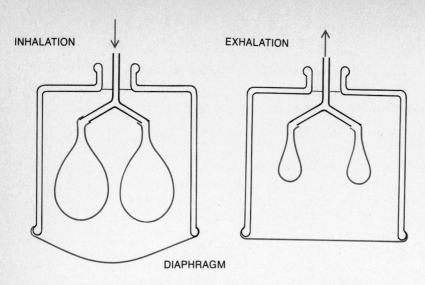

DIAPHRAGM

Mammals have elaborate lungs honeycombed with almost a billion small blind sacs called AL-VEOLI, of diameter 0.1 mm to 1 mm, in which the gas exchange occurs *(Figure 14)*. The total exchange surface in man may be 100 square meters, which is the surface area of a sphere about six meters in diameter. Air is brought in first by a sturdy trachea, dividing into two bronchi, then into about 20 bronchioles, and then into the alveoli. In this way, each ml of air becomes exposed to 300 cm² of exchange surface (in man), as compared with 20 cm² in the frog. A suction method is used to fill the lungs, by expanding its thoracic container (the pleural cavity). This expansion involves principally a contraction of the muscular diaphragm, which pulls it, piston like, down the cylinder formed by the rib cage. The lungs are drawn down behind it. In addition, other muscles lift the rib cage and increase its enclosed volume, so expanding the lungs. Expulsion of air is passive: the above muscles relax, and the elastic lungs return to their contracted volume *(Figure 15)*.

15   **ACTION OF DIAPHRAGM in human respiration is illustrated by a model. Colored bags represent lungs. Diaphragm is pulled downward during inhalation, expanding the lungs and sucking air into them. Relaxation of diaphragm leads to exhalation. Simplified model ignores the movements of the rib cage, which also help to fill and empty the lungs.**

Birds have even greater metabolic needs than mammals, because of the energy demands of flight; they have developed a correspondingly more efficient respiratory apparatus *(Figure 16)*.

In addition to the lungs they have a set of communicating chambers called air sacs, connected to the bronchi and the lungs, and branching throughout the body, even into the bones. By taking up some of the space inside the leg and wing bones that might otherwise be taken up by bone marrow, the air sacs help to make a bird lighter. But weight reduction is not their principal function. Nor is gas exchange, since the sac walls lack the necessary capillary networks. The sacs —or at

fill their buccal cavity with air, then force their air into the lungs; this force-pump method is quite unlike the suction-pump used by most air-breathers. All reptiles have lungs, and only a few aquatic ones (such as soft-shelled turtles) breathe through their skin too. They use a suction pump technique for filling the lungs.

16   **RESPIRATORY SYSTEM OF A BIRD includes both lungs and air sacs. The air sacs are connected to the trachea and branch throughout the body, extending even into the hollow bones.**

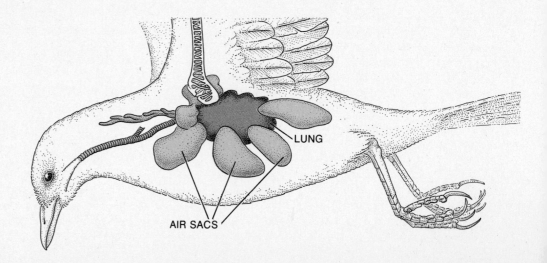

LUNG

AIR SACS

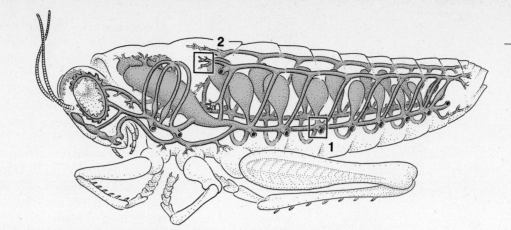

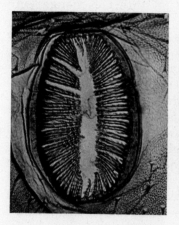
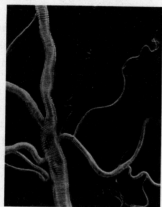

least the larger sacs that lie in the body of a bird —
act as accessory bellows. When a bird inspires air,
it passes first into the posterior air sacs. At the same
time, air previously contained in the lungs passes
forward into the anterior sacs. During the expira-
tion that follows, the anterior sacs empty to the
outside, and the posterior sacs empty their air into
the lungs. The result is that with each breathing
cycle the lungs are completely filled with fresh air.
(In mammals the lungs contain considerable "dead
air" and only renew a portion of their contents
with each breathing cycle.) Moreover, air passes
through the bird's lung in only one direction (from
back to front), which permits much greater efficiency of gas exchange across the lung surfaces. As
one might expect, the alveolar structure of the
mammalian lung is not suitable for such one-way
flow. Instead of alveoli, the birds have air capillaries: tiny tubes that traverse the lung tissue and
serve to channel the flow of air.

**TRACHEAL SYSTEM OF INSECTS is rendered in color 17
in the drawing above. Gases enter and leave through
spiracles (1) on the thorax and abdomen. Spiracles are
often fringed with bristles that filter the air and exclude
water. Highly branched tracheae extend into virtually
every part of the insect body; some are expanded into
air sacs. Gas exchange occurs in thin terminal tracheoles
(2). Photomicrographs at bottom depict details of
spiracle (left), tracheae and tracheoles at low
magnification (center), and a tracheole at a
magnification of about 80,000 × (right). Spiral
thickenings prevent walls from collapsing.**

INSECT TRACHEAE

The lungs of terrestrial vertebrates are an efficient
solution to the problem of gas exchange between
air and blood. An alternative solution is to make
direct use of the gaseous oxygen which terrestrial
life made available. The group that adopted this
strategy was the terrestrial arthropods, including
insects, spiders, and some others. Their tissues

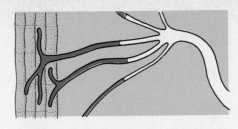

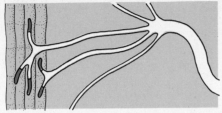

18 GAS EXCHANGE in insects is regulated by the level of liquid in the tracheoles. When the nearby tissue is active, fluid is withdrawn from the ends of the tracheoles (bottom), exposing additional surface area of tracheole for gas exchange.

19 AQUATIC INSECTS include mosquito larva (left) and water beetle Dytiscus (right). Larva breathes by surfacing and thrusting tubelike spiracles (at the end of its abdomen) into the atmosphere. Beetle surfaces briefly and sucks an air bubble into the space beneath its wings, where its spiracles open.

have access to air through a ramifying network of tubes called TRACHEAE, which open to the outside through holes (SPIRACLES) whose opening and closing are controlled by special valves, and which are protected by bristles against the entry of particles or enemies (*Figure 17*). In the smallest insects air simply diffuses into the tracheae's increasingly fine branches, the finest of which is called the TRACHEOLE, about 1 μ in diameter. Where oxygen demand is high, as in muscles, the number of tracheoles is greatly increased. Gas exchange occurs at the tips of the tracheoles (*Figure 18*). Ordinarily there is some fluid in the tips, but this is partly withdrawn during activity, thereby exposing greater surface for direct exchange with tracheolar air.

In larger insects the diffusion path of the narrow tracheae would be too long for adequate gas flow, so the insect has taken part of the outside atmosphere inside him in large air-filled sacs. Air movement in and out of these cavities is promoted by ventilating movements, especially in the abdomen. In one locust, air flows in through anterior spiracles and out through posterior ones at a rate amounting to 1/5 of the insect's body volume each minute.

Some insects evolved into aquatic forms for all or for a part of their lives. This return to the water did

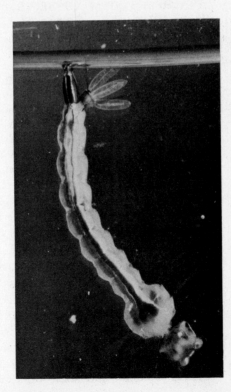

not lead to a return to the conventional watery gill. Instead, special adaptations of the tracheal system evolved *(Figure 19)*. Some aquatic insects return periodically to the surface and thrust spiracles (commonly posterior ones) into the atmosphere; the larvae of common mosquitoes, such as *Aedes*, are a familiar example. Others take the air down with them periodically; the water beetle *Dytiscus* holds a bubble under his wing-covers, and the spiracles open into the bubble. Others that live in mud steal oxygen from water-plants by thrusting a tubular spiracle into the plant's air-filled buoyancy tissue called aerenchyma. Still others such as the nymphs of damsel flies have a network of tracheae in three paddle-like structures (tracheal gills) projecting from their rear. Adaptations for aquatic life may even occur in a truly terrestrial stage of an insect. Many insect eggs have a sculptured water-repellent surface that can sustain a film of air, as shown in the photograph on the opening page of this chapter. If the eggs should be submerged on a rainy day, the trapped air prevents them from drowning.

Lungs, gills and tracheae are only part of the story of gas exchange in animals. The full story of how oxygen gets to the tissues (and CO_2 gets out) will be completed in the next chapter, which discusses the important role of blood in transporting materials to and from the animal cell.

READINGS

M.S. Gordon, G.A. Bartholomew, A.D. Grinnell, C.B. Jorgenson and F.N. White, *Animal Function: Principles and Adaptations*, 3rd Edition, New York, Macmillan, 1977. Chapter 5 of this book (which is essentially a physiological text with a strong comparative slant) presents a very good account of respiratory mechanisms in vertebrates.

P.M. Ray, *The Living Plant*, 2nd Edition, New York, Holt, Rinehart and Winston, 1972. A short and excellent introduction to functional botany with a first rate treatment of water relationships, gas exchange, and stomatal physiology.

F.B. Salisbury and C. Ross, *Plant Physiology*, 2nd Edition, Belmont, CA, Wadsworth, 1978. Contains a fairly detailed treatment of plant water relations and stomatal mechanisms.

K. Schmidt-Nielsen, *Animal Physiology: Adaptation and Environment*, New York, Cambridge Univ. Press, 1975. An excellent, clearly and provocatively written book. Particularly good because it raises new questions as it provides answers. Chapters 1, 2 and 6 are relevant to this chapter.

It is therefore necessary to conclude that the blood in animals is impelled in a circle, and is in a state of ceaseless movement; [and] that this is the act or function of the heart, which it performs by means of its pulse.

—William Harvey (1628)

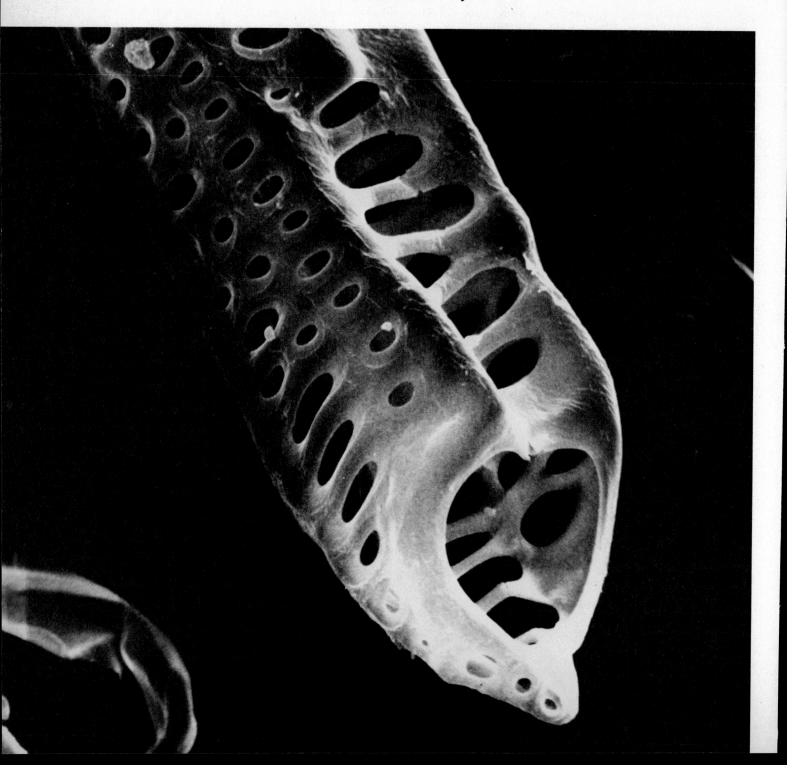

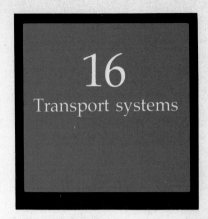

All except the smallest plants and animals need a specialized system to transport substances from cell to cell within the organism. The primary tasks of the system are to deliver nutrients to every cell and to remove waste products. The design of the system depends on the size of the organism and on how active it is. In general large organisms require more extensive systems than small ones, and active organisms require a more rapid system than inactive ones.

The differences between the lifestyles of plants and animals are reflected in the design of their transport systems. Large plants must move materials over long distances, but because they are sedentary it is not important that their biological freight be moved rapidly, nor do they need specialized systems to facilitate the movement of carbon dioxide or oxygen. But sugar synthesized in the leaves must be exported to developing flowers, fruits, roots and even young expanding buds whose need for carbohydrate may temporarily outstrip their own photosynthetic capacity. The stored carbohydrates sometimes found in bulbs, tubers and other storage organs must be mobilized on occasion and transported to growing regions, particularly after a period of winter dormancy. Hormones produced in one part of the plant must be moved to other regions. Ions and water entering the roots from the soil must be distributed throughout the aerial parts of the plant.

A physically active animal must have the machinery to move large quantities of material very quickly, first because working muscles con-

sume enormous quantities of nutrients and generate corresponding quantities of toxic wastes, and second because the tissues and organs of animals do not synthesize their own food as many plant tissues do. All animal nutrients must be imported and then distributed, which imposes additional demands on the transport system.

These design requirements have favored the evolution in animals of an elaborate plumbing system complete with pumps, valves and feedback controls. And because the chemistry of all living cells is essentially "wet" chemistry, all materials must be delivered to the cell either suspended or dissolved in water. Animals have developed a fluid transport vehicle: the blood. In humans and most other higher animals blood transports not only nutrients and wastes, but also the oxygen-carrying pigment hemoglobin, which gives blood its characteristic red color. But in some animals blood serves only as a transport medium; it is unpigmented and therefore plays no special role in oxygenation.

TRANSPORT IN PLANTS

Plants have two major pathways for long-distance transport. The xylem carries water and ions and the phloem carries carbohydrate. It is simplest to trace the transport of water and ions from the terminus of the transport system, the leaf, back to the root where the water and ions enter the plant. The force that drives the system is generated within the leaf. The water content of a spongy or palisade parenchyma cell (*Chapter 14, Figure 3*) at any given instant is determined by three factors: the concentration of dissolved solutes, the local availability of water, and the amount of resistance to stretching (elasticity) of the confining cell wall. The rate at which this cell will show net water uptake is zero when water entering in response to its osmotic po-

XYLEM VESSEL of a geranium plant is depicted in the photograph on opposite page. The xylem tissue was treated to dissolve the intercellular cement, and the separated vessel was then photographed with a scanning electron microscope. Magnification, about 2300X.

337

Plants have two major pathways for long-distance transport. Xylem carries water and ions, and phloem carries carbohydrate.

tential has developed a turgor pressure against the cell wall equal to this osmotic potential *(Chapter 2)*. Should the concentration of dissolved solutes increase —say by active ion accumulation or photosynthetic production of sugars —the osmotic potential of the cell contents is accordingly increased, net water uptake begins, and the process continues until a higher turgor is developed, equal to the new osmotic potential. The elastic properties of the wall then dictate how much water must enter before the new steady state is reached. A thin and highly elastic wall can be stretched far more

before its resistance is equal to a particular turgor pressure than can a thick and inelastic one.

In the cells of the spongy parenchyma, another very important process is at work: water evaporates from moist cell surfaces into intercellular spaces and subsequently diffuses out through the stomata. The evaporation of water from leaves is caused by heat from the sun, and it is this energy that powers the entire water and ion transport system. This evaporative water loss from leaves is called TRANSPIRATION. If water is lost from a cell by evaporation, its volume and turgor pressure decrease, and wall resistance is thus lowered. Now the osmotic potential of the cell contents is no longer fully expressed, and water can enter from an adjacent cell. For the spongy parenchyma cell which we are discussing, this adjacent cell is usually a xylem cell: either a TRACHEID or a VESSEL ELEMENT.

1 XYLEM CELLS and associated parenchyma from the stem of the pipewort plant (Aristolochia) are drawn in cross section (top) and longitudinal section (bottom). At left and center are three types of tracheid. The pair of vessel elements at far right has no end walls, and their sides are perforated by circular-bordered pits.

RINGS IN XYLEM CELLS are clearly visible in photomicrograph of longitudinal section of the stem of a squash (Cucurbita). Rings in cells at left are secondary thickenings of tracheids that matured while stem was still growing. Originally formed much closer together, they became separated as the stem elongated. One complete vessel element and parts of two others are visible at right. The two rings are not thickenings but the vestiges of end walls that became completely perforated when the elements matured, providing an open pipe for the transport of water.

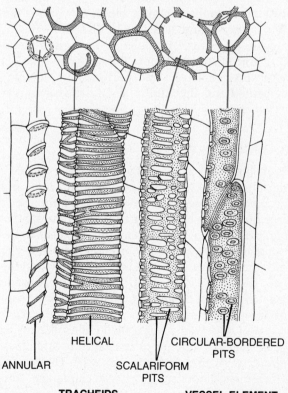

ANNULAR   HELICAL   SCALARIFORM PITS   CIRCULAR-BORDERED PITS

TRACHEIDS   **VESSEL ELEMENT**

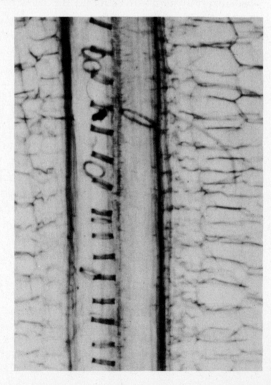

Both of these two types of cell are dead at maturity; their functional part consists only of the shell of primary and secondary wall. The tracheid is the more primitive type, found in the lower vascular plants up through the gymnosperms: the pines, firs, spruces, and their relatives (*Chapter 26*). It is also found among other places in growing regions of the stems and roots of angiosperms (flowering plants). Its primary wall is usually intact, and a secondary wall is deposited inside (*Figure 1*). Annular and helical patterns are laid down while the tissues are still elongating; they can be passively stretched to several times their initial length while still remaining open for water transport (*Figure 2*). In mature regions of a plant one finds tracheids with circular pits. Water passing from one tracheid to the next must pass across the primary wall, a limitation which makes the tracheid less efficient than its more advanced counterpart, the vessel element. The tracheid is in fact a compromise; it serves a dual function, providing both structural support and channels for movement of water and ions.

Like the tracheid, the vessel element has pitted walls; a pair with circular bordered pits appear at far right in Figure 1. Unlike tracheids, however, vessel elements have highly specialized end walls—so specialized, in fact, that they self-destruct as the vessel element reaches functional maturity. Many or a few slits may open up, or, as shown in Figure 1, the entire wall may drop out. The result is the formation of a continuous VESSEL, an open pipe consisting of many vessel elements end-to-end (*Figure 2, right, and Figures 3 and 4*). Such cells are surrounded by extremely thick-walled fibers, which lend structural support but play no role in transport.

Any water which is lost from the photosynthetic cells of a leaf is replaced by water from a xylem cell. If conditions of temperature and humidity are moderate, the system is in dynamic equilibrium: transpirational loss is balanced by osmotic uptake. If temperature is high and humidity low, the rate of evaporative loss from the leaf of a temperate plant may exceed the rate of replenishment. The cell's contents shrink, the turgor is reduced to zero, and the cell collapses. The consequence is the familiar phenomenon of wilting. If shrinkage of the cell contents is excessive, not only does the cell membrane now separate from the wall but the plasmodesmata, delicate intercellular connections, are also broken. Should this happen the wilted tissue will not recover.

If the tracheid or vessel element is full of water

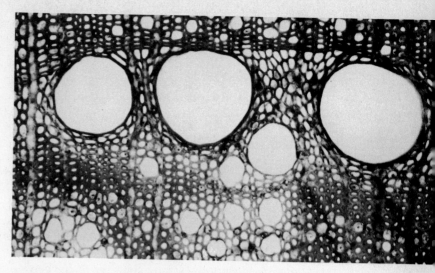

VESSEL ELEMENTS AND FIBERS in a cross section of oak wood. Three large vessel elements and several smaller ones are visible, embedded in a large number of thick-walled fibers. Small fibers at top were formed at the end of a growing season, just before winter dormancy. Adjacent large vessels and larger fibers below them were formed in the following spring. Such transitions form growth rings, which are clearly visible to the unaided eye. **3**

XYLEM VESSELS in a cucumber root. Those shown in this scanning electron micrograph have scalariform-pitted side walls. Magnification, about 1000X. **4**

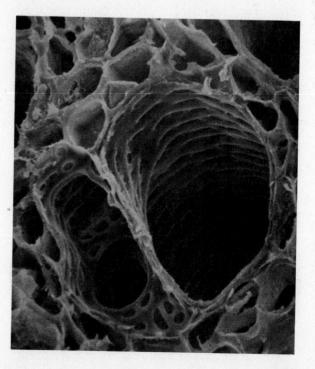

Atmospheric pressure can raise a column of water only about ten meters. A modest redwood tree raises water at least ten times higher.

when the adjacent spongy parenchyma cell begins its removal process, withdrawal of water could cause the xylem element to collapse, except for the extensive deposition of secondary wall material. To empty the cell of its contents one would have to make a space inside by pulling water molecules either away from the wall or away from each other. But because of the strong cohesive forces between water molecules, and the strong adhesive forces between water and the wall material, water removed from one xylem cell is simply replaced by water from an adjacent one. There is a continuous ribbon of water in the xylem from leaf to root, so transpirational loss of water from the leaf pulls water from the underground regions of the plant. This bulk flow is called the transpiration stream.

It is incorrect to think of this process as suction, which is simply removal of air from a system. If the lower end of a suction system is immersed in water, atmospheric pressure simply forces the water in to replace the evacuated air. But atmospheric pressure is only sufficient to raise a column of water about ten meters; clearly trees can move water to much greater heights. A modest redwood tree raises water at least ten times as high. A far better explanation is that of negative pressure.

5   CROSS SECTIONS OF ROOTS. Root of dicot has solid star-shaped mass of xylem but not pith. Root of monocot has separate strands of xylem and phloem. In monocot pith grades into pericycle. Water can enter through root hairs and follow osmotic gradient to xylem.

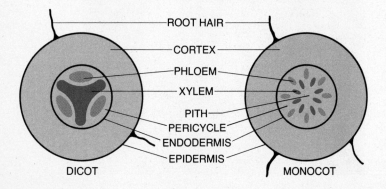

Consider a tree 30 meters tall. If there is no resistance to the movement of water, the equivalent of three atmospheres of pressure is needed to raise water to its crown. Suction is inadequate, but such forces can readily be developed by evaporation. Transpiration can easily develop the necessary three atmospheres of negative pressure to do the job, and the powerful adhesive and cohesive forces prevent the column of water from breaking under tension.

The xylem system of a pine tree, as mentioned above, is an inefficient one. In passing from one tracheid to the next, water must pass through a double layer of primary wall *(Figure 1)*. In the most advanced plants the vessels are far better: the vessel elements of an oak or maple tree, for example, may be close to a millimeter in diameter, and at maturity their end walls are missing. A column of such elements may form a continuous pipe from root to leaf. In any case, the actual tension needed to raise water to the top of our ten-meter tree may be closer to six atmospheres.

Proteins and carbohydrates, among other substances, have a very high affinity for water. If one drops some dry peas into water and confines them, they will absorb water and swell, developing enormous pressure against the walls of their confining vessel. The polar water molecules become associated with local charged sites on the protein or carbohydrate molecules, and become packed closely together. This packing of solvent molecules on the surface of a larger molecule is called IMBIBITION. Imbibed water occupies a smaller volume than water in its free liquid state. Moreover, as a water molecule becomes bound its free movement is reduced and it loses kinetic energy as heat. Thus imbibing peas not only swell, but also become substantially warmer. The magnitude of forces which can be developed by imbibing systems is worth considering: they may be well in excess of 1000 atmospheres. Thus it is not surprising that even a water column under considerable tension in the xylem does not part from the cell wall. Nor is it surprising that badly wilted leaves —from which some water may have been evaporated from the cell walls —show an affinity for water far above their osmotic potential. In fact, they also possess a large imbibitional potential.

Water must be moved from the soil across the root tissue to the root xylem *(Figures 5 & 6)*. It could enter directly into root hair cells following an osmotic potential gradient (solutes in the soil are usually far less concentrated than solutes in the cell). The negative pressures in the xylem are suffi-

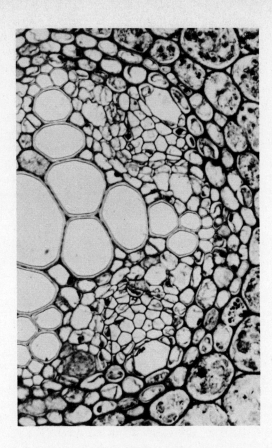

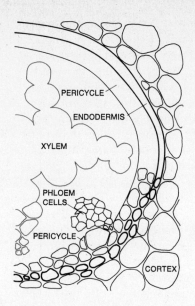

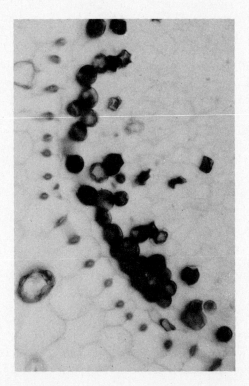

PHOTOMICROGRAPH OF ROOT illustrates another **6** **pathway for entry of water. Photograph at left depicts cross section of root of pea plant. Drawing above identifies principal structures in photo. Water entering root can move between cells, rather than through them, until it reaches the endodermis. Once past that barrier, water can readily enter the xylem.**

cient to remove water from the innermost living cells which in turn could obtain water in replacement from the next cells out, and so on, ultimately removing water from the root hair. A second possible pathway is through the spaces between the root cells. A substantial amount of water probably moves through the root cortex this way. But just before it reaches the xylem it meets an important barrier, a monolayer of cells called the ENDODERMIS. These cells have no intercellular spaces, and their radial walls are all impregnated with suberin—a most effective waxy waterproofing. Thus anything passing from soil to root xylem must pass through the cell membrane and cytoplasm of at least one living cell—the endodermal cell. Endodermal cells are known to be sites of vigorous metabolic activity, and serve as gatekeepers determining what ultimately will or will not find its way into the xylem *(Figures 6 & 7).*

7 **ENDODERMAL CELLS determine what substances enter the xylem. Photomicrograph depicts cross-section of root of the yew (Taxus). Darkly stained pericycle cells contain tannin-like substance. Casparian strips are "gaskets" that ring the endodermal cells and seal the spaces between them.**

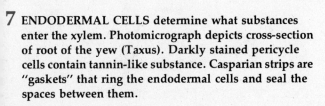

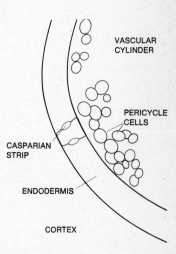

It is clear that water could cross from soil to xylem simply following an osmotic potential gradient. The concentration of solute in the xylem sap is higher than that in the soil, and the endodermis provides an effective semipermeable barrier. Thus the root at least from the endodermis in can be thought of as a sausage casing with higher concentration of solute inside than out. If the root is placed in distilled water, the water enters the xylem and can be raised to substantial heights by simple osmosis. This ROOT PRESSURE certainly plays a role in the elevation of water to higher portions of the plant. However, measured values are far from adequate to move water to the necessary heights of tall trees. Even more important, the rate of the process is far too slow to account for the large volumes of water moved. A tobacco plant at a temperature of 35°C, and relative humidity of 66 percent transpires in excess of 350 ml of water per hour for each square meter of leaf surface. The cut surface above its roots, however, may exude only a few ml per day because of root pressure.

The energy for root pressures is metabolic in origin—the plant uses the energy of ATP to accumulate ions by active transport. The resulting solute gradient is then at least partially responsible for the movement of water in the root. Salts are passed from cell to cell and ultimately secreted across the endodermis and eventually into the xylem. Once the salts reach the xylem, they are simply passively carried in the xylem to the upper portions of the plants, where their uptake by living cells may again be by active transport.

In brief, roots show a substantial water deficit, a deficit of dual origin. First, their own solutes give them an osmotic potential which leads to water uptake. Second and probably more important, transpirational loss of water from the leaves creates a water deficit there which is physically transmitted to the roots by tension on the columns of water in the xylem, and can be added to the local osmotic component. These two forces drive the long distance movement of water and ions.

TRANSPORT OF CARBOHYDRATES

Though xylem tissue may contain some living cells, either as longitudinal strands of parenchyma or as radial bands of tissue—the xylem rays—the cells through which water and ions move are dead and empty at functional maturity. They are simply plumbing. The phloem, on the other hand, through which carbohydrates are moved, contains live, active cells. Individual phloem cells, unlike individual xylem elements, can and do play a role in the movement of materials through this tissue.

It is worth considering the structure of a mature and functional phloem cell. As was the case with xylem, there are somewhat different cell types in the primitive vascular plants and gymnosperms (such as conifers) than in the angiosperms (flowering plants), although the basic features of the plumbing are similar. The more primitive cell type, analogous to the xylem tracheid, is the SIEVE CELL. Elongate and tapered, it has numerous areas of the wall which are especially thin and perforated by numerous minute cytoplasmic strands leading to adjacent sieve cells. These sieve areas provide the only possible transport route from one cell to the next. Each sieve cell is usually in contact with several parenchyma cells. The more advanced cell type, analogous to the xylem vessel element is the SIEVE TUBE ELEMENT. The sieve areas are concentrated on the end walls forming SIEVE PLATES, penetrated by cytoplasmic strands of substantial diameter, which lead from one sieve tube element to the next in a vertical column. These show clearly in cross section in Figure 8. The mature sieve element is singularly uninteresting under the electron microscope; there is no nucleus, rarely a secondary wall, and only a very thin peripheral region of cytoplasm with at most a very few mitochondria.

CROSS-SECTION OF PHLOEM of squash plant shows **8** **two sieve plates (center). The surrounding sieve tube cells were sectioned between end walls and appear clear except for distorted cytoplasm.**

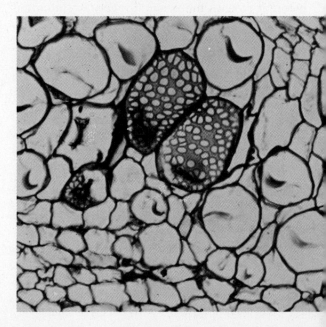

There are no vacuoles, and the interior of the cell is filled with a watery solution containing a rather indistinctive proteinaceous material inelegantly called slime.

Associated with the sieve tube element, is an adjacent COMPANION CELL. It has a normal nucleus, and is packed with mitochondria, endoplasmic reticulum, dictyosomes, and various other organelles. The inner surface of its wall, toward the sieve tube element, is elaborately sculptured, and there is hence a very large surface area of membrane present between the two cells on the companion cell side. The best guess is that the companion cell provides both energy and any needed nuclear information for itself and for the adjacent sieve tube element. In the more primitive case, phloem parenchyma cells play the same role for sieve cells, but the association is clearly not as intimate.

Although it is clear that the products of photosynthesis are translocated through the sieve tubes of the phloem, there is considerable controversy over the mechanism of movement. The most appealing theory has been that there is some sort of mass flow through the sieve tubes, although mass flow through the restrictive sieve plates simply seems unlikely. Work with inhibitors and some ingenius insects has provided at least a reasonable model. In the first place, movement of material through the phloem requires energy. It is readily poisoned with the usual metabolic inhibitors, and is slowed down by lowering the temperature. Movement of xylem sap is unaffected by such treatments. In the second place, mass flow can and does occur. Unfortunately, if one severs a sieve tube, the slime plugs up the cell end nearest the cut, and mass flow of sieve tube contents ceases. Aphids, however, seem to be able to circumvent this problem. They insert their elongated mouth parts (stylets) right through any surrounding tissue and into the sieve tube (*Figure 9*). Evidently they are gentle enough so that slime plugs do not form. If one now removes all of the aphid but the stylet, pure sieve tube contents will continue to exude for a period of hours or even days. Actually, one need not perform such an aphidectomy. The aphid pumps the phloem contents through its digestive tract, extracting only a few amino acids, some sugar, and a few other trivia, and what emerges is at least qualitatively the same as what one obtained from the stylet alone. The first analyses of aphid excrement —euphemistically called honeydew —showed that it contained between 10 and 25 percent sucrose, a far higher concentration than was known to occur in any other plant cell. Moreover the sugar was unphosphorylated, unlike most sugars in photosynthesizing cells.

Most plant physiologists believe that the sugars produced by photosynthesis eventually find their way to the phloem as glucose phosphates. They enter the companion cells where they are converted to the unphosphorylated disaccharide sucrose and actively secreted into the sieve tubes. The extensive area of companion-cell membrane next to the sieve tube element facilitates the process. As the sugar concentration increases in the sieve tubes, their osmotic potential accordingly also increases —reaching values far higher than those of surrounding cells. Water, available either in the adjacent xylem or nearby parenchyma cells, therefore enters causing the development of substantial pressure. Wall resistance prevents excessive stretching, and the consequence of the increased pressure is that the sugar solution is forced out through the sieve plate into the next element down the line. The direction of flow is determined by relative rates of sugar production and utilization in different parts of the plant. In the roots, sugar leaves the phloem, possibly by the reverse process. In many trees, most notably sugar maples, large amounts of sugar are removed from the phloem

STYLET OF AN APHID (center) penetrates a sieve tube. Because aphids can puncture tube so gently that no slime plug forms, experimenters use them to sample the contents of phloem. 9

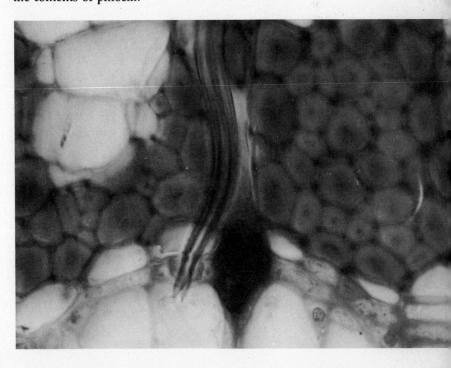

and moved into the living cells of the xylem rays for winter storage. The release of this sugar in large amounts in the spring, both into xylem and phloem, produces the sap from which maple sugar is obtained.

The movement of sugar is a metabolically active process, and the actual driving force is a metabolically generated osmotic potential difference. Movement is from sugar source to sugar sink — normally from leaves to roots, but also frequently from leaves to developing flowers or fruits (consider the rapidly enlarging pumpkin, gaining over 40 grams of dry matter per day) or storage organs *(Figure 10)*.

Though most people accept this pressure flow hypothesis as reasonable, there is still one troublesome fact: the sieve plate or sieve area is eminently unsuited for any kind of bulk flow. The situation is made far worse by the enormous viscosity of the flowing sugar solution. It is clear that at present one must remain open-minded.

Xylem and phloem provide transport pathways for virtually everything in the plant that has to be moved over a long distance. Amino acids can be readily moved along in the phloem, as can hormones such as gibberellins. Nitrogen taken up by the roots as nitrate is reduced to ammonium ion, which then reacts with glutamic or aspartic acids to form glutamine or asparagine, both of which find their way to the xylem for transport to upper parts of the plant. Xylem exudate frequently contains

substantial amounts of cytokinins suggesting that this class of hormone can move upward with the transpiration stream as well. The only serious exception is found with auxin. It may move in the phloem, but in many plants the rates of movement are far too slow, and other living cells are implicated. These hormones are treated in Chapter 18.

TRANSPORT IN ANIMALS

The metabolism of even the most rapidly growing plants proceeds at a leisurely pace compared with that of an active animal. In small invertebrates, dissolved gases, nutrients and waste products are transported throughout the animal by simple diffusion. This is a slow process, and it becomes inadequate for transport distances of a millimeter or more. Most large animals distribute materials internally via the physical movement of bodily fluids. Whether pigmented or not, the fluids are usually called blood, although they may differ greatly in their physiological properties.

Naturally there are exceptions to such generalizations. In jellyfish —even the large species with bell diameters of more than two meters —transport still depends solely on diffusion. The bulk of a jellyfish consists of dead "jelly" (mesoglea); the living tissue is merely a single-celled layer that covers the outside surfaces of the animal. Substances entering or leaving the animal never need to diffuse across more than one layer of cells.

The complexity of true transport systems, such as those of higher animals, depends on how effectively body fluids —mainly blood —are moved, and on the nature of the pathways for the fluids. The simplest transport systems are found in animals such as nematode worms *(Figure 11)*, in which the blood fills all spaces between internal

10 **PRESSURE-FLOW HYPOTHESIS explains transport of substances through xylem and phloem. In this model, dark particles represent sugar molecules and light particles water molecules. Beaker at left represents roots and beaker at right, leaves. Osmotic flow of water into sugar solution creates pressure that drives flow in direction indicated by arrows. Hypothesis is still the subject of debate.**

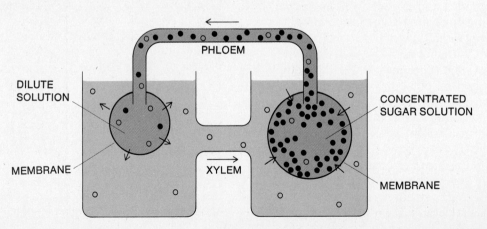

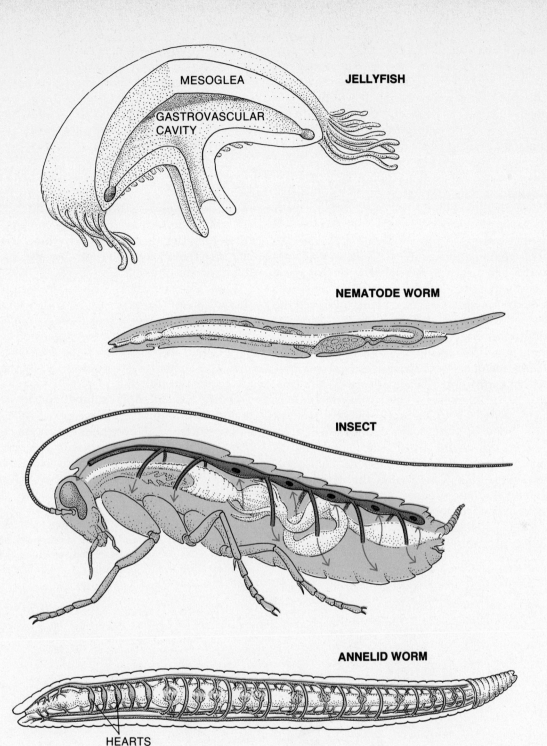

JELLYFISH

MESOGLEA

GASTROVASCULAR CAVITY

NEMATODE WORM

INSECT

ANNELID WORM

HEARTS

ANIMAL TRANSPORT SYSTEMS. Jellyfish have no circulatory system and rely largely on passive diffusion. Nematode worms rely on muscles of body wall to swish around the blood (color) that bathes their organs. Insects also have organs bathed in blood, but blood is circulated by tubular heart (dark color) suspended in it. Annelid worms have closed circulatory system typical of advanced organisms; blood is driven through branched vessels by one or more hearts.

11

organs. The blood simply sloshes around haphazardly when the animal moves. Arthropods (such as insects) have a more complex system. The blood is still an open pool that surrounds all organs, but it is stirred by a tubular heart which fills and empties itself rhythmically. Valves and vessels are often associated with the heart, giving both direction and greater efficiency to the stirring process. Insects are extremely active animals capable of sustaining high metabolic rates; one may wonder how this is possible with such a rudimentary transport system. The answer is that insects do not rely on blood for the transport of oxygen. For this purpose they have a very effective tracheal system *(Chapter 15)*. In arthropods such as crustacea, which lack a tracheal system and are generally less active than insects, oxygen is carried and distributed by the blood.

These open systems contrast with the closed transport systems of other animals, in which the blood is confined to a single set of branching vessels, through which it is driven by one or more hearts. Such closed systems are truly circulatory; that is, the blood travels in a circuit, following the same pathway again and again as it is pumped through the body. Closed circulatory systems occur in annelid worms (earthworms), cephalopod mollusks (octopus, squid), and of course vertebrates. In earthworms the circulatory system consists of a set of dorsal and ventral longitudinal vessels, joined along the sides by dorsoventral vessels. Some of these dorsoventral vessels serve as hearts. Within the tissues the larger vessels branch into networks of fine vessels called capillaries, whose thin walls permit the exchange of nutrients, respiratory gases, and other materials. All closed circulatory systems have hearts, and all depend on capillary networks for the exchange of materials between blood and cells. There are always two principal sets of capillary networks: one set associated with respiratory organs (gills, lungs) whose principal function is oxygen uptake, and one associated with the other tissues and organs of the body, whose function includes not only the delivery of oxygen, but also the exchange of nutrients and other materials.

THE HEART

Hearts as a rule consist of at least one thin walled collecting bag (atrium or auricle) connected to at least one thickly-muscled rhythmic pumping chamber (ventricle), and a set of valves to control the direction of blood flow.

A sensible circulatory system, one would think, should consist of a single circuit with a single heart that pumps the blood first through the lungs or gills, and then through other tissues. The drawback is that the blood has to go through the two principal capillary systems in series. By the time it has gone through the respiratory capillaries its pressure has fallen substantially, and it cannot then pass through the other capillary system with equal efficiency. Nevertheless, fish do precisely this, and they have a heart with but one atrium and one ventricle. Cephalopods have overcome the pressure-drop problem by developing extra

CIRCULATION IN SQUID, an active mollusk with high respiratory demands, depends on three separate hearts: central systemic heart and two booster hearts that force blood through gills.

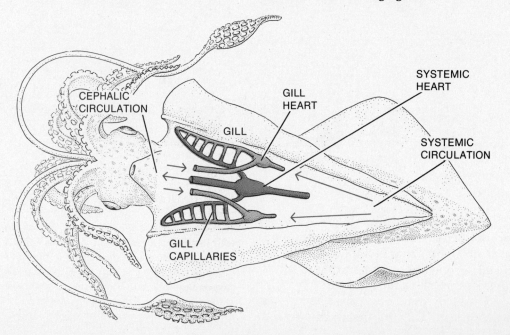

booster hearts (gill hearts) which collect deoxygenated blood, force it through the gills, and thence to the systemic heart which pumps it through the other tissues *(Figure 12)*.

In Amphibia, as in humans, there are two circulatory pathways, one to the lungs (PULMONARY CIRCULATION) and one to the many other organs of the body (SYSTEMIC CIRCULATION). A single strongly-muscled heart pumps the blood through both circuits at once and there are no booster hearts *(Figure 13)*. The heart has two atria, each receiving blood from one of the circuits, but only one ventricle. The disadvantage of a single ventricle is that the deoxygenated and oxygenated blood returning from the two circuits are mixed in the ventricle before they are sent out again by the heart. Ideally, all the "spent" deoxygenated blood returning from the systemic circulation should be sent to the lungs, and the "fresh" oxygenated blood from the lungs should all be sent into the systemic pathway. The circulatory system of amphibians is not as efficient as it might be if the ventricle were divided by a partition that prevented such mixing. Most reptiles have three-chambered hearts comparable to those of Amphibia. Both reptiles and amphibians are "cold blooded" and have metabolic rates that vary with ambient temperature; their demands on circulatory efficiency are therefore not as great as they are in "warm-blooded" animals such as birds and mammals, which maintain high metabolic rates.

In birds and mammals (as well as in crocodilian reptiles) the ventricle is partitioned lengthwise, so that there are essentially two ventricles *(Figure 13)*. Oxygenated blood from the left auricle passes into the left ventricle and is pumped unmixed into the systemic circulation, while deoxygenated blood returning from the systemic circulation passes into the right auricle and ventricle, and is delivered to the lungs.

THE HUMAN CIRCULATORY SYSTEM

Humans possess a typical mammalian circulatory system. The heart is four-chambered, and the blood vessels consist of arteries, veins, and capillaries. Arteries carry blood away from the heart, while veins carry blood toward it. Small vessels are called arterioles and venules.

CIRCULATION IN VERTEBRATES. Heart of fish **13** consists of a single atrium and ventricle. Respiratory and systemic circulation are connected in series. Heart of amphibian has only one ventricle, in which oxygenated arterial blood (dark color) and deoxygenated venous blood (light color) mix to a certain degree. Respiratory and systemic pathways of amphibians are connected in parallel, as in higher organisms. In mammals, birds and some reptiles, mixing of arterial and venous blood is prevented by a partition that divides ventricles into left and right chambers. In effect, partition creates two hearts: one for pumping blood to lungs, the other for pumping blood to the rest of the body.

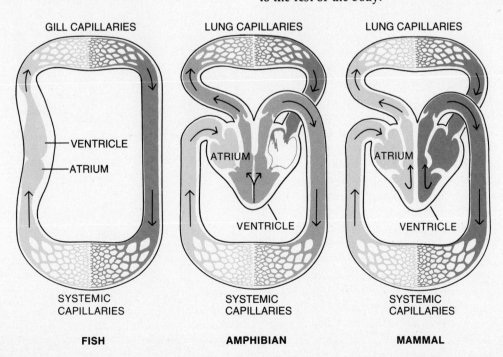

GILL CAPILLARIES   LUNG CAPILLARIES   LUNG CAPILLARIES

VENTRICLE

ATRIUM   ATRIUM

VENTRICLE   VENTRICLE

SYSTEMIC CAPILLARIES   SYSTEMIC CAPILLARIES   SYSTEMIC CAPILLARIES

FISH   **AMPHIBIAN**   **MAMMAL**

The human heart is an indefatigable organ that pumps more than 280 liters (75 gallons) of blood per hour, or about 18 million barrels in a lifetime.

The blood leaves the right ventricle through the pulmonary artery, then passes into the lungs by way of its two branches. In the lungs the branches divide into arterioles and capillaries; the blood becomes oxygenated and loses carbon dioxide. Capillaries emerging from the lungs reassemble into venules and then into pulmonary veins, which return the blood to the left atrium. It then passes into the left ventricle, which forces it out through the aorta, the major artery of the systemic pathway. Many branches lead off from the aorta, bringing oxygenated blood to the various tissues and organs of the body. The first branches are the two coronary arteries that irrigate the heart itself. Other branches lead to the arms and head, the liver, gut, spleen, kidneys and legs *(Figure 14)*. An important vessel is the hepatic portal vein, which conveys freshly absorbed nutrients from the gut to the liver for processing. Deoxygenated blood from the various capillary beds in the systemic circulation eventually flows into two major veins, the anterior and posterior vena cava, which return it to the right atrium of the heart.

The human heart is an indefatigable organ that pumps more than 280 liters (75 gallons) of blood per hour, or about 18 million barrels in a lifetime. The left ventricle is the most powerfully muscled of its four chambers. It supplies the force that pumps the blood through the many capillary networks of the systemic circuit. The right ventricle forces blood only through the lungs, and is comparatively weaker. The two thin-walled atria are weaker still, because their only function is to help fill the ventricles when these expand. Four one-way valves direct the flow of blood within the heart. There are two atrio-ventricular valves, the bicuspid between the left atrium and ventricle, and the tricuspid between the right atrium and ventricle. Blood leaves the ventricles through two semilunar valves *(Figure 15)*. Defects in these valves can lead to backflow or insufficient flow through the valves during the heartbeat, with consequent loss of cardiac efficiency. Such abnormalities are accompanied by audible "hisses" and "whooshes" called

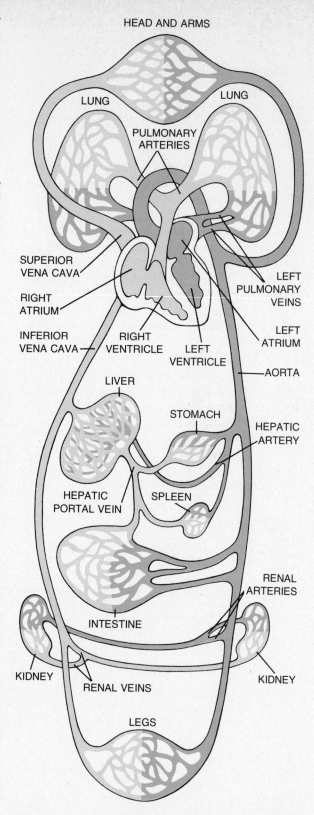

HUMAN CIRCULATORY SYSTEM follows pattern found in other mammals. Arterial blood is shown in dark color, venous blood in light color. This schematic diagram depicts only a few of the major vessels.

14

HEART MURMURS, which a physician can detect with a stethoscope. Rheumatic fever can lead to stenosis (narrowing) of the valves, or to an inability to close properly (valvular insufficiency). The valves are sometimes defective at birth. Surgeons can replace faulty valves with artificial ones. This requires open-heart surgery, but the procedure has become far less risky than it used to be.

The heart of humans, like those of all vertebrates, is MYOGENIC. This means that the heartbeat is initiated within the heart itself, rather than by an outside stimulus. This can be readily demonstrated in the laboratory; the heart of a frog or a turtle, placed in appropriate saline solution, will beat for hours after it has been removed from the animal. The heartbeat is triggered by a small mass of modified muscle tissue called the sino-atrial node (S-A NODE) located in the rear wall of the right atrium. Nodal tissue has the characteristics of both nerve and muscle: it can initiate action potentials and it can contract. The S-A node sets the tempo of the heart. In a resting human being, it emits waves of excitation at the rate of about 70 per minute. They spread over the walls of both atria, inducing them to contract; the right atrium, which receives the signals first, contracts an instant before the left. The signals also reach another node, the atrio-ventricular or A-V NODE, embedded in the partition between the atria. This node is a relay center that transmits the signal from the A-V node to the ventricles. It responds with a signal of its own every time it receives one from the S-A node. It transmits this signal to the muscles of the ventricles by way of a bundle of nodal fibers (Purkinje fibers) whose branches penetrate all parts of the ventricular walls. The ventricles respond by contracting simul-

taneously. The A-V node does not respond immediately to each signal from the S-A node, but does so after a brief delay of about 0.1 second. The delay ensures that the ventricles contract slightly later than the atria, which is essential for proper functioning of the heart. The pacemaker of the heart can become damaged through aging or disease. Artificial pacemakers can stimulate the heart electrically, replacing the damaged nodal centers.

Although the beat itself is initiated in the heart, the rate of the beat is adjusted by nerves and hormones. As anyone knows, exercise or excitement can easily lead to a doubling of the heart rate. Sympathetic and parasympathetic nerves (*Chapter 19*) act on both nodes. Sympathetic stimulation increases the heart rate, while parasympathetic stimulation decreases it. The hormone adrenalin, released under stress, reaches the heart via the bloodstream and causes it to beat faster and more vigorously—the pounding in the chest that one feels when angry or suddenly frightened.

The beating heart alternately contracts (systole) and relaxes (diastole). During systole the heart empties itself and the pressure of the blood reaches a maximum. During diastole the output of the heart drops to zero, and so, one might think, should the blood pressure. But this doesn't happen, because the arteries have thick elastic walls (*Figure 16*). During systole they expand, then recoil slowly during diastole, thus maintaining substan-

HUMAN HEART is supplied with oxygenated blood by coronary arteries (left). Diagrams at right illustrate cardiac cycle. Relaxation of heart (diastole) sucks blood into the two atria at top of heart. Contraction (systole) pumps blood outward to lungs and body. Leaflike valves prevent back flow. 15

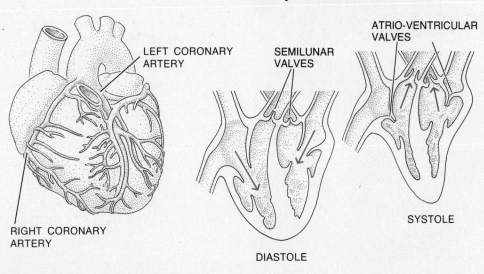

LEFT CORONARY ARTERY

RIGHT CORONARY ARTERY

SEMILUNAR VALVES

ATRIO-VENTRICULAR VALVES

DIASTOLE

SYSTOLE

tial pressure even during the instant when no new blood is being pumped into them. Naturally the pressure is highest in the giant arteries nearest the heart. Because of fluid friction within the vessel walls, blood pressure drops steadily at increasing distances from the heart. In other words, the actual magnitude of the blood pressure depends on where it is measured. Physicians routinely measure it in the upper left arm. The value is expressed as the ratio of systolic to diastolic pressure, in the same units —millimeters of mercury —that weathermen use to indicate atmospheric pressure. For a normal college-age male, blood pressure is about 120/80 mm Hg; for a female, slightly less.

16 **VERTEBRATE BLOOD VESSELS. Photomicrograph at top shows cross-section of a large vein and a small artery. Artery (right) can be distinguished by its firm muscular wall. Below, red blood cells squeeze through capillaries. Cross-section view of capillary appears in Chapter 17.**

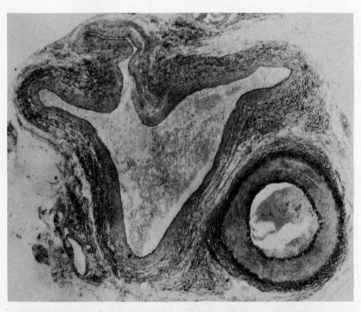

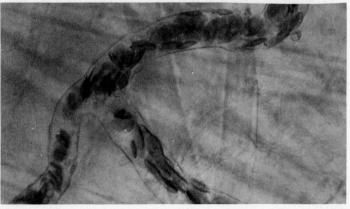

By the time the blood has passed through the capillaries and reaches the veins, its pressure is too low to return to the heart without some additional help. In humans the problem is particularly acute because we stand erect; venous blood from most of the body must overcome gravity to get back to the heart. Veins, unlike arteries, are thin-walled, flaccid and collapsible, as can be seen in Figure 16. Veins must be regularly squeezed and deformed to keep the blood moving. Because veins contain tiny valves at regular intervals to prevent back flow, the squeezing pushes the blood in only one direction: toward the heart. The normal activity of the body muscles provides most of the squeezing force. Breathing movements also help; when the thorax expands during inspiration, the resulting negative pressure causes the major veins of the thoracic cavity to expand and to fill with venous blood from the lower regions of the body. In other words, proper venous return is dependent on the maintenance of at least some activity in the skeletal musculature.

If venous return is impaired, the heart output becomes inadequate; not enough blood reaches the brain, and the person faints. This is precisely what happens, for example, to the soldier who stands at attention too long. Fainting is a biologically adaptive response. When one lies flat, the circulation stands a better chance to recover because it is no longer opposed by gravity.

THE CAPILLARY SYSTEM

While arteries and veins provide for the mass movement of blood, capillaries are the site of the exchange process. These microscopic vessels permeate every tissue of the body, bringing oxygen and nutrients to the cells and carrying away waste products and carbon dioxide. Most capillaries are only wide enough (about .01 mm) to admit red blood cells in single file, as shown in Figure 16. Despite their narrow bore, their collective diameter is larger than that of their parent artery. This is why blood flows slower in a capillary network, like a river in a widened channel. Slow capillary flow

Placed end to end, human capillaries would comprise a thin tube that would encircle the Earth more than twice.

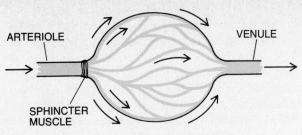

ARTERIOLE · VENULE

SPHINCTER
MUSCLE

CAPILLARY NETWORK

7 FLOW THROUGH CAPILLARY NETWORK is controlled in part by a sphincter muscle that adjusts the diameter of the arteriole leading into the network.

provides more time for the exchange processes to run their course. The number and length of capillaries in the human body are impressive. Placed end to end, it has been estimated, they would comprise a tube that could encircle the Earth more than twice. The surface area of the capillary walls, across which the exchange takes place, is also enormous. In a dog, each cubic centimeter of muscle is provided with a capillary exchange surface of about 500 square centimeters.

When blood enters the capillary network, its pressure is still relatively high. This pressure forces blood fluids, together with dissolved nutrients and oxygen, into the tissues that surround the capillaries. By the time the blood reaches the venous end of the network, its pressure has dropped substantially, and fluid is no longer squeezed out. Osmotic forces then come into play, pulling materials such as CO_2 and wastes, which are at a higher concentration outside the capillaries, into the vessels. Some of the water and other materials lost at the arterial end of the network also re-enter the capillaries by osmosis. But there is always some residual fluid that does not re-enter; it is returned to the blood by way of the lymphatic system.

Blood flow through a capillary network is partly controlled by the action of sphincter muscles in the arterioles *(Figure 17)*. Nerves and hormones regulate the tension of these muscles. By dilating the arterioles in one part and constricting them in another, an organism can adjust the blood flow according to the needs of various regions. After a large meal, for example, extra blood can be shunted to the intestine, and during exercise, to the muscles. During stress adrenalin constricts the blood vessels in the skin (causing an angry or frightened person to turn pale) and dilates the vessels in the skeletal muscles (preparing them for flight or fight). In warm-blooded organisms the regulation of flow through the skin capillaries

plays an important role in heat regulation. In hot weather or after strenuous exercise, when muscular action generates heat, the body is cooled by increased skin circulation; conversely, during cold weather the peripheral circulation is curtailed.

BLOOD

Blood is a fluid tissue. A human being has about five liters of blood. About 50 to 60 percent of it is the yellowish liquid matrix called PLASMA; the remainder is comprised of red blood cells (ERYTHROCYTES), white blood cells (LEUCOCYTES) and PLATELETS. The plasma is mostly water, but contains many other important molecules and ions, including nutrients (fats, glucose and amino acids), hormones, antibodies, nitrogenous waste products (in mammals, mainly urea), and numerous cations and anions.

Erythrocytes are concave disk-shaped cells *(Figure 18)* whose main function is to transport oxygen. They are essentially floating sacs of hemoglobin; each contains about 280 million molecules of this pigment. Their flattened shape provides a greater surface-to-volume ratio which enhances the uptake and delivery of oxygen. Erythrocytes do not divide. They are produced constantly in the bone marrow at a rate of about 140 million per minute, then enter the bloodstream. After about four months the worn-out erythrocytes are destroyed by phagocytic cells in the liver. In mammals the nuclei are extruded from the erythrocytes before

HUMAN BLOOD CELLS. Rounded discs are red blood 18 cells (erythrocytes) that contain hemoglobin. One white blood cell (a leucocyte) is visible at left center, and part of another at lower right corner.

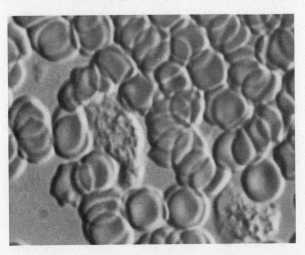

they enter the circulation; the erythrocytes of other vertebrates, however, do have nuclei. Mammalian erythrocytes contain more hemoglobin per unit volume; some workers have suggested that this is the adaptive advantage of the lack of nuclei. A considerable reserve of erythrocytes can be quickly mobilized from the spleen in an emergency. This organ stores blood in large cavities connected to its capillaries. When blood is lost through hemorrhage, or when the demand for erythrocytes rises because of an oxygen shortage in the air (as at high altitudes), the spleen squeezes its blood into the circulation.

Leucocytes are larger and far less numerous than erythrocytes. They have nuclei and are colorless (hence the term white blood cell). They move like amoebas by extending pseudopods and can enter the tissues by squeezing through the junctions between the cells of the capillary wall. Leucocytes are involved in the defense against foreign materials and microorganisms. They are attracted to sites of injury and infection, moving apparently in response to chemicals released at these sites. The leucocyte count in the blood and lymph may rise sharply during infection, providing a useful criterion for the diagnosis of infection.

There are five types of leucocytes; the most abundant are the neutrophils and lymphocytes. Neutrophils are produced in the bone marrow. They mop up bacteria, viruses and scraps of damaged tissue by engulfing them with pseudopods. Lymphocytes play a major role in the body's immune responses. Produced in the lymph nodes, thymus gland, spleen, tonsils and bone marrow, they are involved in antibody production, graft rejection and other immunologic reactions *(Chapter 17).*

Platelets are not whole cells but fragments of the cytoplasm of giant bone-marrow cells called megakaryocytes. They play a crucial role in the formation of blood clots. When platelets encounter a ruptured vessel or damaged tissue, they and the injured tissue release a group of substances called THROMBOPLASTINS. These substances convert a blood protein, PROTHROMBIN, to THROMBIN, which in turn changes FIBRINOGEN into FIBRIN, the insoluble protein that forms the fibrous meshwork of the hardened clot. Fibrin shrinks as it forms, pulling the edges of a wound together and helping to seal it. The details of the clotting process are much more complicated and involve the participation of several other chemical factors, including vitamin K and calcium ions. Agents that remove calcium can serve as anticoagulants, and are usually added to whole blood being stored for transfusions.

THE LYMPHATIC SYSTEM

Earlier in the chapter we mentioned that some of the filtrate from the blood in the capillaries is collected not by the venous capillaries but by the lymphatic system. This fluid, which builds up in the spaces outside the capillaries, contains water, solutes and white blood cells that have left the capillaries, but no red cells. Lymphatic vessels begin as fine capillaries, merging progressively into larger and larger vessels, and ending in a major vessel, the thoracic duct, which empties into the superior vena cava *(Figure 19).* The lymph, as the fluid in the lymphatic system is called, is thus

LYMPHATIC SYSTEM collects lymphatic fluid from the tissue spaces. Lymph capillaries and ducts (dark color) drain into the thoracic duct, which empties into the bloodstream via a junction with the vena cava (light color). Major lymph nodes are shown as swellings along lymph capillaries. **19**

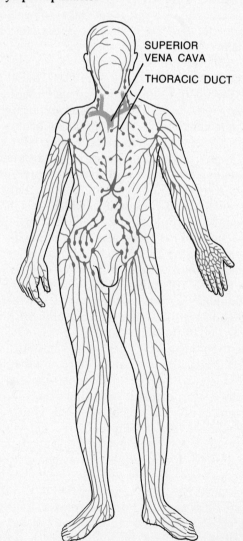

SUPERIOR
VENA CAVA

THORACIC DUCT

dumped back into the blood. Like veins, lymphatic vessels have one-way valves, and as in veins, the fluid within them is pushed along by pressure from adjacent skeletal muscles. Some non-mammalian vertebrates also have special lymph hearts that help with the pumping.

Mammals and birds have lymph nodes along the major lymphatic vessels. Lymph nodes are an important component of the defensive machinery of the body. As noted above, they are a major site of lymphocyte production, and of the phagocytic action that removes microorganisms and other foreign materials from the circulation. The lymph nodes also act as mechanical filters. Soot and dust particles that cannot be digested by the phagocytes are entrapped in the nodes. Lymph nodes tend to swell during infection. Some of them, particularly those on the side of the neck or in the armpit, become noticeable at such times. The nodes also trap metastasized cancer cells: those that have broken free of the original tumor. Because such cells may start additional tumors, physicians often remove the neighboring lymph nodes when they excise a malignant tumor.

It is crucial that there be a correct balance between the rate of lymph formation and its removal by the lymphatic system. The British physiologist, E. H. Starling, proposed early in this century that the balance was dictated by three factors. Starling's hypothesis was that the rate of lymph formation is promoted by the drop in hydrostatic pressure across the capillary wall; that formation is reduced by the greater osmotic pressure of blood as compared with lymph, a result of the greater protein content of blood caused by the filtration effect; and that formation is increased if the leakiness of the capillaries is increased. If any single factor is changed, the rate of lymph formation changes. One example is that high venous blood pressure can cause watery enlargement (edema) of the heart, an important factor in congestive heart disease. Another is that starvation reduces blood proteins, thus lowering the osmotic difference between blood and lymph, and consequently causing a general edema.

Local failure of the lymphatic system can have severe consequences. An especially tragic one occurs in the condition called elephantiasis *(Figure 20)*. It is caused by a parasitic worm *(Wuchereria)* transmitted by mosquitos. They block the lymph flow, causing the accumulation of fluid which distends the limbs to elephant-sized dimensions. The enormous watery swelling is known as a lymphedema.

> Thanks to special pigments, the oxygen-carrying capacity of the blood of many animals is greatly increased (often by more than 50 percent) over what it would be without them.

BLOOD PIGMENTS

Given two systems of equal ability to move blood, if one of them can be made to carry twice as much oxygen per unit volume as the other, and give it up on demand, it will be twice as effective. Thanks to special pigments, the oxygen-carrying capacity of the blood of many animals is greatly increased (often by more than 50 percent) over what it would be without them. In vertebrates the pigment is hemoglobin, contained in the erythrocytes. As noted in Chapter 4, hemoglobin consists of four globular protein chains, each containing a heme group. Every hemoglobin molecule can loosely combine with four molecules of oxygen. The com-

ELEPHANTIASIS results from infection of lymphatic system by a parasitic worm. Lymph drainage is blocked and watery fluid accumulates, causing grotesque swelling. 20

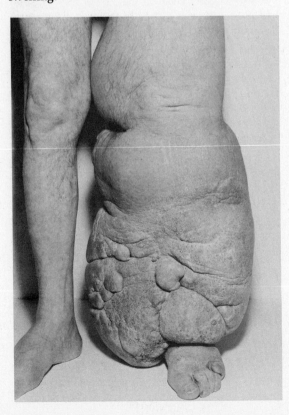

bination involves a bonding of the oxygen molecules to the ferrous ion of the hemes.

The amount of oxygen carried by hemoglobin depends on the partial pressure (essentially the concentration) of oxygen in the blood. When the partial pressure of oxygen (PO_2) is high, as in the lung capillaries, where the blood is exposed to alveolar air, the hemoglobin is loaded with oxygen. Where the PO_2 is low, as in the systemic capillaries, where oxygen is drawn away by the tissues, the hemoglobin unloads its oxygen. When fully loaded with oxygen —that is, when all its available ferrous binding sites are occupied —the hemoglobin is said to be 100 percent saturated. Ideally, hemoglobin should become saturated at the PO_2 prevailing in the lungs, and it should unload oxygen at the PO_2 prevailing in systemic tissues. The relationship between the PO_2 and the degree of saturation of hemoglobin can be depicted as an OXYGEN DISSOCIATION CURVE.

Figure 21 shows such a curve for human hemoglobin. Human hemoglobin becomes saturated when the PO_2 is about 100 mm Hg, which corresponds to the usual PO_2 in the capillaries of the lungs. In the systemic capillaries the PO_2 falls to about 40 mm Hg. The hemoglobin remains almost 70 percent saturated at this pressure, which means that it leaves the capillaries carrying a substantial reserve of oxygen. Tissues such as active muscle, whose oxygen demands are high, can draw upon this reserve. The curve is very steep in the range of 30–40 mm Hg, where oxygen delivery to the tissues takes place. Within this range a slight decrease in PO_2 leads to a substantial release of oxygen by hemoglobin.

Vertebrate hemoglobins are not all alike. The pigments may differ in their amino acid composition and arrangement, and as a consequence, in their oxygen-binding characteristics. The special properties of a particular hemoglobin usually fit the lifestyle of its possessor. Small mammals, for example, have higher metabolic rates and oxygen needs than larger species like humans. The oxygen-dissociation curve of mouse hemoglobin lies to the right of that of human hemoglobin *(Figure 22)*. In other words, mouse hemoglobin gives up oxygen more readily than ours. Conversely the llama of the Andes, a high-altitude mammal, has a curve that lies to the left of ours. Its hemoglobin becomes saturated at a lower PO_2, signifying that it can take up oxygen readily despite the lower concentration of gas in the mountain air.

The human fetus has the special problem of "parasitizing" the oxygen supply of the mother. Its hemoglobin must obtain oxygen from the maternal hemoglobin, and as one might expect, it has a hemoglobin different from the mother's, with an oxygen-dissociation curve that lies to the left of hers *(Figure 22)*. The hemoglobin of the fetus has greater affinity for oxygen than does the maternal hemoglobin. In the capillaries of the placenta, oxygen moves readily from the erythrocytes of the mother to those of the fetus. At birth fetal hemoglobin is replaced by adult hemoglobin.

The muscles of all vertebrates contain myoglobin, a pigment molecule functionally similar to fetal hemoglobin. It consists of only one globular protein chain and one heme group. Like fetal hemoglobin, it has a higher affinity for oxygen than does adult hemoglobin *(Figure 22)*. Myoglobin stores oxygen in muscle. It takes up oxygen from blood, but releases it less readily than hemoglobin; it does not yield its oxygen until the PO_2 is especially low, as during intense exercise. Myoglobin is red like hemoglobin, and is largely responsible for the red color of vertebrate muscles. The whiteness of the meat in the breasts of domestic turkeys and chickens testifies to the low oxygen requirements, and hence the low myoglobin content, of the flight muscles of these nonflying birds.

Many invertebrates also rely on blood pigments for oxygen transport. Hemoglobin occurs in certain annelids (notably the earthworm), mollusks and echinoderms, as well as in a few insects. Some annelids possess a green iron-containing pigment called chlorocruorin. A diversity of species, including many crustaceans and mollusks, have

21 OXYGEN-DISSOCIATION CURVE of human hemoglobin shows saturation at a PO_2 (partial pressure of oxygen) of about 100 mm Hg.

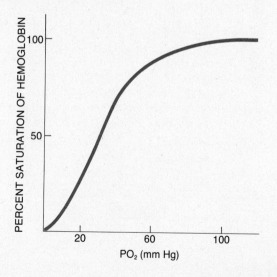

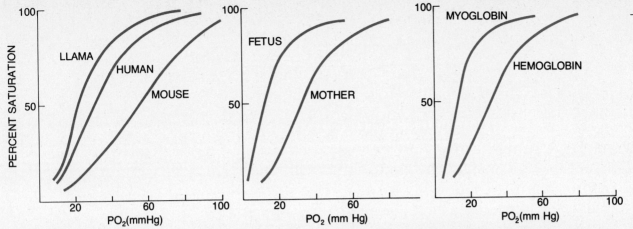

DIFFERENCES AMONG HEMOGLOBINS are evident from their oxygen-dissociation curves. Graph at left compares performance of hemoglobins of three mammals. Center, comparison of curves of fetal and maternal mammalian hemoglobins. Right, the curves of hemoglobin and myoglobin.

hemocyanin, a blue copper-containing compound. These pigments are usually dissolved in the blood rather than being packaged in blood cells.

CARBON DIOXIDE TRANSPORT

Carbon dioxide, the end product of oxidation, is transported from the tissues to the lungs without special pigments. Most of it is carried in the blood plasma as bicarbonate ions. As carbon dioxide is produced in the tissues it diffuses into the blood, where it combines with water to form carbonic acid (H_2CO_3). The reaction is catalyzed by carbonic anhydrase, an enzyme in the erythrocytes. The bicarbonate arises from the dissociation of carbonic acid into hydrogen ions (H^+) and bicarbonate ions (HCO_3^-). Because of their effect on pH, the hydrogen ions are potentially dangerous, but the blood contains substances that can buffer them. As a result, there is only a slight difference between the pH of the blood in systemic capillaries, where carbon dioxide is picked up, and that in lung capillaries, where carbon dioxide is dumped. But even the small difference is important, because the oxygen-carrying capacity of hemoglobin is sensitive to pH. A low pH reduces the affinity of hemoglobin for oxygen, and a higher pH increases it. In the systemic capillaries, where oxygen is needed, the increase in acidity facilitates the release of oxygen from hemoglobin, while in the lungs, where oxygen is plentiful, the decrease in acidity promotes the uptake of oxygen by hemoglobin.

The job of the transport system begins and ends at the gateway to the cell: the cell membrane. Of course to be of any use to an organism a nutrient must cross the membrane and move to the proper site within the cell. Waste products and cellular secretions such as hormones must leave the cell the same way. Here the short-range mechanisms of intracellular transport take over: diffusion, passive and active transport, pinocytosis and cell streaming (*Chapter 2*). This chapter has emphasized the mechanisms of long-distance transport, but the reader should bear in mind that for short-distance transport the individual cells of multicellular plants and animals rely on the same mechanisms as unicellular organisms.

READINGS

K. ESAU, *Anatomy of Seed Plants*, 2nd Edition, New York, John Wiley & Sons, 1977. An excellent treatment of the structural aspects of xylem and phloem.

M.S. GORDON, G.A. BARTHOLOMEW, A.D. GRINNELL, C.B. JORGENSON AND F.N. WHITE, *Animal Function: Principles and Adaptations*, 3rd Edition, New York, Macmillan, 1977. Contains an excellent treatment of animal transport.

P.M. RAY, *The Living Plant*, 2nd Edition, New York, Holt, Rinehart and Winston, 1972. A short and excellent introduction to functional botany with a first rate treatment of water relationships, gas exchange, and stomatal physiology.

K. SCHMIDT-NIELSEN, *Animal Physiology: Adaptation and Environment*, New York, Cambridge University Press, 1975. Physiological solutions to environmental problems are treated in lucid and exciting fashion. Chapters 3 and 4 deal with transport and circulation.

All vital mechanisms, varied as they are, have only one object, that of preserving constant the conditions of life in the internal environment. —Claude Bernard (1878)

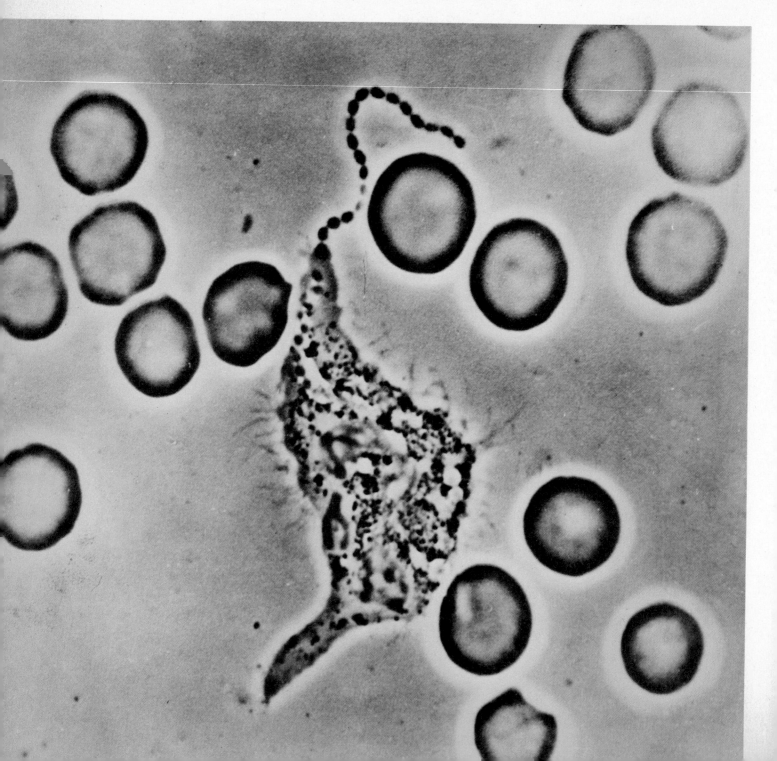

The cells of multicellular organisms are delicate and highly specialized machines which, like high-performance engines, function at their best only under a rather narrow range of conditions. Most of these conditions are dictated by the physics and chemistry of aqueous solutions. Even the hardiest cells, for example, cannot function if the water within them is evaporated or frozen, or if their surroundings are poisonously acidic or alkaline. Cells are especially sensitive to variations in temperature. As a general rule the rate of a chemical reaction doubles for each 10°C rise in temperature. An active cell carries out thousands of reactions per second, and it is easy to see how even a small fluctuation in temperature could disrupt the equilibrium of a network of interlocking reactions.

In the course of evolution multicellular organisms have devised a portable and constant internal environment that insulates most of their cells from the rigors of the outside world. Claude Bernard, the father of physiology, introduced the concept of the internal environment more than a century ago, pointing out that most of the cells of the body live not in contact with the variable external environment, but bathed in body fluids that comprise what he called the *milieu intérieur*. He observed that the stability of the internal environment is a prerequisite of a free and independent existence. In 1930 this concept was extended by the Harvard physiologist Walter Cannon to include HOMEOSTASIS: the tendency to maintain constancy. To some extent the notion of homeostasis may have been somewhat oversold because Cannon worked with vertebrates, especially mammals—organisms that have achieved the highest degree of internal constancy. Other organisms

have to tolerate a good bit of internal inconstancy.

Organisms maintain the constancy of their internal environment by four main types of mechanism: buffering, storage, feedback control and excretion. In warm-blooded animals—mammals and birds—these mechanisms provide extraordinarily stable conditions in which body cells can function at optimum levels. For example, the pH and salt content of their blood and other body fluids remain virtually constant. The virtually weatherproof internal environment of warm-blooded animals is an enormous adaptive advantage because it enables them to survive in a great range of habitats and at a high level of physiological efficiency. But this achievement has its price, notably the need for elaborate sensing and adjusting machinery, and the sensitivity of the tissues to severe accidental changes in factors such as temperature and wetness. Plants and cold-blooded animals operate less efficiently, but they are adapted to tolerate a considerably wider range of internal inconstancy.

Most organisms on Earth have some sensing and timekeeping mechanism that synchronizes their internal activities with the cyclic changes outside, particularly the daily cycle of light and darkness and the yearly cycle of seasonal changes in climate. The workings of these synchronizing mechanisms, known as biological clocks, are one of the most tantalizing mysteries of physiology, and are currently the focus of a massive amount of research. But this is getting a bit ahead of the story. The logical place to begin the study of the internal environment is with the four principal stabilizing mechanisms.

NEUTROPHIL ENGULFS BACTERIUM in the photomicrograph of human blood on opposite page. Part of the bacterium (a chainlike streptococcus) is visible at top center. Dark disks are red blood cells.

BUFFERING

The term BUFFER has long been used by chemists to denote a solution that is in some way resistant to change, usually a change in pH. If a drop of concentrated hydrochloric acid (HCl) is added to a beaker of water, the liquid becomes very acidic, changing perhaps from pH 7 to pH 2.5. If the **beaker contained a buffer, such as a dilute mixture** of sodium acetate and acetic acid, the original pH would be about 4.7. Adding a drop of concentrated HCl would lower the pH very little, perhaps to 4.65; adding a drop of concentrated NaOH would raise the pH only to about 4.75. The solution is said to be buffered against a change of pH, which of course means a change in the concentration of hydrogen ions [H⁺] *(Figure 1)*. All buffers are mixtures of two substances: one that can donate hydrogen ions (such as acetic acid, abbreviated HAc), and one that can accept hydrogen ions (such as acetate ions, Ac⁻). The ratio of the acceptor to the donor determines the pH. If a little HCl is added, some of the acetate will react with it:

$$H^+Cl^- + Ac^- \rightarrow HAc + Cl^-.$$

In the ratio [Ac⁻]/[HAc] the numerator will become smaller and the denominator larger, but so long as both are much larger than the amount of acid or base added, the ratio will not change much.

Many homeostatic mechanisms, both within cells and in extracellular fluids, involve buffering.

1 **BUFFERING is seen when a strong base, sodium hydroxide, is added little by little to a weak acid, one molar acetic acid. Before any base is added, the pH is low, about 2.3. As the first drops of base are added, the pH rises, sharply at first, then levels off as acetate ion accumulates. The pH rises only gradually with further additions of NaOH until nearly all the acetic acid has been neutralized. Then the pH rises very sharply. The nearly horizontal part of this curve is called the region of buffering; the most horizontal part (point of maximum buffering) is at pH 4.7 where equal amounts of acetic acid and acetate ion are present.**

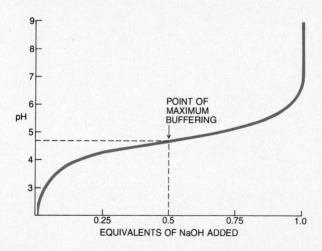

Organisms maintain the constancy of their internal environment by four main types of mechanism: buffering, storage, feedback control and excretion.

Living systems contain many different acceptor-donor pairs, including carbonates, phosphates, and groups that are present in proteins, such as carboxylate and amino groups. Blood contains a particularly subtle and complicated buffering system involving oxygen, hemoglobin and bicarbonate. Buffering systems function simply and automatically, preventing the lethal changes in pH that would occur within tissues whenever substances such as lactic and carbonic acids are produced faster than they can be carried away.

STORAGE

Excess over a certain amount of glucose or fatty acids in the blood is stored away in specialized storage areas, usually as large molecules which are mobilized on demand, when the necessary enzymes are activated. The storage mechanism depends on more complex kinds of equilibria than buffering. Thus a free fatty acid (FFA) undergoes metabolism by a series of enzymes to be deposited as fat. The products (A, B, C and so on) of the enzymes are in equilibrium with each other and with the FFA and fat.

$$FFA \leftrightharpoons A \leftrightharpoons B \leftrightharpoons C \leftrightharpoons fat$$

Just as with buffering, if one raises the FFA concentration all those equilibria are shifted to the right, so that the overall effect is to minimize the change in FFA, and to immobilize the extra FFA as fat. However, the rates at which these equilibria are achieved depend upon the enzymes involved, and the tendency to mobilize reserves on demand can be stimulated by agents such as hormones, that activate suitable enzymes. As with buffering the system is reversible, so that as an organism uses up FFA in its blood, more is mobilized and fat is withdrawn. Similarly, glucose is stored as glycogen in most animals and as starch in most plants. In all cases the effect is homeostatic; the change in circulating free sugars or fatty acids is kept to a minimum.

NEGATIVE FEEDBACK

This form of control was discussed in Part I. Here the reader need only to recall that a room thermostat is an example of a negative-feedback mechanism. When the room gets too warm, the thermostat switches off the heat source. In more general terms, one can say that overproduction triggers compensatory mechanisms which tend to negate the overproduction. For example, an animal that becomes overheated on a hot day is in no position to switch off the heat source; instead he initiates some compensatory behavior, such as sweating or panting. In living systems several such mechanisms are usually integrated to provide a relatively constant internal environment.

Mammals and birds are the only warm-blooded animals (HOMEOTHERMS); they maintain their internal temperature within quite narrow tolerances. In man a few degrees increase of body temperature radically changes nervous co-ordination, as we know from the strange subjective feeling experienced when we have a fever. Normally our oral temperature ranges between 36 and 38°C; severe convulsions occur if the temperature goes to 41°C, and death comes at 45°C. Most other mammals have almost identical body temperatures, but birds are a trifle warmer at about 41°C. Where does the heat supply necessary to maintain such temperatures come from and how is it controlled?

The individual enzymic steps of metabolism are not highly efficient; that is, much of the energy (about 70 percent) tied up in chemical bonds of nutrients is lost as heat as the chemical energy is converted into mechanical or biochemical work. Consequently every living cell generates heat. When we exercise vigorously great amounts of heat are generated and must be dissipated. For instance humans perspire and dogs loll their tongues; in both cases the evaporation of water provides the necessary cooling.

In mammals and birds, detectors of body-temperature appear in two locations. There are receptors in the skin, but the more important system lies in the central brain region called the hypothalamus. It consists of a heat loss center in the anterior part of the hypothalamus, which triggers suitable reactions when its temperature is raised by a fraction of a degree, and a heat conservation center that acts in the opposite way. The responses that these centers evoke vary with species, and are graded according to the extent of the deviation detected. Man adjusts to mild temperature deviations by "vasomotor regulation"; that

is, when the temperature increases, blood vessels under the skin dilate so that more heat is lost by radiation and convection. The opposite occurs when the temperature falls. In some animals, skin areas particularly well supplied with blood vessels serve as radiators that dump excess heat. Good examples are the ears of rabbits and elephants. Conversely, shivering generates extra heat if the temperature is too low for vasomotor compensation. Hair, fur and feathers are important factors in heat control, and their thickness and hence their insulating properties changes with the season. Pilomotor muscles can adjust their effective thickness by erecting them, encasing the animal in an insulating layer of air.

In addition to all these mechanical devices, if cooling is still excessive the homeotherm resorts to hormonal control by producing adrenalin which generally increases glycolysis. An even more direct control is the stimulation (originating again in the hypothalamus) of production of TSH or thyroid-stimulating-hormone from the anterior pituitary gland, which lies very close to the hypothalamus. The TSH is released into the blood and provokes

---

The oxidation of carbohydrates and fats produces only carbon dioxide and water as end products; the carbon dioxide is disposed of by gas exchange, and the water is a valuable acquisition. But the oxidation of protein produces large quantities of toxic ammonia which must quickly be removed.

---

the thyroid to release a series of hormones that elevate the basal metabolism and, by burning stored food reserves, greatly increase the production of body heat. One of the hormones, thyroxin, acts by reducing the efficiency of ATP production; therefore more material will need to be oxidized to produce a mole of ATP. In this way energy is diverted from the production of ATP to the production of heat.

Do these mechanisms fit the category of negative feedback systems? When cold provokes a bird to erect its feathers, it might seem to be a positive rather than a negative response. But in fact all such mechanisms have an on and off character. It is as true to say that a thermostat turns on the furnace when the room gets cold as to say that it turns off the furnace when the room gets too hot. Positive feedback is a quite different and usually disruptive process in which the product of a reaction sequence serves to activate the sequence. Autocatalytic processes are examples of positive feedback. The term negative feedback applies not only to the heating or cooling mechanisms described here, but also to the mechanisms in which glands respond to a chemical signal to turn their regulatory hormones on or off.

EXCRETION

The perpetual intake of food by heterotrophs leads to a perpetual production of byproducts that have to be ejected entirely from the organism if it is to maintain its internal stability. The foods that animals utilize are composed primarily of carbohydrates, proteins and fats. Most of what they ingest is oxidized to produce energy in the form of ATP. The oxidation of carbohydrates and fats produces only carbon dioxide and water as end products; the carbon dioxide is disposed of by gas exchange, and the water is a valuable acquisition. But the total oxidation of proteins produces large quantities of ammonia derived from the NH_2 groups of the amino acids. Unfortunately ammonia is very toxic (rabbits die if their blood ammonia reaches 5 mg per 100 ml) and must be removed efficiently and rapidly.

In protozoans, sponges, coelenterates, and freshwater fish there is extensive movement of water into and out of the body, and the ammonia is simply "washed out" in this water flux. Such animals are called AMMONOTELIC because ammonia is the form of the excreted nitrogen. In the smallest ammonotelic animals, ammonia is lost principally by simple diffusion across the body wall. In freshwater fish, the highly efficient gas-exchange mechanism of the gills is an ideal site of loss. Flatworms have developed special systems to promote water expulsion; water with its dissolved ammonia and carbon dioxide diffuses into tubules, and is expelled to the exterior by the beating of cilia in a specialized flame cell (named after the flame-like flickering of the cilia).

Other animals do not enjoy the luxury of such an ample water flow, and have all met the problem by converting the toxic ammonia to less toxic nitrogen compounds, most often to either urea (ureotelic animals) or uric acid (uricotelic animals). Other less

NH^+

AMMONIUM ION
(AMMONIA IN WATER)

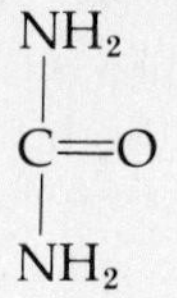

UREA

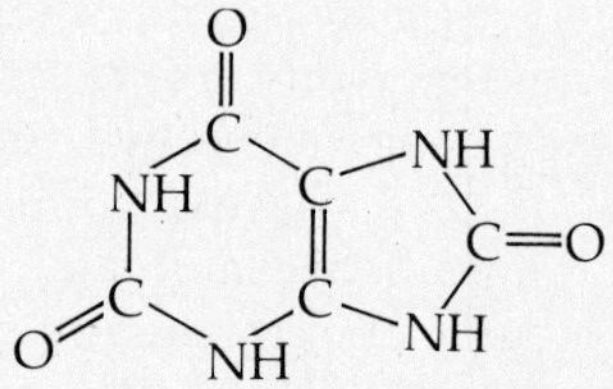

URIC ACID

$(CH_3)_3NO$

TRIMETHYLAMINE OXIDE

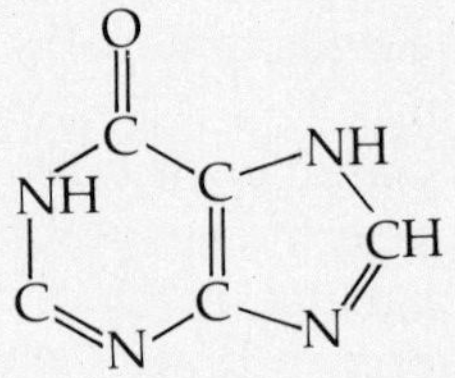

GUANINE

widely used derivatives are trimethylamine oxide, which is common in marine bony fishes; and guanine, which spiders excrete in crystalline form as their major nitrogen product. Animals invariably excrete a mixture of nitrogenous wastes, but one or another of the above usually predominates.

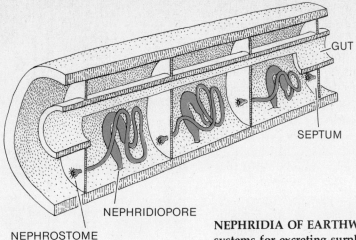

GUT

SEPTUM

NEPHRIDIOPORE

NEPHROSTOME

NEPHRIDIA OF EARTHWORM (color) are simple systems for excreting surplus water and retaining needed salts. Diagram depicts part of an earthworm sliced in half lengthwise. Liquids are drawn into each nephridium through an opening (nephrostome) in the segment ahead, and waste is expelled to the outside through the nephridiopore. Each coiled nephridial tube is associated with a capillary network (not shown) that recovers minerals, preventing their excretion in watery urine.

The relatively simple conversion of ammonia to urea (a very soluble substance of low toxicity), coupled with a specialized system for selectively rejecting it from the blood, helps those animals for whom water conservation is a most important consideration, including mammals and adult amphibia. Urea is soluble and must be disposed of along with a fairly substantial amount of water, making up urine. It is interesting that when the tadpole develops into a frog, it changes from an ammonotelic system (suitable for fresh-water life) to a ureotelic one. Urea production is actually used by elasmobranch fishes (such as sharks and rays) to increase the osmotic pressure of their blood in relation to that of seawater.

The steady production of urea demands a mechanism for its steady removal from blood with only modest water loss—a problem that is complicated by the necessity to regulate water content and salt concentration. Animals that live in fresh-water habitats take up excess water through their skins by osmosis, and get rid of it by urinating copiously. The frog produces urine equivalent to 25 percent of its body weight daily. Earthworms, although terrestrial, live in moist places, and produce 60 percent of their weight of urine daily. By contrast, man produces only two percent of his weight daily. It is clear that there are water-losing and water-conserving animals.

A simple water-losing system is the nephridium of the earthworm. It opens directly into the body cavity, from which the watery contents are drawn into the nephridium by cilia and passed down a long tubule to the exterior (*Figure 2*). Reabsorption of needed salts occurs in the tubule. This process of nonspecific loss, followed by selective reabsorption later in the system, is characteristic also of all the elaborate excretory systems. But clearly the nephridium produces a watery urine that is hypo-osmotic; that is, its osmotic pressure is lower than that of blood, because it is simply blood with salts removed from it by reabsorption.

In higher organisms, the simple nephridium is replaced by a more elaborate nephron; usually thousands of nephrons together comprise a kidney. A simple nephron found in fresh-water fish (*Figure 3*) consists of a tubule which at one end embraces a knot of leaky blood capillaries called a glomerulus (*Figure 4*). In the glomerulus, blood filters through the capillaries to form a filtrate which lacks the cells and proteins of blood. The filtrate passes down the tubule of the nephron where reabsorption of some water and most essential solutes such as glucose and salts occurs. It is collected in a tube called the ureter, and there ejected. A copious, hypo-osmotic urine results. Marine fish cannot afford this copious flow (they are always in danger of desiccation from their salty surroundings) and by switching to urea as their nitrogenous end-products, they can afford to cut down their nephrotic through-put to as little as 0.3 percent of body weight daily in the toadfish. This reduced flow is achieved by greatly reducing the size and number of glomeruli, and in a few fish abolishing them completely; presumably enough filtrate leaks in through the proximal ends of the tubules.

Mammals possess a more elegant and effective kidney with a new feature added to the nephron: a highly efficient device for extracting most of the water as well as the salts from the glomerular filtrate *(Figure 5)*. Despite its efficient high-volume filtration, the nephron produces a concentrated low-volume urine. In man the filtrate amounts to 100 ml per minute, but the urine produced is only 1 ml per minute. The system is also 99 percent efficient in recovering the vital salts lost in filtration, yet the urine is far more concentrated in urea than is blood. The crucial device in this system is the hairpin-like loop of Henle, which, like the circulatory system of a fish gill *(Chapter 15)*, operates on the principle of countercurrent flow. The glomerular filtrate is collected by the proximal tubule of the nephron, and here over half of the water and most essential substances, such as salt and glucose, are reabsorbed by active transport, just as in the sim-

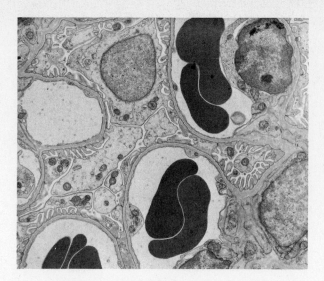

CROSS-SECTION OF GLOMERULUS in kidney of a bat reveals intimate association between capillaries and kidney cells. Dark objects within capillaries are red blood cells. Magnification, 5300X.

4

3 **SIMPLE NEPHRON of fresh-water fish consists of a tubule associated with a glomerulus. Filtrate from glomerulus passes down the tubule, which reabsorbs glucose and salts. Absorbed substances are cycled back into bloodstream via the network of capillaries and veins around the tubule.**

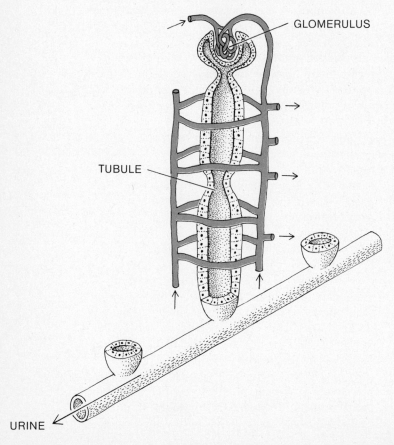

GLOMERULUS

TUBULE

URINE

ple nephron. The filtrate is now fed into the descending arm of the loop of Henle.

The mechanism of water extraction depends on using the loop to create a permanent salt gradient locally, with a high salt concentration at the loop's base and, by simple diffusion, in the interstitial cells surrounding the base *(Figure 6)*. The efflux from the loop is then led (in the collecting duct) along this gradient, which progressively withdraws water from it osmotically (the water permeability of the duct is greater than the salt permeability), the effect being most pronounced at the end of the duct, near the base of the loop. The creation of the local salt gradient depends on the permeability of the ascending arm of the loop to salt but not to water, and the active transport of salt out of this arm. This pumped-out salt diffuses across into the neighboring descending arm of the loop, so that the liquid passing through that arm becomes progressively saltier. The net result is that salt is cycled around the loop without loss or gain, but sets up the desired gradient.

Animals with even greater need to conserve water utilize uric acid as their nitrogenous output. Uric acid is so insoluble that it is osmotically easy to abstract water from a mass of it. Thus birds and reptiles discharge their urine into a cloaca, which also receives the feces. A vigorous water recovery in the cloaca affects both urine and feces, which are finally discharged as a paste, the white part of which is uric acid. Insects, whose small size

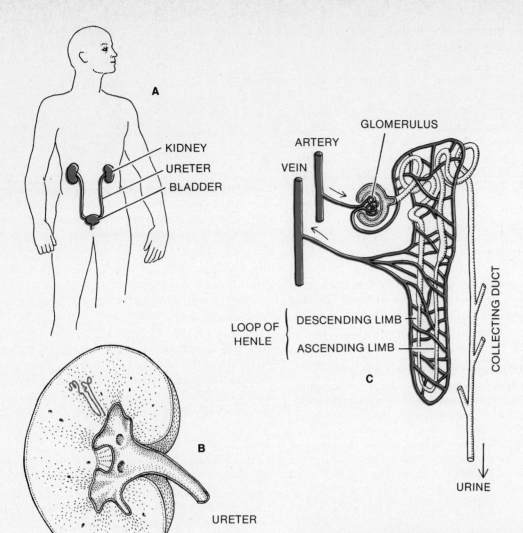

A

KIDNEY
URETER
BLADDER

GLOMERULUS

ARTERY

VEIN

LOOP OF
HENLE

DESCENDING LIMB
ASCENDING LIMB

COLLECTING DUCT

C

B

URETER

URINE

MAMMALIAN EXCRETORY SYSTEM. Each kidney 5
contains thousands of nephrons. Cutaway view of
kidney (B) shows orientation of a single nephron
(color). Diagram of nephron (C) depicts glomerulus and
tubules. Reabsorbed substances are returned to
bloodstream via veins that drain kidney.

exacerbates their water-conservation problem, use a comparable approach. Their excretory apparatus is a fringe of Malpighian tubules around the midgut; these tubules have closed ends projecting into the coelom, from which they collect water and nitrogenous wastes (*Figure 7*). They discharge these tubules into the midgut, converting the nitrogenous material to uric acid. This passes through the gut and into the rectum, along with undigested matter, where the water is extracted. The waste matter is ejected as an almost dry pellet.

There is a curious all-purpose protective system known as microsomal hydroxylase present in the liver of vertebrates and several tissues of insects (and doubtless in many other organisms). It has the ability to hydroxylate almost any apolar (that is, lipid-soluble) compound. Such compounds include natural products such as nicotine and caffeine; synthetic compounds such as DDT and other chlorinated insecticides; and also numerous common drugs such as barbiturates. Apolar compounds are very poorly excreted, and in fact there is a direct correlation between apolarity and retention in the body. In the absence of special mechanisms, such exotic compounds would stay in the blood indefinitely with disastrous consequences. The microsomal hydroxylase, fortunately, is almost totally nonspecific: it will act on virtually any apolar substrate. In this manner it is unlike any of the more familiar enzymes. By hydroxylating these apolar foreign compounds, using atmospheric oxygen and $NADPH_2$, it makes them more polar, and the body can then excrete them. The hydroxylase also has the valuable property of being easily inducible: feeding high levels of apolar compounds causes the enzyme system to multiply to meet the added need.

This same hydroxylase is also involved in hydroxylating the body's own steroids. DDT is thought by some to have ill effects on birds of prey by inducing their livers' hydroxylase, interfering with steroid metabolism, and thus with the ability to control the calcium deposition in their eggs. This may lead to thin-shelled, fragile eggs with little chance of survival.

WATER STABILITY

The problem of water stability is a special one. Water makes up about two-thirds of all tissues, and is the indispensable medium for all biochemical reactions. Terrestrial organisms all share a seri-

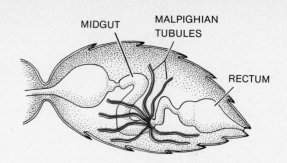

MALPIGHIAN TUBULES of insects collect water and nitrogenous wastes from body cavity, convert the nitrogenous substances to uric acid, and empty into the midgut. Water is recovered in rectum.

7

ous problem: how to maintain enough water despite continual losses from evaporation and from the use of water to flush away the by-products of their metabolism. The threat of desiccation is the price they pay for an ample source of oxygen. Only two phyla, the arthropods and the vertebrates, have made this daring trade-off on a large scale. Many other organisms have left their ancestral seas and lakes but live in wet terrestrial habitats, and die by desiccation if brought into the open. How do mammals and insects (for instance) avoid this fate?

Maintenance of internal wetness in a dry world demands a favorable balance between intake and output. The intake is partly as metabolic water derived from oxidation of carbohydrates and (with an even higher yield) fats. Each gram of glucose produces 0.6 g of water, and a gram of fat produces 1.1 g of water. A human gets about ⅓ liter of water daily by this means. But the bulk of intake is as water in food or drink. Food can be sufficient by itself in some cases; fruit and vegetables contain 75 to 95 percent of water, meats about 50 percent. Should this source prove inadequate, most animals drink bulk water. Their drive to seek such water (which may involve a hazardous expedition from a safe place) is triggered by a reduction in water-content of blood. In mammals, the trigger site has been located as the "thirst center" within the hypothalamus of the brain. If one stimulates the center electrically, the animal shows compulsive drinking, and can literally drink itself to death.

But a land animal that has to be always close to a water source is terribly restricted. To seek emancipation he must reduce his output to the point where he can go for long periods without access to water. Output is in three principal forms: respiratory exchange, excretion, and evaporation from the body's surface. Since respiration always occurs

6 **ABSORPTION OF SODIUM AND WATER by nephron. Sodium ions are removed by active transport (slanted arrows) from ascending limb of loop of Henle. Water is reabsorbed from collecting duct by passive diffusion (right). Numbers indicate sodium concentration in millimoles (thousandths of a mole). Water leaves duct by osmosis because of high concentration of sodium in surrounding tissue. Some sodium from tissues diffuses passively back into descending limb of loop of Henle (left).**

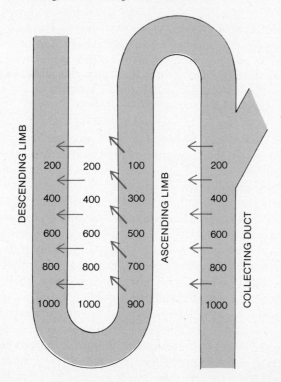

Contrary to legend, the camel does not store water in its hump (it stores fat there), but it can tolerate extreme dehydration. It can lose 40 percent of its body water and survive, whereas most mammals die if they lose 20 percent.

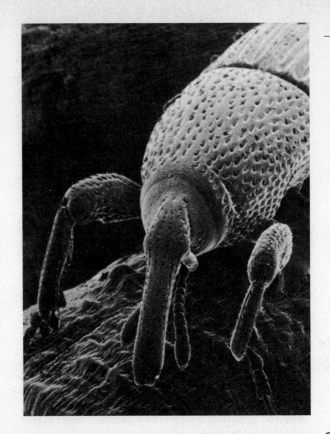

CUTICLE OF INSECT is a waterproof covering of chitin **8** **coated with wax or grease. Photograph depicts granary weevil, a pest that destroys stored wheat. Weevil is only four millimeters long.**

across a wet surface, there is inevitably some water lost in keeping it wet while passing air over it. Land vertebrates have abandoned the external-exchange route (via skin or external gills) by evolving internal lungs; the wet surface is thus exposed only to the minimum volume of air necessary. In insects exchange is also internalized in the tracheae; the external openings (spiracles) are under active control so that they are sealed off to restrict water loss when they are not in use.

Excretory losses of water are minimized by a variety of means. The copious urination of aquatic animals is reduced to a trickle in mammals. A human loses each day a liter of water in urine and 100 ml in feces. The severity of this loss can be partly reduced in mammals if the need arises: the hypophysis (a gland at the base of the brain) responds to a drop in the water content of the blood by secreting ADH or antidiuretic hormone. ADH causes the kidney to withhold water, and can reduce the urine volume by two-thirds.

Evaporation from the body's surface is controlled by a variety of devices in fully "water-emancipated" animals such as mammals and many insects. The insect is especially vulnerable, for its small size involves a high surface to volume ratio — up to 100 times greater than in typical mammals. Insects living in dry air therefore have evolved a thick integument called cuticle, made up in large part of chitin, a polymerized amino-sugar rendered water-impermeable by being tanned with quinones and (like the cutin layer of plants) coated with wax or grease (*Figure 8*). This apolar barrier permits water loss at a rate two thousandfold less, for instance, than from a red blood cell, and about a hundredfold less than from aquatic insects exposed to air.

In mammals the outermost layers of the skin are made up of dying cells (the stratum granulosum) containing a precursor of the horny protein called keratin; and dead flattened cells (the stratum cor-

neum) thickened with keratin and lightly impregnated with wax (*Figure 9*). The dead and dying cells are sloughed off continuously from the innermost layer of the epidermis. The epidermis and the underlying dermis together make up the skin. Its waterproofing property depends on the strata granulosum and corneum. Developmental studies on the guinea pig have shown that after 54 days of fetal development these layers appear and simultaneously the water permeability drops sharply.

Some mammals with unusual water-conservation problems have developed radical stratagems. Contrary to legend, the camel does not store water in its hump (it stores fat there); but it can tolerate extreme dehydration. It can lose 40 percent of its body water and live, whereas most mammals die if they lose 20 percent. In addition the camel can tolerate a higher body temperature than other mammals. Like them, his normal temperature is at about 34°C, but he does not turn on his water-wasting cooling system — his sweat glands — until his temperature reaches 41°, as compared with 37° for man.

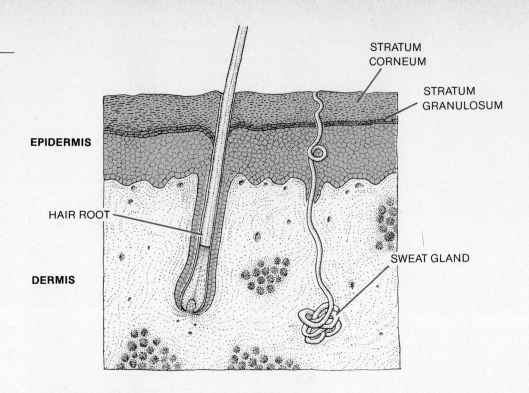

STRATUM
CORNEUM

STRATUM
GRANULOSUM

EPIDERMIS

HAIR ROOT

DERMIS

SWEAT GLAND

9 **HUMAN SKIN** consists of waterproofing layer of dead and dying cells (epidermis) sloughed off from the underlying layer of living cells (dermis). Sweat gland allows controlled release of water.

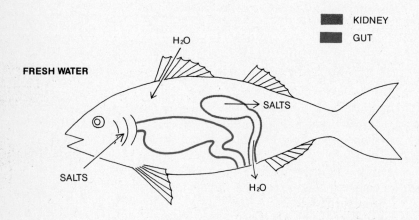

KIDNEY
GUT

FRESH WATER

H₂O

SALTS

SALTS

H₂O

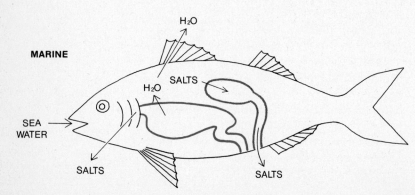

MARINE

H₂O

SALTS

H₂O

SEA
WATER

SALTS

SALTS

Aquatic organisms might seem to be spared from water problems, but many of them must cope with osmotic problems. Marine invertebrates, it is true, enjoy the rare privilege of being isosmotic with their surroundings. But most modern marine fish have evolved from fresh-water ancestors, and as a result the elasmobranches like the shark are about half as salty as the sea, and the bony fishes or teleosts are about a third as salty as the sea *(Figure 10)*. They are thus eternally in danger of being desiccated by water loss. By contrast, all fresh-water animals, vertebrate or invertebrate, are very hypertonic to their virtually salt-free environment, and are in constant danger of being flooded. Obviously all these aquatic animals survive comfortably; they must therefore have strategies to avoid these catastrophes.

A simple mechanical strategy has been developed by clams and oysters to tolerate fresh water that tides may bring out: they close up tight. Most mechanisms are more complex. The elas-

OSMOTIC REGULATION IN FISH. Fresh-water fish are invaded by water that enters through their skins, and eliminate it by urinating copiously. Salts absorbed through gill tissue are conserved by reabsorption in kidney. Salt-water fish lose water osmotically through their skins, and drink sea water continuously to replace it. Resulting excess salts are excreted through gills and kidney.

mobranches compensate for the low osmolarity of the salts in their blood by converting much of their nitrogenous intake to urea $CO(NH_2)_2$ and trimethylamine oxide $(CH_3)_3NO$. These relatively innocuous organic compounds are maintained in the blood at just the concentration to make it isosmotic with the sea. Marine bony fish use another trick: they drink and absorb sea water, but excrete the surplus of salt through their gills, for their kidneys are inadequate for massive salt rejection. The fresh-water fish, by contrast, solve their water problems with their kidneys. They compensate for the steady inundation of water by urinating copiously and continuously, but jealously recovering virtually all the inorganic salts from that urine. Freshwater protozoa have also found a way to rid themselves of the excess water that inevitably invades them. They have contractile vacuoles which pump out the excess, contracting at rates that are a function of the osmotic discrepancy between their cytoplasm and the outer medium (Figure 11).

A few mammals and birds have returned to their ancestral home, the sea, and have had to readapt their essentially terrestrial equipment accordingly. Those that eat fish, such as seals and most whales, are taking in food with a salt concentration that matches their own so they do not have to drink sea water. Mammals, such as baleen whales, that eat marine invertebrates are taking in a too-salty diet, and have had to compensate by evolving special kidneys able to produce unusually concentrated urine. Marine birds (gulls, albatrosses, penguins) and reptiles (sea turtles) have adopted another tactic; the function of getting rid of surplus salt is given to a special salt gland close to the eyes. These animals can all drink sea water, after which they secrete an exceptionally salty fluid, which shows up as a salty nose-drip in the birds and salty tears in the turtles.

INTERNAL INCONSTANCY

Internal constancy is not universal, as shown by the body temperature of POIKILOTHERMS: cold-blooded animals. The terminology is misleading in Greek as well as English, for both terms imply that these animals (all animals except mammals and birds) always have cold blood. On the contrary, on a hot day their blood tends to be hot. Their special feature is that they tend to have INCONSTANT BODY TEMPERATURES that correspond to their environment, hot or cold.

In unduly hot or cold places, poikilotherms may develop behavioral adaptations to lessen the rigors of their environment, for instance by seeking shady places, or operating only at night. In other than tropical regions, such strategems do not permit them to live through the prolonged cold of winter in an active state. They therefore adopt some inactive form to last the winter, such as eggs or pupae in the case of many insects; or they may become metabolically (and therefore behaviorally) almost inert, as in reptiles and amphibia. This taking refuge in some rugged form may not only be triggered by approaching winter, but in many cases by a variety of potentially hazardous conditions. Thus in the little crustaceans called water fleas (Daphnia) a specially thickened egg-case develops around a single egg, the whole package being called an ephippium. These ephippia may form in a variety of disadvantageous conditions, including drying and freezing.

Poikilothermy is not without peril, for virtually all active organisms require relatively constant surroundings. Organisms that lack the ability to maintain constant body temperature must either 1) resort to special behavioral strategies such as migration, 2) resort to physiological strategies such as spending part of their lives in an inactive temperature-resistant state, or 3) avoid extreme environments completely. The relatively recent

1 CONTRACTILE VACUOLE enables Paramecium to bail out excess water that enters it by osmosis. Photomicrograph depicts vacuole when full. Collecting tubules converge toward vacuole, which empties itself through a small central pore barely visible as a circular outline.

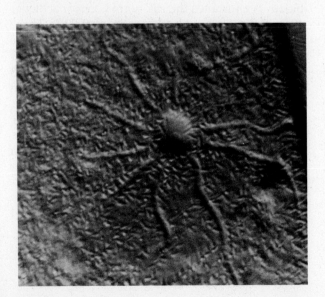

evolution of sophisticated temperature-regulating mechanisms gives homeotherms some distinct biological advantages. Not only does it enhance their overall physiological efficiency, but it opens up environmental niches forbidden to poikilotherms: new geographic locations and new times of the year, particularly winter.

Some homeotherms have also adopted behavioral and physiological strategies to cope with the rugged winters of nontropical regions. Migratory birds, like migratory insects, escape the rigors of winter simply by going somewhere else. Many other insects, as well as many reptiles and mammals, rely on the physiological strategy of hibernation. When cold weather comes they enter a quiescent period during which their body temperature, metabolic rate and body weight drop sharply. In homeotherms the drop in body temperature is striking; within a few hours it drops to a few degrees above air temperature. The temperature of a woodchuck, for example, was observed to drop from 37°C to 13°C, its heart rate from 80 to 4 beats per minute, respiratory rate from 28 to 0.6 breaths per minute, and its basal metabolic rate dropped twentyfold. These changes permit an organism to survive several months without food or water. Compared with poikilotherms, hibernating homeotherms enjoy the competitive advantage of a longer active season —a little earlier in the spring, a little later in the fall —but they are at a disadvantage in year-round competition with homeotherms that possess more sophisticated all-weather temperature regulating systems.

Some organisms survive harsh environmental conditions by altering their biochemical composition. A small tropical fly, *Polybedilum vanderplanki*, has larvae that live in shallow pools which often dry and refill repeatedly. The larvae can tolerate drying up to a remarkable extent: they can reduce their water from a normal 33 percent to a low of 3 percent. When dried up to this latter extent, they can tolerate severe treatments that kill the fully hydrated larvae, including storage over calcium chloride (a dehydrating agent) for seven years, being dropped into liquid helium (−270°C), or being immersed in absolute alcohol for a day. Many other insect species have blood whose water content varies widely in response to environmental changes, and this has been interpreted as a water-storage mechanism that enhances survival through difficult times.

Another useful adaptation occurs in some cold-hardy insects. They have been shown to make glycerol in their blood by dephosphorylating the α-glycerophosphate which all organisms have as an intermediate of glucose metabolism. This happens only as winter approaches, and the insects finally achieve a glycerol level which in some species represents 25 percent of the total weight of the larvae. The glycerol functions as an antifreeze that prevents ice formation. Ice is particularly dangerous to living cells because of its tendency (especially when cooled slowly) to form large crystals that disrupt cellular organization. In the presence of substantial concentrations of glycerol, temperatures can fall to many degrees below freezing (for example, −30°C) without crystal formation; and further cooling leads to "vitrification," the formation of numerous very small crystals to give a glass-like state, generally harmless to tissues. Glycerol is commonly used in the refrigeration of semen for artificial insemination.

Many lower organisms also show a remarkable capacity to tolerate fluctuations in the level of nutrients in their body fluids. Most organisms ingest or synthesize food in a periodic fashion. Consequently, unless some special mechanism exists, there will be a tendency for the blood (or in plants, the cell sap) to be higher in nutrients after a meal (or during daylight), and to fall steadily during growth or other energy-consuming activities. The documentation for such inconstancy is not good, although one would guess it to be very common. One interesting case concerns plant-eating insects. In the vertebrate, the level of blood K^+ is kept rather constant; in man it is 5 mM (5 millimolar), and the $Na^+:K^+$ ratio is 28:1. But plants contain disproportionately high concentrations of soluble K^+, and insects that feed on plants all achieve very high K^+ levels and low $Na^+:K^+$ ratios in their blood. For instance the silkworm *Cecropia* achieves 49 mM K^+ and a $Na^+:K^+$ ratio of 0.09:1. These levels fluctuate dramatically during starvation; the K^+ level of a locust falls to 13 mM from its high of 29 mM (yet its Na^+ level rises a little from 97 to 108 mM). It has been shown that one of the factors that triggers marching behavior in these locusts is the fall of blood K^+, which usually happens when the food supply fails. The uncontrolled drop in K^+ level has become a physiological trigger that evokes a behavioral response —in this case, migration to leafier places. These very high blood K^+ levels would kill a vertebrate whose nerves are extremely sensitive to K^+ *(Chapter 19)*. But insect nerves are relatively insensitive to high K^+ because they are enclosed in a protective sheath. The muscles of insects, however, are quite sensitive to K^+ levels, which explains the action of the trigger.

Another kind of regulatory problem faces aquatic animals whose habitat may subject them to differing salt concentrations. Oysters tolerate the intermittent low-salt conditions in estuaries by simply closing up—a crude kind of regulation which has the drawback of suspending the animal's active life for a time. Similar difficulties in adjusting to changing external conditions are seen in fresh-water invertebrates; when crayfish or fresh-water mussels were exposed to varying Cl^- concentrations, their blood Cl^- levels roughly paralleled these concentrations. Several other invertebrates show a similar inability to compensate for changes in external cations. The levels of Na^+, K^+, Ca^{++} and Mg^{++} in blood can be varied many-fold by exposure of the crab *Eriocheir* or the crayfish *Astacus* to differing salt solutions. In general, aquatic invertebrates have little capacity to compensate for varying salt concentrations, whereas aquatic vertebrates tend to be relatively efficient.

BIOLOGICAL CLOCKS

Certain features of the environment, including temperature and daylength, vary in a predictable way. Most organisms have developed internal clocks to predict such changes, and to modulate their physiology or behavior accordingly, in order to minimize differences between their internal and external environments.

In any part of the world where there are pronounced seasonal changes in climate, organisms appear and disappear at particular times with uncanny precision. The swallows reappear at the Mission San Juan Capistrano in southern California within a week of March 19; the monarch butterflies settle in the trees of Monterey, California, within a few days of mid-October; noxious black flies swarm in the Maine woods in late spring and early summer, then suddenly vanish; blooms of goldenrod cover much of New England and the midwest in August; and emperor penguins breed only in the blackness of the Antarctic winter. These organisms can measure and predict environmental change with considerable precision, and initiate action that may lead to some important biological event many days, weeks or even months later. Biologists have long wondered how they do it, and the answer to this question is even now not entirely clear.

As the Earth spins through its seasonal cycle, there are several environmental parameters that change on an annual basis. Temperatures fluctuate widely, periods of high and low precipitation al-

ternate, and, of course, daylength alternately increases and decreases. Though average seasonal temperatures may vary little from one year to the next, short term variations are apt to be enormous, as a January thaw in Boston followed by a May snowstorm suggests. The time and amount of precipitation in most parts of the world is also highly unpredictable on a week-to-week basis (the Farmer's Almanac notwithstanding) even though seasonal averages may not fluctuate particularly. Unlike temperature and precipitation, however, daylength repeats itself with monotonous precision year in and year out and it is therefore not particularly surprising that it is to this parameter that many organisms respond.

Response to daylength, or PHOTOPERIODISM, is well known among most groups of higher organisms. However, the most detailed studies have been done with flowering plants and insects—partially because they are readily handled in large numbers under laboratory conditions, and partially because many of them have sufficiently short life cycles that their response to photoperiod is quickly evident and hence accessible to experiment. In insects the most carefully studied phenomenon is entry into DIAPAUSE, a dormant state that occurs in different insects at very different developmental stages, from the egg through adulthood. In flowering plants the phenomenon subjected to closest analysis has been the transformation from vegetative to reproductive growth—the initiation and development of flowers—although it has been demonstrated that other phenomena, such as the onset of autumn coloration and entrance into winter dormancy, are also responses to daylength.

Organisms can be broadly fitted into three major groups according to their response to changing daylength. Some undergo a change only if the days become shorter than some critical number of hours. If a day is longer than 15½ hours for the cocklebur *Xanthium*, the plants grow vigorously but produce only new leaves. Only when the period of light is less than 15½ hours do they produce flowers. They are obligate SHORT DAY (SD) plants. Certain aphids reproduce sexually only if daylength is less than 14½ hours. Such short day organisms do not enter sexual reproduction until the daylength is less than some critical value (*Figure 12*). On the other hand, there are both plants and insects in which precisely the opposite phenomenon obtains. Kept on short days, they fail to become sexual (or to enter diapause, or to undergo seasonal changes in color). Only when the day-

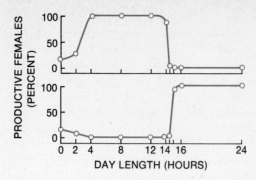

12 REPRODUCTION OF APHIDS is governed by day-length. Short days (top) favor the hatching of females that reproduce asexually. Long days (bottom) favor development of sexually-reproducing females. Days 14.5 hours long—the critical daylength—result in production of both types of female.

length is greater than some critical value does the appropriate response appear. These are the so-called LONG DAY (LD) organisms *(Figure 13)*. Finally there are some organisms which are indifferent to daylength, and are called DAY NEUTRAL (DN); the common dandelion is a particularly obnoxious example. In temperate climates, it will begin flowering almost as soon as the ground is thawed, and will continue flowering until the snow flies. Some organisms show more complex patterns: those which require long and then short days in sequence (or the reverse) and those which must have a period of low temperature that precedes days of a certain length.

Actually the term photoperiodism is misleading,

13 LONG-DAY PLANTS will flower only if exposed to a photoperiod greater than their critical daylength. Curve depicts effect of photoperiod on duckweed (Lemna gibba). Critical daylength for this plant is about 11 hours. White line at right indicates break in time scale of graph.

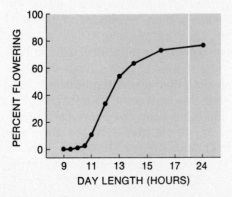

as are terms such as short day and long day organisms, because the organisms are primarily responding not to length of day, but to length of night. Thus an SD plant should be called a LN plant in deference to its responding to the long night. Unfortunately, day-oriented terminology is so deeply rooted in the literature that it has stuck. Two kinds of experiments support the contention that nightlength rather than daylength is critical in determining photoperiodic responses. First, if the experimenter alters the night temperature, there is a substantial change in critical daylength values, but alteration of day temperature has little effect. Second, interruption of a long night by a flash of light under the right conditions makes the organisms behave as though they were on long days. A brief dark period during the day has no effect.

A second feature of the photoperiodic response is that it is an inductive phenomenon, which means that for many organisms a few favorable day-night cycles are sufficient to evoke the photoperiodic response. The cocklebur *Xanthium* provides an extreme example. It can be germinated and grown for months under continuous light. It remains vigorous, producing new leaves, but shows no signs of flowering. But if such plants are exposed to a single long night and are then returned to constant light, flower buds will appear within a few weeks and the plants will then continue to produce new flowers steadily for months. Thus in one or two nights an irreversible change is induced, although it does not express itself until many days later.

A third feature of photoperiodism is that for a species found over a wide range of latitudes, such as the cabbage butterfly (LD) or *Xanthium* (SD), there is a wide variation in critical daylength, with values longer for species found in the more northern latitudes. The adaptive advantage of this sort of variation in potential for daylength response is obvious. Close to the equator where seasonal daylength change is minimal, many organisms are in fact day neutral. However, some which are photoperiodically regulated can discriminate daylength changes of as little as 15 minutes.

If organisms can tell when the light is on and when it is off, what pigment provides them with the appropriate information? With higher plants, the answer is fairly easily obtained by a night-interruption experiment *(Figure 14)*. Two organisms, a long-day plant such as *Spinacea* (spinach) and a short-day plant such as *Xanthium*, are both reared on short days; *Xanthium* flowers, and *Spinacea* remains vegetative. If the experi-

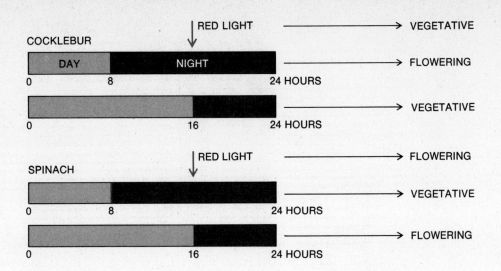

NIGHT-INTERRUPTION EXPERIMENT demonstrates **14**
the opposite responses of a short-day plant (cocklebur)
and long-day plant (spinach). Reared on short days (top
bar of each pair), cocklebur blooms but spinach
remains vegetative. On long days (bottom bar of each
pair) spinach flowers but cocklebur does not. Red flash
during long night makes plants behave as if they were
on long days.

menter administers a brief flash of red light in the middle of the long night, *Xanthium* remains vegetative and *Spinacea* flowers. If he repeats the experiment, but gives a brief exposure to light at the far red portion of the spectrum immediately after the red, the two plants behave as though there had been no night interruption at all; *Xanthium* flowers and *Spinacea* remains vegetative. Phytochrome *(Chapter 12)* is obviously a primary photoreceptor in perceiving the lights-on or lights-off signal.

In animals careful spectral work has lagged somewhat since workers reasonably expected vision to take care of photoperiodic perception. But the British entomologist A. D. Lees has discovered that in at least some aphids certain brain cells, and not the eyes, perceive the photoperiodic information. There is evidence that in lizards and amphibians the pineal gland, rather than the eye, measures the lengths of day and night. M. Menaker and his colleagues have presented powerful evidence that the pineal gland of birds is a photoreceptor. The pigments involved have not been rigorously characterized; but they might be closely related to or identical with the visual pigments, because there is evidence of homology between the centrally located pineal gland and the laterally placed eyes.

Virtually all eucaryotic organisms contain a biological clock, and its presence is demonstrated by circadian rhythms.

Why is there any response to change in daylength in the first place? It is obvious that organisms have some way of measuring time, and that they are remarkably well adapted to a 24-hour day-night cycle. The answer is that photoperiodic phenomena are tied to a fascinating rhythmic system probably found in all eucaryotic organisms. There is without question a biological clock within such relatively advanced organisms, and the outward manifestations of this clock are known as CIRCADIAN RHYTHMS (*circa:* about; *dies* a day).

If an experimenter places certain insects (or, for that matter flying squirrels) in constant conditions of temperature and illumination and monitors their activity he will observe a peak of activity about once every 24 hours, and between these peaks, a period of relative lethargy *(Figure 15)*. If he implants a thermistor, a sensitive temperature-monitoring device, into a hamster and places the hamster under constant environmental conditions with food continuously available, the hamster's temperature will continue to fluctuate up and down on an almost 24-hour cycle for as much as two years. If he measures the level of the enzyme alcohol dehydrogenase in man at hourly intervals throughout several weeks, he will observe that the level is lowest in the early morning and highest near five in the evening—a fact for which many

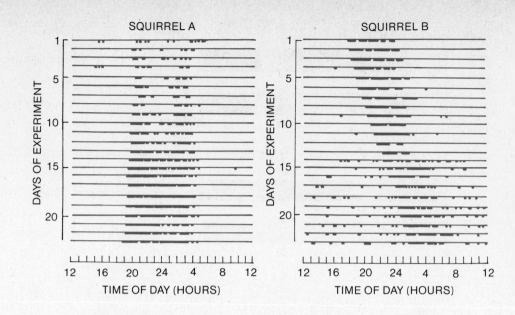

SQUIRREL A

SQUIRREL B

DAYS OF EXPERIMENT

TIME OF DAY (HOURS)

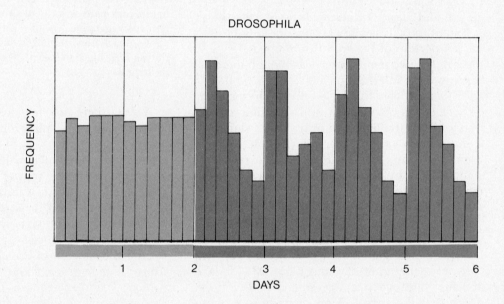

DROSOPHILA

FREQUENCY

DAYS

15 **CIRCADIAN RHYTHMS affect the daily activity of
most eucaryotic organisms. Graphs at top indicate
activity of two flying squirrels confined in a rotating
cage in constant darkness. Heavy horizontal bars
denote intervals of running. Cycle of one squirrel
averaged 23 hours, 58 minutes; the other, 24 hours, 21
minutes. Bar graph below summarizes the emergence
of a population of fruit flies from their pupal cases.
Each vertical bar represents an interval of four hours.
Flies raised in constant darkness (days 1 and 2) emerged
at random throughout the day. Flash of light at the end
of day 2 induced the appearance of a circadian rhythm.
After that, emergence of flies followed a 24-hour cycle.**

cocktail-drinkers should be grateful. The leaflets of
a plant such as clover or the tropical tree *Albizzia*
are normally down and folded at night and up and
expanded in the daytime. The leaves of a number
of plants show such so-called sleep movements.
Under dim light they continue to open and close
on an approximately 24-hour cycle *(Figure 16)*. The
fungus *Neurospora* (bread mold) shows a rhythmic
cycle of spore formation that is sensitive to photoperiod. A colony growing outward from the center
of a petri dish forms spores for about 12 hours,
then enters a 12-hour period of vegetative growth
characterized by simple extension of the edge of
the colony. The spores are produced in clusters on

16 SLEEP MOVEMENTS OF PLANTS such as the tropical tree Albizzia follow a 24-hour cycle. Its leaves are normally folded at night (left) and open during daytime (right). Plant at left was kept in darkness until just before picture was made. Plant at right was left in continuous light.

specialized filaments (conidia, *see Chapter 26*) and the masses of conidia are easily seen with the naked eye. They appear as light-colored bands which form concentric circles in the petri dish — looking like growth rings when viewed from above *(Figure 17).*

There are several important characteristics of all circadian rhythms, independent of the organism chosen for study. First, the period (the time from peak to peak) is remarkably insensitive to temperature. The size or amplitude of the fluctuation may

17 COLONY OF NEUROSPORA growing on a petri dish forms spores for 12 hours, then expands by vegetative growth for the next 12 hours. Masses of spore-forming filaments resemble growth rings.

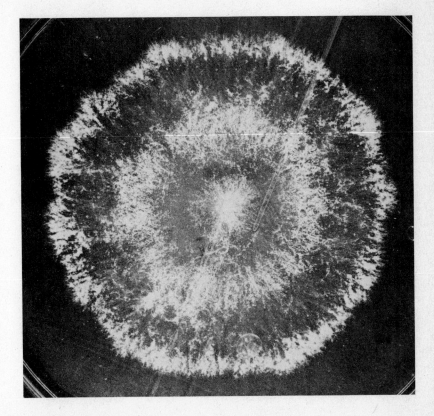

be drastically reduced by lowering the temperature, but the period remains relatively unchanged; about 24 hours. Second, the rhythms are highly persistent; witness the case of the hamster. The continuous cueing provided by light-dark transition is not required for their expression. Third, they can be ENTRAINED, within limits, by light-dark cycles. Note that the period in the above examples was ABOUT 24 hours. If any of these organisms were placed on a day-night cycle totalling EXACTLY 24 hours, the rhythm expressed would show a period of exactly 24 hours. On the other hand, if the experimenter were to use a day-night cycle of, say, 22 hours, then the overt rhythms would show a 22-hour period. If light-dark cycles can entrain the rhythms, it follows that light should also be able to shift the rhythm. If an organism is maintained on constant conditions, with its circadian rhythm expressed on the approximately 24-hour period, a brief exposure to light (or a brief dark period, depending upon the way the experiment was designed) can make the next peak of activity appear either later or earlier than one would have predicted, depending upon when the stimulus was applied. Moreover, the organism does not then return to its old schedule. If the first peak was delayed by six hours, then the subsequent peaks are all six hours late; such phase shifts are hence permanent.

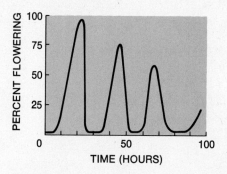

18 24-HOUR CYCLE governs the response of a population of short-day plants (Chenopodium rubrum) to flashes of red light during a 96-hour "night." The light-dark transition at time zero established the rhythm. Flashes that coincided with 24-hour cycle induced flowering; other flashes did not.

There is now ample evidence that interaction of daylength with circadian rhythms is the basis of photoperiodic behavior. A small lawn weed, *Chenopodium rubrum*, provides one sort of evidence. A short day plant, it will flower in response

to a single long night; it shows at least some flowering response to a night as long as 96 hours. A single red light flash during this long night can either enhance or inhibit this response, depending upon precisely when it is administered. The results of such an experiment are diagrammed in Figure 18. The familiar circadian pattern is obvious. Several questions, however, remain unanswered. First, what is the difference between long day and short day organisms? Second, how does the light stimulus affect the biological clock to bring about such profound changes in the growth or behavior in the organisms? Finally, what is the biochemical and biophysical basis for the clock? These questions are currently the subject of considerable research.

INTERNAL RHYTHMS

All these rhythmic activities have their counterparts in the metabolism of the organism. The most obvious counterpart is in the energy use (and therefore oxygen uptake) that accompanies these varied activities. The oxygen uptakes of beans, carrots, mice, crabs and beetles have all been studied, and all show circadian rhythms. It follows that the hundreds of enzymes and intermediates involved in the energy metabolism of the body must show these rhythms.

Other enzymes and their products are geared into these cycles. In the pineal gland of the rat, for instance, several compounds involved in control of the nervous system, including the hormone norepinephrine *(Chapter 18)* show circadian rhythms.

These numerous internal rhythms, accurately cued to a 24 hour period by external light rhythms, give rise to the fluctuations in the activity and physiological state of the whole organism. They **are the intermediate steps between the times of the** biological clock and its expression. But what is the timer? And does the timer get its cue from external factors or is it autonomous?

F. A. Brown of Northwestern University argues that external factors alone constitute the timer of the clock. He points to some metabolic rhythms that do not wander like the activities just described. He recorded for ten years the metabolic rhythms of a set of potato shoots (piling up 1.5 million potato-hours of data) and found sharp peaks at 7 A.M. and 6 P.M. every day. These data were recorded in a system sealed off from external fluctuations in light, temperature, O_2, CO_2, pressure and humidity. Brown's interpretation is that subtle geophysical factors, perhaps rhythmic changes in the earth's magnetic field or the incidence of cosmic rays, synchronize these metabolic cycles despite experimental attempts to control the external conditions. But the inability to pinpoint these factors has made many workers skeptical of their existence, and the prevailing viewpoint is that the clockwork consists of some sort of internal oscillator that is regulated by external phenomena.

The character of the oscillator has been much explored. The numerous metabolic factors which have been examined are not in phase. But that need not prevent one of them from being the controlling factor that drives the other. Is one of the factors the "clock motor" and the others simply the "clock hands" which are driven by the motor? This problem has been rigorously studied in a little marine unicellular plant, the dinoflagellate *Gonyaulax polyedra* (Chapter 26). Populations of these cells show clearcut circadian rhythms in cell division, the capacity to carry out photosynthesis, and the capacity to emit light (luminescence) when being agitated. One approach was to attempt to block just one of the rhythmic activities; if blocking one led to changes in the others, that one could be the clock mechanism. But in fact treatment with specific blockers of photosynthesis, RNA or DNA synthesis, or various other activities, succeeded only in suppressing these specific activities. Upon removal of the inhibitors, the suppressed rhythms reappeared, and no other activities were thrown out of phase. The "mechanism," if there is but one, has therefore not been located. Such experiments, though not conclusive, are in line with the view that the body contains thousands of oscillators, and that entrainment serves to synchronize them with one another by synchronizing them with a simple external oscillator. There is currently great interest in a recently discovered circadian rhythm in the degree of saturation of membrane fatty acids (Chapter 4). Unsaturated fatty acids provide a more fluid and permeable membrane than saturated ones at a given temperature. Changes in saturation could explain a wide range of metabolic circadian rhythms, simply on the basis of alterations in the rate at which small molecules move across membranes. Such a mechanism still does not account for the remarkable insensitivity of the period to temperature change, but it nonetheless provides a testable model.

In other words, it has been as difficult to locate the internal clock mechanism as to detect the exogenous subtle geophysical factors. Although the mechanism of the biological clock is unclear, it is probable that all eucaryotic organisms possess one. The idea of the static nature of the body's composition was overturned in the 1930s and replaced by the notion of a dynamic steady state. But in the last 15 years we have learned that the state is not so steady. The composition of every organism is changing in cyclic fashion throughout each day, and John Doe at noon is literally a different organism from John Doe at midnight.

THE IMMUNE SYSTEM

The relatively constant internal environment is under continuous challenge by invading microorganisms, whose invasion introduces unwanted and often dangerous toxins or worse. Most animals have a potent defense system that protects them from invasion by foreign organisms. It consists of phagocytic cells that engulf and destroy the invaders. In addition all vertebrates, including man, have a second line of defense: an immune system that is highly specific in selecting its target organisms. In a sense the distinction between specific and nonspecific defense systems is artificial, because both systems work together and both depend on the white cells of the bloodstream.

In man, the three principal types of white blood cells are NEUTROPHILS and two kinds of lymphocyte, called B-CELLS (because they originate in bone marrow) and T-CELLS (because they develop in the thymus gland). Unlike red blood cells, white cells can move by themselves, either drifting with the bloodstream or swimming against it; and they can leave the capillaries, squeezing through the walls to reach damaged or infected tissues. Guided by some sort of chemical gradient, neutrophils are the first to arrive at the site of a wound. Assisted by macrophages (giant phagocytic cells found in the tissues), they attack and engulf any foreign matter that forces its way into the body, such as viruses, bacteria or the thorn of a rose. This nonspecific response is the body's first line of defense. Confronted by an infection, the white cells will literally eat themselves to death, engulfing and digesting the foreign matter until the accumulation of toxic breakdown products kills them. A neutrophil can ingest from five to 25 bacteria before it dies. *(See photomicrograph on opening page of this chapter.)*

The second line of defense, specific immunity, is the job of the lymphocytes. Although B-cells and T-cells look identical under the microscope, they respond to foreign matter in different ways. A substance capable of provoking an immune response is called an ANTIGEN. Most antigens are proteins, but certain other large molecules such as polysaccharides are also antigenic. B-cells attack antigens by releasing a quantity of ANTIBODY, protein molecules that circulate in the blood and bind themselves to one specific type of antigen. The production of circulating antibodies by B-cells is sometimes called humoral immunity (from the Latin *humor* = fluid).

Antibodies are proteins consisting of long ("heavy") chains and short ("light") chains strung together by disulfide bonds, as shown in Figure 19.

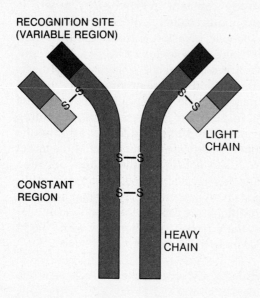

RECOGNITION SITE
(VARIABLE REGION)

CONSTANT
REGION

LIGHT
CHAIN

HEAVY
CHAIN

19 **ANTIBODY MOLECULE consists of two heavy (long) polypeptide chains and two light (short) ones. The gray parts of the chains are the same in all antibodies. The variable recognition sites (color) give each antibody its unique specificity.**

The so-called constant region is the same in all antibodies; the variable region, which contains the site that binds the antigen, differs from one type of antibody to another. The variable region of each type of antibody is tailored to fit one specific antigen. The sequence of amino acids in this region determines the specificity of the antibody. When the antibody encounters an unknown substance, the recognition site senses the shape of the molecule and identifies it as being either "self" (the body's own) or "non-self" (a foreign antigen). Antibodies ignore anything recognized as self, but bind themselves powerfully to foreign antigens, and thus inactivate them. Because invading cells almost always have more than one antigenic molecule on their surface, antigens can bind the

> Once exposed to a particular antigen, the lymphocytes somehow remember it for years afterward, and remain tooled up to produce the corresponding antibody on short notice.

cells into clumps, which are then attacked by phagocytic cells *(Figure 20)*.

T-cells do not release antibodies, but react to antigens on the surfaces of cells. They recognize, engulf and destroy foreign cells, or any of the body's own cells that have been altered by virus infections, cancer, or old age. This mechanism is sometimes called cellular immunity. The important difference between the T-cells and other phagocytic cells is that a particular T-cell will destroy only one specific type of foreign cell, while neutrophils and macrophages will destroy any foreign matter they encounter.

IMMUNOLOGICAL MEMORY

The first time that an animal is exposed to a particular antigen (for example, the organism that causes whooping cough), there is a certain time lag—usually several days—before the number of antibody molecules and the number of activated T-cells circulating in the bloodstream catches up with the number of invaders. But for years afterward, sometimes for life, the immune system somehow "remembers" that particular antigen and remains tooled up to produce antibodies on short notice.

This memory effect explains why IMMUNIZATION has almost wiped out such deadly diseases as smallpox, diphtheria and polio. A very small amount of viral or bacterial protein (often treated to make it harmless) is injected into the body. Later, if very similar disease organisms should attack, the tooled-up cells recognize it and quickly overwhelm it with a massive outpouring of antibodies.

How does this memory work? And why does the system fail to attack the body's own cells and proteins? These questions are the focus of a massive research effort because the answers promise to shed light on cancer, the rejection of transplanted organs, and other medical topics. Several theories have emerged, but the one currently favored by

ANTIGEN

T-CELL

ACTIVATED
T-CELL

KILL FOREIGN OR DISEASED
CELLS ON CONTACT

T-CELL
HELP

PRE-LYMPHOCYTE

B-CELL

ACTIVATED
B-CELL

CIRCULATING
ANTIBODIES

**THE IMMUNE RESPONSE. Developing lymphocyte
(left) differentiates into T-cells and B-cells. Specific
receptors on surface of T-cell are exposed to antigen and
bind it (top). Thus activated, T-cell continues to
multiply (top right). B-cell exposed to antigen interacts
with activated T-cell (bottom right), begins to release
circulating antibodies, and continues to multiply.**

20

most immunobiologists is known as the clonal selection theory.

The biosphere contains an almost endless number of potential antigens. An individual can make specific immune responses to perhaps a few million (roughly 10^6) different antigens. A cell that could make all of them would need 10^6 genes to code for them, according to the rule that one gene codes for one protein. But a human cell contains only about 10^4 genes, and they must code not only for antigens, but for all the other proteins in the body. The theory holds that an individual has a very small number of B-cells and T-cells genetically programmed (in some unknown way) to respond to only one antigen. When they come in contact with that antigen, the cells begin to multiply. After several rounds of multiplication, there is a large clone of them. (Genetically identical descendants of a single cell are known as a clone, thus the name of the theory.) In this way, immunologists believe, the immune system exhibits natural selection; the cells best able to react against a specific antigen have a reproductive advantage and soon greatly outnumber cells unable to react to the antigen.

TRANSPLANTS

If one attempts to transplant an organ or a piece of skin from one human to another, the transplant is recognized as non-self and soon provokes an im-mune response; the tissue is killed or "rejected." But if the transplant is done immediately after birth, or comes from a genetically identical person (an identical twin) the material is somehow recognized as self, and is not rejected. Physicians can overcome the rejection problem for a while by knocking out the immune system with drugs (immunosuppressants) that reduce the action and number of B-cells and T-cells. But this technique leaves the individual wide open to bacteria and viruses, and he must be protected from all such invaders by elaborate isolation procedures.

A slightly different problem arises when one attempts to transplant another important tissue: blood. Early attempts at blood transfusion often killed the patient. The Austrian physician Karl Landsteiner, starting in 1898, found by mixing blood cells and serum from different individuals that only certain mixes were compatible; in others the red blood cells formed clumps that would clog the blood vessels. Landsteiner deduced that there were two red cell antigens, A and B, and two corresponding serum antibodies, α and β. Mixes of A with α or B with β cause clumping. The production of these antigens and antibodies is controlled by genes. A person can be a heterozygote with AB type blood or a homozygote with AA or BB (usually designated simply as A or B). A different allele produces type O blood, which lacks both antigens. An individual who has type A blood can accept a transfusion only from type A donors. An AB individual has no serum antibody for either antigen, and can accept blood from any donor. A type O individual has antibodies for both, and needs type O blood.

The Rh factor, so named because it was first found in Rhesus monkeys, is another blood antigen. In most human populations almost 100 percent of the individuals have this antigen, and their blood is said to be Rh⁺. Among members of the white race, however, only 87 percent are Rh⁺; the others lack this antigen, and their blood is called Rh⁻. In areas where the races have interbred, such as the U.S., the percentage of individuals with the factor lies somewhere between 87 and 100 percent. About 95 percent of American blacks, for example, are Rh⁺. Like antigens A and B, the Rh factor is genetically determined; but the serum antibody to Rh, unlike those for blood type, is not genetically determined. It develops only after exposure to the Rh antigen. If an Rh⁻ woman becomes pregnant with an Rh⁺ child, the embryo's Rh antigen sometimes enters the mother's blood, is recognized as non-self, and the mother makes antibodies against the embryo's blood, often dooming it or the infant to an early death. The problem worsens with subsequent pregnancies. Not many years ago, when an Rh⁺ man married an Rh⁻ woman they would be advised to have only one or at most two children. In recent years this problem has been solved. A substance called Rhogam, derived from the blood of unusually Rh-reactive men and women, is injected into an Rh⁻ mother within 72 hours after her first delivery. It contains antibodies to the Rh⁺ antigens and suppresses the mother's immune attack upon her next baby's blood.

CANCER AND IMMUNITY

In a healthy body no one group of cells becomes dominant. Growth stops when one group has achieved the proper size, as determined by a preset genetic program and the feedback of outside information, such as the degree of cell crowding. An effect called contact inhibition slows or stops the reproduction of cells when they are crowded together. What if a cell should mutate and escape this restriction, continuing to multiply at the expense of its neighbors? This is most likely to happen with epithelial cells (those of the skin and the linings of the body cavities) which wear out and are continously replaced. Is it a coincidence that (based on figures from Denmark over a 25-year period) 92 percent of cancers occur in epithelial cells?

One theory of cancer argues that a principal role of the immune system is to destroy cells that become altered in any way that changes their surface antigens. A key feature of many cancerous cells appears to be a change in their surface properties, so that they become insensitive to contact inhibition. Such a change could be caused by a spontaneous mutation, by introduction of foreign DNA or RNA by a virus, or by the disruption of genes by an environmental agent such as smoking, ultraviolet radiation or pollutants. Most cancers are diseases of old age. When the body gets old, the immune system becomes less effective. Perhaps during youth and middle age the immune system destroys any cancerous cells that might arise; but when it falters cancer cells can multiply unchecked. Humans with a naturally inadequate immune response, or who are under prolonged treatment with immunosuppressants, are particularly prone to cancer. If the immune system does indeed play a major role in the suppression of cancer, the day may come when vaccination will cause cancer to go the way of smallpox, diphtheria and polio.

READINGS

F.A. Brown, J.W. Hastings and J.D. Palmer, *The Biological Clock: Two Views*, New York, Academic Press, 1970. A delightful little paperback describing rhythmic phenomena and the different views of their control. Well illustrated and easy to read.

A.B. Burnett and T. Eisner, *Animal Adaptations*, New York, Holt, Rinehart and Winston, 1964. In 135 pages, this book concentrates on the diversity of the adaptations which animals have evolved to solve basic problems. The section on "Maintenance of the Internal Milieu" is particularly relevant to this chapter.

A.W. Galston and P.J. Davies, *Control Mechanisms in Plant Development*, Englewood Cliffs, NJ, Prentice-Hall, 1970. The first chapter is an excellent discussion of photoperiodism in plants, with brief mention of animal systems.

R.A. GOOD AND D.W. FISHER, Editors, *Immunobiology*, Sunderland, MA, Sinauer Assoc., 1971. A collection of essays on both biological and clinical aspects of immunology. Handsomely illustrated.

G.J.V. NOSSAL, *Antibodies and Immunity*, New York, Basic Books, 1969. Written for the layman by one of the authorities in the field, but by now a bit out of date.

C.L. PROSSER, *Comparative Animal Physiology*, 3rd Edition, Philadelphia, W. B. Saunders Co., 1973. A thorough treatment of regulatory mechanisms (and other matters) in animals. The chapters on water, excretion and temperature are excellent.

F.B. SALISBURY AND C. ROSS, *Plant Physiology*, 2nd Edition, Belmont, CA, Wadsworth, 1978. A comprehensive textbook of plant physiology with a good treatment of circadian rhythms, both in plants and in animals.

An organism like man needs signals that travel between the cells of distant organs. Best known among these, but still mysterious in their modes of action, are the hormones. —S.E. Luria

Hormones are substances that are produced in one part of a multicellular organism and transported to another part, where they induce a change in some biochemical function of the responding cells. In animals these chemical messengers are cast into the bloodstream, the lymphatic system, or some other bodily fluid to reach the target cell *(Chapter 16)*. In plants the path of transport is usually through the xylem or the phloem. Hormones play two major roles in eucaryotes: they control development and regulate physiology. Plant hormones act exclusively in development, governing the direction and timing of growth from germination of the seed or spore through reproduction and aging. Animal hormones serve both functions. They affect not only growth and reproduction, but also digestion, circulation, metabolism, muscle tone and behavior. Together with nerves, hormones serve to maintain the remarkable constancy of a mammal's internal environment *(Chapter 17)*.

In their journey through an organism, hormones may reach many different kinds of cell, but they usually affect only a specific group of "target" cells that have been genetically programmed to detect and respond to a particular hormone. Different groups of target cells may respond differently to the same hormone.

In terms of their biological activity, hormones are stunningly potent substances. Tiny amounts can evoke dramatic responses in the target cells.

"GENERAL" TOM THUMB stands beside the showman P.T. Barnum in photograph on opposite page. Born Charles Stratton in 1838, Tom was exhibited as a midget by Barnum, who invented his stage name and military rank. Because of a hormone deficiency, Tom stood less than 61 cm tall when Barnum discovered him as a boy, and less than 84 cm (33 inches) when fully grown.

The plant hormone indoleacetic acid (IAA), for example, can show activity at concentrations well below one part per million. Hormone-mediated changes qualify as communication in a special sense often attached to the word: the amount of energy involved in the response greatly exceeds that put into the signal. Compared with the response to nerve signals, which typically occur in a fraction of a second, hormonal communication is slow. The swiftest effects are produced by hormones such as adrenalin and noradrenalin, which mediate the stress ("fight or flight") reactions of mammals, but even this dramatic action is sluggish compared with behavioral activity based on nerve conduction.

Hormonal communication has proved entirely adequate to coordinate the rather leisurely activities of plants, but mobile animals obviously need a swifter means of internal communication. Then why do animals still rely heavily on hormones? Why, in the course of evolution, didn't animals discard their sluggish hormone systems entirely, and replace them with nerves? To answer the question, one must look at hormones in an evolutionary perspective.

When the first organisms appeared in the primeval seas billions of years ago they probably communicated with one another during sexual reproduction by means of chemicals passed back and forth in the water. Such chemical messages exchanged between organisms are called PHEROMONES *(Chapter 22)*. This is the prevailing mode of communication among microorganisms, lower invertebrates, and many lower plants alive today, because it utilizes only the simplest sending and receiving apparatus.

When multicellular animals evolved, there arose a need to integrate the activities of many cells. Two kinds of internal communication evolved. The first hormones might have been modified pheromones.

381

When multicellular animals closed off their body cavities, part of the "sea" in which the cells lived became internal, and the pheromones that they used for communication became redefined as hormones. The first nerve cells, which evolved much later, may have been modified hormone-secreting cells. True, a nerve cell is greatly elongated and specialized for the swift propagation of electrical impulses, but its action remains partly chemical: the impulses pass from one nerve cell to another, across the synapse that separates them, when transmitter substances are released like hormones from the tips of the cells *(Chapter 19)*. Hormones and nerves are intimately related and serve similar functions, under the control of the central nervous system. This point will become clearer in the later discussion of the relationship between the pituitary gland and the hypothalamus of the brain.

Plants lack the elaborate systems of regulatory hormones found in higher animals; as a result, they are more susceptible than animals to transient changes in their environment.

In animals hormones are made and released by specialized groups of cells called endocrine tissues. In higher animals the tissues are sometimes compacted into endocrine organs such as the thymus, pituitary and thyroid glands. All of the hormone-producing tissues and organs collectively comprise the ENDOCRINE SYSTEM. In plants less specialized cells from a whole region may make a particular hormone.

As one might expect, hormone molecules do not belong to any single chemical category. They have evolved in many different biological contexts, and many different organic substances have been enlisted to serve as messengers. Some are relatively small polypeptides or proteins; others are steroids (lipids containing the multiple-ring structure shown in Figure 13), amines, or amino acid derivatives. Still others, including some plant hormones, belong to none of these familiar categories. Nevertheless most hormone molecules do have one significant property in common: they are neither very large nor very small. They cannot be very large, because large molecules tend to be in-

soluble and difficult to transport through the body. They cannot be very small, because small molecules are not complex enough to be chemically distinctive and clearly recognizable by the target cells. The largest hormone molecules are proteins such as the human growth hormone, which consists of 188 amino acid units, and the smallest by far is the plant hormone ethylene, which contains only six atoms.

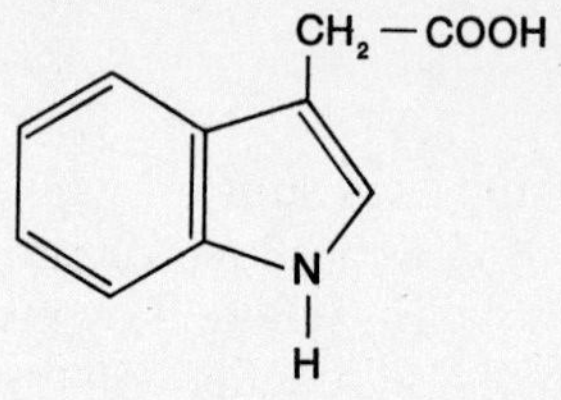

IAA (INDOLEACETIC ACID)

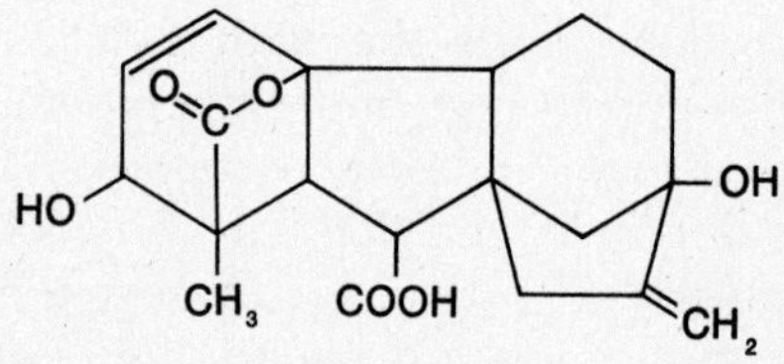

GIBBERELLIC ACID

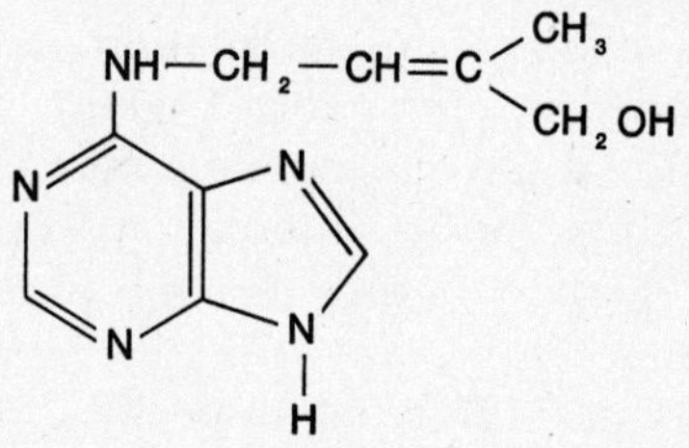

ZEATIN

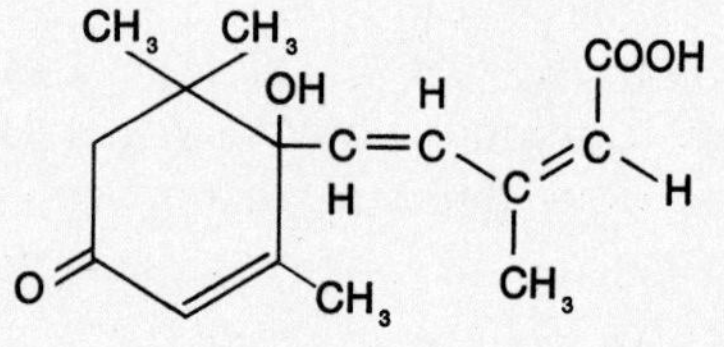

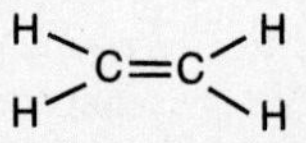

ETHYLENE

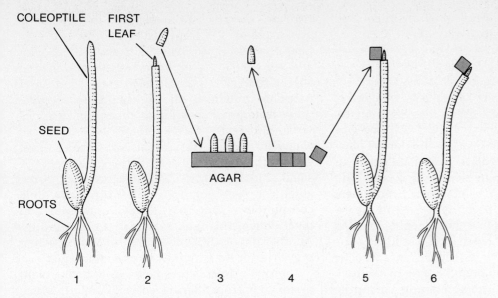

COLEOPTILE
FIRST LEAF
SEED
ROOTS
AGAR

1   2   3   4   5   6

PLANT HORMONES

Plants lack the elaborate systems of regulatory hormones found in higher animals; as a result, they are far more susceptible than animals to transient changes in their environment. Their adaptations to environmental stress follow a different strategy. They rely on developmental changes, often cued by their environment, instead of the swifter hormonal and nervous homeostatic mechanisms found in animals. Such developmental changes in plants are virtually always controlled by hormones.

In higher plants there are six known classes of growth hormones: AUXINS, CYTOKININS, GIBBERELLINS, ABSCISIC ACID and its relatives, ETHYLENE GAS, and FLORIGEN. In lower plants there are a great many other hormones, many of them related to reproductive cycles. There are also many PHEROMONES, substances analagous to hormones, but produced by one organism to influence another. Unlike most insect pheromones, which are released into the air as gases *(Chapter 22)*, the active substances of plants are most often released into water. A chemical system that insures outbreeding in ferns is discussed in Chapter 26.

Structural formulas for five of the six classes of plant hormones are shown in Figure 1. The elusive

THE WENT EXPERIMENT demonstrated the effect of auxin on grasses. Diagram (1) depicts a three-day-old oat seedling. Went removed the coleoptile tips (2) from several seedlings and placed them on a block of agar (3). After an hour or so he removed the tips, divided the agar (4) and placed pieces of it on the decapitated coleoptiles (5). Within about 90 minutes seedlings showed growth curvature away from side touching the agar. Curvature was caused by auxin that diffused into agar.

2

flowering hormone florigen will be discussed later in this chapter. The most important auxin is indoleacetic acid, mentioned in Chapter 12 in connection with the regeneration of xylem in callus tissue. By contrast, there are over 20 known gibberellins, all structural modifications of the gibberellic acid shown. They vary from species to species, and even within a single plant, depending on developmental status —that is, whether or not a plant is flowering. Several may be present at once. There are also several different cytokinins, in addition to the zeatin shown. Likewise, abscisic acid is one of several naturally occurring plant growth inhibitors. Ethylene has no other known physiological homologue.

The first class of plant hormones discovered were the auxins. Building upon research started by Charles and Francis Darwin in 1880, Frits Went first isolated auxin from grass seedlings and showed that it would promote cell elongation when reapplied to the seedling. His classic experiment is shown in Figure 2. In grasses the first portion of the shoot to emerge from the seed is the

1   **PLANT HORMONES. Indoleacetic acid is an auxin, gibberellic acid a gibberellin, zeatin a cytokinin, abscisic acid a growth inhibitor. Ethylene promotes the ripening of fruit and many other processes.**

COLEOPTILE. This hollow cylindrical organ, frequently closed at the tip, protects the shoot apex and the first true leaves as the shoot pushes its way through the soil. The coleoptile tip, the last part of the coleoptile to cease meristematic activity, is the site of auxin synthesis. The auxin is transported to the lower parts of the coleoptile where it promotes extensive growth. Auxin transport is a one-way process, a physiological manifestation of the obvious polarity of the seedling. Auxin transport almost always shows some degree of polarity, but the absolute restriction to downward movement found in the coleoptile probably represents the extreme case. Went excised some coleoptile tips and placed them on moist agar blocks, allowing the auxin to move into the agar. After an hour or so, he removed the tips and placed the agar blocks back asymmetrically onto the remaining coleoptile stumps. Auxin diffusing from the block caused the underlying tissue to elongate extensively, producing a pronounced curvature. Since curvature was proportional to auxin concentration, Went's experiment was the basis for what is still the most sensitive technique for assaying the auxin content of a plant specimen. More than three decades passed before the auxin in coleoptiles was rigorously identified as indoleacetic acid.

In addition to its role in coleoptile elongation, auxin also promotes the elongation of stems and leaf veins. But promotion of cell enlargement is not its only function. Auxin plays a role in inducing differentiation of vascular tissue, not only in the callus system, but in intact plants as well. Although it is a strong inhibitor of root elongation, it induces the mitotic activity that leads to formation of lateral root primordia. Auxin is widely used both in horticulture and agriculture to induce cuttings of valuable plants to form roots. Finally it

regulates the process of ABSCISSION: the separation of mature or senescent leaves or fruit from the parent plant. At the base of many leaf petioles and fruit stalks is found a transverse layer of specialized cells, the abscission zone. Under appropriate conditions, these cells begin to deposit a waxy waterproof covering of suberin in their cell walls, and finally excrete enzymes to digest cell wall polysaccharides. Ultimately the leaf or fruit drops off, leaving a scar protected by suberin. A small amount of auxin flowing down across the abscission zone from fruit or leaf blade will often completely suppress the developmental and biochemical changes that lead to abscission. Another important agricultural use of auxins is to prevent fruit from dropping before it is ripe.

The first cytokinin was isolated, not from plant tissue, but from herring sperm DNA. In 1955, Folke Skoog and his colleagues found that cytokinin had dramatic effects on callus cultures derived from the pith of tobacco stems. On a simple medium of salts and sugar, the tobacco callus remained disorganized, producing irregular parenchyma tissue by rare sporadic mitosis. Cytokinin, however, induced localized pockets of intensive meristematic activity, but with little cell enlargement. Most surprising, however, was the interaction between auxin and the cytokinin. Figure 3 il-

INTERACTION OF AUXIN AND CYTOKININ determines the course of development in culture of tobacco pith callus. Chart summarizes effects of the auxin indoleacetic acid (IAA) and the cytokinin kinetin. Maximum tissue development requires both hormones. High ratio of auxin to cytokinin favors development of roots (upper right); low ratio favors development of shoots (bottom center) and intermediate ratio favors the multiplication of undifferentiated callus cells (middle row).

KINETIN CONCENTRATION (MG/ML)					
0	SMALL CALLUS	SMALL CALLUS	MEDIUM CALLUS ROOTS	MEDIUM CALLUS ROOTS	MEDIUM CALLUS ROOTS
0.2	MEDIUM CALLUS	MEDIUM CALLUS	MEDIUM CALLUS	LARGE CALLUS	LARGE CALLUS
1.0	MEDIUM CALLUS, FEW SHOOTS	MEDIUM CALLUS, MANY SHOOTS	MEDIUM CALLUS, MANY SHOOTS	MEDIUM CALLUS, FEW SHOOTS	LARGE CALLUS
	0	0.005	0.18	1.08	3.0

IAA CONCENTRATION (MG/L)

EFFECT OF GIBBERELLIN on seedlings of mutant dwarf corn. One week before photograph was made seedlings at left were treated with gibberellin. Control seedlings at right were untreated.

4

lustrates the results of varying the concentration of both of these growth regulators. Optimal growth required both substances. A high ratio of auxin to kinetin produced a large number of roots, while a low ratio produced numerous shoots instead. Intermediate ratios merely produced vigorous callus. In other words, it was the interaction of hormones, rather than a single one, that determined the course of development.

The activation of bud meristems in some plants also depends on the interaction of auxin and cytokinin. The reader will recall *(from Chapter 12)* that bud meristems are tiny pockets of potentially meristematic tissue in the leaf axil: the upper angle between leaf and stem. Auxin moving down the stem from the apex keeps these lateral bud meristems dormant. K. V. Thimann and his associates showed that cytokinins were effective auxin antagonists in this system. Small concentrations of kinetin could overcome the auxin inhibition, allowing the bud meristems to begin development.

There are many other cytokinin-mediated processes, mostly involving initiation or maintenance of meristematic activity. One process, however, does not involve cell division at all: the prevention of leaf senescence. If a leaf is detached and kept supplied with ample water and minerals, it nevertheless shortly loses its chlorophyll, turns yellow, and dies. Local application of a cytokinin very much delays the course of senescence. The spot to which the cytokinin was applied remains bright green long after the remainder of the leaf has turned yellow. Senescence involves massive breakdown of proteins and nucleic acids, with the soluble products being exported from the leaf prior to abscission —an excellent conservation tactic. Cytokinins retard this breakdown, by either stimulation of synthesis or inhibition of degradation. Since roots produce cytokinins and export them through the xylem to the shoots, the role of cytokinins in preventing early senescence seems to be logical.

Went and many subsequent workers in plant physiology regarded auxin as the single master compound regulating stem elongation. But work in Japan proved them wrong. The Japanese physiologist Kurosawa was studying a disease of rice seedlings caused by the fungus *Gibberella fujikuroi*. When grown in liquid culture, the fungus produced a mixture of substances that could produce the disease symptoms—seedlings with abnormally long and flimsy stems—in the absence of any infection. In the mid 1950's these experiments finally attracted the attention of workers outside the Orient. The fungal substances turned out to be a mixture of several gibberellins, the natural growth hormones of higher plants.

The potency of gibberellins was dramatically demonstrated in experiments on single-gene mutants of corn and peas. The stems of these mutants failed to elongate normally, and auxins had no effect whatsoever on them. But treatment with microgram quantities of gibberellins completely nullified the effect of the mutant gene. Treated plants were indistinguishable from wild-type seedlings *(Figure 4)*. The mutation was eventually shown to involve a key enzyme in gibberellin synthesis. Like the auxins, gibberellins are synthesized in actively growing regions and exported to other parts of the plant.

Gibberellins play a special role in the germination of seeds of cereal grasses such as oats, wheat, and barley. Figure 5 shows a longitudinal section of an ungerminated barley seed. Within the seed the embryo lies at one end with the coleoptile extending toward the middle and the primary root toward the end. Both are attached to the scutellum, a structure adjacent to the nutritive endosperm. Just beneath the seed coat is a prominent cell layer called the ALEURONE. When the seeds are soaked in water, one of the earliest events in germination is release of gibberellin by the embryo. The gibberellin eventually reaches the aleurone layer, where it induces these cells to begin synthesis and secretion of large amounts of hydrolytic enzymes—amylases for the digestion of starch, proteases, and nucleases. Thus sugars, amino acids, and nucleotides are made available in soluble form for growth of the embryo. They are rapidly absorbed through the scutellum.

Abscisic acid, discussed below in connection with winter dormancy, is in many cases an antagonist of gibberellin. If one applies it to rapidly developing vegetative buds, the buds stop producing normal leaf primordia and produce bud scales instead. They subsequently go dormant. These effects are completely blocked by the application of gibberellin. Conversely, abscisic acid inhibits hydrolytic-enzyme synthesis in the aleurone layer of the seeds of cereal grasses, which is induced by gibberellin. In other cases, abscisic acid may act as an auxin or cytokinin antagonist, though the precise nature of the interaction is still unclear.

The gas ethylene is perhaps as much a pheromone as a hormone. Around the turn of the century kerosene stoves were frequently used to hasten the ripening of lemons and other citrus fruits. Substitution of more modern heating methods, however, proved to be an expensive mistake, since the fruit failed to ripen. Evidently traces

5 **BARLEY SEED contains embryo at one end (bottom). Soaking seed in water triggers germination. Embryo releases gibberellic acid (GA) which reaches the aleurone cells, stimulating them to synthesize hydrolytic enzymes that digest nutrients stored in endosperm. Resulting sugars, amino acids and nucleotides become available for growth of embryo.**

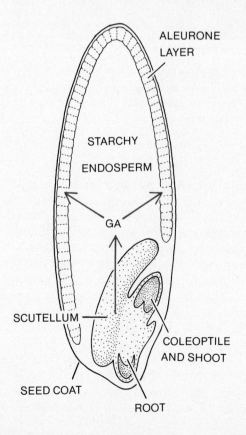

ALEURONE LAYER

STARCHY

ENDOSPERM

GA

SCUTELLUM

COLEOPTILE AND SHOOT

SEED COAT

ROOT

of ethylene from the stoves—not the heat—were promoting the ripening process. It was soon discovered that ripe oranges would hasten the ripening of nearby bananas. The oranges not only responded to ethylene, but eventually began producing it. More than just ripening was involved, however, because it was found that both apples and pears could inhibit the sprouting of nearby potatoes. Many developmental processes, including the dormancy of seeds and the falling of leaves, involve ethylene. In many cases the effects of ethylene resemble those of auxin. Indeed auxin may act in some plants by stimulating ethylene production, but in others ethylene is an auxin antagonist.

The discussion of hormones has thus far emphasized two principles of hormone action: that the target cell is programmed for a specific response, and that two hormones can interact to produce an effect different from those produced by either hormone acting alone. Research on plant growth substances reveals a third: different hormones may limit the same process at different stages of development or in different species. Thus either auxin or gibberellin may limit elongation. Kinetin, which retards senescence in tobacco leaves, is replaced by gibberellin in barley leaves, and by auxin in bean pods. Although the details of hormonal regulation of development are often quite complex, these three underlying principles are simple.

TEMPERATURE, GRAVITY AND LIGHT

With the exception of developmental responses to daylength (photoperiodism, *see Chapter 17*) and temperature, particularly in metamorphosis, animals show little dependence on specific environmental cues. Plants, on the other hand, respond dramatically to such parameters as gravity and changes in temperature or light.

Within limits, temperature changes may alter the rate of animal development, but usually not the pattern. By contrast, temperature changes provide important cues for plants to enter or leave a dormant state. Dormant buds and seeds are relatively dehydrated, with little water to freeze and form damaging ice crystals. Chilling the apices of trees or shrubs of temperate climates causes the shoot apex to begin producing bud scales instead of normal leaves, and eventually to lose water and cease growth. Abscisic acid most certainly plays a role, accumulating as the temperature gradually drops. Other factors, such as decrease in daylength, are involved as well.

The dormant state does not involve cessation of biochemical processes, however. Many seeds and dormant buds must remain below a certain maximum temperature for days or weeks or they will never break dormancy. Blueberry shrubs in New England, for example, must remain below freezing for seven or eight weeks or they will simply remain dormant. Winter rye seeds must be planted in the fall, remaining in the ground through the winter, or they will not germinate. Clearly, even at these low temperatures, some biochemical changes are occurring, changes required for resumption of development. The phenomenon probably involves the effect of temperature on the rates of various synthetic and degradative reactions, perhaps those that involve developmental hormones. The changes that occur during these required cold periods are called VERNALIZATION.

The role of hormones in plant responses to gravity and light direction is much better known. If a plant is placed on its side, the shoot will grow upward directly against the pull of gravity, and the roots will turn down. These responses are called POSITIVE and NEGATIVE GEOTROPISM, respectively. Geotropism of the shoot is reasonably well understood. Auxin, produced by the developing shoot apices and young leaves, is transported in polar fashion down the stem toward the roots (though little may actually go that far). If the shoot is on its side, the auxin is transported laterally to the lower side as well. As a consequence, the cells on the lower side elongate more rapidly than those on the upper, restoring a vertical orientation. In roots the mechanism is less obvious because the auxin transport properties of shoots and roots are quite different. In some roots auxin actually moves toward the root apex. The concentrations found are thousands of times lower than those in shoots. Root geotropism does not occur in the absence of a root cap. There is also some evidence for lateral redistribution of growth inhibitors. Thus lateral transport of auxin is inadequate to account for all of the experimental observations on the geotropism of roots.

The mechanism by which gravity causes lateral displacement of small molecules such as auxin, and perhaps growth inhibitors, is still much under debate. There is some evidence that large organelles such as starch grains are the gravity perceptors, or STATOLITHS, tumbling to the lower side of the cell and in some way altering the transport properties of the lower membrane. Tumbling of starch grains can be observed in both gravity-

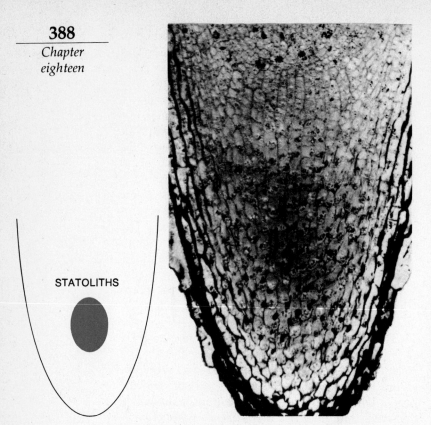

STATOLITHS

6 **ROOT CAP OF PEA PLANT,** shown here in longitudinal section, contains large starch grains called statoliths that may serve as gravity detectors.

sensitive shoots and in root cap cells when tissues are rotated with respect to gravity *(Figure 6).* Moreover, treatments leading to a complete digestion of the starch grains also lead to loss of geotropic sensitivity, though some elongation persists. tips partway up from the base, and collected auxin separately from the illuminated and shaded sides. He found that much more auxin emerged from the side away from the light. He suggested a light-induced lateral transport of auxin away from the light, and together with N. Cholodny presented a unifying hypothesis for both phototropism and geotropism, based on lateral transport of auxin. Though various other proposals were later advanced, the Cholodny-Went hypothesis has withstood experimental test. In coleoptiles, at least, both light and gravity clearly cause a lateral movement of auxin. Figure 7 illustrates one kind of evidence, from the laboratory of W. R. Briggs. Light has no effect on total auxin yields, merely leading to an asymmetric distribution between lighted and shaded sides. The development of the asymmetric distribution, however, requires tissue continuity across the coleoptile. The insertion of a glass barrier that completely isolates the two sides eliminates the differential. Subsequent experiments in several laboratories with radioactive IAA have verified that auxin is indeed laterally transported as a result of unilateral illumination.

Final answers will perhaps emerge from careful studies of the effects of membrane deformation on membrane transport properties.

It is well known that green plants grow toward sunlight, an effect called PHOTOTROPISM. Frits Went did the pioneering experiments on the role of auxin in phototropic curvature of coleoptiles when he first isolated auxin. He split isolated coleoptile

LATERAL TRANSPORT OF AUXIN in response to **7** **light was demonstrated with experiment on coleoptile tips.** Tip at far left was kept in darkness for two hours, and the amount of auxin that diffused into agar block was measured. The same yield of auxin was obtained from agar beneath coleoptile tip growing in light (left center). Assay of agar beneath tip that was partially split by a mica barrier (right center) revealed that the hormone was concentrated on the shaded side. Tip at far right was completely split by barrier that prevented lateral movement of auxin; yield from both sides was the same.

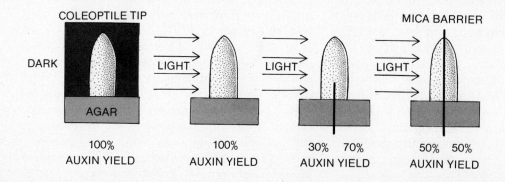

COLEOPTILE TIP   MICA BARRIER

DARK   LIGHT   LIGHT   LIGHT

AGAR

100%   100%   30% 70%   50% 50%
AUXIN YIELD   AUXIN YIELD   AUXIN YIELD   AUXIN YIELD

Under certain illumination conditions, coleoptiles will actually bend away from the light, showing negative phototropism. Under these conditions the distribution of auxin is reversed, with more obtained from the illuminated side of the tip.

Phototropism depends on the wavelength of the light falling on the plant. Both ultraviolet and blue wavelengths elicit a phototropic response from oat coleoptiles, but light beyond 500nm is completely ineffective. There is increasing evidence that the photoreceptor is a flavoprotein, and that an early step is perturbation of an electron transport chain that includes a cytochrome (Chapter 6). The system is probably not mitochondrial, but instead may well be located in the plasma membrane.

ACTION AT THE MOLECULAR LEVEL

How do the various plant hormones act at the molecular level? Despite an impressive mass of experimental work, the exact mechanism of action is unknown for any of them. Some progress has been made, to be sure, particularly in studies of the effects of auxin on cell elongation, and of the effects of gibberellin on the induction of enzyme synthesis in barley seeds.

The first definitive experiments on the mechanism of action of auxin were performed by A. J. N. Heyn in 1939–40. Heyn took sections of oat coleoptiles and incubated some of them in auxin. He supported them horizontally from one end and hung a small weight on the other end, deforming the frail sections downward. After a few minutes he removed the weights and the sections sprang back to a resting position (Figure 8). The degree to which the sharply bent sections straightened out provided a measure of the elastic properties of their cells, and the degree to which the sections remained permanently bent provided

a measure of the plastic properties of the cell walls. The auxin-treated coleoptile sections proved to be more plastic and less elastic than sections treated only with buffer solution. Heyn concluded that auxin was acting to soften the cell wall in some manner, allowing osmotic uptake of water to distend the cell and stretch the wall — an effect typical of elongation growth.

Both softening and elongation presumably involve the synthesis of new proteins. Is the primary target for auxin the cell wall itself? Or is it the chain of reactions between the transcriptions of new mRNA and final protein synthesis?

J. L. Key and his associates have shown that auxin promotes substantial RNA synthesis when applied to soybean hypocotyl sections. A characteristic of most hormones is that they are maximally effective only in a fairly narrow range of concentration. Excess concentrations may be less effective or even inhibitory. Key's experiments showed that the optimal auxin concentration for cell elongation was also optimal for RNA synthesis (Figure 9). It is tempting to postulate that auxin acts like a derepressor in the Jacob-Monod model — a cytoplasmic molecule that can inactivate repressors, permitting operator genes to induce the structural genes to synthesize RNA (Chapter 10).

The processes following gene derepression take time. New messenger and perhaps other classes of RNA must be synthesized and ribosomes must be

EFFECT OF AUXIN on elasticity and plasticity of cells **8**
was determined by hanging a weight on a horizontally suspended section of coleoptile (color). Initial position of coleoptile is shown at left and final position at right. Irreversible bending, the angle between 1 and 3 in the diagram at right, is a measure of the plasticity of the shoot. Reversible bending, the angle between 2 and 3, is a measure of elasticity. Treatment with auxin increases the plasticity and decreases the elasticity.

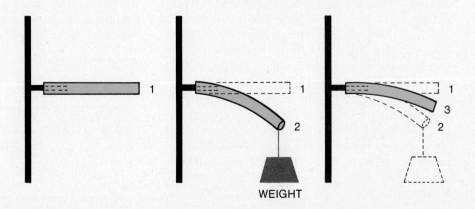

WEIGHT

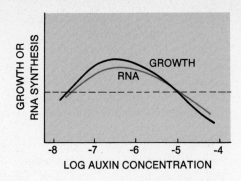

GROWTH OR RNA SYNTHESIS
GROWTH
RNA
-8 -7 -6 -5 -4
LOG AUXIN CONCENTRATION

OPTIMAL CONCENTRATION OF AUXIN for promoting growth in soybean stems coincides with the optimal concentration for promoting the synthesis of RNA. At high concentrations (right) auxin inhibits both processes.

assembled before protein synthesis can begin. The test of the gene-derepression hypothesis is: How soon after application of auxin do the first measureable changes appear? The rates of oxygen uptake and ATP synthesis show increases 30 minutes after the start of auxin treatment, an interval long enough to support the hypothesis. But growth can begin within a minute of auxin application, a period considered too brief to include all of the processes of gene derepression.

It has been known for some years that a drop in pH from say 7 to 4.5 will induce a spurt of growth by coleoptile sections. Recently several workers have demonstrated a rapid excretion of hydrogen ions under the influence of auxin treatment. In addition, there are pronounced electrical changes in coleoptile cells on auxin treatment, perhaps (at least in part) a reflection of the hydrogen ion excretion. There is increasing support for the hypothesis that the earliest growth stimulation by auxin is a consequence of auxin-induced acidification of the cell wall. The exact way in which acidification softens the wall remains elusive, however.

Acid-induced growth does not persist, although auxin-induced growth does. Clearly many long-term effects of auxin must involve the synthesis of new proteins and the derepression of genes. Auxin probably does not act directly on the repressor, but rather at some membrane site. Auxin-induced changes in membrane properties could well release activators or repressors, or influence protein or RNA synthesis in several other ways. The ultimate effect on the cell would depend on which membranes were affected, and what substances were on either side of them at a particular moment.

The effect of gibberellin on the aleurone cells of barley seeds seems a bit more straightforward. The plant embryo produces gibberellins, which induce the aleurone layers to produce several enzymes, including alpha amylase, which degrades starch. The cells synthesize the enzymes from scratch — from amino acids —rather than merely activating

some preformed proenzyme. Secreted into the nutrient-rich endosperm, the enzymes convert (hydrolyze) starch and other nutrients into a water-soluble form. Recent work strongly suggests that transcription is the target of gibberellin action. But despite exhaustive searching, no gibberellin-repressor complex has yet been isolated, nor is there evidence that gibberellin acts directly on the genes. We are still completely in the dark as to the exact way that gibberellin triggers the massive synthesis of only a few hydrolytic enzymes.

HORMONES AND FLOWERING

In flowering plants, hormones play a central role in the flowering process. Almost all studies on flowering have been done on plants that are sensitive to variations in photoperiod (the length of the daylight hours) because in these plants one can exert a very fine control over the flowering process. The effects of photoperiod on flowering were discussed in detail in Chapter 17. Here it is sufficient to say that some plants such as chrysanthemums bloom in autumn, when the daylight hours are short; they are called short-day plants. Long-day plants bloom in summer. What part of the plant detects the variation in day length: the stem, the leaves, or the shoot apex? Many years ago the Russian plant physiologist Mikhail K. Chailakhyan explored this question in detail. To his surprise he found that it was the young but fully expanded leaves, rather than the shoot apex, which measured the length of the light period. Since the developmental changes producing the flower occur in the shoot apex, Chailakhyan proposed the existence of a flowering hormone. Such a hormone would presumably be

---

Although the evidence for the existence of a flowering hormone is detailed and conclusive, florigen has successfully resisted all attempts to isolate and characterize it.

produced in the young leaves only when conditions were favorable for inducing flowering. Once produced, it would then be transported to the shoot apex where it would cause changes in messenger RNA synthesis, and hence in protein synthesis, that lead to the production of the highly modified leaves that comprise the petals and other parts of the flower.

Probably at least three different kinds of substance are involved in the induction of flowering. One of these is almost certainly a gibberellin, discussed earlier in this chapter. Gibberellins are very much involved in stem elongation. Certain long day plants which elongate dramatically from a tight rosette of leaves, in response to a daylength favorable for flowering, will also elongate at unfavorable daylengths if microgram amounts of gibberellin are applied in solution to the apices of their shoots. A second compound is an as yet unidentified inhibitor, and a third is a specific flowering substance called florigen (Chailakhyan's original flowering hormone). High levels of gibberellin and florigen and low levels of inhibitor are prerequisites for flowering. Any one of these three factors may be the one that limits flowering in different species, and the concentration of any one might be determined by photoperiod. Although the evidence for the existence of florigen is detailed and conclusive, florigen itself has resisted all attempts to isolate and characterize it.

One might expect different species to have different florigens, compounding the difficulty of isolation and characterization; or at least that the florigen from long-day plants might differ from that of short-day plants. The evidence, however, is overwhelming in the other direction. Regardless of species or photoperiodic class, a single florigen is involved. The test is relatively simple. One takes a photoperiodically sensitive plant and induces it to flower by exposing it to the appropriate photoperiod. One then returns it to unfavorable photoperiod, removes a portion, and grafts it onto another uninduced plant. The second plant will normally flower if the graft takes *(Figure 10)*. Apparently the only limiting factor is whether or not successful grafts can be made—and they have been made between some quite distantly related species. The donor may be either a long-day or a short-day plant, so long as it has been exposed to the proper photoperiod to induce flowering. Likewise, the recipient may be of either photoperiodic class. A flowering day-neutral plant (one which flowers at a certain size regardless of photoperiod) can act as a donor either to long-day or short-day recipients. Clearly the characterization of florigen remains one of the more important problems facing biologists today.

GRAFTS INDUCE FLOWERING in plants belonging to the same species. Leaf from plant that normally blooms under short-day conditions (left) is grafted onto plant under long-day conditions (center), inducing it to bloom. Grafted leaf from that plant in turn induces flowering in another (right). **10**

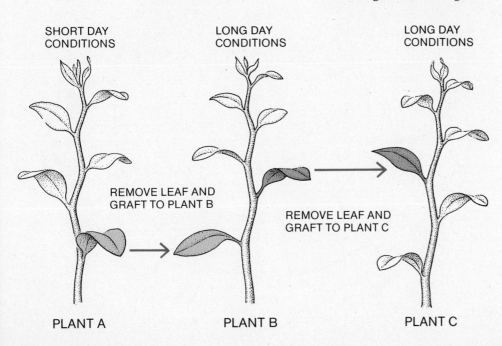

SHORT DAY CONDITIONS

LONG DAY CONDITIONS

LONG DAY CONDITIONS

REMOVE LEAF AND GRAFT TO PLANT B

REMOVE LEAF AND GRAFT TO PLANT C

PLANT A

PLANT B

PLANT C

Traditionally the endocrine and nervous systems of animals have been studied separately, as if they were independent. Actually they comprise a single integrated control system.

ANIMAL HORMONES

In animals as in plants, hormones powerfully influence development. In animals alone, however, hormones also regulate physiology, a function they share with the nervous system. Traditionally the endocrine and nervous systems of animals have been studied separately, as if they were independent. Actually they comprise a single integrated control system. As a general rule, nerves mediate adjustments that must be made swiftly and from moment to moment, while hormones take care of the more gradual, longer-lasting changes in physiology. Of course there are exceptions to this rule, but it remains useful as a general approach to the control strategy of animals.

DEVELOPMENTAL HORMONES

The hormonal control of growth and development in animals involves extensive changes, precisely coordinated. For example, the transformation of a tadpole into a frog includes destructive changes such as the elimination of the tadpole tail, gills and teeth, as well as constructive changes including the full development of limbs, tongue and middle ear. The pituitary gland, the master endocrine organ located next to the brain, begins this process by secreting a hormone that is carried by the bloodstream to the thyroid gland, which responds by secreting into the bloodstream the thyroid hormone that directly triggers metamorphosis. Experimenters have been able to induce complete metamorphosis with thyroxin, one of the thyroid hormones.

11 INTERACTION OF HORMONES during development of cecropia moth. Molting of larva is induced by brain hormone that stimulates prothoracic gland to secrete ecdysone, a hormone that stimulates development. Ecdysone is balanced by juvenile hormone secreted by corpora allata. In pupal state the level of juvenile hormone drops. Shift in balance between two hormones allows pupa to develop into adult.

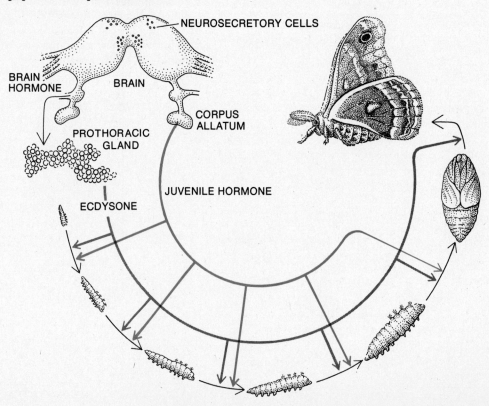

The metamorphosis of the frog also involves extensive changes at the biochemical level. A whole new set of enzymes appears in the liver, and a new kind of hemoglobin in the blood. The areas that resorb the tadpole organs contain high levels of proteases and other degradative enzymes, and a new visual pigment is formed in the retina. These changes are all inducible by thyroxin, which indicates that the specificity of the response depends not on the hormone but on the target cell. A given cell is already programmed to respond in a particular way by its physiological state and its previous development.

Metamorphosis also occurs in many invertebrate animals including sponges, starfish and insects. Carroll M. Williams and his team at Harvard University have studied metamorphosis in the silkworm moth *cecropia*. The early embryo develops into a larva, the familiar caterpillar, which feeds and increases in size. Its growth is physically limited by an exoskeleton of chitin secreted by the epidermal cells. Eventually the entire exoskeleton is shed, a process called molting. The larva grows for a time, forms a new exoskeleton, and molts again. After the fifth molt the new exoskeleton is far more extensive; it encases the entire animal, including its appendages, forming a PUPA. Within the pupa an extraordinary reorganization begins.

Small groups of larval cells, called IMAGINAL DISCS, begin extensive development, reshaping the entire animal, while the larval structures deteriorate. Finally the pupal exoskeleton is molted and the adult moth emerges.

Molting and metamorphosis involve three hormones. Certain brain cells secrete a hormone that causes the PROTHORACIC GLANDS to produce a mixture of two steroids called ECDYSONE. Meanwhile other glands just behind the brain, the CORPORA ALLATA (singular, corpus allatum) are producing a third substance, JUVENILE HORMONE. Throughout development ecdysone is the key substance in balance with juvenile hormone (*Figure 11*). High levels of juvenile hormone favor those synthetic activities associated with a larval molt. As the larva grows, the level of juvenile hormone drops, and ecdysone induces the transformation from larva to pupa. Finally juvenile hormone disppears, and one sees the expression of the full developmental potential of ecdysone: complete metamorphosis and the final pupal molt. But juvenile hormone is not simply an ecdysone antagonist. The British biologist Sir Vincent Wigglesworth has clearly shown that it promotes larval growth. The whole system illustrates the principle of hormonal interaction: two or more hormones acting together may bring about developmental changes quite un-

The paper factor

In the 1950's Karel Slàma, a young Czech entomologist, brought a culture of European insects to the U.S. to study them in the laboratory of Carroll M. Williams at Harvard. The investigators were puzzled because instead of maturing into normal adults, the young nymphs continued to molt and to grow into still larger immature forms. In searching for the reason, Slàma and Williams discovered that the paper towels on the floor of the insects' cages contained a chemical substance that acted very much like insect juvenile hormone. Other investigators showed that this "paper factor" had a molecular structure very similar but not identical to the natural juvenile hormone of the bugs. It turned out to be present in certain kinds of paper (Scottowels, The New York Times and Scientific American) but not in others (the British scientific journal Nature). The source of the substance proved to be hemlock, balsam fir and a few other species of trees that are used to make some kinds of American paper, but not many foreign papers.

Biologists have speculated that a few plants have evolved the ability to manufacture the paper factor or similar substances as a kind of defense against the insects that attempt to feed on them. If young insects consume enough of the chemical, they either die or at least are unable to mature into reproductive adults. This discovery has led entomologists to study the possibility of controlling insects by using preparations containing specific insect hormones (or similar compounds) instead of conventional broad-spectrum insecticides, which are usually toxic to pests and useful organisms.

GLAND	HORMONE	MOST IMPORTANT EFFECTS
HYPOTHALAMUS	RELEASING AND INHIBITING FACTORS (P)	GOVERN SECRETION OF HORMONES BY THE ANTERIOR PITUITARY
	OXYTOCIN (P) & VASOPRESSIN (P)	STORED AND RELEASED BY POSTERIOR PITUITARY (SEE BELOW)
ANTERIOR PITUITARY	GROWTH HORMONE (PP)	STIMULATES GROWTH
	ACTH (PP)	STIMULATES ADRENAL CORTEX
	THYROTROPHIC HORMONE (PP)	STIMULATES THYROID
	FSH (PP)	STIMULATES GROWTH OF OVARIAN FOLLICLES (IN FEMALES) AND SPERMATOGENESIS (IN MALES)
	LH (PP)	STIMULATES CONVERSION OF OVARIAN FOLLICLE INTO CORPUS LUTEUM. STIMULATES SECRETION OF SEX HORMONES BY OVARIES AND TESTES
POSTERIOR PITUITARY	OXYTOCIN (P)	STIMULATES CONTRACTION OF UTERINE MUSCLES AND THE RELEASE OF MILK BY MAMMARY GLANDS
	VASOPRESSIN (P)	CONTROLS EXCRETION OF WATER. STIMULATES CONTRACTION OF SMOOTH MUSCLE IN WALLS OF SMALL ARTERIES
THYROID	THYROXIN (A)	STIMULATES AND MAINTAINS OXIDATIVE METABOLISM
PARATHYROIDS	PARATHORMONE (PP)	MAINTAINS NORMAL CALCIUM METABOLISM AND BONE GROWTH
THYMUS	THYMOSIN (P)	STIMULATES IMMUNE RESPONSE OF LYMPHATIC SYSTEM
PANCREAS	INSULIN (PP)	INCREASES STORAGE OF GLYCOGEN, STIMULATES OXIDATION OF CARBOHYDRATES, LOWERS BLOOD SUGAR
	GLUCAGON (PP)	STIMULATES CONVERSION OF GLYCOGEN TO GLUCOSE IN LIVER
ADRENAL MEDULLA	ADRENALIN (A)	STIMULATES FIGHT-OR-FLIGHT REACTIONS. STIMULATES CONVERSION OF GLYCOGEN TO GLUCOSE
	NORADRENALIN (A)	SUSTAINS BLOOD PRESSURE AND REGULATES REACTIONS TO STRESS
ADRENAL CORTEX	GLUCOCORTICOIDS (S) (CORTISONE, CORTICOSTERONE, HYDROCORTISONE)	CONTROL METABOLISM OF CARBOHYDRATES, PROTEINS AND LIPIDS. REDUCE INFLAMMATION OF TISSUES
	MINERALCORTICOIDS (S) (ALDOSTERONE, DEOXYCORTICOSTERONE)	CONTROL WATER BALANCE AND REGULATE METABOLISM OF SALTS (SODIUM, POTASSIUM AND OTHER MINERALS)
	CORTICAL SEX HORMONES (S)	MAINTAIN SECONDARY MALE SEXUAL CHARACTERISTICS
OVARIES	ESTROGEN (S)	STIMULATES DEVELOPMENT AND MAINTENANCE OF SECONDARY FEMALE SEXUAL CHARACTERISTICS AND BEHAVIOR
	PROGESTERONE (S)	SUSTAINS PREGNANCY AND HELPS TO MAINTAIN SECONDARY FEMALE SEXUAL CHARACTERISTICS
TESTES	TESTOSTERONE (S)	STIMULATES DEVELOPMENT AND MAINTENANCE OF SECONDARY MALE SEXUAL CHARACTERISTICS AND BEHAVIOR

Mammals probably have the most complex endocrine systems of all living organisms. During more than 100 million years of evolution, the trend in mammals has been toward bringing more physiological processes under hormonal control.

like those caused by any single hormone acting alone. Compounds that inhibit these hormones, and hence prevent insects from maturing and reproducing, apparently have evolved as defense mechanisms in certain plants (*Box A*).

MAMMALIAN DEVELOPMENT

Developing mammals do not undergo the spectacular changes in form that are found among amphibians and insects, but mammals probably have the most complex endocrine systems of all living organisms. During more than 100 million years of evolution the trend in mammals has been toward bringing more and more physiological processes under hormonal control. Natural selection has accomplished this in a rather haphazard way, creating an endocrine organ in one convenient position or another, or modifying a steroid or some other available molecule to serve as a hormone. In terms of design the result is a bit of a hodgepodge, but the system as a whole works with exquisite precision.

The position of the endocrine glands of humans is shown in Figure 12, the structure of a few important hormones in Figure 13, and the sources and functions of the major hormones in Table I. In a book of this length it is not possible to consider all of the endocrine organs and hormones. The best approach is to examine a few hormones of particular importance: the growth hormone, the pituitary-hypothalamus system, reproductive hormones, and the hormones that regulate stress reactions.

I ROLES OF PRINCIPAL HORMONES in vertebrates. Abbreviations indicate the chemical nature of each hormone; peptides (P), proteins (PP), amines or amino acids (A), and steroids (S). Table lists only the most important effects of the major hormones. Humans have many more hormones than those listed here.

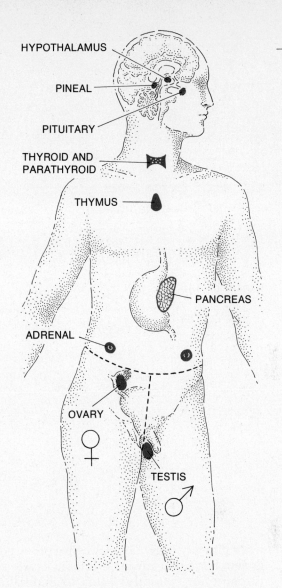

HUMAN ENDOCRINE ORGANS are indicated in color. Such an endocrine system is typical of other mammals as well. **12**

GROWTH HORMONE

The program of mammalian development is under close hormonal control, and errors in this control caused by defects in the endocrine system can lead to tragedy. Such defects include gross errors in the final size to which the organism grows, and disproportionate growth of certain parts of the body.

Growth hormone is a product of the pituitary gland (also called the hypophysis), a round pea-sized organ enclosed in a bony cavity beneath the brain case. It consists of two lobes, which are actually two distinct glands: the anterior and the pos-

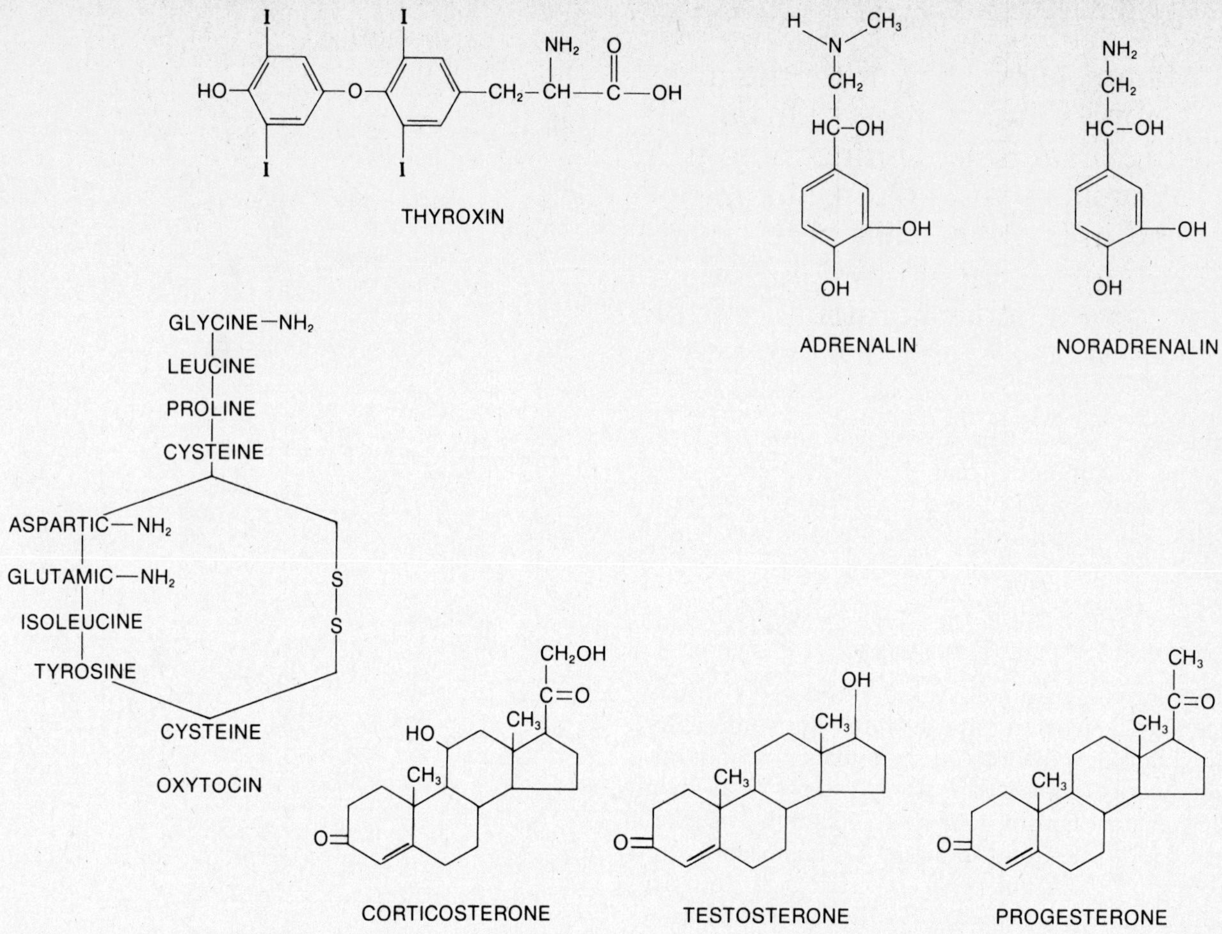

THYROXIN

ADRENALIN

NORADRENALIN

GLYCINE—NH₂
LEUCINE
PROLINE
CYSTEINE
ASPARTIC—NH₂
GLUTAMIC—NH₂
ISOLEUCINE
TYROSINE
CYSTEINE

OXYTOCIN

CORTICOSTERONE

TESTOSTERONE

PROGESTERONE

SOME MAMMALIAN HORMONES. Thyroxin is an amino acid with four atoms of iodine. Adrenalin and noradrenalin are based on the amino acid tyrosine. Oxytocin is a simple polypeptide. Corticosterone, testosterone and progesterone are steroids, all of which have the basic ring structure shown here; they differ in the groups attached to the rings.

terior pituitary (*Figure 14*). Excessive production of growth hormone early in life (caused by a tumor of the anterior pituitary) leads to giantism; inadequate production (caused by atrophy of the gland) leads to dwarfism. The famous American dwarf, General Tom Thumb, was a victim of his pituitary. Could physicians today prevent Tom Thumb's unfortunate growth pattern? No, because growth hormones from other mammals are ineffective in humans; consequently there are several thousand dwarfs living in the U.S. today. When biochemists learn how to synthesize the human hormone, dwarfs may disappear from the population. The reason that many of these dwarfs now go untreated is that treatment of a single individual would require an extract from many human pituitaries per day. If pituitary abnormalities occur after adolescence, the error in growth is one of proportion rather than of overall size. Excessive production of the hormone leads to acromegaly, a fairly rare disease in which bones of the face, hands and feet grow too large.

PITUITARY AND HYPOTHALAMUS

The anterior pituitary is often called the master gland of the endocrine system because it secretes TROPHIC HORMONES: those that control the development and activity of other endocrine organs, including the adrenal cortex, the thyroid and the gonads (internal sex organs). The control involves four different polypeptides: ACTH (adrenocorticotrophic hormone), thyrotrophic hormone, and at least two gonadotrophic hormones. The interaction of the pituitary and its target glands is controlled by both negative and positive feedback loops. The regulation of the thyroid gland is a good example of negative feedback. Thyroxin, mentioned earlier, speeds up oxidative metabolism. (Recall that in frogs it induces metamorphosis, an entirely different function.) When thyrotrophic hormone from the anterior pituitary reaches the thyroid, the production of thyroxin increases. The rising concentration of thyroxin in the blood in turn causes the pituitary to diminish the release of the thyrotrophic hormone. The pituitary hormone speeds up this tightly controlled system, thyroxin slows it down, and together they maintain a finely balanced concentration of thyroxin in the bloodstream. The overall result is a steady control of oxidative metabolism within the body.

The posterior pituitary, or neurohypophysis, is in fact an extension of the hypothalamic region of the brain *(Chapter 19)*. Nerve cells in the hypothalamus produce two hormones, vasopressin and oxytocin, that flow down the stalk that connects the hypothalamus to the posterior pituitary, where they are stored. (This direct secretion of hormones by nerves, called neurosecretion, is common in both vertebrates and invertebrates.) Oxytocin induces contraction of the muscles of the uterus during childbirth. Vasopression acts as the antidiuretic hormone, increasing the retention of water by the kidney tubules and thus decreasing the volume of urine. Vasopressin also causes blood pressure to rise by inducing constriction of the small arteries.

The intimate association between the pituitary gland and the brain raises the question: Is the pituitary truly the master gland? The feedback between the trophic hormones of the anterior pituitary and their target glands is not a closed loop, but is subject to outside influences. The activities of at least some of the target glands fluctuate according to changes in the outside environment. In many mammals the concentration of sex hormones, for example, changes from season to season; there is, in other words, a breeding season and an off season. Similarly the levels of some of the hormones of the adrenal cortex, including cortisone, which assist the body in adjusting to stress, are powerfully influenced by outside events.

The explanation of the variable nature of these interactions is that the brain monitors changes in the environment and, guided by this information, directly controls the anterior pituitary. Information about changes in photoperiod, the warming and cooling of the air, the appearance of an enemy or a mate, is received from the sense organs. After being processed in the forebrain the information is transmitted to the hypothalamus, which responds by secreting special hormones of its own into a set of blood vessels connected directly to the anterior pituitary. These hormones in turn may either induce or inhibit the secretion of the anterior pituitary hormones. Hypothalamic hormones that stimulate the secretion of anterior pituitary hormones are called RELEASING FACTORS; those that act in the opposite way are INHIBITORY FACTORS. Several such hypothalamic factors have now been isolated, identified, and synthesized. All are small peptides. Eventually separate inhibitory and releasing factors may be identified for all anterior pituitary hormones.

In short, the anterior pituitary is not really a master gland in the true sense of the word, but acts under the direction of the hypothalamus. This relationship too is regulated by negative and positive

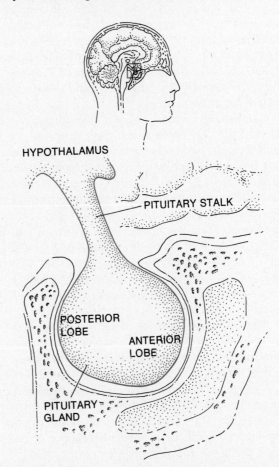

PITUITARY GLAND lies nestled in a bony cavity beneath the brain (top). It is intimately associated with the hypothalamus region of the brain. The two lobes of the pituitary (bottom) secrete hormones that affect many different organs of the body.

14

feedback loops. If small quantities of the female hormone estrogen are implanted directly into certain areas of the hypothalamus, the secretion of female gonadotrophins falls sharply, limiting the further production of estrogen. The same sort of negative feedback probably prevails under natural conditions when estrogen reaches the hypothalamus through the bloodstream. The hypothalamus is probably subject to feedback effects from other hormones as well, with each hormone affecting a different group of target cells.

HUMAN REPRODUCTIVE HORMONES

The hypothalamus and pituitary play a central role in the regulation of human reproduction. At puberty the young male undergoes a series of profound physical and psychological changes. When triggered by releasing factors from the hypothalamus, the anterior pituitary secretes two gonadotrophic hormones: FSH (follicle-stimulating hormone) and LH (luteinizing hormone). LH stimulates the secretion of androgens (male hormones), including testosterone, by special endocrine cells in the testes. FSH, together with LH and testosterone, induces maturation of the testicular germ cells, and the production of spermatozoa *(Chapter 13)*. The androgens are responsible for the characteristic changes of male adolescence, including the growth of hair on the face, the deepening of the voice, and development of a more muscular body.

Parallel physiological events transform the adolescent girl. The same two trophic hormones, FSH and LH, are secreted by the anterior pituitary of the female in response to releasing factors from the hypothalamus. These hormones induce the maturation of the ovaries, which then begin to secrete the two female sex hormones, estrogen and progesterone. The sex hormones in turn initiate the development of the traits characteristic of a sexually mature woman, including enlarged breasts, vagina and uterus, a broad pelvis, pubic hair, and—very suddenly—the onset of the menstrual cycle.

The endocrine control of reproductive physiology seems more complex in women than in men. Mature female mammals, including humans, are subject to an ESTROUS CYCLE. The culmination of the cycle is ovulation, the release of an unfertilized egg from the ovary. The wall of the uterus, properly primed by the effects of estrogen and progesterone, is ready to receive an embryo. In most mammals the female becomes sexually receptive only around the time of ovulation; she is then said to be "in heat" and may even become the aggressive partner. During the rest of the cycle she is unreceptive and seldom if ever attempts to mate. Some species such as sheep and deer ovulate only during a relatively short annual breeding season. In other species the cycle runs continuously; rats and mice ovulate every four or five days.

The human estrous cycle is unusual in two respects. First, as noted in Chapter 13, a woman is never in heat, but remains receptive to some degree throughout most of the cycle. Foreplay and social circumstances have a strong effect on her receptivity. Some students of human evolution ascribe this trait to the development of an unusually strong pair bond in the human species. Instead of approaching females only at the time of heat, human males remain closely affiliated at all times, perhaps because of the long and difficult period of child rearing in the human species. (The origins of human social behavior are examined more closely in Chapter 34.)

The second unusual aspect of the human estrous cycle is menstruation, the periodic flow of blood. In other mammals, as noted in Chapter 13, the cycle is seldom marked by noticeable bleeding. The human cycle lasts about 28 days, hence the term *menses*, which means a month. The hormones that control the menstrual cycle and pregnancy are secreted by the ovary, placenta and the wall of the uterus. Following menstruation, which begins the cycle, the hypothalamus induces the anterior pituitary to secrete FSH and LH, the same hormones that initiate adolescence. These gonadotrophins stimulate the growth of follicles, envelopes of cells that surround a potential egg. The growing follicle cells secrete estrogen, one of the two female sex hormones *(Figure 15)*. The estrogen in turn induces the growth and thickening of the uterine wall. In a positive feedback loop, the rising estrogen concentration activates the LH-stimulating center of the hypothalamus, leading to a sharp rise in the concentration of LH. As the LH level peaks, the best-developed follicle in the ovary bursts and releases the egg, which enters the Fallopian tubes, where it awaits possible fertilization. At ovulation the follicular phase of the menstrual cycle ends and the luteal phase begins.

The luteal phase is named for the corpus luteum (Latin, yellow body), a mass of old follicle cells which have been transformed under the influence of LH into a cluster of yellowish cells richly supplied with blood vessels. The corpus luteum produces the second female sex hormone, proges-

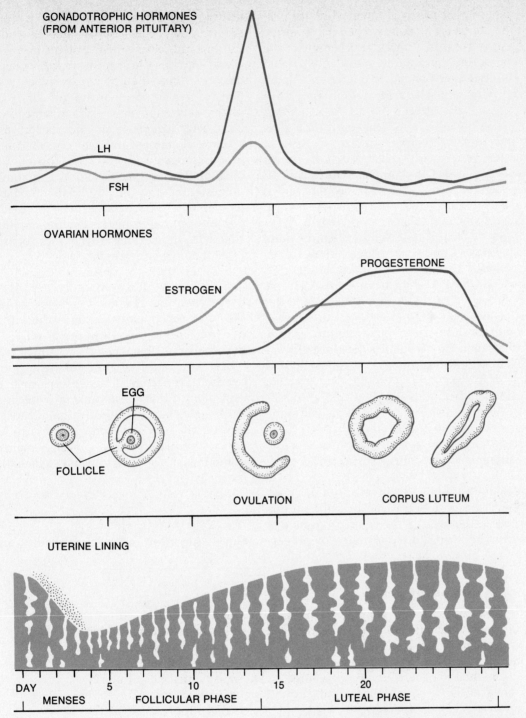

GONADOTROPHIC HORMONES
(FROM ANTERIOR PITUITARY)

LH

FSH

OVARIAN HORMONES

PROGESTERONE

ESTROGEN

EGG

FOLLICLE

OVULATION

CORPUS LUTEUM

UTERINE LINING

DAY   5   10   15   20

MENSES   FOLLICULAR PHASE   LUTEAL PHASE

MENSTRUAL CYCLE of the human female lasts an **15** average of 28 days, and is divided into three phases: the menses (the period of bleeding), the follicular phase and the luteal phase. The two graphs at top depict the varying levels of pituitary and ovarian hormones during the cycle. The diagram (third from top) shows the developing egg in the follicle, its release at ovulation, and the corpus luteum that is left behind. The drawing at bottom indicates the state of the lining of the uterus.

terone, whose principal role is the further preparation of the uterus for nurturing a possible embryo. Progesterone induces a further thickening of the uterine lining and the maturation of glands located in it. Without these changes a fertilized ovum could not implant itself. Progesterone also has a negative feedback effect on the LH-activating center of the hypothalamus, causing the LH output of the anterior pituitary to drop.

Progesterone is rightly called the hormone of pregnancy. So long as the corpus luteum survives, its progesterone prevents the ovary from releasing another egg; in other words, the menstrual cycle is suspended. This inhibition is exploited in birth-control pills, which contain synthetic compounds similar to estrogen and progesterone. They "trick" the hypothalamus into turning off the secretion of gonadotrophins by the anterior pituitary. A woman taking "the pill" does not make fully developed follicles and eggs, nor does she have a corpus luteum and a well-primed uterine wall.

What sustains the corpus luteum during pregnancy? How does it "know" that an embryo is growing in the wall of the uterus? The placenta connecting mother and embryo apparently secretes a hormone similar to LH. This hormone maintains the corpus luteum until birth, then degenerates. In other words, pregnancy sustains the corpus luteum which in turn sustains the pregnancy.

If the egg is not fertilized, no placenta develops and its LH-like hormone is lacking; the corpus luteum regresses and the level of progesterone falls. With the virtual disappearance of proges-

terone, the uterine wall breaks down and menstruation begins as the engorged blood vessels rupture and release their contents into the uterine cavity *(Box B)*. During the first few days of the menstrual flow the body has very low concentrations of female hormones of any kind. This temporary deficiency may be accompanied by feelings of anxiety, irritability and depression, as well as abdominal cramps originating in the uterus. The drop in progesterone level also triggers the beginning of the next cycle. No longer inhibited by progesterone, the gonadotrophin-stimulating centers of the hypothalamus elicit the release of gonadotrophins from the anterior pituitary. This stimulates the development of follicles in the ovary, and a new cycle begins.

Between the ages of 40 and 50 women enter menopause, the post-reproductive period of their lives, when the menstrual cycle comes to a stop. Menopause is apparently caused by a permanent decline in the LH-stimulating activity of the hypothalamus, breaking the chain of events required to complete the menstrual cycle. The ovaries also become less sensitive to the gonad-controlling hormones, causing a decline in the output of estrogen. Although menopause is often accompanied by emotional distress, a woman's basic physical and mental powers remain virtually undiminished, and she can still enjoy a fully satisfying sexual life.

THE ADRENAL SYSTEM

One of the most important regulatory systems of

B A strange window on the uterus

How did biologists study the complex changes in the wall of the uterus during the menstrual cycle? Much of the story was worked out by surgically removing tiny bits of tissue from the uterus on different days of the cycle, then examining stained sections under the microscope. One physiologist, J. E. Markee of Duke University, used an ingenious method to observe the process directly. He transplanted bits of uterine tissue from a female rhesus monkey into the front chamber of the monkey's eye. The tissue survived and was nourished by the blood vessels that supply the eye. The hormones that govern the cycle, broadcast through the bloodstream, reached the eye as well as the uterus, and the transplant responded appropriately. By peering into the eye with a microscope, Markee was able to study the tissue transplant day by day throughout the menstrual cycle. Because the rhesus menstrual cycle is very similar to that of humans, Markee obtained valuable new insights about uterine changes in women. In particular he was able to observe what happens during the luteal phase of the cycle: the events that lead to the breakdown of the uterine lining, the rupture of blood vessels and the onset of menstrual bleeding.

the mammalian body is controlled by the adrenals, pea-sized glands perched on top of each kidney (*Figure 16*). Like the pituitary, the adrenals each consist of two glands, with a distinct set of hormones. The central part, the MEDULLA, secretes adrenalin (sometimes called epinephrine) and noradrenalin. The outer part of the adrenal, the CORTEX, produces several steroid hormones with a different set of functions.

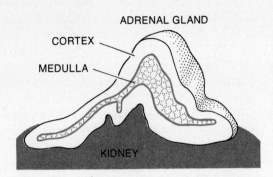

ADRENAL GLAND

6 **ADRENAL GLAND consists of an inner medulla and an outer cortex. Drawing depicts a cross-section through the gland. One adrenal gland is perched on top of each kidney.**

Unlike most other endocrine glands the adrenal medulla is under direct nervous control. It is stimulated by nerves of the autonomic nervous system (*Chapter 19*) that originate in the hypothalamus. This hookup enables the adrenal medulla to respond quickly to outside stresses. When danger threatens, adrenalin and noradrenalin are promptly released into the bloodstream, affecting the body in different but complementary ways. Adrenalin acts quickly in conjunction with the autonomic nervous system to prepare the body for "fight or flight." The heart rate and blood pressure go up, veins dilate, the concentration of sugar in the bloodstream rises, and so does the number of certain white blood cells (eosinophils). The blood flow through skeletal muscle, brain and liver increases by as much as 100 percent; digestive and reproductive functions are inhibited; and one suffers a feeling of anxiety. Adrenalin is secreted whenever one is put in a stressful situation such as cold weather, hostility from others, or a narrow escape from some peril. Adrenalin also promotes the release of ACTH from the anterior pituitary, causing the release of hormones from the adrenal cortex and a more prolonged adjustment of the body to stress. Noradrenalin is released independently in stressful situations and acts mainly to

sustain blood pressure. In other words, adrenalin attends to the emergency while noradrenalin plays a secondary, mainly regulatory role. A curious effect observed in humans is that participation in aggressive encounters induces the release of relatively large quantities of noradrenalin but only moderate amounts of adrenalin; conversely, the anticipation of such events, in the form of anger or fear, favors the release of adrenalin. Hockey players on the bench, for example, might secrete mostly adrenalin, while their teammates on the ice might secrete mostly noradrenalin.

The adrenal cortex has a wide variety of roles, but in contrast with the swift-acting adrenalin and noradrenalin, its hormones generally exert relatively slow sustained effects on overall metabolism, kidney function, electrolyte balance, blood pressure and tissue inflammation. The cortical hormones are known collectively as corticoids, and all are steroid compounds. One group, the glucocorticoids, includes the well-known cortisone. They are released continuously at rates that show a daily rhythm, and promote the conversion of proteins into glucose within the liver, and they also reduce the inflammation of tissues. Another group, the mineralcorticoids (including aldosterone and deoxycorticosterone), regulates the distribution of sodium and other minerals in the tissues.

Under prolonged stress, the appropriate center in the hypothalamus induces the release of ACTH from the anterior pituitary. This substance in turn evokes an outpouring of glucocorticoids from the adrenal cortex. In general these corticoids cause responses opposite to those induced by adrenalin and noradrenalin; they serve as a brake on the body's emergency mobilization. Some quantity of corticoids is required even when the body is not under stress. Animals whose adrenals have been removed experimentally show symptoms identical

In general one can say that hormones released directly by the nervous system are fast-acting and their effects are short-lived. Conversely the hormones of the anterior pituitary and the adrenal cortex mediate effects that are relatively slow and long-lasting.

to those of human patients with Addison's disease (adrenal insufficiency): lowered blood sugar, digestive disturbances, reduced blood pressure and body temperature, kidney failure, and an inability to stand stress of any kind. Without corticoids the body's condition deteriorates in the face of extreme temperatures, prolonged activity, infections, intoxication and the like.

In general one can say that hormones released directly by the nervous system are fast-acting and their effects are short-lived. The hypothalamus itself, for example, makes fast-acting oxytocin, and the adrenal medulla, a neural extension of the hypothalamus, produces fast-acting adrenalin. Conversely the hormones of the anterior pituitary and the adrenal cortex—both non-neural tissues—mediate effects that are relatively slow and long-lasting.

HORMONES AND CELLS

What happens at the molecular level when an animal hormone reaches a target cell? This is one of the key problems in modern biology. Two mechanisms have been proposed, one for protein-like hormones and the other for steroid hormones. Many of the protein, polypeptide and amine-type hormones act via a second messenger (the hormone itself being the first messenger): cyclic AMP (cAMP). Each of these hormones attaches itself to a particular receptor, probably a protein bearing a specific recognition site for the hormone. The receptor lies in the cell membrane and extends outward, making it unnecessary for the hormone to penetrate into the interior of the cell. When the hormone combines with this external receptor it activates an enzyme, adenyl cyclase, which converts ATP to cAMP within the cell. These events were first discovered by the late E. W. Sutherland of Washington University, who was awarded the Nobel Prize in 1971. Working with adrenalin, he found that the cAMP in turn activates a series of reactions that convert the enzyme phosphorylase from an inactive to an active form (*Figure 17*). The active form splits glycogen to form glucose-1-phosphate, which becomes available to the target cell. This glucose derivative can be converted to glucose, which is transported by the bloodstream to other cells.

In the years after Sutherland's early work the list of other systems activated by cAMP seemed to multiply. It is now clear that many hormone effects in vertebrate tissues act via this second messenger (*Box C*). These effects include the action of adrenalin in stimulating muscle and liver cells to convert glycogen to glucose-1-phosphate, and in stimulating the breakdown of lipids in fat cells; the action of ACTH in stimulating the production of glucocorticoids in the adrenals; the action of LH in stimulating the production of progesterone in the ovary; and many more. That such diverse effects can all be mediated by a single compound is explained not only by the specific nature of the receptor on the cell surface, but also by the fact that the target cells have secondary targets within them, targets that activate reaction systems such as those involved in the synthesis of glucocorticoids.

Steroid hormones such as estrogen, progesterone and the hormones of the adrenal cortex do not react with receptors on the cell surface. These hormones are all fat-soluble, and they readily pass

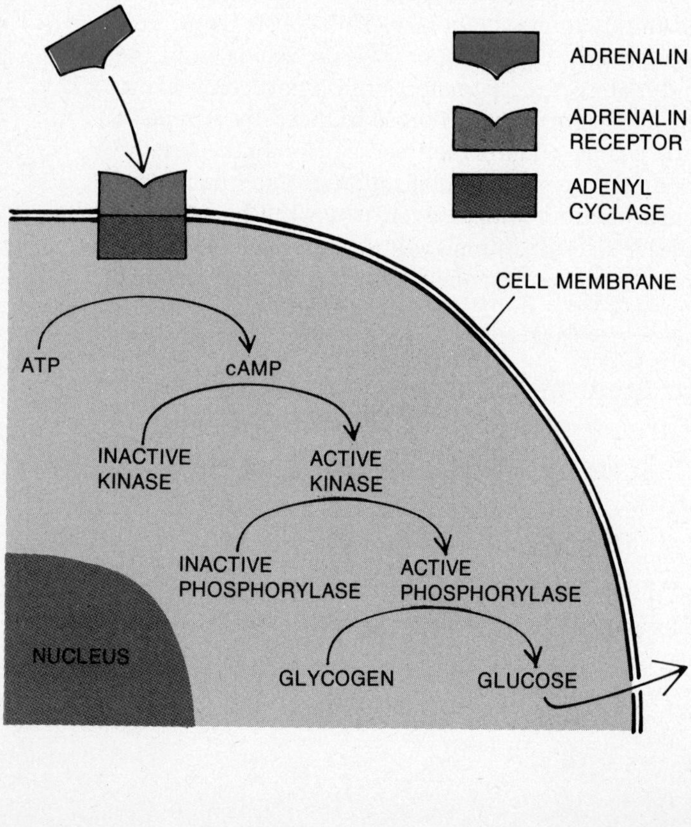

ADRENALIN

ADRENALIN RECEPTOR

ADENYL CYCLASE

CELL MEMBRANE

ATP   cAMP

INACTIVE KINASE   ACTIVE KINASE

INACTIVE PHOSPHORYLASE   ACTIVE PHOSPHORYLASE

NUCLEUS

GLYCOGEN   GLUCOSE

ADRENALIN MOBILIZES GLUCOSE in a series of steps, beginning with the binding of the hormone to a receptor in the cell membrane. Arrows show a series of reactions, beginning with the conversion of ATP to cAMP in the cytoplasm, that result in the conversion of glycogen to glucose. The glucose can be exported from the cell and carried by the bloodstream to other cells. Similar mechanisms are involved in the action of many other hormones.

Maintaining blood sugar between meals

A well-understood case is the action of the hormone glucagon on one of its target organs, the liver. Glucagon, a polypeptide chain of 29 amino acids, is bound to a specific receptor on the outside of liver cells. There it causes the rapid production of cAMP inside the cells which in turn activates the protein kinase of liver. This phosphorylates the enzyme that synthesizes glycogen, which makes it less active. Glucose is then no longer stored, but accumulates in the liver cells and is discharged into the bloodstream. Simultaneously, protein kinase phosphorylates an enzyme that is a mere appendage to another enzyme, called phosphorylase kinase. This enzyme, in contrast to the glycogen-synthesizing one, becomes much more active when it is phosphorylated. The phosphorylase kinase then causes ATP to phosphorylate serines on the enzyme that ultimately attacks glycogen: an enzyme called glycogen phosphorylase, which becomes much more active, causing the production of glucose and its release into the bloodstream. In other words, glucagon touches off a cascade of events that starts with the synthesis of cAMP and ends with a rise in glucose. Both bacteria and higher organisms use cAMP as a sort of distress signal to indicate that glucose is running low and that more should be made from non-glucose sources; but it is odd that this signal is put into action by such different means in the two types of organism.

If cAMP could be produced in a cell but never disappeared, it would be rather like shouting "wolf" continuously; that is, cAMP would not be very useful as a chemical signal. It turns out that phosphodiesterase, an enzyme present in the cytoplasm, hydrolyzes cAMP to an inactive substance. The synthesis of cAMP by adenyl cyclase and the destruction of cAMP by phosphodiesterase go on continuously in cells; only when the synthesis is stimulated by a polypeptide hormone, or when the degradation is inhibited by a drug will the intracellular concentration of cAMP rise. Two drugs that inhibit phosphodiesterase are caffeine and theobromine, which are found in coffee and tea. The "lift" one gets from a cup of coffee is due to the rise in blood sugar caused by the inhibition of cAMP degradation in the liver.

through the lipid-rich cell membranes. Once inside, a steroid hormone must be bound to a RECEPTOR PROTEIN in the cytoplasm if it is to have any effect. This receptor protein is specific for a particular hormone. A receptor for the female sex hormone, estradiol, for example, will not bind male sex hormones such as testosterone and dihydrotestosterone, nor adrenal hormones such as cortisol and aldosterone. The presence of a receptor protein distinguishes a responsive cell from a nonresponsive one. Once the receptor protein has bound estradiol, the complex of the two apparently migrates into the nucleus, though there is some evidence that there is another special receptor protein in the nucleus. In any event, the hormone, bound to a receptor, quickly becomes associated with acidic chromosomal proteins and thus with the DNA of the chromosome (*Figure 18*). There it stimulates the synthesis of certain kinds of messenger RNA, which in turn are exported to the cytoplasm and translated to give specific proteins.

Steroid hormones all act by stimulating the synthesis of new proteins, rather than by altering the activity of protein molecules that were already in the target cells.

Sexual differentiation in humans dramatically demonstrates the roles of hormones and their receptors. For example, a human with an X and a Y chromosome will normally develop into a male. During development an XY embryo produces testosterone, which the target organs convert to dihydrotestosterone, the active form of the hormone. Dihydrotestosterone binds to a cytoplasmic receptor protein and causes characteristic changes in cell metabolism. Some XY males produce very low levels of testosterone and dihydrotestosterone; these males look much more like women, although they have internal testes, not ovaries. Other XY individuals are afflicted by a condition called testicular feminization. They too have internal testes, but consider themselves women, and have the external anatomy of women (*Figure 19*). In such

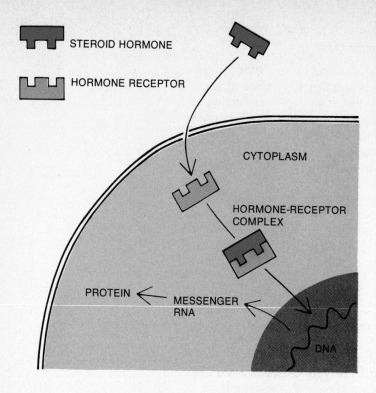

STEROID HORMONE

HORMONE RECEPTOR

CYTOPLASM

HORMONE-RECEPTOR
COMPLEX

PROTEIN ← MESSENGER
RNA

DNA

18 ACTION OF STEROID HORMONES involves a receptor in the interior of the cell. Hormone-receptor complex enters the nucleus and switches on the appropriate genes. Messenger RNA coded by the genes initiates protein synthesis in the cytoplasm.

"females" the amount of circulating testosterone is fully as high as in a normal male, and it is converted to dihydrotestosterone in the target organs. But because of a single gene mutation, they lack the cytoplasmic receptor for dihydrotestosterone, and their tissues are completely unresponsive to it. Normal females, in contrast, have the receptor but make only small amounts of testosterone.

The next chapter describes the nervous system which, together with the endocrine system, orchestrates the responses of animals to their internal and external environments.

READINGS

E. FRIEDEN AND H. LIPNER, *Biochemical Endocrinology of the Vertebrates*, Englewood Cliffs, NJ, Prentice-Hall, 1971. A readable paperback providing a nice overview of vertebrate hormones, their functions, interactions, and modes of action.
A.W. GALSTON AND P.J. DAVIES, *Control Mechanisms*

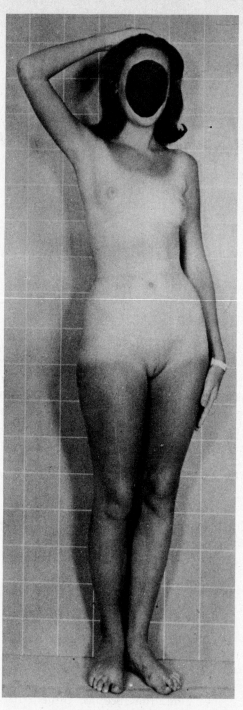

TESTICULAR FEMINIZATION. Despite her female **19** **appearance, young woman in photograph is genetically a male, having both X and Y chromosomes. Instead of ovaries, she has internal testes that produce testosterone. Her vagina ends blindly, with no uterus. The testes in such XY women often become cancerous, so they are usually removed surgically. This has no noticeable physiological effects, because cells of such individuals lack the receptor protein for the activated form of testosterone, and thus are unresponsive to it.**

in Plant Development, Englewood Cliffs, NJ, Prentice-Hall, 1970. A highly readable treatment of phytochrome, plant growth hormones, and hormone-mediated phenomena.

A. LABHART, *Clinical Endocrinology*, New York, Springer Verlag, 1976. An excellent advanced level text on the physiology and biochemistry of hormone action with special emphasis on the clinical aspects of human endocrinology.

J.S. PERRY, *The Ovarian Cycle of Mammals*, Edinburgh, Oliver & Boyd, 1971. Another affordable paperback. The hormonal regulation of female reproduction in mammals is treated concisely, readably and accurately.

H.V. RICKENBERG, Editor, *Biochemistry of Hormones* (MTP International Review of Science, Volume 8), Baltimore, University Park Press, 1974. Provides scholarly articles on the molecular basis of hormone action by some of the top research practitioners in the field.

F.B. SALISBURY AND C. ROSS, *Plant Physiology*, 2nd Edition, Belmont, CA, Wadsworth, 1978. A comprehensive textbook of plant physiology with a complete treatment of plant hormones.

C.D. TURNER AND J.T. BAGNARA, *General Endocrinology*, 6th Edition, Philadelphia, W.B. Saunders, 1976. A comprehensive treatment of animal endocrinology, both vertebrate and invertebrate.

The world is my idea . . . What man knows is not the sun and an earth, but only an eye that sees a sun, a hand that feels an earth. —Arthur Schopenhauer (1818)

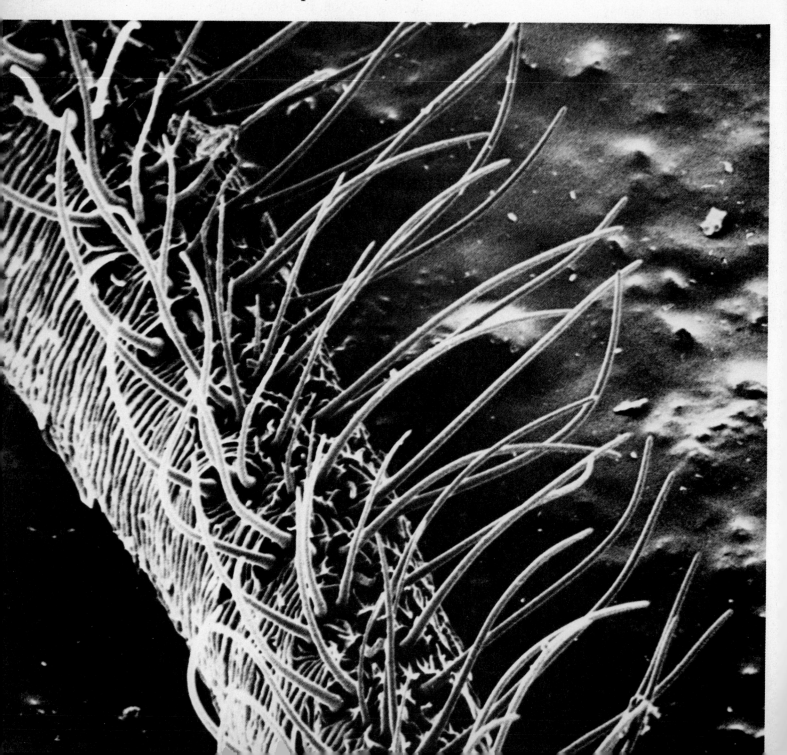

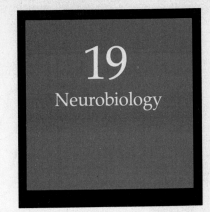

For human beings, the nervous system is what life is about. All emotional and intellectual pleasures are functions of the brain. A person's life, once the short-term needs for adequate food, shelter and sleep are fulfilled, is devoted to modifying the contents of his brain to provide more happiness, satisfaction or fulfillment. In addition, all one knows of the world around him is what his senses perceive. As Schopenhauer's quotation points out, one knows nothing directly, but only through the nervous system.

In all animals, the nervous system performs three functions. It continuously COLLECTS INFORMATION about the external and internal environments. It PROCESSES AND INTEGRATES this information, evaluating it with respect to past experience and future goals. And it ACTS upon this information by co-ordinating the activity of EFFECTORS —muscles and glands. In other words, the nervous system consists of three subsystems. The SENSORY SYSTEM gathers data, converts it into a form that the nervous system can handle, and feeds it to the second subsystem, the CENTRAL NERVOUS SYSTEM (CNS), which in vertebrates consists of the brain and spinal cord. It is essentially a computer containing inherited information (instincts and reflexes) and acquired information (memory and learning). The new input data may be stored or may trigger instructions to activate parts of the third system, the MOTOR SYSTEM. Motor nerves cause muscles or other organs to respond suitably to sensory data.

Almost as a byproduct of all this activity, there emerge for the aspects of life that seem so distinctly human: consciousness, conscience, a sense of personal identity, feelings about other people and things, knowledge and understanding. Perhaps the most intriguing task in modern biology is to try to find the relations between these immensely important subjective aspects and the silent grey-and-white jelly that comprises the brain.

NEURONS

The special cells of the nervous system are the neurons and glia; in addition one finds the cells and materials common to most tissues, such as blood vessels and connective tissue. The NEURON is the only cell involved in actual transmission of the signals, called nerve impulses, which flow through the system. The GLIAL CELLS (from the Greek word for glue) serve unknown functions. Some workers argue that they are nurse cells or transport systems for neurons, providing them with nutrients and perhaps even exchanging RNA with them, exchanges which in some cases might be related to the cellular basis of learning. In some cases they provide an electrically insulating sheath around neuronal processes.

Neurons come in different shapes, sizes and functions *(Figure 1)*. Motor neurons are connected to muscles or glands, and sensory neurons to some kind of sensing cell. Interneurons take impulses from one neuron and pass them to another. The feature that nearly all neurons share is the ability to transmit impulses actively. Under the microscope their cell bodies are distinguished from non-neuronal cells by the presence of so-called Nissl granules. These can be seen under the light microscope only after staining the sample with acidophilic stains (that is, basic stains which react with acidic material) such as thionine. Investigators used to believe that Nissl granules contained some material unique to neurons; actually

SENSORY HAIRS on the antenna of the male Bombyx moth are depicted in the scanning electron micrograph on opposite page. At core of each hair is the partially sheathed dendrite of a chemoreceptive neuron. Antenna is shown at lower magnification in Figure 18.

407

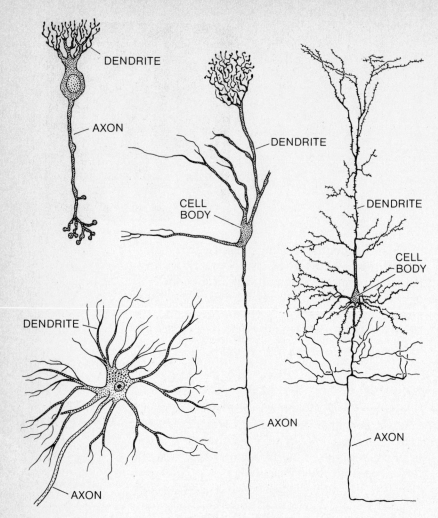

DENDRITE

AXON

DENDRITE

CELL BODY

DENDRITE

DENDRITE

CELL BODY

DENDRITE

AXON

AXON

AXON

SHAPE OF NEURON is related to its function. Bipolar sensory neuron appears at upper left. Two neurons are from the central nervous system: lower left a motor neuron from the spinal cord; right, pyramidal cell from the cerebral cortex. Sensory neuron at center is from olfactory bulb of cat. **1**

they contain merely the RNA of the endoplasmic reticulum. All cells contain such RNA, but in neurons it is present in an unusually condensed state, as electron micrographs clearly show (*Figure 2*).

Most neurons are very long in comparison with their width. An extreme example is the motor neuron that innervates the foot of a giraffe; it is several meters long, but its cell body (which lies in the spinal cord) is only 50 micrometers across. As Figure 1 shows, all neurons have branches or processes. Only one of the many processes (usually one that is much longer and less branched) has the ability to transmit impulses actively; this one, called the AXON, usually carries impulses away from the cell body. The other branched and knobbed processes, called DENDRITES, carry impulses (usually passively) to the cell body.

The junction between two or more neurons is called a SYNAPSE, a term that is also applied to the active junctions between neurons and other cells they may act upon, such as muscles or glands. At a

NISSL GRANULES are clusters of dark-staining material (color) in cytoplasm of neurons (left). Under the electron microscope, granules are revealed as clusters of rough endoplasmic reticulum (right). Neuron in drawing is magnified about 900X; the one in the photomicrograph at right, about 56,000X. **2**

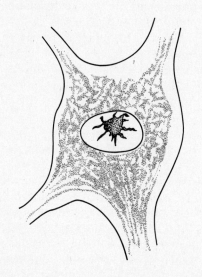

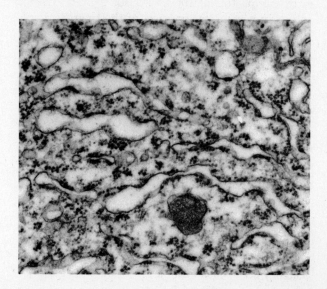

synapse there are actual gaps between the cells, typically 200 Å wide, and we shall see later that conduction across synapses involves processes quite unlike conduction along axons or dendrites. Synapses are usually one-way devices; impulses can flow across them in only one direction. Synapses may occur between an axon and a dendrite, or an axon and a cell body, or even between an axon and an axon. Nearly all neurons have synapses with many other neurons. Neurons can be excitatory or inhibitory, meaning that they either tend to pass an impulse to the neighboring cell, or tend to prevent the neighboring cell from transmitting an excitatory impulse. To illustrate, neuron A in Figure 3 is excitatory for Z, and B is inhibitory for Z. Stimulation of the axon of A causes an impulse to travel to the synapse with Z, in turn exciting Z to send an impulse down its own axon, where it is recorded. But if one stimulates the axons of A and B at the same time, then B's synapse will prevent A from stimulating Z.

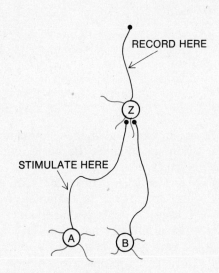

3 **MODEL OF NERVE NET consists of three neurons. Axons and cell bodies are rendered in black, dendrites in color. Axons of neurons A and B synapse with neuron Z. Stimulation of axon of A causes impulse to travel across synapse to the axon of Z, where it can be recorded on an oscilloscope.**

The ability of one nerve cell to turn off the effect of another is extremely important in complex functions. One example is that the brain can select the sensory data to which it wants to pay attention. When being chased by a bull, for example, a person becomes unaware of the beauties of the countryside, and gives his full attention to the problem of escape.

The action potential is the basic unit of communication in all animals, and it has two important features: its velocity depends on the diameter of the axon, and it does not fade out, no matter how long the axon.

MESSAGES ALONG A NEURON

The cell membrane of the axon, like that of all cells, is normally polarized, even when it is resting (not passing a message). In other words, there is a potential difference, called the RESTING POTENTIAL or membrane potential, between the inside and the outside of the cell, just as there is between the two electrodes of a flashlight battery. The potential of a flashlight battery is about 1.5 volts; that across the membrane is typically 0.07 volts, more conveniently described as 70 millivolts (mV). If one could place a wire between the inside and outside of the cell, a current would flow through it because of this potential difference. An easier approach is to poke a tiny sharp electrode into the axon, and to measure the voltage between this electrode and one placed just outside. The sensitive voltmeter used for this measurement is an oscilloscope, which displays the voltage (and any changes of voltage) on a screen.

Information can travel from one part of the body to another along the axon of the neuron, a process called AXONAL CONDUCTION. It can be seen and studied in an axon taken from an isolated nerve (a bundle of axons) —for example, one removed from the leg of a frog or lobster. The message itself consists of a wave of depolarization that rushes down the axon at a rate of say, 10 meters per second. In different axons the rate ranges from 0.6 to 120 meters per second.

If one pokes an electrode into an axon in a living animal and places another electrode outside the axon, one would observe the polarization. If at this moment the animal were to send a message down that nerve, the oscilloscope would show that for a brief moment (about one thousandth of a second: 1 msec) the polarization is reversed *(Figure 4)*. This pulse of depolarization that sweeps along the axon is called an ACTION POTENTIAL; it is the message. Pulses like this tell the brain, when one sits on a hot stove, to get off; and pulses like this, from the

EVENTS AT THE AXON

EVENTS AT THE
OSCILLOSCOPE

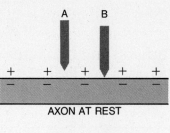

AXON AT REST

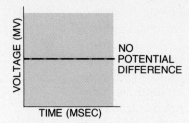

NO
POTENTIAL
DIFFERENCE

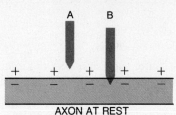

AXON AT REST

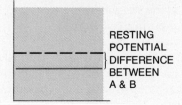

RESTING
POTENTIAL
} DIFFERENCE
BETWEEN
A & B

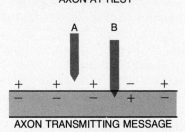

AXON TRANSMITTING MESSAGE

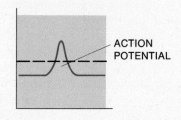

ACTION
POTENTIAL

4 ACTION POTENTIAL is a wave of depolarization that travels along the nerve fiber. Oscilloscope measures the difference in electrical potential between electrodes A and B. When both electrodes are outside the axon there is no electrical potential across them (top). If electrode B penetrates the axon, oscilloscope shows 60–80 millivolt resting potential (middle). As a nerve impulse passes, oscilloscope shows action potential (bottom), which lasts about one millisecond.

5 ION CONCENTRATIONS in giant axon of the squid; concentrations are expressed as millimolar (mM). Because of gradients created by sodium-potassium pump, K^+ ions tend to flow out, and Na^+ and Cl^- tend to flow in. Resting neuron is permeable primarily to K^+; only this tendency is expressed, making the outside of the axon more positive than the inside.

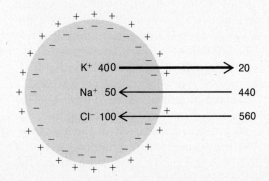

brain to the leg muscles, give the command to move. The action potential is the basic unit of communication in all animals.

The action potential has two important features. First, its velocity depends on the diameter of the axon; axons of large diameter permit greater velocities. Many invertebrates have special "hotline" axons for emergency use, with enormous diameters. These giant axons are favored by physiologists who want to insert electrodes into them. A particular favorite is the axon that instructs the jet-propulsion system of the squid to blast off when danger is detected. This axon, half a millimeter in diameter, transmits action potentials at 30 meters per second—about ten times faster than most other squid axons. The second important feature of the action potential is that it maintains the same shape and size, and travels with the same velocity, no matter how long the axon. Unlike an electrical impulse transmitted down a simple wire, the action potential does not fade out. To maintain the impulse, the axon must continuously supply energy to the pulse as it travels. Its movement, in other words, is active rather than passive. Passive transmission does, however, occur elsewhere in the nervous system, as will be noted later in this chapter.

To understand the action potential, one must first know why the axon is polarized. The explanation lies in the concentration of ions inside and outside the cell. Measurements show that the K^+ concentration is much higher inside the cell than outside; the reverse is true of the concentration of Na^+ (*Figure 5*). Such differences constitute a concentration gradient. Just as a voltage gradient across the ends of a wire causes a current to flow from a high to a low potential, a concentration gradient creates a tendency of ions to flow from a high to a low concentration.

The cause of the Na^+ and K^+ gradients in axons, as in all cells, lies in the sodium pump. The pump consists of one or more proteins embedded in the cell membrane. It uses energy derived from splitting ATP to pump Na^+ out of cells and K^+ into them, thus maintaining the gradients. If the pump is poisoned (for instance by a chemical called ouabain) the Na^+ and K^+ gradients slowly fade away as ions leak across the membrane, tending to equalize the concentration inside and out. Fortunately the leakage rate is very slow in normal cells (including neurons and their axons), so the pump does not have to work constantly at full capacity.

The gradient of K^+ establishes the resting potential. The cell membrane of the resting axon is much more permeable to K^+ than to other important ions such as Na^+ and Cl^-. K^+ tends to diffuse outward, depleting the supply of K^+ ions next to the inner side of the membrane and increasing the supply on the outer side; the outer side, richer in K^+, is therefore more positive than the inside (*Figure 5*). This separation of charges tends to prevent further diffusion, because the departing K^+ ions must now migrate into a relatively positive area, which tends to repel them. The migration continues until the tendency of the K^+ to diffuse outwards is precisely balanced by the repulsive tendency in the opposite direction.

If the resting potential is due to K^+, what about the action potential? The widely accepted view is that of Hodgkin and Huxley of England, who found evidence that the action potential is caused by a wave of sodium permeability. This permeability can be pictured as the brief flinging open of a gate, through which Na^+ rushes in, impelled by its concentration gradient. The Na^+ inflow briefly neutralizes the internal negativity of the axon — that is, depolarizes it. Indeed, the inside becomes briefly more positive than the outside, as shown in Figure 4. But less than a millisecond later, another gate is flung open and the Na^+ gate is shut; K^+ rushes out to restore the original situation. A wave of Na^+ gate-openings flows down the axon, followed by a wave of K^+ gate-openings; the result is the wave of depolarization which we call the action potential.

It is important to distinguish between the Na^+ gate and the Na^+ pump, and to think about their energy needs. The action potential is actively propagated (does not fade out) because of energy stored in the form of ion gradients. These gradients are established by the Na^+ pump (better called the Na^+–K^+ pump) which uses energy from ATP. The Na^+ pump works slowly and steadily whenever the gradient is depleted by leakiness or action potentials. Conversely the Na^+ gate is active only during the action potential. The brief flow of Na^+ (and later K^+) during the action potential does not need ATP; these flows are driven by ion gradients previously created by the Na^+ pump.

CODING IN THE NERVOUS SYSTEM

Figure 6 shows action potentials, not as bell-shaped curves as in Figure 4, but on a compressed

FREQUENCY OF ACTION POTENTIAL increases as the intensity of stimulus increases. Recordings above depict the firing of two nerve cells in the ear of a moth. One cell produces a few large spikes, the other many smaller spikes. In recording A sound level is just above threshold, in B sound is increased by seven decibels, in C by 15 decibels, and in D by 23 decibels. 6

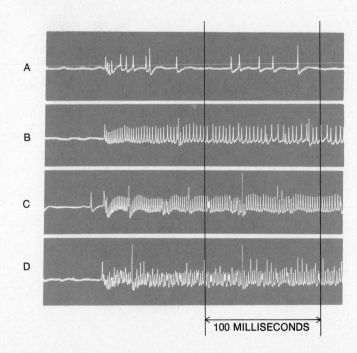

100 MILLISECONDS

time scale that narrows their peak to a spike. Furthermore, instead of recording intracellularly as in Figure 4, the researcher has hooked both electrodes around the outside of the cell, where they record only the transient change in positive charge on the axon's surface.

The function of the action potential is to transmit information. It might be transmitting information about the intensity of pain or odor or light at some particular receptor cell. It cannot signal "more pain is being felt" by sending faster or larger action potentials. It can only shape the pattern of signals, for instance the number per second or the total number in a single burst. Figure 6 shows recordings from two nerves whose signals originate in two sensory cells in the ear of a moth. All the signals in any one axon move at the same rate and have the same height, but the number of them rises as the intensity of the stimulus increases. This sort of frequency modulation is the way that nerves encode quantitative information.

MESSAGES BETWEEN NEURONS

The role of the axon is to get fast accurate data from one place to another. Collected and sorted in the CNS, the data are used to direct bodily actions. When appropriate, the information is combined with other data (for instance, combinations occur of visual, auditory and tactile sense data), and filed for future reference. These complex functions are performed at the synapses, the ultimate integrating units of the nervous system. Integration is the ability to draw some conclusion from a variety of pieces of data. Suppose Z of Figure 3 were a motor neuron governing one's ability to reach out his arm to pick up an apple. Z might well receive excitatory inputs from a brain center; the hunger center might say "yes," the thirst center might say "yes," and perhaps other areas might anticipate the esthetic pleasures of apple-eating and also say "yes."

But there are likely to be inhibitory inputs too. Perhaps a cortical zone says "no" because the last apple eaten led to a stomachache, and another says

7 SYNAPSE between body of one nerve cell and the axons of many others is depicted at progressively higher magnification. Diagram A depicts the end bulbs of many axons in contact with the cell surface. B is an enlarged view of the end bulbs. C shows mitochondria and synaptic vesicles within a single end bulb. D is a view of the synaptic cleft between the membranes of the end bulb (color) and the cell body (gray).

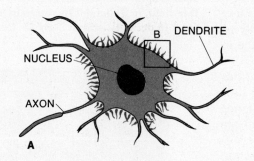

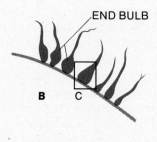

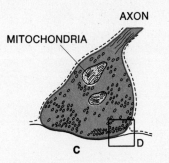

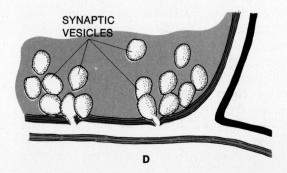

"no" because the apple must be purchased and money is lacking. Neuron Z thus receives both "yes" and "no" inputs, and only if the excitatory impulses are more numerous and sustained than the inhibitory will Z be fired, and then one puts out one's arm to pick up the apple. The essence of Z's action is therefore to respond only if the excitatory influences outweigh the inhibitory.

First consider the simple situation, in which one excitatory neuron (A of Figure 3) fires. A glance at the synapse joining A and Z shows that there is a little swelling (end-bulb) at the end of A's axon (*Figure 7*). The end bulb contains several mitochondria, and clustered around the thickened zone of membrane are hundreds of little bags called synaptic vesicles, about 200 to 400 Å across. These bags contain a chemical substance called a TRANSMITTER. Any one neuron contains only one kind of transmitter; the commonest kinds are acetylcholine, noradrenalin and serotonin. When excitatory neuron A fires, its end bulb releases a pulse of transmitter, probably because the impulse forces several hundred of the vesicles along the presynaptic membrane to empty their contents into the SYNAPTIC CLEFT, a gap approximately 200 Å wide between A and Z. The cell body and the axon of neuron Z are normally polarized, as described above. The transmitter released by neuron A causes partial depolarization of the cell body of Z, because it makes Z suddenly permeable to Na^+. (There is a simultaneous increase in K^+ permeabil-

ity but this only partially offsets the depolarizing Na^+ effect.) The partial depolarization spreads passively over the cell body of Z. If A fires repeatedly, it may cause such a great depolarization of Z that the axon hillock of Z (the area where Z's cell body and axon join), becomes depolarized by some critical amount, typically 30 mV. If this occurs, it triggers an impulse in Z's axon; in other words, Z fires.

Suppose A and B fire simultaneously. B is an inhibitory neuron that releases a different transmitter, which may cause Z to become hyperpolarized: even more polarized than in the resting state. The transmitter could do this by making Z leaky to Cl^- ions, which are more concentrated outside than inside and therefore move in and counteract, by their negative charge, the depolarizing Na^+ effect from A. The outcome is that A cannot depolarize Z enough to trigger the axon hillock. This is the cellular basis of one important kind of inhibition.

Most neurons receive numerous excitatory and inhibitory synapses—up to several thousand in some cases. Whether the neuron fires or not depends upon the ALGEBRAIC SUM OF THE STIMULI; if the excitatory transmitters outweigh the inhibitory ones, the neuron will fire. Thus the neuron can do algebra; it can sum with respect to sign, integrating

REFLEX ARC is a two-neuron circuit, plus an inhibitory neuron called a Renshaw cell. This diagram depicts the circuitry of the knee-jerk reflex. Some components of the muscle-spindle system are omitted.   8

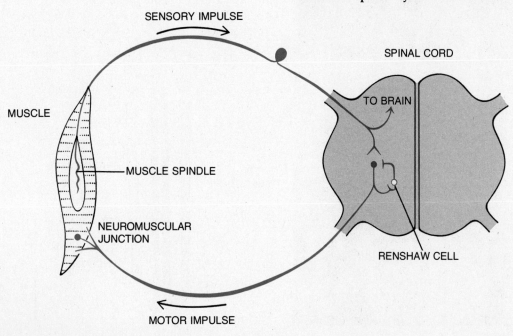

THE KNEE-JERK REFLEX

Nerves are connected to one another in circuits of varying complexity. An extremely simple one is involved in the knee-jerk reflex: the automatic upward kick that occurs when the tendon of the kneecap (patella) is tapped by a neurologist's hammer. The circuit is called a REFLEX ARC, and involves primarily only two cells: a sensory neuron which tells the spinal cord of the blow, and a motor neuron which responds by instructing the muscle to contract *(Figure 8)*. The sensory detector is part of an apparatus designed to monitor stretching. Known as a muscle spindle, it is buried in the muscle which is going to kick, and is attached to the patellar tendon. The tap of the hammer stretches the tendon, causing the sensory neuron to fire a message to the spinal cord. The sensory axon is part of a nerve that enters the cord by a dorsal pathway and synapses inside the cord with the cell body of the correct motor neuron. (It also sends off a branch to tell the brain of the tapping, but the muscle contraction is automatic, or reflex, and does not await instruction from the brain.) This motor neuron now fires off a message, instructing the muscle to contract by exciting the neuromuscular junction where nerve and muscle meet.

Added to this simple and rapid two-neuron circuit is an inhibitory neuron called a Renshaw cell. A branch of the motor neuron excites this little cell, which in turn inhibits the cell body of the motor neuron *(Figure 8)*. This self-inhibition ensures that the motor neuron fires off only one quick volley and then, its job completed, turns itself off.

Most of the circuits of the nervous system are more complicated, involving thousands of interacting neurons. But the way they are hooked together is not different in principle from that found in the simple reflex arc.

SENSORY SYSTEMS

Neurons and their associated structures collect data about the world outside—the sights, smells, touches, sounds, and tastes; and the world inside—including the position and the tension of muscles. These different qualities, or MODALITIES as they are better called, are each detected by appropriate receptors which have the ability to convert (transduce) them into nerve impulses. The impulses are transmitted to the CNS, and constitute the link between the objective world, much of it outside, and the subjective world within.

Not only do these sensory transducers achieve the qualitative effect of converting (for instance)

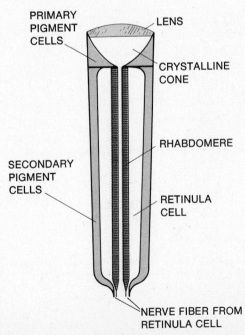

COMPOUND EYE OF INSECT is comprised of many ommatidia. At left, the eye of the housefly magnified 240X. At right, a diagram of a longitudinal section through a single ommatidium. **9**

PRIMARY PIGMENT CELLS

LENS

CRYSTALLINE CONE

RHABDOMERE

SECONDARY PIGMENT CELLS

RETINULA CELL

NERVE FIBER FROM RETINULA CELL

the different "yes" and "no" signals, much as a legislature tabulates a vote to decide if it should act.

The compound eye of the arthropod produces a mosaic image made up of thousands of tiny patches. Although the image is crude and lacks detail, it enables the animal to detect movement extraordinarily well.

sound into electrical pulses, but they achieve a quantitative effect too, for they give information about the intensity of the sound, and about the many different sound components that may be present simultaneously. Full understanding of any sensory system, therefore, requires an understanding of the transduction mechanism, and also of the coding by which variations in the modality are "described" by the patterns of impulses that pass to the brain.

VISUAL SYSTEMS

The simplest "eyes" consist of light-sensitive patches like the eye spot in the protozoan *Euglena*, which do not relay images or other patterned information. They give information that enables the organism to orient with respect to light. Some organisms which have evolved true eyes have also retained simple photoreceptors of this type. Examples are the ocelli which occur in between the compound eyes of many insects, the photoreceptors found in the tails of crayfish and lobsters, and the brain area under the skull of some birds. All these organisms can perceive light even when their true eyes are removed.

The arthropods (such as crustaceans, spiders and insects) have evolved an elaborate compound eye that consists of numerous quasi-independent units called OMMATIDIA; some dragonflies have 20,000 of them in each eye. Each ommatidium has a group of retinula cells grouped around a rodlike RHABDOM *(Figure 9)*. The rhabdom is made up of subunits called rhabdomeres, one derived from each retinula cell. The retinula cell is the true sensory neuron, and the rhabdomere is its photoreceptive part. The rhabdom is surmounted by two lenslike structures that focus light onto it; the crystalline cone and the corneal lens.

This multiple-unit system provides the arthropod with a kind of mosaic image of the outside world built up of tiny patches, each provided by

one ommatidium. Although the image is crude and lacks structural detail, it enables the arthropod to detect movement extraordinarily well. Because an ommatidium recovers rapidly from a light impulse, a blowfly can detect the flickering of a light at a frequency of 250 flashes per second. In contrast, light flickering faster than 50 times a second appears continuous to humans; motion pictures projected at 50 frames per second give the illusion of continuous movement.

The vertebrate eye has a lens protected by a tough cornea; the peripheral part of the lens is covered by a retractable iris *(Figure 10)*. The central part (pupil) can thus be widened in poor light by retracting the iris and letting in more light. The retina is a complicated structure whose photoreceptor cells, called RODS and CONES, are on the outermost layers and so farthest from the incoming light *(Figure 11)*. The rods and cones form synapses

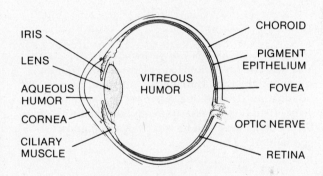

HUMAN EYE has a focusing lens and an iris diaphragm **10** **whose aperture is regulated by ciliary muscle. Interior of eye is filled with transparent gel-like material (aqueous and vitreous humor). Fovea is rodless area that affords acute vision. Point where optic nerve connects with retina is blind spot.**

with the dendrites of fairly simple neurons called BIPOLAR NEURONS because they have only one branched axon and one branched dendrite. The axons of the bipolar cells in turn form a complex web of synapses which eventually connect up with the dendrites of the innermost neuron layer, called GANGLION CELLS. The axons of these ganglion cells are collected to form the OPTIC NERVE that transmits impulses to the CNS.

How does light generate nerve impulses? In 1967 George Wald of Harvard shared a Nobel prize for revealing the details of the first step. All land vertebrates and marine fishes (and at least some insects) have in their eyes a material called rhodopsin, which consists of a protein (opsin) and

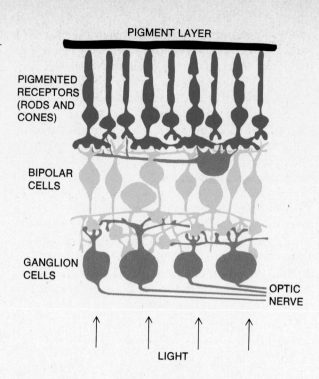

PIGMENT LAYER

PIGMENTED RECEPTORS (RODS AND CONES)

BIPOLAR CELLS

GANGLION CELLS

OPTIC NERVE

LIGHT

11 DIAGRAM OF RETINA of vertebrate eye depicts an area near the fovea. Rods are rendered in dark gray, cones in light gray. Horizontally oriented fibers carry information across retina at two levels, integrating the visual field. Light passes through the whole system before being absorbed by receptors and pigment epithelium.

a derivative of vitamin A called retinaldehyde. The retinaldehyde has a number of possible isomers; one isomer, called the *11-cis* form, is found in rhodopsin. Light causes its conversion to a different isomer called the all-*trans* form (*Figure 12*).

12 RETINALDEHYDE, a component of visual pigments, can be converted by light from the cis form (top) to the trans form (bottom).

11-CIS FORM

CH₂OH

CARBON NO. 11

CH₂OH

ALL-TRANS FORM

Subsequent reactions lead to the splitting of opsin and retinaldehyde. In turn this set of events leads, by a mechanism still unknown in detail, to a hyperpolarization of the membranes of rods and cones, or a depolarization of the membranes of ommatidia. This change in polarization is called the generator potential; if it is large enough it can trigger the nerve cells to transmit action potentials along the axons of the optic nerve.

Rods and cones are special photoreceptors that extend the range of useful light intensities. In dim light, only the rods are functional and all images lack color. In bright light vision shifts to the cones. Color vision is possible because there are three classes of cones, each sensitive to a different range of wavelengths.

A good deal is known about the data which the photoreceptors provide, but two points about the vertebrate eye are particularly important. The first is that the illumination of some receptors leads to the excitation of ganglion cells, while illumination of others has the opposite effect. The surface of the retina consists of hundreds of RECEPTIVE FIELDS, tiny circular areas that affect the firing of a single ganglion cell. The diameters of these fields range from a fraction of a millimeter to more than two millimeters. Some of the fields overlap, and they differ in the way that they respond to light. "On-center" fields, for example, have an inner core of receptors which, when illuminated, lead to the excitation of a ganglion cell, and an outer ring of receptors that react in the opposite way. "Off-center" fields have the reverse arrangement. Acting together, the two types of fields enable the retina to exaggerate the contrast between light and dark images (such as the letters printed on paper).

The second point is that data about the visible world are analyzed in progressively more complex ways between detection and perception. Some of this analysis is done in the retina, quite outside the CNS. The ganglion cells of the retina, in other words, do not simply pass on the crude data received by the receptors, but analyze and shape the data before feeding it into the optic nerve. This "peripheral filtering" is common in many sensory systems.

Hubel and Wiesel of Harvard, working with the cat, have elucidated the next step. The optic nerve feeds into an area on the underside of the brain called the LATERAL GENICULATE BODY. Its cells behave rather like the ganglion cells in that they each respond to a different small disc of retinal area, and they also have on-center and off-center organization. One difference is that the inhibitory periphery of the on-center type is even more pronounced. This exaggerates the tendency, already present in the ganglion cells, to be turned off by light patches which are big enough to provoke peripheral inhibition and so cancel out the central excitation. Discrete points of light are thus dramatically emphasized because they can give excitation without inhibition.

The lateral geniculate body transmits this partially processed data to the back area of the cortex, known as the VISUAL CORTEX. The cells here pick up data that originate in quite widely separated retinal receptors. Different cortical cells therefore respond to different configurations and movements of the retinal image. One kind responds only to lines of light, and only responds well if the line is orientated in a particular direction, so there are "horizontal line detectors" and so on. Others ("edge detectors") detect only the places where sizeable bright areas meet sizeable dark areas. Other more complex edge detectors will only respond if the bright area is to the left, and still others only if the bright area is to the right.

One can imagine the way in which our perception of the world is stitched together from these fragments of analyzed data. The picture of the outside world presented to us by our visual cortex is not a simple point-by-point transcription of the objects in that world, like a television image. Instead it is a collection of statements about contours, lines and movements, much like short radio news bulletins which omit all but selected points of great interest.

AUDITORY SYSTEMS

Only a few invertebrates have ears. These include the cicadas, crickets and a few moths. The moths seem to use their ears exclusively as a warning system for bat attacks, as shown by the brilliant studies of Roeder at Tufts University. The ear of the moth employs only one pair of sensory neurons connected to a sounding board, more properly called a tympanic membrane, on the moth's thorax *(Figure 13)*. In spiders, cicadas, and crickets a similar tympanic system is mounted on the legs; the number of neurons involved is seldom more than 70.

In fishes and reptiles, sound plays a relatively small role, and sounds above a few thousand cycles per second (also known as Hertz, abbreviated Hz) are not detected. It is in the birds and mammals that acoustical communication reaches tremendous complexity, with external structures

EAR OF MOTH is sensitive to the high-frequency sounds of bat "sonar." Arrow indicates the external opening of the ear of Agrotis ypsilon, a moth about 3/4 inch long.

13

(pinnae) to help directionality, and frequency ranges that in the human, for instance, extend from 20 to 20,000 Hz in adults and much more in children. Because the sound waves are in air, but the sensory neurons only respond in liquid, elaborate structures have been developed to transmit the compression waves from the highly compressible gaseous medium to the almost incompressible fluid medium. This is achieved by absorbing the energy over a large disc (the TYMPANIC MEMBRANE or eardrum) and transmitting it via a set of three levers to a very small disc (the OVAL WINDOW of the cochlea). In mammals the three levers, also known as OSSICLES, are picturesquely called the malleus (hammer), incus (anvil) and stapes (stirrup) *(Figure 14)*.

In the human ear, the cochlea is a snail-shaped chamber. A flat membrane (the BASILAR MEMBRANE) is stretched tightly along the course of its canal, separating the tube into two liquid-filled canals *(Figure 15)*. The pressure from the oval window is transmitted via the liquid in the vestibular canal, causing the basilar membrane to bulge; the consequent pressure in the tympanic canal is relieved by allowing outward movement of a sort of pressure-valve called the ROUND WINDOW. The basilar membrane normally presses on a group of sensitive cells (the true sound receptors) embedded in a relatively fixed plate, the tectorial membrane. When the basilar membrane bulges it rubs against these sensitive cells. In some way, the mechanical

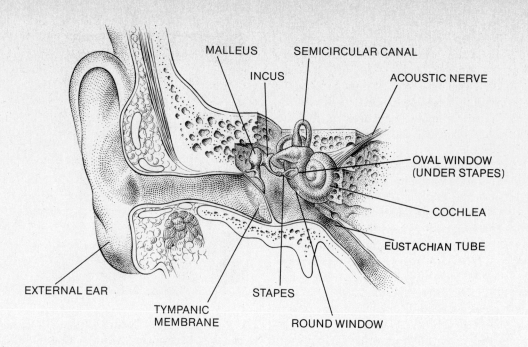

MALLEUS

SEMICIRCULAR CANAL

INCUS

ACOUSTIC NERVE

OVAL WINDOW
(UNDER STAPES)

COCHLEA

EUSTACHIAN TUBE

EXTERNAL EAR

TYMPANIC
MEMBRANE

STAPES

ROUND WINDOW

14 HUMAN EAR. Energy of sound waves in air is absorbed by tympanic membrane and transmitted by three tiny bones (malleus, incus and stapes) to oval window of cochlea. Cochlea is connected to acoustic nerve that leads to brain. Auditory tube opens in the nasal part of the pharnyx.

deformation of these cells causes them to depolarize (presumably by letting in Na^+) and leads to a nerve impulse in the endings of the cochlear nerve which embrace the cell. The system of two membranes and sensitive cells is known as the organ of Corti.

Figure 15 shows how the cochlea narrows at one end, at the apex of the snail shell. Consequently the width of the organ of Corti varies. Von Békésy of the University of Hawaii has shown how the basis of pitch detection depends on this variation at ranges above 60 Hz. The basilar membrane vibrates unequally over its length; each tone gives maximal vibration at one point at which the signal is tuned to the width of the organ. Thus the brain

15 COCHLEA is a coiled tube; each part of it corresponds to a particular pitch. Pressure from oval window causes basilar membrane to bulge, deforming sensitive cells attached to it, as shown in cross-section at right. Deformation generates a nerve impulse.

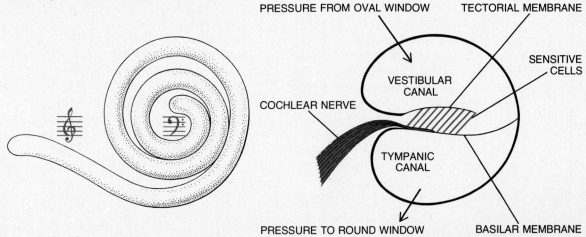

PRESSURE FROM OVAL WINDOW

TECTORIAL MEMBRANE

SENSITIVE
CELLS

COCHLEAR NERVE

VESTIBULAR
CANAL

TYMPANIC
CANAL

PRESSURE TO ROUND WINDOW

BASILAR MEMBRANE

"knows" the pitch by the location of the sensitive cells which are firing. The output from any weakly vibrating cells is inhibited, much as in the center-on phenomenon in the retina, thus sharpening the localization of output from the most sensitive cells.

The term "outer ear" is often used for the pinna and the canal which terminates at the tympanic membrane; "middle ear" is used for the tympanic membrane and the adjacent chamber containing the ossicles; the term "inner ear" is used for the cochlea and its enclosed cells. When we hum, listen to our own voice, or to our crunching of an apple, much of the sound bypasses the outer and middle ears. This explains the big difference between our voice as we think it sounds and as we hear it on a tape recorder.

As in visual systems, some analysis of the auditory input is made peripherally and some in the brain. Capranica of Cornell University has been studying the physiological basis of the ability of cricket-frogs *(Acris crepitans)* to distinguish frog dialects. A female New Jersey frog fails to respond to the mating calls of a male Texas frog. Capranica finds that part of this distinguishing ability depends on the frog's inner ear. The Jersey frog is literally deaf to some of the notes that the Texan uses, notes that Texan females hear perfectly well.

This is a case of peripheral analysis. However, different time sequences as well as different notes make up the various dialects, and these different time sequences are transmitted to the brain, which distinguishes clearly between them. This is a case of central analysis.

CHEMORECEPTION

Chemoreceptors are neurons that detect chemical events. The major external chemical events that humans experience are smell (olfaction) and taste (gustation), and we are most aware of them in the detection and consumption of food. In most mammals, probably including man, olfactory signals play an important part in communicating about sex. In very many animals chemical cues are responsible for elaborate communications about pathways to follow, dangers to avoid, and other important events, as well as playing a major role in

THREE CHEMORECEPTORS. Taste receptor of blowfly consists of two neurons whose dendrites extend upward along sensory hair. Vertebrate taste buds contain hair cells attached to bipolar neuron. Olfactory (smell) receptors in nose of vertebrates are modified epithelial cells.

16

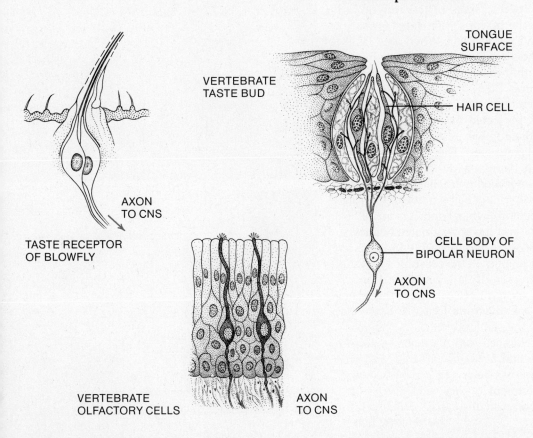

VERTEBRATE TASTE BUD

TONGUE SURFACE

HAIR CELL

CELL BODY OF BIPOLAR NEURON

AXON TO CNS

AXON TO CNS

TASTE RECEPTOR OF BLOWFLY

VERTEBRATE OLFACTORY CELLS

AXON TO CNS

sexual attraction *(Chapter 22)*. There are also internal chemoreceptors, which in man monitor such things as glucose and CO_2 in the blood, and lead to compensations when necessary.

The form and organization of chemoreceptors is much simpler than the complex devices for vision and hearing. In all invertebrates and most vertebrates, chemoreceptors are simply the naked endings of specialized bipolar sensory neurons. One example is the proboscis of the blowfly. It is covered with sensory hairs, each of which contains parts of three sensory neurons. One is a touch receptor. The other two are taste receptors whose cell bodies sit at the base of the hair and whose dendrites pass up the hair to the tip, when they are exposed to the outside *(Figure 16)*.

Vertebrates have developed taste buds on their tongues, which show an important difference from the exposed-dendrite system just described. As Figure 16 shows, a group of cells buds off from the adjacent epithelium (a continuous process —they last only about 250 hours in humans) and develops hairlike processes which are exposed at the tongue's surface. These hair cells are the actual taste receptors, and a single bipolar neuron sends branches of its dendrite to the whole group that comprises the taste bud. The number of possible tastes is still believed to be only four: sweet, sour, bitter, and salt. Specific areas of the tongue are sensitive to each: the tip for sweet, sides for sour, back for bitter, and salt all over. When we "taste" complicated and delicious flavors, we are in fact using olfactory rather than gustatory cells: we are smelling, not tasting.

The olfactory endings of vertebrates lie at the back of the nose, protected by mucus, and quite densely packed; the dog has up to 40 million endings per square centimeter. The dendrite of the sensory neuron touches a chemoreceptor fashioned from a modified epithelial cell. This is a rare case of the receptor not being part of a neuron.

In all chemoreceptors, when contact is made between the receptor and the appropriate chemical a generator potential is created, probably through an effect on Na^+ permeability. One presumes that the chemical combines with a specialized macromolecule of the receptor cell, and that this combination makes the macromolecule change its shape and render the cell leaky to Na^+. This reaction between the chemical and the macromolecule resembles the reaction between substrate and enzyme. The analogy has been stressed by Amoore of the U.S. Department of Agriculture, who argued that differences in the odor of substances

were related to differences in the shape and charge of their molecules. He suggested that for humans, all odors were mixes of seven primary characters (ethereal, camphoraceous, musky, floral, minty, pungent, putrid), and are detected by seven kinds of receptor macromolecules that combine, with varying specificity, with a vast diversity of odorous substances. The theory may prove wrong in its details, but some kind of specific matching along these lines is almost certainly the explanation of variation in odor.

Figure 17 shows a female of the silkworm moth, *Bombyx mori*. How does a potential mate know she is around, and how does he locate her? He knows

FEMALE BOMBYX MOTH is depicted at top. Below, a 1 **magnified view of her scent gland, which releases the pheromone bombykol.**

because she broadcasts, by everting a gland at the tip of her abdomen, a chemical called bombykol. The male *(Figure 18)* detects this with incredible sensitivity, when the airborne bombykol diffuses into the fluid surrounding the partially sheathed dendrites which are at the core of the hairs with which his antennae are covered *(illustrated on opening page of this chapter)*. By piercing the dendrite and recording from it in the presence of different quantities of applied bombykol, Schneider at Seewiesen, Germany found that the system can detect one tenth of a nanogram (a nanogram is 10^{-9} grams) and responds briskly to one nanogram. The male obtains a response whose frequency is clearly related to bombykol concentration over a 10,000-fold concentration range. Thus, sensory data is available for the male to follow a concentration gradient, or to establish if a female scent is faint and far away or rich and, presumably, close.

MECHANORECEPTION

Certain receptors respond to physical forces: twisting, stretching or pressure. It is only for convenience that auditory systems were not considered in this section since the sensory cells of the organ of Corti actually detect a mechanical distortion caused by air waves.

A great many different purposes are served by a kind of all-purpose mechanoreceptor called a hair cell. Hair cells are sensory cells with from one to 50 processes projecting at one end. They form synapses with a neuron which either meets their base end-on, or clasps it in a cup-like way *(Figure 19)*. Deflection of the hairs of these cells in one

MALE BOMBYX MOTH (left) detects bombykol from female with his prominent antennae. Photograph at right depicts part of antenna as seen by the light microscope, revealing sensory hairs. 18

direction causes firing in the associated neuron (once again through unexplained means) while a reverse deflection may cause inhibition of the associated neuron. Thus the cells can indicate directionality.

One use of mechanoreceptors is to inform the organism of the movements it is making. Most vertebrates have the well-known SEMICIRCULAR CANAL SYSTEM which consists of three loops at right-angles to each other, each of which provides information about angular movement in a different plane *(Figure 20)*. These loops, which are chambers in the bone of the skull, and often called the labyrinth, are filled with viscous fluid (endolymph). Each canal has a gelatinous spring-loaded door called a cupula extending part way across it. If one rotates one's head in any plane, the fluid tends to stay in position, thus deflecting the cupula. The consequent bending of the hair cells embedded in the cupula either increases the spontaneous firing of the hair cells (if rotation is in one direction) or inhibits their firing (if rotation is in the opposite direction). The three-plane system is not universal; lampreys have two semicircular canals, and hagfish only one.

The canal system signals changes in angular but not linear velocity. Position in space is signalled by different devices. In vertebrates there is a special chamber (the utriculus) below the semi-circular canals in which a pebble of calcium carbonate

421

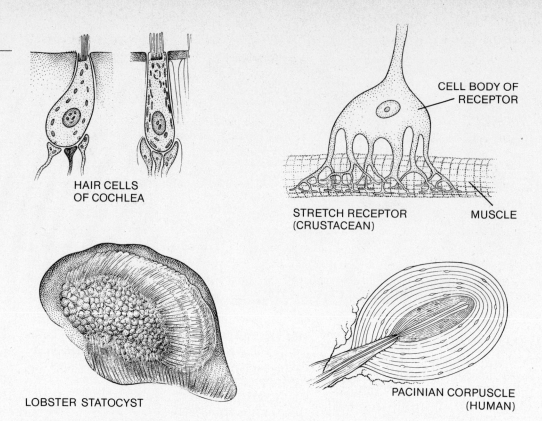

HAIR CELLS
OF COCHLEA

CELL BODY OF
RECEPTOR

STRETCH RECEPTOR
(CRUSTACEAN)

MUSCLE

LOBSTER STATOCYST

PACINIAN CORPUSCLE
(HUMAN)

19 **FOUR MECHANORECEPTORS. Hair cells of cochlea are found on organ of Corti. Hair cell at left forms an end-on synapse with dendrites; cell at right, a cup-like synapse. Stretch receptor tells crustacean when a muscle is stretched or relaxed; inhibitory neuron that switches the receptor off is not shown. Statocyst is a gravity detector. Pacinian corpuscles are touch receptors in the skin of humans.**

20 **SEMICIRCULAR CANALS are fluid-filled chambers in bone of human skull. Drawing depicts a cast of the canals and the cochlea.**

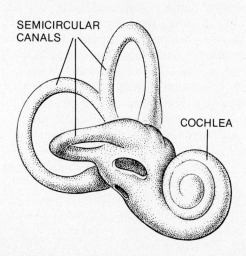

SEMICIRCULAR
CANALS

COCHLEA

called an OTOLITH lies on a bed of hair cells. The position of the pebble, sensed by these cells, tells the animal about the head's position with respect to the Earth's gravitational field. In invertebrates there is a very similar system, a pebble-like STATOLITH lying in a hair-lined sac called a STATOCYST *(Figure 19)*. In 1893 Kreidl showed the importance of these organs by replacing the statoliths of a shrimp with iron filings. When he placed a magnet above them, the shrimps swam upside down.

Fish have a special lateral-line system that tells them of pressure changes caused by objects moving near the fish, as well as its own velocity and direction of locomotion. The system is made up of a series of canals or tunnels whose openings form a row along the side of the fish. Within the epidermis of each canal is a group of hair cells capped by a gelatinous cupula. When water in the canals is moved, the cupula is bent and the hair cells fire.

Other types of mechanoreceptor can respond to touching, stretching and twisting. A very versatile one found in the deep layers of the skin of vertebrates is called the PACINIAN CORPUSCLE *(Figure 19)*. It is shaped like an onion about a millimeter long, and consists of a series of concentric layers of connective tissue around a core that contains the naked terminal of the sensory neuron. Any kind of touch that distorts the skin and its contained "onions" can (if large enough) cause the neuron to fire.

Many other kinds of touch receptor are found nearer the skin's surface. The human has about 640,000 pressure-sensitive spots, especially on the tongue, lips and fingertips. The different roles of these various receptors are by no means clear, but it is apparent that they lie at characteristic skin depths. In mammals one may find very close to the surface a network of free nerve endings that respond to pain, along with little disklike endings ("Merkel's discs") that may signal touch. Somewhat deeper are "Meissner's corpuscles" which also seem to respond to touch; then there are beaded nerve nets supposed to respond to pain; and wrapped around the sheaths of body hairs are nerve terminals which fire when the hair is moved. It is not important to remember these details, but only that there is a great variety of mechanoreceptors in the skin.

Most animals have a variety of kinesthetic receptors or PROPRIOCEPTORS which inform the animal of its position and the stress on its muscles and joints. These receptors continuously feed information back into the CNS and permit the CNS to monitor the success of its instructions to the body. When you decide to pick up something from the floor, you follow the execution of this decision with your eyes, but also with information from all the joints and muscles and touch-detectors involved in the act. In fact several of these pieces of information overlap: you can do the job fairly well with your eyes closed or with anesthetized fingers.

A proprioceptor which has played an important role in neurophysiology is the stretch receptor of crustaceans. The dendrites of this neuron, which is found in the abdominal muscles, wrap around the muscle fibers as Figure 19 shows, and fire when the muscle is stretched. Not shown in the figure is the fact that this neuron is in turn contacted by a completely different inhibitory neuron which can turn off the signal from the stretch receptor. The crayfish, for instance, can tell its stretch receptors that it does not want to know about the state-of-stretch of those muscles at that particular time.

In vertebrates a more elaborate method of signalling the degree of stretch in skeletal muscle has been developed, permitting the animal to achieve highly efficient control of his musculature. At the center of each muscle bundle is a set of modified muscle fibers called the MUSCLE SPINDLE; the modified fibers bulge in their middle, around which is wrapped a spiral sensory nerve ending. When the

muscle is stretched its spindle stretches too, decreasing its bulge and causing the sensory nerve to fire an impulse to the CNS. The CNS can respond with motor impulses to the whole muscle bundle, causing it to contract and thus reducing both the tension on the spindle and the impulses from it. Or the CNS can trigger specialized motor fibers connected only to the spindle, which shorten it and adapt it to the new state of the muscle bundle. In both cases this system tells the CNS about changes in muscle status. If the muscle goes from condition A to condition B, the system only fires while the transition from A to B is occurring. If condition B is sustained, the sensory neuron stops firing; that is, it ADAPTS. Adaptation is a common feature of kinesthetic and many other sensory neurons, and there are many different mechanisms of adaptation.

> Many proprioceptors are adaptive: they fire only when their status changes, then become silent. Thus the CNS is not under eternal bombardment, but is only informed of changes in the status quo.

Proprioceptors also indicate the stresses and movement of the skeleton. In vertebrates the connective tissue of the joints contains receptors sensitive to movement; each one covers a small angle, and together they cover the total span of movement. Arthropods have a variety of proprioceptors. The exoskeletons of insects contain chordotonal organs, usually consisting of a sensory neuron set in an elastic strand; one end of the strand lies in a pliable part of the cuticle. These organs sense changes in one piece of cuticle with respect to another. In addition, dome-shaped sensing systems signal the bending or flexing of the cuticle. Most of these neurons are of the adapting type; they fire when first stimulated, but become silent unless their status changes again. In this way the CNS is not under eternal bombardment, but is only informed of changes in the status quo.

OTHER RECEPTORS

An essay of this length cannot provide details of the many other kinds of neural receptor cells, but the reader should be aware that they exist. There are baroreceptors in mammals that detect changes in blood pressure by following the bulging of the carotid artery. There are thermoreceptors; in mammals and birds there are internal ones in the hypothalamus, as the last chapter mentioned, used in maintaining body temperature between fixed limits. Many invertebrates have thermoreceptors on mouth parts and legs which are useful in seeking prey and in other activities. The pit vipers also have external thermoreceptors, located in pits on their heads, which are used to locate their mammalian prey in the dark.

Electric eels, electric skates and several fresh water fish (gymnotids and mormyrids) have electroreceptors which respond to electric fields generated by the animal. More importantly they detect interruptions of those fields, and so are used to detect other fish (predators or prey) and probably to orient and navigate as well *(Chapter 21)*.

CENTRAL INTEGRATION

How does the CNS use sensory data, integrate it with all other relevant data (incoming or stored) and when necessary undertake some suitable response? Some aspects of nervous systems are rather inflexible, involving a stereotyped response to particular stimuli. Such responses are involved in several basic features of an organism, such as control of heart beat, blood sugar, body temperature, gut movement, blood flow and so on. In complex metazoans these housekeeping functions are performed by a subdivision of the nervous system called the AUTONOMIC NERVOUS SYSTEM, that operates (as its name implies) in virtual independence of the rest of the nervous system. All our glands and smooth muscles are operated by this unobtrusive servant, so that we do not have to consciously transfer food from stomach to duodenum, nor do we consciously narrow our pupils or salivate. We can preset the conditions which will be likely to set off these reactions, but the machinery is not under our direct control.

In vertebrates the autonomic nervous system has two divisions, called SYMPATHETIC and PARASYMPATHETIC, which usually have contrary actions. Thus the sympathetic system releases noradrenalin (also known as norepinephrine) at its nerve endings, which makes the heart beat faster, an action which we can duplicate by injecting noradrenaline. The parasympathetic system releases acetylcholine at its nerve endings, which slows the heart down. Acetylcholine is rapidly destroyed by a blood enzyme, so one has to inject acetylcholine directly into the heart if one wishes to duplicate the effect of parasympathetic stimulation.

Automatic responses are not restricted to the autonomic nervous system. An important kind is REFLEX ACTION, which involves a sensory neuron plus a motor neuron, with few or no interneurons. A well-known example, the knee-jerk reflex, was described above. Similarly the eye will blink if the cornea is touched, and a finger is almost instantly withdrawn from a very hot object. These are safety reactions in which fast action is needed, and take priority over other bodily activities. The response is sometimes said to be wired in; that is, it is not learned. The circuitry is there at birth, like a printed circuit in a TV set.

More complex activities are often wired in, especially in invertebrates. Thus the set of muscular activities that govern flight in insects appears to be automatic; but it is turned on and off by events in the outside or inside world, and in addition, the

CEPHALIZATION of the nervous system is more pronounced in high organisms. Flatworm has a very simple localized system (color), with a few ganglia at the head end. Cockroach, a primitive insect, has several local ganglia and the beginnings of a brain. Ganglia of housefly, a more advanced insect, show a greater degree of consolidation. Cephalization is most extensive in vertebrates, particularly man.

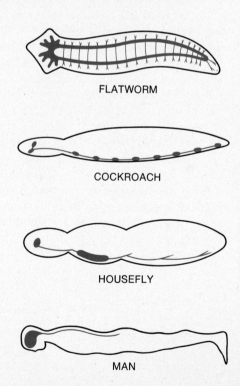

FLATWORM

COCKROACH

HOUSEFLY

MAN

stereotyped activity can be modified in certain limited ways. Thus a flying locust can correct for interfering winds by biasing certain components of its complex flight activity.

This simple life-style of stimulus leading to an automatic response is adequate for many of the simpler invertebrates. But evolution was accompanied by development of more complex behavioral repertories, demanding a kind of activity which took into account a series of factors simultaneously. Consider a mosquito seeking its prey: the mosquito has a whole series of possible activities: flying, settling, walking, sucking, evading, mating, egg-laying. The appropriate behavior is displayed only in the presence of suitable prey which is actively sought and then accurately located. Each operation of landing or sucking or taking off demands an exquisite collaboration of multiple muscles, and a continued awareness of the numerous relevant factors in the outside world, as well as the status of the insect's own body machinery. Such awareness permits the machinery to respond appropriately to the changing world.

The evolution from simple responses to complex behavior is accompanied by movement away from the simple nervous system found for instance in the flatworm, toward an increasing degree of CEPHALIZATION; that is, toward a central conglomeration of nervous tissue to form a brain. The brain receives and transmits data for the whole organism, largely replacing the localized segmental stimulus-and-response machinery, and confers an enormously increased potential for learning (*Figure 21*).

THE CEREBRUM

Within the vertebrates, in which cephalization is very extensive, there is a progressive increase in the relative size of the cerebrum (*Figure 22*). In the codfish the cerebrum is a relatively minor area; in the frog it has become a fairly substantial component; in the goose it is a major area but quite smooth; it has become much convoluted in the

horse; and in the human it has become enormous and very richly convoluted, dominating the underlying regions of the brain. These convolutions add additional surface area per unit volume of cerebral hemisphere; in humans the CORTEX (outermost part) of the hemisphere is the site of associative activity, so the convolutions add to the potential for complex activity. Associative activities include all the most interesting and complex functions of the brain: the ability to put together the sights, smells, and sounds of the world outside, associate them with the learned and innate information already in the cortex, and to respond in an appropriate manner. Man's dominance depends on the immensely rich possibilities for associative processes conferred by this extensive cortex. The abilities to reason extensively, to speak, and to suppress short-term desires in favor of long-term goals, all stem from cortical richness. All this is not to say that there is a simple direct relation between amount of cortex and intelligence; there is evidence that birds, which have very little cortex, use an area in the hypostriatum (which lies below the cortex) in intelligent behavior. Perhaps the evolution towards greater intelligence involved a shift towards the cortex as the location for it, and was accompanied by great enlargement of the cortex.

An important question about the cortex is, to what extent are functions localized within it? One way to explore this has been to see what functions are lost by removing particular pieces, either by surgery or by destruction with heat, chemicals or other means. The second way is to stimulate the local area and see how the animal responds. Work has been done with conscious humans, for exam-

EVOLUTION OF VERTEBRATE BRAIN shows progressive increase in size of the cerebrum (color). Cerebral cortex became convoluted, and older parts of the brain that control coordination, reflexes and instinctive behavior came to lie beneath it.

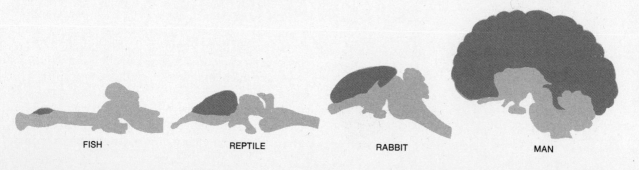

FISH   REPTILE   RABBIT   MAN

22

ple with epileptic problems, with much of the skull removed for exploratory surgery. If one stimulates with an electrode the OCCIPITAL CORTEX, the back of the cortex, the patient sees flashes of light. And if, by surgery or by an accident, the occipital cortex is entirely destroyed, the patient is blind. These effects are limited to a well-defined area, and show that this particular piece of cortex is involved in vision. In mammals the cortex also contains processing zones for other major sensory processes *(Figure 23)*. Just as the hindmost portion is involved in visual events, portions on the side are involved in auditory events, and a strip across the top of the brain in the bodily senses. Just in front of this strip is a cortical area involved in motor activities, with any one piece connected (via numerous relays) to some particular group of muscles.

Substantial areas of the cortex, especially in the frontal portion (frontal lobes) have no simple function; they seem to be involved in complex social senses such as propriety and good and evil. In people obsessed with feelings of guilt and inadequacy, an operation used to be performed in which these areas were destroyed or disconnected, giving a far more tranquil but overly careless personality. Nowadays this lobotomy is seldom done, because tranquillizing drugs perform the role in a less drastic and more reliable way.

Quite apart from housing the points of final analysis of sensory data, the cerebral cortex is the seat of integration of all these sensory modalities and of memory. But these integrative and memory

processes of the cortex are not localized, as we shall see in discussing memory. When the smell of dead leaves evokes a memory of autumns gone by, it takes place in the cortex. But the emotional counterparts of these sensory experiences are probably not cortical; the visual memory of autumn is cortical, but the sad nostalgic feeling is subcortical (since the cortex is anatomically on top, all noncortical events are termed subcortical). The intellectual component is cortical, the emotional component is subcortical.

THE HYPOTHALAMUS

Lying close to the center of the brain is the area known as the HYPOTHALAMUS. The outstanding work of the Swiss neurophysiologist W. R. Hess (who won a Nobel prize in 1949) involved stimulation of this area with deep electrodes. The hypothalamus was thus shown to contain a variety of control centers for basic bodily functions and for certain drives. In Chapter 18 it was mentioned that here are located the temperature-sensing devices that detect fluctuations from an acceptable temperature range, and trigger suitable compensating responses. Nearby centers regulate blood sugar, water balance, carbohydrate and fat metabolism, and blood pressure. The efferent nerves from these centers, which turn on and off the appropriate machinery to accomplish control, operate the sympathetic and parasympathetic branches of the autonomic nervous system. The hypothalamus is thus the point of origin for the systems that maintain the constancy of the internal environment.

These various centers are made up of quite small groups of cell bodies, which are hard to locate since they lie deep within the brain. The procedure used in locating them was to implant long electrodes, stimulate the areas and observe the consequences; and later to remove and slice the brain to discover where the tip of the electrode had been located. In the course of such explorations, Hess and later others have located centers whose stimulation leads to more complex responses than the physiological regulations described above. These complex responses include behavior apparently designed to deal with some need, such as the need to satisfy hunger, thirst or sexual appetite. These behaviors are not stereotyped; they involve a response which meets the need in an appropriate way.

There appears to be a hunger center which when electrically stimulated will cause even an overfed animal to go on eating; and a satiety center, which

23 MAP OF CORTEX indicates the major sensory and motor areas on the surface of the human brain. The symbols OOO and TTT mark locations of olfactory and taste areas that lie beneath the surface.

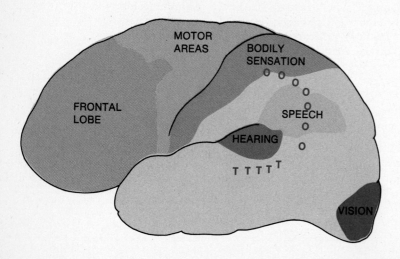

when electrically stimulated will cause even a starved animal to refuse food. There is a thirst center, whose continuous stimulation can cause animals to drink even salty water continously. Even more exciting (but not universally accepted) was the finding by James Olds, then at McGill University, that there is a pleasure center in the hypothalamus. Olds equipped rats with a lever which they could depress at will, and thus stimulate electrodes implanted in their pleasure centers. The rats lost interest in everything except pressing this lever, which they did at rates up to 7000 an hour, and ignored hunger and thirst in their orgy of self-excitement. Such electrodes have been implanted in humans suffering from severe chronic depression. For a time the attempt was a great success, but the stimulation led to progressively less pleasure as time went on. Olds further indicated the existence of a punishment center whose stimulation in rats leads to apparent rage, fright and pain.

The obvious conclusion to be drawn from the above experiment is that important drives, such as hunger, thirst and perhaps pleasure, are mediated through specific circuits of neurons, and perhaps each circuit is localized in a discrete hypothalamic region. If so, much of our lives may be said to be devoted to bringing these neurons to a desirable status. Is the search for happiness merely an effort to produce a particular firing pattern in a few hundred neurons? This conclusion is not universally accepted. Valenstein has claimed that the specificity in these hypothalamic studies is more apparent than real, and is to some extent forced by the experimenter, who sets the scene for the appetitive response he seeks to measure. Valenstein claims that stimulation of a single area can elicit different drives depending upon circumstances, and he believes that the hypothalamic stimulation serves to force reiteration, which can be applied to several quite different behaviors. Clearly the issue is not settled yet; the degree of plasticity (the ability to perform different functions) or of stability of these hypothalamic circuits remains uncertain.

THE CEREBELLUM

Let us consider other localized activities. One fairly well understood system is the CEREBELLUM (*Figure 24*). Put briefly, its function is to modify instructions to motor systems received from the cortex, in the light of the body's status as indicated by proprioceptors. For example, the primary command to pick up a piece of bread on the far side of a table originates in one's motor cortex. But the execution of the command requires not only the stretching out of an arm, but literally dozens of compensating small movements in the trunk, legs, and even toes so that full balance is maintained at all times, in spite of a substantial shift in the center of gravity. These automatic compensations, involving ceaseless small adjustments in the light of ceaseless feedback from proprioceptors, are performed by the cerebellum, which is thus a kind of silent administrator of the motor system.

THE LIMBIC SYSTEM AND THE EMOTIONS

A group of brain structures, collectively known as the LIMBIC SYSTEM, seems to play a large role in emotion (*Figure 24*). The structures form a double loop approximately around the center of the brain and include part of the inner border of the cortex. It was in 1937 that Papez of Cornell proposed that the limbic system, in coordination with the hypothalamus and other subcortical structures, constituted the neural basis of emotion, thus giving neurological support to the popular view that thinking and feeling were fully separable functions. Some of the evidence accumulated in support includes: 1) savage animals such as the agouti and lynx became docile when a particular part of the limbic system, known as the amygdala, was

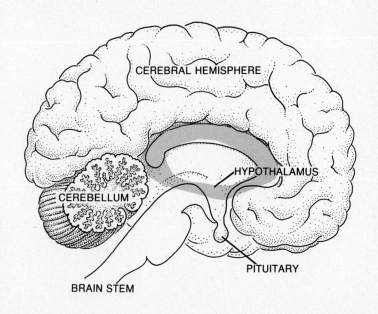

INTERIOR OF HUMAN BRAIN is partially exposed in this saggital section (a vertical cut from front to back). Zone of the limbic system is rendered in color. **24**

removed; 2) stimulation of the amygdala in cats produced aggressive responses such as retraction of ears, hissing, and clawing; 3) in monkeys, removal of parts of the cortex close to the amygdala, along with the underlying amygdala, led to a remarkable lack of response to objects which usually caused fear, such as snakes, strangers, and dogs. But other emotional activities, especially sexual kinds, were greatly intensified, and the animals showed frantic sexual activity.

THE RETICULAR FORMATION

Figure 25 shows a neuronal system which is localized in one sense, but is not a compact area like the cerebellum. Instead it is a fairly diffuse series of neurons which runs as a cylindrical net through the brain stem. Sometimes it is called the brain-stem reticular formation (reticulum means net), sometimes the RETICULAR ACTIVATING SYSTEM or RAS.

The RAS is a system concerned with attention and arousal. A condition of alert wakefulness occurs when the RAS sends impulses to the cortex. Put another way, activation of the RAS results in arousal. But apart from this highly generalized effect, the RAS monitors most of the impulses reaching and leaving the brain, and can quiet some down and energize others. For instance, when you concentrate upon some hard-to-sense event, such

as listening for a small pin to drop, your RAS gives great weight to auditory input, and for a while you may be hardly aware of what you are seeing or touching, or the status of your muscles, for the RAS has dampened these inputs. As well as monitoring inputs, the RAS can modify outputs to emphasize muscular response in one system and deemphasize it in another. If you try to pick up the dropped pin an exquisite degree of attention is given to very precise motor control in the fingers of the active hand, to the neglect of muscular precision in the unused hand.

SLEEPING AND DREAMING

Recently it has become clear that sleep and wakefulness have specific neural bases. The brilliant work of Jouvet in Lyons, France, made it clear that the sleep-wake cycle is not simply an off-on phenomenon. Instead there are three clearly distinguishable active states: wakefulness, light sleep and deep sleep. Each state is associated with the activity of a discrete set of cells in the brain stem. We have already seen that alertness, the full waking state, occurs when the RAS is firing. When sleep occurs, we (and all other sleeping animals) first move into light sleep. The body's muscles still have tone; that is, they still carry their burdens and maintain their flexion, but a different pattern of activity shows up on the electroencephalogram (the EEG or electroencephalogram is a greatly amplified record of the faint electrical waves which the brain generates as a result of the electrical activity in its millions of neurons). After a period of light sleep, most animals move into a new kind of

25 **RETICULAR ACTIVATING SYSTEM, also called the reticular formation, consists of several groups of neurons (light gray) in the brain stem. Drawing also shows the locus coeruleus (black) and nuclei of raphe (color), which regulate sleep. This schematic diagram depicts the brain of the cat as seen from below.**

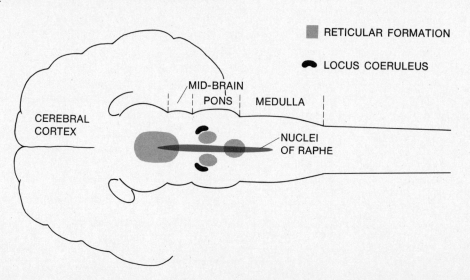

RETICULAR FORMATION

LOCUS COERULEUS

CEREBRAL CORTEX

MID-BRAIN
PONS
MEDULLA

NUCLEI OF RAPHE

sleep called deep sleep, REM sleep or dreaming sleep, in which the EEG pattern returns to a state quite like wakeful EEG. The muscles relax quite suddenly. You can see this in a reader who falls asleep in a chair after dinner; he will doze briefly, then suddenly the head drops and the book is dropped, as muscle tonus disappears. Often the jolt wakes him up. But although tonus is lost, a special muscular activity occurs; the eyeballs roll actively under the lids (hence the term rapid-eye-movement or REM sleep). Often there is flailing of the limbs: a dog lying on its side will start chasing a rabbit. In humans we know that dreaming occurs during REM sleep. Jouvet feels that as one's mind, in dreaming, plays over scenes which are commonly wish-fulfillments, the body accompanies the mental images with roughly appropriate limb movements.

Jouvet showed that the light sleep is due to the activity of a group of neurons whose cell bodies lie in the RAPHE, the center line of the brain stem (Figure 25). The neurons contain serotonin as their chemical transmitter and can be stained histologically. REM sleep is due to the activity of a different group of neurons occurring in the locus coeruleus (blue spot). These neurons can be stained differently, for they contain norepinephrine as their chemical transmitter.

It is entirely wrong to think of the brain as a clump of autonomous sub-organs, with the cortex thinking, the limbic system feeling, and the cerebellum balancing. Brain areas are richly interconnected, and they function with continuous interchanges.

As Figure 25 shows, the raphe neurons, locus coeruleus, and the RAS are anatomically close together. Nevertheless, they can be cut or destroyed individually. If one destroys the raphe neurons of a cat, light sleep disappears. And because one can only enter REM sleep via light sleep, the destruction forbids sleep of any kind. (In actual practice, it is hard to destroy all of this diffuse system, and thus hard to eliminate more than 90 percent of all sleep). If instead one destroys the locus coeruleus,

the animal can have light sleep but not deep sleep. So far no ill consequences have been found in animals deprived of REM sleep.

Although this discussion has stressed localization of function, it would be entirely wrong to think of the brain as a clump of autonomous sub-organs, with the cortex thinking, the limbic system feeling, and the cerebellum balancing. Brain areas are richly interconnected, and they function with continuous interchanges. For example, the hypothalamus contains centers of great importance to the emotions, and so the limbic system is not the sole organ of the emotions. The interplay of these two subcortical systems, and their subordination to the cortex and other areas, must all be taken into account in the neural basis of emotions.

MEMORY AND LEARNING

Memory and learning are so closely related as to be difficult to distinguish experimentally. One tests the learning of an animal by testing its ability to remember particular visual or auditory or other patterns, or recall particular sequences of acts. Yet there are undoubtedly different categories of memory. Two easily demonstrated kinds are short-term and long-term. We can readily remember a new telephone number for about as long as it takes to dial it, but within a few minutes it is lost. Such short-term memory is different from the quasi-permanent memory which stores the telephone number of our own home and a few other familiar numbers.

There seems little doubt that in mammals most or all learning is a function of the cortex. Surprisingly enough, some experiments suggest that specific memories seem not to be located in specific places in the cortex. The most celebrated experiments on this subject were those of Karl Lashley of Chicago. In 1929 he showed that in rats, the memory of how to run a maze was not destroyed by surgical removal of any particular piece of cortex; instead, the rats progressively lost the general ability to remember and to learn as more and more cortex was removed. Lashley called this the principle of mass action. But all learning tasks in all species do not follow this principle. The ability of monkeys to learn that a red square conceals food, but a green circle does not, seems to be localized in the temporal lobes of the cortex.

It is widely believed that memory is a consequence of pathway facilitation. For instance, when one sees a new and striking object it leads to some particular pattern of firing in the visual cortex and association areas. If one can re-play that same pattern precisely, presumably the original sensation will be experienced. The particular pathway (corresponding to the particular pattern just mentioned) is therefore replayed *in toto,* just as playing a recording involves sending the phonograph needle through precisely the original vibrations established when the recording was made. One can imagine that once some particular pathway through a complex network has been followed, the pathway can (when we recall the event) be followed again relatively easily. That is what is meant by pathway facilitation.

Flexner of the University of Pennsylvania found that various antibiotics such as puromycin, which block protein synthesis, when injected into particular brain areas of mice can prevent subsequent learning without affecting things already learned. This suggests that learning involves, at some step, a synthesis of protein. This technique has been elaborated by Agranoff at the University of Michigan, using goldfish. He found that short-term memory was not affected by puromycin, but long-term retention of that same memory was blocked. His findings support the view that the first step of memory formation provides a transient memory trace (short-term memory) which can be fixed in a second step. This second step is the one that requires protein synthesis.

Whether or not the long-term memory trace itself is localized, the fixation site seems to be localized in the hippocampus. It has long been known that a human with damage to the hippocampus acts precisely as one who has lost his ability to fix short-term memory. Such an individual can remember a seven-digit number perfectly well for a few minutes, but cannot recall what he had for breakfast a few hours ago.

EFFERENT SYSTEMS

The systems responsible for carrying out the instructions of the CNS are the EFFERENTS: neurons whose impulses flow from the CNS. Many of them are motor neurons, which control muscles, but others affect glands or modulate sensory inputs. These glands and muscles, collectively called effectors, are considered in the next chapter.

Certain modified neurons secrete materials such as hormones in response to instructions from the CNS. These efferent neurons do not merely control glands, in a sense they are glands. In chromaffin tissue, which plays an important regulatory role by secreting adrenaline and noradrenaline,

one can see how this glandular function is a simple modification of normal neuronal function. The chromaffin tissues are widely distributed, being found in patches all the way down the spinal cord, but especially in a fairly large conglomeration making up the medulla (central portion) of the adrenal gland. Remember that the final terminals of the sympathetic nervous system have noradrenaline as their transmitter substance. In such cases the transmitter acts upon the cells to which the neuronal ending is attached. In chromaffin tissue there is no attached cell; the neuron has become modified to produce such large quantities of noradrenaline (and even more of its methylated derivative, adrenaline) that it can affect cells throughout the body by releasing these hormones into the blood.

In this case the secretory neurons have been so profoundly modified as to be scarcely recognizable as neurons. In invertebrates one finds literally hundreds of NEUROSECRETORY CELLS which are clearly neuronal, but contain additional granules or vesicles producing three types of special hormones: kinetic hormones, which control the actions of muscles or glands; metabolic hormones, which control metabolic processes; and morphogenetic hormones which control growth and development. These neurosecretory cells may be set in the brain itself, as is common in arthropods. Insects have important neurosecretory cells in the middle line of the dorsal surface of the brain. These cells are distinguished from their neighbors by a blue color when suitably illuminated, caused by the secretion which is contained in vesicles of about 2000 Å diameter. When the insect "decides" to moult (a decision based on day length and other data received by the CNS) these cells release brain hormone which then triggers other secretory cells outside the CNS to produce moulting hormone, and so initiates moulting.

Alternatively the neurosecretory cells may be grouped together in secretory nodules outside the CNS, as in the mysteriously named ganglionic-X-organ in the eyestalk of crustacea. This little bundle of neurosecretory cells performs many vital functions: it modifies the blood levels of Ca^{++} and phosphate and blood sugars, it has a restraining influence upon moulting, modifies the adaptation of retinal pigments to changes in light, decreases oxygen consumption of the whole animal, and controls the water uptake that tends to occur during moulting. Probably at least seven distinct hormones are involved in these activities, and there are doubtless others to be discovered.

READINGS

B. KATZ, *Nerve, Muscle, and Synapse*, London, McGraw-Hill, 1966. A magnificent little book dealing with a limited area: transmission of impulses along nerves and across synapses. Rather physical-chemical in approach, very authoritative and reliable.

S.W. KUFFLER AND J.G. NICHOLLS, *From Neuron to Brain*, Sunderland, MA, Sinauer Associates, 1976. A splendid paperback by two outstanding physiologists. Emphasis on neurophysiology, in which it is authoritative and comprehensive. Rather short on biochemistry and pharmacology.

K. OATLEY, *Brain Mechanisms and Mind*, New York, Dutton, 1972. A somewhat popular, richly illustrated paperback that stresses the integrative functions of the brain. Good historical perspective.

T.C. RUCH, H.D. PATTON, J.W. WOODBURY AND A.L. TOWE, *Neurophysiology*, Philadelphia, W. B. Saunders Co., 2nd Edition, 1965. A very thorough treatment from a physiological viewpoint, i.e. stressing function and organization rather than anatomy or biochemistry.

F.O. SCHMITT, *ed.*, *The Neurosciences*, New York, Rockefeller University Press, 1970. A large (1068 pp.) and expensive compendium, with 87 chapters by some of the best neurobiologists in this country. It contains the substance of an intensive study program designed to introduce neurobiology to other scientists, so each chapter is quite intense and detailed. It leans more to the molecular aspects than do most books.

The failing heart cannot pump blood adequately; is it due to failure of the fuel-synthesizing machinery? failure of fuel distribution? failure of fuel consumption? a defect in the contractile machinery? The answers cannot be given until the entire machinery of contractile tissue is understood.
—Phillip Handler

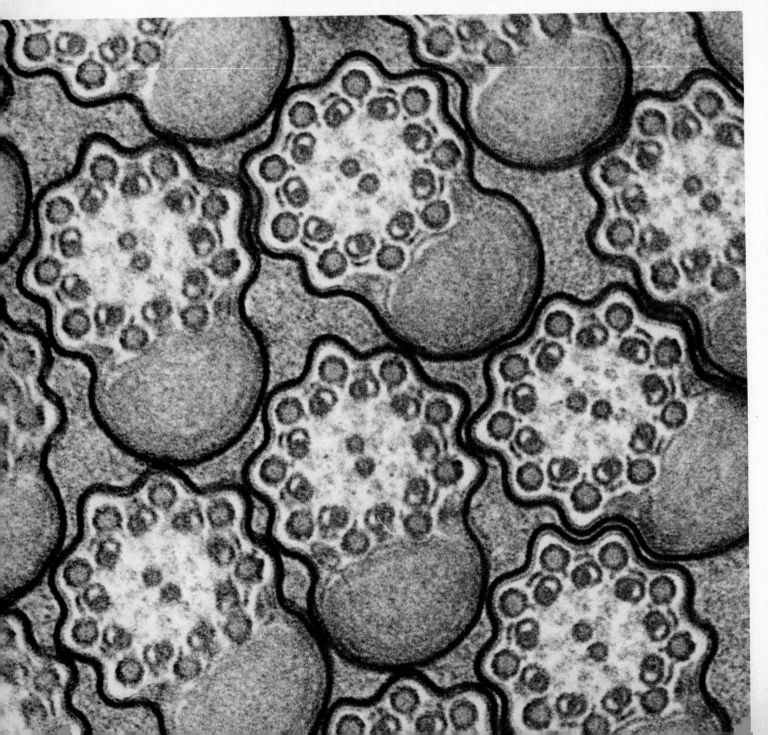

20
Effectors

Multicellular organisms have evolved an astonishing array of specialized "hardware" to carry out the commands of the nervous system, as well as to perform a variety of tasks that are not under direct nervous control. Ciliated cells create mini-currents that sweep fluids from place to place; amoeboid cells engulf bacteria; other cells change the color of an organism, secrete poisons, or emit light. Special organs create sound or generate electric discharges. Such specialized structures are collectively called EFFECTORS. At best it is a very loose category because of the extreme diversity of the structures and of their evolutionary origins. From a human viewpoint perhaps the most important effector is muscle. Because muscle serves so many important functions in man, from breathing and pumping blood to walking and speaking, it has been the subject of intensive research. Recent work has elucidated the mechanism of muscle contraction at the molecular level. A logical place to begin the study of effectors is with ciliated and amoeboid cells —evolutionary carryovers from unicellular life.

CILIATED CELLS

Cilia are common among both unicellular and multicellular organisms. In many unicellular organisms, such as *Paramecium*, they are the only means of locomotion. Flatworms and mollusks use cilia together with the rippling motion of muscles to glide over surfaces. But most multicellular organisms use cilia not to move the cell, but to move a liquid past the surface of a stationary cell. Ciliated cells circulate water through the gills of

SPERM TAILS of an insect are depicted in the electron micrograph on opposite page. Cross-section view reveals the arrangement of filaments within the flagella.

mollusks, and through the bodies of living sponges, which feed on the tiny organisms that they filter from the water *(Chapter 27)*. Many groups of invertebrates feed by trapping microorganisms in mucus and swallowing it, or by sweeping microorganisms directly into the mouth —all with the aid of cilia. In man the respiratory passageways of the lungs and tracheae are lined with mucus-coated cells equipped with hundreds of cilia *(Figure 1)*. The beating of these tiny hairlike structures helps to keep the passages clear by carrying mucus and trapped dust upward toward the throat, where it can be coughed up and spat out or swallowed. Similarly, ciliated cells lining the nasal passages carry dust and mucus downward toward the throat. (Cigarette smoke somehow inhibits normal ciliary action in the respiratory tract, and thus impairs the mechanism for removing foreign matter from the lungs.) Ciliated cells also line the female reproductive tract, creating currents that carry the egg from the ovaries to the uterus.

A cilium pushes against a liquid with the same basic motion as a swimmer's arms during the breast stroke *(Figure 2)*. During the EFFECTIVE STROKE the cilium is held stiffly outward and moved backward through the liquid, propelling the cell forward. During the RECOVERY STROKE the cilium is folded as it returns to the original position. Cilia bend forward slowly and lash backward suddenly. Because the resistance to the motion of an object through a fluid is proportional to the square of its velocity, the resistance to the slow recovery stroke of the cilium is very slight compared with the resistance to the fast effective stroke. Fluids exposed to the beating of cilia are thus propelled in the direction of the effective stroke. Cilia typically beat in coordinated waves. At any particular moment some cilia of a cell are moving through the effective stroke and others are recovering, which assures a steady flow past the cell surface. The mechanism

433

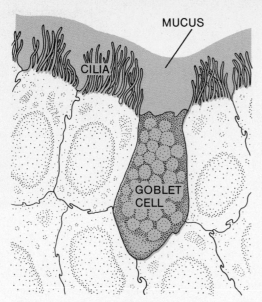

1 **CILIATED CELLS** in the human trachea. Dust and other foreign matter are trapped in film of mucus (color) secreted by goblet cells. Rhythmic beating of cilia sweeps mucus upward toward throat.

that coordinates the beat is not yet established. Suffice to say that the action of these cells appears to be automatic, and is under some form of chemical rather than nervous control.

Most cilia contain two central filaments surrounded by nine pairs of outer ones. The arrangement is virtually identical in all eucaryotes. The only variations involve the central tubules. In some organisms there is only one central tubule, and in others, none at all. A total lack of central tubules is characteristic of certain modified cells which are non-motile, such as those that form the light-sensitive elements of the retina.

2 **STROKE OF A CILIUM** resembles the motion of a swimmer's arms. During the effective stroke (left) cilium is extended stiffly and pushes backward against the fluid medium. During recovery stroke (right) cilium is folded, minimizing resistance to its motion through the fluid. In unicellular organisms, backward movement of cilium propels cell forward. When ciliated cell is fixed in place (as in human trachea), backward lash of cilium pushes fluid in the direction of effective stroke.

> Cigarette smoke inhibits normal ciliary action in the respiratory tract, and thus impairs the mechanism for removing foreign matter from the lungs.

Each pair of outer filaments has two tiny hook-like projections. The hooks, composed of a protein called dynein, are involved in contraction *(Box A)*. The cilium bends when the pairs of filaments on one side slide past their neighbors, which remain extended. The bending of the cilium produces the power stroke that drives it through the surrounding fluid *(Figure 3)*. The central filaments may serve as telegraph lines that conduct impulses along the length of the cilium and coordinate the contraction of the outer filaments.

At the base of the cilia lie basal bodies, or kinetosomes, composed of microtubules arranged differently. Nine triplet tubules form the circular rim of the kinetosome, and there are no central tubules. This arrangement is also found in centrioles, mysterious organelles that play a key role in cell division *(Chapter 2)*. Kinetosomes and centrioles are probably identical in structure. When an unciliated cell develops into a ciliated one, the centrioles may increase in number and migrate toward the cell surface, where they become kinetosomes.

When a cilium occurs by itself or with only a few other cilia on the surface of a cell, it is referred to as a FLAGELLUM (plural, flagella.) Flagella possess the same ultrastructure as other cilia. In most cases they are longer, and employ different forms of stroke. The possession of flagella rather than cilia is a primary criterion for separating two of the largest phyla of protozoans, the Mastigophora and Ciliophora *(Chapter 25)*. Flagella also power the movement of the gametes of many kinds of organisms, including the sperm of man and other vertebrates.

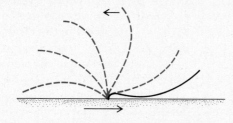

Rigid sperm and displaced hearts

Recent evidence suggests that a peculiar genetic disease of humans is caused by the inability to manufacture the little dynein hooks on the doublet microtubules of cilia and flagella. Men who suffer from this condition make paralyzed sperm, suffer from chronic bronchitis and ear infections, and often have the position of their internal organs inverted, as if their viscera were a mirror image of the usual human body arrangement. How can the absence of dynein hooks account for such diverse symptoms? The hooks play a central role in the apparatus that imparts movement to cilia and flagella. Without hooks, the flagella of the spermatozoa are stiff and immotile. The affected spermatozoa cannot swim like normal ones. A similar paralysis seems to affect the cilia that cover the surfaces of the respiratory tract and the ear ducts. These cilia play an important part in clearing away bacteria and foreign particles that penetrate into these vulnerable areas. Evidently the failure of the ciliary housekeeping apparatus leads to bronchitis and ear infections.

It is much harder to explain the tendency of such individuals to have their innards backwards. One hypothesis suggests that the nonfunctioning cilia change the course of embryonic development. The complex movements of cell layers during embryonic life might be influenced by the cilia on the surfaces of the cells. According to the hypothesis, the beating of cilia imparts a rightward twist to the developing embryo; this twist leads eventually to the placement of the heart to the left, the liver to the right, and so on. In embryos that develop without the beating of cilia, it would be a matter of chance whether the developing viscera turned left or right. Thus about half of such individuals would have visceral reversal.

AMOEBOID CELLS

The one-celled amoeba, a common denizen of pond water, moves about by extending lobe-shaped projections called PSEUDOPODS and seemingly pouring itself into them. This AMOEBOID MOVEMENT characterizes the phylum Sarcodina, to which all amoebas belong *(Chapter 25)*. It also occurs widely in higher organisms, particularly in migrating cells during embryonic development, as well as in blood. Human white blood cells, for example, depend upon amoeboid movement to pursue and capture bacteria and other infectious microorganisms in the blood *(Figure 4)*. The molecular basis of amoeboid movement has yet to be demonstrated with certainty. Of several theories still in contention, all recognize that the movement is accomplished by the forward flow of a relatively liquid phase called ENDOPLASM, located in the core of the organism, followed by its spread to the periphery, where it is converted into a stiff outer layer called ECTOPLASM. The endoplasm moves the pseudopodium forward, and the ectoplasm fixes it in position *(Figure 5)*. But how does the endoplasm move? According to one theory, it consists of rows of molecules that slide forward against the inner surface of the ectoplasm. The chemical structure of

the ectoplasm molecules acts like a ratchet, allowing the ectoplasm to progress forward but not backward.

STRUCTURE OF A CILIUM is depicted in the diagram below. Right, a cross-section cut along the plane indicated in the diagram. Cross-section depicts flagella from tails of guppy sperm, and shows the typical 9 + 2 arrangement of filaments. 3

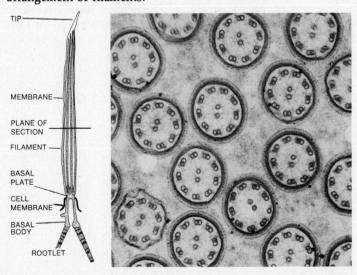

TIP

MEMBRANE

PLANE OF SECTION

FILAMENT

BASAL PLATE

CELL MEMBRANE

BASAL BODY

ROOTLET

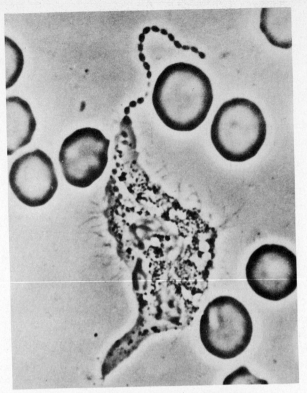

4 HUMAN WHITE BLOOD CELL engulfs a bacterium (the chain of streptococci at top) by amoeboid movements. Dark disks are red blood cells.

A rival theory holds that the endoplasm is squeezed forward by the contraction of rearward molecules, and moves to the front in essentially the same way as toothpaste squeezed from a tube. A third theory, which has received strong experimental support from its author, R. D. Allen of the State University of New York, views the process as

5 AMOEBOID MOVEMENT depends on the streaming of semiliquid endoplasm (horizontal lines) toward the "forward" end of the cell, where the fluid is converted to stiff gel-like ectoplasm (vertical lines). Endoplasm pushes the pseudopod forward and ectoplasm fixes it in place. Ectoplasm is continuously displaced toward rear of cell, where it is converted once again into fluid endoplasm.

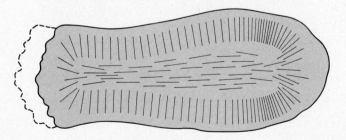

the result of contraction of the ectoplasm at the front of the pseudopod. Allen maintains that the contraction proceeds rearward in short waves from the tip of the pseudopod, an action that pulls the more liquid endoplasm forward to the tip.

THE SKELETON

Organisms larger than amoebas need some sort of rigid or semirigid skeleton to support them against the pull of gravity. The skeleton also provides a firm body part against which muscles can pull. The simplest type of skeleton is found in the earthworm. Called a HYDROSTATIC SKELETON, it consists not of a chain of bones but rather of incompressible liquid trapped in each of its body segments. The walls of each segment are equipped with two sets of muscles, one running lengthwise and the other one encircling the segment *(Figure 6)*. Contraction of the longitudinal muscles shortens the segment, and the incompressible fluid within forces the wall of the segment to bulge outward. (Engineers apply the same principle in designing the hydraulic brakes of automobiles.) Contraction of the circular muscles has the opposite effect, squeezing the segment into a narrow elongated tube. The earthworm moves its body along first by narrowing the segments, so that they extend forward in space, then by anchoring them into the soil with bristles and contracting the segments, thus pulling forward the segments in the rear. By passing alternating waves of contraction and extension steadily along the length of its body, the earthworm is able to make fairly rapid progress through the soil or over its surface.

How can a hydrostatic skeleton be converted into a still more efficient form? One way would be to harden the walls of the segments and to add legs to them. Muscles attached to the walls could then move the legs back and forth. As odd as the idea may seem at first, this is precisely the step taken in the evolutionary origin of the phylum Arthropoda (crustaceans, insects, and related forms) from the phylum Annelida (earthworms and other segmented worms). The hardened walls of the body and appendages are referred to as the EXOSKELETON ("outer skeleton"). In arthropods the exoskeleton consists of CUTICLE, a distinctive layered structure comprised of a thin, waxy EPICUTICLE, which protects the body from drying, and a thicker inner layer, the ENDOCUTICLE, which forms the bulk of the structure. The endocuticle is a tough, pliable material found only in arthropods. It consists of a

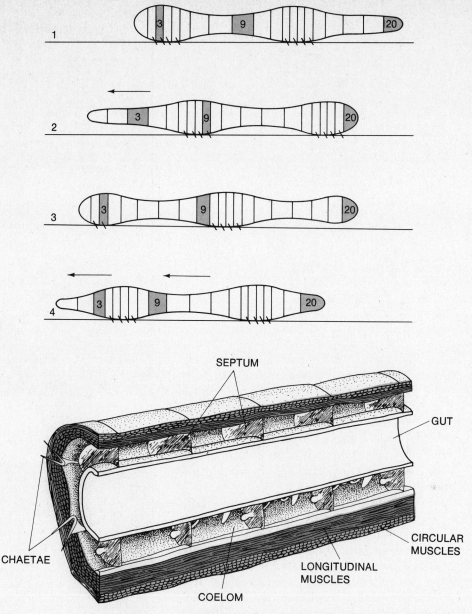

complex of protein and CHITIN, a complex nitrogen-containing polysaccharide. The muscles are attached to the inner surface of the arthropod exoskeleton *(Figure 7).*

THE VERTEBRATE SKELETON

The skeletal arrangement of vertebrates is the exact opposite of that in insects and other arthropods. It is an ENDOSKELETON, an inner scaffolding to which the muscles attach externally. It consists of two kinds of supportive tissue, CARTILAGE and BONE. Cartilage ("gristle" is the more familiar vernacular term) consists of cells widely separated by a rub-

LOCOMOTION OF EARTHWORM. Diagrams at top depict successive stages in locomotion. Contraction of circular muscles squeezes forward segments into a long tube (2). Contraction of longitudinal muscles then shortens the segment (3). Worm anchors it in the soil with bristles (chaetae) and pulls itself forward (4). Shading shows coordinated extension and contraction of selected segments. Longitudinal section of four segments of earthworm (bottom) shows chaetae and muscle layers (color).

6

bery matrix of mixed proteins and polysaccharides. Fibers run in all directions through the matrix, adding to its strength and resiliency. Cartilage occurs in parts of the body that require both stiffness and resiliency, such as the capping material on the

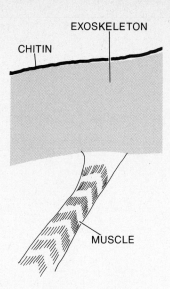

EXOSKELETON

CHITIN

MUSCLE

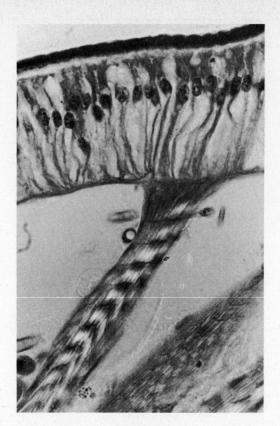

7 **MUSCLES OF INSECT are attached to the inside of its exoskeleton. Drawing at side indicates location of structures in photo.**

ends of bones, the walls of the larynx (voice box), and the protruding parts of the ear and nose. In sharks and rays, often referred to jointly as the cartilaginous fishes, the skeleton is composed entirely of cartilage. It also is the principal component of the embryonic skeleton of man and other vertebrates, being replaced gradually by bone in the course of development.

Bone, despite its dead appearance, contains large numbers of living cells. The calcium in bone is in dynamic equilibrium with that in blood; bone, in fact, is a storehouse of calcium. The bone matrix is rigid because it is impregnated with crystalline calcium phosphate and calcium carbonate. It is most prominently developed in large land-dwelling animals, where its strength is required for the heavier loads moved by their powerful muscles. The architecture of bone varies with its position and function. The shafts of the long bones, for example, consist of cylinders of hard bone surrounding marrow-filled cavities *(Figure 8)*. They are therefore both strong and light, the two prime requirements for efficient locomotion. The cavities

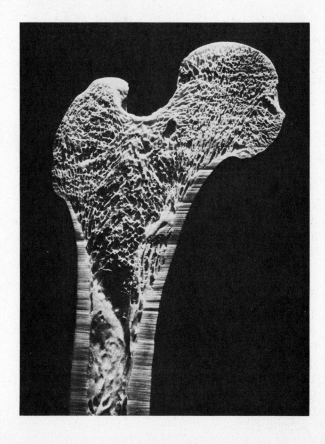

8 **HUMAN FEMUR, the long bone of the upper leg, is depicted in longitudinal section. Cylinder of dense bone (bottom) surrounds the marrow-filled interior of the shaft. Marrow manufactures red blood cells. Spongy bone that comprises the head of the femur combines both strength and lightness.**

are not idle spaces; their marrow is employed in the manufacture of red blood cells for the entire body. The more compact forms of bone are composed of structural units called HAVERSIAN SYSTEMS. Each system is a massive set of concentric bony sheets, or lamellae. Through its center runs a narrow canal containing the blood vessels *(Figure 9)*.

The very rigidity of bones requires the existence of two supplementary forms of binding connective tissue: TENDONS and LIGAMENTS. Both contain large numbers of fibers oriented in the same direction, a property that allows them to be pulled and folded strenuously back and forth without breaking. Tendons bind the muscles tightly to the bone. As the muscle contracts, tendons absorb most of the stress along the lengths of their parallel fibers. Ligaments join bones together. Like leathery hinges, they hold the ends of the bones in position while allowing them sufficient play to bend freely and even to rotate to a limited extent *(Figure 10)*.

MUSCLE

Multicellular animals that rely solely on the beat of cilia for locomotion are confined to a sluggish, creeping existence. The great majority of higher animals, from roundworms to man, employ muscle cells, which are contractile units specialized to move other organs in the body. Three types of muscles can be identified on the basis of differences in the structure of their cells *(Figure 11)*. The simplest are SMOOTH MUSCLES, whose long and spindle-shaped cells are packed with contractile protein which is nearly indistinguishable under the light microscope from the remainder of the cell contents. Smooth muscle cells are evolutionarily the most primitive of the three types. They occur in many of the lower animal phyla, as well as in the walls of the intestine and blood vessels of vertebrates.

In more complexly organized animals, such as vertebrates and insects, muscle cells are often fused into single large units containing many nuclei. Each such unit is referred to as a MUSCLE FIBER, and it is capable of swifter and more powerful contractions than a mass of smooth muscle of comparable size. In some cases the contractile proteins of the muscle fibers are very regularly arranged, forming repetitive sequences of bands, or striations, oriented at right angles to the fiber. Such organs are called STRIATED MUSCLES. Most vertebrate muscles belong to this category, including all those that move the bones and hence provide the means of locomotion.

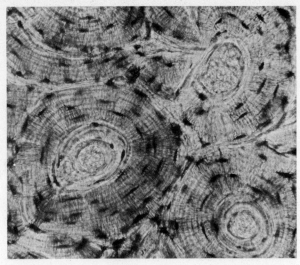

HAVERSIAN SYSTEMS are structural components of dense bone. Cross-section shows concentric lamellae surrounding channels that contain blood vessels. Magnification, about 130X. 9

The third form of muscle that is most commonly recognized is CARDIAC MUSCLE, found in the hearts of vertebrates. To keep the blood circulating evenly, cardiac muscle pulses in rhythmic contractions. It is a peculiar form of striated muscle, composed of profusely branching and interconnected muscle fibers. Each fiber is divided by thin partitions representing the boundaries of the individual cells. Cardiac muscle combines the characteristics of both striated and smooth muscle tissue.

TENDONS AND LIGAMENTS of the human knee 10 **joint. Tendons connect muscles to bone. Ligaments are fibrous "hinges" that bind bones together.**

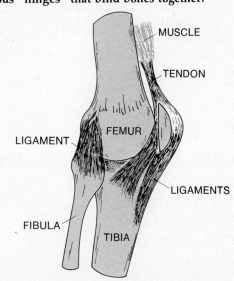

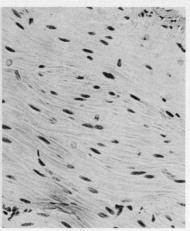

11   THREE TYPES OF MUSCLE. At left, smooth muscle from a snake. Center, striated muscle from a mammal. Right, cardiac muscle. (Photos at left and right, courtesy of Carolina Biological Supply House.)

The evolution of the muscular system is a story of a gradual increase in power, speed, and complexity of action. The invention of muscle cells was a dramatic improvement over the use of cilia, and the origin of muscle fibers represented an advance beyond the exclusive use of smooth muscle cells. In the course of further evolution, as the bodies of animals became larger and more complicated, additional mobility was achieved by alterations in the ways the muscles act. In its most elementary form, muscular movement consists of little more than an alternating contraction and relaxation of sets of muscles. By this means a roundworm can wriggle through a solid medium, a jellyfish can push through the water by expanding and then shutting the umbrella-shaped bell that makes up most of its body, and a snail can slide forward by riding upon the rippling waves of contraction in the muscles of its flattened foot. The next evolutionary step was the addition of a skeleton.

The invention of muscle cells was a dramatic improvement over the use of cilia, and the origin of muscle fiber represented an advance beyond smooth muscle cells.

MUSCLE FUNCTION

One of the basic attributes of muscle cells and fibers is that they are able to exert force in only one direction. Another is that they contract actively and relax passively; that is contract only when they are stimulated. These complex patterns of movement required for locomotion and other behavioral acts consequently require two sets of muscles that move the body parts in opposite directions. Muscles that work against each other in this way are said to be ANTAGONISTIC. The clearest examples of antagonistic action are found in the motion of the legs and other appendages in arthropods and vertebrates. Each joint is bent ("flexed") by one or more FLEXOR MUSCLES and straightened out ("extended") by EXTENSOR MUSCLES, as shown in Figure 12.

The function of muscle is to contract at the appropriate moment, and hence either move some skeletal part to which it is attached (as when one stretches out the arm) or else cause a change in shape or size of some soft tissue (as when the heart beats or the pupil of the eye dilates). In all cases studied it appears that the contraction occurs by the sliding past each other of two kinds of filaments within a cell. H. E. Huxley of Cambridge University has worked extensively with vertebrate striated muscle, and much of the material that follows is the result of his work.

The skeletal striated muscle of vertebrates is under voluntary control, and ordinarily contracts when stimulated by nerves. Such muscle is made up of bundles of muscle fibers, typically 10 to 100 μ in diameter, which often run the whole length of the muscle. Each fiber contains a set of four to 20 filaments called MYOFIBRILS, about one micron in diameter, bathed in cytoplasm (which in muscle is often called SARCOPLASM). Numerous mitochondria are scattered through the sarcoplasm. The banded or striped appearance of the muscle changes when the muscle contracts *(Figures 13 and 14)*. Study of the changing widths of these bands led to the discovery of the contractile mechanism.

The myofibril is the actual contracting unit. In vertebrates it contains two thin filaments of about 50 Å diameter for each thick filament of about 100 Å diameter (1 μ = 10,000 Å). The thin filaments are 2 μ long and made of a protein called ACTIN; the thick ones are 1.5 μ long and made of a protein called MYOSIN. When viewed in cross section, the thin actin filaments are arrayed hexagonally around the thick myosin filaments. The two kinds of filament are connected with each other along the longitudinal axis at 60 Å intervals by bridges.

The contraction of a muscle is caused by the contraction of its myofibrils. The actin and myosin filaments slide over one another, reducing their combined lengths. The thin actin filaments are anchored to a structure that microscopists call the Z

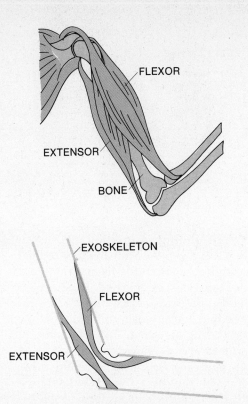

12 FLEXOR AND EXTENSOR MUSCLES in man (top) and insect (bottom). Flexors bend the joint; extensors straighten it. Muscles of vertebrates are attached to outside of skeleton; muscles of insects, to the inside.

STRIATED MUSCLE. Electron micrograph depicts skeletal muscle from the tadpole, magnified about 14,000X. The ribbons extending from left to right are thin longitudinal sections of muscle fibrils. Spaces between fibers contain vessels that comprise sarcoplasmic reticulum. Diagram at side identifies bands. 13

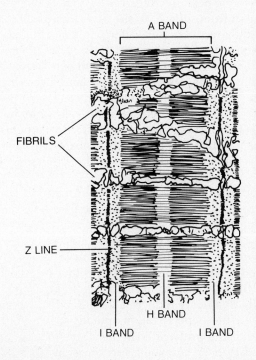

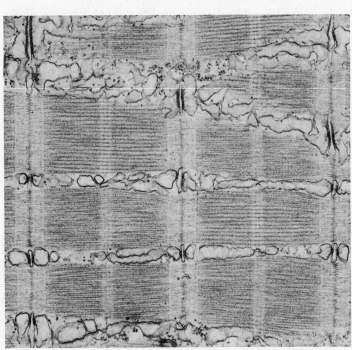

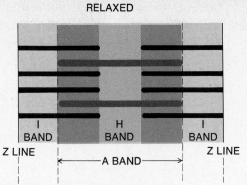

RELAXED

CONTRACTED

I
BAND

H
BAND

I
BAND

Z LINE

A BAND

Z LINE

Z

Z

14 DIAGRAM OF SARCOMERE schematically depicts bands seen in micrographs of striated muscle. Z lines are vertical bands at either end. Broad A band lies in middle of sarcomere, flanked by light I bands. When muscle contracts, I and H bands almost disappear. Horizontal black filaments are actin; horizontal colored filaments, myosin. The sarcomere is the basic contractile unit, bounded by the Z lines.

line (*Figure 14*). By sliding along the unanchored thick myosin filaments, the two sets of thin ones (and hence the attached Z line) are drawn toward each other, shortening the distance between Z lines without shortening any of the filaments. As the diagram indicates, these events lead to several visible changes. The pale I band, which contains actin filaments that do not overlap myosin ones, is reduced to almost nothing by the contraction. The broad dark A band, which contains adjoining actin and myosin filaments, approaches the Z lines, causing the I band to disappear. The pale H band at the center—the gap between the two sets of actin filaments—also virtually disappears in a contracted muscle, because the sets of filaments almost meet.

So much for the structural changes in whole muscle, as observed in studies of thin sections frozen at various stages of contraction, and in x-ray studies of living muscle. What mechanism explains the sliding of filaments during contraction? And what triggers the sliding?

One can extract myosin and actin separately from muscle. If one mixes them small fibrils form. When ATP is added the fibrils contract, and myosin splits a phosphate from the ATP. Evidently the contraction of whole muscle is fueled by ATP. An interesting complication is that unless one uses highly purified actin and myosin in this experiment, Ca^{++} is also required to induce the filaments to contract. Whole muscle also shows this dual need of ATP and Ca^{++} to contract.

The exact way that ATP hydrolysis is geared to

filament sliding is unknown. Presumably the bridges that join the actin and myosin filaments are broken briefly, then the filaments slide and a new set of bridges is established. Huxley has proposed a ratchet model to explain the sliding process. He noted that each myosin unit consists of a globular head (heavy meromyosin) and a stringy tail (light meromyosin), as shown in Figure 15. Sets of such units aggregate tail to tail, forming a myosin filament with bundles of heads extending from each end. The projecting heads form visible bridges that bind to specific sites on the lighter actin filament. Huxley proposed that the two engage each other like a tooth and ratchet; specifically, that the meromyosin bridge is the tooth, and the binding sites along the actin filament are the ratchet. The meromyosin bridge can disengage itself momentarily from the ratchet, allowing the actin filament to slide to the next notch. Figure 15 depicts a single movement from site two to site three (contraction), followed by recovery (relaxation). In a more extensive contraction the filament could slide farther, engaging the bridge at a more distant site. Because energy is necessary for making and breaking bridges—that is, for engaging and disengaging the ratchet—ATP is required for contraction but not for relaxation. In the absence of ATP the filaments lose their sliding ability, which probably explains the phenomenon of *rigor mortis,* the stiffening of muscles in dead animals.

The contractile machinery is set in motion by a nerve impulse. The last chapter described how the arrival of the impulse leads to the depolarization of the muscle membrane immediately adjacent to the nerve ending. Because each nerve fiber is branched, a single one supplies up to a hundred muscle fibers, which comprise a MOTOR UNIT. When the firing of a nerve induces a local depolarization in a fiber, the depolarization spreads throughout the fiber, being propagated in the all-or-none fashion characteristic of axonic transmission (*Chapter*

19). Several milliseconds elapse between the initial depolarization and the contraction of the muscle. The spreading of the message involves the T-SYSTEM, a set of tiny transverse tubes that links

5 RATCHET MODEL is one proposed explanation of muscle contraction. Three drawings at top depict arrangement of myosin and actin filaments. Four drawings below depict the sliding of filaments during contraction (1, 2, 3) and relaxation (4).

FIBER OF MYOSIN, COARSE STRUCTURE

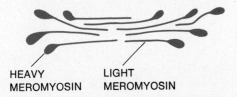

FIBER OF MYOSIN, FINE STRUCTURE

HEAVY MEROMYOSIN   LIGHT MEROMYOSIN

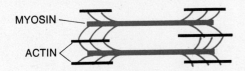

ACTOMYOSIN FIBRIL, COARSE STRUCTURE

MYOSIN

ACTIN

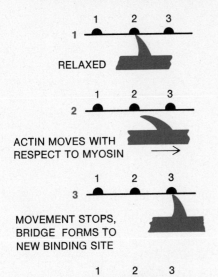

RATCHET MECHANISM OF SLIDING

1   RELAXED

2   ACTIN MOVES WITH RESPECT TO MYOSIN

3   MOVEMENT STOPS, BRIDGE FORMS TO NEW BINDING SITE

4   RECOVERY: ACTIN RETURNS TO STARTING PLACE

together a network called the SARCOPLASMIC RETICULUM that embraces each bundle of myofibrils and contains fluid rich in Ca^{++} *(Figure 16)*. The depolarization initiated by the nerve impulse apparently spreads throughout the T-system and leads to a release of Ca^{++} from the reticulum, allowing the Ca^{++} level in the myofibrils to rise to $10^{-6}M$. The high ionic concentration in turn activates the enzyme ATP-ase, which splits ATP and causes contraction; that is, it causes the actin and myosin filaments to slide.

The relaxation process that returns a muscle to its resting state is essentially a reversal of the sequence that leads to contraction. The reticulum recovers from depolarization and proceeds to reabsorb, with the aid of a biochemical pump, the Ca^{++} just released. When the Ca^{++} level is reduced below $10^{-6}M$ the activity of ATP-ase is turned off, the actin and myosin are easily dissociated, and the muscle can now be passively pulled back to its original position.

FAST AND SLOW MUSCLES

There are notable variations from this basic pattern of muscle contraction. Almost all animals, including vertebrates, have both "fast" skeletal muscles that contract rapidly in response to a nerve impulse, and "slow" muscles that contract about five times more slowly and are important in maintaining posture. The molecular basis of contraction is probably identical in both cases, but the slow muscles have less reticulum and take longer to pump down the Ca^{++} level to which the myofibrils are exposed.

Some flying insects, including flies and bees, have a specialized kind of skeletal striated muscle called asynchronous muscle, which operates their wings. These muscles contract much faster than "normal" skeletal muscles, achieving rates of up to 1000 cycles per second in the midge *Forcipomyia*, compared with up to 35 cycles per second in flying insects with synchronous muscles. There are two striking oddities about asynchronous muscle: although the T-system is present, the sarcoplasmic reticulum is virtually absent. And when the filaments or the muscle are removed and suspended in a solution containing Ca^{++} and ATP, they contract and relax in an oscillatory fashion. Thus the cyclic aspect of their action seems to be regulated by the fibers themselves and not under the control of Ca^{++} uptake by the cytoplasmic reticulum. Presumably the T-system simply initiates and later turns off the cycling behavior, rather than (as in

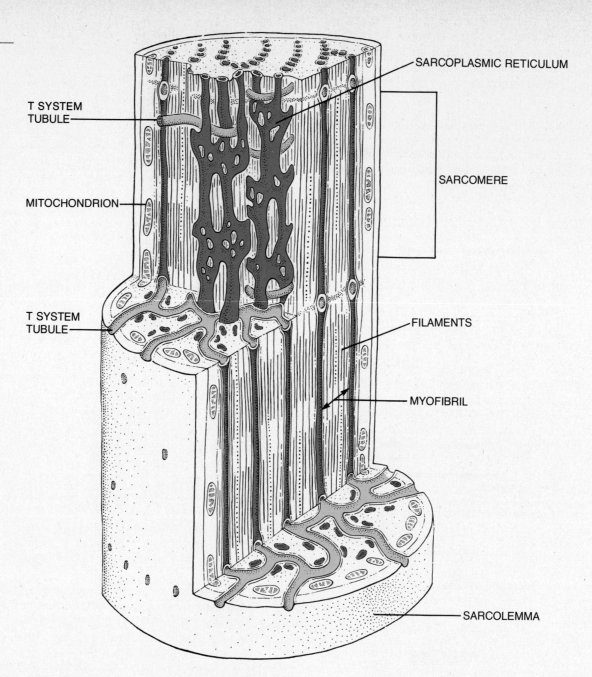

SARCOPLASMIC RETICULUM

T SYSTEM
TUBULE

SARCOMERE

MITOCHONDRION

T SYSTEM
TUBULE

FILAMENTS

MYOFIBRIL

SARCOLEMMA

16 **T-SYSTEM (light color) is a network of transverse
tubules that opens to the outside of muscle fiber.
T-system is joined to the tubules that comprise the
sarcoplasmic reticulum (dark color). Both sets of vessels
are involved in the flow of calcium ions within muscle.
Sarcolemma is membrane around fiber. Other
components of the muscle fiber are identified by labels.
Flow of calcium ions through vessels facilitates muscle
contraction.**

synchronous muscle) relaying information from
the reticulum to control each contraction.

CARDIAC MUSCLE

A special kind of striated muscle is found in the
hearts of vertebrates. Unlike skeletal muscle, cardiac muscle cells are branched, and mixed with
noncontractile cells. Cardiac muscle can contract
rhythmically even in the absence of nerve stimulation; it is "automatic" or "myogenic". Moreover,

the contraction of any one part, such as the ventricle, is almost simultaneous. This synchronous contraction is essential if the ventricle is to expel its content of blood rapidly and completely.

Two special features of heart muscle explain its automatic and synchronous properties. One is that the outer membrane of each muscle fiber lies in very close contact with that of its neighbor at special regions called intercalated discs, and here the fibers are in electrical contact with each other. Consequently, depolarization of any fiber sweeps rapidly throughout neighboring fibers, triggering contraction. The second special feature is the existence of a pacemaker: a small group of modified muscle cells that spontaneously and rhythmically produces a depolarization. The pacemaker of the mammalian heart is called the S-A node, or sino-atrial node, situated on the right atrium (*Figure 17*). The depolarization that it initiates spreads over the atria to another group of modified muscle cells, the A-V or atrio-ventricular node, located between the right atrium and the right ventricle. From the A-V node a bundle of special fibers carries the depolarizing wave swiftly to all parts of both ventricles. The S-A node is not unique in its ability to originate depolarization and hence contractions. If one destroys the S-A node, the A-V node may take over, but its rate is then slower. The ability to contract automatically is a property of all cardiac muscle cells, but in a normal heart the S-A pacemaker imposes its rate upon the whole organ.

The spontaneous beating of the vertebrate heart can be modified by the central nervous system. Nerve endings from the parasympathetic system release acetylcholine and slow the heart rate; and nerve endings from the sympathetic system release norepinephrine and speed the rate. Sites that respond to norepinephrine also respond to epinephrine (also called adrenalin) released into the blood by adrenal glands; that is why excitement or fear or surprise, all of which provoke the adrenals to secrete epinephrine, cause the heart to beat faster.

SMOOTH MUSCLE

Nonstriated or smooth muscle is made up of sets of spindle-shaped cells, each with its own nucleus, and without the banding characteristic of skeletal muscle. There are two types of smooth muscle. MULTIUNIT MUSCLES, found in the iris of the eye, and in the walls of many blood vessels, contract only when stimulated by a nerve or a hormone, and then many fibers contract quite promptly as a unit. The other type, VISCERAL MUSCLE, found in gut walls, contracts slowly and spontaneously. Although Ca^{++} and ATP are probably involved in the contraction, it is not known whether a sliding filament mechanism is involved.

Bivalve mollusks (such as clams and oysters) have developed a special kind of muscle to solve their special problem of how to hold a tremendous tension for very long periods of time, so that their shells can remain closed and protective against all but the most determined predators. The strange feature is that maintenance of this tension involves little more oxygen consumption (and therefore energy utilization) than in the resting state. This phenomenon has been considered in the past to involve some sort of latching mechanism; once the latch is set, no energy is needed to keep it closed. A recent version of this view is that the contracted muscle undergoes "setting"—a change of state, with a great increase of viscosity, rather like the setting of an epoxy glue. The muscles involved contain ribbonlike filaments of a specialized pro-

17 PACEMAKER OF HEART is the sino-atrial node (S-A node). Wave of depolarization spreads from S-A node to atrio-ventricular node (A-V node) and then, via the Purkinje fibers, to rest of heart.

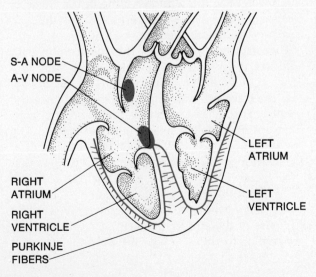

S-A NODE
A-V NODE
LEFT ATRIUM
RIGHT ATRIUM
RIGHT VENTRICLE
LEFT VENTRICLE
PURKINJE FIBERS

Despite its size (about a foot long) the gecko can scramble across ceilings and even hang upside down by one foot from a sheet of glass.

tein called paramyosin with an unusual amino acid composition (no tryptophan and little proline). But nothing is known about the molecular basis of their action.

THE TOE OF THE GECKO

Muscles are an intrinsic part of a vast array of other effectors, including some whose action has puzzled investigators for decades. One of the more unusual is the toe of the green and orange lizard *Gekko gecko,* a native of Malaya and the Philippines *(Figure 18).* Despite its size (about a foot long) the gecko can scramble across ceilings and even hang upside down by one foot from a sheet of glass. Zoologists have long wondered how this agile climber maintains its firm grip on virtually any surface, rough or smooth, including plaster, wood and concrete. Its feet do not have prominent suction cups like those on the toes of some frogs, but instead are ridged with chevron-shaped lamellae *(Figure 19).* Early workers reasoned that some sort of suction device must be present, but studies with the light microscope failed to reveal it.

The studies did reveal some anatomical peculiarities of the gecko's foot. When the gecko lifts its foot from a surface, it bends its toes backward and peels them away from the surface as one would peel a banana. The extensor muscles that open the foot are attached not to the bones, as in other animals, but to the skin at the ends of the

CLOSE-UP OF TOE reveals rows of chevron-shaped **19** lamellae. Surface of each lamella is covered with tiny bristles called setae. Magnification, 35X.

18 GECKO is depicted here about one-third actual size, clinging to the underside of a sheet of glass. Toe of left front foot has been removed for microscopic studies reproduced in Figures 19 to 22.

toes. When the extensor contracts, the lamellae fan outward and release their grip. The flexor muscles are attached to the bases of the lamellae, and are arranged so that the toes of the gecko exert pressure on any flat surface to which the animal clings. Even an amputated gecko limb can be made to adhere to a sheet of glass simply by attaching a small weight to the tendons of the flexor muscles. Only when the extensors are pulled will the toes curl back, releasing the grip.

The structure that gives the toe its powerful adhesive properties was revealed by the scanning electron microscope. Photographs made at low and intermediate magnification showed that the lamellae are covered with tiny double brushes consisting of fibers called setae *(Figures 20 and 21).* At a magnification of 35,000X, which lies at the limit of the instrument's resolving power, two tiny suction cups were visible at the end of each seta *(Figure 22).* Hundreds of thousands of such cups on each toe permit a very close fit between the toes and the contours of the surface. The cups are not set perpendicular to the seta like the suction cup on a dart, but at an angle which makes it easier for the lizard to break the grip of the cups as it peels back its toes. The scanning electron microscope revealed the structure of the cups, and solved an old mystery.

GLANDS

A gland is an organ or group of cells that is specialized to secrete a substance that is released from the cell. The two basic types are ENDOCRINE GLANDS, which secrete hormones more or less directly into the bloodstream and are therefore truly internal organs *(Chapter 18)*; and EXOCRINE GLANDS, which release their products into the digestive tract or outer surface of the body and are therefore to some degree external in their activity. Exocrine glands include the digestive glands *(Chapter 14)*, pheromone glands that produce odor and taste signals for use in communication *(Chapters 21 and 22)*, sweat glands used in the regulation of body temperature *(Chapter 17)*, and poison glands used by a great array of venomous animals, including certain mollusks, annelid worms, spiders, insects, fishes, amphibians, and even mammals (the shrew has a poisonous bite). Some plants also possess exocrine glands. The most familiar are the nectar glands found in flowers (floral nectaries) and scattered over stems and leaves (extrafloral nectaries).

20 SURFACE OF LAMELLA. Scanning electron microscope shows each of the setae as a double brush embedded in the lamella. Photograph is reproduced at a magnification of about 600X.

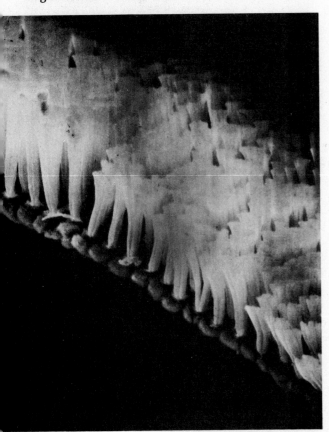

21

FINE STRUCTURE OF SETAE is visible at a magnification of about 3300X. Scanning electron microscope reveals a fringe of tiny cups, two on each of the fibers comprising the setae.

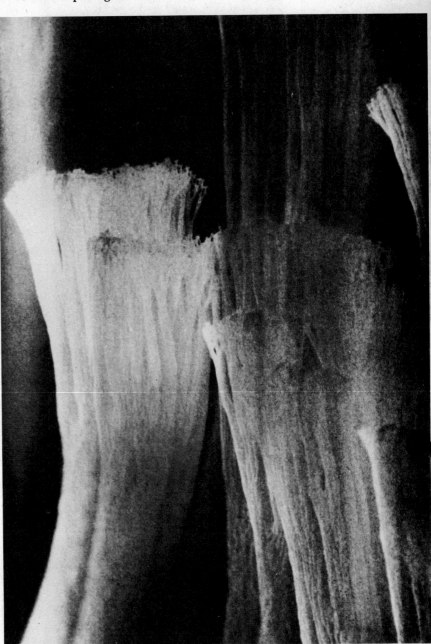

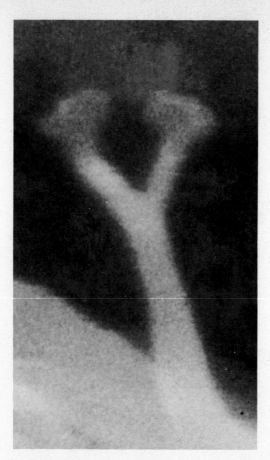

duced only by hydras, jellyfish and other members of the phylum Coelenterata *(Figure 23)*. They are concentrated on the outer surface of the arms of the animal in huge numbers. Each nematocyst is made up of a slender thread coiled tightly within a capsule, which is armed with a spinelike trigger projecting to the outside. When the body of a potential prey organism brushes the trigger the nematocyst fires, turning the thread inside out and exposing little spines along its base. The thread either entangles or penetrates the body of the victim, and a poison is simultaneously released around the point of contact. Once the prey is subdued, it is pulled into the mouth of the coelenterate and swallowed.

NEMATOCYSTS of coelenterates are barbed tubes that lie coiled within specialized cells called cnidoblasts (bottom). Contact with prey triggers the discharge of the nematocyst, which penetrates the tissue of the victim and injects a paralyzing poison. Discharged cnidoblast is shown at top. **23**

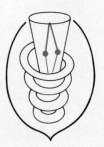

22 **TERMINAL CUPS of a single fiber are magnified 35,000X. Hundreds of thousands of such cups on each toe permit a very close fit between the foot of the gecko and irregular surfaces.**

One remarkable plant glandular system is illustrated in Chapter 14: the carnivorous sundew plant *(Drosera)* captures insects by trapping them in drops of sticky secretion produced by stalked glands that bristle from the surface of every leaf. The insects eventually die and the plant digests their decomposing remains.

TRICHOCYSTS AND NEMATOCYSTS

Some primitive animals possess highly specialized organs that are fired like miniature missiles to capture prey and repel enemies. Ciliated protozoans eject threadlike objects called TRICHOCYSTS from the surfaces of their cells *(Chapter 25)*. The propulsion comes principally from the sudden elongation of the shaft of the trichocyst as it leaves the animal. Trichocysts are employed not only as weapons but also to anchor the organisms to the substratum. NEMATOCYSTS are elaborate cellular structures pro-

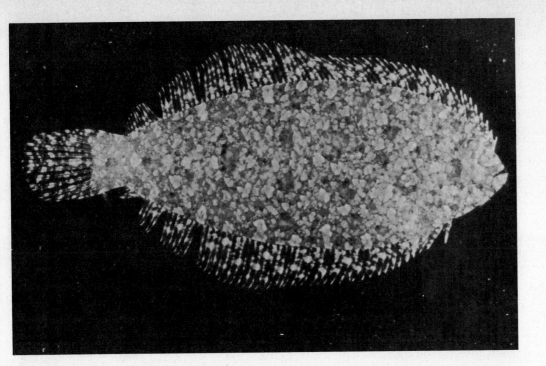

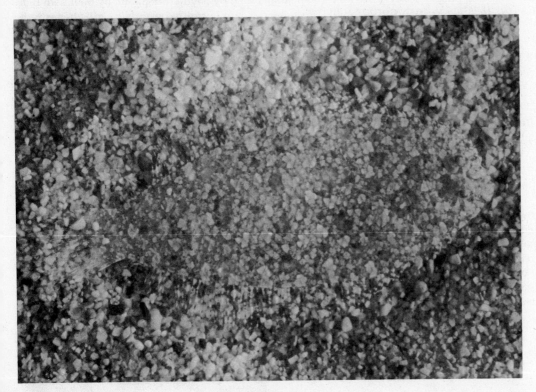

CAMOUFLAGE OF SOLE. Fish in top photograph has not had time to adapt its coloring to the artificial dark background. Below, sole is almost invisible after adapting its chromatophores to match sandy sea bottom. Adaptation to some backgrounds may require as much as a few hours.

CHROMATOPHORES

Chromatophores are pigment-bearing cells in the skin or at least close to the outer surface, which expand or contract to change the color of the organism. They are under nervous or hormonal control or both, and in most cases they can effect the change within minutes or even seconds. In squids, soles, flounders, the famous chameleon (a kind of African lizard) and a few other animals, chromatophores enable the animal to blend in with the background on which it is resting and thus to escape discovery by predators *(Figures 24 and 25)*. In other kinds of fishes and lizards, the color change is used as a signal to communicate with potential mates and territorial rivals belonging to the same species.

BIOLUMINESCENCE

A great many organisms can produce the strange cold light of bioluminescence: bacteria, fungi, radiolarians (a kind of protozoan), dinoflagellates (another kind of protozoan), sponges, corals, coelenterates, nemerteans (marine worms), ctenophores (jellyfish-like animals), clams, snails, centipedes, millipedes, insects, squids, and fishes. The phenomenon is especially frequent among the animals that live in the unlighted depths of the deep sea *(Figure 26)*. It is also disproportionately common in animals that roam shallow water or the land at night. In a few insects, such as the fireflies (which are really beetles), bioluminescence is used in communication between the sexes. The same function is probably served in many of the deepsea fish. But in other kinds of organisms, particularly the bacteria, fungi, and protozoans, the significance of bioluminescence is still very much a mystery *(Figure 27)*. It may in fact be no more than an incidental by-product of peculiar forms of oxida-

25 **CHROMATOPHORE** of a fish is a branched cell containing pigment (shown here as colored dots) in its cytoplasm. Animal is pale when pigment is concentrated in center (left) and turns darker when pigment fills the branches of the cell (right). Some chromatophores contain several pigments, each of which may respond to a different stimulus.

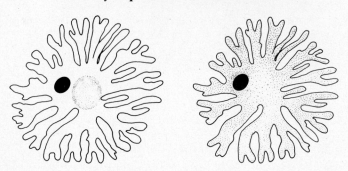

HATCHET FISH (Argyropelecus hemigymnus) is a tiny deep sea species that emits light. More than 100 species of luminous fish are known. Their light probably plays various roles in hunting, mating, and evading predators.

LUMINOUS TOADSTOOLS of the genus Mycena were photographed by daylight (top) and by their own light (bottom). Light is emitted mainly by gills beneath the caps of the fruiting structure.

tion. The light is created when a special substance, LUCIFERIN, is oxidized in the presence of the appropriate enzyme, LUCIFERASE, in specialized cells. The chemical structure of these substances varies in different kinds of animals; in fireflies the luciferin has been identified as a complex aldehyde.

ELECTRIC ORGANS

Members of at least seven families of fish can generate electricity, including the electric eel, the knife fish, the torpedo (a kind of ray), and the electric catfish. The electric organs were usually evolved from muscle, and rely on the same principle of creating an electric potential as nerve and muscle. The electric organs consist of very large disk-shaped cells arranged in long rows like stacks of coins. When discharged simultaneously, these organs can generate far more current than nerve and muscle. The electric eels, for example, produce up to 600 volts with an output of about 100 watts — enough to light a row of light bulbs or temporarily stun a man. In the electric catfish the discharge is apparently used only to repel enemies, but the electric eel uses it to paralyze prey. The eel also generates trains of low-voltage pulses to aid in detecting objects in the water around it, a form of orientation described in Chapter 21.

READINGS

A.L. BURNETT AND T. EISNER, *Animal Adaptation*, New York, Holt, Rinehart & Winston, 1964. A brief treatment of animal behavior and physiology that contains accounts of the evolution of effector systems.

M.S. GORDON, G.A. BARTHOLOMEW, A. GRINNELL, C.B. JORGENSEN AND F.N. WHITE, *Animal Physiology: Principles and Adaptations*, 3rd Edition, New York, Macmillan, 1977. The section on muscle is a comprehensive and unusually thorough review which gives equally good descriptions of the physiological, biochemical and morphological aspects.

D.R. GRIFFIN AND A. NOVICK, *Animal Structure and Function*, 2nd Edition, New York, Holt, Rinehart & Winston, 1970. This is among the best short books on animal physiology, and pays particular attention to effector systems.

H.E. HUXLEY, "The Contraction of Muscle," *Scientific American*, November 1959, and "The Mechanism of Muscular Contraction," *Scientific American*, December 1965. Two outstanding discussions of the mechanism of muscle contraction, written by the man who did much of the work. Together they make a lucid and quite detailed description, which reads like a novel.

But is it not possible that beneath all the variations of individual behavior there lies an inner structure of inherited behavior which characterizes all the members of a given species, genus or larger taxonomic group —just as the skeleton of a primordial ancestor characterizes the form and structure of all mammals today? Yes, it is possible!
—Konrad Lorenz

Guided by the central nervous system, behavior consists of the movements by which organisms survive and reproduce. The more the organisms are required to search through the environment to make their living, the more advanced their behavior. A complex brain (one consisting of a greater number of neurons) permits greater movement and a more precise search of the environment; thus tigers show more complex behavior than flatworms. Behavior is nevertheless not limited to advanced animals, nor even just to animals. Certain carnivorous plants are able to make sudden, directed movements to capture their prey *(Figure 1)*. The Venus flytrap *(Dionaea)*, for example, clamps its modified leaves around insects, while certain fungi are able to snare nematodes by tightening hyphal loops around their bodies. The sensitive mimosas are able to fold and partially retract their leaves when they are touched. These plants accomplish movement without benefit of special nerves or muscles. Instead, they rely on sudden changes in cell turgor. Within the animal kingdom, the most primitive forms of behavior are displayed by those groups which, like plants, lead a wholly sedentary existence. Adult corals and barnacles do little more than extend and retreat their bodies, move their feeding arms, and ingest the small organisms captured as prey. Truly elaborate patterns of behavior occur in closely related animals, such as jellyfish (related to corals) and shrimps and crabs (related to barnacles). But these are the forms that must move from place to place in order to search for food.

Behavior is a biological process basically like digestion, circulation, and other functions with more clear-cut anatomical machinery. Behavior has a genetic basis, and like any genetic trait, it evolves. The potential range of an organism's behavior is wholly controlled by those portions of the DNA allotted to this part of the development of the organism. The development itself is based primarily on the embryology of the sensory and nervous systems and to a lesser extent on that of the endocrine system, which produces hormones with behavioral effects. To an extent that varies enormously among species, development is also influenced by experience. Learning is the modification of behavior patterns by specific experiences. In a broad sense, learning can still be regarded as part of the ontogeny of an individual. Behavior, then, is most easily comprehended if it is examined systematically as a biological process.

SIGN STIMULI

Animals with small brains do not have the equipment to contemplate their environment and to consult their memories before solving each problem encountered. Often the problem is one that they have never experienced in their short lives. Yet in order to survive they must respond with elaborate movements that are precisely timed. No larger-brained, more experienced animal will give them the benefit of the doubt if they make a mistake. How can such relatively simple animals be "programmed" to behave in the right way? The answer is that they respond only to a few key stimuli in the environment. These SIGN STIMULI function like special code words for the animal. (Sign stimuli used in communication within a species are often called social releasers, or simply RELEASERS.) When the animal encounters sign stimuli, providing it is in the right physiological condition to act in the first place, they evoke the

AFRICAN FRESH-WATER FISH Hemichromis fasciatus, depicted in photograph on opposite page, can change its body coloring rapidly. The color changes are displays that express eight different moods.

453

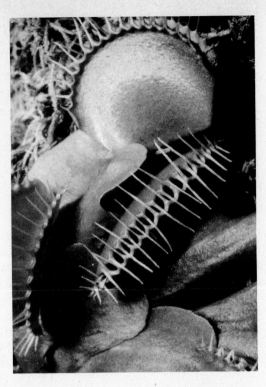

1 **VENUS FLYTRAP** is a carnivorous plant that closes quickly when an insect touches tiny trigger hairs on its specially modified leaves.

2 **TETHERED FEMALE MOSQUITO (left)** attracts flying male (right) who tries to mate with her. Female is glued to wire. Hum of her beating wings is an auditory stimulus that serves as a sexual lure.

appropriate response, called the CONSUMMATORY ACT. In this world of highly abstracted behavior, the odor of a single chemical substance emitted by a female may be the sign stimulus that evokes mating behavior (the consummatory act) on the part of the male. A second chemical substance might identify a prey organism to the male and cause him to eat it. A flash of color can mean a rival male to be challenged; a single peculiar sound, the approach of a predator to be avoided. Characteristically (but not invariably) the sign stimulus consists of relatively few elements, and it is normally encountered only in the object toward which the consummatory act is directed.

These qualities of animal behavior are vividly illustrated in the sexual behavior of *Aedes aegypti*, the mosquito that carries yellow fever. As the female mosquito flies through the air, her wingbeats create a monotonous humming sound. The noise is irritating to human ears, but it is beautiful music to the male mosquito. The beating of the female wings emits sounds varying between 450 to 600 cycles per second. Males are attracted to any steady hum between 300 and 800 cycles per second, and thus are automatically drawn to the females. The effectiveness of this auditory sign stimulus can be demonstrated by introducing tethered females into cages containing males *(Figure 2)*. As long as an individual female buzzes her

The word 'instinct' has had a controversial and confused history. It means too many things to too many people to be valuable as a basic scientific term.

wings in attempts to fly, she attracts males from distances of 25 centimeters or more. The moment she stops, males fly past without noticing her. The great simplicity of the attractive stimulus is disclosed by the fact that tuning forks vibrating in the 300–800 cycles per second range also attract males. If the tuning forks are wrapped in soft material such as cheesecloth, a few males even try to mate with them.

Vertebrates offer just as striking examples of stereotyped behavior. The male of the bullhead (*Cottus gobio*), a common European fresh-water fish, occupies and defends some convenient hiding place such as a crevice beneath a stone. When another creature of comparable size swims in front of its lair, the male darts out and bites it. Taste is apparently then used to recognize the nature of the object. If the object proves to be edible, it is swallowed. If it is inedible the bullhead spits it out. If it is another male, the bullhead lets it go and threatens it with the color change and aggressive posturing characteristic of the species. Finally, if the object turns out to be a female, the bullhead pulls her back into the burrow and attempts to mate with her *(Figure 3)*. Even birds, which we tend to regard as relatively sophisticated animals, display some behavioral patterns that are just as stereotyped as those of the male bullhead. When the male of the European robin establishes its territory in the spring, the sight of another male in the territory causes it to perform threatening songs and postures and even to launch a direct attack. Experiments have shown that the bird does not recognize a rival in the full sense of its "birdness," as a neutral human observer does. What sets it off is the red breast of an adult male robin. Even a simple tuft of red feathers mounted on the end of a wire is sufficient. When a stuffed immature male —lacking a red breast —is placed in the territory, it is seldom attacked.

The concept of the sign stimulus has provided an effective tool for the analysis of much animal behavior. If an animal can be readily fooled by such things as tuning forks and bunches of feathers into performing an entire consummatory act, the experimenter should be able to separate each stimulus normally encountered by the animal and to test its effectiveness as a sign stimulus. By the use of models he can duplicate the stimulus and compare its effectiveness with differing stimuli. An example of the use of models is illustrated in Figure 4. When the eggs of herring gulls (and many other ground-nesting birds) accidentally fall outside the nest perimeter (or are deliberately placed there by the biologist), the parents use their bills to try to roll them back into the nest. This stereotyped response is perfectly suited for the analysis of the sign stimuli by which the birds recognize objects as their own eggs. Models carved from wood and painted in various colors and patterns can be presented to the parents in competition with real eggs. Such experiments have shown that, among other things, the birds prefer the biggest "eggs" they can get. Any such quality that is preferred over the natural stimuli encountered in the life of an animal —in this case abnormally large size —is referred to as a SUPERNORMAL STIMULUS. Many examples of this curious phenomenon have been demonstrated experimentally. No satisfactory general explanation for supernormal stimuli has been produced, but their existence is taken as further evidence of the relative simplicity and automatic nature of much of the behavior of lower animals.

MATING BITE of the male bullhead fish (Cottus gobio). Male captures passing female with his mouth. Recognizing her by taste as a potential mate, he pulls her into lair and attempts to mate.

3

WHAT IS INSTINCT?

So far we have avoided defining the word INSTINCT, a word that has had a controversial and confused history. It means too many things to different persons to be valuable as a basic scientific term. Instinct will no doubt continue to be employed in popular writings to designate in a loose way behavior that is stereotyped—behavior that is highly predictable when the correct stimuli are provided, and rigid in the form of its execution. Egg-rolling in gulls (mentioned above) could be called an instinct. The concept of instinct, however, is clouded by the great variations in the stereotyped patterns of behavior, even within the same species. Instinct is often believed to consist wholly of innate forms of response which are built into the neuronal circuits of the brain from the beginning and are not subject to alteration by experience. But a great deal of new information has accumulated in recent years to show that this feature too is subject to many exceptions.

The complexities in seemingly elementary, "instinctive" behavior are best understood by examining some actual examples. It has long been known that ground-nesting birds such as turkeys, ducks, geese, and pheasants sound alarm calls and crouch down when a hawk or some other bird of prey flies overhead. Experiments performed in 1937 by the pioneer behaviorists Konrad Lorenz and Niko Tinbergen indicated that it is the most basic configuration in the shape of the flying bird, and not any particular anatomical feature, that is crucial in the recognition. When Lorenz and Tinbergen drew cardboard silhouettes of various forms over pens containing game birds, they discovered that a cross-shaped object whose head was shorter than its tail evoked a response. When this identical silhouette was then reversed, so that the head was now longer than the tail, it did not cause the alarm response. The first configuration very roughly imitates a hawk and the second a goose *(Figure 5)*. The result was originally interpreted to be evidence of a strictly inherited response to a more complex sign stimulus, one that entailed the relation of one body part to another. More recent work, however, has shown that the response contains elements of learning. The young of ground-nesting birds are at first afraid of all objects flying overhead, even falling leaves. Eventually they cease reacting to the ones seen repeatedly, such as the falling leaves and more common types of birds. But hawks are relatively scarce in any environ-

4 A SUPERNORMAL STIMULUS. A herring gull attempts to incubate a giant artificial egg presented to it by an experimenter. The bird prefers the model to one of its own eggs (lower right).

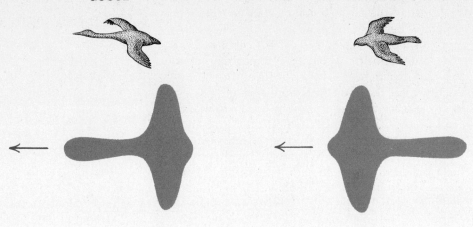

GOOSE   HAWK

NO FEAR   FEAR

DUCKLINGS SHOW FEAR and crouch in grass at the sight of a hawk flying overhead, but not at the sight of a flying goose. Same responses are evoked by sight of a cardboard silhouette (color) passing overhead. When silhouette is moved as shown at left, ducklings respond as if it were a goose. When silhouette is reversed and moved as shown at right, ducklings respond as if it were a hawk. 5

ment, and the birds never become accustomed to their distinctive shape. In this case learning of a special, genetically restricted kind —the identification of the dangerous stimulus by a gradual process of elimination —plays a role in the development of what finally emerges as a stereotyped response.

Many behavioral biologists have found it impossible to make a sharp distinction between instincts and other kinds of behavior. Instead they recognize that there is a broad range of different kinds of behavioral patterns, from highly stereotyped, virtually automatic responses that require no learning process, to very flexible responses that are perfected only during a long learning period. The attraction of the male mosquito to the wingbeat hum of the female is an example of the first extreme, and the mastery of a particular human language is an example of the second. Many, perhaps most, patterns of animal behavior develop within strict constraints imposed by the heredity of the species. But at the same time they add elements of learning before reaching their final form. A second generalization is that most behavior is GOAL-

ORIENTED. This means that the final consummatory act is directed at some specific object in the presence of the animal —a food object, a mate, a territorial rival, or whatever object possesses the sign stimulus. There are typically three periods in the complete unfolding of a behavioral act.

The first is APPETITIVE BEHAVIOR: the animal enters a searching phase appropriate to its physiological requirements. A hungry animal hunts for food, for example, or a sexually mature one primed with reproductive hormones searches for a mate. The tendency to explore in a specific way, combined with a high probability of responding when the appropriate sign stimulus is encountered, is often referred to as a DRIVE. However, like "instinct," the drive is very difficult to define, and its properties even harder to measure, so that most biologists avoid use of the term. It is important to note that appetitive behavior, whether or not one chooses to speak of it as an aspect of "drive," often has the appearance of being automatic and may require no previous experience. A young male of the mourning dove, to take one of the more appealing cases, locates a place to build its nest by a slow trial-and-error process. It assumes the nest-calling position that will later be used to attract a female from within the nest. Its body is thrown forward and its head pulled down, as though its neck and breast already fitted into the nest hollow. During the appetitive phase, however, the body comes down in empty air. The bird now shifts its position and tries again. Eventually it finds a crevice or corner into which it fits, and the experience triggers the next round of behavior, which is the collection of straw for nest building. The appetitive behavior, in other words, brings the male dove to the point where nest building becomes automatic.

Appetitive behavior is followed by a CONSUMMATORY ACT. The appropriate sign stimuli release the appropriate behavioral act. The posturing of the young dove brings it, almost haphazardly, to a naturally formed nest cavity, which serves as the sign stimulus to collect straw.

The final phase is QUIESCENCE. After the consummatory act has been performed, the animal normally slows or halts the appetitive behavior, and it is less likely to perform additional consummatory acts when confronted with sign stimuli a second time. Once started with its nest, the young dove no longer searches for nest cavities and does not respond to additional ones even if given the opportunity.

ORIENTATION

Most behavior is goal-oriented. For this reason it is useful to distinguish between the consummatory act itself and the orientation by which the animal directs the consummatory act at the appropriate object. Orientation is important during some forms of appetitive searching, when the animal organism attempts to move in a constant direction or pattern of search, and during attempts to flee from an enemy or an unpleasant environment. Finally, it attains the greatest precision in the cases of homing from the field to the nest site and in navigation during migrations. Orientation movements can be separated from the consummatory acts with which they are commonly associated, and thus they can be analyzed to a large part as a simpler form of

The most sophisticated guidance systems do not depend on a signal originating from the target, but instead bounce their own signal off the target. Such systems are found in bats, porpoises, and electric eels.

behavior. Several basic types of orientation can be recognized: kineses, taxes, depth perception, and the detection of emitted energy, chiefly sound.

A KINESIS (plural: kineses) is a very elementary form of orientation in which the animal does not specifically direct its body to the stimulus. The stimulus merely causes it to move around to a greater or lesser degree, with the eventual result that it ends up either much farther away from the stimulus or else much closer to it. Consider, for example, the case of the woodlouse searching for a "home." This little isopod crustacean lives in moist places under loose stones, pieces of wood, and other objects on the ground. When a woodlouse finds itself drying out it simply begins to move around a great deal. When it encounters a moist spot it becomes less restless and may halt movement all together. As a consequence individuals tend to congregate in the moist spots found under objects on the ground.

A TAXIS (plural: taxes), in contrast, is a movement in which the animal's body assumes a particular spatial relationship to the stimulus. The nature of the stimulus is often denoted by adding the appropriate Greek prefix to the word taxis: a phototaxis is movement guided by a light, a geotaxis is movement up or down guided by gravity, chemotaxis is movement guided by the smell or taste of some chemical substance, and so forth. A further distinction can be made between a positive taxis (movement toward the stimulus) and a negative taxis (movement away from the stimulus). The flight of a moth toward an electric light, for example, is a positive phototaxis; the retreat of a cockroach from the same light is a negative phototaxis. When a honeybee flies back to its hive after visiting a flower in an open field, it keeps on a straight course by moving at a constant angle to the sun — for example, straight toward the sun, straight away, 15° to the left, or whatever angle leads in the direction of the nest. This SUN COMPASS orientation, as it is called, is another form of phototaxis that is widespread in the animal kingdom. In the older biological literature, taxes used to be called tropisms, and the reader may still occasionally encounter this particular terminology, but the great majority of biologists now use the word tropism to mean only growth of an organism in a particular direction, and not its active movement.

Taxes, or directed movements, can be guided in one or the other of several ways. The simplest method is for the organism to test the intensity of the stimulus at different intervals. Dogs, for example, locate the source of an odor by sniffing repeatedly as they run, changing their direction to move from a weaker to a stronger smell. Humans are able to orient in the same way, but with much less skill. A second, much more efficient form of

taxis is the instantaneous "reading" of direction by means of two receptor organs. Many insects and other lower animals orient toward a light source simply by shifting the body around until an equal intensity of illumination is received in both eyes. An exact balance means that they are now facing the light head on and can move directly toward it or back directly away from it. Several elementary techniques have been devised by biologists to demonstrate the operation of this type of orientation. One is to cover one eye of the animal. The animal now turns its body in the direction of the good eye, since its brain interprets the light to be in that direction. But it can never find a position in which the illumination is balanced in the two eyes, so it continues turning in the same direction indefinitely, making one circle after another. If the animal is negatively phototactic, on the other hand, it attempts to place its body so as to produce a uniform darkness on both eyes. If it is blinded in one eye, it circles in that direction, away from what it believes to be the light source. The water mite *Unionicola* is a parasite of certain fresh-water clams *(Unio)*. When swimming around searching for one of these hosts, it is positively phototactic. This form of orientation keeps it in open, lighted water where it can move unimpeded. But as soon as it smells the odor of a clam, it switches to negative phototaxis and plunges toward the bottom where the clam is almost certain to be found. Mites blinded in one eye can be switched from circular movement toward the good eye to circular movement in the opposite direction simply by adding a few drops of clam juice to the water. A similar and perhaps even more dramatic technique to demonstrate chemotaxis is shown in Figure 6. Honeybees, like most insects, can follow odors to their source by testing the strength of the odor simultaneously with both antennae, their principal organs of smell. They turn in the direction of the antenna that detects the strongest odor. If the experimenter now crosses the antennae and cements them into position, the bee "reads" the odor as coming from the opposite direction, and it accordingly turns away from the true odor source.

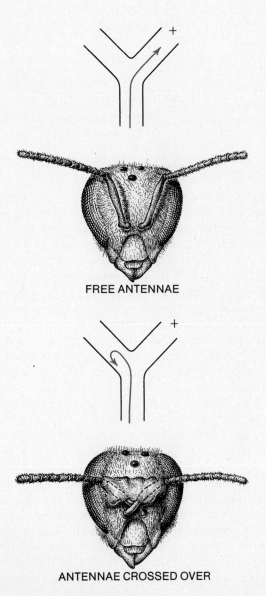

FREE ANTENNAE

ANTENNAE CROSSED OVER

CHEMOTAXIS in the honeybee can be demonstrated by crossing its antennae and gluing them in place. When put in a Y-maze, bees whose antennae have been left in their natural position (top) readily find their way to an odor source (+). Bees with crossed antennae enter wrong channel (bottom). 6

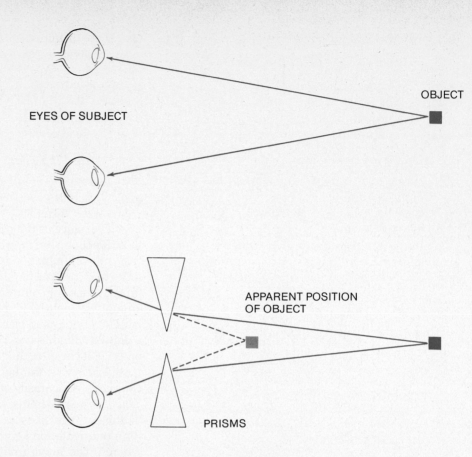

EYES OF SUBJECT

OBJECT

APPARENT POSITION
OF OBJECT

PRISMS

7 **DEPTH PERCEPTION in humans is achieved by
simultaneous focusing of both eyes on an object. Eye
muscles provide brain with information about the angle
of convergence of the lines of vision (arrows). Brain
infers distance to object. Prisms bend light and hence
distort depth perception.**

A further refinement in orientation is the use of
two receptors to achieve DEPTH PERCEPTION, an esti-
mation of the position of the stimulus in space
while it is still far removed. One way humans do
this is by aiming both eyes simultaneously at a
given object. The pull of the eye muscles informs
the brain of the angle at which the eyes are held,
and this information is translated into an intuitive
idea of the distance of the object *(Figure 7)*. A very
similar form of depth perception is achieved by the
barn owl with exceptionally acute hearing. Its large
ears are able to estimate the angle of convergence
to the sounds of a scurrying mouse on the ground
below it. Thus provided with estimates of both the
distance and direction, the bird is able to pounce
on its prey in total darkness.

The most sophisticated guidance systems of all
are those which do not depend on a signal originating from the target but instead bounce their own

signals off the target. Ships at sea locate subma-
rines and other objects underwater by means of
SONAR, which consists of sending out series of
sound pulses and listening for the echoes that
come back. By noting the direction of an echo and
the time required for it to return, an observer on
the ship can pinpoint the position of the object in
the water. Bats use the same principle to guide
their flight around obstacles and to locate and capture insects in the air *(Figure 8)*. They search for
food by continuously emitting a call that consists of
a rapid series of clicking sounds. In the case of the
big brown bat *(Eptesicus fuscus)*, ten clicks are rat-
tled off each second. Although the sounds are
quite loud in terms of decibels, the human ear can
scarcely hear them. This is because most of the
frequencies are 50 to 100 kilohertz per second
(50,000 to 100,000 cycles per second), which is in
the ultrasonic range, above the limit of human
hearing. The bats, of course, hear them very
acutely. When an echo is returned from a flying
insect, the bat rapidly increases the rate of emission of its clicks, permitting it to "zero in" on its
prey. A dipping and swooping bat in the summer
evening's sky is in the pursuit of insects by this

aerial sonar. A bat can be momentarily fooled by a pebble tossed into the air near it. It will swoop at the object, which its echolocation system reports to be an "insect," but soon veers away when it realizes it has been deceived. A very similar echolocation system is used by porpoises to find their way through murky water and to hunt for fish.

A different but equally elaborate form of emitted-energy orientation is employed by the electric eels of South America and certain similar-looking fishes of Africa (*Figure 9*). When searching through the waters of their river homes, these fish generate electrical impulses from special organs located in their tails. The flow forms a dipole field around the body, such that the tail is negative with respect to the head. The lines of flow are perceived by the fish through electric sensory organs which are located mostly on its head. Objects in the vicinity are perceived by the distorting effect they have on the field.

MIGRATION AND HOMING

The annals of natural history contain many astonishing examples of the ability of animals to find their way home after making distant journeys. Salmon, for example, are born in fresh-water streams and soon afterward journey down to sea. Several years later, after they have attained maturity, they swim back upstream to spawn and, in many cases, to die. The particular stream that serves as journey's end is almost invariably the same one in which they were born. It is chosen out of dozens or hundreds of equally suitable streams. The expression "almost invariably" is used advisedly in this case. In one investigation by Cana-

BAT DETECTS MOTH by emitting a series of sonar pulses (color) and homing in on the reflected signals. As bat closes in, it emits pulses more frequently, enabling it to respond rapidly and precisely to the moth's evasive maneuvers.

8

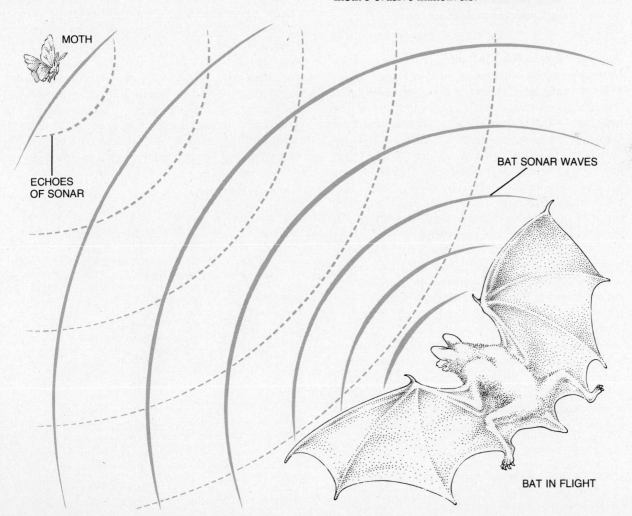

MOTH

ECHOES OF SONAR

BAT SONAR WAVES

BAT IN FLIGHT

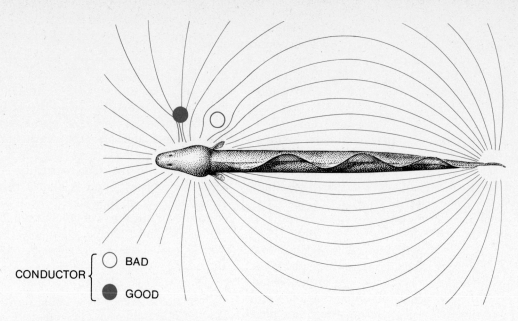

CONDUCTOR { ○ BAD

○ GOOD

9 **ELECTRIC EEL Gymnarchus generates an electric field (colored lines) around its body. Foreign objects, whether good or bad conductors, distort field. Eel detects changes with receptors in its head.**

dian biologists, 469,326 young sockeye salmon were marked in a tributary of the Fraser River. Several years later almost 11,000 were recovered after they had completed a return journey to the very same stream; but not a single one was ever recovered from other streams nearby. What underwater guideposts can these fish possibly follow? It has been discovered by A. D. Hasler and his associates at the University of Wisconsin that the salmon, like many other fish, have an acute sense of smell and are able to remember slight differences in the chemical composition of water. The most reasonable theory to explain salmon homing is that each individual remembers the distinctive "fragrance" of its native stream. As it moves upstream it makes the correct choice each time a new tributary is encountered, until finally it arrives home.

Long-distance migration is especially common in birds, because many species must make annual journeys between their nesting grounds and prime feeding areas far away. Each year over 100,000 sooty terns, an attractive tropical sea bird, travel from the waters off the west coast of Africa all the

way across the Atlantic to Bush Key, a tiny island near the tip of Florida. Here they build their nests and breed *(Figure 10)*. Once the young are fledged, all journey back over the Atlantic. Why do the sooty terns migrate at all? Like many other sea birds, they find protection from cats, foxes, and other predators on isolated islands. It is evidently safer for them to make an entire transoceanic voyage to reach one such haven than it would be to try to nest on the nearby African shores. A somewhat different reason lies behind the north-south migra-

BREEDING AREA
BUSH KEY
(JULY-DECEMBER)

FEEDING AREA
(JANUARY-JUNE)

10 **SOOTY TERN breeds and nests on Bush Key, a tiny island off the Florida coast, then migrates with its offspring to feeding area near Africa.**

tions of birds in the temperate zones. Each spring a legion of migratory forms, from robins, thrushes, and warblers to geese and ducks, make their way north into the greening countryside, where large quantities of food are becoming freshly available. Working rapidly, they are able to rear one or more broods of young. As winter approaches and the food supply declines, all head south again. Some species proceed all the way to Central and South America. The record annual journey in the Western Hemisphere is made by the golden plover, one race of which travels from northern Canada to southern South America. A second race of the same species migrates from Alaska to Hawaii and the Marquesas Islands. Man could never make such journeys unaided by maps and navigational instruments. How do the birds do it? A large part of the answer lies in their ability to use celestial clues. At migration time caged starlings become unusually restless. If permitted to see the sun they begin to fly toward the side of the cage that lies in the direction of their normal migration route. However, when the sky is overcast and the sun is obscured from view, their movements persist, but they are nondirectional. Other migratory birds fly at night and can evidently use the position of the stars to guide them. This surprising fact has been established by several biologists, including S. T. Emlen of Cornell University, who allowed indigo buntings to attempt flights under the artificial night sky of a planetarium. The birds oriented "correctly" with reference to the planetarium sky even when the positions of its constellations did not correspond with the positions of the true constellations outside. Thus other outside influences were eliminated, and it could be concluded that the birds are able to orient to what they believed to be the position of the stars. In one ingenious experiment performed by Emlen, the physiological state of indigo buntings was first altered by exposing them to various artificial day lengths inside closed laboratory rooms, so that some entered the spring migratory condition and others the fall migratory condition. When both groups were then placed in the planetarium, the spring-conditioned birds attempted to fly north and the fall-conditioned birds attempted to fly south. By blotting out selected constellations in the planetarium, Emlen has further learned that the birds apparently find their way by noting the relative positions of constellations, particularly those in the vicinity of the North Star.

Even more impressive than the guidance of migratory movements by celestial clues is the phenomenon of homing. If you were blindfolded, taken to some completely unfamiliar place, and handed a magnetic compass, you could head north, south, or in any direction you arbitrarily chose—but you could not head home. Simple compass reading is also the essential ability demonstrated by migrating birds. But it is not enough in itself to explain how homing pigeons are able to return to their own lofts from as far as 600 miles away in a single day. Nor is it enough to account for such feats as that of one particular Manx shearwater (a kind of sea bird) which, after being carried in an airplane from England to Boston, flew back across the Atlantic and arrived at its nest in England twelve days later. For a bird to home over unfamiliar terrain requires true navigation—that is, the ability to reckon its position on the surface of the earth with reference to the position of the distant goal. One of the currently most attractive hypotheses is that the birds somehow sense the earth's magnetic field, which varies systematically from point to point over the earth's surface. Evidence supporting this idea has come from experiments by William T. Keeton of Cornell University, who attached tiny magnets to the necks of homing pigeons in order to cancel the effects of the earth's magnetic field. Birds thus encumbered lost their homing ability but "control" birds burdened with nonmagnetic metal bars placed at the same position on the neck still managed to home correctly.

CYCLICAL CHANGES IN BEHAVIOR

Few animals display uniform activity around the clock. Most are subject to circadian rhythms (*Chapter 17*) that affect their general levels of activity and even their "willingness" to react to sign stimuli. Sexual behavior in moths, to take one of the most clear-cut examples, is limited to certain hours of the day or night. Each species of giant silkworm moth in the United States has a particular time at which the females release sex pheromones and the males respond. In the promethea moth this is 4–6 P.M. In the cynthia moth it is 10 P.M. to 5 A.M., and in the cecropia moth around 4 A.M. These differences increase the probability that each insect will succeed in finding a mate belonging to its own species.

Reproduction tends to be strongly seasonal. The breeding season in birds, like that of other vertebrates, is heralded by changes in the structure and physiology of the ovaries and testes. These reproductive organs produce the sex hormones, which are mostly androgens in the male and estrogens in the female. The sex hormones in turn induce the development of the characteristic breeding plumage and coloration and alter the behavior of the birds so that they commence territorial and reproductive behavior. The ring dove is a species whose reproductive behavior has been analyzed by the late D. S. Lehrman and his associates at Rutgers University. When males and females of this small relation of the common pigeon are segregated by

11 REPRODUCTIVE BEHAVIOR OF RING DOVE. Male initiates courtship with "bowing coos." Sight of courting male stimulates release of reproductive hormones in female. During nest-building birds copulate, and a week or so later female lays two eggs. Both male and female take turns incubating the eggs. Newly hatched squabs are fed "crop milk," a nutrient fluid regurgitated by both parents.

COURTSHIP   NEST CONSTRUCTION

INCUBATION   "NURSING"

sex in stock cages, they show little evidence of reproductive activity. But if a pair are placed together in a cage, courting begins within a few minutes *(Figure 11)*. The male is the initiator, since his testes are active and probably secreting testosterone, one of the androgen hormones. He faces the female and repeatedly bows and coos. The sight of the displaying male activates centers in the female's brain, which induce secretions from the pituitary gland. The pituitary hormones in turn stimulate the release of the two principal gonadotropins, the follicle stimulating hormone and the luteinizing hormone *(Chapter 18)*. The gonadotropins stimulate the growth of the female's ovaries, which begin to manufacture eggs and to release estrogen into the blood stream. This entire chain of endocrine activity is put into motion within only a day or two. When both birds have thus been primed with hormones, they begin to construct a nest together. During nest building the male continues his courtship, and copulation follows. When the nest is finished, the female becomes closely attached to it and soon lays her first egg, an event that triggers still further endocrine changes in both the female and her partner, leading to the incubation of the egg and care of the young. Step by step,

---

The evolution of mammals shows a trend toward progressive emancipation from hormone-directed behavior, and the substitution of learned responses.

---

as if programmed by a choreographer, hormones lead to behavioral acts, which elicit the release or shutting off of other hormones, which in turn trigger further behavior, and so on until finally the young have been reared and the cycle completed. The ring dove was among the first animals whose reproductive behavior was thoroughly analyzed. By now there is evidence that many other kinds of animals, from cockroaches to higher mammals, also depend upon reciprocal relationships between hormones and behavior to guide their reproductive cycles.

The evolution of mammals shows a trend toward emancipation from hormone-directed behavior, and the substitution of learned responses. When a male rat or mouse is castrated, its sexual

Hormones are very potent in altering the behavioral repertoire of animals. But as controls hormones are relatively crude. Their effects cannot be quickly turned on or off.

behavior rapidly declines to the point of virtual disappearance. Dogs and cats, which have larger brains and display less stereotyped behavior, also suffer a decline of activity following castration. But the loss comes slowly over a period of months or years, and in a few individuals it does not occur at all. Finally, man, apes, and other higher primates are not affected by castration, provided they are sexually mature and experienced at the time of the mutilation.

HUNGER

Earlier we examined the strong inborn tendency of an animal first to explore its environment in a conscious or unconscious search for food, a mate, a nesting place or some other required object; and second, to respond to the object when contact is made. The role of hormones in behavior is to "prime" the animal. These substances affect the intensity of the drives, or to use a more neutral expression, the level of its motivational states. Hormones are very potent in altering the physiology and large parts of the behavioral repertoire of animals. However, as controls they are relatively crude. Their effects cannot be quickly turned on or off. So animals use more direct cues to provide a finer tuning of their motivational states. To put the matter simply, the brain hungers when the body is starved, thirsts when it is desiccated, and lusts when it is sexually deprived. Satiation of these needs is immediately relayed back to the brain without the need for hormones to serve as intermediaries. What, exactly, is the nature of the messages the brain receives?

The psychology of hunger is better understood than that of the other basic appetites. Research has centered on the way that hunger signals are terminated. Until recently there were two competing theories about how hunger is cancelled in the brain through the act of eating. One theory maintained that when the animal takes food in its mouth and chews and swallows it, these muscular actions

satisfy the brain that a meal has been eaten. The other theory held that when the food arrives in the stomach it distends the stomach walls, causing nerve impulses to travel back to the brain; the distention satisfies the brain, and hunger ceases.

The results of a variety of experiments support the second theory. H. D. Janowitz and M. I. Grossman of the University of Illinois surgically modified the throats of dogs so that the esophagus opened to the outside of the throat. When the dogs were fed, the food dropped to the outside without reaching the stomach. Alternatively, the experimenters could fill the stomach with food passed down the esophageal opening, without the dog having performed the acts of chewing and swallowing. It was found that after dogs had undergone the first experience, that is, eating without filling the stomach, they continued trying to eat. In other words, they remained hungry. But when the dogs had their stomachs filled without eating and then were given the chance to eat some more, they refused. It was clear that direct signals from the filled stomach are required before the dog's brain ceases to recognize a state of hunger.

LEARNING

If appetites such as hunger provide a finer adjustment of behavior than the fluctuation of hormone levels, the process of learning permits the most sensitive control of all. Learning is the adaptive change of behavior as a result of experience. The organism alters its behavior to respond to specific stimuli in ways that promote its own survival or that of its offspring. Thus it closely molds its behavior pattern to adapt itself to the particular environment in which it occurs. The same learning potential can be used to fit a great many different environments.

The awkward attempts of young animals to walk, or fly, or find their own way to food are familiar to most of us. The "play" of kittens consists mostly of the constant practice of three maneuvers used by adult cats to catch their prey. They pounce on objects moving along the ground, the method later used to catch mice; they clap their front paws together on a small object in the air, the technique for capturing birds in flight; and they scoop up objects lying on the ground in front of them, the motion employed by adults for catching fish. It is tempting to conclude that such actions are part of the process of learning. In many cases this is demonstrably true. But in others, the animal is not really learning. It is instead undergoing the much more rigid process of MATURATION: the automatic unfolding of the behavioral process as the nervous system and muscles proceed to their full normal development. An instructive example is the flight of birds. Nothing looks more like the process of learning than the "practice" flights of fledging birds, which flap their wings as though testing them prior to the first tentative take-off. But experiments have shown that the process can be achieved purely by maturation. J. Grohmann, a German investigator, reared young pigeons in narrow tubes that prevented them from moving their wings; they could not undertake any form of practice flights. When a control group of pigeons, which were allowed to practice each day, had learned to fly, the experimental birds were released from their bindings. Surprisingly, no difference in the flying ability of the two groups could be detected. The so-called practice flights are evidently merely actions that reflect an incomplete stage in the automatic development of the nervous and muscular systems.

HABITUATION is perhaps the simplest form of true learning, since it involves the loss of old responses rather than the acquisition of new ones. If an animal is repeatedly presented with a stimulus not associated with either a reward or a punishment, it will eventually cease to respond. Habituation is the basis of the "taming" of wild animals. If you carefully handle a freshly captured snake, you will find that after a few hours it will calm down and may eventually lie completely at rest in your hands. This change in behavior is simply habituation to the strange stimuli which your handling first presented it.

Most animals are able to learn to associate two or more stimuli with the same reward or punishment. The great Russian psychologist I. P. Pavlov called this the conditioned reflex, but it is now more accurately called ASSOCIATIVE LEARNING. Pavlov's most famous experiment still provides a simple and instructive example. He first collected and measured the saliva released by a dog when it tasted powdered meat. Then he added a new stimulus, the ticking of a metronome, at the time the meat powder was provided. After the stimuli had been presented jointly in this fashion five or six times the dog was permitted to hear the metronome but was not given meat powder. Nevertheless it still salivated.

Rather than learning to associate new stimuli with old ones that lead to reward or punishment, an animal might explore its environment until it has a particular experience, and then learn the

BAIT

BAIT

INSIGHT LEARNING enables chimpanzee to stack boxes to reach a banana, but raccoon on leash fails to solve the detour problem of reaching a pan of food by first walking away from it.

stimuli associated with the experience. Suppose that Pavlov's hungry dog were allowed to hunt for food until it found a dish of meat powder behind the ticking metronome. Thereafter, the dog would be likely to head directly for the metronome. Learning by trial and error is sometimes alternatively referred to as OPERANT CONDITIONING.

A student enroute to a new classroom might make casual note of such things as the location of a water fountain or alternate exit routes without discernible effort and without benefit of reward or punishment. The information is stored for potential later use. Such LATENT LEARNING also occurs in animals. Like man, they use it most frequently when placed in new surroundings. Bees and wasps, for example, often conduct orientation flights around their new nests, examining and learning enough of the terrain to find their way home on later flights.

INSIGHT LEARNING is the highest form of learning. Referred to as reasoning when it is conducted by man, it consists of recalling a previous experience that involved stimuli different from the current ones and adapting the memory to solve the current problem. Insight sometimes consists of the relatively simple process of GENERALIZATION. Honeybees can be trained to fly to a checkerboard pattern of one color in preference to all other patterns. If this particular checkerboard is now removed and the bees are limited in their choice to a checkerboard of another color, they will still select the substitute over all other patterns. They have generalized the checkerboard pattern and temporarily ignored the color. Much more complex forms of insight learning have been demonstrated in chimpanzees. Perhaps the most striking and humanlike is illustrated in Figure 12. The chimp, with no previous experience of this kind, has reasoned a way to stack boxes on top of each other in order to reach a previously inaccessible banana.

The act of IMPRINTING, as the name itself implies, is a peculiar form of learning that is relatively quick and difficult to reverse, or "unlearn." It is often dramatic in effect, ordinarily occurs only during a limited time during the life of an animal, and does not require a reward or punishment. The best known and most illuminating example comes from a pioneer experiment performed by Konrad Lorenz in Austria. He first divided a clutch of eggs laid by a single graylag goose into two groups. When the goslings hatched from one group, they were permitted to associate with the mother goose. The other goslings were hatched in an incubator, and the first living creature they saw was Konrad Lorenz. In the first few days of their lives, they were allowed to follow Lorenz as though he were their parent. Later, the goslings were marked according to the group to which they belonged and placed together under a box. When released, the two groups separated from each other and ran to their respective adoptive parents (*Figure 13*).

Similar experiments with a variety of birds, mammals, and insects have revealed that imprinting is widespread in the animal kingdom. The period of latency, during which the young animal can be trained in this special way, is often extremely short. The adoption period of ducks, which has been studied in detail by E. H. Hess of the University of Chicago, is illustrated in Figure 14.

THE EVOLUTION OF BEHAVIOR

Circadian rhythms, appetites, hormone levels, maturation, and learning each cause major variations in the behavior of a given animal. The total pattern can be viewed as the program of its behavioral responses. This changing repertory adapts the animal on a moment-to-moment basis to its environment and permits it to survive and

IMPRINTED GOSLINGS follow ethologist Konrad Lorenz as if he were their parent. Lorenz was first living creature that young geese saw during latent period immediately after hatching. 13

reproduce. Now we will consider how the program itself can be altered. Over long periods of time each animal species must adapt by evolution to major changes in the climate and to the shifting composition of the other animals and plants that make up its living environment. The species slowly changes in the hereditary basis of its anatomical and physiological characteristics. It also changes the hereditary control of its behavior program.

For most of the first half of this century, one of the great controversies of biology and psychology raged around the role of instinct in animal behavior. Some psychologists, particularly those who belonged to the "rat psych" school of America (so named because they used white rats extensively in their experiments), believed that behavior

14 **DURATION OF LATENT PERIOD was studied by exposing young ducklings to model of male duck wired for sound (1). Duckling was later tested for imprinting by exposing it to both the male model and a female model that emitted a different sound (2). If duckling followed male, response was scored as positive. Each dot on curve is average test score of ducklings imprinted at that age. Curve indicates that critical age at which ducklings are most strongly imprinted is 16 hours after hatching.**

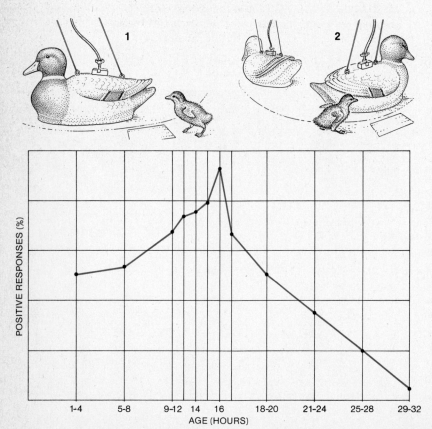

The instinct versus learning controversy is merely another version of the larger nature versus nurture controversy: which is the more important, heredity or environment?

is primarily or even exclusively learned. The young animal, they argued, is born with a brain like a *tabula rasa* —a blank tablet —on which experience is free to write its lessons. Behavior patterns can be developed only by learning. Most biologists regarded this theory as far too extreme. A few propounded the equally extreme theory that in lower animals behavior is usually completely controlled by heredity. The instinct versus learning controversy is but one version of the larger nature versus nurture controversy: which is the more important determining factor, heredity or environment? Which, for example, determines eye color? The answer must be both heredity and environment. Heredity dictates the color, insofar as it differs from one individual to another, but the color itself develops only by a prolonged and exacting interaction of one's genes with the environment.

This evaluation of eye color goes to the heart of the matter. It is meaningful to speak of a "hereditary" trait only if there exist two or more such traits —such as different eye colors —which can be compared between individuals. Thus if one person has brown eyes, and another blue, the difference between the colors is an hereditary difference. The hereditary basis of this particular quantity of human variation has been established beyond doubt by standard genetic analyses. By the same token, the difference between the mating call of a ring dove and that of a rock pigeon is hereditary. The difference is also referred to as innate or "instinctive." To speak of variation in instinctive behavior is to employ a slightly different meaning than the one introduced earlier in this chapter, which stated that if a behavioral pattern varied very little within a species and was therefore highly predictable, it could be referred to loosely as an instinct. Now the term "instinctive" is used for a genetically based difference between two behavioral responses. Both of these crude definitions are valid. The ambiguity of language is unfortunate, and it has been discussed here only because of the importance of the ideas that are so frequently included under the heading of "instinct."

BEHAVIORAL GENETICS

Meanwhile, it is essential to recognize that a great deal of variation in a given behavioral act, whether instinctive or not, does have a genetic basis. Behavioral genetics is a relatively young and rapidly growing field with a significant future. It is the means by which we will eventually come to thoroughly understand the evolution of behavior, in a way that will help us to dispense with troublesome controversies such as the one over the significance of instinct.

That particular behavior patterns are inherited can sometimes be shown in a dramatic fashion by hybridizing two related animal species. W. C. Dilger of Cornell University, for example, used this technique to prove that nest building in lovebirds is subject to genetic control. All species of this group of African parrots build nests from soft materials which they cut into little pieces with scissor-like movements of their bills. In the laboratory, sheets of paper are readily accepted for this purpose. One species, the peach-faced lovebird, carries the scraps of building materials to its nest site by tucking them into the feathers on the lower part of its back *(Figure 15)*. The bird erects the feathers to receive the materials and then compresses them to hold the pieces in place while it flies home. Certain other species, such as Fischer's lovebird, carry the scraps in their beaks, the method used by most kinds of bird. Dilger crossed the two species and waited with great interest to see which carrying technique would be employed by the hybrids. When the young birds first began nest building, they were totally confused. They attempted to tuck the pieces in their feathers like the peach-faced lovebirds, but were unable to raise and lower their back feathers properly, and dropped every piece of building material on the flights to the nest. In six percent of the attempts the birds used their bills to carry pieces in the fashion of Fischer's lovebirds. Gradually, after several months, the hybrids learned to make greater use of their bills. Two years passed, however, before they significantly reduced the amount of effort wasted on feather tucking, and only after three years did they finally switch entirely to the bill-carrying technique.

At least some variation in human behavior also has a genetic foundation. There are quite a few single genes and chromosome abnormalities that

NEST-BUILDING OF LOVEBIRDS is an example of inherited behavior that differs from one species to another. Peach-faced lovebird (top) carries scraps of nest material tucked into its tail feathers. Fischer's lovebird (bottom) carries nest materials in bill, like most other birds.

15

impair intelligence and in some cases cause severe retardation. These conditions, which include Down's syndrome (mongolism) and phenylketonuria (PKU), are fortunately relatively rare and in most cases easily diagnosed. Phenylketonuria

469

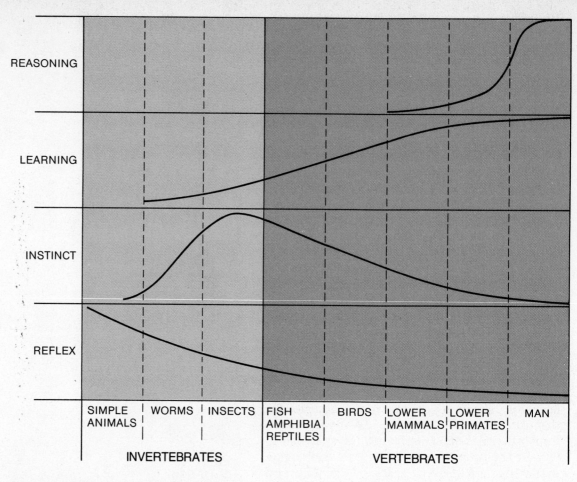

REASONING

LEARNING

INSTINCT

REFLEX

| SIMPLE ANIMALS | WORMS | INSECTS | FISH AMPHIBIA REPTILES | BIRDS | LOWER MAMMALS | LOWER PRIMATES | MAN |

INVERTEBRATES   VERTEBRATES

16 **EVOLUTION OF ADAPTIVE BEHAVIOR can be traced by comparing the differences among primitive and advanced species now alive. Simplest animals rely mainly on inherited behavior: reflexes and instinct. More advanced animals show increased reliance on learning and reasoning.**

can also be prevented if diagnosed in infancy, because it results from a defective utilization of one of the amino acids (*Chapter 29*). Other, normal patterns of behavior are under the control of many genes, and their hereditary components are correspondingly difficult to analyze. The key operational question is the relative contributions of heredity and environment to observed variation in human populations. We all know that talents and personality traits vary enormously. How much of the variation is due to differences in heredity and how much to environment? One crude way of determining whether any genetic influence exists at all is to compare the differences between identical twins with the differences between ordinary twins. Because identical twins originate from the same fertilized ovum (and are therefore termed

monozygotic), they are genetically identical. On the other hand, ordinary twins originate from separate fertilized ova (hence, are called dizygotic), and they are consequently as genetically different from one another as brothers or sisters born in separate years. If variation in a given trait is due in any significant degree to differences in genes, dizygotic twins should differ more from one another than is the case for monozygotic twins—even though the environments in which the twins were raised do not vary more in one case as opposed to the others. The comparisons of monozygotic and dizygotic twins in effect constitute a genetic experiment in which the environment is held "constant." The studies so far have indicated that there may be some genetic influence in the variation of a very wide range of human behavioral traits, including intelligence, psychomotor ability, tendencies toward certain forms of neuroticism, the tendency toward schizophrenia, and others. It must be stressed, however, that the strength of hereditary influences cannot be precisely determined with existing methods, and in each case environmental influences also play a strong role.

Thus it is possible to increase IQ dramatically, or to prevent or at least moderate schizophrenia, by placing individuals in environments designed to accomplish these desirable modifications.

INTELLIGENT BEHAVIOR

One of mankind's most cherished ideas about evolution is that blind instinct has gradually given way to learning and reasoned behavior, and that the trend reaches its highest (but far from perfect) expression in man. This lofty generalization is essentially true. But the evolution of behavior has not been simply the process of animal life striving upward toward a more enlightened view of the world. Behavior has evolved from highly programmed, stereotyped responses to a much more flexible form of adaptation to the environment by means of learning *(Figure 16)*. Stereotyped behavior—or, "instinct"—is the best that can be done with a small number of neurons. An animal with a very small brain is like a carefully programmed automaton keyed to certain principal features of the environment, the sign stimuli, which are vital to its success. It is guided home by one of them, it recognizes a mate by another, it seizes a food item by two more, and so on throughout its life. The large-brained animal is also programmed, but it has enough spare neurons to engage in the luxury of learning. To some extent it can learn to select among the stimuli which in its own particular environment are the most vital to its success. It molds itself to the local environment and responds with greater precision and safety than the small-brained animal. True intelligence and enlightenment have been fortuitous by-products of this trend in human evolution.

READINGS

J. Alcock, *Animal Behavior.* Sunderland, MA, Sinauer Associates, 1975. This is probably the best balanced short textbook on the subject, and one that can be recommended to readers searching for a relatively easy next step in the subject.

L. Ehrman and P.A. Parsons, *The Genetics of Behavior.* Sunderland, MA, Sinauer Associates, 1976. A solid, balanced account of this small but important discipline.

R.A. Hinde, *Animal Behaviour: A Synthesis of Ethology and Comparative Psychology,* 2nd Edition, New York, McGraw-Hill, 1970. The principal, and largely successful, effort to bring together the modern findings of zoology and psychology.

P.H. Klopfer and J.P. Hailman, *An Introduction to Animal Behavior: Ethology's First Century,* Englewood Cliffs, NJ, Prentice Hall, 1967. A sound textbook on animal behavior which stresses the history of ideas.

K. Lorenz, *On Aggression,* New York, Bantam, 1967. This little classic is the source of most of the current ideas (and controversies!) on the evolutionary origin and significance of aggressive behavior in man.

P.R. Marler and W.J. Hamilton III, *Mechanisms of Animal Behavior,* New York, John Wiley & Sons, 1966. The best major textbook on the physiological basis of behavior.

A solitary ant, afield, cannot be considered to have much of anything on his mind; indeed, with only a few neurons strung together by fibers, he can't be imagined to have a mind at all, much less a thought. . . . It is only when you watch the dense mass of thousands of ants, crowded together around the Hill, blackening the ground, that you begin to see the whole beast, and now you observe it thinking, planning, calculating. It is an intelligence, a kind of live computer, with crawling bits for its wits.   —Lewis Thomas

Social behavior alters some of the rules of evolution. Radical changes in adaptation are open to the few animal species that have evolved the capability to cooperate with other members of their species. Evolution can create superbly specialized species of solitary animals, but the achievement of any one species is strictly limited by the capabilities of the individuals that comprise it. A solitary beetle may be able to recognize 100 distinct stimuli in the course of its life. The species as a whole can do no more. If a solitary beetle occupies a typical territory of ten square centimeters, one million beetles of the same species occupy territories totalling ten million square centimeters. Organization of the beetle population ascends no higher.

Social behavior, in contrast, enables a species to transcend the limits of individual capabilities. Using precise forms of communication, males and females can locate one another over long distances and synchronize complex forms of mating behavior. Parents communicate with their offspring to protect and nurture them. Beyond these elementary forms of social behavior, the members of societies can specialize to perform different functions, then form groups and integrate their skills. The complexity and efficiency of the group can exceed that of the individual. These two basic qualities, specialization of individuals and integration of groups, are the criteria for comparing animal societies. Elaborate integration, combined with high levels of individual intelligence, can lead to the next great step in the evolution of behavior: the transmission of learned information from one

generation to another. The biological basis of social behavior is the subject matter of the new—and sometimes controversial—discipline of sociobiology.

COMMUNICATION

The song of a bird, the chirp of a cricket, the flash of a butterfly's wing, the color and scent of a blossom, are all signals being urgently transmitted to other plants or animals. In very general terms, BIOLOGICAL COMMUNICATION can be defined as action on the part of one organism (or cell) that alters the behavior pattern of another organism (or cell) in an adaptive fashion. By adaptive we mean that the signaling and the response contribute in some positive way to the survival or reproduction of one or both of the participants. The acts of communication include some of the most complex and fascinating forms of behavior. The best way to understand them is first to classify them according to the kind of sense organ by which they are received and then to examine the advantages and disadvantages of each sensory system separately. Investigators have studied both communication between members of the same species and communication between different species, sometimes organisms as radically different as plants and animals. Communication involves a variety of signals including chemicals, light and color, sounds, and patterns of movement.

CHEMICAL COMMUNICATION

Chemical signals passed between different organisms of the same species are called PHEROMONES. They are the most widely used signals in both plants and animals and they were probably also the first signals put to service in the early evolution of animal behavior. It can be argued that

LIVING BRIDGE OF ARMY ANTS (Eciton burchelli) spans two logs in the photograph on opposite page. A symbiotic silverfish can be seen crossing the bridge at center. Eciton is a native of the forests of South and Central America.

473

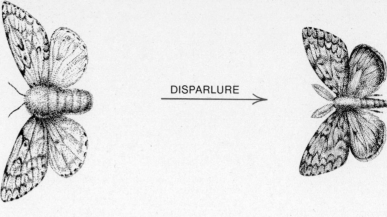

FEMALE

MALE

DISPARLURE

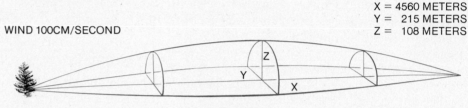

WIND 100CM/SECOND

X = 4560 METERS
Y = 215 METERS
Z = 108 METERS

ACTIVE SPACE

1 **FEMALE GYPSY MOTH releases disparlure from glands in her abdomen. Pheromone evaporates and vapor drifts downwind, where males detect it with their antennae. Diagram below depicts a typical active space downwind from a female in tree. Males within space are attracted to her.**

pheromones even preceded hormones. Consider the fact that communication among single-celled organisms must have come before the origin of the higher organisms, and that this primitive exchange was almost certainly chemical. Since cells in the bodies of higher animals communicate with each other by hormones, pheromones must have preceded hormones and might have given rise to them in a direct evolutionary line.

Pheromones are secreted from cells (usually special glandular cells) as liquids and transmitted either as liquids or gases. Some are normally detected at a distance by smell, others at the surface of the emitting animal by smell or taste. Still others are presented in liquid form at olfactory "signposts" to be smelled by passing animals over long periods of time. The substances function either by evoking immediate behavioral responses, called RELEASER EFFECTS, or by means of PRIMER EFFECTS that work through the endocrine system to alter the physiology and subsequent behavioral patterns of the receptor animal.

Releaser pheromones are widespread in the animal kingdom and serve a great many roles in different kinds of animals: sex attraction, simple aggregation, trail marking, territorial advertisement, individual recognition, and others. Sex attractants constitute an especially large and important category. The females of some kinds of moths, for example, secrete the pheromones from glands in the tips of their abdomens and are able to attract males over long distances *(Figure 1)*. When sexually active males enter the ACTIVE SPACE within which the substance is dense enough to be detected they begin to fly upwind. If they accidentally fly out of the active space, they search back and forth until they strike it again, and then continue their upwind journey. This maneuver automatically brings them to the vicinity of the emitting female. The amounts of pheromone required to create large spaces are very small in many cases, because the feather-shaped antennae of the males are specially constructed to perceive them. In the case of the silkworm moth, for example, it has been estimated by Dietrich Schneider of Germany that only about 40 of the 40,000 hairlike receptors on the male's antennae each need be struck by a single molecule of female pheromone to start sexual activity. The pheromone is so potent that .01 microgram (one hundred millionth of a gram), the minimum average content of a single female moth, would be theoretically adequate to excite more than a billion male moths!

Although pheromones have been most closely analyzed in insects and a few other lower animals, there is little doubt that they are also very important in the communication of fish and mammals. Individuals of the sugar glider (*Petaurus papuanus*), a tree-dwelling marsupial of New Guinea that resembles a little squirrel, are able to discriminate the body odors of other sugar gliders to a remarkably fine degree. In addition to marking their territories with their own distinctive secretions from several glandular sources located over their body, males use specialized movements to smear frontal gland secretion over the bodies of their partners. By this means they recognize their own mates and lay claim to them in the future (*Figure 2*).

Female rhesus monkeys produce sex attractants in their vaginas when they come into heat. These substances, which smell rancid to humans, consist of a simple mixture of fatty acids with low molecular weight. The strongest human body odor is also produced by low-molecular-weight fatty acids secreted by specialized skin glands in the armpits and genital area. These substances appear in quantity only at puberty, and may well serve as subtle pheromones. Body odor does have a sexual significance in many cultures, although the response to it depends strongly on cultural conditioning and on the circumstances under which it is perceived.

The principal function of primer pheromones, in contrast to that of the releaser pheromones just cited, is to trigger a series of physiological changes in the recipient animal. These changes are accompanied by a long-term alteration in the patterns of behavior. Usually the effect is transmitted within the body of the receiver animal by the endocrine glands, which are activated by the pheromones through the brain (or perhaps even directly) and in turn secrete hormones that bring about the ultimate bodily changes. In termites, for example, the queen and the king use primer pheromones to exercise a form of reproductive tyranny. They prevent other members of the colony from developing into their own castes by secreting substances that are ingested and act through the corpus allatum, an endocrine gland controlling differentiation. By this means they remain in full possession of the reproductive role in the colony. Mice produce substances in their urine that alter the reproductive cycles of the females. The smell of the urine from a strange male, to take one of several known examples, can suppress the secretion of prolactin and

SUGAR GLIDERS are squirrel-like marsupial mammals native to New Guinea. Male (top) marks his mate with pheromone secreted by frontal gland, visible as a circular patch on his forehead.

2

block the pregnancy of a female that has been newly impregnated by another male. The interfering male then is free to inseminate his newly acquired mate.

Pheromones are also widespread among protistans and other microorganisms, among fungi, and among the lower plants. In almost all cases they serve a sexual function, bringing gametes together for fertilization. A typically simple system is that of the water mold *Allomyces*, a member of the third major group of higher fungi, the Phycomycetes. *Allomyces* is bisexual, producing distinctive motile

SIRENIN, pheromone released by female gametes of the water mold Allomyces, attracts male gametes.

3

475

male and female gametes, with the female gamete considerably larger than the male. Leonard Machlis of the University of California has shown that the female gamete secretes a pheromone which he named sirenin, an appropriate name for a female sex attractant *(Figure 3)*. By some completely uncharacterized mechanism, the male gamete swims into higher concentrations of this substance, eventually reaching the female gamete and fusing with it to form the zygote. Sirenin is effective in concentrations as low as 10^{-10} molar.

VISUAL COMMUNICATION

Textbooks on animal behavior lay heavy emphasis on visual communication, despite the fact that the phenomenon is less widespread and common than chemical communication. The reason is simple: man has a highly developed visual sense and a relatively poorly developed chemical sense *(see Chapter 34 for the evolutionary explanation of this fact)*. It is natural that biologists would at first pay closest attention to the forms of communication most conspicuous to them. It is also easy to see why certain kinds of animals, notably the birds, fish, and butterflies, have occupied center stage in studies of animal communication. These creatures, like man, rely greatly on vision. For the organisms that have well-developed eyes, visual signals do indeed offer some marked advantages. The visual animal, by adopting a distinctive coloration or pattern of body shading, can project a continuous fixed message with a minimum expenditure of energy. During the breeding season male birds commonly undergo a change in the color of their plumage or exposed fleshy parts of the body to denote that they are in breeding condition. Such signaling can be perceived almost instantly at long distances, and it does not even require that the sender be aware of the existence of the receiver.

Conversely, visual signals can be modified in a way that permits them to be rapidly changed. The rate of information flow is increased, and the messages are kept relevant to shifting circumstances. Cichlid fishes, which are often kept in home aquariums, change their body coloration from one striking pattern to another. *(See illustration on opening page of Chapter 21.)* During territorial contests and sexual encounters it is important for these fishes to be able to transmit the mood, and thereby to suggest the behavior pattern likely to ensue, as each new situation demands. Visual signals can also be employed in creating patterns through time. The male firefly, for example, flashes its light in a pattern that is peculiar to its own species. The human observer can learn to identify the species simply by watching the duration of the flash and the rapidity with which it is repeated. Female fireflies select males of their own species by the same means.

AUDITORY COMMUNICATION

Sound offers certain advantages as a medium of communication. Like visual signaling it can be transmitted efficiently over relatively long distances and altered quickly to fit changing conditions. Unlike visual and chemical signals, it can be used anywhere and at any time. Auditory communication is therefore most commonly encountered among nocturnal animals or those that inhabit dense foliage where vision is of limited usefulness. In the course of his evolution man selected sound as the medium of his remarkable language. The reason for this choice lies in large part in the great variety of signals that can be created by modifying 1) the intensity of the sound ("HELP!" versus "Help me, please"); 2) the duration of the sound ("Heeelp" versus "Help"); and 3) the repetition rate of the sound ("Help! . . . Help!" versus "Help! Help! Help!"). Birds are like human beings, in that they have exploited all three of these qualities in sound communication. Biologists use the sound spectrogram to record and to analyze bird songs.

Intensive research on bird song over the past twenty years has revealed a number of remarkable facts about this seemingly commonplace phenomenon. The most elaborate, and to human

SOUND SPECTROGRAMS of the songs of two male white-crowned sparrows. At top, the rudimentary song of a bird reared in isolation; below, the song of a bird permitted to hear singing of other sparrows. Difference between the two songs reveals the role of learning. Spectrograms record frequencies of sound on the vertical axis, and the sequence in which they were emitted on the horizontal axis.

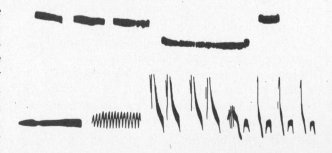

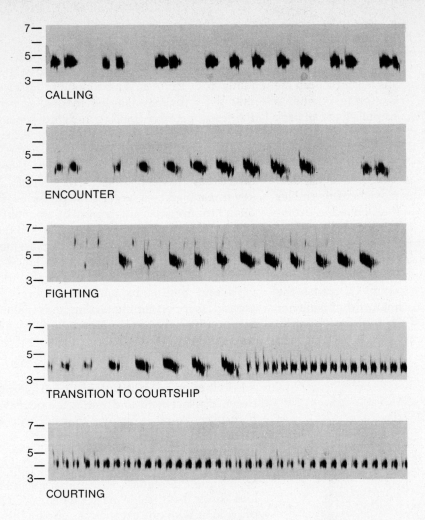

FREQUENCY (KILOCYCLES PER SECOND)

CALLING

ENCOUNTER

FIGHTING

TRANSITION TO COURTSHIP

COURTING

TIME (SECONDS)

CALLS OF A CRICKET are recorded on five spectrograms above. Because insects are relatively insensitive to tone, they vary their messages by changing the intensity and repetition rate of their calls.

5

ears the most beautiful songs are employed by males in the defense of their territories and in attempts to attract females. In at least a few species each male produces its own slight variation of the species theme. This "signature" permits it to be recognized as an individual by its territorial neighbors. A few species transmit the song from generation to generation entirely by heredity, with no learning required. Others, such as the chaffinch, must experience the singing of other individuals to acquire part or all of the song *(Figure 4)*.

The singing of crickets, cicadas, and other insects is much simpler than bird song because insects are relatively insensitive to tone —they do not perceive as many distinctions in pitch as birds. Variety is added primarily by modifying the intensity of the sounds and the rapidity with which they are produced. Examples of two songs that might be distinguished by insects are: "Cheee CHEEE

cheee CHEEE cheee . . ." and "Cheee . . . cheee . . . cheee . . ." This is the reason why insect sounds in general seem so monotonous to human ears *(Figure 5)*.

THE ORIGIN OF SIGNALS

The many remarkable signals used in animal communication did not suddenly spring into existence. They appeared in gradual evolutionary steps, usually as modifications of behavior patterns that originally had nothing to do with communication. Consider the case of the courtship displays of birds. Some have been derived in evolution from INTENTION MOVEMENTS, the preparatory motions that animals go through as they start to run, fly, attack, or engage in some other basic functional response. One example can be seen at the beginning of the take-off leap, when a bird launches its flight. The bird pulls its head back toward the body, bends its legs, and spreads or raises its tail feathers. Many kinds, including ducks, herons, cormorants, and the turkey, have modified the take-off leap to serve as a signal during courtship *(Figure 6)*. The movement resembles the true take-off from which it is derived in evolution, but it is more stereotyped and conspicuous, and no longer leads to flight. This process of the origin of a signal by modification of a behavior pattern originally having nothing to do with communication is called RITUALIZATION. Of course the animal still utilizes the ancestral forms of behavior in addition to its ritualized version. For example, the cormorant still performs a true take-off leap when it flies.

6 **RITUAL TAKE-OFF LEAP of cormorant is part of courtship displays. Stereotyped body movements, including rhythmic wing-flapping, are easily distinguished from bird's true take-off leap.**

A second form of behavior that is often ritualized to create a signal is the DISPLACEMENT ACTIVITY. Animals caught in conflict situations —for example, males confronted by rivals at their territorial boundary, or thwarted in their attempts to court females —sometimes switch to behavior that is totally irrelevant to the circumstances. The animal may suddenly begin to preen itself, or go through the motions of eating or drinking. It has been seriously suggested by many authors that cigarette smoking, a form of nonfunctional "eating," is a case of a human displacement activity. It is certainly true from an anthropomorphic viewpoint that the animals engaging in displacement activities appear to "relieve nervous tension." A similar form of behavior, called a REDIRECTED ACTIVITY, may also occur in conflict situations. In this case the behavioral act is in the right context but directed at the wrong object. For example, a monkey threatened by a larger stronger individual will often turn from this antagonist and threaten a third still smaller monkey. It is sometimes difficult to make a clear distinction between displacement and redirected activities. Consider, for instance, a rooster that turns momentarily from a rival in the barnyard and begins to peck at pebbles on the ground. Is it performing feeding movements in an inappropriate context, in other words engaging in a displacement activity? Or is it aiming its aggressive behavior at an irrelevant object, in a redirected activity?

Whether or not it is possible to distinguish between displacement and redirected activities in a particular case, it is obvious that they occur most frequently in circumstances where there is a need to communicate to other members of the same species. They are therefore convenient patterns to evolve into true signals, and animal species of various kinds have done so repeatedly. Lovebirds provide us with one excellent example. These little parrots, like most birds, tend to scratch the head with the foot when placed in frustrating situations. The response is a seemingly clear-cut case of displacement activity. The male lovebirds belonging to various species show several steps in the progressive ritualization of displacement scratching into a courtship signal. To signal his intention to mate, the male peach-faced lovebird scratches both the feathered portions of his head and his bill. The courtship scratching movements are far more rapid and perfunctory than the "real" head scratching characteristic of normal preening. In more evolved species, such as the Nyasaland and blackcheeked lovebirds, the head-scratching is missing entirely; the movements are directed only at the bill. This "emancipation" of the signal is considered the final step in the evolution of ritualization. Among birds, and other kinds of animals as well, many kinds of signals have evolved through ritualization. Besides flight movements and preening, the basic behavioral patterns involved have included eating, drinking, digging, attacking, and even urination.

ELEMENTARY SOCIETIES

A SOCIETY is a group of individuals that belong to the same species and are organized in a cooperative manner. Reciprocal communication leading to cooperative behavior is the essential ingredient. We can reasonably exclude the simplest AGGREGATIONS, such as masses of ladybird beetles or rattlesnakes that come together to hibernate, because they do not respond to outside stimuli in any fashion as an organized group. A pair of mating animals could be construed as a society only if we stretch the definition to absurd limits. On the other hand, it is useful to regard combinations of adults and offspring as elementary societies providing they are truly bound together by reciprocal communication. There is in fact a special reason for calling them societies: parental care has often served as the evolutionary stepping stone to the most complex forms of society. Some degree of parental care is displayed by a great variety of animal groups, including crustaceans, spiders, centipedes, roaches, beetles, bees, wasps, fishes, lizards, snakes, crocodilians, birds, and mammals. We are accustomed to think of the last two groups, birds and mammals, as the ones that lavish the most attention on their offspring, but this is not invariably the case. *Necrophorus*, one of the most social of beetles, nurtures its young in a very birdlike way *(Figure 7)*. The complexity of behavior displayed by the insect in nest building and in the communication between adults and young is fully comparable with that of birds —a striking example of evolutionary convergence.

Another form of very elementary society is the MOTION GROUP, the most familiar examples of which are swarms of locusts, schools of fish, flocks of birds, and herds of mammals. The members of these assemblages communicate with one another just enough to stay together as they move from place to place in search of food, water, or resting places. There are certain advantages to membership in such a group. Perhaps the most general and obvious benefit is the superior protection it provides against predators. You can witness how this works by quietly approaching a flock of pigeons or other birds that are feeding on the ground. When you come too close, your presence will probably be detected first by the individuals at the edge of the flock, who will then either move away toward the center of the flock, or take flight directly. Both responses will eventually alert the entire group. The important point is that many of the members will be made aware of your presence at a greater distance than if they had been feeding alone. This is why a predator has less chance of catching a particular bird if it is a member of a flock. Many species of social birds have special signals, such as sharp alarm cries, or conspicuous flicking of the wings, that serve to galvanize the flocks more quickly into alertness.

Once a predator launches an attack, membership in a group gives added protection to the individual. Figure 8 shows how flocks of the common European starling thwart the efforts of a predator by presenting a united front that makes assault dangerous. The peregrine falcon, one of their principal natural enemies, catches birds in flight by swooping down from above at enormous speed, in excess of 100 miles per hour. This tactic renders the hawk very vulnerable to injury by collision. It must catch a bird alone and strike it with its talons first. The starlings make this tactic difficult by clumping together in a tight formation. The falcon can hit any one bird easily enough, but it is also likely to crash into others with its head and wings as it plummets down through the flock. The falcon obtains a meal only by making passes to one side of

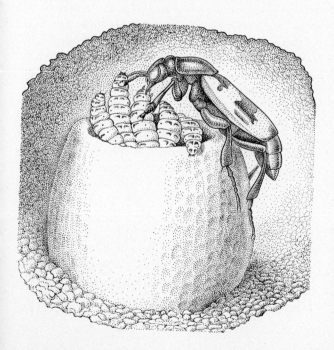

MATERIAL CARE IN BEETLES. In an underground chamber female burying beetle Necrophorus lays her eggs in a depression in a ball of rotting carrion from a dead animal. Hatched larvae beg for food like nestling birds, and mother feeds them with liquid regurgitated from her crop. 7

8 **FLIGHT PATTERN of a group of European starlings changes during attack by peregrine falcon. Starlings usually fly in scattered pattern (left), but assemble in tight formation when attacked (right).**

the flock until it is able to catch a member that has lost contact with the others through inferior maneuvering. Numerous other examples of such group defense exist in birds, fishes, and mammals. Young catfish of the genus *Plotosus* mass together in a solid ball when disturbed, their sharp pectoral fins projecting out in various directions like thorns on a cactus. Adult musk oxen, when confronted by a pack of wolves, form a tight ring with their heads facing outward and the young protected in the center. On the open tundra, where these large animals live, the circle tactic works very well against wolves, but it leaves the musk oxen helpless when hunted by men with guns. As a result, the species has been exterminated over most of its original range during the past 100 years.

Motion groups are believed to serve other functions besides defense against predators. They have been hypothesized to act, according to the species, as a means of discovering food more efficiently, of coordinating reproductive activity and increasing the likelihood of mating, and even, in the case of fish schools, of improving the ease with which the members can move through the water. Most of these ideas, however, remain to be tested by critical research.

HIGHER SOCIAL BEHAVIOR

Parental behavior, aggregations, and motion groups are but short first steps in the evolution of social behavior. How do species move beyond to more complex, highly integrated forms of organization? What we must examine next are not these higher forms of societies themselves, representing as they do the finished product, but rather the elements that have been added to the behavior of individuals to make the societies possible. The first of the ingredients is leadership, action by one or a few individuals that controls the activities of other members and orients the group as it moves from place to place. The existence of a leader is not as important for social behavior as one might at first imagine. Fish schools, for example, are led by the individuals who simply happen to be on the forward fringe at the moment. When the school changes direction a new set of temporary leaders

Reindeer herds are led from one grazing area to another by a vanguard of females, who seem to fall into their roles because they are the most timid and restless members.

takes over *(Figure 9).* Reindeer herds are led from one grazing area to another by a vanguard of individuals, mostly females, who seem to fall into their role simply because they are the most timid and restless members. They are the first to finish eating, the first to rest, and the first to start roving again. In their travels they are followed rather mindlessly by their more contented associates. In wolves, elephants, and some primates such as rhesus macaques, baboons, and gorillas, there exists a leader in the more human sense of the word: a dominant individual, almost invariably a male, who leads the other members from place to place, watches over their activities, occasionally intrudes to settle their disputes, and guards them against enemies. But such a role is developed in only a tiny minority of all the social animal species.

A second important element in the evolution of social behavior is the territory. Biologists define a territory simply as any defended area, or, to be somewhat more technical, an area occupied more or less exclusively by an animal or a group of animals by means of repulsion through aggressive defense or advertisement. We must distinguish such a place from the home range, which is all of the area that is regularly patrolled for food. Often the home range and the territory are the same; the animals defend all of the ground they patrol. But more typically the animals defend only a limited space around their nest, while sharing parts of their home range with their neighbors. Individuals of some species occupy territories only during the breeding season, and they become either solitary wanderers or join social groups at other times.

Territorial behavior varies among different species from the instant, hostile exclusion of intruders all the way to the use of chemical signposts unaccompanied by threats or attacks. The employment of aggressive behavior has been well documented in many different kinds of animals.

Dragonflies of many species, for example, patrol the ponds or lake borders in which their eggs are laid and drive out intruders belonging to the same species by darting flying attacks. A more common and somewhat less direct device of territorial maintenance consists of repetitious vocal signaling. A large part of the bird song heard in spring consists of advertisement by males warning intruders away. Additional examples that are probably familiar to you are some of the monotonous songs of crickets and frogs. Such vocalizing is not directed at individual intruders but is broadcast as a "to whom it may concern" message. There is only one restriction: normally only members of the same species pay attention.

A still more circumspect form of territorial advertisement is seen in the odor signposts laid down at strategic spots within the home ranges of mammals. An excellent example is provided by the European rabbit, *Oryctolagus cuniculus (Figure 10).* Uncontrolled populations of this innocuous appearing species are capable of damaging natural ecosystems by overbreeding and destroying large parts of the vegetation. They are also highly aggressive animals. The adults have been observed to fight with each other over territories and mates. Each territory is defended by a small group of rabbits, who mark the boundaries with little piles of dung pellets. The pellets are impregnated with a smell that originates from the anal gland and is characteristic of the particular animal that produces it. The adult rabbits belonging to a common group also spray urine on each other and their young as a further means of distinguishing friend from foe. A rabbit that intrudes from the outside is alerted to the presence of a resident group by the

SCHOOL OF FISH changes its leadership when it changes direction. Leaders at left (color) are shifted to the flank as the school makes a 90° turn, as shown in drawings at center and right.

9

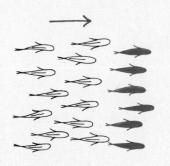

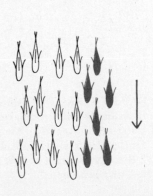

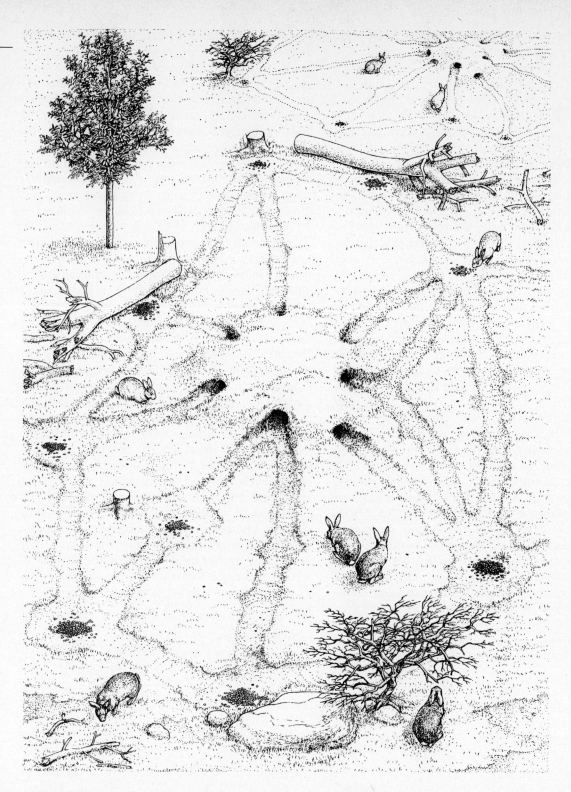

10 **RABBIT TERRITORY is occupied by eight or ten
individuals. Intersecting runs radiate from a central
warren or nest, and on them rabbits place mounds of
fecal pellets scented with secretions from anal gland.
Distinctive scent of pellets warns outsiders that the
territory is occupied.**

pervasive alien smell. If it hesitates to leave it is
attacked and driven out by the natives. A rabbit
that cannot find membership in a group has little
chance to mate and rear young, and, being deprived of a suitable burrow in which to hide, it is

destined for an early death at the hands of predators. The domestic cat utilizes a slight variation of the rabbit method. The hunting ranges of individual cats overlap extensively, and more than one individual usually contributes to the same urine signpost at different times. By smelling the deposits of previous passers-by and judging the duration of the fading odor signals, the foraging cat is able to make a rough estimate of the whereabouts of its rivals. From this information it judges whether to leave the vicinity, to proceed cautiously, or to pass on freely. The domestic dog, as you perhaps have already realized, uses much the same form of communication. The frantic attempts of the family pet to urinate on fireplugs, trees, and other prominent landmarks, and its intense interest in the urine and feces of other dogs wherever it goes, are simply manifestations of a territorial instinct not far removed from that displayed by the rabbits around their warrens.

A pattern of behavior closely akin to territoriality is individual distance, the distance an animal strives to keep between itself and another animal. It is maintained while the animal is both stationary and on the move, and is a kind of floating territory. The constant operation of the effect produces the astonishingly regular spacing patterns often seen in roosting birds and swimming schools of fish. Individual distance is sometimes maintained by overt aggression. For example, a brooding hen begins to show agitation when another hen comes within 20 feet. If the approach is narrowed to under ten feet, the hen lowers her wings and prepares to attack. In other cases, as in schools of fish, the distance is maintained by the simple avoidance of close contact among individuals. The tightness of packing in the school varies among species, and this of course is a direct result of the size of the individual distance that has evolved in each species.

DOMINANCE

The dominance hierarchy is a set of aggressive relationships that develops within a group of animals living together. It can be conveniently viewed as the equivalent of territorial behavior in which the participating animals all coexist within one territory. The result is that one of the group members acts as the territory holder and comes to dominate the others. When food is limited, it eats while the others wait. It takes precedence in the selection of mates and resting places. This "alpha" individual makes frequent use of threats to main-

tain its position, and usually resorts to violent attack if the threats prove inadequate. Below the dominant individual there may exist a second, "beta" animal that controls the rest of the membership in the same way, and beneath it a third-ranking individual, and so on. Because biologists used flocks of chickens in many of the early studies of dominance hierarchies, the hierarchies are sometimes referred to loosely as peck orders. Of course there are many other ways in the animal kingdom of establishing status besides the pecking, wing-flapping, and spur-gouging of the sort practiced by barnyard fowl. Dominance hierarchies are very widespread among social animals of many very different kinds. They are known, for example, in social wasps, bumblebees, hermit crabs, fish, lizards, birds, and mammals. The relationships display one or the other of several geometric patterns. They may consist of simple despotism, in which a single individual is in complete charge and all the subordinate members are of equivalent rank. They may take the form of a straight chain, in which the alpha individual dominates all other members, a second-ranking individual dominates all but the alpha, a third-ranking individual all but the alpha and beta, and so on down the line. Finally, they can include closed loops, in which one member dominates a second, who dominates a third, who in turn dominates the first (*Figure 11*).

Hierarchies are formed in the course of the initial encounters among the group members by means of threats and fighting, but after the issue has been settled, each member gives way to its superiors with a minimum of hostile exchanges. The life of the group may eventually become so peaceable as to hide the existence of such ranking from the observer —until some minor crisis happens to force a confrontation. Troops of the yellow baboon (*Papio cynocephalus*), for example, often go for hours without enough hostile exchanges to reveal the form of the hierarchy. Then in a moment of tension —a quarrel over an item of food is

How can any form of submissive behavior evolve through natural selection? One explanation holds that it is better to be a submissive member of a group than not to belong at all.

sufficient—the ranking is displayed with startling clarity. In its more pacific state the hierarchy is often supported by status signs. The identity of the leading male in a wolf pack is unmistakeable from the way he holds his head, ears, and tail, and the confident, face-forward manner in which he approaches other members of his group. In the great majority of encounters he is able to control his subordinates without any overt display of hostility. Fighting leading to injury or death is very rare.

Control and hostility—these are the harsh ingredients of social organization considered up to now. It is easy to understand how such selfish behavior can be fashioned by straightforward Darwinian natural selection. Numerous experiments with a variety of social animals have confirmed that it pays to be on top. Dominant animals are healthier, live longer, and reproduce more frequently. It is harder to understand, at least at first glance, why animals should subordinate themselves to their dominant group-mates in such a resigned way. How can any form of submissive behavior evolve through natural selection? One

explanation that appears to be emerging is that it is better to be a subordinate member of a group than not to belong at all. Among European rabbits, for example, each group of eight to ten individuals is organized into a dominance hierarchy. The top-ranking males and females enjoy access to the burrows at the center of the warren, and as a result they enjoy longer lives and produce more offspring. The lower ranking animals are not so fortunate, but at least their membership in the group gives them some chance of surviving and reproducing. If they were forced to wander on their own, outside the territorial limits of the established groups, they would soon be caught by a predator or die of starvation. In short, half a loaf is better than no loaf for the social animal.

ALTRUISM AND KIN SELECTION

Compromise within dominance hierarchies is not the whole story behind cooperative behavior. A kind of altruism also exists in some animal societies. Altruism is defined in biology, as in everyday life, as self-destructive behavior for the benefit of others. An altruistic animal is one that reduces its own fitness, its chances of survival and of producing offspring, to increase the fitness of another animal. Altruism is expressed in many forms. A mother bird stays on her nest to resist a predator's attack on her newly hatched young, and

11 PECK ORDERS in chickens. Dominance hierarchy takes three forms. Alpha chicken may dominate all others (left); a second chicken may dominate all but the alpha bird (center); or birds may form a closed loop (right) in which one chicken dominates another, who dominates a third, who dominates first.

risks her own life in the process. A worker bee flies out of the hive to sting an intruder. Because the sting is lined with barbs that anchor it in the enemy's flesh, it remains in place and part of the viscera is pulled out with it when the bee flies off. The result is invariably fatal to the bee. A male baboon falls back to challenge a leopard stalking its troop. The troop escapes, but perhaps at the cost of the life of the male.

How can such acts be explained in ordinary biological terms? Apparently altruistic individuals are less likely to survive and to reproduce their own kind than are the more cowardly and selfish individuals they benefit. According to the elementary Darwinian theory of natural selection, their genes will eventually vanish, and the altruistic behavior programmed by the genes will also disappear. How, then, can altruism come into existence in the first place? The answer generally accepted by biologists is KIN SELECTION, meaning that the altruistic act preserves genes in other individuals that are shared with the altruistic individual by common descent. These individuals are not just offspring. By definition, kin selection affects other kinds of relatives, including brothers and sisters, parents and even more distant categories such as aunts, uncles and cousins. Notice that the baboon does not risk its life in behalf of the leopard, or even baboons in other troops. It does so for individuals that are probably related to itself and therefore more likely to share the genes, including the ones that caused the altruistic behavior. There is a theorem in population genetics which states that an altruistic gene will be favored in natural selection provided the loss it causes to the individual possessing it is compensated for by an improvement in the fitness in the relative that receives the benefit. The more distant the relative, the greater the improvement that is required. For example, about ½ of the genes of a given animal are shared by its brother (or sister). Suppose that an altruistic gene caused the animal to sacrifice its life, or at least to give up having offspring, for the benefit of the brother. In order for the altruistic gene to be favored, the brother would have to have more than twice as many children as a result of the altruistic act. First cousins share ⅛ of their genes on the average. Complete sacrifice for first cousins would thus have to result in more than an eight-fold increase in their fitness for the altruism to gain as a genetic trait. Less than complete sacrifice could be attended by smaller benefits accruing to the relatives. Altruism directed simultaneously at various kinds of relatives, offspring as well as siblings and cousins, would result in a summing of the benefits and an acceleration of the evolution of the altruistic behavior. Kinship, together with the altruism it makes possible, has been one of the essential ingredients in the evolution of the most complex forms of social behavior.

DIFFERENCES

Billions of years of evolution have seen two kinds of organisms, the vertebrates and the insects, climb separate pathways to the highest pinnacles of social behavior. Biologists are fortunate to have both of them for close comparative study. The social insects differ so much from the vertebrates in phylogenetic origin, brain structure, and elemental modes of communication, that for all practical purposes they might as well have originated on some other planet. Perhaps they are the closest we shall ever see to a truly independent form of social being.

The most complex vertebrate societies, and particularly those of birds and mammals, are distinguished from insect societies by a single trait so overriding in its consequences that the other characteristics seem to flow from it. This trait is personal recognition among the members of the group. As a rule each adult animal knows and bears some particular relationship to every other member. Status is extremely important. Where dominance hierarchies exist, elaborate signaling is employed to implement them, as we saw in the case of the wolves. Parent-offspring relationships are specific to individuals, tightly binding, and of relatively long duration. Within primate societies, personal groupings often ascend to the level of the cliques, and, in the case of the chimpanzee, even to the level of child care by "uncles" and "aunts"—adults who form temporary alliances with parents by catering to their offspring. Group members spend large amounts of time and energy in establishing and maintaining these many-sided personal bonds. Associated with the personal form of communication is a prolonged period of socialization of the young. By constant experience, much of it acquired in the form of play, the young animal leaves its personal relationship to its parents and establishes an early status among its age-peers. Some division of labor exists: there are weakly defined specialists in defense, leadership, and foraging. But it is not based upon physiologically different castes. Each member of the group is basically the same, and pursues its own interest as best it can.

In contrast, the higher insect societies, and particularly those of the ants, wasps, bees, and termites, are for the most part impersonal. Limited dominance hierarchies are found in a few species. However, with most social insects personalized relationships play little or no role. The sheer size of the colonies and the short life of the members makes it inefficient, if not impossible, to establish individual bonds. The average adult honeybee, for example, lives for only about one month in the midst of a rapidly changing population of up to 80,000 hive members. An army ant worker has only a few weeks or months to become acquainted with a million or more nestmates. Of course it cannot do this, nor does it even try. Even the relationship of the workers to the queen of the insect colony is impersonal. She is recognized by a small set of pheromones. When little pieces of cork are impregnated with these substances, extracted from the queen's body, the pieces of cork are treated for a while as if they were the queen. Where songbirds have individual calls that are identified by their territorial neighbors, and mammals use individual scent marks, members of an insect colony employ signals that are for the most part uniform throughout the species. The single known exception is the colony odor, which is used to distinguish nestmates, all of them together, from all members of alien colonies. Socialization is minimal in the insects, and play is apparently absent.

If vertebrates had followed the same evolutionary route as insects, the vertebrate society could have evolved into an unimaginably more complex, efficient unit —at the price of independent action on the part of its members.

Yet the insect colony, as a unit, can equal or exceed the accomplishments of the nonhuman vertebrate society —as a unit. The nests of some termites, for example, are more complex in design than those of any mammal or bird. The raids of army ants are more intricate and coordinated than those of a wolf pack. The insects achieve their results with rigid divisions of labor among castes. Colony members are specialized to perform in re- production (the queens), in defense (the soldiers), and in ordinary labor (the workers). The workers often specialize further, as nurses, nest-builders, or foragers. When all of these specialists combine their efforts in the correct way, the colony functions as a superorganism capable of matching the feats of vertebrates. Had vertebrates followed the same route in evolution, the results might have been spectacular. The individual bird or mammal, possessing a brain vastly larger than that of an insect, might have been shaped into a comparably better specialist. The vertebrate society then could have evolved into an unimaginably more complex, efficient unit —at the price, of course, of independent action on the part of its members. Vertebrates have instead remained chained to the cycle of individual reproduction. This forever enhances freedom on the part of the individual at the expense of efficiency on the part of the society.

PRIMATE SOCIETIES

It is among the higher primates that the apogee of the vertebrate form of society is found. If man, with his unique language and revolutionary capacity for cultural transmission is excluded, the primate society can be regarded as a logical extension of the personalized association found in birds and lower mammals. The difference is one of degree and not of kind. An efficient way, therefore, to gain a view of vertebrate societies as a whole is to examine one particular primate species in some depth.

One of the best-studied primates is the rhesus monkey, or rhesus macaque as it is sometimes called, the species of monkey most frequently used in laboratory research. Its scientific name is *Macaca mulatta*. It is but one of 12 species of macaques, comprising the genus *Macaca*, that collectively range from North Africa through the warmer portions of Asia all the way to Japan. The native home of the rhesus monkey is India and the immediately surrounding countries, where it occurs both in forests and in temples and villages. The monkeys rove in bands of from under ten to as many as 78 individuals. They are confined to home ranges of about 8 square kilometers (3 square miles), which overlap broadly with the ranges of other bands. The bands maintain a movable territory: they defend only the ground they occupy at the moment. The defense usually consists of noisy and conspicuous threats, including the peculiar tree-shaking display illustrated in Figure 12. Occasionally fighting breaks out among the groups, and in extreme cases it can lead to injuries. In spite of the isolation that this hostility enforces, a small number of males manage to drift away from their own group from time to time and gain membership in a neighboring group. The result is a continuous gene flow that helps keep the species relatively uniform over its range.

The rhesus society is based to a large degree upon aggression and status. Each adult male occupies a position in a rigid dominance hierarchy that is maintained, virtually moment by moment, through an extraordinarily diverse set of vocal signals, body postures, and facial expressions. The adult females, and to a much lesser extent the juveniles, are also organized into loose hierarchies.

12 **HOSTILE BEHAVIOR** in rhesus monkeys serves to defend their territories and to establish dominance without actual combat. Tree-shaking and ground-slapping are aggressive displays. Fear grimace (sometimes called the fear grin) signals submission and usually causes the attacker to desist.

Some of the forms of communications are shown in Figure 13. In hostile exchanges, which break out frequently, the vocabulary of signals is used to communicate intensity of feeling and, with it, the probability that the signaling animal is about to attack or submit. An individual that is merely expressing a superior attitude toward another may do nothing more than look in the direction of the rival for a brief time. Greater hostility is expressed by a sustained stare. The next time you walk into a room containing caged (and therefore unhappy) rhesus monkeys, notice their response to you. Some will probably hold their bodies rigidly while giving you what appears to be a prolonged, curious stare. The monkey is not expressing curiosity; it is attempting to convey its feeling of hostility, to put fear into you. As aggression mounts among competing monkeys, the next strongest signal is opening the mouth wide. To a human observer the monkey looks astonished, but it is simply angrier.

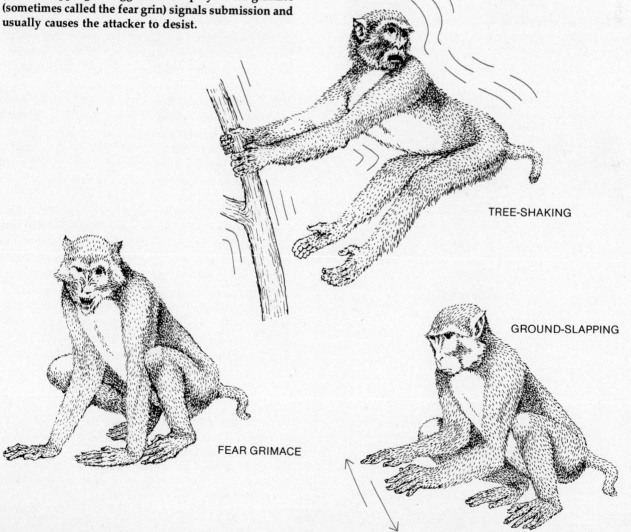

TREE-SHAKING

GROUND-SLAPPING

FEAR GRIMACE

If the opponent does not yield, the monkey may bob its head up and down while continuing to hold its mouth open. A still stronger signal is to lunge forward while slapping the hands on the ground (*Figure 12*). The monkey sometimes accompanies

WALK OF DOMINANT MALE

WALK OF LOW-RANKING MALE

GROOMING

this by shouting "Ho!" in a loud, gruff voice. A highly excited monkey goes beyond these threats and bluffs to chase, hit, and bite its opponent. Hostile actions are matched in frequency by actions of submission. For every serious encounter there must be both winner and loser. The dominated animal is normally able to blunt the aggression of its opponent by simply turning its own head away, the direct opposite gesture to the stare. If this does not work the monkey will grimace, giving the peculiarly mirthless grin illustrated in Figure 12. It may walk away, or crouch down in the sexually submissive posture of the female — even males will adopt the position. A monkey that is being terrorized by a very aggressive opponent will, if all else fails, flee or emit a piercing "Eee!"

Hostile encounters are not random events. They occur principally between individuals of nearly equal rank in the dominance hierarchy or when circumstances cause a momentary sharp conflict of interest between two monkeys of unequal rank. Most of the time, order is maintained by a recognition of status among the monkeys. An individual displays its rank by its posture. High-ranking males are instantly recognizable by their manner. With their heads held high and their tails pointed upward with a little crook in the end, they stroll along slowly and deliberately, regarding each monkey in their path with a seemingly indifferent expression. A low-ranking monkey shows the opposite behavior. When faced by a leading male it moves discreetly out of his line of vision. Its head is kept low, its gaze averted, its tail lowered. Its movements are nervous and furtive (*Figure 13*).

These status signs are used as means of avoiding open conflict with group members of different rank. Rhesus monkeys also have a repertory of conciliatory signals, which include lip-smacking and grooming behavior. Conciliation leads to sexual behavior. It also produces something akin to friendship among members of the same sex, which may produce a relationship that can only be described as an alliance. Pairs of males sometimes work together to enhance their status during hos-

STATUS SIGNS among rhesus monkeys include posture. High-ranking males walk with heads held high and tail hooked upward, in contrast to slouch of low-ranking male. Grooming is conciliatory behavior that leads to mating among members of opposite sex, or friendship with those of same sex.

13

There can be no question as to the advantages won by dominant male monkeys: they have first access to food and mates. Subordinate males mate only with females not claimed at the moment by their superiors.

tile encounters with other members of the troop. So important and pervasive are aggression and rank in the rhesus society that they even dominate the play of the young animals. Juveniles imitate their elders by chasing, wrestling, and biting, as well as by mounting each other in mock sexual activity.

There can be no question as to the advantages won by dominant males: they have first access to food and mates. Subordinate males are able to mate also, but less frequently and only with females not claimed at the moment by their superiors. A female that becomes the sexual partner of a dominant male temporarily acquires his social position and thus gains a greater measure of security during at least part of her pregnancy.

The rhesus society is governed by a communication system that far exceeds in complexity those controlling the simple schools, flocks, herds, and other elementary societies with which we began this review. When the full range of sexual and parent-offspring interactions are added to the codes of territory and dominance, the total repertory of communicative acts known to be utilized by a rhesus society comes to 56. Add to this 27 combinations of these same signals which when presented simultaneously convey still different meanings, and the number of discrete signals is increased to 83. The individual rhesus monkey not only utilizes most of these signals in the course of its life, but it directs them selectively toward specific group members, each of which has a particular significance for its rank within the society and the scope of its daily activities. *Macaca mulatta* is thus a species totally committed to a social existence in the vertebrate mode. A large part of its forebrain is devoted to this part of its life. The forebrain is so large, in fact, because of the requirements placed on it by the exceptionally complex social existence of the species.

The study of primate societies is a relatively new subject. The bulk of our knowledge comes from studies conducted in the field in Africa, Asia, and tropical America only during the past 20 years. Many of the species of monkeys, baboons, and anthropoid apes have been demonstrated to live in societies as complexly organized as those of the rhesus monkey. However, there is a surprising amount of variation among the species in some of the basic details. Territoriality and dominance behavior are particularly subject to variation. The rhesus monkey and its fellow macaque species, together with the baboons, occupy one end of the scale in aggressive behavior. Some anthropoid apes, particularly the chimpanzee and the gorilla, are nearer the opposite, more benign, end of the scale. Bands of gorillas, for example, roam broadly overlapping home ranges but do not interact aggressively when they meet. Each group is typically under the very tolerant control of a single older male. Fights occasionally occur between members of the same band, but they are less frequent and less violent than fights in the rhesus troops. We do not yet have an adequate evolutionary explanation of these differences in the primate societies, which promise to shed light on the origins of our own social behavior.

INSECT SOCIETIES

The typical insect society is a family. The central figure is the queen, the mother of the colony. Her daughters, and in the case of termites her sons as well, make up the worker force. Their sex can be identified, but they are smaller in size than the queen, and they do not normally engage in reproduction. In extreme cases their gonads are completely degenerate. The worker's role is to protect the queen and to rear its brothers and sisters. Their altruism in accomplishing these tasks is nearly total. When the colony has reached a certain size, and is in good health, some of the offspring develop into fully functional males and virgin queens. These individuals mate, travel to new locations, and become the parents of new colonies.

Social life organized in this most advanced fashion, based upon an altruistic worker caste of reproductive neuters, has originated at least 12 times in the insects: at least twice in the wasps, once or twice in the ants, at least eight times in the bees, and once in the termites. The wasps, bees, and ants all belong to the single order Hymenoptera, while the termites comprise the entire order Isoptera. The predilection of the Hymenoptera for altruistic social evolution is a striking curiosity. It can be at least partially explained in the following way.

Sex is determined in the Hymenoptera by haplodiploidy, which means simply that females come from fertilized eggs and males from unfertilized eggs. Males are therefore genetically haploid, and are also homozygous. Sisters are genetically very close to each other, because each one receives identical genetical material from the father. And sharing more genes than is usually the case between sister animals, they are more prone to behave altruistically to each other. The reason is that to favor such an exceptionally close relative is to favor —in large part —one's own genes.

ANT ORGANIZATION

Figure 14 depicts part of a colony of a relatively primitive ant species. Many of the basic features of social life in ants, and social Hymenoptera as a whole, are recorded in this drawing. The queen and worker castes are seen to be anatomically distinct from each other. The queen is a normal female, rather closely resembling the females of the solitary wasps from which the ants evolved. In

14 COLONY OF BULLDOG ANTS (Myrmecia gulosa) shows much of the social behavior characteristic of Hymenoptera in general. Queen stands at left of chamber with a winged male behind her. Workers to the right attend larvae and pupae of queen. Worker at far right doubles up its body as it lays an egg. In a moment it will feed egg to one of the larvae, as another worker is doing at bottom center. Cocoons containing pupae—also offspring of the queen—are piled at the back of the chamber.

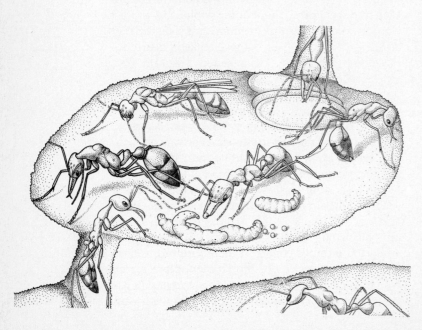

addition to an ordinary and fully functional reproductive system her thorax is constructed for flight. Each mother queen of the kind shown in the drawing is equipped with two pairs of wings when she first emerges from the pupa. She uses them to fly from the parent nest and to participate in a nuptial flight with other young queens and males. After mating she descends to the ground, breaks off her own wings at the base (you can see the dark stumps in the drawing), and sets out to start a colony of her own. This she accomplishes by digging a short burrow in the ground and laying her first small cluster of eggs. When the larvae hatch, they are fed by nutrients metabolized from the queen's wing muscles, together with insects the queen is able to capture in the vicinity of her little nest. All of the members of the first brood are destined to mature into workers. As shown in the drawing, workers are females, but they are smaller than the queen. They never possess wings or wing muscles, and their thoraces are proportionately more slender. Those of a few genera, such as *Myrmecia*, lay eggs, but in most cases feed them promptly to the larvae or another adult member of the colony. Soon after the first group of adult workers emerge from the pupal stage, they take over all of the labor from the queen, from nest building to foraging for food and the care of their immature nestmates. The queen now becomes a quiescent egg-laying machine. Once this point in colony development is reached, the population grows at an accelerating pace. When the colony reaches a mature size, which varies according to species, it begins to produce virgin queens and males. These reproductively competent individuals then depart on nuptial flights of their own to commence a new colony life cycle. The parent colony meanwhile continues to live and grow by the addition of more workers. Each year it produces a new crop of virgin queens and males. So long as the queen lives, which in some species is as long as ten or 15 years, the colony she founded continues to function and to reproduce.

Members of the ant colony coordinate their activities by means of a surprisingly rich communication system. There are signals of touch, including the moment-by-moment jostling, tapping, and licking that can be easily observed by watching a colony through the glass wall of an observation nest. Some ant species also communicate in part by stridulation, the production of sound by rubbing certain body parts together. The effect is a faint squeak, like the sound produced when a piece of metal is scratched lightly against glass or

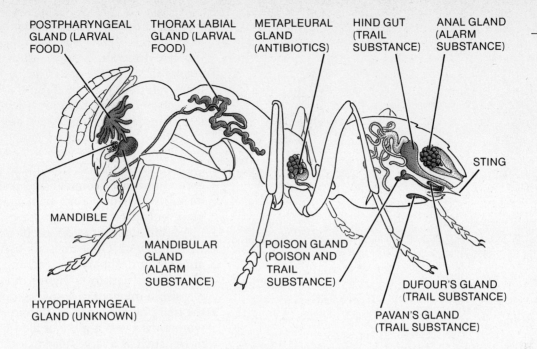

POSTPHARYNGEAL GLAND (LARVAL FOOD)

THORAX LABIAL GLAND (LARVAL FOOD)

METAPLEURAL GLAND (ANTIBIOTICS)

HIND GUT (TRAIL SUBSTANCE)

ANAL GLAND (ALARM SUBSTANCE)

STING

MANDIBLE

MANDIBULAR GLAND (ALARM SUBSTANCE)

POISON GLAND (POISON AND TRAIL SUBSTANCE)

DUFOUR'S GLAND (TRAIL SUBSTANCE)

PAVAN'S GLAND (TRAIL SUBSTANCE)

HYPOPHARYNGEAL GLAND (UNKNOWN)

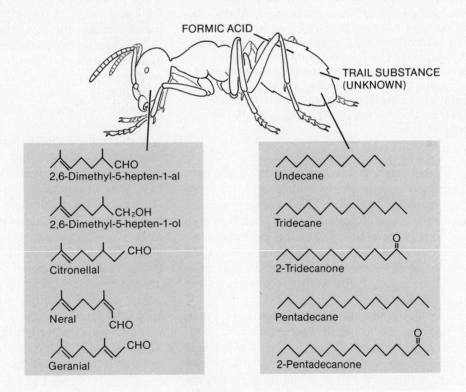

FORMIC ACID

TRAIL SUBSTANCE (UNKNOWN)

2,6-Dimethyl-5-hepten-1-al

2,6-Dimethyl-5-hepten-1-ol

Citronellal

Neral

Geranial

Undecane

Tridecane

2-Tridecanone

Pentadecane

2-Pentadecanone

some other very hard surface. The ants use the signal as a call for help, when they are buried in a cave-in or under attack by an enemy. Most communication, however, is by means of pheromones, chemical signals passed back and forth among members of the same colony. As a rule the pheromones are produced as secretions of

CHEMICAL LANGUAGE OF ANTS consists of pheromones secreted by exocrine glands (color) in drawing at top. Other glands produce special food for larvae or antibiotics to control fungi and bacteria. Species shown at top is Iridomyrmex humilis. Simplified chemical formulas at bottom are alarm pheromones of Acanthomyops claviger. Lines indicate location of glands that secrete them.

15

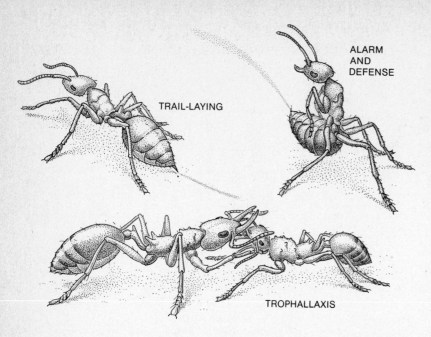

TRAIL-LAYING

ALARM
AND
DEFENSE

TROPHALLAXIS

16 THREE MAJOR TYPES OF SIGNAL are included in the chemical repertory of various ant species. At top left a fire ant worker lays an odor trail from its extruded sting. As the pheromone evaporates its vapor guides other workers. At top right a worker of Formica polyctena sprays a mixture of formic acid and undecane at an enemy. The acid repels the intruder, and the alarm pheromone alerts other ants in the nest. In bottom drawing a Daceton worker regurgitates food to a larger nestmate.

glands that open to the outside of the body, and transmitted to nestmates in liquid or gaseous form. When tasted or smelled, each pheromone induces its own particular response. The vocabulary of ants thus consists of a set of chemical substances released one by one or in combinations to direct the behavior of other colony members. The glandular sources of many of the known pheromones are shown in Figure 15.

Figure 16 shows ant workers engaged in three of the principal forms of chemical communication. ALARM SUBSTANCES are released when a worker is attacked or when it encounters an intruder in the vicinity of the ant nest. Often the pheromones are sprayed out with poisonous secretions used in defense. In ants of the genus *Formica*, as illustrated in the drawing, the worker discharges a stream of formic acid mixed with undecane. The acid serves as a poison to repel invaders, while the undecane serves as a pheromone to alert other worker ants. In still other species, the poison and the pheromone are the same. This is the case for cit-

ronellal and other alarm substances of "citronella ants" (genus *Acanthomyops*) in Figure 15. Because the two functions are so intimately connected among the ants, the use of the secretions is sometimes referred to as an alarm-defense system. At low concentrations, the alarm pheromones are typically simple attractants, which arouse the curiosity of workers and draw them closer to the source of the discharge. As the workers enter zones of ever higher concentration, they become increasingly excited and may commence a true alarm frenzy, during which they rush frantically from place to place and attack any foreign object they encounter.

A more sophisticated kind of pheromone is the TRAIL SUBSTANCE. Odor trails are used by workers to guide their nestmates to food and to new homes. The species shown in Figure 16 is a fire ant *(Solenopsis)*, a member of a species that manufactures the trail pheromone in its Dufour's gland. When a worker discovers a piece of food too large for it to move, such as a large dead insect, it heads homeward dragging its extruded sting behind it. The trail substance is dispersed from the Dufour's gland through the sting onto the ground. As the liquid trace evaporates, it forms a long, narrow cloud of attractive vapor through which other workers move in search of the food. An identical form of trail communication is used when worker ants discover a superior nest site and attempt to recruit other members of the colony to it.

A still more elaborate form of communication is TROPHALLAXIS, the feeding of liquid substances by one worker to another. The most common and basic form of trophallaxis, also illustrated in Figure 16, involves the regurgitation of stored liquid food from the gut. Trophallaxis plays at least two major roles in the organization of ants and other social insects. First, it homogenizes the stomach contents of the colony members. This process was demonstrated in experiments in which radioactive syrup was fed to single workers who were then allowed to return to their nests. At regular intervals afterward, other workers were removed from the nest and their stomach contents examined for radioactivity. It was found that regurgitation rapidly diffuses liquid food throughout the colony. In some species, every worker of the colony received at least some of the material within 24 hours. Rapid trophallaxis ensures that each member of the colony has roughly the same stomach content as every other member. By this means it is made continuously aware of the nutritional condition of the colony as a whole. When the worker acts for itself out of hunger or thirst, it acts for the colony as a whole. Trophallaxis is a form of altruistic behavior that stands in sharp contrast to the selfish maneuvering for status and resources that characterizes the primate society. It also serves a secondary role in the transfer of some of the pheromones, which must be tasted or even eaten in order to produce an effect. The best known of such pheromones is the queen substance of the honeybee, which is discussed later in this chapter.

Since their evolutionary origin about 100 million years ago, the ants have undergone an impressive amount of adaptive radiation *(Chapter 31).* The more than 10,000 species that exist today together range from the arctic circle to the tips of Africa and South America, and they rank among the dominant land animals of the earth. The extreme specialists include the army ants, whose colonies march in packed swarms in search of prey and migrate from one nest site to another at regular intervals. Hundreds of species of army ants are found in various parts of the tropics. The most famous are the driver ants *(Anomma)* of Africa. The

largest colonies contain over 20 million workers, every one of which is the daughter of a single huge queen. These ants attack and consume almost every kind of animal they can trap. However, the largest animals, particularly reptiles, birds, and mammals, usually find it easy to evade the rather slow moving swarms. Another highly evolved group are the fungus-growing ants. These insects are limited to the Western Hemisphere and particularly to the tropics, although one species occurs as far north as New Jersey. The workers collect leaves, flower petals, or insect dung, and raise fungi on these materials deep in the moist chambers of their nests. The hyphae and other growth forms of the fungi form the exclusive food of the ants. The fungi are also highly evolved for the mutualistic association. They are found only in the nests of their ant hosts.

Stranger still is the phenomenon of social parasitism, in which the colony is exploited by some other kind of animal which passes itself off to some extent as a colony member. An especially bizarre form of the symbiosis is slavery. Certain ant species, including those that belong to the genus *Polyergus (Figure 17),* are unable to form independent colonies on their own. Instead, they rely on slave labor for all of the necessities of life. The slave workers, which are former members of colonies belonging to another ant species, are required to build the nests, forage for food, and care for the young of the *Polyergus.* The *Polyergus* cannot even feed themselves—they must be nourished with liquid food regurgitated to them by the slaves. They are, however, superb at fighting and raiding. When a *Polyergus* scout finds a nest of potential slaves, always members of the genus *Formica,* she runs back to her home nest laying an odor trail. This signal brings out an aggressive force of *Polyergus* workers who march quickly to the *Formica* nest. Using their sharp, saberlike mandibles, they dispatch the defending *Formica* workers and carry away their pupae. When the *Formica* pupae hatch later into adult workers, they accept their *Polyergus* captors as sisters and care for them.

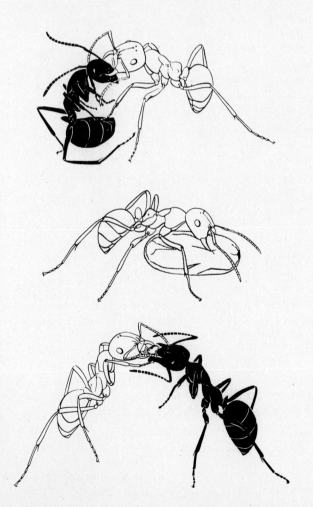

SLAVERY IN ANTS. Workers of the slave-making 17
species Polyergus rufescens have mandibles specially
modified for fighting. At top a Polyergus worker (light
color) kills a Formica worker during a raid on a Formica
colony. At center Polyergus carries a Formica pupa
enclosed in a cocoon back to its own nest. At bottom it
is fed by a Formica adult that has emerged from a
captive pupa.

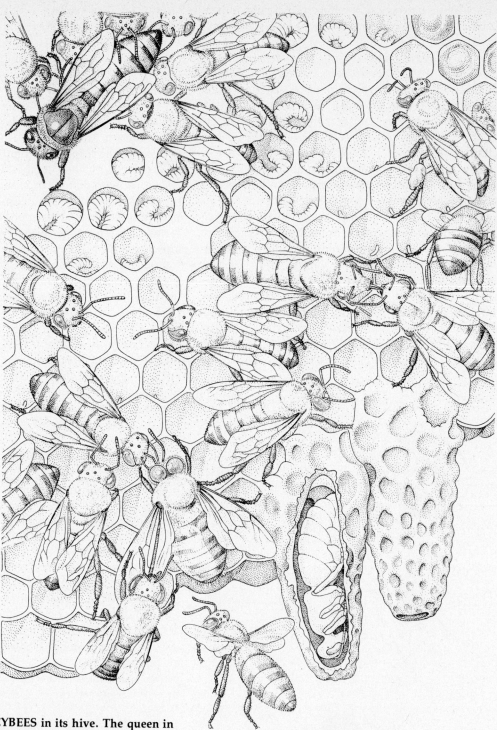

18 COLONY OF HONEYBEES in its hive. The queen in upper left corner is surrounded by attendants. Two bees at right center exchange pollen and nectar. Many of the open cells contain eggs or larvae. Cells at lower right contain pupae of future queens. Male (drone) is visible to left of pupae.

Slavery for them, in other words, is a wholly unconscious and involuntary association.

HONEYBEE ORGANIZATION

Figure 18 illustrates the life of another highly advanced form of social insect, the honeybee. Each hive of bees contains from about 20,000 to 80,000 workers, all of which are females. There is a single mother queen, who exercises a remarkable kind of dominance over her daughters. From her mandibular glands she secretes a continuous supply of a pheromone called queen substance (9-keto-2-decenoic acid). The workers eagerly lick this material up along with other attractive pheromones secreted by the queen's body and pass it among themselves by regurgitation. The queen substance has two inhibitory effects: it prevents the workers from rearing other queens, a psychological effect, and it inhibits the ovaries of the workers from producing eggs, a physiological effect. If the mother queen dies or is deliberately removed by the beekeeper, the resulting shortage of queen substance is felt within hours by the workers. They commence building special, enlarged cells, inside which new queens are reared. A few days later, some of the workers also undergo an enlargement of their ovaries and develop the capacity to lay eggs. The queen substance, then, is a subtle but powerful device which the mother queen uses to prevent the creation of rivals for her supreme reproductive position.

The life cycle of the honeybee is basically similar to that of ants, but with one outstanding difference. When the queen successfully completes her nuptial flight by mating with one or more drones, she does not attempt to start a new colony on her own. Instead, she returns to the parent hive and becomes the new queen there. But what, you may ask, happened to the old queen? The answer is that she left before the new queen emerged from the pupa. She was accompanied by a flying swarm of thousands of workers, who comprised a large fraction of the colony population. Together they searched for a new hive or suitable nesting site. Thus the honeybee colony multiplies in a curious way that is the reverse of the usual procedure in animals generally: it is the parent, not the offspring, who leaves the safety of the established territory and assumes the role of pioneer.

Honeybees are famous for the use of the waggle dance in the recruitment of nestmates to newly discovered food sources and nest sites. The waggle dance of the honeybee is the most famous of all the forms of animal behavior. First "decoded" by Karl von Frisch of Germany in 1945, it is a nearly unique symbolical language by which worker bees communicate the position of food or a new nest site to their sister bees. The waggle dance is easy to understand if one thinks of it as a ritual flight—a scaled-down version of the journey from the nest to the nectar. The essential element in the maneuver is the "straight run," the middle piece in the figure-eight pattern (*Figure 19*). (The remainder of the figure-eight consists of a doubling back to repeat the straight run.) The dancing bee has just returned from several back-and-forth journeys to the target. The straight run it performs is a miniaturized version of the outward flight which it now invites its nestmates to undertake. If the bee

WAGGLE DANCE of honeybee, performed on vertical surface of the comb within the hive, indicates the location of a source of nectar. Direction of straight run (wavy arrow) relative to the vertical indicates direction to food relative to position of the sun. Dance at left indicates a food source that lies directly toward the sun. Dance at right indicates a food source that lies at an angle (Θ) to the right of the sun. The length of the bee's straight run indicates the distance to food source. **19**

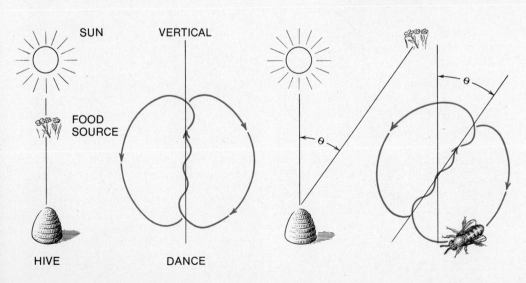

performs on a horizontal surface with a view of the sun or open sky to guide it, the run points directly at the target. Usually, however, the dance is conducted on a vertical comb surface inside the darkened hive. In this case the angle between the straight run relative to the vertical indicates the direction to the target relative to the position of the sun. The duration of the straight run indicates the distance to the source: the longer the straight run takes to complete, the farther away the target. The straight run is emphasized by a rapid waggling motion of the body.

Less well known is the fact that honeybees, like ants, rely primarily on chemical communication to organize the remainder of their social life. In addition to the queen substance just described, there are alarm substances, several kinds of attractants used to recruit nestmates, a peculiar footprint

20   **COLONY OF TERMITES (Amitermes hastatus) consists of interconnecting chambers. In the central chamber the primary queen, her abdomen swollen with eggs, rests with the primary king. Secondary queen occupies chamber at left, while immature queens and males occupy upper chamber. In chamber at lower right corner a soldier (distinguishable by its large mandibles) is accompanied by a nymph that is developing into a soldier. Elsewhere workers attend piles of eggs and larvae wander about.**

The most primitive living termite, Mastotermes darwiniensis of Australia, has reduced entire ranches to dust in one or two years —house, fences and all.

pheromone applied by the tarsi and used as a kind of trail substance, and a hive odor by which nestmates are distinguished from strangers. All of these classes of pheromones have close parallels in ant colonies.

TERMITE ORGANIZATION

Figure 20 shows a section of a termite colony. The most remarkable fact about these insects is that they are so similar in the details of their social organization to the ants, despite having an utterly different phylogenetic origin. Termites can be thought of almost quite literally as social cockroaches. They originated evolutionarily from primitive cockroaches no later than the middle of the Cretaceous Era, over 100 million years ago.

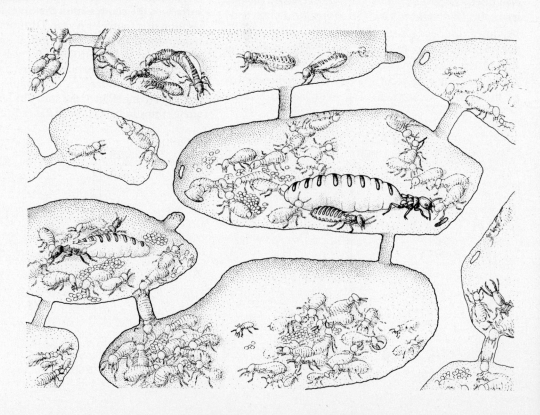

Like the ants, they have become dominant land animals and have produced a wide range of adaptive types. They differ basically from ants, however, in their food habits. From the beginning the termites have been eaters of dead wood and other dead vegetable material. Their great achievement in evolution, next to the invention of society itself, was to enter into symbiosis with protozoans and bacteria capable of digesting cellulose. The termites chew up the vegetable fibers and pass them down their intestine, where the cellulose is broken down by hordes of the little symbionts. In exchange for part of the nutrients the protozoans and bacteria are given protection and a constant supply of new raw materials. The combination is an extremely efficient one, enabling termites to become the chief animal decomposers of wood over much of the world. The most primitive living termite species, *Mastotermes darwiniensis* of Australia, is also one of the most formidable insect pests in the world. Its workers have been observed to attack poles, fences, wooden buildings, wharves, bridges, oil-soaked wood, living trees, crop plants, wool, horn, ivory, vegetables, paper, hay, leather, rubber, sugar, flour, human and animal excrement, billiard balls, and the plastic linings of electric cables. Entire ranches have been reduced to dust in only one or two years—house, fences, and all!

The members of termite colonies communicate primarily by pheromones. A close convergence has occurred between the basic signals of the termite chemical language and those of the ants and other social Hymenoptera. The termites secrete queen substances that inhibit the development of rival queens within the colony. They also have "king substances" that prevent the appearance of new reproductive males. Alarm and trail pheromones and distinctive colony odors are also well developed. Various forms of trophallaxis have been evolved, including both the regurgitation of liquid food and the licking of attractive substances from the body surface. The symbiotic protozoans carried by some species are passed from one termite to another by the unique method of anal trophallaxis, in which one termite extrudes an anal droplet containing the symbionts and feeds it to a nestmate.

READINGS

H. KUMMER, *Primate Societies: Group Techniques of Ecological Adaptation*, Chicago, Aldine Publishing Co., 1971. A short but insightful and thought-provoking book on social evolution in monkeys and baboons.

D. LACK, *Ecological Adaptations for Breeding in Birds*, London, Methuen, 1968. Includes a definitive account of social behavior in birds.

J. VAN LAWICK-GOODALL, *In the Shadow of Man*, Boston, MA, Houghton Mifflin, 1971. A beautifully illustrated account of the behavior of wild troops of the most manlike of the anthropoid apes.

E.O. WILSON, *The Insect Societies*, Cambridge, MA, Belknap Press of Harvard University Press, 1971. A detailed account of the social insects and other arthropods.

E.O. WILSON, *Sociobiology: The New Synthesis*, Cambridge, MA, Harvard University Press, 1975. A comprehensive review of all aspects of social evolution and animal societies.

Part **3**   The diversity of life

The evolution of life on Earth might easily have stopped once the ancient seas were covered with a living scum of primitive green algae. These single-celled autotrophs could perpetuate life forever, capturing energy from the sun, cycling nutrients in and out of the surrounding water, and monotonously replicating themselves by endless cell division. Such organisms could persist indefinitely in an ecological steady state. Their number would remain approximately constant, with the birth rate of new cells equalling the death rate of old ones. The flow of energy too would remain constant, with the amount of radiant energy captured each second equalling the energy lost.

On Earth, of course, the evolution of life has followed a very different scenario. Since the appearance of the first organisms more than three billion years ago, life has relentlessly advanced and diversified. New kinds of plants, some of them comprised of many cells instead of just one, became specialized bottom-dwellers in shallow waters, while others remained floating near the surface. At a very early date the first heterotrophs appeared, including consumers that fed on living plants, predators that fed on the consumers, and scavengers that decomposed the remains of the dead. Perhaps a billion years ago the ranks of the heterotrophs gave rise to "true" animals: multicellular organisms that could move from place to place. Among the multicellular plants and animals there emerged a trend toward the evolution of new and larger organisms with ever more complex anatomy. But at the same time a diversity of small organisms with simple body plans continued to thrive. Eventually radical new "specialists" began to colonize previously uninhabited zones of the planet's surface: the deep sea, the land, and even the air. The result was a steady increase in the biomass —the amount of living matter present on Earth. At the present time the biomass is remarkably more complex and efficient than the minimal green scum that might have existed on a barren planet.

The rest of this book will focus on the emergence of the rich diversity of life on Earth, beginning with the spontaneous appearance of life during the planet's geological youth. Subsequent chapters reconstruct the history of life in terms of the major evolutionary inventions that produced distinctive kinds of organisms. One chapter discusses the most venerable of the biological disciplines: taxonomy, the science of classification. Taxonomy is sometimes regarded as a dull, rather cut-and-dried subject, but we do not intend to let the reader become bored. He will see how easy it is to create his own classifications using only the raw data at hand, and how to use taxonomy to reconstruct evolutionary family trees. The substance of higher classification, meaning the facts about the distinguishing characteristics and relationships of the major groups of plants and animals, should acquire new meaning as one sees how they are interpreted in the context of modern evolutionary theory.

It is mere rubbish, thinking at present of the origin of life;
one might as well think of the origin of matter.
—Charles Darwin

23
Origins

It was only in the last century or two that serious doubts arose about the origin of life. The practical man knew that new life appeared around him all the time: flies and maggots from rotting meat and barnyard manure, lice from sweat, glowworms from rotting logs, eels and fish from sea mud, and frogs and mice from moist earth. No less an authority than Aristotle vouched for such common-sense observations, and for over two thousand years, SPONTANEOUS GENERATION was accepted as a fact of nature.

A few people questioned whether spontaneous generation ever really occurred. Francesco Redi, a Tuscan doctor, demonstrated in 1668 that maggots in meat were only the larvae of flies, and that if the meat were protected so that adult flies could not lay their eggs, no maggots appeared. When the Dutch lens grinder and microscope maker Anthony van Leeuwenhoek discovered micro-organisms in 1676, spontaneous generation received new support. While many people were ready to concede that horses and maggots did not appear spontaneously from nonliving matter, they were less confident about the subvisible creatures that van Leeuwenhoek could find everywhere. The Italian biologist Lazzaro Spallazani (1729–1799) showed that if broths were adequately sterilized and kept from being contaminated by air-borne microorganisms, they remained devoid of life. But he failed to convince his contemporaries, partly because others were performing the same experiments with less care and obtaining

opposite results. Experimental methods were not good enough to rule out spontaneous generation for those who wanted to believe in it.

Louis Pasteur finally laid spontaneous generation to rest permanently in 1862, in response to a competition for a prize set up by the French Academy of Science. Pasteur won the prize for a series of meticulous and conclusive experiments that showed that microorganisms came only from other microorganisms, and that a genuinely sterile broth or solution would remain sterile indefinitely unless contaminated by living creatures. The old aphorism, *Omne vivum ex vivo* (all life from life) became dogma. Pasteur delivered his famous lecture to the French Academy in April, 1864. On the 14th of May a meteorite shower fell near Orgueil, France. A century later, scientists would find these meteorite fragments helpful in studying the evolution of life. They also uncovered a century-old fraud that misfired *(Box A)*.

Pasteur answered an old question, but raised a new and tougher one: If all life comes from preexisting life, where did the first life come from? If one believed in Darwin's new theory of evolution, how could one avoid the logical trap that ultimately the first living organism could not have descended from an earlier one? (Had the Fundamentalists been thinking fast enough, they might have seized upon a combination of the ideas of Pasteur and Darwin as a proof of the initial special creation of life.) Pasteur appeared to leave only two alternatives: Either life was created at a specific time in the past, or life had always existed, somewhere else in the universe if not on Earth. Darwin dismissed the entire controversy as pointless and premature: "It will be some time before we see slime, protoplasm, etc., generating a new animal. But I have long regretted that I truckled to public opinion, and used the Pentateuchal term of *creation*, by which I really meant 'appeared' by some

PRECAMBRIAN ANIMAL in photograph on opposite page looks somewhat like a small jellyfish but does not correspond to any living animal or known fossil species. Named Tribrachidium heraldicum ("three-armed shield") it was found in Ediacara Hills of Australia. About two centimeters in diameter, it lived on sandy sea bottom near shore.

A A humbug meteorite

In the late 18th century, a professor of natural history at Würzburg, Johann Beringer, was the victim of a cruel hoax when his students began seeding a local fossil bed with fake salamanders, lizards, frogs in the act of copulating, and even birds. Beringer published his finds in folios of copperplate engravings which had a great impact on his scientific contemporaries. But his world collapsed when he found fossils with Hebrew characters and finally one with his own name on it. Beringer ruined himself financially trying unsuccessfully to buy back all the copies of his books and destroy them.

Two centuries later, scientists studying one fragment from the Orgueil meteorite shower were astounded to find in it fragments of coal, seed capsules, plant remains, and an optically active water-soluble protein. The meteorite had remained untouched in a sealed glass jar since it was turned in to the Montauban museum in 1864, a few weeks after the shower. Furthermore, the plant fragments were too deeply embedded in the meteorite to be surface contamination from impact with the Earth, and the fusion crust produced by friction heating in the atmosphere was apparently nearly intact. This was evidence for extraterrestrial life several orders of magnitude more definitive than before.

The seed capsules could not at first be matched with those of any known terrestrial species. Then, fortunately, a scientist at the Montauban museum identified them as those of a species of reed common to the countryside around Orgueil. An elaborate century-old hoax rivaling Beringer's fossils began to come to light. The "friction crust" was revealed on closer examination to be faked, and a tiny flake of genuine crust was observed sitting at right angles to the simulated crust. The protein turned out to be an animal glue. Everything could be accounted for by assuming that someone had picked up the newly-fallen meteorite, crushed it, embedded the coal and plant fragments, moistened it and used glue to mold it back together again, and finally heated and polished the surface to simulate a friction crust.

Why would anyone bother with such a hoax? Pasteur had delivered his address to the French Academy of Science disproving spontaneous generation a few weeks earlier, and the issue of the origin of life was widely discussed. Presumably some skeptic with a distorted sense of humor decided to give the public some really impressive evidence for extraterrestrial life, assuming that the meteorite would be examined at the museum and might even reach Pasteur. To his frustration, nothing happened at all. The fragment went into a sealed jar, labeled but unexamined, to sit for another century. The perpetrator presumably went on to other, if not greater, things.

---

wholly unknown process. It is mere rubbish, thinking at present of the origin of life; one might as well think of the origin of matter." With very few exceptions, most serious biologists and chemists accepted Darwin's verdict for more than another 70 years.

In 1924 the Russian biologist A. I. Oparin published a thin monograph in Moscow entitled "The Origin of Life." Although it was never translated from Russian and had no impact on scientific thought at the time, it contained virtually the entire blueprint for what we now believe to be the actual process of evolution of life. Five years later,

the British biologist J.B.S. Haldane arrived at the same ideas independently and published them in *The Rationalist Annual*, again to little effect. It is a comment on scientific attitudes at the time that neither man felt able to propose his ideas in a recognized scientific journal. Not until Oparin expanded his ideas into a book, *Origin of Life*, which was published in 1936 and translated into other languages, did the problem of the appearance of life on Earth receive serious consideration.

Chapter 1 of this text presented essentially the Oparin-Haldane theory. Life, they argued, evolved from organic chemicals in the seas at a time when the Earth's atmosphere contained virtually no free oxygen. These organic chemicals—first simple amino acids, purine and pyrimidine bases, and sugars, and later polypeptides, polynucleic acids, and polysaccharides—were originally synthesized abiologically by the action of ultraviolet radiation, electrical discharges, heat, or radioactivity. They were not rapidly oxidized or broken down, as they would be today, because no free oxygen was available to react with them. Photosynthesis, once it appeared, changed the character of the atmosphere by liberating vast amounts of oxygen. This led to a more efficient life process: respiration. But by hastening the natural oxidation of organic matter and by screening out the Sun's ultraviolet radiation with a high-altitude layer of ozone, the oxygen atmosphere also eliminated the possibility that life could evolve a second time, or that the appearance of new life could be a continual process, as stated by the theory of spontaneous generation.

There is another good reason why the half-way stages toward living organisms could never evolve again now—they would be promptly eaten by older and more efficient competitors. Darwin recognized this in 1871: "It is often said that all the conditions for the first production of a living organism are now present, which could ever have been present. But if (and oh! what a big if!), we could conceive in some warm little pond, with all sorts of ammonia and phosphoric salts, light, heat, electricity, etc., present, that a protein compound was chemically formed ready to undergo still more complex changes, at the present day such matter would be instantly devoured or absorbed, which would not have been the case before living creatures were formed."

The evolution of life attracted the interest of relatively few biologists in the 1940's and early 1950's, but this situation changed drastically after the launching of Sputnik and the beginning of space exploration. With the possibility of finding life on other planets, the beginning of life on our own planet became a nagging question. Paleobiology became interesting, as did exobiology: the study of life on other planets (which one skeptic has described as the only field of science with no subject matter whatsoever).

Any reasonable theory of the origin of life must rest on three kinds of evidence: 1, evidence for a common metabolic heritage in modern organisms; 2, fossil traces of Precambrian life; 3, laboratory experiments simulating presumed early Earth conditions. To these, we should like to be able to add a fourth kind of evidence: comparisons of independent appearance of life on more than one planet.

We shall examine the first three classes of data one at a time, and then assemble a scenario for the beginnings of life. At the present stage of knowledge such a scenario is undoubtedly incomplete, but is unlikely to be wrong in its main points.

A COMMON METABOLIC HERITAGE

The central metabolic processes of modern plants and animals are essentially the same. As explained in Chapter 6, plants and animals alike extract energy from foods by anaerobic fermentation or glycolysis

$$C_6H_{12}O_6 \rightarrow 2\ C_3H_4O_3 + 4\ H\ \text{(on NADH)}$$
GLUCOSE   PYRUVIC ACID

followed by aerobic respiration

$$2\ C_3H_4O_3 + 4\ H\ \text{(from NADH)} + 6\ O_2$$
$$\rightarrow 6\ CO_2 + 6H_2O$$

Green plants add the subtlety of synthesizing foods for later use by the process of photosynthesis

$$6\ CO_2 + 6\ H_2O \xrightarrow[\text{energy}]{\text{light}} C_6H_{12}O_6 + 6\ O_2$$

This pattern persists all the way down the scale of life as far as simple one-celled eucaryotes such as amoebas and green algae. Many other metabolic reactions are identical or closely similar in all eucaryotes, and where they are different, they are often merely variations on the same basic theme.

In terms of metabolism, the real diversity of life is much more evident in procaryotes. Bluegreen algae (not to be confused with green algae, which are plants) and bacteria display a staggering variety of mechanisms for energy storage and food synthesis that have no parallel in the eucaryotes. For example, the combination of carbon and hydrogen with oxygen is not the only way to carry out respiration. Many bacteria use oxygen just as do the higher plants and animals. Others, such as *Desulfovibrio*, can use sulfuric acid for respiration to yield hydrogen sulfide instead of water. Still others can conduct anaerobic respiration by using carbon dioxide as the oxidant, liberating methane gas or acetic acid. The denitrifying bacteria oxidize their foods with nitrate ion and give off nitrous oxide, nitrogen gas, or ammonia. Apparently, from all the possible energy-yielding chemical reactions involving oxidation, the eucaryotes chose the one that reduces oxygen gas to water.

The genuine diversity of life is obscured if one looks only at the eucaryotes. The bacteria help one to see the metabolic pathways that are not common to all organisms.

Even the idea that respiration itself is a universal metabolic heritage is disproved by bacteria. Many bacteria manage quite well with only the energy they receive from fermentation, and do not respire at all. Moreover, several plants, including yeasts, have the option of anaerobic life under certain conditions.

Photosynthesis, a universal mark of plants, also illustrates the diversity of bacterial metabolism. Many bacteria use light energy to generate ATP and NADH, but their photosynthesis does not liberate oxygen. This is because they do not use water as a source of hydrogen atoms as green plants do. Some of the alternative hydrogen atom donors used by bacteria include H_2S, thiosulfate ion, fatty acids, and other organic compounds, or even molecular hydrogen itself. The green and purple sulfur bacteria live in hot springs or other sources of H_2S, take hydrogen from this molecule for photosynthesis, and release sulfate as a waste product. The *Athiorhodaceae* (purple nonsulfur bacteria) need organic compounds, but mainly as a source of hydrogen for photosynthesis, not as a source of energy.

Other bacteria are not photosynthetic at all, but obtain energy for synthesis from chemical reactions. The hydrogen bacteria use the energy of the reaction between hydrogen and oxygen to synthesize carbohydrates from CO_2 and H_2. The colorless sulfur bacteria are not photosynthetic, but obtain their synthesizing energy by oxidizing H_2S to sulfur, or sulfur or thiosulfate to sulfate ions. The iron bacteria, whose ancestors date back to the Precambrian era, oxidize ferrous ions to insoluble ferric hydroxide. The nitrite bacteria oxidize ammonia to nitrite ion, and the nitrate bacteria convert nitrite to nitrate.

The important point to recognize in bacterial metabolism is that the eucaryotes, today's familiar plants and animals, have chosen one particularly efficient set of reactions from among the many possible pathways. The genuine diversity of life is obscured if one looks only at the eucaryotes. The bacteria help one to see what is not common to all organisms.

What remains after all the variation is stripped away is the process of ANAEROBIC FERMENTATION. There is no living form, from bacteria to mammals, that does not extract energy from organic molecules *(recall Chapter 6)*, and all except certain bacteria use glucose. Because comparative biochemistry reveals evolutionary ancestry as clearly as does comparative anatomy, one must conclude that the original ancestor of all organisms, procaryotes and eucaryotes alike, must have absorbed its food from its surroundings and extracted energy from it without the use of molecular oxygen. In short, it must have been an ANAEROBIC HETEROTROPH.

This provides a valuable clue about the conditions under which life began. Life must have emerged when organic compounds were abundant in the seas, not only as sources of energy, but as sources of the materials from which life could evolve in the first place. There was probably no oxygen present, for aerobic respiration is so much more efficient than fermentation alone that it is difficult to imagine that oxygen-using organisms would not have evolved at the beginning, had O_2 been readily available. Another argument against the presence of free oxygen is the very existence of large supplies of organic matter. Had O_2 been present, it would have spontaneously oxidized the supply of organic matter faster than it could have been produced by nonbiological means. One sometimes overlooks the fact that the vast amount of organic matter on the surface of the planet today and the gaseous O_2 around it are both results of the activities of living organisms. If all life were to end today, in a few thousand years all of the organic matter would have been oxidized to CO_2 and water, and the remaining atmospheric oxygen would have been locked up in the Earth's crust in the form of mineral oxides. In short, the atmosphere of the early Earth must have been reducing rather than oxidizing.

THE FOSSIL RECORD

The fossil record of living organisms is plentiful for the last 600 million years. Any recognizable trace of an organic structure preserved from prehistoric times is a fossil. Some of the most dramatic examples are the mammoths that were frozen intact in the icy soil of Siberia. The mineral cast of a bone of a vertebrate and the calcified shell of a mollusk are also fossils. In such specimens the original calcium carbonate or other hard material was gradually washed out and replaced with mineral substances by the infiltration of ground water. A fossil may consist of nothing more than the cast of the impression by the hard surface of a bone or shell on the originally soft soil. Occasionally the only trace of an organism is its footprint or the remains of a burrow that it dug in life. One of the best ways for a particular organism to become a fossil is to die in some favored spot, such as the bottom of a river or the quicksand of a swamp, where it will be quickly covered by mud, sand, peat, or some other material destined to become part of a long-lasting geological formation. As millenia pass, the mud may turn to shale, the sand to sandstone, and the peat to coal. The traces of the organism are then

1 **GEOLOGIC CALENDAR as shown below depicts the history of the Earth as a 30-day month. One day equals 150 million years. First living things appeared very early, perhaps four billion years ago (fourth day). Oldest known fossils lived 3.4 billion years ago, but fossil record is scanty before the opening of the Paleozoic Era (27th day). The four most recent days are depicted in greater detail at bottom. On this scale, first true men appeared only about ten minutes ago, and all recorded history, from ancient Sumer to the present, represents an interval of about 30 seconds.**

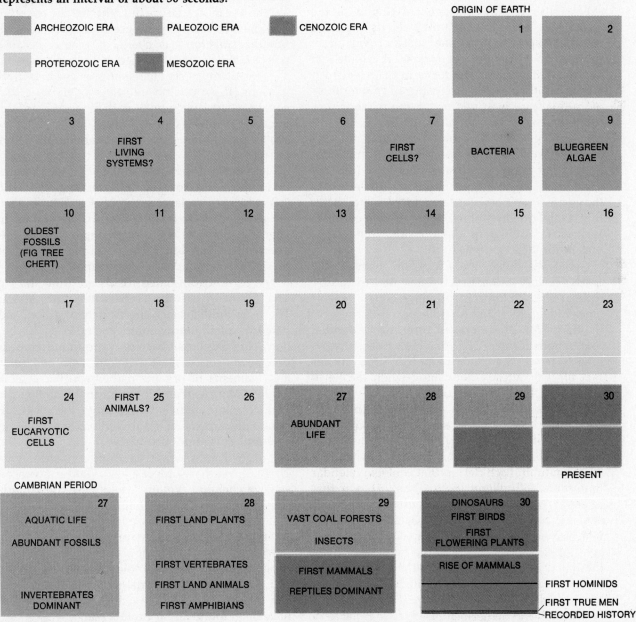

> If we scale down the history of the Earth to a 30-day month, we find that fossil life does not become abundant until the dawn of the Cambrian period, on the 27th day.

locked in and may remain unchanged for millions or even billions of years. When the rocks are later split open by the geologist's hammer, the fossils are exposed for inspection.

Fossils can usually be dated according to the age of the rock strata in which they lie. As a general rule the deepest strata are the oldest, provided that the formation has not been drastically folded or deformed. The age of the rock can be determined by radioisotope techniques using trace quantities of radioactive isotopes of uranium, thorium, rubidium, or potassium. Uranium-238, for example, spontaneously decays into lead at a slow but precisely known rate. By comparing the amount of U^{238} still in the rock with the amount of lead derived from its decay, geophysicists can estimate the age of a sample within an error of about 10 per cent. Different kinds of fossils are used to correlate rock strata in different locations, and to delineate eras in the history of life.

Geologists divide the 4.5 billion years of the Earth's history into five major eras: the Archeozoic (primal life), Proterozoic (primitive life), Paleozoic (ancient life), Mesozoic (middle life), and Cenozoic (modern life). These eras in turn are subdivided into periods (*Chapter 24*). The oldest fossils discovered so far are 3.4 billion years old, which means that life was already flourishing when the Earth was only 1.1 billion years old.

Numbers in the billions are so large that they have little meaning for most readers. To convey a real sense of the relative age of fossil organisms it is useful to scale down the history of the Earth to a 30-day month. Each day on that geologic calendar represents 150 million years (*Figure 1*). On that calendar the Archaean and Proterozoic eras stretch across the first 26 days. The Cambrian period, which opened the Paleozoic era 600 million years ago, marks a great divide in the fossil record. By the dawn of the Cambrian, on the 27th day, the ancient seas teemed with life. Representatives of every modern phylum had appeared, except for vertebrates and higher plants.

PRECAMBRIAN LIFE

Below the earliest Cambrian strata there is a striking dearth of fossils. Various explanations have been suggested for the apparently dramatic explosion of life in the Cambrian era: changes in the radiation level of the sun, climatic changes, an increase in the oxygen content of the atomosphere past a critical point for respiration, or simplest of all, the replacement of softbodied creatures with organisms having shells, armor, and skeletons that would leave more substantial fossil remains. Moreover, in later eras most of the Precambrian rocks have been heated or deformed by geologic processes in a way that would tend to obliterate faint traces in the fossil record.

A few years ago it was thought that the record of Precambrian life was meagre and unimpressive. But the situation has changed recently because paleontologists have learned how and where to look for the microscopic remains of one-celled organisms. If we can rid ourselves of the bias that equates plants and animals with life and regards all microorganisms as pretty much alike, then we shall see that the record of Precambrian life is surprisingly rich.

Over most of the surface of the Earth, Precambrian rocks are covered by later deposits. The "shields," where large areas of Precambrian rock are exposed or at least accessible, are shown in Figure 2. Some of the best examples of Precambrian life are listed in Table I, and the locations of these finds are marked on the map (*Figure 2*).

The earliest well-dated fossils are from the Fig Tree strata in eastern Transvaal, South Africa. These are flintlike black cherts, rich in carbon, and dated by rubidium strontium methods as 3.1 billion years old. Fossil bacteria were found embedded in the rock (*Figure 3*), along with spheroids and filaments of organic matter which resemble modern bluegreen algae. Surprisingly, among the organic compounds extracted from the rocks are two hydrocarbons which are ordinarily considered to be degradation products of the tail of the chlorophyll molecule: pristane and phytane (*Figure 4*). Photosynthesis represents a fairly advanced stage of chemical evolution. Did these compounds come from chlorophyll from the algae or bacteria, and were these organisms already photosynthetic 3.1 billion years ago? At present one can only speculate. Nearby, in the Onverwacht cherts in Swaziland, lie some even older fossils: spheroids and filaments that are certainly older than 3.2 billion years. These are the oldest signs of life that have been found so far on this planet.

Modern algae that inhabit hot springs or geysers lay down limestone deposits in cone-shaped layers called STROMATOLITES. An example of geyserite, an algal limestone from Yellowstone Park, is shown in Figure 5. The cone grows layer by layer, with the apex of the cone downward, like a stack of nested thimbles. Limestone deposits that are almost identical with modern algal stromatolites are found in 2.7-billion-year-old limestone in southern Rhodesia. In other Precambrian deposits of roughly the same age in Australia and Minnesota investigators have found spheroidal and clustered bodies which resemble bacteria and bluegreen algae. Pristane and phytane, the somewhat questionable indications of photosynthesis, also occur in the Minnesota rocks.

One of the most dramatic and varied collections of Precambrian organisms occurs in Gunflint chert, an outcropping of 1.9-billion-year-old rock in the Canadian Shield, exposed near the Minnesota-Ontario border. Thousands of canoeists have followed the Gunflint Trail in an effort to get back to nature and turn back the clock a few years, never

PRECAMBRIAN ROCKS are usually buried under later strata, but in low-lying continental shields they are relatively accessible. Exposed Precambrian strata are rendered in dark color and those overlaid with younger strata in light color. Letters A and L mark fossil sites listed in Table I. Ancient continents were shaped very differently from those of today, and occupied different positions on the face of the globe (Chapter 31). Animal illustrated on opening page of this chapter was found near location L. Bacterium in Figure 3 was found at one of two sites marked A and B.

2

3 OLDEST FOSSIL ORGANISM discovered so far is the rod-shaped Eobacterium isolatum, found in the Fig-tree chert formation. During preparation of sample, fossil was moved from its original position in the rock, leaving its impression visible at bottom. "Yardstick" at far left, one micrometer long, indicates size of the bacterium.

DEPOSIT	BIOLOGICAL REMAINS	AGE (BILLIONS OF YEARS)
ARCHEOZOIC ERA (4.5–2.5 BILLION YEARS AGO)		
A. ONVERWACHT SEDIMENTS SWAZILAND, S. AFRICA	SPHEROIDS, FILAMENTS IN CARBON-RICH CHERTS.	>3.2
B. FIG TREE CHERTS TRANSVAAL, S. AFRICA	FOSSIL BACTERIA ($0.25 \times 0.6\mu$), ALGAE-LIKE SPHEROIDS (PHOTOSYNTHETIC?), FILAMENTOUS ORGANIC STRUCTURES. COMPLEX HYDROCARBONS; PRISTANE, PHYTANE.	3.1
C. BULAWAYO LIMESTONE BULAWAYO, S. RHODESIA	LIMESTONE DEPOSITS RESEMBLING STROMATOLITES LEFT BY ALGAE. LAYERS IN NESTED CONES LIKE MODERN GEYSERITE FROM YELLOWSTONE PARK.	2.7
D. SOUTHERN CROSS QUARTZITE S. CROSS, WESTERN AUSTRALIA	RED HEMATITE AND YELLOW GOETHITE BODIES EMBEDDED IN QUARTZITE. SPHEROIDAL ($\sim 2\mu$). ELONGATED, AND IN RASPBERRY-LIKE CLUSTERS.	2.7
E. SOUDAN IRON FORMATION ELY, MINNESOTA	CARBON-RICH ROCKS CONTAINING PYRITE BALLS WITH MICROSTRUCTURES. SPHEROIDAL (0.8–1.5μ); SINGLE AND CLUSTERED. BACTERIA OR BLUEGREEN ALGAE? HYDROCARBONS; PRISTANE, PHYTANE.	>2.7
LOWER PROTEROZOIC (2.5–1.7)		
F. GUNFLINT CHERT GUNFLINT LAKE TO THUNDER BAY ALONG ONTARIO-MINNESOTA BORDER.	BLUEGREEN ALGAE, FILAMENTOUS AND SPHEROIDAL, FUNGI, ORGANISMS OF NO MODERN PARALLELS. PHOTOSYNTHETIC? BACTERIA ($0.55 \times 1.10\mu$). PRISTANE, PHYTANE, OTHER HYDROCARBONS. STROMATOLITES (UNRELATED TO MICROFOSSILS).	1.9
G. RORAIMA CHERT BRITISH GUIANA	ALGAE RESEMBLING GUNFINT DEPOSITS. UNKNOWN PROTISTS, 160–400μ. SUPERFICIALLY RESEMBLE FORAMINIFERA, SPHERICAL WITH PROJECTIONS.	2.0–1.7
MIDDLE PROTEROZOIC (1.7–1.0)		
H. SAHARAN STROMATOLITES WESTERN SAHARA	LIMESTONE ALGAL DEPOSITS. TWO GENERA, *COLLENIA* AND *CONOPHYTON*. ALTERNATE LAYERS OF $CaCO_3$ AND $Fe(OH)_3$ SUGGEST ASSOCIATION OF PHOTOSYNTHETIC BLUEGREEN ALGAE AND IRON BACTERIA. SAHARAN EARLIEST. LATER STROMATOLITES CONTINUING TO PRESENT. PRECAMBRIAN STROMATOLITES IN AUSTRALIA, CHINA, SIBERIA, URALS, FINLAND, N. EUROPE, CENTRAL AFRICA, TRANSVAAL, INDIA, NORTHERN U.S. AND CANADA.	1.6
I. BECK SPRINGS DOLOMITE SOUTHERN CALIFORNIA	BLUEGREEN ALGAE. POSSIBLE FIRST KNOWN APPEARANCE OF GREEN ALGAE WITH NUCLEATED CELLS (EUCARYOTES).	1.4–1.2
J. NONESUCH SHALES WHITE PINE, MICHIGAN	FILAMENTOUS AND SPHEROIDAL BODIES THE SIZE OF ALGAE. POSSIBLY BIOLOGICAL, SOME ARTIFACTS. PRISTANE, PHYTANE, OTHER HYDROCARBONS.	1.1
K. ALICE SPRINGS, CENTRAL AUSTRALIA; McARTHUR FORMATIONS, AUSTRALIA	POSSIBLE CASTS OF WORM TRACKS. IF VALID, EARLIEST SIGNS OF METAZOANS. POSSIBLE SPONGE SPICULES AND IMPRINTS OF JELLYFISH.	1.5–1.0
LATE PROTEROZOIC (1.0–0.6)		
L. BITTER SPRINGS CHERT NEAR ALICE SPRINGS, AUSTRALIA	BLUEGREEN ALGAE. SPHEROIDS (3–10μ), UNSEGEMENTED AND SEGMENTED FILAMENTS. STRUCTURAL ANATOMY SURPRISINGLY UNDISTORTED. GREEN ALGAE.	1.0–0.9

FIRST APPEARANCE OF ANIMALS AND MULTICELLED ORGANISMS: — 1.0–0.6

SILICEOUS SPONGE SPICULES—USSR, CHINA	POLYCHAETES—GRAND CANYON
SPONGES, RADIOLARA, FORAMINIFERA—N. FRANCE	ARTICULATED ANCESTORS OF TRILOBITES—SWEDEN
SPORES—SCANDINAVIA, RUSSIA, N. FRANCE	MEDUSAS, WORMS, SEA PENS—EDIACARA HILLS, AUSTRALIA
WORM BURROWS (OVER 780 M. Y. OLD)—AUSTRALIA, CHINA, USSR	

PRECAMBRIAN FOSSILS listed in this table stretch across an interval of 3.9 billion years—the first 27 days of the calendar in Figure 1. All Precambrian organisms were aquatic. Letters in column at left are keyed to locations of deposits, as shown on the map of Precambrian outcroppings in Figure 2.

realizing that they could turn the clock back nearly two billion years by examining the rocks at their feet. Unmistakable bluegreen algae, fungi, and bacteria have been found, as well as organisms that have no modern parallels *(Figures 6, 7 & 8)*. One of these unknown organisms, named *Kakabekia umbellata (Figure 9)*, has since been found not to be extinct at all. It is living and thriving in special ammonia-rich soils in Wales *(Box B)*. Pristane, phytane, and other hydrocarbons are present in the Gunflint rocks, along with stromatolites.

In the 1.6 billion year old stromatolites from the western Sahara *(Figure 10)*, calcium carbonate from the bluegreen algae alternates in banded layers with ferric hydroxide, which may have come from iron bacteria. It has been suggested that colonies of algae and bacteria might have lived together, and that seasonal variations in temperature could have made the algae and bacteria alternate in activity and in mineral deposition.

The earliest porphyrins that have been found come from the Nonesuch shales of Michigan. These deposits contain algaelike bodies as well as pristane, phytane, other hydrocarbons, and most significant of all, vanadium porphyrins, typical of the ring compounds that comprise the heart of the chlorophyll molecule. This may be the first solid evidence of the existence of porphyrin compounds and possibly of photosynthesis.

All of the life forms discussed so far appear to have been procaryotes —bacteria or bluegreen algae. The first true plants —green algae —may be present in the Beck Springs dolomite from California and the Bitter Springs chert from Australia. In the California specimens, 1.4 to 1.2 billion years old, can be seen what have been claimed to be eucaryotic cells with nuclei *(Figure 11)*. Both bluegreen and green algae can be seen in the Australian chert, in surprisingly undistorted form, in rocks dated between 900 and 700 million years old. If this interpretation of the fossil microorganisms is correct (other paleobiologists have challenged it), then life finally arrived at the stage of green plant photosynthesis only a few hundred million years before the Cambrian era.

PHYTOL TAIL OF CHLOROPHYLL

PRISTANE

PHYTANE

PRISTANE AND PHYTANE are products of the breakdown of the phytol tail of the chlorophyll molecule. The occurrence of pristane and phytane in fossil-bearing rock suggests the presence of ancient photosynthetic organisms, although this hypothesis is disputed by some paleochemists.

4

LIMESTONE DEPOSITS from Yellowstone National Park were laid down by modern algae living in hot springs and geysers. Deposits resemble fossil stromatolites from Precambrian strata (Figure 10).

5

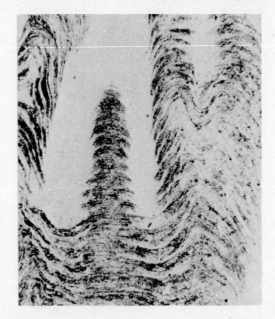

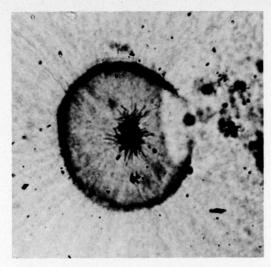

6 COLONY OF FOSSIL ALGAE found in Gunflint chert formation in Ontario resembles modern bluegreen algae of the genus Rivularia, hence its name, Paleorivularia ontarica. It apparently consists of a star of radiating algal filaments surrounded by some sort of membrane or envelope. Diameter of envelope is about 60 micrometers. Fossil is 1.9 billion years old.

7 SPHERICAL ORGANISM with no modern parallel was found in Gunflint chert. Named Eosphaera tyleri, it appears to be a sphere within a sphere, with the inner held away from the outer one by many tiny globules.

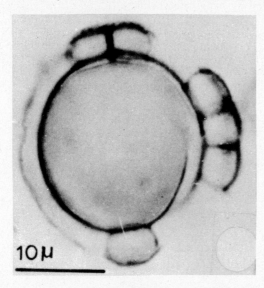

10 μ

The first signs of multicellular life appear in the last half billion years before the Cambrian era, first as worm tracks, and then as impressions in sandstone of softbodied worms, jellyfish, and other creatures with no modern descendants (*see Figure 12 and the photograph that opens this chapter*).

The interval from about 3.5 billion to 800 million years ago —from the 7th to 24th days of the calendar —can be described as the age of procaryotes.

At this point the Precambrian fossil record blends into the Cambrian. The enormous increase in the number of fossils at the beginning of the Cambrian might be explained by a great upsurge in the population of living organisms 600 million years ago, but this population had roots in the earlier eras.

In brief, the interval from approximately 3.5 billion to 800 million years ago —from the 7th to the 24th days of our calendar —can be described as the age of the procaryotes. At the beginning of this period there flourished simple organisms morphologically similar to bluegreen algae and bacteria, which may or may not have been photosynthetic. The evidence of pristane and phytane is too shaky to stand alone, for these compounds have been found in living microorganisms which do not possess chlorophyll. At that time the Earth probably had a reducing atmosphere comprised of hydrogen, ammonia, water, and methane, but no free oxygen. During the course of the following 2.5 billion years, these organisms diversified and developed both photosynthesis and respiration, leading ultimately to eucaryotes and then to multicelled creatures that we would recognize as plants and animals.

TUBULAR FILAMENT from Gunflint chert resembles **8** some living types of bluegreen algaem Unlike modern algae it is branched and lacks septae (perpendicular divisions along the strand). Some tubular fossil species have segments and other structural features characteristic of modern algae.

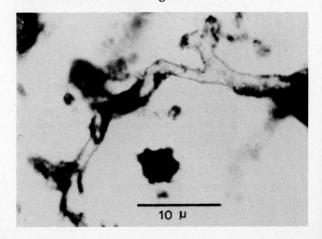

10 μ

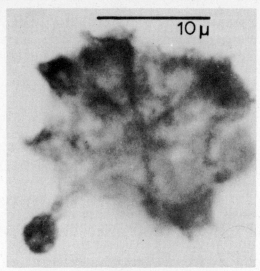

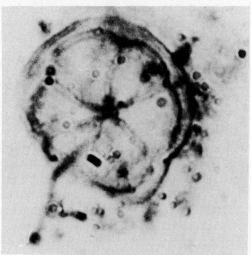

9 UMBRELLA-SHAPED MICROORGANISM from Gunflint chert (top) resembles a previously unidentified microorganism living in soil near Harlech Castle, Wales (bottom). Fossil, named *Kakabekia umbellata*, is about 1.9 billion years old. See Box B.

STROMATOLITE from western Sahara was formed by **10** bluegreen algae living in shallow water 1.6 billion years ago. The vertical section shown here contains layers of calcium carbonate from algae and ferric hydroxide possibly from colonies of iron bacteria that lived alongside the algae.

The main conclusion to emerge from the study of Precambrian microfossils is that life has been present on Earth much longer than anyone previously suspected. Many workers now estimate that the first living systems appeared sometime during the first 500 million years of the planet's existence —first three or four days on our calendar. But even the most primitive Precambrian bacteria are complicated structures compared with nonliving matter. Their discovery sheds no light on the central question of chemical evolution: how did nonliving matter organize itself into a living system?

METEORITES AND EXTRATERRESTRIAL LIFE

One possibility is that life did not originate on Earth, but was "seeded" here by an extraterrestrial object such as a meteor. The Earth is steadily bombarded with showers of meteors, presumably the debris of shattered asteroids, and some of this material contains molecules characteristic of living systems.

Most meteorites are stony or metallic, but a relatively small number are soft and crumbly, with a high carbon content. These meteorites are called CARBONACEOUS CHONDRITES, and the meteorite that

11 GLOBULAR CELLS from stromatolite deposits in Beck Springs, California, are about 1.2 to 1.4 billion years old and resemble green algae. Cells are 14 to 18 micrometers in diameter. Dark spot within each cell is the subject of debate: some workers believe it to be a eucaryotic nucleus, while others think it is an artifact.

fell in a shower around Orgueil belong to this category. They have been studied in the past decade by several research groups, using techniques similar to those for detecting microscopic Precambrian fossils. The results have been inconclusive, and the subject of heated debate. Hydrocarbons have been found, including pristane and phytane. Some of the organic compounds are optical isomers *(Chapter 3)*, which are usually associated with syntheses carried out by living organisms. Spheroids and other organized bodies of fair complexity have been reported, but contamination of the meteorite samples by airborne spores and pollen has confused the issue. In at least one instance, the organized bodies turned out to be ragweed pollen. Most of the complex organized bodies have proven to be terrestrial contaminants, and those that are definitely meteoric in origin are sufficiently

12 SEGMENTED WORM from the late Precambrian is about four centimeters long. Name Spriggina floundersi, it is one of many fossil metazoans found in sandstones of the Ediacara Hills in Australia.

simple that they may be natural mineral formations rather than artifacts of life. Perhaps one should be encouraged by the negative (or at least ambiguous) results of meteorite analysis. These trials serve as a "blank" or a control for the Precambrian fossil analyses. Had the meteorite evidence looked fully as good as the Precambrian rock evidence it would have been necessary to question our apparent ability to find "life" and "organisms" everywhere we looked for them. The fact such well-developed organisms are not found in meteorites is reassuring in regard to what has been found in terrestrial deposits. The presence of pristane, phytane, and other hydrocarbons in the meteorites indicates that at least the first step in molecular evolution—the formation of complex organic compounds—can occur spontaneously even in space. If these meteorites are not evidence for life on some shattered planet, they may be evidence for the universality of the organic-chemical-rich environment in which life could develop.

LABORATORY SIMULATION

By 1952 scientists, among them Harold Urey, were proposing that life evolved on Earth in a primitive reducing atmosphere of hydrogen, water, methane, and ammonia, from organic compounds which had been synthesized by nonbiological means. Possible energy sources for such syntheses would have been ultraviolet radiation from the sun, lightning, and to a lesser extent, radioactive decay in the Earth's crust and heat from volcanos and hot springs *(Table II)*. By far the most plentiful energy source would have been visible light, but it did not become significant in biochemistry until the evolution of photosynthesis. Ultraviolet light, a more concentrated energy source, can break the bonds in carbon compounds. This is why ultraviolet radiation is lethal to microorganisms and is used to sterilize space-probe components.

In 1953, Stanley L. Miller, one of Urey's graduate students, made the first careful experiments to show that organic chemicals, including amino acids, could be synthesized under assumed primitive-Earth conditions. He set up a recirculating system of hydrogen, ammonia, and methane gases and water vapor, and passed these gases over a spark discharge from a high frequency tesla coil, to simulate lightning *(Figure 13)*. After a week of circulation, he opened the flask and analyzed the contents. He found a great variety of organic and amino acids. The electrical discharge produced cyanide compounds and aldehydes from the gases

Care and feeding of a living fossil

The living *Kakabekia (Figure 9)* was discovered during screening to find organisms with high tolerance for ammonia, as part of a study of life under harsh conditions. The organism typically has an umbrella 5 to 10 micrometers in diameter, attached to a 5 to 15 micrometer stalk with a bulb at the other end. The crown of the umbrella has a rim and five to eight ribs. The creature seems to grow the umbrella first, and the stalk and bulb later. Virtually nothing else is yet known about its life cycle. It has appeared only in soil samples from the courtyard and stable areas of Harlech Castle, Wales, and on the fringes of Arctic ice in Alaska, Canada and Iceland, although it has been sought in vain in other samples from Western Europe, American states including New York, Pennsylvania and the Dakotas, and various locations in the Caribbean.

The Harlech Castle samples produce *Kakabekia* when they are cultured on agar plates under an atmosphere of 50 percent ammonia, or in broths which are 5 to 10 molar in ammonium hydroxide. Without the ammonia, no *Kakabekia* appear. It is interesting to speculate that the ancestors of the creature may have evolved in an ammonia-rich atmosphere and found a niche in ammoniacal soils when the character of the atmosphere changed.

Why is Harlech Castle such a good home for *Kakabekia*? The castle is maintained as a symbol of Welsh nationalism, and has stabled horses for over seven centuries. *Kakabekia* has probably thrived on the ammonia from a constant supply of urine from horses and other domestic animals kept in the courtyard. As one biologist remarked, "Hence the umbrella."

MILLER APPARATUS simulated conditions on the primitive Earth. Components of primitive atmosphere—ammonia, hydrogen, methane and water—were exposed to spark discharges that simulated lightning. Condensed gases were recirculated by boiling. After running for a week, apparatus contained mixture of gases and dissolved organic compounds.

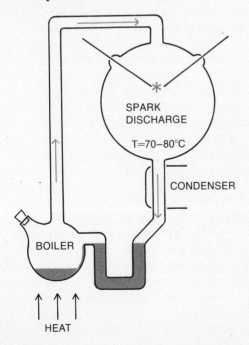

SPARK
DISCHARGE

T=70–80°C

CONDENSER

BOILER

HEAT

POSSIBLE ENERGY SOURCES for nonbiological synthesis of organic compounds on ancient Earth include ultraviolet light from sun, lightning discharges, radioactivity and shock waves from meteors. Despite its enormous total energy content, visible sunlight is too diffuse to be an important factor in nonbiological synthesis.

II

ENERGY SOURCE	ENERGY AVERAGED OVER ENTIRE EARTH (CALORIES PER CM² PER YEAR)
TOTAL SOLAR RADIATION (INCLUDING VISIBLE LIGHT, 7000 Å > λ > 4000 Å).	260,000
ULTRAVIOLET LIGHT	
4000 Å > λ > 2500 Å	570
2500 Å > λ > 2000 Å	85
2000 Å > λ > 1500 Å	3.5
ELECTRICAL DISCHARGE (LIGHTNING)	4
RADIOACTIVITY (TO A DEPTH OF 1 KM BELOW SURFACE)	0.8
VOLCANIC HEAT	0.13
COSMIC RADIATION	0.0015

(Figure 14). These compounds then reacted in water to yield all the substances in Figure 15. The products were not optically active; D- and L- forms of asymmetric compounds were present in equal amounts. In this respect they differ from compounds synthesized by living organisms.

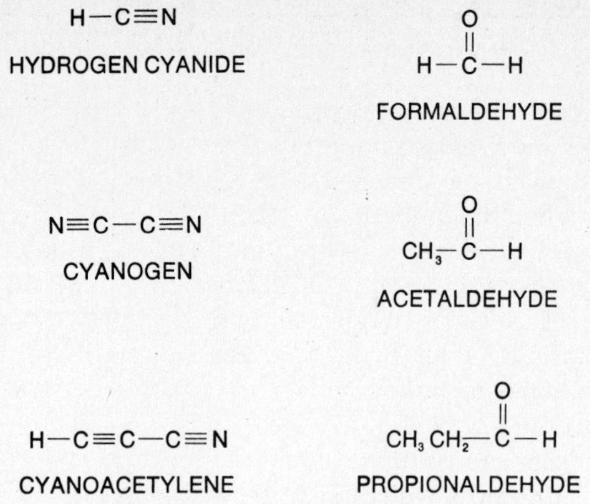

H—C≡N
HYDROGEN CYANIDE

$$H-\overset{\overset{\displaystyle O}{\|}}{C}-H$$
FORMALDEHYDE

N≡C—C≡N
CYANOGEN

$$CH_3-\overset{\overset{\displaystyle O}{\|}}{C}-H$$
ACETALDEHYDE

H—C≡C—C≡N
CYANOACETYLENE

$$CH_3\,CH_2-\overset{\overset{\displaystyle O}{\|}}{C}-H$$
PROPIONALDEHYDE

14 **GASEOUS PRODUCTS of the Miller experiment. These compounds react to form nitriles, which hydrolyze in aqueous solution to yield the products depicted in Figure 15.**

Since Miller's trials many other workers have experimented with electric discharges, ultraviolet light, and radioactive elements, using different gas mixtures which could represent primitive-Earth atmospheres. They have found that it would have been quite possible to produce many amino acids, purines, pyrimidines, sugars such as ribose and 2-deoxyribose, nucleosides, and even complete nucleotides such as ATP, by nonbiological means. The "warm, dilute soup" in which Haldane postulated that life evolved, was quite likely to have been present on the primitive Earth. The easiest nucleotide base to obtain, incidentally, is adenine. Cytosine, guanine, thymine, and uracil must be obtained by more complex reactions, but adenine is simply a pentamer of hydrogen cyanide

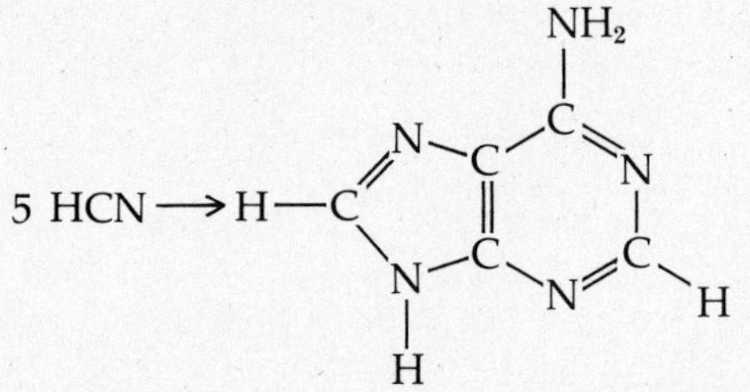

5 HCN ⟶

Irradiation of solutions of adenine, ribose, and phosphate compounds with ultraviolet light of wavelengths 2400–2900 Å has led to the synthesis of AMP, ADP, and ATP. It is likely that these high energy phosphate compounds would have been present when life was evolving, and that the adenosine compounds would have been most common simply because adenine was the most easily synthesized base. The first living organisms (or to reach back one step further, perhaps self-organizing nonliving chemical systems) may have depended on ATP from the seas around them for energy. The pattern that we now see, in which other high free energy organic compounds are broken down in glycolysis and the energy obtained is used to synthesize ATP, may have been a later addition which evolved when the supply of ATP began to run short. The answer to the question, Why do living organisms use ATP to store energy?, may be the same as the answer to the question of why men climbed Mt. Everest: Because it was there.

MICROSPHERES, COACERVATES, AND CELLS

It is a long way from amino acids, nucleotides, and sugars to a living cell. How is the transition to proteins, nucleic acids, and polysaccharides carried out, and how do these lead to a living organism? One idea that is not proposed is that, in some chance fashion, all of the necessary nucleotides to produce the DNA of a bacterium came together one day and started life going. As the British physicist J.D. Bernal has expressed it, "The picture of the solitary molecule of DNA on a primitive seashore generating the rest of life was put forward with slightly less plausibility than that of Adam and Eve in the Garden." Instead, we now picture the gradual evolution of higher and higher levels of complexity in chemical systems, first from monomers to polymers, then to collections of polymers carrying out chemical reactions and perhaps separated from the bulk of their surroundings by a membrane, and finally to isolated chemical systems with the ability to duplicate themselves by means of some template ancestors of our present DNA and RNA. This is the era of CHEMICAL SELECTION—of the constant formation and breakup of localized concentrations of chemicals, and of the survival of those collections of molecules that are best able to grow at the expense of their surroundings without being absorbed by neighboring systems.

The spontaneous formation of polymers is not too difficult to explain. Polymerization of amino acids to protein chains is not spontaneous under standard conditions—that is, concentrations of one mole per liter for each component—but the reaction can become spontaneous if the concentrations of reactants are high enough. It has been proposed that polymerization took place with amino acids concentrated in drying puddles by the

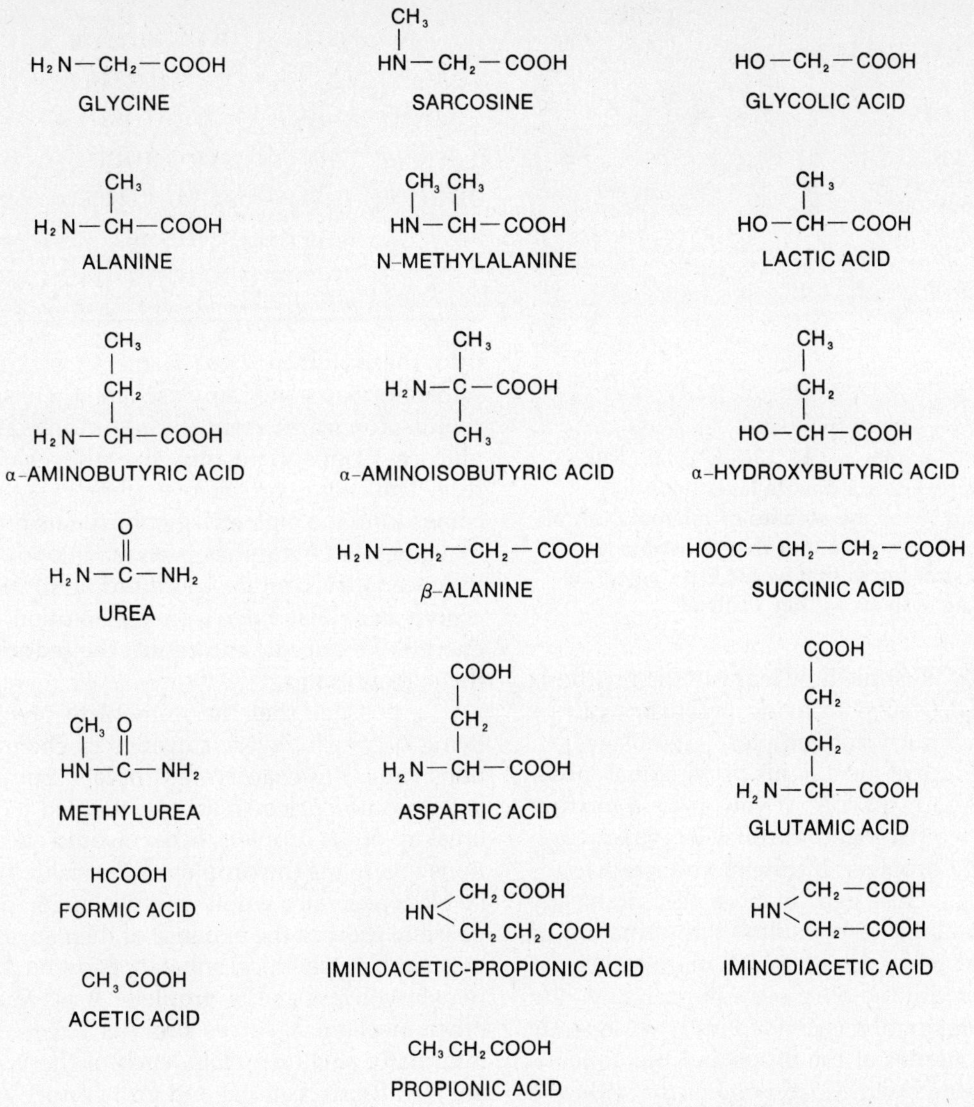

H₂N—CH₂—COOH
GLYCINE

$$CH_3$$
$$|$$
HN—CH₂—COOH
SARCOSINE

HO—CH₂—COOH
GLYCOLIC ACID

$$CH_3$$
$$|$$
H₂N—CH—COOH
ALANINE

$$CH_3 \quad CH_3$$
$$| \quad \; |$$
HN—CH—COOH
N–METHYLALANINE

$$CH_3$$
$$|$$
HO—CH—COOH
LACTIC ACID

$$CH_3$$
$$|$$
$$CH_2$$
$$|$$
H₂N—CH—COOH
α–AMINOBUTYRIC ACID

$$CH_3$$
$$|$$
H₂N—C—COOH
$$|$$
$$CH_3$$
α–AMINOISOBUTYRIC ACID

$$CH_3$$
$$|$$
$$CH_2$$
$$|$$
HO—CH—COOH
α–HYDROXYBUTYRIC ACID

$$O$$
$$\|$$
H₂N—C—NH₂
UREA

H₂N—CH₂—CH₂—COOH
β–ALANINE

HOOC—CH₂—CH₂—COOH
SUCCINIC ACID

$$CH_3 \quad O$$
$$| \quad \;\; \|$$
HN—C—NH₂
METHYLUREA

$$COOH$$
$$|$$
$$CH_2$$
$$|$$
H₂N—CH—COOH
ASPARTIC ACID

$$COOH$$
$$|$$
$$CH_2$$
$$|$$
$$CH_2$$
$$|$$
H₂N—CH—COOH
GLUTAMIC ACID

HCOOH
FORMIC ACID

CH₃COOH
ACETIC ACID

HN$\big\langle$ $\begin{array}{l} CH_2\,COOH \\ CH_2\,CH_2\,COOH \end{array}$
IMINOACETIC-PROPIONIC ACID

HN$\big\langle$ $\begin{array}{l} CH_2—COOH \\ CH_2—COOH \end{array}$
IMINODIACETIC ACID

CH₃CH₂COOH
PROPIONIC ACID

PRODUCTS IN SOLUTION after Miller experiment included the organic compounds shown above. Some of these molecules are not important to the chemistry of life in modern organisms. For example α-alanine is used in proteins but β-alanine is not. Experiment produced both D and L optical isomers. 15

seashore (Haldane), adsorbed on the surface of fine-grained clays and minerals (Bernal), or enclosed in particles called coacervate drops (Oparin). Melvin Calvin has shown that in the presence of certain organic condensing agents, polymerization can even take place in dilute solutions.

Sidney Fox has found that heating a dry mixture of amino acids leads to long chain "proteinoids": polymers having a molecular weight over 10,000. He has suggested that such polymerizations took place in volcanic cinder cones, and that the proteins formed were then washed into the sea. He has also found that his proteinoids have a remarkable tendency to form MICROSPHERES about two micrometers in diameter when hot concentrated solutions are slowly cooled (*Figure 16*). These microspheres show a double-layer boundary resembling a membrane (although without lipids), and swell or shrink as the salt concentration is changed. If allowed to stand for several weeks, the microspheres absorb more proteinoid material from the solution, produce buds and sometimes divide to produce "second generation" microspheres. Cleavage or division can also be induced by changing pH or adding MgCl₂. These microspheres should not be taken to be the ancestors of life; rather, they show some of the remarkable self-organizing properties of relatively simple chemical

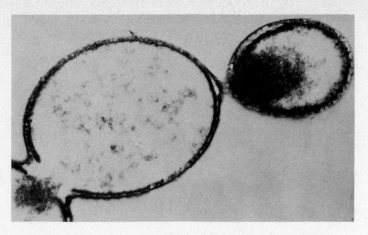

16 MICROSPHERES prepared by adding protein-like material to water show a double-layer boundary roughly similar to the membranes of microorganisms. Microspheres are not alive, but they illustrate the effects of physical forces that would have acted on primitive living systems as they evolved.

systems. They illustrate how many of the functions normally associated with living organisms can be duplicated by much less complex assemblages.

Oparin has spent most of his professional career studying COACERVATE DROPS. If you shake a mixture of a little olive oil in water, the oil will break up into a mass of tiny droplets. If instead you use a long-chain protein such as gelatin and a polysaccharide such as gum arabic, the result is the formation of coacervate drops, which are much more stable. As was the case with the olive oil-water mixture, the coacervate preparation is divided into two separate phases: the interior of the drops and the aqueous solution around them. Coacervate drops will form in solutions of many different kinds of polymers: proteins, nucleic acids, polysaccharides, and various synthetic polymers. Coacervation is simply a matter of physical chemistry, not of life. Yet coacervate drops show several properties relevant to the origin of life. Many substances, when added to a coacervate preparation, are preferentially concentrated within the droplets. Lipids can coat the boundaries of droplets with "membranes" and strengthen the droplets against destruction. If the coacervates are prepared so as to contain enzyme molecules, they can absorb substrates, catalyze a reaction, and let the products diffuse back out into the solution once more. Coacervates containing phosphorylase, for example, will absorb glucose-1-phosphate from the surrounding solution and polymerize it into starch. If a second enzyme, amylase, is also present in the coacervate, it will break up the starch into maltose, which escapes into the solution. The coacervates containing phosphorylase and amylase are then small factories for converting the monosaccharide glucose-1-phosphate into the disaccharide maltose. The energy for the overall process, of course, comes from the high energy phosphate on the original glucose. Oparin has even made photosynthetic coacervates containing chlorophyll, which will absorb an oxidized dye from the solution, use light energy to reduce it, and return the reduced dye to the surroundings.

It is possible that the immediate precursors of living organisms were capsules of chemical reactions similar to coacervate droplets. Some coacervates would enclose reactions that led to the early breakup of the droplets; others would enclose reactions that made the droplets more stable. The more stable coacervates would survive longer, and could possibly grow at the expense of their surroundings by absorbing chemical substances derived from the remains of less stable droplets. If wave action or other mechanical forces broke a large coacervate into many small droplets, each of these might be able to absorb material and grow on its own. This stage of "evolution" would be purely a matter of chemical competition—the kind of competition that decides whether the paper in a book will be oxidized slowly by air and turn brown with age, or oxidized rapidly by a suburban brush fire. Any nonbiological catalysts which accelerate the rates of favorable reactions in a given type of coacervate would give it a great advantage over more slowly-reacting droplets. Chemical selection, therefore, would favor catalyzed reactions. In view of what we know about the thousandfold enhancement of the rate of destruction of hydrogen peroxide when the inorganic catalyst ferric ion is replaced by an organic iron porphyrin, it is not hard to imagine how more and more efficient (and elaborate) catalysts would be developed and retained by chemical selection, until finally the evolving system stumbled onto the ultimate improvement of proteinlike catalysts or enzymes.

Oparin postulated the existence of organized, metabolizing, but nonreproducing systems that he called PROTOBIONTS. According to this reasoning, the breakthrough that led to truly living organisms was the development of REPRODUCTION —the ability of a successful chemical system to ensure its survival by duplicating itself. The molecules in which the instructions for duplication are stored in modern living creatures are DNA or RNA. Yet the LIVING unit of life is not just the nucleic acid, but the entire cell. The organism is as helpless without its DNA as a computer without a program tape, but the DNA alone can no more live than a magnetic tape without a computer can do calculations.

What we have been able to learn about the actual traces of Precambrian life, the heritage of a common chemistry in the biochemistry of modern living organisms, and possible laboratory analogues of certain facets of the evolution of life, makes it possible to construct a scenario for the origin of life. It reflects the speculations and biases of the authors in several places, and would not be accepted in its entirety by any of the scientists most concerned with such matters. But probably no two experts would argue with exactly the same points. The broad features of the scheme are almost certainly correct, but the details take us to the limits of human knowledge, and in some cases a little beyond.

THE SCENARIO

Scene 1. *The solar system condenses from a primeval dust cloud in our part of the Milky Way galaxy 4.5 to 5 billion years ago. The Earth and all of the other planets are formed by the gradual accretion of matter from the dust cloud around the new Sun.* The entire universe is estimated by astronomers to be of the order of 20 billion years old, and our galaxy, the Milky Way, to be 15 billion years old. Our Sun is a second-generation star in this galaxy. The first generation stars were the factories in which the heavier elements were synthesized, elements that were scattered when the first stars exploded. The Sun and the planets were assembled from this stellar debris. This is why the Earth is so much richer in heavy elements than the universe as a whole.

Scene 2. *The inner planets —Mercury, Venus, Earth, and Mars —are too small and too warm to retain their original atmospheres of hydrogen, helium, and the rare gases, in contrast to the larger and colder outer planets of Jupiter and beyond. The gases of the primitive Earth atmosphere escape into space, leaving behind a solid planet.* Radioactive decay and gravi-

tational compression liberate enormous amounts of heat. The smaller planets radiate this heat into space, but the Earth is too large for all of the heat to escape. The interior of the planet melts and the materials separate into an iron-nickel core, an iron and magnesium silicate mantle, and a lighter silicate crust. The core and mantle probably remain semifluid, with convection cells in the mantle that later will cause mountain folding and tectonic processes on the surface of the Earth *(Chapter 1).*

Scene 3. *Gases escaping from the hot interior of the Earth produce a secondary reducing atmosphere composed mainly of hydrogen, water, methane, ammonia, nitrogen, and hydrogen sulfide. No free molecular oxygen is present.* Evidence for this is the fact that all deeply buried rocks are reduced; the oxidized minerals are only a thin layer on the outside. Lavas, basalt and other deep iron-bearing rocks, for example, have the black, grey, and green colors characteristic of ferrous iron. The ferric reds and yellows of sand and sedimentary rock that we see around us are only skin deep.

Scene 4. *Water vapor condenses into seas, and a vast array of organic compounds is produced by the action of ultraviolet light, electrical discharges, radioactive decay in the crust, and volcanic heat. The oceans become a dilute "Haldane soup" of organic matter.*

Scene 5. *Some polymer solutions break up into coacervate droplets, either as a result of wave action or of concentration by evaporation in tide pools. The interior composition of these droplets is often quite different from the surrounding seas. Reactions within coacervate drops differ greatly, and affect the relative stabilities of drops. Droplets form and are broken up, and chemical selection operates between different types of coacervates, leading gradually to particularly stable reaction systems that could be called "protobionts."*

Scene 6. *The first generally available and used energy source for thermodynamically unfavorable chemical syntheses is the supply of naturally-occurring ATP.* This is purely a personal prejudice of the authors because the first thing that all living organisms do with energy extracted from foods is to make ATP with it. All of the locomotor, synthetic, and other activities of a living organism are keyed to the use of ATP as a fuel. Even so universal a process as glycolysis would in this viewpoint be considered as a second-generation metabolic mechanism, tacked onto the ATP-using machinery when the external supply of ATP began to run short.

Scene 7. *Certain of the protobionts develop the ability to employ nucleic acids as templates in the synthesis of useful proteins, which serve both as structural elements and catalysts or enzymes. When this synthetic ability progresses to the point that a protobiont can make a copy of itself, the organism can truly be called living.* The essence of reproduction is copying, but the essence of evolution in the biological sense is that the copying is not absolutely perfect. Copying produces new organisms, but mistakes (mutations) produce variations, and natural selection among variants is the basis of biological evolution. As the protobionts become more stable, chemical selection — the choice between stability or destruction of individuals — is less able to bring about improvements. With the emergence of the hereditary mechanism, a new field for selection arises, and a new kind of impovement, biological evolution, becomes possible.

Scene 8. *Organisms develop synthetic pathways to make substances that they once obtained from their surroundings, but which have been depleted.* This is the beginning of what could properly be called metabolism. It would be closely tied in with the ability to pass on protein-synthesizing machinery from one generation to another, through the hereditary mechanism. Gradually, in a step-by-step fashion, complicated networks of metabolic processes develop and are passed from one individual to its descendants by the hereditary machinery.

Scene 9. *Anaerobic fermentation or glycolysis develops, in which high free energy compounds are used to synthesize and store ATP as the natural supply of ATP runs short.* The new sources of energy are so rich, and organisms that develop glycolysis are so much at an advantage over their rivals, that all traces of the alternate metabolism vanish. Glycolysis becomes a universal heritage of all surviving life. NAD and the flavins develop as hydrogen carriers, and the enzymes of the glycolytic pathway evolve. For such simple catalyses, no structural organization of the enzymes is necessary, and they remain freefloating in the cellular medium.

Scene 10. *Free oxygen from high-altitude photodissociation begins to accumulate in the atmosphere in small amounts and to poison living organisms. Hydrogen peroxide is formed. First iron, then iron porphyrins, and finally the ancestors of catalase are developed to catalyze the destruction of hydrogen peroxide.* In this view, the first use of porphyrins would be in iron-containing heme proteins that eliminate H_2O_2. Photosynthesis, respiration, and oxygen transport in the bloodstream would later all make use of the materials that were already available, the porphyrins.

Scene 11. *Certain organisms begin using magnesium porphyrins to absorb visible light, and to channel the energy obtained into the synthesis of ATP.* Photophosphorylation is chemically simpler and is probably older than complete photosynthesis (see the discussion in Chapter 6). The total solar energy is so much greater than the small fraction available in the ultraviolet region of the spectrum that any organisms that develop the ability to tap the energy of visible light will have a great advantage over their competitors. Chlorophyll as we know it is probably the end product of a long line of less-efficient precursors which were gradually weeded out by natural selection. Ferredoxin and cytochromes *f* and *b* would have developed at this time.

Scene 12. *Bacterial photosynthesis appears, using various donor molecules other than water as sources of hydrogen in the fixation of CO_2.* The green and purple sulfur bacteria use H_2S. Athiorhodaceae use organic matter, more specifically their own waste products. Modern Athiorhodaceae absorb as much as 90 percent of the waste products of glycolysis in building up their own mass by photosynthesis of glucose.

Scene 13. *The bluegreen algae evolve, using water as the hydrogen source in photosynthesis. For the first time oxygen is liberated into the atmosphere from photosynthesis.* The two halves of the reaction, in which protons from water and electrons from excited chlorophyll *a* combine to reduce NADP$^+$, and electrons from the water hydroxyl restore the electron deficiency in chlorophyll *b*, must be kept separated (*Chapter 6*). If not, the photosynthetic process is short-circuited. This means that more elaborate structures must be developed than were formerly necessary. The precursors of chloroplast grana evolve in bluegreen algae.

Scene 14. *Free atmospheric oxygen becomes a serious threat to life, and various organisms develop different defense mechanisms. A layer of ozone, O_3, appears in the upper atmosphere, shielding the surface of the planet from intense ultraviolet radiation, and permitting living creatures to occupy shallower waters.* Free molecular oxygen damages life forms by oxidizing flavins, producing hydrogen peroxide, and destroying the natural supply of organic compounds in the sea. At the same time, the shielding of the surface from ultraviolet radiation greatly decreases the natural synthesis of such compounds. Living creatures become more dependent on photosynthesis and on eating one another for sustenance. Haldane's "soup" becomes pretty thin. Bioluminescence may have originated as a means of eliminating oxygen gas by using it to oxidize organic matter. Similarly, the combination of this dangerous gas with the life form's own waste products as a protection may have been the first step on the road to respiration. (If you must burn your rubbish, you might as well boil water with it.) Alternatively, the combination of oxygen with reduced compounds in the crust of the Earth may have been the first step in developing chemautotrophic nutrition.

Scene 15. *Respirers and chemautotrophs begin to use oxidation reactions as a source of energy for the synthesis of more ATP.* The increased oxygen content of the atmosphere oxidizes the upper layers of the crust and forces the chemautotrophs to the marginal niches in deep wells and underground deposits where they are found today. Simultaneously the increased amount of organic matter produced by living organisms favors respiration. The cytochromes of the terminal respiratory chain develop. A more intricate spatial arrangement is required for these processes, favoring the evolution of the organized membrane-bound enzyme systems that we see today in mitochondria.

Scene 16. *Eucaryotic organisms evolve.* It is possible that the chloroplasts found in eucaryotic plants are the vestigial remains of symbiotic bluegreen photosynthetic algae, and that the mitochondria found in all eucaryotes are the descendants of symbiotic nonphotosynthetic bacteria. The combination of a high oxygen concentration in the atmosphere, the emergence of a really efficient system of respiration, the evolution of eucaryotes, and the appearance of multicellular life forms, all lead to an explosion of life all over the planet, and to the upsurge in number and variety of fossil remains that marks the beginning of the Cambrian era. At this stage the evolution of life as a mechanism is complete, and the further evolutionary development of plants and animals is a familiar story.

THE BERKNER-MARSHALL THEORY

An attempt to attach a time scale to the preceding scenario is shown in Figure 17. This diagram reflects what has become known as the Berkner-Marshall theory: that two of the most dramatic changes in the history of life—the increase in population at the beginning of the Cambrian and the spread of life to the land in the Silurian—were both elicited by critical concentrations of oxygen in the atmosphere. The sudden increase in the living population at the beginning of the Cambrian era may have come about because the oxygen content reached one percent of present levels, making efficient respiration possible. The increased efficiency

How much of this can be believed? Every generation needs its own Creation myths, and these are ours. They are better than any that have come before, but they are subject to revision as we learn more about the history of life.

of energy extraction from a given amount of food would have made it possible for larger and multicellular creatures to exist. Similarly, a rise in oxygen content to 10 percent of present levels would create a high-altitude ozone screen that would shield the surface almost completely from hard ultraviolet radiation. Living organisms could then move from the seas and colonize the exposed surfaces of the Earth. This actually took place for plants in the Silurian period, around 425 million years ago (the 28th day on our calendar).

Scenes 1 through 8 of our scenario probably occurred within the first billion years of our planet's existence. Scene 9, glycolysis, should coincide in time with the word FERMENTERS in Figure 17. The primitive spheroids and filaments in the Onverwacht sediments (*Table I*) may be the remains of such protodecomposers, with a crude anaerobic fermentation metabolism. Photosynthesis (scenes 11–13) is likely to have arisen around 3.1 to 2.7 billion years ago. The Fig Tree bacteria and the Bulawayo algal stromatolites may date from this period. If the Bulawayo deposits do represent bluegreen algae, with O_2-liberating photosynthesis, then the first steps toward our present oxygen-rich atmosphere would have been taken 2.7 billion years ago. The era from that time until the late Precambrian would have been one of gradual evolutionary improvement of the photosynthetic machinery, increasing oxygen concentration, and slow development of respiratory metabolism (scenes 14 & 15). Scene 16, the appearance of eucaryotes, evidently occurred slightly less than one billion years ago, judging from the presence of green algae in the fossils at Bitter Springs, Australia.

The Archeozoic Era (4.5–2.5 billion years ago, *Table I*) would therefore be the era of the evolution of anaerobic, heterotrophic life. The Lower and Middle Proterozoic (2.5–1.0 billion years ago) would see the evolution of photosynthesis, the

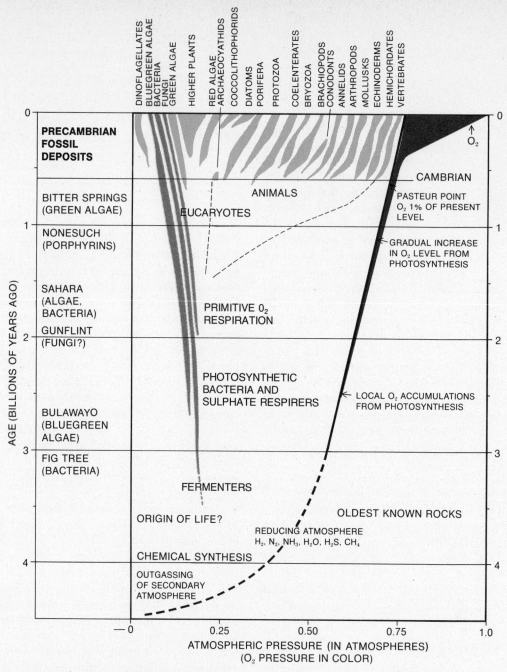

17 **BERKNER-MARSHALL THEORY holds that the
proliferation of life at the beginning of the Cambrian
Era and the invasion of land by living organisms
depended on concentration of oxygen in ancient
atmosphere. Diagram shows evolutionary tree
superimposed on graph of evolving atmosphere. Width
of branches indicates relative abundance of different
types of organisms. Principal Precambrian fossil
deposits are dated at left. Curve at right shows
progressive increase of atmospheric pressure, with
partial pressure of oxygen gas in color. Oxygen pressure
rose sharply in last 0.5 billion years.**

change in character of our atmosphere, and the
development of respiration. The Late Proterozoic
(1.0–0.6 billion years ago) would date approximately from the appearance of eucaryotes, include
the rise of multicellular organisms, and end with
the rapid rise in population brought on by an accumulation of many improvements in the machinery of life.

How much of this can be believed? Every generation needs its own creation myths, and these are ours. They are probably more accurate than any that have come before, but they are undoubtedly subject to revision as we find out more about the nature and the history of life. The best that can be said for any scientific theory is that it explains all the data at hand and has no obvious internal contradictions.

READINGS

The following two papers, written independently of one another by a Russian and an English biochemist, were the starting points for modern studies of the origin of life. They are both clear, short, and elegant. They are also so far ahead of their time that one gets the uneasy feeling that all we have done since 1924 is to fill in the technical gaps in Oparin and Haldane's schemes. Both papers are reprinted as appendices to Bernal's *Origin of Life:*

A.I. OPARIN, "The Origin of Life," Moscow, 1924.
J.B.S. HALDANE, "The Origin of Life," in *The Rationalist Annual*, London, 1929.

The articles and books listed below are particularly good introductions to the evolution of life for newcomers to the subject:

J.D. BERNAL, *The Origin of Life,* Cleveland, World Publishing Company, 1967. A clear and non-technical introduction by one of the best minds in the field. Raises philosophical as well as scientific questions. Well illustrated.

S. L. MILLER AND L. E. ORGEL, *The Origins of Life on Earth,* Englewood Cliffs, NJ, Prentice-Hall, 1974. The best general introduction to the subject by two active investigators. A good key to the research literature.

I.S. SHKLOVSKII AND C. SAGAN, *Intelligent Life in the Universe,* San Francisco, Holden-Day, 1966. Chapters 14–17 contain a good introductory discussion of the appearance of life on Earth. Excellent color plates of Precambrian fossils.

GEORGE WALD, "The Origins of Life," in *The Scientific Endeavor*, New York, Rockefeller University Press, 1964. This paper, given at the Centennial celebration of the National Academy of Science in 1963, is also published in *Proc. Nat. Acad. Sci. (U.S.),* 52:595 (1964). Ranks with the papers by Oparin and Haldane for clarity and brevity.

A.G. FISCHER, "Fossils, Early Life, and Atmospheric History", and L.V. BERKNER and L.C. MARSHALL, "History of Major Atmospheric Components", in *Proc. Nat. Acad. Sci. 53,* 1205 and 1215 (1965). Two excellent discussions of the evolution of the atmosphere and the development of life.

J.W. SCHOPF, "Precambrian Paleobiology: Problems and Perspectives", in *Ann, Rev. Earth and Planet. Sci. 3,* 213 (1975). A skeptical reassessment of the early fossil record of Precambrian life, by a very conservative scientist. When is a filament an organism, and when is it a mineralogical defect? Excellent tabulations of fossil finds, and a guide to the scientific literature.

For readers who want to go more deeply into the evolution of life, the books listed below are more advanced but particularly good:

A.I. OPARIN, *Genesis and Evolutionary Development of Life,* New York, Academic Press, 1968. Approximately every seven years from 1936, Oparin has written a book on the origin of life summarizing the state of the field at the time. This book now supersedes *Life: Its Nature, Origin and Development* (1961) and the various editions of *The Origin of Life on Earth*. This book includes a good historical introduction to what people thought about spontaneous generation of life since the time of the Greek philosophers.

A.I. OPARIN, A.G. PASYNSKII, A.E. BRAUNSHTEIN, AND T.E. PAVLOVSKAYA (editors), *The Origin of Life on the Earth,* New York, Pergamon Press, 1959. The proceedings of the First International Symposium on the Origin of Life, held in Moscow in 1957. Scientists from all over the world were represented. Some of the discussions and arguments (reprinted in full) are particularly interesting. Do not confuse this book with Oparin's early books with almost the same title.

S.W. FOX (editor), *The Origins of Prebiological Systems,* New York, Academic Press, 1965. Proceedings of a conference held in Florida in 1963. A sequel to the Moscow symposium, with much the same cast of contributors.

It must not be imagined that because we place one genus or one family before another we would consider them as more perfect, as superior to others in the system of nature. He alone has this pretension who pursues the chimerical project of placing the organisms in a single line; we have long ago renounced this scheme. —Georges Cuvier (1828)

"I know *who* I am but I don't know *what* I am!"

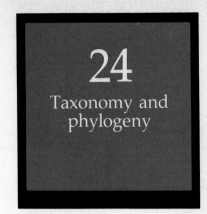

When man encounters the unknown, his first rational response is to classify. Strange objects, people, or phenomena must be compared with one another, collected into abstract groups, and named. Once classified, they can be subjected to a more sophisticated form of scrutiny. It was natural that the first efforts of biology, dating back to the time of Aristotle, were devoted to classification. We call the science of classification of organisms TAXONOMY. Other words for the same discipline are systematics and biosystematics. Since almost all taxonomists believe in evolution, a study of the similarity among organisms is normally elaborated in some fashion into evolutionary hypotheses. These interpretations are expressed in PHYLOGENETIC TREES, which represent the lines of descent from one species or other group of organisms to another. In the present chapter we shall consider first the basic rules of taxonomic procedure, including the rules of nomenclature (the assignment of technical names), and then show how biologists go about constructing classifications and phylogenetic trees.

The basic unit of classification is the SPECIES. Later, in Chapter 30, the definition and full biological meaning of this unit will be examined. For the moment, it is enough to define the species loosely as a group of organisms that closely resemble one another, either because they interbreed freely or because they have descended from common ancestors in the not too distant past. In either case, the members of the species are genetically similar; they resemble one another more than they do the members of other species.

According to the internationally accepted rules of nomenclature, each species is assigned two names, one identifying the species itself and the other the GENUS to which the species belongs. The genus is a group of species that resemble one another (the plural of the word genus is genera

and its adjective form is generic). In many cases the name of the taxonomist who first proposed the species name is added on the end. Thus *Homo sapiens* Linnaeus is the name of the modern human species. *Homo* is the genus to which the species belongs, *sapiens* identifies the species, and Linnaeus was the first author to have used the species name *sapiens*. You can think of the generic name *Homo* as being equivalent to your surname and the specific name *sapiens* as being equivalent to your first name. This two-name system, referred to as binomial nomenclature, is universally employed throughout biology.

The list of names used dates back to the works of Carolus Linnaeus (1707–1778), the great Swedish biologist who first applied the binomial system on a rigorous and worldwide scale to both plants and animals. To avoid confusion, a law of priority is observed —where two or more names have been used for the same species, providing neither predates the work of Linnaeus, the oldest name is retained and the others discarded. Thus when the ant species *Lasius pallitarsus* Provancher was recently shown to be the same species as *Lasius sitkaensis* Pergande, the older name *L. pallitarsus* took precedence over the newer name *L. sitkaensis*. *L. sitkaensis*, to use the common expression, was "synonymized" or "placed in synonymy" under *L. pallitarsus*. It is the junior synonym and *L. pallitarsus* is the senior synonym, a fact that usually is expressed in taxonomic writings by the following convention: *Lasius pallitarsus* Provancher (= *Lasius sitkaenis* Pergande). Common names of animals in use for more than 50 years are exempted from the change resulting from the discovery of an obscure older name. This rule prevents taxonomists from indulging in the gamesmanship of upsetting the classification by searching through long-forgotten literature.

Notice that the generic name is capitalized and

523

the specific name is not. In botany the specific name is sometimes capitalized, when the name of a person or place is incorporated in it, for example *Vernonia Blodgetti*, one of the iron-weeds. The generic and specific names are always italicized. These scientific names, or "technical names" as they are often called, are usually derived from more or less appropriate Greek or Latin words. Thus *Homo sapiens* is Latin for "wise man," *Lasius pallitarsus* means "hairy one with pale feet," and *Lasius sitkaensis* means "hairy one from Sitka (Alaska)," the place where the first specimens were collected. If the name chosen by the taxonomist is not Greek or Latin in origin, it is usually Latinized by adding a suitable ending. Occasionally taxonomists become frivolous in their choice of

The name Zyzzyx was applied to a genus of wasps by an entomologist for the express purpose of contributing the last name at the end of taxonomic catalogs. Nomenclature can sometimes be whimsical.

names and have to be curbed. For example, the generic names *Peggichisme* (pronounced Peggy kiss me), *Marichisme*, *Polychisme*, and so on, given originally to a group of hemipteran bugs, have been invalidated through a special ruling of the International Commission of Zoological Nomenclature. So has the crustacean name *Cancelloidokytodermogammarus (Loveninuskytodermogammarus)*, for obvious reasons. On the other hand, *Zyzzyx*, which was

1 **IMAGINARY PLANT SPECIES can be clustered into genera in several ways. Classification scheme at top might be followed by a taxonomist who is a "lumper"; scheme at bottom, by a "splitter."**

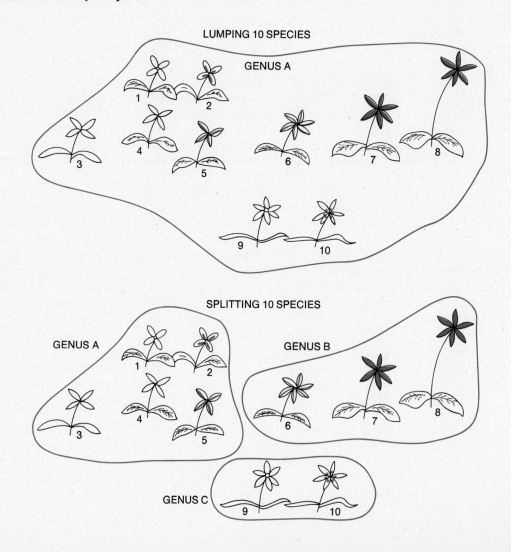

LUMPING 10 SPECIES

GENUS A

SPLITTING 10 SPECIES

GENUS A

GENUS B

GENUS C

applied to a genus of wasps by an entomologist for the express purpose of contributing the last name at the end of taxonomic catalogs, has been allowed to stand. Nomenclature can sometimes be whimsical.

Suppose a taxonomist undertook to revise a group of species, meaning that he was examining the similarities among the species and attempting to classify and to name them, as well as to deduce something about their phylogeny. He would begin by deciding which species are represented and what criteria he will use to distinguish between them. The preferred criterion is whether or not the organisms interbreed freely with one another in nature. Then, by a careful study of the available information on morphology, physiology, and behavior, he would decide on a measure of resemblance among the species. There are easy-to-follow statistical procedures for taking such measures. But the procedures still contain many arbitrary features, with the result that most taxonomists still try to judge the matter subjectively. Whichever measure is used, the next problem is CLUSTERING the species into higher taxonomic categories.

Here the taxonomist is faced with a choice: he must decide the limits of each genus. A precedent he might consider, one of many available, is the placement of all of the oak species in the genus *Quercus*. The question is: Which of the various species can be clustered together to form the genera? Figure 1 shows that this is basically an arbitrary choice. By studying the position of the plants, each representing one species, one can see that it is equally valid in this particular case to combine all of the species into a single genus, or to split them into two or more genera. The lower of the two arrangements shows the species divided into three genera, but it could have been four, if we decided to put species 3 in a genus of its own, or even five, if species 6 were also separated. Taxonomists who like to combine species into a small number of large genera are often described lightly as "lumpers," while their colleagues who prefer a large number of small genera must be called, of course, the "splitters." Everyone recognizes that the choice is at least partly a matter of taste. But taxonomy is not a subject to be taken casually. The two most important criteria of good taxonomic work are first, the judgment used in defining species, or other "operational taxonomic units" such as local populations; and second, the clarity with which the relationships are described.

Similar subjective techniques are required in the classification of categories higher than the genus. Once the limits of the genera have been decided upon, the taxonomist clusters them into FAMILIES. Figure 2 shows how the similarities among species comprising different genera form the basis for grouping particular genera into single families or setting them apart in different families. An example of a family is the Formicidae, which includes all of the genera and species of ants in the world. At this point the reader should take note of an important fact about the clustering procedure that often escapes beginners: It is not necessary to include many species in each group belonging to higher categories. A genus may contain only a single species, and a family may contain only a single genus; it is possible for a family to be composed of only one species. This extreme case is exemplified by family II in Figure 2.

In animal classification the names of families are identified by -idae endings. Thus Formicidae is the family that contains all ant species, while Hominidae contains man and his fossil relatives. The family names are based on a member genus; hence Formicidae is based on *Formica* (just remove the -a and add -idae), and Hominidae is based on

SEVEN GENERA, A to G, containing a total of 20 species (denoted by numerals), are classified into families by a lumper (top) and a splitter (bottom). Both arrangements are equally valid because a family, like a genus, is an arbitrarily defined cluster of similar species.

2

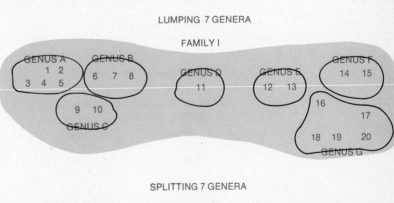

LUMPING 7 GENERA

FAMILY I

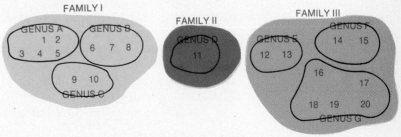

SPLITTING 7 GENERA

FAMILY I   FAMILY II   FAMILY III

Homo (from the base of the genitive form hominis, plus -idae). Plant classification follows the same procedure, except that the ending -aceae is used instead of -idae. Thus the Rosaceae is the family that includes the genus of roses *(Rosa)* and all its relatives. Notice that unlike the generic and species names, the name of the family is not italicized.

To demonstrate how a taxonomic key is designed and used we have created four species of an imaginary insect called Hypotheticus.

And so it goes on up into still higher categories. The taxonomist groups families into orders, orders into classes, classes into phyla, and phyla into kingdoms according to the same procedures. The hierarchy of units to which our own species *(Homo sapiens)* belongs is shown in Box A.

Turning to the plant kingdom, a parallel set of categories can be listed for the common rose, *Rosa gallica (Box B)*.

Once the classification and nomenclature are settled, the taxonomist must worry about making it possible for other biologists to identify the species he has revised. He accomplishes this in three ways: by placing identified specimens in museum collections for reference purposes, by publishing illustrations and verbal descriptions in articles, and by constructing TAXONOMIC KEYS. To see how a taxonomic key is designed and used, consider the four species of the imaginary insect genus *Hypotheticus (Figure 3)*. We use an arbitrary example only to be able to simplify and clarify without distorting reality. We start with four *Hypotheticus* species that can be separated easily from one another by characteristics that are obvious in the drawing. The characteristics, or CHARACTER STATES as they are usually called, are also listed in Table I and marked as either present (+) or absent (−) in each of the four species.

Using the information concerning the character states, the reader can construct a simple taxonomic key *(Table II)*. By using the key to identify one or two of the species from the drawings alone, the reader can quickly see how the keys are constructed. Start with the first couplet of traits, make a choice between the two alternatives given there, and proceed on down. By examining the multiple character states that distinguish each of the four species, it is clear that the key to *Hypotheticus* can follow other sequences of characters than the one chosen here.

Now proceed to the construction of a hypothetical phylogeny, still using the convenient set of four *Hypotheticus* species. A first step is to deduce the BRANCHING SEQUENCES in the history of the genus. The basic, simple question is, which species

---

A KINGDOM: **Animalia** (animals)
   PHYLUM: **Chordata** (chordates)
      SUBPHYLUM: **Vertebrata** (vertebrates)
         CLASS: **Mammalia** (mammals)
            ORDER: **Primates** (primates)
               FAMILY: **Hominidae** (man and close relatives)
                  GENUS: *Homo* (modern man and precursors)
                     SPECIES: *Homo sapiens* (modern man).

B KINGDOM: **Plantae** (plants)
   PHYLUM: **Tracheophyta** (vascular plants)
      SUBPHYLUM: **Pteropsida** (ferns and seed plants)
         CLASS: **Dicotyledoneae** (dicots)
            ORDER: **Rosales** (saxifrages, psittosporums, sweet gum, planetrees, roses, and relatives)
               FAMILY: **Rosaceae** (cherry, plum, hawthorn, roses, and relatives)
                  GENUS: *Rosa* (roses)
                    SPECIES: *Rosa gallica* (domestic rose).

SPECIES	BODY ENTIRELY WHITE	HORNS PRESENT ON HEAD AND THORAX	WINGS VERY SHORT	EXACTLY TWO SPOTS ON ABDOMEN
HYPOTHETICUS ALBUS	+	−	−	−
HYPOTHETICUS CORNUTUS	−	+	−	−
HYPOTHETICUS BRACHYPTERUS	−	−	+	−
HYPOTHETICUS BIMACULATUS	−	−	−	+

I CHARACTER STATES of the species that comprise the genus Hypotheticus. Only the presence (+) or absence (−) of each character state is indicated.

evolved from which other species? Or, alternatively, which species share immediate common ancestors? Suppose we had at our disposal the following bits of information, of the sort that is often available in taxonomic studies: All early fossils of *Hypotheticus* and similar genera are completely white and hornless and possess normally developed wings. A few fossil specimens of *Hypotheticus* from a later geologic period have two spots on the abdomen (like the modern species *H. bimaculatus*). The four-spotted condition, so far as is known, is limited to the modern species *H. brachypterus*.

We conclude that white bodies, hornlessness, and normal wings are the primitive (original)

A KEY TO THE SPECIES OF HYPOTHETICUS		
1.	BODY ENTIRELY WHITE	H. ALBUS
	HEAD AND THORAX BLACK	GO TO ITEM 2
2.	HEAD AND THORAX BEARING A TOTAL OF FIVE LARGE SPINES OR HORNS	H. CORNUTUS
	HEAD AND THORAX LACKING SPINES AND HORNS	GO TO ITEM 3
3.	WINGS EACH MUCH SHORTER THAN THE WIDTH OF THE BODY	H. BRACHYPTERUS
	WINGS EACH MUCH LONGER THAN THE WIDTH OF THE BODY	H. BIMACULATUS

II TAXONOMIC KEY of the genus Hypotheticus. Key refers to characteristics illustrated in Figure 3.

character states and all the other character states we listed were derived in evolution. If this is indeed the case, it is possible to study how the various character states are combined and then to come up with the simple phylogenetic diagram shown in Figure 4. The reader should try this himself to make sure he understands it. Notice that the diagram shows only the branching sequences. No attempt is made to make the time scale precise or absolute; an exact chronology can be determined only with information derived from fossil specimens. Once the fossil data are at hand, we can construct a more detailed phylogenetic diagram *(Figure 5)* in which the distance separating the end points of the branches indicates the degree of difference between the species. This difference is usually the best yardstick for estimating the amount of evolution that has occurred since the branches diverged.

HOMOLOGY

The phylogenetic analysis of the imaginary genus *Hypotheticus* was based on a fundamental but unspoken assumption. When two species of

IMAGINARY GENUS OF INSECTS includes four **3** species. Early primitive species (albus) has white body, no horns, and normal wings. One advanced species (bimaculatus) has two spots on abdomen. Two other advanced species (cornutus and brachypterus) have four spots. Cornutus has horns on thorax, brachypterus has short wings. Genus was created to show principles of identification and phylogeny.

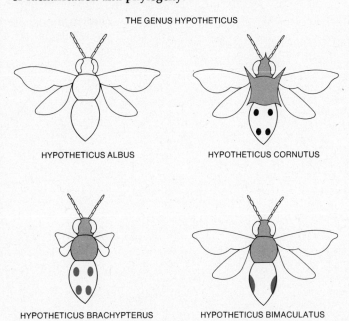

THE GENUS HYPOTHETICUS

HYPOTHETICUS ALBUS   HYPOTHETICUS CORNUTUS

HYPOTHETICUS BRACHYPTERUS   HYPOTHETICUS BIMACULATUS

Hypotheticus were found to share similar character states, such as black coloration or horns on the body, it was implicitly assumed that these states were acquired from a similar or identical state in some ancestral species. If this assumption were true, a taxonomist would be justified in saying that possession of similar character states is evidence of a phylogenetic relationship between the two species. When the character states found in any two species owe their resemblance to a common ancestry, taxonomists say the states are HOMOLO-GOUS, or are HOMOLOGUES of each other. HOMOLOGY is defined as correspondence between two structures due to inheritance from a common ancestor.

Homologous structures can be identical in appearance and, at least in theory, can even be based on identical genes. Conversely, such structures can diverge in evolution until they become quite different in both appearance and function. But homologous structures always retain certain basic features that betray their common ancestry. The classic example of homology combined with di-

4 ELEMENTARY PHYLOGENY of Hypotheticus is summarized in this diagram. Branching sequence is based on assumption that species having similar traits descended from a common ancestor.

The classic example of homology combined with evolutionary divergence is found in the comparative anatomy of the forelimbs of vertebrates.

vergence is found in the comparative anatomy of the vertebrate forelimb. When one compares the forearm of a man with that of a monkey, for example, it is easy to make a detailed, bone-by-bone, muscle-by-muscle comparison and to conclude thereby that the forearms, as well as their various parts, are homologous. But if one compares a human forearm with the foreleg of a dog, strong differences are obvious in both structure and function of the forelimb. This organ is used for locomotion by the dog but for grasping and manipulation by the man. Even so, all of the bones can still be matched. It is reasonable to conclude that these structures, and therefore the forelimbs as a whole, are homologous in man and the dog. More extreme cases exist within the vertebrate animals.

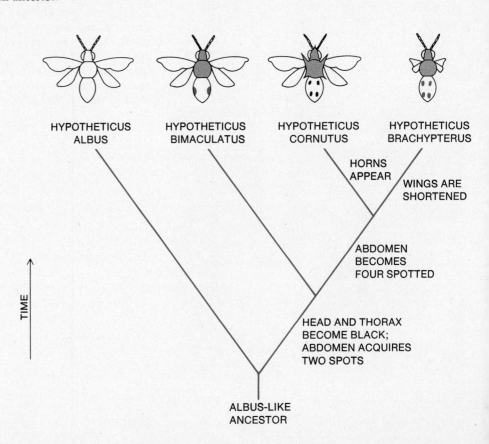

HYPOTHETICUS ALBUS   HYPOTHETICUS BIMACULATUS   HYPOTHETICUS CORNUTUS   HYPOTHETICUS BRACHYPTERUS

HORNS APPEAR

WINGS ARE SHORTENED

ABDOMEN BECOMES FOUR SPOTTED

HEAD AND THORAX BECOME BLACK; ABDOMEN ACQUIRES TWO SPOTS

TIME

ALBUS-LIKE ANCESTOR

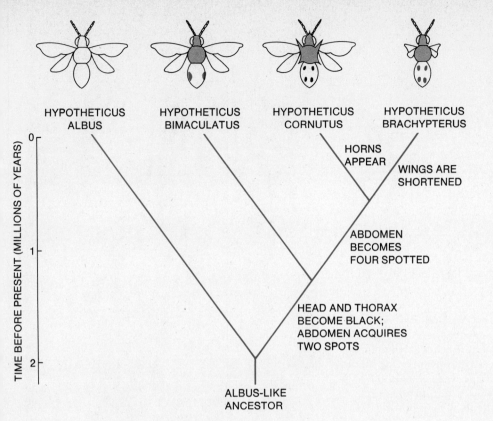

The wing of a bird and the flipper of a seal at first glance appear to be radically different from each other or from the forearm of a man, yet they too are constructed around bones that can be matched on a nearly perfect one-to-one basis.

The wing of a fly shares the same function as a bird's wing, and the two organs even resemble each other to a slight degree in external form. Yet close examination shows that the two organs are completely different in their basic structure. The entire wing of the fly is a membranous outgrowth of the external skeletal covering of the body. Its main supports are not bones but rather columns of hardened protein combined with a complex polysaccharide called chitin. The bird and fly wings are said to be ANALOGUES of each other. ANALOGY is defined as a resemblance in function which is based on CONVERGENT EVOLUTION rather than descent from a common ancestor *(Figure 6)*.

The taxonomist constructs his classifications from similarities and differences that he can directly observe in the relatively small number of specimens at his disposal. To add the element of phylogeny to these classifications, he must then decide whether the characteristics shared by the various groups are homologous or analogous — whether they are due to common ancestry or to

DETAILED PHYLOGENY of Hypotheticus includes information in Figure 4, plus fossil data (which establishes time scale) and a rough estimate of amount of evolution since branches diverged.

5

convergent evolution. Such decisions are almost always subjective and difficult to make. Typically they can be put to a test only by consulting fossils, which provide the only direct record of the ancestry of living organisms.

For resolving questions of phylogeny the most useful fossils are those that preserve the features actually used by the taxonomist. Living gastropods (snails), for example, are classified on the basis of variations in their soft bodies. Fossil gastropods are known mainly on the basis of their shells; the soft parts are rarely preserved. Thus there is an element of uncertainty in establishing lines of inheritance of the important taxonomic features.

Few fossils are composed of the actual material of the once-living animal or plant. More commonly the fossil vertebrate bone or invertebrate shell is a mineral replacement of the original *(Figure 7)*. Frequently, as with petrified wood, the original structure of the organism is preserved — replaced molecule by molecule as the original material is removed by decomposition.

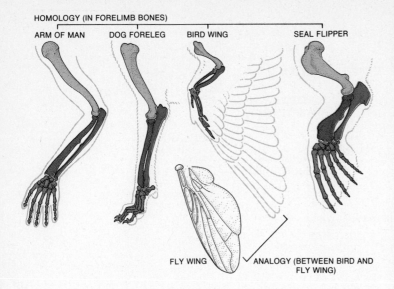

HOMOLOGY (IN FORELIMB BONES)

ARM OF MAN DOG FORELEG BIRD WING SEAL FLIPPER

FLY WING | ANALOGY (BETWEEN BIRD AND FLY WING)

6 HOMOLOGY AND ANALOGY. Forelimb bones of various vertebrates are homologous, even though they sometimes serve very different functions. Homologous bones in drawing are rendered in corresponding shades of gray and color. Wing of the fly is analogous to that of a bird, meaning that it is similar in function and to some extent in form, despite its wholly different evolutionary origin.

Sometimes images of the soft parts are preserved as casts retaining some of the original molding of the organism. A particularly favorable process is the deposition of carbonized films *(Figure 8)*. The films are created when the volatile components of the decomposing organism are released and fixed directly onto the surface of a smooth, finely structured surrounding material. The finest of the smaller fossils are those preserved in amber, the fossilized gum of trees. When a tree is injured or for some other reason exudes gum, the material serves as a natural flypaper that traps pollen, bits of leaves, and the smallest animals, especially insects. In some of the amber fossils the anatomical details are nearly as complete as those in specimens freshly fixed by the biologist and embedded in clear balsam for microscopic examination. Every hair and wrinkle can still be studied under magnification. One of the oldest and most significant amber specimens appears in Figure 9. This worker ant, discovered in 1967, was the earliest social insect of its kind known at the time, and it proved that social life originated as long as 100 million years ago (a termite of similar age has been uncovered since then). It also constitutes one of the missing links of evolution, which connects the ants with the nonsocial wasps that gave rise to them.

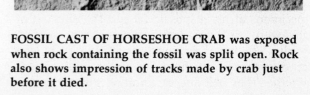

FOSSIL CAST OF HORSESHOE CRAB was exposed **7** **when rock containing the fossil was split open. Rock also shows impression of tracks made by crab just before it died.**

8 FOSSIL OF A FROG (Eopelotes beyeri) was laid down about 25 million years ago. Soft parts of body were preserved as a film of carbon on rock.

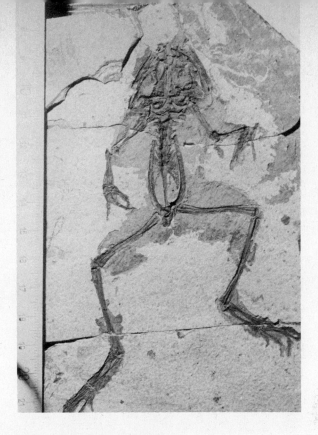

With the new information provided by this fossil it was possible to state with some assurance that the ants originated from one particular family of wasps, the Tiphiidae. Accordingly, the new genus to which the fossil ant belongs was named *Sphecomyrma*, which means "wasp ant."

THE EVOLUTIONARY TIME SCALE

In the early 19th century paleontologists began the arduous task of reconstructing the history of life on Earth by the study of fossils. By the latter part of the century it was clear that the sequence of fossils reflected many of the evolutionary changes predicted by Darwin. But an urgent question remained. What was the time scale of the fossil-dated geological ages? How many thousands or millions of years did major evolution require? Geologists, by various indirect methods such as estimating the time required to lay down the total deposits of rocks, judged that the Earth must be at least hundreds of millions of years old. This, the biologists agreed, was consistent with the time required for the sweeping changes that evolution has

100-MILLION-YEAR-OLD ANT depicted in **9** photograph below is exquisitely preserved in amber. Ant was trapped in the gum of a sequoia tree in an ancient forest near the Atlantic shore of New Jersey. This specimen is an evolutionary link between ants and their ancestors, the nonsocial wasps.

III **GEOLOGICAL ERAS since the dawn of the Cambrian period. Table corresponds to the last four days of the geological calendar in Chapter 23, and lists the major biological events of each period. Curve at right indicates the evolutionary trend toward a greater diversity of organisms. Dip in middle of curve reflects the widespread extinctions around the Permian period. Dates were obtained from measurements of radioactivity in rocks.**

wrought among living organisms. In the last two decades an absolute time scale was finally made possible by the radioactive dating techniques discussed in Chapter 23. Those techniques are based on decay processes that take millions of years, and therefore serve to date the older geological deposits. For the dating of younger deposits, a powerful tool is RADIOCARBON DATING, which is based on the decay of carbon-14, the isotope familiar from

ERA	PERIOD	BIOLOGICAL EVENTS	YEARS AGO	NUMBER OF ANIMAL FAMILIES
CENOZOIC	QUATERNARY	MODERN MAN	11 THOUSAND	
		EARLY MAN: NORTHERN GLACIATION	5 TO 3 MILLION	
	TERTIARY	LARGE CARNIVORES	13 ± 1 MILLION	
		FIRST ABUNDANT GRAZING MAMMALS	25 ± 1 MILLION	
		LARGE RUNNING MAMMALS	36 ± 2 MILLION	
		MANY MODERN TYPES OF MAMMALS	58 ± 2 MILLION	
		FIRST PLACENTAL MAMMALS	63 ± 2 MILLION	
MESOZOIC	CRETACEOUS	FIRST FLOWERING PLANTS; CLIMAX OF DINOSAURS AND AMMONITES FOLLOWED BY EXTINCTION	135 ± 5 MILLION	
	JURASSIC	FIRST BIRDS, FIRST MAMMALS; DINOSAURS AND AMMONITES ABUNDANT	180 ± 5 MILLION	
	TRIASSIC	FIRST DINOSAURS; ABUNDANT CYCADS AND CONIFERS	230 ± 10 MILLION	
PALEOZOIC	PERMIAN	EXTINCTION OF MANY KINDS OF MARINE ANIMALS, INCLUDING TRILOBITES. GLACIATION AT LOW LATITUDES	280 ± 10 MILLION	
	CARBONIFEROUS	GREAT COAL FORESTS, CONIFERS. FIRST REPTILES. SHARKS AND AMPHIBIANS ABUNDANT. LARGE AND NUMEROUS SCALE TREES AND SEED FERNS	345 ± 10 MILLION	
	DEVONIAN	FIRST AMPHIBIANS AND AMMONITES; FISHES ABUNDANT	405 ± 10 MILLION	
	SILURIAN	FIRST TERRESTRIAL PLANTS AND ANIMALS	425 ± 10 MILLION	
	ORDOVICIAN	FIRST FISHES; INVERTEBRATES DOMINANT	500 ± 10 MILLION	
	CAMBRIAN	FIRST ABUNDANT RECORD OF MARINE LIFE. TRILOBITES DOMINANT, FOLLOWED BY EXTINCTION OF ABOUT TWO-THIRDS OF TRILOBITE FAMILIES	600 ± 50 MILLION	
	PRECAMBRIAN	FOSSILS EXTREMELY RARE, CONSISTING OF PRIMITIVE AQUATIC PLANTS AND SOME SIMPLE ANIMALS. EVIDENCE OF GLACIATION. OLDEST DATED ALGAE AND BACTERIA, 3.1 BILLION YEARS		

(NUMBER OF ANIMAL FAMILIES axis: 0 200 400 600 800 1000)

tracer experiments *(Chapter 3)*. The C^{14} atoms in nature are eventually oxidized to CO_2 and are incorporated into plants during photosynthesis. So long as the C^{14} circulates through living plants and animals and back into the atmosphere or water, the proportion of C^{14} to C^{12} (the much more abundant nonradioactive isotope) remains constant. But as soon as the organism dies the amount of C^{14} declines due to a decay into C^{12}. Measurement of the residual radioactivity can be used to date samples up to 50,000 years old, which makes the technique particularly useful in obtaining precise dates of early man and other relatively recent fossils.

Chapter 23 covered the history of the Earth to the beginning of the Cambrian period, 600 million years ago. Table III covers the interval from then to the present—the last four days on the 30-day calendar in Chapter 23. The curve that accompanies Table III indicates a trend toward increasing diversity among living things. Between the earliest recorded beginnings of life and the first true animal fossils, which occur in the late Precambrian, stretches an interval of no less than two billion years—about 13 days on our geological calendar. Once evolution entered its multicellular phase, 600 million years ago, the number of different organisms increased steadily to the present remarkable level. Right now life on Earth is in a period of unfolding. But this expansion of diversity has not always been a smooth upward progression. Widespread extinctions, accompanied by rapid evolutionary change and important shifts in the composition of life, took place around the close of the Cambrian, Ordovician, Devonian, Permian, Triassic and Cretaceous periods. The most serious perturbation occurred at or near the close of the Permian period, and its effect is clearly evident in the dip in the curve of animal diversity of Table III. At this time no fewer than 24 orders, and a great many of the animal families throughout the world, became extinct. Several theories have been advanced to explain these crises in the history of life. Some geologists have blamed them on sudden episodes of mountain building, associated with a general uplift of the land, or on the drifting apart of the continents that began at this time. Others blame drastic changes in climate, resulting in cooling and even glaciation of much of the Earth's surface. Still others have blamed massive outbursts of cosmic radiation from outer space. The truth is that no single kind of cataclysmic event has been consistently associated with all of the crises. Possibly more than one cause was at work and there may have been others that remain undiscovered.

THE CLASSIFICATION OF ORGANISMS

This chapter closes with a table of the major groups of organisms, including the five kingdoms (monerans, protists, fungi, plants, and animals), most of the phyla, and many of the most important classes *(Box C)*. The phylogenetic relationships of the kingdoms are roughly indicated in Figure 10. For simplicity, groups belonging to lower categories, such as orders and families, are mostly omitted. There is no need to try to memorize these names at this time. They are presented here so that the reader can scan them and obtain a preliminary view of the total diversity of life. For later chapters that deal with the individual phyla, the table will provide a convenient reference for comparisons among the groups.

FIVE KINGDOMS OF ORGANISMS and their relationships are represented by this phylogenetic tree. For simplicity, lower categories such as orders and families have been omitted. **10**

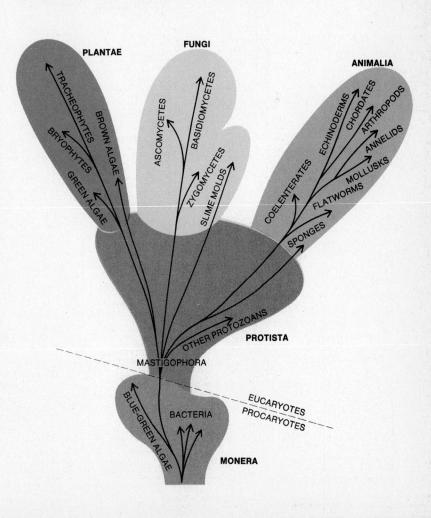

	KINGDOM	PHYLUM	SUBPHYLUM	CLASS	SUBCLASS	ORDER
THE PROCARYOTES (CHROMOSOMES NOT LOCATED WITHIN A NUCLEUS)	**MONERA**	**SCHIZOPHYTA OR SCHIZOMYCETES:** BACTERIA (UNKNOWN NUMBER OF SPECIES)				
		CYANOPHYTA: BLUE-GREEN ALGAE (2500 SPECIES)				
THE EUCARYOTES (CHROMOSOMES LOCATED WITHIN A NUCLEUS)	**PROTISTA**	**MASTIGOPHORA OR FLAGELLATA:** FLAGELLATES		**PHYTOMASTIGINA:** PLANT-LIKE FLAGELLATES		
				ZOOMASTIGINA: ANIMAL-LIKE FLAGELLATES		
		SARCODINA OR RHIZOPODA: AMOEBAS, FORAMINIFERANS, HELIOZOANS, RADIOLARIANS				
		SPOROZOA: SPOROZOANS, INCLUDING MALARIAL PARASITES				
		CILIOPHORA OR CILIATA: CILIATES (6000 SPECIES)				
	FUNGI	**ZYGOMYCOTA** (500 SPECIES): TUBE FUNGI, INCLUDING SOME RUSTS, BREAD MOLDS, WATER MOLDS, AND OTHERS				
		ASCOMYCOTA (35,000 SPECIES): SAC FUNGI, INCLUDING YEASTS, POWDERY MILDEWS, CUP FUNGI, BLUE AND GREEN MOLDS, SOME BREAD MOLDS, AND OTHERS				
		BASIDIOMYCOTA (25,000 SPECIES): CLUB FUNGI, INCLUDING MOST MUSHROOMS, TOADSTOOLS, BRACKET FUNGI, SMUTS, AND RUSTS				
		MYXOMYCOTA: SLIME MOLDS (450 SPECIES): SOMETIMES PLACED IN PROTISTA				

C

THE CLASSIFICATION OF ORGANISMS (CONTINUED)

THE EUCARYOTES (CHROMOSOMES LOCATED WITHIN A NUCLEUS)

KINGDOM	PHYLUM	SUBPHYLUM	CLASS	SUBCLASS	ORDER
FUNGI CONTINUED	**FUNGI IMPERFECTI:** UNDER THIS LOOSE CATEGORY ARE INCLUDED A WIDE VARIETY OF FUNGUS SPECIES WHICH CANNOT BE PLACED TO ONE OF THE CLASSES ABOVE, BECAUSE THE SEXUAL PART OF THE LIFE CYCLE HAS BEEN LOST IN EVOLUTION OR ELSE EXISTS BUT SIMPLY IS NOT YET ELUCIDATED. MOST FUNGI IMPERFECTI PROBABLY BELONG TO EITHER THE ASCOMYCOTA OR THE BASIDIOMYCOTA.—LICHENS: THESE ARE COMPOUND ORGANISMS, EACH OF WHICH CONSISTS OF ALGAE LIVING WITHIN THE BODY OF A FUNGUS, USUALLY A MEMBER OF THE ASCOMYCOTA OR BASIDIOMYCOTA.				
PLANTAE	**PYRROPHYTA:** DINOFLAGELLATES (1100 SPECIES)				
	CHRYSOPHYTA: DIATOMS AND RELATED ALGAE (10,000 SPECIES)		**XANTHOPHYCEAE:** YELLOW-GREEN ALGAE **CHRYSOPHYCEAE:** GOLDEN-BROWN ALGAE **BACILLARIOPHYCEAE:** DIATOMS		
	PHAEOPHYTA: BROWN ALGAE (1000 SPECIES)				
	RHODOPHYTA: RED ALGAE (3000 SPECIES)				
	CHLOROPHYTA: DESMIDS AND OTHER GREEN ALGAE (6000 SPECIES)				
	BRYOPHYTA: LIVERWORTS, MOSSES, AND RELATED FORMS (250,000 SPECIES)		**ANTHOCEROPSIDA:** HORNWORTS **HEPATICAE, OR HEPATICOPSIDA:** LIVERWORTS **BRYOPSIDA** OR MUSCI: MOSSES		
	TRACHEOPHYTA: VASCULAR PLANTS (250,000 SPECIES)	**PSILOPSIDA:** PSILOTUM AND OTHERS	**PSILOPHYTALES:** PSILOPHYTON AND OTHER PRIMITIVE EXTINCT PLANTS **PSILOTALES:** PSILOTUM AND RELATED FORMS		
		LYCOPSIDA: LYCOPODS **SPHENOPSIDA:** HORSETAILS (EQUISETUM) AND RELATED FORMS			

	KINGDOM	PHYLUM	SUBPHYLUM	CLASS	SUBCLASS	ORDER
	PLANTAE CONTINUED	**TRACHEOPHYTAE** CONTINUED	**PTEROPSIDA:** FERNS AND SEED PLANTS	**FILICINEAE:** FERNS		
				GYMNOSPERMAE: PINES, FIRS, GINKGOS, AND OTHER GYMNOSPERMS		
				ANGIOSPERMAE: ANGIOSPERMS OR FLOWERING PLANTS. THIS GROUP CONSISTS OF THE MONOCOTS (ORDER MONOCOTYLEDONAE), WHICH INCLUDE THE GRASSES, SEDGES, LILIES, PALMS, AND ORCHIDS; AND THE DICOTS (ORDER DICOTYLEDONAE), WHICH CONSISTS OF THE BULK OF THE REMAINING HIGHER PLANTS		
	ANIMALIA	**PORIFERA:** SPONGES (10,000 SPECIES)				
		SNIDARIA OR COELENTERATA: HYDRAS, JELLYFISH, AND RELATED FORMS (9000 SPECIES)		**HYDROZOA:** HYDRAS AND HYDROIDS		
				SCYPHOZOA: JELLYFISH		
				ANTHOZOA: SEA ANEMONES, CORALS		
		CTENOPHORA: COMB JELLIES, SEA GOOSEBERRIES (90 SPECIES)				
		PLATYHELMINTHES: FLATWORMS (13,000 SPECIES)		**TURBELLARIA:** "PLANARIANS" AND OTHER FREE-LIVING FLATWORMS		
				TREMATODA: FLUKES (ALL PARASITIC FORMS)		
		MESOSOMA: MESOZOANS (SMALL, EXTREMELY SIMPLIFIED PARASITES OF MARINE INVERTEBRATES: (50 SPECIES)		**CESTODA:** TAPEWORMS (ALL PARASITIC FORMS)		
		NEMERTINEA OR RHYNCHOCOELA: NEMERTEANS OR RIBBON WORMS (750 SPECIES)				
		ROTIFERA: ROTIFERS (1500 SPECIES)				
		GASTROTRICHA: GASTROTRICHS (175 SPECIES)				
		KINORHYNCHA: KINORHYNCHANS (64 SPECIES)				
		NEMATODA: ROUNDWORMS OR NEMATODES (10,000 SPECIES)				
		NEMATOMORPHA: HAIR WORMS (100 SPECIES)				
		ACANTHOCEPHALA: ACANTHOCEPHALAN WORMS (ALL PARASITES OF ARTHROPODS: 300 SPECIES)				
		ENTOPROCTA: MOSS ANIMALS OR ENTOPROCTS (60 SPECIES)				

THE EUCARYOTES
(CHROMOSOMES LOCATED WITHIN A NUCLEUS)

	KINGDOM	PHYLUM	SUBPHYLUM	CLASS	SUBCLASS	ORDER
	ANIMALIA CONTINUED	**PRIAPULIDA:** PRIAPULIDS (CUCUMBER-SHAPED MARINE ANIMALS: 8 SPECIES)				
		SIPUNCULA: SIPUNCULIDS (CYLINDRICAL MARINE WORMS: 250 SPECIES)				
		MOLLUSCA: MOLLUSKS (80,000 SPECIES)		**AMPHINEURA:** CHITONS		
				MONOPLACOPHORA: MONOPLACO-PHORANS		
				GASTROPODA: SNAILS, CONCHS, SLUGS		
				BIVALVIA, OR PELECYPODA: CLAMS, OYSTERS, AND OTHER BIVALVE MOLLUSKS		
				SCAPHOPODA: SCAPHOPODS OR TUSK SHELLS		
				CEPHALOPODA: SQUIDS, OCTOPODS		
		ECHIURA: ECHIURIDS (CYLINDRICAL MARINE WORMS; 60 SPECIES)				
		ANNELIDA: SEGMENTED WORMS (7000 SPECIES)		**POLYCHAETA:** POLYCHAETES (EXCLUSIVELY MARINE WORMS)		
				OLIGOCHAETA: OLIGOCHAETES, INCLUDING EARTHWORMS AND FRESHWATER WORMS		
				HIRUDINEA: LEECHES		
		TARDIGRADA: WATER BEARS (180 SPECIES)				
		PENTASTOMIDA: TONGUE WORMS (65 SPECIES)				
		ONYCHOPHORA: ONYCHOPHORANS (65 SPECIES)				
		ARTHROPODA: ARTHROPODS (900,000 SPECIES)		**TRILOBITA:** TRILOBITES (PRIMITIVE EXTINCT FORMS)		
				CHELICERATA: SCORPIONS, SPIDERS, MITES, HORSESHOE CRABS		
				PYCNOGONIDA: SEA SPIDERS		
				CRUSTACEA: CRUSTACEANS, INCLUDING WATER FLEAS, CRABS, SHRIMPS, AND LOBSTERS		
				INSECTA OR HEXAPODA: INSECTS		**ODONATA:** DRAGONFLIES
						BLATTARIA: COCKROACHES
						ISOPTERA: TERMITES
						ORTHOPTERA: GRASSHOPPERS, CRICKETS, WALKING STICKS
						HEMIPTERA: STINK BUGS, ASSASSIN BUGS, BEDBUGS, WATER BOATMEN

THE EUCARYOTES
(CHROMOSOMES LOCATED WITHIN A NUCLEUS)

	KINGDOM	PHYLUM	SUBPHYLUM	CLASS	SUBCLASS	ORDER
	ANIMALIA CONTINUED	ARTHROPODA CONTINUED		INSECTA CONTINUED		**HOMOPTERA:** APHIDS, SCALE INSECTS, CICADAS **ANOPLURA:** LICE **COLEOPTERA:** BEETLES **LEPIDOPTERA:** BUTTERFLIES, MOTHS **DIPTERA:** FLIES, MOSQUITOES, GNATS **HYMENOPTERA:** SAWFLIES, WASPS, BEES, ANTS **SIPHONAPTERA:** FLEAS
				DIPLOPODA: MILLIPEDES **CHILOPODA:** CENTIPEDES **PAUROPODA:** PAUROPODS **SYMPHYLA:** SYMPHYLANS		
THE EUCARYOTES (CHROMOSOMES LOCATED WITHIN A NUCLEUS)		**PHORONIDA:** PHORONIDS (WORMLIKE MARINE ANIMALS; 15 SPECIES) **BRYOZOA OR ECTOPROCTA:** BRYOZOANS OR MOSS ANIMALS (4000 SPECIES) **BRACHIOPODA:** BRACHIOPODS OR LAMP SHELLS (260 SPECIES) **CHAETOGNATHA:** ARROW WORMS (50 SPECIES) **ECHINODERMATA:** ECHINODERMS (5300 SPECIES)		**CRINOIDEA:** SEA LILIES AND FEATHER STARS **ASTEROIDEA:** STARFISH OR SEA STARS **OPHIUROIDEA:** BRITTLE STARS, BASKET STARS **ECHINOIDEA:** SEA URCHINS, SAND DOLLARS **HOLOTHUROIDEA:** SEA CUCUMBERS		
		POGONOPHORA: POGONOPHORANS (80 SPECIES) **HEMICHORDATA:** ACORN WORMS (80 SPECIES) **CHORDATA:** CHORDATES (39,000 SPECIES)	**VERTEBRATA:** VERTEBRATES	**AGNATHA:** AGNATHS OR JAWLESS FISHES, INCLUDING LAMPREYS **PLACODERMI:** PLACODERMS (EXTINCT ARMORED FISHES) **CHRONDRICHTHYES:** ELASMOBRANCHS, INCLUDING SHARKS, RAYS, AND CHIMAERANS **OSTEICHTHYES:** BONY FISHES (MOST LIVING FORMS OF FISHES) **AMPHIBIA:** AMPHIBIANS, INCLUDING FROGS, TOADS, SALAMANDERS, AND CAECILIANS		

	KINGDOM	PHYLUM	SUBPHYLUM	CLASS	SUBCLASS	ORDER
C **THE CLASSIFICATION OF ORGANISMS (CONTINUED)**	ANIMALIA CONTINUED	CHORDATA CONTINUED	VERTEBRATA CONTINUED	**REPTILIA:** REPTILES **AVES:** BIRDS (SOME RECENT AUTHORS PLACE BIRDS WITH DINOSAURS IN A SEPARATE CLASS. THE DINOSAURIA) **MAMMALIA:** MAMMALS		
					PROTOTHERIA: MONOTREMES (EGG-LAYING MAMMALS, INCLUDING THE DUCKBILL PLATYPUS AND SPINY ANTEATERS, LIMITED TO AUSTRALIA AND NEW GUINEA) **METATHERIA:** MARSUPIALS (POUCHED MAMMALS) **EUTHERIA:** PLACENTAL MAMMALS	
						INSECTIVORA: SHREWS, MOLES, TENRECS
						CHIROPTERA: BATS
						CARNIVORA: DOGS, CATS, BEARS, SEALS, SEA LIONS
						RODENTIA: MICE, RATS, SQUIRRELS, PORCUPINES, BEAVERS
						LAGOMORPHA: RABBITS, HARES
						PRIMATES: LEMURS, LORISES, TARSIERS, MONKEYS, APES, MAN
						ARTIODACTYLA: EVEN-TOED UNGULATES—CATTLE, DEER, SHEEP, PIGS, CAMELS, HIPPOPOTAMUSES
						PERISSODACTYLA: ODD-TOED UNGULATES—HORSES, ASSES, ZEBRAS, RHINOCEROS
						PROBOSCIDEA: ELEPHANTS
						CETACEA: WHALES, PORPOISES
						SIRENIA: SEA COWS (MANATEES)

THE EUCARYOTES (CHROMOSOMES LOCATED WITHIN A NUCLEUS)

READINGS

B. KUMMEL, *History of the Earth: An Introduction to Historical Geology*, 2d ed., San Francisco, W.H. Freeman & Co., 1970. One of the best introductions to paleontology, strong in both geological and biological aspects.

G.H.M. LAURENCE, *Taxonomy of Vascular Plants*, New York, Macmillan, 1951. The excellent introductory chapters are recommended as a botanical supplement to Mayr's book.

E. MAYR, *Principles of Systematic Zoology*, New York, McGraw-Hill, 1969. This is the best available introduction to the principles of classification and is recommended as the next book to read for the student interested in exploring this growing branch of biology. Although intended for zoologists, most of the ideas and techniques are equally useful in botany.

R. SOKAL, "Numerical Taxonomy," *Scientific American*, December 1966. A brief, lucid introduction to some of the recently developed techniques for measuring similarity of taxonomic units and testing phylogenetic hypotheses.

R.H. WHITTAKER, "New Concepts of Kingdoms of Organisms," *Science*, 163: 150–160 (1969). An authoritative account of the higher categories of organisms, their evolutionary relationships, and the history of the subject of the classification of kingdoms and phyla.

What is a microorganism? There is no simple answer to this question. The word 'microorganism' is not the name of a group of related organisms, as are the words 'plants' or 'invertebrates' or 'frogs'. The use of the word does, however, indicate that there is something special about *small* organisms; we use no special word to denote large animals or medium-sized ones. —W. R. Sistrom

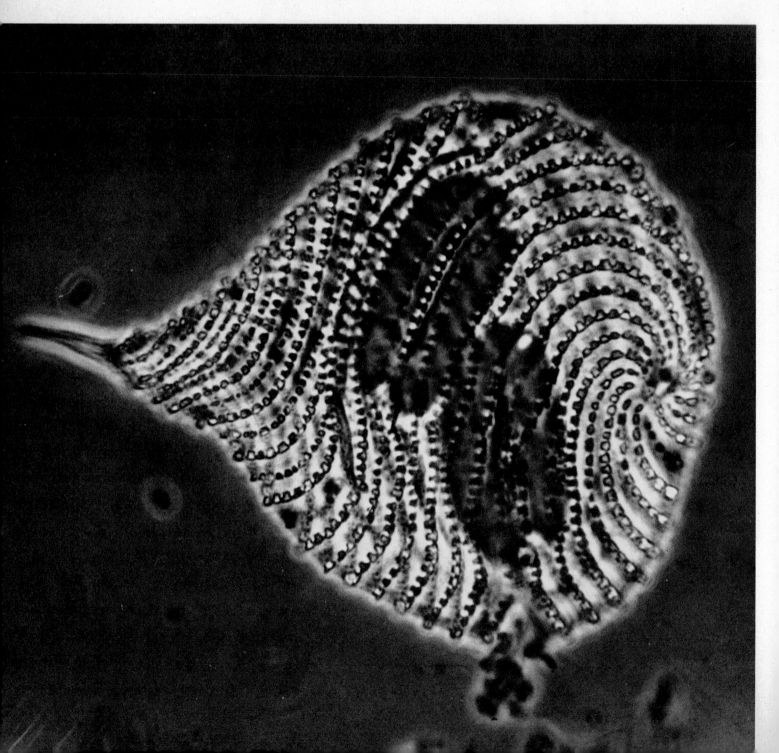

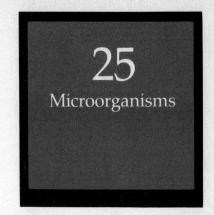

The human being, a giant organism, is inordinately aware of other large organisms. We are very familiar with the differences that separate dogs from cats and lilies from pine trees. But we are only dimly aware of the vast and strikingly diverse world of the microorganisms—the viruses, bacteria, protozoans, and other creatures that are nearly or wholly below the range of unaided vision. This bias is reflected vividly in most classification schemes, including the one outlined in the previous chapter. With considerable confidence taxonomists draw limits around such phyla of big organisms as the Mollusca (snails, clams, squids, and their relatives) and Tracheophyta (flowering plants), certain in their knowledge that the species enclosed are of common ancestry and share many basic anatomical features.

When sorting out microorganisms, however, taxonomists lack this sure touch. Until very recently most recognized the phylum Protozoa, a unit embracing all of the single-celled eucaryotes. Now it is recognized that the differences among the "classes" of protozoans, namely the Mastigophora, the Sarcodina, the Sporozoa, and the Ciliophora, are at least as great as those that distinguish most phyla of the higher animals and plants. The Mastigophora, for example, include certain organisms that are animal-like in structure and nutrition and others that are plant-like. Consequently, many (but not all) biologists prefer to raise the classes to the rank of full phyla, as in Chapter 24. An even more serious area of uncertainty is the division of microorganisms into king-

doms, the highest of the taxonomic categories. Bacteria and bluegreen algae are procaryotes, and a strong case can be made for distinguishing them from eucaryotes as a separate kingdom, the MONERA (pronounced mon-eé-ra). Currently the classification of microorganisms is in a state of flux. There is a great amount of new information, and the classification scheme is likely to undergo some major changes before it is finally stabilized.

THE KINGDOM MONERA (PROCARYOTES)

The architecture of procaryotic and eucaryotic cells is compared in Figure 1. Although considered the simplest of living organisms, the procaryotes are enormously more complex than single proteins or other nonliving molecules. The basic unit of these organisms is the procaryotic cell, which contains a complete complement of genetic and protein-synthesizing machinery, including DNA, RNA, and all of the enzymes needed to translate the code into protein. The cell also contains at least one system for generating the ATP needed to keep the machinery running. In earlier chapters it was pointed out that the procaryotic cell differs from the eucaryotic cell in four important ways. First, its genetic material is not organized within a distinctive nucleus, and the DNA is not part of a DNA-protein complex called chromatin. The elaborate **mechanism of mitosis** *(see Figure 16 in Chapter 2)* is missing; the cells divide by simple fission, after replicating their DNA. Second, there are none of **the familiar membrane-bounded organelles** — mitochondria, chloroplasts, golgi apparatus, endoplasmic reticulum—that occur in the cells of higher organisms. Third, there is no subcellular structure that can carry out the entire process of aerobic respiration, or, in the case of the photosynthetic procaryotes, photosynthesis. The cell itself is the smallest entity that can carry out these

EUGLENA SPIROGYRA is a unicellular flagellate. Photomicrograph on opposite page depicts the flexible pellicle that covers the organism, emptied of its cellular contents. Spiral ridges on pellicle are beset with small beads whose function is unknown. Magnification, about 2500X. Flagella are not visible in this picture.

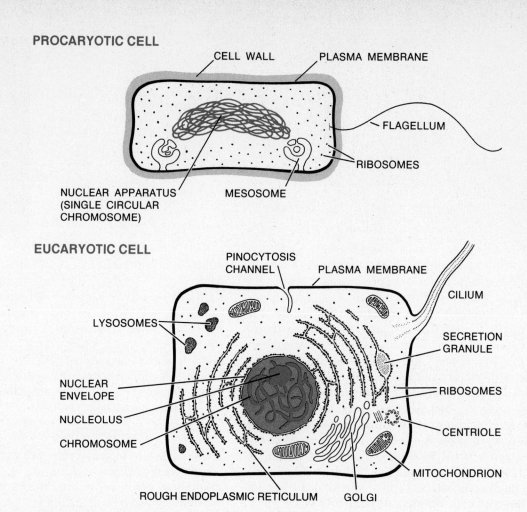

PROCARYOTIC CELL

CELL WALL • PLASMA MEMBRANE • FLAGELLUM • RIBOSOMES • MESOSOME • NUCLEAR APPARATUS (SINGLE CIRCULAR CHROMOSOME)

EUCARYOTIC CELL

PINOCYTOSIS CHANNEL • PLASMA MEMBRANE • CILIUM • LYSOSOMES • SECRETION GRANULE • NUCLEAR ENVELOPE • RIBOSOMES • NUCLEOLUS • CENTRIOLE • CHROMOSOME • MITOCHONDRION • ROUGH ENDOPLASMIC RETICULUM • GOLGI

1 **PROCARYOTIC AND EUCARYOTIC CELLS show marked differences in complexity and internal organization. Schematic diagram depicts a typical bacterium and a typical animal cell.**

2 **STRUCTURAL FORMULAE for the two unusual nitrogen-containing sugar derivatives found in the polymers of bacterial cell walls.**

N–ACETYLMURAMIC ACID   N–ACETYLGLUCOSAMINE

reactions. Finally, almost all procaryotes have a cell wall whose composition is unique. It consists of a backbone molecule of polymerized sugars, N-acetyl muramic acid and N-acetyl glucosamine *(Figure 2)*, together with characteristic amino acids that cross-link the adjacent chains. The molecule N-acetyl muramic acid is found in the cell wall of all bacteria. Although this wall material, called the MUCO-COMPLEX SUBSTANCE, may in fact account for only a fraction of the total wall substance (proteins, lipids, and perhaps true polysaccharides make up the rest), it is thought to be responsible for the structural integrity of the wall. This substance is not found in the cells of any other organisms.

Absence of characteristic eucaryote organelles should not be construed as an absence of any membranous structures inside the procaryote cell. Membranous structures called mesosomes are frequently associated with formation of new cell walls during cell division. Many aerobic bacteria have rather elaborate internal membrane systems to which respiratory enzymes are bound, as in

**BLUEGREEN ALGAE are procaryotes that contain 3
photosynthetic pigments bound to elaborate internal
membranes. Electron micrograph depicts Spirulina,
which is a bluegreen alga found in the warm sulfur
springs of Yellowstone National Park. The genetic
material in center of the cell is ringed by four or five
layers of photosynthetic membranes. Magnification,
40,000X.**

mitochondria. Similarly, the bluegreen algae and photosynthetic bacteria may have elaborate internal membranes to which photosynthetic pigment systems are bound *(Figure 3)*.

In the difficult task of classifying the procaryotes, we follow the arbitrary practice of dividing them into two phyla: the SCHIZOPHYTA, which includes the myxobacteria, spirochetes, and true bacteria; and the CYANOPHYTA, which includes only the bluegreen algae. Botanists inevitably include the bluegreen algae in their surveys of the plant kingdom, and may even include most or all of the bacteria among the lower fungi. Some authors even consider the higher algae and fungi to be "higher protists." Nevertheless, man's compulsion to classify requires a choice, and our scheme makes at least as much sense as calling bluegreen algae plants because they were studied primarily by botanists, and bacteria fungi because they are frequently included in surveys of the lower plants.

THE BACTERIA (PHYLUM SCHIZOPHYTA)

Several major groups of bacteria can be identified on the basis of striking characteristics. Cells of the MYXOBACTERIA are short rods that glide from place to

> The eubacteria encompass an enormous number of species, a wide variety of structural forms, and a truly incredible array of biochemical tricks for obtaining energy.

place. The physical basis of this gliding movement is not yet known. There are no readily definable motor structures such as flagella, though the cell walls are unusually small and flexible. They form remarkable fruiting structures similar to those of the cellular slime molds. A group of cells aggregates, and from the cell mass there arise characteristic fruiting bodies. These may be simple globes over a millimeter in diameter, or they may be somewhat more complex branched structures *(Figure 4)*. Then either single cells transform themselves into thick-walled spores, or whole clusters of cells form a dessication-resistant cyst. Both spores and cysts can germinate under favorable conditions to yield the typical gliding vegetative cells.

Like the myxobacteria, the SPIROCHETES also have rather thin and flexible walls. They possess a unique structure called the AXIAL FILAMENT. The cells themselves are long rods coiled helically around the filament, which is anchored to the cytoplasm at both ends, and extends the length of the cell. This axial filament is thought to be responsible for motility of these organisms, but the actual mechanism by which it works is unknown. Many spirochetes are parasites in man, including the organism that causes syphilis *(Figure 5)*.

The EUBACTERIA, sometimes called true bacteria, unlike the other two groups possess a thick and

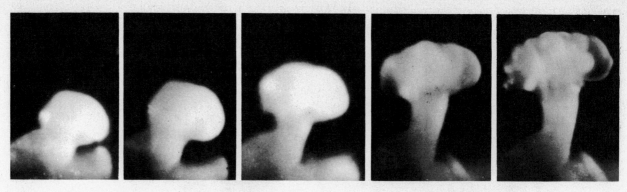

4 FRUITING BODY of a myxobacterium (genus Chondromyces). This series of photomicrographs shows the development of the branched reproductive structure from an aggregated mass of individual bacterial cells. The tree-like structure is usually a little less than one millimeter high. Tips of the branches develop into microscopic cysts containing hundreds or thousands of rod-shaped cells.

5 SPIROCHETES are long rod-shaped bacteria whose bodies are coiled into a helix around an axial filament. Photograph depicts Treponema pallidum, which causes syphilis. Two organisms are shown.

relatively stiff cell wall. Though many are not motile, some can move by means of flagella, whip-like filaments that extend singly or in tufts from one or both ends of the cell, or all around it *(Figure 6)*. Remember that the flagella of higher organisms —plant, animal, or protozoan —are monotonously alike: they consist of a hollow cylinder of nine pairs of fibrils surrounding two central fibrils. By contrast, the bacterial flagellum consists of a single protein subunit, flagellin. Thus flagellum architecture provides another major difference between procaryotic and eucaryotic cells *(Figure 7)*.

The eubacteria encompass an enormous number of species, a wide variety of structural forms, and a truly incredible array of biochemical tricks for ob-

FLAGELLATED BACTERIA include Pseudomonas (upper left), which has a single flagellum; a spirillum (upper right) with tufts of flagella at both ends; and Proteus vulgaris, which has many flagella. Bacteria are not drawn to same scale. **6**

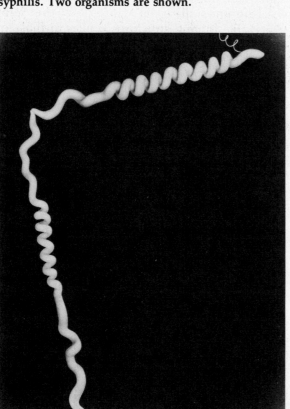

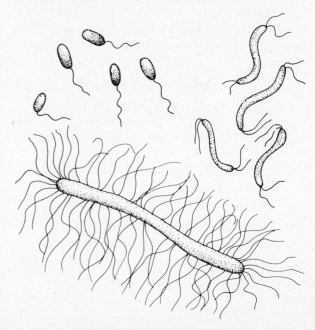

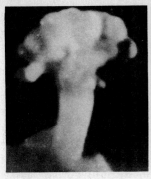

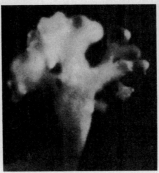

taining energy. Some of these tricks have been mentioned in previous discussions of autotrophy and photosynthesis, and it is necessary here merely to note the extraordinary diversity of organic molecules that can serve both as carbon and energy sources for some eager bacterium. At one time bacteriologists smugly stated that there was perhaps no carbon compound in existence that could not be metabolized by at least one strain of bacteria. Only recently have they found, to everyone's distress, that chlorinated hydrocarbons like DDT and related insecticides, and chlorinated derivatives of benzene like the weed killer, 2,4-D, are in fact relatively poor carbon and energy sources for either bacteria or fungi, and thus can accumulate in toxic amounts in the landscape.

Certain motile bacteria have evolved a remarkable chemoreception system in which certain proteins in the cell membrane serve as chemodetectors for such things as sugars. When a sugar gradient exists, the consequence is that flagellar movement is sufficiently directed to move the bacterium in the direction of the higher concentrations. Some groups of bacteria can even accomplish the remarkably difficult task of fixing atmospheric nitrogen *(Chapter 14)*. There are, however, some limitations. Although many bacteria can obtain energy either by respiration or fermentation *(Chapter 6)*, and are therefore FACULTATIVE ANAEROBES, others can live only by fermentation, and, indeed oxygen is a poison. These are called OBLIGATE ANAEROBES. The obligate anaerobe *Clostridium botulinum*, the cause of a particularly deadly kind of food poisoning, has already been mentioned in Chapter 6. Finally, at least among the true bacteria, one cannot find the process of aerobic photosynthesis *(Chapter 6)*; however, if one agrees with microbiologist Roger Stanier that the bluegreen algae are simply a neglected group of bacteria, this limitation does not apply.

Eubacteria characteristically exist in one or another of three different shapes *(Figure 8)*. There

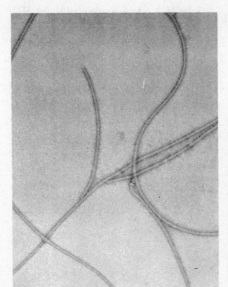
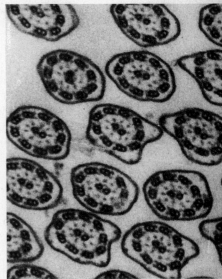

BACTERIAL FLAGELLA (left) consist of a single fibril of the protein, flagellin. Flagella of eucaryotes, shown in cross-section at right, contain two central fibrils surrounded by nine pairs. **7**

SHAPES OF BACTERIA. Bacterial cells occur in three basic forms, either singly or in groups. Rod shape is most common. **8**

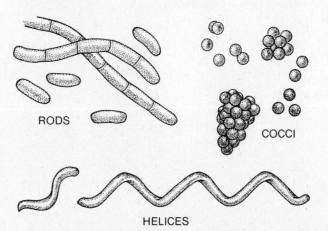

RODS

COCCI

HELICES

are spheres (coccus, plural cocci), rods (bacillus, plural bacilli), and helical forms (spirillum, plural spirilla). The cocci can occur singly, or in two or three dimensional arrays of chains, plates, or blocks of cells. Bacilli and spirilla occur singly or in chains, but the chains do not really signify multicellularity; each cell is fully viable and independent. Most bacteria reproduce simply by the fission of one cell into two. Chains arise merely by the accidental adhesion of cells after fission.

In addition to these basic shapes, bacteria exhibit several other structural features. Some bacteria form stalks by which they are attached to their substrate. The stalk may be an extension of the cell wall or an extracellularly secreted product. In either case it can lead to the formation of small clusters of cells all attached to the same spot. Still other eubacteria comprise filaments that approach but do not attain true multicellularity. The cells occur in chains and when cell division occurs, it occurs simultaneously in all cells of the filament. The filament is enclosed in a thin and delicate sheath, and these bacteria reproduce by releasing flagellated cells from the open end of the sheath.

Another odd and interesting group of eubacteria are the ACTINOMYCETES. Just as the myxobacteria bear a remarkable resemblance to the cellular slime molds, the actinomycetes resemble the filamentous fungi *(Chapter 26)*. They develop vegetatively into elaborately branched systems of filaments that lack any cross walls whatsoever. Such a mat of filaments is called a MYCELIUM. Presumably each branching system contains many copies of the

genetic material, just as the branched filamentous fungi contain large numbers of nuclei. Actinomycetes reproduce usually by forming chains of spores at the tips of the filaments, a process resembling asexual spore formation in many fungi. Nevertheless, the actinomycetes are true procaryotes, closely related to the unicellular cocci or bacilli. Frequently the branched filamentous growth ceases and the structure breaks up into typical spheres or rods. The actinomycetes contain several important genera, including *Mycobacterium*, the organism that causes tuberculosis, and *Streptomyces*, the organism that produces streptomycin and several other antibiotics.

The last group of nonphotosynthetic eubacteria to be considered are the RICKETTSIAE (singular, rickettsia) and related organisms. These extremely small parasites have never been cultured outside of living cells. They are more complex than mere proteins and other large organic molecules, because they contain both kinds of nucleic acid, a number of enzymes, and a wall that contains the mucocomplex substance. Rickettsiae are agents of several serious diseases in man, including Rocky Mountain spotted fever and typhus, and are frequently carried by arthropods, particularly ticks and fleas. They do not seem to cause any disease symptoms in their arthropod hosts.

This survey of the eubacteria appropriately closes with two groups of PHOTOSYNTHETIC BACTERIA: green and purple. Many photosynthetic bacteria can use hydrogen sulfide (H_2S) as a reducing agent under strictly anaerobic conditions *(Chapter 6)*. Unlike the higher photosynthetic organisms and bluegreen algae, they lack chlorophyll *a* as their key photosynthetic pigment, having instead either bacteriochlorophyll (purple bacteria) or chlorobium chlorophyll (green bacteria). Both of these pigments are sufficiently modified within the cells to allow the organisms to use for photosynthesis

9 ABSORPTION SPECTRA of a green alga (sea lettuce) and green and purple photosynthetic bacteria. Bacteria have no difficulty in growing in stagnant water beneath fairly dense layers of algae because wavelengths of light that they need are not appreciably absorbed by the algae (1 nm = 10^{-9} meter).

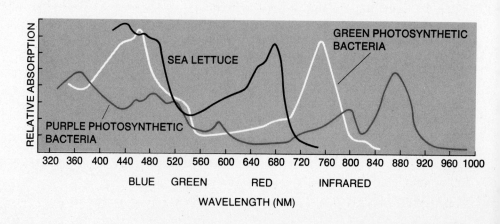

light of longer wavelength than that used by all other photosynthesizing organisms. They have no difficulty in growing in stagnant water beneath fairly dense layers of algae, because the wavelengths of light that they need are not appreciably absorbed by the algae. The comparative absorption spectra of green algae and purple and green photosynthetic bacteria are shown in Figure 9. The green bacteria are all nonmotile rods, and most of the purple bacteria are motile rods or spirilla. One purple bacterium reproduces by budding, the rest by simple fission. Since these organisms can fix nitrogen, they may be very important ecological components of stagnant water.

BLUEGREEN ALGAE (PHYLUM CYANOPHYTA)

Of all known organisms, these procaryotes are without question the most nutritionally independent. They perform aerobic photosynthesis (using chlorophyll *a*), they can respire aerobically, and many can fix nitrogen on a large scale. They require only a few mineral elements, water, nitrogen gas, oxygen, and carbon dioxide. Despite their ability to do the kind of photosynthesis otherwise characteristic only of eucaryotic photosynthesizers, they are true procaryotes. They contain none of the familiar membrane-limited organelles of eucaryotic cells, they do not have a discrete nucleus, and their cell walls contain the mucocomplex substance. Even so, the bluegreen algae contain by far the most elaborate internal membrane system found in procaryotes, the photosynthetic lamellae. They are also the only procaryotic photosynthesizers which contain chlorophyll *a*.

Bluegreen algae are either free-living, colonial, or filamentous. Some filamentous forms show differentiation into at least three different cell types: vegetative cells, spores, and heterocysts (*Figure 10*). Heterocysts are important structures for two reasons. First, when filaments reproduce by fragmentation, the weak point may be the heterocyst. More important, however, heterocysts evidently lack the oxygen-evolving portion of photosynthesis, and therefore are more anaerobic. As a consequence, the extremely oxygen-labile nitrogenase enzyme required for nitrogen fixation survives, and it is in the heterocysts that nitrogen fixation occurs. All of the known bluegreen algae with heterocysts fix nitrogen, and the forms that lack heterocysts do not.

Although sexual reproduction is encountered among bacteria, it is absent in the bluegreen algae.

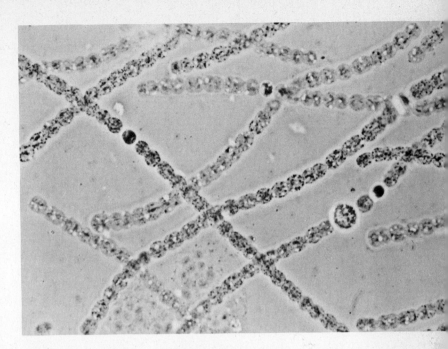

BLUEGREEN ALGA Anabaena appears under the microscope as filaments of cells embedded in a gelatinous matrix. Single large cell at right center is a heavy-walled spore. Dark cells are nitrogen-fixing heterocysts.

10

Reproduction is by simple fission either to produce a new vegetative cell or to produce a resistant spore. Recently viruses have been discovered that can infect bluegreen algae and then transfer genetic material from one alga to another by transduction, but true sexuality has never been observed.

THE VIRUSES

As pointed out in Chapter 2, this group of very important "organisms" causes the most difficult problems in the entire art of classification, and challenges the best ideas about the dividing line between the living and nonliving. Are viruses alive? In Chapter 1 we stated that living organisms are generally considered to share certain funda-

---

It is perhaps best to consider viruses not as organisms, but as interesting and frequently dangerous biochemicals. They can even be crystallized like common table salt.

---

mental common properties. First, whether by anaerobic or aerobic respiration, or by some sort of chemosynthetic or photosynthetic reaction, they make ATP. Second, whether they are bacteria or elephants, they have the complete system for transcribing and translating the messages encoded in their DNA, a system which requires the several kinds of RNA. Third, the cell of a living organism invariably arises directly from a preexisting cell, whether by simple fission, as in the bacteria, or by the complicated processes of mitosis or meiosis found in higher organisms. Fourth, when a living cell divides, most or all of its components are increased in amount and then parceled out between the two daughter cells. Fifth, living cells are invariably bounded by a membrane, a structure which offers at least a modicum of regulation of the intracellular environment.

11 **STRUCTURE OF VIRUSES shows a symmetry that reflects their chemical composition. Coats of polyhedral viruses are molecular "boxes" fashioned from identical protein subunits. Rod-shaped tobacco mosaic virus is a helical rod (helix not shown in drawing). Internal components of mumps and influenza viruses also show helical symmetry. Other viruses display a more complex symmetry. Spiked outlines of mumps, herpes and influenza viruses indicate their asymmetrical protein coats.**

COMPARISON OF VIRUSES AND CELLS reveals that **I** **viruses lack the characteristics generally used to define living systems.**

	VIRUSES	CELLS
NUCLEIC ACIDS	DNA OR RNA	DNA AND RNA
ATP SYNTHESIS	NO	YES
ORIGIN OF NEW VIRION OR CELL	SYNTHESIS BY HOST	FISSION, MITOSIS, OR MEIOSIS
LIMITING MEMBRANE	ABSENT	PRESENT

Viruses fail on all five counts (*Table I*). In the spectrum of biological objects (note the avoidance of the expression "living things"), they are extraordinarily simple. They contain either DNA or RNA, but never both, and they have in addition only one or a few different kinds of protein molecules. They have no enzymatic machinery for synthesizing ATP. Because they contain only one kind of nucleic acid, viruses are incapable of synthesizing their own proteins. And because a whole virus never arises directly from a preexisting virus, a virus cannot synthesize the full complement of biochemicals to parcel out to two daughter viruses. Finally, viruses have no membranes. It is perhaps best to consider viruses not as organisms, but as interesting and frequently dangerous

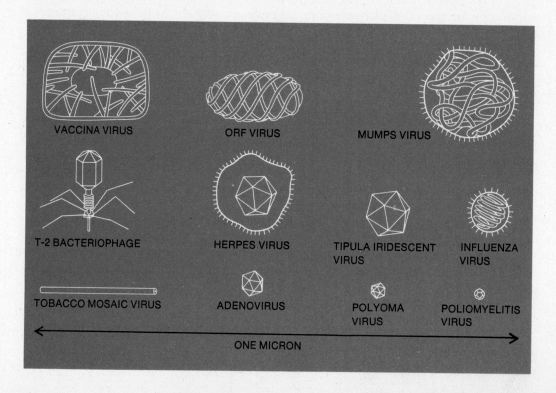

VACCINA VIRUS ORF VIRUS MUMPS VIRUS

T-2 BACTERIOPHAGE HERPES VIRUS TIPULA IRIDESCENT VIRUS INFLUENZA VIRUS

TOBACCO MOSAIC VIRUS ADENOVIRUS POLYOMA VIRUS POLIOMYELITIS VIRUS

ONE MICRON

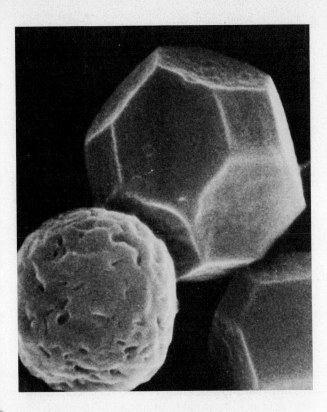

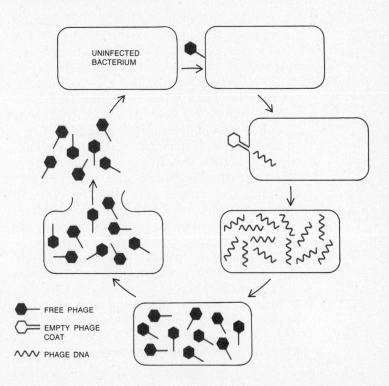

functioning as an enzyme that attacks the bacterial cell wall. The nucleic acid is then injected into the bacterium, but the protein coat remains outside. Once inside the bacterium, the viral DNA becomes inserted into the bacterial DNA and is then indistinguishable from a linear series of bacterial genes. Subsequent events may proceed in either of two directions. If the phage is VIRULENT, its nucleic acid replicates much more rapidly than the bacterial DNA, and is rapidly transcribed and translated into viral coat protein and a number of viral enzymes (*Figure 13*). These processes are mediated both by bacterial and viral-coded enzymes, and the protein synthesis is accomplished using ATP, rRNA and tRNA from the bacterial host. Large numbers of new virions assemble themselves, as **mentioned in Chapter 11. Ultimately the bacterial** cell bursts open or LYSES, releasing perhaps several hundred new virions. Conversely, if a phage is TEMPERATE, its DNA replicates only when the bacterial DNA replicates, but is apparently not transcribed into messenger for new coat protein (*Figure 14*). In this condition, it is said to be in the PROPHAGE

POLYHEDRAL INCLUSION BODIES of an insect virus were photographed with scanning electron microscope. This virus infects the caterpillar of the silkworm moth Bombyx mori (illustrated in Chapter 19). Each polyhedron contains several hundred virus particles. Rounded object is an incompletely formed inclusion body. Picture is reproduced at a magnification of about 9000X.

biochemicals (*Figures 11 and 12*). They can even be crystallized like common table salt.

What, then, is the "life history" of a virus? Viruses are obligate parasites, if indeed a complex biochemical can really be called a parasite. The most widely studied and perhaps most complex viruses known are those that infect bacterial cells: the bacteriophages, or phages for short. The cycle of phage infection and reproduction has been mentioned in several previous chapters. Here the process will be described in somewhat more detail.

The fundamental particle of the virus, perhaps analogous to the cell, is the VIRION, which consists of a central core either of DNA or RNA surrounded by a coat composed of one or at most a few different kinds of proteins. The bacteriophage T4, for example, consists of a head region of nucleic acid (DNA) surrounded by protein, and a hollow and rather complex tail. It becomes attached to the bacterial wall tail-first, the tail protein apparently

BACTERIOPHAGE CYCLE begins when phage attaches itself to wall of bacterium (top) and injects its DNA. After replicating within host cell, phage DNA directs synthesis of viral coat proteins. Viruses assemble themselves within the bacterium, and are released when cell disintegrates. **13**

UNINFECTED BACTERIUM

FREE PHAGE

EMPTY PHAGE COAT

PHAGE DNA

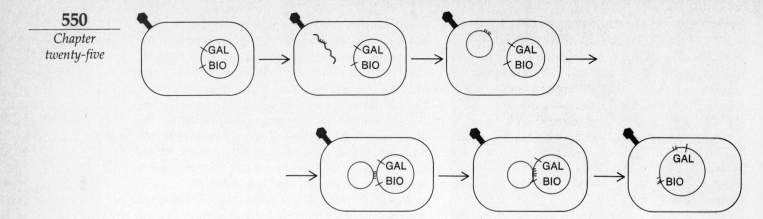

14 FORMATION OF VIRAL PROPHAGE. Viral nucleic acid (color) is injected into bacterium and becomes circular (top). Homologous regions of viral and bacterial genomes pair (bottom center) and become integrated (far right). Attachment site for viral nucleic acid lies between markers gal and bio on bacterial chromosome (see four tickmarks). Separation of marks is due to crossover.

stage, and the bacterium is said to be LYSOGENIC. Occasionally the prophage begins the rapid replication and mRNA synthesis characteristic of virulent strains, and viral replication and the lytic process take place. Curiously, a bacterium containing a prophage is immune to infection by another virion of the same strain. The prophage apparently codes for a repressor molecule which prevents a second infection.

In this description of virus infection, replication and release, a serious puzzle remains: how do the RNA viruses function? It is clear that some can serve directly as messenger RNA for their own protein synthesis, utilizing the RNA polymerase system of the host cell. This mechanism, however, does not provide for new strands of virus RNA. Recently a team of investigators headed by Howard W. Temin and David Baltimore have provided an important clue. They have shown that in tissue culture animal cells infected with certain tumor-producing RNA viruses carry with them an enzyme awkwardly named REVERSE TRANSCRIPTASE. This enzyme, coded by the viral RNA, reverses the usual transcription process by serving as a template for DNA synthesis. The resulting DNA acts just as the DNA of the phage described above, entering into the host cell and coding for new viral RNA and coat protein. This would assure the continuation of the "life cycle" of the virus.

Viruses that infect higher plants and animals, including those that cause poliomyelitis, influenza,

and other diseases in man, do not show the clear differentiation into head and tail regions. They are rodshaped, or perhaps cuboidal, like the geometrically symmetrical heads of the phages. Frequently the entire virion enters the host cell, including the viral protein. Although the coat protein may play a role in virus replication, it is not essential because in both plant and animal viruses the isolated nucleic acid is fully capable of initiating infection.

The distribution of viruses in terms of host organisms is odd. Viral diseases of flowering plants (angiosperms) are very common, but they are rare in the cone-bearing seed plants (gymnosperms), lower vascular plants (such as ferns), as well as algae and fungi. Almost all vertebrates are susceptible to viral infection, but among invertebrates such infections are common only in arthropods. A group of viruses called ARBOVIRUSES (short for arthropod-born viruses) causes serious diseases such as encephalitis in man and other mammals. Though carried within the arthropod's cells and transmitted to the other host by a bite (certain ar-

Because of their extraordinary versatility in nutrition, the flagellates are said to bridge the gap between plants and animals at the unicellular level. And because some of them form large and well-organized colonies, they also span the gap between unicellular and multicellular organisms.

boviruses are carried by mosquitoes) they apparently do not affect the insect host severely —just the bitten and infected mammal. No satisfactory explanation has yet been found for the curious selectivity of viral infection, but one might eventually emerge from an hypothesis concerning the origin of viruses. The suggestion is that viruses arose from cells as detached pieces of genetic material. These pieces, while still within the cells, somehow acquired the facility for more rapid replication than the remainder of the genetic material. Upon release, perhaps at cell death, such pieces might enter or be engulfed by neighboring cells, there to repeat their replicative cycle. If this hypothesis is correct the biological specificity of viral infection is no longer a mystery; only those organisms whose detached genetic material became viruses would be subject to reinfection by a nucleic acid fragment bearing some resemblance to their own. In the absence of a clear demonstration of viral origin by this mechanism, however, such reasoning must be regarded as highly speculative.

THE KINGDOM PROTISTA

In an important sense the phylum MASTIGOPHORA, the flagellates, can be regarded as the most fundamental of all the eucaryotic phyla. If every trace of life on Earth were removed except the members of this one group, a large percentage of the species would probably survive —or to be more precise, they would survive so long as a supply of fixed nitrogen was available (there are no nitrogen-fixers in the phylum). Not only would they sustain life entirely by themselves, but they would provide a favorable starting point for a renewal of evolutionary diversification of both the plants and animals. No other single phylum outside the procaryotes has the right combination of self-reliance, primitive traits, and diversity of adaptive types among its members to approach such a potential.

Figure 15 depicts a *Euglena*. Like most other members of its phylum, this common freshwater form possesses a relatively elementary cell plan and a well-formed nucleus. It propels itself through the water with one of its two flagella, which sometimes doubles as an anchor to hold the organism in place. Each flagellum provides power by means of a wavy motion that spreads from base to tip, as explained in Chapter 20. *Euglena* reproduces by mitosis —the simplest and most direct way possible. It has very flexible nutritional requirements. In sunlight it is fully autotrophic, using its chloroplasts to synthesize organic compounds through photosynthesis. When kept in the dark, the organism loses its photosynthetic pigment and begins to feed exclusively on dead organic material floating in the water around it.

Because of their extraordinary versatility in nutrition, the flagellates are said to bridge the gap between plants and animals at the unicellular level. Some relatively large green colonial flagellates appear quite similar to green algae or the motile aquatic stages of some of the higher plants. Other flagellates, in contrast, strongly resemble tiny animals. Some species related to *Euglena* are

EUGLENA GRACILIS, a photosynthetic heterotroph, **15** **propels itself through the water with the longer of its two flagella. Rudimentary second flagellum does not extend beyond canal at rear of organism. Starch manufactured by photosynthesis is stored as paramylon granules.**

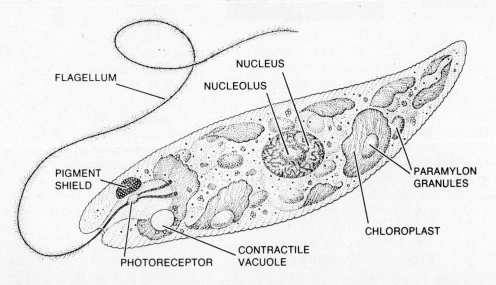

FLAGELLUM   NUCLEUS   NUCLEOLUS   PARAMYLON GRANULES   PIGMENT SHIELD   CHLOROPLAST   PHOTORECEPTOR   CONTRACTILE VACUOLE

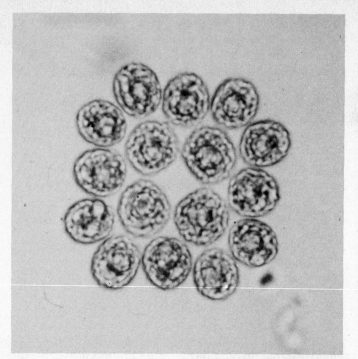

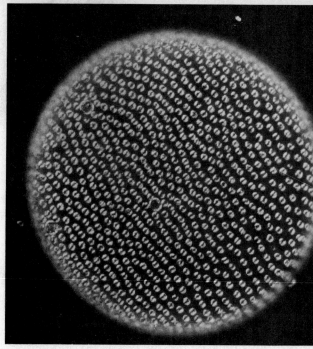

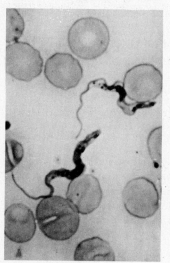

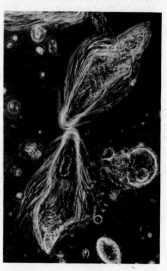

DIVERSITY OF FLAGELLATES is illustrated by organisms pictured above. Top left: Gonium, a loosely colonial flagellate whose separate cells are held together by a gelatinous matrix. Top right: Volvox, a colonial flagellate consisting of a hollow ball of cells. Bottom left: human blood containing Trypanosoma gambiense, which causes African sleeping sickness. Bottom right: cellulose digesting flagellate from hindgut of woodeating cockroach Cryptocercus. Photomicrographs are not reproduced to the same scale.

with the green algae in the following chapter, while the *Trypanosoma* and the cellulose-digesting flagellate from the cockroach *Cryptocercus* are normally considered animals!

The flagellates also span the gap between single-celled and many-celled organisms. Surprisingly large and well-organized colonies of cells are formed in such fresh-water groups as the genus *Volvox*. The cells are not differentiated into tissues and organs as in higher plants and animals, but the colonies show vividly how the preliminary step to this great evolutionary advance might have been taken. In addition, the intermediate stages between the one-celled state of *Euglena* and the extreme colonial state of *Volvox* are preserved in such loosely colonial forms as *Gonium* and *Pandorina*.

PHYLUM SARCODINA, the amoebas and their relatives, have for generations been portrayed in popular writing as blobs of glop—the simplest

devoid of chloroplasts and make their living by preying on other protozoans, including *Euglena*. An impressive diversity of other "zooflagellates," so labeled in order to distinguish them from the plantlike "phytoflagellates," live as internal parasites of larger animals, including man. Within the guts of certain wood-eating roaches and termites live an array of huge zooflagellates possessing some of the most bizarre and complicated body forms found anywhere among the protists. Some idea of the diversity of this phylum is conveyed by Figure 16. *Gonium* and *Volvox*, top, are both photosynthetic, and might well have been treated along

552

form of animal life imaginable. An examination of the typical amoeba, illustrated in Figure 17, shows why such a conclusion was reached. The animal consists of a single cell with no definite shape. It feeds on small organisms and organic particles by engulfing them with PSEUDOPODS —extensions of the constantly changing body mass. A pseudopod is formed when, in response to an internal stimulus, the cytoplasm at some point near the edge of the cell becomes more liquid than the surrounding material. Pressure from inside causes the cell to bulge outward at that point. The exact physical basis of the movement is still a matter of controversy. Particles of food are simply engulfed by pseudopods and sealed off in food vacuoles within the cytoplasm. The material is then slowly digested and assimilated into the main body of the organism. Excess water accumulated by freshwater species is periodically voided by means of contractile vacuoles, which literally squeeze the material out of the body into the surrounding medium. The pseudopods are also the organs of locomotion. The animal extends a pseudopod in the direction it "wishes" to go, anchors it, and perhaps, by a prolonged contraction of long protein chains in the cytoplasm, the amoeba pulls itself forward.

Despite its apparently primitive characteristics, the amoeba is probably not a primeval organism. Compelling evidence points to the conclusion that its simplicity is a secondarily derived condition in evolution. The phylum Sarcodina apparently originated from ancestors within the Mastigophora. Some intermediate forms still exist (Figure 18). The example shown, *Mastigamoeba aspera,* is in fact such an exactly intermediate link that it could equally well be placed in either phylum.

The amoeba is in fact a rather advanced form of protozoan specialized for life on the bottoms of lakes, ponds and other bodies of water. Its creeping form of locomotion and manner of engulfing particles require it to remain close to a relatively rich supply of sedentary organisms and organic particles. The remaining Sarcodina form an astonishing array of even more specialized forms (Figure 19). All are animal-like, existing as predators, parasites, or scavengers. There are SHELLED AMOEBAS that live in casings of sand grains glued together, or in spiny or scaly shells secreted by the animal itself. FORAMINIFERANS are marine creatures that secrete shells of calcium carbonate. Their pseudopodia are long, threadlike, branched, and interconnected with one another to create a sticky net that the animals use to catch smaller plankton.

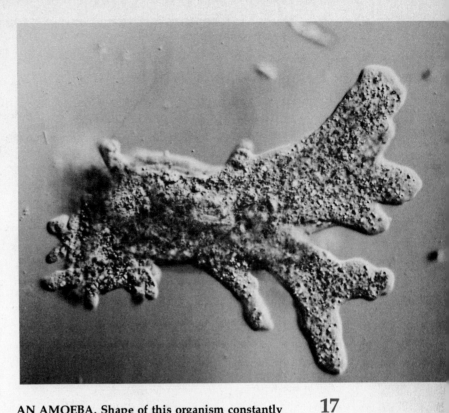

AN AMOEBA. Shape of this organism constantly changes as it extends pseudopods (right) to move from place to place or to engulf food. 17

MASTIGAMOEBA has pseudopods like the Sarcodina and a flagellum like the Mastigophora. It can legitimately be assigned to either phylum. 18

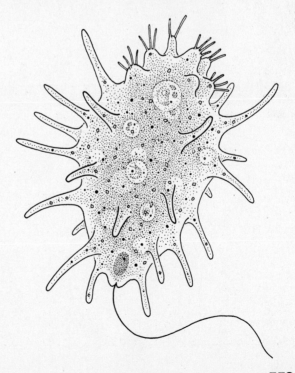

553

RADIOLARIA AND FORAMINIFERA are shelled, amoeba-like members of the phylum Sarcodina. Siliceous skeletons of radiolarians (left) display an extraordinary diversity of architecture. Tiny pores in surface are openings through which the living organisms extended pseudopods. Magnification, about 100X. Foraminifera secrete chalky shells (right). Magnification, about 50X.

The shells of individual foraminiferan species have distinctive shapes, and they are easily preserved as fossils in marine sediments. Each geological period had its own distinctive species. For this reason, plus the fact that they are so abundant, foraminiferan remains are especially valuable as indicators in the classification and dating of sedimentary rocks, and they also serve as indicators in oil prospecting. HELIOZOANS are fresh-water forms surrounded by a bristling array of long pseudopods.

LIFE CYCLE OF MALARIAL PARASITE (Plasmodium) involves two hosts. Sporozoites enter human bloodstream with saliva of Anopheles mosquito (left center) and develop into merozoites, primarily in cells of liver and lymphatic system (1). Merozoites infect red blood cells and reproduce (2); some develop into gametes (3). When infected human is bitten by mosquito (4), gametes enter insect's stomach, differentiate further, and fuse in pairs (5). Resulting zygotes bore into gut lining (6) and form cysts in which sporozoites proliferate asexually (7). When cysts burst (8), sporozoites bore into salivary glands and accumulate in saliva. When mosquito bites human, cycle begins anew.

IN MOSQUITO

5 6

8

IN MAN

4

GAMETES

♂
♀

3

SPOROZOITES

1

IN LIVER AND IN
RETICULOENDOTHELIAL
SYSTEM

2

IN BLOOD
(48 HOUR CYCLE)

MEROZOITES

MEROZOITES

Like the foraminiferans, they drift in the water and use the pseudopodia to trap smaller organisms. A third group with the same feeding technique are the RADIOLARIANS. Found exclusively in the sea, they are perhaps the most beautiful of all microorganisms. Almost all radiolarian species secrete siliceous skeletons from which needlelike pseudopods project. The skeletons of the different species are as varied as snowflakes, and many of them have elaborate geometrical designs. A few of the radiolarians are among the largest of the protozoans, with skeletons measuring several millimeters across.

The SPOROZOANS (Phylum Sporozoa) are exclusively parasitic forms which derive their name from the fact that some of them produce sporelike infective stages. Sporozoans generally have an amoeboid body form, but this in no way indicates a relationship to the Sarcodina. The development of the trait is a common evolutionary event in parasitic protozoans. This mark of degeneracy has appeared, for example, even in parasitic dinoflagellates. The sporozoans, like many obligate parasitic forms among the higher animals, display elaborate life cycles featuring asexual and sexual reproduction by a series of very dissimilar life stages. Often

1 **BLOOD OF MALARIA VICTIM. Several red blood cells contain merozoites. The cell at right is replete with merozoites and close to bursting. Crescent-shaped cell is a gamete of the parasite which infects Anopheles mosquito.**

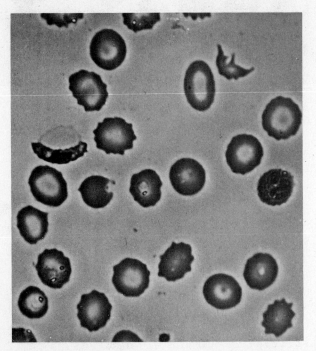

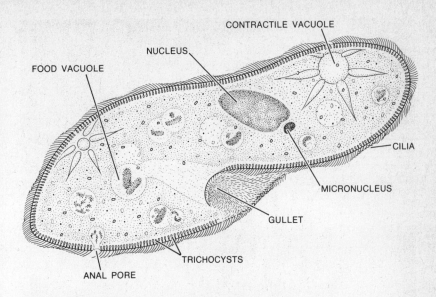

ANATOMY OF PARAMECIUM, as seen under the light microscope, exhibits the complexity of structure characteristic of ciliates. Undischarged trichocysts are embedded in the pellicle, between the cilia. Micronucleus carries the genetic information exchanged during conjugation. **22**

these stages are associated with two kinds of host organisms. The best known sporozoan is the malaria parasite *Plasmodium*, a highly specialized organism that spends part of its life cycle within vertebrate red blood cells (*Figures 20 and 21*).

PHYLUM CILIOPHORA, the ciliates, ranks with the flagellates as the most diverse and ecologically important of the protozoan phyla. Ciliates are all animal-like in nutrition, and are much more specialized in body form than most flagellates and other protists. They are characterized by the possession of hairlike CILIA which have the same cable-like ultrastructure as flagella and are believed to have evolved from them. A second characteristic of ciliates is the possession of two types of nuclei: a large MACRONUCLEUS and, within the same cell, from one to as many as 80 MICRONUCLEI. The little micronuclei are the essential carriers of the genetic information. The macronucleus is sometimes called the vegetative nucleus because it is not essential in reproduction but does play a vital role in the expression of genetic information in the phenotype.

The ciliates represent the zenith of evolution among unicellular organisms. Their astonishing complexity of structure and behavior is exemplified by *Paramecium*, the most famous and thoroughly studied member of the phylum (*Figures*

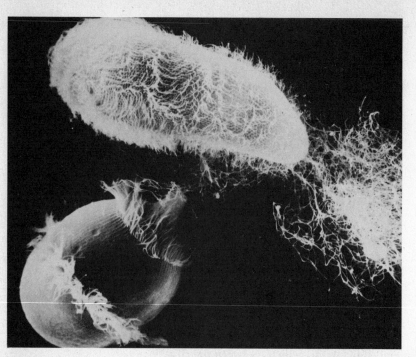

22 *and 23)*. Its slipper-shaped body is covered by an elaborate pellicle, a skin composed principally of an outer membrane and an inner layer of closely packed, kidney-shaped structures called alveoli that embrace the cilia. Also present as a layer of the pellicle are the unique defensive organelles called trichocysts. Expelled by a microscopic explosion in a few thousandths of a second, the trichocysts emerge as sharpened darts driven forward at the tip of a long expanding shaft *(Figure 24)*.

The cilia provide a form of locomotion that is generally superior to that made possible by flagella and pseudopods. The paramecium can direct the beat of these organelles to propel itself forward or backward. When it encounters a barrier or a negative stimulus it can back off swiftly. (Some of the larger ciliates hold the speed record for protists — over 2 millimeters per second.) A few of the cilia of the paramecium are sensory in function, and they

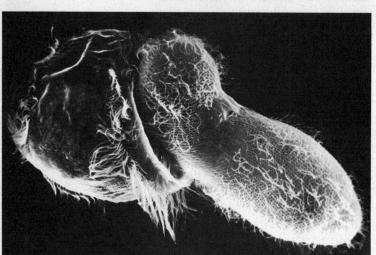

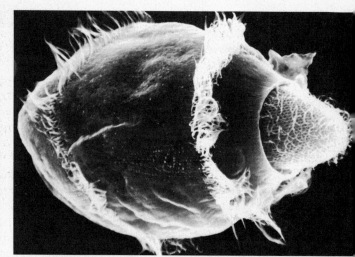

23 DIDINIUM DEVOURS PARAMECIUM in this series of scanning electron micrographs. Top, globular Didinium closes in on the Paramecium, which fires a barrage of trichocysts, visible at right of picture. Didinium makes contact (above) and engulfs the Paramecium (right). Trichocysts do not deter Didinium, which paralyzes prey with tiny darts of its own.

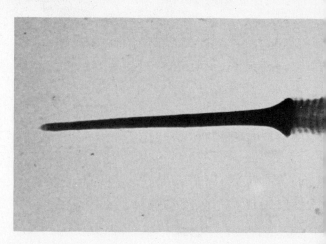

24 DISCHARGED TRICHOCYST consists of needlelike tip and banded filament. In the laboratory, Paramecium ejects trichocysts when disturbed. In nature, Paramecium probably uses these "harpoons" to defend itself against predators and to anchor itself in place. Magnification, 27,000X.

are somehow able to transmit stimuli back through the remainder of the cytoplasm in a way that permits rapid coordinated movements on the part of the entire organism. Paramecia usually reproduce by cell division. But an elaborate form of sexual behavior called CONJUGATION occurs when two paramecia line up tightly against each other and fuse in the oral region of the body *(Figure 25)*. During the next several hours there is an extensive reorganization and exchange of nuclear material. The macronuclei break up and disappear. Each of the micronuclei divides, and one of the two daughter nuclei migrates over to the other cell, where it fuses with its counterpart. The exchange is perfectly reciprocal—each of the two paramecia gives and receives an equal amount of genetic material. The two organisms now pull away and depart, having been genetically refreshed. Finally, the macronucleus is reconstituted under the direction of the newly organized micronuclei. As a rule, each familial line of paramecia must periodically go through the process of conjugation. It has been shown by laborious experimentation that if some species are not permitted to conjugate, the asexual lines can live through no more than about 350 cell divisions before they die out.

Most ciliates possess all of the traits just described for *Paramecium*. Some, however, are especially notable for the exceptional degree of development of individual organelle systems *(Figure 26)*. Certain of the hypotrich ciliates, for example, have the equivalent of legs. Fused cilia called CIRRI move in an independent but coordinated fashion, enabling the animal to walk over surfaces. This degree of coordination is made possible by nerve-like neurofibrils that lead to individual cirri. When these are cut in laboratory experiments, the coor-

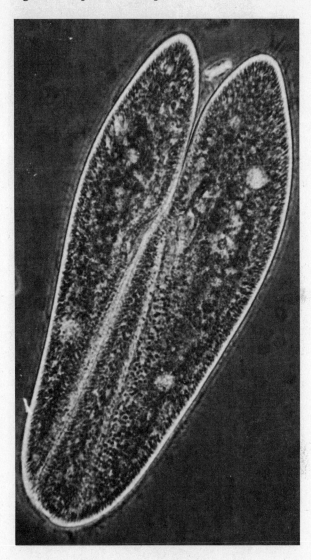

CONJUGATING PARAMECIA align themselves tightly side by side, and their oral grooves fuse. After exchanging genetic material (micronuclei), the two organisms separate and depart.

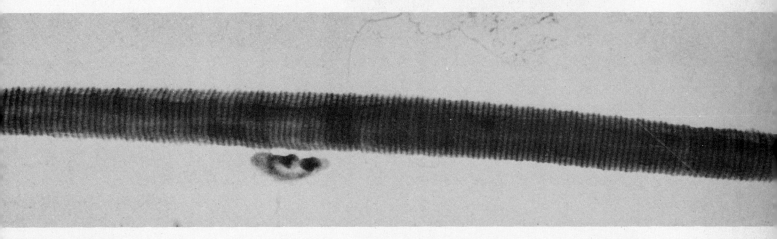

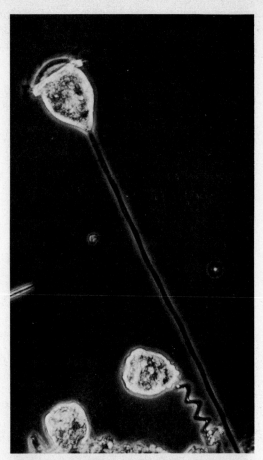

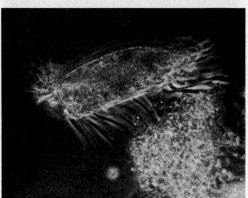

dination is lost. Many kinds of ciliates possess myonemes, musclelike fibers within the cytoplasm. A contraction of myonemes causes an astonishingly quick retraction of the stalk of forms such as *Vorticella* when the animal is disturbed. Possibly the ultimate in cytoplasmic organization is displayed by highly specialized ciliates that live in the digestive tracts of cows and many other hooved animals (*Figure 27*). They possess not only myonemes (musclelike fibers), neurofibrils, and elaborately fused cilia, but also a cytoplasmic "skeleton" and a "gut" complete with mouth, esophagus, and anus. The ciliates represent the zenith of evolution among unicellular organisms. When examining the intricate structure of one of these organisms, the reader may have to stop and remind himself that he is looking at only one cell.

DIPLODINIUM DENTATUM, a ciliate that lives in the 27 rumen of cattle, is one of the most complex ciliates. Surface view of the cell appears at left. In longitudinal section at right, micronucleus lies at far right, with macronucleus immediately to its left. Plates of cytoplasmic "skeleton" are visible at left of section. One group of myonemes retracts the mouth, and another ring-shaped group closes it. "Gut" of Diplodinium is differentiated into a mouth, an esophagus and an anus.

26 OTHER CILIATES include Vorticella (top) and Eoplotes (bottom). Muscle-like myonemes in stalk enable Vorticella when disturbed to contract rapidly, like the one at lower right. Cirri on ventral surface of Eoplotes enable it to walk across surfaces. Cirri consist of closely compacted cilia whose action is coordinated by neurofibrils.

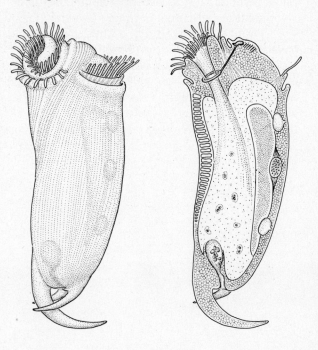

READINGS

R. D. BARNES, *Invertebrate Zoology*, Philadelphia, Saunders, 3rd Edition, 1974. One of the best textbooks in the field, containing an excellent brief review of the protozoans.

T. D. BROCK, *Biology of Microorganisms*, Englewood Cliffs, NJ, Prentice-Hall, 2nd Edition, 1974. An attractively produced and very readable account of microorganisms, with emphasis on the bacteria.

H. CURTIS, *The Marvelous Animals: An Introduction to the Protozoa*, Garden City, NY, The Natural History Press, 1968. An excellent popular account of the protozoans, with enough detail to satisfy almost anyone but a research specialist.

A. JURAND, AND G. G. SELMAN, *The Anatomy of Paramecium Aurelia*, London, Macmillan, 1964. A modern treatise of the anatomy of a protist. Remarkable photographs of the ultrastructure of organelles offer a view of the phenomenal complexity that prevails even at this microscopical level of biological organization.

G. F. LEEDALE, *Euglenoid Flagellates*, Englewood Cliffs, NJ, Prentice-Hall, 1967. An authoritative treatise of this fascinating group of plant-animals, including summaries of much contemporary experimental work.

W. R. SISTROM, *Microbial Life*, New York, Holt, Rinehart, & Winston, 2nd Edition, 1969. A brief and well-written survey of bacteria and viruses.

R. Y. STANIER, E. A. ADELBERG, AND J. L. INGRAHAM, *The Microbial World*, Englewood Cliffs, NJ, Prentice-Hall, 4th Edition, 1976. Perhaps the best introductory textbook on microbiology, containing a review of the broad principles of classification and the biology of each of the major groups.

We have not to deal with a directly ascending series of organic forms which advance from the lower and simpler to the higher and more complex, but we must rather conceive of the Vegetable Kingdom . . . as a copiously branching tree, whose boughs have their origin in a common stem, but stand in no direct communication with each other. —K. Goebel (1882)

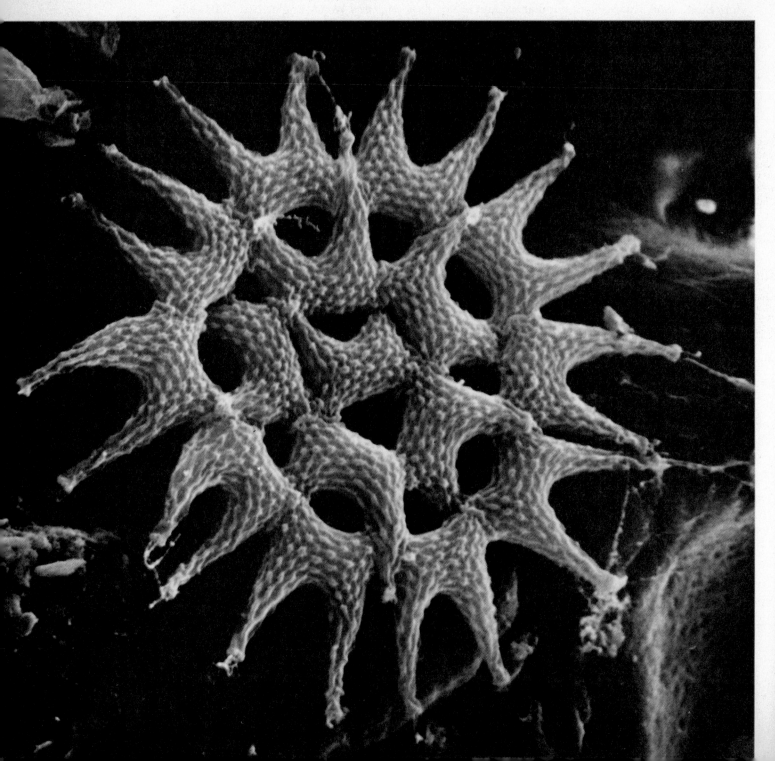

26
Fungi and plants

The evolutionary steps that led from the single-celled organisms to multicellular fungi and plants eventually produced whole new strategies of existence. The true significance of the difference between the two great levels of organisms is that the single-celled monerans and protistans diversified to exploit the opportunities open to microscopic organisms. The second great assemblage, the fungi and plants, diversified to exploit the opportunities open to large organisms. The fungi are today among the important degraders of dead organic matter; they probably never possessed the ability to photosynthesize or to live autotrophically in any other way. Plants, including algae, stand in sharp contrast. All but a few are capable of photosynthesis. They make up the overwhelming bulk of autotrophic organisms on the planet today.

THE KINGDOM FUNGI

To most people the word fungus evokes a rather vague, unpleasant image. Fungi are thought of as sprawling globs of living matter that occupy a very low place in the evolutionary order of things. The truth is nearly at the opposite extreme. Many fungi are exquisitely constructed, and their life cycles are among the most complex to be found anywhere in nature. Moreover, their great diversity in anatomy and life cycles makes the kingdom Fungi difficult to define as a taxonomic unit. Some workers consider the fungi simply as a category of non-photosynthetic plants. But we feel that the large group that includes slime molds, molds, yeasts, rusts, and mushrooms is sufficiently coherent and

COLONIAL ALGA Pediastrum boryanum is a freshwater green alga. Photomicrograph on opposite page, made with the scanning electron microscope, depicts a colony of cells at a magnification of 6100X.

sufficiently different from plants to comprise a separate kingdom.

The higher fungi all follow a characteristic pattern of filamentous growth, with individual filaments or HYPHAE becoming enmeshed into a cottony mass called a MYCELIUM, and the mycelium sometimes becoming organized into elaborate fruiting structures such as puff balls or mushrooms. Their walls contain a variety of polysaccharides, but usually also contain CHITIN, the same polymer that hardens the exoskeleton of insects and other arthropods, and/or cellulose, characteristic of the cell walls of higher plants. Chitin is a polymer of N-acetyl glucosamine *(Figure 2 in Chapter 25)*, a modified amino sugar found in bacterial cell walls. Lower fungi may lack true cell walls in all phases of their life cycle except for the spore stage; the presence of characteristic fruiting structures distinguishes these fungi from protozoans or plants.

The protistan phyla can be readily separated from one another on the basis of cell structure. There is certainly no difficulty in distinguishing a flagellate such as *Euglena* from a ciliate such as *Paramecium* or from an *Amoeba*. But such variation in cell structure simply does not occur in the fungi, except perhaps in the slime molds, and the phyla are instead based on the method and characteristic structures associated with sexual reproduction. Other criteria, such as presence or absence of cross walls in the hyphae, are of more limited use.

One note on classification: Most botanists prefer the word DIVISION to the word PHYLUM. In this book the term phylum is used throughout. Also, a few authors still follow the older practice of treating the phyla of fungi as classes of a single phylum (phylum Fungi). We have placed the names of these classes, for example class Myxomycetes which is synonymous with the phylum Myxomycota, in parentheses to help the reader

561

make cross references to textbooks that still use the older system.

The SLIME MOLDS belong to the phylum Myxomycota (class Myxomycetes). If the nucleus of an amoeba began rapid mitotic division, accompanied by a tremendous increase in its cytoplasm and organelles, the organism would resemble the vegetative phase of the true acellular slime molds, one of the two major groups of the phylum Myxomycota. During most of its life history, an acellular slime mold exists as a wall-less mass of protoplasm which streams over its substrate in a remarkable network of strands called a PLASMODIUM (*Figure 1*). A degree of structural rigidity is provided by the outer protoplasm, which is normally in a gel state rather than being a fluid sol. A myxomycete such as *Physarum* provides a dramatic example of cytoplasmic streaming. As with pseudopodium formation in *Amoeba*, the outer protoplasmic gel becomes more fluid in places and there is a rush of cytoplasm creating a new bulging mass of plasmodium. In some manner this streaming reverses its direction every few minutes as cytoplasm rushes into a new area, draining away from an older one. As the plasmodium spreads over its substrate it engulfs food particles, predominantly bacteria and other small organisms. Sometimes an entire wave of plasmodium moves across the substrate, leaving strands behind. A contractile

SPORANGIA of the slime mold Physarum arise from heaped masses of plasmodium. Nuclei of cells on outside of knobs develop into spores. 2

1 PLASMODIUM OF SLIME MOLD on a decaying log. Slime mold spreads across its substrate by the process of cytoplasmic streaming. Plasmodium engulfs bacteria and other small organisms.

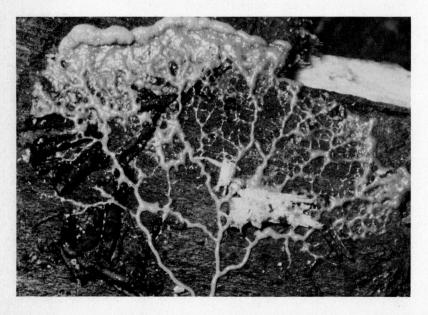

protein called myxomyosin (analogous to the contractile protein of muscle, actomyosin) is involved in the streaming movement. Minute fibers called microfilaments actually mediate the movement in conjunction with a myosin-like molecule.

So long as the food supply is adequate, and other conditions such as moisture and pH are optimal, *Physarum* can grow almost indefinitely in its plasmodial stage. But if conditions become harsher, one of two things can happen: the plasmodium can form a resistant structure, a rather irregular mass of hardened cell-like compartments called a sclerotium, which rapidly becomes a plasmodium again upon return of favorable conditions; or the plasmodium can transform itself into spore-bearing fruiting structures (*Figure 2*). Rising from heaped masses of plasmodium, these stalked or branched structures derive their rigidity from the deposition of cellulose or chitin at the surfaces of their component cells.

The plasmodium is a diploid structure, and during the development of the fruiting structure called the SPORANGIUM (*Figure 2*) its cells divide by meiosis. One or more knobs, variously colored and shaped, may develop on the end of the stalk, and the outer nuclei become surrounded by walls and become spores. Eventually, as the spore-bearing sporangium dries, the spores are shed. They then germinate into wall-less flagellated haploid cells which can manage on their own quite nicely, becoming walled and resistant cysts when conditions become unfavorable. Upon the return of more favorable conditions the cysts release flagellated swarm cells that can either divide mitotically to produce more swarm cells, or can function as gametes. Two swarmers fuse to form a diploid zygote, which divides by mitosis, but without wall formation, to form a new plasmodium.

The second principal group of Myxomycota is the cellular slime molds. One member of this small but interesting group, *Dictyostelium discoideum*, has already been discussed (*see Figure 27 in Chapter 12*). Where *Physarum* and other acellular slime molds have the plasmodium as their basic vegetative unit, an amoeboid cell serves this role in the cellular slime molds. Large numbers of uninuclear cells called myxamoebas engulf bacteria and other food particles and reproduce by mitosis and fission. So long as food and moisture are available, this simple developmental stage persists indefinitely. Should conditions become stringent, however, the cellular slime molds, like their acellular counterparts, form fruiting structures. The apparently independent myxamoebas aggregate into an irregular mass called a PSEUDOPLASMODIUM. Unlike the true plasmodium of the acellular slime molds, this structure is not simply a giant lump of cytoplasm: the individual myxamoebas retain their membranes and therefore their identity. The pseudoplasmodium may migrate over the substrate for a period of time before ultimately constructing a delicate, stalked fruiting structure (reported to contain cellulose or chitin). During this period some of the amoeba and their nuclei fuse. Then meiosis occurs, probably still during the pseudoplasmodium stage, although the nuclei and chromosomes are so incredibly small that the precise details of alternation of generations remain to be elucidated in most species.

FILAMENTOUS FUNGI (phylum Phycomycota or class Phycomycetes) have characteristics that enable the taxonomist to say, conclusively, "These plants are obviously fungi." The filamentous fungi are distinguished by a cell wall composed either of chitin, or cellulose, or both. The organisms have long and elaborate branched hyphae and may form enormous mycelial mats. But despite these characteristic fungal traits, phycomycetes are sometimes completely lacking in cross walls, which suggests that they are not truly cellular organisms. Even when cross walls exist, they contain large perforations through which extensive cytoplasmic streaming occurs. Thus, except for certain reproductive stages, there is no single structural unit with its single nucleus, confining walls, and appropriate organelles, that can be called a cell. Is this organizational plan really different from that of the acellular slime molds? After all, the slime molds frequently make cellulose or chitin, at least during the fruiting stage. Furthermore, many of the phycomycetes form motile flagellated cells, either vegetative spores or gametes at some stage in their life history.

Both phycomycetes, which are fungi, and actinomycetes, which are bacteria, reproduce by forming spores at the tips of filaments. The similarity is an elegant example of convergent evolution.

The actinomycetes, discussed with the Eubacteria in Chapter 25, provide some interesting analogies to the phycomycetes. Recall that the actinomycetes are walled procaryotes that form filamentous masses without any dividing walls. Presumably each individual contains many copies of the genetic material, just as in the phycomycetes. Reproduction is frequently by formation of spores at the tips of the filaments, also just as in many of the phycomycetes. The similarity between these two types of organism is an elegant example of convergent evolution.

The simplest of the phycomycetes are the chytrids. These organisms, many of which are parasites on aquatic organisms, may have no more than one or a few nuclei. Short filamentous processes digest their way into the substrate, which is either dead organic detritus or living cells. The chytrids

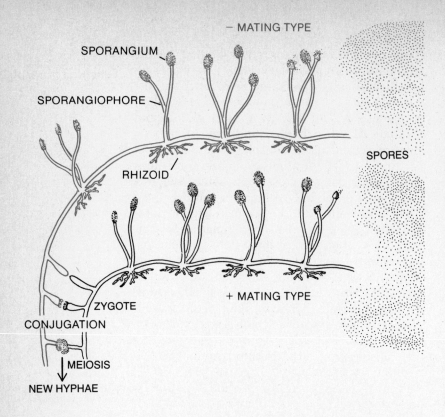

SPORANGIUM

SPORANGIOPHORE

– MATING TYPE

RHIZOID

SPORES

ZYGOTE

+ MATING TYPE

CONJUGATION

MEIOSIS

NEW HYPHAE

3   **BLACK BREAD MOLD (Rhizopus stolonifer) spreads across bread by extending hyphae like the two above. Rhizopus reproduces asexually by shedding spores, or sexually by conjugation.**

reproduce vegetatively by forming a swollen cell in which mitosis produces numerous flagellate spores, called ZOOSPORES. (A large number of fungi and algae reproduce vegetatively from zoospores, both from haploid and diploid phases of the life cycle.) Sexual reproduction, which has been observed in only a few species, involves conjugation of a kind similar to that observed in bacteria and protozoans. One organism produces a slender tube which grows toward and then fuses with

another. A nucleus migrates across, and nuclear fusion produces a zygote.

An example of a phycomycete with a more elaborate life cycle is the common water mold *Allomyces*. The motile female gamete of *Allomyces* produces the chemical pheromone sirenin that attracts the swimming male gamete, leading to fertilization and zygote formation. *Allomyces* is also one of those organisms that has an isomorphic alternation of generations *(Chapter 13)* in which the diploid and haploid phases of the life cycle are indistinguishable except on the basis of their chromosome number and their reproductive products. The diploid sporophyte produces zoospores both by mitosis and by meiosis in discrete SPORANGIA, cell-like units delimited by distinct cross walls. Diploid zoospores simply produce new sporophytes, while haploid zoospores produce gametophytes. A single gametophyte produces both male and female gametes at the tips of hyphae (usually with the male produced terminally and female just below) in specialized walled off structures called gametangia.

The filamentous fungi just described are all aquatic and saprophytic —they obtain nutrients by breaking down dead organic matter. Many others are terrestrial, and a few are serious plant parasites. *Albugo* is a well-known parasitic genus causing a mealy blight on sweet-potato leaves, morning glories, and numerous other plants. It is an obligate parasite, and has never been grown on any medium other than its plant host.

FORMATION OF AN ASCUS begins when heterocaryon filament (far left) walls off a pair of dissimilar nuclei and forms a hook-shaped structure known as a crozier. The filament develops, step by step, into an ascus containing eight ascospores (far right).   **4**

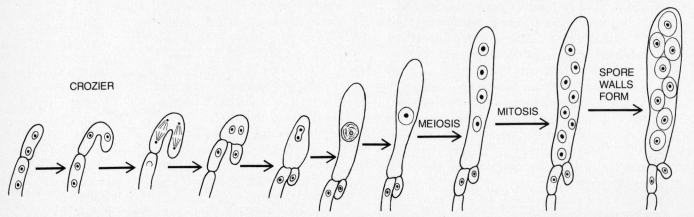

CROZIER

MEIOSIS   MITOSIS   SPORE WALLS FORM

564

A phycomycete which the reader has certainly seen at one time or another is *Rhizopus stolonifer*, the black bread mold. The mycelium creeps over the substrate, growing forward by means of specialized hyphae *(Figure 3)*. Large numbers of SPORANGIOPHORES (sporangium-bearing stalks) are produced, each of which bears a single sporangium containing many hundreds of minute spores. As in other phycomycetes, the spore-forming structure is cut off from the rest of the hypha by a wall. Sexual reproduction is by conjugation, with hyphae of adjacent different strains coming together to fuse and form heavy-walled zygotes. Several pheromones are involved in the process, as in *Allomyces (see Chapter 22)*. Since *Rhizopus* is terrestrial, it is perhaps not surprising that at least two of the pheromones are passed as a gas from one organism to another.

ASCOMYCETES (phylum Ascomycota) comprise a large and diversified group of fungi distinguished by a unique saclike structure called the ASCUS. Both nuclear fusion and subsequent meiosis take place within individual asci. The meiotic products become delimited as ASCOSPORES, and are ultimately shed to begin the new gametophytic generation. The ascomycete hyphae are broken into segments by cross walls, but a pore in each cross wall permits extensive movement of cytoplasm and organelles (including the nuclei) from one "cell" to the next.

The ascomycetes can be divided into two broad groups, depending on whether or not the asci are contained within a specialized fruiting structure. Those species having a fruiting structure are collectively called the euascomycetes, while those without it are called the hemiascomycetes. The sexual cycle of the euascomycetes involves the formation of an important and distinctive structure known as the HETEROCARYON. The heterocaryon (meaning "different nuclei") is defined as the sometimes-extended mycelial stage in which both parent nuclei are present in the cytoplasm. It occurs when two compatible strains of a particular fungus fuse as though they were conjugating. Nuclei are passed from one mycelium to the other, but then proceed to divide synchronously with the host nuclei for a considerable period of time. Only with the formation of asci do the nuclei finally fuse. Sometimes the fusing hyphae form well-differentiated fruiting structures, as shown in Figure 5 in Chapter 13.

Ascus formation itself is a peculiar process. One might expect that a hyphal tip containing a pair of dissimilar nuclei would simply become walled off,

the nuclei would fuse, and then meiosis would proceed as usual. What actually happens is illustrated in Figure 4. A hook or crozier forms at the tip of the heterocaryotic hypha, and the paired nuclei come to lie on either side of it. Both nuclei then undergo mitosis simultaneously with their mitotic axes longitudinal to the hyphal axis. New walls are laid down as shown, and the tip, containing a single nucleus, fuses with the hyphal wall it has come to touch, releasing that nucleus back into the lower portion of the hypha. The other two nuclei finally fuse, and meiosis begins ascospore formation.

This generalized ascomycete life cycle should be kept in mind in the following special consideration of the first of the two major groups, the hemiascomycetes. These organisms are very small in general, and frequently are unicellular. Perhaps the best known are the yeasts, especially baker's or brewer's yeast *(Saccharomyces cerevisiae)*. Other yeasts occur naturally on fruits such as figs and, in particular, grapes, and play an important role in the making of wine. Of all of the fungi, the yeasts are undoubtedly the most important domesticated ones. The yeasts multiply either by simple fission or, in the better known genera, by a process of budding; the outgrowth of a new cell from the surface of an old. The single cells are haploid. Conjugation occurs only occasionally between two adjacent compatible cells, and nuclear fusion is followed immediately by meiosis and a single mitosis, so that the entire structure becomes an ascus. There is no heterocaryon stage.

Among the euascomycetes, the second major group of ascomycete fungi, are several common molds: *Aspergillis*, *Penicillium*, and *Neurospora*, the

CONIDIOSPORES of euascomycetes are tiny chains of spores. Those of Aspergillus are tightly clumped, those of Penicillium more loosely branched. 5

ASPERGILLUS

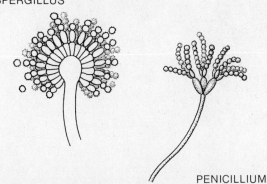

PENICILLIUM

brown, green, and pink molds, respectively. All three are ubiquitous, frequently occurring on old bread, and are common laboratory contaminants. *Neurospora* is already familiar to the reader as one of the favorite organisms of experimental genetics *(Chapter 8)*. Asexual reproduction occurs through extensive production of CONIDIOSPORES, illustrated in Figure 5. These small chains of spores are produced literally by the millions, and are sufficiently resistant to survive for weeks. When heterocaryon formation occurs, the fruiting structure is rather small and inconspicuous.

Penicillium is the organism which produces the antibiotic penicillin, presumably to defend itself from competing bacteria. The two species *Penicillium camemberti* and *Penicillium roqueforti* are the organisms responsible for the characteristic flavor of Camembert and Roquefort cheeses. Not surprisingly, people who are hypersensitive to penicillin may react violently if they make the mistake of eating one of these cheeses.

A large number of the euascomycetes are serious parasites on higher plants. The powdery mildews are a group that infects cereal grains, lilacs, and roses, to name but a few host plants. They may also be a serious problem to grape growers, and a great deal of research has been done on ways to control these serious agricultural pests.

6 BRACKET FUNGI grow on tree trunks and dead logs, and produce large fruiting structures like those below. Spore-forming basidia fill pores on one surface of the fruiting structure.

The euascomycetes also include the cup fungi. In these organisms, the largest fruiting structures are small cups several centimeters across. The inner surfaces of the cups are covered with a mixture of sterile filaments and asci, and produce huge numbers of spores. Although these fleshy structures appear to be composed of distinct tissue layers, microscopic examination shows that their basic organization is still filamentous. Such fruiting structures are formed only by the heterocaryon mycelium, however.

BASIDIOMYCETES are all members of the phylum Basidiomycota. Some produce the most spectacular fruiting structures found anywhere in the fungi. These are the puffballs which may be over half a meter in diameter, the mushrooms of all kinds (including the poisonous toadstools), and the giant bracket fungi often encountered on trees and fallen logs in a damp forest *(Figure 6)*. Some basidiomycetes are among the most serious plant pathogens, including the wheat rust *(Puccinia graminis)* and the smut fungi that parasitize cereal grains. The basidiomycete wall, like that of the phycomycetes and ascomycetes, contains chitin. Basidiomycete hyphae are characterized by being completely septate (walled off), with the result that nuclei cannot migrate throughout the mycelium.

The BASIDIUM is the site of nuclear fusion and meiosis in the basidiomycetes, playing the same role as the ascus in the ascomycetes. All known basidiomycetes possess the heterocaryon stage in their sexual cycle. Strictly speaking, this stage should be called a DICARYON, since in the septate mycelium at this stage, each cell (it is finally safe to use this word with reference to a fungal hypha) contains two nuclei, one from each of the original haploid partners.

The organization of the elaborate fruiting structure of a higher basidiomycete is illustrated in Figure 7, a diagrammatic cross section of one of the gill mushrooms. There is a cap or pileus, on the under side of which are found the gills. It is on the sides of the gills that basidia are found in enormous numbers between sterile filaments. The basidia discharge their spores into the air spaces between adjacent gills, and they sift down into air currents for dispersal to start new haploid mycelia. The exact pattern of the gills and the spore color are useful criteria for determining mushroom species. One simply places a mature cap on a piece of white paper for a few hours in a quiet place. The ejected basidiospores settle from between the gills, leaving on the paper an elegant replica of the gill pattern, and also clearly revealing spore color.

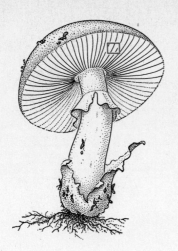

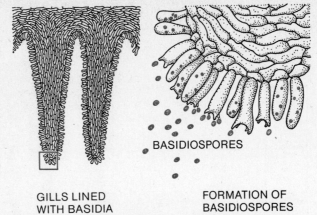

BASIDIOSPORES

GILLS LINED
WITH BASIDIA

FORMATION OF
BASIDIOSPORES

7

GILL MUSHROOMS are basidiomycetes. Gills on underside of fleshy cap are lined with basidia. Tiny spores are dispersed by wind. Small colored squares at left and center indicate area of next section to right.

The three preceding phyla of higher fungi (Phycomycota, Ascomycota, Basidiomycota) are most readily distinguished by their manner of sexual reproduction. But a large number of fungi, both saprophytes and parasites, have no sexual stages; presumably these stages have been lost in evolution. It thus becomes difficult to classify them with any of the three major phyla. A basidiomycete can normally be spotted by its completely septate hyphae, but distinguishing between phycomycetes and ascomycetes is frequently difficult or impossible. Fungi without a taxonomic home are simply dumped into the orphanage known as the FUNGI IMPERFECTI.

LICHENS

This unusual group of organisms defies insertion into any general classification of plants or fungi. The reason is very simple: a lichen is a meshwork of two radically different organisms, a fungus and an alga. Together they can survive some of the harshest environments on Earth. The fungus is usually an ascomycete, but occasionally it is either a basidiomycete or a member of the fungi imperfecti (one phycomycete has been reported); and the alga may be either a blue-green from the Monera, or a green, from the phylum of plants called the Chlorophyta, which will be described later in this chapter. Little experimental work has been done with the lichens, perhaps because they grow so slowly. Thus it is only recently that workers have been able to culture the fungal and algal partners separately and then to reconstruct a lichen from the two. The reassembly does not work if the growth medium is too rich.

Lichens are found in all sorts of exposed habitats: tree bark, open soil, or bare rocks. The reindeer ''moss,'' actually not a moss at all but the lichen *Cladonia,* covers vast areas in the arctic and subarctic regions. Both the fungal partner and the alga can readily grow alone, and even within the lichen, algal growth may outstrip fungal or vice versa, depending upon environmental conditions. The most widely held interpretation of the relationship is that it represents a type of mutually beneficial symbiosis. Filaments of the fungal mycelium are tightly pressed against the algal cells, and sometimes even invade them. The algae not only survive these indignities, but continue their growth and photosynthesis.

A cross section of a typical lichen is shown in Figure 8. There is a tight upper region of fungal filaments alone, an algal layer, a looser filamentous fungal layer, and finally filaments that attach the whole structure to its substrate. The whole meshwork has properties that enable it to hold water fairly tenaciously. Certain nutrients for the algae must arrive in part through the fungal hyphae, the meshwork must provide a suitably moist environment for algal growth, and the fungi must derive fixed carbon from algal photosynthesis.

Lichens can reproduce simply by fragmentation

Photosynthetic plants, the primary producers of organic matter in all food chains, probably outweigh all other organisms by a factor of ten.

SOREDIUM

UPPER CORTEX

ALGAL LAYER

MEDULLA

LOWER CORTEX

8 CROSS SECTION OF LICHEN THALLUS shows fungal components in gray and algal cells in black and white. Upper cortex is protective surface. Algal layer is site of photosynthesis. Medulla apparently serves as food-storage area. Rhizines on lower cortex anchor lichen in place. Soredia are specialized reproductive structures consisting of one or a few algal cells surrounded by fungal hyphae. Once detached from the thallus, they are dispersed by air currents.

of the plant body, which is called the THALLUS, or else by specialized structures called SOREDIA. The soredia consist of one or a few algal cells surrounded by fungal hyphae. They become detached, move in air currents, and on arriving at a favorable location, once again set up the partnership. In cases in which the fungal partner is an ascomycete or a basidiomycete, the fungus may go through its sexual cycle, producing either ascospores or basidiospores. When these are discharged, however, they leave the lichen unaccompanied by the algal partner, and thus are not thought to be essential to lichen reproduction. Nevertheless, many lichens produce characteristic fruiting structures in which the asci or basidia are located.

THE KINGDOM PLANTAE

The enormous kingdom of organisms known as the plants, starts with the algae and ends with the seed plants. The word enormous refers to biomass, rather than to number of species. Photosynthetic plants, the primary producers of organic matter in all food chains, probably outweigh all other organisms by a factor of ten. This observation is consistent with the simple fact that transfer of chemical energy from the eaten to the eater, or from host to parasite, is only about ten percent efficient. (The

energetics of food chains will be treated in detail in Chapter 33.)

In general, plants are defined as eucaryotic photosynthetic organisms. Those few plants that are not photosynthetic are clearly closely related to other plants that are. We exclude the bluegreen algae as procaryotes that are probably little more than modified bacteria, and we exclude fungi because none of them is photosynthetic.

In drawing lines within the plant kingdom, the criteria used for separating groups of algae are different from those used to separate groups of higher plants. The phyla of algae may be distinguished by their photosynthetic pigments, primary photosynthetic storage product, and to a lesser extent, their cell-wall composition. The higher phyla are distinguished by their life cycle, by the absence or presence of xylem, and by the type of xylem *(Chapter 16).*

ALGAE

Algae probably carry out 50 to 90 percent of all of the photosynthesis occurring on the planet, with higher plants accounting for most of the rest. The overall contribution of the bluegreen algae and photosynthetic bacteria is relatively minute. The algae exhibit a remarkable range of growth forms. Some are simply unicellular; others, known as COENOCYTES, are filaments comprised either of distinct cells or multinuclear structures without cross walls. A few are multicellular and intricately branched or arranged in multicellular, leaflike extensions. In extreme cases the masses are even subdivided into tissues and organs. Almost all these types can sometimes be found within a single phylum —for example, the green algae,

PHYLUM	PRINCIPAL PHOTOSYNTHETIC PIGMENTS	STORAGE PRODUCT	PRINCIPAL CELL WALL MATERIALS*
CYANOPHYTA (BLUEGREEN ALGAE)	CHLOROPHYLL A, β-CAROTENE, PHYCOBILINS	CYANOPHYTE STARCH	CELLULOSE, PECTIC COMPOUNDS, MUCOCOMPLEX SUBSTANCE
PYRROPHYTA (INCLUDES DINO-FLAGELLATES)	CHLOROPHYLLS A, C, β-CAROTENE	STARCH, FATS, OILS	CELLULOSE, PECTIC COMPOUNDS
CHRYSOPHYTA (INCLUDES DIATOMS)	CHLOROPHYLLS A, C, β-CAROTENE AND FUCOXANTHIN	CHRYSOLAMINARIN, OILS	CELLULOSE, PECTIC COMPOUNDS, SILICA
PHAEOPHYTA (BROWN ALGAE)	CHLOROPHYLLS A, C, β-CAROTENE AND FUCOXANTHIN	CHRYSOLAMINARIN, OILS, MANNITOL	CELLULOSE, PECTIC COMPOUNDS, ALGINIC ACIDS (CALCIUM CARBONATE)
RHODOPHYTA (RED ALGAE)	CHLOROPHYLL A, D?, β-CAROTENE, PHYCOBILINS	FLORIDEAN STARCH	CELLULOSE, PECTIC COMPOUNDS, MUCILAGES (E.G., AGAR), (CALCIUM CARBONATE)**
CHLOROPHYTA (GREEN ALGAE)	CHLOROPHYLLS A, B, β-CAROTENE	STARCH	CELLULOSE, PECTIC COMPOUNDS (CALCIUM CARBONATE)**
BRYOPHYTA (MOSSES AND ALLIES)	CHLOROPHYLLS A, B, β-CAROTENE	STARCH	CELLULOSE, PECTIC COMPOUNDS, LIGNIN***
TRACHEOPHYTA (VASCULAR PLANTS)	CHLOROPHYLLS A, B, β-CAROTENE	STARCH	CELLULOSE, PECTIC COMPOUNDS, LIGNIN***

*When present
**Some species
***Some cells

I BIOCHEMICAL CHARACTERISTICS of algae. Algae are frequently classified according to their photosynthetic pigments, food-storage compounds and cell wall materials. Bluegreen algae (which are not really plants) and higher plants are included in table for comparison.

phylum Chlorophyta. Life cycles also show extreme variation, but all except the Rhodophyta (red algae) have forms with flagellated motile cells in at least one stage of their life cycle, and some (for example, the Pyrrophyta) are unicellular and motile throughout most of their existence.

Table I shows some of the principal biochemical characteristics of the phyla of algae. Except for mannitol (a sugar alcohol), fats, and oils, all of the storage compounds listed, in particular chrysolaminarin and various forms of starch, are polymers of glucose. These polymers differ among themselves in the kind of chemical linkage between adjacent glucose molecules, in the degree of branching of the polysaccharide chains, and in chain size. The pectic compounds found in the cell walls are all variously modified polymers of galacturonic acid. The Cyanophyta (bluegreen algae) and higher plants are included for comparison.

The DINOFLAGELLATES (phylum Pyrrophyta), a group of predominantly unicellular algae, are of

interest for two reasons. First, the dinoflagellates are probably second in importance only to the diatoms (members of the phylum Chrysophyta) as primary photosynthetic producers of organic matter in the oceans. Second, some of the close relatives of the dinoflagellates, although still true members of the Pyrrophyta, superficially resemble amoebas: they lack cellulose walls, possess contractile vacuoles, and occasionally feed on other organisms. Although the Pyrrophyta, as this evidence suggests, are perhaps closely related to the protistans (they are sometimes included with the flagellates) we include them with the plants both because they photosynthesize and because most species can (and do) make a cellulose cell wall.

The dinoflagellates themselves *(Figure 9)*, the most abundant of the Pyrrophyta, are notably peculiar cells. They are biflagellate, with one flagellum in an equatorial groove around the cell, and the other starting at the same point and passing down a longitudinal groove before it extends free into the surrounding medium. Dinoflagellates are mostly marine, and sometimes reproduce in enormous numbers in warm and somewhat stagnant waters. Certain species produce a potent nerve toxin, and when they reach high numbers

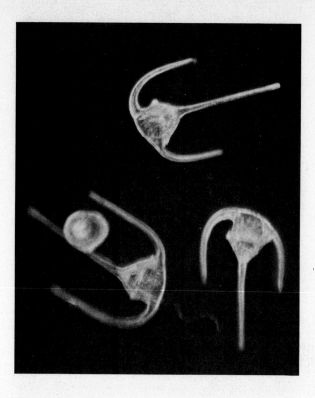

9 **CERATIUM TRIPOS is an anchor-shaped marine dinoflagellate.**

would seem to justify this choice. Many have sufficient carotenoids in their chloroplasts to give them a yellow or brownish color; all make chrysolaminarin, a special form of carbohydrate, and oils as photosynthetic storage products; members of all classes can deposit silicon in their cell walls; and the cell wall of some members of all classes is constructed in two pieces like the top and bottom of a box of stationery, with the walls of the top overlapping the walls of the bottom.

Architectural magnificence on a microscopic scale is the hallmark of the diatoms. These unicellular organisms, marine or fresh-water inhabitants, produce silicon-impregnated cell walls which show intricate patterns, and in fact the taxonomy of the diatoms is entirely based on the wall patterns *(Figure 10)*. Looking down on the top of the "box," one sees one or the other of two basic shapes: diatoms that are radially symmetrical or those that are bilaterally symmetrical. Asexual reproduction is by simple cell division, for which the siliceous cell wall presents a serious problem: the wall can increase in size but only to a limited extent. The consequences of this problem are illustrated in Figure 11. Both the top and bottom of the "box" become tops of new "boxes" and, as a result, the new cells which started as bottoms must be smaller. If the process continued indefinitely,

(called a bloom) the resulting "red tide" can kill enormous numbers of fish. A particularly severe red tide in the Gulf of Mexico in the summer of 1971 killed tons of fish along the west coast of Florida. Likewise the genus *Gonyaulax* produces a severe toxin which can become accumulated in shellfish in amounts which while not fatal to the shellfish may kill the person who eats it.

Many dinoflagellates contain a remarkable biochemical system for generating light. In complete darkness, cultures of these organisms emit a faint glow. If one suddenly disturbs the culture physically, by stirring it or bubbling air through it, the organisms each emit a number of bright flashes, perhaps a thousand-fold brighter than the dim glow of the quiet culture. The flashing then rapidly subsides. A ship passing through a tropical ocean containing a rich growth of these species produces a bow wave and a wake that glow eerily as millions of dinoflagellates discharge their light system.

DIATOMS and their relatives comprise the phylum Chrysophyta. Both unicellular and filamentous forms are known. There is disagreement as to whether or not both types should be included in the same phylum, but several characteristics

DIATOMS. Despite their splendid diversity of form, all diatoms show either radial or bilateral symmetry. Magnification about 1200X. **10**

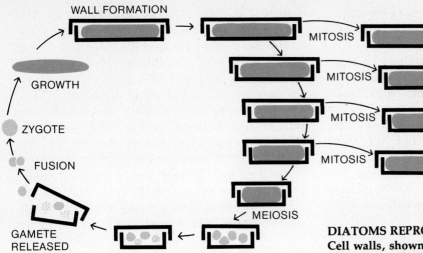

WALL FORMATION

MITOSIS

MITOSIS

MITOSIS

MITOSIS

GROWTH

ZYGOTE

FUSION

MEIOSIS

GAMETE RELEASED

DIATOMS REPRODUCE both sexually and asexually. Cell walls, shown edge-on in this schematic diagram, are two-part "boxes." In asexual reproduction (right) the parts separate and each becomes the top of a new box. In the process the offspring within the shell become progressively smaller. Zygotes produced by sexual reproduction (left) grow and lay down new full-size shells. **11**

one cell line would simply vanish. Sexual reproduction largely solves the problem. A given cell will form a pair of gametes (sometimes one is flagellated), the cell walls are shed, and fusion of gametes occurs. The zygote is sometimes called an auxospore, a name whose Greek origin (*auxein:* increase) refers to the fact that it increases substantially in size before a new wall is laid down.

Diatoms are probably the most important photosynthesizers in the open ocean. They are ubiquitous, and frequently occur in enormous numbers. The walls of dead cells are very resistant to breakdown, with the result that certain sedimentary rocks are composed almost entirely of these siliceous skeletons that sank to the sea floor. Diatomaceous earth, obtained from such rocks, has many industrial uses, from insulation and filtering to metal polishing.

Dinoflagellates and diatoms are among the tiniest algae. Conversely, the BROWN ALGAE (phylum Phaeophyta) are undoubtedly the largest. Some giant kelps, such as *Macrocystis*, may be almost 60 meters long (a smaller relative *Laminaria*, appears in Figure 2 of Chapter 14). Although they exhibit a variety of life cycles, the brown algae are always multicellular, composed either of branched filaments or leaflike growths called thalli. These organisms are almost exclusively marine. Some are found floating in the open ocean; the most famous example is *Sargassum*, which forms dense mats of vegetation in the Sargasso Sea in the mid-Atlantic. Most, however, are attached to rocks near the shore. A few brown algae seem to thrive only where they are regularly exposed to heavy surf; a notable example is the sea palm, *Postelsia*, of the Pacific coast (*Figure 12*). The attached forms all de-

velop a specialized structure called a HOLDFAST, which literally glues them to the rocks. The plant may show extensive differentiation into stemlike stalks and leaflike blades, and gas-filled cavities or bladders are also frequent. For biochemical reasons that are only poorly understood, these gas cavities

SEA PALM (Postelsia) is a brown alga that flourishes along the Pacific coast. Resembling a miniature palm tree about one meter tall, these plants can withstand pounding of heaviest surf. **12**

often contain as much as five percent carbon monoxide —a concentration high enough to kill a human.

In addition to organ differentiation, the larger brown algae also exhibit considerable tissue differentiation. Many of the giant kelps have photosynthetic filaments only in the outermost regions of the various organs. Inside the photosynthetic region lie filaments of long cells which very much resemble the phloem of higher plants. Called trumpet cells because they have flaring ends, they are sieve tubes *(Chapter 16)* that rapidly conduct the products of photosynthesis (mostly mannitol) through the body of the plant.

The brown algae exemplify the extraordinary diversity found in the algal superphylum. One example of a common simple brown alga is *Ectocarpus.* Its branched filaments a few cm in length are commonly found growing on shells and stones, and the gametophyte and sporophyte phases of the alternation of generations can only be distinguished by chromosome number (haploid or diploid) or reproductive products (zoospores or gametes). Thus the generations are isomorphic — haploid and diploid plants have the same form. In contrast, the kelps (genus *Laminaria*) and other brown algae show the more complex heteromorphic alternation of generations. The large and obvious plant is the sporophyte. Meiosis in special fertile regions of the leaflike fronds produces haploid zoospores. These germinate to form a tiny filamentous gametophyte. The gametophytes bear either eggs or sperm. The rockweed *(Fucus)* carries gametophyte reduction still further: there is no longer any multicellular haploid phase, although there is still a multinuclear haploid stage. The gametes themselves are formed directly by meiosis.

The brown algae obtain their characteristic color from the carotenoid fucoxanthin *(Figure 13)*, present in large amounts in the plastids. The combination of this yellowish-orange pigment with the green of chlorophylls *a* and *c* yields the dirty brown color. The cell walls may contain as much as 25 percent alginic acid, a gummy polymer of mannuronic and glucuronic acids (both sugar acids). The substance serves to cement cells and filaments together, and also provides good holdfast glue. It is used commercially as an emulsifier in ice cream.

The RED ALGAE (phylum Rhodophyta) include plants which grow in the shallowest tide pools, but also includes the plants found deepest in the ocean (as deep as 170 meters where nutrient conditions are right and the water is clear). Very few red algae inhabit fresh water. Most grow attached to some substrate by a holdfast. Almost all of them are multicellular. The characteristic color of the red algae is a result of the pigment phycoerythrin, found in relatively large amounts in the chloroplasts of most species in addition to phycocyanin, carotenoids, and chlorophyll. In a sense, red algae, along with several other groups of algae, are misnamed. They have the capacity to change the relative amounts of the various photosynthetic pigments depending upon the light conditions where they are growing. **Thus the leaflike *Chondrus crispus*, a common North** Atlantic species, may appear bright green when it is growing unshaded in a tide pool, reddish brown when it is growing under other vegetation, and deep red when it is growing at extreme depths. The pigmentation reflects to a remarkable degree the wavelengths of the light which reach the plants. In deep water, the photosynthetically effective light penetrating the ocean is mostly in the blue-green part of the spectrum, and the plants accumulate large amounts of phycoerythrin, the pigment which absorbs in that region *(Figure 14)*. Chlorophyll, central to photosynthesis in all plants, is not as sensitive to light quality. Variation in pigmentation dependent on light color is known as CHROMATIC ADAPTATION, and it is well illustrated in the red algae.

In addition to being the only true algae with phycoerythrin and phycocyanin among their pigments, the red algae have two other unique characteristics. They store the products of photosynthesis in the form of floridean starch, which is composed of very small branched chains of about fifteen glucose units, and they produce no motile flagellated cells at any stage in their life cycle. Curiously, sex-

PHOTOSYNTHETIC PIGMENTS of algae include fucoxanthin, a yellowish-orange pigment found in brown algae and the Chrysophyta. 1

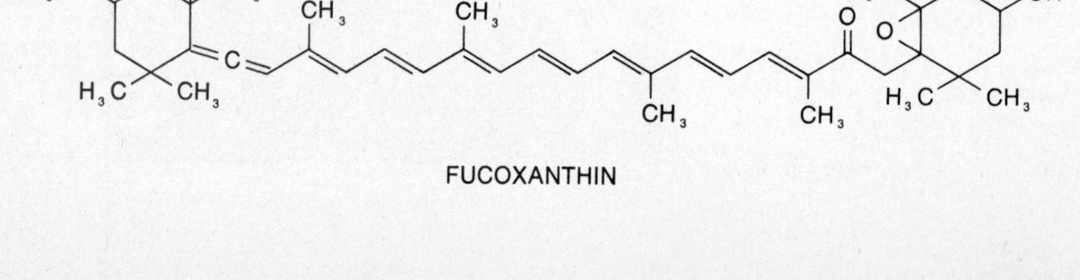

FUCOXANTHIN

ual reproduction is not by conjugation, as in those fungi that lack motile reproductive cells. The male gametes are naked and ameboid, while the female gametes are completely immobile. The branched filamentous genus *Nemalion* provides a good example. The cell destined to be a female gamete develops a long slender walled projection to which a male gamete becomes attached. The male nucleus eventually penetrates the projection, and nuclear fusion occurs to form the zygote. The new zygote in *Nemalion* then divides by meiosis followed by mitosis to produce a fairly large number of specialized, wall-less haploid spores, each of which is liberated and can form a new gametophyte. Note that the *Nemalion* plant is itself haploid, and the only stage of the life cycle which is diploid is the zygote. Other red algae, however, may have quite different patterns, either isomorphic or heteromorphic.

Like coral animals, some red algal species are very important in the formation of tropical reefs. They share with these primitive animals the biochemical machinery for depositing calcium carbonate as a precipitate in and around their cell walls. After the death of the cells the calcium carbonate persists, sometimes forming substantial rocky masses.

Some red algae also produce large amounts of mucilaginous substances, polysaccharide in nature, and based mostly on the sugar galactose with a sulfate group attached. This material readily forms solid gels and is the source of agar, so widely used in the laboratory for making a solid aqueous medium on which tissue cultures and bacteria are grown for experimental work.

The GREEN ALGAE (phylum Chlorophyta) and the photosynthetic flagellates such as *Euglena* and *Volvox* (*Figures 15 and 16 in Chapter 25*) which most botanists include with the green algae are the only groups which contain the photosynthetic pigments characteristic of the higher plants. Chlorophyll *a* predominates, and a major pigment is chlorophyll *b*, which occurs in none of the other algae. The carotenoids, predominantly β-carotene and certain xanthophylls (carotenoids with one or more hydroxyl groups) are likewise those characteristic of higher plants. The principal photosynthetic storage product is also familiar: long straight chains of glucose (*amylose*) or long branched chains (*amylopectin*) which together make up what we call starch.

Uniformity of pigmentation and photosynthetic storage product is combined in the green algae with an incredible variety in shape, construction of

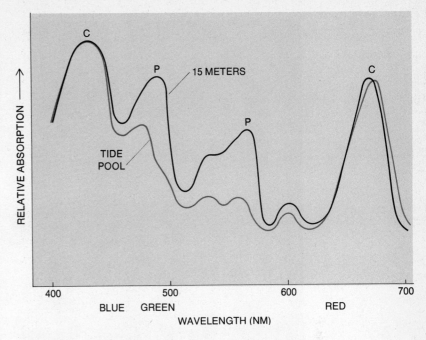

ABSORPTION SPECTRA of the red alga Chondrus. **14** Two spectra are shown, one for a plant growing at a depth of 15 meters, the other for one growing near the surface of a tide pool. Peaks labeled (P) are due to phycoerythrin; those labeled (C), to chlorophyll. Plant growing at greater depth contains substantially more phycoerythrin but about the same amount of chlorophyll as surface plant.

plant body, and life cycle. *Chlorella* is an example of the simplest type: single-celled, immobile, and apparently asexual. *Chlamydomonas* is more complex: unicellular, flagellated, and sexual. Both *Pediastrum* (*illustrated on the opening page of this chapter*) and *Hydrodictyon* (*Figure 15*) are colonial but lack motility. *Oedogonium* is filamentous and sexual, with each cell uninucleate. *Cladophora* is multicellular, but each cell is multinucleate. *Bryopsis* is tubular and coenocytic, forming cross walls only with the formation of reproductive structures. *Acetabularia* (*see Figure 6 in Chapter 10, and accompanying discussion*) is a single giant uninucleate cell with remarkable morphology, becoming multinucleate only at the reproductive stage. *Ulva* (*Figure 4 in Chapter 13*) is a membranous sheet two cells thick whose unusual appearance justifies its common name of sea lettuce. Finally, there are the remarkable unicellular DESMIDS (*Figure 16*) with their elaborately sculptured cell walls.

Sexual reproduction also runs a long gamut, from forms such as *Spirogyra*, a filamentous alga with one or more spiral chloroplasts in each cell,

573

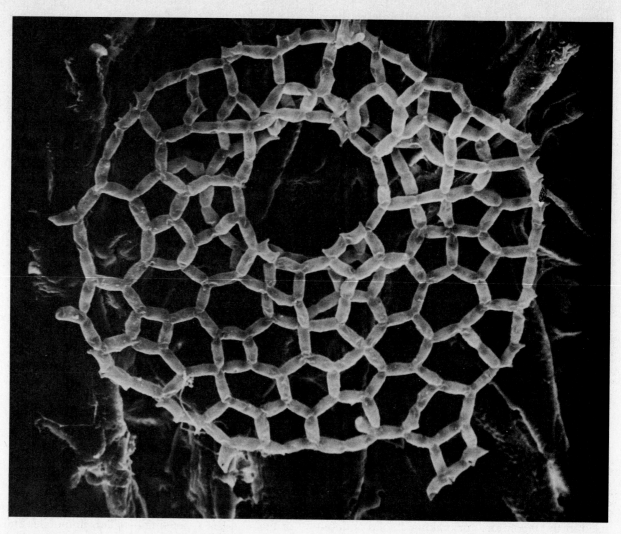

15 **HYDRODICTYON is a colonial green alga that forms an elaborate network of cells. This photograph, made with the scanning electron microscope, is reproduced at a magnification of 600X. Some colonies are large enough to be seen by the unaided eye.**

which produces no motile gametes and reproduces by conjugation *(Figure 17),* through *Oedogonium,* in which the female gamete is immobile and the male gamete free-swimming and flagellated, to *Ulva,* in which both gametes are motile by flagella. The plants may be haploid, with meiosis occurring as the first division of the zygote *(Chlamydomonas),* or diploid, with meiosis occurring at the time of gamete formation *(Bryopsis).* Still other forms show the isomorphic alternation of generations, with diploid and haploid plants morphologically indistinguishable from each other *(Ulva).*

Green algae have been extremely important to studies of photosynthesis. It was in *Chlorella* that

Calvin and his colleagues first worked out the details of CO_2 fixation. Mutants of *Chlamydomonas* have been elegantly used to work out the pathways of electron transport in illuminated chloroplasts.

The Chlorophyta have been widely used for other special research purposes as well. *Acetabularia* is uniquely suited for certain types of studies concerning the interaction of nucleus and cytoplasm *(Chapter 10).* Coenocytic forms such as *Valonia,* a sphere which may grow to be over two centimeters in diameter, provide elegant experimental material for studying movement of ions across membranes, since it is so easy to get electrodes inside. *Nitella,* divided into distinct nodes and internodes, has also been useful in both membrane studies and in studies of the deposition and growth of cell walls. The internodes, which may be over a millimeter in diameter and four centimeters in length, have no cross walls. An electric stimulus applied at one end of an internodal cell propagates rapidly to the other end in a way that has fascinated neurophysiologists for many years; this plant cell is the nearest botanical counterpart to the neuron.

THE HIGHER PLANTS

Virtually all green multicellular land plants belong to two major phyla: *Bryophyta* (mosses, liverworts, and hornworts) and *Tracheophyta* (vascular plants). Tracheophytes represent one of the triumphs of evolution. They include virtually all the familiar plants, trees, and shrubs that now dominate the landscape. Probably descended from ancestral green algae, the higher plants first invaded the land in the Silurian period of the Paleozoic era, about 415 million years ago (the 28th day of our geologic calendar).

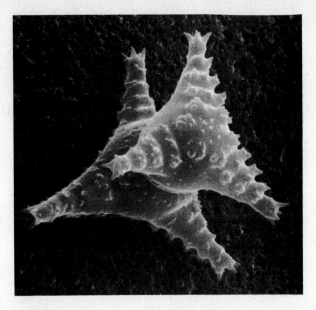

Most of the characteristics that distinguish higher plants from algae are adaptations to life on land.

DESMIDS have a highly sculptured cellulose wall. **16** This scanning electron micrograph depicts a single desmid. The nucleus normally lies in the isthmus between the two three-cornered halves. After mitosis a nucleus migrates into each half, the halves separate, and each half regenerates its missing counterpart. Magnification, about 500X.

Most of the characteristics that distinguish the higher plants from the algae are evolutionary adaptations to life on land. A woody skeleton may provide support against gravity in the absence of the buoyancy of water. No longer bathed, higher plants evolved specialized absorbing tissues, such as roots that project into the soil, specialized conducting tissues to carry minerals and nutrients throughout the plant, and protective coverings to prevent drying out. The reproductive patterns of higher plants also show modifications for life on land, culminating in the evolution of the seed.

Two ancestral groups of higher plants probably evolved separately from the ancient green algae. One group culminated in the bryophytes, the other in tracheophytes. These two phyla are only indirectly related to each other, either through a single common ancestor or through two separate ancestors among the fresh-water green algae.

PHYLUM BRYOPHYTA

If the reader were to skim botany texts published during the last 100 years or so, he would discover that the classification of plant phyla has been notoriously unstable. Even a brief comparison of

CONJUGATING FILAMENTS OF SPIROGYRA. After **17** conjugation tubes (at left and right) fuse, the walls between them disintegrate and a nucleus moves from one cell to the other, where it fuses with the host nucleus to form a zygote. Upon germination, zygote nucleus undergoes meiosis and the daughter cells eventually form new filaments. Dark spots are starch grains in the spiral chloroplasts (barely visible) for which organism was named.

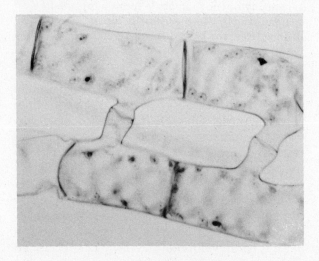

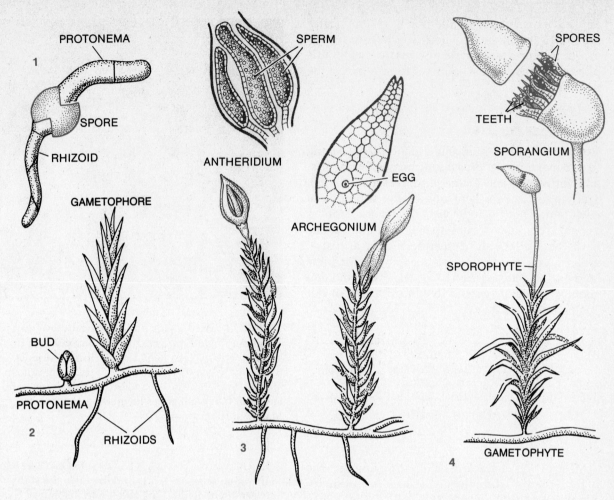

18 **LIFE CYCLE OF MOSSES begins with the sprouting of a spore (1). Shoots of gametophyte plant (2) develop sex organs at their tips, either male antheridia or female archegonia (3). In some species, both sex organs occur on the same plant; in other species, on separate plants. After fertilization the egg develops into a sporophyte plant, which is always attached to the tip of the gametophyte. The sporophyte forms a capsule-like sporangium (4), which eventually bursts. The cap disintegrates, and curved "teeth" around the rim fling the spores outwards.**

modern texts reveals substantial disagreement among the specialists as to where to draw the lines between phyla. Only the bryophytes have survived this lumping and splitting (mostly splitting) relatively unscathed. A glance at the life cycle in the phylum reveals the reason *(Figure 18)*. Alternation of generations is heteromorphic, as in many algae, and both gametophyte and sporophyte may be many centimeters in length. However, no matter how large and complex the sporophyte is, it begins its development within a specialized

gametophytic organ called the ARCHEGONIUM, and never becomes independent. Although the bryophyte sporophyte may be photosynthetically self-sufficient, its only contact with the substrate is indirect. Any water and minerals it needs it must obtain from underlying gametophytic tissue.

The bryophytes have achieved an important evolutionary advance in their sexual cycle. The female gamete is never motile, but develops within the multicellular gametophytic archegonium. The **male gametes are also borne within multicellular** gametophytic organs called ANTHERIDIA (singular, antheridium). They are motile, and must swim to the archegonium and down its neck to effect fertilization. Unlike the algal zygote, which is wholly on its own following fertilization, the bryophyte zygote divides and goes through a distinctive embryo phase while being protected and almost entirely nourished by the surrounding gametophytic tissue. These added forms of protection for the gametes and zygote undoubtedly helped to enable

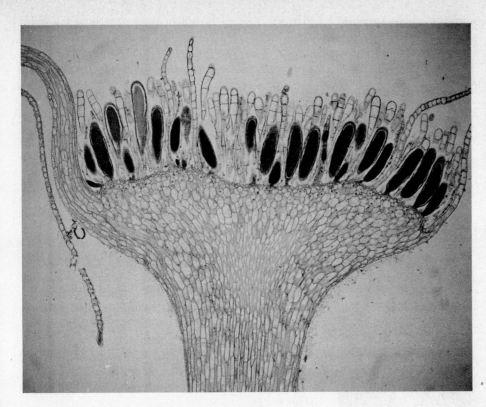

ANTHERIDIA AND ARCHEGONIUM of the moss 19
Mnium (shown at left and right, respectively) are borne
on the tips of separate gametophores. Dark masses
within antheridia are clusters of mature sperm. Large
egg is visible within the archegonium, at end of canal
leading to tip of the organ. Both antheridia and
archegonium are surrounded by sterile filaments of
cells. Magnification: antheridia, 25X; archegonium,
110X.

the bryophytes to be among the first plants to conquer the land.

The most familiar bryophytes are, of course, the mosses (class Musci). It is worth looking closely at the reproductive cycle of a typical moss such as *Mnium* or *Polytrichum*. Spore germination (which frequently requires light and is phytochrome mediated) produces a branched filamentous plantlet or PROTONEMA (plural protonemata; *see Figure 18*). However, certain filaments with odd diagonal cross walls are nonphotosynthetic, and serve to anchor the plant to the substrate. These filaments, called RHIZOIDS, are the bryophyte counterpart of the root hairs of higher plants. After a period of filamentous growth, cells close to the tips of the photosynthetic branches begin rapid cell division in three dimensions to form buds. These buds eventually differentiate a distinct apex and produce the familiar leafy moss plant with the leaves spirally arranged. These leafy GAMETO-PHORES, as they are called, produce both antheridia and archegonia at their tips *(Figure 19)* and fertilization ensues. (Sometimes a gametophyte may produce both antheridia and archegonia, sometimes only one or the other.) In most mosses, sporophyte development then follows a remarkably precise pattern, resulting ultimately in the for-

mation of an absorptive foot, a stalk, and, at the tip, a swollen sporangium or capsule.

During this rigidly programmed development, the archegonial tissue itself is also growing rapidly, and for a time it keeps pace with the rapidly expanding sporophyte. Finally, however, the archegonium loses out, and it is split apart around its middle. The top portion frequently persists on top of the rapidly elevating capsule as a little pointed cap, the CALYPTRA. The top of the capsule is ultimately shed, after meiosis has led to the creation of numerous mature spores within. Groups of cells just below the lid become partially lignified and dried, forming a series of teeth surrounding the opening *(Figure 20)*. Highly responsive to humidity, the teeth arch into the mass of spores and then out again as the atmosphere is first dry and then moist. The spores are thus dispersed when the

20 SPORE CAPSULE of moss plant (Funaria hygrometrica) is ringed with pointed teeth that fling the spores outward when the humidity of the surrounding air rises. Magnification, about 150X.

21 LIVERWORTS. Plants of the aquatic liverwort Ricciocarpus float on the surface. Antheridia and archegonia are buried in central grooves. The liverworts are about two or three cm long. Tiny oval plantlets between the liverworts are Lemna, one of the tiniest flowering plants.

surrounding air is moist, under the most favorable conditions for subsequent germination.

Only a few mosses lack this pattern of sporophyte development, and a familiar exception is *Sphagnum*. In terms of biomass, *Sphagnum* probably outweighs all other mosses put together, being found in tremendous quantities in northern bogs and tundra, and extending high into the arctic. The *Sphagnum* sporophyte has a very simple capsule with an air chamber in it. In some unresolved way, air pressure is built up in this air chamber, so that eventually the capsule literally blows its lid, dispersing the spores with an audible pop.

The LIVERWORTS (class Hepatici) are readily distinguished from the mosses on two grounds: they never produce a filamentous protonema, and their sporophyte capsules are very simple—a globular capsule wall surrounding a mass of spores. Also, the gametophyte architecture is much more variable. The simplest gametophytes are nothing but flat plates of cells perhaps a centimeter or so in length, producing antheridia and/or archegonia on their upper surfaces and rhizoids on their lower *(Figure 21)*.

Spore dissemination in liverworts is likewise variable, and sometimes downright peculiar. In some species spores simply are not released until the surrounding capsule wall rots. In others, however, the spores are disseminated by bizarre structures called elaters, which are located within the capsule. The elaters are long cells with a helical thickening of secondary wall. As they lose water, the whole cell body shrinks longitudinally to a fraction of its former length, thus compressing the helical thickening just like a spring. When the stress becomes sufficient, a vapor bubble forms and the compressed "spring" snaps back to its resting position, tossing spores in all directions.

The HORNWORTS (class Anthocerotae) appear at first glance to be liverworts with very simple gametophytes. These latter are simply flat plates of cells, frequently only a few cells thick. However, they have two characteristics found nowhere in any other bryophytes. First, the archegonia, instead of being borne on stalks, are embedded in the gametophytic tissue. Second, of all the bryophytic sporophytes, those of the Anthocerotae come closest to being capable of indefinite growth. Consider either the liverwort or moss capsule—its growth is apical, and formation of a capsule terminates any further potential for elongation. With a plant such as the hornwort *Anthoceros*, however, a region of the stalk below the capsule remains capable of indefinite meristematic activity, continuously producing new spore-bearing tissue above *(Figure 22)*. Sporophytes of hornworts growing in mild and continuously moist conditions have been reported to reach as much as 20 centimeters in length.

Vascular plants are an extraordinarily large and diverse group, but they were launched by a single evolutionary event: the invention of the tracheid.

The bryophytes are an ancient group that probably arose in or before the Devonian period. Exactly what the ancestral algae that gave rise to them might have looked like is anyone's guess, since the only hard evidence comes from photosynthetic pigmentation.

VASCULAR PLANTS

The VASCULAR PLANTS (phylum Tracheophyta) are an extraordinarily large and diverse group, yet they can be said to have been launched by a single evolutionary event. Sometime during the Paleozoic era, probably well before the Silurian period, the sporophyte generation of some long extinct plant produced a wholly new cell type: the TRACHEID. This event was probably almost as important for the eventual evolution of terrestrial life as was the earlier appearance of aerobic photosynthesis for life in general. As noted in Chapter 16, the tracheid is the principal element of the xylem of all vascular plants except the angiosperms, where it nevertheless persists along with the more specialized and efficient system of vessels and fibers. The evolutionary appearance of tracheary tissue (composed of tracheids) had two important consequences: first, it provided a pathway for long-distance transport of water and mineral nutrients from a source of supply to regions of need; second, it provided something almost completely lacking —and unnecessary— in the largely aquatic algae —rigid structural support. The tracheid set the stage for the complete and permanent plant invasion of the land masses of the world.

We have decided to include all vascular plants in the phylum Tracheophyta on the grounds that

members of every single group possess one peculiar type of cell in the xylem, the SCALARIFORM-PITTED TRACHEID, a tracheid with ladder-like rows of pits on the sides *(see Figure 1 in Chapter 16)*.

The life cycle of the vascular plants which lack seeds is remarkably uniform. Alternation of generations is uniformly heteromorphic, with both haploid and diploid phases being free-living and independent at maturity. The fern life cycle, illustrated later in this chapter, can serve with minor modifications for all of these seedless organisms. Unlike the bryophytes, in which the sporophyte is attached to and completely dependent upon the gametophyte, the tracheophyte sporophyte is the large and obvious plant that one normally sees in nature. The gametophytes are rarely more than a centimeter or two in length and seldom live longer than a few weeks at most. By contrast, the sporophyte of a tree fern may be 15 or 20 meters high and live for years. As in the fungi, algae, and for that matter the bryophytes, the most prominent resting stage in the life cycle is the single-celled spore. The shoot apices of the sporophytes may show winter or (in arid regions) summer dormancy, but upon return to favorable conditions, they merely resume the general growth pattern which they had arrested. The seedless vascular plants still must have an aqueous environment at one stage of their life cycle because fertilization is accomplished by a motile flagellated sperm.

PSILOPSIDS, the first vascular plants, belong to a now-extinct subphylum (Psilopsida). In the Silurian and Devonian periods of the Paleozoic era, the psilopsids appear to have been the only vascular plants, and they dominated the landscape. Some of the psilopsid plants of that time anticipated structural features of all of the other subphyla of the Tracheophyta, and they strengthen the case for a common origin of all vascular plants from some common green algal precursor.

In 1917, R. Kidston and W. H. Lang first reported some remarkably well preserved fossils of vascular plants embedded in Devonian rocks not far from Rhynie, Scotland. Considering the age of the rocks (over 395 million years), the preservation was remarkable. These plants all had a simple vascular system consisting of well-preserved tracheids

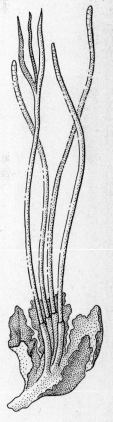

ANTHOCEROS is a hornwort whose elongated sporophytes grow upward from a flat gametophyte. The inconspicuous gametophyte tends to die back as the sporophytes mature. Plant is drawn slightly larger than life size.

22

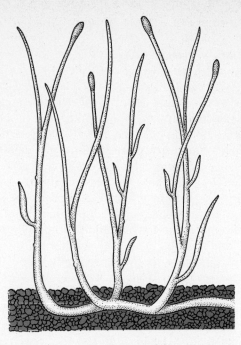

23 RHYNIA. Reconstruction of this extinct psilopsid is based on fossil found near Rhynie, Scotland. Its aerial shoots were less than half a meter tall.

and poorly-preserved phloem, surrounded by a cortex. Many had no central pith. Furthermore, although flattened scales were found on the axes of some of them, these appendages entirely lacked vascular tissue, and thus were not comparable with the true leaves of any other vascular plants. Roots were absent, and the plants were apparently attached to the soil by horizontal portions of stem (rhizomes) that bore rhizoids. These horizontal stems also bore what were evidently aerial branches, and sporangia —homologous with the bryophyte capsule —were found at their tips. Branching was dichotomous, with the shoot apex seemingly dividing to produce two equivalent new stems, each pair diverging at approximately the same angle from the original stem *(Figure 23)*. Scattered fragments of such plants had been found earlier but never in such profusion or so well preserved.

The presence of tracheids clearly indicated that these were vascular plants. But were they sporophyte or gametophyte? Close inspection of thin sections of sporangia revealed tetrads of spores. In virtually all of the living seedless vascular plants (with no evidence to the contrary from fossil forms) the four products of a meiotic division remain attached to one another during their development into spores. Only when the spores are mature do they ultimately separate, and even after separation, their walls reveal the exact geometry of the way in which they were attached. Thus a tetrad of four closely packed spores is inevitably found only immediately after meiosis, and the plant that produced the tetrads must have been the diploid sporophyte. No identifiable gametophytes have ever been observed.

There is still some debate about whether psilopsids are entirely extinct. In numerous textbooks there are lengthy discussions of two rootless and spore-bearing plants, *Psilotum* and *Tmesipteris*, as the only living relics of this early Paleozoic phylum *(Figure 24)*. But David W. Bierhorst has presented powerful arguments based on anatomy, morphology, and development, that these two genera, which together contain only a handful of species, are very probably either primitive ferns or reduced

PSILOPSID OR FERN? Psilotum nudum may be either **2** a living survivor of a subphylum that became extinct in the Devonian period, or it may be a primitive or a reduced type of fern. Sporangia occur in clusters of three. Plant has no roots, and tiny thornlike scales instead of leaves.

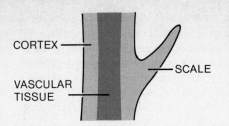

CORTEX

VASCULAR TISSUE

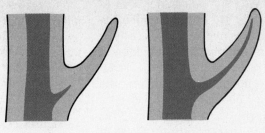

SCALE

ferns. There are indeed no roots, but odd ferns such as *Stromatopteris*, from New Caledonia, may form roots only quite late in development. Although *Psilotum* has no true leaves, only minute scales, *Tmesipteris* has respectable flattened photosynthetic organs with well developed vascular tissue. Finally, there is an enormous hole in the geological record to explain: no psilopsid fossils appear anywhere after the Devonian period.

FERNS AND FERN ALLIES

The true ferns (subphylum Pteropsida, class Filicinae) comprise one of the major living groups of vascular plants. One key to understanding the evolution of ferns and the other higher plants is close attention to the origin and modification of the true LEAF. Up to this point the word leaf has been used rather loosely. We spoke of moss leaves and leafy liverworts; we also commented on the absence of true leaves in psilopsids. In the strictest sense, a leaf is a flattened photosynthetic structure emerging laterally from a main axis or stem and possessing true vascular tissue. Although various authors have tried to invent other names for the bryophyte structure, use of the word leaf has unfortunately persisted. In our remaining discussion, however, we shall use it in the restricted sense, and hence dismiss the Bryophyta and Psilopsida as leafless.

This tight definition allows a closer look at true leaves in the Tracheophyta, and it turns out that there are probably two different types, of different origin. The first type of leaf is usually small and only rarely has more than a single vascular strand, at least in plants alive today. It occurs in two subphyla related to ferns: the Lycopsida (club mosses) and Sphenopsida (horsetails or scouring rushes), of which only a few genera still survive. Its evolutionary origin is thought to be the progressive development of vascular tissue within small scalelike outgrowths of the stem, and it is therefore called a MICROPHYLL (*Figure 25*). The principal characteristic of such a leaf is that its vascular

MICROPHYLL might have evolved from scales such as those on the stems of psilopsids. Diagram shows progression from a scale without vascular tissue (left) to one with some vascular tissue (center), to true microphyll (right). All three types appear in fossil plants. 25

strand departs from the vascular system of the stem in such a way that there is scarcely any perturbation in the conformation of the stem vascular cylinder. Even in the fossil lycopsids and sphenopsids of the Carboniferous, many of which had microphylls several centimeters in length, the vascular arrangement is the same.

The other type of leaf, called the MEGAPHYLL, is first encountered in the subphylum Pteripsoda (*Figure 26*). It is thought to have arisen from the flattening of a dichotomously branching stem system, with the development of extensive photosynthetic tissue between the branch members. Another feature of the megaphyll leaf type is that its vascular system creates a major alteration in the architecture of the stem vascular system where it departs for the leaf base. This alteration is called a leaf gap (*Figure 27*). The vascular tissue of the stem is a hollow cylinder of xylem with phloem and then cortex outside and pith inside, both com-

PROPOSED ORIGIN OF MEGAPHYLL. Branching stem system (left) became progressively reduced (left center) and flattened (right center). Flat plates of photosynthetic tissue evolved between small end branches (right). The end branches evolved into the veins of leaves. 26

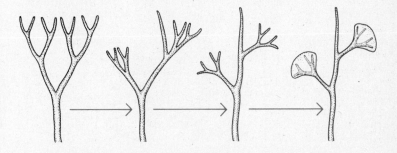

posed of parenchyma cells. All of the vascular tissue of a given region arches out to the leaf petiole, and there is a clear parenchymatous connection in the stem between the pith and the cortex. Shortly apical from the leaf the gap is closed again, and the next gap up appears just above the next leaf.

The TRUE FERNS comprise the class Filicinae, as mentioned above. In the chapter on reproduction (*Chapter 13*), the fern life cycle was presented briefly, and has already been mentioned in connection with characterization of the seedless vascular plants in the present chapter. It is illustrated in detail in Figure 28. It is clearly heteromorphic, with entirely different and completely independant gametophytes and sporophytes. The specialized **structures in which the gametes develop are further** illustrated in Figures 29 and 30.

Only two additional aspects require comment here. First, there are a few genera of ferns which produce a tuberous fleshy gametophyte instead of the characteristic flattened photosynthetic structure described previously. They depend upon a symbiotic fungus for nutrition, and in some cases (*Psilotum*) even the embryo sporophyte must be-

27 **LEAF GAPS are discontinuities in cylinder of vascular tissue (color) in stem. Schematic diagram depicts stem of typical vascular plant. Gaps appear where leaves diverge from stem.**

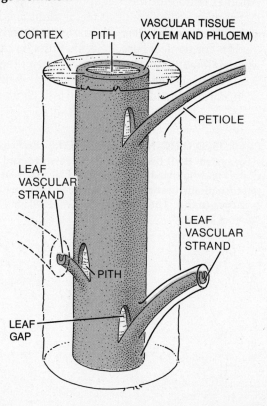

LIFE CYCLE OF FERN involves a large sporophyte plant and an inconspicuous gametophyte. Swimming sperm shed by antheridia of one gametophyte fertilize eggs within archegonia of another. Zygote develops into embryonic sporophyte within the archegonium and eventually sprouts leaves and roots. Structure of antheridia and archegonia is illustrated in more detail in Figures 29 and 30.

come associated with the fungus before any extensive development will proceed. Second, there are two orders of aquatic ferns, the Marsileales and the Salviniales, which have evolved heterospory: separation of sexes in the sporophyte. Male and female spores (which will germinate to produce male and female gametophytes, respectively) are produced in different sporangia, and the male spores are always much smaller and greater in number. All other ferns are homosporous.

As briefly mentioned in the chapter on reproduction (*Chapter 13*), ferns have evolved an elegant system for getting the sperm from one gametophyte to the egg of a different one. As a fern spore germinates, it first forms a small filamentous gametophyte. Soon lateral cell divisions lead to the formation of a heart-shaped structure whose lower surface is covered with cells called rhizoids, which, like root hairs, serve both to anchor the plant and to absorb nutrients. Eventually archegonia appear near the notch of the heart-shaped gametophyte (*Figures 28 and 29*). At about the same time, the young gametophyte begins to produce and release a potent pheromone called antheridogen. The pheromone diffuses into the environment, inducing nearby spores to germinate and to form antheridia (*Figure 30*). The induced antheridia mature not long after the archegonia of the original gametophyte, and shed antherizoids (swimming sperm) that fertilize the eggs in the archegonium. Only spores that germinate far from other gametophytes will go on later to form archegonia. Moreover, as an isolated gametophyte begins to form archegonia, it loses its sensitivity to antheridogen.

The story does not end there. Germinating spores themselves produce small amounts of antheridogen, inducing neighboring spores to germinate. Thus a whole gametophyte colony on a damp log may be a highly coordinated population with respect to successive production of the male and female gametes on one individual, and their simultaneous production on adjacent individuals, insuring a measure of outbreeding in these haploid hermaphroditic plants. The evidence to date

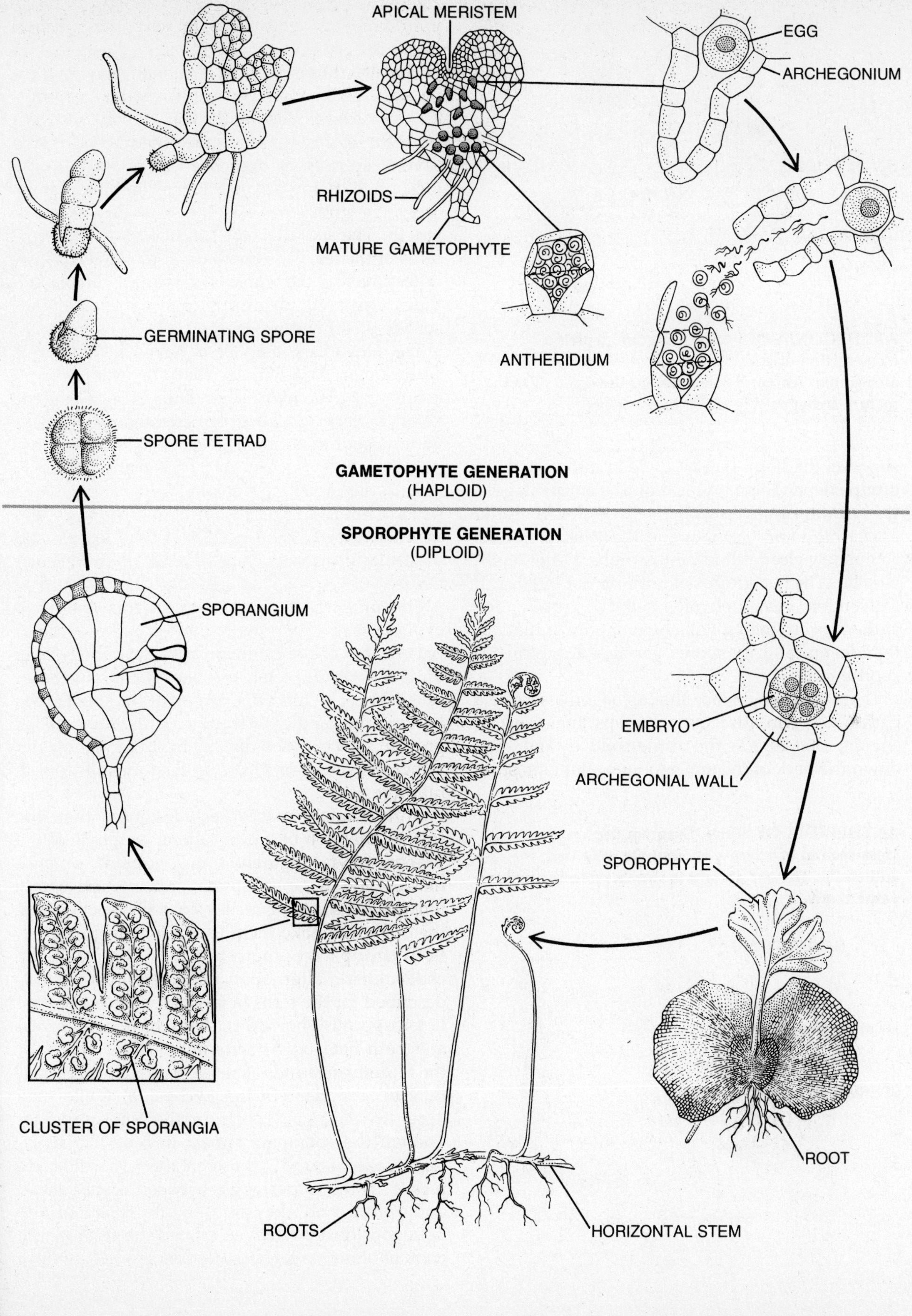

APICAL MERISTEM

EGG

ARCHEGONIUM

RHIZOIDS

MATURE GAMETOPHYTE

ANTHERIDIUM

GERMINATING SPORE

SPORE TETRAD

GAMETOPHYTE GENERATION
(HAPLOID)

SPOROPHYTE GENERATION
(DIPLOID)

SPORANGIUM

EMBRYO

ARCHEGONIAL WALL

SPOROPHYTE

CLUSTER OF SPORANGIA

ROOT

ROOTS

HORIZONTAL STEM

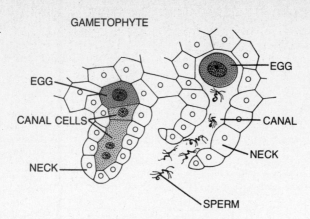

GAMETOPHYTE

EGG

CANAL CELLS

NECK

EGG

CANAL

NECK

SPERM

29 **ARCHEGONIA OF FERN. Diagram depicts a
cross-section of a fern gametophyte with two
archegonia. Antherizoids swim up the canal of the
mature archegonium (right) to reach egg.**

suggests that ferns can be divided into two major
groups depending on which of two antheridogens
they produce. One of the two is a gibberellin
(*Chapter 18*) and the nature of the other is not yet
completely clear although it is probably not a gibberellin. The two groups of ferns do not interact. If
a given fern gametophyte is found to produce one
antheridogen, then it will respond only to that antheridogen, and its spores germinate only in response to it.

There is also evidence that archegonia secrete a
chemical sperm attractant, perhaps analogous to
sirenin, that directs the antherizoid to and then
down the neck of the archegonium. But almost no

30 **ANTHERIDIA OF FERN. Diagram depicts a
cross-section of a fern gametophyte with two
antheridia. The one at right is shown releasing mature
antherizoids.**

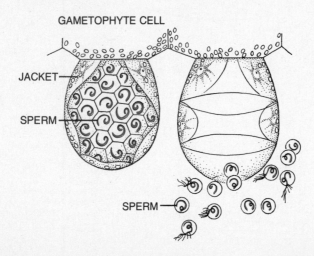

GAMETOPHYTE CELL

JACKET

SPERM

SPERM

information is available as to what the chemical
might be.

The usual familiar fern is a relatively small inhabitant of shaded moist woodland and swamps.
The fern frond is in fact the leaf; moving outward
from the base, one must trace the main axis and
several subsets of branches before finding any
mesophyll. Some leaves become climbing organs,
and may grow to be as much as 30 meters in
length. The sporangia are found on the undersur-
faces of the leaves, sometimes at the edges, sometimes covering the whole undersurface, and sometimes clustered in groups called SORI (singular,
sorus), illustrated in Figure 31.

The Devonian fossil beds have yielded forms
with some characteristics that are psilopsid and
some that resemble other subphyla. Thus the
genus *Protopteridium* had flattened branch systems
with mesophyll, but on the same axes bore terminal sporangia (*Figure 32*). This plant evidently
lacked true roots. During late Paleozoic times, the
ferns underwent some wild evolutionary experimentation in the structure of their leaves, and
particularly in the arrangement of their vascular
tissue.

The modern ferns still reflect this extensive
evolutionary experimentation with the vascular
pattern, but a true cambium is still extremely rare.
Thus even a 20-meter tree fern is constructed of
cells which are direct derivatives of the shoot apex.
Lacking the rigidity of woody plants, they do not
grow in sites exposed directly to strong winds, but
rather in ravines or in the understories of tropical
rain forests.

Unfortunately space precludes more than the
briefest mention of the remaining subphyla of the
Tracheophyta, namely the Lycopsida or club
mosses, and the Sphenopsida, or horse-tails (sometimes called scouring rushes since silica deposits in
the cell walls make them useful for this purpose).
Both groups are spore-bearing, and have the
characteristic alternation of generations already
described for the ferns (a few of the Lycopsida are
heterosporous, however). Both have true roots,
and both bear only microphylls. Here, however,
the resemblance ends. The leaves are arranged spirally on the stem in the Lycopsida, while they
occur in distinct whorls in the Sphenopsida.
Though the sporangia appear in cone-like structures called STROBILI, those in the Lycopsida are
tucked in the upper angle between a specialized
microphyll and the stem, while those of the
Sphenopsida are recurved toward the stem on the
ends of short stalks (SPORANGIOPHORES). Growth in

the Lycopsida is entirely from apical meristems, while that in the Sphenopsida comes to a large extent from discs of meristematic tissue just above each whorl of leaves; each disc is called an INTER-CALARY MERISTEM.

Though only trivial elements of the vegetation of today, these two groups appear to have provided the dominant vegetation during the Carboniferous. One kind of coal, called cannel coal, is formed almost entirely from fossilized spores of a tree lycopsid named *Lepidodendron*, an indication of the importance of this genus in the forest of that time.

SEED PLANTS

With the GYMNOSPERMS (class Gymnospermae)—the pines, firs, cedars, and their relatives—one finally comes to plants that have developed a true seed habit. In true seed plants such as gymnosperms, the production of male gametophytes as pollen (in fact the whole mechanism of pollination) frees the organisms once and for all from a need for liquid water for fertilization. The female sporangium is enclosed in a special layer of sporophytic tissue called the INTEGUMENT, destined eventually to develop into the seed coat. The integument, together with its enclosed female sporangium and the tissue that attaches it to the parent sporophyte, forms the OVULE. The small opening in the integument at the apex of the ovule, through which pollination eventually occurs, is called the MICROPYLE. The female gymnosperm gametophyte usually produces eggs enclosed in archegonia within the integument *(Figure 33)*.

Although there are probably fewer than 750 species of living gymnosperms, these plants take a back seat only to the angiosperms in their dominance of the land masses. The great Douglas fir and cedar forests of the Pacific Northwest, and the massive forests of pine, fir, and spruce that clothe the northern continental regions and upper slopes of mountain ranges, rank among the great vegetation formations of the world. All of these trees belong to one particular order, the Coniferales—the CONIFERS or cone-bearers *(Figure 34)*. The sporophylls of the conifers are borne in apical cones or strobili. All are heterosporous, and the sexes are separated at the sporophyte level. Male and female sporophylls—specialized leaves bearing sporangia—are found in separate male and female cones.

The life cycle of the pine shown in Figure 33 illustrates several important features. First, it illustrates the transformation of a female sporangium

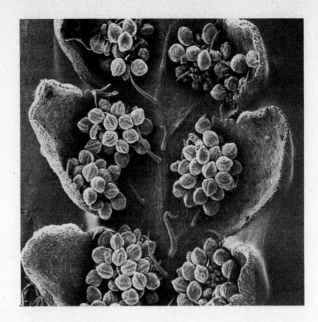

SORI are clusters of sporangia on the undersides of leaves of ferns. Photograph depicts leaves of the fern *Hypolepis tennifolia*. **31**

or MEGASPORANGIUM into an ovule by the addition of an integument, and the subsequent development of the fertilized structure into a seed. It also illustrates the separate development of the extremely small male gametophyte. Beginning as a MICRO-SPORE in a MICROSPORANGIUM, it germinates to produce a tube which digests its way through parent

FOSSIL FERN (Protopteridium) had leaflets and sporangia on the same branches. This genus dates from Devonian period, and lacks true roots. **32**

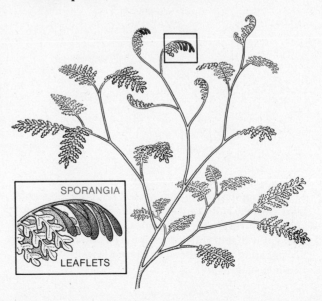

SPORANGIA

LEAFLETS

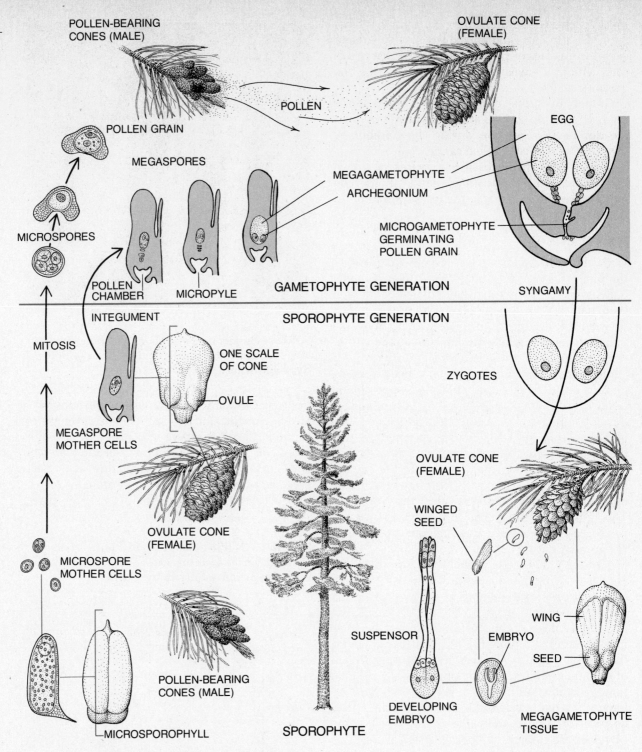

POLLEN-BEARING
CONES (MALE)

OVULATE CONE
(FEMALE)

POLLEN

POLLEN GRAIN

EGG

MEGASPORES

MEGAGAMETOPHYTE

ARCHEGONIUM

MICROSPORES

MICROGAMETOPHYTE
GERMINATING
POLLEN GRAIN

POLLEN
CHAMBER

MICROPYLE

GAMETOPHYTE GENERATION

SYNGAMY

INTEGUMENT

SPOROPHYTE GENERATION

MITOSIS

ONE SCALE
OF CONE

OVULE

ZYGOTES

MEGASPORE
MOTHER CELLS

OVULATE CONE
(FEMALE)

OVULATE CONE
(FEMALE)

WINGED
SEED

MICROSPORE
MOTHER CELLS

WING

SUSPENSOR

EMBRYO

SEED

POLLEN-BEARING
CONES (MALE)

DEVELOPING
EMBRYO

MEGAGAMETOPHYTE
TISSUE

MICROSPOROPHYLL

SPOROPHYTE

33 **LIFE CYCLE OF A PINE is typical of gymnosperms.
Mature sporophyte (bottom center) has both male and
female cones, and gametophytes develop within them.
Ovule matures after fertilization and becomes the seed.
Suspensor disintegrates during development, finally
disappearing by the time the embryo within the seed
becomes mature.**

sporophyte tissue, eventually releasing a gamete
nucleus near the egg. Primitive gymnosperms
still produce a swimming sperm at the very
end, though the sperm must swim only a few
micrometers—a relic of an earlier stage of evolution.

The word gymnosperm means, literally, naked-

seeded. In the conifers, the ovules (which develop into seeds upon fertilization) are borne exposed on the upper surfaces of the sporophylls. Any protection from the environment derives merely from the fact that the sporophylls are tightly pressed against each other within the cone. Indeed, some pines have such tightly closed female cones that normally only fire suffices to split them open and to release the seeds. One important example is the lodgepole pine, which covers vast fire-ravaged areas in the Rocky Mountains and elsewhere.

All living gymnosperms have an active cambium, and all but a few produce tracheids as the sole type of water-conducting cell found in the xylem. Despite this apparently relatively inefficient design for a water transport system, gymnosperms are among the tallest trees known. The coastal redwoods of California are the record holders—the largest are well over 100 meters high. The xylem produced by the gymnosperm cambium is the principal resource of the lumber industry. A single redwood tree is reputed to contain enough lumber for an entire small housing development, a fact that has caused heavy cutting and brought this majestic species to extinction over large parts of its former range.

It is once again to the Devonian rocks that we must turn for the first fossil evidence of the group in hand. The early gymnosperm story is an especially interesting one, illustrating the difficulties under which students of fossil plants must frequently labor. Many years ago a relatively rare and poorly preserved plant fossil, *Archaeopteris*, was described (*Figure 35*). From its thin carbonized films, it appeared to many workers to be a primitive heterosporous fernlike plant similar to *Protopteridium*. However, another fairly common Devonian fossil, *Callixylon*, remained a puzzle. It consisted only of well-petrified logs, some over a meter in diameter and more than 20 meters long. Among the delicate and herbaceous psilopsids, it seemed entirely out of place. Its wood was composed of peculiar characteristically pitted tracheids, and it obviously had possessed a highly successful functional vascular cambium. In 1960, Charles Beck of the University of Michigan resolved the puzzle. He found fronds of *Archaeopteris* clearly attached to and part of *Callixylon* logs. Since the name *Archaeopteris* had been published first, the name *Callixylon* was dropped, and the "rare" *Archaeopteris* was suddenly recognized to be a common Devonian plant. It had both psilopsid and fernlike characteristics; but its woody tissue, based on tracheids, was clearly that of a gymnosperm.

By Carboniferous time, several new lines of gymnosperms had evolved, including one group

FOSSIL GYMNOSPERM Archaeopteris was once believed to be two separate organisms. Petrified logs of this tree were named Callixylon, and carbonized films of fronds, Archaeopteris. Discovery of one specimen with fronds attached to log revealed it to be a common plant of Devonian period.

35

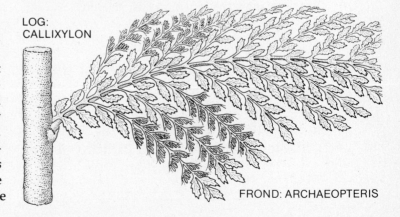

LOG: CALLIXYLON

FROND: ARCHAEOPTERIS

that possessed fernlike foliage but with characteristic gymnosperm seeds attached to the leaf margins. The first true conifers also appeared about the same time. They were either not dominant trees, or else they did not grow where conditions were right for fossilization. In fact gymnosperms are apparently either at their peak right now, or else they are still an emerging group—despite the extremely long history already behind them.

FLOWERING PLANTS

At the summit of plant evolution stand the FLOWERING PLANTS (class Angiospermae). In earlier chapters, when "plants" were mentioned in discussing processes such as water balance, long distance transport in the xylem and phloem, leaf structure, or hormonal regulation of development, we usually meant this very particular group of plants—the angiosperms. Although a great deal of angiosperm (plant) biology has already been covered, one aspect was deliberately postponed for this chapter: the details of sexual reproduction. The reason for waiting was that sexual reproduction in angiosperms presents unique characteristics found nowhere else in the plant kingdom—characteristics which in fact provide coherence for the entire class.

The reproductive organ of the angiosperm is the flower *(Figure 36)*. It is a strobilus, like that of many other groups of vascular plants, consisting of a compact axis with modified sporangium-bearing leaves. The leaves bearing the microsporangia are called STAMENS, and those bearing the megasporangia are called CARPELS. In addition, there frequently exist a number of specialized sterile leaves found below the sporophylls, the upper being called the PETALS (collectively the COROLLA), and the lower the SEPALS (collectively the CALYX). An ideal flower (for which there is undoubtedly no exact counterpart in nature) is shown in Figure 37. From base to apex, the sepals, petals, stamens, and car-

> Gymnosperms are either at their evolutionary peak right now, or else they are still an emerging group, despite an extremely long history stretching back to the middle of the Paleozoic era.

ALYSSUM FLOWER was photographed with the scanning electron microscope. Anthers at tips of the six stamens are open and beginning to shed pollen. Pistil dominates center of picture; ovary lies within its base. Petals are visible in background. Magnification, about 75X.

pels are arranged in whorls and attached to a central RECEPTACLE. Towards the tip of each stamen are located four microsporangia which produce pollen and hence the male gametophytes. Folded within the carpels are found the ovules, each with its enclosed female sporangium. Note that unlike the ovule of the gymnosperms, that of the angiosperms has not one but two integuments. The car-

REPRODUCTIVE CYCLE of flowering plants. An ideal flower is depicted at center, with female cycle at left and male cycle at right. Each ovule (bottom left) encloses a female sporangium. Meiosis of mother cell produces four megaspores. Three degenerate, and remaining one divides mitotically, producing an embryo sac with eight nuclei (upper left). Male sporangia are fused into anthers at tip of filament (lower right). Microspore produced by meiosis divides mitotically before becoming a walled pollen grain. When pollen reaches the female stigma, pollen tube grows downward toward ovary and eventually penetrates the embryo sac (top center). Meanwhile the generative nucleus of pollen grain divides mitotically to produce two sperm nuclei, which fuse with nuclei in embryo sac to yield a diploid zygote and a triploid endosperm nucleus.

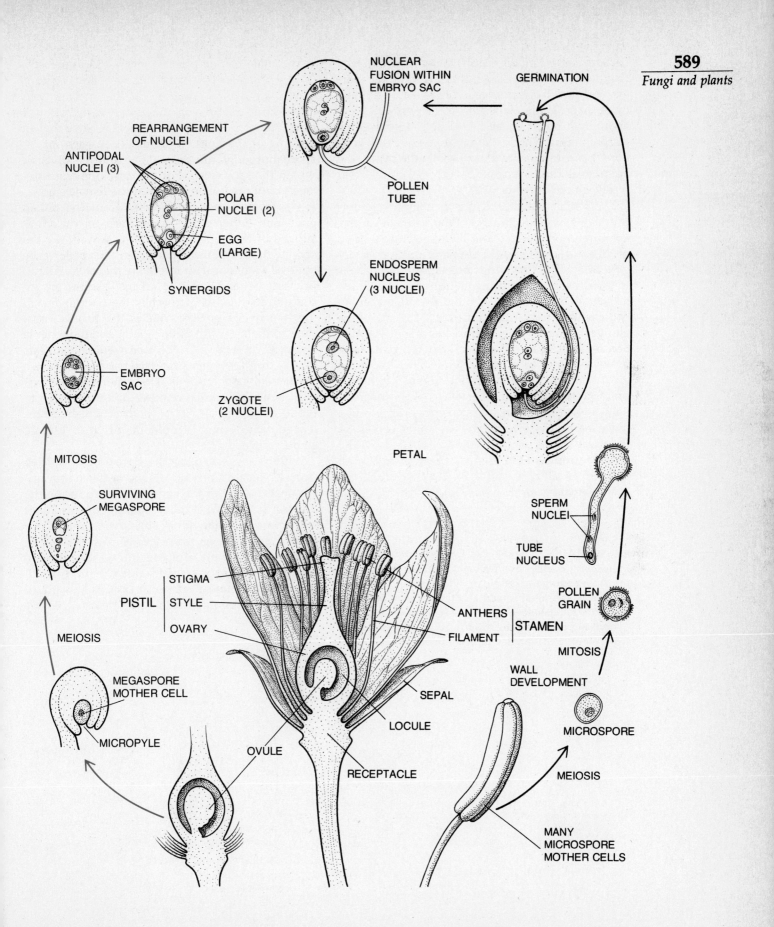

NUCLEAR FUSION WITHIN EMBRYO SAC

GERMINATION

REARRANGEMENT OF NUCLEI

ANTIPODAL NUCLEI (3)

POLAR NUCLEI (2)

EGG (LARGE)

SYNERGIDS

POLLEN TUBE

ENDOSPERM NUCLEUS (3 NUCLEI)

EMBRYO SAC

ZYGOTE (2 NUCLEI)

MITOSIS

PETAL

SURVIVING MEGASPORE

SPERM NUCLEI

TUBE NUCLEUS

MEIOSIS

POLLEN GRAIN

PISTIL

STIGMA

STYLE

OVARY

ANTHERS

FILAMENT

STAMEN

MITOSIS

WALL DEVELOPMENT

MEGASPORE MOTHER CELL

SEPAL

MICROPYLE

LOCULE

MICROSPORE

OVULE

RECEPTACLE

MEIOSIS

MANY MICROSPORE MOTHER CELLS

pel consists of a modified leaf, folded in such a way that the ovules are inside and fused along the margin and the structure is called the PISTIL. Thus the ovule is no longer exposed on the surface of the sporophyll but is entirely enclosed in sporophyll (carpel) tissue. The swollen base of the pistil, containing one or more ovules, is called the OVARY, the apical stalk the STYLE, and the terminal surface that receives the pollen the STIGMA.

The female gametophyte is a singularly unimpressive object *(Figure 37)*. One cell within the sporangium divides meiotically to produce a chain of four megaspores. All but one of these then degenerate. The surviving megaspore undergoes mitosis, usually producing eight nuclei, which are all bounded by a single membrane. The nuclei become rearranged so that three are at the end of the elliptical gametophyte nearest the micropyle, two become paired in the center, and the remaining three are grouped at the far end. Since there are eight in all, and they fall in this strange but highly predictable placement, they have all been given names: the three nearest the micropyle are the egg nucleus plus two synergids, those in the center are the polar nuclei, and those at the far end are the antipodal nuclei. The eight-nucleate structure all together is called the EMBRYO SAC; it constitutes the entire female gametophyte.

The male gametophyte is even less impressive *(Figure 37)*. Meiosis occurs within the four sporangia (fused into two anthers at the tip of a stalk called the filament), sometimes very early in flower development. Before the pollen is shed by an opening of the anthers (dehiscence), each microspore normally undergoes one mitotic division within the spore wall. Further development is delayed until the pollen arrives at the female stigma. A POLLEN TUBE then germinates from the pollen grain and either grows downward on the inner surface of the style or digests its way down the spongy tissue of this female organ. The pollen tube follows a chemical gradient in the style that is probably at least partially an increase in calcium concentration, until it reaches the micropyle. Of the two nuclei present, one is the TUBE NUCLEUS, close to the tip of the pollen grain, and the other is the GENERATIVE NUCLEUS. The tube eventually digests its way through megasporangial tissue and reaches the female gametophyte. The generative nucleus meanwhile has undergone one mitotic division to produce two sperm nuclei, both of which are released into the cytoplasm of the female gametophyte.

Within the embryo sac, there are now ten nuclei: eight female and two male. One of the sperm nuclei next fuses with the egg nucleus producing the diploid zygote. The other sperm nucleus fuses with the two polar nuclei to form a triploid nucleus. While the zygote nucleus begins cell division to form the new sporophyte embryo, this triploid nucleus undergoes rapid mitosis to form a specialized tissue called ENDOSPERM. At this point, the female antipodal nuclei and synergids simply degenerate.

38 **DEVELOPMENT OF SEED in Angiosperms.**
Endosperm nucleus divides many times to form nutritive endosperm, while zygote divides to form embryo. Note initial formation of suspensor, which elongates to push developing embryo into endosperm. Everything including the ovary wall comprises the fruit. Double coat marks the outside of the seed.

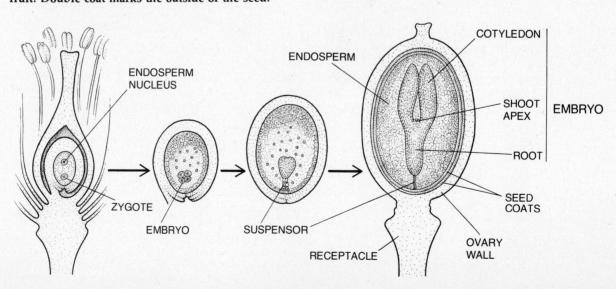

Unquestionably the flowering plant marks the summit of plant evolution. Angiosperms are the dominant vegetation of the planet.

Shortly after fertilization, a highly coordinated growth and development of embryo, endosperm, integuments, carpel, and even sometimes receptacle tissue ensues *(Figure 38)*. As large amounts of nutrient are moved in from other parts of the plant, the endosperm begins accumulating both starch and protein. The embryo becomes differentiated into characteristic root and shoot regions, and the first leaves (COTYLEDONS) appear. The integuments develop into a double-layered seed coat, sometimes fleshy, sometimes heavily lignified and hard, and the carpel (sometimes plus the receptacle) becomes ultimately the wall of the FRUIT which encloses the seed.

This maturing structure contains three different generations in one system. The integument represents the original parent sporophyte, the triploid endosperm was derived two-thirds from the female gametophyte plus one-third from the male, and of course the embryo represents the next sporophyte generation. Angiosperms also frequently involve the megasporophyll itself (the carpel) very extensively in fruit formation, an ingenious device for seed dissemination by animals who eat the fruits and then either spit out the seeds or else pass them undigested through their guts.

Of all the characteristic traits of the angiosperms, including the presence of vessels in the xylem, companion cells in the phloem, a closed megasporophyll (carpel) which participates in fruit formation, and specialized sterile leaves such as petals and sepals associated with the sporophylls, only one trait—the strange double fertilization mechanism just described—is found in all angiosperms and only in angiosperms. For example, the xylem of certain primitive woody angiosperms contains only scalariform-pitted tracheids, and several gymnosperms form true vessels. With double fertilization, the angiosperms have finally and conclusively abandoned both the archegonium, the ancient device that protected the egg and zygote, and the swimming sperm. The origin of this process in geological time is wholly unknown. Although it is very complicated, the double fertilization process has nevertheless been highly successful. It has produced the dominant land vegetation of the planet.

Something more must be said about the ideal flower which was invented for the above discussion. Because it produces both megasporangia and microsporangia it is said to be PERFECT, meaning that it contains both male and female parts. Many angiosperms produce two types of flower on the same plant, one type with only megasporangia and the other with only microsporangia. Consequently either the carpels or the stamens are nonfunctional or absent in a given flower. Plants of this kind are known as MONOECIOUS, meaning "one-housed," but, it must be added, with separate rooms. In some other species of angiosperms separation of the sexes is complete at the whole-plant level. A given plant produces either male or female sporophylls, but never both. This situation is called DIOECIOUS —"two-housed." In other words, there are truly female plants and truly male plants. In the ideal flower we also illustrated distinct petals and sepals, arranged in distinct whorls, whereas sometimes there is no distinguishing between the two, and they are spirally arranged. In such a case, these appendages are called TEPALS. Sometimes appendages of any sort —petals, sepals, or tepals —are completely absent.

The most primitive flower, from an evolutionary point of view, has a large number of tepals (or sepals and petals), carpels, and stamens, all spirally arranged *(Figure 39)*. Evolutionary change within the angiosperms involved a number of striking changes from this early condition: reduction in the number of each type of organ, differentiation of petals from sepals, stabilization of each type of organ to a fixed number, arrangement in whorls, and finally change in symmetry from radial (as in a lily) to bilateral (as in a sweet pea or orchid) often accompanied by an extensive fusion of parts. The most primitive carpels were clearly modified leaves, appearing as folded but incompletely closed structures —really intermediate between the gymnosperms and advanced angiosperms in this condition *(Figure 40)*. In the more highly specialized flowers, the carpels become fused and then progressively more and more buried in receptacle tissue, so that the most highly evolved have the other flower parts attached at the very top of the ovary rather than at the bottom. Primitive stamens, like primitive carpels, were also leaflike, little resembling those of the ideal flower illustrated in Figure 37 *(Figure 40)*.

Angiosperms are divided into two distinct subclasses, the MONOCOTS (subclass Monocotyledonae)

and the DICOTS (subclass Dicotyledonae). The names derive respectively from the existence of but a single embryonic cotyledon in monocots, versus a pair of embryonic cotyledons in dicots. There are, however, other major differences between the two groups. Monocots, which include among other things grasses, cattails, lilies, orchids, and palm trees, have leaves with parallel veins *(see Figure 4 of Chapter 14)*. Unlike dicots and gymnosperms, they almost never have a cambium. Bundles of vascular tissue, each containing xylem and phloem, are found throughout the stem, instead of in a ring surrounding the pith. In the root, instead of the solid central core of xylem with phloem outside, monocots have a ring of alternating bundles of xylem and phloem, surrounding a pith *(Chapter 16)*.

The absence of a cambium means that a palm tree, like a tree fern, is entirely the direct product of the apical meristem. On the other hand, grasses make extensive use of intercalary meristems, located just above the point of attachment of each leaf. The undifferentiated and tender meristematic tissue forms a weak point, which is why grass stems can sometimes be pulled apart into sections so easily. A final distinguishing feature of monocots is that the various floral parts exist in threes or sixes, when the number is fixed at all, instead of fours and fives, as with dicots.

The dicots include the vast bulk of familiar seed plants: most of the herbs, weeds, vines, trees, and shrubs that cover the earth. The venation of dicot leaves is usually branched or reticulate (net-like in pattern). The presence of a cambium is common—hence there are many woody species. Oaks, willows, violets, sunflowers, and chrysan-

39 PRIMITIVE FLOWER has many stamens and carpels arranged spirally. Photograph depicts flower of the tulip tree (Liriodendron). Central column consists of many separate carpels, tightly packed. Elongated rod-shaped structures are the stamens.

40 EVOLUTION OF CARPEL AND STAMEN is depicted in side view (top) and cross section (bottom). Carpel began as a modified leaf with sporangia near edges (far left). As evolution progressed, edges rolled inward and fused (left center). Three carpels fused to form a three-chambered ovary (center). Three stamens at right illustrate progression from primitive to advanced form. Primitive stamen (right center) was leaflike structure bearing four long sporangia (color). Diagram depicts stamens from actual plants: Austrobaileya (center), magnolia (right) and lily (far right).

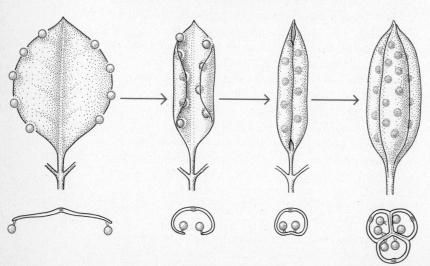

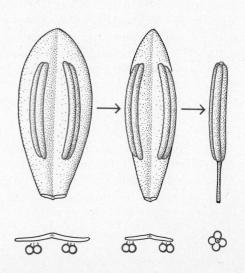

themums are examples chosen almost at random from the immense diversity of dicots. The sunflowers and the chrysanthemums are representatives of the Compositae, the most specialized dicot family known *(Figure 41)*. Their "blossom" is actually a tightly packed mass of flowers on a common receptacle. The outermost flowers are bilaterally symmetrical, each producing a single strap-shaped structure which one could easily mistake for an individual petal of a single flower, and the central ones are very inconspicuous and radially symmetrical. Frequently the outer flowers of the composite blossom are only female, while the inner ones are perfect, having both male and female parts.

There is not space here to discuss the tremendous variation in the structure of flowers or of the fruits and seeds which develop from them within the angiosperms, and particularly within the dicots. One can only point out that the taxonomy of the angiosperms depends heavily upon characters of these structures, and that there is an enormous technical terminology to deal with them.

The origin of the angiosperms is at the present time a paleobotanical mystery. The first clearly angiosperm pollen is found in Cretaceous sediments. By the end of the Mesozoic era, approximately 65 million years ago (half a day on our calendar), the angiosperms were well launched. They underwent an enormous evolutionary burst, which is probably still underway, starting in the early Cenozoic era. While various gymnosperms, from the seed ferns to the cycads, have been proposed as possible angiosperm ancestors, one is always left with the curious problem of where and how that one universal and distinctive angiosperm trait, the double fertilization, originated in evolution. No living gymnosperm has anything approaching it, and fossil gymnosperms are not likely to be particularly helpful. Direct evidence must be obtained from cytological studies of ovules, and unfortunately the fossil gymnosperm ovules are not well enough preserved for this kind of study.

READINGS

D. W. BIERHORST, *Morphology of Vascular Plants,* New York, Macmillan, 1971. A detailed and extremely up-to-date treatment of the vascular plants, both living and fossil.

BLOSSOM OF SUNFLOWER (Helianthus) is actually a **41** **large cluster of small flowers. Each of the long outer "petals" is formed by the fusion of five petals of a miniature flower. Center of blossom consists of many tiny "flowerlets," each having functional anthers and pistil.**

H. C. BOLD, *Morphology of Plants,* New York, Harper & Row, 3rd Edition, 1973. An excellent general text on the entire plant kingdom.

P. H. RAVEN, R. F. EVERT AND H. CURTIS, *Biology of Plants,* New York, Worth, 2nd Edition, 1976. An excellent introductory botany text with particularly strong treatment of the life histories of fungi and plants.

R. F. SCAGEL, R. J. BANDONI, G. E. ROUSE, W. B. SCHOFIELD, J. R. STEIN, AND T. M. C. TAYLOR, *An Evolutionary Survey of the Plant Kingdom,* Belmont, CA, Wadsworth, 1966. An unusual and complete survey of the plant kingdom, by authors who are specialists in each of the various groups.

Anyone looking into the pages of the present handbook
will soon find out that the zoology of dreams is far poorer
than the zoology of the Maker. —Jorge Luis Borges

ORNITHISCHIAN

SAURISCHIAN

PTEROSAUR

PLESIOSAUR

ICHTHYOSAUR

27
Animals

Plants exist on energy received directly from the sun. Because of the abundance of this resource, plants are able to anchor themselves to one spot, and, thus rooted, they can afford to adopt a body form that is circular or loose and sprawling. They need neither muscles nor the sensory receptors and nervous systems required for locomotion. Animals, in contrast, feed on plants and other animals, or on the decomposing remains of these organisms. As a result, animals cannot remain motionless like plants. In a later chapter on ecology (*Chapter 33*) the reader will learn how it is that only about ten percent of the energy used by the plants is available to the animals that eat them, and only about one percent of the original energy is available to the animals that feed on the plant-eaters. Therefore, animals must extract energy from an area much broader than that occupied by their own bodies. Two alternative strategies are available: one is active searching; the other is to remain SESSILE—fixed in place—while drawing in quantities of a medium, usually water, from which food can be extracted. It will be profitable to keep the distinction in mind as we review each phylum. Much of the anatomy and behavior of animals can be interpreted as adaptation to one or the other of these two feeding methods. The animals that search for food are the ones that are the most animal-like in character. They possess the best developed locomotion, sensory organs, and nervous system—and hence the most complicated and precise behavior.

MESOZOIC REPTILES depicted on opposite page include Triceratops, an herbivore (top); a flying carnivore (left center); Tyrannosaurus, a terrestrial carnivore (right center); an aquatic plesiosaur and an ichthyosaur (bottom). Plesiosaurs had long necks; ichthyosaurs were more fishlike.

What, then, is an animal? It is easy enough to distinguish a moose from a willow tree, but it is an exercise in philosophy when one tries to label the most primitive groups of organisms. An animal is, in final analysis, a member of a group of phyla that biologists have chosen to call animal. Of course, rational criteria are used. Certain traits are shared by all these phyla: mobility, well developed sensory and nervous systems, and responsiveness to external stimuli by behavior. But the unavoidable arbitrariness of our definition was already established when the line was drawn between protists and animals simply on the grounds that the former are unicellular and the latter multicellular. In fact, there are many ciliates with more complex structure and behavior than the simplest sponges and coelenterates—both of which are unequivocally considered animals.

The change from protistans to "real" animals nevertheless represents a quantum jump in evolution. The essential part of the change is the beginning of the DIVISION OF LABOR among cells. Within the ancestral colonies of cells—perhaps similar to those still existing in *Volvox* and other colonial protistans—some cells began to differentiate into somatic types and others into gametes and gamete precursors. Once this step was taken, it was possible for the units to become increasingly differentiated as specialized types, but all the while improving their coordination as working groups. These groups comprised ever larger and more complex organisms: animals. Figure 1 summarizes the prevailing opinion about the origin of the most primitive animal groups from protozoan types. The reader should bear in mind that this phylogenetic scheme is no more than an educated guess based upon careful study of living forms. The greatest part of the evolution it represents took place in Precambrian time, and the few fossils preserved from that time, such as those from the Ediacara

595

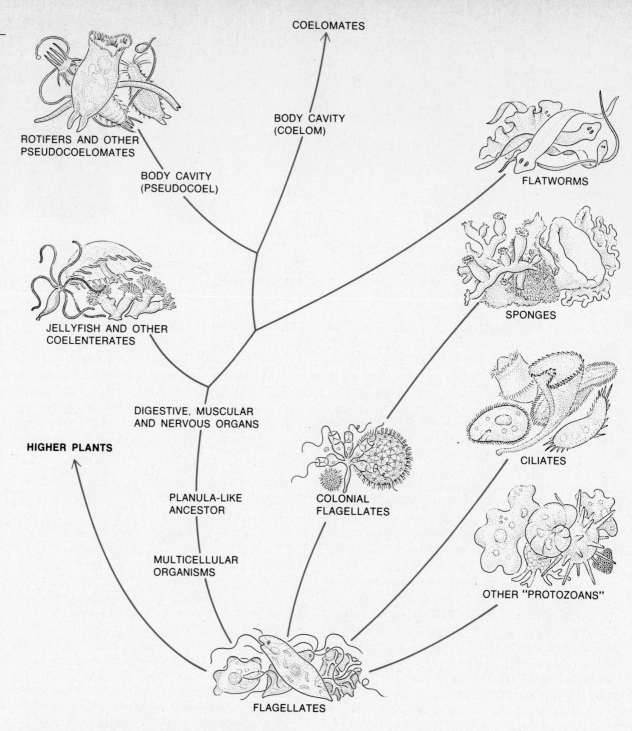

COELOMATES

ROTIFERS AND OTHER
PSEUDOCOELOMATES

BODY CAVITY
(COELOM)

BODY CAVITY
(PSEUDOCOEL)

FLATWORMS

JELLYFISH AND OTHER
COELENTERATES

SPONGES

DIGESTIVE, MUSCULAR
AND NERVOUS ORGANS

HIGHER PLANTS

CILIATES

COLONIAL
FLAGELLATES

PLANULA-LIKE
ANCESTOR

MULTICELLULAR
ORGANISMS

OTHER "PROTOZOANS"

FLAGELLATES

1 **PHYLOGENETIC TREE OF LOWER ANIMALS**
summarizes current thinking about their evolutionary
relationships. Flagellates, ciliates and other
"protozoans" are classified as protists. Multicellular
forms are classified as true animals. Drawings show a
few modern members of each group.

Hills of South Australia, have not yielded much
new information concerning the origins of the
modern fauna. But they do reveal that a great deal
of evolutionary experimentation took place in
these earliest of times. In addition to representatives of phyla still living, there are some strange
unidentifiable animal forms quite possibly belong-

ing to phyla that became extinct before the Cambrian period *(Figure 2)*.

About one million living species of animals are now known to science, and estimates of the true number range between two and ten million. They are the products of over a billion years of evolution, much of which has not been preserved in the fossil record and hence must be reconstructed indirectly. The essential facts of animal diversity do not lend themselves to a streamlined, easily assimilated presentation. This chapter simplifies the material as much as possible without distorting it, but it should be read selectively. The reader should obtain an overview of animal evolution while concentrating on the special topics that he finds most important and interesting.

Animals are classified into three main types—acoelomate, pseudocoelomate, and coelomate—according to the anatomy of their soft tissues and organs, and into two other categories—vertebrate and invertebrate—according to the nature of their supporting tissue and skeleton. The reader should familiarize himself with the vocabulary in Box A. The acoelomate animals include sponges, jellyfish, and flatworms. Pseudocoelomates include rotifers and roundworms. And the coelomates include all of the so-called higher animals, from invertebrate mollusks and arthropods to chordates such as reptiles, birds, and man.

PRECAMBRIAN ANIMAL (Parvancorina) is an offshoot from the evolutionary mainstream, an evolutionary experiment that failed. Fossil of this aquatic animal, which resembles no known living species, is depicted about three times its actual size.

> Sponges look so much like plants that for centuries they were called zoophytes (animal-plants). Their bodies consist of a loose republic of cells built around a water-canal system.

THE ACOELOMATE ANIMALS

The SPONGES (phylum Porifera from the Latin, meaning "pore bearer") are an ancient group of extremely primitive animals. They are so distinct that many specialists place them in a subkingdom of their own, the Parazoa. Almost all are marine animals, with only a few, relatively scarce species being found in fresh water. Through most of its life cycle the individual sponge is immobile, attached tightly to the substrate. It feeds by drawing water to itself and filtering out the small organisms and nutrient particles that flow past the walls of its inner cavities. Sponges look so much like plants that for centuries they were called zoophytes (animal-plants), and their exact position in nature was a matter of dispute. Only in the middle of the 19th century was their capacity to create currents of water recognized as a diagnostic animal trait. One of the characteristics of sponges is the possession of flagella by which the cells lining the body chambers set up the water currents. It is easy to suppose, as do a majority of invertebrate zoologists, that the first sponges originated from colonial flagellates. The feeding cells even have collars, consisting of modified cilia, that surround the bases of the flagella, a trait shown by flagellates of the order Choanoflagellida *(Figure 3)*. Whether they descended directly from colonial species of this particular group is uncertain. But at the very least, the sponges are obviously a group that diverged very early from protistan stock. They represent an evolutionary dead end, having given rise to no other known phylum. Their gross structure is unique among animals: there is no mouth or true digestive cavity, nor muscles, nor nervous system. For that matter, there are no organs at all in the usual sense of the word.

The body consists of a loose republic of cells built around a unique water-canal system. The sponge depicted in Figure 3 is typical of the simpler species. Water currents are set up by the flagel-

A Categories of animals

The most primitive animals are ACOELOMATES, animals whose bodies lack a coelom or any other internal cavity (see diagram). Pseudocoelomates have a false coelom, a kind of body cavity found only in a small number of relatively simple phyla such as rotifers. The PSEUDOCOEL is derived directly from the blastocoel of the embryo: the first cavity formed inside the proliferating ball of cells. It provides a liquid-filled space in which many of the body organs float. Coelomate animals have a true COELOM: a special kind of body cavity found in the most evolved animal phyla. The coelom is derived from the embryonic mesoderm, and it is lined with a special mesodermal lining called the peritoneum. The internal organs of coelomate animals hang down into the coelom, but do not float free within it; they are slung in pouches of the peritoneum.

The VERTEBRATE animals are those in which the nerve cord is enclosed in a backbone composed of bony segments, or vertebrae. The vertebrates are vastly outnumbered by the invertebrates both in number of species and in number of individuals, but because of the extraordinary interest they hold for us (as we are vertebrates ourselves) a basic distinction is usually made between the disciplines of VERTEBRATE ZOOLOGY as opposed to INVERTEBRATE ZOOLOGY. Zoology courses at the intermediate level are divided this way in most college biology departments in the United States and Canada.

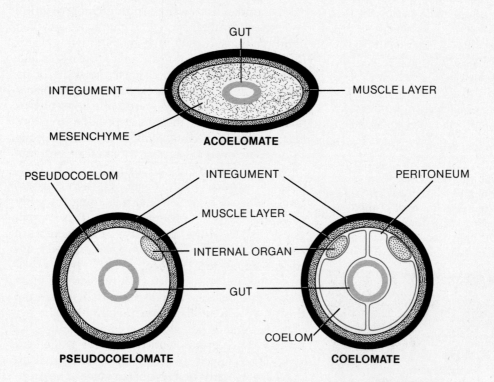

BODY CAVITIES of animals are illustrated in these schematic cross sections. Acoelomates such as sponges, coelenterates and flatworms have no body cavity; spaces between body wall and gut are filled with a jellylike matrix containing cells. Pseudocoelomates such as rotifers and nematodes have true fluid-filled body cavities, but the cavities lack a specialized cellular lining (peritoneum). All higher animals, such as mollusks, insects and vertebrates, have a true coelom lined with a peritoneum, although in some adult arthropods the coelom is almost obliterated, being replaced by a secondarily-formed body cavity called a hemocoel.

598

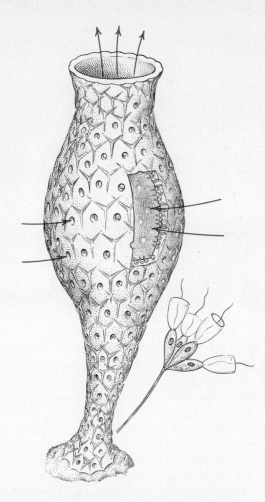

SIMPLE SPONGES have body cavities lined with flagellated cells that resemble free-living choanoflagellates (bottom). Current created by these cells pulls water into the sponge through pores. Mineral spicules strengthen the wall of the sponge. Arrows indicate flow of water.

ganism. If given the opportunity, the cells reassemble into new sponges. Tests have shown that they are attracted to each other by "aggregation substances," a protein-polysaccharide mixture located in the cell surfaces. For generations, sponge fishermen in the Mediterranean and Caribbean have used the trick of cutting up individuals of commercial species and waiting until they grow to marketable size. The fact is that a sponge, insofar as it can be distinguished as a separate organism, is virtually indestructible in a favorable environment. About the only way it can die is for all of its cells to be consumed by predators or by disease organisms.

Structural complexity has been acquired by a few kinds of sponges, but nevertheless in a curiously limited way. Size cannot be increased without improving the flow of water through the porocytes. Some sponges have made this improvement without adding new tissues or cell types. They have merely increased the number of choanocytes by folding up the layer bearing these cells back into the thickening body walls.

JELLYFISH, SEA ANEMONES, CORALS and related forms belong to the phylum Coelenterata. The coelenterates have advanced beyond the sponges by acquiring authentic organs and tissues. They possess tentacles, epithelial cells with muscle fibers, cells that discharge unique stinging organs called NEMATOCYSTS, and nerve nets. There is a true mouth and a blind digestive sac called the GASTROVASCULAR CAVITY. The digestive apparatus, plus the row of tentacles that normally surrounds the mouth, enable the animal to capture and swallow a much wider range of food particles than is available to sponges. The nematocysts play an important role in paralyzing and anchoring the prey. They are responsible for the nettling sting that some jellyfish can inflict on human swimmers. In extreme cases, involving the Portuguese man-of-war *(Physalia)* and the tropical Pacific sea wasp *(Chironex),* the injuries can be fatal. The coelenterate body possesses RADIAL SYMMETRY —seen from above it displays a perfectly circular outline. If you cut the animal in half through any point including the center, the two halves will be mirror images of each other. Put another way, the animal has a top and bottom but no front or rear. Most other animals, including all of the vertebrates and higher invertebrates, are characterized by BILATERAL SYMMETRY, meaning that there is a head and a rear, and only a bisection down the exact midline of the body can produce halves that are mirror images of each other. The radial, flowerlike form is often

lated cells (choanocytes) that line the inside of the body cavity. The water flows into the animal by way of minute pores perforating special epidermal cells. It passes up into a chamber of the body and out of a large terminal opening called the osculum. Between the thin epidermis and the choanocytes lies a zone of colloidal gel, the mesenchyme. Inside the layer are found AMOEBOCYTES, wandering amoeba-like cells responsible for whatever low levels of communication occur among the primary cells. Also present in the mesenchyme are the SPICULES, thin secreted objects that stiffen and support the body of the sponge.

So loosely organized are all these units that it is possible to squeeze the entire body of some kinds of sponges through a fine mesh, separating all the cells from one another, without killing the or-

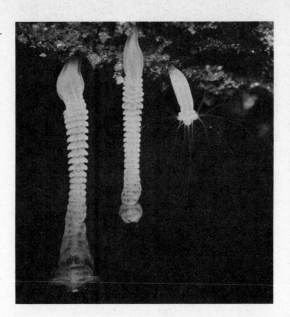

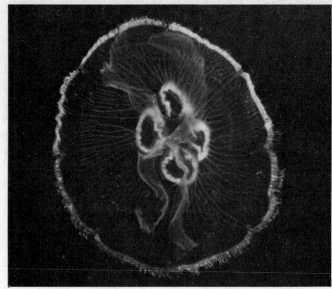

4 **AURELIA**, a common jellyfish, passes through alternation of generations typical of other coelenterates. At left, three sessile polyps in various stages of development hang from the underside of a rock. Right, the free-swimming medusa has four clearly visible lobes lined with cilia that sweep food into the central mouth. Polyps are reproduced about 3X actual size; medusa, about ½X.

combined with brilliant coloration—brilliant reds, purples, and blues, sometimes iridescent and shimmering—to rank some coelenterates among the most beautiful of all organisms. A few coelenterates are also extraordinarily large. The record is held by the giant jellyfish of the northern oceans, the lion's mane *(Cyanea)*, whose bell sometimes exceeds two meters in diameter.

5 **PLANULA** is the free-swimming larva that develops from the fertilized eggs of coelenterates. Larva settles on sea bottom and becomes a polyp. Magnification, about 50X.

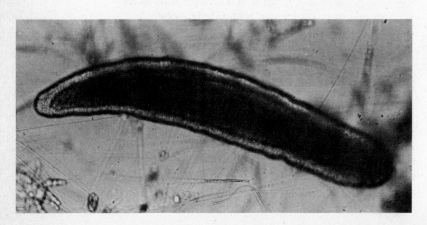

Most of the basic coelenterate features are lucidly displayed by the fresh-water animal *Hydra*, familiar from high-school biology labs *(Chapter 14)*. *Hydra* is considered a specialized coelenterate. In particular, it has a shortened life cycle. The majority of other coelenterate species, including those considered primitive, pass through an alternation of generations essentially like that of *Aurelia (Figure 4)*. The POLYP is the sessile (fixed) stage, in which the body consists of a cylindrical stalk with the end bearing the mouth and tentacles facing upward. The MEDUSA is the alternate, free-swimming stage. Its body is that of the typical jellyfish, shaped like a bell or an umbrella with the mouth and tentacles facing downward. In the completely alternating coelenterate life cycle, the polyps give rise to medusae by asexual budding, and the medusae generate the polyp stage by sexual reproduction. In the latter process, the medusae release reproductive cells into the water. When an egg is fertilized it develops into a free-swimming ciliated larva called a PLANULA (or planula larva), which settles to the bottom and transforms into a polyp. Because the planula possesses such an elementary body plan *(Figure 5)* it is widely believed to resemble the ancestral multicellular animal that arose from protozoans and gave rise in turn to the coelenterates and other metazoans. Although the polyp and medusa stages are radically different from one another in outward appearance, they actually share a closely similar body plan. The medusa can be envisioned as a polyp that lost its stalk, turned over, and swam away. Conversely, a polyp can be viewed as a medusa that turned over,

PORTUGUESE MAN-OF-WAR (Physalia) is a colonial coelenterate. A modified medusa serves as a float, and modified polyps as tentacles. Armed with stinging nematocysts, tentacles can capture and paralyze small fish, as shown here. Physalia is depicted about ⅔ actual size. 6

grew a stalk, and attached itself. Most of the outward difference between these stages is due to the development of the MESOGLEA, the middle body layer. The mesoglea of polyps is usually thin, but in medusae it is very thick and constitutes the bulk of the animal. In fact, the jellylike consistency of the mesoglea is the reason that the medusae are commonly called jellyfishes. It is an adaptation that permits these animals to float easily through the open water.

This distinction between the polyp and medusa stages must be kept in mind while considering di-versity within the coelenterates. The complex "organism" of the Portuguese man-of-war, for example, is actually a large colony of polyps, each of which is specialized to contribute to the locomotion, feeding, or reproduction of the assemblage as a whole *(Figure 6)*. The corals also consist of polyp

colonies. In most coral species the individual polyp lays down a skeletal container of calcium carbonate that surrounds and protects its soft body. This skeletal cup is highly regular in form. Coral colonies grow by the budding of individual polyps, with the skeletal cups being added one to the other in a set geometric pattern that varies among species *(Figure 7)*. As a consequence, each coral formation consists of a beautiful but bewildering array of skeletal forms. The vernacular names of the groups of species — horn corals, brain corals, staghorn corals, organ pipe corals, sea fans, sea whips, and others — convey the impression accurately. As the colony grows, old individuals die, leaving their calcareous skeletons intact beneath. In time the living members form a layer on top of a growing coral reef of skeletal remains. The reefs, which are often thousands of years old, play a major role in the formation of tropical islands, particularly the atolls. They also provide the physical basis of the most stable and diverse of all the marine ecosystems *(Chapter 33)*.

7 CORALS are coelenterates that live in warm shallow seas. Polyps tipped with feathery tentacles emerge from their stony skeletons to feed. Polyps of this colony of Galaxea are largely retracted.

FLATWORMS (PHYLUM PLATYHELMINTHES)

To most people the term worm conjures up a picture of a long, squirming creature, living in some hidden place without benefit of eyes, head, or legs. The truth is that many animals possess approximately this body form, and they are very difficult to classify. Some are brightly colored, a few have large eyes and legs, and many live in the open. In the early days of zoology only a single phylum, Vermes (Latin for worm), was recognized for all. Now it is understood that worms belong to many phyla and display a wide range of anatomical complexity. The simplest of these animals are the Platyhelminthes. As their Greek name literally states, they are the flat worms, distinguished by the horizontal flattening of their body which gives them the appearance of a piece of ribbon or tape *(Figure 8)*. The best known members of the phylum are parasitic forms, the tapeworms (class Cestoda) and flukes (class Trematoda), some of which cause schistosomiasis and other serious diseases in man. The third group, the turbellarians (class Turbellaria) are all free-living; most are marine, but a few live in fresh water or moist habitats on land.

The fresh-water turbellarian *Dugesia*, known

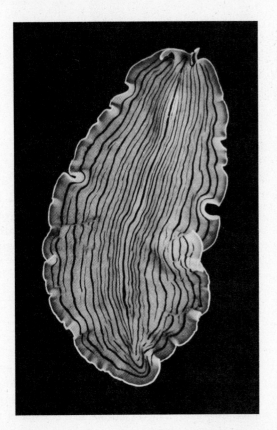

POLYCLAD FLATWORM (Prostheceraeus vittatus) is a **8** marine species about one inch long. Head end of worm lies at top of picture.

better by the popular name planaria, has been a favorite subject of generations of physiologists. It can be taken as typical of the more primitive, free-living members of the phylum. In one crucial respect the planarian is more advanced than a coelenterate: it has a head, with which it advances during locomotion; and with the existence of this body part come all the characteristics of bilateral symmetry. The sensory apparatus is concentrated at the front, where it is most needed, and consists of a chemoreceptor organ and very simply constructed eyes, called ocelli (singular, ocellus). The head also contains a primitive brain, comprised of little more than anterior thickenings in the longitudinal nerve cords. The intestinal tract of the planarian, like that of the coelenterate, consists simply of a mouth opening into a blind sac. However, the intestine is often highly branched, forming intricate geometric patterns.

Planarians and other flatworms are hampered by certain anatomical restrictions that stem from their simple body plan. The primitiveness of the circulatory and excretory systems dictates that each cell must be near a surface in order to respire. A flattened body form is therefore a necessary design feature. In the largest free-living flatworms, members of a mostly marine group called triclads, the animal has the shape of an oval, slightly crumpled sheet of paper. The flattened body form in turn precludes a well-developed muscular system. Turbellarians, the only flatworms whose life habits require them to move around a great deal, depend primarily on the use of broad layers of cilia to glide over surfaces.

Among the turbellarians are found the acoels (order Acoela), tiny marine worms that appear to be the most primitive members of the phylum. They lack a digestive cavity (acoel means no gut), distinct reproductive organs, or any trace of excretory organs. In fact, the acoels are not very much advanced over the planula larva of the coelenterates, and they are regarded by a few zoologists as being close to the ancestry of all the metazoan animals.

THE PSEUDOCOELOMATE ANIMALS

Phylum Rotifera, the rotifers or wheel animalcules, includes many of the animals visible in a drop of pond water at 100× magnification *(Figure 9)*. Most are tiny, not much larger than ciliate protozoans and easily mistaken for them. Each rotifer actually contains between 500 to 1000 cell nuclei—the number is a fixed species characteristic—and a

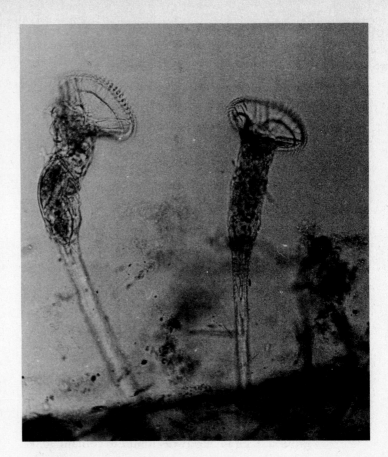

ROTIFERS in a drop of pond water. When not swimming freely, the rotifer anchors itself to the substrate with an adhesive that is produced by a gland in its "foot."

9

surprisingly highly organized organ structure. Rotifers possess a complete gut, one that passes from a mouth at the anterior end to an anus at the posterior end. There is also a body cavity, the pseudocoel defined at the beginning of this chapter, within which the internal organs are suspended. Rotifers are easily recognized by the possession of two distinctive organs connected with the ingestion of food. First is the CORONA, a con-

If all matter except roundworms were to vanish, a ghostly outline of the surface of the Earth could still be seen in the form of the countless nematodes that live in the soil, the water, and in plants and animals.

spicuous ciliated organ surmounting the head of some species. The coordinated beat of the cilia sweeps particles of organic matter from the water into the mouth of the animal and down to its MAS-TAX, a complicated skeletal structure, bearing teeth, that grinds the food up. A few rotifer species prey on protozoans and small metazoans. Their mastax is toothed and can be everted through the mouth to seize small objects.

A few rotifer species are marine, but most live in fresh water. A small number, loosely referred to as terrestrial, actually occupy an unusual aquatic habitat. They rest on the surfaces of mosses and lichens in a desiccated, inactive form until a rain occurs, then swim about in the films of water that temporarily cover the plants. Rotifers are remarkable in another respect. Each species is found over most of the entire earth, flourishing wherever it can find a suitable habitat.

ROUNDWORMS (PHYLUM NEMATODA)

Ubiquity and abundance are cardinal features of this pseudocoelomate phylum. Humans unintentionally eat and drink large numbers of nematodes in their lifetime. A single rotting apple from the ground of an orchard was found to contain 90,000 roundworms belonging to several species. One square meter of mud from off the coast of Holland yielded 4,420,000 individuals, and the soil of rich farmland has up to 3 billion to the acre. It has been said that if all matter except roundworms were suddenly to vanish, a ghostly outline of the surface of the earth could still be seen in the form of the countless nematodes that live mostly as scavengers through the upper layers of the soil, in the bottoms of lakes and streams, and in plants, and as parasites in the bodies of most kinds of animals. Nematodes are pseudocoelomate animals with a body organization at approximately the same grade of complexity as the rotifers *(Figure 10)*. One feature that sets them apart is their CUTICLE: a tough, smooth, elastic "skin" composed of several layers with differing chemical compositions. This organ no doubt gives the nematodes superior protection from their environment, but it also strictly limits their mobility. It forms a kind of straightjacket that prevents the animals from gorging themselves on large meals and forces them to feed more or less continuously. As a result, nematodes are animals adapted to living within their food, chiefly as scavengers, plant feeders, and parasites. Many nematodes are also predators but depend on smaller animals, including protozoans and other

nematodes, which are available in large numbers. In either case, nematodes can be characterized as depending on the conspicuous success of other kinds of organisms.

THE ANNELID SUPERPHYLUM

In Figure 11 a major branch has been added to the expanding phylogenetic tree of the animal phyla. Under the broad and informal grouping labeled annelid superphylum have been placed three strikingly distinct phyla, the annelid worms, the mollusks, and the arthropods. What is the justification for making such an improbable taxonomic cluster? The answer lies in largely hidden details of the

NEMATODE WORM. Diagram of male nematode indicates the presence of a gut (light color) with mouth and anus, a nervous system with a ganglion that serves as a rudimentary brain, and a testis (dark color) that opens into the anus.

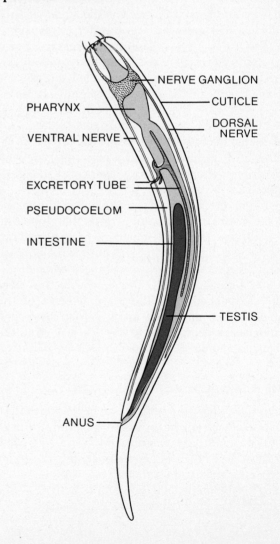

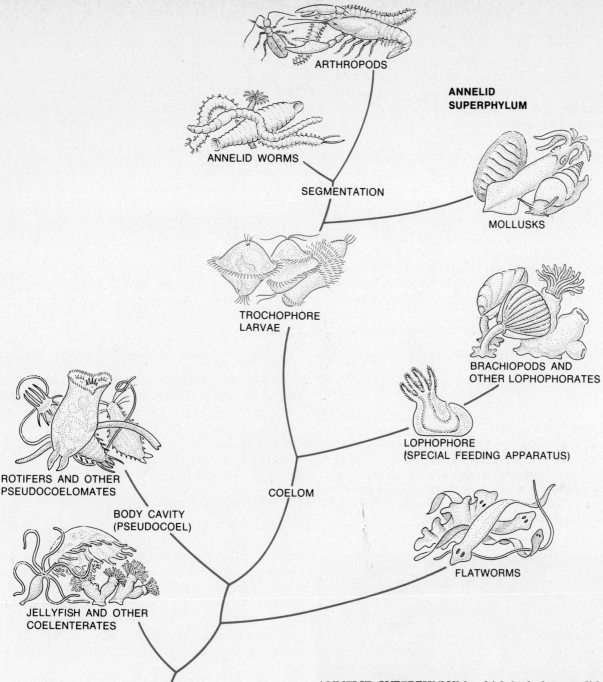

ARTHROPODS

ANNELID
SUPERPHYLUM

ANNELID WORMS

SEGMENTATION

MOLLUSKS

TROCHOPHORE
LARVAE

BRACHIOPODS AND
OTHER LOPHOPHORATES

ROTIFERS AND OTHER
PSEUDOCOELOMATES

LOPHOPHORE
(SPECIAL FEEDING APPARATUS)

BODY CAVITY
(PSEUDOCOEL)

COELOM

FLATWORMS

JELLYFISH AND OTHER
COELENTERATES

ANNELID SUPERPHYLUM, which includes annelid **11**
worms, mollusks, and arthropods, is a major addition
to the phylogenetic tree depicted in Figure 1. Members
of superphylum are linked by common features of
development, particularly by the presence of a
trochophore larva (Box B).

early development of these animals, where certain
points of correspondence link them with flatworms, rotifers, and others of the lower animal
phyla, while separating them from other higher
forms of animals. Of equal significance is the possession, in certain marine annelids and mollusks,
of the TROCHOPHORE LARVA *(Figures 12 and 24)*, a distinctive form propelled by a row of cilia that encircles the middle of the body (the name means

"wheel bearer"). Only some of the marine annelids and mollusks go through a trochophore
stage during their development, but enough do to
suggest that the trochophore is a primitive rem-

The biogenetic law

A venerable theory known as the biogenetic law states that "ontogeny recapitulates phylogeny." In its original 19th-century sense the law meant that each major embryonic stage represents an ancestral adult form from the remote past. The earlier the stage, the more ancient the ancestor. Thus the fertilized egg of some higher animal was thought to represent, in a modified form of course, a protozoan forebear: the blastula a planulalike descendant of the protozoan, the gastrula perhaps the flatworm descendant of the planula, and so on. Evolution was thought to proceed by the addition of new, more advanced adult stages to the older ones, which were pushed back one step in the developmental sequence. In this simplistic model, the trochophore was viewed as being a fairly close approximation of the adult invertebrate that gave rise to both the annelids and the mollusks.

The great majority of biologists today reject the strict interpretation of the biogenetic law, and substitute the hypothesis that the various life stages tend to evolve at different rates in response to the needs imposed by their differing ways of life. In some phyla the adult stages outstrip the larvae in diverging from ancestral forms. In other phyla the larvae outpace the adults. In the case of annelids and mollusks, the adults have evolved more rapidly. The trochophore is probably close to the original larval form of the common ancestor, a relict that continues to bear witness to the common ancestry long after the adult stages have ceased to do so.

12 **TROCHOPHORE LARVA of the marine annelid Polygordius. Ciliated larva has a complete gut and an excretory protonephridium, as well as an embryonic mesoderm. Such larvae are thought to be a primitive survival of the extinct common ancestor of the annelid superphylum. (See also Figure 24.)**

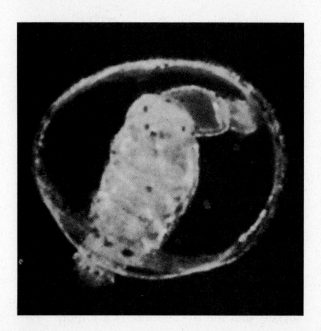

nant of the remote ancestor that gave rise to these now very divergent groups. No arthropod possesses this stage, but other evidence strongly suggests that arthropods evolved directly from annelid forebears, perhaps from a line that had already lost the trochophore (Box B).

THE COELOMATE ANIMALS

The evolution of the coelom made possible the development of the physically largest and most complex organ systems. The mesodermal lining of the coelom, the peritoneum, provides the means to suspend the internal organs in the fluid-filled body cavity while at the same time containing and separating them. The annelid superphylum, together with the others discussed in the rest of this chapter, are collectively called the higher animals.

SEGMENTED WORMS (PHYLUM ANNELIDA)

Compared with rotifers and flatworms, the earthworm is definitely a higher animal (Figure 13). Its many organ systems are completely formed and it is capable of relatively quick, accurately directed movements during locomotion, feeding, and reproduction. In fact, improved locomotion can be taken as the key to those aspects of annelid biology which are the most distinctive. The earthworm is a segmented animal constructed of a number of muscular doughnut-shaped rings separated by thin partitions or septa (singular, septum). Segmentation is essential to the crawling movement of the worm. Each segment can be shortened or lengthened by its own set of muscles, or else shortened at one side and lengthened at the other to provide a curve in the body. The segments can be expanded and contracted in a coordinated fashion to produce waves of bulges moving up and down the length of the body. The result is the rhythmic, shuffling locomotion that characterizes earthworm movement and enables the animal to push its way rapidly through the soil. Now consider the full con-

DIAGRAM OF EARTHWORM shows well-developed **13** **organ systems, particularly for locomotion and reproduction. Colored frame at upper right indicates location of cross-section view at middle left. Earthworms are hermaphrodites having both male and female sex organs.**

1 HEARTS
2 BRAIN
3 VENTRAL NERVE CORD
4 SPERM RECEPTACLES
5 TESTES AND SPERM SACS
6 OVARY
7 OVIDUCT
8 SPERM DUCT
9 SEPTUM
10 NEPHRIDIA

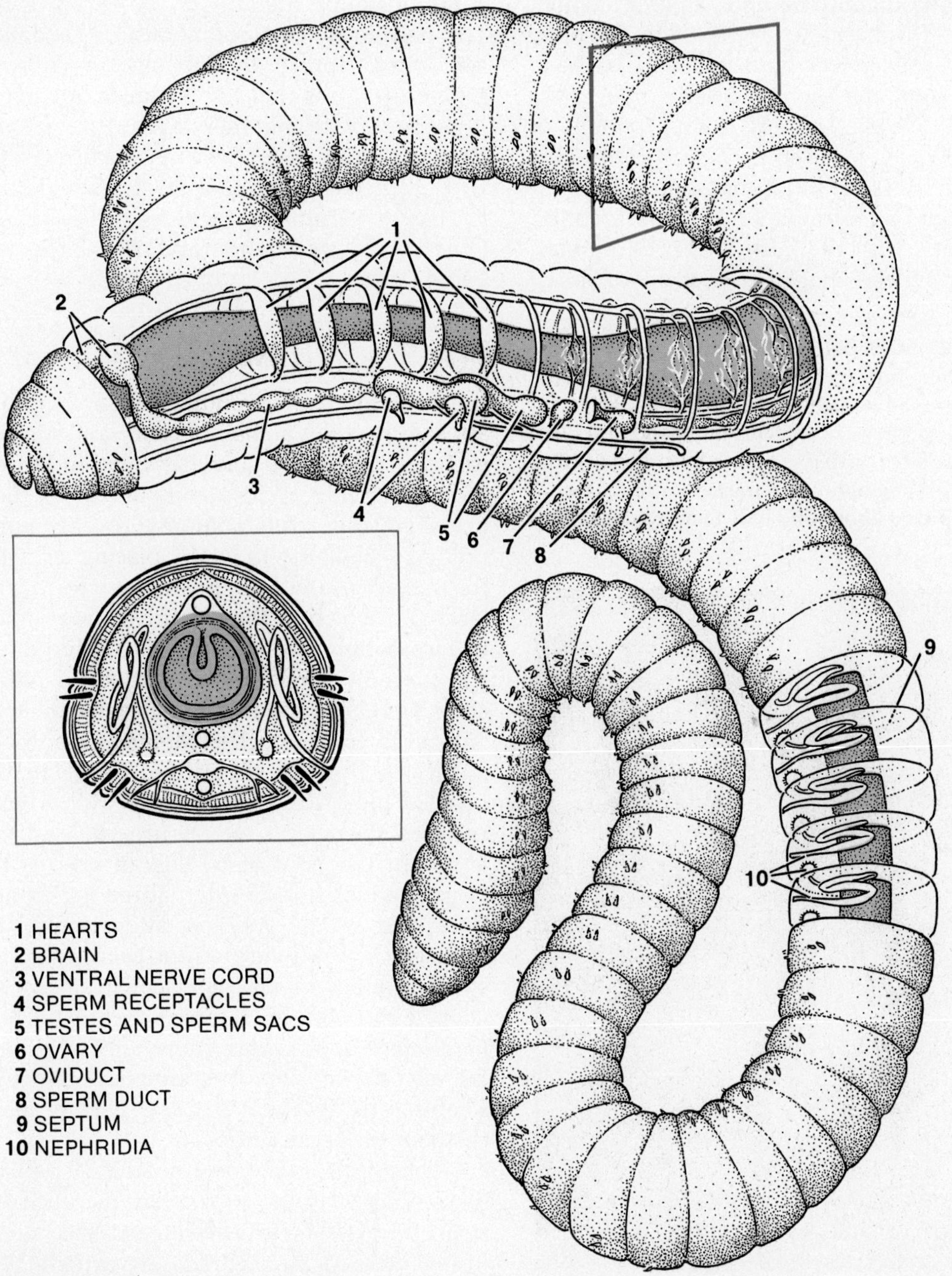

sequences of the adoption of this form of movement on the earthworm's body plan. The musculature must be segmented in order to permit the waves of contraction. Separate nerve centers (ganglia) are needed in each segment to give it independence of operation; but the centers must also be connected by the nerve cords in order to permit their operation to be coordinated. The septa close off the coelom in each segment, creating separate liquid-filled chambers that are capable of exerting pressure against the ground as the muscles surrounding them contract and expand. Each chamber, as a result, also becomes a potential pool of stagnation. This hazard is overcome by the presence of separate excretory units (NEPHRIDIA) in each segment. In sum, the segmentation of the earthworm seems to make sense as a total pattern of adaptation toward a more efficient form of locomotion through the soil and submarine mud and sand. It provides an example of the generalization that the groups with the most advanced animallike traits are those that are required to move around the most in the course of their daily existence.

Although the earthworm possesses the fundamental annelid body plan, it is unusual in several other respects. The vast majority of other annelids are marine and aquatic. A large number of the marine forms in turn, comprising the class Polychaeta, bear lateral appendages (PARAPODIA) on each segment. The major adaptive forms within the Annelida are impressively diverse. There are scavengers and soil feeders such as the earthworms and their aquatic equivalents which together make up the bulk of the class Oligochaeta. The polychaetes include voracious predators that seize and capture animals their own size, and sedentary forms that live in tubes and trap small organisms with plumelike tentacles (*Figure 14*). Parasitic forms occur within both the Polychaeta and Oligochaeta, while a third annelid class, the leeches or Hirudinea, includes predators of other invertebrates and temporary external parasites on vertebrates.

14 PEACOCK WORM (*Sabella*) is a marine polychaete that lives in a long tube fashioned from mucus, sand or lime. Only the head, which bears fanlike tentacles, protrudes from the tube. Tentacles trap food and also serve as gills. Food is carried to mouth by action of cilia. One worm in photograph has retracted its tentacles. Worms are depicted about one-half their actual size.

MOLLUSKS (PHYLUM MOLLUSCA)

It is hard to imagine animals less alike in outward appearance than a snail, a clam, and a squid. Yet each respresents but one class within the phylum Mollusca. Only a close comparison of their embryology and adult anatomy reveals numerous points of similarity to justify placing all of them in the same phylum. Each possesses a FOOT, a muscular organ which in the primitive state serves both in locomotion and as a sturdy underpinning for the viscera. The foot of the squid and other cephalopods is modified into a head, equipped with eyes and tentacles. Mollusks also possess a MANTLE, a sheet of specialized tissue that covers most of the viscera like a body wall. The mantle secretes the SHELL, which provides an external armor in the bulk of molluscan groups but has been modified into an internalized backbone in the slugs and squids and lost altogether in the octopuses. Within the mantle cavity are found a small number of uniquely constructed, feather-shaped GILLS, often referred to by the technical name ctenidia (singular, ctenidium) in order to distinguish them from other kinds of animal gills. A scraping organ, the RADULA, is usually located in the floor of the mouth.

Perhaps the single most notable fact about mollusks is the way the body plan, incorporating the basic, primitive features just listed, has been varied in evolution to permit radically different adapta-

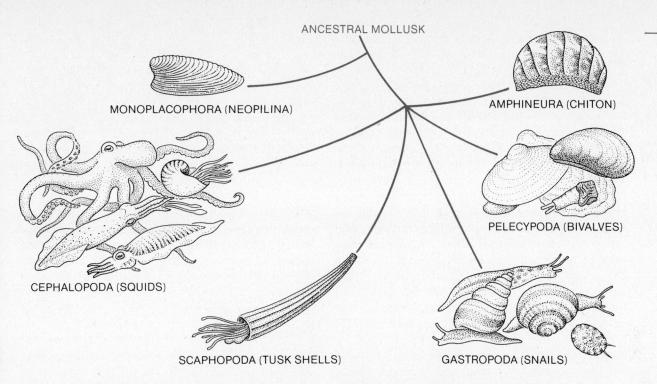

ANCESTRAL MOLLUSK

MONOPLACOPHORA (NEOPILINA)

CEPHALOPODA (SQUIDS)

SCAPHOPODA (TUSK SHELLS)

AMPHINEURA (CHITON)

PELECYPODA (BIVALVES)

GASTROPODA (SNAILS)

tions to the environment *(Figure 15)*. The chitons (class Amphineura) are the most primitive major group. The body is symmetrical, and the internal organs, particularly the digestive and nervous systems, are simply constructed. Development proceeds through a trochophore larva almost indistinguishable from that found in certain primitive annelids. The adult chiton spends most of its life clamped tightly to rock surfaces by means of its large, muscular foot. It is capable of moving slowly by means of series of rippling waves through the foot. The body of the chiton might, in fact, have been derived from that of some ancestral flatworm. Its foot could represent a modification of the flatworm's lower surface. The gills and shell plates of the chiton are divided into repeating units down the length of the body in a way vaguely reminiscent of the segmentation of the Annelida. However, the condition is not at all comparable to the complete septum formation that characterizes the annelid body, and it has a different embryonic origin.

In 1952 an oceanographic vessel dredged up ten specimens of an unusual little mollusk resembling a limpet from the deep Pacific just off the coast of Costa Rica. The finding of these animals, placed in the genus *Neopilina*, created a sensation, because they turned out to be monoplacophorans (class Monoplacophora), a group previously known only from fossils dating from the Cambrian period. In

SIX CLASSES OF MOLLUSK are depicted in this diagram, which suggests the diversity of variations on the basic body plan. Features common to all mollusks include a muscular foot, mantle, shell, gills and radula. Chitons are the most primitive mollusks and cephalopods the most advanced. **15**

other words, an entire major group had been discovered alive that had previously been known only from fossils. The *Neopilina* are remarkable in possessing multiple gills, muscles, and nephridial units that are repeated down the length of the body *(Figure 16)*. At first it was argued that they are much more primitive in this regard than the chi-

NEOPILINA is a living fossil, the only known survivor of a class of mollusks that flourished during the Cambrian period. **16**

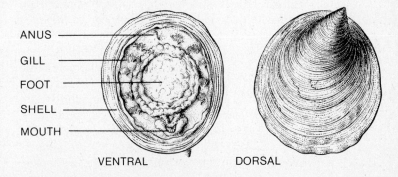

ANUS

GILL

FOOT

SHELL

MOUTH

VENTRAL   DORSAL

tons, and that they form a link to the common ancestor of the mollusks and annelids. However, an equally strong argument has since been made that the trait was derived secondarily in evolution and is not the same as the segmentation of annelids. It now still seems more likely that mollusks are basically "lump-shaped," that they have never passed through a segmented stage in evolution. Still, this conclusion is a tentative one that could easily be overturned by the discovery of additional fossil forms from the Cambrian or Precambrian, or even new living fossils in addition to *Neopilina*.

The BIVALVES (class Bivalvia or Pelecypoda) are the familiar clams, oysters, scallops, mussels, and other important edible shellfish, together with a host of similar, lesser known forms. They are characterized by the great reduction in the head. An interesting exercise for you is to try to pick out the front and rear ends of a freshly opened oyster or clam. The foot is compressed laterally (Pelecypoda means hatchet foot) and is used by many clams for burrowing into mud and sand and only secondarily for moving from place to place.

The GASTROPODS (class Gastropoda) are relatively primitive mollusks that use their foot primarily for locomotion. For reasons that have not yet been satisfactorily explained, most of the body has undergone a 180-degree counterclockwise turn in evolution, with the result that the digestive tract and nervous system are twisted into a U-shape, and the anus, the opening of the mantle, and the gills are moved to the front of the body just behind the head. As improbable as this change might seem, the Gastropoda have gone on to become the most successful and diverse of the molluscan classes. Among the crawling forms are a bewildering variety of snails, whelks, limpets, slugs, abalones, drills, and the often brilliantly ornamented nudibranchs. Other mollusks, the limpets, are sedentary and superficially resemble chitons. Still others, the sea butterflies and fishlike pteropods, have modified the foot into swimming organs and move easily through the open water of the sea.

Finally, the CEPHALOPODS (class Cephalopoda) are by purely human standards very close to the apex of invertebrate evolution. As exemplified by the squids, octopuses, and chambered nautilus, they possess a well-formed head with a relatively large brain, and are capable of extensive learned behavior. The cephalopod looks at the world with large, bright eyes, superficially similar to our own but radically different in embryonic origins. Its arms and tentacles, another distinctive cephalopod invention, are able to manipulate small objects

We are now living in the Age of Arthropods, or more precisely, the Age of Insects. The total number of insects on Earth is estimated to be 10^{18} —a billion billion —almost one billion insects for every human being.

with impressive skill. The cephalopods never attained the level of intelligence of advanced vertebrates, but of all the invertebrates they came closest to assembling the necessary anatomical ingredients to make such an evolutionary step possible.

ARTHROPODS

The arthropods are one of the two or three most successful phyla on Earth. The phylum Arthropoda includes trilobites, crustaceans, spiders, centipedes, insects, and related forms. The original arthropods evolved from annelid ancestors in Precambrian times. They acquired a unique armor, the EXOSKELETON or CUTICLE, composed of layers of protein and a strong, flexible polysaccharide called CHITIN. In some groups of arthropods, especially the crustaceans, the exoskeleton is impregnated and toughened with mineral salts such as calcium carbonate and calcium phosphate. The reason why the exoskeleton evolved in ancient times is unknown. We can only guess that it originally afforded protection from other predatory animals, a function that it still serves today. Once this particular strategy was adopted, the subsequent evolution of the phylum was channeled into new directions. In order to maintain agility, the armor must be hinged; and so arthropods retained and elaborated upon the original annelid segmentation. In order to move rapidly it was necessary to have hinges in the exoskeleton of the legs; and so the arthropods evolved this unique feature also *(Figure 17)*. The musculature was adapted to the special needs of armored existence. Instead of the simple repetition of segmental muscles found in the annelids, muscles became specialized to operate particular segments of the bodies and the appendages attached to them. Within most of the evolving lines of the Arthropoda, certain of the appendages were modified from their original walking or swimming functions to aid in eating and reproduction. Others assumed a primarily sensory function. This division of labor provided a versatility to the individual arthropod not available to its annelid ancestors.

The adoption of an exoskeleton and its accessories had another curious and profound influence on the later stages of arthropod evolution. Encasement within armor does more than just protect an animal from predators. It provides the body with the support it needs for walking on land and, of equal importance, it keeps the animal from drying out quickly when it chooses to leave the water. Arthropods were, in short, among the best candidates among the invertebrates to colonize the land. This they accomplished repeatedly, often with spectacular success. Crustaceans, in particular sowbugs and crabs, have numerous terrestrial representatives. Scorpions and spiders are the products of other, very successful invasions of the land, and none of their immediate water-dwelling ancestors survives. But all of these groups are completely overshadowed in numbers and diversity by the insects. The data in Table I indicate that in terms of species numbers it makes sense to say that we are now living in the age of arthropods—or more precisely, the age of insects, which comprise the great majority of arthropod species.

Why have the insects undergone such a disproportionate evolutionary diversification? The following is at least a partial answer. The insects originated from centipedelike ancestors at least as far back as the Devonian period, over 350 million years ago (on the 28th day of our evolutionary calendar). It appears that this early start gave them

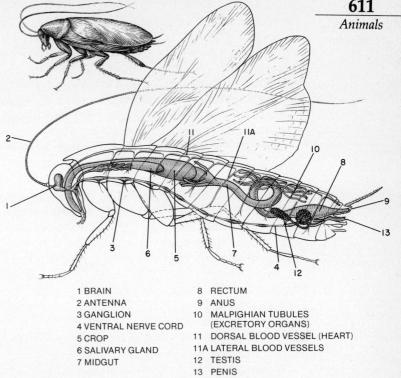

1 BRAIN
2 ANTENNA
3 GANGLION
4 VENTRAL NERVE CORD
5 CROP
6 SALIVARY GLAND
7 MIDGUT
8 RECTUM
9 ANUS
10 MALPIGHIAN TUBULES (EXCRETORY ORGANS)
11 DORSAL BLOOD VESSEL (HEART)
11A LATERAL BLOOD VESSELS
12 TESTIS
13 PENIS

MALE COCKROACH exhibits the principal features of **17** **the arthropod body plan. Armored exoskeleton is segmented and hinged. Various segments have evolved scores of specialized appendages, including those for grasping, chewing, walking, and (in advanced insects) flying. Respiratory system, which was discussed in detail in Chapter 15, is omitted from this diagram.**

I **DIVERSITY OF ARTHROPODS is evident from this tabulation of species in the major phyla of living animals. Number of species in some lesser-known phyla (such as Nematoda) is undoubtedly higher than the number given here, because only a fraction of the species have been classified.**

PHYLUM	APPROXIMATE NUMBER OF KNOWN LIVING SPECIES (IN THOUSANDS)
PROTOZOA	30
PORIFERA	5
COELENTERATA	5
PLATYHELMINTHES	13
NEMATODA	10
MOLLUSCA	110
ANNELIDA	9
ARTHROPODA	900
ECHINODERMATA	6
CHORDATA	45
OTHER PHYLA	20

an advantage in exploiting the newly formed forests and other primitive forms of land vegetation. By Carboniferous times, a great diversity of types already swarmed over the land, and winged insects had appeared—the first animals to fly. The new environments penetrated by the insects were like a new planet, an all but empty ecological world, comparable in size and complexity to the surrounding seas that gave birth to the earlier forms of life. Viewed in this light, it is not too surprising that insects have become at least as diverse as all the marine invertebrates put together.

There still exists a group, the ONYCHOPHORANS, that links the annelids to the primitive arthropods. The phylogenetic position of the Onychophora is so close to being intermediate that they are sometimes listed as a class within the Arthropoda and sometimes as a phylum by themselves. In the systematic list of Chapter 24, we adopted the latter choice, in accordance with the position taken by an increasing number of specialists. Onychophorans look like slugs with legs *(Figure 18)*. The internal structure is closer to that of the annelids, although the open circulatory system resembles that of the arthropods. The body is covered by a thin, flexible cuticle, which contains chitin like that of the arthropods but is not divided into articulating plates. Onychophorans are exclusively terrestrial, and today are found only in a few places in the tropics and southern temperate zone.

The most primitive undoubted arthropods were the TRILOBITES (class Trilobita), a group that flourished in the seas of the Cambrian and Ordovician periods but became extinct by the close of the Paleozoic era *(Figure 19)*. Trilobites were heavily armored, and the segmentation of their bodies and appendages followed a relatively simple, repetitive plan. Another ancient group, still present with us as living fossils, are the horseshoe crabs (class Xiphosura). These large animals are common in shallow waters along the east coast of North America and southeast Asia, where they scavenge and prey on mollusks and other bottom-dwelling invertebrates.

A selection of arthropods is depicted in Figure 20. The ARACHNIDS (class Arachnida) are relatives of horseshoe crabs that invaded the land and, like the insects, enjoyed an early and lasting success. All have six pairs of appendages, of which all posterior four are legs and the anterior two are modified for feeding. The body is divided into an anterior por-

FOSSIL TRILOBITE belongs to a class of arthropod that flourished in the seas of the Cambrian and Ordovician periods. Traces of the shell are visible at the top of this fossil impression, which is reproduced about actual size. 19

18 **PERIPATUS has a soft body like annelids and segmented legs like arthropods. Once thought to be an evolutionary link between the two groups, onychophorans are probably an offshoot from ancestral annelids, rather than being in direct line of descent. Peripatus gives birth to living young.**

tion, which bears all of the appendages, and a posterior abdomen. The most diverse and ecologically important members are the scorpions, harvestmen (daddy longlegs), spiders, mites, and ticks. The spiders have evolved into superb pedators—some make use of excellent vision to chase and seize their prey, while others rely principally on elaborate silken webs. Silk, incidentally, is produced by at least a few members of most of the other arthropod classes. Spiders are distinguished by the great efficiency with which they have put it to use.

The CRUSTACEANS (class Crustacea) are the dominant arthropods of the sea. One group alone, the alga-feeding copepods, are so dense in the plankton that they may well be the most abundant of all groups of animals. Although crustaceans are distinctive in many ways, they can be characterized most simply as arthropods with two pairs of antennae. In addition to the familiar shrimps, sowbugs, sand fleas, lobsters, crayfish, and crabs, all of which belong to phylogenetically advanced

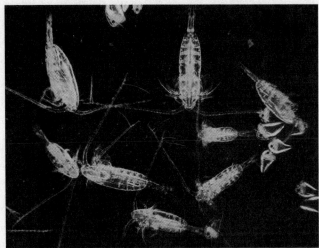

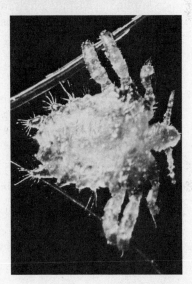

20 ARTHROPODS. Fisher spider (top left) shown eating minnow, is an arachnid. Copepods at upper right, one of the main components of zooplankton, are microscopic crustaceans. Coral shrimp (center left) tears away at a starfish, a member of a group (echinoderms) that has virtually no natural enemies. Center right, a crab louse (Phthirius pubis) clings to human pubic hair. Barely two millimeters long, "crabs" have once again become common in North America. Bottom, the mite Dichocheles infests one ear of moths. Mite lays a chemical trail that prevents another mite from infesting the other ear, which would deafen moth to bat sonar, jeopardizing lives of both arthropods.

groups, there is a vast array of more primitive forms, many bearing a superficial resemblance to shrimps, that are abundant but too small to attract popular notice. One group, the barnacles (subclass Cirripedia) have developed a wholly sedentary existence. The unique calcareous shells of these animals cause them to resemble mollusks; but, as the great zoologist Louis Agassiz remarked over a century ago, a barnacle is "nothing more than a little shrimp-like animal, standing on its head in a limestone house and kicking food into its mouth."

The MYRIAPODS are an assemblage of four classes of terrestrial animals: the centipedes ("hundred legs"; various species really have 15 to 173 pairs of legs), the millipedes ("thousand legs"; the true number varies between 40 and 200 pairs), the symphylans, and the pauropods. All are superficially similar in appearance. They are characterized by the possession of a well-formed head and an elongate, flexible, segmented body. Many of the body segments bear legs. The name myriapod, once applied to a single phylum embracing all four groups, means many legs. The centipedes, pred-

21 **METAMORPHOSIS OF INSECTS. Some orders undergo incomplete metamorphosis (hemimetabolous development). Nymph that hatches from egg becomes an adult after growth and molting. Other orders undergo complete metamorphosis (holometabolous development). Egg hatches into larva which grows, molts, and becomes an inactive pupa, often encased in a cocoon. During pupal stage new adult develops from disks of cells present in larva, using tissues of larva as raw materials.**

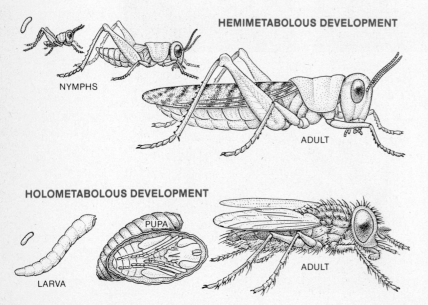

NYMPHS

HEMIMETABOLOUS DEVELOPMENT

ADULT

HOLOMETABOLOUS DEVELOPMENT

PUPA

LARVA

ADULT

ators of insects and other small animals, and the millipedes, scavengers and plant eaters, are the most abundant and diverse of the four groups. The symphylans, which resemble small white centipedes and are relatively abundant in the soil, are most notable as the probable ancestors of the insects.

INSECTS

The INSECTS (class Insecta), the little conquerors of the land and air, differ from other arthropods in the possession of three basic body parts (head, thorax, abdomen), a single pair of antennae on the head, and three pairs of legs originating from the thorax. Respiration is accomplished by means of air sacs and channels, called TRACHEAE (singular, trachea) that extend from external openings inward to tissues throughout the body. The higher insects are distinguished by the power of flight. The adults, but not the immature stages, possess two pairs of stiff, membranous wings attached to the thorax.

The number of described insect species is rapidly approaching one million. One authoritative estimate based on statistical analysis places the actual number of species alive on the earth at present at approximately three million. The number of individuals, incidentally, is believed to be in the neighborhood of 10^{18}, or a billion billion, roughly one billion insects for every human being. It is easy to see by these numbers alone why entomology is regarded as a full science within itself. Entomologists divide the known living species into about 26 orders. One can make some immediate sense out of this bewildering variety by recognizing the existence among the living species of four major adaptive levels, or EVOLUTIONARY GRADES. At the lowest level are the APTERYGOTES, or wholly wingless forms. These include the springtails (order Collembola) and silverfish (order Thysanura), smallish insects that mostly live hidden lives in the soil and humus, beneath rocks, and in the crevices of tree trunks. Some, particularly the thysanurans, are not far removed from the ancestral myriapods. The winged insects, or PTERYGOTES, encompass three grades. The more primitive forms have wings that cannot be folded back against the body. They are often excellent flyers but require a great deal of open space in which to maneuver. Foremost among these paleopterous insects are the orders Odonata (dragonflies and damselflies) and Ephemeroptera (mayflies). The third and next highest grade consists of the neopterous orders:

Orthoptera (grasshoppers, crickets, roaches, mantids, walking sticks), Isoptera (termites), Plecoptera (stoneflies), Dermaptera (earwigs), Thysanoptera (thrips), Hemiptera ("true" bugs), Homoptera (aphids, cicadas, leafhoppers, etc.), Neuroptera (lacewings, etc.), Coleoptera (beetles), Trichoptera (caddisflies), Lepidoptera (butterflies and moths), Diptera (flies), Hymenoptera (sawflies, bees, wasps, ants), and other, lesser orders. The secret of the enormous success of many of these groups is believed to lie to a large extent in the neopterous condition—the ability to fold the wings over the back when the insect is not using them. This trick bestows new versatility on the insects that possess it. They are able to fly from one place to another, then upon landing tuck the wings out of the way, and crawl into crevices and other tight places just as wingless insects. The meaning of this adaptation will become especially clear to anyone who tries to catch a flying roach with his hands. Several orders, including the Anoplura (lice) and Siphonaptera (fleas), have returned secondarily to a wholly flightless condition.

The fourth and final evolutionary grade occurs within certain of the neopterous insects. It is the acquisition of holometabolous development, or COMPLETE METAMORPHOSIS *(Figure 21)*. The consequence of this form of life cycle is the extreme specialization of life stages. The larvae became totally adapted to feeding and growing, while the adults became more specialized for reproduction and dispersal. Often the adult acquires different food habits as well, so that the larvae and adults are able to exploit a larger part of the environment between them. Not surprisingly, the holometabolous insects include several of the most successful and diverse of all the insect orders: Neuroptera, Coleoptera, Trichoptera, Lepidoptera, Diptera, and Hymenoptera.

THE LOPHOPHORATE ANIMALS

There exists a small group of coelomate animals, collectively called lophophorate phyla, which are closest to the annelid superphylum but are distinctive enough to deserve separate mention. Their diagnostic trait is the possession of a LOPHOPHORE, a specialized form of food-catching organ that superficially resembles the crown of tentacles of a coelenterate polyp. A closer examination shows the lophophore to be a U-shaped fold of the body wall encircling the mouth and giving rise to numerous hollow ciliated tentacles. The cilia, through coordinated beating, pull a current of water into the lophophore, where small invertebrates are sifted out and swallowed.

The BRACHIOPODS or lamp shells (phylum Brachiopoda) are lophophorate animals that resemble little clams. Like these bivalve mollusks, they possess a mantle and a calcareous shell with two opposing valves that can be pulled shut to protect the soft body inside. The shell differs from that of the mollusks, however, in that the two valves move up and down rather than from side to side. The brachiopods are an ancient group that reached its peak in Paleozoic and Mesozoic times. Among the few types still living in marine waters are the species of the genus *Lingula*, living fossils whose history extends all the way back to the Ordovician period *(Figure 22)*.

LINGULA, a brachiopod, lives in burrows on sea bottom, anchored by its stalk. Its genus has changed very little in the past 400 million years.   22

The BRYOZOANS or MOSS ANIMALS (phylum Bryozoa or Ectoprocta) are by all reasonable criteria one of the most important animal groups, despite the fact that they traditionally receive little attention from biologists and the general public. Most of the 4000 species live in the sea, where they are sometimes dominant elements among the bottom-dwelling organisms *(Figure 23)*. Some species settle thickly on the bottoms of ships and constitute, along with the barnacles, the principal elements of the fouling community that reduces speed and increases the cost of world shipping by hundreds of millions of dollars annually. All bryozoans are colonial. Superficially they resemble the colonial corals, but they differ by their coelomate body cavity, by their use of lophophores to catch food, and by their overall more complex body structure. In some bryozoans the individual colony members are specialized in anatomy and function, some being concerned solely with feeding and others with reproduction or defense.

THE ECHINODERM SUPERPHYLUM

At first glance a starfish and a bird seem even less related than our earlier pairing of squids and but-

23 **FLUSTRELLA is a bryozoan. Ciliated tentacles resemble those of corals, and create currents that drive food particles into the mouth.**

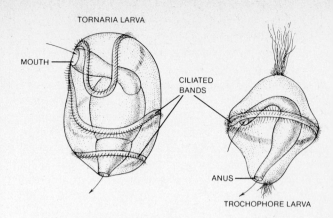

TORNARIA LARVA of echinoderms and primitive chordates is compared with the trochophore larva of annelids and mollusks in diagram above.

terflies. Yet in this case also the evidence compels the comparison. The primitive members of the Echinodermata and Chordata, the phyla to which starfishes and birds respectively belong, are very similar in their early stages of development. Many species in this echinoderm superphylum share a characteristic form of larva, in this case the TORNARIA LARVA, easily recognized by the winding, longitudinal bands of cilia by means of which it swims through the open sea *(Figure 24)*. The echinoderms and chordates are distinguished from other animals by equally fundamental properties in the earliest stages of embryonic development. First, the cleavage of the fertilized egg is INDETERMINATE. If you allow one or a few divisions to occur and then separate the cells with a fine needle, each cell may still develop into a full embryo. In many other animal phyla, by contrast, cleavage is DETERMINATE. Separated cells only develop into parts of embryos corresponding to their original position in the fertilized egg, and the excision of a cell causes the loss of a structure. Second, the cleavage pattern of the echinoderms and chordates is radial, which means that the cells divide along a plane either parallel or at right angles to the long axis of the fertilized egg. In other animals the cleavage is spiral: the plane of division is oblique to the long axis, causing the cells to be arranged in a spiraling pattern. Finally, the mouth of the embryo of echinoderms and chordates originates at some distance from the blastopore. The two phyla are accordingly referred to as the deuterostomes (second mouths). In most other animal groups, referred to as the protostomes (first mouths), the mouth arises next to the blastopore. These three striking embryonic characteristics indicate that ancestors of

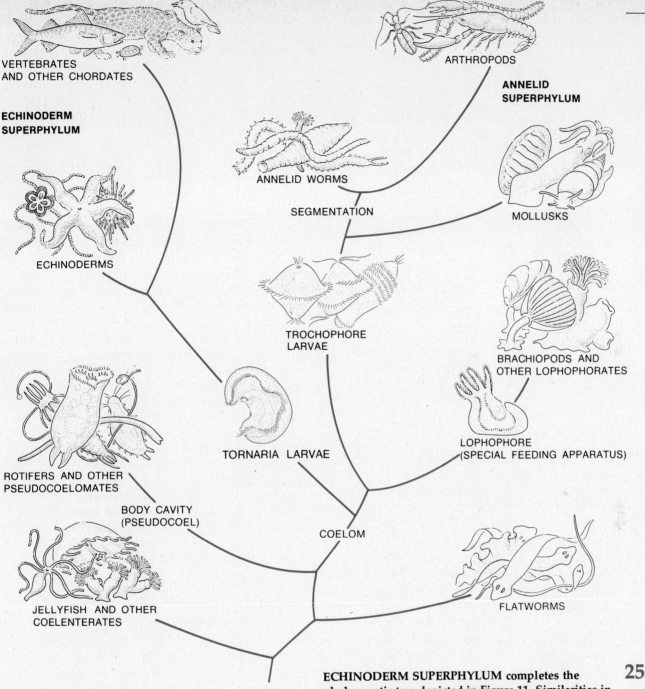

VERTEBRATES
AND OTHER CHORDATES

ARTHROPODS

**ECHINODERM
SUPERPHYLUM**

**ANNELID
SUPERPHYLUM**

ANNELID WORMS

SEGMENTATION

MOLLUSKS

ECHINODERMS

TROCHOPHORE
LARVAE

BRACHIOPODS AND
OTHER LOPHOPHORATES

TORNARIA LARVAE

LOPHOPHORE
(SPECIAL FEEDING APPARATUS)

ROTIFERS AND OTHER
PSEUDOCOELOMATES

BODY CAVITY
(PSEUDOCOEL)

COELOM

JELLYFISH AND OTHER
COELENTERATES

FLATWORMS

**ECHINODERM SUPERPHYLUM completes the
phylogenetic tree depicted in Figure 11. Similarities in
embryonic development, and between the larvae of
echinoderms and those of certain primitive chordates,
provide strong evidence that the two groups arose from
a common ancestor.**

25

the echinoderms and chordates split off from the
rest of the coelomate animals at a very early stage
in their evolution *(Figure 25).*

ECHINODERMS

The phylum Echinodermata includes the sea stars,
sea urchins, sea cucumbers, crinoids, and related
forms. The echinoderms are truly different ani-

mals, so distinctive in body plan from other higher
animals that if they were not already so familiar as
sea animals they might seem to have originated
from some other world. Their single most characteristic feature is the WATER-VASCULAR SYSTEM, an

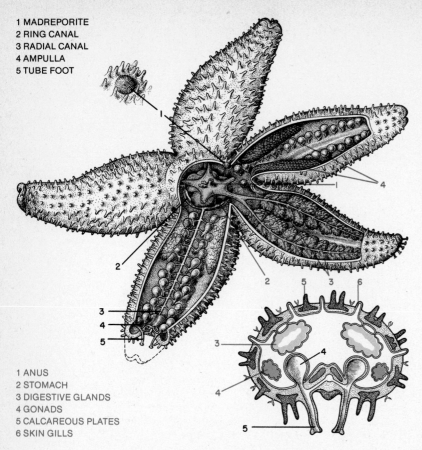

1 MADREPORITE
2 RING CANAL
3 RADIAL CANAL
4 AMPULLA
5 TUBE FOOT

1 ANUS
2 STOMACH
3 DIGESTIVE GLANDS
4 GONADS
5 CALCAREOUS PLATES
6 SKIN GILLS

26 **CUT-AWAY VIEW OF STARFISH reveals the water-vascular system, digestive glands (light color), and sex organs (dark color). Function of water-vascular system is explained in text.**

array of canals and tubelike appendages that serve several functions simultaneously, the principal ones of which are locomotion and the capture of food. The structure of this system can be roughly grasped by examining a diagram such as that of the starfish shown in Figure 26. Although the total functioning of the system has never been fully understood, the mechanical operation of the basic unit, the TUBE FOOT, is clear enough. Each tube foot is a little adhesive organ that creates suction by hydraulic expansion and contraction. When the bulb (ampulla) attached to it contracts, the tube foot elongates. On touching a surface, the terminal sucker on the end of the tube foot draws back slightly, creating a small vacuum. Adhesiveness is simultaneously increased by the secretion of a sticky substance around the sucker. With hundreds of tube feet acting simultaneously a starfish can exert enormous force. It can grasp a clam in its arms, anchor the arms with tube feet, and by

steady contraction of muscles in the arms gradually pull the shell apart. The tube feet serve starfish and other echinoderms in a variety of other ways. They are used in locomotion and as organs of touch. Because of their thin walls, they also function as the equivalent of lungs or gills during respiration.

Echinoderms are distinctively armored. They possess an internal skeleton of calcareous plates that are either articulated, permitting flexibility, or else fused to form a rigid skeletal box. Warty or spiny projections extend outward as extra protective devices—the word echinoderm in fact means spiny skin. Most members of the phylum are radially symmetrical, but embryological studies show the condition to have been derived secondarily from the bilateral symmetry of ancestral forms.

The echinoderms are a very ancient and diverse group. In addition to the starfish (class Asteroidea) there are four major living groups. The brittle stars (class Ophiuroidea) resemble starfish but have long flexible arms that allow them to move rapidly over the bottom or even in a few species to swim. The sea urchins and sand dollars (class Echinoidea) are sluggish, round or ovoid animals whose skeletons are fused into a single box. They lack arms and progress sluggishly over the bottom in search of all forms of organic matter, plant and animal. The sea cucumbers (class Holothuroidea) are also sluggish omnivores. Their cucumber-shaped bodies lack arms and spines, and the skeleton is reduced to tiny plates scattered through the thick, leathery skin. The sea lilies (class Crinoidea) are usually sessile forms attached to the ocean floor by a stalk. A crown of feeding tentacles at the oral end completes their flowerlike appearance.

CHORDATES (PHYLUM CHORDATA)

Most of the chordates in the sea, and all on the land, are vertebrates. The remaining species, the INVERTEBRATE CHORDATES, belong to two subphyla that are extremely dissimilar in outward appearance. They share certain similarities in their embryonic development and display, at some point of their life cycle, the three diagnostic characteristics of the phylum Chordata: 1, a dorsal, hollow nerve cord; 2, clefts in the wall of the throat region, usually referred to as gill slits, which function to circulate water during feeding and respiration; and 3, a notochord, a unique stiffening rod located along the back. The tunicates (subphylum Urochordata) resemble tadpoles in the larval stage. The technical name of the subphylum (tailchorded) derives from

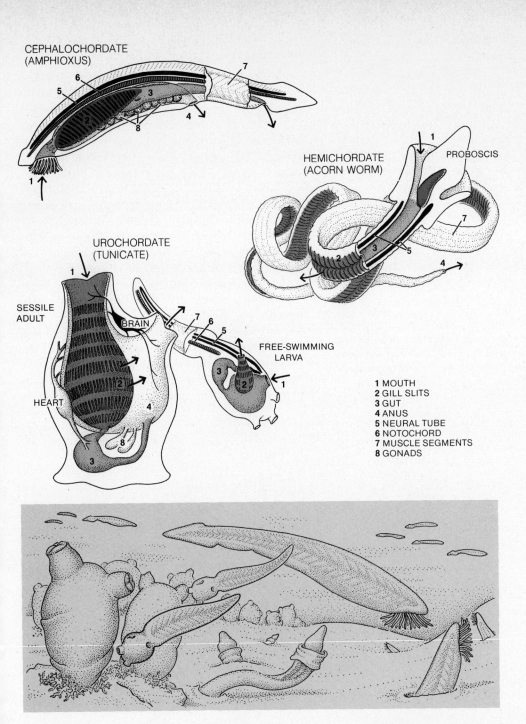

CEPHALOCHORDATE
(AMPHIOXUS)

HEMICHORDATE
(ACORN WORM)   PROBOSCIS

UROCHORDATE
(TUNICATE)

SESSILE
ADULT

BRAIN

FREE-SWIMMING
LARVA

HEART

1 MOUTH
2 GILL SLITS
3 GUT
4 ANUS
5 NEURAL TUBE
6 NOTOCHORD
7 MUSCLE SEGMENTS
8 GONADS

the fact that the notochord is confined to the tail in this stage. The adults are sedentary animals that superficially resemble sponges but have much more complex body structures. Many of the species are colonial. The lancets or cephalochordates (subphylum Cephalochordata), which means head chorded because the notochord extends all the way through the head, are small fishlike animals that burrow in the sand and feed by filtering food parti-

INVERTEBRATE CHORDATES have a notochord, dorsal neural tube, and pharyngeal gill slits. Amphioxus is fishlike, but spends much of its time buried in sands beneath coastal waters, feeding by sifting microorganisms from water passing through its mouth and gills. Acorn worms and tunicates feed in much the same way. Like amphioxus, acorn worms live partially buried, but adult tunicates are exposed and sessile. Tunicate larva is free-living and resembles a tadpole. The drawing below the diagrams depicts these primitive chordates in their marine habitat.

27

cles from the water. The most familiar examples belong to the genus *Branchiostoma* and are better known by their old technical name amphioxus. A third group, usually associated with the invertebrate chordates, are the acorn worms (phylum Hemichordata). These bizarre marine creatures live in burrows and draw suspended food particles to themselves by means of cilia that cover their large proboscis. Because they have gill slits and an organ in the head that used to be considered a short notochord, the acorn worms were formally placed as a subphylum of the Chordata. However, the "notochord" is now known to be an outpocketing of the gut just anterior to the mouth. The absence of this most diagnostic chordate trait is considered ample reason to place the hemichordates in a phylum of their own (*Figure 27*).

28 THE FROG exemplifies the characteristics of a typical vertebrate: segmented vertebrae surrounding the nerve cord; closed circulatory system; and a well-developed nervous system with advanced sensory organs. In advanced vertebrates the notochord is present only in embryos.

VERTEBRATES

The VERTEBRATES (subphylum Vertebrata) rank with the arthropods as dominant animals of sea, land, and air. The fishes in particular are the principal large-bodied carnivores and scavengers of the sea and fresh water, while amphibians, reptiles, and many of the birds and mammals occupy the same position on the land. Other kinds of birds, and a large array of mammals, from mice and rabbits to kangaroos, antelopes, and elephants, are among the most important plant feeders. Although low in numbers of individual organisms, the vertebrates rival the major invertebrate phyla in diversity and ecological significance. But, of course, the single fact of overriding consequence is that the vertebrates produced us, and thus deserve our closest attention.

A vertebrate is an animal with a series of segmented bones, vertebrae, which surround the notochord and nerve cord (*Figure 28*). In the evolutionarily more advanced groups the notochord is present only during embryonic stages

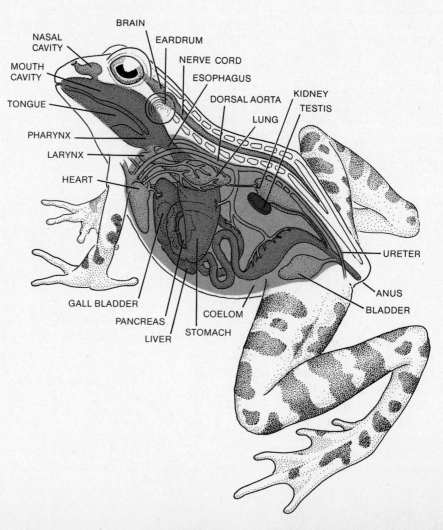

BRAIN
NASAL CAVITY
EARDRUM
MOUTH CAVITY
NERVE CORD
ESOPHAGUS
TONGUE
DORSAL AORTA
KIDNEY
TESTIS
LUNG
PHARYNX
LARYNX
HEART
URETER
ANUS
BLADDER
GALL BLADDER
COELOM
PANCREAS
STOMACH
LIVER

and is supplanted in its protective role by the vertebrae during most of the life of the organism. The vertebrate organism is further characterized by a complicated, closed circulatory system in which the blood, containing hemoglobin-filled red blood cells, is pumped by a chambered heart through dense systems of capillaries that penetrate deeply into all of the active tissues of the body. This better-designed circulatory system makes it possible for animals to grow large in size while remaining physically very active. A second trait associated with increased size and activity is the better-developed nervous system. The brain of even the most primitive vertebrates is superior to that of the invertebrates. In the birds and mammals it has advanced to a wholly new level of complexity and organization. A third concurrent development is the improved sensory apparatus, including large eyes capable of sharp image perception, ears that serve in primitive forms as organs of equilibrium and in higher forms for both equilibrium and hearing, and, in the more primitive, aquatic forms, a lateral-line system that provides a sixth sense capable of detecting slight changes in water pressure and currents.

Where did the vertebrates come from? The living invertebrate chordates are the logical animals to consider as possible ancestors. At first glance these strange little creatures seem like very improbable candidates. Yet one group, the tunicates, does possess the requisite traits. The larvae, which superficially resemble minute tadpoles, have long tails containing a notochord and a large part of the nerve cord. After hatching from the egg each larva swims about for a short time, then settles down to the bottom, loses its tail and notochord, and is transformed into a spongelike adult. If the tunicates are the ancestors of the vertebrates, it might seem that they gave rise to these advanced organisms while their entire lives were spent in the tadpole stage; then they went on to evolve the spongelike adult stage. This is an appealingly straightforward guess, but it is equally conceivable that the opposite happened. According to one popular current theory, the vertebrates arose by neoteny—the takeover of the reproductive function by the larval forms and the elimination of the original adult form from the life cycle. The tadpolelike larvae, in other words, evolved directly into the first vertebrate adults by eliminating the spongelike adult stage.

Neither version of the theory of the tunicate origin of vertebrates can be tested directly, because the primordial vertebrates have not yet been found

LAMPREY EEL, shown clinging to a catfish, belongs to a group of jawless fishes (agnaths) that have survived for more than 425 million years. Sea lampreys are blood-sucking parasites. Other lampreys feed on dead or dying fish. Modern lampreys are very different from their Paleozoic ancestors.

29

in the fossil record. They are, however, closely approached by the jawless fishes, or AGNATHS (class Agnatha), a group that dates back at least to Ordovician times. A diversity of agnaths, many of them possessing strangely shaped, armored bodies, flourished over a period of tens of millions of years, then disappeared almost completely as more advanced kinds of fishes became abundant. Only a few relict forms—the lampreys, hagfishes, and slime eels—survived beyond the Devonian period down to the present (Figure 29). All of these creatures are degenerate predators, making their living either by sucking blood from living fish or eating the flesh of dead or dying fish. Their eel-like bodies have completely lost the bones that characterized the early agnaths.

In the Devonian period, generally referred to as the Age of Fishes, an immense variety of more advanced types came to throng the seas and fresh water in company with the agnaths (Figure 30). Among the most primitive were the archaic placoderms (class Placodermi). These jawed fishes

Sharks and their relatives "chose" speed and mobility over the sluggish existence of most earlier fishes.

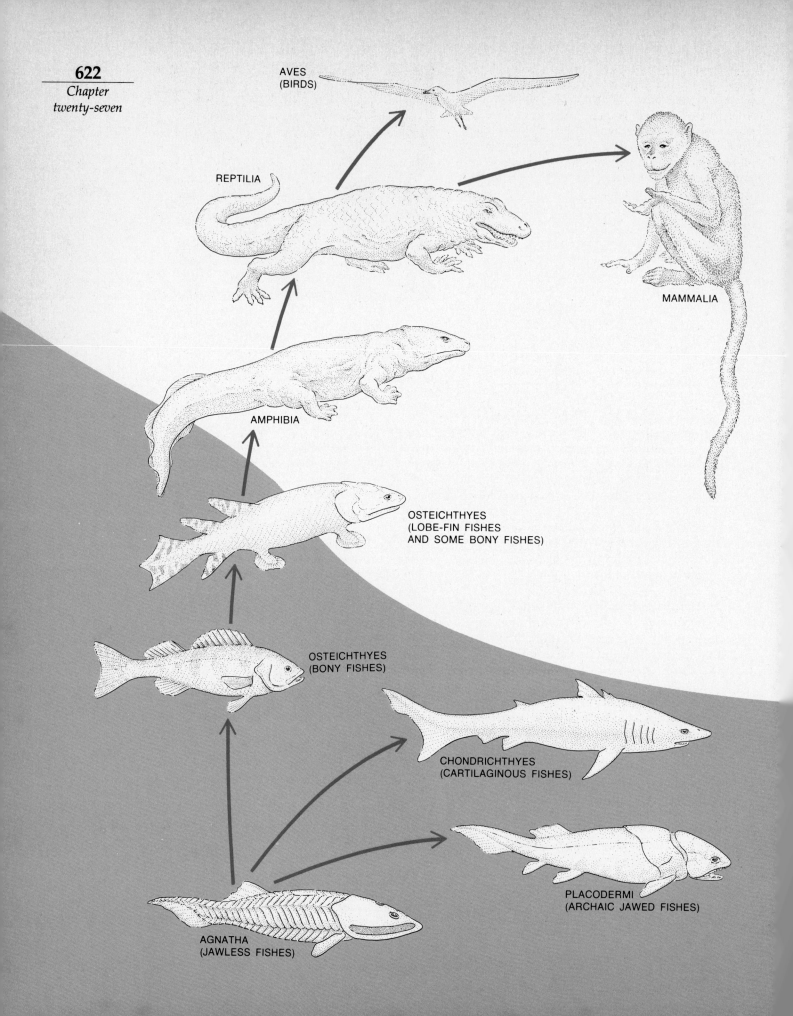

AVES
(BIRDS)

REPTILIA

MAMMALIA

AMPHIBIA

OSTEICHTHYES
(LOBE-FIN FISHES
AND SOME BONY FISHES)

OSTEICHTHYES
(BONY FISHES)

CHONDRICHTHYES
(CARTILAGINOUS FISHES)

PLACODERMI
(ARCHAIC JAWED FISHES)

AGNATHA
(JAWLESS FISHES)

EVOLUTION OF VERTEBRATES. The first vertebrates are thought to have resembled the jawless fishes (agnaths) that later virtually disappeared as more advanced fishes became abundant. Lobe-fin fishes gave rise to amphibians, which marked the beginning of the transition to land. Reptiles were first true land animals, and they in turn gave rise to both birds and mammals.

attained a distinctly higher evolutionary grade than the agnaths. Their jaws were derived from some of the cartilaginous or bony HYOID ARCHES that support the gill region *(Figure 31)*. Many also developed elaborate sets of swimming fins and sleek body forms that must have improved their maneuverability in the open water. A few attained huge size and were probably the top-level predators of the world at that time. The placoderms, however, were a short-lived group. All but a very few disappeared by the close of the Devonian period, and none survived to the end of the Paleozoic.

During the Devonian two other major groups began careers that were to be crowned with success down to the present time. The first were the sharks, skates, and chimaeras—the chondrichthyans (class Chondrichthyes). Their bodies possess a single type of stabilizing system which they share with all other higher fishes: a pair of pectoral fins just behind the gill slits, and a pair of pelvic fins just in front of the anal region. By observing a fish swimming in a tank, one can quickly appreciate the role of these appendages in providing balance and mobility. The chondrichthyans are distinguished from other higher fishes by the fact that their internal skeleton is composed entirely of cartilage, without any trace whatever of bony elements. The skin, unlike that of many early agnaths and placoderms, is not armored. For the most part it is flexible and leathery, sometimes bearing bristly projections that give it sandpaperlike consistency. The sharks and their relatives "chose" speed and mobility over the more sluggish, turtlelike existence of most earlier fishes. Another notable fact about the chondrichthyans is that they originated in the sea, and they have only rarely been able to penetrate bodies of fresh water. They employ a unique device to cope with the problem of water loss through osmosis. Their blood contains large quantities of dissolved urea which, in combination with trimethylamine oxide and the ordinary blood salts, keeps the body in osmotic balance with the sea water.

The "real" fish, in the minds of most persons, are the BONY FISHES or osteichthyans (class Os-

teichthyes). They are indeed the most abundant and diverse of the fish classes in both fresh and salt water. Their variety is so great that no concise definition of the class they comprise is possible. As their name implies they have a more extensive internal bony skeleton than all the other fish groups. The gill slits, which open directly to the outside in sharks and other chondrichthyans, open into a single chamber in the osteichthyans. The chamber is covered by a flap, usually bony in nature, called the operculum. The osteichthyans possess the two pairs of lower fins just described in the chondrichthyans. But they are basically different because they originated in fresh water. The numerous marine groups that we know today are secondary invaders of the oceans who did not adopt the chondrichthyan method of using blood urea to cope with osmosis. Instead, they make up for the loss of water through the gills and other permeable membranes by drinking sea water continuously and urinating sparingly. The excess salt that is unavoidably absorbed with the sea water is then excreted through the gills. The primitive bony fish possessed lungs, which supplemented the gills in respiration. Perhaps they used these extra organs to overcome occasional oxygen short-

HYOID GILL ARCHES evolved into the jaws of the 31 placoderms that flourished in Devonian seas. Gill arches of agnaths are shown for comparison.

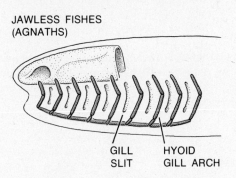

JAWLESS FISHES
(AGNATHS)

GILL SLIT   HYOID GILL ARCH

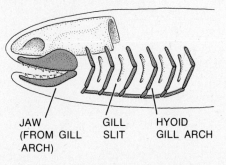

PRIMITIVE JAWED FISHES

JAW (FROM GILL ARCH)   GILL SLIT   HYOID GILL ARCH

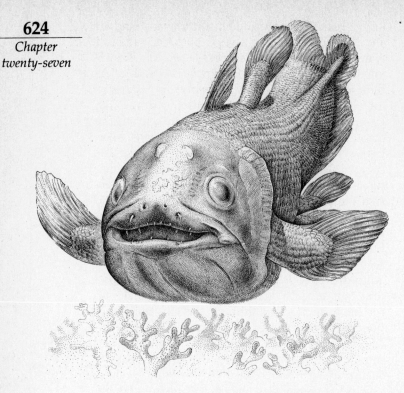

32 LATIMERIA is a lobe-fin fish, a member of a group thought to be extinct for 75 million years. Many have been caught in the last three decades. Presence of lungs marks it as a link between life in the sea and life on land. Bony "stalks" attach lobed fins to body.

33 LEGS OF AMPHIBIANS evolved from the bony fins of lobe-fin fishes. The fishes were probably able to crawl from stream to stream or from pond to pond on their stubby fins.

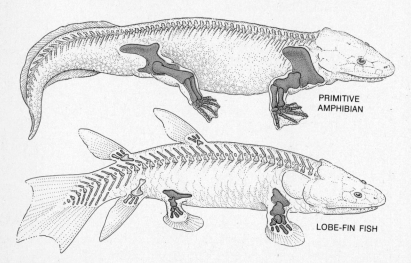

PRIMITIVE AMPHIBIAN

LOBE-FIN FISH

ages encountered in the fresh-water habitats. The great majority of later bony fish converted the lungs into swim bladders, which now serve as organs that help keep the fish suspended in the water. Some of the bottom-dwelling forms went on to eliminate the swim bladders. Today, only a tiny group of species, the lungfishes, still use the lungs for the original purpose of respiration. These are all fresh-water forms that occupy limited ranges in the tropical portions of South America, Africa, and Australia.

The invention of lungs by the early bony fishes set the stage for the invasion of land by their descendants. The osteichthyans which accomplished this feat were the crossopterygians, or LOBE-FIN FISHES, close allies of the lungfishes. This group flourished from Devonian into Mesozoic times. It used to be thought that they had become extinct about 75 million years ago. Then, in 1939, a living crossopterygian was caught by a commercial fisherman from waters off the coast of East Africa. Since that time diligent search has yielded many more specimens of this extraordinary fish, which was given the name *Latimeria (Figure 32)*. Anatomists have taken full advantage of the unexpected and rare opportunity to conduct a detailed study of a genuine "missing link."

The first land vertebrates, the AMPHIBIANS (class Amphibia), arose from crossopterygians in Devonian times. This epochal event *(Figure 33)* did not require a new design of the respiratory system, for the ability to breathe air had been achieved earlier when the ancestral bony fishes evolved lungs at the beginning of their history. Instead, the crucial step was the evolution of the stubby fins of the crossopterygians into the walking legs of the amphibians—legs whose basic design has since been bequeathed all the way on down to us in their basic design. Quite likely the Devonian crossopterygians were able to crawl from one pond or stream to another by pulling themselves along on their fins. The earliest amphibians, which remained very fishlike in most of their body structure, merely perfected this particular locomotory ability.

Most of the modern amphibians remain chained to the water for at least part of their life cycle. The bulk of the anurans (frogs and toads) and salamanders spend part or all of their adult lives on the land, typically in a moist habitat, but return to fresh water to lay their eggs. The primitive structure of the egg requires this reversion to the ancestral home. The eggs are small, surrounded by delicate membranes, and contain only limited supplies of yolk. Usually they give rise to an aquatic larva, such as the tadpole of the frogs and toads, which pursues a fishlike existence for some period of time before transforming into the terrestrial adult form.

To understand the vulnerability of the amphibians is to understand a great deal about the advance in evolutionary grade that brought about the origin of the REPTILES (class Reptilia). The primitive reptiles arose from primitive amphibians in the Upper Carboniferous Period at the latest, some 310 million or more years ago. The reptiles became the first true, fully liberated land animals. Those that live mostly in water, such as the majority of turtles, entered this habitat later as a secondary adaptation. The liberation of the reptile life cycle from the water was achieved by the AMNIOTIC EGG *(Chapter 13)*. This type of egg is intimately familiar to everyone. It is preserved in hens and other birds which are the direct descendants of reptiles. It has a leathery or brittle, calcium-impregnated shell that resists evaporation of the precious fluids inside. Within the shell and surrounding the embryo are several membranes, including the amnion from which the egg gets its name, that give added protection from desiccation and assist the embryo in excretion and respiration. Finally, the embryo is supplied with large quantities of yolk that permit it to attain a relatively advanced state of development before it must break the shell and face the outside world. Such an egg need not be laid in the water. It can be deposited on the land, even in a dry place, a circumstance that permits the adult

> During the Mesozoic era reptiles became the dominant large animals of the land. By the end of the Mesozoic most of the great assembly had disappeared, to be replaced in the Cenozoic by an equally impressive group of mammals.

reptile to move for unlimited distances away from bodies of fresh water.

Reptiles are advanced beyond the amphibians in other ways. Fertilization is achieved within the body of the female by copulation, sparing the adults from the necessity of shedding their gametes in water and permitting mating on land. The ventricle of the heart is partially divided into chambers that separate the freshly oxygenated blood from unoxygenated blood and permit it to be pumped to needy tissues more efficiently (for more details, see page 347). Respiration is improved by a

NILE CROCODILE. Crocodiles are the closest living relatives of dinosaurs on one hand and birds on the other. Photograph reveals two of the features that distinguish crocodiles from alligators: narrow snout and prominent fourth tooth of lower jaw, which is exposed even when jaw is closed. Closable "lid" on nostril is an adaptation to a secondarily aquatic existence.

34

bellows-like movement of the ribs. Even the brain is more advanced. It is in the reptiles that we find the first small cerebral hemispheres, which in the mammals later became the centers of the higher mental faculties. The total result of all these changes was the transformation of the reptile into a more alert and adaptable animal *(Figure 34)*.

During the Mesozoic era, often called the Age of Reptiles, the reptiles went through an extraordinary diversification and became the dominant large animals of the land. During this time two orders, the Ornithischia and Saurischia, popularly called the dinosaurs, were the prevalent large

35 FEATHERS. Drawing at top depicts wing of the herring gull. Only the principal feathers are shown, and the cartilage that supports them has been omitted to reveal bones of bird's upper arm (far right), forearm (right center), wrist and hand (center). Outer half of wing, from wrist to wingtip, is adapted for control and propulsion; inner half, for lift. Drawing at center shows one of the primary feathers from wingtip. Vane consists of parallel barbs on either side of shaft (rachis), as shown in detailed view at lower right. Overlapping barbules are engaged by tiny hooks that allow them to slide relative to one another, which explains the flexibility of the vane. The intricate meshwork also gives vane a light and relatively airtight surface, well adapted for flight and insulation.

ARRANGEMENT OF FEATHERS IN WING

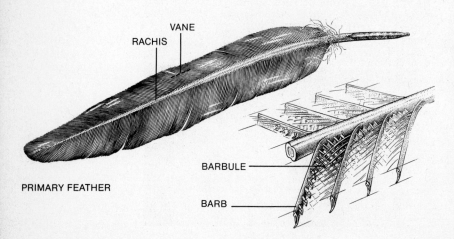

VANE
RACHIS
PRIMARY FEATHER
BARBULE
BARB

reptiles. Recent evidence suggests that many of the dinosaurs were as warm-blooded and agile as modern mammals. Some are even believed to have had coats of fur. But many other forms flourished, from the familiar and relatively humble turtles, snakes, and lizards to extreme aquatic and marine groups such as the plesiosaurs, ichthyosaurs, and mosasaurs. By the end of the Mesozoic, most of the great assembly disappeared, to be replaced in the Cenozoic by an equally impressive group of mammals.

Well into the Mesozoic, at least three separate lines of reptiles achieved the capacity for sustained flight by a flapping movement of wings. Two of the groups, referred to collectively as pterosaurs (winged reptiles), still retained a basically reptilian anatomy. They became extinct by the close of the Mesozoic era. The third group advanced so far as to deserve the distinction of being placed in a class of their own. These are of course the BIRDS (class Aves), who survived the demise of the ruling reptiles to become one of the dominant vertebrate groups of the Cenozoic era. Zoologists sometimes lightly refer to birds as "glorified reptiles" or "hot lizards," and there is an important truth embodied in the jest. Virtually all of the important differences between the birds and their reptilian ancestors have arisen as adaptations to flight. The single most characteristic feature is their feathers, which are highly modified versions of the old reptilian scales *(Figure 35)*. The flying surface of the wing is created by large quills that arise from the forearm and from the reduced, stubby fingers. Other strong feathers sprout like a fan from the shortened tail and serve as a stabilizer during flight. Still other feathers, the contour feathers and down feathers, cover the body like a tight-fitting garment and provide the insulation needed in the control of body heat. The body skeleton was extensively modified to serve the requirements of flight. One of the most conspicuous changes was in the shape of the sternum, which has been transformed into a large vertical keel for the attachment of the breast muscles. These muscles (the familiar white or breast meat of chicken and other fowl) function to pull the wings downward during the main propulsive movement in flight. Flying also requires a constant high temperature, so birds have become warm-blooded. It requires a highly efficient circulation, which has been achieved by the complete division of the ventricle into two chambers. One chamber pumps "used" blood to the lungs; the other receives the freshly oxygenated blood from the lungs and pumps it to the rest of the body. The brain has been greatly enlarged over that of reptiles. The change is not in the cerebral cortex, the principal seat of intelligence in mammals, but rather in the centers of sight and muscular coordination. The beaks of modern birds completely lack teeth (truly nothing is scarcer than a hen's tooth), but primitive forms from the Mesozoic, such as *Archaeopteryx (Figure 36)*, still had teeth. *Archaeopteryx* was intermediate between the ancestral reptiles and modern forms in other ways as well. Although it was covered by feathers and had well-developed wings, the fingers of its forearm had not been much reduced, and it still possessed a long tail. Also, its brain case and sternum were not significantly advanced beyond the reptilian condition.

Some recent authors, the same ones who consider many dinosaurs to have been warm-blooded, regard the birds to be little more than flying members of this group. They place the two kinds of animals in a single class, the Dinosauria. In other words, the dinosaurs did not become extinct at the end of the Mesozoic Era. They are still singing outside our windows!

MAMMALS

In a real sense the highest evolutionary grade of all animals is the MAMMAL (class Mammalia). About 63 million years ago —only 11 hours ago on our calendar —the Mesozoic era, the Age of Reptiles, gave way to the Cenozoic era, the Age of Mammals. The change was not one simply of mammals arriving on the scene to displace an obsolete set of reptiles. On the contrary, these animals appeared in the early part of the Mesozoic and coexisted as lesser partners of the ruling reptiles for tens of millions of years. When the dinosaurs and most other dominant reptile groups disappeared at the close of the Mesozoic, the mammals increased dramatically in numbers and diversity. Why the reptiles declined is not really known, and it remains the subject of perennial speculation. It has often been suggested that the mammals gained enough momentum to assist in the process. Whether or not they did play the role of exterminators, they were the inheritors of that part of the ecological world earlier controlled by the reptiles.

We should not hesitate to designate ourselves, and other mammals, as superior animals, physically and mentally *(Figure 37)*. Virtually everything that is distinctively mammalian is geared to an increase in speed, in alertness, in intelligence, and in protection given the young. As the name implies

ARCHAEOPTERYX was a primitive bird of the Mesozoic era. This cast of a fossil impression reveals obvious reptilian features, particularly the long bony tail. The head is bent back, so jaws and teeth are not clearly visible. Birdlike features include feathers and well-developed wings.

36

(from *mammae*, or milk glands), only the mammals suckle their young with nutritive fluid. Except in the most primitive forms, the embryos are not deposited in shelled eggs but are nurtured within the body of the mother until they reach an advanced stage of development. The jaw is strengthened, and teeth are differentiated into types variously specialized for cutting, chewing, and grinding. A muscle wall, the DIAPHRAGM, completely separates the chest cavity from the abdominal organs and increases the depth and efficiency of breathing. The heart, like that of birds, is improved by a complete division of the ventricle into two chambers. The legs are swung beneath the body, permitting a clean back-and-forth motion of the legs and greater speed than is possible with the sprawling pos-

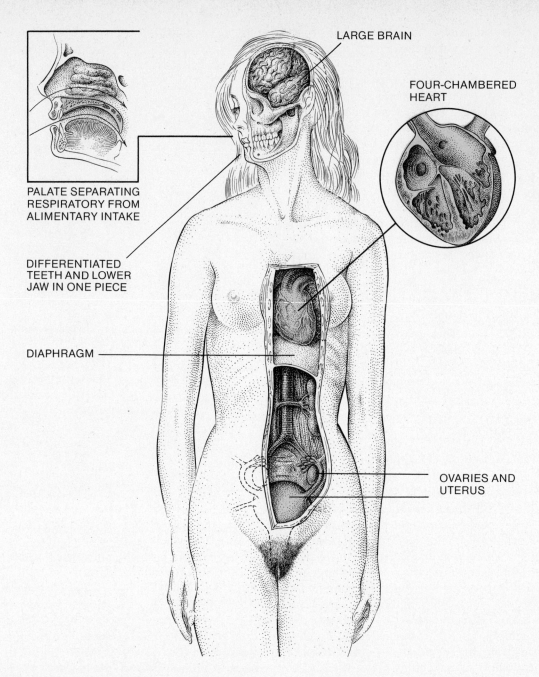

LARGE BRAIN

FOUR-CHAMBERED
HEART

PALATE SEPARATING
RESPIRATORY FROM
ALIMENTARY INTAKE

DIFFERENTIATED
TEETH AND LOWER
JAW IN ONE PIECE

DIAPHRAGM

OVARIES AND
UTERUS

37 **WOMAN AS A MAMMAL. In addition to mammary glands, several other anatomical traits are characteristic of the class Mammalia. Some of them are identified above; others are discussed in the text. Uterus is somewhat enlarged, as it would appear early in pregnancy. Features illustrated here are typical of placental mammals; marsupials have an external pouch for carrying young.**

ture of the reptile. The mammal is warm-blooded; its temperature is kept high and relatively constant, allowing it to be active in a wide range of temperatures and climates. Its skin is typically covered with hair, which provides mechanical protection and aids in heat conservation. The brain is greatly enlarged, chiefly in the cerebral hemispheres. The key to mammalian success is internal activity guided by intelligence. Almost every living mammal is capable of more learned behavior than any living amphibian or reptile.

READINGS

R.M. Alexander, *The Chordates*, New York, Cambridge University Press, 1975. A comprehensive and easily comprehended account of the biology of vertebrates and other members of the phylum Chordata.

R.D. Barnes, *Invertebrate Zoology*, 3rd Edition, Philadelphia, Saunders, 1974. One of the best general textbooks, covering all groups from the Protozoa to the invertebrate chordates.

F.M. Bayer and H.B. Owre. *The Free-Living Lower Invertebrates*, New York, Macmillan, 1968. Beautifully written and illustrated, this is an authoritative and esthetic introduction to the acoelomate animals.

D.J. Borror, D.M. DeLong, and C.A. Triplehorn, *An Introduction to the Study of Insects*, 4th edition, New York, Holt, Rinehart, and Winston, 1976. Workmanlike and clearly written introduction to insect diversity.

W.D. Russell-Hunter, *A Biology of Higher Invertebrates*, New York, Macmillan, 1969. A somewhat more advanced treatment of functional anatomy and evolution. It gives a scholar's view of the current "state of the art" of invertebrate zoology.

Part The strategy of evolution

Evolution is the ultimate existential game, a game most species are doomed eventually to lose. Defeat — extinction —is final. Victory brings only the meager privilege of staying for another round, a round in which nature is both the opponent and the umpire, selecting the winners by rules that change unpredictably. In one generation a species may have to survive prolonged drought; in the next, unexpected cold, followed perhaps by a food shortage or a new disease.

The species alive today, from *E. coli* to man, comprise a small group of temporary winners. Shaped by the interaction of life and Earth for roughly four billion years, modern organisms are formidable competitors, genetically adapted to survive in harsh and variable environments. At times they seem to follow a collective strategy for survival. Strictly speaking, however, living organisms do not consciously evolve "in order to" meet a particular environmental challenge. The teleological idea that creatures shape their evolutionary responses is biologically unsound. Species evolve because certain combinations of genes possessed by some organisms enable those organisms to meet the test of the environment better than others, and hence these organisms are able to survive and reproduce when others die. In retrospect we can observe certain trends in evolution —patterns of play common to all living things — and for verbal convenience we term them a strategy. In this sense, then, we can say that life has staked its survival on a strategy of expansion and diversity. Each species tends to multiply and occupy new territory, expanding aggressively until halted by natural conditions: a food shortage, a geographic barrier, predators, and so on. Diversity enables life to hedge its bets in the contest with nature. The more kinds of organisms that exist, the smaller the chance that all life

631

might be destroyed by any one environmental disaster.
And an assortment of organisms can exploit the resources
of an environment more fully than can a single species.

Diversity is a result of genetic variability. The most successful players of the game of evolution somehow strike a
balance between adaptiveness (the physiological capacity
to cope with their present environment) and adaptability
(the capacity to produce offspring with new combinations
of genes, combinations that may be better suited to future
environments).

Mutation and sexual reproduction are the mechanisms
that create genetic variability, mutation by creating new
genes and sexual reproduction by rearranging genes into
new combinations. Imagine that two individuals in a
species both possess *Aa* genotypes for a single locus. If
they could only reproduce asexually their offspring would
all be *Aa*. But if they could mate, their offspring would be a
mixture of *AA*, *Aa* and *aa*. Collectively the sexual offspring
could adjust to environmental conditions that favored any
of the three genotypes. They are more adaptable than the
otherwise equivalent asexual offspring.

The more variable a species, the more heterozygous
combinations it contains, the faster it can evolve to adjust
to new conditions in the environment. But the greater its
variability, the less perfect the adaptation to any particular
environment at a given moment. The genotype *AA* may be
superior in one environment and *aa* in another. Adaptiveness and adaptability are antagonistic requirements, and
each species must maintain a degree of genetic variability
that represents a compromise between the two.

Evolution is not a game of individuals but of groups of
organisms. The basic unit of evolutionary studies is the
population —a group of individuals of the same species
who exchange genes by interbreeding. For this reason the
study of evolution is to a large extent a branch of population biology.

Individual mortality is part of the price that a species
pays for long-term adaptability. Each individual is doomed
to perish regardless of whether or not his species survives.
The introduction to Part II suggested that if individual organisms developed the potential to live forever, and ceased
reproduction to avoid the risk of being replaced by their
offspring, there would be no evolution. A species of im-

mortals would, in effect, be staking its existence on an all-or-nothing solution to the problems of survival. The odds against success would be overwhelming. Living organisms must continually invent new solutions to the challenge of the environment. Nature selects the best solutions and in the process chooses the players for the next round of evolution.

Chapter 28, *The Process of Evolution,* introduces the basic theories of evolution. Several decades ago ideas about evolution were based largely on the fossil record of changes in the gross anatomy of plants and animals. Today geneticists recognize that evolution begins with a change in a molecule of DNA. Detection of the resulting alteration in the structure of a cell protein (often an enzyme) enables investigators to trace evolutionary relationships at the molecular level. Such studies provide an independent perspective on earlier hypotheses about the branching points of the tree of evolution *(Chapter 29).*

The remaining chapters in this section provide a glimpse of the game in action, beginning with the multiplication of species *(Chapter 30).* The factors that govern the distribution of animals and plants are described in the chapter on biogeography *(Chapter 31)* and in the following two chapters, which treat populations and ecosystems. The section ends with a special review of the origin of man. This final chapter is based on the proposition that our place in the universe will not be grasped until the human species is fully understood as a product of biological evolution.

There is grandeur in this view of life, with its several powers, having been originally breathed into a few forms or into one; and that . . . from so simple a beginning, endless forms most beautiful and most wonderful have been, and are being, evolved. —Charles Darwin

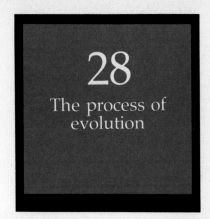

28

The process of evolution

Evolution—is it a fact or a theory? This question echoes an old and virulent controversy. It is important not only historically but also because of the light it can still shed on the distinction scientists make between fact and theory.

The process of evolution is a fact. It occurs. Biologists have watched and measured its progress at the level of the gene. They have created new species in the laboratory and in the experimental garden. They have collected a very large amount of fossil evidence, in many cases so complete that it cannot be rationally explained by any nonevolutionary hypothesis. On the other hand, *how* evolution occurs is a complex matter subject to theory. Of the two leading theories of the nineteenth century, Lamarckism and Darwinism, the Darwinian explanation has increasingly gained acceptance until now it is almost universally used to account for evolutionary phenomena. Moreover, the modern version of Darwinism has been aligned so consistently with genetics, paleontology, systematics, and other branches of biology, that it must be regarded as one of the most firmly grounded and reliable explanatory systems in all of science.

LAMARCKISM

To understand Darwinism fully, one needs to look back briefly in history to see how it contrasts with its chief rival. In 1809 Jean Baptiste de Lamarck, the great French systematist, proposed a very ingenious theory of evolution. He suggested that organisms evolve into new kinds of organisms be-

H.M.S. BEAGLE, depicted on opposite page, was the ship that carried young Charles Darwin on round-the-world voyage that led to his theory of evolution. Ship was painted in 1841 by Owen Stanley as she lay at anchor in Sydney harbor.

cause, in a nearly literal sense, they want to. They try to do something new, or they try to improve on what their ancestors were doing before them, and the effort results in a change in the size of the particular organ or capacity that is exercised. If this change is passed on in some degree to the offspring of the organism that made the effort, the result is evolution. Lamarck cited some arresting examples from nature that he believed could be explained by his theory. Giraffes have long necks, he argued, because for many generations they have stretched their necks in an attempt to reach the more succulent branches and leaves high up in trees. As the members of each generation succeeded in lengthening their necks, they passed some or all of this acquisition to their offspring, who were then in a position to add to it still further on their own. The stork, to take a second of Lamarck's hypothetical examples, has a long neck because it has attempted over many generations to catch fish in ever deeper water. To this Lamarck added an extra bit of speculation of the kind that often got him into trouble with his contemporaries: the long legs of the stork can be explained by the fact that the bird does not like to get its belly wet.

Lamarckism, also known as the theory of evolution by the inheritance of acquired characteristics, actually has some good features as theories go: it is simple, clear provocative, and easy to test. But of course it is also wrong. It is clear from a modern understanding of genetics that DNA directs the formation of the phenotype—the length of the giraffe's neck, for example—but does not accept instruction back from the phenotype. Of necessity something else directs the modification of DNA.

DARWINISM

The directing force was described independently in 1858 by Charles Robert Darwin and Alfred Russell Wallace in the *Journal of the Linnean Society of London.* In 1859 Darwin published the book *On The Origin of Species by Means of Natural Selection.* This classic work became the means by which the theory gained support and, in the process, came to be called Darwinism.

The theory of Darwin and Wallace can be expressed as follows: Every population of individual organisms contains genetic variability. Some of the hereditary traits permit individuals to survive and reproduce better than others. As a consequence, these superior traits become more prevalent in later generations. It follows that evolution has occurred. As long as there is genetic variability in the population evolution can continue.

Table I presents the approximate form of the original 1858 argument. This concise statement contains certain expressions used by Darwin and Wallace and other writers during this period in the history of biology. Two key phrases, "struggle for existence" and "survival of the fittest," have found their way into everyday speech, and contributed an extraordinary amount of emotional heat to popular discussions about evolution down to the present day. The fundamental difference between Lamarckism and Darwinism can be deduced from the table. Lamarckism is a primitive conception based on the idea that change is transmitted in a directed manner from individual to individual. Darwinism, in contrast, is based on the concept of the population as the unit of evolutionary change. In the Darwinist view it is the hereditary content of

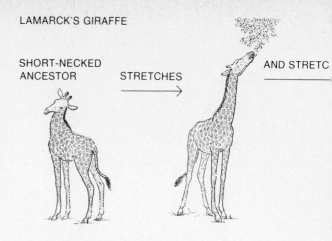

LAMARCK'S GIRAFFE

SHORT-NECKED ANCESTOR   STRETCHES →   AND STRETC

RIVAL EXPLANATIONS OF EVOLUTION were advanced by Lamarck and Darwin. According to Lamarck, the long neck of the giraffe arose as a result of generations of stretching to reach food. According to Darwin, giraffes with longer necks arose as a result of natural selection.

the population as a whole, and not just that of the single individual, which evolves *(Figure 1).*

Natural selection, as conceived in Darwinism, embraces a vast array of phenomena that affect the survival and reproduction of organisms. It occurs whenever some genotypes produce more representatives in the next generation than others. This greater fitness can come from a superior resistance to parasites and predators, or to cold, heat, drought, and other extremes of weather. It can result from a greater capacity to invade habitats not previously occupied by the species. It can also result, of course, from the mere capacity to breed faster. Any or all of these qualities in combination can serve as the components of fitness. With each

I **DARWIN-WALLACE THEORY** of natural selection is a logical system of postulates and deductions. Brief synopsis in this table illustrates the conceptual framework of the theory.

POSTULATE	DEDUCTION
EACH POPULATION OF PLANTS OR ANIMALS TENDS TO GROW GEOMETRICALLY—THE MORE INDIVIDUALS THAT EXIST, THE FASTER THEIR NUMBER INCREASES. BUT THE SPACE AND FOOD THEY HAVE AVAILABLE TO LIVE ON INCREASE SLOWLY OR NOT AT ALL.	THERE IS A CONTINUING STRUGGLE FOR EXISTENCE AMONG THE MEMBERS OF THE GROWING POPULATION.
HEREDITARY DIFFERENCES EXIST AMONG MEMBERS OF THE POPULATION THAT AFFECT THEIR ABILITY TO SURVIVE AND TO REPRODUCE.	THE RESULT IS A CONTINUING PROCESS OF THE SURVIVAL OF THE FITTEST (NATURAL SELECTION).
NEW HEREDITARY VARIATION CONTINUES TO APPEAR IN THE POPULATION INDEPENDENTLY OF THE SELECTION PROCESS.	ORGANIC EVOLUTION OCCURS.

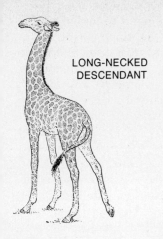

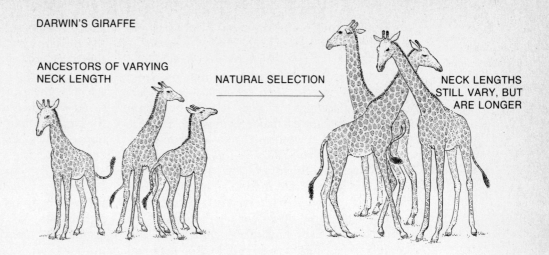

DARWIN'S GIRAFFE

LONG-NECKED
DESCENDANT

ANCESTORS OF VARYING
NECK LENGTH

NATURAL SELECTION

NECK LENGTHS
STILL VARY, BUT
ARE LONGER

passing generation the superior genotypes will become more numerous relative to the less fit ones.

To complete this process of evolution one must imagine, with Darwin and Wallace, that new genotypes appear spontaneously in the population. They are then tested against the other genotypes by natural selection. Some will prove superior, and increase in numbers. Others will prove inferior, and disappear. If new genetic variation originates spontaneously and is subject to natural selection as a continuing process, evolution can go on indefinitely. In time each population will replace a large part of its genetic material. The result of this process is the creation of a very different kind of organism.

THE MODERN CONCEPT

The fundamental idea of the Darwin-Wallace theory is still regarded as essentially correct by biologists. It is now possible, however, to redescribe the evolutionary process in the language of modern genetics. Evolution can be broadly defined as a change in the heredity of a population. Population genetics permits an even more precise definition: evolution is any change in gene frequency in a population. More precisely, it is the change in relative frequency of two or more alleles occupying the same chromosome locus within a given population. Suppose that we are following the change of frequency of two such alleles, one of which can be labeled A and the other a. Each organism in the population will therefore consist of one of the following three diploid combinations: AA, Aa, or aa. The allele A can be dominant over a, or not; this particular distinction does not matter at this point. Suppose further that in a certain population under

study, 60 percent of all the alleles at the locus were found to be A, and 40 percent a. The GENE FREQUENCIES are therefore 60 and 40 percent respectively; let us use the decimal notation to relabel them as 0.60 and 0.40. Employing the usual symbols of genetics, the first frequency is denoted p and the second q. Then $p + q = 0.60 + 0.40 = 1.00$. Suppose that we followed the frequencies p and q over three successive generations and recorded that they changed from one generation to the next as follows: $0.60 + 0.40 = 1.00$; to $0.59 + 0.41 = 1.00$; to $0.57 + 0.43 = 1.00$. In this case p, the frequency of A, steadily decreased while q, the frequency of a, increased by the same amount. We have observed evolution at the lowest possible level.

In such an elementary case it is easy to conceive of a reversal in the trend, as in the following imaginary example: $0.57 + 0.43 = 1.00$; to $0.59 + 0.41 = 1.00$; to $0.60 + 0.40 = 1.00$. In practice, however, evolution of the whole population in all of its traits is seldom reversed. The reason is simple. The heredity of most species of organisms is based not on a single locus but on thousands or tens of thousands of loci, many of which are represented in the population by more than just one or two alleles. Try to imagine most or all of this huge assemblage of allele systems being allowed to undergo various complicated changes in relative frequency, then reversing themselves and returning, all in concert, back to the starting point. Such an event is extremely unlikely. Even without the aid of this particular argument from genetics, biologists long ago recognized that evolution very seldom if ever reverses itself to any significant extent. This generalization is called DOLLO'S LAW, after Louis Dollo, a Belgian paleontologist who first derived it from his studies of fossils. Actually, Dollo's

637

Law is not a true law in the sense of physics and chemistry, but rather a strong inference based on empirical evidence from the fossil record combined with some persuasive theoretical deductions from population genetics. Even so, it has some far-reaching implications. A well-known application of the law is that once a major anatomical structure has been lost or transformed into another structure, it can never be regained in its original form. The hind legs of seals, for example, which evolved into flippers as part of a marine adaptation in the remote past, can never return to a walking form identical with that of land-dwelling mammals. The eyes of blind cavefish, having been genetically lost, cannot be regained. It is also true that no mammal can ever evolve back into the ancestral reptile, no reptile can evolve into the ancestral amphibian, and so on down the phylogenetic tree.

A corollary of Dollo's Law states that as species evolve —and all species do evolve, almost all of the time —they always diverge from one another. All bird species have always diverged from each other and, as a group, from all mammal species. Occasionally we witness some amount of convergence in life habits or in body form. Bats, for example, are mammals that acquired birdlike wings and the power of flight. But the resemblance is strictly superficial. Modern species of bats are genetically more different from birds than were their flightless ancestors.

The problems of the science of population genetics can be reduced to two fundamental questions stemming from the original Darwin-Wallace model of evolution. First, what is the origin of the basic units of genetic variation at the levels of the gene and chromosome? Second, what are the causes of changes in their frequencies in populations? Answers to these questions can be formulated in a concise manner as follows. New units of genetic variation originate both by gene mutations and grosser alterations in the structure of the chromosomes. Mutations and chromosome aberrations thus create the raw materials of evolution. But by themselves they do not cause more than trivial changes in gene frequencies, except in the few cases where their rates are abnormally high. The bulk of evolution, consisting of changes in the gene frequencies of entire populations, is caused by a complex array of other causes. Of these by far the most important is natural selection.

Even without the benefit of modern genetic knowledge the early Darwinian notion of random fluctuation turned out to be an adequate approximation of mutations and chromosome aberrations

as these processes have come to be understood in the present century. The development of evolutionary biology since about 1920 is often referred to as the MODERN SYNTHESIS, or NEO-DARWINISM, by which is meant that Mendelian genetics has been fused with the theory of natural selection, creating the basic discipline of population genetics. Population genetics has in turn been applied with great success to the reshaping of such subjects as systematics, speciation theory, ecology, and biogeography, which will be explored in some detail in subsequent chapters. The centennial of the publication of *The Origin of Species*, celebrated around the world in 1959 by conferences and by the publication of memorial books, saw the modern synthesis still in full progress.

THE HARDY-WEINBERG LAW

The keystone of modern evolutionary theory is the Hardy-Weinberg Law, named after the British mathematician (G. H. Hardy) and German biologist (W. Weinberg) who developed it independently in 1908. The law states that the processes of sexual reproduction do not of themselves change the frequencies of genes (for example, of A and a) or of the diploid genotypes (AA, Aa, and aa). There is an important formula connected with this law that allows us to predict gene frequencies from genotype frequencies and vice versa. Before turning to the formula it will be well to have the implication of the central conclusion clearly in mind, that evolution cannot be caused by sexual reproduction. At first this may seem an obvious conclusion to draw. But no regular cellular event is more complex or disruptive than gametogenesis, which is the very first step of sexual reproduction. In diploid organisms a crucial event is the first meiotic division, during which the homologous chromosomes come together in synapsis, exchange segments, and then separate into different daughter cells. Repeated over many generations, this process increasingly fragments and disperses the original genotype. Consider your own genotype. One-half of your genes come from each of your two parents, one-fourth on the average from each of your four grandparents, one-eighth from each of your eight great-grandparents, and so on back through time. You are separated from those ancestors who lived during the years of the American Revolution, for example, by approximately seven generations. There were no fewer than $2^7 = 128$ ancestors in this generation. To make the matter more interesting, let us suppose that one of them was a veteran of the Revolutionary War. You can be proud of him for that, but bear in mind that he contributed only about 1/128 of your own genes, or the equivalent of less than one-half of one of the 46 chromosomes in each of your cells. Where did the rest of him go? Providing he founded a lineage of average size, his genes have been dispersed to a very large array of people now living. If he and his descendants followed the American pattern of large families, this number is likely to be more than 128 persons (Figure 2).

From this example one can see why population geneticists like to think of evolution occurring in a GENE POOL, all of the genes of the population taken together, rather than down separate lineages of organisms. When evolution occurs over more than a few generations (it often occurs over thousands) it is almost impossible to think in terms of individual organisms, and of the fate of individual genotypes. The truth is that the individual genotype has no integrity over long stretches of time. Within several generations the "chromosome chopper" of meiosis has begun to dissolve it into the gene pool. The Hardy-Weinberg formulation, then, is expressed in terms of the frequencies of genes and genotypes in the gene pool. It says that for the simplest possible case of two alleles A and a on the same locus, in each and every generation the frequencies of these two alleles, p and q, will remain constant. It says further that the frequencies of the genotypes will remain constant and related to the allele frequencies as follows:

IN WORDS

Frequency of AA + Frequency of Aa + Frequency of aa = 1,

IN SYMBOLS

$p^2_{(AA)} + 2pq_{(Aa)} + q^2_{(aa)} = 1$.

To understand why this must be so, first examine the idealized life cycle of a diploid organism shown in Figure 3. Since the basic Mendelian laws of heredity are followed, the set of interbreeding individuals is referred to as a MENDELIAN POPULATION. It is necessary to add the condition that breeding occurs at random within the population. Up to this point we have been referring to the concept of a population loosely, leaving it to your intuitive feel of the idea. Now we must be more specific and say that the population is a group within which any adult is just as likely to select any member of the other sex as a mate, regardless of its location, as any other. Such a randomly mating population is called a DEME. It is the basic unit of

evolution. A deme can be all of the mice infesting a small house, for example, or all of the robins nesting in a woodlot. The geographic limits of demes and the numbers of organisms comprising them differ widely among different kinds of plants and animals.

In a deme, the gametes are mixed at random. As illustrated in the example of Figure 3, each gamete carries either A or a. In order to see how the Hardy-Weinberg formulation is derived from a simple knowledge of p and q, the frequencies of the two alleles, we need only utilize one quite elementary theorem of probability theory from mathematics: the probability of the union of independent events is equal to the product of their separate probabilities. In the Hardy-Weinberg formulation the probabilities are identical to the gene frequencies. If the frequency (p) of A genes is 0.6, to take an arbitrary example, then the probability of any given sperm or egg bearing an A allele as opposed to an a allele is also 0.6. (Similarly, 0.5 of tossed coins come up heads, and the probability of a given coin coming up heads is therefore also 0.5.) Six out of ten random selections of a sperm or egg will bring up an A allele. And since $p + q = 1$, the probability of drawing an a allele is $1 - 0.6 = 0.4$. Now the probability of joining two A-bearing gametes together at fertilization is (according to the theorem just cited) $p \times p = p^2 = 0.36$. Therefore 0.36 or 36 percent of the offspring in the next generation should have AA genotypes. Also, the probability of bringing two a-bearing gametes together at fertilization is $q \times q = q^2 = 0.16$. Table II shows that there are two ways of producing a heterozygote: A sperm with a egg, the probability of which is $p \times q$; and a sperm with A egg, the probability of which is also $p \times q$. Consequently the probability of obtaining a heterozygote from both ways is $2pq$.

From the Hardy-Weinberg formula,

$p^2_{(AA)} + 2pq_{(Aa)} + q^2_{(aa)} = 1$,

it is easy to show that the gene frequencies p and q remain constant each generation. Notice that out of the sum on the left side of the formula, a total of $p^2 + pq$ consists of A alleles. The fraction that this comprises of all the alleles is

$$\frac{p^2 + pq}{p^2 + 2pq + q^2} = p^2 + pq = p^2 + p(1 - p) = p.$$

By a parallel procedure it can be shown that the frequency of a in the next generation will be q. Thus the original gene frequencies are preserved.

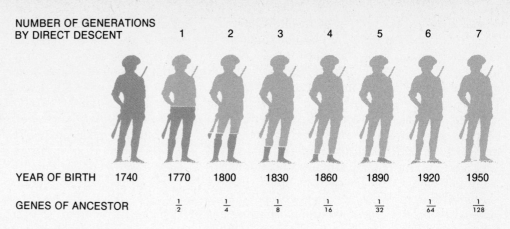

NUMBER OF GENERATIONS BY DIRECT DESCENT	1	2	3	4	5	6	7
YEAR OF BIRTH	1740	1770	1800	1830	1860	1890	1920	1950
GENES OF ANCESTOR		$\frac{1}{2}$	$\frac{1}{4}$	$\frac{1}{8}$	$\frac{1}{16}$	$\frac{1}{32}$	$\frac{1}{64}$	$\frac{1}{128}$

2   DISPERSION OF GENOTYPE in successive generations is a result of meiosis and sexual recombination. Living man at far right, seven generations removed from ancestor who lived during American Revolution (here depicted as Minuteman), would carry only 1/128 of ancestor's original genes.

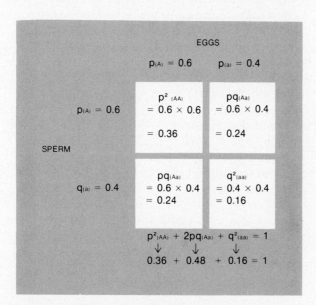

II   HARDY-WEINBERG FORMULA can be derived by calculating the probability of each possible combination of two alleles in sexual reproduction. Table shows worked-out example in which p = 0.6 and q = 0.4. Sum of probabilities is always 1.

A fundamental trait of Mendelian demes is that no matter what the proportions of the diploid genotypes are at the beginning, they will come to fit the Hardy-Weinberg distribution in the next generation and remain true to it every generation afterward. For example, suppose we started a new population by mixing 6000 *AA* individuals with 4000 *aa* individuals. No *Aa* individuals at all are included in this first generation. In the mixture $p = 0.6$ and $q = 0.4$. Applying the Hardy-Weinberg formula we can predict that in the next (second) generation approximately 36 percent of the individuals will be *AA*, 48 percent *Aa*, and 16 percent *aa*. This will be the distribution for every generation afterward.

The Hardy-Weinberg formula is the starting theorem of evolutionary genetics. In the study of real populations it is used to test hypotheses of both random mating and of evolutionary change. If a local population is really a deme —if its members are breeding randomly —and if it is undergoing no evolutionary change, the diploid frequencies will give an approximate fit to the Hardy-Weinberg formula based on the observed gene frequencies. However, if the diploid frequencies deviate significantly from the expected Hardy-Weinberg values, it is necessary to examine for evidence of (1) nonrandom mating among the genotypes, such as a preference of individuals for mates of their own genotypes; or (2) a change in the frequencies due to evolution occurring from one generation to the next. In some cases both phenomena exist. In other words, the Hardy-Weinberg formula is a null hypothesis. When proposed, it is the equivalent of suggesting that no preferential mating and no significant evolution are occurring.

As species evolve —and all species do evolve, almost all the time —they always diverge from one another.

ORIGINAL POPULATION

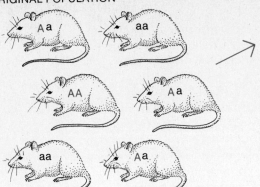

GAMETES

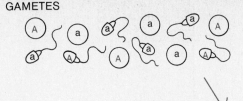

OFFSPRING

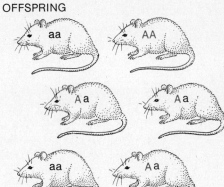

3 HARDY-WEINBERG LAW states that frequencies of two alleles remain constant from generation to generation in a Mendelian population. Diagram shows two generations of mice. Despite meiosis and sexual recombination, frequencies of alleles A and a remain unchanged in the second generation.

At this point you should solve some problems using the Hardy-Weinberg formula to make sure you really understand it *(see Box A).*

The genetic effect of normal sexual reproduction is to create new diversity in the diploid stage at each generation—without altering the frequencies of the genes involved. Sex is immensely effective in this, its primary role. Among asexual organisms in which no recombination occurs, the only way for a population to increase its variability is through new mutations or immigration of new types from the outside. Where normal sexual reproduction exists, on the other hand, new combinations of genes can be assembled on the same chromosome through crossing over at each gametogenesis *(see Chapter 9)*. If the organisms happen to be diploid, even more new combinations of chromosomes can be created at each fertilization. The collaboration of crossover, which generates new combinations of genes on the same chromosome, and independent assortment, which changes the combinations of chromosomes, produces a virtually endless genetic diversity.

To see the last principle clearly, consider the simplest possible case, in which two alleles on one locus produce just three diploid genotypes, *AA*, *Aa*, and *aa*. Addition of a second locus with two alleles, *B* and *b*, makes nine genotypes possible: *AABB, AABb, AAbb, AaBB, AaBb, Aabb, aaBB, aaBb, aabb.* The number of possible genotypes goes up

The genotype of an individual has no integrity over long stretches of time. In several generations the chromosome chopper of meiosis dissolves it into the gene pool.

swiftly with the addition of more such loci. In fact, the number in a population containing n of the loci is 3^n. This remains true where each locus contains only two alleles each. The number is usually higher in actual cases. In general, if the number of genotypes possible at the first locus is m_1, and the number possible at the second locus is m_2, and so on, the total number of genotypes that can be put together from all of the n genotypes is simply the product of all these numbers; it is $m_1 \times m_2 \times m_3 \times \cdots \times m_n$. In most Mendelian populations, both the number of loci and the number of alleles per locus are large. For example, in both *Drosophila melanogaster* and *Homo sapiens*, it is estimated that at least 10,000 loci exist, many of which carry numerous alleles. The total number of conceivable diploid genotypes in such species is astronomical; in fact, in these two examples it is greater than the number of all the atoms in the visible universe!

A Problems in population genetics

PROBLEM. Twenty-five percent of the individuals of a population of cavefishes are found to be albino, a trait known to be controlled by a single recessive gene. Estimate the frequency of the gene in the population as well as the frequency of its heterozygotes.

ANSWER. Let us arbitrarily label the albino allele a and the "normal," non-albino allele as A. It was stated that $q^2_{(aa)} = 0.25$. Then $q = 0.5$, the frequency of a in the population. And from this $p = 1 - 0.5 = 0.5$. The frequency of the heterozygotes is then estimated to be $2pq = 0.50$.

PROBLEM. A species of flowering plants contains two alleles, a_1 and a_2, which control the color of the seed coats. In a breeding experiment a new population is started by mixing 5000 homozygotes of the allele a_1 with 95,000 homozygotes of the allele a_2. Predict the genotype frequencies in the next generation and in each generation thereafter.

ANSWER. From the data given, $q = 0.05$ and $p = 0.95$. It can be predicted that in the next generation, and in each succeeding generation, the frequency of a_1a_1 individuals will be $q^2 = (0.05)^2 = 0.0025$; the frequency of a_2a_2 individuals will be $p^2 = (0.95)^2 = 0.9025$; and the frequency of a_1a_2 individuals will be $2pq = 0.0950$.

PROBLEM. The alleles at one locus in a population of moths is known to consist initially of 70% M and 30% m, where m controls a color trait recessive to that controlled by M. After a severe frost, and before the moths have a chance to breed again, the genotype frequencies are found to be 55% MM, 44% Mm, and 1% mm. Interpret the result.

ANSWER. If $p = 0.70$ and $q = 0.30$ as given, we should expect the following genotype frequencies: $p^2_{(MM)} = 0.49$, $2pq_{(Mm)} = 0.42$, and $q^2_{(mm)} = 0.09$. But after the frost the population contains far fewer mm individuals than predicted by the Hardy-Weinberg formula, somewhat more Mm individuals, and even more MM individuals. The most reasonable hypothesis is that a higher proportion of individuals possessing the M gene survive than those possessing the m gene. If the explanation is true, the moth population can be said to have undergone a small amount of evolution.

Thus, changing its heredity in a kaleidoscopic fashion, the sexual species exposes a new array of genotypes to the environment at each generation. At the same time the species holds the frequencies of its alleles nearly constant. As a result, sexually reproducing organisms enjoy an adaptability in the face of a changing environment far beyond the range of asexual species with the same amount of allele diversity. This adaptability seems to be the reason why sexuality is nearly universal in nature. It has been relinquished only in a few groups of organisms where peculiarities in the form of adaptation, such as the need to reproduce rapidly by means of budding, set asexual reproduction at a special premium.

AGENTS OF EVOLUTION

Sexual reproduction creates diversity but does not force changes in gene frequencies. It is therefore the agent of adaptability but not of evolution. In other words, the population uses sex to increase the diversity it needs to evolve, but sexual reproduction alone does not cause the population to evolve. Instead, the essential changes in gene frequencies are caused by four evolutionary agents: mutation pressure, genetic drift, gene flow, and natural selection.

Gene flow and selection are the most important in controlling microevolution—the simple changes in gene frequencies in local populations that have been discussed so far. Selection alone is the guiding agent in the larger progressive changes of evolution, such as the origin of new structures and new major taxonomic groups. Such creative changes are based on hundreds or thousands of separate microevolutionary events, which are assembled and focused by the selective pressures exerted by the environment.

Let us now consider each of the four evolution-

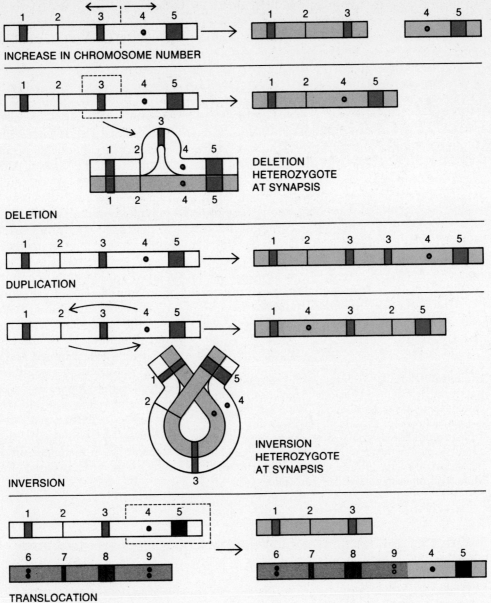

INCREASE IN CHROMOSOME NUMBER

DELETION
HETEROZYGOTE
AT SYNAPSIS

DELETION

DUPLICATION

INVERSION
HETEROZYGOTE
AT SYNAPSIS

INVERSION

TRANSLOCATION

4 **MODEL OF CHROMOSOME depicts five types of aberration.** Chromosome resembles putty stick that can be broken and reassembled in a limited number of ways. Bands and numbers are included simply as visual reference points. When an altered chromosome (color) pairs with a normal one, the two must undergo contortions to match up during synapsis and meiosis.

ary agents separately to understand their mode of action. Afterward, we will review several actual cases of microevolution to learn how biologists analyze changes in gene frequency with reference to these agents. Finally, at the end of this chapter

and in subsequent chapters, we will examine the transition from microevolution to the more complex evolutionary events.

MUTATION PRESSURE

Mutations are of two basic kinds: POINT MUTATIONS and CHROMOSOME ABERRATIONS. The former involve changes too minute to be directly observed by any existing form of microscopy. Chapter 8 pointed out that genetic and biochemical analyses utilizing bacteria have confirmed that point mutations are molecular events involving substitutions of some nucleotide pairs for others in the DNA molecule. At the other end of the scale, chromosome aberrations constitute major structural changes that can sometimes be observed under the light microscope. Such large alterations encompass not one but hundreds or thousands of nucleotide pairs. Chromosome aberrations are often regarded by students as rather difficult to master, but there is an easy way for you to comprehend them in a few minutes. With the aid of Figure 4, think of a chromosome as a stick of putty and imagine all of the ways one can modify it without twisting or pulling it out of shape. This list, if complete, will form a catalog of the chromosome aberrations. The putty stick, serving as a model of the chromosome, can be broken up to create a larger number of smaller sticks (increase in chromosome number). It can be joined to another putty stick (fusion, leading to reduction in chromosome number). One can take a piece out of the stick (deletion); double an existing piece (duplication); remove a piece, flip it over,

and reinsert it (inversion); or transfer a piece to another stick (translocation). It is also possible simply to duplicate the entire stick. If the full set of chromosomes is duplicated—if the chromosome number per cell is doubled, tripled, or more—the result is called polyploidy. In their genetic consequences chromosome aberrations can be thought of in much the same way as point mutations. They have similar effects on the phenotype, in that their presence commonly alters such traits as size, color, fecundity, and behavior. All but polyploidy segregate and recombine according to the same Mendelian laws. Because of its powerful effect on meiosis, polyploidy can result in the instantaneous creation of new species, a process which will be examined more closely in a later section. Inversions do not create such extraordinary effects, but they have played an important role in the history of population genetics. The reason is that both their homozygotes and heterozygotes are easy to detect with the light microscope in some species (*Figure 5*) and therefore can be quickly counted in populations in the laboratory and the field. Instead of using point mutations, which are often difficult to detect in the heterozygous state, geneticists often turn to inversions to conduct basic studies on such processes as natural selection and genetic drift.

Mutations, both point mutations and chromosome aberrations, create all of the raw material on which evolution is based. But can they also *direct*

5 INVERSION MUTATION spans the segment numbered 3–5 in this schematic diagram of the third chromosome of Drosophila pseudoobscura (top). This particular inversion is called the arrowhead arrangement. When mutant chromosome synapses with a noninverted or "standard" chromosome it must loop before pairing can occur, as shown at lower right. Mutant chromosome is rendered in color.

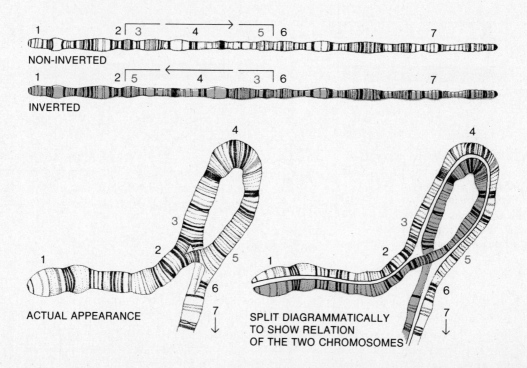

1   2 ⌐ 3   4 →   5 ⌐ 6   7
NON-INVERTED

1   2 ⌐ 5   ← 4   3 ⌐ 6   7
INVERTED

ACTUAL APPEARANCE

SPLIT DIAGRAMMATICALLY
TO SHOW RELATION
OF THE TWO CHROMOSOMES

ORGANISM	PHENOTYPIC CHARACTER AFFECTED	RATE	UNIT MEASURED
BACTERIOPHAGE T2	ATTAINS LYSIS INHIBITION, R → R$^+$	1×10^{-8}	PER GENE PER REPLICATION
BACTERIA *ESCHERICHIA COLI*	ACQUIRES ABILITY TO FERMENT LACTOSE, *LAC* ⇌ *LAC*$^+$	2×10^{-7}	PER CELL PER DIVISION
	ACQUIRES ABILITY TO UTILIZE HISTIDINE, *HIS*$^-$ → *HIS*$^+$	4×10^{-8}	PER CELL PER DIVISION
	LOSES ABILITY TO UTILIZE HISTIDINE, *HIS*$^+$ → *HIS*$^-$	2×10^{-6}	PER CELL PER DIVISION
HIGHER PLANTS CORN *(ZEA MAYS)*	SHRUNKEN SEEDS *Sh* → *sh*	1×10^{-5}	PER GAMETE PER GENERATION
	PURPLE SEEDS *P* ⇌ *p*	1×10^{-6}	PER GAMETE PER GENERATION
INSECTS *DROSOPHILA MELANOGASTER*	WHITE EYE *W* ⇌ *w*	4×10^{-5}	PER GAMETE PER GENERATION
	BROWN EYE *Bw* ⇌ *bw*	3×10^{-5}	PER GAMETE PER GENERATION
RODENTS HOUSE MOUSE *(MUS MUSCULUS)*	PIEBALD COAT COLOR *S* ⇌ *s*	3×10^{-5}	PER GAMETE PER GENERATION
MAN *(HOMO SAPIENS)*	NORMAL ⇌ HEMOPHILIAC	3×10^{-5}	PER GAMETE PER GENERATION
	NORMAL ⇌ ALBINO	3×10^{-5}	PER GAMETE PER GENERATION

II **RATES OF SPONTANEOUS MUTATION, as measured in populations of several different organisms, indicate that mutation is a rare event. Albino mutant, for example, turns up in only one of every 300,000 human gametes. Long-term evolutionary effect of such a rate is virtually negligible.**

evolution? Can mutations alone progressively change the frequencies of alleles or chromosome types (for example, a set of four such alleles or types: a_1, a_2, a_3, and a_4) that can mutate one into the other and back again:

$$a_1 \rightleftarrows a_2 \rightleftarrows a_3 \rightleftarrows a_4$$

Mutations are usually reversible, as suggested by the arrows that lead backward as well as forward.

Selection alone guides the larger progressive changes of evolution, such as the origin of new structures or of major taxonomic groups.

Suppose that the forward rates were very much higher than the backward rates. If no other evolutionary agent, such as natural selection, intervened to reverse the process, a population composed originally of a_1 alleles might well end up composed mostly of a_4 alleles. Such a change would be evolution by MUTATION PRESSURE.

However, the possibility that much evolution occurs by mutation pressure in any particular case is remote. The reason is easy to establish from the data in Table III, which exemplify "typical" mutation rates that have been measured in real populations. In this table, read the numbers as follows: 10^{-1} means 1/10 or 0.1, 10^{-2} means 1/100 or 0.01, 10^{-3} means 1/1000 or 0.001, and so on downward; 10^{-6} per cell per division therefore means 1/1,000,000 mutations per cell per division or (in more readily visualized form) one cell out of every million cells that divide; 2×10^{-6} means two cells out of every million; 4×10^{-8} means four cells out of every hundred million; and so forth.

The most important fact about these mutation rates and others that have been measured is that they are so low. Rates ranging as high as one in a thousand per generation are seldom encountered. One in a million is a much more typical number. The low rates are sufficient to create the variability needed to permit evolution to occur, but they are not great enough to drive evolution along by mutation pressure. If a mutation is occurring at the rate of one organism per million, this means a change in the gene frequency of only one in a million (say, 0.264318 to 0.264319), a negligible alteration that will almost certainly be offset by other, more powerful agents such as natural selection and gene flow.

GENETIC DRIFT

Genetic drift is the alteration of gene frequencies through chance alone. Genetic drift is most likely to be effective in very small populations: those containing fewer than 100 individuals. Suppose you had just completed a cross of $Aa \times Aa$ individuals of *Drosophila*, and you selected four individuals from among the offspring to start a new population. If the ones taken were chosen at random, would you expect them to yield the exact 1:2:1 Mendelian ratio? By reflecting on this for just a moment, you should be able to see that when only four individuals are involved, such a ratio would be obtained only part of the time. On other occasions, chance alone would dictate that the ratios would be one *AA* and three *Aa*, or two *AA* and two *aa*, or all *AA*, or all *aa*, and so on. To take a close analogy, when two coins are flipped they sometimes come up as two heads, sometimes as two tails, and sometimes as one head and one tail. If a population is small enough to have its gene frequencies altered by chance alone in this fashion, the result is evolution by genetic drift. Notice that it is even possible for an allele (or chromosome type) to be lost entirely from a population in one short step.

Genetic drift always influences gene and chromosome frequencies to some degree, for the reason that no matter how large the population there is always some chance deviation from the exact numerical result predicted by Mendel's laws. Even Mendel recorded such deviations. A famous experiment he reported, in which a 3:1 ratio was obtained in the F_2 generation of pea plants from crosses between the dominant yellow and recessive green seed coat colors, did not really yield this exact ratio. The actual numbers recorded were 6,022 yellow and 2,001 green, or a ratio of 3.0095:1. When the numbers from seven other confirmatory experiments published since 1866 are added,

the total is 153,902 yellow and 51,245 green, or a ratio of 3.003:1. In fact, a slightly different result can be expected each time the same experiment is performed. Notice that as the population size was increased in this case, the statistical error, or deviation from the statistically expected result, was reduced. In general, changes in gene frequency due to genetic drift become negligible as population sizes become very large.

Genetic drift, like mutation pressure, is probably of minor significance in most populations, most of the time. However, the process is likely to become important to populations when the number of breeding individuals drops to a few tens or hundreds. Then there is a good chance of some alleles becoming lost while others become simultaneously fixed. To use our familiar two-allele case again, the population might come to consist entirely of *AA* or *aa* organisms rather than of a mixture of organisms bearing *AA*, *Aa*, and *aa*. Geneticists have found that such a reduction in genetic variability not only lowers the capacity of a population to adapt to changes in the environment, but it also tends to reduce the overall fitness of the population as well. The reason is that certain of the alleles produce impaired fertility or survival ability when in the homozygous condition. If species become too small in population size, they may reach the point where the accidental fixation of less adaptive genes speeds their decline still more. This is one explanation that has been suggested for the final decline of the heath hen, a bird species that became extinct in the eastern United States in the 1930's. It also appears to explain the reduction of fertility to dangerously low levels in the last surviving herds of the wisent, the European equivalent of the North American bison. Inbreeding has the same effect as small population size in promoting genetic drift. No matter how large the total population size, so long as the average deme (the group of individuals that breed freely among themselves) is small, chance fluctuations in gene frequency can occur.

Another way in which genetic drift can operate is through the FOUNDER EFFECT. New populations in new places are usually started by a very small number of pioneering individuals lucky enough to reach a place where they can survive and reproduce. Because they contain only a small fraction of the alleles found in the source population, they must be different from the source population at the very beginning. The new populations will also differ among themselves because of chance differences in the genes of their founders. Geneticists

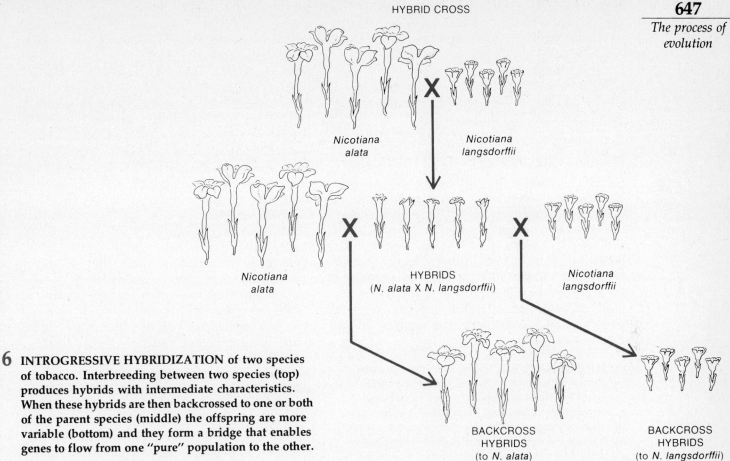

*Nicotiana
alata*

*Nicotiana
langsdorffii*

*Nicotiana
alata*

HYBRIDS
(*N. alata* X *N. langsdorffii*)

*Nicotiana
langsdorffii*

BACKCROSS
HYBRIDS
(to *N. alata*)

BACKCROSS
HYBRIDS
(to *N. langsdorffii*)

6 **INTROGRESSIVE HYBRIDIZATION** of two species
of tobacco. Interbreeding between two species (top)
produces hybrids with intermediate characteristics.
When these hybrids are then backcrossed to one or both
of the parent species (middle) the offspring are more
variable (bottom) and they form a bridge that enables
genes to flow from one "pure" population to the other.

have often used the founder effect in an attempt to
explain the small, seemingly meaningless, differences that occasionally distinguish populations belonging to the same species.

GENE FLOW

Most demes are only partially isolated from other
demes of the same species. Some organisms are
still able to migrate from one to the other. If the
migrants are then able to survive and to breed in
their new home, they add to the gene pool of the
deme. This injection of new genes by migration is
referred to as gene flow. In many cases the immigration rate is relatively low, adding perhaps no
more than one in a million individuals per generation. At this level it is equivalent in its effect to
mutation pressure, capable of adding new genetic
material but not of contributing significantly to the
alteration of gene frequencies. In other cases the
rate of immigration is high, as much as one immigrant for every hundred individuals in the host

population or higher. Gene flow of this magnitude
can serve both as a source of new genetic variability and as a prime mover of evolution, on a par
with natural selection.

Of equal importance to the amount of gene flow
is the degree of genetic difference that separates
the immigrants from their host population. If the
two are genetically identical, then of course no
evolution occurs, no matter how intense the gene
flow. But if they are very different, a small amount
of immigration can result in a great deal of evolution, especially if some of the introduced genes
prove to be adaptively superior in the new environment. The extreme case of such exchange occurs during HYBRIDIZATION —the interbreeding between populations that are different enough to be
ranked as species. If the hybrid products themselves are able to breed back with one or both of
the parent populations, the process is called INTRO-
GRESSIVE HYBRIDIZATION. It results in the production

of radically new genetic combinations within the parent species. Most of these products are less well adapted to the local environment and destined for early extinction in the process of natural selection. But some, acting like daring new mutations, may cause the parent populations to take new directions in evolution. An example of introgressive hybridization is given in Figure 6.

NATURAL SELECTION

Natural selection is the only one of the agents of evolution that specifically adapts populations to their immediate environment. Its result is always the same: some genotypes gain in the population at the expense of others. The selective force can act on the variability of a population in one or another of several radically different patterns, producing the array of effects illustrated in Figure 7. It is important to understand the consequences of each of these patterns separately. Let us begin by briefly

considering the significance of the normal distribution used in the example. Many but not all biological traits vary according to this mathematical form. In simplest terms the normal distribution, which produces a bell-shaped frequency curve of the kind shown in Figure 7, arises commonly when many factors contribute jointly and independently to the trait under consideration. In the case of population genetics, the factors are usually multiple genes that operate together to influence the same phenotypic trait. Of course, genes (or more precisely, alleles) that happen to be on the same locus still segregate and recombine in a Mendelian fashion.

STABILIZING SELECTION involves a disproportionate elimination of the extremes, that is, the tapered edges of the normal distributions. This process occurs repeatedly in all populations. In all but the smallest and most isolated populations the variability is increased each generation by mutation and gene flow. Stabilizing selection then "pulls in the skirts" of population variation by reducing or eliminating all but the optimal phenotypes (and the genotypes controlling them) that are best suited to the local environment. An important special form of stabilizing selection makes possible the phenomenon of BALANCED POLYMORPHISM. This is the coexistence of two or more phenotypes not connected by intermediate phenotypes. Sexual differences, and diffences between developmental stages of the kind that occur in the life cycle of a single individual, are excluded from this definition of balanced polymorphism. Examples include the blood types of human beings. Each condition of

7 **EFFECTS OF NATURAL SELECTION on a population are plotted on two sets of curves. Graphs illustrate the impact of the three principal types of selection— stabilizing, directional and disruptive—on the size of an imaginary organism. In this particular species, most individuals are roughly from 65 to 85 millimeters long, as indicated by the normal distribution curves at left. Height of curve corresponds to the number of individuals having a given size. In the set of curves at right, arrows pointing upward indicate phenotypes (in this case, sizes) favored by natural selection; downward arrows, sizes disfavored by selection. Selection alters the range of sizes, as shown by curves at bottom right.**

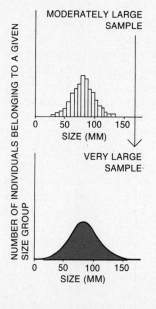

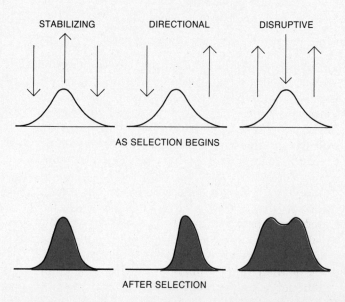

the A, B, and O series, as well as the Rh positive and Rh negative factors and other, less familiar series, are controlled by single alleles. An example of balanced polymorphism involving *Drosophila* is given in Figure 8. In this case we follow the practice of the original investigator of labeling the alleles *E* and *e*, although for our purposes they could be equally well labeled *A* and *a*. Experiments have shown that the heterozygote (*Ee*) is superior in fitness, that is, in ability to survive and reproduce, to either of the two homozygotes (*EE* and *ee*). Since the superior heterozygote contains both of the alleles, *E* and *e*, neither one can eliminate the other in evolution. A great deal of the natural genetic variability of populations is based on the coexistence of such multiple alleles. It is believed that heterozygote superiority is an important, and perhaps the most important, mechanism perpetuating genetic variability after it has been created by mutation.

DISRUPTIVE SELECTION is a rarer phenomenon, or at least one less well documented at the present time. Its greatest significance is that it has the potential to create balanced polymorphism, by breaking the population up into two or more distinct types, and it can therefore serve as an alternative mechanism to heterozygote superiority. In extreme cases populations subjected to disruptive selection might even divide into two or more species, although such a process has not yet been demonstrated outside the laboratory.

Figure 9 summarizes, in simplified graphic imagery, the effects of the four principal agents of evolution —mutation, gene flow, genetic drift and natural selection —on imaginary populations. These agents are the mainsprings of evolution. The most important is the dominant process that guides progressive evolution: DIRECTIONAL SELECTION. Rather than attempt to characterize this crucial process in general theoretical language, we shall provide several case histories that reveal the diverse ways in which it can operate.

EE · Ee · ee

ABDOMEN OF FRUIT FLY illustrates the effects of balanced polymorphism. Differences in color shown here are due to a pair of alleles (E and e) on the same locus. Heterozygote (Ee) has superior fitness to both homozygotes, so neither allele eliminates the other by natural selection.

8

FOUR AGENTS OF EVOLUTION and their effects on imaginary populations. The whole range of possible phenotypes and genotypes in a population can be represented by an abstract two-dimensional "map." Arrows indicate direction of evolutionary forces. Original population is mapped in dark color; evolutionary changes, in lighter colors. The larger the shaded area, the greater the amount of variation in the population; and the more the original areas are shifted, the greater the amount of evolution.

9

MUTATION PRESSURE

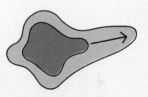

GENE FLOW

GENETIC DRIFT

OR

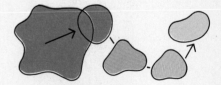

GENETIC DRIFT

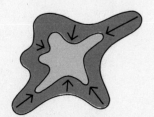

STABILIZING SELECTION

DIRECTIONAL SELECTION

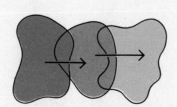

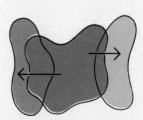

DISRUPTIVE SELECTION

MICROEVOLUTION

The following two case histories were selected from among many in the literature because they have been the subject of unusually meticulous research and for this reason can be examined the most critically. In each example you will notice that natural selection, and particularly the directional forms of natural selection, is the principal agent involved. This reflects the general rule, well established during the past 40 years, that when microevolutionary change is closely analyzed, natural selection is almost always found to be operating and is usually of overriding importance.

INDUSTRIAL MELANISM

Two centuries ago, before the burgeoning industries of England began dumping black soot and other airborne pollutants on the countryside, the vegetation was everywhere green and fresh. Even near the cities this was true. In addition, the trunks of the trees had a whitish color for the most part, caused by the luxuriant growth of lichens. Many insect species, moths among them, lived among the lichens or at least rested there during the day.

The larger species were mostly light colored and difficult to see when they sat motionless on a lichen background. In the late 1840's an increasing number of very dark (melanic) individuals began to appear in the moth populations near the cities. In one species that has been intensively studied, the peppered moth *Biston betularia*, the melanic form increased in numbers until it comprised 98 percent of the population in and around the city of Manchester. Genetic studies have shown that the dark coloration is under the control of a single dominant gene that arises spontaneously by mutation from the "typical" color gene. It thus appears that within 50 years the local moth populations of such industrial cities as Manchester had undergone a nearly complete gene substitution.

The selective force in these cases is, strange to say, not the pollution itself but rather birds that live in the polluted areas. Insect-eating birds obtain a good part of their food by searching for the insects that rest on the tree trunks. Insects whose coloration blends in with the background are more difficult to find, and they are eaten less often. Thus in the unpolluted rural regions of England, where

PEPPERED MOTH is a case history of evolution in action. Moth exists in two forms, one light, one dark. Photograph at left shows both forms at rest on a lichen-covered tree trunk. The light form is almost invisible. Photograph at right shows both forms on soot-blackened bark, where dark form is better camouflaged. In only a century, dark moth has replaced light one as the predominant form.

	LIGHT MOTHS	DARK MOTHS
DORSET, ENGLAND:		
WOODLAND NOT POLLUTED BY SMOKE (LIGHT BACKGROUND)		
RELEASED BY INVESTIGATOR	496	473
RECAPTURED LATER BY INVESTIGATOR	62	30
PERCENT RECOVERED	12.5	6.3
BIRMINGHAM, ENGLAND:		
WOODLAND POLLUTED BY SMOKE (DARK BACKGROUND)		
RELEASED BY INVESTIGATOR	137	447
RECAPTURED LATER BY INVESTIGATOR	18	123
PERCENT RECOVERED	13.1	27.5

SURVIVAL VALUE OF PIGMENTATION was measured by releasing both light and dark forms of peppered moth and later recapturing the survivors. In unpolluted woodland (top) survival rate of light moths was about twice that of dark ones. In polluted woodland (bottom) ratio was reversed.

lichens are still light in color, the non-melanic form of *Biston betularia* predominates in the populations to the near exclusion of the melanic form. But in urban areas, where air pollutants have darkened the lichens, the melanic forms prevail *(Figure 10)*.

Experiments have been performed in which non-melanic and melanic moths were released together in both polluted and unpolluted woodland, and their subsequent fates observed. One set of data, gathered by the British entomologist H. B. D. Kettlewell, is presented in Table IV. It shows that the form whose color matches its background is strongly favored over the other. This selection is intense enough to cause a significant change in gene frequency within a single generation. Extended over many tens of generations, it would result in a nearly complete substitution of the favored gene for the other. This is what did happen in England in the late nineteenth century. Since

---

As an application of Darwinian theory, the story of sickle-cell anemia contains an ironic twist: protection against one deadly disease was bought at the price of another.

---

these data were published, Kettlewell and other observers have succeeded in watching several species of birds as they selectively picked off the moths of contrasting coloration. The industrial melanism example reveals that evolution can occur rapidly at the level of the single gene. It further shows that such evolution occurs in natural populations in ways that are most difficult to predict in advance.

THE SICKLE-CELL TRAIT

The case of sickle-cell anemia in human beings is the classic textbook example of balanced genetic polymorphism. Its cause has been traced all the way down to the molecular level, and the reason for the high frequency of the gene in some human populations has apparently been explained in full by classical population genetic theory.

Sickle-cell anemia is a severe hereditary disease that afflicts a large percentage of the native populations of Africa. It is caused by an abnormal form of hemoglobin, called hemoglobin S, the molecules of which have the peculiar tendency in the deoxygenated state of linking up to form long rods. The molecular formations are sufficiently rigid to distort the red blood cells from their normal disc shape to a twisted, sickle-like form *(Figure 11)*. These cells tend to clog the smaller blood vessels. The automatic response of the body is then to destroy them, so that a shortage of red blood cells, anemia, is the result. The sickle-cell condition is under the control of a single gene. This gene produces hemoglobin S by directing the substitution of a single amino acid in the approximately 600 residues that make up the hemoglobin molecule *(Chapter 6)*. Persons who are heterozygous for the

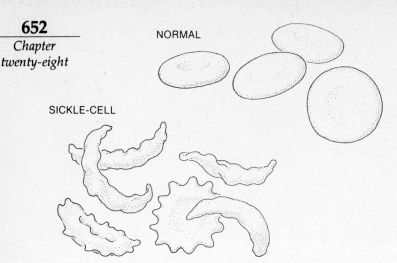

NORMAL

SICKLE-CELL

11 **RED BLOOD CELLS** of humans change from normal disc shape (top) to sickle shape (below) when blood of a person who is homozygous for the sickle-cell gene is deoxygenated during respiration.

gene, by virtue of carrying one normal gene and one sickle-cell gene, manufacture both normal hemoglobin and hemoglobin S. They show signs of anemia only under conditions of stress, such as breathing at high altitudes, when an unusual demand is placed on the hemoglobin. Homozygous carriers, on the other hand, are much more severely hit. Most die in childhood. The remainder suffer from chronic anemia with periodic crises when blood is temporarily cut off from various body organs.

Although sickle-cell anemia is virtually lethal in the homozygous state, the gene is abundant in many parts of Africa *(Figure 12)*. As many as 40 percent of the members of some of the tribes carry the trait, mostly in the heterozygous condition. Among black Americans approximately nine percent are heterozygous for the gene, and about 0.25 percent are homozygous. How can a gene imposing such lowered fitness on its homozygotes rise to these levels? Whenever population geneticists encounter a situation of this kind, they immediately consider the possibility of stabilizing selection by the mechanism of balanced polymorphism. If sickle-cell homozygotes are being eliminated at a high rate by anemia, something else must be eliminating the "normal" homozygote as well, with the result that the heterozygote enjoys a superior fitness over both homozygotes. The something else appears to be malignant tertian malaria, a particularly virulent form of this blood

disease caused by the protozoan *Plasmodium falciparum*. Not only do the ranges of the sickle-cell gene and *Plasmodium falciparum* roughly coincide, but field studies have indicated that persons carrying hemoglobin S are more resistant to the parasite than those lacking it.

This final example of application of Darwinian theory contains an ironic twist. Protection against one deadly disease is revealed to have been bought at the price of another. Nothing could illustrate more clearly that it is the population, not the individual, that constitutes the ultimate unit of evolution. For individuals do not become hemoglobin heterozygotes by any kind of personal design or physiological adaptation, nor are they able to confer their valuable heterozygosity to particular offspring by any exercise of personal choice. Some of their children will be heterozygotes by chance and others homozygotes. Among the homozygotes chance will also decree who will be susceptible to malaria and who will die from anemia. In a sense the anemics can be regarded as losers in the evolutionary game, the population's mindless response to the selective pressure exerted by the malarial parasites.

If you do not find this example provoking, consider the one of schizophrenia. There is abundant evidence that this very common mental disease has a strong hereditary component. According to one hypothesis it is controlled by a single domi-

MAP OF AFRICA shows distribution of sickle-cell gene. The range roughly coincides with that of the malaria-causing protozoan Plasmodium falciparum. Darker color indicates higher frequencies.

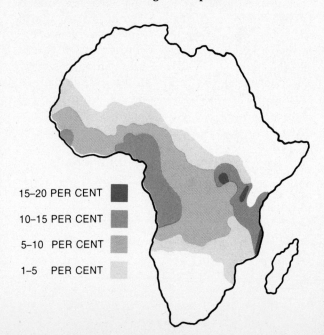

15–20 PER CENT

10–15 PER CENT

5–10 PER CENT

1–5 PER CENT

nant gene with low penetrance or at most a small number of such genes —meaning that only a small minority of the carriers develop the disease to any perceptible degree. The high frequency of schizophrenia in populations, despite the relatively high mortality and low reproductive replacement of those afflicted by it, suggests the operation of a hidden selection pressure favoring its survival in less overt ways. This adaptive factor could be the well-known greater ability of schizophrenics to withstand shock from wounds and operations. [See Huxley, J. S., E. Mayr, H. Osmond, and A. Hoffer, *Nature*, **204:** 220–221 (1964). The hypothesis presented here is still the subject of controversy and continued research.] In considering the deeper meaning of such an evolutionary trade-off it is useless to ask whether complete sanity is preferable to resistance to wound shock, just as it is useless to ask whether it is better to resist malaria than to avoid anemia. Nature, in permitting the operation of balanced genetic polymorphism, does not ask such questions.

MACROEVOLUTION

Each of the examples of microevolution mentioned, involving shifts in the frequencies of small numbers of genes, could be multiplied a hundredfold from reports in the scientific literature. Biologists have been privileged to witness the beginnings of evolutionary change in many kinds of plants and animals and under a variety of situations, and they have used this opportunity to test the assumptions of population genetics that form the foundations of modern evolutionary theory. The question that should be asked before we proceed to new ideas is whether more extensive evolutionary change, macroevolution, can be explained as an outcome of these microevolutionary shifts. Did birds really arise from reptiles by an accumulation of gene substitutions of the kind illustrated by the industrial melanism of moths?

The answer is that it is entirely plausible, and no one has come up with a better explanation consistent with the known biological facts. One must keep in mind the enormous difference in time scale between the observed cases of microevolution and macroevolution. Under natural conditions the nearly complete substitution of the melanic gene of the peppered moth took 50 years. Evolution of the magnitude of the origin of the birds usually, perhaps invariably, takes many millions of years. As paleontologists explore the fossil record with increasing care, transitions are being documented between increasing numbers of species, genera, and higher taxonomic groups. The reading from these fossil archives suggests that macroevolution is indeed gradual, paced at a rate that leads to the conclusion that it is based upon hundreds or thousands of allele substitutions no different in kind from the ones examined in our case histories.

READINGS

P.R. Ehrlich, R.W. Holm, and P.H. Raven, *eds., Papers on Evolution*, Boston, Little, Brown, 1969. Many of the basic articles from modern evolutionary biology are reprinted here, with commentaries by the editors. This is a useful adjunct to the textbooks suggested below.

L.E. Mettler and T.G. Gregg, *Population Genetics and Evolution*, Englewood Cliffs, NJ, Prentice-Hall, 1969. A brief textbook comparable to that by Stebbins but with much greater emphasis on population genetics. It would be useful as a stepping stone for the student who wishes to undertake advanced reading in the latter subject.

R.E. Ricklefs, *Ecology*, Newton, MA, Chiron Press, 1973. The chief virtue of this book is that it is one of the few at an elementary level that solidly connects evolutionary theory with population ecology. It is also valuable for its imaginative approach to the theoretical ideas of the two subjects.

G.L. Stebbins, *Processes of Organic Evolution*, 3rd Edition, Englewood Cliffs, NJ, Prentice-Hall, 1977. Perhaps the best short textbook on general evolutionary biology and a good book to read next. Well-balanced presentation of population genetics and speciation theory, with both plant and animal examples.

E.O. Wilson and W.H. Bossert, *A Primer of Population Biology*, Sunderland, MA, Sinauer Associates, 1971. This brief self-teaching textbook is designed to form the bridge between your elementary textbook and more advanced treatments of population genetics, evolutionary theory, and ecology.

[Natural selection] is the composer of the genetic message,
and DNA, RNA, enzymes, and the other molecules in the
system are successively its messengers. — George Gaylord Simpson

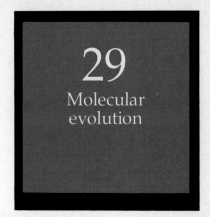

When biologists began to study fossils seriously in the eighteenth century, almost everyone agreed that the ancient bones were significant, but disagreed about what they signified. One proposal, widely held, was that they were the remains of life before the biblical flood. In 1726, when the Swiss physician Johann Scheuchzer unearthed the remains of a giant salamander in a quarry in Baden, he mistakenly took it to be the skeleton of a drowned sinner, and named it *Homo diluvii testis* ("human witness to the flood"). Another opinion, after the battle lines hardened, was that fossils were placed in the earth by God as though they were the relics of ancient life, in order to test the faith of the faithful. The view finally prevailed that God does not set obstacle courses any more than, in Einstein's famous phrase, he plays dice with the universe. These fossils are the legitimate remains of earlier life on Earth, and are one basis for constructing a history of the evolutionary process.

Another basis, already examined in Chapter 23, is the comparative anatomy of living creatures. Although the fossil evidence was a strong stimulus to thinking about evolution, the family relationships that we see today would alone be enough to construct a reliable outline of the process. Today, more often than not, the fine points of relatedness between two species are settled by examining the species at all stages of their life cycles, rather than by looking for a common ancestor in the necessarily fragmentary and incomplete fossil record.

CYTOCHROME *C* **is an ancient protein found in all plants and animals. Drawing on opposite page shows a front view of the molecule. Heme group (white) is seen edge-on. Interior of molecule is filled with hydrophobic side chains packed around the heme. All of the charged side chains lie on the outside. Large spheres indicate locations of α-carbon of amino acids.**

Another tool for studying evolution has been developed in the past few years: the structures of macromolecules. Fred Sanger won the Nobel Prize in 1956 for his discovery of the way to determine the sequence of amino acids in a protein chain, and for his application of the method to insulin. Five years later, Max Perutz and John Kendrew were awarded the Nobel Prize for the first successful x-ray analyses of the three-dimensional folding of proteins. These two methods, protein sequence and protein structure analysis, have opened up a new field which can be called molecular anatomy. This area, when developed further, will have as much or more to tell us about family relationships among living organisms than macroscopic anatomy.

The history of every living organism is written in its enzymes. As generation follows generation in the line of living creatures, point mutations occur in the bases of the DNA that specify the amino acid sequence of a particular enzyme. These mutations show up as changes in the protein sequence, and affect the functioning of the enzyme. The codon AUA for isoleucine, for example, has three other possible bases at each of three positions (*Figure 1*), and can therefore mutate to nine other triplet codons. Two of these, UUA and CUA, are translated into isoleucine as before. These are invisible mutations whose presence cannot be detected in the resulting enzyme. Four others cause the substitution of chemically similar hydrophobic amino acids: valine, leucine, or methionine. These are called conservative mutations. It is quite possible that these new amino acids would be as good as the old one, and that the enzyme's operation would be unimpaired. However, mutations at the second position to ACA, AAA, or AGA, are more radical. They lead to amino acids with different properties: polar but uncharged in the case of threonine, or

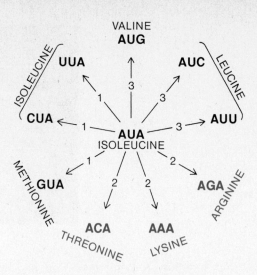

1 POINT MUTATIONS can produce nine new codons from the isoleucine codon AUA (center). Codons printed in black are hydrophobic like isoleucine. Those in color have sharply different properties.

positively charged for lysine or arginine. If it happens that the enzyme cannot tolerate a positive charge at a particular point in its structure, then the mutation of AUA to AGA would knock out the enzyme. If the enzyme is sufficiently important, its deficiency might be lethal. The unfortunate individual who receives this defective DNA as his genetic inheritance will not be viable, but will die and take the record of the lethal mutation with him (*Box A*).

In most cases, the lines are not so sharply drawn. The altered enzyme may work slightly less well or perhaps even slightly better. If the enzyme is ruined, the host may find an alternative pathway for carrying out the same metabolic process, or may obtain some needed compound from the environment rather than by synthesizing it. The result will be a small selection pressure in a population for or against the individuals with the mutation in sequence. Eventually the entire population, except for a small minority, may have enzymes with a favorable new sequence. Molecular evolution proceeds exactly like the more familiar macroscopic evolution—the introduction of variation by random mutation, followed by selection from among the variants according to their efficiency in coping with the environment. Out of this interplay of variation and selection comes evolutionary change.

In fact, molecular evolution and macroscopic evolution are two views of the same process.

Large-scale evolution deals with changes in visible traits: eye color, wing or limb structure, ability or inability to metabolize a given substance or to synthesize it if need be. All of these, if we really understood them, could be reduced to a pattern of chemical reactions mediated by enzymes. Geneticists often study a mutation, such as an unusual eye color in the fruit fly *Drosophila*, without worrying about its biochemical origin. But it would be equally appropriate to study the reactions and enzymes responsible for the abnormal eye pigmentation. Most of the macroscopic traits that have been studied, however, involve several enzymes operating in concert. When we look just at the changes in a single enzyme, we are getting closer to the site of mutation and the source of variation, although perhaps farther away from the impact of selection pressures and the test of fitness. Mutation acts on molecules of DNA, and selection acts on populations of individuals. Protein structures are one step removed from the DNA, but still many steps removed from populations.

This chapter will summarize several examples of molecular evolution that have been studied to date, beginning with the most straightforward case, cytochrome *c*. The multiple chains of the hemoglobins will suggest the ways in which new proteins can develop. The reader will see that cytochrome *c* and the hemoglobins evolved at different rates, and why this occurred. The serine proteases —chymotrypsin, trypsin, and elastase— are somewhat further along the same road. They are independent but obviously related enzymes in the same individual. These enzymes and subtilisin provide a striking example of convergent evolution. And finally, the fibrinopeptides and histones illustrate the upper and lower limits found so far for the rates of molecular evolution.

SIMPLE VARIATION AND SELECTION: CYTOCHROME *C*

Cytochrome *c* is one component of the respiratory chain of mitochondria (*see Figure 14 in Chapter 6*). It and the other proteins of the citric acid cycle and respiratory chain are part of the common heritage of all eucaryotic organisms —plants and animals alike. Cytochrome *c* is therefore one of the most widespread of the easily accessible proteins. It is easily accessible because it is not bound tightly to the mitochondrial membrane as are many of the other cytochromes and respiratory enzymes. Cytochrome *c* is small, being a polymer of around 104 amino acids, with a molecular weight 12,400,

A molecular disease

Tyrosine is manufactured in humans with the aid of the enzyme phenylalanine hydroxylase, which catalyzes the reaction by which oxygen is added to phenylalanine. Approximately one person in 160 carries a flaw in the DNA that should make this enzyme. This defective gene is recessive —that is, if a person inherits a functioning gene for phenylalanine hydroxylase from either parent he can make the enzyme, even though the gene from the other parent is faulty. But the unfortunate one person in 25,000 who is homozygous for the defective gene is unable to make the enzyme, and unable to convert phenylalanine to tyrosine in the normal way.

This alone is not too harmful; other pathways for synthesizing tyrosine exist. However, the unused phenylalanine is converted to phenylpyruvate, as much as one or two grams per day being made by a normally little-used process. Phenylpyruvate first builds up in the blood and then spills over into the urine, where it can be detected easily. The high phenylpyruvate concentration of the blood is severely damaging to the developing brain and nervous system of a young child. If the condition persists, the result is progressive and irreversible mental retardation, minor epileptic fits, skin eczema, and lack of skin and hair pigmentation.

Fortunately, phenylketonuria can be detected by routine testing of the urine of newborn infants. At present there is no way to repair the defective DNA or to supply the child with artificial enzyme. However, if his phenylalanine intake can be kept low by special diets during his developing years, then brain damage can be avoided and other side effects minimized.

Several other "molecular diseases" are known which are caused by mutated DNA in defective genes transmitted via the hereditary mechanism just as normal genes are. These include albinism (absence of pigmentation) and sickle-cell anemia (defective hemoglobin molecules). In a sense these hereditary and inborn errors in metabolism caused by defective DNA are the price we pay for the long-range advantages of molecular evolution, and thus for evolution as a whole.

---

whereas the other mitochondrial proteins have molecular weights of 50,000 to several hundred thousand. Like myoglobin and hemoglobin, it has a heme group, an organic ring structure surrounding an iron atom. The bright red color because of the heme makes it easy to follow during extraction and purification in the laboratory. All of these advantages: smallness, extractability in quantity, wide distribution, and great antiquity, have caused more attention to be paid to cytochrome *c* sequences than those of any other protein. We now know the amino acid sequences of the cytochromes from over 70 species, from man, horse, and fruit fly, through yeast and bread mold, to sunflower seed and wheat germ. In addition, the three-dimensional structures of horse, tuna, and bonito (a relative of tuna) cytochromes have been worked out, and we can match primary sequences to molecular structure. Cytochrome *c* provides a picture in miniature of the evolutionary process.

The amino acid sequences for 33 cytochromes *c* are compared in Figure 2. A noticeable feature is the great similarity in proteins from different species. Hydrophobic, acidic, basic residues, and glycines have been distinguished by color in the figure in order to emphasize these similarities. It is obvious that the different cytochromes are variations on a common theme. Many positions along the chain appear to have been occupied by one amino acid throughout the entire history of the eucaryotes. It is hard to believe that mutations never occurred at the corresponding places in the DNA. Instead, these positions in the protein must be so vital that all variants are ruthlessly weeded out by selection. This is the case, for example, for cysteines 14 and 17, histidine 18 and methionine 80, all of which are totally invariant.* Chemical tests have demonstrated, and x-ray analysis con-

*Cysteine 14 has been replaced by alanine in two microorganisms: Euglena and Crithidia. Evidently this heme connection can be dispensed with, as long as the other one is there.

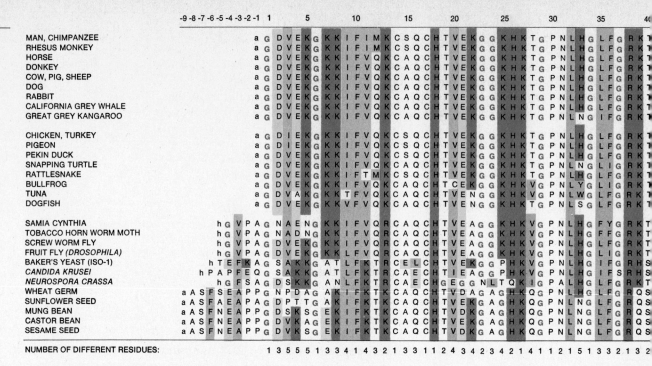

2 **AMINO ACID SEQUENCES** of cytochromes c from 33 species of organism are listed in the diagram on these two pages. The one-letter amino acid symbols are explained in Table I. In addition, h indicates free amino end of polypeptide chain, a the acetylated amino end, and x trimethylated lysine. Hydrophobic residues appear against a grey background, basic side chains against a dark color background, and acid side chains against a light color. Glycine (G) is printed in colored type. Pattern of vertical bands indicates how chemical nature of side chains has been preserved during evolution.

firmed, that all these side chains interact with the heme group. The two cysteines make covalent bridges between heme and polypeptide. The heme supplies four of the six octahedral ligands around the iron atom, and the fifth and sixth are provided by histidine 18 and methionine 80. Any mutation which touches these amino acids is fatal.

Other sections of the sequence are invariant, most notably segment 70–80 leading up the heme-linked methionine. Yet other regions, although not invariant, can tolerate only conservative changes in amino acids. Region 80–85 is always hydrophobic, or at least free of charged groups. Region 86–93 is charged, with a predominant but not absolute separation of positive and negative charges. Region 94–98 is never charged, and 99–104 is never hydrophobic.

The simplest interpretation is that all of these cytochrome *c* molecules are folded in the same way, and what we are observing are the constraints on a successfully folded and operating molecule. The inference is that these proteins are all descendants of a common ancestral cytochrome *c*. There is good chemical evidence for this inference in that all of these cytochromes are more or less interchangeable in reactions with cytochrome oxidases from other species. The x-ray analyses of horse and fish cytochromes have shown that for the past 400 million years at least, there has been no change in the folding of the cytochrome *c* molecule. These molecules are interchangeable parts even though there is no likelihood, outside of the laboratory, that interchangeability would be an advantage.

The x-ray analysis of cytochrome *c* has shown

X-ray analyses of horse and fish cytochromes have shown that at least for the past 400 million years there has been no change in the folding of the cytochrome *c* molecule.

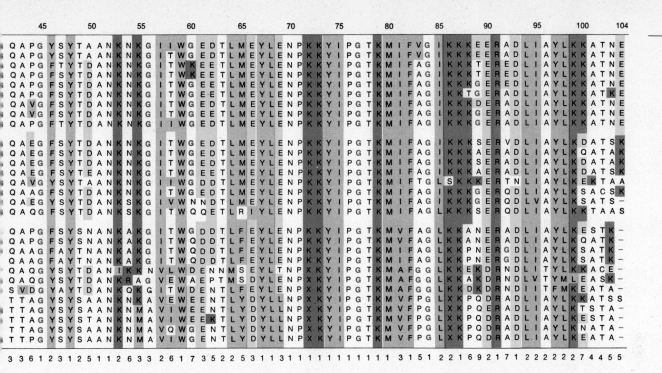

the reasons for most of the similarities of Figure 2. The front view of the molecule, with the heme group seen edge-on, appears on the opening page of this chapter. The regions in which a hydrophobic character is conserved are packed around the heme, giving it a local nonaqueous environment. To a good first approximation, the heme is enclosed in hydrophobic side chains, the main polypeptide chain is wrapped around this package, and the outer surface of the molecule is covered with charged groups.

These charged groups themselves are subject to constraints. The acidic and basic groups are not scattered at hazard over the surface, as they appear to be in most other globular proteins. Instead the negative charges are on the back hemisphere of the molecule as seen in the frontispiece to this chapter, while several positive charges are always found around the exposed edge of the heme. This probably results from the necessity for cytochrome *c* to interact with two other macromolecular assemblages larger than itself, the reductase and the oxidase. (The reductase contains cytochrome *b* of Figure 14 in Chapter 6; the oxidase contains cytochrome *a*.) It is known from chemical studies, for example, that the oxidase interaction involves positive charges on cytochrome *c*; and is now believed that the electron transferred to and from cytochrome passes to the iron and back out again

TWENTY AMINO ACIDS are listed in this table, together with the symbols used in the illustrations that accompany this chapter. Column at right suggests ways to remember the symbols.

AMINO ACID	THREE-LETTER SYMBOL	ONE-LETTER SYMBOL	MNEMONIC DEVICE
HYDROPHOBIC			
ALANINE	ALA	A	INITIAL LETTER
VALINE	VAL	V	INITIAL LETTER
LEUCINE	LEU	L	INITIAL LETTER
ISOLEUCINE	ILE	I	INITIAL LETTER
METHIONINE	MET	M	INITIAL LETTER
PHENYLALANINE	PHE	F	"**F**ENYLALANINE"
TYROSINE	TYR	Y	"T**Y**ROSINE"
TRYPTOPHAN	TRP	W	"T**W**YPTOPHAN"
BASIC			
LYSINE	LYS	K	NEAREST UNUSED LETTER TO L
ARGININE	ARG	R	"**R**-GININE"
HISTIDINE	HIS	H	INITIAL LETTER
ACIDIC			
ASPARTIC ACID	ASP	D	"ASPAR**D**IC ACID"
GLUTAMIC ACID	GLU	E	"GLU**E**"
OTHERS			
CYSTEINE	CYS	C	INITIAL LETTER
GLYCINE	GLY	G	INITIAL LETTER
PROLINE	PRO	P	INITIAL LETTER
SERINE	SER	S	INITIAL LETTER
THREONINE	THR	T	INITIAL LETTER
ASPARAGINE	ASN	N	"AS**N**"
GLUTAMINE	GLN	Q	"**Q**-TAMINE"

Figure 2 annotations (mean differences in cytochrome sequences of major categories):

- PRIMATES VS. OTHER MAMMALS: 10.1 ± 0.8
- (mammals): 5.1 ± 1.3
- MAMMALS VS. BIRDS: 9.7 ± 1.3
- REPTILES VS. MAMMALS: 14.8 ± 4.3
- BIRDS VS. REPTILES: 12.8 ± 5.1
- AMPHIB. VS. HIGHER VERT.: 13.5 ± 2.8
- FISH VS. LAND VERT.: 18.6 ± 2.2
- INSECTS VS. VERTEBRATES: 26.4 ± 2.5
- LOWER PLANTS VS. ANIMALS: 47.4
- HIGHER PLANTS VS. ANIMALS: 45.0 ± 1.8
- YEASTS VS. MOLD: 41.5 ± 0.5
- LOWER VS. HIGHER PLANTS: 48.5 ± 3.2

	Man, Chimpanzee	Rhesus Monkey	Horse	Donkey	Cow, Pig, Sheep	Dog	Rabbit	California Grey Whale	Great Grey Kangaroo	Chicken, Turkey	Pigeon	Pekin Duck	Snapping Turtle	Rattlesnake	Bullfrog	Tuna	Dogfish	Silkworm Moth	Tobacco Hornworm Moth	Screwworm Fly	Fruit Fly (Drosophila)	Baker's Yeast	Candida (a Yeast)	Bread Mold (Neurospora)	Wheat Germ	Sunflower Seed	Mung Bean	Castor Bean	Seasame Seed
Man, Chimp / Monkey	1																												
Horse	12	11																											
Donkey	11	10	1																										
Cow	10	9	3	2																									
Dog	11	10	6	5	3																								
Rabbit	9	8	6	5	4	5																							
Whale	10	9	5	4	2	3	2																						
Kangaroo	10	11	7	8	6	7	6	6																					
Chicken	13	12	11	10	9	10	8	9	12																				
Pigeon	12	11	11	10	9	9	7	8	11	4																			
Duck	11	10	10	9	8	8	6	7	10	3	3																		
Turtle	15	14	11	10	9	9	9	8	11	8	8	7																	
Rattlesnake	14	15	22	21	20	21	18	19	21	19	18	17	22																
Bullfrog	18	17	14	13	11	12	11	11	13	11	12	11	10	24															
Tuna	21	21	19	18	17	18	17	17	18	17	18	17	18	26	15														
Dogfish	24	23	16	15	16	17	17	16	20	19	19	17	19	26	20	20													
Silkworm M.	31	30	29	28	27	25	26	27	28	28	27	27	28	31	29	32	32												
T.H.W.M.	31	30	28	27	27	25	26	27	28	28	26	27	29	33	30	30	31	5											
S.W. Fly	27	26	22	22	22	21	21	22	24	23	23	22	24	29	22	24	25	14	12										
Fruit Fly	29	28	24	24	24	23	23	24	26	25	25	24	24	31	22	25	26	15	14	2									
Yeast	45	45	46	45	45	45	45	45	46	46	46	46	49	48	47	47	48	47	45	45	45								
Candida	51	50	51	50	50	49	50	50	51	50	50	50	52	52	51	47	52	47	46	47	47	28							
Mold	48	46	46	46	46	46	46	46	49	47	46	46	49	48	49	48	49	47	46	41	41	41	42						
Wheat	43	43	46	45	45	44	44	44	47	46	46	46	46	47	48	49	49	45	42	45	47	47	50	54					
Sunflower	42	42	46	45	45	44	44	44	44	46	44	44	44	45	47	48	48	44	44	44	46	48	42	54	15				
Mung Bean	45	45	48	47	47	46	46	46	46	48	46	46	46	45	49	50	50	46	47	47	49	50	52	53	16	11			
Castor	41	41	41	40	43	42	42	42	43	45	43	42	42	43	45	45	48	46	45	44	46	46	50	43	13	11	8		
Sesame	40	40	45	44	43	42	43	43	42	46	44	43	43	42	47	48	49	45	45	46	48	47	50	42	13	12	6	6	

II DIFFERENCES IN CYTOCHROME SEQUENCES

indicate the evolutionary distance between various species. To find the number of amino acid residues that differ in the cytochromes of say, a bullfrog and a dog, find the bullfrog line at left and the dog column at bottom. Answer is the number at the intersection of line and column—in this case, 12. Headings at top of columns indicate mean differences in cytochrome sequences of important categories of organisms, such as insects and vertebrates.

via the exposed edge of the heme. In addition to these interactions, the protein may also be bound loosely to the mitochondrial membrane.

Another distinctive feature of the sequences of Figure 2 is the constancy of glycines throughout evolution. Glycine, alone among the amino acids, has no side chain. The x-ray analysis has shown that most of the glycines occur in tight corners where the polypeptide chain must make a sharp bend. Glycine is necessary in many places because there is no room for a side chain.

The selection pressures which maintain a working cytochrome *c* molecule, therefore, are severe. Glycine and hydrophobic, basic, and acidic side chains all play essential roles and are maintained throughout the course of evolution.

If one compares any two species in Figure 2, he will see that in general the farther apart the species are on the evolutionary tree, the more unlike their cytochromes are. In fact, one can use the differences in sequences to construct a phylogenetic tree. The differences between species are listed in Table II. The species fall into natural groups which agree with the traditional phyletic groupings. The mammals differ among themselves by roughly five amino acids out of 104 (with the primates being less like the other mammals). Mammals differ from birds and reptiles by roughly 10–15 residues (the low figure for mammals *vs.* birds will be reconsidered later). Fish differ from all of the land vertebrates by 18–20 residues, and insects differ from vertebrates by an average of 26 residues. Plants and animals differ by an average of 45–48 residues, but although nearly half the amino acids are different, there is no mistaking the essential similarity of sunflower and kangaroo cytochrome *c.*

We could construct a phylogenetic tree from the data in Table II, drawing branches and junctions so that the distance from one species back to a junction and forward to a second species was proportional to the amino acid difference between species. Contradictions would be encountered in individual species, and we would have to choose a minimal-error tree which minimized the overall discrepancy between all real and calculated species differences at once. There are some advantages to working with DNA codons rather than amino acids, the most obvious being the fact that mutations take place in DNA bases, not in amino acids. The disadvantage is that the real DNA base sequences are not known*; but if enough data are available, the most likely set of base sequences, which permits point mutations to produce the observed amino acid changes, can be deduced.

A typical phylogenetic tree constructed solely from protein sequence information is shown in Figure 3. In this tree, the vertical distance down to a junction and back up to another species represents the degree of difference in DNA base sequences between species. A computer program was used to find the branch lengths that best minimize the overall errors. There are some obvious flaws. The branches between kangaroo and the other non-primate mammals, and between tuna and the land vertebrates, involve backward steps. The mammalian anomaly comes about because the primates are

*Rapid DNA sequencing methods developed by W. Gilbert at Harvard in 1975 may change this situation in a few years.

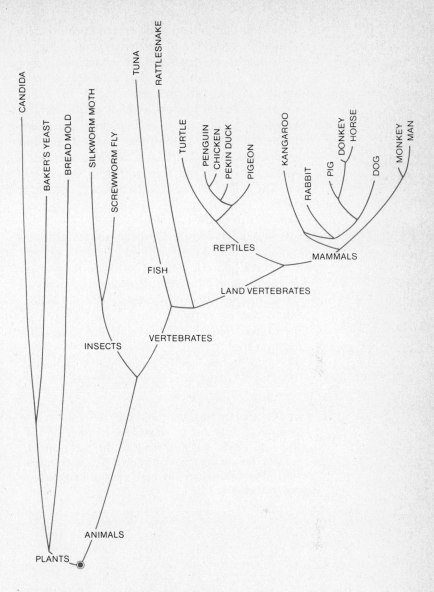

PHYLOGENETIC TREE in computer-plotted diagram is 3 based solely on sequence of amino acids in cytochrome c from many species. Vertical axis indicates degree of difference among sequences.

less like the other placental mammals than the marsupial kangaroos are. Primates are very much creatures apart, on the evidence of Table III. Another flaw is the offshoot of the rattlesnake from the line of land vertebrates before the reptile/mammal separation, whereas it should split off from the line leading to turtle. This flaw is detectable in Table II; rattlesnake is less like the mammals than are turtle, bullfrog, or fish. Part of the trouble is poor statistics from few data. Is this strangeness of rattlesnake cytochrome *c* typical of all snakes or even most reptiles? We simply do not know yet.

661

One important lesson to be drawn from the data in Table II is that the hierarchy picture of life, in which insects are higher than yeasts and other microorganisms, fish are higher than insects, reptiles higher than fish, mammals higher than reptiles, and primates highest of all, is totally wrong. If you compare all animals or all higher plants with the yeasts and mold in the Table, you will find that they are equally far away. A silkworm moth and a dogfish are just as far removed from yeast as is man. The collection of living creatures is not a ladder, but a shrub, of which all branch tips are equally far from the roots.

So far we have implicitly been equating differences in sequences with separation in evolutionary history and hence in time. This assumes that the rate of evolution of cytochrome *c* has not changed radically in the course of evolution, or else the correlation between sequence differences and phylogenetic trees would not be valid. But is this assumption true?

The dates for most of the branch points indicated in Table II and Figure 3 are known from radioisotope dating of the fossil record. One can therefore plot the average number of differences between two branches against the time in the past at which their evolutionary histories diverged. The slope of the resulting curve at any time in the past will indicate the rate of evolution of the protein. The data for this comparison are given in Table III. The figures have been normalized to a standard chain length of 100 amino acids. Another correction must be made to compensate for the fact that a single amino acid difference between horse and duck, for example, may be the result of two or more successive changes at that position in the sequence. The longer ago that two lines separated, the greater the probability that we are missing some mutations by virtue of their having occurred at the same amino acid position.

The dates of divergence are somewhat earlier than might be expected, because what is important is not the era in which a given family flourished or dominated, but the time when the ancestors of the line separated from their relatives which would continue in the older patterns. The great age of the mammals, for example, is the Cenozoic, the last 63 million years. Mammals first appear in the fossil record in the Jurassic, 180 million years ago, but even this is not good enough. The ancestors of

III **RATE OF CYTOCHROME EVOLUTION can be determined by plotting the number of changes per 100 amino acid residues against the time that has elapsed since two lines of organisms diverged. The data in this table and in Table II are the basis for the straight-line curve in Figure 4.**

COMPARISON	CHANGES	CHAIN LENGTH	CHANGES PER 100 RESIDUES	CHANGES CORRECTED FOR REPEATED MUTATIONS	ERA OF DIVERGENCE	MILLIONS OF YEARS SINCE DIVERGENCE
PRIMATES *VS.* OTHER MAMMALS	10.1 ± 0.8	104	9.7 ± 0.8	10.2 ± 0.9	LATE CRETACEOUS	90
HORSE, DONKEY *VS.* OTHER NONPRIMATE MAMMALS	5.1 ± 1.3	104	4.9 ± 1.3	5.0 ± 1.3	LATE CRETACEOUS	90
MAMMALS *VS.* BIRDS	9.7 ± 1.3	104	9.3 ± 1.3	9.8 ± 1.5	LATE CARBONIFEROUS	300
MAMMALS *VS.* REPTILES	14.8 ± 4.3	104	14.2 ± 4.1	15.3 ± 4.8	LATE CARBONIFEROUS	300
BIRDS *VS.* REPTILES	12.8 ± 5.1	104	12.3 ± 4.9	13.2 ± 5.6	LATE PERMIAN	240
AMPHIBIANS *VS.* HIGHER VERTEBRATES	13.5 ± 2.8	104	13.0 ± 2.7	14.0 ± 3.1	LATE DEVONIAN	360
FISH *VS.* LAND VERTEBRATES	18.6 ± 2.2	104	17.9 ± 2.1	19.7 ± 2.5	DEVONIAN	400
INSECTS *VS.* VERTEBRATES	26.4 ± 2.5	108	25.4 ± 2.3	29.4 ± 3.0	EARLY CAMBRIAN	600
LOWER PLANTS *VS.* ANIMALS	47.4 ± 2.0	112	42.3 ± 1.8	54.9 ± 3.4	(PRECAMBRIAN)	(~1200)
HIGHER PLANTS *VS.* ANIMALS	45.0 ± 1.8	112	40.1 ± 1.6	51.0 ± 3.0	(PRECAMBRIAN)	(~1200)
LOWER *VS.* HIGHER PLANTS	48.5 ± 3.2	112	43.4 ± 2.9	56.9 ± 5.3	(PRECAMBRIAN)	(~1200)
YEASTS *VS.* MOLD	41.5 ± 0.5	112	37.1 ± 0.5	45.7 ± 0.9	(PRECAMBRIAN)	(900)

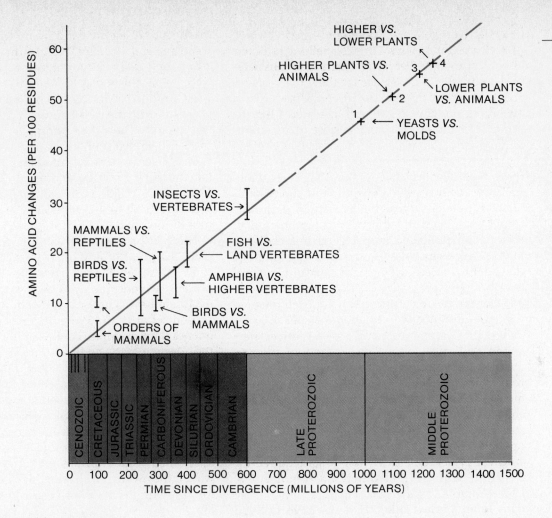

4 **RATE OF EVOLUTION** of cytochrome c. Mean differences between cytochromes c in divergent lines of evolution are plotted against the time when the two lines diverged. Vertical bars indicate mean deviations from average. Data are from Tables II and III. Crosses at upper right indicate divergences for which no good time estimates are available. Broken line indicates extrapolated curve.

these first mammals were the therapsid reptiles of the middle Permian and late Triassic, and these derived from the anapsid reptiles of the Permian and ultimately the cotylosaurs of the late Carboniferous. The separation of those early reptiles which would ultimately produce the mammals, from those which are the ancestors of modern reptiles, must therefore be pushed back as far as 300 million years ago. By the same kind of sleuthing, the separation between birds and reptiles is placed 240 million years ago, in the late Permian.

The resulting plot of the degree of sequence differences against time since separation is shown in

Figure 4. The lengths of the vertical lines indicate the spread in the average figures as obtained from Tables II and III. A straight line can be drawn within these limits of uncertainty, indicating that, to a good first approximation, the rate of evolution of cytochrome *c* has not altered appreciably since the beginning of the Cambrian. Two exceptions, alluded to earlier, are the primates and the bird/mammal comparison.

Why should the rate of evolution of cytochrome

The properly chosen set of 10 to 15 invertebrate cytochrome sequences could settle once and for all the question of the origin of the chordates from the invertebrates.

---

c be effectively constant throughout a period which saw the evolution of man from something as different as marine invertebrates? The reason illustrates one advantage of molecular anatomy over macroscopic anatomy in the study of the process of evolution: proteins are so well-insulated from the actual selection process, and so close to the source of mutations, that what one observes is the rate of mutation dampened down by a general selection for suitability which changes little over the geological time periods. Mitochondria, and the respiratory chain, evolved in the Precambrian. From the similarity of plant and animal mitochondria today, there is no reason to suspect that mitochondria have evolved since the Cambrian to anything like the extent that cellular organization and external morphology have evolved. The efficient power-pack for the cell, once arrived at, has been retained without substantial modification for over 600 million years. As new base mutations produce new amino acid substitutions, these random changes are measured against a relatively unvarying set of requirements for a working cytochrome *c* molecule. The rate at which acceptable amino acid changes accumulate in a random manner is a measure of the strictness of the specifications for an operative protein. These specifications are quite strict for cytochrome *c*, in view of its required interaction with several other macromolecules. Cytochrome *c*, as a result, changes only slowly with time. Certain other proteins have evolved much more rapidly, and one histone makes cytochrome *c* look like a will-o'-the-wisp. These examples will be discussed later in the chapter.

Why should birds and mammals be more alike in cytochrome structure than their evolutionary history warrants? It is suggestive that these are the two warm-blooded groups, whose temperature is maintained in defiance of the environment. Both groups would need more efficient sources of energy than cold-blooded creatures, and the similarity of mammalian and avian cytochrome *c* may be an example of convergent evolution in the face of similar selection pressures.

The striking dissimilarity between primates and other mammals is harder to explain. It is tempting to think that the more intellectually active primates have evolved a better (or at least different) mitochondrial respiratory system in the face of different energy needs. But this is speculation, and tainted with anthropocentrism as well. At the moment, the primate anomaly is a quandary that is worth mulling over at the end of a hard day at the laboratory.

One intriguing dividend from Figure 4 is the extrapolation of the plot back to the divergence of plants and animals. According to this linear extrapolation, this took place on the order of 1100 to 1300 million years ago. The earliest signs of eucaryotes in the fossil record are from roughly this period. The Bitter Springs green algae date from 700 to 900 million years ago, but the Beck Spring dolomite deposits of eastern California contain putative fossil eucaryotes that are probably 1200 to 1400 million years old. Molecular anatomy, therefore, adds to the evidence supporting our picture of the origin of life in Chapter 23.

The cytochrome *c* story, although far more complete than that of any other protein, is still woefully underdeveloped. The full possibilities of protein sequence analysis as a tool for studying evolution have hardly been touched. The properly chosen set of 10 to 15 invertebrate cytochrome sequences, for example, could settle once and for all the question of the origin of the chordates from the invertebrate phyla. Many other relationships between invertebrate phyla could similarly be resolved, and invertebrate evolution placed on a firm footing. The advantage of molecular anatomy is that the results obtained from cytochrome *c* can be checked with any other protein that is as widely distributed. The enzymes of the glycolytic pathway, if they were not so large, would be ideal tools for studying even earlier stages in the history of life.

THE EVOLUTION OF NEW PROTEINS

A study of evolution could in principle be carried out with any globular protein. But at present the only other protein to be studied to a comparable extent is hemoglobin, mainly because, like cytochrome *c*, it is widespread, easily obtainable in pure form, and not too large. The hemoglobins are oxygen carriers, and were not needed until metazoans became too large for the transportation of oxygen by simple diffusion through body fluids. The hemoglobins evolved more recently than the cytochromes and are therefore not as widespread in the living kingdom. They are found in all vertebrates and in many invertebrates, in some species only in larval stages.

Hemoglobin has the added complication of more than one type of chain, which shows us the beginnings of the process by which new proteins evolve. Hemoglobins from the adults of all vertebrates except a few primitive fish have four polypeptide chains per molecule: two α (alpha) chains and two β (beta) chains, as noted in Chapter 5 (see Figure 18 in Chapter 5). The two α chains each have 141 amino acids, and the two β chains, 146. X-ray crystal structure analysis has revealed that each of these chains is folded almost exactly like the chain of myoglobin, which has 153 amino acids (Figure 16, Chapter 4). Furthermore, the amino acid sequences show that the α, β, and myoglobin chains are related, although less so than the cytochromes. The α and β chains in man differ by 84 amino acids, which is roughly the same proportional change as that between the cytochromes of man and bakers' yeast.

Hemoglobin is the carrier of oxygen in the blood, and myoglobin is the storage molecule for O_2 in the tissues. They are similar in folding, in sequence, and function. Can we interpret these similarities as arising from a common evolutionary ancestry, and can we learn anything about the evolutionary process from them? It is easy to see how evolutionary divergence can arise in a given protein in different lines of descent, but how can divergence take place within an individual organism?

The story is a little more complicated. Many of the higher vertebrates have a fetal hemoglobin, which disappears shortly after birth as the adult form begins to be produced. This fetal hemoglobin has the two α chains of the adult form, but uses a third type of chain, the γ (gamma) chain, in place

of the β. The conversion from fetal to adult hemoglobin is a changeover from synthesis of γ to synthesis of β. In addition, primates have a small fraction of yet another hemoglobin, in which the β chains are replaced by δ (delta) chains. (There are other minor variants that can be ignored in this discussion.) The primates, at least, can manufacture four hemoglobin chains: α, β, γ, and δ. How do these fit into an evolutionary pattern?

The amino acid differences in the hemoglobins of man and horse, as well as sperm whale myoglobin, are compared in Table IV. Over 20 species have been examined, but these few sequences illustrate the important trend. The most similar chains are the human β and δ, with only 10 differences. The second most similar, however, are not these two and another human chain, but the α chains from horse and man, with 18 changes. Furthermore, horse β is more like human β and δ than is human γ. The α chains from horse and man are really quite unlike any of the β, δ, or γ chains from these two species, differing by a fairly uniform 84 to 89 amino acid residues. Whale myoglobin, not unexpectedly, differs from these by considerably more, and differs from all the hemoglobin chains by the same amount (allowing for statistical fluctuations).

From these data one can construct an evolutionary history and a tree, which the other known sequences only serve to confirm. The tree obtained is shown in Figure 5, and the history is the following: At some time in the distant past, oxygen-using organisms had a one-chain globin which served both as a carrier and storage agent for oxygen. Well before the radiation of invertebrate phyla that we associate with the Cambrian, this protein differentiated into two separate oxygen-binding heme proteins, one for storage (myoglobin) and one for transport (hemoglobin). Primitive fish such as the lamprey still have a one-chain hemoglobin, as do many marine worms and insects. At some time during the evolution of the higher fish, a radically new step was taken. The hemoglobin line of descent differentiated into two lines, producing two slightly different proteins that we now call types α and β. These two chains associated in a tetramer to produce the four-chain hemoglobin molecule pictured in Figure 18, Chapter 5. Carp, for example, have such a hemoglobin, which became the universal pattern for subsequent vetebrates. The β chain line divided again during mammalian evolution to produce β and γ chains, and a molecule with extra oxygen-binding power useful to the embryo or fetus. The β line split yet again in pri-

V DIFFERENCES IN HEMOGLOBINS of man and horse, and myoglobin from sperm whale. As in Table II, number at intersection of horizontal row and vertical column indicates how many amino acids differ between the two polypeptide chains being compared. The α chain of human hemoglobin has a total of 141 residues; whale myoglobin, 153.

	HORSE α	HUMAN α	HORSE β	HUMAN β	HUMAN δ	HUMAN γ	WHALE MB
HORSE α	0	18	84	86	87	87	118
HUMAN α	18	0	87	84	85	89	115
HORSE β	84	87	0	25	26	39	119
HUMAN β	86	84	25	0	10	39	117
HUMAN δ	87	85	26	10	0	41	118
HUMAN γ	87	89	39	39	41	0	121
WHALE MB	118	115	119	117	118	121	0

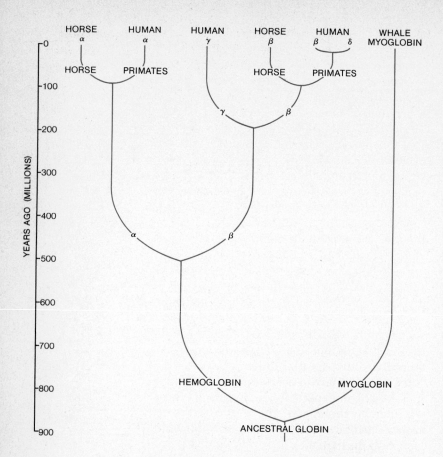

5 EVOLUTION OF GLOBINS is the basis for this phylogenetic tree. Tree was constructed primarily from the sequence data in Table IV. Dates of the branching points came from Figure 8.

mates, to produce a δ chain of unknown utility at present.

The way in which this differentiation of one chain into several can take place at the genetic level is shown in Figure 6. It is apparently not uncommon for genetic material to be repeated accidentally during cell division. The gene for an enzyme is usually envisioned as a particular sequence of DNA that produces only one polypeptide chain. Relatively little study has been made yet of the phenomenon of polymorphism—the presence of several genes producing the same protein with minor variations in sequence. Yet studies of *Drosophila* and the mouse indicate that polymorphism may be more widespread than is now realized. *Drosophila* has several variants of many of its enzymes, indistinguishable by activity, and detectable only by careful electrophoretic methods. Each of these variants must be coded by

its own segment of DNA, and these segments could hardly have arisen other than by accidental duplication of an original gene.

These duplicated genes are the raw materials of molecular evolution. If, in the simplest case, two genes both make the same enzyme, then a mutation in one gene which would otherwise be lethal can be accepted because the other gene continues to make the essential enzyme. The mutated gene may produce an enzyme which is less effective, or which is not enzymatically active for the original reaction at all. If as the result of several mutations, this aberrant gene produces a protein that is catalytically effective in some other reaction (probably closely related to the original one), then natural selection will favor those individuals with this new gene, and a new enzyme will have been born. By some such mechanism, the gene for the ancestral globin must have duplicated. The two loci then went their own way evolutionarily, one producing a protein that was better at binding oxygen, the other producing a protein that was better at picking it up in an oxygen-rich environment and delivering it to the first protein. Myoglobin and hemoglobin had been born. The later differentiation of α, β, γ, and δ chains must have had a similar origin: accidental gene duplication, random mutation in one or both genes, and natural selection from among the proteins produced by these genes. Mutation without gene doubling could never produce new proteins, and gene doubling without mutation would not lead to evolution at all.

A one-chain hemoglobin from the midge, *Chironomous thummi*, has been crystallized and its structure found by x-ray diffraction methods. As indicated in Figure 7, this globin is folded the same way as the myoglobins and all of the chains of the vertebrate globins. The same folding has been found in the marine worm *Glycera dibranchiata*. This particular "myoglobin fold" is evidently very old. It was developed before the ancestral globin differentiated into hemoglobin and myoglobin, and before separation of the evolutionary lines leading to insects, marine worms, and chordates.

Enough hemoglobin sequences are known so that we can perform the same kind of comparison that was done for cytochrome *c* in Tables II and III. The results appear in Figure 8. Hemoglobin, like cytochrome *c*, appears to have evolved at a roughly uniform rate since the Cambrian, although at a faster rate than cytochrome *c*. If we define the Unit Evolutionary Period as the length of time required for a one percent change in sequence to show up between two diverging lines, then the U.E.P. for cytochrome *c* is approximately 20 million years, while the U.E.P. for hemoglobin is only 5.8 million years. Why should hemoglobin change so much more rapidly?

The answer is that hemoglobin reacts with small molecules such as O_2 rather than other macromolecules, and as a result the specifications for an operative hemoglobin molecule are less stringent. Although the rate of appearance of mutations in hemoglobin and cytochrome genes is presumably comparable, the rate of appearance of acceptable mutations is slower in cytochrome *c*. If one is going to hit a machine at random with a hammer, he will

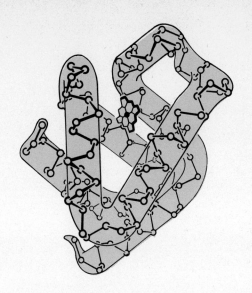

INSECT HEMOGLOBIN. Larva of midge has single-chain hemoglobin with same three-dimensional folding as vertebrate hemoglobins and myoglobin. (Compare with Figure 16 in Chapter 4.) **7**

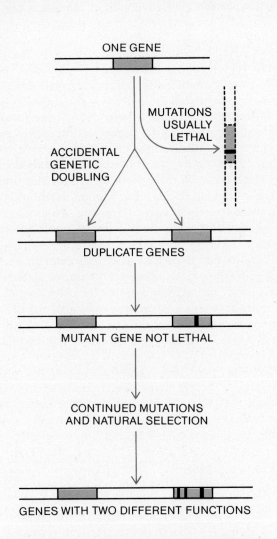

ONE GENE

MUTATIONS USUALLY LETHAL

ACCIDENTAL GENETIC DOUBLING

DUPLICATE GENES

MUTANT GENE NOT LETHAL

CONTINUED MUTATIONS AND NATURAL SELECTION

GENES WITH TWO DIFFERENT FUNCTIONS

6 ACCIDENTAL DUPLICATION of a gene may account for existence of several slightly different types of polypeptide chain in hemoglobin molecule. Mechanism shown here postulates gene doubling, followed by mutation (dark band) in one of the doubled genes.

have a smaller probability of ruining it with a given set of blows if it is a simple gasoline engine than if it is a precision lathe.

The spread between orders of mammals in comparing α chains and β chains in Figure 8 illustrates this idea. From one mammalian order to another, the α chains vary by 16.3 ± 2.8 residues, whereas the β chains vary by 22.2 ± 3.2 residues. Why should the α chains be more fixed in sequence? The answer probably is that each α chain has to be able to fit efficiently with a γ chain in fetal hemoglobin and a β chain in the adult protein, whereas the requirements for a β chain are only that it mesh well with α. The β and γ chains are just different enough that the constraints on α are slightly more rigid than on β, and the α chains vary slightly less than β.

Figure 8 indicates some new lines of reasoning which can be tested against other evidence. Since the α, β split took place during the evolution of the higher fish, this point is dated at the beginning of the Ordovician period. We do not know enough about the distribution of fetal hemoglobin to pinpoint the appearance of the γ chain on historical grounds, but the degree of difference between γ and β chains suggests that the split took place in the late Triassic, or when the mammals were evolving (*point 3*). This at least makes very good sense, for γ chains, where they are known, are found

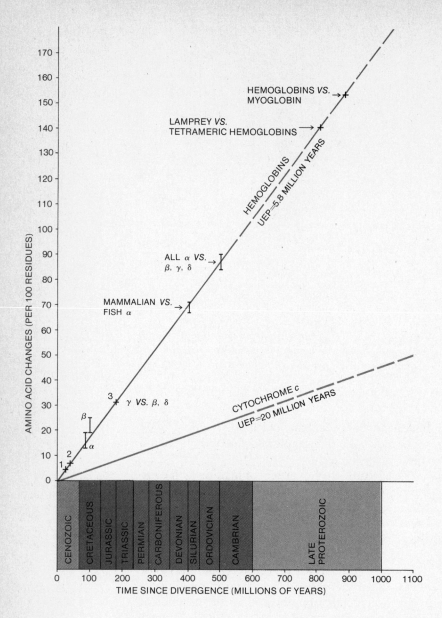

only in mammals. The β, δ split is similarly dated as having occurred roughly 40 million years ago in the primate line *(point 2)*, and the differentiation of man and gorilla from the other primates is dated around 28 million years ago. This last date is currently relevant to the disagreement among anthropologists as to whether genus *Homo* had separated from the other great apes and gone his lone way as early as 15 to 20 million years ago. Many maintain that this was so from the evidence of the fossil *Ramapithecus (see Chapter 34)*. But both the hemoglobin evidence and that from other proteins such as serum albumin support the more traditional idea: that man and the great apes did not separate until four to five million years ago.

THE LIMITS OF EVOLUTION: FIBRINOPEPTIDES AND HISTONES

Several other families of proteins are well-enough studied today to tell us something about the evolution of macromolecules. These include the family of digestive enzymes tryspin, chymotrypsin, and elastase, and the family of dehydrogenase enzymes involved in glycolysis and the citric acid cycle. References on these are found in the Readings at the end of the chapter. But to end this discussion we want to look at two contrasting proteins with very fast and very slow rates of evolutionary change.

Proteins that evolve as slowly as cytochrome *c* are of little help in working out the fine structure of evolution. Once it is known that the cytochromes *c* of man and chimpanzee are identical, no more phylogenetic information can be obtained from them. A more rapidly changing protein is more informative.

The fibrinopeptides A and B are among the most rapidly evolving proteins, largely because they are among the most useless and therefore subject to the least selective pressure against change. Fibrin of blood clots is a highly aggregated, α-helical pro-

8 RATE OF HEMOGLOBIN EVOLUTION is more rapid than that of cytochrome c. Rates are measured in terms of Unit Evolutionary Period (U.E.P.), the time required for a one percent difference to appear between amino acid sequences of diverging lines. As in Figure 5, broken lines indicate extrapolated curve and crosses mark points that cannot be dated from independent evidence. Point 1 (lower left) marks separation of man and gorilla from monkey β chains, point 2 separation between primate β and δ chains, and point 3 separation of γ from β. At 90 million years, bar marked α indicates differences between α chains in different orders of mammals, and β the differences in β chains between orders.

Both the hemoglobin evidence and that from other proteins such as serum albumin support the traditional idea that man did not separate from the great apes until four to five million years ago.

tein. Before the clot forms, the protein chain is held apart in a nonaggregating molecule called fibrinogen. When clotting occurs, an enzyme (thrombin) chops out two short peptides from the fibrinogen molecule, permitting it to collapse into fibrin and aggregate. The peptide fragments, having fulfilled their role, are discarded.

The discarded peptides have been used to study the evolution of mammals, especially the artiodactyls, which include pigs, deer, and cattle, The family tree that is produced (Figure 9) agrees with the traditional phylogenetic tree and clears up one or two uncertainties in details. The fibrinopeptides illustrate the validity of protein sequence comparison as a method of taxonomy. But of more interest in the scope of this chapter is the rate at which the fibrinopeptides evolve. This is shown in Figure 10. The fibrinopeptides evolve much more rapidly than hemoglobin or cytochrome c: their U.E.P. is approximately 1 million years. The reason is obvious. If the only purpose of these peptides is to hold fibrinogen molecules in an unfolded configuration until they are cut out, then the requirements for

The fibrinopeptides are among the most rapidly evolving proteins, largely because they are among the most useless and hence subject to little selective pressure against change.

workable fibrinopeptides are unlikely to be very strict. A greater proportion of the random mutations in sequence will be acceptable, and changes will accumulate faster.

Exactly the opposite is found in one of the histones, or basic proteins that wrap around DNA in the nucleus. It is not clear exactly what role the histones play, but is likely that they contribute to turning genes on and off. They are therefore at the very heart of the living organism, and could be expected to be quite critically specified. Histone fraction IV is a polypeptide chain of 102 amino acids. Amino acid sequences have been determined for histone IV from calf thymus and pea seedlings, and only two changes are found, both conservative ones involving the substitution of one hydrophobic side chain for another. If we accept

9 **FAMILY TREE OF ARTIODACTYLS was deduced from comparison of fibrinopeptides. It agrees in every respect with the accepted phylogenetic tree. The close resemblance illustrates the validity of sequence analysis as a tool for studying the details of evolutionary history.**

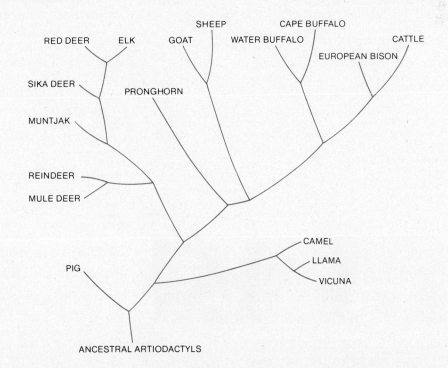

the cytochrome *c* date of roughly 1200 million years since the divergence of animals and plants, then the U.E.P. for histone IV is on the order of 600 million years!

The entire discussion of the relative rates of molecular evolution is summarized in Figure 11. In the words of Louis Sullivan, the great architect and teacher of Frank Lloyd Wright, "Form follows function." This dictum is as true for macromolecules as for buildings. The function dictates the form, and also determines what proportion of random alterations is likely to be acceptable. Out of

this selection process comes the difference in rates of evolution. It is possible, therefore, to select a protein with the most suitable rate of change in studying evolution. If answers to broad questions affecting the entire sweep of living history are sought, then a slowly-evolving protein like cytochrome *c* is required. If what is desired is fine-structure information about a small group of related organisms, then a rapidly changing fibrinopeptide is called for. The tools for the study of phylogeny and evolution are there, and can be selected for the problem at hand.

10   **RATE OF FIBRINOPEPTIDE EVOLUTION has remained roughly constant during the divergence of the mammals in the Cretaceous period and the Cenozoic era. Rate is far faster than those of hemoglobin and cytochrome c. The Unit Evolutionary Period (U.E.P.) is only about one million years.**

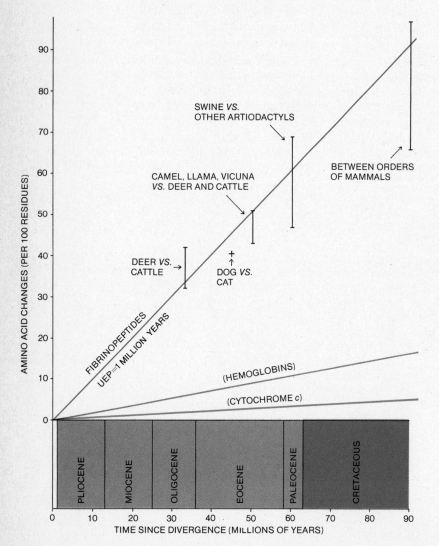

READINGS

Most of the story of molecular evolution must be dug out of the original literature, and elementary accounts are hard to come by. The following are the best starting points:

M.O. Dayhoff, "Computer Analysis of Protein Evolution," *Scientific American*, July 1969. Construction of phylogenetic trees from sequence data.

R.E. Dickerson, "Cytochrome *c*: The Structure and History of an Ancient Protein," *Scientific American*, April 1972. The evolution of the protein in the light of its three-dimensional structure.

R.E. Dickerson and I. Geis, *The Structure and Action of Proteins*, New York, Harper & Row, 1969. A continuation of the material in this chapter, and at about the same level. Extensively correlated with three-dimensional protein structures.

H. Neurath, "Protein-Digesting Enzymes," *Scientific American*, December 1964. Predates the x-ray analyses of the trypsin family, but a good summary of the sequence evidence at the time.

R.M. Straud, "A Family of Protein-Cutting Proteins," *Scientific American*, July 1974. The three-dimensional story of the trypsin family.

E. Zuckerkandl, "The Evolution of Hemoglobin," *Scientific American*, June 1965. Old, but still an excellent introduction. Clear and readable.

The following reviews or symposium volumes are the next step if you want to go beyond the *Scientific American* level:

M.O. Dayhoff, *Atlas of Protein Sequence and Structure, 1969*, Silver Spring, MD, National Biomedical Research Foundation, 1969. A standard reference compendium of amino acid sequences in proteins, with introductory chapters on molecular evolution. Very little on three-dimensional structure.

R.E. Dickerson, "The Structure of Cytochrome *c* and the Rates of Molecular Evolution," *Journal of*

FIBRINOPEPTIDES
(1.1 MILLION YEARS)

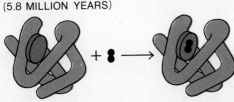

FIBRINOGEN   FIBRIN

GLOBINS
(5.8 MILLION YEARS)

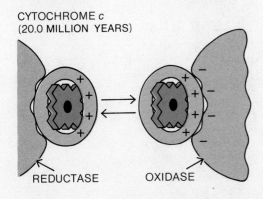

CYTOCHROME *c*
(20.0 MILLION YEARS)

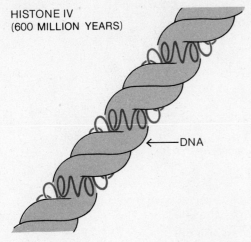

REDUCTASE   OXIDASE

HISTONE IV
(600 MILLION YEARS)

← DNA

Molecular Evolution, 1:26, 1971. The evidence behind the cytochrome story.

R.E. DICKERSON, R. TIMKOVICH AND R.J. ALMASSY, "The Cytochrome Fold and the Evolution of Bacterial Energy Metabolism, *Journal of Molecular Biology*, 100:473, 1976. Evidence that respiration may have evolved from bacterial photosynthesis.

T.H. JUKES AND C.R. CANTOR, "Evolution of Protein Molecules," *in Mammalian Protein Metabolism*, Vol. III, New York, Academic Press, 1969.

R.J. POLJAK, "X-Ray Diffraction Studies of Immunoglobulins." *Advances in Immunology* 21:1, 1975.

M.G. ROSSMANN, A. LILJAS, C-I BRANDEN, AND L.J. BANASZAK, "Evolutionary and Structural Relationships among Dehydrogenases," *The Enzymes* (Paul Boyer, ed.) 3rd Ed.; Vol. XI, p. 62. New York, Academic Press, 1975.

E.L. SMITH, "Evolution of Enzymes," in *The Enzymes*, (Paul Boyer, ed.) 3rd Edition, Vol. I, p. 267. New York, Academic Press, 1970.

1 **INTERACTION OF MACROMOLECULES determines the rate of molecular evolution. Fibrinopeptides do not interact with other molecules and hence evolve rapidly. Globins interact with O₂ and other small molecules. Cytochromes must mate with the cytochrome oxidase and cytochrome reductase complexes. Histone IV binds to DNA within the nucleus. Its specifications are the most rigid of all, and it is the slowest to evolve.**

One might ask: 'Why not simply ignore the species problem?' This also has been tried, but the consequences were confusion and chaos. The species is a biological phenomenon that cannot be ignored. —Ernst Mayr

Virtually all of the classification of plants and animals is based on the concept of species —the next level of organization above the individual organism. With the specimen assigned subjectively to a particular species according to degrees of similarity with other individuals, the taxonomist's task becomes a matter of constructing a simple hierarchy. Species are subdivided into geographical sets of populations, called subspecies, and clumped into sets of similar species called genera (singular: genus). The genera are in turn clumped to form still higher taxonomic categories *(see Chapter 24)*.

Of greater contemporary significance, the species is the basic unit in almost all modern ecological and evolutionary studies. In describing the process by which populations and communities of organisms interact and evolve, the biologist resorts almost unconsciously to the species as the lowest operational unit.

It is curious that although Charles Darwin entitled his book *On The Origin of Species*, by our present understanding he did not write on the origin of species at all. Darwin identified natural selection as the mechanism by which evolution mostly occurs, revealing how a given species can change its genetic constitution through time. But he failed to show how one species can split into two or more. The reason for this omission is quite simple. Darwin was preoccupied with proving that species are

COURTSHIP FLIGHT of male sulfur butterfly as it would look to an insect. Photograph on opposite page, made with ultraviolet light, shows male (right) approaching female, flashing the bright upper surface of his wings. The flash area is conspicuous to the insect, whose eyes are sensitive to ultraviolet, but not to humans. Wing-flashing is one of the mechanisms that enable sulfur butterflies to recognize members of their species, and thus prevent interbreeding with other species.

flexible, arbitrary units that natural selection alters relentlessly through time. He said, in opposition to the prevailing opinion of the day, that "no clear line of demarcation has as yet been drawn between species and subspecies —that is, the forms which in the opinion of some naturalists come very near to, but do not quite arrive at the rank of species; or, again, between sub-species and well-marked varieties, or between lesser varieties and individual differences. These differences blend into each other in an insensible series; and a series impresses the mind with the idea of an actual passage." In other words, species cannot exist as discrete units.

THE CONCEPT OF SPECIES

Today biologists take an intermediate position in assessing the problem. They agree that species are indeed changeable and encompass a great deal of variation, as Darwin was anxious to point out. But biologists also recognize that sharp discontinuities exist between populations. The gaps are based on the absence of free interbreeding among the populations that coexist geographically. Since this negative criterion of reproductive isolation is a natural one and not based on some arbitrary degree of difference selected by the taxonomist, it is referred to as the BIOLOGICAL SPECIES CONCEPT. The concept defines the species as a population or series of populations within which free gene flow occurs under natural conditions. This means that all of the normal, physiologically competent individuals in a given place at a given time are capable of breeding with all of the other individuals of the opposite sex belonging to the same species, or at least that they are capable of being linked genetically to them through chains of other breeding individuals. By definition they do not breed freely with members of other species.

673

Natural conditions are a basic part of this definition. In establishing the limits of a species it is not enough merely to prove that genes of two or more populations can be exchanged under experimental conditions. The populations must be demonstrated to interbreed fully in the free state. To illustrate the point, let us consider a familiar case with some surprising implications. Lions and tigers are genetically closely related, despite their obvious differences in outward appearance. They are frequently crossed in zoos to produce hybrids, called "tiglons" (tiger as father) and "ligers" (lion as father). But this accomplishment does not prove them to belong to the same species. The ability to hybridize under a suitable experimental environment can be said to be a necessary condition by our definition, but it is not sufficient. The important question is whether the two forms cross freely where they occur together in the wild. Lions and tigers did coexist over most of India until the 1800's, when lions began to be reduced by hunting *(Figure 1)*. Now they are nearly extinct, limited to a few hundred individuals in the Gir Forest in the state of Gujurat. There is no doubt that lions and tigers were fully isolated reproductively during their coexistence, for no tiglons or ligers have ever been found in India. Suppose that lions and tigers had been shown to be wholly intersterile under experimental conditions. This could have reasonably been interpreted to mean that they are distinct species, because the condition could be assumed to hold in nature also. But the opposite evidence means nothing, since many other genetic devices in addition to mere intersterility might (and obviously do) operate to isolate them in nature. In fact the lion and the tiger differ strongly in their behavior and the habitats they prefer to live in. The lion is more social, roaming in small groups called prides, and it prefers open country. The tiger is solitary, and is found more frequently in forested regions. These differences between the two forms could be great enough to account fully for their failure to hybridize.

The biological species concept has proven useful because it accounts for a large percentage of the discontinuities that occur in natural phenotypic variation. Its validity can be further substantiated by comparing species concepts across human cultures. In 1927 Ernst Mayr, an evolutionist and a student of birds who has specialized in the study of species, led an expedition to collect and identify

1 **RANGES OF LION AND TIGER were once far more extensive than they are today. Until the nineteenth century the two species overlapped throughout much of India (darkest shading).**

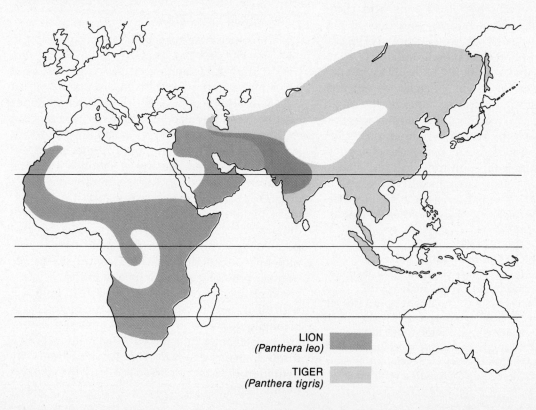

LION
(Panthera leo)

TIGER
(Panthera tigris)

the birds of the remote Arfak Mountains of northwestern New Guinea. In the course of his study he came to recognize 138 species by means of the biological criterion. His taxonomic judgment was based mostly on the phenotypic variation of the birds collected, but all of the evidence indicated that the entities he came to recognize are true biological species. The native Arfak people among whom he lived depend in good part on hunting for their existence. At the time of his visit, Mayr found that they distinguished and assigned special names to 137 of the 138 forms which he had recognized as species and labeled with technical names derived from Greek and Latin. The Arfak people missed the full roster of 138 only because two of the species were so similar in outward appearance that it required the scrutiny of a trained ornithologist to tell them apart. Biologists do not in fact always perform as well as aboriginal peoples. Six tribes of Amerindians living in the rain forests of the Amazon-Orinoco basin use a total of about 5,000 names to distinguish more than half of the tree species that grow within their hunting grounds. The explanation of this botanical skill is that the people depend on the trees for their livelihood and also refer to them constantly in their folklore and religion. Each of several individual tribes knows over 1,000 species, which would be an impressive accomplishment even for a professional botanist. Not only do the entities usually match the genetic concepts of the botanists, but in some cases the Indians have called the attention of the botanists to obscure species previously overlooked by the professionals.

SPECIES AND PHENOTYPE

When a biologist distinguishes a species on the basis of the modern biological concept, he does more than simply point out the existence of a state of reproductive isolation. The rank of species normally carries with it two other basic attributes that repay closer study:

The definition of the species as a closed gene pool is the best devised thus far, but compared with other scientific concepts it has some glaring weaknesses.

---

1. The species, in order to be reproductively isolated, has probably enjoyed a relatively long evolutionary history during which it adapted to the environment in ways peculiar to itself.

2. Consequently, the species will be found to differ from other species in many phenotypic characters in addition to the ones first noticed by the taxonomist. These characters will probably involve morphology, physiology, biochemistry, and (in the case of animals) behavior. Further, the species will be found to occupy a distinctive geographic range of its own.

The power of the species concept is most effectively illustrated in examples of the so-called SIBLING SPECIES. These are valid biological species that are difficult to tell apart by ordinary means. Sometimes taxonomists have stumbled upon the existence of groups of sibling species by discovering gaps in what was previously considered to be continuous variation within relatively obscure characteristics. When detailed analyses were then conducted, the gaps were seen to be maintained by reproductive isolation between biological species. Afterwards other previously unsuspected phenotypic gaps were brought to light. An important example of a complex of sibling species unveiled in this manner is that of the malaria mosquitoes of Europe. Prior to 1934 malaria in Europe was thought to be carried by a single species, called *Anopheles maculipennis*. Entomologists were puzzled, however, by the fact that in some parts of Europe the mosquito was abundant but malaria was absent. They noticed that in a few places the mosquito ignored man while continuing to feed on domestic animals. In some localities it bred in brackish water and elsewhere was limited to fresh water. All of this variation had no particular meaning until finally it was shown to be correlated with constant differences in the egg color and shape of the egg masses. Further analysis then revealed that most of the variation, including the ability to transmit malaria versus the lack of the ability, can be used to separate genuine biological species. The modern view of the *Anopheles maculipennis* complex of sibling species is summarized in Table I. This understanding has helped greatly in the control of malaria in Europe.

WEAKNESSES OF THE CONCEPT

The definition of the species as a closed gene pool is the best devised thus far, but in comparison with other scientific concepts it has some glaring weaknesses. Compare it, for example, with a strong

CHARACTERISTIC COMPARED	A. MELANOON & SUBALPINUS	A. MESSEAE	A. MACULIPENNIS	A. ATROPARVUS
EGG COLOR	ALL BLACK OR (SUBALPINUS) WITH DARK CROSS BARS	TRANSVERSE BARS PART OF A DIFFUSE DARK PATTERN	TWO BLACK CROSS BARS ON LIGHT BACKGROUND	DAPPLED OR WITH WEDGE-SHAPED BLACK SPOTS
EGG MASS	LARGE AND SMOOTH	LARGE AND ROUGH	LARGE AND ROUGH	SMALL AND SMOOTH
X-CHROMOSOME	STANDARD	EXTENSIVE REARRANGEMENT	STANDARD	STANDARD
THIRD CHROMOSOME	INVERSION IN RIGHT ARM	INVERSION IN RIGHT ARM	INVERSION IN RIGHT ARM	STANDARD
HABITAT	OFTEN RICE FIELDS	COOL, STANDING FRESH WATER	COOL, RUNNING FRESH WATER	COOL, SLIGHTLY BRACKISH WATER
HIBERNATION	NO	YES	YES	NO
FEEDING ON MAN	UNKNOWN	RARELY	NO	YES
RANGE	MEDITERRANEAN	CONTINENTAL AND NORTHERN EUROPE	MOUNTAINS OF EUROPE	NORTHERN EUROPE
MALARIA CARRIER	NO	NO (RARELY)	NO	SLIGHTLY

I **MALARIA MOSQUITOES of the Anopheles maculipennis group in Europe can be told apart by the characteristics listed in this table. One of the distinguishing characteristics is the ability to transmit malaria. Only the two species that feed mainly on man (far right) are dangerous.**

concept of the physical sciences such as the electron. This ultimate physical unit is actually a summary term for a set of exact measurables, namely 4.8×10^{-10} units of negative charge, 9.1×10^{-28} grams of mass, and so forth. The value of such a physical concept is determined by how often it appears in descriptions and laws and by the predictive value of the theory that uses it. Physicists recognize that belief in the reality of electrons is based on the fact that the concept is needed in the explanation of so many and varied phenomena — cathode rays, photoelectric effects, currents in solids and liquids, radioactivity, chemical bonds, and so on. It is probable that any other culture developing a science of physics would create the concept of the electron.

Now consider the biological species concept. It looks very good in cases like the Arfak Mountain bird fauna and the *Anopheles* mosquitoes of Europe. In general, where the biotas (faunas and floras taken together) are considered at just one place and over a period of time too brief to permit evolution, the species concept is strong. We must stress the importance of one place and one time — the smaller the place and the shorter the time the easier the concept is to use. For this reason it is often said that the ideal species to which the definition can be applied is the nondimensional species, a population observed over too small an area and through too short a time for these dimensions to play a significant role. For it is the dimensions of space and time, when their effects on variation are fully considered, that cause the greatest difficulty in the biological species concept. These difficulties can be enumerated as follows:

1. Populations that do not occur in the same places cannot be evaluated precisely, because we do not know whether they would interbreed freely with each other if they came together under natural conditions.

2. Populations that do not occur together in time cannot be evaluated for the same reason. Figure 2 reveals the gradual evolution of one species of snail, as shown by fossil shells from successive Pliocene strata. The snails at the beginning and at the end of the progression would almost certainly be considered distinct species solely on the strength of the morphological differences that separate them. But if this proposition is accepted we are forced to ask: which of the connecting populations should be separated as additional species? The answer is that the biological species

A. LABRANCHIAE	A. SACHAROWI
SIMILAR TO *ATROPARVUS* BUT PALER, DARK SPOTS SMALLER	GRAY WITHOUT PATTERN
VERY SMALL AND ROUGH	NONE
STANDARD	SMALL INVERSION
STANDARD	INVERSION IN LEFT ARM
MOSTLY WARM, BRACKISH WATER	SHALLOW, STANDING WATER, OFTEN BRACKISH
NO	NO
YES, WITH PREFERENCE	ALMOST EXCLUSIVELY
CHIEFLY SOUTHERN EUROPE	EASTERN MEDITERRANEAN AND NEAR EAST
VERY DANGEROUS	VERY DANGEROUS

the simple reason that the concept is based wholly on the exchange of genes during sexual reproduction. Where sexual reproduction is lacking, arbitrary species definitions based on the degree of difference in the phenotypes must be used.

4. Populations that exchange genes with other populations to an intermediate degree are common in some groups, especially plants. The populations cannot make up their minds whether to interbreed freely with each other, in which case they would all form a single species, or whether to maintain strict reproductive isolation, in which case they could easily be classified as distinct species. Instead they maintain an intermediate degree of gene flow. Such populations are often referred to as semispecies. They provide unusually difficult problems for the taxonomist attempting to separate and classify them. Several well-known genera of plants containing semispecies are listed in Table II. Here they are contrasted with the single plant genus *Asclepias* (the milkweeds), which is well known for its fully isolated biological species.

REPRODUCTIVE BARRIERS

How do two species coexist in the same place and time and yet not interbreed? This reproductive iso-

concept is quite meaningless in this case. It is necessary to turn to a different, quite arbitrary species definition based on the amount of difference in the phenotypic characters preserved in the fossils. This criterion is admittedly selected more for convenience than for its anticipated theoretical value.

3. Populations that are either completely asexual or obligatorily self-fertilizing also lie outside the domain of the biological species concept, for

FOSSIL SNAIL SHEELS from successive rock strata of the Pliocene epoch demonstrate how a macroevolutionary change can occur in a series of microevolutionary steps. The ten snails depicted below were fossilized in fresh-water deposits in Yugoslavia sometime between three and ten million years ago. Specimens at top left and bottom right scarcely seem to belong to same species. **2**

GENUS	COMMON NAME	AREA COVERED IN STUDY	FAMILY	NUMBER OF WELL-DEVELOPED SPECIES	NUMBER OF SEMISPECIES
PINUS	PINE	PACIFIC COAST	PINACEAE	12	6
QUERCUS	OAK	PACIFIC COAST	FAGACEAE	10	7
DIPLACUS	MONKEY FLOWER	PACIFIC COAST	SCROPHULARIACEAE	3	4
AQUILEGIA	COLUMBINE	NORTHERN HEMISPHERE	RANUNCULACEAE	2	3?
ASCLEPIAS	MILKWEED	NORTH AMERICA	ASCLEPIADACEAE	108	0

II **SEMISPECIES IN PLANTS create difficult problems of classification. In many taxonomic groups there is an intermediate amount of gene flow between coexisting populations. Such populations are called semispecies to distinguish them from more sharply differentiated species.**

lation is a secondary result of genetic differences that exist between them. The differences are generally referred to as INTRINSIC ISOLATING MECHANISMS, in order to contrast them with such environmental (extrinsic) barriers as mountain ranges, rivers, and straits, that separate some of the populations physically. Intrinsic isolating mechanisms are the set of all the genetic differences between species which, regardless of the origin and the primary adaptive significance of the differences, reduce the chance of interbreeding. In other words, they are the sum of everything that can go wrong in reproduction between two populations.

To take one concrete example, if one species has evolved so that it is active only during the day, and the other so that it is active only during the night, this difference in adaptation to the environment also serves to isolate the two species reproductively. It is an intrinsic isolating mechanism. Biologists who study speciation (the process of species multiplication) have found it convenient to classify the isolating mechanisms into two kinds:

3 **ISOLATING MECHANISMS OF BIRDS include song, habitat and color of plumage. These characteristics, which enable ornithologists in the field to distinguish between species, also serve to keep different species from interbreeding. Flycatchers shown are natives of eastern U.S.**

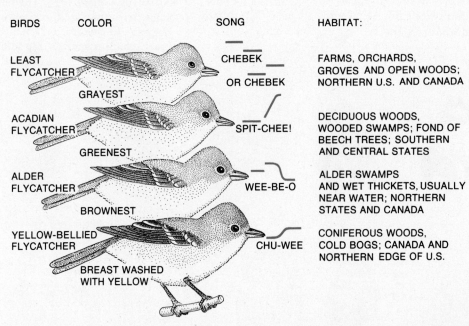

BIRDS   COLOR   SONG   HABITAT:

LEAST FLYCATCHER   GRAYEST   CHEBEK OR CHEBEK   FARMS, ORCHARDS, GROVES AND OPEN WOODS; NORTHERN U.S. AND CANADA

ACADIAN FLYCATCHER   GREENEST   SPIT-CHEE!   DECIDUOUS WOODS, WOODED SWAMPS; FOND OF BEECH TREES; SOUTHERN AND CENTRAL STATES

ALDER FLYCATCHER   BROWNEST   WEE-BE-O   ALDER SWAMPS AND WET THICKETS, USUALLY NEAR WATER; NORTHERN STATES AND CANADA

YELLOW-BELLIED FLYCATCHER   BREAST WASHED WITH YELLOW   CHU-WEE   CONIFEROUS WOODS, COLD BOGS; CANADA AND NORTHERN EDGE OF U.S.

those that operate to keep the species apart before mating occurs (premating isolating mechanisms) and those that prevent hybrid offspring from developing or breeding in case the premating isolating mechanisms fail (postmating isolating mechanisms). A more complete classification is given in Table III.

Intrinsic barriers tend to act in sequence. For example, if there are no ecological barriers, or if they exist but are insufficient, temporal differences might suffice as barriers. If these also are lacking, behavioral differences might exist that keep the species apart. And so on down the list. Sexual reproduction consists of an elaborate series of exceedingly delicate steps. Anything that can go wrong at any place in the sequence qualifies as an intrinsic isolating mechanism. Species, in the course of their evolution, obey Dollo's Law (see Chapter 28) and always continue to diverge from each other in genetic composition. As a result, they will tend always to augment and to strengthen the intrinsic isolating mechanisms that separate them, for the simple reason that the greater the genetic difference between them, the more things there are that can thwart them if they attempt to interbreed.

INTRINSIC ISOLATING MECHANISMS keep two species separate by preventing them from mating or by preventing hybrid offspring from developing or breeding.

PREMATING ISOLATING MECHANISMS	
1.	ECOLOGICAL: THE SPECIES OCCUPY DIFFERENT HABITATS OR AT LEAST BREED IN DIFFERENT HABITATS. THESE DIFFERENCES, LIKE ALL OF THE OTHER INTRINSIC MECHANISMS LISTED BELOW, HAVE AN HEREDITARY BASIS.
2.	TEMPORAL: THE SPECIES BREED AT DIFFERENT SEASONS OR AT DIFFERENT TIMES OF THE DAY.
3.	BEHAVIORAL: IN THE CASE OF ANIMALS, THE COURTSHIP BEHAVIOR DIFFERS.
4.	MECHANICAL: MORPHOLOGICAL OR PHYSIOLOGICAL DIFFERENCES PREVENT NORMAL MATING.
POSTMATING ISOLATING MECHANISMS	
5.	MECHANICAL: THE SPERM ARE UNABLE TO REACH OR TO FERTILIZE THE EGGS.
6.	MORTALITY: THE HYBRID DIES AT SOME STAGE PRIOR TO MATURITY.
7.	STERILITY: THE HYBRID MATURES BUT IS STERILE TO SOME DEGREE.
8.	FITNESS: THE HYBRID IS FERTILE BUT THE F$_2$ HYBRID GENERATION HAS LOWER FITNESS.

A great many familiar facts of natural history can be usefully interpreted with reference to this particular aspect of speciation. Consider, for example, the case of the three species of birds called flycatchers that are found in eastern North America. They are nearly identical in size, plumage, and other outward characteristics. In other words, they are sibling species. In Roger Tory Peterson's *A Field Guide to the Birds*, birdwatchers

---

Intrinsic isolating mechanisms are the sum of everything that can go wrong in reproduction between two populations.

---

are advised to take note of differences in habitat preference and song when attempting to identify flycatchers to the species level (*Figure 3*). These nonanatomical characteristics are indeed strong. In fact, they are the means by which the species themselves are kept apart. That is, the segregation into the major habitats, combined with radically different songs used during territorial defense and courtship, are probably sufficient to prevent free interbreeding of the species. As might be expected, many of the diagnostic characteristics by which we classify species are the intrinsic isolating mechanisms by which the organisms themselves maintain the separateness of the species.

One more example, selected from among many in the recent biological literature, will illustrate how a systematic investigation of species barriers is conducted. The butterflies shown in Figure 4 are representatives of two of the common European species of "Whites" in the family Pieridae. *Pieris napi* is a lowland dweller, *Pieris bryoniae* is restricted to the mountains. The populations overlap slightly at an elevation of approximately 1000 meters. Here on the mountainsides they hybridize occasionally, but over the greater part of their ranges they maintain their integrity as species. Entomologists have analyzed the differences that separate the two species, producing the list shown in Figure 4. The peculiar combination of isolating mechanisms that operate in this case would have been impossible to predict in advance. Because of their secondary nature and the virtually random manner in which they arise, each set of isolating mechanisms tends to be different from the next.

HOW SPECIES MULTIPLY

How do separate species arise? There are two basically different mechanisms. Geographic or allopatric speciation involves the intervention of external barriers that split populations, allowing them to diverge to species level. Sympatric speciation involves internal factors that subdivide populations on the spot and without the aid of external barriers. (Allopatric means literally "belonging to different countries," while sympatric means "belonging to the same country.")

4 ISOLATING MECHANISMS OF INSECTS were studied in two species of white European butterfly. Their ranges overlap on mountainsides and the two species occasionally interbreed.

PIERIS NAPI

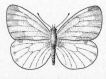

PIERIS BRYONIAE

PREMATING ISOLATING MECHANISMS		
1 ECOLOGICAL	FLIES MOSTLY IN THE VALLEYS	FLIES MOSTLY IN THE MOUNTAINS OVER 1000 METERS ELEVATION
2 TEMPORAL	STARTS BREEDING EARLIER IN THE YEAR	STARTS BREEDING LATER IN THE YEAR
3 BEHAVIORAL	PREFERS TO MATE IN SUNNY WEATHER	HAS A STRONGER TENDENCY TO MATE IN CLOUDY WEATHER
4 MECHANICAL	NONE KNOWN	NONE KNOWN

POSTMATING ISOLATING MECHANISMS	
5 MECHANICAL	NONE AT THIS LEVEL; SPERM CAN REACH THE EGGS
6 MORTALITY	THE HYBRIDS ARE LESS LIKELY TO SURVIVE. THE LARVAE ARE MORE SUSCEPTIBLE TO VIRUS DISEASE. ALL SURVIVING FEMALES DIE WHEN THEY EMERGE FROM THE CHRYSALIS. MALES ARE NORMAL.
7 STERILITY	NO INFORMATION ON EITHER SPECIES
8 FITNESS	NO INFORMATION ON EITHER SPECIES

In geographic speciation, a single population (or series of populations) is first divided by an extrinsic barrier —a river, mountain range, an arm of a sea. The isolated populations then diverge from each other in evolution because of the inevitable differences of the environments in which they find themselves. Since all populations evolve when given enough time, divergence between all extrin-

SIMPLE FISSION is one biogeographic process that leads to the formation of new species. Drawing 1 at upper left shows a single species of butterfly living in its habitat. In drawing 2 (lower left) the intrusion of an uninhabitable habitat—in this case, a river valley—divides the species into two isolated populations. Eventually they evolve into two distinct species (3). If the barrier then disappears—for example, if the river dries up or changes its course—the two new daughter species can coexist in the same habitat without interbreeding, as shown in the drawing at lower right.

sically isolated populations must eventually occur. By this process alone the populations can acquire enough differences (intrinsic isolating mechanisms) to reduce gene flow should the extrinsic barrier be removed and the populations again come into contact. If sufficient differences have accumulated, the populations can coexist as newly formed species. If it has not occurred yet, the populations will resume exchanging genes when the contact is renewed.

Sympatric speciation can take one of two very different forms: polyploidy or disruptive selection. When there is an accidental doubling in the number of chromosomes in some of the offspring, the resulting polyploid individuals will usually not be able to interbreed with the normal population that gave rise to them. If they are able to interbreed among themselves, the polyploids constitute an instantaneously created new species. In disruptive selection a new selective force removes intermediate individuals from a previously freely interbreeding population, thus splitting the population as effectively as if some external barrier had been placed between the two new populations.

Geographic speciation appears to be the most frequent process in both plants and animals. Polyploidy is responsible for the origin of approximately half of the living species of higher plants, but it is a relatively insignificant process in animal speciation. Finally, disruptive selection has not been demonstrated with complete certainty in nature. It remains a good possibility, one that will be very difficult to observe if it does occur but potentially of importance as a mechanism for the rapid multiplication of species.

The principal modes of speciation can now be examined in fuller detail, with some of the theoretical difficulties added. Figures 5 and 6 show the two most common ways in which geographic speciation occurs. They represent, in a deliberately oversimplified form, the process of speciation by fission and by multiple invasion respectively.

6
▼

MULTIPLE INVASION also leads to the formation of new species. Drawing 1 (far left) depicts a single population of snails on one side of an unchanging geographic barrier. Somehow a few individuals manage to cross the barrier (2), starting a new and largely isolated population. Eventually the two populations evolve into separate species (3). If a second colonization follows, the two daughter species will overlap without interbreeding (4).

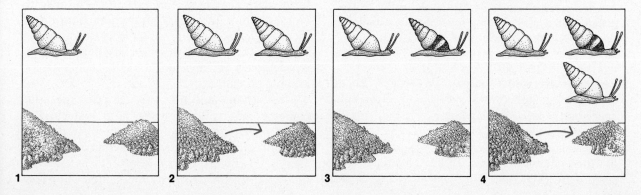

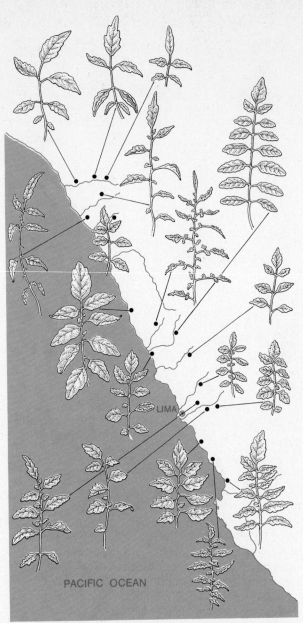

7 **GEOGRAPHIC SPECIATION has been thoroughly studied in the wild tomato Lycopersicon peruvianum, which grows along the coasts of Peru and Chile. Isolated populations have become noticeably differentiated, as is evident from the wide variation in the shapes of their leaves.**

Diagrams exemplify complex processes with deceptive ease, and certain facts about geographic speciation must be borne in mind. First, all populations will evolve given enough time. Second, virtually all geographically separated populations live in different environments. This last statement may seem extreme at first. But recall that the physical environment does shift from point to point on the globe, often dramatically and over relatively short distances. The total climate and chemistry of the soil change in going inland from a coast, or up from the coastal plain to the mountains, or out across the sea from a continent to an island, and so on all around the world. Even sharper changes occur in the vegetation and in the composition and relative abundance of the animal species living on the vegetation. It follows that geographically isolated populations are probably always under different selective pressures. Consequently, when they evolve, they will diverge from each other. Another generalization to keep in mind is that when the populations have diverged enough so that effective intrinsic isolating mechanisms have come into existence, they must be ranked as species even though they have not yet come into contact. The coming together of the newly formed daughter species is the final step of speciation, but it is not the crucial one.

Most histories of geographic speciation must be deduced. This is accomplished by comparing sets of living populations that show various stages in evolutionary divergence. An especially informative example is provided by the wild tomato species *Lycopersicon peruvianum (Figure 7)*. This plant occurs in the natural state along the coasts of Peru and Chile and along the Pacific foothills inland. Its populations show extreme differences in some morphological characteristics, including leaf shape, as illustrated in Figure 7. When samples from various populations were grown and crossbred under standard experimental conditions, they revealed a bewildering variety in the degree of interfertility. The species has differentiated in such a way that if we were to select samples from particular localities and match them in pairs in breeding experiments, we would have to define some pairs as belonging to the same species (freely interbreeding), some as completely distinct species (no interbreeding), and some as semispecies (showing intermediate degrees of interfertility). The way the crosses come out appear random and are unpredictable. *Lycopersicon peruvianum* is not at all unusual in this regard. Its complex pattern is just what should be expected from the theory of geographic speciation.

Most understanding of geographic speciation comes from analyses of the kind just cited, in which different stages of the process are deduced by comparing different sets of living populations. This is essentially the same procedure by which the process of mitosis in a root is inferred by com-

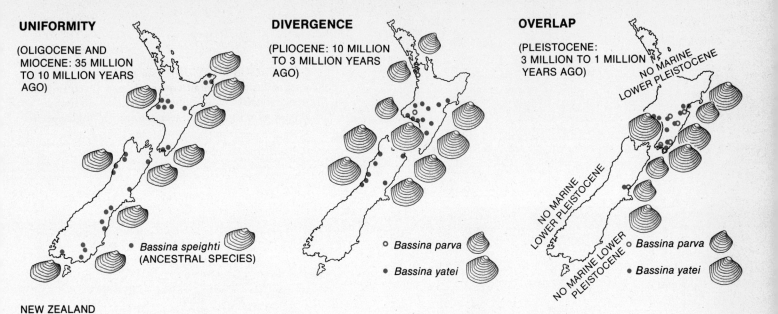

NEW ZEALAND

UNIFORMITY

(OLIGOCENE AND MIOCENE: 35 MILLION TO 10 MILLION YEARS AGO)

Bassina speighti (ANCESTRAL SPECIES)

NEW ZEALAND

DIVERGENCE

(PLIOCENE: 10 MILLION TO 3 MILLION YEARS AGO)

○ *Bassina parva*

● *Bassina yatei*

OVERLAP

(PLEISTOCENE: 3 MILLION TO 1 MILLION YEARS AGO)

NO MARINE LOWER PLEISTOCENE

NO MARINE LOWER PLEISTOCENE

NO MARINE LOWER PLEISTOCENE

○ *Bassina parva*

● *Bassina yatei*

NEW ZEALAND FOSSILS reveal the gradual emergence of two new species of cockle. Dots in map at left indicate fossil beds of the ancestral species. Eventually it diverged into two populations with separate ranges (center). By the early Pleistocene, the populations had become separate species with overlapping ranges (right). Because of changes in the coastline, some areas show no marine fossils from the Pleistocene epoch.

paring cells in different stages of division. If we had a complete enough fossil record, however, it should be possible to trace the process through time as it actually happened. This has actually been done in the remarkable example illustrated in Figure 8. The cockle genus *Bassina*, like many mollusks that live in shallow marine water, leaves numerous fossil remains, making it possible to trace some of the geographic variation of populations through time. As late as the Miocene Epoch, about 10 to 35 million years ago, there was a single, relatively uniform species occupying the New Zealand coasts. By Pliocene time, 3 to 10 million years ago, the populations had undergone strong differentiation, and they still occupied different ranges. There is no information on the barrier that separated them, or whether by this period the populations had already reached the level of distinct species. In any case, by lower Pleistocene time, 1 to 3 million years ago, the populations had become distinct species with overlapping ranges. This example is considered typical in that it required millions of years to complete. However, evidence exists to indicate that in some cases geographic speciation can occur in a much shorter period of time, within a few thousand years or even less. For example, certain native Hawaiian moths of the genus *Hedylepta*, whose caterpillars feed on leaves of the banana, are believed to have originated since man came to the islands about 2000 years ago.

When populations come together after a period of geographic isolation, one of three events can occur:

1. If the populations have attained the level of full species and are very different from each other, they may coexist with no significant interaction.

2. If the populations have failed to attain the level of species, they will interbreed to some extent. They may fuse completely into a single species, or else maintain an intermediate level of gene flow, producing that most perplexing category we considered earlier, the semispecies.

3. The populations may have attained the level

> The ability to displace increases the number of species that can be fitted into an ecological community, just as the compressibility of clothing increases the number of items that can be packed into a suitcase.

GALAPAGOS
ISLANDS

0°   0°

ECUADOR

PERU

GALAPAGOS ISLANDS lie on the equator, 600 miles west of Ecuador. Small square in map at left indicates area of detailed map below it. Here Darwin found 14 species of finch, all descended from a single species.

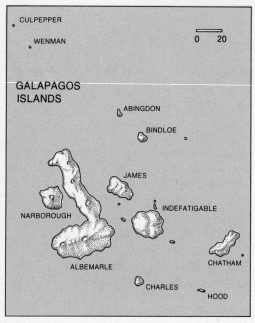

CULPEPPER

WENMAN

0   20

GALAPAGOS
ISLANDS

ABINGDON

BINDLOE

JAMES

INDEFATIGABLE

NARBOROUGH

CHATHAM

ALBEMARLE

CHARLES

HOOD

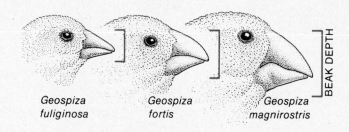

Geospiza fuliginosa   *Geospiza fortis*   *Geospiza magnirostris*

BEAK DEPTH

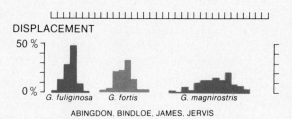

DISPLACEMENT

50 %

0 %
G. fuliginosa   G. fortis   G. magnirostris

ABINGDON, BINDLOE, JAMES, JERVIS

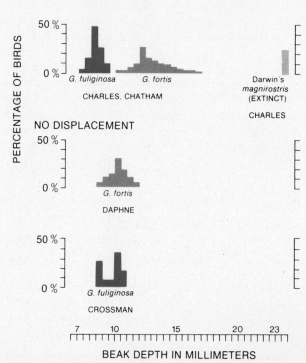

50 %

0 %
G. fuliginosa   G. fortis

CHARLES, CHATHAM

Darwin's
magnirostris
(EXTINCT)

CHARLES

NO DISPLACEMENT

50 %

0 %
G. fortis

DAPHNE

PERCENTAGE OF BIRDS

50 %

0 %
G. fuliginosa

CROSSMAN

7   10   15   20   23

BEAK DEPTH IN MILLIMETERS

10 VARIATION IN BEAK DEPTH of Galapagos finches is a classic example of character displacement. Drawings at top compare beaks of three species. Thicker beak enables a finch to feed upon seeds too hard for competing species to crack. Bar graphs show variation in beak depth among populations on several islands. Vertical scale shows percentage of birds having beaks of the depth indicated on horizontal scale. On each of two islands inhabited by a single species (two bottom rows) finches have beaks of intermediate depth. On four islands (top row) fortis is squeezed by two other species, displacing the beak depths of all three. On Charles and Chatham islands (second row) fortis and fuliginosa compete with each other. Beak size of fortis is displaced toward the right of the graph—that is, toward the larger beak size of magnirostris, a species that has become extinct on Charles island.

of species but interfere with one another in such a way as to cause one or both of them to diverge still further in evolution. This peculiar form of post-contact evolution is called CHARACTER DISPLACEMENT. There are two ways in which interference can occur. First, the two newly formed species may still be so similar in food habits, habitat preference, and other ecological traits, that they compete with each other. This means that where one species is present, there is less food, nesting space, or other necessities of life for the other to utilize. It will therefore be advantageous for the two species to be less alike, and genotypes that resemble the competing species less will be favored in evolution. As a result, the two species will diverge in evolution more rapidly where their populations come into contact. The same pattern of character displacement can be obtained when the populations coexist as semispecies. The presence of semispecies means that interbreeding is not complete. Attempts to interbreed between the two newly formed species can result in gamete wastage — fewer offspring are produced by a given amount of effort to mate and reproduce. Either the potential mate fails to respond to the mating attempt, or else hybrid offspring are produced but are less successful at developing and breeding on their own. As a consequence, the genotypes of either species that avoid such attempts at interbreeding will be favored in natural selection over those that do attempt breeding. The two semispecies will then diverge from each other and eventually become full species with more nearly perfect intrinsic isolating mechanisms.

DARWIN'S FINCHES

Figures 9 and 10 illustrate a classic example of character displacement associated with geographic speciation by the multiple invasion of islands. The birds are finches, a group of species found only on the Galapagos Islands. When Darwin visited this archipelago, located astride the equator 600 miles off the coast of Ecuador, the differences he observed among the populations of plants and animals in going from island to island convinced him of the truth of evolution, that "species are not immutable." This realization came first from a rather casual study of the mockingbirds found on several of the Galapagos Islands. It was cemented later by examinations of other kinds of animals, including the giant tortoises and remarkably diverse finches, that are found only on the Galapagos. Approximately a century later David Lack, a British or-

nithologist, undertook a more detailed study of Darwin's finches, and it was he who discovered the example of character displacement presented here.

The Galapagos Islands are a series of more or less rounded volcanic peaks that project up through the deep waters of the Pacific. All of the native species of plants and animals originated from a few lucky individuals that were accidentally transported as stragglers or waifs over the ocean.

Difficulty in distinguishing subspecies lies at the heart of the problem of defining race in the human species.

Most of the colonists originated from the coast of South America, but a very few others came from Polynesian islands far to the west. Because the Galapagos Islands are permanently separated from each other by ocean water, speciation has occurred chiefly by means of multiple invasions from one island to the other. All of the 14 living species of finches, comprising an entire subfamily of their own, are believed to have evolved from a single species that reached the islands sometime in the last several million years and then divided many times over through the process of multiple invasion. As newly formed species accumulated, they faced the problem of interference through competition and hybridization, and character displacement may have occurred frequently. The example shown in Figure 10 is in an early and unusually well marked stage. Here we see three seed-eating species undoubtedly descended from a common ancestor. The larger the bill, the harder the seed that can be cracked open and eaten. Thus the three species probably avoid competition by feeding on different kinds of seeds. They may also use the bill shape during courtship as a premating isolating mechanism. Thus differences in bill size (measured here as the depth) are important in preventing interference between the species, and this is the character that has undergone displacement.

Character displacement has a significance that goes well beyond its role in speeding up evolution after the completion of speciation. It is also the means by which species are packed into ecological communities. Evolution by displacement is rather like the deformation that occurs when a full suitcase is closed. The ability to displace increases the number of species that can be fitted onto an island or other ecological community, just as the ability of clothing to be folded and squeezed increases the number of items that can be forced into a suitcase. The question of how many species can coexist in one place is called the species packing problem. Studies of character displacement have provided an important means of partially solving this fundamental problem. It has been found, for example, that where displacement has occurred in a single characteristic, such as beak depth in the species of *Geospiza*, the species usually evolve apart until they differ by approximately 30 to 50 percent. This is exemplified in the two species, *fuliginosa* and *fortis*, that show the most clear-cut pattern of displacement. Obviously, if there is a minimal degree of difference that species must attain in order to coexist, there is a corresponding maximal number that can be packed into any one place. For example, the species number is one to three in the *Geospiza* of the Galapagos Islands, with the larger islands supporting the larger number. The exact mathematical relation between the minimal character difference and the maximum number of species that can coexist has not yet been solved, but it is a problem on which theoretical biologists are now actively working.

SPECIATION BY POLYPLOIDY

Polyploidy, the exact multiplication of the number of chromosomes in the cell, is the means by which species can be created in only one or two generations. This is the case because multiplying the number of chromosomes interferes with the process of meiosis, and makes breeding between the diploid species and its polyploid offspring difficult or impossible. A little reflection on the mechanics of meiosis will make the basis of the process clear. Consider (to take an actual case, from the genus *Galax*, a group of woodland-dwelling evergreen herbs) a species with a diploid chromosome number of 12. This means that in each cell there are two copies of 6 different chromosomes (the haploid number). At meiosis the two chromosomes of each kind must pair with each other in order to complete the process of gamete formation. Each gamete will end up containing one of each kind of chromosome, a total of 6 chromosomes. Now suppose that due to an accident during meiosis some new plants are produced with exactly twice that many chromosomes, 4 of each kind or 24 in all. These polyploid plants can produce gametes. The polyploid gametes will contain 12 chromosomes (2 of each kind), instead of just 6, and they can combine to form fertile offspring with the polyploid number of 24. They can also combine with haploid gametes (1 of each kind) to form hybrids with a total of 18 chromosomes (6 + 12 = 18). However, these 18-chromosome hybrids cannot produce normal gametes. The reason is that when meiosis begins and the time comes for the chromosomes to pair, there will be three of each of the 6 kinds, two from the polyploid parent and one from the diploid parent. Two chromosomes of a kind can pair easily, but three run into severe mechanical difficulties, causing uneven distribution of chromosomes in the daughter cells. The meiosis is often interrupted at this point, and the hybrid is thereby rendered partially or wholly sterile. If diploids can breed freely with other diploids, and

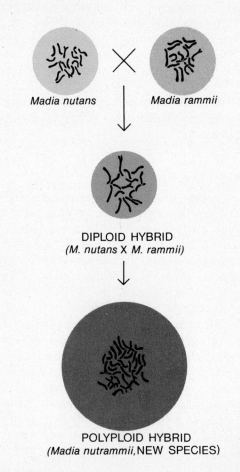

NEW POLYPLOID SPECIES can be created from a diploid hybrid in one generation. Diagram shows how a new species of California tarweed arose spontaneously from two diploid species growing in an experimental garden.

Madia nutans ✕ *Madia rammii*

DIPLOID HYBRID
(M. nutans X *M. rammii)*

POLYPLOID HYBRID
(Madia nutrammii, NEW SPECIES)

CLINE in size of two Australian birds. Larger birds are found in colder climates. Average size of southern specimens of Rhipidura is 11 percent larger than northern ones. In Seisura, which is divided into three populations, difference is 22 percent.

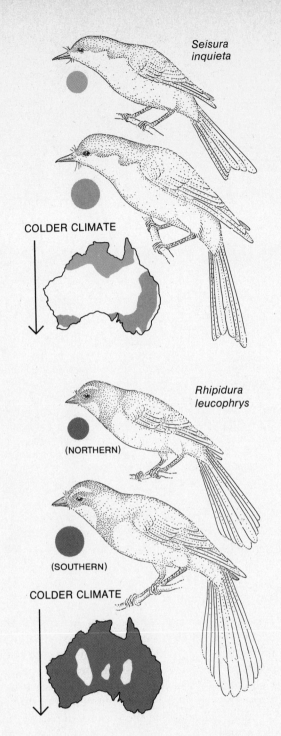

polyploids with other polyploids, but the hybrids between the two are sterile, the diploids and polyploids are two different species. The new polyploid species has thus been created in one step.

Another and even more important way of creating polyploid species is by multiplying the number of chromosomes of the diploid hybrids of species. It happens that two distinct plant species occasionally produce hybrid offspring which prove sterile because the chromosomes from the two parent species are too different to pair successfully. However, if the hybrid undergoes an exact doubling of its chromosome number, it has no difficulty at meiosis. There are two of each kind of chromosome from each parent instead of just one, and these two pair in the usual fashion to permit the completion of meiosis. The hybrid polyploid is a new species, isolated from its diploid parents for the same reason that a non-hybrid polyploid is isolated. A typical example of speciation by hybrid polyploidy is given in Figure 11.

GEOGRAPHIC VARIATION AND SUBSPECIES

Earlier we considered the chaotic pattern of geographic variation and intersterility in the wild tomato *Lycopersicon peruvianum*. The theory of geographic speciation predicts that just such confusing examples will be commonplace. But the fact that the process is by nature so complex and unpredictable makes it one of the least understood and most challenging subjects in modern evolutionary biology. We are made aware of the profundity of the problem in trying to devise a classification of populations below the level of the species.

A population that differs significantly from other populations belonging to the same species is referred to as a geographic race or subspecies. We can restate the generalization just made about geographic variation by saying that subspecies show every conceivable degree of differentiation from other subspecies. At one extreme are the populations that fall along a CLINE—a simple gradient in the geographic variation of a given character. In other words, a character that varies in a clinal pattern is one that changes in a gradual manner from

one end of the range of the species to the other. The closer together in space the populations of the species are throughout the range, the more even the clinal variation is from population to population. Figure 12 shows two examples of clines that exemplify this generalization. They also illustrate another feature of clines, namely that the trend in the variation can often be correlated with climatic differences that occur within the range of the

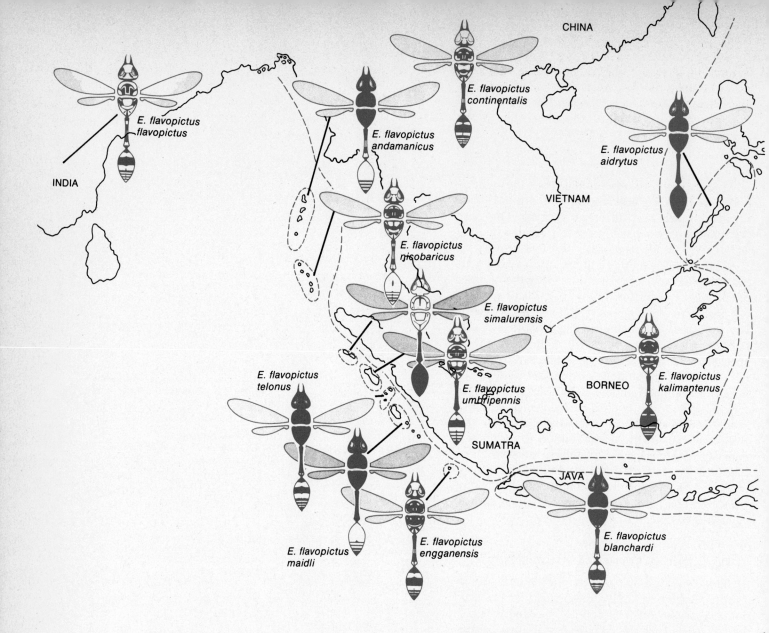

CHINA

*E. flavopictus
continentalis*

*E. flavopictus
flavopictus*

INDIA

*E. flavopictus
andamanicus*

*E. flavopictus
aidrytus*

VIETNAM

*E. flavopictus
nicobaricus*

*E. flavopictus
simalurensis*

*E. flavopictus
telonus*

*E. flavopictus
umbripennis*

BORNEO

*E. flavopictus
kalimantenus*

SUMATRA

JAVA

*E. flavopictus
maidli*

*E. flavopictus
engganensis*

*E. flavopictus
blanchardi*

species. In the case of the two Australian bird species illustrated, the cline is consistent with Bergmann's Rule, which states that the larger subspecies of warm-blooded vertebrates occur in colder climates. The colder climate in this case is of course the southern, more temperate portion of Australia. The adaptive significance of Bergmann's Rule appears to be the greater ability of larger animals to conserve heat, because they have a lower ratio of body surface exposed relative to the total body weight. The same explanation holds for Allen's Rule: that in warm-blooded animal species the subspecies living in colder climates tend to have shorter protruding body parts, such as bills, tails, and ears. A third such climatic rule (out of a great many others that could be cited) is Gloger's Rule, that subspecies living in warm and humid

POTTER WASP, a native of tropical Asia, is differentiated into many subspecies. The most strongly differentiated populations are isolated on islands. Subspecies are designated by three-part names.

areas are usually more darkly pigmented than those living in cool and dry areas. These rules, which are named after the nineteenth century zoologists who first proposed them, are not laws in any strict sense but at most only statistical generalizations based on the study of a great many species.

Most geographic variation is far too irregular to be classified into clines. Figure 13 is an example of the sharp, chaotic patterning of the kind that has long attracted the attention of evolutionists. The tropical wasp species illustrated here shows the

688

greatest amount of differentiation among its populations that occupy small islands, where the degree of isolation is greatest. Taxonomists often identify such distinctive populations by formally designating them as subspecies and giving them a third technical name. The entire species, comprised of all the populations together, is referred to by a binomial in the usual manner. In this case the species is *Eumenes flavopictus*. Each population within the species that can be distinguished as a subspecies is given a three-part name, for example *Eumenes flavopictus flavopictus* of India or *Eumenes flavopictus continentalis* of southeastern Asia.

Another important generalization about geographic variation is that it can occur in virtually every kind of characteristic—from skin color and head shape, to take some arbitrarily selected examples, to body temperature, urine composition, and patterns of innate behavior. Any trait subject to genetic variation (and hence capable of evolutionary change) can show geographic variation. Species often differ drastically from each other in the particular traits by which they vary and in the intensity by which these traits vary. One species, for example, may show great geographic variation in skin color and none in head shape, while a second, otherwise similar species varies significantly in head shape but not in skin color.

Despite the seeming clarity of the patterns exhibited in the geographic variation of many species, and the convenience of labeling their geographic divisions as subspecies, the subspecies has some inherent weaknesses that make it very difficult to use in analysis and classification. These weaknesses stem directly from our ignorance of some of the most basic processes of microevolution. The first difficulty comes from the definition of the population. Few species are actually broken up into populations as we ideally conceive them—isolated or semi-isolated groups of randomly breeding individuals. Most species are structured in more complex, ambiguous patterns. Take for example that famous animal species, the gorilla of Africa. Approximately 10,000 individuals of *Gorilla gorilla*

exist in the eastern part of the range of the species. All are confined to a central portion of the continent *(Figure 14)*. These individuals are grouped into what look superficially to be about 60 populations that occupy 10 to 100 square miles of mountain country each. In the center of the range there is a large area in which the species appears to be sparse but continuously distributed. The truth is that the exact limits of these "populations" are unknown, since the rate at which gorillas move from one area to another to breed is not known. Put in the language of population genetics, we do not know the rate of gene flow. Lacking that crucial parameter, very little more can be concluded about the population structure of gorillas. *Gorilla gorilla* is not at all unusual in this regard. In fact, it is much better known at the present time than the vast majority of animal and plant species.

The second major difficulty of the subspecies concept is the tendency of various characteristics to display differing patterns of geographic variation within the same species. The redbacked salamander *Plethodon cinereus (Figure 15)* is quite typical in this regard. You can see that if you decide to define

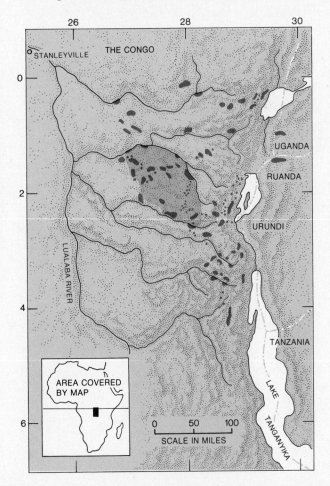

RANGE OF MOUNTAIN GORILLA lies in East Africa. Tiny black square in miniature map at lower left indicates area covered by larger detailed map. Dark colored spots designate location of gorilla populations, comprising a total of about 10,000 animals. Lighter colored area near center of map marks a region of continuous but sparse occupation. Rates of movement between areas, and hence the true population limits, are largely unknown.

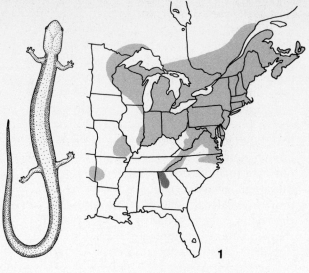

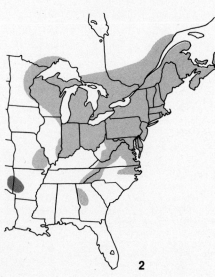

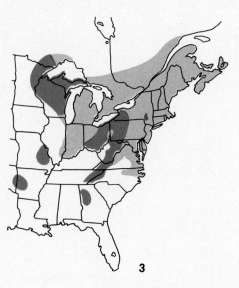

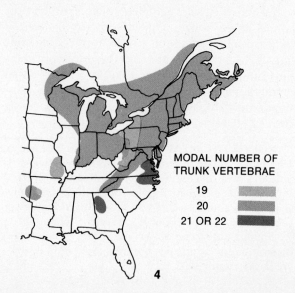

the subspecies of this very abundant American vertebrate on the basis of any one of the four morphological characters chosen, you will have no trouble. Next try to define the subspecies with two of the characters. This is difficult, but still possible. When all four characters are used as criteria, however, the task becomes so difficult as to raise the question of whether subspecies are worth the effort. The discordance of patterns becomes an overpowering consideration when it is realized that in many species tens or even hundreds of characters show discordant variation of a comparable degree. In fact, the more characters that are added to the analysis, the more subspecies it would be necessary to distinguish, and the less practical the classification.

Difficulty in distinguishing subspecies lies at the heart of the problem of defining race in the human species. Skin color alone follows one pattern of geographic variation, nose shape follows another pattern, blood types still another, and so on down an unending list. Anthropologists who attempt to classify subspecies, or distinct races, or whatever they choose to call them, run into predictable difficulties. Within the past 25 years various anthropologists, using their own favorite characters

GEOGRAPHIC VARIATION in four traits of the North American salamander Plethodon cinereus. Maps 1, 2 and 3 show ranges of subspecies that differ in stripes and coloration. Map 4 shows range of salamanders with a particular number of vertebrae. Distribution pattern varies from trait to trait, a common occurrence that makes classification into subspecies difficult and often misleading.

MODAL NUMBER OF
TRUNK VERTEBRAE

19

20

21 OR 22

as the criteria, have produced widely varying estimates of the number of geographic races in the entire human species. One recognized 5 races, another 6, another 30, and still another 37 plus about 30 "subraces." To reject this fallacy is not to deny that mankind is a geographically variable species. On the contrary, he shows a respectable amount of such variation when compared with other animal and plant species. But he is quite typical in the discordant nature of his patterns. It is necessary to realize that subspecies are not a realistic unit of classification in most cases. As a consequence, more and more biologists have begun to analyze geographic variation in terms of the individual character patterns instead of attempting to distinguish populations on the basis of geographical ranges.

READINGS

V. GRANT, *Organismic Evolution*, San Francisco, W.H. Freeman, 1977. A review of speciation in the broad context of evolutionary theory, up-to-date and with examples from both plants and animals.

E. MAYR, *Populations, Species, and Evolution*, Cambridge, MA, Belknap Press of Harvard University Press, 1970. An abridged and very readable version of the definitive work on modern speciation theory as it applies to animals. It is, however, more of a scholarly review than a textbook.

G.L. STEBBINS, *Processes of Organic Evolution*, 3rd Edition, Englewood Cliffs, NJ, Prentice-Hall, 1977. This intermediate-level text gives the best short review of speciation.

To do science is to search for repeated patterns, not simply to accumulate facts; and to do the science of geographical ecology is to search for patterns of plant and animal life that can be put on a map. —Robert H. MacArthur

Why did the lion, which once ranged from Africa to India, and the tiger, which ranged from Persia to Siberia, not expand their ranges to cover all of Africa and Asia, or even the entire earth? On a larger scale, why are the rats and mice (family Muridae), common frogs (family Ranidae), and pond turtles and their relatives (family Emydidae) now dominant groups that are spreading over almost the entire earth; while at the same time other groups, such as the rhinoceros (family Rhinocerotidae) and tapirs (family Tapiridae), have given up their wide ranges and are retreating to odd corners of the tropics? This kind of question is the concern of biogeography.

The task of the biogeographer is formidable: he must collect the piecemeal data concerning the distribution of species provided to him by taxonomic studies and somehow explain this information in terms of the principles of evolutionary theory, ecology, geology, and climatology. Biogeography has two disparate major branches: historical biogeography, in which the biologist attempts to account for the histories of the species of plants and animals, from the moment of their origin to the time they propagate other species or else go extinct; and ecological biogeography, in which the biologist deals with the adaptations the species have acquired in evolution and the position which they occupy, in the literal physical sense, in local habitats. In the past, biologists have often pursued these two subdisciplines separately, as though they were different subjects. But they really treat

TYPICAL AFRICAN ANIMALS in photograph on opposite page include plains zebras, Grant's gazelles, yellow baboons and a single bustard (bird at lower left). Photo was made at a watering place near Lake Amboseli in Kenya.

two aspects of the same phenomenon and must be joined into a single science.

ADAPTIVE RADIATION AND CONVERGENCE

Once a population has acquired intrinsic isolating mechanisms, thus becoming a full fledged species, it is destined to diverge forever in its genetic composition from all other species. But this does not mean that it will evolve without reference to other species. The direction in which a species evolves is determined by the opportunities open to it, and these in turn are set by the other species of animals and plants with which it shares its environment. Obviously, an insect species cannot become specialized for eating pine seeds if no pine trees grow within its range. Less obviously, the specialization it chooses depends in part on the kinds of other specialists already present. If a species is given a choice between pine seeds and fir seeds, where only the pine seeds are being utilized by other insect species, the species will be more likely to evolve toward specialization in fir seeds. In this case competition is the force that moved it in one direction as opposed to another. Each animal species survives by eating other species —plant and/or animal —and competes with other species for the food. And each species is eaten by some other. Thus each species is tightly constrained wihin an ecological community, and there is a definite limit on the adaptive pathways it can follow in evolving.

Again and again, on different continents and islands and in different geological times, evolution has created ecological communities that resemble each other in their broadest features. The origins of the species that make up these separated communities are often drastically different, but the final products are usually remarkably similar. The

693

proliferation of species into different niches of a community is called ADAPTIVE RADIATION. The similarity acquired by species that come to occupy approximately the same ecological position but in different communities is called CONVERGENCE. Figure 1 indicates the complementary relationship of these two phenomena. Here we have an example of two adaptive radiations that occurred at quite different geological times.

WAVES OF EVOLUTION

A common characteristic of evolution in general is the dynastic succession of groups, each of which flowered into adaptive radiation as it supplanted the group preceding it. According to conservative estimates, over 99 per cent of all evolutionary lines that existed in the past have become extinct. The vast majority of these lines were products of adaptive radiation that gave way to succeeding radiations. In some cases, as in the amphibian-reptile-

mammal lines, the ancestors of the new radiations were products of the previous ones (*Figure 2*). But in many other cases entire groups were supplanted by lines phylogenetically remote from their own, sometimes invaders from other continents.

Adaptive radiation not only occurs repeatedly through time, it also occurs simultaneously on different, isolated parts of the earth. Perhaps the most dramatic example of simultaneous radiation is the origin of the modern mammalian fauna of Australia. The continent of Australia has been a giant island, isolated from Asia by a seaway in the vicinity of modern Indonesia, since before the beginning of the Cenozoic Era approximately 63 million years ago. At a very early period in this long spell of isolation, or even before it began, Australia was colonized by primitive egg-laying mammals called monotremes. These strange little animals make up only a minute fraction of the present Australian fauna; the echidnas and the duckbill platypus are the only surviving representatives. The early colonists also included the marsupials. These are mammals somewhat more advanced evolutionarily than monotremes, distinguished by their peculiar means of reproduction. The young

1 **ADAPTIVE RADIATION of reptiles during the Mesozoic Era produced an assortment of animals that are more or less convergent with forms produced by adaptive radiation of modern mammals.**

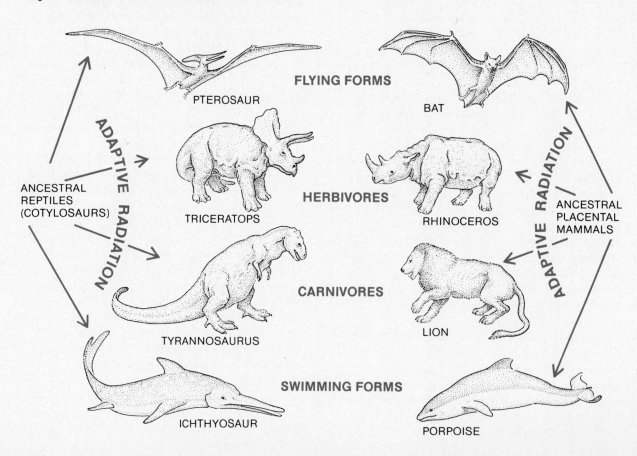

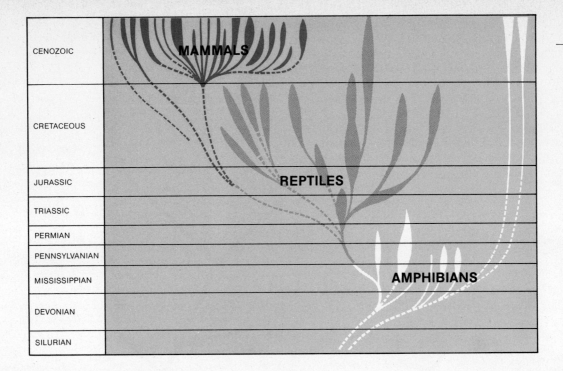

CENOZOIC	MAMMALS
CRETACEOUS	
JURASSIC	REPTILES
TRIASSIC	
PERMIAN	
PENNSYLVANIAN	
MISSISSIPPIAN	AMPHIBIANS
DEVONIAN	
SILURIAN	

2 DYNASTIC SUCCESSION of amphibians, reptiles and mammals is depicted in this highly schematic diagram. Each group held a dominant position for a time and then was replaced by another group. Width of branches indicates abundance of species. Dashed lines indicate unknown ancestral forms.

are born while still very small and undeveloped; they then crawl into a pouch (the marsupium) located on the mother's belly and attach themselves firmly to the mother's teats to complete their development. The marsupials have been an outstanding success in Australia. Today they comprise about two-thirds of the native mammal species of the continent. During the past 70 million years the marsupials have undergone an adaptive radiation of the first magnitude, filling most of the niches open to land-dwelling mammals *(Figure 3)*. In other lands equivalent radiation was achieved by placental mammals. The degree of convergence that has occurred during the radiations of the two groups is astonishing. In some instances, for example the flying squirrel (placental) *versus* the sugar glider (marsupial), the woodchuck (placental) *versus* the wombat (marsupial), and the placental *versus* the marsupial versions of the wolf and the mole, the external resemblances are so close that special instruction is needed to place a given species in the proper group. In addition, unique forms have been produced on both sides. For example, the kangaroos are the large herbivores

(plant-eaters) of Australia. Some species live in trees and browse on the vegetation, while others are specialized for grazing grass and low herbiage like sheep or cattle. But the body form of the kangaroos is unique. While filling the niches for large herbivores in Australia, these animals have developed their own peculiar means of locomotion and a body form that goes with it. The placental mammals of the remainder of the world have also produced unique specialists, from the horses and other hooved herbivores, to the aerial bats and exclusively aquatic and marine forms such as the manatees, porpoises, and whales.

Adaptive radiation is produced by speciation. It is possible for a single ancestral species, originating from a few pioneering organisms, to invade a new continent or group of islands and to generate species that spread out and fill multiple niches in the radiative pattern. To see more clearly how such an event can occur, let us next consider a case of adaptive radiation and convergence on a smaller scale. One of the most striking examples known is illustrated in Figures 4 and 5. The Hawaiian honeycreepers, comprising a distinct family of their own (Drepanididae), are believed to have originated from a single species of small, goldfinch-like birds that colonized Hawaii from the temperate zone of Asia or North America. When the ancestral drepanidids landed some million years ago, they found a rich new environment devoid of most of the kinds of bird that are dominant ele-

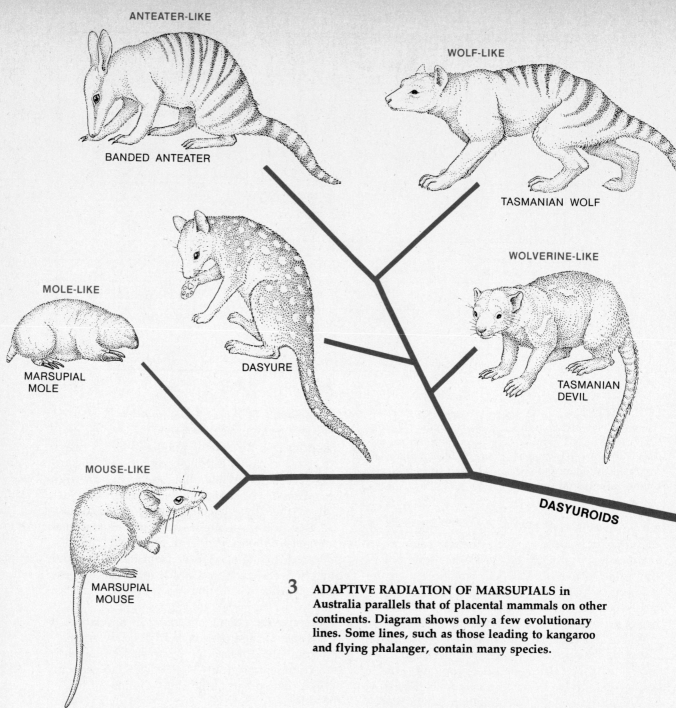

ANTEATER-LIKE

BANDED ANTEATER

WOLF-LIKE

TASMANIAN WOLF

WOLVERINE-LIKE

TASMANIAN DEVIL

MOLE-LIKE

MARSUPIAL MOLE

DASYURE

MOUSE-LIKE

MARSUPIAL MOUSE

DASYUROIDS

3 ADAPTIVE RADIATION OF MARSUPIALS in Australia parallels that of placental mammals on other continents. Diagram shows only a few evolutionary lines. Some lines, such as those leading to kangaroo and flying phalanger, contain many species.

ments of the faunas in the rest of the world. Hawaii contained no finches, parrots, woodpeckers, or hummingbirds. Even today the drepanidids share the Hawaiian islands with only about 17 other native land and fresh-water bird species, such as the native crow and the Nene, or Hawaiian Goose. The reason Hawaii had such a poor fauna even earlier is of course its isolation from other land masses. Only a very few plant and animal species have succeeded in reaching and colonizing Hawaii. Those that do make the landfall are presented with unusual opportunities for the exploitation of unfilled niches. Speciation occurs easily among the expanding populations, because the water gaps separating the islands are powerful barriers to gene flow between the populations inhabiting different islands. In the case of birds generally, and drepanidids in particular, speciation has occurred through the process of multiple invasion (*see Chapter 30*). As

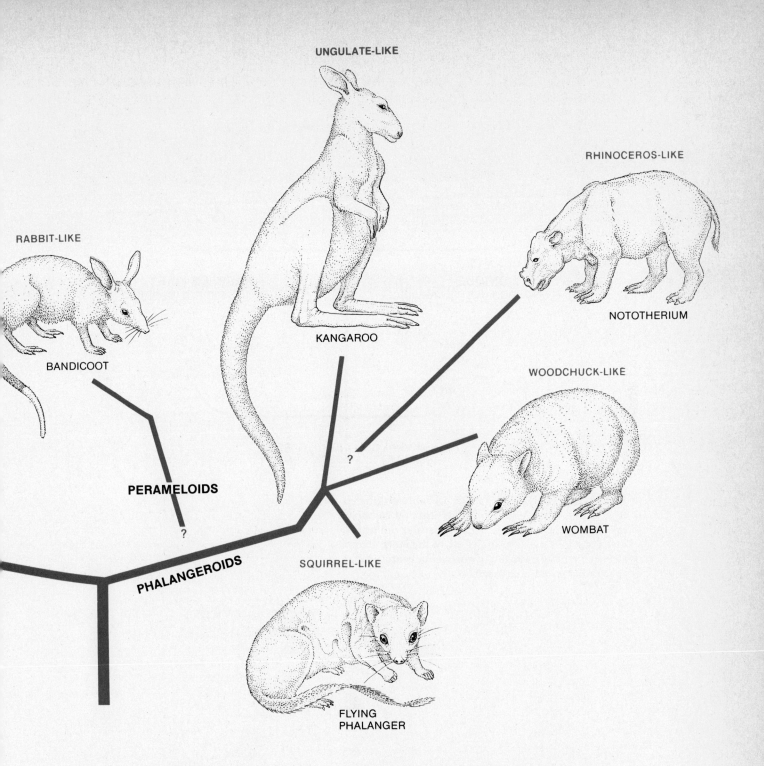

UNGULATE-LIKE

RHINOCEROS-LIKE

RABBIT-LIKE

NOTOTHERIUM

BANDICOOT

KANGAROO

WOODCHUCK-LIKE

PERAMELOIDS

?

WOMBAT

?

PHALANGEROIDS

SQUIRREL-LIKE

FLYING
PHALANGER

the species diverged from each other, they tended to expand quickly into the unfilled ecological positions that were so abundantly available around them. The result was the achievement of an unusually broad radiation with the production of a relatively small number of species. When Europeans first settled the Hawaiian Islands in the early 1800s, they found the bird fauna in a relatively intact condition. The original Polynesian settlers who had preceded them by some two thousand years had not yet destroyed enough of the native habitats, or introduced sufficient numbers of harmful animal and plant species, to damage the natural living environment of Hawaii. Among the drepanidids were species that physically resembled finches, warblers, and parrots, and possessed similar food habits. There was a "woodpecker", *Hemignathus wilsoni*, that chiseled open dead wood with its stiff lower bill and picked out insects with its sharp, curved upper bill. The

697

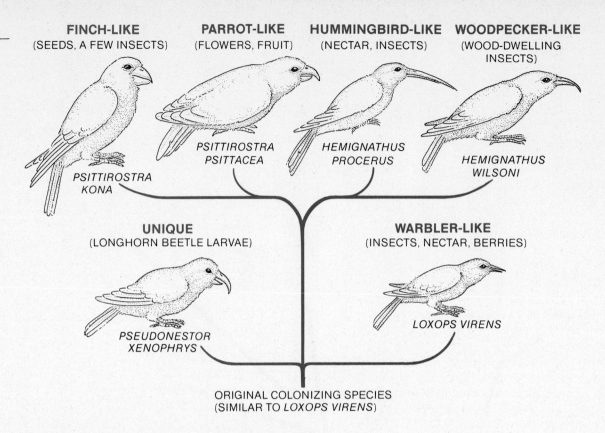

FINCH-LIKE
(SEEDS, A FEW INSECTS)

PARROT-LIKE
(FLOWERS, FRUIT)

HUMMINGBIRD-LIKE
(NECTAR, INSECTS)

WOODPECKER-LIKE
(WOOD-DWELLING INSECTS)

PSITTIROSTRA KONA

PSITTIROSTRA PSITTACEA

HEMIGNATHUS PROCERUS

HEMIGNATHUS WILSONI

UNIQUE
(LONGHORN BEETLE LARVAE)

WARBLER-LIKE
(INSECTS, NECTAR, BERRIES)

PSEUDONESTOR XENOPHRYS

LOXOPS VIRENS

ORIGINAL COLONIZING SPECIES
(SIMILAR TO *LOXOPS VIRENS*)

4 **HAWAIIAN HONEYCREEPERS** probably originated from a single species of finch-like bird that colonized the islands and later radiated by means of multiple invasion. This drawing shows six of the more extreme adaptive forms. In the early nineteenth century 22 species inhabited the islands.

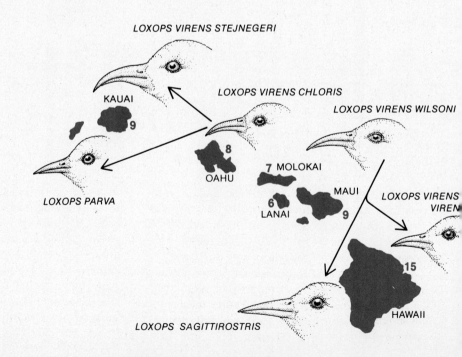

LOXOPS VIRENS STEJNEGERI

KAUAI

LOXOPS VIRENS CHLORIS

9

LOXOPS VIRENS WILSONI

8

OAHU

7 MOLOKAI

LOXOPS VIRENS VIREN

6
LANAI

MAUI

9

15

HAWAII

LOXOPS PARVA

LOXOPS SAGITTIROSTRIS

familiar true woodpeckers of the family Picidae accomplish the same goal in other parts of the world by chopping into wood with the entire bill and retrieving the insects inside with a long, sticky tongue. There was also at least one unique adaptive form: *Pseudonestor xenophrys*, which made its living by tearing open dead wood and trees with a parrot-like beak and feeding on the large longhorn beetle larvae that are abundant in such places. This bird, like many of the most specialized and interesting of the drepanidids, became extinct when its environment was altered by the European settlers of Hawaii—an irreplaceable loss to evolutionary biology and to human esthetics.

Adaptive radiations like those in the Australian marsupials and Hawaiian honeycreepers have occurred repeatedly in many other phylogenetic groups of plants and animals in isolated places all around the world. One lesser known but comparable example is provided by the plants of St. Helena *(Figure 6)*. This tiny island, located midway between Africa and South America in the South Atlantic Ocean, has been reached by very few species of plants and animals. The higher plants that did succeed were presented with the same degree of opportunity as the Drepanididae of Hawaii. In the near absence of competitors they were free to radiate in many directions. In particular, no true tree species had colonized St. Helena, and thus at first there were no forests. The plants that arrived were small, herbaceous forms belonging to the sunflower family Compositae. Their achievement in evolution was to make a forest of their own, by evolving into tree-like forms.

BIOGEOGRAPHIC REGIONS

The world is not a continuous sphere of ecologically uniform terrain. Much of the land is divided into islands, while the six continents are separated from each other by oceans, seas, and straits. The continents themselves are extremely diverse, being comprised of broad deserts, forests, grasslands, and rivers and lakes. The consequence of this elementary generalization is that from the point of view of dispersing organisms the world is a great

5 **MAP OF HAWAIIAN ISLANDS shows number of honeycreeper species originally present on each island. Species and subspecies of the genus Loxops originated in isolated populations on different islands. In the course of their divergence they developed bills of different shapes.**

and complicated system of Australias, Hawaiis, and St. Helenas. Wherever enduring barriers exist, the floras and faunas separated by them tend to evolve in their own directions, to produce their own adaptive radiations, and thus to create self-contained communities. As a result, the world can

---

A few pioneers belonging to a single ancestral species can invade a new continent or group of islands and generate new species that spread out and fill multiple niches.

---

be divided into biogeographic regions, each of which contains a distinctive assemblage of plant and animal species. Figure 7 is a map of the classic zoogeographic regions of the world, which are the most distinctive areas based on the animal species alone. A similar (but not identical) map could be drawn for the phytogeographic regions, based on the plants.

The mammals are a typical animal group. Australia and the immediately adjacent islands, comprising the Australian Region, are distinguished by kangaroos, bandicoots, the koala, the wombat, and other products of the marsupial radiation, as well as the echidnas and the platypus. Africa south of the Sahara Desert, the Ethiopian Region, has the African elephant, a large array of distinctively African antelopes, two species of rhinoceros, the chimpanzee, the gorilla, and many others. Tropical Asia, the Oriental Region, is distinguished by the Indian elephant, the Malay tapir, the water buffalo, and the tiger. The Neotropical Region, comprised of tropical Mexico and Central and South America, has the llama, the spider monkey, marmosets, and the great anteater. The Nearctic Region (North America) is distinguished by the pronghorn antelope and grizzly bear, among others, while the Palaearctic Region (Europe, North Africa, and temperate-zone Asia) has the hedgehog, the wild ass, and the chamois. These lists of diagnostic species could be enormously lengthened with the names of additional species that represent virtually every other major group of plant and animal. Each species comprises, to a greater or lesser extent, one product of an adaptive radiation that was hemmed in by the great physical barriers dividing the continents. Similar but less distinctive biogeographic regions can be distinguished in the life of the sea. Here the barriers consist primarily of the continents.

It must be added that the biogeographic regions serve only for a first, crude description of the collective distribution of life. The regions are not separated from one another by sharp boundaries. Most of the diagnostic species occur only within a small part of the region they represent, and no two have exactly equal ranges. As a result, biogeographers often attempt to subdivide the regions into biotic provinces and other, lesser divisions. Many of the older textbooks on biogeography and evolution are filled with this kind of information. But the effort is seldom useful, because the limits of each such subdivision depend entirely on the set of species the author chooses to define it. The limits therefore change as the lists of diagnostic species change. There exists a close analogy between this shifting quality of bio-

6 TREES OF ST. HELENA have all evolved from about four species of the sunflower family that colonized the island as small herbaceous plants. Drawing shows details and overall forms of four of these trees. Originally island had no true trees and no forests. Virtual absence of competitors allowed invaders to radiate in many directions, particularly into tree-like forms.

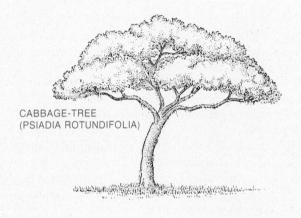

CABBAGE-TREE
(PSIADIA ROTUNDIFOLIA)

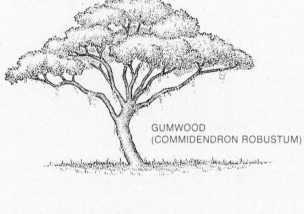

GUMWOOD
(COMMIDENDRON ROBUSTUM)

geographic units and the difficulties that accompany the delimitation of geographic subspecies *(see Chapter 30)*. The geographic limits of the subspecies usually depend on the choice of the characteristics used to define them. The variable characteristics, like the species themselves within the biogeographic region, tend to be independent in their patterns of distribution. It is further true that not all of the species obey the limits of the biogeographic region. Many are found on both sides of the lines drawn by biogeographers to separate the regions. This is the reason that the "line" is more accurately represented as a band or zone within which the representatives of a region become increasingly prevalent as one travels in the direction of the region. The only precise line that can be drawn is the one that represents the shift from a majority of species derived from one region to a majority derived from the other region.

THE BALANCE OF SPECIES

Balance of nature is not just a figure of speech. It can be measured precisely in the numbers of species of plants and animals that occupy a given island or other piece of the earth's surface over a specified period of time. These numbers tend to remain constant for long intervals. As new species colonize the area old species become extinct, in approximately a one to one relationship. The

SHE-CABBAGE-TREE
("LACHANODES" SPECIES)

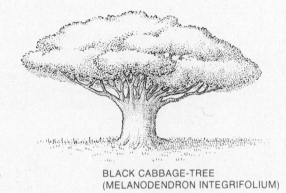

BLACK CABBAGE-TREE
(MELANODENDRON INTEGRIFOLIUM)

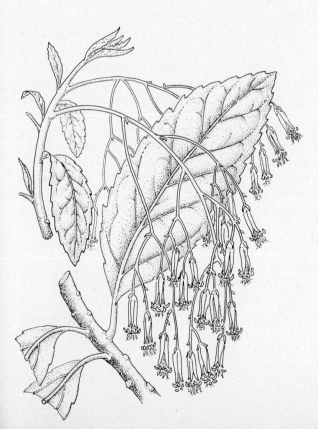

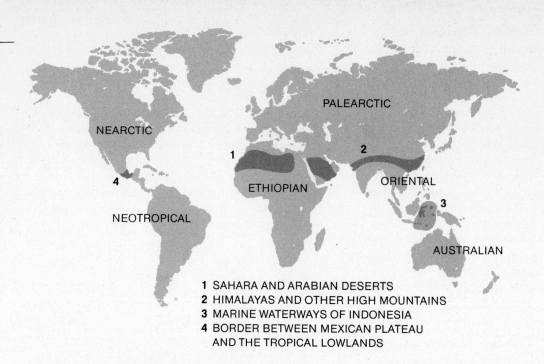

1 SAHARA AND ARABIAN DESERTS
2 HIMALAYAS AND OTHER HIGH MOUNTAINS
3 MARINE WATERWAYS OF INDONESIA
4 BORDER BETWEEN MEXICAN PLATEAU
AND THE TROPICAL LOWLANDS

7 **MAJOR ZOOGEOGRAPHIC REGIONS of the world.**
Numbers refer to the barriers that separate them.
Barriers are shown as broad bands to indicate gradual
change from one fauna to the next.

equilibrial number can be rather quickly moved up or down by changing the rate at which species are added to the system, or by altering the environment of the area.

One of the most reliable indicators of the existence of a balance in the number of species is the area-species curve. One example is given in Figure 8. The relationship between area and the number of species is most clearly expressed when the faunas and floras of different islands are compared with one another. Only species that are members of some well-defined taxonomic group can be reliably used for this purpose. For example, it is meaningful to include all of the land and fresh-water birds, as shown in Figure 8, or all of the birds belonging to some lesser subdivision of the birds such as the family Drepanididae, or the snails, or the higher plants, and so forth. Very roughly, the number of species belonging to such a taxonomically uniform group increases according to the formula

$$S = CA^z,$$

where S is the number of species present, A is the area of the island, C is the value of S when $A = 1$ (its value is not important for our purposes), and z is a number that varies from one taxonomic group to another and from place to place. In the great majority of cases z falls somewhere between 0.25 and 0.35. In the case of the birds of the East Indies, for example, z is 0.28. Another way of expressing this formula is to say that the number of species increases as the fourth to third root of the area. Still another very rough way of stating the relationship is to note that when the area is increased tenfold, the number of species approximately doubles. An island with an area of 100 square kilometers supports about twice as many species as one with an area of ten square kilometers. A great advantage of the area-species formula is that it allows the prediction of the number of species present on a given island before the island is even explored. The formula is useful in dealing with any isolated biologic community. It can be applied to lakes, ponds, and streams—islands of water in a sea of land. It is further applicable to habitat islands—patches of habitat surrounded by other kinds of habitats. To

An island with an area of 100 square kilometers supports about twice as many species as an island with an area of ten square kilometers.

the animals preferring spruce forest, for example, a patch of spruce forest in the middle of grassland is an island. A single tree standing alone in a grassy field can serve as an island to the insects and other small creatures that are strongly dependent on the tree.

Several factors can cause a group of species on an island (or habitat island) to deviate from the area-species curve. One is failure to attain equilibrium. If the island is young, or has undergone a recent catastrophe that eliminated much of the life existing on it, the species may still be in the act of colonizing it. This means that the rate at which new species are arriving is greater than the rate at which they are going extinct. The number of species present on the island is thus increasing toward its equilibrial level on the area-species curve. A second cause of deviation is the distance effect: the number of species present on an island is smaller than expected because the island is remote from the sources of immigrant species. Because fewer colonists are arriving, a smaller number of

species are present. This continues to be true even when the number reaches equilibrium. Also, fewer genera and higher taxonomic groups will be represented on the island. In extreme cases, islands lack such groups as fresh-water fishes, snakes, ants, and (as on St. Helena) even conventional trees. Such faunas and floras are referred to as DISHARMONIC. Ireland has a disharmonic flora and fauna. There are no snakes, as well as many other basic life forms that live elsewhere in Europe. The snakes were not banished by St. Patrick, as folklore tells it. If they ever lived on Ireland in the first place, they were eliminated by glaciers during the ice ages of the Pleistocene, and they have not succeeded in recolonizing this relatively isolated European island since that time.

The process of the reduction in numbers of species and higher taxonomic groups during dispersal to remote places is called FILTERING. Filtering is most clearly observed along a string of islands, where more distant islands harbor progressively fewer kinds of plants and animals. An excellent example from the insect fauna of the Pacific region is given in Figure 9. Since the islands form a sequence of stepping stones leading from one place to another they are often called filter bridges.

8 **AREA-SPECIES CURVE of the fresh-water and land birds of the East Indies. Straight-line curve indicates that the numbers of species on individual islands are approaching theoretical limit.**

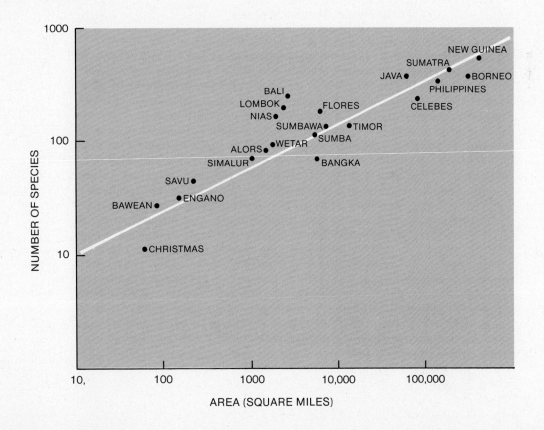

These biogeographic highways are especially important at the boundaries between the biogeographic regions, because they permit the steady diffusion of species from one region to the other. In other words, filter bridges ensure that the boundaries are not absolute barriers.

Figures 10 and 11 show some of the basic ideas of species equilibrium theory, a relatively new branch of biogeography that accounts for such basic phenomena as the area-species curve, the distance effect, and filtering. The theory starts with two generalizations that have been verified in field studies. First, the smaller the island —either a true island or a habitat island —the higher the extinction rate on it. If ten species of birds colonize two islands, one with an area of a thousand square miles and the other with an area of only one square mile, each species will be able to build up a much larger population on the larger island. Each is therefore much less likely to be completely wiped out by later fluctuations in population size on the large island than on the small island. The second generalization is that the more remote the island, the fewer the colonists that are able to reach it. Consequently, a smaller number of new species

will colonize it in a given period of time. In Figures 10 and 11 the unit of time used is the year, but the time scale is unimportant; we could equally well use millenia, or months, or even days.

By studying these simple diagrams you should be able to see clearly the causes of the upward slope of the area-species curve and of the distance and filtering effects. Another important result of species equilibrium theory is the turnover equation. Although the derivation of this equation is mathematically complex, the result is quite straightforward and simple:

$$\text{Species extinction rate at equilibrium (species/year)} \doteq \frac{\text{The number of species at equilibrium (S)}}{\text{The time (years) required for the island to fill from zero to 90\% of equilibrium}}$$

The dot over the equation sign means that the relation is only an approximate one. Instead of years (which are used here for clarity), other time units such as months or centuries could have been inserted. The turnover equation has some fascinating applications.

Krakatoa is a volcanic island located in the Sunda Straits between Sumatra and Java. In August, 1883, it erupted in the greatest explosion in recent history, one which covered it with red-hot pumice and destroyed all life on its surface. After the rocks cooled, Krakatoa was rapidly recolonized by plants and animals from nearby islands. The birds regained an equilibrial number of about 30

9   FILTERING OF WEEVILS across the islands of the Pacific. Islands act as filter bridges. Number of genera belonging to the subfamily Cryptorhynchinae progressively decreases on islands farther from New Guinea, the large island at upper left that serves as a principal source area.

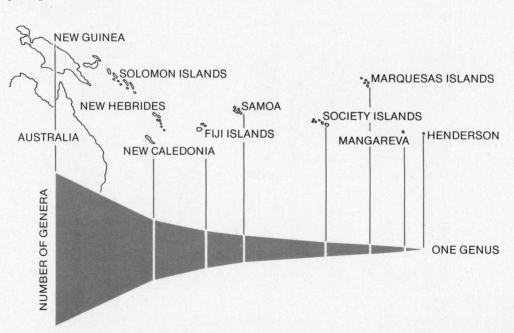

species in 25 to 36 years. If we take 30 years as the time required to reach 90 per cent of equilibrium, the turnover equation predicts that

$$\text{Species extinction rate at equilibrium (species per year)} \doteq \frac{30 \text{ species}}{30 \text{ years}} \doteq 1 \text{ species per year}$$

This is an unexpectedly high extinction rate. Yet censuses made in 1919–1921 and 1932–1934 indicate that at least one species was going extinct every five years on Krakatoa, and the true figure was probably much closer to the predicted rate of one species per year. As equilibrium was reached, each species that became extinct was replaced on the average by one new colonizing species from nearby islands.

Under some circumstances it is possible to make a direct test for the existence of a species equilibrium. The coast of southern Florida, for example, is dotted by thousands of small islands consisting entirely of red mangrove trees rooted in shallow water. Those islands that are about 40 feet in diameter each contain 20 to 45 species of insects, spiders, mites, and other arthropods. The constancy of this number from island to island suggests that it is in equilibrium, but such evidence is not by itself conclusive. An experiment was recently performed to observe the entire colonization process, in much the way the colonization of Krakatoa had been studied after the 1883 explosion. Six of the little mangrove islands were covered with nylon tents and fumigated with methyl bromide. The fumigation treatment destroyed all of the insects and other arthropods but left the red mangrove trees intact. Subsequent recolonization of the islands was very rapid, as shown in Figure 12. All of the islands reattained species numbers that were close or equal to the original level. Equally impressive is the fact that the more distant islands, which started with the smallest number of species (in accordance with the distance effect) ended with the smallest number. Although the numbers of species were preserved in the end, the species compositions were not. The species present at the conclusion of the experiment on each island were different for the most part from those that lived there before the fumigation. Furthermore, there was a high rate of extinction throughout the colonization process, in accordance with the turnover equation. Equilibrium was achieved when the extinction rate had reached such a high level that it balanced the immigration rate of species onto the islands.

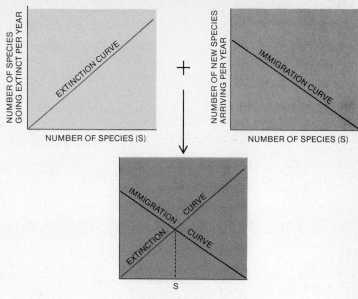

SPECIES EQUILIBRIUM THEORY holds that the extinction rate of species increases as more species crowd onto an island (graph at top left). Number of immigrating species gradually decreases (top right) because there are fewer potential colonizers left in the source area. In the combined graph at bottom, S represents the equilibrium number of species when the rate of extinction equals the rate of immigration. From then on, the number of species inhabiting the island remains constant. 10

EFFECTS OF SIZE AND REMOTENESS on the equilibrium number of species on an island. On smaller islands (left) extinction occurs more rapidly, raising slope of extinction curve and decreasing equilibrium number of species. Isolated islands (right) are hard for immigrant species to reach, which lowers both the immigration curve and the equilibrium number of species on the island. 11

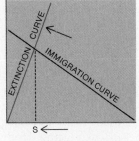

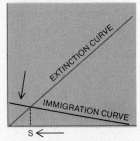

705

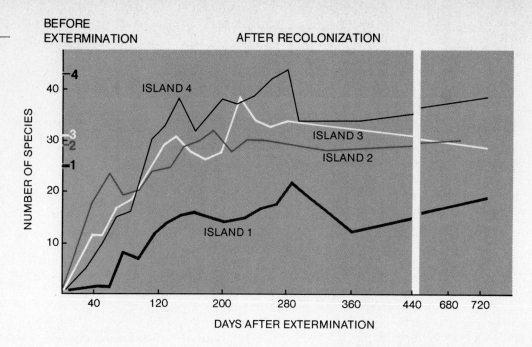

BEFORE EXTERMINATION   AFTER RECOLONIZATION

NUMBER OF SPECIES

ISLAND 4

ISLAND 3

ISLAND 2

ISLAND 1

DAYS AFTER EXTERMINATION

RECOLONIZATION of four small islands in the Florida Keys is plotted as four curves. Horizontal bars at far left indicate the original number of species on each island. Bars are color coded to correspond to curves. Within a year islands had almost reattained original number of species.

FAUNAL DOMINANCE

The principle of species equilibrium can be extended from islands such as Krakatoa and the Florida Keys to the large islands and continents of the world. The faunas of these much greater land masses still conform to the area-species curves. There is reason to believe that if they were to be monitored over long periods of time they would be found to be in equilibrium. But the process is complicated by the fact that large islands and continents also tend to be the theaters of adaptive radiation. They are the centers of the biogeographic regions, each of which has a unique assemblage of animals that fill its major niches. Thus two continents cannot be compared with the same assurance as, say, two small islands in southern Florida. What would happen if two of the unique continental faunas were to be mixed? Such mixing does happen, gradually and over long periods of time, by dispersal across the boundaries of the biogeographic regions. And as it occurs, the elements of certain faunas tend to move in and to replace those in other faunas. This pattern is a relatively consistent one, and it is referred to as faunal dominance.

As a rule, the dominant animal species arise on the largest land masses, which contain the greatest, most diverse assemblages of species. Throughout the Cenozoic Era, extending from about 63 million years ago to the present, North America, Europe, and Asia have been closely connected to form what is in effect a single land mass, through which the major groups of plants and animals dispersed relatively easily. The effectiveness of this northern area in creating dominant elements is dramatically illustrated in the case of its exchanges with South America. For most of the Cenozoic Era a seaway connected the Caribbean Sea with the Pacific Ocean. South and North America were separated, and Central America existed as a group of islands scattered through the seaway. During its period of isolation South America developed a mammalian fauna almost

> Much of the old South American fauna has vanished, evidently the victims of competition and replacement by invaders from the World Continent. No one will ever ride a litoptern or feed peanuts to a toxodont in a zoo.

wholly different from that in the rest of the world. Forms evolved that strikingly resembled horses; others were reminiscent of rhinoceroses, great cats, and other basic mammalian types *(Figure 13)*. This adaptive radiation closely paralleled radiation occurring in Australia and the World Continent at the same time. Prior to the emergence of the Panama Isthmus 3 to 5 million years ago, there

CONVERGENT TYPES OF MAMMALS existed in North and South America before the emergence of the Isthmus of Panama linked the two regions. Eventually many of the original South American species became extinct. Of the mammals shown here, all but the two marsupials are placentals.

13

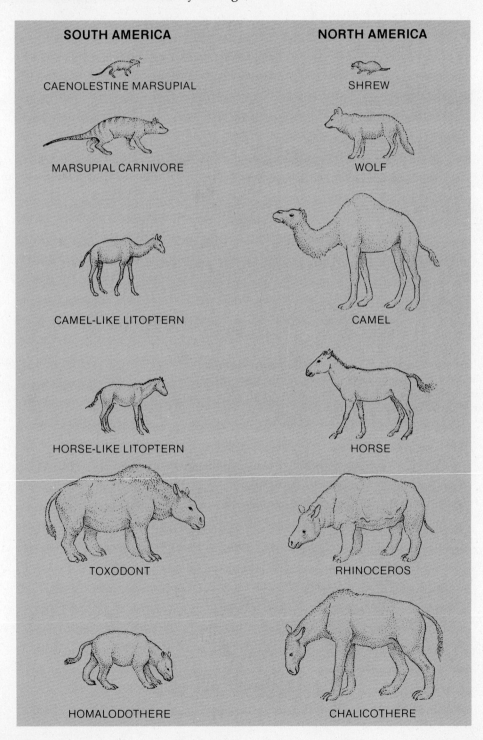

SOUTH AMERICA | NORTH AMERICA

CAENOLESTINE MARSUPIAL | SHREW

MARSUPIAL CARNIVORE | WOLF

CAMEL-LIKE LITOPTERN | CAMEL

HORSE-LIKE LITOPTERN | HORSE

TOXODONT | RHINOCEROS

HOMALODOTHERE | CHALICOTHERE

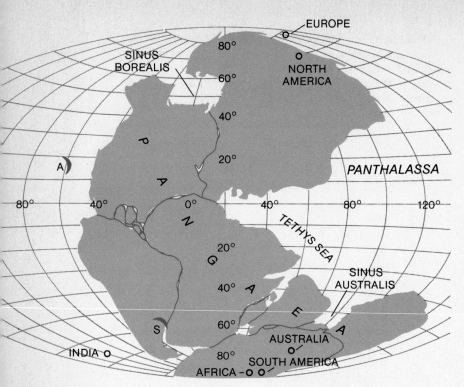

SUPERCONTINENT OF PANGAEA stretched from pole to pole in the early Paleozoic Era, about 200 million years ago. Small black circles indicate varying positions of the magnetic poles. Colored crescents A and S are included as geographic reference points. Today A is site of Antilles in the Caribbean and S the Scotia arc in the South Atlantic, off tip of South America. Panthalassa was the ancestral Pacific Ocean, and the Tethys Sea was the ancestral Mediterranean. Central meridian in this and the two following maps lies 20 degrees east of present zero (Greenwich) meridian.

15 **20 MILLION YEARS LATER,** early in the Mesozoic Era, Pangaea had split into two parts. Laurasia was comprised of land that is now North America, Europe and Asia. Gondwana later became South America, Africa, Australia and Antarctica. Arrows indicate direction of continental drift.

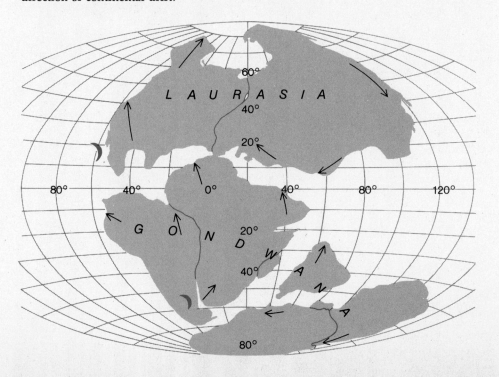

were 23 families of mammals in South America and 27 in North America. None of the families was shared by the two continents. When the Isthmus rose slowly from the sea, it separated the Caribbean Sea from the Pacific Ocean and created a land bridge over which many of the species of North American mammals could spread south into South America, and some South American mammals could spread north. After a relatively brief period of faunal enrichment in both continents, the total number of families began to decline. Today there are 30 families in South America and 23 in North America. The numerical picture is therefore not drastically different from the original one before the creation of the Isthmus. But now 17 of the families occur on both continents, and most of these are of northern origin. A large part of the old South American fauna has vanished, evidently the victim of competition and replacement by the invaders from the World Continent. The reader will never ride a litoptern or feed peanuts to a toxodont in the zoo.

CONTINENTAL DRIFT

For decades most interpretations of biogeography have been based on the assumption that the continents have always been anchored in their present positions. Geologists realized that shorelines gradually changed as shallow seas advanced and

> Early in the Paleozoic Era all, or nearly all of the land on Earth was joined together in a single supercontinent that geologists have named Pangaea.

retreated over the low areas, but they believed that the centers of the continents remained stationary. But in the last decade or so geologists have accepted the theory that the land masses of Earth have been imperceptibly but steadily moving—a process called continental drift. Early in the Paleozoic Era all, or nearly all, of the emergent land was joined together in a single supercontinent which geologists have named Pangaea *(Figure 14)*. By early Mesozoic times Pangaea had broken into two gigantic pieces. To the north was Laurasia, comprised of what is today North America, Europe, and Asia. To the south was Gondwana,

110 MILLION YEARS LATER, at the close of the Mesozoic, the modern continents had broken apart and were drifting away from one another. North America and Europe were still joined to Greenland (top), and Australia remained connected to Antarctica (bottom). India lies off African coast.

16

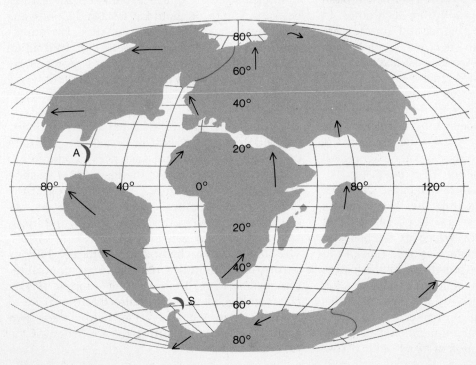

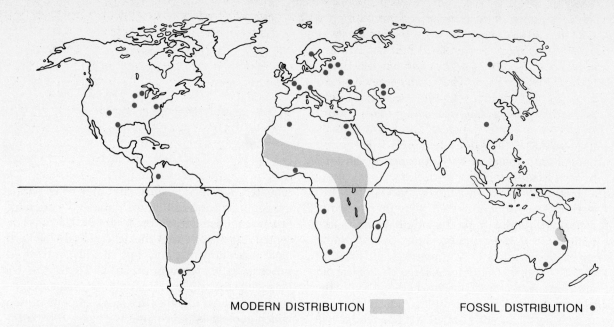

MODERN DISTRIBUTION ▢   FOSSIL DISTRIBUTION •

17 **DISTRIBUTION OF FOSSIL LUNGFISHES suggests
that their range was established before the continents
separated completely. Modern distribution suggests
that lungfish may have spread between Africa and
South America before continents were separated by the
formation of the Atlantic Ocean.**

destined later to form South America, Africa, Australia, and Antarctica (*Figure 15*). By the close of the Mesozoic Era, approximately 65 million years ago, most of the modern continents had broken apart and were drifting away from each other. In the far north only North America and Europe were still solidly joined by what is now Greenland. On the opposite side of the world Australia and Antarctica remained connected (*Figure 16*).

The broad outlines of this astonishing view of the history of the earth were first proposed in 1912 by the German meteorologist Alfred Wegener. The recent acceptance of the theory is due largely to the discovery of new forms of geological evidence. All of the relevant information now seems to fall neatly into place: the jigsawlike fit of the shapes of the present continents into the larger pieces theorized to exist in past eras (especially true of South America and Africa, which are not believed to have separated until late Mesozoic times); the matching of geological formations at the places where the junctions were believed to have existed; the orientations exhibited by magnetic particles fixed in rocks laid down during successive stages of movement (a phenomenon called paleomagnetism); and the distribution of many kinds of fos-

sils deposited over the several supercontinents during different ages. Taken together, these diverse facts can be satisfactorily explained only by the theory of continental drift.

How does the concept of continental drift affect biogeography? Obviously, the global patterns of dispersal in past eras must have differed greatly from those displayed by modern organisms. Dinosaurs and other reptiles of the Mesozoic Era, for example, appear to have experienced little difficulty ranging throughout Laurasia and Gondwana and even back and forth between these two supercontinents. In general, dispersal of individual groups of land animals was much more nearly global in earlier times than it is now. It would have been much more difficult to define regions of animals and plants.

Some living groups of birds and mammals, evolved during the Cenozoic Era, after the present continents had come into existence. It continues to make sense to reconstruct the histories of existing groups in terms of modern world geography. However, many other groups of organisms, including some insects, freshwater fishes, and frogs, have histories that extend back into the Mesozoic. They may have reached their present distributions before continental breakup was complete. Lungfishes, to take one striking example (*Figure 17*), might well have spread directly from Africa to South America, or vice versa, before these two continents separated during the formation of the Atlantic Ocean.

READINGS

S. CARLQUIST, *Island Life, A Natural History of the Islands of the World*, Garden City, NY, The Natural History Press, 1965. A superb general book on biogeography that goes far beyond its main topic of islands to treat continents as well. It analyzes most of the basic topics of biogeography in considerable depth, yet in clear, simple language and with a wealth of illustrations. This book is strongly recommended as the next book to read on biogeography after you finish this chapter.

P.J. DARLINGTON, *Zoogeography: The Geographical Distribution of Animals*, New York, John Wiley & Sons, 1957. A classic in its field, written in clear language with many examples.

R.S. DIETZ AND J.C. HOLDEN, "The breakup of Pangaea," *Scientific American*, October 1970. A clear account of the geological evidence for continental drift.

B. KURTEN, "Continental drift and evolution," *Scientific American*, March 1969. A speculative but interesting and provocative essay on the significance of continental drift in biogeography.

R.H. MACARTHUR, *Geographical Ecology: Patterns in the Distribution of Species*, New York, Harper & Row, 1972. For students with a good mathematical background, this is the best introduction to general quantitative theories of biogeography.

Ecological explosions differ from some of the rest by not making such a loud noise and in taking longer to happen. That is to say, they may develop slowly and they may die down slowly; but they can be very impressive in their effects, and many people have been ruined by them, or died, or forced to emigrate. —Charles Elton

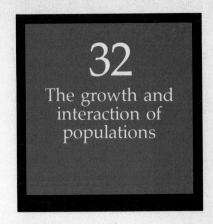

Imagine that one could select a single bacterium at random from the surface of this book, and could somehow endow it and all of its descendants with the power to grow and to reproduce without any restriction. In a month this bacterial colony would weigh more than the visible universe, and would be expanding outward at the speed of light. Similarly, a single pair of Atlantic cod and their descendants reproducing without hindrance would in six years fill the Atlantic Ocean with their packed bodies. After a few more decades they, too, would weigh as much as the visible universe and be expanding at the speed of light.

Man is one of the most slowly maturing and breeding of all organisms. But if the existing human population could somehow achieve the impossible feat of continuing to increase at its present rate, which is less than its maximum potential, it would come to weigh as much as the entire earth in 2000 more years. Four thousand years later it would weigh as much as the visible universe — and, of course, be expanding at the speed of light.

All populations have this same potential for explosive growth. The reason is quite simple. Since all, or nearly all, mature individuals in the population can produce offspring, the rate at which a whole unrestricted population increases is exactly proportional to the number of individuals in the population. Therefore the larger the population, the faster it grows. And the faster it grows, the sooner it becomes still larger, and so on upward at an accelerating pace. In other words, the size of a population continuously accelerates when restraints are removed. This form of increase is called

SCHOOL OF BAITFISH (Stolephorus purpureus) in the underwater photograph on opposite page scatters before the attack of a predatory member of the tuna family (Euthynnus affinus).

EXPONENTIAL GROWTH, and it is expressed mathematically in the following way:

$$\text{Rate of increase in number of individuals} = \left(\begin{array}{c}\text{Average birth rate}\end{array} - \begin{array}{c}\text{Average death rate}\end{array}\right) \times \begin{array}{c}\text{Number of individuals}\end{array}$$

The difference between the average birth rate and average death rate, the term enclosed in parentheses in the above equation, is called the INTRINSIC RATE OF INCREASE. To see this relationship more clearly, consider the exponential growth of mankind. At the present time, as everyone knows who has heard of the population crisis, the number of human beings is increasing at an accelerating rate. In South America, to take one example, the total population is now approximately 200 million. The birth rate, according to one authoritative estimate, is 42 children born for every 1000 persons every year, or 4.2 percent. The death rate is 19 persons out of every 1000 persons every year, or 1.9 percent, which less than balances the birth rate. The birth rate minus the death rate is the number of new persons added to every 1000 every year; this figure is 42 − 19 = 23, or 2.3 percent. Translating the percentages into decimals we have a birth rate of 0.042 per person per year, a death rate of 0.019 per person per year, and an intrinsic rate of increase of 0.023 per person per year. In South America therefore, the population is increasing at approximately the following rate:

$$\text{Rate of increase in number of individuals} = \begin{array}{c}\text{Intrinsic rate of increase}\end{array} \times \begin{array}{c}\text{Number of individuals}\end{array}$$

$$= (0.042 - 0.019) \times 200,000,000$$
$$= 0.023 \times 200,000,000$$
$$= 4,600,000 \text{ per year}$$

713

Ecologists often refer to the normal long-range condition of populations — including those of man —as one of zero population growth.

When South America acquires a population of 500 million (this will occur around the year 2010 if the present trend continues), its growth rate will be up to 11,500,000 additional persons per year. At one billion (projected time: 2037) it will be adding 23,000,000 per year. Both estimates are made simply by multiplying the expected population size in that year times the present intrinsic rate of increase (0.023). These figures are only approximate, since they are based on data collected around 1960. As living conditions change and the proportions of individuals belonging to different age groups shift with respect to one another, the rate of increase also changes. For this reason we can use the expression intrinsic rate in only a loose sense when describing the particular case of South America.

The intrinsic rate of increase also varies enormously from species to species. In the laboratory rat, for example, it is 5.4 (540 percent) per year. In most bacteria it is between 10 and 100 (1000 to 10,000 percent) per day. The intrinsic rate of increase also varies among populations belonging to the same species, depending on the physiological condition of the organisms and the particular environment in which they exist. In extreme cases the intrinsic rate drops to zero. This is merely the formal way of saying that if organisms are living in excessively harsh circumstances, they cannot reproduce.

Of course, no population can maintain exponential growth indefinitely. At any given time, most populations have a steady size, or else their numbers are fluctuating up and down around a constant average value. In other words, the actual rate of increase in the number of individuals is zero. For this reason, ecologists often refer to the normal long-range condition of populations —including those of man —as one of zero population growth. Once in a while, a population is temporarily reduced by catastrophe to a very low level. Or, as in the case of man, it finds a new way to exploit the environment and thus enters temporarily unfilled areas. Then its numbers begin to increase rapidly. At first the growth conforms approximately to the exponential type expressed by our equation. But this condition is short-lived —no population can weigh as much as the visible universe! What happens is that the growth curve soon bends over to assume an S shape, as shown in the right-hand side of Figure 1. The number of individuals increases to the point where the environment in which they live can support no more. At this population size, called the CARRYING CAPACITY of the environment (for the particular population at that particular time), the death rate equals the birth rate. The rate of increase in the number of individuals is then zero. Temporary deviations from zero will probably occur in the future, causing the popula-

1 POPULATION GROWTH CURVES. Beginning with a small number of individuals, the size of a population can increase in either exponential (left) or logistic fashion (right).

EXPONENTIAL (UNRESTRICTED) GROWTH   LOGISTIC (RESTRICTED) GROWTH

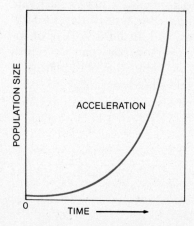
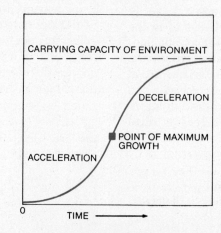

tion to grow for a short time, or to decline, but the average value over long periods of time will be zero. As a result the population will fluctuate up and down around the carrying capacity of the environment for most of its existence. For most of its history, to take a familiar example, the human species had an actual rate of increase at or close to zero. Only within the last several thousands of years have technological and medical advances permitted a temporary exponential increase.

Why does expansion up to the carrying capacity of the environment follow a logistic (S-shaped) form? The reason is that as numbers increase, the birth rate slows down, or the death rate increases, or both occur together. Eventually the death rate comes to equal the birth rate, and the carrying capacity of the environment is reached. This more realistic model of logistic growth must be substituted for the exponential growth model given earlier. By adding the carrying capacity of the environment to the equation, one arrives at an equation for LOGISTIC GROWTH of a population:

$$\text{Rate of increase in number of individuals} = \left(\begin{array}{c}\text{Average} \\ \text{birth} \\ \text{rate}\end{array} - \begin{array}{c}\text{Average} \\ \text{death} \\ \text{rate}\end{array}\right)$$

$$\times \left(\frac{\begin{array}{c}\text{Carrying} \\ \text{capacity of} \\ \text{the environment}\end{array} - \begin{array}{c}\text{Number of} \\ \text{individuals}\end{array}}{\begin{array}{c}\text{Carrying capacity} \\ \text{of the environment}\end{array}}\right) \times \begin{array}{c}\text{Number of} \\ \text{individuals}\end{array}$$

The only difference between the logistic growth equation and the exponential growth equation is the introduction of the second term in parentheses, the one containing the number for the carrying capacity of the environment. A little thought about this term and about its effects as the number of individuals goes from near zero to the carrying capacity, will clarify how it gives rise to the S shape of the logistic curve. When the number of individuals in the population is very low, the term is nearly equal to the carrying capacity divided by the carrying capacity (nearly one). If one substitutes the number one or some low number close to it for the "number of individuals" in the equation, the exponential growth equation results. In other words, when the population is very small and still far from approaching the carrying capacity, it grows in a way not very different from the idealized exponential form. However, if you make the population size equal to the carrying capacity, the term in parentheses is zero. Zero multiplied against the other terms in the right-hand side of

the equation makes the entire expression zero. The population starts growing at nearly the rate predicted by the exponential form, but whenever the population reaches the carrying capacity the rate slows down steadily until it stops entirely. It has reached zero population growth. The total result is the logistic growth curve (*Figure 2*). A more sophisticated discussion of the equation appears in Box A.

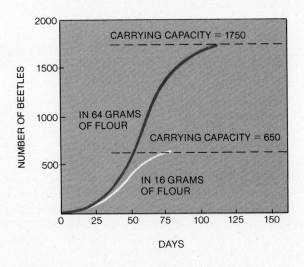

LOGISTIC GROWTH CURVES for two populations of **2** **the flour beetle Tribolium confusum. Carrying capacity was varied by altering the amount of flour in which beetles lived. After equilibrium is reached, number of beetles in each population continues to fluctuate slightly.**

SURVIVAL AND OPTIMAL YIELD

There are two fundamental ideas in ecology to be gained from a study of the logistic growth curve and the equation on which it is based. The first is that the carrying capacity of the environment is independent of the intrinsic rate of increase. How fast a population can grow has no bearing on the size it can finally attain, and vice versa. Slow-breeding elephants are capable of filling up the entire world (given a little extra time), but the fastest breeding population of bacteria will soon become extinct—if it is limited to a drying pool of water. Ecologists have used this principle to show that when attempting to control a pest species, such as rats infesting dumps or seagulls endangering aircraft around landing strips, it is much more efficient to reduce the breeding sites or food supply of the population than it is to try to exterminate the animals themselves. Efficient garbage removal gets rid of rats better than rat traps. Taking away the resources of the species reduces the carrying capacity of its environment, a permanent alteration, but killing off part of the population is a temporary measure which the animals swiftly counteract through their awesome power of exponential increase. The same principle is important in a different way in the planning of conservation policies. When trying to protect a rare animal species, the most important step is to provide it with a preserve, an area containing its favorite habitat, to ensure that it will have a sufficiently high upper population limit. Once this is accomplished, it is possible to remove part of the population by hunting or capturing specimens for exhibition without endangering the species or even significantly reducing its numbers.

The second basic point about population growth is that the greatest amount of increase does not occur either when the population is just beginning to grow or when the population has reached the carrying capacity of the environment. It occurs at one particular point in between. Look again at the idealized logistic curve of Figure 1. The steepest rise of the S-shaped curve is in the center of the S, exactly halfway on the vertical axis between the baseline and the carrying capacity of the environment. At this point the greatest number of new individuals is added to the population in a given amount of time. This maximum rate of population growth is referred to as the OPTIMAL YIELD. Because not all growth curves fit the ideal logistic form, the "optimal yield problem" can seldom be solved so simply, and ecologists must resort to more sophisticated techniques. But almost without exception the optimal yield point occupies an intermediate position on the growth curve. Suppose a trapper were interested in getting the maximum number of foxes from a forest, or a fisherman in trying to extract the maximum number of fish from a pond, lake, or ocean fishing bank. The ideal economic strategy for either to follow would be to "crop" the population down to the point of optimal yield. If left near the maximum that the environment can support, or if exploited to the point where the or-

---

A Mathematical description of population growth

The reader may wish to see how the principles of population growth presented in this chapter are expressed in the mathematical language used by population ecologists. The equation for exponential growth, which is given in the text in words, can be written more concisely as follows:

$$\frac{dN}{dt} = rN$$

where dN represents the amount of growth in an infinitesimally small amount of time, dt, and dN/dt, which is pronounced "dee N dee t", is the rate of increase in the number of individuals in the population. N is the number of individuals present at the moment the count is taken, and r is the intrinsic rate of increase. The "solution" of this rate equation, from which we draw the curve of Figure 1, is

$$N = N_o e^{rt}$$

Here N is expressed as a function of time. N_o is the population at the starting point, t is the amount of time that passed since the starting point, and e is the base of natural logarithms ($e = 2.71828 \ldots$). Since t, the independent variable, is in the exponent of this equation, the growth is referred to as exponential growth.

Logistic growth can be expressed as

$$\frac{dN}{dt} = r\left(\frac{K - N}{K}\right)N$$

where K is the carrying capacity of the environment. This rate equation is identical to the one given in words in the text. It also has a solution, one which yields the S-shaped curve of Figure 1, but the form is complicated and will not be given here.

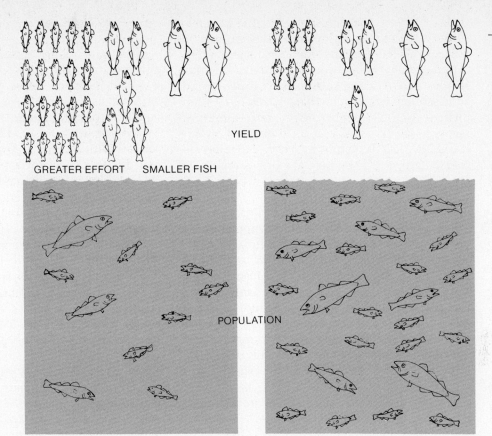

YIELD

GREATER EFFORT   SMALLER FISH

POPULATION

3 FISH POPULATIONS at point of optimal yield (right) can be harvested with much less effort than populations reduced by overfishing (left), and proportion of large fish caught is higher.

ganism becomes scarce, the yield will decline. One of the most important activities of applied ecology, therefore, is the solution of optimal yield problems. One theoretical but realistic example is given in Figure 3. This case, incidentally, illustrates a third principle: that the heavier the exploitation, the more the population will come to consist of younger and smaller individuals, which are also usually the least valuable commercially.

Already many of the great fishing areas of the world have had their populations driven below the point of optimal yield by excessive harvesting. The Georges Bank off the coast of New England, source of the cod, halibut, and other prime food fishes, is one example of acute concern to the United States. As the fish populations are reduced, more expensive and sophisticated ships and trawling techniques are required to produce the same amount of yield, and prices rise accordingly. Another example of excessive harvesting, one of

truly tragic proportions, is provided by the whaling industry. The blue whale, the largest creature that ever lived on the land or the sea, has now been hunted to the point of near extinction. The same fate has overtaken the humpback whale, while others among the great whales, notably the sperm whale and the sei whale, are being decimated and their existence will eventually be threatened. The senselessness of this destruction is deepened by the fact that the whaling industry has carried its activities to the point that whaling can now be only marginally profitable and is in danger of wiping itself out. If the whaling nations, and particularly Japan and the Soviet Union as the chief offenders, could agree to limiting their catches in such a way as to allow the whale populations to approach the optimal yield level, the species could be saved —and whaling could again become a very profitable industry. But ignorance and greed have prevented such action. As D. W. Ehrenfeld has pointed out: "Because of their status as an exclusively international resource, blue whales could have been protected only by *all* whaling nations acting in concert. Yet with the exception of Norway, which instituted strong unilateral conservation measures pertaining to its own industry, the record of the member nations of the International Whaling Commission is a sickening testimonial to corporate and national greed, shortsightedness, and inertia. The Whaling Commission itself had neither inspection nor enforcement powers, and any of its majority decisions could, in effect, be vetoed by a single member nation.''

AGE AND THE PROBABILITY OF DEATH

The intrinsic rate of increase of a population depends on two statistical qualities: the ages of the organisms comprising the population, and the reproductive performance of the organisms in each age group. Obviously, a population consisting entirely of individuals too old to reproduce has a zero rate of increase and is doomed to early extinction. A population made up mostly of immature individuals has a low rate of increase but is destined to improve it in a short time, and it probably has a secure future.

The age distribution of a population is the proportion of individuals in each age group. For example, the population of the United States in 1950 could be analyzed as follows: 29.0 percent were 0–19 years old, 45.7 percent were 20–49 years old, 16.2 percent were 50–64 years old, and 9.1 percent were 65 years old or older. When a population is allowed to exist in a constant environment for several generations or more, so that its birth and death rates become constant, its age distribution becomes stable. That is, an age distribution is reached which is maintained without change from that time forward. The stable age distribution, like the intrinsic rate of population increase, differs greatly from species to species and depends to some extent upon the environment in which the population happens to find itself. It further resembles the intrinsic rate of increase in being an ideal condition seldom maintained for long in nature. Because the environment is not constant, most populations may only approach stable age distributions or at best hold them for brief periods of time.

One of the elements that determines the form of the age distribution is the survivorship curve. Examples of the three basic shapes this curve can take are shown in Figure 4. The oyster is an example of a species in which vast numbers of young stages are produced, and the majority quickly die. Only a tiny fraction succeed in attaching themselves to a rock or to some other support, the necessary step for completing the life cycle. The survival rate among oysters reaching this point is much higher. The hydra exemplifies species with constant mortality rates. An individual is just as likely to die when it is one year old as when it is one day old. Man and fruit flies kept in the laboratory, in contrast to the oyster and hydra, are species with a definite period of senescence. Provided with a good environment, most individuals live to a certain age in reasonably sound health. In man the age of senescence is approximately the biblical "three score and ten" (70) years. Then the diseases and infirmities of old age begin to set in, and death becomes increasingly probable with each passing year.

No human being has been reliably certified to have lived beyond 120 years. At this point it is worth asking a peculiar sort of question as part of our review of ecology. Why do senescence and death occur at all? To the evolutionary biologist no question having to do with any regular process of life, or its termination, is out of order. Each process has an evolutionary history, and is susceptible to explanation by means of the theory of natural selection. The explanation for the origin of senescence and scheduled death most generally accepted among biologists is the broken test tube theory of Peter Medawar, the distinguished British physiologist. Consider, Medawar said, a row of test tubes that are in constant use in a laboratory. They are sure to be broken at a regular rate just through everyday accidents. After an equally predictable period of such exposure, most will have perished and been replaced. A survivorship curve can be constructed for the test tubes, although they are inanimate objects, just like that for certain organisms—for example hydras and sparrows. Now consider organisms. If they were exactly like

---

Man and fruit flies, in contrast to the oyster and hydra, are species with a definite period of senescence. Senility is the price paid by individuals who manage to survive beyond the usual time allotted by accidental death. They succumb to "play now, pay later" genes.

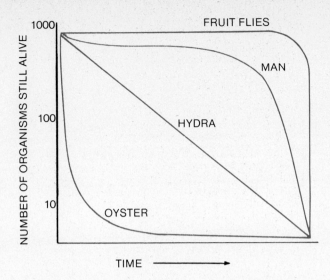

4 **SURVIVORSHIP CURVES for four species of animals. Starting with equal populations of 1000 individuals, each species follows a characteristically shaped curve. Time scale is adjusted to cover same distance for each species.**

test tubes and did not age, organisms still would die at a steady rate from accidental causes. For most individuals an accidental death consists of being eaten by predators, killed by disease, or starving. If by age x (in primitive man x might have been 70 years) the great majority have been eliminated, natural selection will favor individuals that are vigorous and able to breed at earlier ages. Genes that cause senility and increase the probability of death after age x would not be penalized — what difference would they make if nearly everyone has already been eliminated by accidents? If these same genes also added vigor and reproductive ability to individuals younger than x, they would be favored through natural selection. Senility is therefore the price paid by individuals who manage to live beyond the usual time allotted by accidental death. They succumb to "play now, pay later" genes. Senility becomes a common condition only when populations are sufficiently well fed and protected to reduce mortality by predation, disease, and other external causes. Senility used to be rather rare in human populations. In the earliest days of civilization, the average life expectancy was only about 15 years. Mortality was very high in infancy and childhood, and only a few individuals reached what would be today regarded as middle age. By Roman times life expectancy had risen to 25 years, and by 1900 it was still only 50 years or less. In some European countries today it surpasses 70 years.

REGULATION OF POPULATION GROWTH

In logistic growth, what is it exactly that slows the population down and finally brings it to a halt? Almost anything can. The factors of population control are exceedingly diverse, and differ from species to species. They are sometimes difficult to identify except through careful, lengthy examination. Darwin expressed the matter trenchantly when he said, "We behold the face of nature bright with gladness, we often see superabundance of food; we do not see, or we forget, that the birds which are idly singing round us mostly live on insects or seeds, and are thus constantly destroying life; or we forget how largely these songsters, or their eggs, or their nestlings, are destroyed by birds and beasts of prey; we do not always bear in mind, that though food may now be superabundant, it is not so at all seasons of each recurring year."

To be eaten, to starve, to fall victim to disease — these and all other kinds of ill fortune met by some members of the population, will increase the overall death rate of the population, or depress its birth rate, or both. But not every one of the effects regulates the population, in the sense that it helps to hold the population at the carrying capacity of the environment. To regulate population growth, a process must increase in its intensity as the population grows larger. When it does this, a process is known as a DENSITY DEPENDENT EFFECT (some textbooks use the less precise term density dependent factor). In Figure 5 are presented the principal categories of density dependent effects. As a population of organisms grows more dense, its predators have an easier time finding prey, and the mortality rate of the population therefore increases. Infectious disease is more easily transmitted (epidemics among humans occur most often in the crowded cities). Competition increases, food and space grow scarcer. Mortality goes up and the average birth rate declines. Emigrations may also increase as the organisms become more restive, and the population growth rate declines still further. In the end, one or more of these density dependent effects drives the mortality high enough and the birth rate low enough until the two come to be equal, and there is no longer any net growth in the population size. The population has attained zero population growth. Several actual examples of density dependent effects are illustrated in Figure 6.

Sometimes density dependent effects are mediated by striking physiological changes that

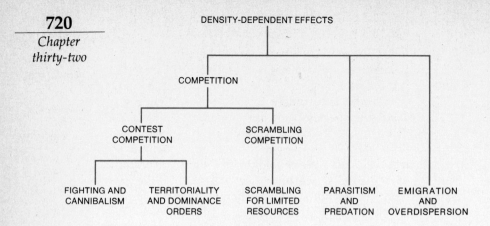

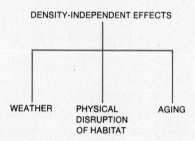

5 FACTORS THAT LIMIT LOGISTIC GROWTH include both density dependent and density independent effects. Density dependent effects tend to stabilize population size. Independent effects also affect size of population, but cannot hold it constant. Diseases are included under parasitism.

are interesting to the biologist in their own right. As laboratory mice are crowded together, for example, the average birth rate declines. Recent studies have shown this change to be due in large part to the presence of a pheromone released with the urine. As the density of mice increases, the smell of the pheromone intensifies. At high levels the odor alters the balance of sex hormones in the bodies of the females and in many of them ovulation is suppressed. There is also a sharp increase in the occurrence of pseudopregnancy, an abnormal swelling of the uterus that blocks true pregnancy. A third effect of the urinary pheromone is to increase the production of adrenal hormones (corticosteroids), which secondarily reduce the reproductive capacity of the mice. Biologists have only begun to explore this pheromone system and the many other kinds of physiological mechanisms responsible for population control in animal and plant species.

In contrast to the governing role played by den-

sity dependent effects, DENSITY INDEPENDENT EFFECTS operate without reference to population size. A flash flood in the desert destroys a certain fraction of the populations it touches, without reference to whether they are sparse or dense at the time. The same kind of random destruction is caused by many other adverse events that occur periodically in the physical environment, from landslides and volcanic eruptions to unfavorable changes in the weather. The physiological process of aging is also largely independent of the population density. Density independent effects can assist in moving population growth upward or downward, but they cannot hold the population size at a constant level.

COMPETITION

Competition, as ordinary experience teaches us, is what happens when there is not enough of something desirable to go around. The ecologist defines it in a manner closely akin to this ordinary usage: competition is the active demand of two or more organisms for a common vital resource. An animal that aggressively challenges another over a piece of food is obviously competing. So is a plant that absorbs phosphates through its root system at the expense of its neighbor, or cuts off its neighbor from sunlight by shading it with its leaves.

The techniques of competition are extraordinarily diverse. As indicated in Figure 5, some ecologists make a rough distinction between contest competition, in which the organisms confront one another and actively contend for the resource, and scramble competition, in which the winning organism is the one that is able to get to the resource first or else proves more efficient at using it up. In extreme contest competition, direct aggression is employed. When barnacles of the species *Balanus balanoides* invade rock surfaces occupied by a second barnacle species, *Chthamalus stellatus*, they eliminate these competitors by direct physical removal. In one population studied in Scotland, ten percent of the individuals in a colony of *Chthamalus* were overgrown by the shells of the *Balanus* within a month, and another three percent were undercut and lifted off during the same period. A few others were crushed from the side by the expanding shells of the *Balanus*. By the end of the second month 20 percent of the *Chthamalus* had been eliminated, and eventually all disappeared. Individuals of *Balanus* also destroy each other, but at a slower rate than they do members of the competitor species. Thus competition for space in this case occurs both between and within each species of barnacle.

Ant colonies are notoriously aggressive toward each other, and colony warfare both within and between species has been witnessed by many entomologists. The colonies often establish their territorial limits in this fashion, and sometimes they fight to the end. A striking example is presented in Figure 7.

Contest competition does not always take the form of open combat. Threats and mutual avoidance of animals serve the same purpose in many species. And in the scramble forms of competition we encounter an array of complex and subtle phenomena that are often uncovered only by the closest study. Flies provide some of the purest examples of scramble competition. An analysis of populations of the fruit fly *Drosophila melanogaster* revealed that when food is short the fly larvae most likely to survive are the ones that 1, feed most rapidly; 2, are best adapted to the particular medium at hand; 3, have the greatest weight at the

DENSITY DEPENDENT EFFECTS in four animal populations. As population density increases, so do parasitism (top left), mortality due to food shortage (top right), and emigration (bottom left). Birth rate decreases (bottom right). Each species has its own set of density dependent controls. **6**

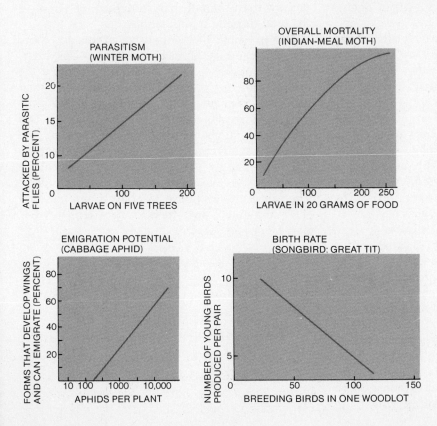

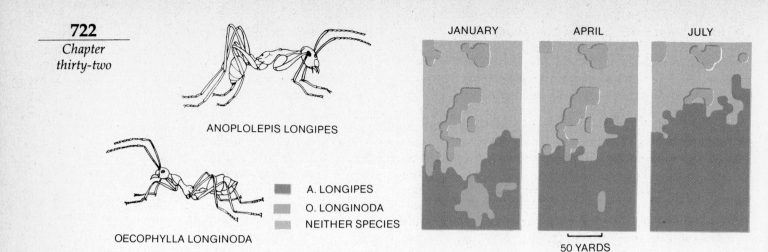

ANOPLOLEPIS LONGIPES

OECOPHYLLA LONGINODA

■ A. LONGIPES
■ O. LONGINODA
■ NEITHER SPECIES

JANUARY   APRIL   JULY

50 YARDS

7 **WARFARE AMONG ANTS is one form of competition. Three "maps" show progress of territorial struggle between colonies of ants on a coconut plantation in Tanzania. Here Anoplolepis replaces Oecophylla, but the reverse often occurs where soil is less sandy and vegetation is thicker.**

beginning of the competition; and 4, are most resistant to the mechanical and chemical changes due to crowding. Combat and other forms of aggression have nothing to do with the interactions of the larvae, but severe competition occurs nonetheless.

What keeps species from completely eliminating their less successful competitors? Why haven't the *Chthamalus* barnacles and one or the other of the ant species mentioned earlier gone extinct? The answer is that if a pair of species are too similar to each other in their requirements, one of them does indeed become extinct. This generalization is usually referred to as Gause's principle, or the PRINCIPLE OF COMPETITIVE EXCLUSION. It can be stated in broad but concise terms as follows: No two species that are ecologically identical can long coexist. The idea of ecological identity and difference can be more easily grasped by considering the concept of the niche. In Figure 8 we have presented a simplified version of the way ecologists view the niches of species. Each species, or more precisely each local population, has a temperature range in which it can successfully live and reproduce itself. It also has an array of kinds of food on which it can subsist. In the case of plants, each population has a certain list of required nutrients it must obtain from the soil, in addition to the basic requirement of radiant solar energy. Furthermore, there exists a particular range of humidity within which the

population can succeed. So far we have listed three dimensions of the niche. Figure 8 shows that up to three can be conveniently represented by a simple graph. But of course in nature there exist many more than three dimensions. One must add, for example, the time of day or year in which the species is active, the major habitat in which it lives, the place in the habitat where it lives, and so on. These additional components cannot be added to this particular graph, but it should be clear how the biologist analyzes them, component by component, with the aim of characterizing much or all of the total niche of a population.

The principle of competitive exclusion, then, states that unless the niches of two species differ, they cannot coexist. The implication is that if the two species are so genetically similar that their niches are the same, they cannot occupy the same geographical range. It is also true that the two species can be genetically quite different but find themselves forced to live in the same place and do the same things in a particular environment. In this case, too, one species will replace the other. The pair of species can coexist only if new pieces of the environment are added to the extent that one part favors one species and the other part favors

---

If a pair of species are too similar to each other in their requirements — that is, if they occupy identical ecological niches —one of them is doomed to become extinct.

the second species. An example of this aspect of the principle of competitive exclusion is provided in Figure 9.

The concept of the niche provides only a first, crude understanding of competitive exclusion. There is more to the story than just the tolerance limits of the species. The flour beetle example in Figure 9 illustrates that there is also a connection between competition and logistic population growth. In the experiment the two species are permitted to increase upward from low population levels. Each tends to grow along the familiar lines of the S-shaped logistic curve, finally reaching the carrying capacity of the environment. Competitive exclusion occurs if one species produces enough individuals to prevent the population of the other

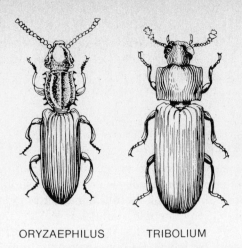

ORYZAEPHILUS   TRIBOLIUM

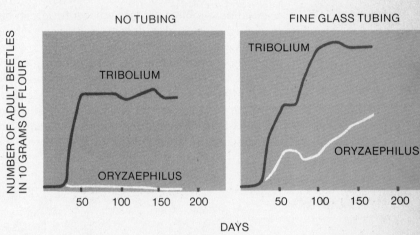

COMPETITIVE EXCLUSION in two species of flour beetle. In pure flour Tribolium defeats Oryzaephilus (left), but when fine glass tubing is added to the flour—creating tiny shelters to which Oryzaephilus can escape—the two species can coexist, as shown by graph at right.

9

from increasing. By adding a new piece to the environment, in this case the glass tubing that favors the *Oryzaephilus*, the losing species can sometimes be taken out of the reach of its competitor. Another way of accomplishing the same thing is for the losing species to change its niche through evolution. The environment remains the same, but the losing species has now been able to take itself far enough from the reach of its competitor to permit it to coexist. This is one form of the evolutionary process of character displacement, described in Chapter 30. Figure 10 presents an imaginary example to illustrate the relationship, as ecologists see it at the present time, between logistic growth, specialization into different niches, and competitive exclusion. Here the dimension of the niche that has been diversified is the place where the moths live

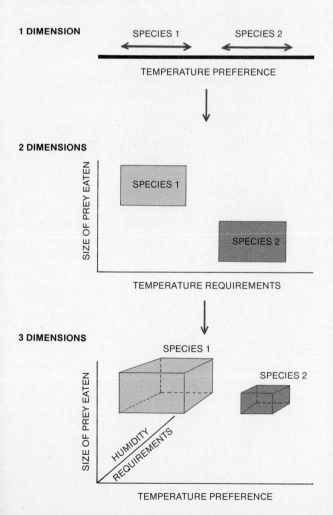

8 **ECOLOGICAL NICHE can be represented by a one-, two- or three-dimensional graph. Niche may consist of more than three variables, and variables may differ from the three plotted here.**

and breed. We could have just as easily made it the time in which they live and breed, or the part of the forest they favor, or many other ecological properties, utilized singly or in combination.

EMIGRATION, MIGRATION, AND POPULATION CYCLES

One of the most effective devices operating in the control of population size is emigration, the simple departure of individuals from the mother population. Emigration of people to the New World, and later to Australia and Africa, was of course a principal means by which Europe abated its human population problem for several centuries. When the emigration is also density dependent, it can serve as one of the most efficient and finely tuned regulators of population size. The behavior of animals prone to density dependent emigration is typically different from that of other members of the population. They are more restless, and they tend to travel long distances in a single direction. The trip is evidently designed not to move them from

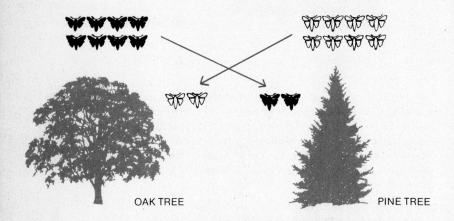

OAK TREE   PINE TREE

10 COMPETING SPECIES COEXIST by changing their niches through evolution. In diagram above, black moth has specialized for life on oak tree, and white moth for life on pine. Each species can attempt to invade trees of the other, but carrying capacity of "home" trees is too small to support a horde of invaders large enough to compete successfully with the rival species on rival's home trees.

one particular spot to another, but only to transport them away from the crowded mother population. In short, emigration is in many cases not merely an accidental departure from the range of the population but a distinctive behavior pattern designed to disperse the population. The indi-

vidual organism that emigrates is gambling that somewhere a habitat will be found that is more favorable to life than the crowded one left behind. The gamble pays off frequently enough to make emigration a favored behavioral trait in natural selection.

Animals prone to density-dependent emigration are more restless, and they tend to travel long distances in one direction. The trip is not intended to move them from one spot to another, but to transport them away from the crowded mother population.

The most dramatic case of emigration behavior known is that displayed by the plague locusts. From time to time immense swarms of grasshoppers, containing millions of individuals, materialize in remote, arid parts of the world and sweep over nearby grassland and farm country. They consume vegetation while on the move, and they are capable of destroying crops over thousands of acres. One of the plagues of ancient Egypt was the grasshoppers, or locusts as they are called when in the migratory phase. Even today locusts are a principal economic problem throughout the drier parts of Africa and the middle East. The plague locust phenomenon occurs in many species of grasshoppers. It always originates in peaceable populations that permanently occupy certain restricted areas and in this respect do not differ from the populations of other kinds of grasshoppers. But as crowding develops in the populations of this solitary phase of the plague species, many of the individuals begin to undergo a transformation. Slowly, over a period of three generations, they develop longer wings, a more slender body shape, and a darker color. The fully transformed gregarious phase is more active than its solitary parents and grandparents. The individuals begin to band together and to fly beyond the usual feeding areas. If enough members of the gregarious phase assemble in this fashion, they become a plague swarm. Once on the move, the swarms travel impressive distances *(Figure 11)*. They have

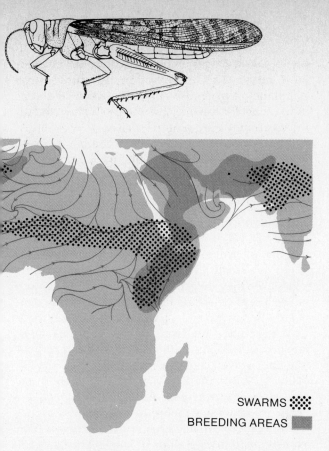

SWARMS ⠿
BREEDING AREAS ▨

PLAGUES OF LOCUSTS occur when populations in home breeding areas (dark tint) become too dense. Migratory swarms fly far from home and tend to descend in areas where windflow patterns converge. Arrows indicate direction of wind. Shown here is desert locust, one of many plague species.

been observed to cross from Somalia, on the northeast corner of Africa, to the island of Socotra in the Arabian Sea, a distance of 200 miles. Individual locusts from Africa have even landed on the Azores, a group of mid-Atlantic islands 900 miles from the nearest point of land in Africa.

Many other examples of density dependent movement exist. Some aphid populations, for example, begin to produce winged individuals in larger numbers as the populations become crowded. These migratory forms then fly away to start new populations. In other animal species, the migratory phase is automatic and not controlled by population density. Many winged insect species pass through a brief period in the life cycle, usually shortly after the emergence of the adult, in which they undertake determined migratory movement. Emigration, whether density dependent or not, always has the same two results: lessening of the population pressure in the home area, and an increase in the likelihood of starting new populations

elsewhere. In the long term, benefits are bestowed on both the individuals who remain behind and the pioneers who succeed in finding new places to live. The phenomenon, however, is not to be confused with migration in the strict sense. As practiced by birds, migration is the seasonal, back-and-forth flight between two favorable living areas. The migratory bird species come to the north temperate zones in the spring, when the exceptional bursts of vegetation and insect life make the rearing of their young especially favorable. They remain there through the summer and fall until both the weather and the food supply begin to deteriorate. Then they migrate southward to warmer climates to spend the winter. The long lives and great flying power of individual birds have made possible the evolution of this very special kind of oriented movement.

Another peculiar phenomenon associated with the regulation of population growth is the POPULATION CYCLE. As exemplified in the famous case of the lemming *(Figure 12)*, the population cycle is a more or less regular oscillation in the population size. It differs only in degree from the irregular fluctuation in numbers displayed by most animal and plant populations. Biologists have always been fascinated by the great sweep and regularity of some of these cycles, which are sometimes accompanied, as in the lemming, by mass emigrations during the times of greatest population density. Efforts to explain the causes of population cycles have not met with success. No single explanation has been found to be satisfactory in all cases. According to one hypothesis concerning the brown lemming cy-

POPULATION CYCLE of the brown lemmings near Barrow, Alaska. Number of individuals oscillates widely through a cycle that lasts three or four years. Reason for cycle is unknown.   **12**

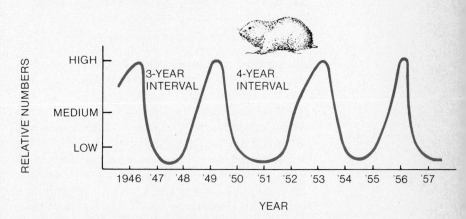

cle, the population "crash" occurs when the population becomes so dense that it consumes all of the arctic vegetation on which it depends for a living. This may occur at some places (for example, Barrow, Alaska) but not at others. According to current theory, other populations fluctuate cyclically either because of density dependent hormone changes of the kind described earlier for the mouse, or because of short-term genetic changes that occur in the populations as they increase in size. This is one of the many fields of ecology which is in an early and very active stage of investigation.

PREDATORS AND PREY

Predation is the act of consuming another organism. Predators are either herbivores (devourers of plants), carnivores (devourers of other animals), or omnivores (devourers of both). Predators and prey can exercise a reciprocal control of each other's population size in the following way. When the prey become too numerous, they are cropped back in a density dependent manner by the predators. When the predators become too numerous, they crop the prey down to a low level, which causes them to run out of food and to suffer a population decline of their own.

A simple and instructive example of the balance between predator and prey is that of the wolves and moose of Isle Royale. Isle Royale is a 210-square-mile island located in Lake Superior near the Canadian shore. It is kept in its primitive condition by the U.S. National Park Service. Early in this century moose colonized Isle Royale, probably by walking over the 15-mile stretch of ice from Canada during the winter. In the absence of timber wolves and other predators, the moose increased rapidly. By the mid 1930's the herd had increased to between 1,000 and 3,000 animals. At this point the moose population far exceeded the carrying capacity of the island for moose, and the low vegetation on which they depend for existence was soon consumed. A population crash ensued, reducing the herd to well below the carrying capacity. As the vegetation grew back, the herd expanded rapidly again—and crashed again in the late 1940's. In 1949 timber wolves crossed the ice from Canada to Isle Royale. Their appearance had a marked and beneficial effect on the Isle Royale environment. The wolves reduced the number of moose to between 600 and 1,000, somewhat below the carrying capacity. The browse vegetation has returned in abundance, and the moose now have

plenty to eat. Their numbers are controlled by predation rather than by starvation. The timber wolf population remained steady at between 20 and 25 individuals for a while, then began a slow and almost certainly temporary increase.

What controls the number of timber wolves? Why don't they just keep eating moose until none of these prey are left, then suffer a population crash of their own? The answer is very simple. The wolves catch all of the moose they possibly can, and their effort keeps the moose population down to 600 to 1,000 individuals. It is very hard work to trap and kill a moose. The wolves travel an average of 15 to 20 miles a day during the winter. Whenever they detect a moose they try to capture it. Most of their efforts fail. During one study conducted by L. David Mech the Isle Royale wolfpack was observed to hunt various moose on 131 separate occasions. Fifty-four of the moose escaped before the wolves could even get close. Of the remaining 77 that the wolves were able to confront, only six were overcome. All this effort yielded a "crop" of about one moose every three days. That was enough to provide each of the wolves with an average of 10 to 13 pounds of moose per day. Apparently the wolves simply cannot increase the yield beyond this point, and their number has consequently stabilized. The moose, by unwillingly supplying the wolves with one of their members about every three days, have stabilized their own population. The predator-prey system is in balance. As a curious side effect, the moose herd is kept in good physical condition, since the wolves catch mostly the very young, the old, and the sickly individuals. And, finally, because the moose population is not permitted to increase to excessive levels, the vegetation on which they feed remains healthy.

The Isle Royale example is one of the best documented cases of a balanced predator-prey system. It is generally true that herbivorous animals separated from their predators tend to increase to excessive levels and to strip the landscape of their food supply, often with disastrous effects on the environment. Insect pests are usually nothing more than species that have been introduced into new countries or new environments without their predators. Unshackled from these density dependent controls, the populations grow explosively. They can then destroy the crops, or the shade trees, or whatever plants they are adapted to feed upon. Entomologists are sometimes able to solve the problem by means of biological control, which is usually the introduction of predators or disease organisms that specialize on the herbivore back in its native country. One of the most successful applications of the technique occurred during attempts to combat the cottony-cushion scale, *Icerya purchasi*, a pest of citrus trees. Scale insects are relatives of aphids. Like aphids, they feed by sucking the sap from plants with their long, tubular mouth-parts. The cottony-cushion scale, which derives its common name from the tufts of white waxy material that cover its back, is a native of Australia. In the late 1800's the species was accidentally introduced into California on small citrus trees. Its populations increased rapidly, to the point that it singlehandedly threatened the California citrus industry. In a pioneering experiment in biological control, an entomologist, A. Koebele, traveled to Australia in search of some of the natural enemies of the cottony-cushion scale. He discovered a lady-bird beetle, a species of the genus *Rodalia*, which serves as an especially effective predator. When introduced into California, the beetle multiplied rapidly and brought the cottony-cushion scale down to tolerable levels. The scale insect is not extinct. It still increases to pest levels in isolated places where the *Rodalia* has not reached, or where indiscriminate use of chemical insecticides has killed off the *Rodalia*. But through most of California, both the cottony-cushion scale and the *Rodalia* ladybird beetles exist at relatively low population densities. The citrus trees, upon which the scale insects still depend for food, remain in good condition. The close parallels between this system and the somewhat more natural Isle Royale system should be readily apparent.

Biological control has not been limited to insect pests. When the European rabbit increased to excessive numbers following its introduction into Australia, it was brought under partial control by the deliberate addition of the virus that causes myxomatosis, a usually fatal disease of rabbits. Plants often become serious pests when introduced into new countries without their natural herbivores. An especially dramatic example is that of the prickly pear *(Opuntia)*, a cactus introduced into Australia as an ornamental plant sometime prior to 1839. Some of the plants escaped from cultivation and spread rapidly to cover over 60 million acres in Queensland and the warmer parts of New South Wales. Much of this land was solidly covered by groves of the cactus. Biologists then found a moth, *Cactoblastis cactorum*, that is an effective natural enemy of such plants in South America. When freed in Australia, *Cactoblastis* multiplied and spread rapidly. It destroyed the cactus over much of its range by literally eating it up. Today both the cactus and the *Cactoblastis* coexist at relatively low densities. In other words, the herbivore-plant system has attained the same kind of balance as the animal predator-prey systems described earlier.

Two important principles of applied ecology have emerged from experience in pest control. The first is that species of animals and plants cannot be introduced to new countries without a real danger of their swiftly multiplying to pest levels. The fate of individual immigrant populations is unpredictable. Often they become extinct or remain as scarce, inconspicuous elements of the host fauna and flora. But in too large a percentage of cases, species that were minor elements in their home country rise to pest levels in the new countries. This is why most nations in the world today maintain strong quarantine regulations on the import of exotic plants and animals. The second principle to be learned is that the most beneficial way to control pests is to add predators to the system. When it works, the procedure is invariably better in the long run than the use of poisons or mechanical means. The predators normally affect only the pest species and do not harm the rest of the environment; and being self-sustaining, they are far less expensive.

SYMBIOSIS

Symbiosis translated literally from the Greek means life together. Biologists define symbiosis as the living together of two species in a prolonged and intimate ecological relationship, ordinarily involving frequent or permanent bodily contact. Symbiosis can take one or the other of the following three basic forms: in PARASITISM one species is benefited at the expense of the other; in COMMENSALISM one species is benefited while the other species is neither benefited nor harmed; in

Generally the deadliest of the pathogens are those that are most poorly adapted to the host species. The ideally adapted parasite is one that can flourish without reducing its host's ability to grow and reproduce.

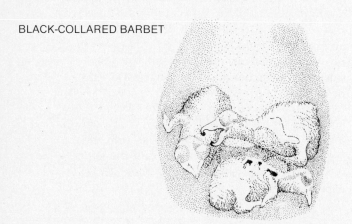

LESSER HONEYGUIDE

BLACK-COLLARED BARBET

list of such parasites that live in human beings, although perhaps only by reading and not by experience. Into this group fall the many kinds of viruses (measles, German measles, mumps, smallpox), rickettsiae (typhus, Rocky Mountain spotted fever), *Corynebacterium diptheriae* (a bacterium causing diphtheria), *Pasteurella pestis* (a bacterium causing the plague), *Vibrio cholerae* (a bacterium causing cholera), several species of *Plasmodium* (protozoans causing malaria), *Ancylostoma duodenale* (hookworm), *Diphyllobothrium latum* (tapeworm), and so forth. A large part of the practice of medicine is simply applied parasitology. Few persons realize that for every one of the parasitic species that cause serious disease in man and other organisms, there are many others that give their hosts little or no trouble. It is generally true that the deadliest of the pathogens are the ones that are the most poorly adapted to the species affected. Frequently, the organism that succumbs is a secondary host that has picked up the disease from the primary host species by accident. The ideally adapted parasite is one that can

13 BROOD PARASITISM by the Lesser Honeyguide of Africa. Female lays her eggs in nest of host such as barbet. Newly hatched honeyguide uses its hooked beak to kill the host's young, as shown in drawing at bottom. It will then take the place of its victims and be raised by their parents.

MUTUALISM both species are benefited by the relationship.

Because of the complexity of the subject, the study of symbioses is virtually a science unto itself. Symbioses of one kind or another occur in all the major groups of organisms, from protists to mammals. They are extraordinarily diverse in kind, and the more advanced types employ bizarre adaptations that completely transform the life cycles and even the physical form of the participants. Furthermore, the life cycles are the most complex found in nature.

The most familiar parasites are the pathogens, the organisms that cause disease in their hosts. You are probably already acquainted with a long

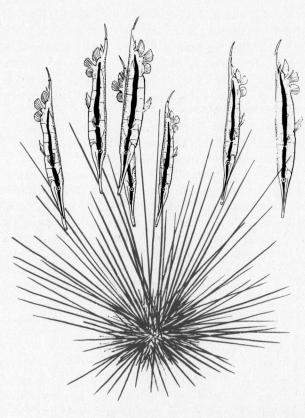

SYMBIOSIS between tropical fish Aeoliscus strigatus and a sea urchin is an example of commensalism. Fish is protected by sheltering spines without affecting sea urchin one way or the other.

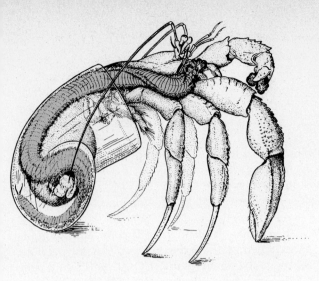

5 NEAR-PARASITIC SYMBIOSIS is involved in the association between a hermit crab and the polychaete worm Nereidepas fucata. Drawing depicts a crab forced to live in an artificial glass shell instead of usual mollusk shell. Worm (color) has moved in with the crab and steals some of its food.

flourish without reducing its host's ability to grow and reproduce.

Most parasites make their living by consuming part of their host's tissue. A somewhat antic but still accurate characterization of such a parasite is a predator that consumes its prey in units of less than one. But there are other ways of living at the expense of another species. One of the most bizarre is BROOD PARASITISM, substitution of the parasite's young in place of the host's young. Providing the substitution goes undetected by the host adults, the parasitic brood then eat the provisions intended for the brood it replaced. The phenomenon has a fairly widespread occurrence in the birds and certain groups of insects, particularly the wasps. Among the more familiar parasitic birds are the cowbirds and the cuckoos. A less familiar but striking example from the honey-guides of Africa is illustrated in Figure 13.

Perhaps the most familiar examples of organisms that live in commensalism, or neutral symbiosis, are the bryophytes, mosses, bromeliads, orchids, and other plants that grow on the trunks and branches of trees. They flourish at no visible expense to the trees because they occupy the surface of what is in effect protective tissue, which is not thwarted in its primary role by the relatively lightweight symbionts. Commensalism is widespread throughout the animal kingdom, and is especially common among marine invertebrates.

The host organisms are typically slow-moving or sessile, housed in structures, such as shells or burrows, which can be readily shared by the smaller commensal species. One excellent example involving a fish and an echinoderm is shown in Figure 14. In a second example involving two invertebrates *(Figure 15)*, we see the fine line that separates commensalism from parasitism. The polychaete worm lives by taking some of the food collected by the hermit crab. In this sense it is mildly parasitic, despite the fact that it does not harm its host by a direct attack in the manner usually employed by parasites. In any given case, it is very difficult to say that a species is a pure commensal, because this is equivalent to concluding that it has no effect on the host organisms whatever.

Mutualism, like parasitism and commensalism, occurs widely through most of the principal plant and animal groups and includes an astonishing diversity of physiological and behavioral adaptations. Some of the most advanced, and ecologically most important, examples occur among the plants. Nitrogen-fixing bacteria of the genus *Rhizobium* live in special nodules in the roots of legumes. In exchange for protection and a constant environment, the bacteria provide the legumes with substantial amounts of nitrates which aid in their growth *(see Chapter 14)*. Corals are animals that gain much of their energy from single-celled algae, called zooxanthellae, which are sheltered within their tissues. In exchange, they provide the algae with nutrients from crustaceans and other small animals which they capture with their tentacles. Lichens are perhaps the ultimate mutualistic symbionts. They are actually compound organisms consisting of highly modified fungi that harbor blue-green and green algae among their hyphae. Together the two components form a compact and highly efficient unit. In general, the fungus absorbs water and nutrients and forms most of the supporting structure, while the alga conducts the photosynthesis. This improbable combination has proven especially efficient at occupying habitats that are inhospitable to most other plants —rock surfaces, the bark of trees, and bare, hard ground. The lichens, including the so-called reindeer moss, are among the dominant plants of the treeless arctic tundra, and they are common among the pioneer organisms that colonize newly exposed rock and soil *(see Chapter 26)*.

A radically different form of mutualism is illustrated in Figure 16. Many kinds of ants depend partly or wholly upon aphids and scale insects for their food supply. They "milk" these inoffensive little creatures by stroking them with their fore legs and antennae. The "cattle" respond by excreting droplets of honeydew, which is simply partly digested plant sap that has passed all the way through their guts. In return for this sugar-rich food, the ants protect their charges from parasitic wasps, predatory beetles, and other natural enemies.

FLOWERS AND INSECTS

A very special kind of symbiosis exists between flowering plants and the butterflies, bees, and other insects that visit them. The flower is a signal to the insects upon which the plant depends to carry its pollen to other plants and thus achieve cross-fertilization. The shape of the flower is such that when the insect alights, it has no trouble finding the spot—usually the center of the blossom—where nectar and pollen are located. In exchange for receiving nectar, the insect inadvertently carries pollen from flower to flower. In this way the plant has adopted the insect as its sperm carrier.

The details of this symbiotic communication, as exemplified by the honeybee and the flowers it attends, make a marvelous story. Experiments have proven that bees can see color, but they see a visual spectrum different from ours. They can perceive ultraviolet, which man cannot, but they are far less sensitive than man to light at the opposite (red) end of the spectrum. When the human observer looks at a flower through a transformer that shifts the ultraviolet energies to the visible spectrum, he sees the flower for the first time the same way as the bee. The results can be striking. What previously had seemed to be plain white or yellow blossoms now are found to possess (in the eye of the bee) strongly contrasting patterns. These ultraviolet colors permit the bees to distinguish species of plants that are superficially the same to the human eye. The geometric patterns also provide geometric clues, called nectar guides, that direct the bee to the position in the flower where the nectar is located *(Figure 17)*. No example illustrates so well the difference between the sensory world of man and that of other organisms.

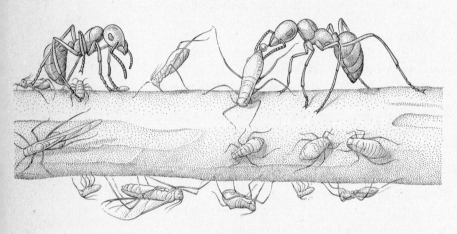

16 MUTUALISM between ants and aphids. When stroked by ant's antennae, aphid excretes sugar-rich "honeydew" (top center). In return, ants protect aphids from parasitic wasps and other enemies.

17 BEE'S-EYE VIEW of the marsh marigold (Caltha palustris) at left is remarkably different from view seen by human eye (right) because eye of bee is sensitive to ultraviolet light. Ultraviolet photograph at left reveals pattern of nectar guides that direct bee to location of food, and incidentally ensure that she will pollinate flower. To humans, flower appears to be a bright, even yellow.

ANTI-PREDATION

Organisms possess equally sophisticated devices to avoid predators. Various plants, for example, are protected by formidable armaments: cacti and roses have spines, magnolias have leathery coats

18 **BIRD IS STARTLED** when the sphinx moth flashes the fake eyespots on its wings. Such eyespots are found in several species of moth, and serve to discourage predators such as birds.

on their leaves, grasses are stiffened by harsh silica inclusions, and so on. Some tropical plants, such as the bull's horn acacias of Latin America, live in close symbiosis with swarms of stinging ants that repel both insects and browsing mammals. In addition various plants store an array of poisons, including nicotine and other alkaloids, fragrant terpenoid substances such as limonene, and others. A few plant species have even gone so far as to produce chemicals closely similar to the growth hormones of insects. When insects swallow an excess of these hormone mimics their development is altered in ways that eventually cripple or kill them.

Various animals avoid detection through the use of CRYPTIC APPEARANCE: they resemble pieces of the environment on which they live. Some species employ only shading or color. The flounder, for example, alters its color to blend in with the patch of ocean floor on which it happens to be resting at the moment. Certain insects are also shaped to resemble inedible or neutral objects such as twigs and leaves. These animals typically hold their bodies in positions that increase the resemblance still further. In order to observe the phenomenon of cryptic appearance yourself, all you have to do is closely inspect a few tree trunks in warm weather. You will soon encounter a variety of moths, barklice, and other insects whose color closely resembles that of the bark on which they are resting. Some will elude your attention until your activity forces them to move. The great majority of examples of cryptic animals are in the act of avoiding predators. Conversely some spiders, predatory bugs and mantids resemble bark or flowers and use this deception to ambush unwary prey.

WARNING APPEARANCE is one of the most common of all biological attributes. Some examples are familiar to the layman. The boldly striped body color of the hornet, the brilliant red-and-black bands of the deadly coral snake and the fake eyespots of certain moths are cases in point *(Figure 18)*. A good rule to remember when entering a strange habitat anywhere in the world is: leave the most conspicuous animals alone, especially if they are beautiful! Many of them are transmitting a warning to all would-be enemies. Their appearance says in effect, "I am poisonous to eat or venomous. I don't need to hide or to run because my potential enemies learn to avoid me in ways that they will remember."

Perhaps the next best thing to having unpleasant traits and warning signals to advertise them is to resemble some other species that does. In this situation, called BATESIAN MIMICRY, the predator is not able to distinguish the harmless mimic from the dangerous model and leaves both alone. The mimicry typically involves color, form, and behavior. The scarlet king snake, for example, is a

BATESIAN MIMICRY in butterflies. Monarch butterfly (top) is distasteful to birds, who learn to avoid it. Protection is thereby granted to the closely similar viceroy butterfly (bottom), although it is quite edible. **19**

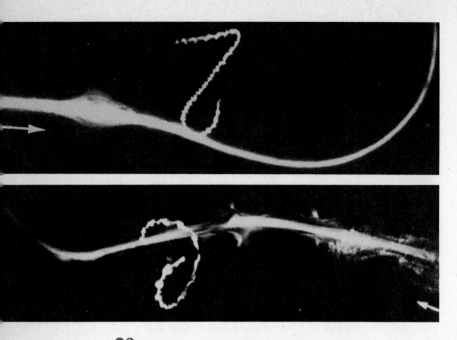

sonic cries of bats, so that those predators can be avoided in time *(Figure 20)*. Many insects and other kinds of animals possess elaborate glandular systems, some equipped with aiming devices, that are used to spray repellant substances at enemies. One of the most spectacular examples is the explosive "chemical gun" of the bombardier beetles, illustrated in Figure 21.

READINGS

S.M. HENRY, *ed., Symbiosis. Volumes I, II.* New York, Academic Press, 1966, 1967. The most thorough and authoritative review of all aspects of symbiosis. It is unfortunately somewhat detailed and advanced for the beginning student.

E.J. KORMONDY, *Concepts of Ecology,* Englewood Cliffs, NJ, Prentice-Hall, 1969. A brief but well-written exposition of most of the elementary concepts of population and community ecology, with strong emphasis on human applications.

E.P. ODUM, *Fundamentals of Ecology,* 3rd Edition,

20 **MOTH AVOIDS BAT in the streak photograph at top by going into a spiral dive when it detects the bat's sonar. Moth in bottom picture was less fortunate. Arrows indicate direction of bat's flight.**

nonvenomous form which expertly mimics the coral snake. A second case involving two well-known butterflies is shown in Figure 19. A second form of mimicry is MULLERIAN MIMICRY, in which two or more unpleasant species resemble each other. Each, then, serves as both a model and a mimic. They "team up" in a way that insures more frequent exposure and quicker learning on the part of the predator than would be the case if they were advertising their repellant qualities separately.

Some of the most precise acts of communication occur in the recognition and avoidance of predators. Here are several of the many remarkable examples that have been discovered in recent years: Sea anemones are very sedentary, flower-shaped animals that spend almost all of their lives attached to rocks on the sea bottom. However, when they detect the presence of a starfish or one of a very few other natural enemies, they pull loose and swim away—a very startling sight to the human observer. The little wheel animalcule *Brachionus* serves as prey for another wheel animalcule, *Asplancha.* When young *Brachionus* are reared in water containing chemical substances from *Asplancha,* they grow long spines that make them immune from attack. The hearing sense of moths is especially adapted to listen for the ultra-

BOMBARDIER BEETLE (genus Brachinus) is the insect counterpart of the skunk. Photograph shows a beetle spraying forward, in response to forceps pinching its front leg. Beetle ejects spray at temperature of boiling water, and can fire more than 20 times in succession, making it almost invulnerable to attack. **21**

Philadelphia, W.B. Saunders Company, 1971. A very good general introductory textbook of ecology, clearly written and rich in ideas and examples from population ecology.

R.E. RICKLEFS, *Ecology*, Newton, MA, Chiron Press, 1973. Probably the most comprehensive and best organized of all general ecology texts; it is especially strong on evolutionary aspects.

E.O. WILSON AND W.H. BOSSERT, *A Primer of Population Biology*, Sunderland, MA, Sinauer Associates, 1971. A self-teaching textbook that covers the basic principles of population ecology.

It is interesting to contemplate an entangled bank, clothed with many plants of many kinds, with birds singing on the bushes, with various insects flitting about, and with worms crawling through the damp earth, and to reflect that these elaborately constructed forms, so different from each other, and dependent on each other in so complex a manner, have all been produced by laws acting around us.

—Charles Darwin

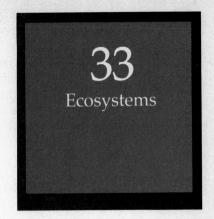

Each organism on Earth is a member of an ecosystem, a unit that consists of the other organisms that affect it (the community of species, comprising the biotic environment) plus the nonliving matter and radiant energy that make up its physical environment. Because these numerous elements act on each other in a reciprocal manner, the ecosystem is an almost endlessly ramifying and complicated unit. In this respect it accurately reflects the state of ecology, the science that studies it. Ecology in fact differs from the other disciplines of biology in that it does not treat the organism simply as an isolated unit, with a narrowly defined input of radiant energy, nutrients, and stimuli. Ecology alone attempts to account for the effects the organism has on the remainder of the ecosystem, and vice versa. The movements of the organism, its discharge of waste products, and ultimately its contribution of material through the death and decay of its own body, all change the environment to some degree.

It is easy to think of a pond or an island in the middle of the ocean as a single ecosystem. But it is often equally useful to deal with far less discrete units: a patch of pine forest, for example, or the grassy border of a highway. The limits drawn around ecosystems are always arbitrary and selected for convenience only, because ecosystems are never entirely closed. All patches of nature are linked tightly to the surrounding environment. Even a pond receives its water from precipitation plus the drainage from nearby land. Much of the organic matter in a pond is leached from the surrounding soil or is conducted to it in the bodies of immigrating organisms.

LAND BIOMES

The most superficial classification of ecosystems is into biomes and biome types. The BIOMES are the great communities of species that occupy the major patches of the environment, assemblages that have a strikingly different physical appearance from each other. Thus the grassland of the western United States is one biome, and the nearby desert is a second biome. A grassland and a desert also exist in southern South America, but the species that comprise these biomes are almost wholly different from their counterparts in the United States. We speak of the biomes that resemble one another in physical appearance but differ in species composition as comprising a worldwide BIOME TYPE. All of the grasslands of the world are said to comprise the grassland biome type.

Ecologists often differ in their classifications of the major patches of the environment. The biomes, like the subspecies and biogeographic regions described in earlier chapters, are strictly defined by the elements put into them. The borders of the biome shift as we add or subtract species that are utilized as the diagnostic elements. The great boreal forests, for example, are usually designated as one of the biomes of North America. They consist of dense stands of coniferous (cone-bearing) trees such as spruce, fir, and larch. The boreal forest biome is sometimes referred to lightly as the "spruce-moose biome" by ecologists, who thus follow the practice of including one of the principal animal inhabitants in naming the biome. The ecologists are well aware that parts of the coniferous forest lack the moose, other parts lack the

TROPICAL CORAL REEF is the most complex marine environment and contains the richest community of organisms. Photograph on opposite page depicts reef covering a vertical underwater cliff off Sharmesh Sheikh in the Red Sea. Among the many kinds of living coral visible in the picture are two species of staghorn coral that form the large mass at bottom. Projecting upward from it at left center is a lace-like fire coral.

735

spruce, while still other parts have neither spruce nor moose. The point is that the biomes seldom exist as sharply defined patches. They have broad borders, and the species that comprise them have weakly correlated geographical ranges.

In spite of this basic limitation, the biome concept is very useful for pointing out two important generalizations about life on this earth. The first is that the physical environment is all-important in determining the gross appearance of the organisms that exist in an ecosystem, especially the height, the profile, and the arrangement of the vegetation. The second generalization is a corollary of the first: given a particular physical environment in different parts of the world, the species of plants and animals will adapt to it by acquiring much the same outward body forms, regardless of their phylogenetic origins. The last statement should seem vaguely familiar. The classification of biome types is in fact simply another way of describing convergent evolution (*see Chapter 31*).

In Figure 1 are represented four of the biome types of the world. A simplified view of each of the plant formations is given, together with a characteristic "climograph" that documents the year-round changes in temperature and precipitation. Because temperature and precipitation are among the most crucial physical factors in determining the structure of the vegetation, in a large percentage of cases it is possible to predict the biome type at a given locality from the information provided in such climographs alone. Other physical factors are important, however. The structure and chemistry of the soil can be equally vital. There are, for example, certain trace elements known to be essential for the full development of plants. These include boron, chlorine, cobalt, copper, iron, manganese, molybdenum, sodium, vanadium, and zinc. When one or more of these substances is present in too low concentration, land which according to the temperature-precipitation climograph should carry a forest will instead be covered with shrubby vegetation or be grassland bearing a few scattered trees. Where grassland was indicated, there may exist sparse vegetation more closely resembling that of a desert. Thus a single factor, such as temperature, humidity, or the concentration of trace elements, is capable of determining the presence or absence of species and thus the character of the entire biome. The generalization is sometimes expressed as the Law of the Minimum: the factor that is most deficient is the one that determines the presence and absence of species. It does not matter, for example, how right temperature and sunlight are in a given locality, or how rich the nutrients and trace elements in the soil—if the precipitation is very low, the result will still be a desert. Only a careful study of all of the climatic and soil characteristics of a region can begin to reveal the reasons for the existence of cer-

1 FOUR TYPES OF BIOME are illustrated on these two pages. Accompanying climographs indicate year-round changes in temperature and precipitation. Numbers on climographs refer to months, with January as month number 1, February number 2, and so on. Place names on climographs indicate stations where readings were taken. Drawings depict generalized biome types, not specific places.

TROPICAL RAIN FOREST

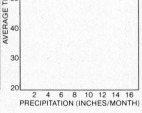

BARRO COLORADO, PANAMA

DECIDUOUS FOREST

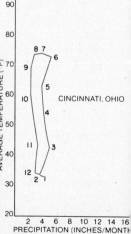

CINCINNATI, OHIO

tain plant formations, together with the animal species that depend on them.

Figure 2 depicts the distribution throughout the world of all of the principal biome types recognized in a recent classification. The tundra (from the Finnish *tunturi* —a treeless plain) is the cold treeless land that encircles the arctic and elsewhere occupies the highest mountain tops above the treeline. The vegetation of the tundra resembles grassland but is actually made up of a mixture of lichens, mosses, grasses, sedges, and low-growing willows and other shrubs. A permanent layer of frozen soil, the permafrost, lies from a few inches to a few feet beneath the surface. It prevents the roots of trees and other deep-growing plants from becoming established, and it slows the drainage of surface water. As a result the flat portions of the tundra are dotted with shallow lakes and bogs, and the soil between them is exceptionally wet.

The taiga (from a Siberian word for the coniferous forest) is the great boreal forest of the north temperate zone, and the biome type that lies next to the tundra. It is a humid, dense formation comprised of a relatively few species of coniferous (cone-bearing) trees. In many localities there are just two dominant trees: one fir and one spruce species. This kind of monotony in both structure and species composition is typical of the taiga.

At the opposite extreme is the tropical rain forest, the biome type known to the popular imagination as the teeming jungle. A lowland tropical rain forest in prime condition is actually a glorious sight. The highest trees reach 100 feet, with a few "emergents" soaring to 120 feet or more. Beneath the highest canopy are many lower layers of trees.

Vines and palm trees are abundant, and dense clusters of orchids and other commensalistic plants plaster many of the trunks and branches. The tight, multiple canopies of the trees allow little sunlight to reach the floor of the forest. As a result, few shrubs and herbaceous plants grow there, and it is relatively easy for a man to walk through a rain forest. Decomposition of fallen leaves and dead wood is so rapid that humus is thin, and it is even missing in spots on the forest floor. The diversity of life is the greatest found anywhere on Earth, on the land or in the sea. In a single square mile of the richest forests can be found hundreds of species of trees, and hundreds more of birds, reptiles and amphibians, butterflies, and ants, and dozens of species of mammals. The environment is divided up to an astonishing degree by specialists among the plants and animals. For example, there is an entire little flora of lichens, mosses, and other small plants that grow as commensals on the leaves of the forest trees. In this microvegetation are hidden a little fauna of insects, mites, nematode worms, and other small invertebrates. In the mountain rain forests of New Guinea a similar miniature flora and fauna flourishes on the backs of large plant-eating beetles. Such diversification is probably a result of the great age and stability of the moist tropical regions of the world. It also reflects the fact that through most of the geological past the tropical rain forest covered a much greater part of the world than it does today. Sixty million years ago, such forests grew as far north as the southern United States and the British

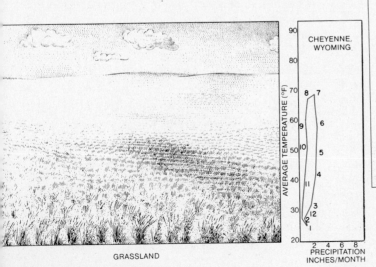

GRASSLAND

CHEYENNE, WYOMING

HOT DESERT

YUMA, ARIZONA

2 **DISTRIBUTION OF LAND BIOMES throughout the
world. Maps show all the major biome types
recognized in a recent classification. Highly productive
biomes such as tropical rain forest are rendered in color.
Comparatively unproductive biomes such as polar ice
cap, tundra and taiga are shown in light shades of gray.
Map reveals scarcity of prime agricultural land.**

of traveling 1° latitude northward at sea level. This
apparent telescoping of space is due, of course; to
the rapid decrease in temperature with altitude,
usually (but not always) accompanied by an increase in precipitation and humidity. Even tundra,
or biomes closely approaching it, can be found at
the tops of the highest mountains in the United
States.

isles. The tropical rain forests, and perhaps also
the rich adjacent savannas and thorn scrubs, have
served as the headquarters from which dominant
groups of vertebrates have repeatedly arisen and
from which they have then spread to other biome
types around the world *(see Chapter 31).*

The next time you have a chance to climb a high
mountain, whether on foot or by automobile, take
the opportunity to observe the changes in biome
types over relatively short distances as you progress upward. The Sierra of California, the high
Rockies, and the White Mountains of New Hamp-
shire are especially suitable for this purpose. You
will find, as suggested in Figure 3, that a rise of
about 200 feet in elevation is the rough equivalent

The diversity of life in a tropical rain
forest is the greatest found anywhere
on Earth. In a single square mile of
the richest forests can be found
hundreds of species of trees;
hundreds more of birds, reptiles,
amphibians, butterflies and ants; and
dozens of species of mammals.

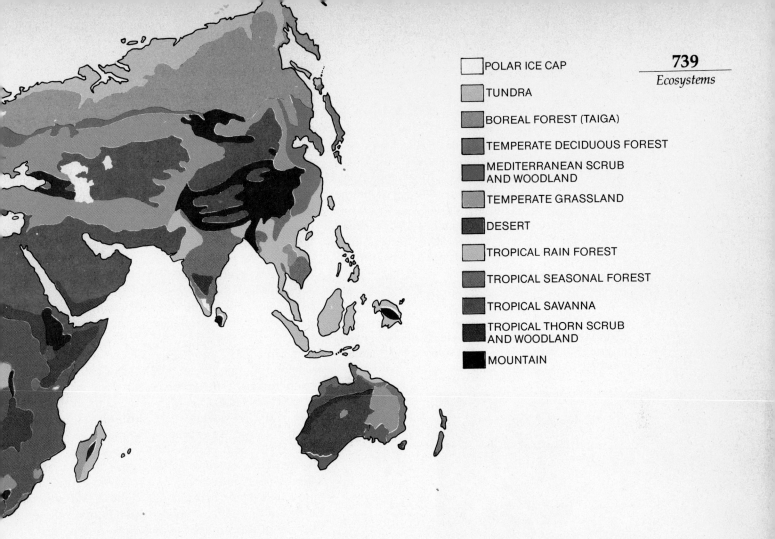

POLAR ICE CAP

TUNDRA

BOREAL FOREST (TAIGA)

TEMPERATE DECIDUOUS FOREST

MEDITERRANEAN SCRUB AND WOODLAND

TEMPERATE GRASSLAND

DESERT

TROPICAL RAIN FOREST

TROPICAL SEASONAL FOREST

TROPICAL SAVANNA

TROPICAL THORN SCRUB AND WOODLAND

MOUNTAIN

EFFECT OF ELEVATION AND LATITUDE on biome types. In North America mean temperature falls as one travels north or climbs upward from sea level. On mountainsides a rise of about 200 feet in elevation corresponds to a northward shift of roughly one degree of latitude at sea level.

3

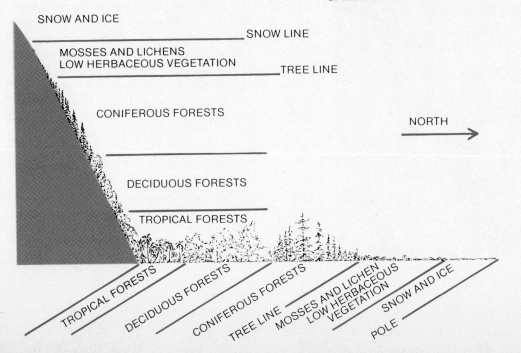

SNOW AND ICE

SNOW LINE

MOSSES AND LICHENS
LOW HERBACEOUS VEGETATION

TREE LINE

CONIFEROUS FORESTS

NORTH

DECIDUOUS FORESTS

TROPICAL FORESTS

TROPICAL FORESTS

DECIDUOUS FORESTS

CONIFEROUS FORESTS

TREE LINE

MOSSES AND LICHEN LOW HERBACEOUS VEGETATION

SNOW AND ICE

POLE

THE SEA

The oceans and seas cover 70.8 percent of the earth's surface, but, except for the shallow water of the margin, they are far less complex in structure and productivity than the land. Instead of forests and grasslands, the green plants of the open sea consist mostly of phytoplankton, the microscopic diatoms and other plant cells that drift in the lighted layers of the water *(Figure 4)*. Consequently, as one looks down into open ocean water signs of plant life are seldom seen. It is nevertheless there in abundance. The phyto-

4 PLANKTON consists of microscopic plants such as diatoms and dinoflagellates (phytoplankton, top) and tiny animals such as copepods (zooplankton, bottom). Photos are not reproduced to same scale.

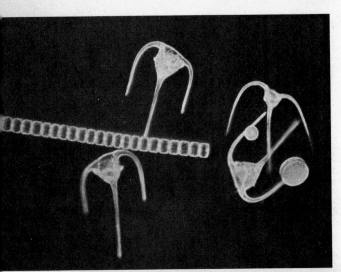

plankton are eaten by a great array of invertebrate animals, ranging in size from protozoans to large medusae. These are in turn consumed by other, larger invertebrates and vertebrates. Like the phytoplankton, some of the invertebrates also drift with the water currents and are therefore collectively called zooplankton. The drifting plants and animals together are simply called the PLANKTON. Other animals, the NEKTON, swim actively in search of prey. From the point of view of the biogeographer an ocean or sea can be thought of as a large basin, with shallow water lapping the rim and sunlight penetrating for a limited distance into the upper layers *(Figure 5)*.

Because of the vastness and uniformity of this basin, and the relatively rapid mixing of the organisms within it, ecologists do not attempt to divide the oceans into biome types like those on land. Instead, they distinguish the waters of the shallow continental shelves (the NERITIC PROVINCE) from those of the main part of the basin (the OCEANIC PROVINCE). Next, they distinguish the bottom of the ocean at all depths (the BENTHIC DIVISION) from the open water above the bottom (the PELAGIC DIVISION). Finally, they distinguish the upper, lighted zone of the water from the deep, lightless zone —often referred to as the ABYSSAL ZONE. The exact boundary between the lighted and abyssal zones at any given locality depends upon the intensity of the sunlight, which in turn depends on the latitude, and upon the turbidity of the water. Usually, no light penetrates below 600 meters. Beyond this depth the phytoplankton cannot grow. The animal life of the deep sea depends instead on the rain of dead and dying organisms that drift down from the lighted layers above.

The physical environment of the ocean is most diverse in the shallow water along its margin. Here there exists the greatest variation in temperature, salinity, light intensity, and water turbulence. Also, the ocean floor receives much more energy from sunlight than from any other source. And here is the maximum luxuriance of both plant and animal life. The shoreward portion of this marginal strip, a subdivision of the neritic province, is often referred to as the LITTORAL ZONE. It extends from the uppermost line of tidal wave action out into the water to the depths at which the water is no longer thoroughly stirred by tides and waves. Figure 6 depicts one of the typical communities that occur in the littoral zone. Here, in contrast to the deeper water, the plant life includes many larger forms of multicellular algae (the "seaweed") and a few higher plants such as the ubiquitous eel grass

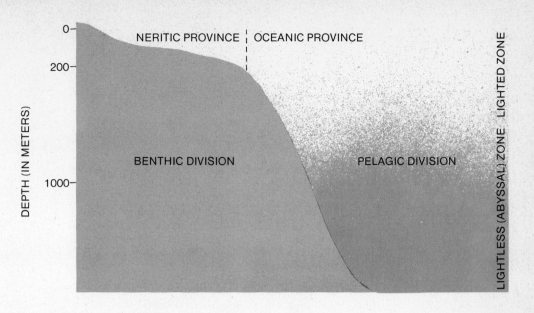

OCEANIC ECOSYSTEM is classified not into biomes but into provinces, divisions and zones. Special local communities such as coral reefs and intertidal zones are not included in diagram.

(Zostera). But most plant growth is out of sight, in the form of microscopic algae.

The most diverse —and most interesting —of all marine formations is the coral reef (*Figure 7*). There are many parallels between the coral reef and the tropical rain forest. The reefs are limited to shallow tropical waters to depths of about 60 meters. They are built up principally from the lime skeletons of millions of coral polyps. At any given moment the living generation of polyps forms a thin, growing layer on the tops of the massed skeletons of their ancestors. Some other organisms, such as lime-secreting species of plants (especially algae), single-celled foraminiferans, bryozoans, mollusks, and serpulid worms, add to the material and in a few localities are among the principal contributors. The individual reefs are relatively stable and often very old —hundreds or thousands of years old. They are also complex in form, like tropical rain forests. The reason is that many kinds of coral add their own distinctive skeletal forms —the staghorn corals, the organ corals, the brain corals, the fire corals, and many others. These supporting structures present other kinds of animals with a diversified landscape within which they can specialize and radiate. In the reefs of Port Galera, in the Philippine Islands, to take one typical locality, are found 111 species of corals, 70 of chaetopod worms, 10 of sipunculid worms, and between 200 and 250 species of crustaceans, as well as thousands of other species of brittle stars, crinoids, holothurians, mollusks, and other representatives of virtually every animal phylum.

SUCCESSION

The tropical rain forest and the coral reef do not spring full-blown from the ground or sea floor. They take possession of a patch of land or sea bottom by a long process called succession, in which empty space is first filled by a simple community of pioneer species, then gradually by more complex and bulky assemblages, and finally by the fully developed community that characterizes the local environment. The final community, or at least the most stable and longest-lived one, is referred to as the CLIMAX COMMUNITY. It is the end of the succession, and usually can be identified as belonging to one of the biome types or marine communities of the kind just described. An example of a succession is presented in Figure 8. Here the empty space is provided in the form of a new beach created by the waves of Lake Michigan. The first plants to take hold in the bare sand are grasses. As these bind the sand together and add humus, the next group of plants is able to take root. The transition flora, chiefly pine, oak, and cottonwood, add still more humus. Finally, after a few hundreds or thousands of years, the climax forest, comprised chiefly of beech and maple, takes over. Many other ways exist in which new space can be created and a succession set in motion. Destruction of existing

741

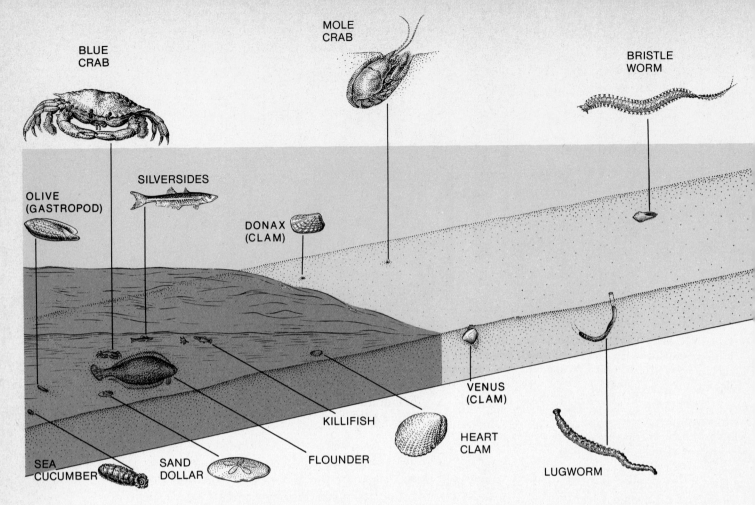

BLUE CRAB

MOLE CRAB

BRISTLE WORM

SILVERSIDES

OLIVE (GASTROPOD)

DONAX (CLAM)

VENUS (CLAM)

KILLIFISH

HEART CLAM

FLOUNDER

SEA CUCUMBER

SAND DOLLAR

LUGWORM

6 **BEACH COMMUNITY** along Atlantic coast harbors a rich abundance of life. Littoral zone extends from high tide line to deep still water. Distribution of species changes rapidly as one passes from area that lies permanently under water (dark color at left) to high tide level (light color). Dry land is at far right. Plant life, not shown, consists of microscopic algae, seaweed and eel grass.

ecosystems by fire, severe wind storms, flooding, and landslides are among the most frequent. Since these are the recurrent catastrophes of nature, they ensure that no spot on earth supports a climax community forever. At any given time, part of the natural environment is always at or near a climax condition, while other parts are passing through successions. This dynamic equilibrium on a grand scale ensures that species specialized for successional stages always have habitats available to them, and consequently they do not go extinct. Successions also occur within the lesser ecosystems. When a tree falls, its wood forms space for decomposition species, and as it decays and crumbles, these species succeed one another in a highly

regular pattern. When a rock or piece of metal or glass is dropped into the sea, its surface is colonized by a regular progression of bacteria, algae, coelenterates, bryozoans, polychaete worms, tunicates, barnacles, and other invertebrates. You can observe this particular succession yourself merely by placing some empty bottles in a tidal pool or some other quiet, convenient body of marine water and checking them periodically over a period of a few weeks or months.

Although many successions are predictable and regular, one should not think of them as the replacement of whole groups of species that come and go in unison. Figure 9 gives a picture of the process as it really appears. One can see that the bird species participating in this particular succession colonize the habitat and disappear from it in an irregular sequence, and their tenures in the habitat are broadly overlapping. The same is true of the plant species, the histories of which are not included in this diagram.

Why do successions occur? Why don't the first

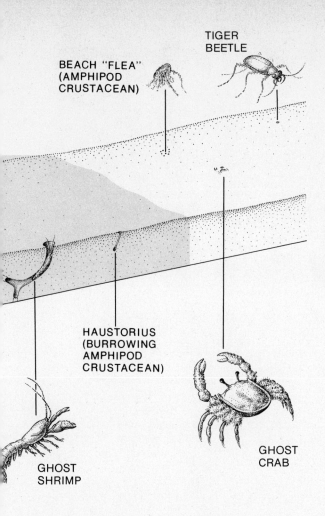

BEACH "FLEA"
(AMPHIPOD
CRUSTACEAN)

TIGER
BEETLE

HAUSTORIUS
(BURROWING
AMPHIPOD
CRUSTACEAN)

GHOST
SHRIMP

GHOST
CRAB

truders reach their full size, they cast too much shade for the *Eucalyptus* to reproduce themselves. The *Eucalyptus* finally die out as the rain forest takes over the land completely.

Not all successional stages prepare the way for their own decline and fall. In the eastern United States, mixed forests of pine and oak modify the soil in a way that makes it more favorable for the growth of their own seedlings than for those of their competitors. Succession in such cases occurs simply because slower-growing trees rise to dominance at a later time, and alteration of the environment therefore may have nothing to do with the replacement. In fact, ecologists recognize two broad classes of species involved in this automatic kind of succession. Opportunistic species, sometimes referred to more colorfully as fugitive species, are able to disperse widely, and they grow and breed rapidly. Many annual weeds belong to this category. The opportunistic species fill the pioneer and early successional positions. In the course of their evolution they have adopted the strategy of finding and utilizing empty space before other species pre-empt it. They are prepared to reproduce quickly and get out (send out their seeds or other dispersal stages) before other species take over. Stable species, on the other hand, specialize in competitive superiority. Forest trees of most kinds obviously belong to this category. They grow and disperse more slowly, but in their encounters with opportunistic species they are able to win the ground and consequently enjoy tenancy of it for longer periods of time.

FOOD WEBS AND ENERGY FLOW

Almost all of the energy utilized by living organisms comes from the sun. Even the fossil fuels—coal, oil and natural gas—upon which so much of the economy of modern civilization is based, represents reserves of captured solar energy that were locked up by organisms millions of years ago. The amount of life on the Earth has always been strictly limited by the quantities of solar energy available to it. Each year this planet receives 5×10^{24} gram calories of energy from the Sun, of which 4×10^{20} is captured in photosynthesis. This primary solar energy cannot be increased. It is utilized to construct and to run the elaborate chemical machinery of life. But energy is lost at each step as it is used by organisms. In time all of it is frittered away as heat, the form of energy that cannot be recovered.

The Earth is thus an open system with respect to

colonists simply take the space and hold it against subsequent intruders? There are two forces at work to overcome such pre-emption. In some instances the species alter the environment in a way to make it more favorable for other species than for themselves. For example, the first insects to attack a dead tree are specialized for boring into hard wood. As they crumble the wood, in collaboration with fungi and bacteria also specialized for this early stage of decay, the wood becomes less favorable to them and their offspring and more favorable to the insect species that prefer trees in an advanced state of decay. As the species specialized for each stage of succession eat themselves literally out of house and home, the ruins they leave behind provide an excellent habitat for the next set of species. In a parallel manner, plants sometimes foreclose future reproduction by the process of their own growth. In Australia, for example, the *Eucalyptus* trees of the open, sunny savannas provide shade in which young trees from the nearby rain forests can sprout and grow. When these in-

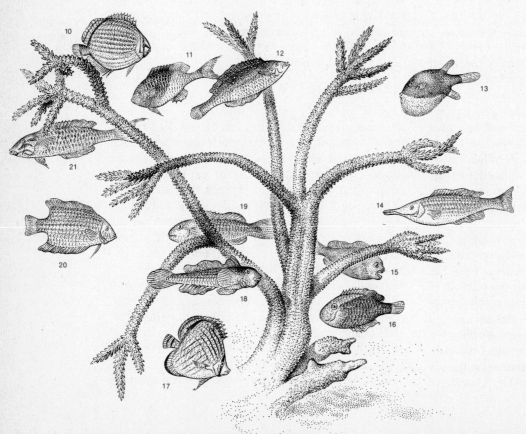

1 RHINECANTHUS ACULEATUS (TRIGGER FISH)
2 BALISTAPUS UNDULATUS (TRIGGER FISH)
3 AMANSES (FILE FISH)
4 CANTHIGASTER (PUFFER)
5 AROTHRON (PUFFER)
6 CIRRHITIDAE (HAWKFISH)
7 CHAETODON AURIGA (BUTTERFLY FISH)
8 HOLOCENTRUS SPINIFERUS (SQUIRREL FISH)
9 CEPHALOPHOLIS (GROUPER)
10 CHAETODON (BUTTERFLY FISH)
11 CHROMIS (BLUE CORAL FISH)
12 STETHOJULIS AXILLARIS (WRASSE)
13 AROTHRON HISPIDUS (PUFFER)
14 GOMPHOSUS (WRASSE)
15 CARACANTHUS
16 CENTROPYGE (BUTTERFLY FISH)
17 CHAETODON (BUTTERFLY FISH)
18 PARAGOBIODON (GOBY)
19 GOBIODON (GOBY)
20 EPIBULUS INSIDIATOR (WRASSE)
21 THALASSOMA (WRASSE)

CORAL-DWELLING FISH show a wide range of specialization in the positions they occupy while resting and feeding on the reef. Drawing depicts two kinds of branching coral found in the Marshall Islands of Micronesia, together with a few of the fish that inhabit them. Corals with long branches, such as the one at bottom, usually grow on parts of reef that are sheltered from rough waves.

energy. Energy passes quickly through the living organisms that cover the Earth. They cannot hold on to it. The best they can do is pass it on in ever diminishing quantities to their offspring and to the organisms that consume their bodies. In order to understand fully this basic principle of ecological energetics, it is necessary to consider first the phenomenon of the FOOD CHAIN. A food chain is most commonly a sequence of species of organisms that are related to each other as prey and predators: one species is eaten by another, which is eaten in turn by a third, and so on. Examples of typical food chains from terrestrial and marine environments are shown in Figures 10 and 11. Ecosystems are vastly more complex than these deliberately simplified diagrams suggest. They contain a great many more chains, and most of the chains are tied together by cross-connectives, as exemplified by the three chains leading to the Great Horned Owl. For this reason the complete ensemble of chains representing an entire ecosystem is often referred to as the FOOD WEB.

Each species forms a step, or link as it usually is called, in one or more food chains. The position located on the chain is referred to as the TROPHIC LEVEL. Thus the green plants, which are the producers for the entire community, comprise the first trophic level. The second trophic level is formed by the herbivores, which are the consumers of the green plants, the third trophic level by the carnivores, which eat the herbivores, the fourth trophic level by the secondary carnivores, which eat the carnivores, and so on. In almost all ecosystems there are top carnivores, one or more large, specialized animal species that browse on the animals in the lower trophic levels but are not ordinarily consumed by predators themselves. The larger whales enjoy this status, as do lions, wolves, and man, the most voracious of all the top carnivores. In addition to the producer-to-carnivore chains there are parasite chains, in which small animals feed on their larger animal hosts, and decomposer chains, in which bacteria, fungi, and a huge diversity of animal species feed on the dead bodies of organisms from all trophic levels.

As energy flows through the various food chains, it is being constantly divided into three channels *(Figure 12)*. Some of it goes into PRODUCTION, which is the creation of new tissue by growth and reproduction, as well as the manufacture of energy-rich storage products in the form of fats and carbohydrates. Some of the energy is lost from the ecosystem by EXPORT, the emigration of organisms coupled with the passive transport of dead organic material out of the ecosystem by the actions of wind and water. The rest of the energy is lost permanently to the ecosystem and all other ecosystems by means of RESPIRATION. The leakage due to respiration is very high. In fact, only a small fraction of the energy is transferred successfully from one trophic level to the next. Ecologists put the figure at about ten percent. The exact mea-

■ ORIGINAL SAND   ■ SAND WASHED UP BY THE WAVES AND BLOWN BY THE WIND   ■ HUMUS ADDED BY PLANTS AND ANIMALS

SUCCESSION OF PLANTS has altered the beach dunes on the Indiana shore of Lake Michigan. First plants to invade sand are grasses. The final climax vegetation belongs to the deciduous forest biome.   **8**

surement used to make this important generalization is ecological efficiency, the ratio of production available to one trophic level to the rate of production available to the next trophic level, on which it depends for its food. Consider the following very simple ecosystem: a field of clover, the mice that eat the clover, and the cats that eat the mice. According to the "ten percent rule" of ecological efficiency, we would expect that for every 1,000 calories of new clover grown, about 100 calories of

9 SUCCESSION OF BIRDS accompanies the progression from grassland to oak and hickory forest in Georgia. Succession resembles that shown in previous illustration, although here the starting point is an abandoned field instead of a lakeshore beach. The climax forest harbors not only the greatest diversity of species, but also supports the greatest number of individual organisms.

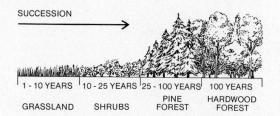

SUCCESSION →

	1 - 10 YEARS	10 - 25 YEARS	25 - 100 YEARS	100 YEARS
	GRASSLAND	SHRUBS	PINE FOREST	HARDWOOD FOREST

GRASSHOPPER SPARROW
MEADOWLARK
FIELD SPARROW
YELLOWTHROAT
YELLOW-BREASTED CHAT
CARDINAL
TOWHEE
BACHMAN'S SPARROW
PRAIRIE WARBLER
WHITE-EYED VIREO
PINE WARBLER
SUMMER TANAGER
CAROLINA WREN
CAROLINA CHICKADEE
BLUE-GRAY GNATCATCHER
BROWN-HEADED NUTHATCH
WOOD PEWEE
HUMMINGBIRD
TUFTED TITMOUSE
YELLOW-THROATED VIREO
HOODED WARBLER
RED-EYED VIREO
HAIRY WOODPECKER
DOWNY WOODPECKER
CRESTED FLYCATCHER
WOOD THRUSH
YELLOW-BILLED CUCKOO
BLACK AND WHITE WARBLER
KENTUCKY WARBLER
ACADIAN FLYCATCHER

NUMBER OF COMMON SPECIES	2	8	15	19
DENSITY (PAIRS PER 100 ACRES)	27	123	113	233

mice and 10 calories of cats would be added. Notice that the calories present in the clover are the same ones that the cats use. The cats, however, are specialized carnivores. They have sharp teeth that are adapted for shearing meat. The mice have teeth that are adapted for chopping and grinding seeds and other vegetable material. From the point of view of the cat population, the mouse population is a device for converting clover calories into an edible form. Our generalization states that the best the mice can do is to make about ten percent of the clover calories available to the cats. If cats were preyed upon, the predator would find these animals about equally efficient at converting mouse calories. Measurements in diverse ecosystems have shown that the ecological efficiencies that exist between different links in food chains actually vary from about five to 25 percent. Most are close enough to ten percent to make this figure useful for rough first approximations.

A major complicating factor in the analysis of energy flow is the fact that species cannot always be sorted neatly into trophic levels. Individual species often play more than one role. The crow, for example, is both a predator of insects and small animals and a scavenger of dead birds and mammals—and therefore a decomposer. Some other birds feed on fruit and nectar, which makes them herbivores, and also on a wide variety of insects, which makes them first and second level carnivores. Yet even with these qualifications, the ten percent rule of ecological efficiency can be used to predict accurately a major consequence in the organization of ecosystems: food chains seldom have more than four or five links. The reason is that a 90 percent reduction (approximately) in productivity with each step results in only $^1/_{10} \times \, ^1/_{10} \times \, ^1/_{10} \times \, ^1/_{10} = \, ^1/_{10,000}$ as much productivity at a trophic level four links removed from the green plants. In fact, the top carnivore that is producing only one-ten-thousandth as many calories as the plants on which it ultimately depends must be both sparsely distributed and far-ranging in its activities. Wolves must travel as much as 20 miles a day to find enough energy *(Chapter 32)*. The territories of individual tigers and other great cats often cover hundreds of square kilometers. Such organisms are simply too sparse to support predators on their own. No animal species preys on tigers—not because tigers are so formidable, but because they produce too few calories per square kilometer to make it worthwhile. For this reason the great white shark, which feeds to a large extent on seals and other marine mammals, is a rare species.

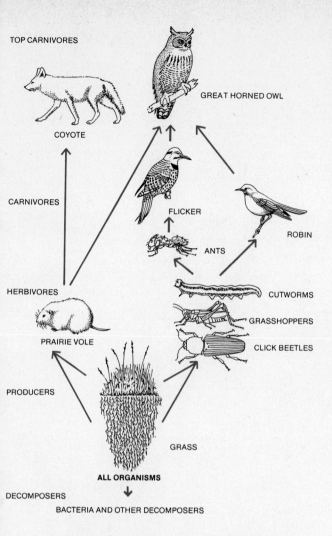

TOP CARNIVORES

GREAT HORNED OWL

COYOTE

CARNIVORES

FLICKER

ROBIN

ANTS

HERBIVORES

CUTWORMS

GRASSHOPPERS

PRAIRIE VOLE

CLICK BEETLES

PRODUCERS

GRASS

ALL ORGANISMS

DECOMPOSERS

BACTERIA AND OTHER DECOMPOSERS

FOOD CHAIN IN GRASSLAND of Canada. Diagram shows only a few of the chains and branches in this biome. Chains vary from one to five links. Trophic levels are indicated at far left.

Let us next consider the matter of production by entire ecosystems. The rate of calorie manufacture varies enormously among both the major biome types and the marine and aquatic habitats. As Figure 13 shows, communities in the great majority of natural systems have balanced production and respiration rates. This, of course, is what we should expect of any system that is both stable and relatively independent of other systems. A discrete community that is producing more than it is respiring is closely similar to a population that is actively growing; in fact it is probably composed of such populations. It cannot continue in this direction for long. It must stabilize, which means that its rate of respiration must come to equal its rate of produc-

tion. Conversely, a community that is respiring more than it is producing is a dying community. The algae, bacteria, and other inhabitants of swamps and polluted waters always tend to decline. They are sustained only because they receive new energy through the import of materials from the outside.

The reader should note that there is only a loose relationship between the amount of living material present in an ecosystem, usually referred to as the BIOMASS, and its rate of production. It is quite possible for a grassland, with a relatively low biomass, to outproduce a forest with a biomass hundreds or

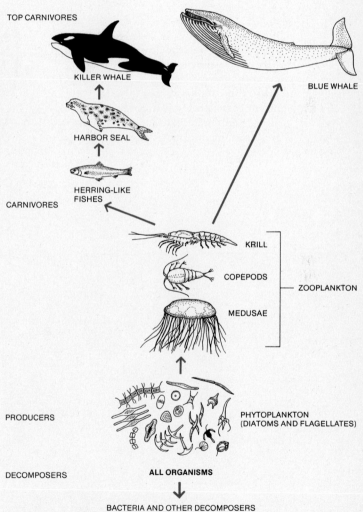

TOP CARNIVORES

KILLER WHALE

BLUE WHALE

HARBOR SEAL

CARNIVORES

HERRING-LIKE FISHES

KRILL

COPEPODS

ZOOPLANKTON

MEDUSAE

PRODUCERS

PHYTOPLANKTON (DIATOMS AND FLAGELLATES)

DECOMPOSERS

ALL ORGANISMS

BACTERIA AND OTHER DECOMPOSERS

MARINE FOOD CHAINS are represented in this highly simplified diagram. As in preceding illustration, trophic levels are indicated at far left. Most of the organic matter that feeds marine organisms is synthesized by phytoplankton living in the sunlit surface waters.

11

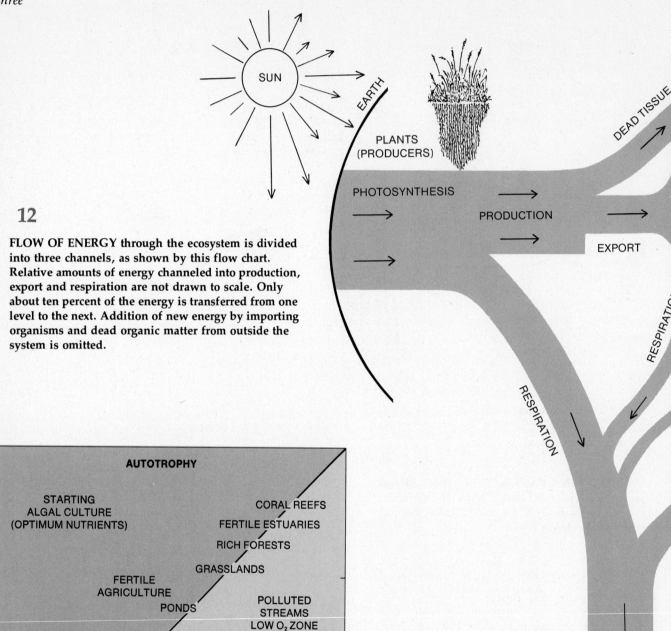

12

FLOW OF ENERGY through the ecosystem is divided
into three channels, as shown by this flow chart.
Relative amounts of energy channeled into production,
export and respiration are not drawn to scale. Only
about ten percent of the energy is transferred from one
level to the next. Addition of new energy by importing
organisms and dead organic matter from outside the
system is omitted.

13

COMMUNITY METABOLISM. In nature the
respiration of an entire community usually balances its
production. Such communities lie along diagonal line
in the graph. Communities that produce more than they
consume lie above the line; those that consume more
than they produce lie below it.

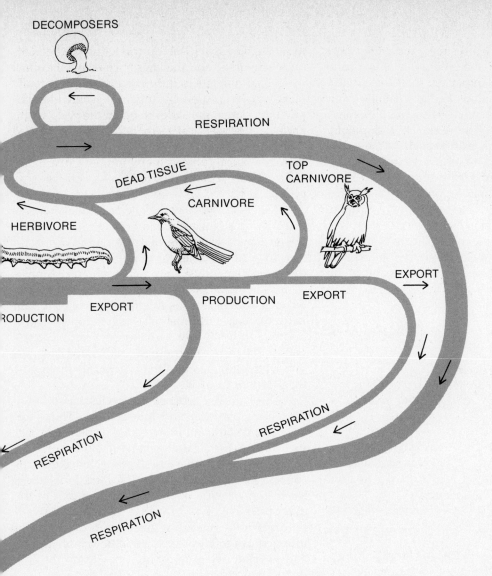

DECOMPOSERS

RESPIRATION

DEAD TISSUE

TOP CARNIVORE

CARNIVORE

HERBIVORE

EXPORT

EXPORT

PRODUCTION

EXPORT

RODUCTION

EXPORT

PRODUCTION

RESPIRATION

RESPIRATION

RESPIRATION

thousands of times greater. All that is required is that the forest have a lower production when measured on a per gram basis. In general, early successional stages have a lower biomass and a higher production than the climax stages that replace them. Once again it is possible to compare this quality of communities with a similar quality in single populations: the optimal yield of populations, you will recall, occurs at an intermediate stage of population density and not at the maximum level permitted by the environment. The loose connection between the biomass and production is portrayed with particular vividness when the same trophic levels in different kinds of communities are compared. Figure 14 presents

No animal species preys on tigers —not because tigers are so formidable, but because they produce too few calories to make it worthwhile.

examples of the kind of diagram used by ecologists to summarize the quantitative properties of whole communities. The biomass pyramid shows the total weight of organisms belonging to the different trophic levels that can be found at any instant in a circumscribed area, in this case one square meter. The energy pyramid shows the rate of production per unit time in the different levels. As one might expect from the ten percent rule of ecological efficiency, energy pyramids are very similar from one ecosystem to the next. However, biomass pyramids vary greatly. The reason is that plants, the producer organisms, exhibit such extreme variability in their ability to photosynthesize. The algae of the ocean can greatly outproduce most land plants, including grasses, on a per gram basis. Consequently, they are able to support a proportionately much larger biomass of herbivores. Note that the production of ocean herbivores is still only about ten percent that of algae. But their biomass is greater, because the algae are growing and being eaten at an exceptionally high rate.

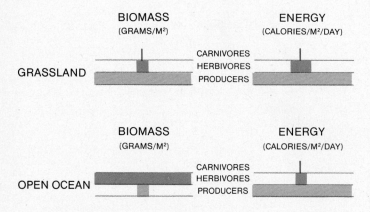

14 BIOMASS AND ENERGY PYRAMIDS for two very different ecosystems. The form of energy pyramids is similar from one system to another, but form of biomass pyramids varies considerably.

CYCLING OF MATERIALS

Although the world is an open system with respect to energy, it is a closed system with respect to materials. Water, carbon, nitrogen, phosphorus, and the other elementary building materials of living tissue are cycled perpetually through organisms to the environment and back through organisms again. The carbon and nitrogen atoms of which we are constructed at this moment are the same atoms that composed dinosaurs, insects, and trees in the Mesozoic Era. Most of the atoms are build into complex molecules by the organisms, only to be released in the form of small molecules during the process of death and decomposition. For a period of time that varies from a few seconds

to millions of years —according to the kind of substance involved and the laws of chance —the small molecules circulate in the environment. Eventually most are recaptured by organisms and incorporated into large molecules to begin the cycle again. Because the chemical elements are being passed back and forth between biological organisms and their geological (physical) environment, the material cycles are usually referred to as BIOGEOCHEMICAL cycles.

Between 30 and 40 of the chemical elements participate in the biogeochemical cycles. Some are the obscure trace elements that are required only in tiny amounts. Iodine, for example, is essential for the endocrine physiology of vertebrates. Molybdenum is essential for nitrogen-fixing organisms; when it is deficient, as it is in the soils of certain parts of Asia and Australia, vegetation is sparse and scrubby. Other substances, such as the radioactive form of strontium (Sr^{90}), have been put into circulation in abnormally large quantities by the activities of man.

Each of the biogeochemical cycles has its own distinctive pathway and rate of flow. Those followed by carbon and nitrogen are displayed in Figures 15 and 16. Several general qualities of cycles are exemplified in these two cases. Notice that circulation occurs chiefly when the carrier molecules are in the gaseous phase and become part of the atmosphere. Notice, too, that there are two very different routes which a given atom of carbon or nitrogen can follow. One leads through organisms and is relatively quick. The other leads through decay, entombment in geological deposits, and liberation after very long periods of time by exposure and weathering or through the intervention of man. A third general principle to note is that man has added new pathways to the biogeochemical cycles. He has also altered to some degree the lengths of each of the cycles, with far-reaching consequences.

POLLUTION

The word pollution has many meanings and implications for different people. The ecologist, with special insight, defines it as the misplacement of resources by alteration of the biogeochemical cycles. Only very recently has man become aware of his own profound influence on these cycles. Previously we thought the world to be a bottomless sink into which waste materials could be poured forever. They would disappear, we believed, into the uncharted depths of the soil and of the sea.

The carbon and nitrogen atoms of which we are composed are the same atoms that composed dinosaurs, insects and trees in the Mesozoic Era.

When they returned, if ever, they would come back to us transformed into the pure, "natural" products that we first exploited. Now we know better. The evidence shows that mankind is an inordinately demanding passenger on this spaceship Earth. He has drastically altered the direction and rate in which certain materials are entering the biogeochemical cycles. Many of these substances are now changing the complexion of entire ecosystems as large as forests, estuaries, rivers, lakes,

and bays; and some are beginning to act as outright poisons.

No matter what is done to the biogeochemical cycles, they will still be cycles. Eventually the rates of input and output of each of the substances, for example nitrogen into nitrates and nitrates back into nitrogen, will strike a balance or at least begin to approach a balance. Why, then, should we worry about perturbations in the system? The answer is that both the perturbations and the new balances that follow them are unhealthy for man and most species of plants and animals. They are unhealthy for the reason that we belong to ecosys-

THE CARBON CYCLE. Carbon is removed from the CO_2 pool in the atmosphere by the process of photosynthesis. Animals obtain carbon by eating plants or other animals that feed on plants. Animal respiration, the decay of organic matter, and the combustion of fossil fuels return CO_2 to atmosphere.

15

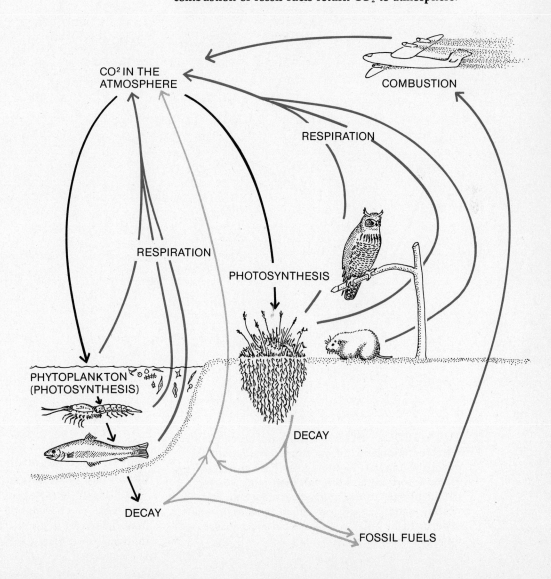

CO^2 IN THE ATMOSPHERE

COMBUSTION

RESPIRATION

RESPIRATION

PHOTOSYNTHESIS

PHYTOPLANKTON (PHOTOSYNTHESIS)

DECAY

DECAY

FOSSIL FUELS

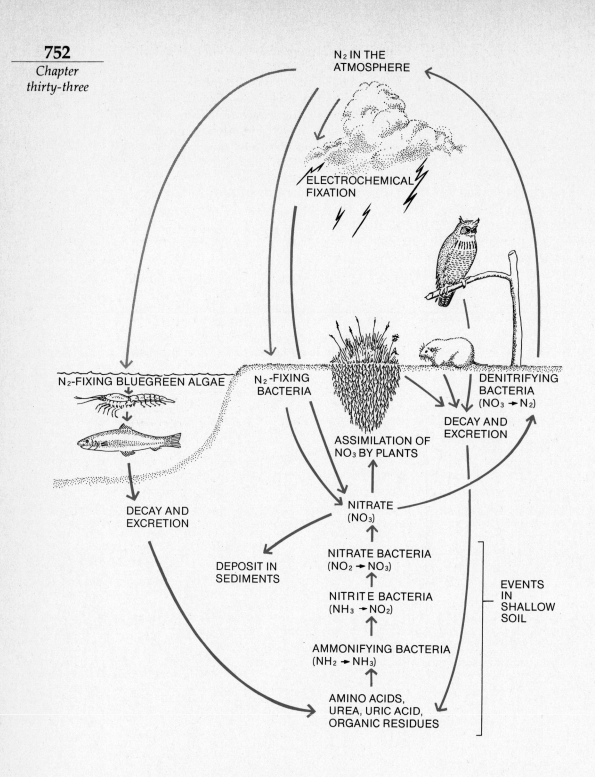

N₂ IN THE
ATMOSPHERE

ELECTROCHEMICAL
FIXATION

N₂-FIXING BLUEGREEN ALGAE

N₂-FIXING
BACTERIA

DENITRIFYING
BACTERIA
(NO₃ → N₂)

DECAY AND
EXCRETION

ASSIMILATION OF
NO₃ BY PLANTS

DECAY AND
EXCRETION

NITRATE
(NO₃)

DEPOSIT IN
SEDIMENTS

NITRATE BACTERIA
(NO₂ → NO₃)

NITRITE BACTERIA
(NH₃ → NO₂)

EVENTS
IN
SHALLOW
SOIL

AMMONIFYING BACTERIA
(NH₂ → NH₃)

AMINO ACIDS,
UREA, URIC ACID,
ORGANIC RESIDUES

16 THE NITROGEN CYCLE. Bacteria and bluegreen algae
convert atmospheric nitrogen to a form that other
organisms can use. Lightning has the same effect.
Plants absorb nitrates through their roots. Animals
obtain nitrogen by eating plants or other animals. Dead
organic matter and waste products of animals return
nitrogen to the soil. Denitrifying bacteria recycle it back
into the atmosphere.

tems that evolved for millions of years in the old,
unpolluted environment.

Pollution alters the distribution of materials in
two ways. It changes the rate of flow of the basic
materials, and it introduces some of the materials
in the form of new, man-made molecules with

which decomposer species cannot cope. One of the most pervasive consequences of pollution belonging to the first category is EUTROPHICATION, the addition of excess nutrient materials to water. Man, the great and wasteful consumer, injects excess phosphates, nitrates, ammonia, and diverse forms of organic residues into the lakes, streams, and estuaries. Most are passed in sewage and industrial waste. Other substantial amounts come from the drainage of farmlands, where manure and synthetic fertilizers are spread on the ground during the winter and flushed into streams during the spring thaws and rains. Considerable additional quantities of nitrates are added from the combustion of coal and other fossil fuels. Although these materials are mostly distasteful excretory products for us, they are vital nutrients to aquatic plants. The plants are thus provided with a eutrophic (richly nourished) environment. Algae multiply until they form "blooms" that turn the water green and coat the surface with a scum resembling green paint. As eutrophication proceeds, the original dominant elements of the freshwater phytoplankton, the desmids, are replaced by diatoms, which give way to flagellates and other green algae. Finally, the blue-green algae take over. As the masses of algae and other aquatic vegetation decay, the water is depleted of oxygen. Anaerobic bacteria become the prevalent decomposers. These organisms cannot break organic molecules all the way down to CO_2. Instead, the end products of their respiration are such substances as hydrogen sulfide and propionic and butyric acids. The water becomes offensive in smell and taste to human beings. At the height of its eutrophication Lake Monona, in Wisconsin, was typical when it was compared by one observer to "a foul and neglected pig sty." Finally, as the oxygen is used up, much of the original fauna declines, to be replaced by midges, certain types of "trash fish," and other resistant animals.

The recent history of Lake Erie is an especially instructive example of the downward spiral of eutrophication. When it was opened to water traffic and commerce in the early 1800's, this was one of the beautiful lakes of America, teeming with wildlife and game fish. Today over 10 million people live in the Lake Erie basin. Each year Detroit, Toledo, Cleveland, Buffalo, and nearby towns and cities pour over 40 billion gallons of raw sewage and industrial waste into Lake Erie. The Canadian side is intensively farmed and contributes additional waste nutrients in the form of excess fertilizers. Some of the consequences of the pollution are shown in Figures 17 to 19. As dissolved solids increased in the lake water through the years, algae proliferated. At the water filtration plant at Cleveland, for example, these phytoplankton organisms increased from 81 per milliliter in 1929 to 2,423 per milliliter in 1962. Algal blooms began to turn patches of the water green. Populations of bacteria also grew dramatically. At the western end of Lake Erie, the numbers of *Escherichia coli* increased from 175 to 449 per 100 mil-

The history of Lake Erie is typical to a greater or lesser degree of the other Great Lakes, as well as most of the other bodies of fresh water near centers of human population.

liliters during the period 1913 to 1946. *E. coli* originates in the human intestine, and it is used as an index of the degree of pollution from sewage and thus of the probable occurrence of disease-causing microorganisms. As the *E. coli* level became too high, public health regulations required the closing of as many as three-fourths of the beaches along the southern shore of Lake Erie.

For the original aquatic organisms living in Lake Erie pollution literally has been a threat to survival. As the oxygen levels dropped, particularly along the lake bottom, many of the native species began

17 DISSOLVED SOLIDS in Lake Erie have increased by almost 50 percent in the last 70 years. Higher concentration of solids led to a drastic increase in populations of algae and bacteria.

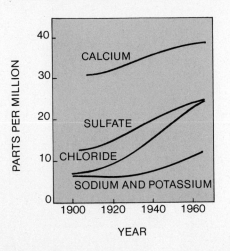

to decline, and pollution-tolerant species of clams, snails, and midges sharply increased. Nymphs of the mayfly genus *Hexagenia* were replaced by oligochaete worms as the dominant organisms of the lake bottom. Fishes have been affected just as drastically. In 1899, before the onset of the heaviest pollution, the dominant fish species were lake herring (cisco), blue pike, carp, yellow perch, sauger, whitefish, and walleye. Lake trout were common in deeper waters. By 1925 the lake herring industry collapsed when the species became too scarce for

exploitation to be commercially feasible. Since 1945 blue pike, sauger, and whitefish have almost disappeared from the lake, and the lake trout has evidently become totally extinct. In 1965 commercial fishing had become dependent on the less valuable yellow perch, smelt, sheepshead, whitebass, carp, catfish, and walleye. Lake Erie, in short, is not yet a dead lake. New efforts are now being conducted to reverse its downward spiral. But it is still biologically less diverse than it used to be, and from a human standpoint it provides a far less pleasant and useful environment. Furthermore, it is continuing to deteriorate. The history of Lake Erie is typical to a greater or lesser degree of the other Great Lakes, as well as most of the other bodies of fresh water located near centers of human populations around the world.

Man, while thoughtlessly flushing excessive nutrients into the waters all around him, has also been adding abnormal quantities of metals and other elements that were previously scarce in ecosystems. Lead and mercury are two familiar substances that have been increased to hazardous levels in many places. Perhaps the most insidious of all additions is strontium 90, a radioactive isotope that originates in the fallout from atomic tests and also as a by-product of fission in nuclear reactors. When strontium 90 is absorbed by plants and then passed along the food chains, it is concentrated in a stepwise manner in the bodies of organisms. This enrichment is especially pronounced in vertebrates, where strontium tends to replace calcium in the construction of bones (*Figure 20*). There it lodges for the lifetime of the organism,

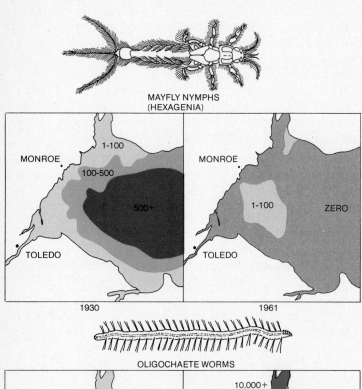

MAYFLY NYMPHS
(HEXAGENIA)

1-100

MONROE
100-500

500+

TOLEDO

1930

MONROE

1-100

ZERO

TOLEDO

1961

OLIGOCHAETE WORMS

MONROE

UP TO
1000

TOLEDO
10,000+

1930

MONROE

10,000+

1000—
10,000

UP TO
1000

TOLEDO

1961

18 **EFFECT OF POLLUTION at western end of Lake Erie. Original mayfly population decreased (top) as oligochaete worms increased (bottom). Contours indicate numbers of organisms per square meter.**

EXTINCTION OF LAKE TROUT in Lake Erie is a result of oxygen depletion aggravated by overfishing and predation by sea lamprey.

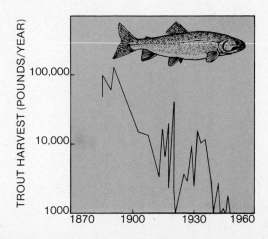

TROUT HARVEST (POUNDS/YEAR)

100,000

10,000

1000

1870   1900   1930   1960

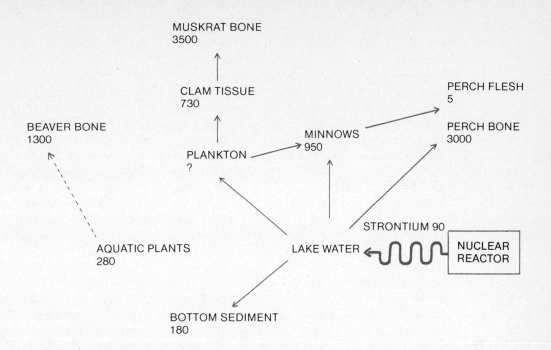

MUSKRAT BONE
3500

CLAM TISSUE
730

PERCH FLESH
5

BEAVER BONE
1300

MINNOWS
950

PERCH BONE
3000

PLANKTON
?

STRONTIUM 90

AQUATIC PLANTS
280

LAKE WATER

NUCLEAR
REACTOR

BOTTOM SEDIMENT
180

STRONTIUM-90 CONTAMINATION of Perch Lake in Ontario was caused by seepage of radioactive waste from Atomic Energy of Canada, Ltd. Strontium became concentrated in bones of vertebrates. Numbers indicate relative concentration of strontium at various levels of food web compared with the concentration in lake water, which has been arbitarily defined as 1.

emitting a steady bombardment of "hard" (ionizing) radiation into the surrounding tissues. A second menace originating in atomic fallout is cesium 137. Because this radioactive isotope behaves chemically like potassium, it is absorbed easily by organisms and is distributed quickly throughout their bodies. Like strontium 90, it is concentrated with each link in the food chain. Cesium 137 has proved especially dangerous in the simple ecosystems of the arctic tundra. There, during the height of the above-ground atomic weapons testing in the 1950's and early 1960's, cesium 137 was picked up rapidly by lichens. Caribou live principally on lichens during the winter, and eskimos and Laplanders live principally on caribou meat. Before the overall cesium 137 levels began to decline in the arctic following the nuclear test ban treaty, the concentrations of the isotope in these northernmost peoples had reached perilous levels.

Of equal significance to ecosystems are the new organic molecules produced and discarded by man. So long as these substances are BIO-DEGRADABLE, capable of being broken down to sim-

pler components by organisms in the ecosystem, they pose no greater threat than the natural products, such as phosphates and nitrates, that contribute to pollution. But when substances cannot be easily degraded by organisms, they create a wholly new kind of problem. Fifteen years ago artificial detergents of the kind used daily in households were accumulating at an alarming rate in the freshwater systems around the cities and towns of the United States. Although not poisonous while at low to moderate levels, they became a nuisance because of the large masses of foam they generated in streams, filtration plants, and other places where the water is stirred vigorously. These early detergents built up for the simple reason that the organic portions of their molecules had been synthesized in a form the microorganisms found extremely difficult to metabolize. When manufacturers deliberately switched to the production of biodegradable forms of detergents, that particular problem disappeared.

Unfortunately, no easy way out exists in the case of the hard pesticides, the chemicals which are effective in killing insects and other pests but which are also difficult for organisms to digest. The roll call of these persistent poisons is a familiar list of household words: aldrin, chlordane, dieldrin, DDT, heptachlor, lindane, and others. DDT is by far the most important in its impact on the environment. One of the first of its class of insecticides, it was introduced with spectacular success during the early years of World War II. By eliminating insect vectors such as mosquitoes, flies, and lice, DDT permitted the control of malaria, typhus, and other diseases over large parts of the world. It is credited with having saved over half a billion human lives. It also saved agriculture from uncounted millions of dollars damage from a variety of plant-eating insect pests. DDT in fact has been one of the great benefactors of mankind, and it justly won a Nobel Prize for its discoverer, Paul Mueller of Switzerland. But because it also degrades slowly and is relative toxic, it has now turned into an ecological nightmare. The reason for this transformation is made more explicit by the diagram in Figure 21. DDT has a strong affinity for fatty tissues, and it is moved efficiently through food chains in all kinds of ecosystems. DDT now occurs in easily detectable levels almost everywhere in the world. It is a principal component of dust in air samples from remote Pacific islands. It exists in the fatty tissue of penguins in the Antarctic. It occurs at hazardous levels in the Cahow, or Bermuda Petrel (*Pterodroma cahow*), a species consisting of only about 100 individuals all of which nest on the island of Bermuda. The nearest source of DDT for the Cahow is the American mainland, 650 miles away. DDT is particularly dangerous to birds because of the way it is concentrated during its passage through food chains. In its early career, before it was indicted by Rachel

> The industrial nations of the world are creating trash and liquid pollutants faster than these materials can be recycled by microorganisms. This trend must somehow be reversed unless we wish to see the biosphere gradually covered by a detritosphere of refuse.

Carson's book *Silent Spring*, DDT was used indiscriminately in cities in attempts to control the little bark beetles that transmit Dutch elm disease. The insecticide reached the soil of yards and gardens in high concentrations. When earthworms consumed the contaminated soil, they concentrated the DDT by a factor of more than ten. Robins and other songbirds that ate the earthworms concentrated the DDT again—often above their own level of tolerance. The bird numbers were temporarily reduced by 30 to 90 percent. Some of the species of top carnivores among the birds appear to have been affected even more drastically. In the past 25 years two of our most magnificent birds of prey, the Peregrine Falcon and the Bald Eagle—the latter our national symbol—have disappeared from a large portion of their former range in North America and are in real danger of eventual extinction. There is strong evidence to suggest that these species are losing their power of reproduction because DDT interfers with calcium deposition in their eggs. The eggs are too fragile to support the weight of the parent birds, and they usually break before hatching.

There is irony—and eventual hope—in the conception of pollution as the misplacement of resources. If man has witlessly altered the biogeochemical cycles to his own disadvantage in the past, he can, with appropriate planning and effort, change them back to his advantage in the future. Nutrients can be recovered before they enter streams to continue the process of eutrophication. Or they can be diverted to places where eutrophication is desirable, for example commercial fish ponds, and bays and estuaries where high productivity of crustaceans, mollusks, and food fish is desired. Poisonous minerals, such as lead and mercury, can be filtered out more effectively and perhaps even to economic advantage. Some substances, such as copper, must be recovered and recycled more effectively, because they are now being lost to the sea in forms that are not economically retrievable by any current technology. The organic molecules synthesized by industry for the manufacture of plastics, pesticides, detergents, and other disposable materials must always be quickly biodegradable. At the present time the industrial nations of the world are creating trash and liquid pollutants faster than these materials can be metabolized and recycled by microorganisms. This trend must somehow be reversed, unless we wish to see the biosphere of the earth gradually covered over with a detritosphere of refuse.

One additional form of pollution about to become a menace is thermal pollution. Already the discharge of heated effluents into rivers near major industrial cities sometimes raises the summer temperature of the water to above 90 degrees Fahrenheit. The highest temperatures that most fish in North America can tolerate are between 77° and 97°F. They cease breeding at still lower maximum temperatures. With the proliferation of nuclear reactors, which are envisioned as one of man's primary energy sources for the future, the problem will soon become acute. It is projected that by the year 2000, nuclear plants in the United States will be producing as much as one million megawatts of energy. The use of natural waters to cool the condensers would require the substantial heating of an amount of water equivalent to a third of the yearly freshwater runoff in the entire nation. It is obvious even to nonecologists that such an effect would be catastrophic for all forms of aquatic life and create a hazard for terrestrial and marine life as well. Consequently, alternative ways of dissipating heat from nuclear power plants must be found, perfected, and utilized.

THE FRAGILE ENVIRONMENT

Ecologists are fond of quoting the following epigram about the environment: "You cannot do just one thing." You cannot exterminate one species, or dam one river, or cut down one forest, without altering other parts of the environment—often drastically. Consider, for example, the lesson taught by the Aswan High Dam. This great structure was built by the Egyptians with the help of the Soviet Union. The idea was to control the flow of the Nile River. To replace the unchecked floods that have covered the lower Nile Valley since before recorded history, there was to be a continuous release of the stored water to permit year-round artificial irrigation. Like most such schemes the program has worked—at least at the beginning. But unforeseen side effects are now appearing that threaten the value of the whole enterprise. First, the flood plains of the lower Nile have been deprived of the annual load of fertile silt that was deposited on them each year in the past. It is now going to be necessary to import artificial fertilizers to maintain the desired level of agricultural production. Second, the silt is being deposited behind the dam, which therefore may in time have to be abandoned altogether. Third, the employment of controlled irrigation without periodic flooding and flushing of the soil increases the level of salinity, with adverse effects on agriculture. Fourth, the loss of the nutrients that were previously carried into the eastern Mediterranean Sea by the flooding Nile has caused a drastic reduction in the sardine population. The annual catch has dropped from 18,000 tons a year to 500 tons. Fifth, the reduction of flow of fresh water from the Nile has increased the salinity in the eastern Mediterranean, causing

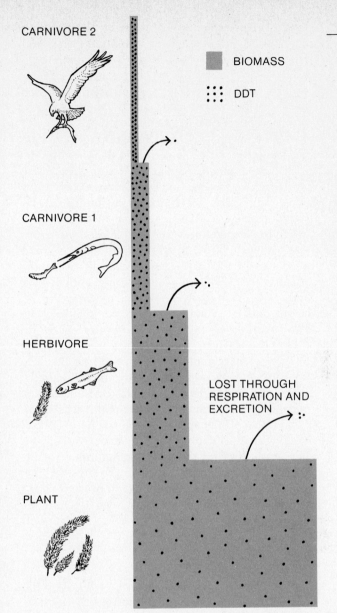

DDT IS CONCENTRATED by the ecosystem. At each level of the food chain, more than half the biomass is lost through excretion, respiration and decay. But most of the DDT is retained, and as a result becomes highly concentrated in carnivores.

21

an increase of fish species introduced from the Red Sea (via the Suez Canal) at the expense of native species. Sixth, there has been an increase in the incidence of the dread disease schistosomiasis. The pathogen, a blood fluke, depends on snails to complete its life cycle, and freshwater snails are much more abundant in irrigation canals than in natural river systems. In reviewing the results of such poorly planned modifications of the environment, Garrett Hardin of the University of California at Santa Barbara has remarked, "The effects of any sizable intervention in an ecosystem are like ripples spreading out on a pond from a dropped pebble; they go on and on."

To initiate an ecological catastrophe man does not have to go so far as to poison the environment with industrial waste or to build some mammoth structure like the Aswan High Dam. He can sometimes do it with the innocent introduction of a single species of plant or animal in a new land. When the Canadian water weed, *Elodea canadensis*, was established in England in the 1880's, it multiplied explosively. For years it clogged rivers and streams to the extent of interfering with fishing and barge transport. Then, inexplicably, it declined and is today a common but economically unimportant element of the freshwater flora. When the African malarial mosquito *Anopheles gambiae* was accidentally introduced into Brazil in the 1930's, it spread rapidly through populated areas and soon was responsible for hundreds of malarial deaths annually. In the end it was eradicated by an all-out effort of the Rockefeller Foundation with the help of the Brazilian government. The application of the needed amounts of insecticides throughout the range of the mosquito required the efforts of three thousand persons over a period of three years and cost over two million dollars.

The greatest destroyer of ecosystems in history, next to man himself, is the humble, winsome European rabbit.

The reader may be surprised that the greatest destroyer of ecosystems in history, next to man himself, is the humble and winsome European rabbit, *Oryctolagus cuniculus*. The European settlers of Australia and New Zealand deliberately introduced rabbits into these countries during the 1800's, in the belief that they would provide a ready supply of meat that could be harvested from the wild countryside. The rabbits exceeded this expectation to a tragic degree. Protected from predators and harsh weather by the labyrinthine retreats of their warrens, they developed abnormally dense populations that stripped vegetation and

loosened topsoil over wide areas. Both farming and sheep herding were made impracticable in the hardest hit areas, and many farmers were forced to abandon their land —or to try to make a living by farming rabbits! The native marsupial fauna of Australia has been diminished and pushed back over part of their native ranges by the habitat destruction caused by this single alien species. On small remote islands where the European rabbit has been introduced the effects can be even more severe. Consider the sad history of Laysan Island, a tiny coral reef (area: 2 square miles) in the northern portion of the Hawaiian chain. In its original state Laysan supported attractive groves of sandalwood trees, thickets of bushes, and scattered palm fans. There was a small but distinctive fauna of land birds found nowhere else in the world: a race of drepanidid honey-eaters (*Himatione sanguinea freethii*), a species of warbler (*Acrocephalus familiaris*), and a species of flightless rail (*Porzanula palmeri*). Around the year 1903, domestic rabbits were deliberately introduced onto the island from Europe. These individuals found themselves in a very congenial environment, totally lacking in their natural enemies. They did what any species does under the circumstances: multiplied to the utmost limit permitted by the living space and food. By 1912 there were over 5,000 rabbits present, and they were rapidly destroying the vegetation. An attempt to eradicate them through systematic shooting failed. By 1923 Laysan was reduced to a barren strip of sand containing only a few stunted trees. Twenty-two of the 26 original plant species had been eliminated. The *Acrocephalus* warbler had vanished, and the last three *Himatione* honeyeaters died during a sandstorm while an expedition studying them was on the island. The last of the *Porzanula* rails died within two more years. The rabbits themselves had been reduced to a state of starvation. The population was down to a few hundred individuals, which were finally hunted down and totally eliminated. But it was too late. This one poorly adapted animal species had virtually destroyed the entire island ecosystem.

READINGS

P.R.EHRLICH, A. EHRLICH, AND J.P. HOLDREN, *Ecoscience: Population, Resources, Environment*, San Francisco, W.H. Freeman, 1977. A definitive and excitingly written source book on population and environmental problems.

E.P. ODUM, *Fundamentals of Ecology*, 3rd Edition,

Philadelphia, W.B. Saunders Company, 1971. See comments at the end of Chapter 32.

E.J. KORMONDY, *Concepts of Ecology,* 2nd Edition, Englewood Cliffs, NJ, Prentice-Hall, 1976. This is a good, brief textbook that stresses ecosystems and cites many examples from the literature of the past ten years.

R.H. WHITTAKER, *Communities and Ecosystems,* 2nd Edition, New York, Macmillan, 1975. Similar to the Kormondy text but stronger on theoretical topics.

D.W. EHRENFELD, *Biological Conservation,* New York, Holt, Rinehart & Winston, 1970. An excellent little book for the student who wants to pass directly into readings on applied aspects of ecology.

K.E.F. WATT, *Ecology and Resource Management,* New York, McGraw-Hill, 1968. A provocative exposition of the possible methods of environmental engineering that can be developed from a knowledge of population and community ecology.

R.E. RICKLEFS, *Ecology,* Newton, MA, Chiron Press, 1973. See comments at the end of the previous chapter.

G.A. ROHLICH, *ed., Eutrophication: Causes, Consequences, Correctives,* Washington, DC, National Academy of Sciences, 1969. The National Academy of Sciences-National Research Council report on this major form of pollution, recommended to the student who would like to make his own evaluation of the source materials.

If I were forced to call the human species apes, I should at least show that they are culture-apes, the culminating-tradition apes, the ideal conception of all apes. That man is naked is irrelevant; he might just as well be furry.

—Konrad Lorenz

Fifty million years ago Africa was an island continent. Together with Arabia it was cut off from Europe to the north and India and Asia to the east by the great Tethys Sea, a body of shallow tropical water that stretched between the Atlantic and Indian Oceans *(Chapter 31)*. During its isolation Africa, like all great islands, evolved a distinctive fauna and flora. The early colonists of many groups of mammals in particular gave rise to spectacular adaptive radiations, parallel and simultaneous with the developments on the island continents of Australia and South America *(see Chapter 31)*. Elephants and their kin originated on Africa, together with a great assemblage of less familiar types—elephant shrews, hyracoids, tenrecs, arsinotheres, barytheres, moeritheres, and others. The Tethys Sea did not form a total barrier, and occasional groups were able to cross from Europe or Asia at various times and to colonize the island continent. During the Oligocene epoch, primitive forerunners of the modern pigs, buffalos, antelopes, and lions invaded Africa and began secondary radiations that rivaled, and in some instances replaced, the older ones. Africa, in short, became a major center of mammalian evolution.

One of the earliest imports were the primates (order Primates). The exact place of origin of this key group has not been established by fossil discoveries. The most primitive primates were similar to the living tree shrew illustrated in Figure 1. They descended some time in the Cretaceous period from small animals that probably resembled living shrews of the primitive mammalian order Insecti-

vora. By Eocene time, at the latest, the primitive primates had established themselves in Africa. By the end of the Miocene, a few tens of millions of years later, they had completed a remarkable adaptive radiation that produced lemurs and similar primitive forms, baboons and other kinds of Old World monkeys, anthropoid apes, and man himself. Meanwhile, in South and Central America, an independent radiation had produced the New World monkeys, including the wooly monkeys, squirrel monkeys, spider monkeys, howlers, marmosets, tamarins, and others.

Simply to define a primate is to understand a great deal of the evolution that made the origin of man possible. The primates are a group of mammals that are fundamentally adapted to life in trees by means of improved balance, leg and arm flexibility, and vision. The primitive tree shrews, lemurs, monkeys, and apes were apparently all tree-dwelling. They retained generalized limb structures, with five digits terminating each limb. In early stages of evolution the limbs acquired the ability to rotate and to move in many directions, so that the body could be moved or supported while the animal was in any of a wide variety of positions in the tree canopies. The fingers and toes became capable of grasping the trunks and branches. The claws changed into flat nails, and the inner surfaces of the hands and feet were sculpted into sensitive, hairless pads. Balance and motor control were correspondingly improved. Surrounded as they were by the open spaces of their habitat, and depending completely on the judgment of distances simply to move about, the primates' sense of smell was deemphasized while simultaneously their vision was improved. The eyes were enlarged and moved forward in the head, providing overlapping images and instant three-dimensional perception of the environment.

During the African radiation, some primates

BUSHMAN MOTHER in the photograph on opposite page prepares the skin of a freshly killed animal. Hunter-gatherers such as the African bushmen are modern survivals of an early stage in the evolution of human society.

TREE SHREW

TARSIER

LEMUR

MAN

NEW
WORLD
MONKEY

OLD WORLD MONKEY

APE (CHIMPANZEE)

1 **PRIMATES** are mammals adapted to life in trees.
Illustration depicts living representatives of the seven
major groups. The earliest primates probably
resembled the tree shrew at upper left.

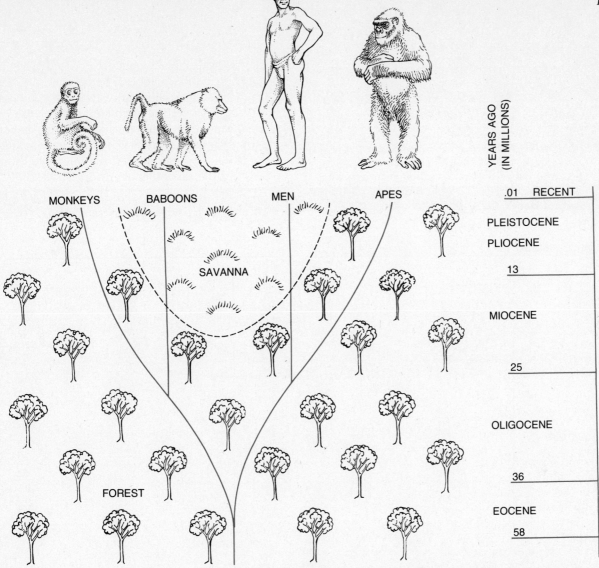

YEARS AGO
(IN MILLIONS)

MONKEYS   BABOONS   MEN   APES

SAVANNA

FOREST

.01	RECENT
	PLEISTOCENE
	PLIOCENE
13	
	MIOCENE
25	
	OLIGOCENE
36	
	EOCENE
58	

ADAPTIVE RADIATION produced the higher primate groups of Africa. Baboons and man have adapted to life on the savanna, the tree-studded grassland adjacent to the tropical forests.

2

were able to enlarge their size by one or the other of two very different means. The first was brachiation—movement by grasping branches over the head and swinging along beneath them. This is the only means by which a large animal can travel easily through the tree tops. The true apes—the chimpanzee, gorilla, gibbon, and orang-utan—become highly modified for brachiation. They have long, powerful arms and shoulders and relatively heavy, deep chests. Their hind legs are short and their feet shaped for grasping. As a consequence locomotion on the ground is usually on all fours, with most of the weight being supported on the knuckles of the feet and hands. It

is important to realize that these several traits, which in the popular mind are the essence of the ape's nature, are not primitive characteristics. They are advanced specializations of large animals for life in trees. Man never passed this way in evolution. Although his ancestors in the early and middle Cenozoic could be called primitive apes, he did not descend in a direct line from any creature closely resembling one of the modern apes.

The second way large size could be achieved by

> The bipedal stride is unique in the animal kingdom. It freed the hands, giving the man-apes the potential for carrying weapons and tools.

3 **MAN AND HIS ANCESTORS comprise the family Hominidae. Breakthrough in human evolution occurred between 2 and 4 million years ago when Australopithecus or a similar form gave rise to the earliest Homo, the true ancestral man.**

primates was simply to abandon an arboreal existence and return to the ground, where bigness is no impediment to locomotion. In Africa two principal lines of primates made this transition independently—the baboons and man *(Figure 2)*. The baboons have regained a close approach to the quadrupedal (four-footed) locomotion of their ancient ground-dwelling ancestors. They run swiftly on all fours over the open land of the savanna. At night they often climb trees in order to sleep more safely. The ancestors of man, the man-apes as they are usually called, also left the forests to hunt and sleep in the savannas of Africa. In so doing, they

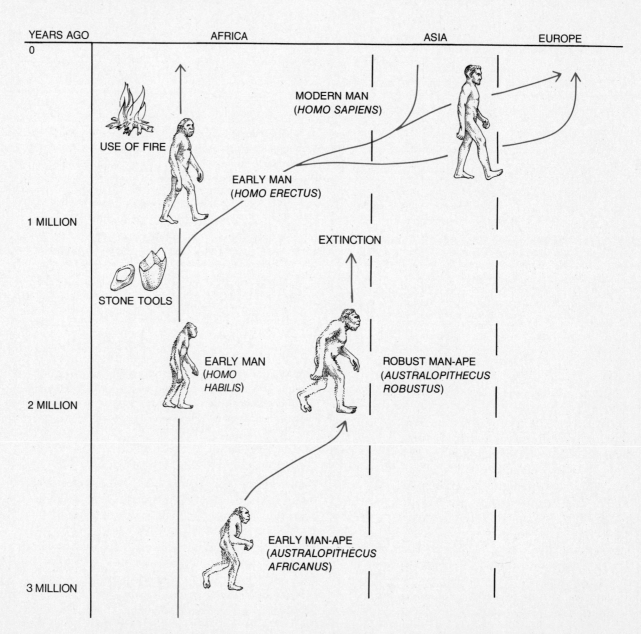

| YEARS AGO | AFRICA | ASIA | EUROPE |

USE OF FIRE

MODERN MAN
(*HOMO SAPIENS*)

EARLY MAN
(*HOMO ERECTUS*)

STONE TOOLS

EXTINCTION

EARLY MAN
(*HOMO HABILIS*)

ROBUST MAN-APE
(*AUSTRALOPITHECUS ROBUSTUS*)

EARLY MAN-APE
(*AUSTRALOPITHECUS AFRICANUS*)

0

1 MILLION

2 MILLION

3 MILLION

adopted a form of locomotion destined to have enormous secondary consequences in their subsequent evolution. This is the bipedal stride, in which the body is held in a fully erect position and moved exclusively by the stiff swinging motions of the hind legs beneath it. The bipedal stride is unique in the animal kingdom. It freed the hands and gave the man-apes the potential for carrying weapons and tools.

FROM MAN-APES TO MEN

Figure 3 depicts the phylogeny of the family Hominidae —man and his immediate ancestors —which has been deduced from fossil material discovered in the past one hundred years throughout Europe, Asia, and Africa. In this book we do not have space to describe the story of the arduous search for those remains, which has provided the piece-by-piece solution of the puzzle of man's origin. Suffice it to say that it has been, and continues to be, one of the most exciting chapters in the history of science, and we recommend further reading on the subject if for no other reason than sheer pleasure. Our discussion will be confined to several of the key biological generalizations that have emerged.

The first generalization is that man's erect posture and bipedal locomotion, the qualities which at a distance make an individual look human, appeared full blown long before the great enlargement of the brain that truly characterizes our species. Three to four million years ago man-apes of the genus *Australopithecus* had a human posture combined with a brain not much larger than that of a gorilla. The man-apes were only moderately impressive. They used stones as weapons and possibly also as crude cutting instruments. The man-apes may have also wielded sticks and bones, but chimpanzees do no less. They strip small branches of their leaves and use them to root out termites from their burrows; they also crumple leaves in their hands and use them to sponge up water from narrow recesses. The great leap forward in human evolution began about four to two million years ago when *Australopithecus* or an earlier similar form gave rise to the earliest "true" men of the genus *Homo* (Figure 4). From 1,500,000 to 500,000 years ago a distinctly more advanced form, *Homo erectus*, appeared and began to spread from Africa through

EARLIEST TRUE MEN included Homo habilis of Africa. In this reconstruction of a Tanzanian grassland two million years ago, a band of hunters drives rival predators from a fallen dinothere, an extinct relative of the modern elephant. In the left foreground are spotted hyenas; at right, a female sabertooth cat and her two cubs. The men, less than 1.5 meters (5 feet) tall were individually no match for the large carnivores; a high degree of cooperation was required to handle such large prey. This social behavior evolved together with higher intelligence and the superior ability to use tools.

4

large parts of Europe and Asia. These men had a cranial capacity between that of *Australopithecus* and modern man, and used fire and tools as advanced as stone axes.

What caused the man-ape *Australopithecus* to cross the threshold to *Homo?* A growing amount of circumstantial evidence suggests that early *Australopithecus africanus* was a modest hunter, subsisting at least in part on birds, reptiles, and rodents and other small mammals. He also probably consumed fruits and vegetables such as tubers. Later when he acquired the capacity to hunt antelopes and other large mammals, a rich new source of protein was added to his diet. At about this time men probably began to hunt in groups; this was at least partially responsible for the evolution of the genus *Homo.*

Hunting groups were primarily or exclusively male. An elaborate language was developed to coordinate the actions of the hunters and the complicated social relations necessitated by division of labor between the sexes at home.

Pack-hunting is rare in nature; it occurs, mainly among wolves, wild dogs and killer whales. It requires not only cooperation but considerable communication and division of labor. The more these qualities can be refined, the more successful the hunt. Early man evolved a complex social structure, in which the hunting groups were primarily or exclusively male; the women and children may have remained secluded in caves and bivouacs while the hunters were abroad. An elaborate language, flexible yet precise, was developed to coordinate the actions of the hunting groups and the complicated social relations necessitated by the division of labor between the sexes at home. The brain grew larger, providing a physical basis for the evolving language. The larger brain also permitted a dramatic increase in other dimensions of intelligence, leading to the evolution of true culture: the handing of knowledge and traditions from one generation to the next. All of these traits were advanced as *Homo* arose from the early *Australopithecus* or their ancestors.

In previous chapters evolution was described as opportunistic. Each important adaptive radiation produces some types that are "thrown away" — doomed to early extinction; others survive and give rise to the next evolutionary grade. Human evolution was no exception. As *Australopithecus africanus* progressed, it gave rise to a second man-ape species, *Australopithecus robustus*. The massive jawbones and teeth of *A. robustus* were best suited for grinding roots and seeds, and suggest that it became more vegetarian at the same time *A. africanus* and *Homo* were turning increasingly to the pursuit of animals. This remarkable creature became extinct a little less than a million years ago without having contributed to the ancestry of *Homo.* A similar radiation of short-lived races occurred during the last 100,000 years of the evolution of modern man *(Homo sapiens)*. The most famous example is Neanderthal Man *(Homo sapiens neanderthalensis)*, who had a full-sized brain but heavy skull bones and brutish features dominated by a thick brow ridge. The Neanderthal race lived from about 100,000 to 35,000 years ago around the Mediterranean and northward for some distance into Europe. Other relatively primitive races that have since passed into extinction were Solo Man of Java and Rhodesian Man of Africa.

Where did the *Australopithecus* man-apes originate? This question is the equivalent of asking when it was that the ancestors of the man-apes began to diverge from those of the chimpanzee, gorilla, and other modern apes. Few fossils have been discovered that shed light on this crucial episode. The best candidate for the primal ancestor is *Ramapithecus punjabicus*, known from only a few teeth and jaw fragments collected at two localities in India and Kenya. The specimens, which date from the transition of Miocene to Pliocene times approximately 14 million years ago, are more man-like in structure than apelike. *Ramapithecus* has accordingly been placed in the Hominidae. It is very tentatively judged to be an ancestor of man in a direct line. But it is also so similar to the primitive apes that it is considered to fall near the point at which the Hominidae first diverged as a separate group.

ADAPTATIONS OF MODERN MAN

The evolution of the man-apes and early man has shaped the biology of the modern human species. All of the uniquely human traits are adaptations to environments in which man evolved in the millions of years before history. These adaptations in-

Man lost most of his hair. Perhaps the reason that he became a "naked ape" was to cool the body during the pursuit of prey on the African plains.

clude profound modifications of anatomy, physiology, and behavior.

The form of the skeleton is adapted to fully erect posture and bipedal locomotion *(Figure 5)*. The spine is curved in a way that distributes the weight of the trunk more evenly down its length. The chest has been flattened, a geometrical change that moves the center of body weight back toward the spine. The pelvis is broadened to provide adequate attachment for the powerful striding muscles of the upper leg, and realigned to convert it into a lower, basinlike support for the viscera. The tail has disappeared and its shortened vertebrae now curve down and in to form the coccyx, which makes up part of the floor of the pelvis. This transformation of the tail vertebrae is unique among the vertebrates. The occipital condyles, those bony points by which the skull pivots on the neck vertebrae, have rotated far beneath the lower surface of the skull until the weight of the head is nearly (but not exactly) balanced upon them. The leg is straight, aligned in such a way that it transmits the entire weight of the body through the knee joint. The feet are specialized to accommodate the bipedal stride. They have been narrowed and lengthened, and the toes have been shortened to become little more than flexible extensions of the feet. The result is that the entire weight of the body can be placed on one foot at a time and passed from the heel to the ball and toes as the foot rocks forward during the walking motion. The opposable thumb of the hand, a primate trait, attains its maximum development in man. It is relatively very long and powerfully muscled, providing the hand as a whole with a strong, precise grip. The face has retreated back into the skull, presumably to assist the shift of gravity of the head in the direction of the occipital condyles. As the brain grew in size, the forehead expanded, completing the evolution of the main structure of the human face.

Man lost most of his hair. The reason why he became a "naked ape," to use the phrase in the title of a popular book on human evolution, is still a matter of conjecture. Perhaps the most plausible explanation is that nakedness was originally a de-

vice to assist in the cooling of the body during the strenuous pursuit of prey in the hot sunlight of the African plains. It is correlated with man's heavy reliance on sweating to reduce body heat. The human body is uniquely adapted for producing large quantities of sweat. It contains between two and five million sweat glands, a number far in excess of that found in any other primate.

HUMAN SKELETON reveals modifications for upright posture and bipedal stride. Straight leg, curved spine, flattened chest and positioning of head all serve to align weight of body along a vertical axis. Broad pelvis provides attachment for powerful striding muscles, and together with curved coccyx, supports the viscera. Shortened toes facilitate heel-and-toe walking motion.

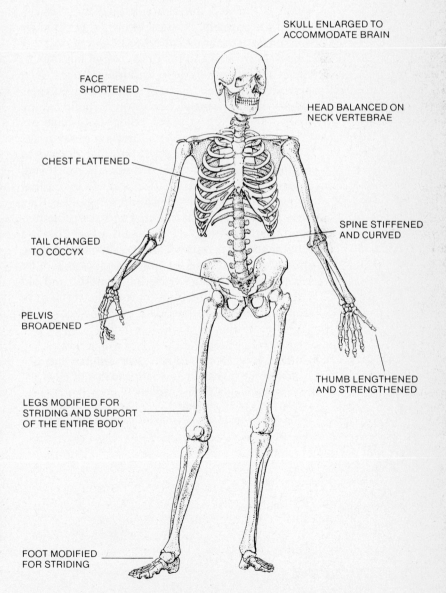

5

SKULL ENLARGED TO ACCOMMODATE BRAIN

FACE SHORTENED

HEAD BALANCED ON NECK VERTEBRAE

CHEST FLATTENED

SPINE STIFFENED AND CURVED

TAIL CHANGED TO COCCYX

PELVIS BROADENED

THUMB LENGTHENED AND STRENGTHENED

LEGS MODIFIED FOR STRIDING AND SUPPORT OF THE ENTIRE BODY

FOOT MODIFIED FOR STRIDING

The reproductive biology of man has undergone extensive changes. The estrous cycle of the female has been altered in two ways that affect sexual behavior. Menstruation, the loss of blood which occurs at the end of the cycle whenever fertilization fails to occur, has been intensified. The females of many primates experience slight bleeding at this time, but only in women is there a heavy loss of blood from the sloughing uterine wall. The estrus itself, the period of "heat" during which ovulation occurs and the female actively seeks copulation, has been virtually erased. Instead of the intensive bouts of copulation during estrus encountered in other primates, there is more or less continuous sexual activity throughout the human cycle. And in place of the sexual signals of estrus, such as changes in skin color surrounding the female sexual organs or the release of distinctive odors, copulation is initiated through extended foreplay between the partners. Physical attraction is based on conspicuous anatomical features including the breasts of women and the pubic hair of both sexes. As a consequence of these multiple changes, men and women form intimate, long-lasting partnerships. Human sexual evolution is believed by many zoologists and anthropologists to have occurred as part of the unique division of labor between the sexes that arose when man turned to hunting in groups. Continuous sexual liasons were needed to maintain the support of individual women and their young children, because they were sequestered in the camps and not permitted to join the chase. The period of childhood lengthened to permit the more elaborate learning procedures required to join the societies. Life beyond the reproductive period—middle and old age in the peculiarly human sense of these stages—was greatly extended, probably because of

6 EVOLUTION OF THE BRAIN. Extraordinary increase in size of the human brain relative to that of other vertebrates is due almost entirely to growth of cortex. Older parts of the brain that control coordination, reflexes and instinctive behavior came to lie beneath the cortex.

> The expansion of the cortex was associated with the most important development of all: the human form of language.

the adaptive advantage of ensuring the care of the last offspring produced.

The brain was enormously increased by the growth of the cortex, which contains the centers for memory and complex computation *(Figure 6)*. During the approximately one million years required for the transition from *Australopithecus* to *Homo*, the brain doubled in weight, and intelligence increased to a degree unprecedented in all earlier animal evolution.

The expansion of the cortex was associated with the most important development of all: the development of the human form of language. You will recall from Chapter 22 that animal communication consists of the exchange of a limited repertory of signals which are hereditarily fixed and relatively invariable within species. Almost all of these signals pertain to immediate circumstances. They may consist of a threat, a plea for food, or an invitation to "follow me," in other words single statements that are simple, urgent, and direct. Human language, in contrast, has been liberated in time and space. It can refer not only to immediate circumstances but also to past and future times and to distant places. The basis of this change, like the one that transformed human sexuality, has been speculated to lie in the necessities imposed by the group hunting and division of labor adopted by early human societies. The increase in information was made possible by several new qualities in the vocal exchanges. Words became arbitrary in meaning—different sounds could be applied to

FISH   REPTILE   RABBIT   MAN

different objects (and later, to different ideas) according to the desires of the speakers. True syntax was added — the meaning of words changed according to the other words with which they were spoken and the order in which all appeared. These two traits, arbitrary symbolism and syntax, made possible a productive language, one in which the number of messages can be multiplied without limit, simply by adding newly coined words or combining old ones in new combinations. Man is the only species that succeeded in creating a truly productive language, and it is the scaffolding of human culture and civilization.

READINGS

B. CAMPBELL, *Human Evolution: An Introduction to Man's Adaptations*, 2nd Edition, Chicago, Aldine Publishing Co., 1974. Easily the best written and most authoritative account of the adaptive basis of human evolution. This is the ideal book to read after concluding the present chapter.

D. MORRIS, *The Naked Ape*, New York, McGraw-Hill, 1969. A popular and entertainingly written version of the subjects covered by Campbell's work. This is the book to suggest to friends who do not have the biological background equivalent to an elementary college course.

D. PILBEAM, *The Ascent of Man: An Introduction to Human Evolution*, New York, Macmillan, 1972. This book covers much the same ground as Campbell's but lays more emphasis on the modern races of man. It also provides a fascinating account of the history of the discovery of human fossils.

C. SAGAN, *The Dragons of Eden*, New York, Random House, 1977. A sound and entertaining popular account of early human evolution with special attention to the origin of intelligence.

E. O. WILSON, *Sociobiology: The New Synthesis*, Cambridge, MA, Belknap Press of Harvard University Press, 1975. Contains a reconstruction of early human social evolution in comparison with the social systems of other primates.

Part **5** Alternative futures

The Romans named the first month of the year after Janus, a god with two faces who could look backward in time as well as forward. Man is the first animal to have the ability to look backward, in the sense that he can study the process of evolution that produced him. Even more important, he has the ability to look forward, aware that he now has the power to shape not only his own evolution but also that of other species. Man will decide —either consciously or by default —the next steps in the future of life on this planet.

It is a grave responsibility. In the words of biologist Clifford Grobstein, ''Man has long assumed purpose in the universe —a purpose that he has personalized in a Divine Being. Man is now challenged to provide purpose himself, since science is placing in his hand increasing power to control events and hence to implement purpose.''

Is our species equal to the task? Man's performance to date raises doubts about his competence as overlord of Earth. He has squandered its resources as if there were no tomorrow, and has driven one species after another to extinction. He has just begun to realize that overconfidence in his mastery and an arrogant disregard of his place in nature jeopardize his own chances for survival. One biologist observed that it is not at all certain that superior intelligence confers a long-term survival advantage on the species that possesses it.

Restoring the equilibrium of the Earth is a worldwide problem. Mankind as a whole will be required to act with a degree of cooperation and altruism heretofore more characteristic of insect than of human societies. Human nature changes slowly if at all, and so do human institutions. Can the reaction time of our culture be speeded up enough to cope with the series of crises now rushing toward us?

Prophecy is a risky game, but the task can be simplified

770

by projecting present trends into the future, preferably the near future. This process leads us to predict that most of the ecological crises discussed in the next two chapters will be resolved by the end of the 21st century, either by man's reason or by nature's indifference. We have attempted to make some educated guesses about the outcome of a few of these crises, particularly the most urgent one: the population explosion. But our guesses are colored by our individual, personal and largely nonscientific judgments of the nature of man. We found ourselves in disagreement on this subject, and our views of man's prospects spread across a spectrum ranging from tempered optimism to gloom.

The optimists among us believe that our species possesses the flexibility and generosity, or at least the enlightened self-interest, to find a reasonable solution to our predicament; and that although the future may hold some difficult readjustments for man, catastrophe is neither inevitable nor probable. Chapter 35 reflects the view that the biological sciences will lead man to a new harmony with nature.

The pessimists argue that there is no precedent in human history for the required degree of global altruism or even global cooperation; that it may already be too late for whole nations to reverse the disastrous momentum of rising population, soaring expectations, and commitment to economic growth —in short, that man's cultural reaction time is too slow to solve so many urgent problems so quickly. The view that man is about to enter one of the Dark Ages of human history is the theme of Chapter 36. The reader can decide for himself which of our alternative futures he prefers to believe; and as a member of the next generation, he can help to shape it.

Have you descended to the springs of the sea, or walked in the unfathomable deep? Have the gates of death been revealed to you? Have you ever seen the doorkeepers of the place of darkness? Have you comprehended the vast expanse of the world? Come, tell me all of this, if you know. —Book of Job, The New English Bible

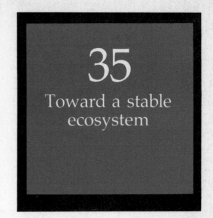

Devoid of science as well as impoverished of philosophy, the human species for the greater part of its existence has occupied a fragile and subordinate position in nature. Near the beginning of recorded history, God, according to the biblical account, could select Job as a representative of humanity and mock him for his ignorance.

The rise of modern science and technology has utterly changed man's position. He has personally descended to the deepest trenches of the sea and can orbit the Earth at just beneath escape velocity. He controls nuclear power and transmits messages through space at the speed of light. He understands the chemistry of the gene and has taken the first steps toward the artificial synthesis of life in the laboratory. He has walked on the moon, an achievement that would surely have strained even Job's capacity for dreaming.

But man has not freed himself. Like the mythical Greek hero Prometheus, who tricked Zeus and stole fire from the gods, man's success brings with it the risk of cruel punishment. The control of atomic power is accompanied by the danger of atomic war. Man's eminence as a biological species has temporarily abolished the old population controls and launched him into the population explosion, an unchecked growth that can eventually consume the resources of the world and cancel all the benefits of science and technology. It has been argued that even if we escape atomic war we will achieve essentially the same result from the popula-

MEGASTRUCTURE on opposite page was envisioned by architect Paolo Soleri as a component of a Utopian city of the future. A cluster of these enormous multi-level structures, taller than the World Trade Center and much more massive, would house all the activities of a city. By eliminating suburbs and other one-level forms of urban sprawl, Soleri's high-density city would conserve the surrounding countryside.

tion explosion —if we are foolish enough to continue reproducing at the same rate for another hundred years. For this reason the population factor must weigh heavily in all attempts at prophecy.

THE POPULATION EXPLOSION

The alarming facts concerning human population growth can be quickly comprehended by studying the growth curve reproduced in Figure 1. Human numbers remained almost constant for hundreds of thousands of years. If the entire history of *Homo*, roughly the period of the Old Stone Age, were put in scale, the base line in Figure 1 would extend to the left for 35 feet. It has been estimated that 77 billion babies have been born during this time. No fewer than 65 billion, or 94 percent, have lived since 6000 B.C., and 3 billion, or 4 percent, are alive today. In only about 30 more years, if the current growth rate is sustained, the world population will double again, to about 6 or 7 billion. The biological basis of the problem is disclosed by the dip in the curve seen in the 14th Century. This temporary decline, in 1348–1350 to be exact, was caused by the Black Death (the Germans called it the Great Dying) which destroyed a quarter of the population of Europe. The disease responsible was plague, the pathogen of which is a single bacterial species (*Pasteurella pestis*) transmitted by fleas from rats to man. Those of us today who take for granted the hygienic comforts of western civilization find it hard to appreciate the threat imposed by such infectious diseases in earlier times. Typhus, malaria, syphilis, cholera, yellow fever, and a long deadly list of other diseases collaborated with famine and war to make early death highly probable. This was the reason there was no population problem throughout most of man's history. Man's enemies were the density dependent controls in whose presence he had first evolved. A

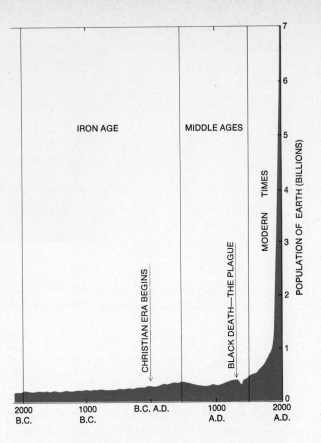

IRON AGE   MIDDLE AGES

MODERN TIMES

CHRISTIAN ERA BEGINS

BLACK DEATH—THE PLAGUE

POPULATION OF EARTH (BILLIONS)

2000 B.C.   1000 B.C.   B.C. A.D.   1000 A.D.   2000 A.D.

1 **HUMAN POPULATION has remained virtually constant in size during most of man's existence on Earth. In 1800 A.D. world population reached one billion. By 1900 it had doubled. At present rate of growth (about 70 million people per year) it will approach seven billion by the year 2000.**

high natural birth rate was required to maintain the population at its constant low level. The population biology of mankind, in short, was typical of an animal species. With the advent of agriculture and later of more careful hygiene and modern medical practice, the death rate was drastically reduced in all the societies that could make use of these technologies. But the birth rate remained the same. As a result, a net annual gain in the citizenry of between one and three percent was achieved in most countries, and the populations have grown as inexorably as bacteria released into a flask of nutrient broth.

Mankind, you must now well realize, has already passed the point where it places a severe strain on the resources of the globe. This one species already consumes more food than all other land vertebrates combined. We have stripped most of the forests and grasslands or altered them to our purpose, and the fish of many of the richest banks of the sea have been harvested below their point of optimal yield. Although human populations are expanding exponentially, the size of the earth remains finite. In the time of Christ the density was approximately one person per square kilometer. Today it is approximately 20/km² and in less than 30 years it will exceed 40/km². In the year 2600, if the present growth is somehow miraculously preserved, there will be just enough land for each person to stand shoulder to shoulder. The food energy available can be increased slightly by technological improvements, but in the long run it too will be a finite resource. All of the algae and other plant growth of the oceans together produce somewhere from 5.0 to 13.5×10^{16} calories per year, and all the plants of the land produce between 5.0 and 7.2×10^{16} calories per year. Thus a maximum total of 1 to 2×10^{17} calories are annually made available to animals over the entire Earth in the form of plant food. A single human being requires an average of 2,200 calories per day or, 8×10^5 calories per year, and the entire human race requires over 10^{15} calories per year. It follows that mankind already pre-empts at least one percent of all the energy available, and will continue to do so, even if every person were to go on to a strictly vegetarian diet! It is in fact already too late for the whole world to eat steaks, because, as mentioned in Chapter 33, cows and other herbivores are very inefficient devices for gathering and packaging energy. In a few more decades of unabated population growth, it may be too late for the whole world even to eat grass.

ENERGY AND RESOURCES

Man is altering the world in other ways than by simply consuming more and more of the available food. The technologically advanced nations are also demanding energy for industrial and household use at an accelerating pace. In 1970 the United States generated approximately 1.5 trillion kilowatt-hours of electrical energy by tapping hydroelectric power and consuming gas, oil, coal, and nuclear fuels. In the year 2000, according to conservative estimates, this nation will be generating 9 trillion kilowatt-hours per year. The needs of the developing countries will increase even faster. A variety of nonrenewable resources, from such fossil fuels as oil to copper and other metals, and even so vital an agricultural material as phosphate, are being depleted. Eventually we must find substitutes for them or revert, in ways painful to contemplate, to much lower levels of consumption. The great American dream of the nineteenth century, that this country contains land enough for our descendants to the thousandth and thousandth generation, has been cut cruelly short by the elementary mathematics of population growth and rising economic expectations.

Assessments of the population explosion are conventionally limited to the question of whether man can survive, and in what degree of comfort. We should never forget that the rest of the living world is being endangered as well, with uncertain ultimate consequences to ourselves. Dozens of species of mammals and birds have become extinct through man's activities during the past 200 years alone. We can only guess the number of other, less conspicuous kinds of plants and animals that have met the same fate. Many species must surely have been eliminated without our being aware that they ever existed. The principal human influence on the environment is generally to simplify it —to reduce it to uniform conditions which, for the moment at least, seem most favorable to sustain high densities of human beings. The result is a steady reduction in the number of species of plants and animals that can exist on the planet. Of course, a very few agricultural and pest species benefit. Corn, cattle, rats, and cockroaches are able to build abnormally large populations alongside those of man. But the vast majority of species are reduced in number and increasingly restricted in distribution. The exact result of all this attrition can only be guessed, but there is every reason to believe that our more enlightened descendants will come to regret it deeply.

All human problems are subordinate to the population crisis. This means that if we continue on our present course, the problem will surely finish us. In the case of atomic war, it is necessary to perform the drastic act of starting the war in order to suffer its terrible effects. But in the case of the population crisis, nothing more is required than mindlessly to continue what we are already doing. Fortunately, an awareness of the problem has begun to grow among educated people, and there are signs from India to the United States that political leaders are willing to try to do something about it. It is a cardinal principle that every nation must be brought through the demographic shift (Figure 2). In the most primitive human societies, such as the bushmen of Africa and the highland peoples of New Guinea, which are still essentially in the Stone Age, a high death rate compensates for the naturally high birth rate. The populations consequently grow slowly or not at all. In the developing countries, the death rate has been re-

duced but the people still cling to the old custom of producing large families. The result is rapid population growth. Only when birth rates are brought down by contraception and deliberate population planning can the demographic shift be completed. Then low birth rates match low death rates, and the population size again stabilizes. At the present time only the most advanced industrial countries have made any progress toward completing the demographic shift. Few of these have actually reduced birth rates all the way to the point of halting population growth. To do so requires some degree of altruism and a willingness to discuss what remains an emotional subject hedged by religious taboos. Population control will certainly be discussed with ever greater freedom and maturity in the future. It is also likely that the subject of optimal population size, already a favorite in academic groups, will soon be debated more publicly. The reader will recall from Chapter 32 that the final population size (K, the carrying capacity of the environment) is independent of the rate of population growth by which it was reached. In other words, the demographic shift can be terminated at any population level. A decision must eventually

DEMOGRAPHIC SHIFT is necessary to attain population stability. In primitive societies high birth rate is offset by high death rate (left) and population size remains constant. Transition to modern technology (center) slashes the death rate. Birth rate remains high, causing sharp increase in population. To restore equilibrium, birth rate must be reduced to correspond to death rate (right).

2

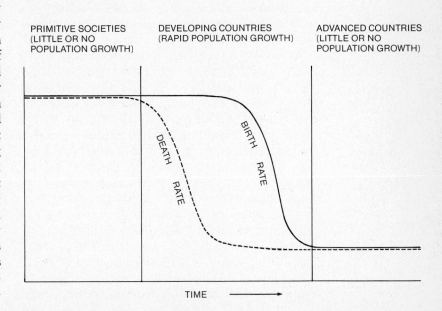

PRIMITIVE SOCIETIES (LITTLE OR NO POPULATION GROWTH)   DEVELOPING COUNTRIES (RAPID POPULATION GROWTH)   ADVANCED COUNTRIES (LITTLE OR NO POPULATION GROWTH)

DEATH RATE

BIRTH RATE

TIME

be made concerning the final level. Will our descendants settle for one billion Americans, saving whatever might remain of the countryside by compressing the population into strip cities of megastructures extending for hundreds of miles along the coasts and rivers? Will they choose the present population size and encourage citizens to move into smaller, more livable communities? Or will they prefer to bring birth rates temporarily below death rates, in order to reduce the population size and enjoy more space than is available at the present time? We cannot say which course will be chosen, only that the choice will have to be deliberately and carefully made in the not too distant future.

MAN'S ANIMAL NATURE

There are pessimists who claim that man is fundamentally too irrational to arrive at such decisions and will not find his way to a permanently secure future. You have no doubt heard all your life the argument that human nature can't change. Or, as the seventeenth century dramatist Molière acidly expressed it for all the cynics in this world, "Man, I can assure you, is a nasty animal." This pessimistic view has gained new credence in recent years with the revelation of man's evolutionary beginnings in Africa. It is widely believed among experts on the subject that the massive enlargement of the brain during the past two million years of evolution was associated with an increasing shift to a carnivorous diet. When the ancestral man-apes took up hunting in groups, they enriched their communication system and divided labor among the group members. This theory envisions our distant ancestors as pack apes or killer apes, specialized for the pursuit and destruction of large animals. Perhaps the man-apes were also aggressively territorial. From such postulates it is easy to form a rationalization for our own bad behavior. Raymond Dart, the South African who participated in the early discoveries of the man-ape fossils, put it this way in 1953: "The blood-bespattered, slaughter-gutted archives of human history from the earliest Egyptian and Sumerian records to the most recent atrocities of the Second World War accord with early universal cannibalism, with animal and sacrificial practices or their substitutes in formalized religions and with the world-wide scalping, headhunting, body-mutilating and necrophiliac practices of mankind in proclaiming this common blood lust differentiator, this mark of Cain that separates man dietetically from his anthropoidal relatives and allies him rather with the deadliest Carnivora."

Dart has undoubtedly painted too lurid a picture. But the opinion persists among many scientists that the beastliness in human nature is just that —beastliness held over from our man-ape ancestry, during which aggression, territorial instinct, and blind tribal loyalty became valuable adaptive traits. A few popular writers, including Robert Ardrey, Konrad Lorenz, and Desmond Morris, have developed variations on the theme in the last few years. They interpret men as naked apes, to use Morris' catchy phrase, peculiar bipedal primates that are ill-adapted for the complexity of their own hastily built civilizations. These authors offer the curiously comforting proposition that our sins are only animal sins ("Original Sin" if you wish), the result of instincts that originated long ago as stereotyped, strictly inherited traits.

The killer-ape theory has not been accepted by all serious students of the subject. There are many, especially psychologists and sociologists trained in the liberal humanist tradition, who totally reject the idea that modern man is slave to the residual instincts of the man-apes. Instead, they view the shortcomings of human behavior as the result of the imperfections of society. The human mind is interpreted to be an extremely flexible, imprintable instrument, which goes wrong only when it is subjected to abnormal stress or is trained incorrectly in the first place.

The great debate about human nature reveals both the shortcomings of the social sciences, particularly human psychology and sociology, and their critical importance for the future. Perhaps you have recognized that this debate is a continuation of the old nature-versus-nurture controversy which has troubled philosophers since the seventeenth century. Man is obviously higher than a man-ape and nowhere close to being an angel. But exactly where he stands, in what dimensions of innate morality and intelligence, is an extraordinarily difficult matter to assess. The crucial question

---

Biology is now entering its golden age. Gaining daily in depth and competence, it will be used to direct the coming stabilization of the world ecosystem.

---

is, To what degree is man's nature flexible by education and the constraints of law, and conversely to what extent should his governments and institutions be shaped to conform to his instinctive nature? Biology can find no higher goal than to join the social sciences in the search for a workable answer.

Future historians may judge that science in the remaining part of the twentieth century consisted of a turning inward, an increasing appreciation of the uniqueness of the Earth as opposed to all other accessible worlds, and of the living as opposed to the nonliving. In the first chapter of this book we explained why doubts exist that any life form, at least any exhibiting advanced organization, will be encountered elsewhere in the solar system. Life can exist on the planets of other stars, to be sure, but their distance makes it doubtful that we will explore them in the foreseeable future. The reaches of space are so immense that the human mind falters in trying to visualize them. The nearest star, Alpha Centauri, is 4.3 light-years away. It would take many human generations to travel just to that one place and back. A yard of earth then, with its teeming organisms, seems incomparably more valuable and interesting than a yard of Moon or of Venus, or inhabited worlds that for the moment can be reached only in the pages of science fiction.

Biology is now entering its golden age. Gaining daily in depth and competence, it will be used to direct the coming stabilization of the world ecosystem. Population growth must be halted at a level that permits a pleasant life for every human being. Material cycles must be carefully assayed and managed to ensure the steady supply of the essential resources in perpetuity. We confidently predict that within a few more decades medical science will be perfected to control the remaining diseases and to postpone senility to a maximum age. Genetic surgery, the transplanting of healthy genes for defective mutant alleles, will probably be attempted. If it succeeds, it will open the door to the eventual full genetic reconstitution of the human species; and this in turn will create a new moral dilemma. Shall we control our own evolution? If so, who will choose the "superior" genes with which to endow future generations? Meanwhile, the technique of transplanting entire organs, such as the heart and liver (and perhaps ultimately the brain), will be steadily improved. The search will continue for a technique to induce organs and limbs to regenerate when they have been lost by disease or accident. It is conceivable that such methods can be advanced to the point where death

by natural causes is circumvented altogether. This achievement would produce a second moral dilemma. If the death rate is drastically reduced, the birth rate must be correspondingly reduced. The result would be a second, unnatural demographic shift, resulting in a world of ancients kept alive by innumerable anatomical and biochemical crutches. Birth and childhood would become rare episodes. As unappealing as this prospect may seem to members of our own, tumultuous, youth-oriented culture, it is nevertheless the goal toward which we are striving —because death must always seem to be the enemy.

Primitive societies were joined intimately with nature. The hunter and gatherer had to have expert knowledge of the plants and animals on which his daily existence depended. His tribe was a modest component of the local ecosystem. With the industrial revolution and rapid urbanization, men drew temporarily away from the land, comfortable in the belief that they had "conquered nature," and entered a wholly new form of existence. The illusion took less than a century to shatter, as the problems created by overpopulation, pollution, and resource shortages became critical. Future societies will return to nature, but in a vastly more enlightened way. Man will again become a component of the stable ecosystem, perhaps even a modest one. He will look upon the millions of **species of plants and animals on Earth, himself in**cluded, as the ultimate wonder of his existence, deserving of unending study and of esthetic appreciation.

READINGS

P.R. Ehrlich, A.H. Ehrlich, and J.P. Holdren, *Ecoscience: Population, Resources, Environment*, San Francisco, W.H. Freeman & Company, 1977. A brilliant and authoritative account of the ramifying problems created by runaway population growth in the human species, the innumerable ecological consequences, and the steps that must be taken to save our future.

G. Feinberg, *The Prometheus Project*, New York, Doubleday, 1969. A clear-eyed prophecy of science in the future service of mankind, with due attention to the recent achievements and promises of biology.

P. Handler, ed., *Biology and the Future of Man*, New York, Oxford University Press, 1970. The best thoughts of 185 leading American biologists are presented in this survey sponsored by the National Academy of Sciences.

The world is being carried to the brink of ecological disaster
not by a singular fault which some clever scheme can
correct, but by the phalanx of powerful economic, political,
and social forces that constitute the march of history.
Anyone who proposes to cure the environmental crisis
undertakes thereby to change the course of history.
                                        —Barry Commoner

In *The Limits to Growth*, Dennis L. Meadows and his colleagues cite a French riddle for children: A farmer's pond is gradually covered by a giant water lily that doubles in size every day. If allowed to grow unchecked it would cover the pond in 30 days, choking off all other life in the water. The farmer decides not to worry about cutting back the lily until it covers half the pond. On what day will that happen? The answer, of course, is the 29th day. He has only one day to save his pond.

Meadows uses the riddle to illustrate the suddenness with which exponential growth approaches a fixed limit. The story is also a parable of man's plight on spaceship Earth. At the present point in human history, population, pollution and the consumption of nonrenewable resources are increasing exponentially; that is, they are increasing like compound interest. World industrial output, for example, has been rising at a rate of seven percent a year, which may sound modest until one considers what it means. One hundred dollars invested at seven percent compound interest will double every ten years. Exponential growth is consuming billions of tons of resources and adding billions of tons of pollutants to the environment every year. Clearly this growth cannot long be sustained on a planet with limited resources. The choice that faces mankind is whether to try to live within the limits of the Earth's resources by accepting a self-imposed limit on growth, or to keep on growing until stopped by nature, in ways that are likely to be much harsher than those that society might choose for itself.

GRAFFITI are children's reaction to the squalid anonymity of life in urban slums. Photograph on opposite page depicts names and street numbers scrawled on wall of a building in New York.

In theory it is possible to stabilize the human population and the economy at a level that the Earth can sustain for many millenia. But such a decision would involve drastic changes in virtually every aspect of human activity. It would mean limiting the world birth rate —by coercion if necessary; diverting capital from industry to the uneconomic production of food; and perhaps redistributing the wealth of the rich nations. As Barry Commoner has pointed out, such a decision would be no less than a commitment to change the course of history.

The solution to the ecological crisis is not primarily technological but political and social. Most of the necessary technology already exists; what is missing is the foresight and the will to apply it. Exponential growth confronts mankind with the need to make profound changes in human society in a very short time, probably within only two or three generations. And these decisions must involve a degree of global cooperation for which there is no precedent in human history.

Can human institutions react energetically enough, and swiftly enough, to avert disaster? Faced with difficult choices, human beings and human institutions usually find it easier to deny for as long as possible that a problem exists, and then to assume that if left alone, things will somehow work themselves out. Every year that the decisions are postponed, the greater the chance that the accumulated insults to the environment will become irreversible.

Recent history provides scant evidence that human society can restructure its priorities so quickly. Chapter 35 concluded with an affirmation that man will somehow find the necessary solutions simply because he MUST. No such optimism will be found here. Although we should like to believe otherwise, the authors of this chapter be-

> Those who believe that the Earth is already overcrowded will derive no comfort from the thought that within the lifetime of infants now alive, there will be four people on Earth for every one alive today.

lieve that individuals, corporations and nations will pursue their short-sighted self interest as they have in the past, until disaster shocks them out of their complacency.

POPULATION

Chapter 35 noted that the population of the Earth will approximately double in the next three decades, approaching seven billion by the year 2000. Most of this growth will occur in the Third World nations of Asia, Africa and Latin America, which contain two-thirds of the Earth's present population. In most industrialized nations population is increasing from 0.5 to 1.0 percent a year, and shows signs of leveling off. In the Third World, population is multiplying at rates of from two to three percent a year, and in many countries the **rates are still rising. Between 40 and 45 percent** of their population is less than 15 years old, which means that even if the birth rate is somehow stabilized at the replacement level, the absolute numbers of people will continue to increase for many decades.

If the developed nations can stabilize their populations at the replacement level by the year 2000, and the Third World nations by 2040 (an optimistic assumption), then about a century from now the population of the Earth will be 15.5 billion. Those who believe that the Earth is already overcrowded will derive no comfort from the thought that within the lifetimes of infants now alive, there will be four people on Earth for every one alive today.

One of the great unanswered questions of demography is how to achieve population stability. The desire for children is one of the mainsprings of human behavior. Obviously the first step toward stabilizing population is to put safe, reliable, easy-to-use methods of contraception into the hands of every couple in both developed and undeveloped countries. This will enable them to have the number of children they want. The next step is

somehow to persuade them to want only the number of children that Earth can support. But if voluntary methods continue to fail, what is the next step? In India, after assuming dictatorial powers, the government of Indira Gandhi decreed that any government employee who added a fourth child to his family after 1977 would be fired. Her successors promptly retreated from this unpopular position. Some workers have suggested that the only foolproof method of population control would be one that was not under the control of the couple—for example, an oral contraceptive dissolved in the public water supply, with the antidote available only upon application to the government. At birth each person might be issued a coupon entitling him or her to have one child, just as present law restricts a man to one wife.

In a democracy such coercive measures do not appear to be politically feasible in the foreseeable future. Any government that suggested them would be committing political suicide. But the alternative to these distasteful proposals is even more painful: the restoration of population equilibrium by starvation, disease, or global war. Society must make a choice; there is no way to escape that. To do nothing, or to take halfway measures, is to decide by default.

FOOD PRODUCTION

There is general agreement that perhaps one-third of the people on Earth today are undernourished. Perhaps 10 to 20 million deaths each year can be attributed directly or indirectly to malnutrition. Data from the United Nations Food and Agriculture Organization indicate that the diets in most of the developing countries do not supply the basic protein and caloric requirements. Although total world food production is rising, the gains have been wiped out by population growth.

Agronomists remain guardedly optimistic that the food supply can be doubled to feed the seven billion people who will be on Earth at the turn of the century, but many doubt that the Earth can sustain that population for long. Some workers argue that in terms of food supply, in the long run the Earth can sustain only 1.5 billion people, and that if one also takes into account the depletion of nonfood resources, plus the impact on the environment, the optimum long-term population of the Earth is only one billion—less than one-third the present level.

Within the past decade agronomists have introduced high-yield varieties of rice and wheat, new techniques of farming, and new ways to cope with plant diseases and insect pests. Sometimes called the Green Revolution, this "package" of seeds and techniques has enabled some farmers in countries such as Pakistan and Mexico to increase the productivity of their land, sometimes doubling local yields. The Green Revolution is not the answer to overpopulation but it might buy enough time, perhaps two decades, to work out a long-term solution.

The Green Revolution also has its dark side. It depends on enormous increases in the use of pesticides and nitrate fertilizers. Intensive use of pesticides is "addictive" because the pesticide selects for resistant strains of insects and destroys many of the beneficial parasites and predators that would otherwise limit the pest population. More and more spraying becomes necessary to maintain a given level of crop yields. The American agricultural system is already "hooked" on pesticides, and the Green Revolution is spreading this dependence to the poorer countries.

The present world production of pesticides is more than one million tons a year. To double world food production by the end of the century perhaps six million tons a year will be necessary because of their diminishing effectiveness. Pesticides ultimately accumulate in the oceans. As much as 25 percent of the DDT produced so far may have already found its way to the sea, and has begun to show up in the cells of marine organisms. A sixfold increase in the use of pesticides threatens to disrupt the marine ecosystem, perhaps reducing or contaminating the fish harvest upon which the world now depends for 20 percent of its protein.

Some of the worst pest problems are still to come. In the developing countries vast areas are being planted in single crops of the new cereal grains, creating splendid opportunities for pest epidemics. Much has been written about new approaches to pest control, particularly about narrow-spectrum insecticides and biological control, but so far these methods have proved to be expensive and of limited effectiveness.

The new high-yield strains of wheat and rice require enormous applications of fertilizer —from ten to 27 times more than the traditional varieties. At present the global use of fertilizers totals about 62 million tons a year, almost all of it in the advanced countries. U.S. farmers use nitrates at the rate of 70 kilograms per arable hectare (about 2½ acres), compared with eight kilograms in India and five in Africa. In 25 years the world consumption of fertilizers could exceed 700 million tons a year, with

Because cities must absorb the first shock of the population explosion, the quality of life in urban areas has deteriorated appallingly in the past two decades.

serious impact on inshore waterways. Not only does the runoff of excess nutrients threaten to choke lakes and streams with blooms of foul-smelling, fish-killing algae, but nitrates also pose a serious health hazard when they seep into drinking water. Nitrates are relatively harmless to humans but they can be converted by intestinal bacteria, especially in infants, to poisonous nitrites.

As with pesticides, there is also a point of diminishing return in the use of fertilizer. Illinois farmers used about 100,000 tons of nitrates to obtain an average yield of about 70 bushels of corn per acre in 1958. Eight years later they used 400,000 tons to obtain a yield of 95 bushels per acre. The extra 25 bushels required the use of four times as much fertilizer. There is mounting evidence that intensive application of nitrates eventually damages the soil, particularly the poorer soils found in many of the developing countries, reducing its response to additional fertilizer and impairing its future food-producing capacity. Moreover, the price of fertilizers and pesticides has increased sharply in the last few years. To buy them, the impoverished nations of the Third World must use their meager supply of foreign exchange. Despite the Green Revolution, over-population confronts the Third World with a dismal choice of risks: ignore the new agricultural technology and starve now, or apply it and starve later.

THE SOCIAL IMPACT

In little more than a decade, the Green Revolution has begun to tear at the social fabric of the undeveloped nations. The thrust of the Green Revolution is toward efficient large-scale farming, which favors large landowners who can afford the increased investment in fertilizers, pesticides and machinery. In the Third World as in the U.S., more food is being produced by fewer hands. In one village in India, for example, tenant farmers and sharecroppers are being squeezed off the land by landowners who prefer to cultivate the land with hired labor and pocket the profits. In short, the

Green Revolution has aggravated the classical political tensions between the haves and the have-nots. And throughout the world unemployed rural laborers are migrating to cities.

Because the city must absorb the first shock of the population explosion, the quality of life in urban areas has deteriorated appallingly in the last two decades. Growth is fastest around the most crowded metropolises of the Third World. Right now in the Third World there are about 100 cities with populations over one million. One-third of all Argentinians now live in greater Buenos Aires, and the population of Jakarta has soared from 550,000 to 4.5 million since the end of World War II. By the end of the century, according to U.N. projections, Mexico City will be one of the largest cities in the world, with a population of more than 30 million. Despite the efforts of many governments to encourage the pioneering of undeveloped regions, the migration to the cities continues. Ironically, even as living conditions become more desperate, the tide of aspiration throughout the world continues to rise. For some the city holds the promise of excitement and glamor; for most, at least the hope of a job and relief from loneliness and rural drudgery —a hope that for the unskilled worker all too often withers to despair in a decaying slum.

The fate of American cities is a portent of the fate of all cities. Analysis of the latest census data reveals what is happening to New York. In the last decade almost a million middle class whites fled the city, to be replaced by about the same number of blacks, Puerto Ricans and immigrants —legal and illegal —from the Caribbean and Latin America. Many of these people are from rural areas and have no marketable skills. The welfare rolls quadrupled to 1.3 million; roughly one of every six New Yorkers is now on welfare. Unemployment, drug addiction, crime and delinquency skyrocketed. The middle class fled to the suburbs, followed by many large corporations which found it increasingly difficult to attract young executives to the city. Hundreds of city blocks deteriorated into slums, particularly in Brooklyn and the Bronx, as the middle class moved out. The exodus of taxpaying citizens and businesses, and the abandonment of slum buildings by landlords (thus removing them from the tax rolls) were major factors in driving the city to the edge of bankruptcy. The story is being repeated in most of the older cities of the U.S. The American city has become, in the words of *The New Republic*, the domain of "the very rich, the very poor, the old, and the peculiar."

THE AUTOMOBILE

The sprawl of suburbs has been facilitated by the automobile. In Los Angeles, a collection of suburbs in search of a city, the tyranny of the automobile has become almost absolute. In the absence of even the rudiments of a mass transit system, the private automobile (or two or three of them) has become a necessity. Other cities in America are well along the road to Los Angelization. Between 1960 and 1970 the automobile population of the U.S. increased twice as fast as the human population, and the growth was concentrated in the suburbs. Even the Arab oil embargo of the mid-1970's, with the resulting shortage of gasoline, produced no more than a pause in the upswing of car sales. Highways, gas stations and parking lots now cover more of the surface area of the U.S. than do homes, stores and schools.

Automobiles burn more than 35 percent of the total annual energy budget of the nation, and collectively they are a major source of air pollution. Most of the gains from the new emission-control devices have been wiped out by the increase in the number of cars; each car pollutes less, but there are more cars to do the polluting. Even in death a car is a polluter. Unprofitable as scrap, it is abandoned on the street or stacked in hideous junkyards. It is probably the greatest single contributor to the trashing of America.

The automobile provides a case history of the difficulty of working out political solutions to environmental problems. British economist Ezra J. Mishan considers the invention of the private automobile "one of the great disasters to have befallen the human race." He argues that for a fraction of the money the nation is currently spending on the maintenance of private cars and the government services necessary to keep traffic moving, the country could provide comfortable, frequent and highly efficient public transport service (bus, train or subway) in all the major population areas; restrain and gradually reverse suburban sprawl, the futile attempts of commuters to get away from it all; restore quiet and dignity to cities, enabling people to stroll unobstructed by traffic and to enjoy once more the charm of towns and villages.

Mishan believes that the U.S. is almost hopelessly locked into its commitment to the automobile as the primary means of transit. One of every six Americans earns his living from an industry directly related to automobiles. Of the top 25 U.S. corporations (ranked according to their 1972 revenue), 17 derive a major portion of their income from products or services related to automobiles. Millions of jobs and billions of dollars add up to enormous political clout, and so nothing is done. So far no American city has dared to ban private automobiles, and the highway lobby has succeeded in blocking legislative attempts to channel highway funds into mass transit.

The rest of the world, which once mocked America's obsession with the automobile, now seems hellbent to catch up. Other nations are repeating our mistakes with the automobile, dooming themselves in turn to repeat our mistakes with highways, pollution and suburban sprawl. A political solution to the automobile problem appears to be nowhere in sight. Even a dictator might cringe at the thought of the fierce reactions that would be stirred up by an attempt to curtail the use of automobiles in America. It seems unlikely that if we lack the foresight and the will to get out of our cars and onto bicycles, buses and trains, we are not going to be able to make the much tougher decisions that are necessary to restore environmental sanity.

ECONOMIC GROWTH

The industrial output of the world is increasing at an even faster rate than population, and most economists see this as a good thing. Most of the growth is occurring in advanced countries, widening the economic gap between rich and poor nations. The Japanese economy, with a gross national product (GNP) of $1,190 per capita, is growing at almost ten percent a year; the Soviet Union, with a GNP of $1,100 per capita, is growing at 5.8 percent a year; and the U.S. with a GNP of more than $4,000 per capita, at 3.4 percent a year. Conversely the economy of India, with a GNP of $100 per capita, is growing at one percent a year.

The exponential economic growth of the industrial nations makes enormous demands on the nonrenewable resources of the planet, and one can only wonder how long it will be before the rest of the world catches on to what this means. The U.S., for example, consumes 44 percent of the total annual world output of coal, 28 percent of its iron output, 33 percent of its petroleum, and 63 percent of its natural gas. If the present rates of consumption persist, the known reserves of coal will be exhausted in 300 years, oil in 70 years, most metals in 50 years, and natural gas in 30 years. The diminishing supply has already begun to push prices upward. The price of mercury has increased 500 percent in the last 20 years and the price of lead has increased 300 percent in the last 30. In the early 1970s, as most readers will recall, the price of oil increased 400 percent.

An industrial economy not only depletes resources, but creates massive quantities of pollutants, some of them highly toxic. One of the most ominous pollution threats is radioactive waste from nuclear power plants. By the year 2000, according to current plans, roughly half of the electric power needs of the U.S. will be generated by nuclear plants. If the present safety standards are still in effect, these plants will release into the environment each year some 25 million curies of nuclear waste, mostly in the form of krypton gas and tritium (H^3) in coolant water. A curie is an enormous amount of radiation, equivalent to that emitted by one gram of radium. The "hottest" wastes from reactors, too dangerous to release, must be stored in sealed containers in abandoned mines, on the sea bottom, or in giant tank farms until their radioactivity decays, which will take centuries. Satisfactory sites for storage of radioactive waste are already in such short supply that a former AEC commissioner suggested rocketing the wastes into orbit around the sun. By the year 2000, U.S. nuclear plants will produce roughly a trillion curies of stored waste per year.

At present no one knows how much mercury, lead, radioactivity or other pollutants the world ecosystem can absorb without irreversible damage, or how one exotic pollutant may interact with others to produce nasty surprises. But without doubt there is an upper limit, and the danger of reaching that limit is increased by the long lag between the time a pollutant is dumped into the environment and the first appearance of biological damage. By then a "reservoir" of the pollutant may have accumulated in the environment, as was the case with DDT. With the continued expansion of industry and agriculture, the 15 billion people on the planet by the middle of the 21st century might well push the pollution burden to 100 times its present level. Technological developments might allow the expansion of industry with decreasing pollution, but only at high cost. The U.S. Council on Environmental Quality has estimated that even a partial cleanup of American air, water and solid-waste pollution would require an expenditure of $105 billion. A nation can defer such expenses only if it is willing to risk the health of its citizens and irreversible damage to the environment.

The notion of a zero-growth economy is heresy to both capitalist and socialist societies. Because most of the economic growth of the next few decades will occur in industrial countries, the primary focus of any effort to limit resource depletion must be the corporations of the U.S., Europe and Japan, and, to a lesser extent, the socialist-industrial nations of eastern Europe.

CORPORATION VERSUS PUBLIC

It is an axiom of classical economics that the primary goal of a business venture is to maximize profit, but Harvard economist John Kenneth Galbraith argues that the primary aim of the modern corporation is growth. Skeptics might well ask themselves what corporation president would boast of a DECREASE in sales. The present trend toward the formation of giant conglomerate corporations further supports Galbraith's contention. With so powerful a commitment to growth the corporation stands as the most serious obstacle to the attainment of a steady-state economy.

In theory the role of regulating corporations falls to the government, but in the words of ecologist William W. Murdoch, "large corporations not only have the power to pollute, they have the economic and political power to prevent, delay and water down regulatory legislation. They also have the power and connections to ensure that the regulatory agencies don't regulate as they ought to."

Because of the danger that the corporate system poses to the environment, several political theorists have maintained that the corporation as we know it is obsolete. Unfortunately the performance of the socialist nations—with the exception of China—does not promise a viable alternative. The industrial-minded bureaucracy that governs the Soviet Union appears to be as unresponsive to the welfare of the biosphere as any giant corporation. Protests from Soviet biologists failed to halt the pollution of Lake Baikal, and the Caspian Sea has become a sink for industrial waste, threatening the caviar-producing sturgeon with extinction. Pollution in the socialist countries is not yet so severe as in the West, but with expanding industrialization it is rapidly catching up.

In Mishan's words, "The spreading suburban wilderness, the near traffic paralysis, the mixture of pandemonium and desolation in the cities, a sense of spiritual despair scarcely concealed by the frantic pace of life" are part of the price one pays for an expanding economy. To men born in a century shaped by the assumption that growth is good, the idea of an equilibrium population and a spaceship economy in which resources are con-

> The ecological crisis will eventually lead to severe limitation of freedoms that individuals and businesses now take for granted. In the 21st century the citizens of even the richest nations will live under restrictions that would seem intolerable by 20th century standards.

tinually recycled may seem revolutionary, and indeed it is. The idea of a zero-growth economy has only recently begun to be discussed seriously, and as one might expect, it has encountered almost unanimous opposition from business. It seems unlikely that either the advanced industrial nations or the underdeveloped ones will voluntarily abandon their drive for growth and "progress"—the advanced countries because their political structure is dominated by business, and the poor ones because of the fear that a halt to expansion might lock them into perpetual poverty.

INDIVIDUALISM AND AUTHORITY

Political theorists have recently recognized that the ecological crisis will eventually lead to severe limitations of freedoms that individuals and businesses now take for granted. The freedom to be selfish exacts a terrible price in overpopulation, pollution, the plundering of resources, and human misery. As a political goal, the halt of economic growth in the advanced nations and population growth in the poor nations may be unattainable under democratic rule. In a steadily worsening situation, the need for speedy and drastic solutions will favor the rise of increasingly authoritarian governments. One Third World government has attempted to dictate how many children a couple can have (India), and another has decreed where its citizens must live (the revolutionary regime in Cambodia deported virtually the entire population of Phnom Penh to rural areas). Throughout the world, private ownership of land—the ultimate resource—is likely to become a thing of the past, as it already has in most of the socialist nations. Other scarce resources will be rationed, particularly such "necessities" as gasoline and meat. In the 21st century the citizens of even the richest nations will live under restrictions that would seem intolerable by 20th century standards. Whether these restrictions are imposed by the economics of scarcity or by the decree of government planners is a matter of debate, but they are unlikely to be accepted voluntarily by a citizenry accustomed to thinking primarily in terms of short-sighted self interest.

Harvard psychologist B. F. Skinner looks at the problem in a different way. In his book *Beyond Freedom and Dignity,* he argues that human survival depends on reshaping the attitudes and behavior of mankind, particularly the substitution of altruism for individualism. Aggressive individualism, an effective form of behavior for taming a frontier or resisting the domination of tyrants, becomes dangerous to the species on a crowded and civilized planet.

Skinner maintains that human behavior is a product of long and subtle social conditioning. He advocates the development of a technology of behavior: the engineering of psychological tools to manipulate human behavior toward socially desirable patterns. Such techniques have already been used successfully in a limited way with mental patients and juvenile criminals. But whether behavioral engineering can control the conduct of all mankind is open to doubt. The prospect stirs uneasy thoughts of George Orwell's *1984,* and fears about how such a powerful tool might be used by political leaders such as the Chairman of the People's Republic of China, a Latin American dictator, or even the President of the United States. Can man be programmed to act with the altruism characteristic of an insect society? Should he be? And to what ends?

WAR

The most immediate threat to the future of life on Earth is annihilation by thermonuclear weapons or by chemical or biological warfare. Oddly enough, in recent years this fact seems to have been pushed below the threshhold of public attention. Each of the perennial confrontations between major powers —in Berlin, Hungary, Czechoslovakia, Cuba, Laos, Vietnam and the Middle East —carries the risk that a strategic escalation of threat and counterthreat will trigger a war that would destroy hundreds of millions of people and contaminate the entire planet. Five or ten more confrontations between the great powers might leave us only a 50-50 chance of surviving until the year 2000.

The impoverished hordes of the Third World are erupting as a political force that could endanger the precarious balance between the superpowers. The ability of a few fanatics to disrupt the activities of advanced nations has been demonstrated by skyjackers, terrorists and guerrillas. How much greater will be the disruptive power of tomorrow's wretched billions in Asia, Africa and Latin America —desperate men with nothing to lose? In Vietnam a tiny Asian nation defeated France and stalemated America. The nations of the Third World (except China) lack hydrogen bombs and intercontinental missiles, and none seems likely to develop them. But atomic bombs are another matter. Within a few decades many poor nations will have at least one nuclear reactor capable of manufacturing enough plutonium to assemble a bomb. A missile is not necessary to deliver such a bomb; the hold of a freighter would serve just as well. A tankerful of nerve gas or a few canisters of botulinum toxin could decimate a city no matter how carefully it is guarded.

We might ask ourselves how long the nations of the Third World will tolerate the widening gap between their wretchedness and the prosperity of the advanced nations. Our survival depends on their co-operation, and theirs on ours. Will we be willing to lower our standard of living so that they can rise from squalor? An international redistribution of wealth would require a degree of altruism that does not seem to be characteristic of our species, or at least not of our political system. It seems more probable that in the next few decades the advanced nations will try to segregate themselves in privileged enclaves, responding only half-heartedly to the misery around them.

LIFEBOAT ETHICS

A few ecologists have concluded that the situation in the Third World has already become hopeless, and that only a policy of enlightened selfishness can prevent the advanced industrial nations from being dragged down with it. The most eloquent spokesman for this viewpoint is Garrett Hardin of the University of California, Santa Barbara. Two of his essays have become focal points of controversy: *The Tragedy of the Commons,* published in *Science* in 1968, and *Living on a Lifeboat,* published in *BioScience* in 1974. In a sense, his viewpoint is the opposite of Skinner's.

Hardin compares each wealthy nation to a 60-man lifeboat containing 50 people. The people of the poor nations, packed into dangerously overcrowded lifeboats, continually fall out and swim about, hoping to be admitted to a rich lifeboat.

Seeing 100 others struggling in the water, what should the occupants of the rich lifeboat do?

Hardin outlines three alternatives: the occupants can follow the Christian-Marxist ethic and take all of the needy into the boat, jamming 150 people into a craft that can hold no more than 60. The result is that "The boat is swamped, and everyone drowns. Complete justice, complete catastrophe." The second alternative is to admit only ten more to the boat, loading it to capacity and eliminating any safety factor, an action for which "we will sooner or later pay dearly. Moreover, *which* ten do we let in?. . . . The best ten? The neediest ten? How do we discriminate?" The third alternative, which Hardin clearly favors, is to admit no more to the boat, preserve the small safety factor, and to guard against boarding parties (such as illegal immigrants). "The last solution," Hardin concedes, "is abhorrent to many people. 'It is unjust,' they say. Let us grant that it is. 'I feel guilty about my good luck,' say some. The reply to this is simple: Get out and yield your place to others. Such a selfless action might satisfy the conscience of those who are addicted to guilt, but . . . the person to whom a guilt-addict yields his place will not himself feel guilty about his sudden good luck (if he did he would not climb aboard)." The lifeboat, Hardin concludes drily, "purifies itself of guilt."

It is difficult to argue with Hardin's metaphor. Several biologists have taken issue with his underlying assumptions, maintaining that, with proper management, the Earth's resources can, at least temporarily, support the hordes that will inhabit the planet at the turn of the century and beyond. Unquestionably Hardin is raising grim issues that most people would rather not think about. If Hardin proves to be right, in the near future the prosperous nations of both East and West will be forced, in the interest of survival, to re-examine the ethical systems on which their societies, at least in theory, are built.

What then are the prospects for the future of mankind? In the long run, most of us do not believe that the human species will be destroyed, because it is extraordinarily resilient. But in the words of C. P. Snow, "there is likely to be a prolonged period of hardship, sporadic famine rather than widespread, commotion rather than major war. To avoid worse," he adds, "will require more foresight and will than men have shown themselves capable of up to now." In short, we believe that mankind is about to enter one of the Dark Ages of human history. Most of the ecological problems that threaten us will be resolved by the end of the 21st century, either by human intelli-gence or by nature's ruthless indifference. This much seems certain: the civilization that emerges from that century will be vastly different from the one we know today.

READINGS

B. COMMONER, *The Closing Circle*, New York, Bantam Books, 1972. A lucid account of what has gone wrong and what must be done about it. Commoner points out that industrial society ignores one of the cardinal rules of ecology: everything must go somewhere.

D.H. MEADOWS, *et al.*, *The Limits to Growth*, New York, Universe Books, 1972. A team of scientists at MIT extrapolated present trends into the future, using a computer model of the world ecosystem. This easy-to-read paperback summarizes the results. The book has provoked fierce controversy. *Science* called it "a computer view of doomsday." People just don't want to believe it, which is part of the problem.

E.J. MISHAN, *The Costs of Economic Growth*, Harmondsworth (England), Penguin Books, 1969. In the tradition of Thorstein Veblen, Mishan is both an economist and a witty critic of society. He is one of the pioneer advocates of a zero-growth economy. A hardcover edition is available in the U.S. from Praeger, New York.

W.W. MURDOCH, *ed.*, *Environment*, 2nd Edition, Sunderland MA, Sinauer Associates, 1975. Amateurs spin a lot of wooly talk about ecology. This is a source-book by a score of professionals.

C.P. SNOW, *Public Affairs*, New York, Scribners, 1971. Lord Snow, novelist and scientist, was praised for his essay, *The Two Cultures and the Scientific Revolution*. This book is a collection of essays on the interface between science and politics, with emphasis on the future of man.

L.S. STAVRIANOS, *The Promise of the Coming Dark Age*, San Francisco, W.H. Freeman, 1976. A distinguished historian envisions the immediate future as a Dark Age comparable to the one that followed the fall of the Roman Empire. He seems curiously undisturbed by that prospect, maintaining that it will be a dynamic period of change leading to a new Golden Age.

K.E.F. WATT, *The Titanic Effect: Planning for the Unthinkable*, Sunderland MA, Sinauer Associates, 1974. Written by an outstanding ecologist who points out that the magnitude of disasters can be decreased by anticipating them and planning intelligently. The final two chapters, "Scenarios of the Future" and "Political Remedies," are pertinent to this chapter.

Abscisic acid (ab sighs' ik) [L. *abscissio:* the act of breaking off]. A plant growth hormone having growth-inhibiting action. Frequently antagonizes other plant growth hormones such as auxins or gibberellins.

Absolute temperature scale. A temperature scale in which the degree is the same size as in the centigrade scale, and the zero is the state of no molecular motion. Absolute zero is −273° on the centigrade scale.

Acclimation (a cli ma' shon). A change in an organism which improves its ability to tolerate a changed "climate" (i.e. environmental situation).

Acrosome (a' krow soam) [Gr. *akros:* highest or outermost + *soma:* body]. The structure at the forward tip of an animal sperm, which first fuses with the egg membrane and enters the egg cell.

Actin [Gr. *aktis:* a ray]. One of the two major proteins of muscle; it makes up the thin filaments.

Action potential. An impulse in a nerve which is propagated nondecrementally, taking the form of a wave of depolarization or reverse polarization imposed on a polarized cell surface.

Activating enzyme. An enzyme that couples a low energy compound with ATP to yield a high-energy derivative. As used in this book, an enzyme that condenses ATP with a specific amino acid to give enzyme-bound AMP-amino acid.

Activation energy. The energy barrier which blocks the tendency for a set of chemical substances to react. A reaction is speeded up if this energy barrier is surmounted by adding heat energy or is lowered by finding a different reaction pathway with the aid of a catalyst. Designated by the symbol, E_a.

Active site. The region on the surface of an enzyme where the substrate binds, and where catalysis occurs. Active sites have been found which are shallow depressions, grooves (for binding chains), and deep cavities (for trapping side chains or the end of a polypeptide chain).

Active transport. The transport of a substance across the plasma membrane against a concentration gradient, that is, from a region of low concentration to a region of high concentration. Active transport requires the expenditure of energy and is a "saturable" process (see carrier-facilitated diffusion and free diffusion).

Adaptation (a dap tay' shun). In evolutionary biology, a particular structure, physiological process, or behavior that makes an organism more fit to survive and reproduce. Also, the evolutionary process that leads to the formation of such a trait.

Adaptive radiation. The division of a single species into many species specialized to diverse ways of life.

Adenosine triphosphate. (see ATP)

Afferents [L. *ad:* to + *ferre:* to bear]. Neurons which carry impulses to the central nervous system.

Aleurone layer (al' yur own) [Gr. *aleuron:* wheaten flour]. In grass seeds, a specialized cell layer just between the seed coat and the endosperm, synthesizing hydrolytic enzymes under the influence of gibberellin, and thus helping mobilize reserves for the developing embryo.

Alga (al' gah). Any one of a wide diversity of plants belonging to the phyla Cyanophyta, Pyrrophyta, Chrysophyta, Phaeophyta, Rhodophyta, and Chlorophyta. Most live in the water, where they are the dominant plants; most are unicellular, but a minority are multicellular (the "seaweeds" and similar plants).

Allele. A particular form of a gene, distinguishable from other forms or alleles of the same gene.

Allopatric (al' lo pa' trick) [Gr. *allos:* other + *patria:* homeland]. Pertaining to populations that occur in different places.

Allopolyploid (al' lo pol' lee ploid) [Gr. *allos:* different + *poly:* many + *ploos:* fold]. A polyploid constructed of sets of chromosomes that originated from two or more species.

Allostery (al' lo ster y) [Gr. *allo:* different + *stereos:* structure]. Regulation of the activity of an enzyme by the binding, at a site other than the catalytic active site, of an effector molecule which does not have the same structure as any of the enzyme's substrates. If the catalyzed reaction is an early step in the ultimate synthesis of the effector molecule, then the effector molecule regulates its own synthesis in a form of feedback control.

Alternation of generations. The succession of haploid and diploid phases in a sexually reproducing organism. In higher animals, the haploid phase consists only of the gametes. In plants, however, the haploid phase may be the more prominent phase (as in fungi and mosses) or may be as prominent as the diploid phase (see the life cycle of *Ulva* in text). In higher plants, the diploid phase is the more prominent.

Alveolus (al ve' o lus) (plural: alveoli) [L. *alveus:* a cavity]. A small bag-like cavity, especially of the kind which constitute the blind sacs of the lung.

Amino acid (a mee' no). An organic compound of the general formula: $H_2N—CHR—COOH$, where R can be one of twenty or more different side chains. An amino acid is so named because it has both a basic amine group, $—NH_2$, and an acidic carboxyl group, $+COOH$.

Amphi- [Gr. both]. A prefix used to denote a character or kind of organism that occupies two or more states. For example, amphibian (an animal that lives both on the land and in the water).

Amphibian (am fib' ee-an). A member of the vertebrate class Amphibia, such as a frog, toad, or salamander.

Ammonotelic (amon o tel' ic) [Gr. *telos:* end]. An organism in which the final product of breakdown of amino compounds (primarily proteins) is ammonia.

Amoeba (a mee' bah) [Gr. *amoibe:* change]. Any one of a large number of different kinds of unicellular animals belonging to the phylum Sarcodina, characterized among other features by its ability to change shape frequently through the protrusion and retraction of soft extensions of cytoplasm called pseudopodia.

Amoeboid (a mee' boid) [Gr. *amoibe:* change]. Referring to any cell that behaves more or less like an amoeba, constantly changing its shape by the protrusion and retraction of soft extensions of cytoplasm called pseudopodia.

Amplification (gene). Cellular mechanism for increasing the dosage of specific genes by making numerous DNA copies of the gene. In particular, the genes that are responsible for producing ribosomal RNA have been shown in some cells to undergo numerous replications.

A.M.U. (atomic mass unit). The basic unit of mass on an atomic scale, defined as one twelfth the mass of a carbon-12 atom. There are 6.023×10^{23} a.m.u. in one gram. This number is known as Avogadro's number.

Anaerobic (an air row' bic) [Gr. *an:* not + *aer:* air + *bios:* life]. Occurring without the use of oxygen; as "anaerobic fermentation."

Analogy (a nal' o jee) [Gr. *analogia:* resembling]. A resemblance in function, and often appearance as well, between two structures which is due to convergence in evolution rather than to common ancestry. (Contrast with homology.)

Anaphase (an' a phase) [Gr. *ana:* indicating upward progress]. The stage in cell division at which the separation of sister chromosomes (or, in the first meiotic division, of paired homologues) occurs. Anaphase lasts from the moment of first separation to the time at which the moving chromosomes converge at the poles of the division.

Angiosperm (an' jee o spurm) [Gr. *angion:* vessel + *sperma:* seed]. One of the "higher plants"; literally one whose seed is carried in a "vessel," which is the fruit.

Anion (an' ey on) [Gr. *ana:* back + *ienai:* to go; from anode, the pole in a battery at which electrons flow back into the external circuit]. A negatively charged ion. (Contrast with cation.)

Anisogamy (an' eye sog' ah mee) [Gr. *aniso:* unequal + *gamos:* marriage]. The existence of two dissimilar gametes.

Annelid (ann' el id). A member of the phylum Annelida; one of the segmented worms, such as an earthworm or a leech.

Annual. Referring to a plant whose life cycle is completed in one growing season. (Contrast with perennial.)

Anther (an' ther) [Gr. *anthos:* flower]. A pollen-bearing portion of the stamen of a flower.

Antheridium (an' ther id' i um) [Gr. *antheros:*

blooming]. The multicellular structure that produces the male gamete in Bryophyta and seedless Tracheophyta.

Anticodon. A "triplet" of three nucleotides in transfer RNA that is able to pair with a complementary triplet in messenger RNA (a codon), thus aligning the transfer RNA on the proper place on the messenger.

Antipodals (an tip' o dal) [Gr. *anti:* against + *podus:* foot]. Cells (usually three) of the mature embryo sac of a flowering plant, located at the end opposite the egg (and micropyle).

Apex (a'peks). The tip or highest point of a structure, as the apex of a growing stem or root.

Apical (a' pi kul). Pertaining to the apex or tip, as the apical meristem, which is the actively growing tissue at the tip of a stem or root.

Appetitive behavior. Searching behavior, often of a specific nature, which terminates when the animal contacts the appropriate signal from the environment (sign stimulus) and performs the corresponding instinctive act (consummatory act).

Archegonium (ar' ke goe' nee um) [Gr. *archegonos:* first of a kind]. The multicellular structure that produces eggs in Bryophyta and seedless Tracheophyta.

Archenteron (ark en' ter on) [Gr. *archos:* beginning + *enteron:* bowel]. The earliest primordial animal digestive tract, it first appears during gastrulation.

Area-species curve. The graphical representation of the relation between the area of an island (or any other circumscribed part of the earth's surface) and the number of species that inhabit it.

Arthropod (arth' row pod). A member of the phylum Arthropoda, such as a crustacean, a spider, a centipede, or an insect.

Artifact [L. *ars, artis:* art + *facere:* to make]. Something made by human effort or intervention. In biology, something that was not present in the living cell, but was unintentionally produced by an experimental procedure.

Ascospore (ass' ko spor). A fungus spore produced within an ascus.

Ascus (ass' cuss) [Gr. *askos:* bladder]. In higher fungi belonging to the group ascomycetes, the club-shaped sporangium within which spores are formed by meiosis (usually of a single zygote nucleus).

Asexual reproduction. A form of reproduction, such as budding or simple fission, that does not involve the fusion of gametes.

Associative learning. "Pavlovian" learning, in which an animal comes to associate a previously neutral stimulus (such as the ringing of a bell) with a particular reward or punishment.

Assortment (genetic). The random separation during meiosis of nonhomologous chromosomes and of genes carried on nonhomologous chromosomes. For example, if genes A and B are borne on nonhomologous chromosomes, meiosis of diploid cells of genotype *AaBb* will produce haploid cells of the following types in equal numbers: *AB, Ab, aB, ab.*

Atomic mass unit. (see a.m.u.)

Atomic number. The number of protons in the nucleus of an atom, also equal to the number of electrons around the neutral atom. Important in determining chemical properties of the atom.

Atomic weight. The weight of an atom on the a.m.u. scale. Approximately equal to the total number of protons and neutrons in its nucleus.

ATP (adenosine triphosphate). A compound containing adenine, ribose, and three phosphates. It is the "common currency" of energy for most cellular processes.

Auricle (or' i cal) [L. *auris:* an ear]. The thin-walled chamber, also called "atrium," of the heart which contains blood and feeds it to the ventricle for pumping; also, the outer ear.

Autonomic nervous system. The system (which in vertebrates is made up of sympathetic and parasympathetic subsystems) which controls such involuntary "housekeeping" functions as that of the guts and glands.

Autopolyploid (au' tow pol' lee ploid) [Gr. *auto:* self + *poly:* many + *ploos:* fold]. A polyploid constructed entirely of sets of chromosomes that originated from the same species.

Autoradiography. The detection of a radioactive compound or organelle in a cell by putting it in contact with a photographic emulsion and allowing the compound to "take its own picture." The emulsion is developed and the location of the radioactivity in the cell is seen by the presence of silver grains in the emulsion.

Autosome. (see sex chromosome)

Autotrophic (au' tow trow' fik) [Gr. *autos:* self + *trophe:* food]. Capable of living exclusively on inorganic materials, water, and some energy source such as sunlight; for example, most plants and bluegreen algae and many bacteria. (Contrast with heterotrophic.)

Auxin (awk' sin) [Gr. *auxein:* increase]. In plants, a substance that regulates growth by affecting cell elongation or other alterations.

Auxotroph (awks' oh trofe') [Gr. *auxanein:* to grow + *trephein:* to nourish]. A mutant form of an organism that requires a nutrient (or nutrients) not required by the wild-type, or reference, form of the organism (see prototroph).

Avogadro's number, N. The conversion factor between a.m.u. and grams. More usefully, the number of atoms in that quantity of an element which, expressed in grams, is numerically equal to the atomic weight in a.m.u. N = 6.023×10^{23} atoms.

Axon [Gr. *axon:* an axle]. Process of the neuron which can carry action potentials; is often the longest and least branched process of the cell body; and usually carries impulses away from the cell body of the neuron.

Bacteriophage (bak teer' e o faj') [Gr. *bakterion:* little rod + *phagein:* to eat]. One of a group of viruses that infect bacteria and ultimately cause their disintegration.

Bacterium (plural: bacteria) [Gr. *bakterion:* little rod]. A procaryote—that is, a cell with a genome consisting of a simple DNA molecule not contained in a nuclear envelope—that has a cell wall and that does not carry out photosynthesis at all, or, if it does photosynthesize, that does not produce molecular oxygen as a product of photosynthesis.

Basidium (bass id' ee yum) [Gr. *basis:* base, L. diminutive]. In higher fungi belonging to the group basidiomycetes, the characteristic sporangium in which four spores are formed by meiosis, and then borne briefly externally before being shed.

Batesian mimicry. Mimicry by a relatively harmless kind of organism of a more dangerous one, by which the mimic enjoys protection from predators that mistake it for the dangerous model. (Contrast with Mullerian mimicry.)

Behavioral biology. The scientific study of all aspects of behavior, including neurophysiology, ethology, and sociobiology.

Bilateral symmetry (bye lat' e ral sym' me tree). The condition in which only the right and left sides, divided exactly down the back, are mirror images of each other. (Contrast with radial symmetry.)

Binomial (bye nome' e al). Consisting of two names; for example the binomial nomenclature of biology which gives the name of the genus followed by the name of the species.

Biogeochemical cycle. The route of passage of particular materials, such as water or carbon, from the physical environment into organisms and back out again, in an unending alternation.

Biogeography. The scientific study of the geographic distribution of organisms. Ecological biogeography is concerned with the habitats in which organisms live, historical biogeography with the complete geographic ranges of organisms and the historical circumstances that determine the ranges.

Biological species concept. The conception of the basic unit of classification as a population or series of populations of organisms capable of freely interbreeding with each other and reproductively isolated from other species.

Bioluminescence. The production of light by biochemical processes in an organism.

Biomass (bye' oh mass). The total weight of all the living organisms, or some designated group of living organisms, found in a given area.

Biome (bye' ome). A major portion of the living environment of a particular region, such as a fir forest or grassland, characterized by its distinctive vegetation and maintained by local conditions of the climate.

Biome type. One of a broad category of biomes, such as all of the grasslands of the world taken together.

Biota (bye oh' tah). All of the organisms, including fauna, flora, and microorganisms, found in a given area (see fauna and flora).

Biotic (bye ah' tik). Pertaining to any aspect of life, especially to characteristics of entire populations or ecosystems.

Blastocoel (blass' toe seal) [Gr. *blastos:* sprout + *koilos:* hollow]. The hollow central cavity of a blastula.

Blastodisc (blass' toe disk) [Gr. *blastos:* sprout + disc]. A disc of cells forming on the surface of a large yolk mass, comparable to a blastula, but occurring in forms in which the massive yolk restricts cleavage to one side of the egg only.

Blastula (blass' chu luh) [Gr. *blastos:* sprout]. An early stage in animal embryology, usually a hollow sphere of cells surrounding a central cavity.

Bohr effect (Bor effect). The reduction of hemoglobin's affinity for oxygen caused by acid conditions, usually as a result of CO_2.

Bud primordium [L. *primordium:* the beginning]. In plants, a small mass of potentially meristematic tissue found in the angle between the leaf stalk and the shoot apex. Will give rise to a lateral branch under appropriate conditions.

Budding. Asexual reproduction in which a more or less complete new organism simply grows from the body of the parent organism and eventually detaches itself.

Buffering (buf' er ing). A process by which a system resists changes, particularly in pH, in which case added acid or base is partially converted to a less active form.

Bulb. An underground storage organ composed principally of enlarged and fleshy leaf bases.

Callus [L. *calleo:* to be thick skinned]. In plants, wound tissue, or relatively undifferentiated proliferating cell mass, frequently maintained in tissue culture.

Calorie. The amount of heat required to raise the temperature of one gram of water by one degree Celsius (1°C). In many physiological measurements, including those used in nutrition studies, the "calorie" refers to the kilocalorie, which is the amount of heat required to raise one kilogram of water by 1°C.

Calyptra (ka lip' tra) [Gr. *kalyptra:* covering for the head]. A hood or cap found partially covering the apex of the sporophyte capsule in many moss species, formed from the expanded wall and neck of the archegonium.

Calyx (kay' licks) [Gr. *kalyx:* cup]. All of the sepals of a flower collectively.

Cambium (kam' bee um) [L. *cambiare:* to exchange]. A meristem which gives rise to radial rows of cells in stem and root, increasing them in girth; commonly applied to the vascular cambium which produces wood and phloem, and the cork cambium, which produces bark.

cAMP (cyclic AMP). A compound, formed from ATP, that mediates the effects of numerous hormones in animals. It is also needed for the transcription of catabolite-repressible cistrons or operons in bacteria.

Capillaries (kap' il eryz) [L. *capillaris:* hair]. Very small tubes, especially the smallest blood-carrying vessels of animals between the termination of the arteries and the beginnings of the veins.

Carbohydrates (kar bo hi' drate) [E. carbon + hydrate]. Organic compounds with the general formula: $(CH_2O)_n$. The most common examples are sugars, starch, and cellulose.

Carboxylic acid (kar box sill' ik). An organic acid containing the carboxyl group: —COOH, which dissociates to the carboxylate ion: —COO .

Cardiac (kar' dee ak) [Gr. *kardia:* heart]. Pertaining to the heart and its functions.

Carotene (ka' ro teen) [L. *carota:* carrot]. A yellow or orange pigment commonly found as an accessory pigment in photosynthesis, but also found in fungi; contains only carbon and hydrogen.

Carotenoid (ka rot' e noid) [L. *carota:* carrot]. A class of photosynthetic accessory pigments, including xanthophylls and carotenes.

Carpel (kar' pel) [Gr. *karpos:* fruit]. The organ of the flower that contains one or more ovules.

Carrier-facilitated diffusion (passive transport). Transport of a substance across the plasma membrane by carrier molecules, but without a "pump." This process is "saturable," but cannot cause the net transport of a substance from a region of low concentration to a region of high concentration (see free diffusion and active transport).

Carrying capacity. In ecology, the largest number of organisms of a particular species that can be maintained indefinitely in a given part of the environment.

Caryokinesis (carry oh kin ee' sis) [Gr. *karyon:* kernel + *kinein:* to move]. The division of the nucleus of a dividing cell. As opposed to cytokinesis.

Caste. In social insects, any set of individuals of a particular anatomical type, or age group, or both, that performs specialized labor in the colony.

Catabolite repression. The decreased synthesis of many enzymes that tend to provide glucose for a cell caused by the presence of excellent carbon sources, particularly glucose.

Catalyst (kat' a list) [from Gr. *cata-*, implying the breaking down of a compound]. A chemical substance which accelerates a reaction, without itself being consumed in the overall course of the reaction. Enzymes are biological catalysts.

Cation (kat' ey on) [Gr. *cata:* out + *ienai:* to go; from cathode, the pole in a battery from which electrons flow out into the solution]. A positively charged ion. (Contrast with anion.)

Cell wall. A relatively rigid structure that encloses cells of plants, fungi, and most bacteria. The cell wall gives these cells their shape and limits their expansion in hypotonic media.

Cellulose (sell' you lowss) [E. cell + ose, suffix indicating a sugar or other carbohydrate]. A straight-chain polymer of glucose molecules, used by plants as a structural supporting material.

Central nervous system. That part of the nervous system which is condensed and centrally located, e.g. the brain and spinal cord of vertebrates; the chain of cerebral, thoracic and abdominal ganglia of arthropods.

Centrifuge [L. fleeing from center]. A device in which a biological preparation can be spun around a central axis at high speed, creating a centrifugal force that mimics a very strong gravitational force. Suspensions will tend to settle out under these conditions, the larger, denser particles sedimenting more rapidly.

Centriole (sent' ree ole) [L. *centrum:* center]. A cylindrical organelle of animal cells (and some plant cells) containing microtubules in a characteristic arrangement consisting of nine triplets. A centriole is at each pole of the spindle apparatus in dividing cells, but the exact role of the centriole in the elaboration of the spindle is not known (see kinetosome).

Centromere (sent' row mere) [L. *centrum:* center + Gr. *meros:* part]. The point on a chromosome to which spindle fibers attach at cell division. Generally seen as a constriction. Not necessarily central in its location on the chromosome.

Cephalopod (sef' a low pod). A member of the mollusk class Cephalopoda, such as a squid or an octopus.

Chiasma (kie as' muh) (plural: chiasmata) (kie as' muh tuh [Gr: cross]. An "x"-shaped connection between paired homologous chromosomes at meiosis. A chiasma is the visible manifestation of crossing-over between homologous chromosomes.

Chitin (kye' tin) [Gr. *chiton:* tunic]. The characteristic flexible organic component of the exoskeleton of insects and other arthropods, consisting of a complex nitrogen-containing polysaccharide.

Chloroplast [Gr. *chloros:* green + *plast:* a combining form meaning "a particle"]. An organelle bounded by a double membrane and containing the enzymes of photosynthesis. Chloroplasts occur only in eucaryotes.

Chromatic adaptation [Gr. *chroma:* color]. The ability of some algae to modify the kind and amount of different photosynthetic accessory pigments, producing those which absorb the wavelengths of light maximally available in their environment.

Chromatid (kro' ma tid'). Each of a pair of new sister chromosomes from the time at which the molecular duplication occurs until the time

at which the centromeres separate at the anaphase of cell division.

Chromatin [Gr. *chroma:* color]. The nucleic acid–protein complex found in eucaryotic chromosomes.

Chromatophore (krow mat' o fore) [Gr. *chroma:* color + *phoreus:* carrier]. A pigment-bearing cell that expands or contracts to change the color of the organism.

Chromosome (krome' oh zome) [Gr. *chroma:* color + *soma:* body]. Complex structure found in the nucleus of a eucaryotic cell, composed of nucleic acids and proteins (primarily basic histones), and bearing part of the genetic information (genes) of the cell.

Chromosome aberration. Any larger change in the structure of the chromosome, including duplication or loss of chromosomes (aneuploidy) or duplication of complete sets of chromosomes (polyploidy), usually gross enough to be detected with the light microscope.

Ciliate (sil' ee ate). A member of the protistan phylum Ciliophora, unicellular organisms that propel themselves by cilia.

Cilium (sil' ee um) (plural: cilia) [L. *cilium:* eyelash]. Hairlike organelles used for locomotion by many unicellular organisms and for moving water and mucus by many multicellular organisms.

Circadian rhythm (sir kade' ee an) [L. *circa:* approximately + Gr. *dies:* day]. A rhythm in behavior, growth, or some other activity that recurs about every 24 hours under constant conditions. (The prefix *circa-* refers to the lack of precision in the timing.)

Cistron. The genetic unit of function, often considered loosely equivalent to a gene. Generally each cistron contains the genetic information for a single polypeptide chain.

Class. In taxonomy, the category below the phylum and above the order; a group of related, similar orders.

Codon. A "triplet" of three nucleotides in messenger RNA that directs the placement of a particular amino acid into a polypeptide chain.

Coelom (see' lome) [Gr. *koiloma:* cavity]. The body cavity of higher metazoan animals, which is lined with epithelium of mesodermal origin.

Coenocyte (seen' a sight) [Gr: common cell]. A "cell" bounded by a single plasma membrane, but containing many nuclei.

Cofactor (ko' fak tor). An organic molecule bound to an enzyme, and necessary for its catalytic activity. Flavin molecules derived from riboflavin (vitamin B$_2$) are cofactors for many enzymes.

Coitus (koe′ i tus) [L. *coitus:* a coming together]. The act of sexual intercourse.

Coleoptile (koe′ lee op′ tile) [Gr. *koleos:* sheath + *ptilon:* feather]. A pointed sheath covering the shoot of grass seedlings; part of the cotyledon.

Collagen [Gr. *kolla:* glue]. A fibrous protein found extensively in bone and connective tissue.

Commensalism. The form of symbiosis in which one species benefits from the association, while the other is neither harmed nor benefited.

Communication. Action on the part of one organism (or cell) that alters the pattern of behavior in another organism (or cell) in an adaptive fashion.

Community. Any ecologically integrated group of species of microorganisms, plants, and animals inhabiting a given area.

Companion cell. Specialized cell found adjacent to each sieve tube element in flowering plants. Probably provides the energy for driving materials through the phloem.

Complementation. The production of a wild-type phenotype through the combined activities of two chromosomes in a common cytoplasm with each chromosome bearing at least one recessive mutation. If, under these conditions, a mutant phenotype is observed, the two recessive mutations on the chromosomes are said to be in the same functional unit or cistron.

Conditional lethal mutation. A mutation that causes death only under certain conditions. For example, "temperature sensitive" lethal mutations cause death of mutant individuals at elevated temperatures, but not at low temperatures.

Conidium (koh nid′ ee um) [Gr. *konis:* dust]. An asexual fungus spore borne singly or in chains either apically or laterally on a hypha.

Conifer (kon′ e fer) [Gr. *konos:* cone + *phero:* carry]. One of the cone-bearing plants, mostly trees, such as pines and firs.

Conjugation (kon′ jew gay′ shun) [L. *conjugare:* yoke together]. The close approximation of two cells during which they exchange genetic material, as in *Paramecium* and other ciliated protozoans.

Conservative amino acid change. The substitution, during the course of evolution of a protein, of one amino acid at a given point in the chain by another amino acid of very similar chemical character. Substitution of lysine for arginine, aspartic acid for glutamic acid, and leucine for valine are examples of conservative changes.

Constitutive enzyme. An enzyme that is present in approximately constant amounts in a system, whether substrates for it are present or absent. Thus it is part of the "constitution" of the cell (see inducible enzyme).

Consummatory act. One of a set of very specific, stereotyped patterns of motor responses. The behavior of most animal species consists largely or wholly of such consummatory acts and is loosely referred to as "instinctive."

Continental drift. The gradual drifting apart of the world's continents that has occurred over a period of hundreds of millions of years.

Contractile vacuole. An organelle, often found in protozoa, which pumps excess water out of cells and keeps them from being "flooded" in hypotonic media.

Co-repressor. A low-molecular-weight compound that unites with a protein (the apo-repressor) to form a functional repressor or "holo-repressor."

Corolla (koh rol′ lah) [L. *corolla,* diminutive of *corona:* wreath, crown]. All of the petals of a flower collectively.

Corpus [L. *corpus:* body]. The portion of the shoot apical meristem which gives rise to the bulk of the stem tissues, e.g. cortex, vascular tissue, pith.

Cortex [L. *cortex:* bark or rind]. In plants, the tissue between the epidermis and the vascular tissue of a stem or root.

Cotyledon (kot′ ee lee′ don) [Gr. *kotyledon:* a hollow space]. The first leaf or leaves of the seed which unfold as the seedling sprouts.

Covalent bond. A chemical bond which arises from the sharing of electrons between atoms. Usually a strong bond.

Cristae. Small shelf-like projections of the inner membrane of the mitochondrion. The cristae are the site of oxidative phosphorylation.

Cross-pollination. The fertilization of one plant by pollen from another plant. (Contrast with self-pollination.)

Crossing-over. The mechanism by which linked markers undergo recombination. In general, the term refers to the reciprocal exchange of corresponding segments between two homologous chromatids. However, the reciprocality of crossing-over is problematical in procaryotes and viruses and, even in eucaryotes, very closely linked markers often recombine by a nonreciprocal mechanism.

Crustacean (crus tay' see-an). A member of the arthropod class Crustacea, such as a crab, shrimp, or sowbug.

Cryptic appearance. The resemblance of an animal to some part of its environment, which helps it to escape detection by predators.

Cupula. A cap, such as the gelatinous cap which extends across each semicircular canal in the ear's labyrinth, and responds to fluid flow in the canal.

Cutin (cue' tin) [L. *cutis:* skin]. A mixture of long straight chain hydrocarbons and waxes secreted by the plant epidermis, providing an impermeable coating on aerial plant parts.

Cyclic AMP. (see cAMP)

Cytochromes (sy' to chromes) [Gr. *cyto:* cell + *chroma:* color or pigment]. Iron-containing red proteins, components of the electron-transfer machinery in photosynthesis and respiration.

Cytokinesis (site' oh kin ee' sis) [Gr. *kytos:* vessel + *kinein:* to move]. The division of the cytoplasm of a dividing cell.(Contrast with caryokinesis.)

Cytokinin (site' oh kye' nin) [Gr. *kytos:* vessel + *kinesis:* movement]. A member of a class of plant growth hormones playing a role in cell division, retarding of senescence, bud expansion, and so on.

Cytoplasm. The contents of the cell, excluding the nucleus.

Darwinism. The theory of evolution by natural selection, as propounded by Charles Darwin.

Deciduous (de sid' you us) [L. *decidere:* fall off]. Referring to a plant that sheds its leaves at certain seasons. (Contrast with evergreen.)

Decomposer. An organism, such as a bacterium, fungus, or carrion beetle, that consumes dead organic matter, recycling it through the ecosystem.

Degeneracy. The situation in which a single amino acid may be represented by any one of several different nucleotide "triplets" or codons in messenger RNA. Most of the amino acids can be represented by more than one codon.

Deletion (genetic). A mutation resulting from the loss of a continuous segment of a gene or chromosome. Such mutations never revert to wild-type and act in genetic crosses as if they occupied several clearly distinct sites marked by other mutations. (See point mutation.)

Dendrite [Gr. *dendron:* a tree]. Process of the neuron which often cannot carry action potentials; is commonly much branched and relatively short compared with the axon; and commonly carries impulses to the cell body of the neuron.

Density gradient. A solution, usually contained in a centrifuge tube, that varies continuously from a higher density at the bottom to a lower density at the top.

Deoxyribonucleic acid. (see DNA)

Determination. Process whereby a group of cells, a single cell, or even part of an embryonic cell becomes fixed into a predictable developmental pathway.

Dicaryon (die kare' e on) [Gr. *di:* two + *karyon:* kernel]. An organism, typically a fungus, with paired (but not fused) nuclei derived from different parents. (Contrast with zygote.)

Dicot (short for dicotyledon) [Gr. *dis:* two + *kotyledon:* a cup-shaped hollow]. Any member of the angiosperm class Dicotyledonae, plants in which the embryo produces two leaves prior to germination. Leaves of most dicots have major veins arranged in a branched or reticulate pattern.

Differentiation. Process whereby originally similar cells follow different developmental pathways. The actual expression of determination.

Dioecious (die eesh' us) [Gr: two houses]. Organisms in which the two sexes are "housed" in two different individuals, so that eggs and sperm are not produced in the same individuals. Examples: humans, fruit flies, oak trees, date palms (see monoecious).

Diploid (dip' loid) [Gr. *diploos:* double]. Having a chromosome complement consisting of two copies (homologues) of each chromosome. A diploid individual (or cell) usually arises as a result of the fusion of two gametes, each with just only copy of each chromosome. Thus, the two homologues in each chromosome pair in a diploid cell are of separate origin, one derived from the female parent and one from the male.

Displacement activity. The performance of a behavioral act, usually in conditions of frustration or indecision, which is not directly relevant to the situation at hand.

Diverticulum (di ver tic' u lum) [L. *divertere:* turn away]. A small cavity or tube that connects to a major cavity or tube.

DNA (deoxyribonucleic acid). The fundamental hereditary material of all living organisms. Except for bacteria and bluegreen algae, stored in the cell nucleus. A nucleic acid polymer using deoxyribose rather than ribose.

Dominance. In genetic terminology, the ability of one allelic form of a gene to determine the phenotype of a heterozygous individual, in which the homologous chromosome carries a different allele. For example, if *A* and *a* are two allelic forms of a gene, *A* is said to be dominant to *a* if *AA* diploids and *Aa* diploids are phenotypically identical and are distinguishable from *aa* diploids. The *a* allele is said to be recessive.

Dominance hierarchy. The set of relationships within a group of animals, usually established and maintained by aggression, in which one individual has precedence over all others in eating, mating, and so on; a second individual has precedence over all but the highest-ranking individual, and so on down the hierarchy.

Drive. A loose term used to describe the tendency of an animal to seek out an object (for example, a mate, a particle of food, or a nest site) and to perform the appropriate response toward it.

Duplication (genetic). A mutation resulting from the introduction into the genome of an extra copy of a segment of a gene or chromosome.

Ecdysone (eck die' sone) [Gr. *ek:* out of + *dyo:* to clothe]. In insects, a hormone inducing molting.

Echinoderm (e kine' oh durm). A member of the phylum Echinodermata, such as a starfish or sea urchin.

Echolocation. (see sonar)

Ecology (ee kol' oh jee) [Gr. *oikos:* house + *logos:* discourse, study]. The scientific study of the interaction of organisms with their environment, including both the physical environment and the other organisms that live in it. In recent years, ecology is also often used in a loose manner to indicate the environment itself.

Ecosystem (eek' o sis tum). The organisms of a particular habitat, such as a pond or forest, together with the physical environment in which they live.

Ecto- (ek' toh) [Gr: outer, outside]. A prefix used to designate a structure on the outer surface of the body. For example, ectoderm. (Contrast with endo- and meso-.)

Ectoderm [Gr. *ektos:* outside + *derma:* skin]. The outermost of the three embryonic tissue layers first delineated during gastrulation. Gives rise to the skin, sense organs, nervous system, etc.

Effector. Any organ or cell that moves the organism through the environment or else alters the environment to the organism's advantage. Examples include muscle, bone, and a wide variety of exocrine glands.

Effector, allosteric. A molecule, usually quite different from the substrate, which binds to an effector site on an enzyme and increases or decreases its catalytic activity.

Efferents [L. *ex:* out + *ferre:* to bear]. Neurons which carry impulses from the central nervous system.

Electron (e lek' tron) [L. *electrum:* amber (static electricity), from Gr. *elektor:* bright sun (color of amber)]. One of the three most fundamental particles of matter, with mass approximately 0.00055 a.m.u. and charge −1.

Embryo sac. In angiosperms, the female gametophyte. Found within the ovule, it consists of eight or fewer cells, membrane bounded, but without cellulose walls between them.

Emigration. The deliberate and usually oriented departure of an organism from the habitat in which it has been living.

Endo- [Gr: within, inside]. A prefix used to designate an innermost structure. For example, endoderm, endocrine. Often contrasted with meso- (in the middle) and ecto- (on the outside).

Endocrine gland (en' do krin) [Gr. *endon:* inside + *krinein:* to separate]. Any gland, such as the adrenal or pituitary gland of vertebrates, that secretes certain substances, especially hormones, into the body through the blood or lymph.

Endocytosis. A collective word for two very similar processes, pinocytosis and phagocytosis, involving the uptake of liquids or solids, respectively.

Endoderm [Gr. *endon:* within + *derma:* skin]. The innermost of the three embryonic tissue layers first delineated during gastrulation. Gives rise to the digestive and respiratory tracts and structures associated with them.

Endodermis [Gr. *endon:* within + *derma:* skin]. In plants, a specialized cell layer marking the inside of the cortex in roots and some stems. Frequently a barrier to free diffusion of solutes.

Endoplasmic reticulum, rough [Gr. *endon:* within + L. *plasma:* form; L. *reticulum:* little net]. A system of folded membranes with ribosomes attached, found in the cytoplasm of eucaryotic cells. The ribosomes attached to the rough endoplasmic reticulum engage primarily in the synthesis of secretory proteins.

Endoplasmic reticulum, smooth. A system of folded membranes found in the cytoplasm of eucaryotic cells. Smooth endoplasmic reticulum is devoid of associated ribosomes. It plays a role in the packaging of secretory products and in many other cellular processes.

Endosperm (en' do sperm) [Gr. _endon:_ within + _sperma:_ seed]. A specialized triploid seed tissue found only in angiosperms, which contains stored food for the developing embryo.

Endothermic reaction (en' do therm ik) [Gr. _endo:_ in + _therme:_ heat]. A chemical reaction which absorbs heat.

Enthalpy (en' thal py) [Gr. _en:_ in + _thalpein:_ to heat]. The heat flow in or out of a chemical system reacting in constant overall pressure, as in our atmosphere. Designated by the symbol H. In an endothermic reaction, the enthalpy of the reacting system rises; in an exothermic reaction, it falls.

Entropy (en' tro py) [Gr. _en:_ in + _tropein:_ to change; "concerned with chemical changes"]. A measure of the degree of disorder in any system. A perfectly ordered system has zero entropy; increasing disorder is measured by positive entropy. Spontaneous reactions in a closed system are always accompanied by an increase in disorder and entropy. Designated by the symbol S.

Enzyme (en' zime) [Gr. _en:_ in + _zyme:_ yeast]. A globular protein, on the surface of which are found chemical groups arranged so as to make the enzyme a catalyst for a chemical reaction.

Epi- [Gr: upon, over]. A prefix used to designate a structure located on top of another. For example, epidermis, epiphyte.

Epidermis [Gr. _epi:_ upon + _derma:_ skin]. In plants and animals, the outermost cell layers. Only one cell-layer thick in plants.

Epinephrine (ep i nef' rin) [Gr. _epi:_ upon + _nephros:_ a kidney]. The hormone secreted principally by the medulla of the adrenal gland, which is situated on the kidney; epinephrine is also called adrenaline. It is secreted as a result of fear or excitement and produces effects on the circulatory system, on glucose mobilization, etc.

Epiphyte (ep' e fyte) [Gr. _epi:_ upon + _phyton:_ plant]. A specialized plant that grows on the surface of other plants but without parasitizing them.

Epoch (ep' ok). A lesser division of geological time; for example, the Miocene epoch, which lasted about 12 million years.

Equilibrium, chemical. A state in which forward and reverse reactions are proceeding at counterbalancing rates, so there is no observable change in the concentrations of reactants and products.

Equilibrium constant. A particular kind of ratio of the concentrations of products to reactants when equilibrium has been reached in a chemical reaction. Designated by the symbol, K_{eq}.

Era (ehr' ah). One of the major divisions of geological time; for example, the Mesozoic era, which lasted about 170 million years.

Estrus (es' truss) [L. _oestrus:_ frenzy]. The period of heat, or maximum sexual receptivity, in the female. Ordinarily the estrus is also the time of the release of eggs in the female.

Estrous cycle. The repeated series of changes in reproductive physiology and behavior in the female, culminating in estrus.

Ethology (ee thol' o jee) [Gr. _ethos:_ habit, custom + _logos:_ discourse]. The study of whole patterns of animal behavior in natural environments, stressing the analysis of adaptation and evolution of the patterns.

Eucaryotes (you car' ry ots) [Gr. _eu:_ true + _karyon:_ kernel or nucleus]. Organisms whose cells contain their genetic material inside a nucleus. Includes all life above the level of bacteria and bluegreen algae.

Evergreen. A plant that retains its leaves through all seasons. (Contrast with deciduous.)

Evolution. Any gradual change. Organic evolution, often referred to as evolution for short, is any genetic change in organisms from generation to generation.

Evolutionary biology. The collective branches of biology that deal with the evolutionary process and the characteristics of populations of organisms, as well as ecology, behavior, and systematics.

Exo- (eks' oh). Same as ecto-.

Exobiology. The study of the probable existence of life on other planets.

Exocrine gland (eks' oh krin) [Gr. _exo:_ outside + _krinein:_ to separate]. Any gland, such as the salivary gland, that secretes to the outside of the body or into the alimentary tract.

Exoskeleton (eks' oh skel' e ton) [Gr. _exo:_ outside + skeleton]. A hard covering on the outside of the body; the exoskeleton of insects and other arthropods has essentially the same functions as the bony internal skeleton of vertebrates.

Exothermic reaction (eks' oh therm' ik) [Gr. *exo:* outside + *therme:* heat]. A chemical reaction that gives off heat.

Exponential growth. Growth, especially in the number of organisms of a population, which is a simple function of the size of the growing entity: the larger the entity, the faster it grows.

Extrinsic isolating mechanism. Any barrier in the environment, such as a river bed, desert, or ocean, that isolates one population from another.

Family. In taxonomy, the category below the order and above the genus; a group of related, similar genera.

Fauna (faw' nah). All of the animals found in a given area. (Contrast with flora.)

Faunal dominance. The ability of species originating from one part of the world to spread their range to other parts of the world and displace other species already living there.

Feedback control. Control of a particular step of a multi-step process, induced by the presence or absence of a product of one of the later steps. A thermostat regulating the flow of heating oil to a furnace in the home is a feedback control device.

Fermentation (fer men ta' shun) [L. *fermentum:* yeast]. The degradation of a molecule such as glucose to smaller molecules with the extraction of energy, without the use of oxygen (i.e., anaerobically).

Fertilization membrane. A membrane surrounding an animal egg which becomes rapidly raised above the egg surface within seconds after fertilization, serving to prevent entry of a second sperm.

Fiber. An elongated and tapering cell of vascular plants, usually with substantial secondary wall. The wall may or may not be lignified, and the mature fiber may or may not contain a living protoplast.

Flagellate (flaj' e late). A member of the phylum Mastigophora, which are unicellular organisms that propel themselves by flagella.

Flagellin (fla jell' in). The protein from which procaryotic but not eucaryotic flagella are constructed.

Flagellum (fla jell' um) (plural: flagella) [L: whip]. Long, whip-like appendage that propels pro-

caryotic or eucaryotic cells.

Flora (flore' ah). All of the plants found in a given area. (Contrast with fauna.)

Florigen. A plant hormone involved in the conversion of a vegetative shoot apex to a flower.

Flower. The total reproductive structure of an angiosperm; its basic parts include the calyx, corolla, stamens, and carpels.

Food chain. A portion of a food web, most commonly a simple sequence of prey species and the predators that consume them.

Food web. The complete set of food links between species in a community; a diagram indicating which ones are the eaters and which are consumed.

Fossil. Any recognizable structure originating from an organism, or any impression from such a structure, that has been preserved from prehistoric times.

Frame-shift mutation. A mutation resulting from the addition (or deletion) of a single base-pair in the DNA sequence of a gene. As a result of this, mRNA transcribed from such a gene is translated normally until the ribosome reaches the point at which the mutation has occurred. From that point on, codons are read out of proper register and the amino acid sequence bears no resemblance to the normal sequence.

Free diffusion. Diffusion directly across the plasma membrane without the involvement of carrier molecules. Free diffusion is not "saturable," and cannot cause the net transport from a region of low concentration to a region of higher concentration (see carrier-facilitated diffusion and active transport).

Free energy. That energy which is available for doing useful work, after allowance has been made for the increase or decrease of disorder. Designated by the symbol G, and defined by: $G = H - TS$, where H = enthalpy, S = entropy, and T = absolute temperature.

Fruit. In angiosperms, a ripened and mature ovary (or group of ovaries) containing the seeds. Sometimes applied to reproductive structures of other groups of plants, and includes any adjacent parts which may be fused with the reproductive structures.

Gametangium (gam i tan' gee um) [Gr. *gamos:* marriage + *angeion:* vessel or reservoir]. Any plant structure within which a gamete is formed.

Gamete (gam' eet) [Gr. *gamete:* wife or *gametes:*

husband]. The mature sexual reproductive cell: the egg or the sperm.

Gametocyte (ga meet' oh site) [Gr. *gamete:* wife, *gametes:* husband + *kytos:* cell]. The cell that gives rise to sex cells, either the eggs or the sperm. (See oocyte and spermatocyte.)

Gametogenesis (ga meet' oh jen' e sis) [Gr. *gamete:* wife, *gametes:* husband + *gignomai:* be born]. The specialized series of cellular divisions that leads to the production of sex cells (gametes). (See oogenesis and spermatogenesis.)

Gametophyte (ga meet' oh fyte). In plants with an alternation of generations, the haploid phase that produces the gametes. (Contrast with sporophyte.)

Gastropod (gas' troh pod). A member of the molluscan class Gastropoda, such as a snail or conch.

Gastrovascular cavity. Serving for both digestion (gastro-) and circulation (vascular); in particular the central cavity of the body of jelly fish and other Coelenterata.

Gastrula (gas' true luh) [Gr. *gaster:* stomach]. An embryo forming the characteristic three cell layers, ectoderm, mesoderm, and endoderm, which will give rise to all of the major tissue systems of the adult animal.

Gate. A system which permits a membrane to be transiently permeable to an ion, which therefore flows in the direction determined by its concentration gradient.

Gel. A mixture that is either semisolid or solid, but which has a large amount of liquid trapped within the solid component.

Gene. A unit of heredity. Often used as the rough equivalent of the cistron, the unit of genetic function which carries the information for a single polypeptide.

Gene flow. The exchange of genes between different species (an extreme case referred to as hybridization) or between different populations of the same species. A principal driving force of evolution.

Gene pool. All of the genes in a population.

Genetic drift. Evolution (change in gene proportions) by chance processes alone.

Genophore (jean' oh for) [Gr. *gen:* to produce + *phorain:* to bear]. The DNA molecule that carries the genes of a virus or procaryote. As opposed to the more complex chromosomes of the eucaryotes.

Genotype (jean' oh type) [Gr. *gen:* to produce + *tupos:* impression]. An exact description of the genetic constitution of an individual, either

with respect to a single trait or with respect to a larger set of traits. (Contrast with phenotype.)

Genus (jean' us) (plural: genera) (jen' e rah) [Gr. *genos:* stock, kind]. A group of related, similar species.

Geotropism [Gr. *ge:* the earth + *trope:* a turning]. A directed plant growth response to gravity.

Gibberellin. One of a class of plant growth regulators playing a role in stem elongation, seed germination, flowering of certain plants, etc. Named for the fungus *Gibberella* from which gibberellins were first isolated.

Gizzard (giz' erd) [L. *gigeria:* cooked chicken guts]. A very muscular part of the stomach of birds that grinds up food, sometimes with the aid of fragments of stone.

Glia (Gr: glue). Cells found only in the nervous system, which do not conduct nerve impulses.

Gluconeogenesis (glue' ko neo gen' esis) [Gr. *neos:* new + *gen:* become]. The formation of glucose from components derived from the breakdown of other metabolites.

Glucose (glue' kose) [Gr. *gleukos:* sweet wine mash for fermentation]. The most common carbohydrate sugar, with the formula $C_6H_{12}O_6$.

Glycogen (gly' ko gen). A branched-chain polymer of glucose molecules, similar to starch but of higher molecular weight. Exists mostly in liver and muscle; the principal storage carbohydrate of most animals and fungi.

Glycogenolysis (gly' ko ge nol' i sis) [Gr. *lusis:* loosening]. The enzymatic breakdown of glycogen to make glucose or phosphate esters of glucose.

Glycolysis (gly kol' li sis) [from glucose + lysis]. The enzymatic breakdown of glucose during the process of anaerobic fermentation. One of the oldest of energy-yielding mechanisms in living organisms.

Golgi complex (goal' gee). A system of concentrically folded membranes found in the cytoplasm of eucaryotic cells. Plays a role in the production and release of secretory materials such as the digestive enzymes manufactured in the acinar cells of the pancreas. First described by Camillo Golgi (1844 – 1926).

Gonad (goh' nad) [Gr. *gone:* seed, that which produces seed]. An organ that produces sex cells in animals: either an ovary (female gonad) or testis (male gonad).

Growth. Irreversible increase in volume (probably the most accurate definition, but at best a dangerous oversimplification).

Growth ring. A growth layer of xylem or phloem produced by the vascular cambium, as seen in transverse section. Reflects periodic environmental changes that affect rate of cambial activity and growth of the cells produced—alternating summer and winter or wet and dry periods, for example.

Gymnosperm (jim' no sperm) [Gr. *gymnos:* naked + *sperma:* seed]. A plant, such as a pine or other conifer, whose seeds do not develop within an ovary (hence, the seeds are "naked").

Habituation (ha bich' oo ay shun). The simplest form of learning, in which an animal presented with a stimulus without reward or punishment eventually ceases to respond.

Haploid (hap' loid) [Gr. *haploeides:* single]. Having a chromosome complement consisting of just one copy of each chromosome. This is the normal "ploidy" of gametes or of asexual spores produced by meiosis or of organisms (such as the gametophyte generation of plants) that grow from such spores without fertilization.

Hardy-Weinberg Law. The law that the basic processes of Mendelian heredity (meiosis and recombination) do not alter either the frequencies of genes or their diploid combinations. The Law also states how the percentages of diploid combinations can be predicted, on the basis of a simple formula, from a knowledge of the percentage of genes.

Hemoglobin (hee' moh glow' bin) [Gr. *haima:* blood + L. *globus:* globe]. The colored protein of the blood which transports oxygen.

Hermaphroditism (her maf' row dite' is-m) [Gr. *hermaphroditos:* a person with both male and female traits]. The coexistence of both female and male sex organs in the same organism.

Hertz (Abbreviated as Hz). Cycles per second.

Hetero- [Gr: other, different]. A prefix used in biology to mean that two or more different conditions are involved; for example, heterotroph, heterozygous.

Heterocaryon (het' er oh care' ee ahn) [Gr. *heteros:* different + *karyon:* kernel]. A cell or organism carrying a mixture of genetically distinguishable nuclei. A heterocaryon is usually the result of the fusion of two cells without fusion of their nuclei.

Heterochromatin (het' er oh chrome' a tin) [Gr. *heteros:* different + *chroma:* color]. Chromosomal material that appears to be dense and darkly stained in interphase, when the bulk of the chromosomal material is diffuse and lightly stained.

Heteromorphic (het' er oh more' fik) [Gr. *heteros:* different + *morphe:* form]. Having a different form or appearance, as two heteromorphic life stages of a plant. (Contrast with isomorphic.)

Heterospory (het' er os' poh ree) [Gr. *heteros:* different + *spora:* seed]. The condition in plants of producing two kinds of spores, the small microspores and large megaspores. (Contrast with homospory.)

Heterotrophic (het' er oh trow' fik) [Gr. *heteros:* different + *trophe:* food]. Requiring organic materials for food. (Contrast with autotrophic.)

Heterozygous (het' er oh zie' gus) [Gr. *heteros:* different + *zugotos:* joined]. Of a diploid organism having different alleles of a given gene on the pair of homologues carrying that gene (see homozygous).

Hfr (h-f-r: for "high frequency of recombination"). Donor bacteria (specially *E. coli* or *Salmonella*) in which the F-factor has been integrated into the genophore. This produces a bacterium that transfers its chromosomal markers at a very high frequency to recipient cells.

Holdfast. In many large attached algae, specialized tissue attaching the plant to its substratum.

Homeostasis (home' ee o sta' cis) [Gr. *homoios:* like, sameness + *stasis:* position]. The maintenance of a steady state, such as a constant temperature or a stable social structure, by means of physiological or behavioral feedback responses.

Homeotherm (home' ee o therm) [Gr. *homos:* same + *therme:* heat]. An animal which maintains a constant body temperature by virtue of its own heating and cooling mechanisms.

Homologue (home' o log') [Gr. *homos:* same + *logos:* word]. One of a pair, or larger set, of chromosomes having the same overall genetic composition and sequence. In diploid organisms, each chromosome inherited from one parent is matched by an identical (except for mutational changes) chromosome—its homologue—from the other parent.

Homology (ho mol' o jee) [Gr. *homologi(a):* agreement]. A similarity between two structures which is due to inheritance from a common ancestor. The structures are said to be homologous. (Contrast with analogy.)

Homospory (home os' poh ree) [Gr. *homoios:* like +

spora: seed]. Producing just one kind of spore. (Contrast with heterospory.)

Homozygous (home' o zie' gus) [Gr. *homos:* same + *zugotos:* joined]. Of a diploid organism having identical alleles of a given gene on both homologous chromosomes. An organism may be a "homozygote" with respect to one gene and, at the same time, a "heterozygote" with respect to another (see heterozygous).

Hormone (hore' mone) [Gr. *hormon:* excite, stimulate]. A substance produced in one part of a multicellular organism and transported to another part where it exerts its specific effect on the physiology or biochemistry of the target cells.

Hydrogen bond. A chemical bond which arises from the attraction between the slight positive charge on a hydrogen atom, and a slight negative charge on a nearby fluorine, oxygen, or nitrogen atom. Weak bonds, but found in great quantities in proteins and other biological macromolecules.

Hydrogenase. An enzyme found in a few bacteria and algae which can transfer electrons to H^+ to generate molecular hydrogen, or from molecular hydrogen to some other electron acceptor of appropriate redox potential, leaving H^+ ions.

Hydrolyze (hi' dro lize) [Gr. *hydro:* water + *lysis:* cleavage]. To break a chemical bond, such as a polypeptide bond, with the insertion of the components of water, $-H$ and $-OH$, at the cleaved ends of a chain. The digestion of proteins is a hydrolysis.

Hydrophobic [Gr. *hydro:* water + *phobia:* fear; "disliking water"]. Molecules and amino acid side chains which are mainly hydrocarbons— compounds of C and H with no charged groups or polar groups—have a lower energy when they are clustered together than when they are distributed through an aqueous solution. Because of their attraction for one another and their reluctance to mix with water they are called "hydrophobic." Oil is a hydrophobic substance; phenylalanine is a hydrophobic amino acid in a protein.

Hymenopteran (hy' men op' ter an). A member of the insect order Hymenoptera, such as a wasp, bee, or ant.

Hypertonic [Gr: higher tension]. A medium with a higher concentration of osmotically active particles than is present in cells. In hypertonic media, water will flow out of cells, causing shrinkage or plasmolysis (see hypotonic and isotonic).

Hypha (high' fuh) [Gr. *hyphe:* web]. In the fungi, any single filament. May be multinucleate (phycomycetes, ascomycetes) or multicellular (basidiomycetes).

Hypotonic [Gr: lower tension]. A medium that has a lower concentration of osmotically active particles than does a cell. In hypotonic media, water will tend to flow into cells (see hypertonic and isotonic).

Imbibition [Lat. *imbibo:* to drink]. The binding of a solvent to another molecule. Rubber will imbibe gasoline, dry starch and protein will imbibe water. Binding forces can be very large, leading to extensive swelling of the imbibing system.

Imprinting. A rigid form of learning, in which an animal comes to make a particular response only to one other animal or object.

Individual distance. The fixed distance an animal strives to keep between itself and other members of the same species.

Inducer. A small molecule which, when added to a system of growth medium, causes a large increase in the level of some enzyme. Generally it acts by binding to repressor and changing its shape so that the repressor does not bind to the operator.

Inducible enzyme. An enzyme that is present in much larger amounts when a particular compound (an inducer) has been added to the system (see constitutive enzyme).

Inhibitor. A substance which binds to the surface of an enzyme and interferes with its action on its substrate molecules.

Insight learning. The most complex form of learning, in which the brain puts together separate memories and reasons out new relationships and solutions to problems.

Instinct. Behavior that is relatively highly stereotyped, more complex than the simplest reflexes, and usually directed at particular objects in the environment. Learning may or may not be involved in the development of the behavior; the important point is that the behavior develops toward a narrow, predictable end-product.

Insulin (in' si lin) [L. *insula:* an island]. An animal hormone synthesized in islet cells of the pancreas, which promotes the conversion of glucose to the storage material, glycogen.

Integument [L. *integumentum:* covering]. In gymnosperms and angiosperms, a layer of tissue around the ovule which will become seed coat. Gymnosperm ovules have one integument, angiosperm ovules two.

Intention movement. The preparatory motions that animals go through prior to a complete behavior response; for example, the crouch before flying, the snarl before biting, etc.

Intercalary meristem. A meristematic region in plants which occurs not apically, but between two regions of mature tissue. Intercalary meristems occur in the nodes of grass stems, for example.

Interphase. The period between successive cell divisions during which the chromosomes are diffuse and the nuclear envelope is intact. It is during this period that the cell is most active in transcribing and translating genetic information. However, because the nucleus of an interphase cell is less interesting to look at than the nucleus of a dividing cell, this phase has been misnamed "the resting stage," a term that is still sometimes used.

Intrinsic isolating mechanism. A genetically based trait that helps prevent members of one species from breeding with members of other species.

Inversion (genetic). A rare mutational event that leads to the reversal of the order of genes within a segment of a chromosome, as if that segment had been removed from the chromosome, turned 180°, and then reattached.

Invertebrate. Any animal that is not a vertebrate, that is, whose nerve cord is not enclosed in a backbone of bony segments.

In vitro [L: in glass]. In a test tube, rather than in a living organism (see *in vivo*).

In vivo [L: in the living state]. In a living organism. Contrast with *in vitro*. Many processes that occur *in vivo* can be reproduced *in vitro* with the right selection of cellular components.

Ion (ey' on) [Gr. *ion:* wanderer]. An atom with electrons added or removed, giving it a negative or positive charge. (See anion and cation.)

Ionic bond. A chemical bond which arises from the electrostatic attraction between positively and negatively charged groups of ions. Usually a strong bond.

Isogamy (ey sog' ah mee) [Gr. *isos:* equal + *gamos:* marriage]. A kind of sexual reproduction in algae and fungi in which the gametes (or gametangia) are not distinguishable on the basis of size or morphology.

Isomorphic (ey' so more' fik) [Gr. *isos:* equal + *morphe:* form]. Having the same form or appearance, as two isomorphic life stages. (Contrast with heteromorphic.)

Isotonic [Gr: same tension]. A medium that has the same concentration of osmotically active particles as a cell, so that there is no net inflow or outflow of water (see hypotonic and hypertonic).

Isotope (ey' so tope) [Gr. *isos:* equal + *topos:* place—i.e. same place in the chemical periodic table]. Two isotopes of the same chemical element have the same number of protons in their nuclei, but differ only in the number of neutrons.

Juvenile hormone. In insects, a hormone maintaining larval growth.

Keratin (ker' a tin) [Gr. *keras:* horn]. A protein which contains sulfur, and is part of such hard tissues as horn, nail, and the outermost cells of the skin.

Key. In taxonomy, a device for quickly identifying a specimen down to the species or a higher category to which it belongs. The key consists of a series of choices made according to whether the specimen possesses one trait as opposed to another.

Kinesis (ki nee' sis) [Gr: movement]. Orientation behavior in which the organism does not move in a particular direction with reference to a stimulus but instead simply moves at an increasing or decreasing rate until it ends up farther from the object or closer to it. (Contrast with taxis.)

Kinetosome (ki nee' to zome) [Gr. *kinein:* to move + *soma:* body]. A cylindrical, microtubular organelle found at the base cilia and flagella of eucaryotic cells. In structure, the kinetosome is twin to the centriole.

Lamarckism (Lah mark' iz-um). The theory of evolution by acquired characteristics, as propounded by Jean Baptiste de Lamarck.

Larva [L. *larva:* ghost, early stage]. An immature stage of any invertebrate animal that differs strongly in appearance from the adult.

Latent learning. Learning that is not put to immediate use but is "stored" for possible future use.

Leaf axil. The upper angle between a leaf and the stem, site of axillary or lateral buds which under appropriate circumstances become activated to form lateral branches.

Leaf primordium [L. *primordium:* the beginning].

A small mound on the flank of a shoot apical meristem; will give rise to a leaf.

Life cycle. The entire span of the life of an organism from the moment of fertilization (or asexual generation) to the time it reproduces in turn.

Linkage (genetic). Association between markers on the same chromosome or genophore such that they do not show random assortment. Linked markers recombine with one another at frequencies less than 0.50; the closer the markers on the chromosome, the lower the frequency of recombination.

Lipids (lip' id) [Gr. *lipos:* fat]. The substances in a cell which are easily extracted by organic solvents. Fats, oils, waxes, steroids, and other large organic molecules, including those which with proteins make up the cell membranes.

Logistic growth. Growth, especially in the size of an organism or in the number of organisms that constitute a population, which slows steadily as the entity approaches its maximum size. (Contrast with exponential growth.)

Lumen (loo' men) [L. *lumen:* light]. The cavity inside any tubular part of an organ, such as a piece of gut or a kidney tubule.

Lumper. A taxonomist who prefers to classify organisms into relatively few groups. (Contrast with splitter.)

Lymphatics (lim fat' ics) [L. *lympha:* water]. The system in animals of (1) spaces between the capillaries and the body cells which contain filtered blood (lymph) and (2) the vessels which collect the lymph and return it to the blood system.

Lysosome (lie' so zome) [Gr. *lusis:* a loosing + *soma:* body]. A membrane-bounded inclusion found in eucaryotic cells. Lysosomes contain a mixture of enzymes that can digest most of the macromolecules found in the rest of the cell.

Macro- (mack' roh) [Gr. *makros:* large, long]. A prefix commonly used to denote something large; contrast with micro-.

Macroevolution (mack' ro ev o loo' shun). A large amount of evolutionary change, involving many elementary changes in gene proportions or chromosome structure. (Contrast with microevolution.)

Mammal (mam' el) [L. *mamma:* breast, teat]. Any animal of the class Mammalia, characterized by the production of milk by the female mammary glands and the possession of hair for body covering.

Marine. Pertaining to the sea. (Contrast with terrestrial.)

Marsupial (mar soo' pee al). A mammal belonging to the mammal subclass Metatheria, most of which, like the opossums and kangaroos, have a pouch (the marsupium) that contains the milk glands and serves as a receptacle for the young.

Maternal inheritance (or, cytoplasmic inheritance). Inheritance in which the phenotype of the offspring depends on factors, such as mitochondria or chloroplasts, that are inherited from the female parent through the cytoplasm of the female gamete.

Maturation. The automatic development of a pattern of behavior, which becomes increasingly complex or precise as the animal matures. Unlike learning, the development does not require experience to occur.

Mega- [Gr. *megas:* large, great]. A prefix often used to denote something large. (Contrast with micro-.)

Megaphyll [Gr. *megas:* large + *phyllon:* leaf]. In vascular plants, a leaf thought to be derived from a flattened branch system, its vascular tissue forming a leaf gap where it attaches to the stem vascular tissue.

Megasporangium. The special structure (sporangium) that produces the megaspores.

Megaspore [Gr. *megas:* large + *spora:* seed]. In plants, a haploid spore that produces a female gametophyte. In many cases the megaspore is larger than the male-producing microspore.

Meiosis (my oh' sis) [Gr: diminution]. Cell division of a diploid cell to produce four haploid daughter cells. The process consists of two successive cell divisions with only one cycle of chromosome replication.

Menses (men' sees) [L. *menses:* plural of month]. The "period" of a sexually mature girl or woman; the days in each reproduction cycle during which blood and mucosal tissue flow from the uterus.

Meristem [Gr. *meristos:* divided]. A plant tissue made up of actively dividing cells.

Mesenchyme (mes' en kyme) [Gr. *mesos:* middle + *enchyma:* infusion]. Embryonic or unspecialized cells derived from the mesoderm.

Meso- (mes' oh) [Gr: middle]. A prefix often used to designate a structure located in the middle, or a stage that appears at some intermediate time. For example, mesoderm, Mesozoic.

Mesoderm [Gr. *mesos:* middle + *derma:* skin]. The middle of the three embryonic tissue layers first delineated during gastrulation. Gives rise to skeleton, circulatory system, muscles, excretory system, and most of the reproductive system.

Messenger RNA (mRNA). A disposable copy of one of the strands of DNA, it carries information for the synthesis of one or more proteins.

Meta- [Gr: between, along with, beyond]. A prefix used in biology to denote a change or a shift to a new form or level; for example, as used in metamorphosis.

Metamorphosis (met' a mor' fo sis) [Gr. *meta:* beyond + *morphe:* form, shape]. A strong change occurring from one developmental stage to another, as for example from a tadpole to a frog or an insect larva to the adult.

Metaphase (met' a phase) [Gr. *meta:* between]. The stage in cell division at which the centromeres of the highly condensed chromosomes are all lying on a plane (the metaphase plane or plate) perpendicular to a line connecting the division poles.

Metazoan [Gr. *meta:* between, among + *zoon:* animal]. Pertaining to any or all of the multicellular animals with the exception of the sponges, which are often distinguished as the "parazoan" animals.

Micro- (mike' roh) [Gr. *mikros:* small]. A prefix often used to denote something small; often contrasted with macro- or mega-.

Microbiology (mike' roh bye ol' o jee) [Gr. *mikros:* small + *bios:* life + *logos:* discourse]. The scientific study of microscopic organisms, particularly bacteria and other monerans, unicellular algae, and protistans.

Microevolution (mike' roh ev o loo' shun). A small amount of evolutionary change, consisting of minor alterations in gene proportions, chromosome structure, or chromosome numbers. (A larger degree of change would be referred to as macroevolution or simply as evolution.)

Microfilament. Minute fibrous structures found in the cytoplasm of eucaryotic cells. They appear to play a role in the motion of cells as if they were small muscles that could produce contractions of the cell surface wherever such contractions might be needed to produce the desired alteration in shape.

Microorganism (mike' roh or' ga niz' um). Any microscopic organism, such as a bacterium or one-celled alga.

Microphyll [Gr. *mikros:* small + *phyllon:* leaf]. A leaf possibly derived from a scale-like outgrowth of the stem. Its vascular tissue does not form a leaf gap where it attaches to the stem vascular tissue.

Micropyle (mike' roh pile) [Gr. *mikros:* small + *pyle:* gate]. Opening in the integument(s) of a seed plant ovule through which pollen grows to reach female gametophyte within.

Microsporangium. The special structure (sporangium) that produces the microspores.

Microspore (mike' roh spore) [Gr. *mikros:* small + *spora:* seed]. In plants, a haploid spore that produces a male gametophyte. In many cases the microspore is smaller than the female-producing megaspore.

Microtubule: Minute tubular structures found in centrioles, spindle apparatus, cilia, flagella, and other places in the cytoplasm of eucaryotic cells. These tubules play a role in the motion and maintenance of shape of eucaryotic cells.

Middle lamella. A thin layer of pectic compounds (polymers of galacturonic acid) cementing together two adjacent plant cell walls.

Mimicry (mim' ik ree). The resemblance of one kind of organism to another, which serves the function of discouraging potential enemies. See Batesian mimicry and Mullerian mimicry.

Mitochondrion (plural: mitochondria) [Gr. *mitos:* thread + *chondros:* cartilage, or grain]. An organelle that occurs in eucaryotic cells and contains the enzymes of the Krebs tricarboxylic acid cycle, the respiratory chain, and oxidative phosphorylation. A mitochondrion is bounded by a double membrane.

Mitosis (my toe' sis) [Gr. *mitos:* thread]. Cell division in eucaryotes leading to the formation of two daughter cells each with a chromosome complement identical to that of the original cell.

Mole. A quantity of a compound whose weight in grams is numerically equal to its molecular weight expressed in atomic mass units. Avogadro's number of molecules, 6.023×10^{23} molecules.

Mollusk (mol' lusk). A member of the phylum Mollusca, such as a snail, clam, or octopus.

Moneran (moh neer' an). A member of the Kingdom Monera which consists of bacteria and bluegreen algae.

Monocot (short for monocotyledon) [Gr. *monos:* one + *kotyledon:* a cup-shaped hollow]. Any member of the angiosperm class Monocotyledonae, plants in which the embryo produces but a single leaf prior to germination. Leaves of most monocots have their major veins arranged parallel to each other.

Monoecious (mo neesh' us) [Gr: one house]. Organisms in which both sexes are "housed" in a single individual, which produces both eggs and sperm. In plants, these are found in different flowers. Examples: corn, peas, earthworms, hydras (see dioecious, perfect flower).

Monotreme (mon' oh treem). An egg-laying mammal, in particular a platypus or an anteater, belonging to the subclass Prototheria.

Morphogenesis (more' fo jen' e sis) [Gr. *morphe:* form + *genesis:* origin]. The development of form. Morphogenesis is the overall consequence of determination, differentiation, and growth.

Morphology (more fol' o jee) [Gr. *morphe:* form + *logos:* discourse]. The scientific study of organic form, including both its development and function.

mRNA. (see messenger RNA)

Morula (mor' you la). A solid ball of cells formed by the first few cell divisions of the mammalian zygote.

Muco-complex substance. A characteristic material found in bacterial cell walls, composed of polymerized amino sugars cross-linked by amino acids.

Mullerian mimicry. The resemblance of two or more unpleasant or dangerous kinds of organisms to each other; the mimicry gives each added protection because its potential enemies find it easier to recognize as a member of a group. (Contrast with Batesian mimicry.)

Multicellular [L. *multus:* much + *cella:* chamber]. Consisting of more than one cell, as for example a multicellular organism. (Contrast with unicellular.)

Muscle cell. The basic cellular unit of muscle, a highly specialized cell packed with contractile protein which functions to move other organs in the body.

Mutagen (mute' ah jen) [L. *mutare:* change + Gr. *gignomai:* causing]. An agent, especially a chemical, that increases the mutation rate.

Mutation. In the broad sense, any discontinuous change in the genetic constitution of an organism. In the narrow sense, the word usually refers to a "point mutation": a change along a very narrow portion of the nucleic acid sequence.

Mutation pressure. Evolution (change in gene proportions) by different mutation rates alone.

Mutualism. The type of symbiosis, such as that exhibited by fungi and algae in forming lichens, in which both species benefit from the association.

Mycelium (my seal' ee yum) [Gr. *mykes:* mushroom]. In the fungi, a mass of hyphae.

Myoglobin (my' o globe' in) [Gr. *mys:* muscle + L. *globus:* sphere]. The special hemoglobin found in muscle, of lower molecular weight and carrying less oxygen than blood hemoglobin.

Myosin [Gr. *mys:* muscle]. One of the two major proteins of muscle, it makes up the thick filaments.

Natural selection. The differential contribution of offspring to the next generation by various genetic types belonging to the same population. The mechanism of evolution proposed by Charles Darwin.

Negative control. The situation in which a regulatory macromolecule (generally a repressor) functions to turn transcription off. In the absence of a regulatory macromolecule, the structural genes are turned on.

Nematocyst (ne mat' o sist) [Gr. *nema:* thread + *kystis:* cell]. An elaborate threadlike structure produced by cells of jellyfish and other coelenterates, used chiefly to paralyze and capture prey.

Nematode (nem' ah toad). A member of the phylum Nematoda; a roundworm.

Neoteny (nee ot' e nee) [Gr. *neo:* new + *tein:* stretch]. The capacity of the larval stage to become sexually mature.

Nephridium (nef rid' ee um) [Gr. *nephros:* kidney]. An organ which is involved in excretion, and often in water control, involving a tube which opens to the exterior at one end.

Nephron (nef' ron) [Gr. *nephros:* kidney]. The basic component of the kidney, which is made up of numerous nephrons. Its form varies in detail, but always has at one end a device for receiving a filtrate of the blood, and then a tubule which absorbs selected parts of the filtrate back into the bloodstream.

Neural crest [Gr. *neuron:* nerve]. Cells which become detached from the forming neural tube and wander great distances to fulfil a wide range of developmental roles.

Neural plate. A thickened strip of ectoderm along the dorsal side of the early vertebrate embryo; gives rise to the central nervous system.

Neurophysiology (nure' oh fiz' ee ol oh jee). The scientific study of the functioning of the nervous system.

Neurula (nure' yu la) [Gr. *neuron:* nerve]. Embryonic stage during formation of the dorsal nerve cord by two ectodermal ridges.

Neutron (new' tron) [E: neutral]. One of the three most fundamental particles of matter, with mass approximately 1 a.m.u. and no electrical charge.

Nitrogenase. In nitrogen-fixing organisms, an enzyme complex which mediates the stepwise reduction of atmospheric N_2 to ammonia.

Nomenclature. The method of assigning names in the classification of organisms.

Nonsense (chain-terminating) mutations. Mutations that change a codon for an amino acid to one of the codons (UAG, UAA, or UGA) that signal termination of translation. The resulting gene product is a shortened polypeptide that begins normally at the amino-terminal end and ends at the position corresponding to the altered codon.

Nuclear envelope. The surface, consisting of two layers of unit membrane, that encloses the nucleus of eucaryotic cells.

Nucleic acid (new klay' ik) [E: nucleus of cell]. A long-chain alternating polymer of deoxyribose or ribose sugar rings and phosphate groups, with organic bases (adenine, thymine or uracil, guanine, cytosine) as side chains. DNA and RNA are nucleic acids.

Nucleoid (new' klee oyd). The region, not bounded by any membrane, that harbors the DNA genophore of a procaryotic cell.

Nucleolar organizer (new klee' o lar). A region on a chromosome that appears to be associated with the formation of a new nucleolus following cell division. Presumably the site of the genes that correspond to ribosomal RNA.

Nucleolus (new klee' oh lus) [L: little kernel]. A small, clear body, generally spherical, found within the nucleus of eucaryotic cells. It is the site of synthesis of ribosomal RNA.

Nucleotide. The basic chemical unit in a nucleic acid. A nucleotide in RNA consists of one of four nitrogenous bases linked to the sugar, ribose, which in turn is linked to phosphate by ester bonds. In DNA, deoxyribose is present instead of ribose, and the base thymine is present instead of uracil.

Nucleus (new' klee us) [from L. diminutive of *nux:* kernel or nut]. (1) The dense central portion of an atom, made up of protons and neutrons, with a positive charge. Surrounded by a cloud of negatively charged electrons. (2) Centrally located chamber of eucaryotic cells that is bounded by a double membrane and contains the chromosomes. The information center of the cell.

Nyctinasty (nik' tee nasty) [Gr. *nyctos:* night + *nastos:* pressed close]. The night folding and morning opening of the leaves of certain higher plants.

Ommatidium [G. *omma:* an eye]. One of the units which, collected into groups of up to 20,000, make up the compound eye of arthropods.

Ontogeny (on toj' e nee) [Gr. *onto:* from "to be" + *gignomai:* be born, produce]. The development of a single organism in the course of its life history. (Contrast with phylogeny.)

Oocyte (oh' oh site) [Gr. *oon:* egg + *kytos:* cell]. The cell that gives rise to eggs in animals.

Oogenesis (oh' oh jen e sis) [Gr. *oon:* egg + *gignomai:* be born]. Female gametogenesis, leading to production of the egg.

Operator. The region of an operon that acts as the binding site for the repressor.

Operon. A genetic unit of transcription, typically consisting of several structural genes or cistrons that are transcribed to give a single messenger RNA molecule; the operon contains at least two transcriptional control regions: the promoter and the operator.

Optimal yield. The largest rate of increase that a population can sustain in a given environment. There exists a particular population size, less than the carrying capacity, at which this yield is realized.

Order. In taxonomy, the category below the class and above the family; a group of related, similar families.

Organ. A formed body, such as the heart, liver, brain, root, or leaf, composed of different tissues integrated to perform a distinct function for the body as a whole.

Organelles (or' gan els') [L: little organ]. Organized structures that are found in or on cells. Examples: ribosomes, nuclei, mitochondria, chloroplasts, cilia, contractile vacuoles.

Organic. Pertaining to any aspect of living matter: to its evolution, structure, or chemistry. The term is also applied to any chemical compound that contains carbon.

Organism. Any living creature.

Osmosis. The tendency of water to move from a

region in which it is more concentrated (e.g., pure water or a hypotonic solution) to a region in which it is less concentrated, that is, in which the concentration of dissolved molecules or ions is higher.

Ovary (oh' var ee). Any female organ, in plants or animals, that produces an egg.

Oviduct [L. _ovum:_ egg + _duco:_ to lead]. In mammals, the tube serving to transport eggs to the uterus or outside.

Ovule (oh' vule) [L. _ovulum:_ little egg]. In plants, an organ that contains a gametophyte and, within the gametophyte, an egg; when it matures, an ovule becomes a seed.

Ovum (oh' vum) [L. _ovum:_ egg]. The egg, the female sex cell.

Oxidation (ox i day' shun). Relative loss of electrons in a chemical reaction; either outright removal to form an ion, or the sharing of electrons with substances having a greater affinity for them such as oxygen. Most oxidations, including biological ones, are associated with the liberation of energy.

Paleobiology. The attempted reconstruction, from fossil evidence, from knowledge of the properties of chemical substances, and from theories about primitive Earth conditions, of the early stages in the evolution of life.

Paleobotany (pale' ee oh bot' ah nee) [Gr. _palaios:_ ancient, old]. The scientific study of fossil plants and all aspects of extinct plant life.

Paleontology (pale' ee on tol' oh jee) [Gr. _palaios:_ ancient, old + _logos:_ discourse]. The scientific study of fossils and all aspects of extinct life.

Palisade parenchyma. In leaves, one or several layers of tightly packed columnar photosynthetic cells, frequently found just below the upper epidermis.

Parasitism. The form of symbiosis in which one species lives at the expense of the other, but not ordinarily to the point of killing its host (such destruction would then be designated predation).

Parazoan [Gr. _para:_ near, close by + _zoon:_ animal]. Pertaining to the sponges (phylum Porifera), which are so different in basic structure from the rest of the multicellular animals (the metazoans) that they are often distinguished by this special name.

Parenchyma (pair en' kyma) [Gr. _para:_ beside + _enchyma:_ infusion]. A plant tissue composed of relatively unspecialized cells without secondary walls.

Parthenogenesis (par then' oh jen' e sis) [Gr. _parthenos:_ virgin + _gignomai:_ be born]. The production of an organism from an unfertilized egg.

Passive transport. (see carrier-facilitated diffusion)

Pathogen (path' o jen) [Gr. _pathos:_ suffering + _gignomai:_ causing]. An organism that causes disease.

Penis (pee' nis). The mammalian male organ used in coitus (sexual intercourse).

Peptide bond. The connecting group in a protein chain, $-CO-NH-$, formed by the removal of water during the linking of amino acids, $-COOH$ to $-NH_2$. Also called an amide bond.

Perennial (pe ren' ee al) [L. _per:_ through + _annus:_ a year]. Referring to a plant that lives from year to year. (Contrast with annual.)

Perfect flower. A flower with both stamens and carpels, therefore hermaphroditic.

Pericycle [Gr. _peri:_ around + _kyklos:_ ring or circle]. In plant roots, tissue just within the endodermis, but outside of the root vascular tissue. Meristematic activity of pericycle cells produces lateral root primordia.

Period. A division of time of intermediate duration, less than an era and greater than an epoch; for example, the Triassic period, which lasted for about 50 million years.

Permease. A protein in membranes that specifically transports a compound or family of compounds across the membrane.

Petal. In an angiosperm flower, a sterile modified leaf, nonphotosynthetic, frequently brightly colored, and often serving to attract pollinating insects.

Phage (faj). Short for bacteriophage.

Phagocytosis [Gr: eating cell]. The uptake of a solid particle by forming a pocket of cell membrane around the particle and pinching off the pocket to form an intracellular particle bounded by membrane (see pinocytosis and endocytosis).

Phenotype (fee' no type) [Gr. _phaenein:_ to show + _tupos:_ impression]. The observable properties of an individual as they have developed under the combined influences of the genetic constitution of the individual and the affects of environmental factors. (Contrast with genotype.)

Pheromone (fer' o mone) [Gr. _phero:_ carry + _hormon:_ excite, arouse]. A chemical substance used in communication between organisms of the same species.

Phloem (flo' im) [Gr. *phloos:* bark]. In the more advanced plants, the food-conducting tissue, which consists of sieve cells or sieve tubes, fibers, and other specialized cells.

Phospholipid. Cellular materials that contain phosphorus and are soluble in organic solvents. An example is lecithin (phosphatidyl choline). Phospholipids are important constituents of cellular membranes (see lipid).

Photoperiod (foe' tow peer' ee od). The duration of a period of light, such as the length of time in a 24-hour cycle in which daylight is present.

Photosynthesis (foe tow sin' the sis) [literally, "synthesis with light"]. Metabolic processes by which visible light is trapped and the energy used to synthesize energy-rich compounds such as ATP or glucose.

Phototropism [Gr. *photos:* light + *trope:* a turning]. A directed plant growth response to light.

Phycoerythrin (fy' ko ai rith' rin) [Gr. *phykos:* seaweed + *erythros:* red]. In bluegreen and red algae, a photosynthetic accessory pigment, an open chain tetrapyrrole bound to protein, bright red or orange.

Phycocyanin (fy' ko sie' an inn) [Gr. *phykos:* seaweed + *kyanos:* dark blue]. In the bluegreen and red algae, a photosynthetic accessory pigment, an open chain tetrapyrrole bound to protein, deep blue.

Phylogeny (fy loj' e nee) [Gr. *phyle:* tribe, race + *gignomai:* be born, produce]. The evolutionary history of a particular group of organisms; also, the diagram of the "family tree" that shows which species may have given rise to others. (Contrast with ontogeny.)

Phylum [Gr. *phylon:* tribe, stock]. In taxonomy, a high-level category just beneath the kingdom and above the class; a group of related, similar classes.

Physiology (fiz' ee ol' oh jee) [Gr. *physis:* natural form + *logos:* discourse, study]. The scientific study of the functions of living organisms and the individual organs, tissues, and cells of which they are composed.

Phytochrome (fy' to krome) [Gr. *phyton:* plant + *chroma:* color]. A plant pigment regulating a large number of developmental and other phenomena in plants; can exist in two different forms, one of which is active and the other of which is not. Different wavelengths of light can drive it from one form to the other. An open chain tetrapyrrole bound to protein.

Phytoplankton (fy' to plangk' ton) [Gr. *phyton:* plant + *planktos:* wandering]. The autotrophic portion of the plankton, consisting mostly of algae.

Pinocytosis [Gr: drinking cell]. The uptake of liquids by engulfing a sample of the external medium into a pocket of the plasma membrane followed by pinching off the pocket to form an intracellular vesicle (see phagocytosis and endocytosis).

Pistil [L. *pistillum:* pestle]. The female structure of an angiosperm flower, within which the ovules are borne. May consist of a single carpel, or of several carpels fused into a single structure. Usually differentiated into ovary, style, and stigma.

Pit. In botany, a small cavity in a cell wall which is not a complete perforation.

Pith. In plants, relatively unspecialized tissue found within a cylinder of vascular tissue.

Placenta (pla sen' ta). The organ, found in most mammals, that provides for the nourishment of the fetus and elimination of the fetal waste products. It is formed by the union of membranes of the mother's uterine lining with membranes from the fetus.

Placental (pla sen' tal). Pertaining to mammals of the subclass Eutheria, a group which is characterized by the presence of a placenta and which contains the great majority of living species of mammals.

Plankton (plangk' ton) [Gr. *planktos:* wandering]. The free-floating organisms of the sea and freshwater that for the most part move passively with the water currents. Consisting mostly of microorganisms and small plants and animals.

Planula (plan' yew la) [Diminutive of L. *planum:* something flat]. The free-swimming, ciliated larva of the coelenterates.

Plaque (plack) [Fr: a metal plate or coin]. A circular clearing in a turbid, confluent layer of bacteria growing on the surface of a nutrient agar gel. Produced by successive rounds of infection initiated by a single bacteriophage.

Plasma membrane (plasmalemma, or cell membrane). A lipid-containing membrane that regulates the entry and exit of molecules and ions. Every cell has a plasma membrane.

Plasmodesma (plural: plasmodesmata) [Gr. *plasma:* formed or molded + *desmos:* band]. A cytoplasmic strand connecting two adjacent plant cells, occurring predominantly in primary pit fields.

Plasmodium [Gr. *plasma:* mold or form]. In the non-cellular slime molds, a multinucleate mass of protoplasm, surrounded by a membrane, characteristic of the vegetative feeding stage.

Plasmolysis (plaz mol' y sis). Shrinking of a cell by outflow of water in a medium of high osmotic pressure. Plasmolysis refers to cells that have a rigid cell wall. The plasma membrane and the cytoplasm pull away from the cell wall as the cytoplasm shrinks.

Plastid. Organelle in plants which serves for food manufacture by photosynthesis and/or storage; bounded by a double membrane.

Pneumatophore (new mat' o for) [Gr. *pneuma:* air, wind, breathing + *phoresis:* a being borne]. A specialized root structure growing up into the air from roots in a very low oxygen environment, permitting gas exchange. Found on plants such as mangrove and cypress, which grow in stagnant swampy areas.

Poikilotherm (poy' kill o therm) [Gr. *poikilos:* varied + *therme:* heat]. An animal which is not homeothermic, and whose body temperature therefore tends to vary with the surrounding temperature.

Point mutation. A mutation that results from a small, localized alteration in the chemical structure of a gene. Such mutations can give rise to wild-type revertants as a result of reverse mutation. In genetic crosses, a point mutation behaves as if it resided at a single point on the genetic map (see deletion).

Polar body. A nonfunctional nucleus produced by meiosis, accompanied by very little cytoplasm. The meiosis which produces the mammalian egg produces in addition three polar bodies.

Polar nucleus. One of two nuclei derived from each end of the angiosperm embryo sac, both of which become centrally located. They fuse with a male nucleus to form the primary triploid nucleus which will produce the endosperm tissue of the angiosperm seed.

Pollen [L. *pollen:* fine powder, dust]. The fertilizing element of flowering plants, containing the male gametophyte and the gamete, at the stage in which it is shed.

Pollination. Process of transferring pollen from the anther to the receptive surface (stigma) of the ovary in plants.

Polycistronic messenger. A messenger RNA molecule containing information for the synthesis of more than one polypeptide chain. Generally such messengers are transcribed from an operon.

Polymorphism (pol' lee mor' fiz um) [Gr. *polys:* many + *morphe:* form, shape]. (1) In genetics, the coexistence in the same population of two distinct hereditary types based on different alleles. (2) In social organisms such as colonial coelenterates and social insects, the coexistence of two or more functionally different castes within the same colony.

Polyploid (pol' lee ploid). A cell or an organism in which the number of complete sets of chromosomes is greater than two.

Polyribosome. A complex consisting of a thread-like molecule of messenger RNA and several (or many) ribosomes. The ribosomes move along the mRNA, synthesizing polypeptide chains as they proceed.

Polysome. (see polyribosome)

Polytene (pol' lee teen) [Gr. *polloi:* many + *taenia:* ribbon]. An adjective describing giant interphase chromosomes, such as those found in the salivary glands of fly larvae. The characteristic, reproducible pattern of bands and bulges seen on these chromosomes has provided a method for preparing detailed chromosome maps of several organisms.

Population. Any group of organisms, capable of interbreeding for the most part, and coexisting at the same time and in the same place.

Positive control. The situation in which a regulatory macromolecule is needed to turn transcription of structural genes on. In its absence, they will be off.

Postmating isolating mechanism. Any intrinsic isolating mechanism, for example hybrid sterility, that operates after mating.

Predator. An organism that kills and eats other organisms. Predation is usually thought of as involving the consumption of animals by animals, but in the broad usages of ecology it can also mean the eating of plants.

Premating isolating mechanism. Any intrinsic isolating mechanism that operates prior to the completion of mating.

Primate (pry' mate). A member of the order Primates, such as a lemur, monkey, ape, or man.

Primary cell wall. The first formed plant cell wall, a fabric of cellulose microfibrils embedded in a gelatinous matrix of pectic compounds and hemicelluloses. The cellulose microfibrils may have a preferred orientation, but not a very strict one.

Primary pit fields. Thin area in the primary walls of two adjacent plant cells through which very fine cytoplasmic strands or plasmodesmata connect the cytoplasm of one cell to the next.

A membrane barrier may or may not exist across the plasmodesmata.

Primer pheromone. A pheromone (chemical signal) that acts through the endocrine system to alter the physiology of an organism in some way and eventually causes the organism to respond in a different way. (Contrast with releaser pheromone.)

Primitive streak. A line running axially along the blastodisc, the site of inward cell migration during formation of the three ply embryo.

Primordial germ cells [L. *primus*: first]. Cells determined early in embryonic growth of animals to become gametes (or to produce gametes on their ultimate arrival at the gonads).

Pro- [Gr: first, before]. A prefix often used in biology to denote a developmental stage that comes first or an evolutionary form that is more primitive than another. For example, procaryote, prophase.

Procambium [Gr. *pro*: before + L. *cambiare*: to exchange]. In plants, the embryonic tissue which will give rise to the first xylem and phloem.

Procaryotes (pro car' ry otes) [L. *pro*: before + Gr. *karyon*: kernel or nucleus]. Organisms whose genetic material is distributed throughout the cell, rather than being isolated in a nucleus. Includes bacteria and bluegreen algae. Considered an earlier stage in the evolution of life than the eucaryotes.

Prolamellar granule [L. *pro*: before + *lamina*: thin plate]. A highly ordered structure found within a mature proplastid, containing some carotenoids and protochlorophyll, but no chlorophyll. In the light, will undergo massive reorganization plus synthesis to form chloroplast lamellae.

Promoter. The region of an operon that acts as the initial binding site for RNA polymerase.

Prophage (pro' faj'). The noninfectious phage units that are linked with the chromosomes of the host bacteria and multiply with them but which do not cause the dissolution of the cell. Prophage can later enter into the lytic phase to complete the virus life cycle.

Prophase (pro' phase) [Gr. *pro*: before]. The first stage of cell division, during which chromosomes condense from diffuse, thread-like material to discrete, compact bodies.

Proplastid [L. *pro*: before + Gr. *plastos*: molded]. A plant cell organelle which under appropriate conditions will develop into a plastid, usually the photosynthetic chloroplast. If plants are kept in the dark, proplastids may become quite large and complex.

Proprioreceptor (L. *proprius*: own). A receptor which provides information about the position or orientation or stress in an animal's own tissue.

Protease (pro' te ase). A proteolytic enzyme.

Protein (pro' teen, less often pro' te in) [Gr. *protos*: first]. One of the most fundamental building substances of living organisms. A long-chain polymer of amino acids with twenty different common side chains. Occurs with its polymer chain extended, in fibrous proteins, or coiled into a compact macromolecule in enzymes and other globular proteins.

Proteolytic enzyme (pro te o lit' ik) [from protein + Gr. *lysis*: cleavage]. An enzyme whose main catalytic function is the cleavage and digestion of a protein or polypeptide chain. The digestive enzymes trypsin, pepsin, and carboxypeptidase are all proteolytic enzymes.

Protistan. Pertaining to the kingdom Protista, which embraces most of what used to be included in the old phylum Protozoa (and still is by some biologists), including the flagellates, amoebas, ciliates, and a few other eucaryotic, unicellular organisms.

Proton (pro' ton) [Gr. *protos*: first]. One of the three most fundamental particles of matter, with mass approximately 1 a.m.u. and charge + 1.

Protonema (pro' tow nee' mah) [Gr. *protos*: first + *nema*: thread]. The hair-like growth form which constitutes an early stage in the development of the moss gametophyte.

Protoplasmic streaming. The metabolically driven motion of the contents of a eucaryotic cell. This process of "mixing" allows transport of materials, over distances that are too great to allow mixing by diffusion alone.

Protoplast. A cell which would normally have a cell wall, but from which the wall has been removed by enzymatic digestion or by special growth conditions.

Prototroph (pro' tow trofe') [Gr. *protos*: first + *trephein*: to nourish]. The nutritional wild-type, or reference form, of an organism. Any deviant form that requires for growth nutrients not required by the prototrophic form is said to be a nutritional mutant, or auxotroph.

Protozoa. A group of single-celled organisms classified by some biologists as a single phylum; includes the flagellates, amoebas, and ciliates. This textbook follows most modern classifications in elevating the protozoans to a distinct kingdom (Protista) and each of their major subgroups (for example, the amoebas) to the rank of phylum.

Pseudoplasmodium [Gr. *pseudes:* false + *plasma:* mold or form]. In the cellular slime molds such as *Dictyostelium,* an aggregation of single amoeboid cells. Occurs prior to formation of fruiting structure.

Pseudopod (soo' do pod) [Gr. *pseudes:* false + *podos:* foot]. A temporary, soft extension of the cell body which is used in locomotion, attachment to surfaces, or engulfing particles.

Pupa (pew' pa) [L. *pupa:* doll, puppet]. In higher insects (the Holometabola), the inactive developmental stage that intervenes between the larva and the adult.

Quiescent zone. In root or shoot apical meristems, a central group of cells which divide only infrequently, and have relatively low levels of protein and nucleic acid synthesis.

Radial symmetry (ray' dee al sym' me tree). The condition in which two halves of a body are mirror images of each other regardless of the angle of the cut, providing the cut is made along the center line. Thus a cylinder cut lengthwise down its center displays this form of symmetry. (Contrast with bilateral symmetry.)

Radical amino acid change. The substitution, during the course of evolution of a protein, of one amino acid at a given point in the chain by another amino acid of quite different character. Substitution of aspartic acid for phenylalanine, or glycine for tryptophan, would normally be considered radical changes. Such radical changes are usually weeded out by natural selection if they occur in important parts of the molecule.

Recapitulation. The repetition of stages of evolution (phylogeny) in the stages of development in the individual organism (ontogeny).

Receptacle [L. *receptaculum:* reservoir]. In an angiosperm flower, the end of the stem to which all of the various flower parts are attached.

Recessiveness. (see dominance)

Recombinant. An individual, meiotic product, or a single chromosome in which genetic materials originally present in two individuals end up in the same haploid complement of genes. The reshuffling of genes can be either by independent segregation, or by crossing over between homologous chromosomes. For example, a human may pass on genes from both parents in a single haploid gamete.

Redirected activity. The direction of some behavior, such as aggression, away from the primary target and toward another, less appropriate object.

Reduction (re duk' shun). Gain of electrons; the reverse of oxidation. Most reductions lead to the storage of chemical energy, which can be released later by an oxidation reaction. Energy-storage compounds such as sugars are highly reduced compounds.

Reductionism (re duk' shun ism). An overly simplified attitude which holds that the properties of a complex system can be explained completely from a knowledge of the properties of only its components. Usually ascribed by psychologists to biologists, by biologists to chemists, and by chemists to physicists.

Reflex [L. *reflexus:* reflected]. An automatic action, involving only a few neurons (in vertebrates, often in the spinal cord) in which a motor response swiftly follows a sensory stimulus.

Region. In biogeography, a major division of the world distinguished by its peculiar animals or plants. For example, Africa south of the Sahara is recognized as constituting the Ethiopian Region.

Regulator gene. A gene that contains the information for making a regulatory macromolecule, often a repressor protein.

Releaser. A sign stimulus used in communication. Often the term is used broadly to include any sign stimulus.

Releaser pheromone. A pheromone (chemical signal) that is quickly perceived and causes a more or less immediate response. (Contrast with primer pheromone.)

Repressible enzyme. One whose synthesis can be decreased or prevented by the presence of a particular compound.

Repressor. A protein coded by the regulator gene. The repressor can bind to a specific operator and prevent transcription of the operon.

Respiration (res pi ra' shun) [literally, "breathing"]. The oxidation of the end products of glycolysis to produce much more energy than the simple glycolytic process can liberate. The oxidant in the respiration of all higher organisms is oxygen gas. Some bacteria can use nitrate or sulfate ion instead of O_2.

Reversion (genetic). Mutational event that restores wild-type phenotype to a mutant.

Revision. In taxonomy, the reclassification of a group of species as the result of additional study.

Rhizine (ry' zeen) [Gr. *rhiza:* root]. Structure on a lichen which is composed of fungal hyphae and anchors the lichen to its substratum.

Rhizoids (ry' zoyds) [Gr. *rhiza:* root]. Hair-like extensions of cells in mosses, liverworts, and a few vascular plants that serve the same function as roots and root hairs in higher plants. The term is also applied to branched root-like extensions of some fungi and algae.

Rhizome (ry' zome) [Gr. *rhizoma:* mass of roots]. A special underground stem (as opposed to root) that runs horizontally beneath the ground.

Ribonucleic acid. (see RNA)

Ribose (rye' bose). A sugar of chemical formula: $C_5H_{10}O_5$, one of the building blocks of nucleic acids.

Ribosome. A small organelle that is the site of protein synthesis.

Ritualization. The evolutionary modification of a behavior pattern that turns it into a signal used in communication, or at least improves its efficiency as a signal.

RNA (ribonucleic acid). A nucleic acid polymer using ribose. Serves as the genetic storage material in some viruses. In other organisms, used in the copying and translation of genetic information into protein structure.

Root pressure. Movement of water and ions up the stem, driven by osmotic entry of water into the roots. Rate and pressure insufficient to account for massive water movement up stems of trees, a movement which requires transpiration.

Saltatory conduction [L. *saltare:* to jump]. A fast mechanism of impulse transmission down a nerve, in which the impulse jumps from one node of Ranvier to another.

Sap. An aqueous solution of nutrients, minerals, and other substances that passes through the xylem.

Saprophyte (sa' pro fight) [Gr. *sapros:* putrid + *phyton:* plant]. An organism (normally a bacterium or fungus) which obtains its carbon and nitrogen directly from dead organic matter.

Secondary cell wall. A later formed cell wall in some plant cells. The cellulose microfibrils are tightly packed and show a high degree of orientation; may be impregnated with lignin or a waxy substance like suberin.

Segregation (genetic). The separation of alleles, or of homologous chromosomes, from one another during meiosis so that each of the haploid daughter cells produced by meiosis contains one or the other member of the pair found in the diploid mother cell, but never both.

Self-pollination. The fertilization of a plant by its own pollen. (Contrast with cross-pollination.)

Semen (see' men) [L: seed]. The thick, whitish liquid produced by the male reproductive organ in mammals, containing the sperm.

Semi-conservative replication. The common way in which DNA is synthesized. Each of the two partner strands in a double helix acts as a template for a new partner strand. Hence, after replication, each double helix consists of one old and one new strand.

Semispecies. Populations that are reproductively isolated from one another to an intermediate degree. They are neither fully isolated, as true species, nor freely interbreeding along their boundaries, which would make them subspecies belonging to the same species.

Sepal (see' pul). One of the outermost structures of the flower, usually protective in function and enclosing the rest of the flower in the bud stage.

Sex chromosome. In organisms with a chromosomal mechanism of sex determination, one of the chromosomes involved in sex determination. One sex chromosome, the X-chromosome, is present in two copies in one sex and only one copy in the other sex. The autosomes, as opposed to the sex chromosomes, are present in two copies in both sexes. In many organisms, there is a second sex chromosome, the Y-chromosome, that is found in only one sex—the sex having only one copy of the X.

Sex linkage. The pattern of inheritance characteristic of genes located on the sex chromosomes of organisms having a chromosomal mechanism for sex determination. The sex that is "diploid" with respect to sex chromosomes can assume three genotypes: homozygous wild-type, homozygous mutant, or heterozygous carrier. The other sex, "haploid" for sex chromosomes, is either hemizygous wild-type or hemizygous mutant.

Sexuality. The ability, by any of a multitude of

mechanisms, to bring together in one individual genes that were originally carried by two different individuals. The capacity for genetic recombination.

Sieve area. Portion of a sieve tube element or sieve cell wall with clusters of pores through which the protoplasts of adjacent phloem cells are interconnected.

Sieve cell. Specialized cell found in the phloem of gymnosperms and spore-bearing vascular plants, through which organic matter, principally carbohydrate, moves from leaves to the rest of the plant. Lateral walls have local areas of perforations leading to adjacent sieve cells.

Sieve plate. In sieve tubes, the highly specialized end walls in which are concentrated the clusters of pores through which the protoplasts of adjacent sieve tube elements are interconnected.

Sieve tube. A column of specialized cells found in the phloem, specialized to conduct organic matter, principally carbohydrate from photosynthesizing leaves to other plant parts. Found principally in flowering plants.

Sieve tube element. A single cell of a sieve tube, containing some cytoplasm but few organelles, with highly specialized perforated end walls leading to elements above and below.

Sigma factor. A polypeptide subunit of RNA polymerase that helps to determine the specificity of binding of the enzyme to particular promoters in DNA.

Sign stimulus. The single stimulus, or one out of a very few stimuli, by which an animal distinguishes key objects, such as an enemy, or a male, or a place to nest, etc.

Social insect. One of the kinds of insect that form colonies with reproductive castes and worker castes; in particular, the termites, ants, social bees, and social wasps.

Society. A group of individuals belonging to the same species and organized in a cooperative manner; in the broadest sense, includes parents and their offspring.

Sociobiology. The scientific study of animal societies and communication.

Sodium pump. A system, usually requiring ATP, which transports sodium, against its concentration gradient, from the inside of cells to the outside.

Sol. A liquified gel.

Somite (so' might) [Gr. *soma:* body]. One of the segments into which an embryo becomes divided longitudinally, leading to the eventual segmentation of the animal as illustrated by the spinal column, ribs, and associated muscles.

Sonar. The mode of orientation used by bats, porpoises, and a few other animals in which the positions of objects are estimated by emitting sounds and listening for the echoes that bounce back from them.

Spawning. The direct release of sex cells into the water.

Specificity site. The region on the surface of an enzyme where a part of the substrate binds which helps the enzyme to recognize the proper substrate.

Speciation (spee' shee ay' shun). The processes of diversification of populations and the multiplication of species.

Species (spee' shees). The basic lower unit of classification, consisting of a population or series of populations of closely related and similar organisms. The more narrowly defined "biological species" consists of individuals capable of interbreeding freely with each other but not with members of other species.

Species equilibrium. A condition in which the number of species going extinct in a given area per unit time equals the rate at which new species are arriving; thus the number of species in the area remains constant.

Sperm [Gr. *sperma:* seed]. A male reproductive cell.

Spermatocyte (spur mat' oh site) [Gr. *sperma:* seed + *kytos:* cell]. The cell that gives rise to the sperm in animals.

Spermatogenesis (spur mat' oh jen' e sis) [Gr. *sperma:* seed + *gignomai:* be born]. Male gametogenesis, leading to the production of sperm.

Spermatozoon (spur' ma toh zoe' an). A sperm.

Sphincter (sfingk' ter) [Gr. *sphigkter:* shut tight]. A ring of muscle which can close an orifice, for example at the anus.

Spindle apparatus. An array of microtubular fibers stretching from the metaphase plate to the poles of a dividing cell and playing a role in the movement of chromosomes at cell division. Named for its shape.

Splitter. A taxonomist who prefers to divide organisms into relatively many groups. (Contrast with lumper.)

Spongy parenchyma. In leaves, a layer of loosely packed isodiametric photosynthetic cells with extensive intercellular spaces for gas diffusion. Frequently found between the palisade parenchyma and the lower epidermis.

Spontaneous generation. The idea that life evolves continually from non-living matter. Usually distinguished from the current idea that life evolved from non-living matter under primordial conditions at an early stage in the history of our planet.

Spontaneous reaction. A chemical reaction which will proceed on its own, without any outside influence. A spontaneous reaction need not be rapid.

Sporangiophore [Gr. *phore:* to bear]. Any branch bearing one or more sporangia.

Sporangium (spor an' gee um) [Gr. *spora:* seed + *angeion:* vessel or reservoir]. In plants, any specialized structure within which one or more spores are formed.

Spore [Gr. *spora:* seed]. Any asexual reproductive cell, capable of developing into an adult plant, without gametic fusion. Haploid spores develop into gametophytes, diploid spores into sporophytes. In procaryotes, a resistant cell capable of surviving unfavorable periods.

Sporophyll. Any leaf or leaflike structure which bears sporangia; refers to carpels and stamens of angiosperms, sporangium-bearing leaves in ferns, and their allies.

Sporophyte (spo' roh fyte). In plants with an alternation of generations, the diploid phase that produces the spores. (Contrast with gametophyte.)

Stamen (stay' men). A male (pollen-producing) unit of a flower, usually composed of an anther, which bears the pollen, and a filament, which is a stalk supporting the anther.

Starch [O.E. *stearc:* stiff, from Gr. *stereos:* solid]. A branched-chain polymer of glucose molecules, used by plants as a means of storing energy.

Statoliths [Gr. *statos:* positioned + *lithos:* stone]. In plants, subcellular organelles (presumably starch grains) that sense the direction of gravity, helping to orient the plant.

Steady state. An apparently unchanging condition which is due to balanced synthesis and degradation of all components in the system.

Stigma [L. *stigma:* mark, brand]. The part of the pistil, at the apex of the style, which is receptive to pollen, and on which pollen germinates.

Stoma (plural: stomata) [Gr. *stoma:* mouth]. Small openings in the plant epidermis which permit gas exchange; bounded by a pair of guard cells whose osmotic status regulates the size of the opening.

Strobilus (strobe' a lus) [Gr. *strobilos:* a cone]. The cone, or characteristic multiple fruit, of the pine and other gymnosperms. Also, a cone-shaped mass of sporophylls found in club mosses and certain fern allies. Same as strobile.

Structural gene. A gene that encodes the primary structure of a protein (usually an enzyme).

Style [Gr. *stylos:* pillar or column]. In flowering plants, a column of tissue extending from the tip of the ovary, and bearing the stigma or receptive surface for pollen at its apex.

Sub- [L: under]. A prefix often used to designate a structure that lies beneath another or is less than another. For example, subcutaneous, subspecies.

Subspecies (sub' spee' shees). A subdivision of a species. Usually defined more narrowly as a geographical race: a population or series of populations occupying a discrete range and differing genetically from other geographical races of the same species.

Substrate (sub' strayte). The molecule or molecules on which an enzyme exerts catalytic action.

Succession. In ecology, the gradual and predictable series of changes in species composition of organisms, from the time of colonization of an empty space to the maturing of the final, stable, climax community.

Supernormal stimulus. Any stimulus, or any intensity of a variable stimulus, that is preferred by animals over the natural sign stimulus.

Suspensor. In plants, a cell or group of cells derived from the zygote, but not actually part of the embryo proper, which in some higher plants pushes the young embryo deeper into nutritive gametophyte tissue or endosperm by its growth.

Symbiosis (sim' bee oh' sis) [Gr. *symbiosis:* to live together]. The living together of two or more species in a prolonged and intimate ecological relationship. (See parasitism, commensalism, mutualism.)

Symmetry. In biology, the property that two halves of an object are mirror images of each other. (See bilateral symmetry and radial symmetry.)

Sympatric (sim pa' trick) [Gr. *syn:* together + *patria:* homeland]. Referring to populations whose geographic ranges overlap at least in part.

Synapsis (sin ap' sis). The highly specific parallel alignment (pairing) of homologous chromosomes during the first division of meiosis.

Synergids (si nur' jids). Two cells found close to the egg cell in the angiosperm embryo sac; they disappear shortly after fertilization.

Systematics. The science of classification of organisms.

Taxis (tak' sis) [Gr. *taxis*: arrange, put in order]. The movement of an organism in a particular direction with reference to a stimulus. A taxis usually involves the employment of one sense and a movement directly toward or away from the stimulus, or else the maintenance of a constant angle to it. Thus a positive phototaxis is movement toward a light source, negative geotaxis is movement upward (away from gravity), and so on.

Taxonomy (taks on' oh me) [Gr. *tasso*: arrange, classify]. The science of classification of organisms.

Telophase (teel' oh phase) [Gr. *telos*: end]. The final phase of mitosis or meiosis during which chromosomes become diffuse, nuclear envelope reforms, and nucleoli begin to appear in the daughter nuclei.

Tepal. In an angiosperm flower, a sterile modified leaf. This term is used to refer to such flower parts when one is unable to distinguish between petals and sepals.

Terrestrial (ter res' tree al). Pertaining to the land. (Contrast with marine.)

Territory. A fixed area from which an animal or group of animals excludes other members of the same species by aggressive behavior or display.

Testicle (tes' ti kul) [L. *testiculus*: diminutive of *testis*]. Same as testis.

Testis (tes' tis) (plural: testes) [L. *testis*: witness]. The male gonad, that is, the organ that produces the male sex cells.

Tetrad (te' trad) [Gr. *tetras*: foursome]. Paired homologous chromosomes during the first prophase and metaphase of meiosis. Each chromosome at these stages will be a double structure, consisting of two chromatids joined at a not-yet-divided centromere.

Thallus (thal' us) [Gr. *thallos*: sprout]. Any plant body which is not differentiated into root, stem or leaf.

Thyrotropic hormone. A hormone which is produced in the pituitary gland of amphibia such as the frog, transported in the blood stream to the thyroid gland, inducing the thyroid gland to produce the thyroid hormone that regulates

metamorphosis from tadpole to adult frog.

Tissue. A group of similar cells organized into a functional unit and usually integrated with other tissues to form part of an organ such as a heart or leaf.

Tonoplast [Gr. *tonos*: something stretched + *plasma*: anything formed or molded]. In plant cells, the membrane surrounding a large central vacuole.

Tornaria (tor nare' e ah) [L. *tornus*: lathe]. The free-swimming ciliated larva of certain echinoderms and hemichordates; its existence indicates the evolutionary relationship of these two groups.

Trachea (tray' key a) [Gr. *trakhoia*: a small rough artery]. A tube which carries air to the bronchi of the lungs of vertebrates, or to the cells of arthropods.

Tracheid (tray' key id). A distinctive conducting and supporting cell found in the xylem of nearly all vascular plants, characterized by tapering ends and walls that are pitted but not perforated.

Transcription. The synthesis of RNA, using one strand of DNA as the template.

Transduction. Transfer of genes from one bacterium to another with a bacterial virus acting as the carrier of the genes.

Transfer RNA (tRNA). A category of relatively small RNA molecules (about 75 nucleotides). Each kind of transfer RNA is able to accept a particular activated amino acid from its specific activating enzyme, after which the amino acid is added to a growing polypeptide chain.

Transformation. Mechanism for transfer of genetic information in bacteria in which pure DNA extracted from bacteria of one genotype is taken in through the cell surface of bacteria of a different genotype and incorporated into the genophore of the recipient cell. By extension, the term has come to be applied to phenomena in other organisms in which specific genetic alterations have been produced by treatment with purified DNA from genetically marked donors.

Transition mutation. A base substitution mutation in which a purine (A or G) replaces a purine, and a pyrimidine (T or C) replaces a pyrimidine. (Contrast with transversion mutation.)

Translation. The synthesis of a protein (polypeptide). This occurs on ribosomes, using the information encoded in messenger RNA.

Translocation (genetic). A rare mutational event

that moves a portion of a chromosome to a new location, generally on a nonhomologous chromosome.

Transpiration [L. *spirare:* to breathe]. The evaporation of water from plant leaves and stem, driven by heat from the sun, and providing the motive force to raise water (plus ions) from the roots.

Transversion mutation. A base substitution mutation in which a purine (A or G) replaces a pyrimidine (T or C) and *vice versa.* (Contrast with transition mutation.)

Trichocyst (trike' o sist) [Gr. *trichos:* hair + *kystis:* cell]. A threadlike organelle ejected from the surface of ciliate protozoans, used both as a weapon and an anchoring device.

Triplet. (see codon)

tRNA. (see transfer RNA)

Trochophore (troke' o fore) [Gr. *trochos:* wheel + *phoreus:* bearer]. The free-swimming larva of some annelids and mollusks, distinguished by a wheel-like band of cilia around the middle, and indicating an evolutionary relationship between these two groups.

Trophallaxis. The exchange of liquid substances from one member of an insect society to another.

Trophic level. The position of a species in a food chain, indicating which species it consumes and which consume it.

T-system. A set of transverse tubes that penetrates skeletal muscles and terminates in blind sacs (cisternae). The T-system transmits impulses to the sacs, which then release Ca^{++} to initiate muscle contraction.

Tuber [L. *tuber:* swelling]. A short fleshy underground stem, usually much enlarged, and serving a storage function, as in the case of the potato.

Tunica [L. *tunica:* a garment]. The outermost cell layers of the shoot apical meristem, giving rise to leaf primordia on its lateral flanks.

Turgor pressure. The pressure which a plant cell exerts against its constraining wall, a function of the osmotic concentration of the cell, water availability, and the relative elasticity and plasticity of the wall itself.

U.E.P. (Unit Evolutionary Period). The time (in millions of years) for a one percent difference in amino acid sequences to show up between two divergent lines of evolution of a protein.

Unicellular (yoon' e cel' yew ler) [L. *unus:* one +

cella: chamber]. Consisting of a single cell; as for example a unicellular organism. (Contrast with multicellular.)

Unit Evolutionary Period. (see U.E.P.)

Unit membrane. The boundary layer that separates compartments of cells and that separates the cell from its surroundings. Under the electron microscope, the unit membrane appears to be made up of two closely-spaced layers. Chemically, the membrane is composed mainly of phospholipids, which appear to form a double-layer, and proteins, which appear to be embedded in the phospholipid layers. Some organelles, such as mitochondria and nuclei, are surrounded by double membranes composed of two unit membranes.

Uterus (yoo' ter us) [L. *uterus:* womb]. The womb, a specialized portion of the female reproductive tract in certain mammals, which receives the fertilized egg and nurtures the embryo in its early development.

Vacuole (vac' you ole) [Fr: small vacuum]. A liquid-filled cavity in a cell, enclosed within a unit membrane. Vacuoles play a wide variety of roles in cellular metabolism, some being digestive chambers, some storage chambers, some waste bins, some cellular "bladders," and so forth.

Vascular (vas' kew lar). Pertaining to organs and tissues, such as blood vessels in animals and phloem and xylem in plants, that conduct fluid.

Vernalization [L. *vernalis:* belonging to spring]. Events occurring during a required chilling period, leading eventually to the breaking of dormancy of certain buds and seeds. Vernalization may require many weeks of below-freezing temperatures.

Vertebrate. An animal whose nerve cord is enclosed in a backbone of bony segments, called vertebrae. The principal groups of vertebrate animals are the fishes, amphibians, reptiles, birds, and mammals.

Vessel [L. *vasculum:* a small vessel]. In botany, a tube-shaped element of the xylem which consists of hollow cells (vessel elements) placed end to end and connected by perforations. Together with the tracheid, it conducts water and minerals in the plant.

Villus (vil' us) (plural: villi) [L. *villus:* shaggy hair]. A hair or a hair-like projection from a membrane, for example from many gut walls.

Virion. The virus particle, the minimum unit capable of infecting a cell. The equivalent of the cell in true organisms.

Virus [L: poison, slimy liquid]. Any of a group of ultramicroscopic, infectious particles constructed of nucleic acid and protein that can reproduce only in living cells and do not contain enough distinct structures themselves to be ranked as living cells.

Vitamin [L. *vita*: life]. Any one of several structurally unrelated organic molecules which an organism cannot synthesize itself, but nevertheless requires in small quantity for normal growth and metabolism.

Waggle dance. The running movement of a working honeybee on the hive, during which the worker traces out a repeated figure-eight. The dance contains elements that transmit to other bees the location of food and new nest sites.

Wall pressure. The physical resistance of the plant cell wall to turgor pressure.

Warning appearance. A conspicuous trait, such as bright coloration, serving to warn potential predators that the animal is poisonous or otherwise dangerous.

Wild-type. Geneticists' term for standard or reference type. Deviants from this standard, even if the deviants are found in the wild, are said to be mutant.

Xanthophyll (zan' tho fill) [Gr. *xanthos:* yellowish-brown + *phyllon:* leaf]. A yellow or orange pigment commonly found as an accessory pigment in photosynthesis, but found elsewhere as well; contains one or more oxygens in addition to carbon and hydrogen.

Xylem (zy' lem) [Gr. *xylon:* wood]. In evolutionarily more advanced plants, the woody tissue that conducts water and minerals; xylem consists of tracheids, vessels, fibers, and other highly specialized cells.

Yolk. The stored food material in animal eggs, usually rich in protein and lipid.

Zoology (zoe ol' o jee) [Gr. *zoon:* animal + *logos:* discourse]. The scientific study of animals.

Zooplankton (zoe' o plangk' ton) [Gr. *zoon:* animal + *planktos:* wandering]. The animal portion of the plankton.

Zoospore (zoe' o spohr) [Gr. *zoon:* animal + *spora:* seed]. In algae and fungi, any swimming spore. May be diploid or haploid.

Zygote (zye' gote) [Gr. *zygotos:* yoked]. The cell created by the union of two gametes, in which the gamete nuclei are also fused. The earliest stage of the diploid generation.

818

819

822

Cup fungi, fruiting structures of, 566
Cuticle
  arthropod, 610, 611
  exoskeleton, composition of, 436–437
  insect, function of, *365*
  nematode, 604
Cutin, leaf structure and, 293
Cuttings, auxins and, 384
Cyanide, stomata and, 323
Cyanophyta, *see also* Bluegreen algae
  biochemical characteristics, *569*
  members of, *543*
Cyclic adenosine monophosphate
    (cAMP)
  catabolite repression and, 213
  cell aggregation and, 262
  as second messenger, 402
Cyclic phosphorylation, function of, 122
Cycloheximide, ribosomes and, 114
Cynthia moth, cyclical behavior of, 463
Cysts, slime mold, 563
Cysteine
  cytochromes *c* and, 657, 658
  disulfide bonds and, 66, 67
Cytidine triphosphate, aspartate trans-
    carbamylase and, 94
Cytochrome(s), respiration and, *113*
Cytochrome *a*, DNA and, 114
Cytochrome *b*₆, photosynthesis and,
    122
Cytochrome *b*₅₅₉, photosynthesis and,
    123
Cytochrome *c*
  amino acid sequences, 657, *658–659*
    differences, *660*
  charge distribution in, 659
  drawing of, *654*
  evolution, rate of, 662–664, *663*
  phylogenetic tree and, 661–662
  properties of, 656–657
  simple variation and selection of,
    656–660
Cytochrome *f*, photosynthesis and, 122,
    123
Cytogenetics, gene maps and, 197–199
Cytokinesis, cell division and, 34
Cytokinin
  isolation of, 384
  plants and, 383, 384–386
Cytosine, nucleic acids and, *70*, 71

Damsel flies, nymphs, gas exchange by,
    335
Dandelion, day length and, 370
Daphnia, resistant stage of, 367
Darwinism, evolution and, 635–637
Darwin's finches, speciation and, 685–
    686
Darwin's theory, genetics and, 177
Darwin-Wallace theory, postulates and
    deductions, 636
Dating, radiocarbon and, 41, *42*, 532–
    533

Daylength, predictability of, 369
DDT
  benefits and disadvantages of, 755–
    756, *757*
  degradation of, 545
  microsomal hydroxylase and, 363
Death
  postponement, consequences of,
    777
  probability, age and, 718–719
Degeneracy, genetic code and, 128
Degradation, control and, 214
Deletions
  detection of, 199
  mutations and, 139, 141–142, 159
Deme, definition of, 639
Demographic shift, populations and,
    775
Dendrites, function of, *408*
Dendrolasia, leaves of, *320*
Denitrification, need for, 300–301
Density dependent effects, population
    and, 719–720, *721*
Deoxycorticosterone, effects of, 401
Deoxynucleoside triphosphates, struc-
    tures of, *135*
Deoxyribonucleic acid (DNA), 62, *126*
  amount per cell, 216
  autoradiography of, 27
  bacterial maleness and, 166–167
  cells and, 11–12
  chemistry of, 70–72
  chloroplasts and, 25, 114, 232
  chromosomes and, 30
  duplication of, 71–72, 131–133
  enzymes and, 55–56
  foreign, growth in bacteria, 169–171
  function of, 55, 65, 70–71
  genetic analysis of, 154
  genophore and, 27
  information and, 127–128
  Meselson-Stahl experiment and,
    133–135
  mitochondria and, 24, 114, 115
  nucleoli and, 34
  perpetuation of, 226
  procaryotic, 541
  recombinant, controversy and, 171
  reverse transcriptase and, 550
  ring-shaped, occurrence of, *28*
  sequencing of, 152–153
  steroid hormones and, 403
  strands of, *217*
  synthesis
    cell cycle and, 34
    enzymes and, 135–138
  transformation and, 130
Deoxyribonucleic acid ligase, function
    of, 136, 170
Deoxyribonucleic acid polymerase, func-
    tion of, 135–137
Deoxyribose, structure of, *62*
Depolarization, muscle contraction and,
    443

Deposit feeders
  digestion in, 313
  examples of, 305, 307
Depth perception, orientation and,
    460–461
Desert
  extent of, *738–739*
  plant adaptations to, 119, 295
Desmids, *575*
Detergents, pollution by, 755
Determination, definition of, *242*
Deuterostomes, mouth of, 616
Development
  indeterminate, 616
  mammalian, 249–250
  order and, 229
DFP, *see* Diisopropyl phosphoro-
    fluoridate
Diabetes, treatment of, 200
Diamond pipes, Earth's mantle and,
    3
Diapause, photoperiodism and, 369
Diaphragm
  action of, 332
  contraception and, 286
  mammals and, 627
Diastole, heartbeat and, *349*
Diatoms, *570*
  characteristics of, 570–571
Dicots
  characteristics of, 592–593
  examples of, 592–593
Dictyostelium discoideum, 262
  aggregation of, 262
  life cycle of, 563
Didinium, feeding by, *556*
Differentiation
  definition of, *242*
  reversibility, 236–239
  RNA and, 234–235
  sexual, hormones and, 403–404
Difflugia corona, pattern inheritance in,
    202–203
Diffraction pattern, lysozyme, *74*
Diffusion
  ammonia excretion and, 360
  carrier-facilitated, 15–*16*
  cell membrane and, 15–16
  gas exchange and, 319, 320
  stomata and, 321
  transport and, 344
Digestion
  in carnivores, 312–313
  control of, 310
  early experiments on, *308*
  in herbivores, 310–312
  macromolecules and, 302
  products, uptake of, 309–310
Dihydrotestosterone, formation of,
    403–404
Dihydroxyacetone phosphate,
    glycolysis and, *104*
Diisopropyl phosphorofluoridate, inhi-
    bition by, 82–83

823

825

828

830

833

838

Credits

CHAPTER 1

Opener: NASA   1. NASA   2. After J. Tuzo Wilson   3. Verne Shoemaker   4. NASA   5. NASA

CHAPTER 2

Opener: L. Orci & A. Perrelet   Box B *(left)*: Louis W. Ladbaw *(right)*: T. F. Anderson   3. J. David Robertson   7. David B. Slautterback   10. Keith R. Porter   11. Myron C. Ledbetter   12. Keith R. Porter   13. Myron C. Ledbetter   14. Walter Stoeckenius   17. Walter Plaut

CHAPTER 3

Opener: Erwin W. Mueller

CHAPTER 4

Opener: E.S. Machlin & James J. Burton   3. J. David Robertson   4. After "The Structure and Function of Histocompatibility Antigens" by Bruce A. Cunningham, © October 1977 by *Scientific American Inc.* All rights reserved.   14. © 1969, R.E. Dickerson & I. Geis   21. From R.E. Dickerson & I. Geis, *Chemistry, Matter and the Universe*, W.A. Benjamin, 1976

CHAPTER 5

Opener: R.E. Dickerson   Box C *(left)*: E. Margoliash *(right)*: © 1969, R.E. Dickerson & I. Geis   13. © 1969, R.E. Dickerson & I. Geis   15. © 1969, R.E. Dickerson & I. Geis   18. © 1969, R.E. Dickerson & I. Geis

CHAPTER 6

Opener: Space Environment Services, SEL/ERL/ NOAA   2. S. Mudd & T.F. Anderson   3. Hilda A. Agar & Howard C. Douglas   4. Keith R. Porter   5. L.K. Shumway   19. From E.T. Weier, R.C. Stocking, & M.G. Barbour, *Botany*, 5th Edition, John Wiley & Sons, 1975   23. Norbert Pfennig

CHAPTER 7

Opener: O.L. Miller Jr. & B.R. Beatty   1. After D.M. Bonner   6. M.S. Meselson & F.W. Stahl

CHAPTER 8

Opener: L. Caro & R. Curtiss   3. A.K. Kleinschmidt   4. Millard Susman   7. Thomas Seale

CHAPTER 9

Opener: J.R. McIntosh   5. Walter Plaut   6. Walter Plaut   7. Walter Plaut   11. R.L. Phillips & A.M. Srb, *Canadian J. Genet. Cytol.*, 9:768, 1967   16. *(top)*: Walter Plaut *(bottom)*: Hal Krider   17. Adapted from "Ovarian Development in *Drosophila Melanogaster*" by R.C. King, Academic Press, 1970. Photo: A.S. Fox & R.A. Kreber   18. After D.L. Nanney

CHAPTER 10

Opener: O.L. Miller Jr., B. Hamkalo & C.A. Thomas Jr.   7. O.L. Miller Jr. & B.R. Beatty   8. Elizabeth Kaveggia

CHAPTER 11

Opener: Jerome Gross   1. Jerome Gross   2. Manfred E. Bayer   3. Barbara Panessa   4. David A. Stetler   5. Ulrich Clever   6. F.C. Steward   7. After "Transplanted Nuclei and Cell Differentiation" by J.B. Gurdon, © December 1968 by *Scientific American Inc.* All rights reserved.

CHAPTER 12

Opener: Chester F. Reather, courtesy of Carnegie Institution of Washington   4. Lloyd M. Beidler   11. Chester F. Reather, courtesy of Carnegie Institution of Washington   16. F.C. Steward   17. E.H. Newcomb   20. Ray Evert   22. H.A. Phillips   23. Thomson Associates by J.V. Herbert   27. John Tyler Bonner

CHAPTER 13

Opener: D.W. Fawcett & E. Anderson   1. J.A.L. Cooke   2. Thomas Eisner   7. W.H. Hodge   8. Alice F. Tryon   11. Michael J. Reber from Chesapeake Biological Laboratory   12. Stanley & Kay Breeden   13. J.A.L. Cooke   15. Ederic Slater   16. J.A.L. Cooke   17. Richard C. Kern   18. Virginia Page

CHAPTER 14

Opener: Penelope Jenkin   1. Ursula Goodenough   2. Walter H. Hodge   3. J.H. Troughton & F.B. Sampson   5. J.H. Troughton & L.A. Donaldson   6. Walter H. Hodge   7. Walter H.

Hodge   8. After Scagel *et al.*   9. Allan H. Bennett   10. Harold J. Evans   12. Harold J. Evans   13. Thomas Eisner   16. Charles Walcott   17. E.W. Strauss   21. Thomas Eisner   22. H. Heusser   23. *(top, middle)*: Thomas Eisner *(bottom)*: Penelope Jenkin   24. *(left)*: Ralph Buchsbaum *(top right)*: William A. Wimsatt *(bottom right)*: Hans Banziger   25. Thomas Eisner

CHAPTER 15

Opener: Julius H. Comroe, Jr.   1. William C. Steere   2. After G.M. Smith   4. B.E. Juniper   5. J.H. Troughton & L.A. Donaldson   7. G.D. Humble & K. Raschke   8. Walter H. Hodge   9. Thomas Eisner   11. Philadelphia Zoological Society   12. Thomas Eisner   17. *(left, center)*: Thomas Eisner *(right)*: D.S. Smith & P.L. Miller   19. Thomas Eisner

CHAPTER 16

Opener: P. Dayanandan   1. Katherine Esau   2. Thomson Associates by J.V. Herbert   3. Thomson Associates by J.V. Herbert   4. J.H. Troughton & L.A. Donaldson   6. Thomson Associates by J.V. Herbert   7. Thomson Associates by J.V. Herbert   8. Thomson Associates by J.V. Herbert   9. M.H. Zimmerman   16. Thomas Eisner   18. Thomas Eisner

CHAPTER 17

Opener: Pfizer Inc.   4. Keith R. Porter   8. B.K. Filshie & C.D. Beaton   11. Thomas Eisner   16. John Lupo   17. Stuart Brody

CHAPTER 18

Opener: The Bettmann Archive   4. C.F. Cleland   6. H.A. Phillips   19. From R.H. Williams, Ed., *Textbook of Endocrinology*, 5th Edition, W.B. Saunders. Courtesy of Judson J. Van Wyk

CHAPTER 19

Opener: R.A. Steinbrecht   1. From S.W. Kuffler & J.G. Nicholls, *From Neuron to Brain*, Sinauer, 1976   2. M.M. Salpeter   6. K.D. Roeder   9. T.J. McDonald   13. Thomas Eisner   17. R.A. Steinbrecht   18. R.A. Steinbrecht

CHAPTER 20

Opener: David Phillips   3. D.W. Fawcett   4. Pfizer Inc.   7. Thomas Eisner   8. Gjon Mili   9. Thomas Eisner   11. *(left & right)* Carolina Biological Supply Co. *(center)*: Thomas Eisner   13. Keith R. Porter   18. Joseph F. Gennaro   19. Joseph F. Gennaro   20. Joseph F. Gennaro   21. Joseph F. Gennaro   22. Joseph F. Gennaro   24. Hans R. Haefelfinger   26. Yata Haneda   27. Yata Haneda

CHAPTER 21

Opener: Hermann Kacher   1. William M. Harlow

2. Thomas Eisner   13. Hermann Kacher   15. William C. Dilger

CHAPTER 22

Opener: Carl W. Rettenmeyer   2. Thomas G. Schultze-Westrum   4. Peter Marler   10. From "Territorial Marking by Rabbits" by R. Mykytowycz, © May 1968 by *Scientific American Inc.* All rights reserved.   14. From E.O. Wilson, *The Insect Societies*, Belknap Press, Harvard University Press, 1971   18. From E.O. Wilson, *The Insect Societies*, Belknap Press, Harvard University Press, 1971   20. From E.O. Wilson, *The Insect Societies*, Belknap Press, Harvard University Press, 1971

CHAPTER 23

Opener: Martin F. Glaessner   3. E.S. Barghoorn & J.W. Schopf   5. E.S. Barghoorn & S.A. Tyler   6. E.S. Barghoorn & S.A. Tyler   7. E.S. Barghoorn & S.A. Tyler   8. E.S. Barghoorn & S.A. Tyler   9. *(top)*: E.S. Barghoorn & S.A. Tyler *(bottom)*: S.M. Siegel *et al.*   10. N. Menchikoff   11. G. Licari   12. Martin F. Glaessner   13. After S.L. Miller   16. Sidney Fox   17. After A.G. Fischer

CHAPTER 24

Opener: Joseph G. Farris, © 1976, *The New Yorker* Magazine Inc.   7. Bernhard Kummel, *History of the Earth*, 2nd Edition, W.H. Freeman & Co., 1970   8. F.M. Carpenter   9. Richard D. Estes

CHAPTER 25

Opener: Gordon F. Leedale   3. Norma J. Lang   4. John Tyler Bonner   5. HEW, Center for Disease Control   7. Steven H. Larsen   8. After W.R. Sistrom   10. Ray Evert   11. After "The Structure of Viruses" by R.W. Horne, © January 1963 by *Scientific American Inc.* All rights reserved.   12. B.K. Filshie & C.D. Beaton   16. *(top left)*: John Kingsbury *(top right)*: Thomas Eisner *(bottom left)*: Eric V. Gravé *(bottom right)*: Gordon F. Leedale   17. R.D. Allen   19. Eric V. Gravé   21. Eric V. Gravé   23. Gregory Antipa   24. R. Jakus   25. Eric V. Gravé   26. Eric V. Gravé

CHAPTER 26

Opener: Harvey J. Marchant   1. Ross E. Hutchins   2. Ross E. Hutchins   6. Walter H. Hodge   9. Douglas P. Wilson   10. John Troughton   12. C.F. Cleland   15. Harvey J. Marchant   16. Jeremy D. Pickett-Heaps   17. Thomson Associates by J.V. Herbert   19. Ray Evert   20. John Troughton   21. William C. Steere   22. William C. Steere   24. Walter H. Hodge   31. John Troughton   34. Walter H. Hodge   36. J.H. Troughton & L.A. Donaldson   39. Reed C. Rollins   41. W.E. Boggs

CHAPTER 27

2. Martin Glaessner   4. Douglas P. Wilson   5. Ralph Buchsbaum   6. Douglas P. Wilson

7. Lev Fishelson   8. Douglas P. Wilson
9. L.W. Mullinger   12. J.A.L. Cooke
14. Douglas P. Wilson   18. Thomas Eisner
19. American Museum of Natural History
20. *(top left)*: John Gerard *(top right)*: Douglas P. Wilson *(center left)*: Hermann Kacher *(center right)*: R.D. Silberglied *(bottom)*: K.D. Roeder & Helen Ghiradella
22. Ralph Buchsbaum   23. J.A.L. Cooke
29. American Museum of Natural History
34. Jane Burton   36. American Museum of Natural History

CHAPTER 28

Opener: National Maritime Museum, Greenwich, England   10. H.B.D. Kettlewell

CHAPTER 29

Opener: © 1972, R.E. Dickerson & I. Geis   3. After W.M. Fitch & E. Margoliash   6. After R.E. Dickerson & I. Geis   9. After G.A. Mross & R.F. Doolittle

CHAPTER 30

Opener: Robert E. Silberglied   2. From T. Dobzhansky, *Evolution, Genetics and Man*, 1955. By permission of John Wiley & Sons Inc.   13. After J. van der Vecht

CHAPTER 31

Opener: Irven DeVore/AnthroPhoto   14. After "The Breakup of Pangaea" by Robert S. Dietz and John C. Holden, © October 1970 by *Scientific American, Inc.* All rights reserved.   15. After "The Breakup of Pangaea" by Robert S. Dietz and John C. Holden, © October 1970 by *Scientific American Inc.* All rights reserved.   16. After "The Breakup of Pangaea" by Robert S. Dietz and John C. Holden, © October 1970 by *Scientific American, Inc.* All rights reserved.

CHAPTER 32

Opener: Eugene L. Nakamura   14. From D. Davenport, in S.M. Henry, Ed., *Symbiosis*, Vol. 1, Academic Press, 1966   15. From M. Caullery, *Parasitism and Symbiosis*, Sigdwick & Jackson Ltd., 1952   17. Thomas Eisner   18. Thomas Eisner   19. Thomas Eisner   20. Frederick A. Webster   21. Thomas Eisner

CHAPTER 33

Opener: Lev Fishelson   4. Douglas P. Wilson
8. After Ralph Buchsbaum   21. After "Toxic Substances and Ecological Cycles" by George M. Woodwell, © March 1967 by *Scientific American Inc.* All rights reserved.

CHAPTER 34

Opener: Irven DeVore/AnthroPhoto   4. From E.O. Wilson, *Sociobiology: The New Synthesis*, Belknap Press, Harvard University Press, 1975

CHAPTER 35

Opener: From Paolo Soleri, *Arcology*, 1969. By permission of the M.I.T. Press, Cambridge, Mass.   1. After A. Desmond

CHAPTER 36

Opener: Steven Balkin

EDWARD O. WILSON is Baird Professor of Science at Harvard University. Born in Birmingham, Alabama, he earned his doctorate from Harvard in 1955 and joined the Harvard faculty the following year. For four years he taught a general education course in elementary biology, and currently is in charge of the first half of the introductory course for majors in biology. His research interests span the general theory of systematics, evolution and biogeography, and the social behavior of insects. A member of the National Academy of Sciences and the American Philosophical Society, he is the author of *The Theory of Island Biogeography*, with the late Robert H. MacArthur, *A Primer of Population Biology*, with William H. Bossert, *The Insect Societies*, and *Sociobiology*. In 1977, Wilson was a recipient of the National Medal of Science.

THOMAS EISNER is Jacob Gould Schurman Professor of Biology at Cornell University. Born in Berlin, he earned his Ph.D. in biology at Harvard in 1955 and moved to Cornell in 1957. He has published extensively on the behavior and chemical ecology of insects, particularly on chemical communication. Eisner has taught the introductory biology course at Cornell, as well as courses in neurobiology and behavior, and chemical ecology. A member of the National Academy of Sciences, he has also served on the editorial board of *Science*.

WINSLOW R. BRIGGS is Director of Plant Biology of the Carnegie Institution of Washington located in Stanford, California. A native of Minnesota, he did his graduate work at Harvard, receiving his Ph.D. in 1955. He moved to Stanford University the following year, and remained there until he returned to Harvard in 1967. At Harvard he taught plant physiology and introductory botany. He assumed his present position in 1973. Briggs' chief research interests are plant growth

and development, and the biophysics and biochemistry of plant responses to light.

RICHARD E. DICKERSON is Professor of Chemistry at the California Institute of Technology. He earned a Ph.D. in physical chemistry from the University of Minnesota in 1957. He worked in England with J. C. Kendrew on the first high-resolution x-ray analysis of the structure of a crystalline protein. In 1959 he joined the faculty of the University of Illinois, and in 1963 moved to Caltech. His main research interests are the structure and function of proteins, and the evolution of protein molecules. He is the author or co-author of four other books, including *Chemistry, Matter and the Universe* with Irving Geis.

ROBERT L. METZENBERG is Professor of Physiological Chemistry at the University of Wisconsin Medical School. He earned a Ph.D. in biochemistry from Caltech in 1956. Since then he has remained more or less continuously at the University of Wisconsin, where he has taught medical students, medical technologists, graduate students, and first-year undergraduates in the Biology Core Curriculum. His current research interest is the problem of how genes regulate metabolism in eucaryotes.

RICHARD D. O'BRIEN holds three posts at Cornell University. He is director of the Division of Biological Sciences, Professor of Neurobiology, and Richard J. Schwartz Professor of Biology and Society. He earned his Ph.D. from the University of Western Ontario in 1954. After six years in a research institute of the Canadian Department of Agriculture, he moved to Cornell in 1960. Before assuming his current duties, he served as Professor of Entomology, Chairman of Biochemistry, and Chairman of the Department of Neurobiology and

Behavior. His main research interests are the molecular basis of chemical transmission and the action of nerve poisons.

MILLARD SUSMAN is Chairman of the Departments of Genetics and Medical Genetics at the University of Wisconsin at Madison. He received a Ph.D. in genetics from Caltech in 1962. After a year's postgraduate work in London, he went to the University of Wisconsin. His research specialty is the genetics and development of bacteriophage viruses. He teaches microbial genetics and was a member of the team that created and taught the Biology Core Curriculum.

WILLIAM E. BOGGS is an author and editor. A native of Pennsylvania, he graduated from the University of Pittsburgh in 1954. He began his career with McGraw-Hill in 1957, and two years later became one of the editors at *Scientific American*. Eventually he returned to McGraw-Hill as biology and chemistry editor. His by-line has appeared in a number of publications, particulary *The New Republic*, where he was one of the earliest opponents of the supersonic transport. In 1977, as a consultant to the American Federation of Information Processing Societies, he created the pilot issue of *Abacus*, a new magazine for the computing profession.

GEOLOGIC CALENDAR on these two pages depicts the history of the Earth as a 30-day month. One day equals 150 million years. First living things appeared very early, perhaps four billion years ago (fourth day). Oldest known fossils lived 3.1 billion years ago, but fossil record is scanty before the opening of the Paleozoic Era (27th day). On this scale, first true men appeared only about ten minutes ago, and all recorded history, from ancient Sumer to the present, represents an interval of about 30 seconds.

ARCHEOZOIC ERA

PROTEROZOIC ERA

PALEOZOIC ERA

MESOZOIC ERA

CENOZOIC ERA

3

4 FIRST LIVING SYSTEMS?

5

10 OLDEST FOSSILS (FIG TREE CHERT)

11

12

17

18

19

24 FIRST EUCARYOTIC CELLS

25 FIRST ANIMALS?

26